OXFORD IB DIPLOMA PROGRAMME

2014 EDITION

CHEMISTRY

COURSE COMPANION

Sergey Bylikin
Gary Horner
Brian Murphy
David Tarcy

OXFORD
UNIVERSITY PRESS

OXFORD

UNIVERSITY PRESS

Great Clarendon Street, Oxford, OX2 6DP, United Kingdom

Oxford University Press is a department of the University of Oxford. It furthers the University's objective of excellence in research, scholarship, and education by publishing worldwide. Oxford is a registered trade mark of Oxford University Press in the UK and in certain other countries

British Library Cataloguing in Publication Data

Data available

978-0-19-839212-5

17

Paper used in the production of this book is a natural, recyclable product made from wood grown in sustainable forests. The manufacturing process conforms to the environmental regulations of the country of origin.

Printed in Great Britain by Bell and Bain Ltd, Glasgow

Acknowledgements

The publishers would like to thank the following for permissions to use their photographs:

Cover image: Pasieka/Science Photo Library

p5a: Laguna Design/Science Photo Library; **p5b:** Jerry Mason/Science Photo Library; **p6a:** Charles D Winters/Science Photo Library; **p6b:** Africa Studio/Shutterstock; **p6c:** Geoff Tompkinson/Science Photo Library; **p10:** Getty Images; **p13:** AFP/Stringer/Getty Images; **p17:** Laguna Design/Science Photo Library; **p19:** Science Photo Library; **p23:** Science Photo Library; **p25:** Science Photo Library; **p28:** Charles D Winters/Science Photo Library; **p41:** One-Image Photography/Alamy; **p42:** A Barrington Brown/Science Photo Library; **p46:** Gianni Tortoli/Science Photo Library; **p51b:** Giphostock/Science Photo Library; **p52:** Physics Department, Imperial College/Science Photo Library; **p69a:** Science Photo Library; **p69b:** Science Photo Library; **p69c:** Science Photo Library; **p69d:** Science Photo Library; **p89:** Charles D Winters/Science Photo Library; **p107a:** Laguna Design/Science Photo Library; **p107b:** Fundamental Photographs; **p108a:** Fundamental Photographs; **p108b:** Fundamental Photographs; **p108c:** Library of Congress/Science Photo Library; **p116:** John Cole/Science Photo Library; **p117:** Russell Knightley/Science Photo Library; **p118:** Andrew Lambert Photography/Science Photo Library; **p119a:** Russell Knightley/Science Photo Library; **p119b:** Russell Knightley/Science Photo Library; **p120:** Andrew Lambert Photography/Science Photo Library; **p123:** Thomas Fredberg/Science Photo Library; **p130:** Clive Freeman/Biosym Technologies/Science Photo Library; **p144:** Paul Vinten/iStock; **p147:** Danicek/Shutterstock; **p149:** Incamerastock/Alamy; **p150:** Chien-min Chung/In Pictures/Corbis; **p156:** NASA/Science Photo Library; **p168:** NASA/Science Photo Library; **p180a:** Charles D Winters/Science Photo Library; **p180b:** Charles D Winters/Science Photo Library; **p180c:** Charles D Winters/Science Photo Library; **p185:** Andrew Lambert Photography/Science Photo Library; **p192:** Charles D Winters/Science Photo Library; **p196a:** Charles D Winters/Science Photo Library; **p196b:** Charles D Winters/Science Photo Library; **p197:** Andrew Lambert Photography/Science Photo Library; **p200:** Andrew Lambert Photography/Science Photo Library; **p202:** Charles D Winters/Getty Images; **p203a:** Andrew Lambert Photography/Getty Images; **p203b:** Andrew Lambert Photography/Getty Images; **p211:** AJP/Shutterstock; **p212:** Realimage/Alamy; **p220:** Tyler Olson/Shutterstock; **p223:** Richard Wareham Fotografie/Alamy; **p228:** Andrew Lambert Photography/Science Picture Library; **p230a:** Jaxa; **p230b:** Martin Bond/Science Photo Library; **p231:** Tim Graham/Getty Images; **p242a:** Kenneth Eward/Biografx/Science Photo Library; **p242b:** Professor K Seddon and Doctor T Evans, Queen's University, Belfast/Science Photo Library; **p247:** Science Photo Library; **p252:** Andrew Lambert Photography/Science Photo Library; **p255:** Andrew Lambert Photography/Getty Images; **p257:** Science Photo Library; **p262a:** Martyn F Chillmaid/Science Photo Library; **p262b:** Charles D Winters/Science Photo Library; **p285a:** National Institute of Advanced Industrial Science and Technology; **p360:** Charles D Winters/ Science Photo Library; **p285b:** Dennis Schroeder, NREL/US Department of Energy/ Science Photo Library; **p285c:** National Institute of Advanced Industrial Science and Technology; **p306a:** Royal Society of Chemistry; **p306b:** Royal Society of Chemistry; **p307:** Science Photo Library; **p308:** Charles D Winters/Science Photo

Library; **p310:** Patrick Aventurier/Getty Images; **p315:** Valua Vitaly/Shutterstock; **p317:** Sheila Terry/Science Photo Library; **p332:** Chemical Education Digital Library; **p346:** Bob Adelman/Corbis; **p352:** Andrew Lambert Photography/Science Photo Library; **p353:** Laguna Design/Science Photo Library; **p365:** Andrew Lambert Photography/Science Photo Library; **p366:** AdStock/Universal Image Group/Getty Images; **p368:** Gabriel Sperandio/Getty Images; **p378:** Sam Ogden/Science Photo Library; **p391:** Getty Images; **p398:** Patrick Landmann/Science Photo Library; **p404:** Marytn F Chillmaid/Science Photo Library; **p409:** Andrew Lambert Photography/Science Photo Library; **p414:** Dr Morley Read/Science Photo Library; **p415:** Photostock-Israel/Science Photo Library; **p462:** Du Cane Medical Imaging Ltd/Science Photo Library; **p464:** Jon Wilson/Science Photo Library; **p472:** NASA/Science Photo Library; **p482a:** Brian Young/Virginia Tech Chemistry Department; **P482b:** Science Photo Library; **p487:** Biosym Technologies Inc/Science Photo Library; **p488a:** Kletr/Shutterstock; **p488b:** Clive Freeman/Biosym Technologies/Science Photo Library; **p491:** Martyn F Chillmaid/Science Photo Library; **p495:** Andrew Lambert Photography/Science Photo Library; **p497:** Martyn F Chillmaid/Science Photo Library; **p503:** Victor Habbick Visions/Science Photo Library; **p506:** Digital Instruments/Vecco/Science Photo Library; **p517:** David Parker/IMI/University of Birmingham High TC Consortium/Science Photo Library; **p519:** Stefano Torrione/Hemis/Alamy; **p530:** Laguna Design/Science Photo Library; **p544a:** Microfield Scientific Ltd/Science Photo Library; **p544b:** Herve Conge, ISM/Science Photo Library; **p551a:** Pascal Goetgheluck/Science Photo Library; **p551b:** Gusto Images/Science Photo Library; **p557:** Pasieka/Science Photo Library; **p559a:** Laguna Design/Science Photo Library; **p559b:** Laguna Design/Science Photo Library; **p559c:** Laguna Design/Science Photo Library; **p559d:** Laguna Design/Science Photo Library; **p559e:** Steve Gschmeissner/Science Photo Library; **p559f:** Fotoedgaras/iStock; **p575:** Pasieka/Science Photo Library; **p576:** Jacopin/Science Photo Library; **p577:** Jesse Grant/Stringer/WireImage/Getty Images; **p588:** Power and Syred/Science Photo Library; **p591:** US National Library of Medicine/Science Photo Library; **p593:** Ingram/OUP; **p598:** Photodisc/OUP; **p600a:** Alamy Creativity/OUP; **p600b:** Charles D Winters/Science Photo Library; **p600c:** Power and Syred/Science Photo Library; **p601a:** REX/KPA/Zuma; **p601b:** Clive Freeman, The Royal Institution/Science Photo Library; **p610:** Charles D Winters/Science Photo Library; **p620:** White/OUP; **p624:** A Barrington Brown/Science Photo Library; **p638:** Lynn McLaren/Science Photo Library; **p642:** Klaus Guldbrandsen/Science Photo Library; **p649:** Kenneth Eward/Biografx/Science Photo Library; **p656:** Picture Garden/Getty Images; **p658a:** Frank Khramer/Getty Images; **p658b:** E.O/Shutterstock; **p662:** Paul Rapson/Science Photo Library; **p664:** Ashley Cooper/Visuals Unlimited Inc/Getty Images; **p666a:** Science Museum/Science and Society Picture Library; **p666b:** Maximilian Stock Ltd/Science Photo Library; **p668:** Vaughn Melzer/JVZ/Science Photo Library; **p672a:** Karen Kasmauski/Science Faction/SuperStock; **p672b:** Sheila Terry/Science Photo Library; **p672c:** Rev Ronald Royer/Science Photo Library; **p675:** OUP; **p677a:** Cate Gillon/Getty Images; **p677b:** Steigers Corporation; **p685a:** Joe Amon/Denver Post/Getty Images; **p685b:** Tom Stoddart/Getty Images; **p688a:** Getty Images; **p688b:** Sheila Terry/Science Photo Library; **p690:** Dorling Kindersley/Getty Images; **p692:** Mark Sykes/Science Photo Library; **p694:** Martin Bond/Science Photo Library; **p695:** Volker Steger/Science Photo Library; **p696:** Lawrence Berkeley National Laboratory/Science Photo Library; **p697:** Andrew Lambert Photography/Science Photo Library; **p699a:** Trans-Ocean/Emilio Segre Visual Archives/American Institute of Physics/Science Photo Library; **p699b:** Charles D Winters/Getty Images; **p700a:** Derek Lovley/Science Photo Library; **p700b:** Volker Steger/Science Photo Library; **p704:** Christopher Groenhout/Getty Images; **p706:** US Department of Energy/Science Photo Library; **p713:** US Air Force/Science Photo Library; **p718:** US National Library of Medicine; **p721:** Stevie Grand/Science Photo Library; **p730:** John Durham/Science Photo Library; **p732:** Dr Jeremy Burgess/Science Photo Library; **p745:** NASA; **p746a:** Jean-Yves Sgro/Visuals Unlimited Inc/Science Photo Library; **p746b:** Library of Congress/Science Photo Library; **p748a:** Thomas Deerinck, NCMIR/Science Photo Library; **p748b:** Jacopin/Science Photo Library; **p754:** Patrick Landmann/Science Photo Library; **p758:** David Nunuk/Science Photo Library; **p762:** Phillipe Benoist/Look at Sciences/Science Photo Library; **p765:** CNRI/Science Photo Library; **p769:** Astier-Chru Lille/Science Photo Library; **p778:** Science Photo Library; **p780:** Science Photo Library

Artwork by Six Red Marbles and OUP

Contents

Course book definition

The IB Diploma Programme course books are resource materials designed to support students throughout their two-year Diploma Programme course of study in a particular subject. They will help students gain an understanding of what is expected from the study of an IB Diploma Programme subject while presenting content in a way that illustrates the purpose and aims of the IB. They reflect the philosophy and approach of the IB and encourage a deep understanding of each subject by making connections to wider issues and providing opportunities for critical thinking.

The books mirror the IB philosophy of viewing the curriculum in terms of a whole-course approach; the use of a wide range of resources, international mindedness, the IB learner profile and the IB Diploma Programme core requirements, theory of knowledge, the extended essay, and creativity, action, service (CAS).

Each book can be used in conjunction with other materials and indeed, students of the IB are required and encouraged to draw conclusions from a variety of resources. Suggestions for additional and further reading are given in each book and suggestions for how to extend research are provided.

In addition, the course books provide advice and guidance on the specific course assessment requirements and on academic honesty protocol. They are distinctive and authoritative without being prescriptive.

IB mission statement

The International Baccalaureate aims to develop inquiring, knowledgeable and caring young people who help to create a better and more peaceful world through intercultural understanding and respect.

To this end the organization works with schools, governments and international organizations to develop challenging programmes of international education and rigorous assessment.

These programmes encourage students across the world to become active, compassionate and lifelong learners who understand that other people, with their differences, can also be right.

The IB Learner Profile

The aim of all IB programmes to develop internationally minded people who work to create a better and more peaceful world. The aim of the programme is to develop this person through ten learner attributes, as described below.

Inquirers: They develop their natural curiosity. They acquire the skills necessary to conduct inquiry and research and snow independence in learning. They actively enjoy learning and this love of learning will be sustained throughout their lives.

Knowledgeable: They explore concepts, ideas, and issues that have local and global significance. In so doing, they acquire in-depth knowledge and develop understanding across a broad and balanced range of disciplines.

Thinkers: They exercise initiative in applying thinking skills critically and creatively to recognize and approach complex problems, and make reasoned, ethical decisions.

Communicators: They understand and express ideas and information confidently and creatively in more than one language and in a variety of modes of communication. They work effectively and willingly in collaboration with others.

Principled: They act with integrity and honesty, with a strong sense of fairness, justice and respect for the dignity of the individual, groups and communities. They take responsibility for their own action and the consequences that accompany them.

Open-minded: They understand and appreciate their own cultures and personal histories, and are open to the perspectives, values and traditions of other individuals and communities. They are accustomed to seeking and evaluating a range of points of view, and are willing to grow from the experience.

Caring: They show empathy, compassion and respect towards the needs and feelings of others. They have a personal commitment to service, and to act to make a positive difference to the lives of others and to the environment.

Risk-takers: They approach unfamiliar situations and uncertainty with courage and forethought, and have the independence of spirit to explore new roles, ideas, and strategies. They are brave and articulate in defending their beliefs.

Balanced: They understand the importance of intellectual, physical and emotional balance to achieve personal well-being for themselves and others.

Reflective: They give thoughtful consideration to their own learning and experience. They are able to assess and understand their strengths and limitations in order to support their learning and personal development.

A note on academic honesty

It is of vital importance to acknowledge and appropriately credit the owners of information when that information is used in your work. After all, owners of ideas (intellectual property) have property rights. To have an authentic piece of work, it must be based on your individual and original ideas with the work of others fully acknowledged. Therefore, all assignments, written or oral, completed for assessment must use your own language and expression. Where sources are used or referred to, whether in the form of direct quotation or paraphrase, such sources must be appropriately acknowledged.

How do I acknowledge the work of others?

The way that you acknowledge that you have used the ideas of other people is through the use of footnotes and bibliographies.

Footnotes (placed at the bottom of a page) or endnotes (placed at the end of a document) are to be provided when you quote or paraphrase from another document, or closely summarize the information provided in another document. You do not need to provide a footnote for information that is part of a 'body of knowledge'. That is, definitions do not need to be footnoted as they are part of the assumed knowledge.

Bibliographies should include a formal list of the resources that you used in your work. 'Formal' means that you should use one of the several accepted forms of presentation. This usually involves separating the resources that you use into different categories (e.g. books, magazines, newspaper articles, internet-based resources, CDs and works of art) and providing full information as to how a reader or viewer of your work can find the same information. A bibliography is compulsory in the Extended Essay.

What constitutes malpractice?

Malpractice is behaviour that results in, or may result in, you or any student gaining an unfair advantage in one or more assessment component. Malpractice includes plagiarism and collusion.

Plagiarism is defined as the representation of the ideas or work of another person as your own. The following are some of the ways to avoid plagiarism:

- words and ideas of another person to support one's arguments must be acknowledged
- passages that are quoted verbatim must be enclosed within quotation marks and acknowledged
- CD-Roms, email messages, web sites on the Internet and any other electronic media must be treated in the same way as books and journals
- the sources of all photographs, maps, illustrations, computer programs, data, graphs, audio-visual and similar material must be acknowledged if they are not your own work
- works of art, whether music, film dance, theatre arts or visual arts and where the creative use of a part of a work takes place, the original artist must be acknowledged.

Collusion is defined as supporting malpractice by another student. This includes:

- allowing your work to be copied or submitted for assessment by another student
- duplicating work for different assessment components and/or diploma requirements.

Other forms of malpractice include any action that gives you an unfair advantage or affects the results of another student. Examples include, taking unauthorized material into an examination room, misconduct during an examination and falsifying a CAS record.

Using your IB Chemistry *kerboodle* Online Resources

What is Kerboodle?

Kerboodle is an online learning platform. If your school has a subscription to IB Chemistry Kerboodle Online Resources you will be able to access a bank of resources and assessments to guide you through this course.

What is in your Kerboodle Online Resources?

There are three main areas on the IB Chemistry Kerboodle: planning, resources, and assessment.

Resources

There a hundreds of extra resources available on the IB Chemistry Kerboodle Online. You can use these at home or in the classroom to develop your skills and knowledge as you progress through the course.

- Hundreds of worksheets – read articles, perform experiments and simulations, practice your skills, or use your knowledge to answer questions.

- Find out more by looking at links to recommended sites on the Internet, answer questions, or do more research.

- Plus more to come in regular updates to Kerboodle!

Planning

This area is for your teacher so you won't have access to material in here.

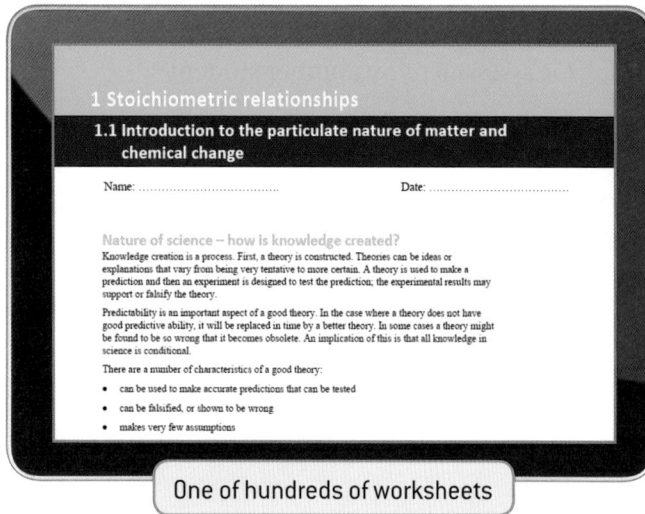

One of hundreds of worksheets

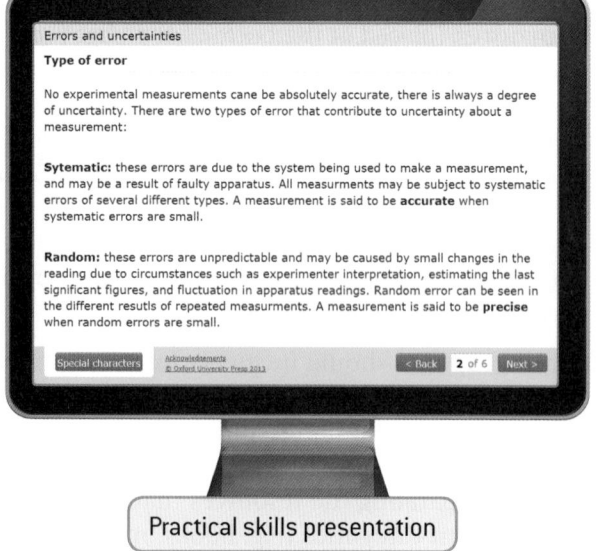

Practical skills presentation

Assessment

Click on the assessment tab to check your knowledge or revise for your examinations. Here you will find lots of interactive quizzes and exam-style practice questions.

○ Formative tests: use these to check your comprehension. Evaluate how confident you feel about a sub-topic, then complete the test. You will have two attempts at each question and get feedback after every question. The marks are automatically reported in the markbook, so you can see how you progress throughout the year.

○ Summative tests: use these to practice for your exams or as revision. Work through the test as if it were an examination – go back and change any questions you aren't sure about until you are happy, then submit the test for a final mark. The marks are automatically reported in the markbook, so you can see where you may need more practice.

○ Assessment practice: use these to practice answering the longer written questions you will come across when you are examined. These worksheets can be printed out and performed as a timed test.

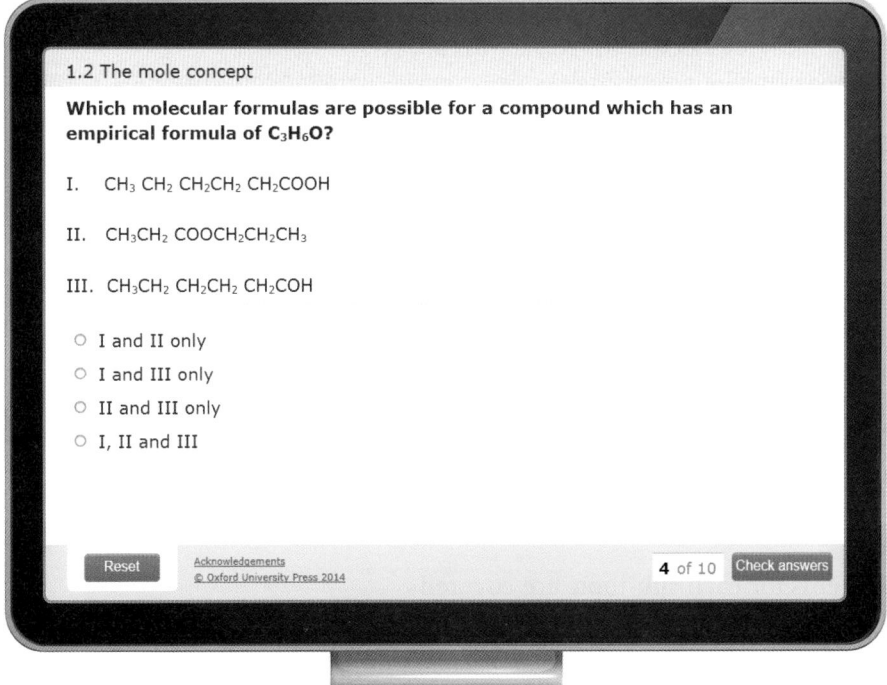

1.2 The mole concept

Which molecular formulas are possible for a compound which has an empirical formula of C_3H_6O?

I. $CH_3\ CH_2\ CH_2CH_2\ CH_2COOH$

II. $CH_3CH_2\ COOCH_2CH_2CH_3$

III. $CH_3CH_2\ CH_2CH_2\ CH_2COH$

○ I and II only

○ I and III only

○ II and III only

○ I, II and III

Reset   Acknowledgements   © Oxford University Press 2014   **4** of 10   Check answers

Don't forget!

You can also find all the textbook answers on our free website
www.oxfordsecondary.co.uk/ib-chemistry

Introduction

This book is a companion for students of Chemistry in the International Baccalaureate Diploma Programme.

Chemistry is one of the pivotal science subjects of the IB Diploma Programme. It is an experimental science that combines academic study with the acquisition of laboratory and investigational skills. Chemistry is often called the central science, as chemical principles underpin both the physical environment in which we live and all biological systems. Apart from being a subject worthy of study in its own right, chemistry is also a prerequisite for many other disciplines such as medicine, biological and environmental sciences, materials and engineering. A study of chemistry invariably involves fostering of a wide range of additional generic, transferable skills, such as analytical skills, problem-solving, data-handling, IT and communication skills, critical-thinking, numeracy and scientific literacy skills. During the two years of an IB Diploma Programme Chemistry Course, students are encouraged to develop knowledge of chemistry and an understanding of the nature of scientific inquiry. With its focus on understanding the nature of science (NOS), IB Chemistry learners will develop a level of scientific literacy that will better prepare them to act on issues of local and global concern, with a full understanding of the scientific perspective.

The structure of this book closely follows the chemistry programme in the Subject Guide.

Topics 1 - 11 explain in detail the core material that is common to both SL and HL courses. Topics 12 - 21 explain the AHL (additional higher level material). Topics A, B, C and D cover the content of the options. The optional topics cover four of the major domains in Applied Chemistry: Materials, Biochemistry, Energy and Medicinal Chemistry. Each option has a number of common strands – quantitative aspects, analytical techniques, environmental perspectives and integrated organic chemistry linkages.

All topics in the book include the following elements:

Understandings

The specifics of the content requirements for each sub-topic are covered in detail. Concepts are presented in ways that promote enduring understanding.

Applications and skills

These sections help you to develop your understanding by considering a specific illustrative example, often following a step-by-step working method approach or by considering a particular chemical experiment, involving key laboratory techniques.

Nature of science

Here you can explore the methods of science and some of the knowledge issues, theories, hypotheses and laws that are associated with scientific endeavour. This is done using carefully selected examples, including chemical research that led to paradigm shifts in our understanding of the world. NOS underpins each topic presented and throughout the book

there are a wide range of NOS based questions and exercises to challenge your chemical understanding and draw on your scientific perspectives. NOS is an assessable component of the programme and sample NOS style questions are integrated throughout the book.

Theory of Knowledge

These short sections have headings that are equivocal 'knowledge questions'. The text that follows often details one possible answer to the knowledge question. We encourage you to draw on these examples of knowledge issues in your TOK essays. Of course, much of the material elsewhere in the book, particularly in the NOS sections, can be used to prompt TOK discussions.

TOK provides a space for you to engage in stimulating wider discussions about questions such as whether there should be ethical constraints on the pursuit of scientific knowledge. It also provides an opportunity for you to reflect on scientific methodologies, and how these compare to the methodologies of other areas of knowledge. TOK is not formally assessed in the IB Chemistry programme, but it plays a pivotal role in the teaching of IB science.

Activities and quick questions

A variety of short topics or challenging questions are included with a focus on active learning. We encourage you to research these topics or problems yourselves using information readily available in textbooks or from the Internet. The aim is to promote an independent approach to learning.

End -of-topic questions

At the end of each topic you will find a wide range of questions (multiple-choice, data-base exercises, extended response, NOS style problems and hypothesis style questions).

Answers can be found at www.oxfordsecondary.co.uk/ib-chemistry

Meet the authors

Sergey Bylikin was awarded a PhD in Chemistry from Moscow State University in 1998 and, one year later, received the State Prize of the Russian Federation in Chemistry. Until 2009, he was assistant professor at Russian State Medical University, after which point he took up a role at the Open University in the UK. Sergey is an author of several textbooks. He has been associated with the IB since 2007 and was involved in the latest IB Chemistry curriculum review.

Gary Horner, a graduate of the University of Queensland, has taught Chemistry since 1986 in Australia, Switzerland and Hong Kong. In his International School career Gary has held various leadership positions, including that of CAS coordinator and Head of Science. Since 2000, he has had significant involvement with the IBO, attending workshops across Europe and Canada and leading workshops in India, Hong Kong and Japan. In 2010, he began advising on the IB Chemistry curriculum review and is a member of the team developing the latest DP science course. Gary is currently teaching at King George V School in Hong Kong.

Brian Murphy graduated with a PhD in Inorganic Chemistry from University College Cork. Following postdoctoral and teaching posts in the UK and Ireland, he moved to the United Arab Emirates to take up a position at UAE University, where he became Head of the Department of Chemistry and associate professor of Inorganic Chemistry. After 8 years he moved back to Ireland to take up a post at Athlone Institute of Technology, where he is currently a senior lecturer. Brian has been associated with the IB since 1998 and was involved in the design of the latest IB Chemistry curriculum.

David Tarcy graduated *cum laude* with a degree in Science Education from Whitworth College and has done graduate work in sciences and information technology in the Northwest USA and Queensland, Australia. He has taught in the USA, Australia, Europe, and Southeast Asia and has been involved in curriculum writing, moderation, and question setting for various exam boards and institutions. David is active in many chemistry education discussion boards, is an IB Diploma Programme Chemistry Workshop Leader and Field Representative and was involved in the design of the latest IB Chemistry curriculum.

A project of this size would not have been possible without support and encouragement. To the greatest extent, the authors would like to thank their families for their love and patience. In particular, special appreciation goes to:

Brian Murphy - to my wife Mary, for all her love, understanding and unremitting support (míle buíochas!), parents, Teresa and Joe (RIP) who instilled in me an appreciation of internationalisation from an earlier age, sister, Lorraine and her family; **Gary Horner** - to my parents Dennis and Myrtle for their devotion, vision and unwavering support of their children's happiness, my sister Susan for her eternal friendship, selflessness and professional expertise; **David Tarcy** - to Tina Walton, my brothers Gary and Brian, for their input and support, as well as the many friends and professional colleagues I have met through my teaching career for their support, advice, and friendship; **Sergey Bylikin** - to Natasha for her patience, support and invaluable comments.

1 STOICHIOMETRIC RELATIONSHIPS

Introduction

There is a broad community of people working within a wide variety of scientific disciplines and approaching their inquiry with common methodology, terminology and reasoning processes. Chemistry can be regarded as the central science, and mathematics the language of science. In this chapter we begin to lay down many of the foundations on which an understanding of chemistry is based. From the classification of matter to the IUPAC organization of the nomenclature of organic and inorganic compounds and the representations of chemical reactions by equations, this chapter discusses the comprehensive language of chemistry.

For chemists, the mole concept is of fundamental importance. Its definitions in relation to the number of particles, mass and the volume of a gas elicit universal understanding and stoichiometry, the quantitative method of examining the relative amounts of reactants and products in a particular chemical reaction is developed. Treatment of the gas laws and the application of volumetric analysis complete this introductory chapter.

1.1 Introduction to the particulate nature of matter and chemical change

Understandings

→ Atoms of different elements combine in fixed ratios to form compounds, which have different properties from their component elements.

→ Mixtures contain more than one element and/or compound that are not chemically bonded together and so retain their individual properties.

→ Mixtures are either homogeneous or heterogeneous.

Applications and skills

→ Deduction of chemical equations when reactants and products are specified.

→ Application of the state symbols (s), (l), (g), and (aq) in equations.

→ Explanation of observable changes in physical properties and temperature during changes of state.

Nature of science

→ Making quantitative measurements with replicates to ensure reliability – definite and multiple proportions.

The atomic theory

A universally accepted axiom of science today is that all matter is composed of atoms. However, this has not always been so. During the seventeenth century the *phlogiston theory* was a widely held belief. To explain the process of combustion it was proposed that a fire-like element called **phlogiston**, said to be found within substances, was released during burning. Quantitative investigations of burning metals revealed that magnesium in fact gains rather than loses mass when it burns in oxygen, contradicting the phlogiston theory.

Scientists use a wide range of methodologies, instruments, and advanced computing power to obtain evidence through observation and experimentation. Much of the technology commonly used today was not available to scientists in the past, who often made ground-breaking discoveries in relatively primitive conditions to feed their appetite for knowledge. Over time, theories and hypotheses have been tested with renewed precision and understanding. Some theories do not stand the test of time. The best theories are those that are simple and account for all the facts.

The **atomic theory** states that all matter is composed of atoms. These atoms cannot be created or destroyed, and are rearranged during chemical reactions. Physical and chemical properties of matter depend on the bonding and arrangement of these atoms.

States of matter

Matter is everywhere. We are made up of matter, we consume it, it surrounds us, and we can see and touch many forms of matter. Air is a form of matter which we know is there, though we cannot see it. Our planet and the entire universe are made up of matter and chemistry seeks to expand our understanding of matter and its properties.

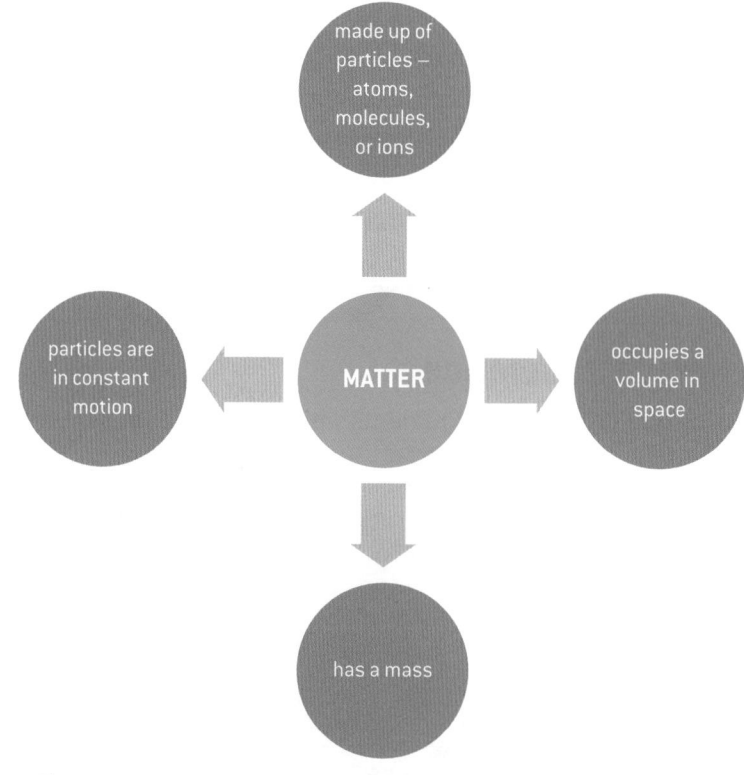

▲ Figure 1 The characteristics of matter

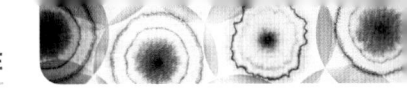

The properties of the three **states of matter** are summarized below.

Solid	Liquid	Gas

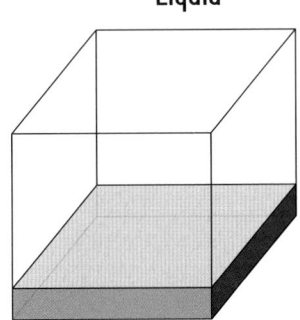

- fixed volume
- fixed shape

- cannot be compressed
- attractive forces between particles hold the particles in a close-packed arrangement

- particles vibrate in fixed positions

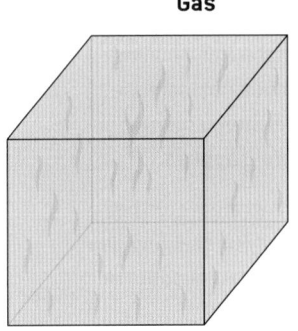

- fixed volume
- no fixed shape – takes the shape of the container it occupies
- cannot be compressed
- forces between particles are weaker than in solids

- particles vibrate, rotate, and translate (move around)

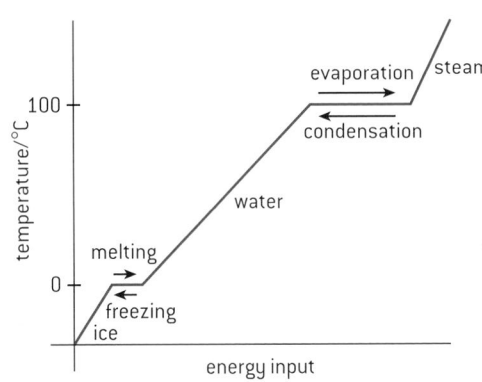 *(Gas diagram)*

- no fixed volume
- no fixed shape – expands to occupy the space available

- can be compressed
- forces between particles are taken as zero

- particles vibrate, rotate, and translate faster than in a liquid

The way the particles of matter move depends on the temperature. As the temperature increases the average kinetic energy of the particles increases – the particles in a solid vibrate more. The particles in liquids and gases also vibrate, rotate, and translate more.

> SI (Système International) units are a set of standard units that are used in science throughout the world. This will be discussed in great detail in sub-topic 1.2.

Temperature

There are a number of different temperature scales. The most commonly used are the Fahrenheit, Celsius, and Kelvin scales. All three are named in honour of the scientist who developed them.

The SI unit for temperature is the **kelvin** (K). The Kelvin scale is used in energetics calculations (see topic 5).

Absolute zero is zero on the Kelvin scale, 0 K (on the Celsius scale this is –273 °C). It is the temperature at which all movement of particles stops. At temperatures greater than absolute zero, all particles vibrate, even in solid matter.

You can convert temperatures from the Celsius scale to the the Kelvin scale using the algorithm:

temperature (K) = temperature (°C) + 273.15

Changes of state

If you heat a block of ice in a beaker it will melt to form liquid water. If you continue heating the water, it will boil to form water vapour. Figure 2 shows a heating curve for water – it shows how its temperature changes during these **changes of state**. We shall look at the relationship between temperature and the kinetic energy of particles during these changes of state.

> When describing room temperature, we might say '25 degrees Celsius (25 °C)' or '298 kelvin (298 K)' (to the nearest kelvin). Note that we use just the word kelvin, not degrees kelvin. The boiling point of water is 100 °C or 373 K, and the melting point of water is 0 °C or 273 K.

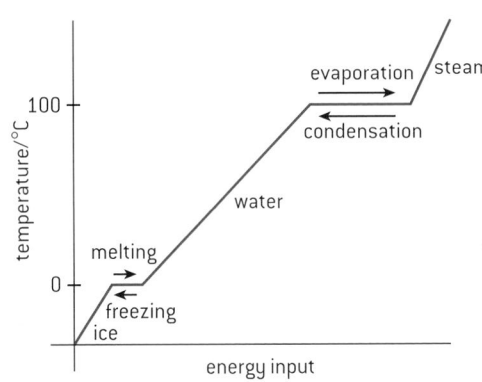

▲ Figure 2 The heating curve for water

Freeze-drying is a food preservation technique that uses the process of **sublimation**. Foods that require dehydration are first frozen and then subjected to a reduced pressure. The frozen water then sublimes directly to water vapour, effectively dehydrating the food. The process has widespread applications in areas outside the food industry such as pharmaceuticals (vaccines), document recovery for water-damaged books, and scientific research laboratories.

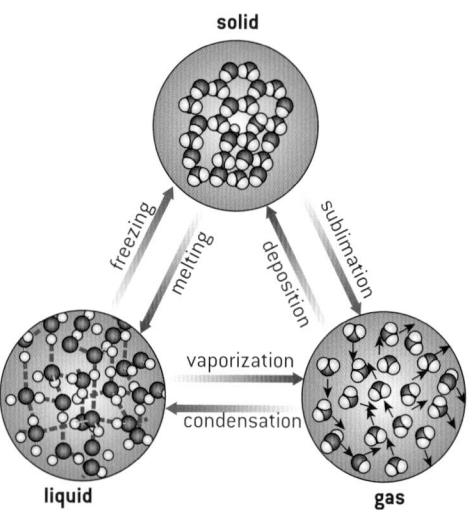

▲ Figure 3 Changes of state for water

What happens to the particles during changes of state?

- As a sample of ice at –10 °C (263 K) is heated, the water molecules in the solid lattice begin to vibrate more. The temperature increases until it reaches the melting point of water at 0 °C (273 K).

- The ice begins to **melt** and a solid–liquid equilibrium is set up. Figure 2 shows that there is no change in temperature while melting is occurring. All of the energy is being used to disrupt the lattice, breaking the attractive forces between the molecules and allowing the molecules to move more freely. The level of disorder increases. (The nature of the forces between molecules is discussed in sub-topic 4.4.)

- Once all the ice has melted, further heating makes the water molecules vibrate more and move faster. The temperature rises until it reaches the boiling point of water at 100 °C (373 K), and the water starts to **boil**.

- At 100 °C a liquid–gas equilibrium is established as the water boils. Again the temperature does not change as energy is required to overcome the attractive forces between the molecules in the liquid water in order to free water molecules from the liquid to form a gas. (Equilibrium is covered in sub-topic 7.1.)

- The curve in figure 2 shows that while the water is boiling its temperature remains at 100 °C. Once all the liquid water has been converted to steam, the temperature will increase above 100 °C.

- Melting and boiling are **endothermic** processes. Energy must be transferred to the water from the surroundings to bring about these changes of state. The potential energy (stored energy) of the molecules increases – they vibrate more and move faster.

- Cooling brings about the reverse processes to heating – the **condensation** of water vapour to form liquid water, and the **freezing** of liquid water to form a solid.

- Condensation and freezing are **exothermic** processes. Energy is transferred to the surroundings from the water during these changes of state. The potential energy of the molecules decreases – they vibrate less and move slower.

- **Vaporization** is the change of state from liquid to gas which may happen during boiling, or by **evaporation** at temperatures below the boiling point. In **sublimation** matter changes state directly from the solid to gas phase without becoming a liquid. **Deposition** is the reverse process of sublimation – changing directly from a gas to a solid.

Elements and compounds

An **element** contains atoms of only one type. Atoms of elements combine in a fixed ratio to form **compounds** composed of molecules or ions. These rearrangements of the particles of matter are the fundamental cornerstone of chemistry, represented in formulae and balanced chemical equations. (Atoms are covered in detail in sub-topic 2.1.)

Chemists study how elements and compounds react with one another, the many different chemical and physical properties of the substances created in these reactions, and how they can be used in many important applications.

The compound sodium chloride, NaCl, is made up of the elements sodium and chlorine.

The group 1 alkali metal sodium is a soft metal that undergoes rapid oxidation in air and violently reacts with water, creating alkaline solutions. Sodium is stored under oil to prevent these reactions. It is the sixth most abundant element on the planet, (2.26% by mass).

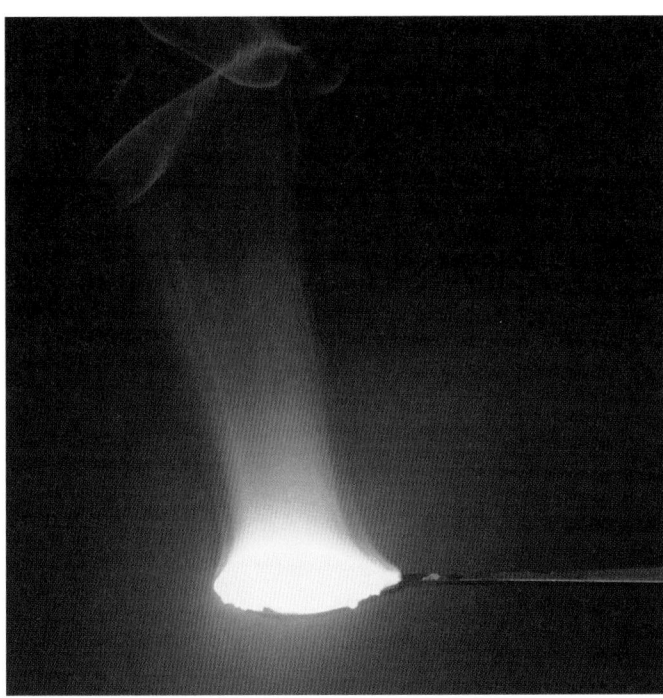

▲ Figure 4 Elemental sodium is a reactive alkali metal

▲ Figure 5 The structure of sodium chloride. It consists of a crystalline lattice of sodium ions (purple) and chloride ions (green)

The halogen chlorine is a gas at room temperature. Chlorine, Cl_2, is highly irritating to the eyes, skin, and the upper respiratory tract.

The highly reactive elements sodium and chlorine combine to form the ionic crystalline compound sodium chloride, commonly called table salt and consumed daily in the food we eat. The properties and uses of sodium chloride are very different from those of its constituent elements.

Mixtures

A **pure substance** is matter that has a constant composition. Its chemical and physical properties are distinct and consistent. Examples include the elements nitrogen, N_2 and argon, Ar and compounds such as water, H_2O, table salt, NaCl, and glucose, $C_6H_{12}O_6$.

Pure substances can physically combine to form a **mixture**. For example, sea water contains mainly sodium chloride and water. Pure substances can be separated from the mixture by physical techniques such as filtration, fractional distillation, or chromatography. The

Proust's law of constant composition
(1806) stated that compounds have
distinct properties and the same
elemental composition by mass.

▲ Figure 6 Chlorine reacts vigorously
with sodium metal

▲ Figure 7 Table salt is the compound sodium
chloride, NaCl(s). It has very different properties
from those of its constituent elements

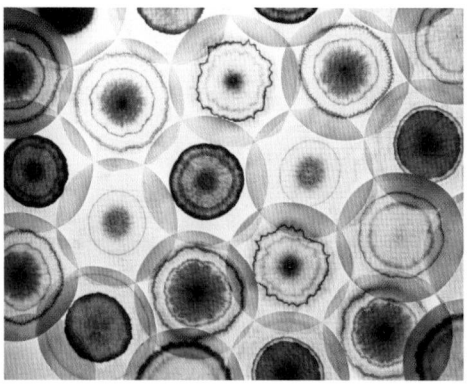

▲ Figure 8 Paper chromatography is used to
investigate industrial dyes by separating them
into their pure constituent components

elements or compounds that make up a mixture are not chemically
bound together.

Homogeneous mixtures have both uniform composition and uniform
properties throughout the mixture. Examples include salt water or a
metal alloy such as brass. **Heterogeneous** mixtures have a non-uniform
composition and hence their properties vary throughout the mixture.
Examples include foods such as tom yum goong (Thai hot and sour
prawn soup) or Irish stew (a mixture of cubed meat and vegetables).

Figure 9 summarizes the classification of matter into elements,
compounds, and mixtures.

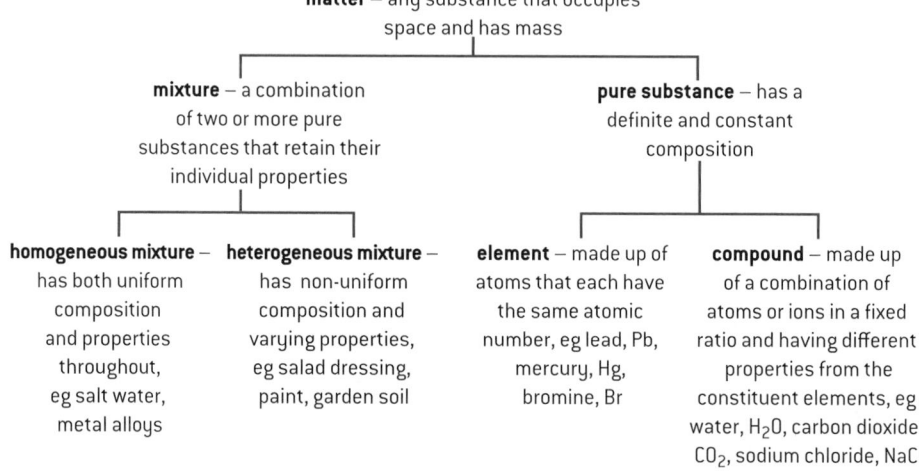

▲ Figure 9 Elements, compounds, and mixtures

🌐 The language of chemistry

Chemistry has a universal language that transcends borders and
enables scientists, teachers, and lecturers, students, and citizens of
the wider community to communicate with each other. Chemical
symbols and equations are a language that requires no translation.
Knowledge of the symbols for elements and compounds and their
relationship to one another as displayed in a balanced equation
unlocks a wealth of information, allowing understanding of the
chemical process being examined.

Chemical symbols are a way of expressing which elements are present
and in which proportions, in both organic and inorganic compounds.
The International Union of Pure and Applied Chemistry (**IUPAC**) is
an organization that develops and monitors a system of standardized
nomenclature for both organic and inorganic compounds. IUPAC's role
is to provide consistency in the naming of compounds, resulting in a
language of symbols and words that require no translation from one
country or culture's language to another.

Useful Resource

The IUPAC Gold Book (http://goldbook.iupac.org/index.html) is IUPAC's
compendium of chemical terminology.

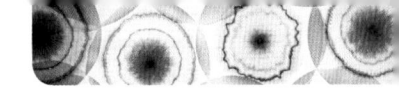

TOK

Language is a crucial component in the communication of knowledge and meaning. Does the language of chemistry with its equations, symbols, and units promote or restrict universal understanding? What role does linguistic determinism play?

For example, the concept of equilibrium is often initially misinterpreted. Preconceived ideas focus on a 50:50 balance between reactants and products. It requires an understanding that equilibrium means that both the forward and reverse reactions are occurring at the same rate before we can see that an equilibrium reaction might favour the formation of products or reactions, or that such a reaction could be non-spontaneous.

Name of polyatomic ion	Formula	Name of polyatomic ion	Formula
ammonium ion	NH_4^+	phosphate(V) ion	PO_4^{3-}
carbonate ion	CO_3^{2-}	phosphonate ion	PO_3^{3-}
hydrogencarbonate ion	HCO_3^-	sulfate(VI) ion	SO_4^{2-}
hydroxide ion	OH^-	sulfate(IV) ion	SO_3^{2-}
nitrate(V) ion	NO_3^-	ethanedioate ion	$C_2O_4^{2-}$
nitrate(III) ion	NO_2^-	peroxide ion	O_2^{2-}

▲ Table 1 Common polyatomic ions

Common combinations of elements: Background to writing equations

An **ion** is a charged species. **Anions** are negatively charged and **cations** are positively charged.

There are a number of common polyatomic ions that exist in many of the substances you will study and work with. You need to be familiar with the names and formulae of these ions, shown in tables 1 to 3.

Name of acid	Formula
hydrochloric acid	HCl
nitric(V) acid	HNO_3
phosphoric(V) acid	H_3PO_4
sulfuric(VI) acid	H_2SO_4
ethanoic acid	CH_3COOH

▲ Table 2 Common acids

 ## Writing and balancing equations

An ability to write equations is essential to chemistry and requires a full understanding of the language of equations. At the most fundamental level, formulae for the reactants are put on the left-hand side along with their state symbols (s), (l), (g), (aq), and those for the products on the right-hand side. The arrow represents a boundary between reactants and products. State symbols can be deduced by referring to the solubilities of ionic salts and the state of matter of the element or compound at a given temperature.

A reaction may be described in terms of starting materials and products. The process of transforming these words into a balanced chemical equation starts with the construction of chemical formulae. Writing ionic and covalent formulae will be discussed in depth in topic 4.

Name of anion	Formula	Naming suffix
sulfide ion	S^{2-}	-ide
sulfate(VI) ion	SO_4^{2-}	-ate
sulfate(IV) ion	SO_3^{2-}	-ate

▲ Table 3 Naming anions. The prefix identifies the element present and the suffix the type of ion (eg element or polyatomic ion)

Quick questions

Write equations for the following chemical reactions, including state symbols. Refer to the working method on the next page on balancing equations if you need to.

1 Zinc metal reacts with hydrochloric acid to form the salt zinc chloride. Hydrogen gas is evolved.

2 Hydrogen gas and oxygen gas react together to form water.

3 At a high temperature, calcium carbonate decomposes into calcium oxide and carbon dioxide.

Worked example

Magnesium burns in oxygen to form a white powder known as magnesium oxide. Write a chemical equation to represent this change, including state symbols.

Solution

The reactants are the metal magnesium, a solid at room temperature, and the diatomic molecule, oxygen, which is a gas. The product is the oxide of magnesium, magnesium oxide which is a solid substance.

$$2Mg(s) + O_2(g) \rightarrow 2MgO(s)$$

Working method: how to balance chemical equations

The examples below involve reactions of metals. Figure 10 reminds you that metals are below and to the left of the metalloids in the periodic table.

Remember that to balance an equation you change the coefficient of a formula (add a number in front of the formula). You do not change the formula itself.

Step 1: First balance the metallic element on each side of the equation – add a number in front of the symbol on one side if necessary so that there is the same number of atoms of this element on each side.

Step 2: Balance any elements that occur in only one formula on the reactant and products side. Sometimes polyatomic ions remain unchanged in reactions and they can be balanced easily at this stage.

Step 3: Balance the remaining elements if necessary.

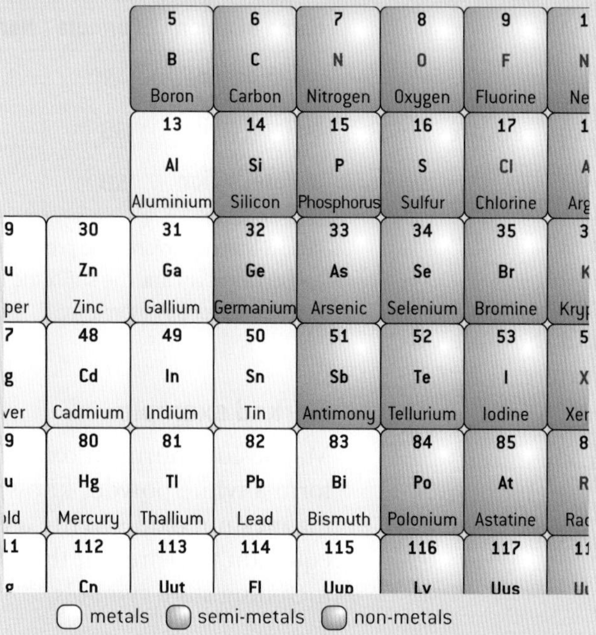

▲ Figure 10 Metals are below and to the left of the metalloids in the periodic table

Example 1

The alkaline earth metal calcium reacts with water to produce an alkaline solution. Balance the following equation.

> **Step 1:** Balance the metal Ca first. It is balanced.

$$Ca(s) + H_2O(l) \rightarrow Ca(OH)_2(aq) + H_2(g)$$

> **Step 2:** Balance O next, as it occurs in only one formula on each side. (H occurs in both products.) Multiply H_2O by 2 to balance O.

$$Ca(s) + 2H_2O(l) \rightarrow Ca(OH)_2(aq) + H_2(g)$$

> **Step 3:** You can now see that hydrogen has been balanced by step 2, which often happens. Always check to make sure.

The equation is now balanced overall.

Example 2

Potassium hydroxide is a soluble base that can neutralize the diprotic acid sulfuric acid. Diprotic acids produce two hydrogen ions when they dissociate. Balance the following equation.

> **Step 1:** Balance K by doubling KOH on the reactant side.

$$H_2SO_4(aq) + KOH(aq) \rightarrow K_2SO_4(aq) + H_2O(l)$$

> **Step 2:** Both O and H occur in two compounds on both sides of the equation. The sulfate ion is unchanged in the reaction and is balanced, so the coefficient for H_2SO_4 will stay the same. There are 4 H atoms on the reactant side, so multiply H_2O by 2.

$$H_2SO_4(aq) + 2KOH(aq) \rightarrow K_2SO_4(aq) + H_2O(l)$$
$$H_2SO_4(aq) + 2KOH(aq) \rightarrow K_2SO_4(aq) + 2H_2O(l)$$

The equation is now balanced.

Some types of reaction

Combination or **synthesis** reactions involve the combination of two or more substances to produce a single product:

$$C(s) + O_2(g) \rightarrow CO_2(g)$$

Decomposition reactions involve a single reactant being broken down into two or more products:

$$CaCO_3(s) \rightarrow CaO(s) + CO_2(g)$$

Single replacement reactions occur when one element replaces another in a compound. An example of this type of reaction is a redox reaction (topic 9):

$$Mg(s) + 2HCl(aq) \rightarrow MgCl_2(aq) + H_2(g)$$

Double replacement reactions occur between ions in solution to form insoluble substances and weak or non-electrolytes, also termed **metathesis** reactions:

$$HCl(aq) + NaOH(aq) \rightarrow NaCl(aq) + H_2O(l)$$

This example is an acid-base reaction discussed further in topic 8.

> The names and symbols of the elements can be found in section 5 of the *Data booklet*.

 # Some applications and reactions of butane

Fuels and refrigerants

Butane, C_4H_{10} is mixed with other hydrocarbons such as propane to create the fuel liquefied petroleum gas (LPG). This is used in a wide variety of applications.

Methylpropane (also called isobutane) is an isomer of butane. Isomers have the same chemical formula but their atoms are arranged structurally in a different way. Methylpropane is used as a refrigerant, replacing the CFCs that were previously used for this purpose.

Ozone occurs naturally in the stratosphere, in the upper atmopshere. Ozone filters out most of the harmful ultraviolet rays from the sun. Without this protection the ultraviolet radiation would be harmful to many forms of life, causing skin cancer in humans and other problems.

> ### CFCs and the impact of science and technology
>
> The process of refrigeration involves the energy changes of a condensation–evaporation cycle using volatile liquids. Chlorofluorocarbons (CFCs) were traditionally used in refrigerators and air-conditioning units. They cause depletion of the ozone layer in the atmosphere, which protects us from the harmful effects of ultraviolet radiation in sunlight.
>
> CFCs are now banned in many countries, and non-halogenated hydrocarbons such as propane are more commonly used instead. There is more about this in sub-topic 5.3.

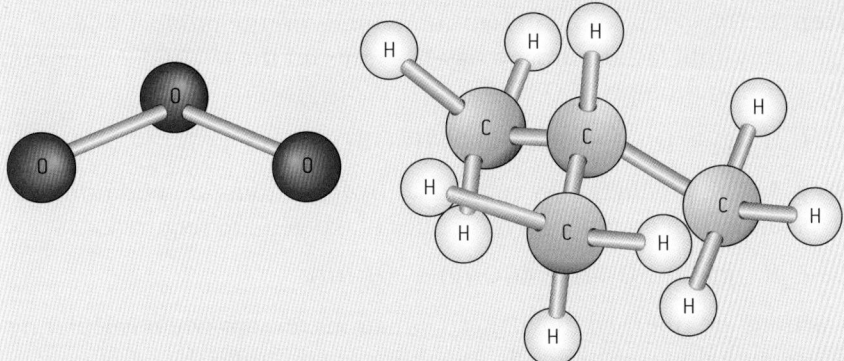

▲ Figure 11 Ozone, O_3 ▲ Figure 12 Methylpropane is used as a refrigerant

CFCs undergo reactions with the ozone in the stratosphere, causing it to break down. The 'ozone hole' is a thinning of the ozone layer that appears over the polar regions of the Earth each spring. The use of CFCs has caused this depletion of the ozone layer, so they have now been replaced by methylpropane.

The combustion of hydrocarbons, C_xH_y produces carbon dioxide and water.

Since 1997, taxis in Hong Kong have been powered by liquefied petroleum gas (LPG). Today there are over 18 000 LPG taxis and 500 LPG light buses operating there. LPG, consisting of butane and/or propane, undergoes combustion to release energy to power the vehicle. The reaction produces carbon dioxide and water (sub-topic 10.2). LPG burns much more cleanly than petrol or diesel.

▲ Figure 14 Rush hour in Hong Kong

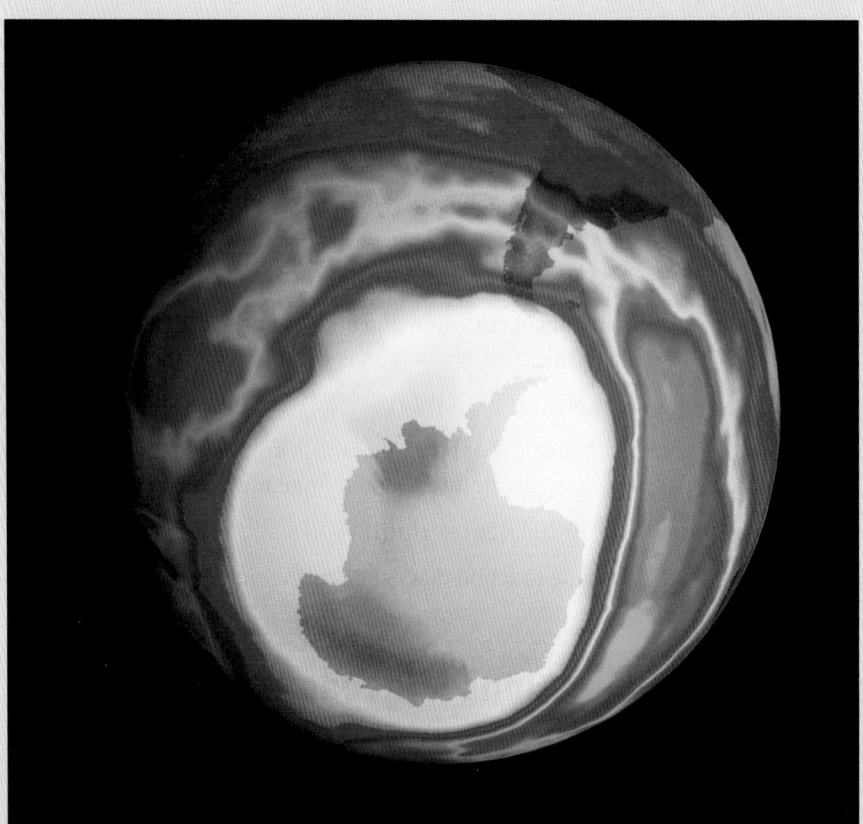

▲ Figure 13 The ozone 'hole' was first noticed in the 1970s and is monitored by scientists worldwide

Balancing the equation for the combustion of butane

The combustion of butane is an exothermic reaction.

$$C_4H_{10}(g) + O_2(g) \rightarrow CO_2(g) + H_2O(l)$$

Step 1: There are no metal atoms to balance, so balance the carbon atoms first by multiplying CO_2 by 4.

$$C_4H_{10}(g) + O_2(g) \rightarrow 4CO_2(g) + H_2O(l)$$

Step 2: Oxygen is found in two compounds on the product side so leave this until last. Hydrogen has 10 atoms on the left and 2 atoms on the right, so multiply H_2O by 5.

$$C_4H_{10}(g) + O_2(g) \rightarrow 4CO_2(g) + 5H_2O(l)$$

Step 3: The products now contain 13 oxygen atoms, an odd number. To balance the equation 6.5 molecules of oxygen are required.

$$C_4H_{10}(g) + 6.5O_2(g) \rightarrow 4CO_2(g) + 5H_2O(l)$$

Fractions are not used in balanced equations, except when calculating lattice enthalpy (see topic 15). We therefore multiply the whole equation by 2.

$$2C_4H_{10}(g) + 13O_2(g) \rightarrow 8CO_2(g) + 10H_2O(l)$$

The complex coefficients in this example show why the method of balancing equations on page 8 is more efficient than just trial and error.

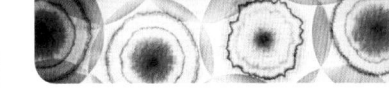

The atom economy

The global demand for goods and services along with an increasing world population, rapidly developing economies, increasing levels of pollution, and dwindling finite resources have led to a heightened awareness of the need to conserve resources. Synthetic reactions and industrial processes must be increasingly efficient to preserve raw materials and produce fewer and less toxic emissions. Sustainable development is the way of the future.

To this end the **atom economy** was developed by Professor Barry Trost of Stanford University Stanford, CA, USA. This looks at the level of efficiency of chemical reactions by comparing the molecular mass of atoms in the reactants with the molecular mass of useful compounds.

$$\frac{\text{percentage}}{\text{atom economy}} = \frac{\text{Molecular mass of atoms of useful products}}{\text{Molecular mass of atoms in reactants}} \times 100\%$$

The atom economy is important in the discussion of Green Chemistry, which we will discuss later in this book. In an ideal chemical process the amount of reactants = amounts of products produced. So an atom economy of 100% would suggest that no atoms are wasted.

Activity

a) Suggest why even if a chemical reaction has a yield close to 100%, the atom economy may be poor. Carry out some research into this aspect.

b) Discuss some other ways a chemical process may be evaluated other than the atom economy, eg energy consumption etc.

c) Deduce the percentage atom economy for the nucleophilic substitution reaction:

$$CH_3(CH_2)_3OH + NaBr + H_2SO_4 \rightarrow CH_3(CH_2)_3Br + H_2O + NaHSO_4$$

Quick questions

Identify the type of reaction and then copy and balance the equation, using the smallest possible whole number coefficients.

1   $SO_3(g) + H_2O(l) \rightarrow H_2SO_4(aq)$

2   $NCl_3(g) \rightarrow N_2(g) + Cl_2(g)$

3   $CH_4(g) + O_2(g) \rightarrow CO_2(g) + H_2O(g)$

4   $Al(s) + O_2(g) \rightarrow Al_2O_3(s)$

5   $KClO_3(s) \rightarrow KCl(s) + O_2(g)$

6   $C_3H_8(g) + O_2(g) \rightarrow CO_2(g) + H_2O(g)$

7   $Ni(OH)_2(s) + HCl(aq) \rightarrow NiCl_2(aq) + H_2O(l)$

8   $AgNO_3(aq) + Cu(s) \rightarrow Cu(NO_3)_2(aq) + Ag(s)$

9   $Ca(OH)_2(s) \rightarrow CaO(s) + H_2O(l)$

1.2 The mole concept

Understandings

→ The mole is a fixed number of particles and refers to the amount, *n*, of substance.

→ Masses of atoms are compared on a scale relative to ^{12}C and are expressed as relative atomic mass (A_r) and relative formula/molecular mass (M_r).

→ Molar mass (M) has the units $g\,mol^{-1}$.

→ The empirical formula and molecular formula of a compound give the simplest ratio and the actual number of atoms present in a molecule respectively.

Applications and skills

→ Calculation of the molar masses of atoms, ions, molecules and formula units.

→ Solution of problems involving the relationships between the number of particles, the amount of substance in moles and the mass in grams.

→ Interconversion of the percentage composition by mass and the empirical formula.

→ Determination of the molecular formula of a compound from its empirical formula and molar mass.

→ Obtaining and using experimental data for deriving empirical formulas from reactions involving mass changes.

Nature of science

→ Concepts – the concept of the mole developed from the related concept of 'equivalent mass' in the early 19th century.

SI: the international system of measurement

Throughout history societies have developed different forms of measurement. These may vary from one country and culture to another, so an internationally agreed set of units allows us to understand measurements regardless of the language of our culture.

Units of measurement are essential in all walks of life. The financial world speaks in US dollars, the resources industries use million tonnes (MT), precious metals are measured in ounces, agricultural manufacturing uses a range of measures including yield per hectare, and environmental protection agencies, amongst others, talk about parts per million (ppm) of particulate matter. Which units do chemists use?

The desire for a standard international set of units led to the development of a system that transcends all languages and cultures – the **Système International d'Unités** (SI). Table 1 shows the seven base units of the SI system. All other units are derived from these seven base units.

Accuracy and SI units

Continual improvements in the precision of instrumentation used in the measurement of SI units have meant that the values of some physical constants have changed over time. **The International Bureau of Weights and Measures** (known as **BIPM** from its initials in French) monitors the correct use of SI units, so that in all applications of science, from the school laboratory to the US National Aeronautics and Space Administration (NASA), SI units are used and are equivalent in all cases.

Property	Unit	Symbol
mass	kilogram	kg
temperature	kelvin	K
time	second	s
amount	mole	mol
electric current	ampère	A
luminosity	candela	cd
length	metre	m

▲ Table 1 The seven base units of the SI system

Table 2 shows two quantities that are used throughout the study of chemistry, along with their units. Table 3 is a list of standard prefixes used to convert SI units to a suitable size for the application you are measuring.

Avogadro's constant (N_A)	Molar volume of an ideal gas at 273 K and 100 kPa
6.02×10^{23} mol^{-1}	2.27×10^{-2} m^3 mol^{-1} ($= 22.7$ dm^3 mol^{-1})

▲ Table 2 Useful physical constants and unit conversions

Amount of substance: The mole

Chemists need to understand all aspects of a chemical reaction in order to control and make use of the reaction. From large-scale industrial processes such as electrolytic smelting of aluminium and industries involved in processing of food and beverages, to pharmaceutical companies synthesizing medicines and drugs, the ability to measure precise amounts of reacting substances is of crucial importance.

All chemical substances are made up of elements that are composed of their constituent atoms, which vary in the number of protons, neutrons, and electrons (topic 2). Chemists use a system to measure equal amounts of different elements regardless of how big their atoms are, which allows them to calculate reacting quantities. The **mole** is an SI unit, symbol **mol**, defined as a fixed amount, n, of a substance. This

▲ Figure 1 A platinum–iridium cylinder at the National Institute of Standards and Technology, Gaithersburg, MD, USA, represents the standard 1 kg mass

Study tips

Physical constants and unit conversions are available in section 2 of the *Data booklet*. The value of Avogadro's constant (L or N_A) will be provided in Paper 1 questions, and may be referred to in the *Data booklet* when completing both Papers 2 and 3.

Prefix	Abbreviation	Scale
nano	n	10^{-9}
micro	µ	10^{-6}
milli	m	10^{-3}
centi	c	10^{-2}
deci	d	10^{-1}
standard	–	1
kilo	k	10^3
mega	M	10^6
giga	G	10^9

▲ Table 3 Useful prefixes, their abbreviations and scales

Stoichiometry uses the quantitative relationships between amounts of reactants and products in a chemical reaction. These relationships depend on the law of conservation of mass and definite proportions. They allow chemists to calculate the proportions of reactants to mix, and to work out expected yields, from the ratios of reactants and products according to the balanced chemical equation.

definition can be applied to atoms, molecules, formula units of ionic compounds, and electrons in the process of electrolysis.

This fixed amount is a number of particles called **Avogadro's constant** (symbol L or N_A) and it has a value of 6.02×10^{23} mol^{-1}. Avogadro's constant enables us to make comparisons between chemical species. A mole of any chemical species always contains an identical number of representative units.

Relative atomic mass, relative formula mass, and molar mass

Isotopes are atoms of the same element that have the same number of protons in the nucleus but different numbers of neutrons (see sub-topic 2.1). Isotopes of an element have different mass numbers. The **relative abundance** of each isotope is a measure of the percentage that occurs in a sample of the element (table 4).

The masses of atoms are compared with one another on a scale in which a single atom of carbon-12 equals 12 units. The **relative atomic mass** A_r of an atom is a weighted average of the atomic masses of its isotopes and their relative abundances. The existence of different isotopes results in carbon having an A_r of 12.01. The **relative molecular mass** or **relative formula mass** M_r for a molecule or formula unit is determined by combining the A_r values of the individual atoms or ions. A_r and M_r have no units as they are both ratios.

The **molar mass** is defined as the mass of one mole of a substance. It has the unit of grams per mole, g mol^{-1} (figure 2).

Isotope	Relative abundance	Atomic mass
^{35}Cl	75%	35.0
^{37}Cl	25%	37.0
Relative atomic mass A_r		35.5

▲ Table 4 The relative atomic mass of chlorine is the weighted average of the atomic masses of its isotopes and their relative abundance

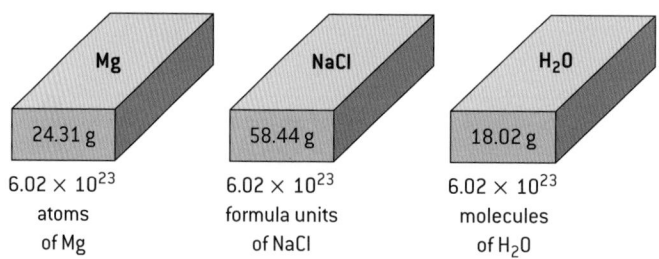

▲ Figure 2 The molar mass of a substance contains Avogadro's number of representative particles (the particles may be atoms, molecules, or ions)

TOK

Scientific discoveries are the product of many different ways of knowing (WOK). To construct knowledge and understanding, scientists can use intuition, imagination, reasoning, and even emotion, as well as detailed investigation and analysis of large volumes of data that either support or disprove observations and hypotheses. Sometimes it can just be a matter of serendipity. The scale of Avogadro's constant (602 000 000 000 000 000 000 000) passes beyond the boundaries of our experience on Earth. The population of the planet is dwarfed by this number. How does this experience limit our ability to be intuitive?

13
Al
26.98

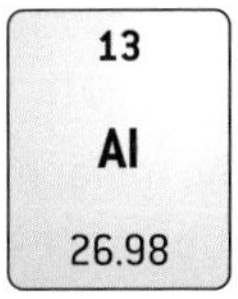

▲ Figure 3 The element aluminium as represented in the periodic table

 # Worked examples: A_r and M_r

Example 1

State the relative atomic mass A_r of aluminium.

Solution

Figure 3 shows the periodic table entry for aluminium.

$A_r(\text{Al}) = 26.98$

Example 2

Calculate the molar mass M_r of sulfuric acid, H_2SO_4.

Solution

Table 5 shows the data needed to answer this question.

Element	Relative atomic mass A_r	Number of atoms	Combined mass/g
hydrogen	1.01	2	2.02
sulfur	32.07	1	32.07
oxygen	16.00	4	64.00

▲ Table 5

$M_r(H_2SO_4) = (2 \times 1.01) + (1 \times 32.07) + (4 \times 16.00)$

$M_r(H_2SO_4) = 98.09 \text{ g mol}^{-1}$.

Example 3

Calculate M_r of copper(II) sulfate pentahydrate, $CuSO_4 \cdot 5H_2O$.

Solution

Many transition metal complexes (sub-topic 13.1) contain water molecules bonded to the central metal ion. The formula $CuSO_4 \cdot 5H_2O$ shows that 5 mol of water combines with 1 mol of copper(II) sulfate.

Element	Relative atomic mass A_r	Number of atoms	Combined mass/g
copper	63.55	1	63.55
sulfur	32.07	1	32.07
oxygen	16.00	4	64.00
oxygen	16.00	$5 \times 1 = 5$	80.00
hydrogen	1.01	$5 \times 2 = 10$	10.10

▲ Table 6 Calculating the molar mass of copper(II) sulfate pentahydrate

$M_r(CuSO_4 \cdot 5H_2O) = 249.72 \text{ g mol}^{-1}$.

Negative indices and units

An **index** or **power** is a mathematical notation that shows that a quantity or physical unit is repeatedly multiplied by itself:

$m \times m = m^2$

A **negative index** shows a reciprocal:

$\dfrac{1}{x} = x^{-1}$

$\text{dm}^{-3} = \dfrac{1}{\text{dm}^3}$

Concentration (molarity): units may be written as mol dm^{-3}, M, or mol L^{-1} (US).

Enthalpy of neutralization: units are kJ mol^{-1}.

Initial rate of reaction: units are mol dm^{-3} s^{-1}.

Study tips

- When adding and subtracting numbers, always express the final answer to the same number of decimal places as the least precise value used.

- When dividing or multiplying, always express the answer to the same number of significant figures as the least precise value used.

Quick question

Calculate the molar mass of the following substances and ions.

a) $Mg(NO_3)_2$

b) Na_2CO_3

c) $Fe_2(SO_4)_3$

d) S_8

e) $Zn(OH)_2$

f) $Ca(HCO_3)_2$

g) I_2

h) $MgSO_4 \cdot 7H_2O$

i) $[Al(H_2O)_6]^{3+}$

j) P_2O_5

Primary standards

A **primary standard** is any substance of very high purity and large molar mass, which when dissolved in a known volume of solvent creates a primary standard solution.

Primary standard solutions are used in acid–base titrations to improve the accuracy of the final calculation. The concentration of a primary standard can be determined accurately.

Mole calculations

All chemists, whether in the scientific community, manufacturing industries, or research facilities, work every day with reacting quantities of chemical substances and so need to perform stoichiometric calculations. The relationship between the amount (in mol), number of particles, and the mass of the sample is summarized in figure 4.

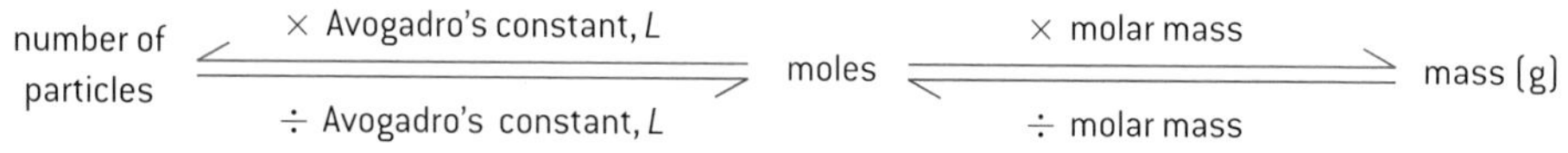

▲ Figure 4 The relationship between amount, mass, and number of particles

🌐 Worked examples: mole calculations

Example 1

Calculate the amount (in mol) of carbon dioxide, $n(CO_2)$ in a sample of 1.50×10^{23} molecules.

Solution

$$\text{amount (in mol) } n = \frac{\text{number of particles}}{\text{Avogadro's constant, } L}$$

Rearranging and substituting values:

$$n(CO_2) = \frac{1.50 \times 10^{23}}{6.02 \times 10^{23}\,\text{mol}^{-1}}$$

$$= 0.249 \text{ mol}$$

Example 2

Calculate the number of carbon atoms contained in 1.50 mol of glucose, $C_6H_{12}O_6$.

Solution

→ 1 molecule of glucose contains 6 atoms of carbon, 12 atoms of hydrogen, and 6 atoms of oxygen.

→ 1 mol of glucose contains 6 mol of C atoms.

→ 1.50 mol of glucose contains 9 mol of C atoms.

$$\text{number of atoms} = \text{amount (in mol) } n \times \text{Avogadro's constant, } L$$

$$= 9 \text{ mol} \times 6.02 \times 10^{23}\,\text{mol}^{-1}$$

$$= 5.42 \times 10^{24}\,\text{C atoms}$$

Study tip

The answer is recorded to 3 significant figures, as this is the precision of the data given by the examiner (1.50 mol).

Example 3

Calculate the amount (in mol) of water molecules in 3.01×10^{22} formula units of hydrated ethanedioic acid, $H_2C_2O_4 \cdot 2H_2O$.

Solution

→ For every 1 formula unit there are 2 molecules of water.

→ 1 mol of a substance contains Avogadro's number of particles.

Therefore,

$$\text{amount (in mol) } n = \frac{\text{number of particles}}{\text{Avogadro's constant, } L}$$

$$n(H_2C_2O_4 \cdot 2H_2O) = \frac{3.01 \times 10^{22}}{6.02 \times 10^{23}} = 0.0500 \text{ mol}$$

$$n(H_2O) = 2 \times 0.0500 \text{ mol} = 0.100 \text{ mol}$$

Units

Amount of substance n has the units mol

$$n = \frac{m}{\text{molar mass}}$$

Mass m has the units g; molar mass has the units $g\,mol^{-1}$.

Example 4

Calculate the amount (in mol) in 8.80 g of carbon dioxide, CO_2.

Solution

$$n(CO_2) = \frac{m}{\text{molar mass}}$$

$$= \frac{8.80 \text{ g}}{12.01 + 2(16.00) \text{ g mol}^{-1}}$$

$$= 0.200 \text{ mol}$$

Example 5

Calculate the mass in g of 0.0120 mol of sulfuric acid, H_2SO_4.

Solution

Calculate the molar mass of H_2SO_4 and substitute into the equation:

$$\text{mass (g)} = n(H_2SO_4) \times M_r(H_2SO_4)$$

$$= 0.0120 \text{ mol} \times [2(1.01) + 32.07 + 4(16.00)] \text{ g mol}^{-1}$$

$$= 1.18 \text{ g}$$

Example 6

Calculate the number of chlorine atoms in a 6.00 mg sample of the anti-cancer drug cisplatin, *cis*-diamminedichloroplatinum(II), $Pt(NH_3)_2Cl_2$.

Solution

→ First convert the mass in mg to g.

→ Next find the amount in mol by calculating the molar mass.

→ Finally remember that there are 2 mol of chlorine atoms in every mol of cisplatin.

$$6.00 \text{ mg} = 6.00 \times 10^{-3} \text{ g}$$

$$n[Pt(NH_3)_2Cl_2]$$

$$= \frac{6.00 \times 10^{-3}\,g}{195.08 + 2(14.01) + 6(1.01) + 2(35.45)}$$

$$= 2.00 \times 10^{-5} \text{ mol}$$

$$n(Cl) = 2 \times 2.00 \times 10^{-5} \text{ mol} = 4.00 \times 10^{-5} \text{ mol}$$

$$\text{number of atoms (Cl)} = 4.00 \times 10^{-5} \text{ mol} \times 6.02 \times 10^{23} \text{ mol}^{-1}$$

$$= 2.41 \times 10^{19}$$

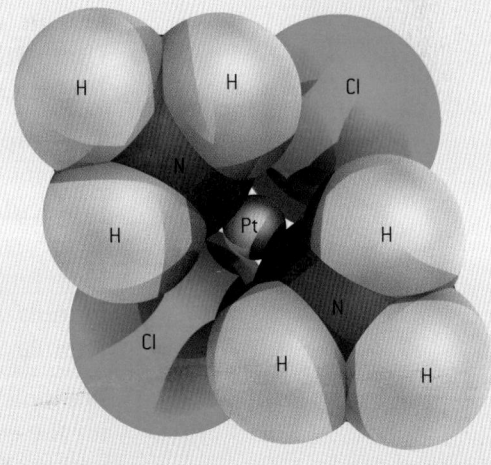

▲ Figure 5 The anti-cancer drug cisplatin

Quick questions

1 Calculate the amount (in mol) in each of the following masses:

   a) 8.09 g of aluminium

   b) 9.8 g of sulfuric acid

   c) 25.0 g of calcium carbonate

   d) 279.94 g of iron(III) sulfate.

> **2** Calculate the mass (in grams) in each of the following:
>
> **a)** 0.150 mol of nitrogen, N_2
>
> **b)** 1.20 mol of sulfur dioxide, SO_2
>
> **c)** 0.710 mol of calcium phosphate, $Ca_3(PO_4)_2$
>
> **d)** 0.600 mol of ethanoic acid, $C_2H_4O_2$.
>
> **3** Calculate the number of particles present in the following:
>
> **a)** 2.00 mol of vanadium, V
>
> **b)** 0.200 mol of sodium chlorate(VII), $NaClO_4$
>
> **c)** 72.99 g of iron(III) chloride, $FeCl_3$
>
> **d)** 4.60 g of nitrogen(IV) oxide.

Experimental empirical and molecular formula determination

The term "empirical" describes information that is derived through observation and/or investigation, using scientific methods. Chemical laboratories involved in medical research and development, manufacturing, or food production will often carry out analyses of the composition of a compound in processes that may be either qualitative or quantitative in nature.

Qualitative analysis focuses on determining which elements are present in a compound. It could also verify the purity of the substance. **Quantitative analysis** enables chemists to determine the relative masses of elements which allows them to work out their exact composition.

The **empirical formula** of a compound is the simplest whole-number ratio of atoms or amount (in mol) of each element present in a compound. The **molecular formula** is the actual number of atoms or amount (in mol) of elements in one structural unit or one mole of the compound, respectively. Therefore the molecular formula is a whole-number ratio of the empirical formula. Sometimes the empirical formula is the same as the molecular formula. Table 7 shows some examples.

For ionic compounds the empirical formula is the same as the formula for the compound, since the formula represents the simplest ratio of ions within the structure (figure 6).

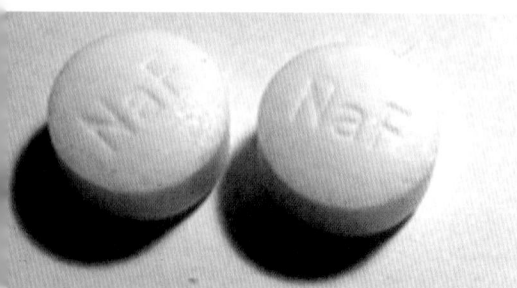

▲ Figure 6 Sodium fluoride, NaF has a 1:1 ratio of ions in its empirical formula. It is used in some countries to enhance the health of teeth

Substance	Molecular formula	Empirical formula
ethane	C_2H_6	CH_3
water	H_2O	H_2O
hydrogen peroxide	H_2O_2	HO
butanoic acid	$C_4H_8O_2$	C_2H_4O
glucose	$C_6H_{12}O_6$	CH_2O

▲ Table 7 Some examples of molecular and empirical formulae

Worked examples: percentage composition by mass

You can use your understanding of how to calculate the molar mass of a compound to calculate the percentage by mass of elements in a compound.

Example 1

Calculate the percentage by mass of sulfur in sulfuric acid, H_2SO_4.

Solution

$$\% \text{ sulfur} = \frac{A_r(S)}{M_r(H_2SO_4)} \times 100\%$$

$$= \frac{32.07}{2(1.01) \times (32.07) \times 4(16.00)} \times 100\%$$

$$= 32.69\%$$

If you have a compound of unknown formula but you know the percentage composition by mass of the elements present, you can calculate the empirical formula and, in some cases, the molecular formula.

Example 2

Determine the empirical formula of an organic compound that contains 75% carbon and 25% hydrogen by mass.

Solution

The first step is to determine the ratio of $n(C)$ to $n(H)$:

$$\text{relative amount of substance} = \frac{\% \text{ composition}}{\text{molar mass}}$$

$$n(C) = \frac{75}{12.01} = 6.24$$

$$n(H) = \frac{25}{1.01} = 24.75$$

Now take the smallest quotient (6.24). Use this as the divisor to determine the lowest whole-number ratio of the elements:

carbon $\quad \dfrac{6.24}{6.24} = 1$

hydrogen $\quad \dfrac{24.75}{6.24} = 3.97$

Because the percentage composition is experimentally determined it is acceptable to round to the nearest whole number if the number is close to a whole number. Therefore the simplest whole-number ratio of carbon to hydrogen is 1:4 and the empirical formula is CH_4.

Sometimes multiplication is needed to convert the ratio to whole numbers:

example 1   1:1.25   Multiply each side by 4:
$$4(1):4(1.25) \approx 4:5$$

example 2   1:1.33   Multiply each side by 3:
$$3(1):3(1.33) \approx 3:4$$

> **Study tip**
>
> Empirical formulae are based on experimental data; those for example 2 would likely have been determined by a combustion reaction. The value of 3.97 rather than 4 for hydrogen comes from experimental error.

Example 3

Upon analysis, a sample of an acid with a molar mass of 194.13 g mol⁻¹ was found to contain 0.25 g of hydrogen, 8.0 g of sulfur, and 16.0 g of oxygen. Determine the empirical formula and the molecular formula.

$$n(S) = \frac{8.0}{32.07} = 0.25 \qquad \frac{0.25}{0.25} = 1$$

$$n(O) = \frac{16.0}{16.00} = 1.0 \qquad \frac{1.0}{0.25} = 4$$

$$n(H) = \frac{0.25}{1.01} = 0.25 \qquad \frac{0.25}{0.25} = 1$$

Therefore the empirical formula is HSO_4.

To calculate the molecular formula, calculate the empirical formula mass and determine how many empirical formulae make up the molar mass.

$$\frac{\text{molar mass}}{\text{empirical formula mass}}$$

$$= \frac{194.13}{1.01 + 32.07 + 4(16.00)} = \frac{194.13}{97.08} = 2$$

The molecular formula of the acid is $2(HSO_4)$ or $H_2S_2O_8$. This compound is called peroxodisulfuric acid (figure 7).

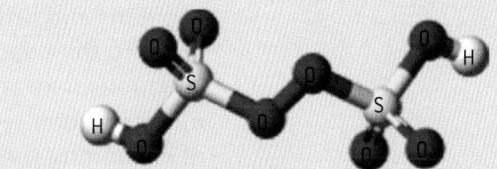

▲ Figure 7 Molecular model of peroxodisulfuric acid

1.3 Reacting masses and volumes

Understandings

→ Reactants can be either limiting or excess.

→ The experimental yield can be different from the theoretical yield.

→ Avogadro's law enables the mole ratio of reacting gases to be determined from volumes of the gases.

→ The molar volume of an ideal gas is a constant at specified temperature and pressure.

→ The molar concentration of a solution is determined by the amount of solute and the volume of solution.

→ A standard solution is one of known concentration.

Applications and skills

→ Solution of problems relating to reacting quantities, limiting and excess reactants, theoretical, experimental, and percentage yields.

→ Calculation of reacting volumes of gases using Avogadro's law.

→ Solution of problems and analysis of graphs involving the relationship between temperature, pressure, and volume for a fixed mass of an ideal gas.

→ Solution of problems relating to the ideal gas equation.

→ Explanation of the deviation of real gases from ideal behaviour at low temperature and high pressure.

→ Obtaining and using experimental values to calculate the molar mass of a gas from the ideal gas equation.

→ Solution of problems involving molar concentration, amount of solute, and volume of solution.

→ Use of the experimental method of titration to calculate the concentration of a solution by reference to a standard solution.

Nature of science

→ Making careful observations and obtaining evidence for scientific theories – Avogadro's initial hypothesis.

Stoichiometry

A balanced chemical equation provides information about what the reactants and products are, their chemical symbols, their state of matter, and also the relative amounts of reactants and products. Chemical equations may also include specific quantitative data on the enthalpy of the reaction (see topic 5). **Stoichiometry** is the quantitative method of examining the relative amounts of reactants and products. An understanding of this is vital in industrial processes where the efficiency of chemical reactions, particularly the **percentage yield**, is directly linked to the success and profitability of the organization.

From a balanced chemical equation the coefficients can be interpreted as the ratio of the amount, in mol, of reactants and products. This is the equation for the reaction used for the manufacture of ammonia in the Haber process (see topic 7):

$$N_2(g) + 3H_2(g) \rightleftharpoons 2NH_3(g) \qquad \Delta H = -92.22 \text{ kJ}$$

It shows that one molecule of nitrogen gas and three molecules of hydrogen gas combine in an exothermic reaction to produce two molecules of ammonia. However, when setting up a reaction the reactants may not always be mixed in this ratio – their amounts may vary from the exact stoichiometric amounts shown in the balanced chemical equation.

The limiting reagent

Experimental designers of industrial processes use the concept of a **limiting reagent** as a means of controlling the amount of products obtained. The limiting reagent, often the more expensive reactant, will be completely consumed during the reaction. The remaining reactants are present in amounts that exceed those required to react with the limiting reagent. They are said to be **in excess**.

It is the limiting reagent that determines the amount of products formed. Using measured, calculated amounts of the limiting reagent enables specific amounts of the products to be obtained. The assumption made here is that the experimental or actual yield of products achieved is identical to the theoretical or predicted yield of products. This is rarely the case. Much effort is focused on improving the yield of industrial processes, as this equates to increased profits and efficient use of raw materials.

 ## Worked example: determining the limiting reagent

In the manufacture of phosphoric acid, molten elemental phosphorus is oxidized and then hydrated according to the following chemical equation:

$$P_4(l) + 5O_2(g) + 6H_2O(l) \rightarrow 4H_3PO_4(aq)$$

If 24.77 g of phosphorus reacts with 100.0 g of oxygen and excess water, determine the limiting reagent, the amount in mol of phosphoric(V) acid produced (the **theoretical yield**) and the mass, in g, of phosphoric acid.

Solution

The amount in mol of phosphorus and oxygen is determined using the working method from sub-topic 1.2:

$$n(P_4) = \frac{m}{M}$$

$$= \frac{24.77 \text{ g}}{4(30.97) \text{ g mol}^{-1}} = 0.2000 \text{ mol}$$

$$n(O_2) = \frac{m}{M}$$

$$= \frac{100.0 \text{ g}}{2(16.00) \text{ g mol}^{-1}} = 3.125 \text{ mol}$$

$$P_4(l) + 5O_2(g) + 6H_2O(l) \rightarrow 4H_3PO_4(aq)$$

	$P_4(l) + 5O_2(g) + 6H_2O(l) \rightarrow 4H_3PO_4(aq)$			
M(g mol⁻¹)	123.88	32.00		
m/g	24.77	100.0	excess	
n_i/mol	0.200	3.125	excess	0
n_f/mol				

To determine the amount of oxygen that will react with the phosphorus we can use a cross-multiplication technique:

$$P_4 : O_2$$

$$1 : 5$$

$$0.200 : \alpha$$

$$1 \times \alpha = 0.2000 \times 5$$

$$\alpha = 0.2000 \times \frac{5}{1}$$

$$\alpha = 1.000 \text{ mol}$$

Therefore 0.2000 mol of phosphorus requires 1.000 mol of oxygen to completely react. There is 3.125 mol of oxygen available so this is in excess and phosphorus is the limiting reagent. All the phosphorus will be consumed in the reaction and $3.125 - 1.000 = 2.125$ mol of oxygen will remain after the reaction comes to completion.

The limiting reagent dictates the amount of phosphoric acid produced. The mole ratio is used to determine the amount of product, in mol. Four times the amount in mol of phosphoric acid will be produced compared with the amount of phosphorus:

$$P_4(s) + 5O_2(g) + 6H_2O(l) \rightarrow 4H_3PO_4(aq)$$

M(g mol^{-1})	123.88	32.00	
m/g	24.77	100.0	excess
n_i/mol	0.2000	3.125	excess	0
n_f/mol	0.0	2.125	excess	0.8000

The mass of phosphoric acid, H_3PO_4 produced can be determined by multiplying n_f by M_r:

$$m = M \times n$$

$$= [3(1.01) + 30.97 + 4(16.00)] \text{ g mol}^{-1}$$
$$\times 0.8000 \text{ mol} = 78.40 \text{ g}$$

This value represents the theoretical yield of phosphoric acid. Theoretical yields are rarely achieved in practice.

Quick questions

1 Butane lighters work by the release and combustion of pressurized butane:

$$2C_4H_{10}(g) + 13O_2(g) \rightarrow 8CO_2(g) + 10H_2O(l)$$

Determine the limiting reagent in the following reactions:

a) 20 molecules of C_4H_{10} and 100 molecules of O_2

b) 10 molecules of C_4H_{10} and 91 molecules of O_2

c) 0.20 mol of C_4H_{10} and 2.6 mol of O_2

d) 8.72 g of C_4H_{10} and 28.8 g of O_2

2 Two aqueous solutions, one containing 5.3 g of sodium carbonate and the other 7.0 g of calcium chloride, are mixed together. A precipitation reaction occurs:

$$Na_2CO_3(aq) + CaCl_2(aq) \rightarrow 2NaCl(aq) + CaCO_3(s)$$

Determine the limiting reagent and the mass, in g, of precipitate formed (the theoretical yield).

3 The oxygen required in a submarine can be produced by a chemical reaction. Potassium superoxide, KO_2 reacts with carbon dioxide, CO_2 to produce oxygen and potassium carbonate, K_2CO_3.

a) Write the balanced chemical equation for this reaction.

b) 28.44 g of KO_2 reacts with 22.00 g of CO_2. Deduce the limiting reagent.

c) Calculate the mass, in g, of K_2CO_3 produced.

d) Calculate the mass, in g, of O_2 produced.

4 A solution of 155 g of potassium iodide, KI is added to a solution of 175 g of nitric acid, HNO_3. The acid acts as an oxidizing agent.

$$6KI(aq) + 8HNO_3(aq) \rightarrow 6KNO_3(aq) + 2NO(g)$$
$$+ 3I_2(s) + 4H_2O(l)$$

a) Deduce which reagent is in excess.

b) Determine how many grams of this reactant will remain unreacted.

c) Determine how many grams of nitrogen monoxide, NO will be produced.

5 Chlorine gas is produced by the reaction of hydrochloric acid, and the oxidizing agent manganese(IV) oxide, MnO_2:

$$MnO_2(s) + 4HCl(aq) \rightarrow MnCl_2(aq) + Cl_2(g) + 2H_2O(l)$$

At 273.15 K and 100 kPa, 58.34 g of HCl reacts with 0.35 mol of MnO_2 to produce 7.056 dm^3 of chlorine gas.

a) Deduce the limiting reagent.

b) Calculate the theoretical yield of chlorine.

Theoretical and experimental yields

The balanced chemical equation represents what is theoretically possible when a reaction is carried out under ideal conditions. It allows the expected amount of products to be calculated – the **theoretical yield**.

Scientists in industry work to maximize the yield of reactions and maximize profits. However, under experimental conditions and especially in large-scale processes, many factors result in a reduced yield of products. These factors could include:

- loss of products from reaction vessels
- impurity of reactants

- changes in reaction conditions, such as temperature and pressure
- reverse reactions consuming products in equilibrium systems
- the existence of side-reactions due to the presence of impurities

To calculate the **percentage yield** a comparison is made between the theoretical yield and the actual amount produced in the process – the **experimental yield**:

$$\% \text{ yield} = \frac{\text{experimental yield}}{\text{theoretical yield}} \times 100\%$$

Worked example: determining theoretical yield

Respirators are being used increasingly with concern for workplace safety and rising levels of environmental pollution. Iodine(V) oxide, I_2O_5 reacts with carbon monoxide, CO and can be used to remove this poisonous gas from air:

$$I_2O_5(s) + 5CO(g) \rightarrow I_2(g) + 5CO_2(g)$$

100.0 g of I_2O_5 reacts with 33.6 g of CO. Calculate the theoretical yield of carbon dioxide and given an experimental yield, in mol, of 0.900 mol CO_2, calculate the percentage yield.

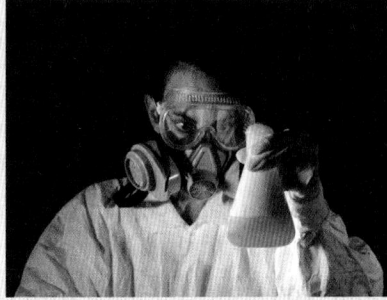

▲ Figure 1 A chemist wearing a respirator for safety

Solution

Step 1: Calculate the initial amount in mol of reactants and determine the limiting reagent:

$$n(I_2O_5) = \frac{m}{M}$$
$$= \frac{100.0 \text{ g}}{2(126.90) + 5(16.00) \text{ g mol}^{-1}}$$
$$= 0.2996 \text{ mol}$$

$$n(CO) = \frac{m}{M}$$
$$= \frac{33.6 \text{ g}}{12.01 + 16.00 \text{ g mol}^{-1}}$$
$$= 1.20 \text{ mol}$$

Step 2: Using mole ratios, determine the limiting reagent.

$$I_2O_5 : CO$$
$$1 : 5$$
$$0.3000 : \alpha$$

$$1 \times \alpha = 0.3000 \times 5$$
$$\alpha = 0.3000 \times \frac{5}{1}$$
$$\alpha = 1.500 \text{ mol}$$

The reaction of 0.3000 mol of I_2O_5 requires 1.50 mol of CO for completion. However, only 1.20 mol of CO is available; therefore this is the limiting reagent.

The ratio of limiting reagent CO to product CO_2 is 5:5 or 1:1. The number of mol of CO_2 that is theoretically possible is therefore 1.2 mol.

It was found that 0.90 mol or 39.61 g of CO_2 was produced. This is the **experimental yield**.

To determine the percentage yield of CO_2 we first need to calculate the theoretical yield of CO_2:

$$m = M \times n$$
$$= [12.01 + 2(16.00)] \text{ g mol}^{-1} \times 1.20 \text{ mol}$$
$$= 52.8 \text{ g}$$

Then:

$$\% \text{ yield} = \frac{\text{experimental yield}}{\text{theoretical yield}} \times 100\%$$
$$= \frac{39.61 \text{ g}}{52.8 \text{ g}} \times 100\% = 75.0\%$$

Worked examples: Avogadro's law

Example 1

Calculate $n(O_2)$ found in a 6.73 dm^3 sample of oxygen gas at STP.

> 1 mol O_2 occupies 22.7 dm^3 at STP

Solution

$$n(O_2) = \frac{6.73 \text{ dm}^3}{22.7 \text{ dm}^3} = 0.296 \text{ mol}$$

Example 2

The hydrogenation of ethyne, C_2H_2 involves reaction with hydrogen gas, H_2 in the presence of a finely divided nickel catalyst at 150 °C. The product is ethane, C_2H_6:

$$C_2H_2(g) + 2H_2(g) \rightarrow C_2H_6(g)$$

When 100 cm^3 of C_2H_2 reacts with 250 cm^3 of H_2, determine the volume and composition of gases in the reaction vessel.

Solution

According to Avogadro's law, for every 1 molecule of ethyne and 2 molecules of hydrogen, 1 molecule of ethane will be formed. Looking at the volumes reveals that only 200 cm^3 of the hydrogen is required, and that 100 cm^3 of ethane will be formed. The final mixture of gases contains both ethane and unreacted hydrogen:

$$C_2H_2(g) + 2H_2(g) \rightarrow C_2H_6(g)$$

initial volume, V_i/cm^3	100	250	0
final volume, V_f/cm^3	0	50	100

After reaction there will be 150 cm^3 of gases in the vessel comprising 50 cm^3 of H_2 and 100 cm^3 of C_2H_6.

The gas laws

The gas laws are a series of relationships that predict the behaviour of a fixed mass of gas in changing conditions of temperature, pressure, and volume.

You have seen that Avogadro's law states that the molar volume (22.7 dm^3 at STP) is independent of the composition of the gas.

Boyle's law

Robert Boyle (1627–1691) discovered that when the temperature remains constant, an inverse relationship exists between pressure and volume. Gases contained in smaller volumes will have an increased number of collisions with the surface of the container, so exert a higher pressure. The relationship between pressure p and volume V can be expressed as:

$$p \propto \frac{1}{V} \qquad \text{or} \qquad V_1p_1 = V_2p_2$$

where V_1 and p_1 represent the initial volume and pressure and V_2 and p_2 the final volume and pressure, respectively.

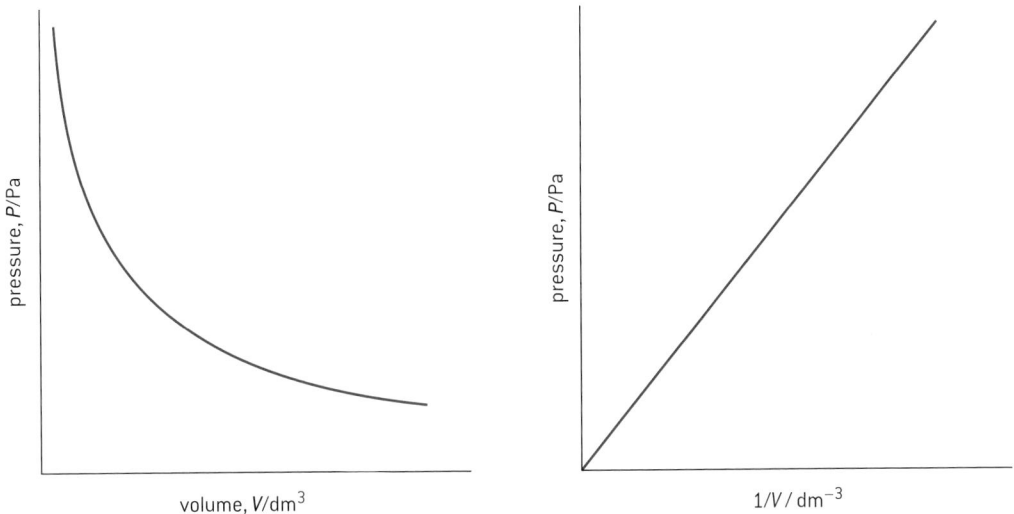

▲ Figure 5 Boyle's law: the pressure of a gas is inversely proportional to the volume at constant temperature

 Worked example: Boyle's law

A helium-filled weather balloon is designed to rise to altitudes as high as 37 000 m. A balloon with a volume of 5.50 dm³ and a pressure of 101 kPa is released and rises to an altitude of 3500 m where the atmospheric pressure is 68 kPa. Calculate the new volume, in dm³. It is assumed that the temperature and amount, in mol, remain constant.

Solution

First make a summary of the data:

$p_1 = 101$ kPa

$V_1 = 5.50$ dm³

$p_2 = 68$ kPa

$V_2 = \alpha$ dm³

Making V_2 the subject of the expression:

$$V_2 = V_1 \times \frac{p_1}{p_2}$$

$$= 5.50 \text{ dm}^3 \times \frac{101 \text{ kPa}}{68 \text{ kPa}}$$

$$= 8.17 \text{ dm}^3$$

Charles's law

Jacques Charles (1746–1823) investigated the relationship between the temperature and volume of a gas. He discovered that for a fixed mass of gas at a constant pressure, the volume V of the gas is directly

We saw in sub-topic 1.1 that absolute zero is zero on the kelvin scale, 0 K (−273.15 °C). The idea of negative temperatures and the existence of a minimum possible temperature had been widely investigated by the scientific community before Lord Kelvin's time (1824–1907). Kelvin stated that absolute zero is the temperature at which molecular motion ceases. According to Charles's law, if the temperature of a system was to double from 10 K to 20 K, the average kinetic energy of the particles would double and the volume would correspondingly double.

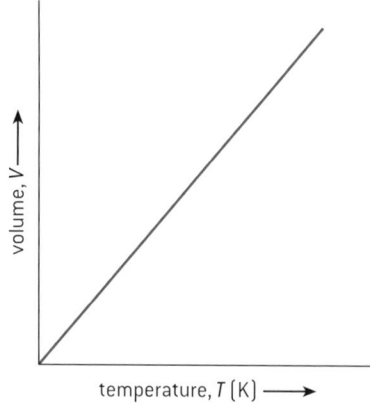

▲ Figure 7 Charles's law: the volume of a gas is directly proportional to absolute temperature at constant pressure

proportional to the absolute temperature T in kelvin. This relationship can be expressed as:

$$V \propto T \quad \text{or} \quad \frac{V_1}{T_1} = \frac{V_2}{T_2}$$

When an inflated balloon is placed into a container of liquid nitrogen (boiling point −196 °C), the average kinetic energy of the particles decreases. The gaseous particles collide with the internal wall of the balloon with less frequency and energy and it begins to deflate – the volume reduces. If the balloon is then removed from the liquid nitrogen and allowed to return to room temperature the balloon will reinflate.

▲ Figure 6 Reducing the temperature reduces the average kinetic energy of the particles of a gas, and the volume reduces

🌐 Worked example: Charles's law

A glass gas syringe contains 76.4 cm³ of a gas at 27.0 °C. After running ice-cold water over the outside of the gas syringe, the temperature of the gas reduces to 18.0 °C. Calculate the new volume, in cm³, occupied by the gas.

Solution

$$V_1 = 76.4 \text{ cm}^3$$

$$T_1 = 27.0 + 273.15 = 300.15 \text{ K}$$

$$V_2 = \alpha \text{ cm}^3$$

$$T_2 = 18.0 + 273.15 = 291.15 \text{ K}$$

$$\frac{V_1}{T_1} = \frac{V_2}{T_2}$$

$$V_2 = \frac{V_1 \times T_2}{T_1}$$

$$= 74.1 \text{ cm}^3$$

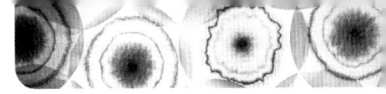

Gay-Lussac's law

Having established gas laws stating that pressure is inversely proportional to volume at constant temperature and that volume is directly proportional to temperature at constant pressure, the remaining relationship involves pressure and temperature, at constant volume.

Gay-Lussac's (1778–1850) work with ideal gases led him to the understanding that when the volume of a gas is constant, the pressure of the gas is directly proportional to its absolute temperature. The relationship can be expressed as:

$$p \propto T \quad \text{or} \quad \frac{p_1}{T_1} = \frac{p_2}{T_2}$$

Figure 8 demonstrates that when the temperature reaches absolute zero (0 K), the kinetic energy of the ideal gas particles is zero and it exerts no pressure. As the temperature increases, the particles collide with the walls of the container with increased force and frequency, causing increased pressure.

The combined gas law

The three gas laws, Charles's law, Boyle's law, and Gay-Lussac's law, are combined in one law called the **combined gas law**. For a fixed amount of gas, the relationship between temperature, pressure, and volume is:

$$\frac{p_1 V_1}{T_1} = \frac{p_2 V_2}{T_2}$$

The ideal gas equation

The **ideal gas equation** describes a relationship between pressure, volume, temperature, and the amount, in mol, of gas particles. Having established that pressure and volume are inversely proportional and that both pressure and volume have a direct relationship with the temperature of a gas and the amount of gas particles, the ideal gas equation combines these interrelationships:

$$pV = nRT$$

Collaboration

The scientific community is highly collaborative. Evidence that is fundamental to understanding is often challenged, tested, and utilized by other scientists to develop new understanding and investigate the possibility of developing new general laws.

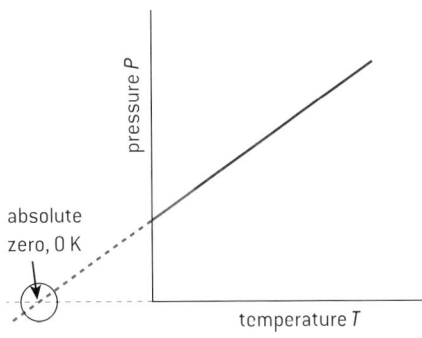

▲ Figure 8 Gay-Lussac's law: the pressure of a gas is directly proportional to absolute temperature at constant volume

The gas constant and the units of the ideal gas equation

R is called the **gas constant** and it has a value of 8.31 J K^{-1} mol^{-1}. This value is provided in section 2 of the *Data booklet*.

The inclusion of R in the ideal gas equation requires the following units: p (Pa), V (m^3), and T (K). Note that 1 Pa = 1 J m^{-3}; this allows you to see how the units in the ideal gas equation are balanced:

$$p(\text{J m}^{-3}) \times V(\text{m}^3) = n(\text{mol}) \times R(\text{J K}^{-1}\text{mol}^{-1}) \times T\,(\text{K})$$

1 dm^3 = 1 × 10^{-3} m^3.

13 Aspirin, one of the most widely used drugs in the world, can be prepared according to the equation given below.

salicylic acid   ethanoic anhydride   *aspirin*   ethanoic acid

A student reacted some salicylic acid with excess ethanoic anhydride. Impure solid aspirin was obtained by filtering the reaction mixture. Pure aspirin was obtained by recrystallization. Table 9 shows the data recorded by the student.

Mass of salicylic acid used	3.15 ± 0.02 g
Mass of pure aspirin obtained	2.50 ± 0.02 g

▲ Table 9

i) Determine the amount, in mol, of salicylic acid, $C_6H_4(OH)COOH$, used. [2]

ii) Calculate the theoretical yield, in g, of aspirin, $C_6H_4(OCOCH_3)COOH$. [2]

iii) Determine the percentage yield of pure aspirin. [1]

iv) State the number of significant figures associated with the mass of pure aspirin obtained, and calculate the percentage uncertainty associated with this mass. [2]

v) Another student repeated the experiment and obtained an experimental yield of 150%. The teacher checked the calculations and found no errors. Comment on the result. [1]

IB, May 2009

14 Brass is a copper-containing alloy with many uses. An analysis is carried out to determine the percentage of copper present in three identical samples of brass. The reactions involved in this analysis are shown below.

Step 1: $Cu(s) + 2HNO_3(aq) + 2H^+(aq) \rightarrow$
$\qquad\qquad Cu^{2+}(aq) + 2NO_2(g) + 2H_2O(l)$

Step 2: $4I^-(aq) + 2Cu^{2+}(aq) \rightarrow 2CuI(s) + I_2(aq)$

Step 3: $I_2(aq) + 2S_2O_3^{2-}(aq) \rightarrow 2I^-(aq) + S_4O_6^{2-}(aq)$

A student carried out this experiment three times, with three identical small brass nails, and obtained the following results.

Mass of brass = 0.456 g ± 0.001 g

Titre	1	2	3
Initial volume of 0.100 mol dm^{-3} $S_2O_3^{2-}$ (± 0.05 cm^3)	0.00	0.00	0.00
Final volume of 0.100 mol dm^{-3} $S_2O_3^{2-}$ (± 0.05 cm^3)	28.50	28.60	28.40
Volume added of 0.100 mol dm^{-3} $S_2O_3^{2-}$ (± 0.10 cm^3)	28.50	28.60	28.40
Average volume added of 0.100 mol dm^{-3} $S_2O_3^{2-}$ (± 0.10 cm^3)	28.50		

▲ Table 10

i) Calculate the average amount, in mol, of $S_2O_3^{2-}$ added in step 3. [2]

ii) Calculate the amount, in mol, of copper present in the brass. [1]

iii) Calculate the mass of copper in the brass. [1]

iv) Calculate the percentage by mass of copper in the brass. [1]

v) The manufacturers claim that the sample of brass contains 44.2% copper by mass. Determine the percentage error in the result. [1]

IB, May 2010

2 ATOMIC STRUCTURE

Introduction

Australian-born British physicist William Lawrence Bragg (1890–1971) shared the 1915 Nobel Prize in Physics with his father, Sir William Henry Bragg, for their analysis of crystal structures using X-rays, which led to the development of X-ray crystallography. William Bragg is the youngest Nobel laureate on record, having received the prize at the age of only 25. In a tape-recorded interview in 1969 Bragg said:

A wrong theory is always so much better than no theory at all.

William Lawrence Bragg

Chemistry is sometimes described as the "central science" and at the centre of chemistry lies atomic theory. Every chemical reaction can be explained in terms of atoms. In this topic we shall examine the various theories and models that have led to our current understanding of the structure of the atom.

2.1 The nuclear atom

Understandings

→ Atoms contain a positively charged dense nucleus composed of protons and neutrons (nucleons).

→ Negatively charged electrons occupy the space outside the nucleus.

→ The mass spectrometer is used to determine the relative atomic mass of an element from its isotopic composition.

Applications and skills

→ Use of the nuclear symbol notation $_Z^A X$ to deduce the number of protons, neutrons, and electrons in atoms and ions.

→ Calculations involving non-integer relative atomic masses and abundance of isotopes from given data, including mass spectra.

Nature of science

→ Evidence and improvements in instrumentation — alpha particles were used in the development of the nuclear model of the atom that was first proposed by Rutherford.

→ Paradigm shifts — the subatomic particle theory of matter represents a paradigm shift in science that occurred in the late 1800s.

Background to atomic theory

Two Greek philosophers, Leucippus and Democritus stated around 440 BC that matter was composed of indivisible particles termed *atomos*. However, no concrete scientific evidence was given to support this hypothesis and so it was not accepted to any great degree by the scientific community at the time.

Dalton's atomic theory

In 1808 the English schoolteacher John Dalton developed an atomic model of matter that was supported by experimental data. This model formed the origin of atomic theory that underpins much of modern science. We shall see in this topic how this model was progressively refined and replaced over time.

Dalton called the indivisible building blocks that comprise matter "atoms". Dalton's theory can be summarized as follows.

- *Postulate 1:* All matter (materials) consists of very small particles called **atoms**.

- *Postulate 2:* An **element** consists of atoms of one type only.

- *Postulate 3:* **Compounds** consist of atoms of more than one element and are formed by combining atoms in whole-number ratios.

- *Postulate 4:* In a chemical reaction atoms are not created or destroyed.

The simple "laws of chemical combination" were known to the scientific community in the 1800s and Dalton's theory explains a number of these.

- *The law of definite proportions*: This was proposed by a French scientist, Joseph Proust, in 1799. The law states that a compound always has the same proportion of elements by mass. For example, if you measure the mass of sulfur and oxygen in sulfur trioxide, SO_3 it will always contain 1 part sulfur and 3 parts oxygen by mass.

- *The law of conservation of matter*: Matter cannot be created or destroyed. The total mass of matter following a chemical reaction is equal to the total mass of matter before the start of the reaction.

For a scientific theory to be accepted it should not only provide an explanation of known observations but should be able to predict correctly the outcomes of future experiments.

Dalton used his theory to deduce another law:

- *The law of multiple proportions*: If two elements X and Y combine in different ways to form more than one compound, the masses of X that combine with a fixed mass of Y can be expressed as a ratio of small whole numbers.

Example: The law of multiple proportions

Chemists in Dalton's time did not use the mole as a measure of the amount of substance. As an example of the law of multiple proportions, consider measuring the mass of carbon and oxygen in forming the two compounds carbon monoxide, $CO(g)$, and carbon dioxide, $CO_2(g)$. An experiment might measure that 3 g of carbon combines with 4 g of oxygen to form carbon monoxide, whereas 3 g of carbon combines with 8 g of oxygen to form carbon dioxide. Carbon and oxygen have combined in *different ratios to give different compounds*:

$CO(g)$:	C:O ratio = 3:4
$CO_2(g)$:	C:O ratio = 3:8

The ratio of the masses of oxygen that combine with the same mass of carbon to form the two compounds is 1:2 (a simple ratio of whole numbers).

Study tip

In science, a **law** can be considered a summary of several observations.

TOK

John Dalton was a brilliant scientist. He never married and said: "My head is too full of triangles, chemical properties, and electrical experiments to think much of marriage!" He was an multidisciplinary scientist, who worked in the disciplines of physics, mathematics, biology, and philosophy, as well as chemistry. Do you think philosophy still has a place in modern scientific thinking? Debate this question in class and consider why scientists should always try to embrace an interdisciplinary approach in their thinking.

Dalton was colour blind and saw himself as being dressed in grey clothes. His only known pastime was bowling. Could the wooden balls on the bowling green have influenced his theories of the atom? How important is the work–life balance for the scientific practitioner or indeed for society as a whole?

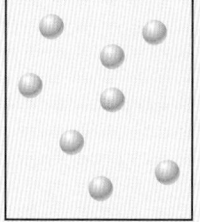 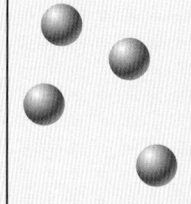 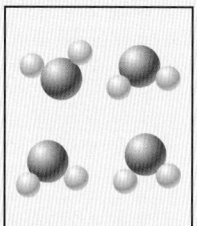

atoms of element X   atoms of element Y   compound consisting of elements X and Y

▲ Figure 1 Schematic showing some of the principles of Dalton's theory. Examine each of the four postulates and discuss each one in relation to the three representations shown here

Quick question

Deduce the ratio of the mass of oxygen per gram of sulfur in the compounds sulfur dioxide, $SO_2(g)$, and sulfur trioxide, $SO_3(g)$.

Thomson's "plum-pudding" model of the atom

Although Dalton's 1808 postulates had merit, his theory did not answer one fundamental question: "What is the atom composed of?"

It was almost another 100 years before scientists began to gather evidence to answer this question. One of the first leaders in the field was the English physicist J.J. Thomson (1856–1940), who worked at the Cavendish laboratory at the University of Cambridge, UK. In 1906 Thomson won the Nobel Prize in Physics for the discovery of the electron. Thomson worked on cathode rays, which he suggested consist of very small negatively charged particles called **electrons**.

> The term "electron" was originally proposed by the Irish scientist George Johnstone Stoney in 1891.

Thomson proposed what is now termed the **"plum-pudding"** model of the atom – he said that the atom was similar to a plum pudding (a dessert eaten on Christmas day in the UK and Ireland), with negatively charged particles (like raisins) embedded in a positive region (the "pudding") of the atom.

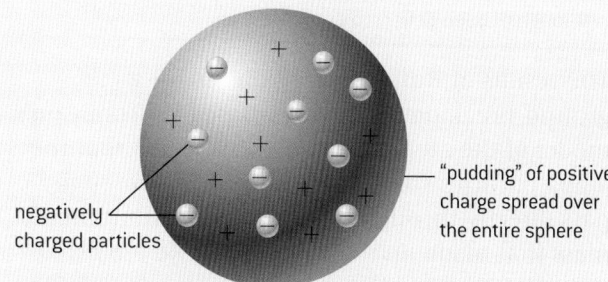

negatively charged particles

"pudding" of positive charge spread over the entire sphere

▲ Figure 2 Thomson's "plum-pudding" model of the atom. In the analogy, raisins represent negatively charged particles embedded in a pudding of positive charge. Overall there is a balance between the positive and negative charges since the atom is electrically neutral

Rutherford's gold foil experiment

Thomson's model raised a number of questions. Because matter is electrically neutral, the presence of negatively charged particles in atoms implies that they must also contain positively charged particles. The search for these particles and for a more detailed model of the atom led New Zealand physicist Ernest Rutherford (1871–1937) and co-workers to conduct the gold foil experiment in 1909. Published in 1911, this experiment tested Thomson's "plum-pudding" model by placing a thin gold metal foil in an evacuated chamber and bombarding it with alpha particles (figure 3). Alpha (α) particles are high-energy, positively charged He^{2+} ions emitted from naturally occurring radioactive elements such as radium.

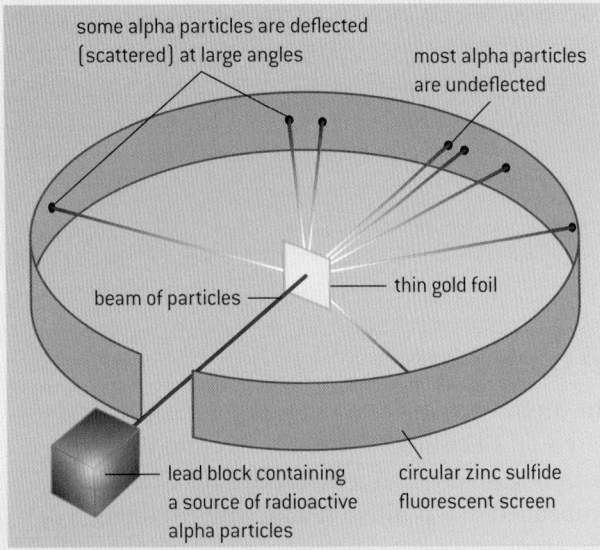

▲ Figure 3 Rutherford's experiment. The zinc sulfide fluorescent screen was used to detect alpha particles that had passed through or been deflected by the gold foil

The results of Rutherford's experiment were ground-breaking at the time. Based on Thomson's model of the atom, Rutherford expected that the alpha particles would have sufficient energy to pass directly through the uniform distribution of mass that made up the gold atoms. He predicted that the alpha particles would decelerate and that their direction on going through the gold foil would involve only a minor deflection. However, his results were astonishing.

Most of the alpha particles went through the gold foil and some were deflected slightly as expected. But some particles were deflected by very large angles and some even bounced straight back towards the source. These particles had collided head-on with what we now know to be the **nucleus** in the gold atom (figure 4). Rutherford described this result by commenting:

> It was as incredible, as if you had fired a 15 inch artillery shell at a piece of paper and it came straight back and hit you!

Rutherford based his explanation on the fact that the gold foil consists of thousands of gold atoms. When the beam of positively charged alpha particles bombarded the foil the majority of the particles passed through undeflected, since the atom consists mainly of empty space. However, at the core of the atom lies a dense region of positive charge called the nucleus. When an alpha particle came close to the nucleus of a gold atom it deflected through a large angle, and when it hit the nucleus it reflected back along its initial path.

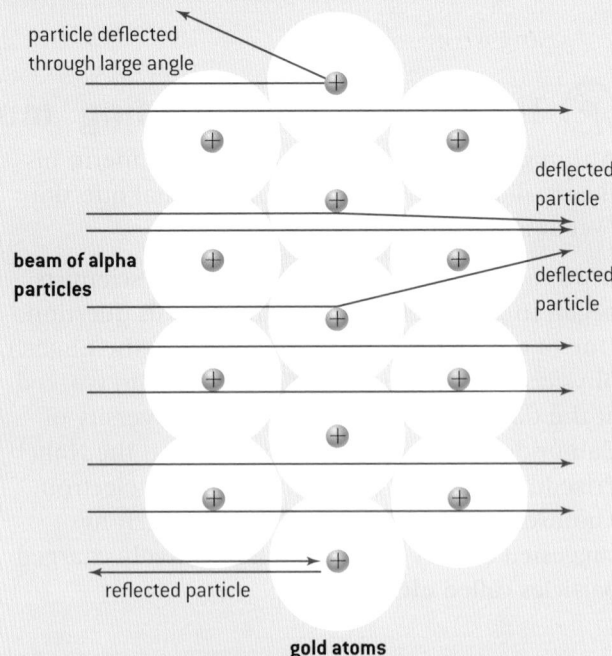

▲ Figure 4 Rutherford's model, which explains his findings in the gold foil experiment

The scale of the atom

Rutherford's work has formed the basis of much of our thinking on the structure of the atom. Rutherford is rumoured to have said to his students:

All science is either physics or stamp-collecting!

The vast space in the atom compared to the tiny size of the nucleus is hard to fully appreciate. Rutherford's native New Zealand is a great rugby-playing nation. Imagine being at Eden Park stadium (figure 6) and looking down at the centre of the pitch from the top row of seats on the upper tier of the stand. If a small grape were placed at the centre of the field, the distance between the grape and you would represent the distance between the electron and the nucleus.

The relative volume of open space in the atom is vast and our simple representation of Rutherford's atomic model in figure 4 is obviously unrealistic. The nucleus occupies a tiny volume of the atom and the diameter of an atom is approximately 100000 times the diameter of the nucleus. We shall return to the idea of the scale of the atom in sub-topic 2.2.

Atoms themselves are extremely small. The diameter of most atoms is in the range 1×10^{-10} to 5×10^{-10} m. The unit used to describe the dimensions of atoms is the **picometre**, **pm**:

$$1 \text{ pm} = 10^{-12} \text{ m}$$

In X-ray crystallography a commonly used unit for atomic dimensions is the **angstrom**, symbol **Å**:

$$1 \text{ Å} = 10^{-10} \text{ m}$$

For example, the atomic radius of the fluorine atom is quoted in section 9 of the *Data booklet* as 60×10^{-12} m (60 pm). To convert this to Å we can use **dimensional analysis**, using the conversion factors given above:

$$60 \text{ pm} \times \frac{10^{-12} \text{ m}}{1 \text{ pm}} \times \frac{1 \text{ Å}}{10^{-10} \text{ m}} = 0.60 \text{ Å} = 6.0 \times 10^{-1} \text{ Å}$$

Can we see atoms and are they real?

All the models we have discussed have assumed that atoms are real. However, for many people objects are only "real" when they can be seen. In 1981 two physicists, Gerd Binnig and Heinrich Rohrer, working at IBM in Zurich, Switzerland invented the **scanning tunnelling microscope (STM)**, an electron microscope that generates three-dimensional images of surfaces at the atomic level. This allowed scientists the ability to observe individual atoms directly. The Nobel Prize in Physics in 1986 was awarded to Binnig and Rohrer for their ground-breaking work.

The nucleus in science

The word "nucleus" means the central and most important part of an object. The word is used in both chemistry (the nucleus of an atom) and biology (the nucleus of a cell).

▲ Figure 5 The 100 New Zealand dollar note, issued in 1999, shows a picture of Lord Rutherford, reflecting his immense contribution to science. Do any bank notes in your own country have pictures of famous scientists?

▲ Figure 6 Eden Park, Auckland, New Zealand

TOK

The American theoretical physicist Richard Feynman (1918–1988) said:

> If … all of scientific knowledge were to be destroyed, and only one sentence passed on to the next generation… I believe it is that all things are made of atoms.

Are the models and theories that scientists create accurate descriptions of the natural world, or are they primarily useful interpretations for the prediction, explanation, and control of the natural world?

No subatomic particles can be directly observed. Which ways of knowing do we use to interpret indirect evidence, gained through the use of technology?

Subatomic particles and descriptions of the atom

After Rutherford's experiment in 1909 a number of experiments followed in the period to approximately 1935, culminating in scientists having a much more detailed picture of the structure of the atom.

Atoms consist of three types of subatomic particle:

- the proton
- the neutron
- the electron.

Section 4 of the *Data booklet* gives the mass, in kg, and the charge in coulombs, C, of each of these subatomic particles. The masses given are very small and the **atomic mass unit**, amu, is a convenient unit for these masses (table 1).

$$1 \text{ amu} = 1.660539 \times 10^{-24} \text{ g}$$

Subatomic particle	Charge	Mass/amu	Location
proton	+1	~ 1	nucleus
neutron	0	~ 1	nucleus
electron	−1	$\dfrac{1}{1836}$	outside the nucleus in the electron cloud

▲ Table 1 A comparison of the subatomic particles

The **neutron** was discovered by British physicist James Chadwick in 1932 (figure 7).

Chadwick's discovery of the neutron was based on an experiment in which beryllium, Be, placed in a vacuum chamber was bombarded with alpha particles, He^{2+}, emitted from polonium. The beryllium was found to emit neutrons and based on Chadwick's mass calculations he was able to prove categorically that the particles were in fact neutrons and not gamma rays as had been previously thought:

$$_2^4\alpha + {}_4^9\text{Be} \rightarrow {}_6^{12}\text{C} + {}_0^1n$$

▲ Figure 7 British physicist Sir James Chadwick (1891–1974), who was awarded the Nobel Prize in Physics in 1935 for discovering the neutron

The discovery of the neutron was at the time the last piece of the jigsaw puzzle of atomic structure. Rutherford always postulated the existence of the neutron but had no conclusive evidence until Chadwick's discovery.

The atomic number, Z

The atoms of each element have an individual **atomic number**, Z:

- The **atomic number** is the number of protons in the nucleus of an atom of an element. Different elements have different atomic numbers.

For a neutral atom the number of electrons equals the number of protons, for example:

- Z for oxygen, O, is 8. Therefore the oxygen atom has 8 protons and 8 electrons.
- Z for copper, Cu, is 29. Copper atoms have 29 protons and 29 electrons.

The mass number, A

The mass of the atom is concentrated in the nucleus, which contains both protons and neutrons.

- The **mass number**, A, is the number of protons ⏐ the number of neutrons in the nucleus of an atom.

For example:

- Z for fluorine, F, is 9. Therefore fluorine has 9 protons and 9 electrons.
- A for fluorine-19 is 19. Therefore fluorine-19 has $19 - 9 = 10$ neutrons.

 ## The nuclear symbol

The **nuclear symbol** includes both A and Z for a particular element X and is represented like this:

$$_Z^A X$$

Isotopes

As you saw in sub-topic 1.2, **isotopes** are different forms of the same element that have the same atomic number, Z, but different mass numbers, A, because they have different numbers of neutrons in their nuclei.

For example, hydrogen has three isotopes:

$_1^3 H$ (tritium)

1 proton, 1 electron, 2 neutrons

$_1^2 H$ (deuterium)

1 proton, 1 electron, 1 neutron

$_1^1 H$ (hydrogen)

1 proton, 1 electron, 0 neutrons

In nature most elements exist as mixtures of isotopes. For example, boron contains the two naturally occurring isotopes boron-10 (natural abundance 19.9%) and boron-11 (natural abundance 80.1%).

Isotope enrichment: Nuclear energy and nuclear weapons

Uranium found in nature consists of three isotopes with the relative abundances and atomic compositions shown in table 2.

Isotope	Relative abundance	Number of protons	Number of electrons	Number of neutrons
^{234}U	0.0055%	92 protons	92 electrons	142 neutrons
^{235}U	0.7200%	92 protons	92 electrons	143 neutrons
^{238}U	99.2745%	92 protons	92 electrons	146 neutrons

▲ Table 2 Isotopes of uranium

Uranium-235 is used in nuclear reactors where it undergoes **fission** (splitting) with the release of a large amount of energy. Natural uranium has a much higher abundance of U-238 than U-235 so uranium ore may be **enriched** to increase the proportion of U-235. The separation of natural uranium into enriched uranium and depleted uranium is the physical process of **isotope separation**.

Because they are the same element (same Z) isotopes have the same chemical properties but they show different physical properties due to their different mass numbers, A.

The difference in mass between U-235 and U-238 can be used to enrich a fuel with U-235. In some nuclear reactors natural uranium is used as the fuel but uranium used for nuclear weapons needs to be of higher grade and is usually enriched.

Activity

1 In class, discuss the pros and cons of nuclear energy and debate the issue of countries developing nuclear weapons programmes.

2 **a)** Deduce the number of protons, electrons, and neutrons in the isotopes $^{37}_{17}Cl$ and $^{35}_{17}Cl$.

   b) Deduce the number of protons, electrons, and neutrons in the ion, $^{37}_{17}Cl^-$.

Radioisotopes

As well as boron-10 and boron-11, boron also has a number of **radioisotopes** (radioactive isotopes). Examples are boron-8, boron-9, boron-12, and boron-13. Radioisotopes are used in nuclear medicine for diagnostics, treatment, and research, as tracers in biochemical and pharmaceutical research, and as "chemical clocks" in geological and archaeological dating.

Iodine radioisotopes as medical tracers

The thyroid gland in the neck releases thyroxine and triiodothyronine into the bloodstream. These **hormones** or chemical messengers control the body's growth and metabolism. An overactive thyroid gland produces an excess of these two hormones and this accelerates the metabolism of the body leading to symptoms such as high levels of anxiety, goitre (swelling of the thyroid gland) and weight loss.

Iodine is concentrated in the thyroid gland. The radioisotope iodine-131 emits gamma (γ) rays which are high-energy (short-wavelength) photons. Iodine-131 is used in the treatment of thyroid cancer and also in diagnostics, to determine whether the thyroid gland is functioning normally. In hospital, a patient is given radioactive iodine-131 and an image of the thyroid gland can be obtained, for example using a gamma camera. In contrast, iodine-125 is used to treat prostate cancer and brain tumours.

Positron emission tomography (**PET**) scanners give three-dimensional images of tracer concentration in the body, and can be used to detect cancers (see sub-topic D.8). **Single-photon emission computed tomography** (**SPECT**) **imaging** can be used to detect the gamma rays emitted from iodine-131.

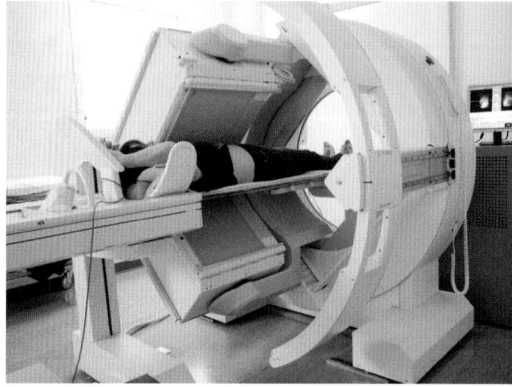

▲ Figure 8 A single-photon emission computed tomography scanner can be used to detect the gamma rays from iodine-131

Cobalt-60 in radiotherapy

Cobalt-60 also emits gamma rays and is used to treat cancer.

Carbon-14 in cosmic, geological, and archaeological dating

Radioisotopes are often used as radioactive clocks for the dating of cosmic, geological, and archaeological matter. The American scientist Professor Willard Libby won the Nobel Prize in Chemistry in 1960 for his method that uses carbon-14 for age determination in archaeology, geology, geophysics, and other branches of science.

Nitrogen is present in the Earth's atmosphere as the isotope nitrogen-14. The atmosphere is constantly bombarded by highly penetrating cosmic rays from outer space and this neutron bombardment causes radioactive carbon-14 to form, along with hydrogen, according to the nuclear equation:

$$^{14}_{7}N + ^{1}_{0}n \rightarrow ^{14}_{6}C + ^{1}_{1}H$$

This neutron bombardment results in a constant supply of carbon-14 in the atmosphere, as it is continuously formed from nitrogen-14. Nitrogen gas consists of 78% of the Earth's air by volume.

The **half-life**, $t_{1/2}$ is the time it takes for an amount of radioactive isotope to decrease to one-half of its initial value. The half-life for the carbon-14 decay process is 5730 years.

Carbon-14 can be oxidized to form carbon dioxide. Living plants absorb carbon dioxide for photosynthesis and assimilate the carbon into other compounds in their bodies. Animals consume plants, taking in their carbon compounds, and they exhale carbon dioxide. In all living organisms the ratio between carbon-12 and carbon-14 found in the atmosphere is essentially constant at any given time, since carbon is continually exchanged with the atmosphere in the processes of life. When a living organism dies however, its carbon is no longer exchanged with the atmosphere or with other organisms.

The carbon-14 isotope may then undergo decay to form nitrogen, emitting beta particles (electrons) in the process:

$$^{14}_{6}C \rightarrow ^{14}_{7}N + ^{0}_{-1}e^{-}$$

The net result is that there is a gradual decrease in the ratio of carbon-14 to carbon-12 in the organism's body. The amount of carbon-14 in the body of a plant or animal that was once living can be measured. Scientists can use this method to determine the age of artefacts such as wood, paintings, papyrus, ancient manuscripts, and scrolls.

The Shroud of Turin

The Shroud of Turin is a linen cloth believed by many people to be the one used to wrap the body of Jesus Christ after his death. The cloth shows the image of a person who appears physically traumatized and many believe that it represents the crucifixion of Jesus. In 1988 the Vatican in Rome commissioned three independent analytical laboratories based at the University of Oxford, UK, the Swiss Federal Institute of Technology, and the University of Arizona, USA, to carry out carbon-14 dating on the Shroud. All three results confirmed that the samples taken from the cloth originated between 1260 and 1390 AD, suggesting that the Shroud was not the burial cloth of Jesus. Nevertheless, the controversy and debate about the Shroud continues amongst scientists, theologians, and historians to this day.

In July 2013 Giulio Fanti and co-workers from the University of Padua, Italy, published research in the journal *Vibrational Spectroscopy* which shows a two-way relationship between age and a spectral property of ancient flax textiles. The media reported their findings worldwide, claiming that the results dated the Shroud of Turin between 300 BC and 400 AD, which could date from the time of Christ.

▲ Figure 9 The Shroud of Turin

Useful resource

An app (application) has been developed by the Diocese of Turin in Italy and the International Centre of Sindonology (scientific study of the Shroud). The app is named *Shroud 2.0* and using this you can explore the various images, scientific and theological interpretations.

Activity

In class, consider and debate the aspects of hypothesis, theory, technology, and analytical evidence surrounding the Shroud of Turin.

Relative atomic mass

The mass of the electron is negligible $\left(\frac{1}{1836}\,\text{amu}\right)$. The mass of the atom is concentrated in the nucleus in the protons and neutrons. However, the mass of a single atom is tiny, as seen in table 1 of this sub-topic and section 4 of the *Data booklet*, and it is more convenient to use a system of *relative* atomic masses. The **atomic mass unit** (more correctly termed the **unified atomic mass unit** according to IUPAC) and relative atomic mass are defined as follows:

- The **unified atomic mass unit** is a non-SI unit of mass and is defined as one-twelfth of the mass of a carbon-12 atom in its ground-state. This unit is used to express masses of atomic particles: 1 amu or 1 u = $1.6605402 \times 10^{-27}$ kg.

- The **relative atomic mass**, A_r, is the ratio of the average mass of the atom to the unified atomic mass unit.

As mentioned in sub-topic 1.2, the average mass of the atom is a weighted average of the atomic masses of its isotopes and their relative abundances.

The mass spectrometer

The **mass spectrometer** is an instrument used to determine the relative atomic mass of an element. It can also show its isotopic composition.

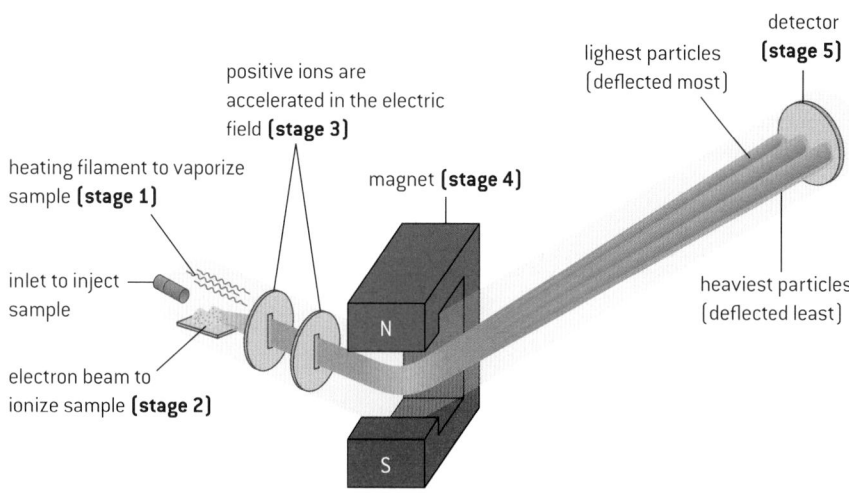

▲ Figure 10 Schematic diagram of a mass spectrometer

There are five stages in this process:

- **Stage 1** (vaporization): The sample is injected into the instrument where it is heated and vaporized, producing gaseous atoms or molecules.

- **Stage 2** (ionization): The gaseous atoms are bombarded by high-energy electrons, generating positively charged species:

$$X(g) + e^- \rightarrow X^+(g) + 2e^-$$

- **Stage 3** (acceleration): The positive ions are attracted to negatively charged plates and accelerated in the electric field.

- **Stage 4** (deflection): The positive ions are deflected by a magnetic field perpendicular to their path. The degree of deflection depends on the mass-to-charge ratio (the m/z ratio). The species with the smallest mass, m, and the highest charge, z, will be deflected the most. Particles with no charge are not deflected in the magnetic field.

- **Stage 5** (detection): The detector detects species of a particular m/z ratio. The ions hit the counter and an electrical signal is generated.

The instrument can be adjusted so that only positive ions of a single charge are detected. The deflection will then depend only on the mass.

The mass spectrum is therefore a plot of relative abundance (of each isotope) versus m/z or the mass number, A. The height of each peak indicates the relative abundance of the respective isotope.

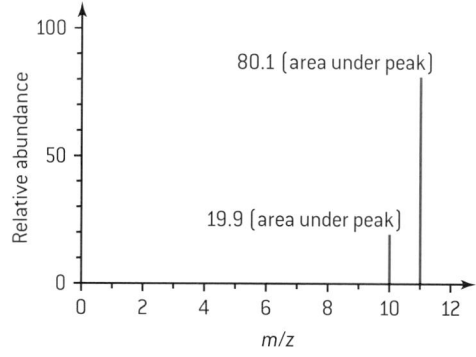

▲ Figure 11 Mass spectrum of boron. The two peaks correspond to two isotopes

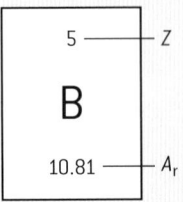
 Worked examples: calculations involving non-integer relative atomic masses and abundances of isotopes

Example 1

Boron has two naturally occurring isotopes with the natural abundances shown in table 3.

Isotope	Natural abundance/%
^{10}B	19.9
^{11}B	80.1

▲ Table 3 Isotopes of boron

Calculate the relative atomic mass of boron.

Solution

The relative atomic mass is the weighted average of the atomic masses of the isotopes and their relative abundance:

$$\text{relative atomic mass} = \left(10 \times \frac{19.9}{100}\right) + \left(11 \times \frac{80.1}{100}\right) = 10.8$$

Example 2

Rubidium has a relative atomic mass of 85.47 and consists of two naturally occurring isotopes, ^{85}Rb (u = 84.91) and ^{87}Rb (u = 86.91). Calculate the percentage composition of these isotopes in a naturally occurring sample of rubidium.

Solution

Note that in this example exact u values are given correct to two decimal places so you need to use this information in your answer. In Example 1 no such precise information was given.

Take a sample of 100 atoms. Let x = number of ^{85}Rb atoms and $(100 - x)$ = number of ^{87}Rb atoms in the sample.

$$A_r = 85.47 = \frac{84.91x + 86.91(100 - x)}{100}$$

cross-multiplying:

$$84.91x + 86.91(100 - x) = 8547$$

$$84.91x + 8691 - 86.91x = 8547$$

solve by making x the subject of the expression:

$$-2.00x = -144$$

$$x = 72.00$$

The sample contains 72.00% ^{85}Rb and 28.00% ^{87}Rb.

Example 3

Deduce the relative atomic mass of the element X from its mass spectrum in figure 13 and identify X from the periodic table.

Solution

- The mass spectrum shows two isotopes, X-69 and X-71.

- In theory the area under each peak is proportional to the number of atoms of each isotope. In calculations the peak height can be taken as an approximation of the relative numbers of atoms. The peak heights are X-69 = 27 units and X-71 = 41 units.

- The naturally occurring isotopes must sum to 100%.

- The total height of both peaks is 68 units. To deduce the relative atomic mass of X we need to determine the relative abundance of each isotope:

$$\text{X-69:} \quad \left(\frac{27}{68}\right) \times 100 = 40\%$$

$$\text{X-71:} \quad \left(\frac{41}{68}\right) \times 100 = 60\%$$

- The relative atomic mass of X can now be determined using the procedure from worked example 1:

$$\text{relative atomic mass} = \left(69 \times \frac{40}{100}\right) + \left(71 \times \frac{60}{100}\right)$$
$$= 70.2 \text{ (or 70 correct to 2 SF)}$$

- From the periodic table in section 6 of the *Data booklet*, X must be Ga ($Z = 31$), which is quoted as having $A_r = 69.74$. The value of 70.2 from this calculation is closest to this value. In this calculation if you use peak heights instead of peak areas, the precision of the calculations will be 2 SF at best, so this is the reason why all figures were expressed to 2 SF.

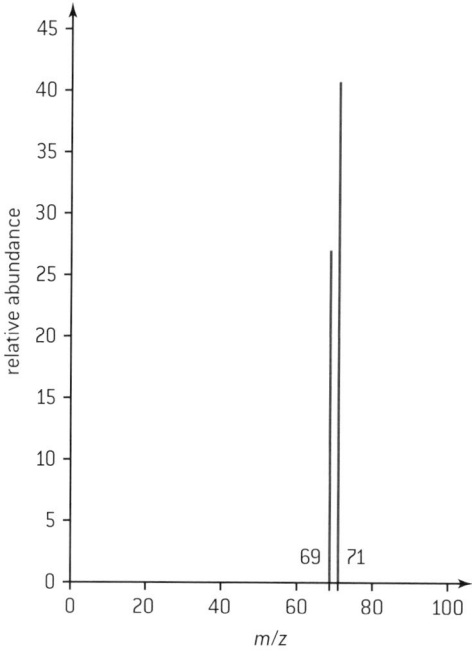

▲ Figure 13 Mass spectrum of X showing the relative abundances of its naturally occurring isotopes

2.2 Electron configuration

Understandings

→ Emission spectra are produced when photons are emitted from atoms as excited electrons return to a lower energy level.

→ The line emission spectrum of hydrogen provides evidence for the existence of electrons in discrete energy levels, which converge at higher energies.

→ The main energy level or shell is given an integer number, n, and can hold a maximum number of electrons, $2n^2$.

→ A more detailed model of the atom describes the division of the main energy level into s, p, d, and f sublevels of successively higher energies.

→ Sublevels contain a fixed number of orbitals, regions of space where there is a high probability of finding an electron.

→ Each orbital has a defined energy state for a given electron configuration and chemical environment and can hold two electrons of opposite spin.

Applications and skills

→ Description of the relationship between colour, wavelength, frequency, and energy across the electromagnetic spectrum.

→ Distinction between a continuous spectrum and a line spectrum.

→ Description of the emission spectrum of the hydrogen atom, including the relationships between the lines and energy transitions to the first, second, and third energy levels.

→ Recognition of the shape of an s atomic orbital and the p_x, p_y, and p_z atomic orbitals.

→ Application of the Aufbau principle, Hund's rule, and the Pauli exclusion principle to write electron configurations for atoms and ions up to $Z = 36$.

Nature of science

→ Developments in scientific research follow improvements in apparatus – the use of electricity and magnetism in Thomson's cathode rays.

→ Theories being superseded – quantum mechanics is among the most current models of the atom.

→ Use theories to explain natural phenomena – line spectra explained by the Bohr model of the atom.

The electromagnetic spectrum

What visions in the dark of light!

Samuel Beckett (1906–1989), Irish novelist, poet, and playwright who won the Nobel Prize in Literature in 1969

The developments that have led to much of our understanding of the electronic structure of the atom have come from experiments involving light. Visible light, the light we see, is full of scientific intrigue.

Visible light is one type of **electromagnetic radiation**. Other examples include radio waves, microwaves, infrared radiation (IR), ultraviolet radiation (UV), X-rays, and gamma rays. The **electromagnetic spectrum** (EMS) is a spectrum of wavelengths that comprise the various types of electromagnetic radiation.

The energy, E, of electromagnetic radiation is inversely proportional to the wavelength, λ:

$$E \propto \frac{1}{\lambda}$$

High-energy radiations such as gamma rays and X-rays have small wavelengths, and low-energy radiations such as radio waves have long wavelengths.

Wavelength is related to the frequency of the radiation, ν, by the expression:

$$c = \nu\lambda$$

where c is the speed of light (3.00×10^8 m s^{-1}).

The SI unit of energy is the joule, J; for wavelength the metre, m; and for frequency the hertz, Hz.

> **Absorption, emission and continuous spectra**
>
> A white-hot metal object such as an incandescent light bulb filament emits the full range of wavelengths, producing a **continuous spectrum** including all the colours of the rainbow from red to violet.
>
> If a pure gaseous element such as hydrogen is subjected to an electrical discharge the gas will glow – it emits radiation. The resultant **emission spectrum** consists of a series of lines against a dark background.
>
> If a cloud of a cold gas is placed between a hot metal and a detector, an **absorption spectrum** is observed. This consists of a pattern of dark lines against a coloured background. The gaseous atoms absorb certain wavelengths of light from the continuous spectrum.
>
> Absorption and emission spectra are widely used in astronomy to analyse light from stars.

▲ Figure 1 The aurora borealis in Lapland, Sweden. The aurora borealis (or Northern Lights) is a display of coloured light visible in the night sky at high latitudes. It occurs when charged and energetic particles from the sun are drawn by the Earth's magnetic field to the polar regions. Hundreds of kilometres up they collide with gaseous molecules and atoms, causing them to emit light

Emission spectra and Bohr's theory of the hydrogen atom

In the 1600s Sir Isaac Newton (1642–1727) showed that if sunlight is passed through a glass prism the visible light is separated into different colours generating a continuous spectrum. This spectrum contains light of all wavelengths and so appears as a continuous series of colours, each colour merging into the next with no gaps. The familiar example of a continuous spectrum is a rainbow. The wavelengths of visible light range from 400 to 700 nm.

Many sources of radiation produce a line spectrum rather than a continuous spectrum. If a pure gaseous element is subjected to a high voltage under reduced pressure, the gas will emit a certain characteristic colour of light. For example, sodium emits yellow light. If this light is

▲ Figure 2 White light as perceived by the human eye consists of many colours or wavelengths of light. Shown here is the continuous spectrum of white light emitted by an incandescent light bulb filament

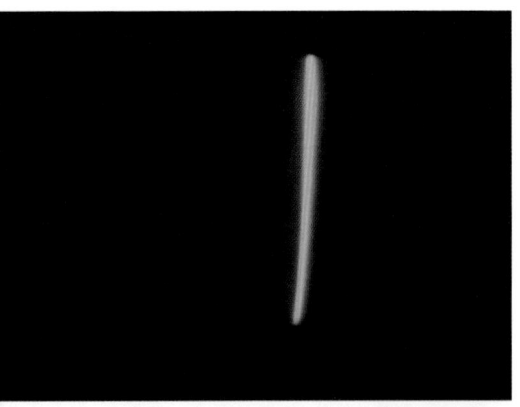

▲ Figure 3 Line emission spectrum of sodium. The spectrum looks like a single bright yellow line but at high resolution it is possible to see two lines very close together corresponding to the wavelengths 589.0 nm and 589.6 nm

passed through a prism, the resultant spectrum is not continuous but consists of a black background with a small number of coloured lines each corresponding to a characteristic wavelength.

Each element has its own characteristic line spectrum which can be used to identify the element. For example, in the visible region of the line emission spectrum of sodium two distinct yellow lines, corresponding to the wavelengths 589.0 nm and 589.6 nm, can be seen on a black background (figure 3).

🌐 Flame tests

Flame tests are often used in the laboratory to identify certain metals. The colour of the flame varies for different elements and can be used to identify unknown substances. The colours are due to the excitation of electrons in the metals by the heat of the flame. As the electrons lose the energy they have just gained, they emit photons of light.

Analogy

You might think of a line emission spectrum as being analogous to a barcode. Every product in a shop has its own unique barcode which gives it an identity, and the same is true of the line emission spectra of the elements. Each line emission spectrum is different and is characteristic of a specific element.

Quantization of energy

The precise lines in the line emission of an element have specific wavelengths. Each characteristic wavelength corresponds to a discrete amount of energy. This is the basis of **quantization**, the idea that electromagnetic radiation comes in discrete "parcels" or quanta. A **photon** is a quantum of radiation, and the wavelength, λ, and energy, E, of a photon are related by the equation:

$$E = h\nu = \frac{hc}{\lambda}$$

where:

h = Planck's constant = 6.63×10^{-34} J s

ν = frequency of the radiation

c = speed of light = 3.00×10^{-8} m s^{-1}

This equation can be found in section 1 of the *Data booklet*. It shows that E is inversely proportional to λ: the greater the energy of the photon, the smaller the wavelength, and vice versa.

In 1913 the Danish physicist Neils Bohr (1885–1962) examined the line emission spectrum of the hydrogen atom. Bohr proposed a theoretical explanation for the spectrum based on classical mechanics. His model proposed the following:

- The hydrogen atom consists of a positively charged particle called the proton at its centre, around which a negatively charged particle called the electron moves in a circular path or **orbit**, similar to the way that planets orbit the sun. Although there is an inherent attraction between the two oppositely charged species, this force of attraction is balanced by the acceleration of the electron moving at high velocity in its orbit.

- Bohr suggested that each orbit has a definite energy associated with it: the energy of the electron orbiting the positively charged centre in a particular orbit is fixed or **quantized**. The energy of the electron in a particular orbit is given by the expression:

$$E = -R_H\left(\frac{1}{n^2}\right)$$

where:

R_H = Rydberg constant = 2.18×10^{-18} J

n = principal quantum number, with positive integer values 1, 2, 3, 4, ... depending on the orbit or energy level the electron occupies

- When an electron in its **ground-state** is excited (for example, by subjecting it to an electrical discharge), it moves to a higher energy level and stays in this **excited-state** for a fraction of a second.

- When the electron falls back down from the excited-state to a lower energy level it emits a **photon**, a discrete amount of energy. This photon corresponds to a particular wavelength λ, depending on the energy difference between the two energy levels (figure 4).

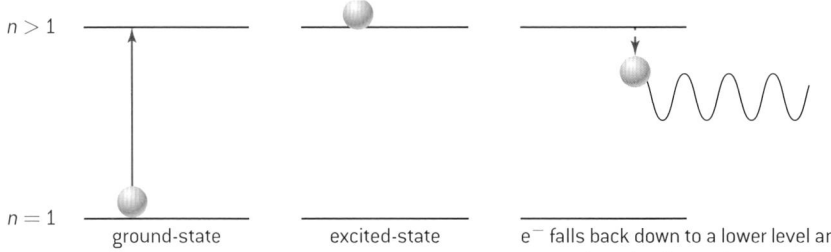

$n > 1$

$n = 1$

    ground-state   excited-state   e⁻ falls back down to a lower level and energy is emitted as a photon of light of wavelength, λ, corresponding to the energy difference between the two energy levels

▲ Figure 4 Principles of the Bohr model of an atom when an electron is excited. n is the principal quantum number

Note that an electron can be excited to any energy level higher than its current level: in figure 4 instead of being excited to $n = 2$ it could be excited to $n = 3$, $n = 4$, etc. The electron can also fall back down to any lower energy level.

The difference in energy between the two energy levels can be expressed as follows, where i represents the initial state and f represents the final state:

$$\Delta E = E_f - E_i$$

$$= h\nu = \frac{hc}{\lambda}$$

We can rearrange this expression noting that:

$$E = -R_H\left(\frac{1}{n^2}\right)$$

$$\Delta E = E_f - E_i = \left[-R_H\left(\frac{1}{n_f^2}\right)\right] - \left[-R_H\left(\frac{1}{n_i^2}\right)\right]$$

$$= \left[R_H\left(\frac{1}{n_i^2}\right)\right] - \left[R_H\left(\frac{1}{n_f^2}\right)\right]$$

$$\Delta E = \left[R_H\left(\frac{1}{n_i^2} - \frac{1}{n_f^2}\right)\right] = h\nu = \frac{hc}{\lambda}$$

Analogy

Think about standing on the bottom step of a flight of stairs. You could jump to the second step, or you could jump higher to the third or fourth step. Suppose you jump from the first step to the fifth step. You stay there for a few seconds and then jump back down. You might jump down to the first step, or jump two steps down to the third step, or jump three steps down to the second step. This is analogous to the way excited electrons can jump from a higher energy level to a lower one.

You always jump to a step, not to some place between steps. This shows the idea of *quantization* – each step is analogous to an energy level, which has a definite, discrete energy. Jumping up steps requires an amount of energy, and jumping down steps releases discrete amounts of energy.

Why the negative sign?

The negative sign in the expression for E is an arbitrary convention. It means that the energy of the electron in the atom is less than its energy if the electron was located an infinite distance away from the nucleus.

Conventions are often used in chemistry. Another example of an arbitrary convention is always placing the cathode on the right-hand side in a cell diagram (see topic 9). Can you think of any other conventions that we use in chemistry?

The hydrogen line emission spectrum consists of a series of lines of different colours (violet, blue, blue–green, and red) in the visible region of the spectrum. The series of lines shown in figure 5 is called the **Balmer series**, which comprises lines associated with electronic transitions from upper energy levels back down to the $n = 2$ energy level.

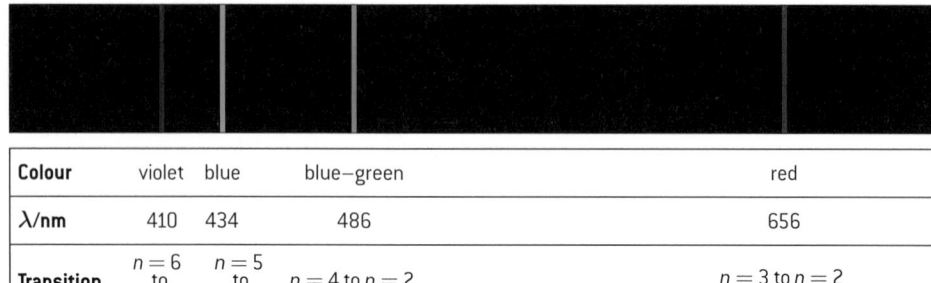

Colour	violet	blue	blue–green	red
λ/nm	410	434	486	656
Transition	$n = 6$ to $n = 2$	$n = 5$ to $n = 2$	$n = 4$ to $n = 2$	$n = 3$ to $n = 2$

Study tip

You are not required to know the names of the individual series of spectral lines. However, you are required to know which transition corresponds to which region of the EMS, eg the transition $n = 4$ to $n = 1$ will be seen in the UV region, etc.

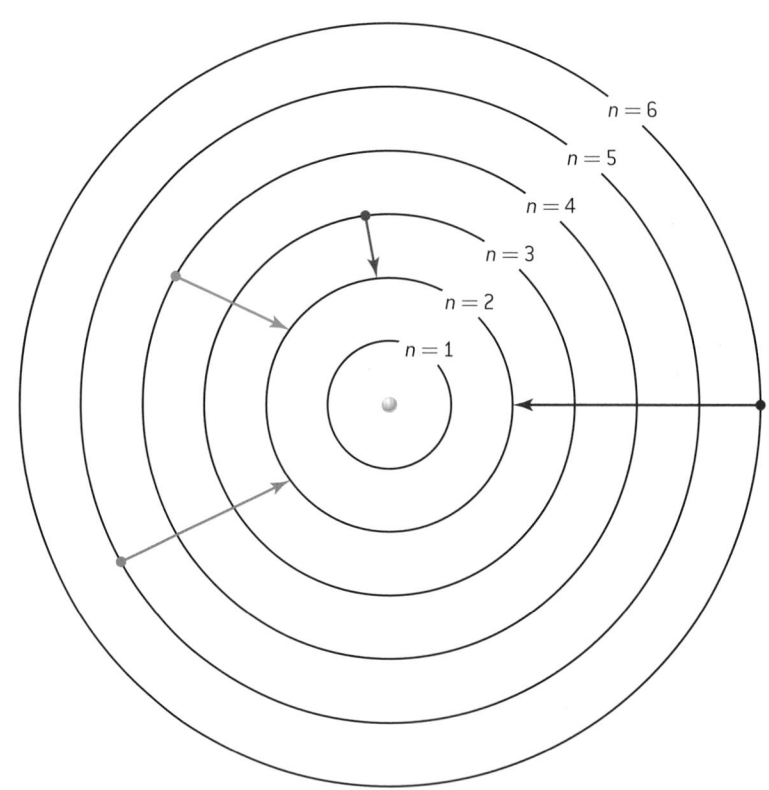

▲ Figure 5 Line emission spectrum of the hydrogen atom. Four lines are seen in the visible and ultraviolet regions of the spectrum; these make up the Balmer series

Other series of lines exist corresponding to transitions to the $n = 1$ and $n = 3$ energy levels (table 1). These are observed in the ultraviolet and infrared regions of the EMS.

Series	n_f	n_i	Region of EMS
Lyman	1	2, 3, 4, 5, ...	UV
Balmer	2	3, 4, 5, 6, ...	visible and UV
Paschen	3	4, 5, 6, 7, ...	IR

▲ Table 1 Different series of lines in the hydrogen line emission spectrum

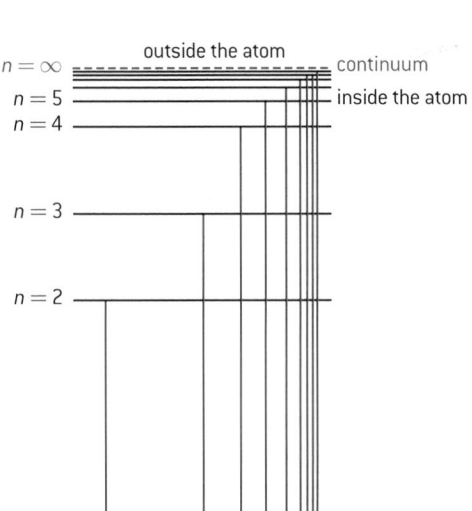

▲ Figure 6 Some transitions to the $n = 1$ level from higher levels for the Lyman series (in the UV region) of spectral lines that occur in the emission spectrum of the hydrogen atom

Quantization and atomic structure

The line emission spectrum of hydrogen provides evidence for the existence of electrons in discrete energy levels, which get closer together (they are said to **converge**) at higher energies. At the limit of this convergence the lines merge, forming a **continuum**. Beyond this continuum the electron can have any energy; it is no longer under the influence of the nucleus and is therefore outside the atom. Such an electron may be referred to as a **free electron**.

 Models of the atom and electron arrangements

The Bohr theory of the atom is a basis for writing **electron arrangements**. An electron arrangement gives the number of electrons in each shell or orbit, for example:

electron arrangement of H: 1
electron arrangement of P: 2, 8, 5
electron arrangement of Ca: 2, 8, 8, 2

Electron arrangements are a very useful tool for explaining and predicting the chemical properties of an element.

In the Bohr model of the atom the energy levels are often drawn as concentric circles, as shown in figure 7 for phosphorus.

Limitations of the Bohr theory

This model has now been superseded and is associated with a number of misconceptions:

- It assumes that the positions of the electron orbits are fixed. This is incorrect; in fact orbits do not actually exist (we shall shortly introduce the idea of an orbital).

- It assumes that energy levels are circular or spherical in nature. This is also incorrect.

- It suggests an incorrect scale for the atom – remember from sub-topic 2.1 that the atom is made up of mainly empty space.

There were some fundamental theoretical problems pertaining to the Bohr model:

- Bohr limited his calculations to just one element, namely hydrogen. The model did not explain the line spectra of other elements containing more than one electron.

- Bohr suggested that the electron is a subatomic particle orbiting the nucleus.

Nevertheless, Bohr made a significant contribution to our understanding of electronic structure and in particular, some of the merits of his theory are the following:

- It was based on the fundamental idea of quantization – the fact that electrons exist in definite, discrete energy levels.

- It incorporated the idea of electrons moving from one energy level to another.

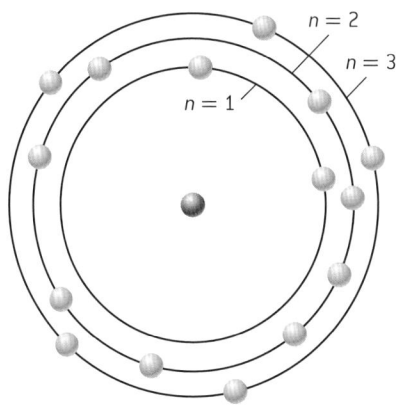

▲ Figure 7 Electron arrangement for phosphorus according to the Bohr model

The quantum mechanical model of the atom

The Bohr theory provided a first approximation of atomic structure, and in particular the arrangement of electrons. It has since been replaced by more sophisticated mathematical theories from the field of **quantum mechanics**, which incorporates the wave-like nature of the electron. Some of the key ideas are described below.

Heisenberg's uncertainty principle states that it is impossible to determine accurately both the momentum and the position of a particle simultaneously (topic 12). This means that the more we know about the position of an electron, the less we know about its momentum, and vice versa. Although it is not possible to state precisely the location of an electron in an atom and its exact momentum along a trajectory at the same time, we can calculate the *probability* of finding an electron in a given region of space within the atom.

Schrödinger's equation was formulated in 1926 by the Austrian physicist Erwin Schrödinger (1887–1961). His sophisticated mathematical equation integrates the dual wave-like and particle nature of the electron. This ground-breaking work led to the birth and subsequent development of the field of quantum mechanics. In 1933 Schrödinger received the Nobel Prize in Physics with Paul Dirac.

The solution to Schrödinger's equation generated a series of mathematical functions called **wavefunctions** describing the electron in the hydrogen atom and associated possible energy states the electron can occupy. Each wavefunction is represented by the symbol, ψ. The square of the wavefunction, ψ^2, represents the probability of finding an electron in a region of space at a given point a distance, r, from the nucleus of the atom. ψ^2 is termed the **probability density**. The equations are very complex but at this level all we need to consider are the basic principles underpinning the results.

The wavefunctions of electrons in an atom are described by atomic orbitals:

- An **atomic orbital** is a region in space where there is a high probability of finding an electron.

Any orbital can hold a maximum of two electrons. There are several types of atomic orbital: s, p, d, and f, etc. Each type has a characteristic shape and associated energy.

Study tip

Atomic orbitals have different shapes. For SL you need to be familiar with the shapes of the s and p atomic orbitals, while for HL you need to know the shapes of the s, p, and d atomic orbitals. We shall return to the shapes of the d orbitals in topic 13 when we discuss crystal field theory.

Analogy

Imagine that you are a student in an IB chemistry class in Quito in Ecuador, waiting for your teacher to arrive at 8.00 am. At 8.15 am there is no sign of your teacher and your class decide to go looking for him. You decide first to define the most probable places the teacher is likely to be. Suggestions from the class include:

The teacher:

- is possibly in the staff room, the chemistry laboratory, or the library
- may be in the school principal's office or in the school car park
- could be at his house in Quito

- could perhaps be at the airport
- might even have gone home to South Africa!

If the class went looking for the teacher they would most likely start looking in the most probable locations closest to the classroom. But at 8.15 am they do not know with any degree of certainty precisely where the teacher is.

A three-dimensional graph could be drawn with a cluster of dots showing areas where there is a high probability of finding him. This is the idea of an **orbital**. A boundary surface could be drawn around this cluster of dots to define a region of space where there is a 99% chance of finding the teacher. This might be the school perimeter, or Quito where he lives.

If you were also asked to measure the distance from the classroom to the exact location where the teacher is you could not do this at 8.15 am, as you do not know his exact location with absolute certainty.

What aspects of quantum mechanics does this analogy capture?

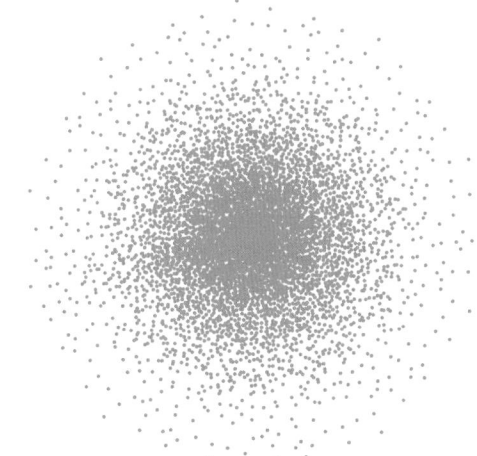

▲ Figure 8 An orbital is a three-dimensional graph with a cluster of dots showing the probability of finding the electron at different distances from the nucleus

The s atomic orbital

An **s orbital** is *spherically symmetrical*. The sphere represents a boundary surface, meaning that within the sphere there is a 99% chance or probability of finding an electron (figure 9).

The p atomic orbital

A **p orbital** is *dumbbell shaped*. There are three p atomic orbitals, p_x, p_y, and p_z, all with boundary surfaces conveying probable electron density pointing in different directions along the three respective Cartesian axes, x, y, and z (figure 10).

Energy levels, sublevels, orbitals, and electron spin

The Bohr model introduced the idea of a main energy level, described by n, which is called the **principal quantum number**. This can have positive integer values 1, 2, 3, etc. In the quantum mechanical model, as n increases, the mean position of an electron is further from the nucleus. The energies of the orbitals also increase as n increases. Each main energy level or shell can hold a maximum number of electrons given by $2n^2$. So the electron capacity for $n = 1$ is 2, for $n = 2$ is 8, for $n = 3$ is 18. That is why we have two elements in the first row of the periodic table, eight elements in the second, etc.

The energy levels are split up into **sublevels**, of which there are four common types: s, p, d, and f. Each sublevel contains a number of orbitals, each of which can hold a maximum of 2 electrons (table 2).

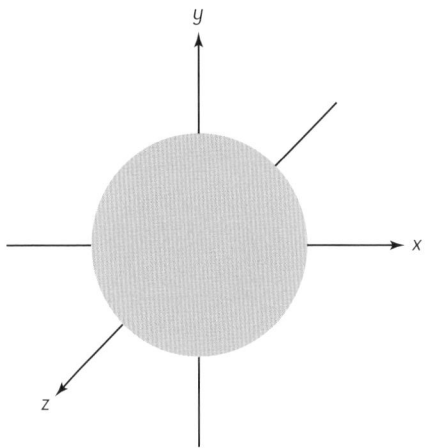

▲ Figure 9 The s atomic orbital is spherically symmetrical

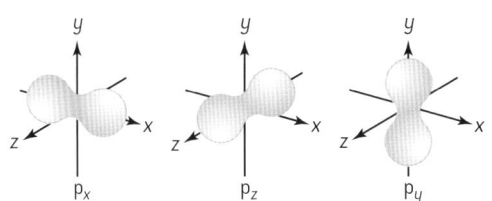

▲ Figure 10 The three p atomic orbitals are dumbbell shaped, aligned along the x, y, and z axes

Sublevel	Number of orbitals in sublevel	Maximum number of electrons in sublevel
s	1	2
p	3	6
d	5	10
f	7	14

▲ Table 2 Sublevels of the main energy levels in the quantum mechanical model

For convenience, an "arrow-in-box" notation called an **orbital diagram** is used to represent the electrons in these atomic orbitals (figure 11). We shall use orbital diagrams to represent electron configurations.

s sublevel (one box representing an s orbital)

p sublevel (three boxes representing the three p orbitals p_x, p_y, and p_z)

d sublevel (five boxes representing the five d orbitals)

f sublevel (seven boxes representing the seven f orbitals)

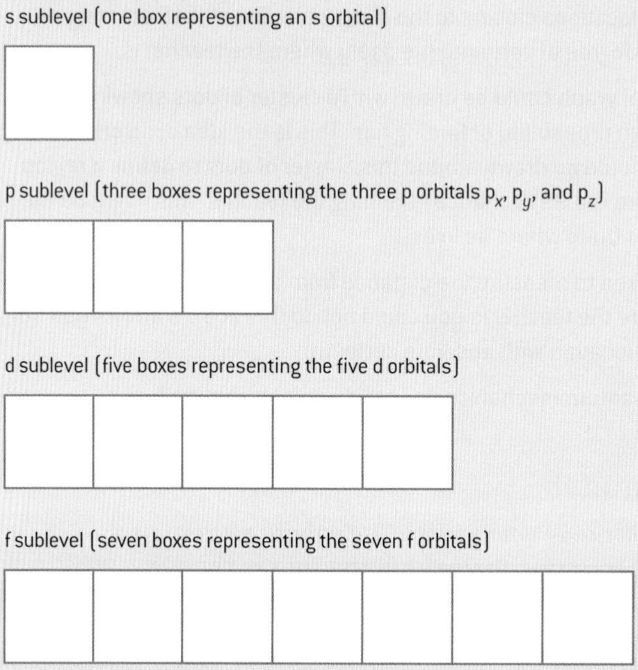

▲ Figure 11 Orbital diagrams are used to represent the electron configurations for atoms. Arrows are drawn in the boxes to represent electrons, a maximum of 2 electrons in each box (orbital)

Two electrons in the same orbital have opposite values of the **spin magnetic quantum number, m_s**. The sign of m_s $\left(+\frac{1}{2} \text{ or } -\frac{1}{2}\right)$ indicates the orientation of the magnetic field generated by the electron. A pair of electrons in an orbital behaves as two magnets facing in opposite directions and therefore is commonly represented by two arrows in a box (figure 12).

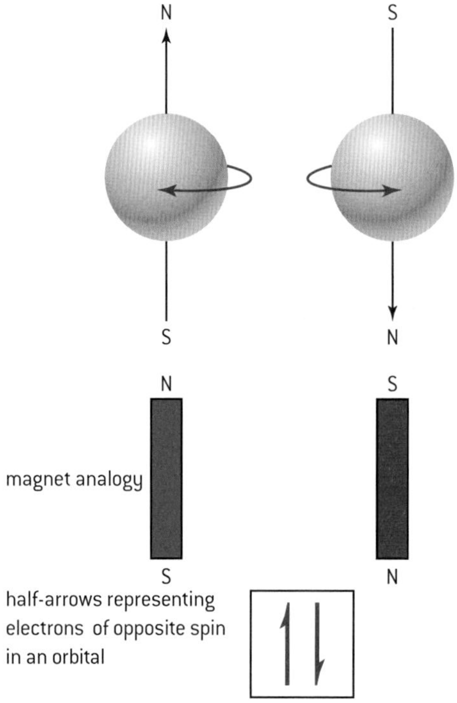

magnet analogy

half-arrows representing electrons of opposite spin in an orbital

▲ Figure 12 Electron spin is represented by arrows in orbital diagrams

Quantum numbers

In this mathematical model of the electronic structure of the atom there are four **quantum numbers**. The first is the **principal quantum number**, n, which represents the energy level. The second quantum number, the **azimuthal quantum number**, l, describes the sublevel, and the third quantum number, the **magnetic quantum number**, m_l, the atomic orbital. The fourth quantum number, the **spin magnetic quantum number**, m_s, describes the spatial orientation of the electron spin. Quantum numbers are not formally examined in the IB Chemistry Diploma, but you need to know the principles of energy levels, sublevels, atomic orbitals, and electron spin.

You might think of the four quantum numbers as an electronic postal address. The country represents the energy level, the province the sublevels, the town the orbitals, and the street number or postal code the spin of the electron.

Writing electron configurations

We shall now develop these ideas further by writing **electron configurations** for atoms and ions.

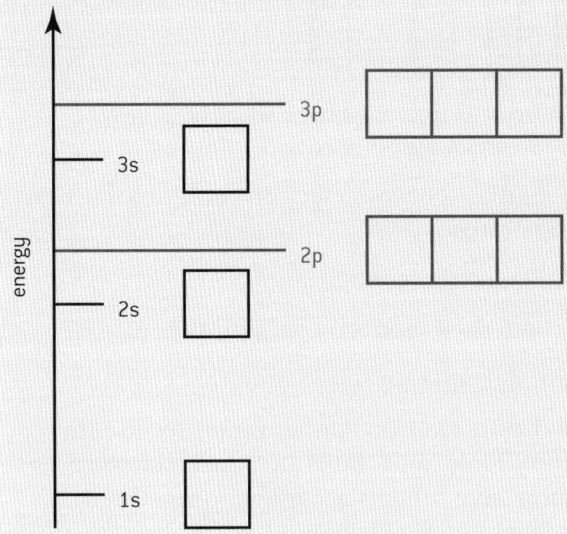

▲ Figure 13 This is the order of energy levels of the first few sublevels

There are three principles that must be followed when representing electron configurations.

1 The **Aufbau principle** states that electrons fill the lowest-energy orbital that is available first. Figure 13 shows the sublevels for the first few energy levels.

Up to Ca ($Z = 20$) the Aufbau principle correlates precisely with experimental data and the 4s level is filled first before the 3d level since it is lower in energy. The condensed electronic configuration for Ca is written as $[Ar]4s^2$. However, for Sc ($Z = 21$), the two levels are comparable in energy with the 4s level now slightly higher in energy than the 3d level and hence the 3d is filled first. The condensed configuration for Sc therefore is correctly written as $[Ar]3d^14s^2$. This trend continues along the 3d sublevel. For Zn ($Z = 30$), the 4s level now is much higher in energy than the 3d and the condensed electron configuration for Zn is best written as $[Ar]3d^{10}4s^2$ for this reason. This is consistent with experimental data which shows that when the 3d-block elements are ionized, the electrons are removed from the 4s before the 3d levels, which makes sense since the 4s is higher in energy than the 3d for this block of elements. The situation overall is quite complex as in the case of Sc the filling of the last three electrons does not continue in the 3d level, and experimental data does not provide evidence for an $[Ar]3d^3$ electron configuration for Sc. The reason for this is that the 3d orbitals are more compact than the 4s orbitals and hence electrons entering the 3d orbitals will experience a much greater

Study tips

- For the IB Chemistry Diploma you need to be able to deduce the electron configurations for the atoms and ions of the elements up to and including $Z = 36$ (Kr).

- The periodic table showing atomic number, Z, is provided in section 6 of the *Data booklet*.

mutual repulsion. In an excellent article written by E. Scerri, Department of Chemistry and Biochemistry, at the University of California, USA and published in *Education in Chemistry*, 7th November 2013, the reason is explained as follows: "The slightly unsettling feature is that although the relevant s orbital can relieve such additional electron-electron repulsion, different atoms do not always make full use of this form of sheltering because the situation is more complicated than just described. One thing to consider is that nuclear charge increases as we move through the atoms, and there is a complicated set of interactions between the electrons and the nucleus as well as between the electrons themselves".

2 The **Pauli exclusion principle** states that any orbital can hold a maximum of two electrons, and these electrons have opposite spin.

3 **Hund's rule of maximum multiplicity** states that when filling **degenerate** orbitals (orbitals of equal energy), electrons fill all the orbitals singly before occupying them in pairs. This is illustrated in figure 14.

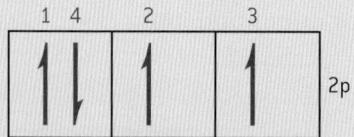

▲ Figure 14 Electrons fill each orbital singly before occupying them in pairs

There are three ways electron configurations can be illustrated:

1 full electron configuration

2 condensed electron configuration

3 orbital diagram representation.

To write an electron configuration we use the periodic table, and "build up" the electrons in successive orbitals according to the three principles described above.

The periodic table can be shown as four blocks corresponding to the four sublevels s, p, d, and f (figure 15).

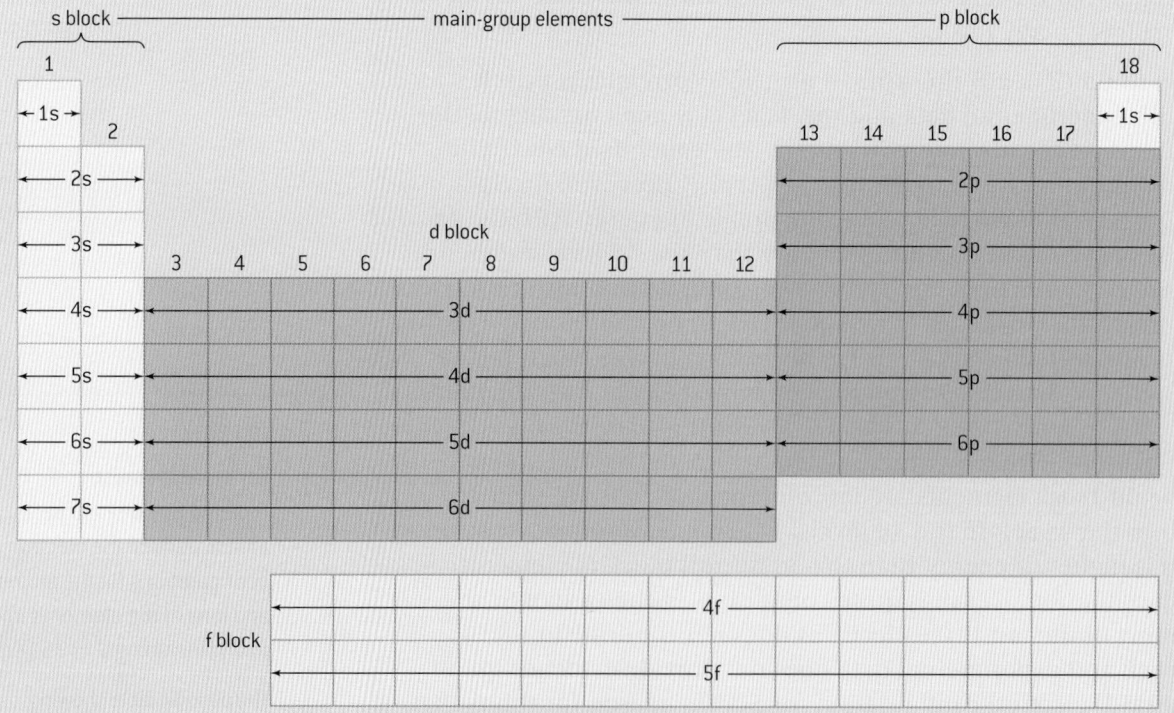

▲ Figure 15 The blocks of the periodic table correspond to the sublevels s, p, d, and f

Full electron configurations

Table 3 shows the **full electron configurations** for some of the first 36 elements.

Element	Z	Electron configuration
Period 1 elements:		
H	1	$1s^1$
He	2	$1s^2$
Period 2 elements:		
Li	3	$1s^2 2s^1$
Be	4	$1s^2 2s^2$
B	5	$1s^2 2s^2 2p^1$
C	6	$1s^2 2s^2 2p^2$
N	7	$1s^2 2s^2 2p^3$
O	8	$1s^2 2s^2 2p^4$
F	9	$1s^2 2s^2 2p^5$
Ne	10	$1s^2 2s^2 2p^6$
Period 3 elements: continue with the same filling pattern, for example:		
Na	11	$1s^2 2s^2 2p^6 3s^1$
Mg	12	$1s^2 2s^2 2p^6 3s^2$
Al	13	$1s^2 2s^2 2p^6 3s^2 3p^1$
Ar	18	$1s^2 2s^2 2p^6 3s^2 3p^6$
Period 4 elements: After Z = 30 the 4p sublevel is filled:		
K	19	$1s^2 2s^2 2p^6 3s^2 3p^6 4s^1$
Ca	20	$1s^2 2s^2 2p^6 3s^2 3p^6 4s^2$
Sc	21	$1s^2 2s^2 2p^6 3s^2 3p^6 3d^1 4s^2$
Ni	28	$1s^2 2s^2 2p^6 3s^2 3p^6 3d^8 4s^2$
Zn	30	$1s^2 2s^2 2p^6 3s^2 3p^6 3d^{10} 4s^2$
Ga	31	$1s^2 2s^2 2p^6 3s^2 3p^6 3d^{10} 4s^2 4p^1$
Br	35	$1s^2 2s^2 2p^6 3s^2 3p^6 3d^{10} 4s^2 4p^5$
Kr	36	$1s^2 2s^2 2p^6 3s^2 3p^6 3d^{10} 4s^2 4p^6$

▲ Table 3 Full electron configurations for some of the first 36 elements

Condensed electron configuration

You can see above that full electron configurations become quite lengthy and cumbersome with increasing atomic number. An element's chemistry is dictated by its outer **valence electrons** (as opposed to the inner **core electrons**), and a more convenient way of representing electron configurations is as the **condensed electron configuration**:

[nearest noble gas core] + valence electrons

Some exceptions: Chromium and copper

Two of the first 36 elements have electron configurations that differ from what you may predict. These two elements are Cr ($Z = 24$) and Cu ($Z = 29$):

Cr $1s^2 2s^2 2p^6 3s^2 3p^6 3d^5 4s^1$

Cu $1s^2 2s^2 2p^6 3s^2 3p^6 3d^{10} 4s^1$

In these two elements electrons go into the 3d orbitals before completely filling the 4s orbital. Chromium has a half-filled 3d sublevel of 5 electrons and copper has a completely filled 3d sublevel of 10 electrons. Half-filled and completely filled 3d sublevels reduce the overall potential energy of an atom, so the electron configurations $3d^5 4s^1$ and $3d^{10} 4s^1$ are more stable than $3d^4 4s^2$ and $3d^9 4s^2$, respectively.

For example:

He   [He]

O   [He]$2s^22p^4$

Ne   [He]$2s^22p^6$ or simply [Ne]

P   [Ne]$3s^23p^3$

Orbital diagrams

Orbital diagrams make use of the arrows-in-boxes notation described in figures 11 and 13, with arrows representing electrons and boxes representing orbitals. Degenerate orbitals are represented by boxes joined together to show their energy equivalence.

Orbital diagrams may show all the orbitals as in the full electron configuration, or just the orbitals beyond the nearest noble gas core as in the condensed electron configuration. Orbital diagrams may have steps showing the energy levels or may be represented on one line. For example, figure 16 shows two types of orbital diagrams that can be used to represent fluorine:

F $1s^22s^22p^5$

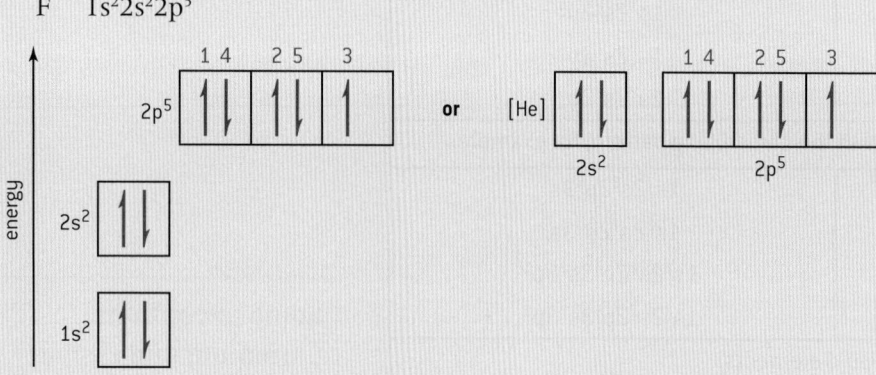

▲ Figure 16 Orbital diagrams showing the electron configuration for fluorine

The condensed version is more convenient and will be used in this book. For example, the orbital diagrams for the elements chromium, cobalt, and bromine are represented as follows:

Cr   [Ar]$3d^54s^1$

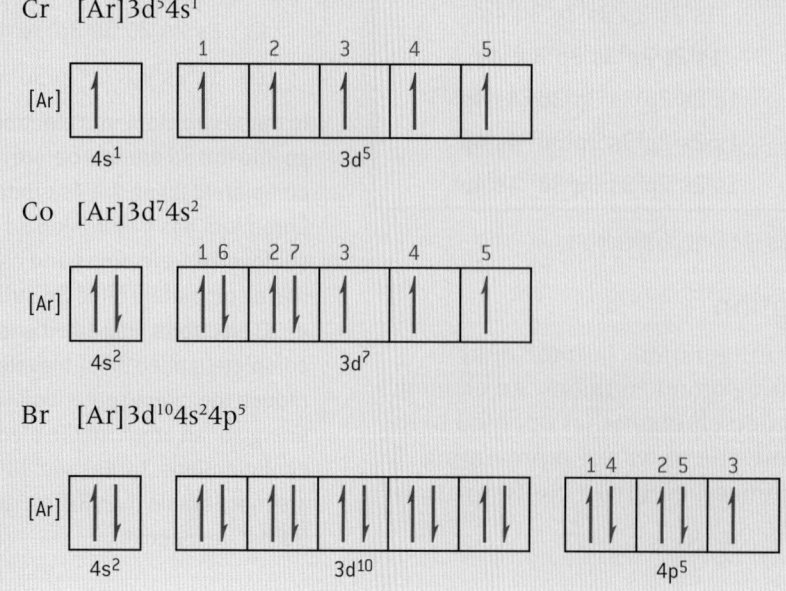

Co   [Ar]$3d^74s^2$

Br   [Ar]$3d^{10}4s^24p^5$

Worked examples: electron configurations

Example 1

Deduce the full electron configurations for Mg, Mg^{2+}, O, and O^{2-}.

Solution

- From table 3:

  Mg $1s^22s^22p^63s^2$

  To write the electron configuration for the Mg^{2+} cation, 2 electrons must be removed. These are taken from the orbital of highest principal quantum number n; in this case, the $3s^2$ orbital:

  Mg^{2+} $1s^22s^22p^6$

- From table 3:

  O $1s^22s^22p^4$

  To write the electron configuration for the O^{2-} anion, two electrons must be added according to the same principles as before:

  O^{2-} $1s^22s^22p^6$

  Notice that the electron configurations for the species Mg^{2+} and O^{2-} are identical: they contain the same number of electrons and are said to be **isoelectronic**. Na^+, F^-, and Ne are also isoelectronic with Mg^{2+} and O^{2-}. However, each of these species has a different number of protons (atomic number Z table 4):

Species	Atomic number, Z (number of protons)	Number of electrons
O^{2-}	8	10
F^-	9	10
Ne	10	10
Na^+	11	10
Mg^{2+}	12	10

▲ Table 4 Isoelectronic species

Example 2

Deduce the condensed electron configurations of S, S^{2-}, Fe, Fe^{2+}, Cu, and Cu^+.

Solution

- S $[Ne]3s^23p^4$

  For the S^{2-} anion we add 2 electrons:

  S^{2-} $[Ne]3s^23p^6$ or simply [Ar]

- The electron configuration for Fe ($Z = 26$) can be deduced as:

  Fe $[Ar]3d^64s^2$

  For the Fe^{2+} cation 2 electrons are removed from the orbital of highest n; in this case, the 4s orbital:

  Fe^{2+} $[Ar]3d^6$

- The copper electron configuration is one of the two exceptions that you must remember:

  Cu $[Ar]3d^{10}4s^1$

  To form the Cu^+ ion, again the electron is removed from the orbital of highest n; in this case, the 4s level:

  Cu^+ $[Ar]3d^{10}$

Example 3

Deduce the orbital diagrams for Ni, Ni^{2+}, and Se.

Solution

First write the condensed electron configuration for the species. Then draw the orbital diagram, remembering that two electrons in the same orbital have opposite spin quantum numbers:

Ni $[Ar]3d^84s^2$

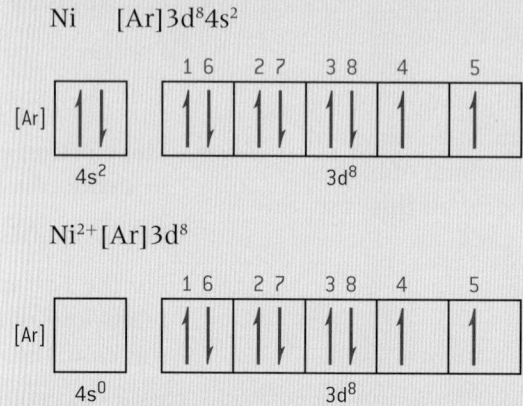

$Ni^{2+}[Ar]3d^8$

Notice that in the orbital diagram for the Ni^{2+} cation there are no electrons in the 4s orbital – the box should be left blank.

For selenium:

Se $[Ar]3d^{10}4s^24p^4$

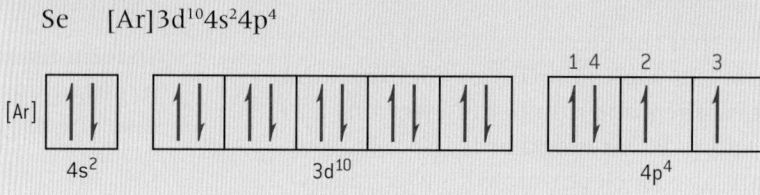

🧬 Experimental evidence for electron configurations

Direct evidence of the electron configuration for an element can be found from magnetic measurements. There are different types of magnetism, including **paramagnetism** and **diamagnetism**. A paramagnetic material has at least one unpaired electron and hence can be attracted by a magnetic field. The greater the number of unpaired electrons, the greater the force of attraction in a magnetic field. In contrast, a diamagnetic material has all its electrons paired and can be repelled by a magnetic field.

Developments in scientific research over the past 50 years have led to a number of improvements in instrumentation which have allowed scientists to determine the number of unpaired electrons in an atom.

Questions

1 What is the number of protons, electrons, and neutrons in boron-11?

   A. 5 protons, 5 electrons, and 11 neutrons

   B. 5 protons, 5 electrons, and 10.81 neutrons

   C. 5 protons, 5 electrons, and 6 neutrons

   D. 11 protons, 11 electrons, and 5 neutrons

2 What is the number of protons, electrons, and neutrons in $^{34}_{16}S^{2-}$?

   A. 18 protons, 16 electrons and 18 neutrons

   B. 16 protons, 18 electrons and 34 neutrons

   C. 16 protons, 18 electrons and 18 neutrons

   D. 16 protons, 16 electrons and 18 neutrons

3 Which statements about the isotopes of chlorine, $^{35}_{17}Cl$ and $^{37}_{17}Cl$, are correct?

   I. They have the same chemical properties.

   II. They have the same atomic number.

   III. They have the same physical properties.

   A. I and II only

   B. I and III only

   C. II and III only

   D. I, II and III   [1]

   IB, May 2011

4 A sample of element X contains 69% of ^{63}X and 31% of ^{65}X. What is the relative atomic mass of X in this sample?

   A. 63.0

   B. 63.6

   C. 65.0

   D. 69.0   [1]

   IB, May 2010

5 What is the relative atomic mass of an element with the mass spectrum shown in figure 17?

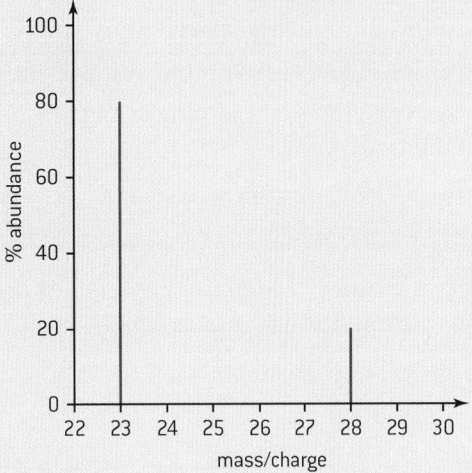

▲ Figure 17

   A. 24

   B. 25

   C. 26

   D. 27   [1]

   IB, May 2009

6 Which is correct for the following regions of the electromagnetic spectrum?

	Ultraviolet (UV)		Infrared (IR)	
A.	high energy	short wavelength	low energy	low frequency
B.	high energy	low frequency	low energy	long wavelength
C.	high frequency	short wavelength	high energy	long wavelength
D.	high frequency	long wavelength	low frequency	low energy

       [1]

   IB, May 2009

7 In the emission spectrum of hydrogen, which electronic transition would produce a line in the visible region of the electromagnetic spectrum?

   A. $n = 2 \rightarrow n = 1$

   B. $n = 3 \rightarrow n = 2$

   C. $n = 2 \rightarrow n = 3$

   D. $n = \infty \rightarrow n = 1$ [1]

   IB, May 2011

8 Which describes the visible emission spectrum of hydrogen?

A. A series of lines converging at longer wavelength

B. A series of regularly spaced lines

C. A series of lines converging at lower energy

D. A series of lines converging at higher frequency [1]

IB, May 2010

9 What is the order of increasing energy of the orbitals within a single energy level?

A. $d < s < f < p$

B. $s < p < d < f$

C. $p < s < f < d$

D. $f < d < p < s$ [1]

IB, May 2009

10 What is the condensed electron configuration for Co^{3+}?

A. $[Ar]4s^23d^7$

B. $[Ar]4s^23d^4$

C. $[Ar]3d^6$

D. $[Ar]4s^13d^5$

11 Draw and label an energy level diagram for the hydrogen atom. In your diagram show how the series of lines in the ultraviolet and visible regions of its emission spectrum are produced, clearly labelling each series. [4]

IB, May 2010

12 a) List the following types of electromagnetic radiation in order of **increasing** wavelength (shortest first).

I. Yellow light

II. Red light

III. Infrared radiation

IV. Ultraviolet radiation [1]

b) Distinguish between a continuous spectrum and a line spectrum. [1]

c) The thinning of the ozone layer increases the amount of UV-B radiation that reaches the Earth's surface (table 5).

Type of radiation	Wavelength / nm
UV-A	320–380
UV-B	290–320

▲ Table 5

Based on the information in table 5 explain why UV-B rays are more dangerous than UV-A. [3]

IB, Specimen Paper

13 a) Deduce the full electron configuration for Mn and Mn^{2+}.

b) Deduce the condensed electron configuration for Cu^{2+}.

c) Draw orbital diagrams for Co^{2+} and As.

14 ⊘ Atoms are often drawn as spheres. Comment on the use of this representation as a model.

15 ⊘ Developments in scientific research follow improvements in apparatus. Discuss this statement with regard to the use of electricity and magnetism in Thomson's experiments with cathode rays.

16 ⊘ In many textbooks the electronic configuration for vanadium is written as $[Ar]4s^23d^3$. This is common practice and widely accepted by the chemical community. However, suggest why this way of writing the electronic configuration for vanadium may be at odds with experimental evidence. You might like to read the following article: http://www.rsc.org/eic/2013/11/aufbau-electron-configuration to guide you in your answer.

3 PERIODICITY

Introduction

Science is full of factual information. However, some of the greatest scientific discoveries have resulted from scientists being able to interpret vast amounts of data and deduce clear patterns emerging from it. In 1869 the Russian chemist Dmitri Mendeleev recognized that if elements were arranged in order according to their atomic weight (relative atomic mass), a definite pattern could be seen in the properties of the elements. This led ultimately (after some refinement of the theory) to the development of the most important tool available to chemists, the periodic table of elements, which lies at the core of chemistry.

As the table developed it became clear that the chemical and physical properties of the elements are a periodic function of Z, the atomic number. In this topic we shall examine the nature of the periodic table, establish what information can be extracted from it, and explore how repeated (**periodic**) patterns can be linked to the properties of the elements.

3.1 Periodic table

Understandings

→ The periodic table is arranged into four blocks associated with the four sublevels — s, p, d, and f.

→ The periodic table consists of groups (vertical columns) and periods (horizontal rows).

→ The period number (n) is the outer energy level that is occupied by electrons.

→ The number of the principal energy level and the number of the valence electrons in an atom can be deduced from its position on the periodic table.

→ The periodic table shows the positions of metals, non-metals and metalloids.

Applications and skills

→ Deduction of the electron configuration of an atom from the element's position on the periodic table, and vice versa.

Nature of science

→ Obtain evidence for scientific theories by making and testing predictions based on them – scientists organize subjects based on structure and function; the periodic table is a key example of this. Early models of the periodic table from Mendeleev, and later Moseley, allowed for the prediction of properties of elements that had not yet been discovered.

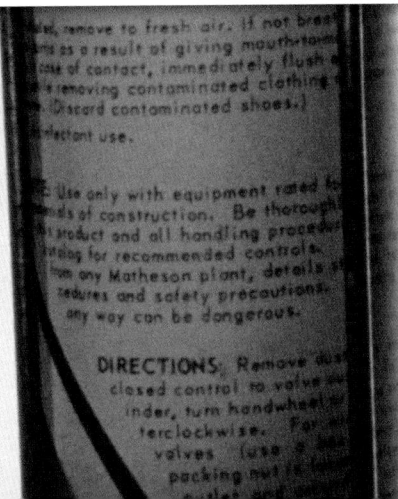

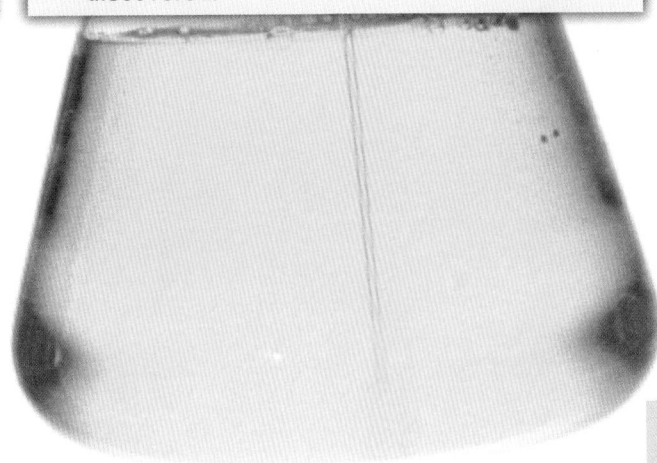

The development of the periodic table

Evidence for scientific theories is obtained by making predictions and then testing them against proposed theories. Scientists often try to classify their subject based on structure and function, and the periodic table of elements is a good example of this.

The development of the periodic table took place over a number of years and has involved scientists from different countries building on the foundations of each others' work and ideas.

Four key scientists contributed to the development of the modern periodic table, as summarized below.

Döbereiner

In 1817 the German chemist **Johann Döbereiner** (1780–1849) discovered that the elements calcium, strontium, and barium had similar properties and that the atomic weight (relative atomic mass using today's terminology) of strontium was approximately the mean of the sum of the atomic weights of calcium and barium. He classified this trio of elements as a **triad**. Döbereiner also recognized other triads – one involving chlorine, bromine, and iodine, and another involving sulfur, selenium, and tellurium. This discovery was called the **law of triads**. Surprisingly the scientific community at that time did not pay much attention to this law and the classification of the elements into triads was limited to just a few elements. However, Döbereiner's hypothesis suggesting there was an inherent link between atomic weight and the properties of elements was an important stepping stone in the development of the periodic table of elements.

Newlands

In 1864 the English chemist **John Newlands** (1837–1898) discovered that when elements were arranged in order of atomic weight, there appeared to be evidence of a pattern with the properties of the elements repeated in **octaves** consisting of seven elements, such that each element had properties similar to the eighth element above or below it. This term was named based on the analogy of an octave in music – the same note is repeated at intervals of eight on the musical scale. In 1865 Newlands published this idea of the **periodicity** of elements (that is, a repeated pattern) when arranged in order of atomic weight. This became known as the **law of octaves**.

If the elements are arranged in order of their equivalents with a few slight transpositions, it will be observed that elements belonging to the same group appear on the same horizontal line. It will also be seen that the numbers of analogous elements differ by seven or by some multiples of seven. Members stand to each other in the same relation as the extremities of one or more octaves of music. Thus in the nitrogen group, between nitrogen and phosphorus there are seven elements; between phosphorus and arsenic, fourteen; between arsenic and antimony, fourteen; and lastly, between antimony and bismuth, fourteen also. This peculiar relationship I propose to provisionally term The Law of Octaves.

J.A.R. Newlands, 'a letter to the editor', *Chemical News*, 12 (18th August 1865).

Newlands's idea of octaves applied to only a limited number of known elements. He tried to apply this principle to the known elements (about 60 at the time). However, they did not all neatly fit this type of pattern: highly reactive metals such as lithium, sodium, potassium, rubidium, and caesium became grouped with very unreactive metals such as silver and copper. One idea Newlands had was to place two elements together, in one box of a periodic table, to take account of this. Newlands presented his law to the Chemical Society in England but his ideas were not accepted. His presentation to the Chemical Society of this work in 1866 was not published. As a result Newlands felt ridiculed and returned to his position of chief chemist at a sugar plant.

Mendeleev

In 1869, four years after Newlands's ideas were first mooted, the Russian chemist **Dmitri Mendeleev** (1834–1907) discovered, like Newlands, that if the elements were arranged in order of atomic weight, a repeated pattern of their properties could be identified. This was termed the **periodic law**. The main difference between Newlands's and Mendeleev's work was that Mendeleev considered the properties of the elements very carefully and grouped together only elements that had similar properties. In 1869 Mendeleev published his first periodic table of elements.

Mendeleev improved the table over time and left gaps for undiscovered elements, so that each element fell into the correct group. Using this approach Mendeleev was able to predict the existence and properties of undiscovered elements. However, some elements did not obey Mendeleev's version of the periodic law. For example, iodine (atomic weight 126.90) had to be placed in the table *after* tellurium, despite the fact that tellurium had a higher atomic weight (127.60).

Moseley

It soon became apparent to the scientific community that arranging the elements in order of atomic weight was problematic. In 1913 the British physicist **Henry Moseley** (1887–1915) arranged the elements in the periodic table in order of atomic number, Z, instead of atomic weight. This is the basis for the modern periodic table of elements.

Figure 1 summarizes the contributions of some of the various scientists who developed the periodic table over time.

J.W. Döbereiner (1780–1849)	**J.A.R. Newlands** (1837–1898)	**Dimitri Mendeleev** (1834–1907)	**Henry Moseley** (1887–1915)
Law of triads (1817) – link between atomic weight and different elements in groups of threes.	**Law of octaves** (1865) – when elements were arranged in order of atomic weight there appeared to be evidence of a pattern with the properties repeated in **octaves** consisting of seven elements.	**The periodic law** (1869) – when the elements were arranged in order of atomic weight a repeated pattern of their properties was found.	**The modern periodic law** (1913) – when the elements were arranged in order of increasing atomic numbers (Z), their properties recurred periodically.

▲ Figure 1 Scientists who contributed to the development of the periodic table of elements

Activity

In modern science do you think that theoretical research has a much greater chance of acceptance by the scientific community if it is supported by empirical evidence? Discuss this in class.

A **hypothesis** is a proposal that tries to explain particular phenomena. A **theory** results from testing a hypothesis and may subsequently replace the hypothesis. A hypothesis can therefore be considered a tentative explanation that can be tested through investigation and exploration whereas a theory is an established array of ideas or concepts which may then be used to make predictions.

In science there are two ways of arriving at a particular conclusion – **inductive reasoning** and **deductive reasoning** (figure 2). Inductive reasoning is a "bottom-up" approach whereas deductive reasoning may be described as a "top-down" approach. With inductive reasoning definite measurements and observations can lead scientists to establish the existence of possible trends or a pattern. From such a pattern a hypothesis can be formulated that can ultimately lead to a theory based on certain conclusions. In deductive reasoning, the starting point involves the theories themselves. These are tested based on experimental (empirical) work.

What role did inductive and deductive reasoning play in the development of the periodic table? What role do inductive and deductive reasoning have in science in general?

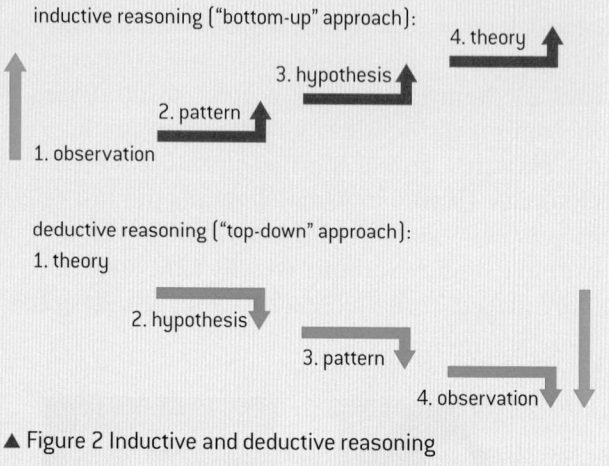

▲ Figure 2 Inductive and deductive reasoning

The periodic table today

In the modern periodic table the elements are arranged in order of increasing atomic number, **Z**, with elements having similar chemical and physical properties placed underneath each other in vertical columns called **groups**. The groups are numbered from 1 to 18; certain groups have their own names (table 1).

Group number	Recommended name
1	alkali metals
2	alkaline earth metals
15	pnictogens
16	chalcogens
17	halogens
18	noble gases

▲ Table 1 Names of groups recommended by IUPAC in the periodic table of elements

Useful resource

Much information on each element can be found on the "WebElements" periodic table website. This resource was compiled by Professor Mark Winter at the University of Sheffield, UK.
http://www.webelements.com/

The current periodic table consists of 118 elements and is shown in figure 3. Each group is characterized by a number of distinct properties. For example, the noble gases in group 18 are very unreactive (though there are known compounds containing noble gases, such as XeF_4). Helium, the lightest of the noble gases, is used for filling balloons and has many industrial applications because it is non-flammable and does not typically form chemical compounds with any elements.

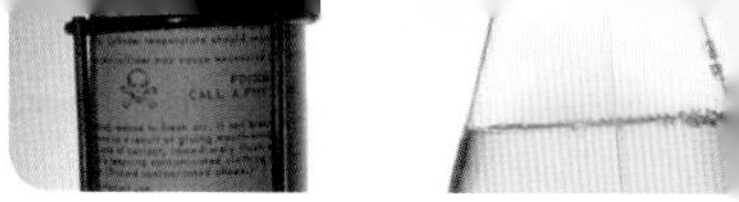

▲ Figure 3 The modern periodic table of elements

The horizontal rows of elements numbered from 1 to 7 are termed **periods**. The **period number** is equal to the principal quantum number, n, of the highest occupied energy level in the elements of the period. For example, calcium (Ca), $Z = 20$, is in period 4 so has four energy levels with $n = 1$, 2, 3, and 4.

Metals, non-metals, and metalloids

The periodic table is also split broadly into **metals** and **non-metals**; these are separated by a stepped diagonal line. The elements to the left of this line are the metals (excluding non-metallic hydrogen which is a gas) and the non-metals lie to the right.

Metals:

- are good **conductors** of heat and electricity

- are **malleable** (capable of being hammered into thin sheets)

- are **ductile** (capable of being drawn into wires)

- have **lustre** (they are shiny).

Mercury, Hg, $Z = 80$, is a liquid and can dissolve many other metals. The solutions formed in this way are called **amalgams**; for example, Ag−Sn−Hg can be used as a filling for teeth. We shall discuss metals further in sub-topic 4.5.

Non-metals

Non-metals are poor conductors of heat and electricity. Typically non-metals gain electrons in chemical reactions (they are **reduced**), whereas metals lose electrons (they are **oxidized**).

Metalloids

Some of the elements close to the stepped diagonal line have both metallic and non-metallic properties. The elements boron, B, $Z = 5$, silicon, Si, $Z = 14$, germanium, Ge, $Z = 32$, arsenic, As, $Z = 33$, antimony,

Quick question

Suggest **two** reasons why authorities in Sweden banned the use of mercury dental fillings since 2008.

71

Electron configurations and the periodic table

Sub-topic 2.2 showed that the electron configuration of an element can be expressed in three ways:

- full electron configuration
- condensed electron configuration
- orbital diagram.

For example, for fluorine, F, $Z = 9$:

- full electron configuration: $1s^2 2s^2 2p^5$
- condensed electron configuration: $[He]2s^2 2p^5$
- orbital diagram:

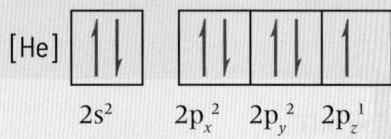

Figure 4 can be a powerful tool when writing electron configurations: the position of an element in the periodic table can be used to deduce the electron configuration, as the following worked example shows.

Worked example: deduction of the electron configuration from the element's position in the periodic table

1   Consider the element selenium, which has the chemical symbol Se.

   a) State the number of protons and electrons in an atom of Se.
   b) State in which group of the periodic table selenium belongs.
   c) State the number of valence electrons in an atom of Se.
   d) State the number of protons and electrons in the anion, Se^{2-}.
   e) Deduce the full electron configuration of Se.
   f) Deduce the condensed electron configuration of Se.
   g) Draw the orbital diagram for Se.

Solution

a) $Z = 34$, so Se has 34 protons and 34 electrons (atoms are neutral).

b) Se is in group 16 (the chalcogens).

c) Group 16 elements have 6 valence electrons.

d) For Se^{2-} the number of protons equals Z for Se, namely 34. However, since it is an anion carrying two negative charges it has gained two electrons, so it has a total of 36 electrons.

e) The full electron configuration for Se is $1s^2 2s^2 2p^6 3s^2 3p^6 3d^{10} 4s^2 4p^4$.

f) The condensed electron configuration for Se is $[Ar]3d^{10}4s^2 4p^4$.

g) The orbital diagram for Se is given below:

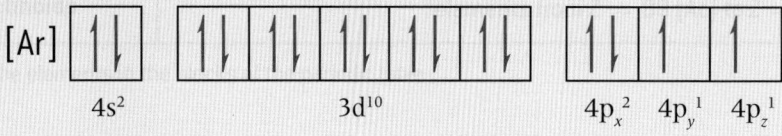

3.2 Periodic trends

Understanding

→ Vertical and horizontal trends in the periodic table exist for atomic radius, ionic radius, ionization energy, electron affinity, and electronegativity.

→ Trends in metallic and non-metallic behaviour are due to the trends above.

→ Oxides change from basic through amphoteric to acidic across a period.

Applications and skills

→ Prediction and explanation of the metallic and non-metallic behaviour of an element based on its position in the periodic table.

→ Discussion of the similarities and differences in the properties of elements in the same group, with reference to alkali metals (group 1) and halogens (group 17).

→ Construction of equations to explain the pH changes for reactions of Na_2O, MgO, P_4O_{10}, and the oxides of nitrogen and sulfur with water.

Nature of science

→ Looking for patterns – the position of an element in the periodic table allows scientists to make accurate predictions of its physical and chemical properties. This gives scientists the ability to synthesize new substances based on the expected reactivity of elements.

Trends in physical and chemical properties

Electron configurations (topic 2), which can be explained through quantum mechanics, help us understand many aspects of atomic properties such as atomic radius, ionization energy, electron affinity, and electronegativity. These properties, described in this topic, in turn provide a better understanding of chemical reactions. At the same time, properties are peppered with patterns and trends, and these patterns are mirrored in chemical properties.

Patterns lie at the heart of the periodic table of elements – elements show trends in their atomic and chemical properties across periods and down groups. The position of an element in the periodic table allows scientists to make accurate predictions about its behaviour in chemical reactions and therefore facilitate the synthesis of new compounds.

Atomic radius

The radius of a circle, R_c, is the distance from the centre of the circle to a point on the circumference. It is easily measured and has a definite value.

In the Bohr model of the hydrogen atom (sub-topic 2.2) the core of the atom is the nucleus while the single electron lies in a fixed **orbit**. Based on this model it would appear that the radius of the atom, R_e, can also be measured, as according to Bohr the electron is in a fixed position within a defined orbit.

However, as described in topic 2 we now know that the Bohr model of the atom is highly simplistic and electrons are in fact located in **atomic orbitals**, which are regions of space where there is a high probability of finding an electron. This means that the position of the electron is not fixed, so we cannot measure the radius of the atom in the same way as we measure the radius of a circle. When looking at atomic models, we need to move away from the simplistic Bohr model where atoms are often represented as spheres. Based on quantum mechanics we know that atoms cannot be represented as spheres with fixed boundaries. The **boundary surface** (i.e. the atomic orbital) in fact represents a 99% probability of finding an electron in that region of space.

One way of overcoming this problem and finding the radius of an atom is to consider two non-metallic atoms chemically bonded together, that is, consider an X_2 diatomic molecule. The distance between the two nuclei of the X atoms is given by d, and the **bonding atomic radius**, R_b, is defined as:

$$R_b = \tfrac{1}{2}d$$

This is shown in figure 1, using the example of iodine. The bonding atomic radius is sometimes termed the **covalent radius**.

For metals the bonding atomic radius is $\tfrac{1}{2}d'$ where d' now represents the distance between two atoms adjacent to each other in the crystal lattice of the metal.

An alternative atomic radius is the **non-bonding atomic radius**, R_{nb}. Consider a group of gaseous argon atoms. When two argon atoms collide with one another there is very little penetration of their electron cloud densities. Argon does not form a diatomic species. If argon is frozen in the solid phase the atoms would touch each other (topic 1) but would not be chemically bonded. In this case the distance between the argon

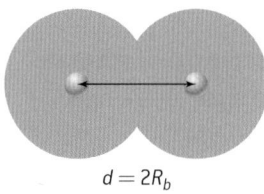

$d = 2R_b$

▲ Figure 1 The iodine diatomic molecule, I_2. The bonding atomic radius, R_b, for iodine is 136 pm ($d = 272$ pm), where 1 pm $= 10^{-12}$ m

Useful resource

The "Periodic Table of Videos" website, developed by Professor Martyn Poliakoff, CBE and co-workers at the University of Nottingham in the UK provides videos for all 118 elements. Professor Poliakoff is a research professor and is also a pioneer in the field of **green chemistry** which is discussed at several points in the IB Chemistry Diploma programme. http://www.periodicvideos.com/

TOK

- We saw in sub-topic 3.1 that Mendeleev examined the properties of elements in minute detail and grouped elements with similar properties together. When Mendeleev published his first periodic table of elements in 1869 he left gaps in the table for as yet undiscovered elements, and hence elements fell into their correct groups. Mendeleev was therefore able to predict the properties of yet undiscovered elements at the time. The predictive power of Mendeleev's periodic table illustrates the "risk taking" nature of science. What is the distinction between scientific and pseudoscientific claims?

- The periodic table is an excellent example of classification in science. It classifies elements in several ways – metals, non-metals, and metalloids; main-group and transition elements; groups and periods; elements with acidic, basic, and amphoteric oxides; and s, p, d, and f sublevels. How do classification and categorization help and hinder the pursuit of knowledge? For example, scandium will be discussed further in topic 13. Why is it incorrect to classify scandium as a non-transition element?

atoms could be measured and hence R_{nb} could be found (figure 2). The non-bonding atomic radius is often termed the **van der Waals' radius**.

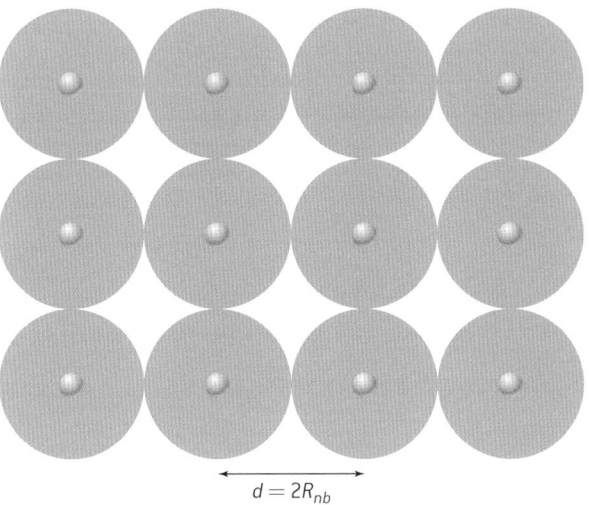

$$d = 2R_{nb}$$

▲ Figure 2 Atoms of argon in the solid phase. The atoms are touching but not chemically bonded. The non-bonding atomic radius of argon R_{nb} is 188 pm ($d = 376$ pm)

Section 9 of the *Data booklet* provides data for the covalent atomic radii of the elements. The general term "atomic radius" is used to represent the mean bonding atomic radius obtained from experimental data over a wide range of elements and compounds. Note that the bonding atomic radius is always smaller than the non-bonding atomic radius. The approximate bond length between two elements can also be estimated from their atomic radii.

For example, for the interhalogen compound BrF:

    atomic radius of bromine = 117 pm

    atomic radius of fluorine = 60 pm

    bond length of Br−F = 177 pm

Compare this with the experimental bond length of Br−F in the gas phase (176 nm).

> **Quick question**
>
> Predict the bond lengths in:
>
> **a)** iodine monobromide, IBr
>
> **b)** trichloromethane (chloroform), $CHCl_3$.

Effective nuclear charge and screening effect

In an atom the negatively charged electrons are attracted to the positively charged nucleus. A valence or outer-shell electron is also repelled by the other electrons in the atom. The **core electrons** in the inner non-valence energy levels of the atom reduce the positive nuclear charge experienced by a valence electron. This effect of reducing the nuclear charge experienced by an electron is termed **screening** or **shielding**.

The net charge experienced by an electron is termed the **effective nuclear charge**, Z_{eff}. This is the nuclear charge, Z, (representing the atomic number) minus the charge, S, that is shielded or screened by the core electrons:

$$Z_{eff} = Z - S$$

where Z = actual nuclear charge (atomic number) and S = **screening or shielding constant**.

Z_{eff} can be worked out using **Slater's rules**. You can read about these rules in advanced textbooks on inorganic chemistry, but you are not required to calculate Z_{eff} using Slater's rules as part of the IB Chemistry Diploma programme. You do need to understand the principle of screening and for our purposes you can consider S as a parameter related to the number of core electrons in an atom.

Worked example: estimating nuclear charge

Estimate the effective nuclear charge experienced by the valence electron in the alkali metal potassium.

(For comparison, using Slater's rules Z_{eff} for potassium is calculated as 2.2.) As chemists we need to be aware of the limitations of many of our assumptions, equations, and rules.

Solution

Potassium, K has the electron configuration $1s^2 2s^2 2p^6 3s^2 3p^6 4s^1$ and $Z = 19$.

K has a total of 19 electrons and one valence electron (it is in group 1). This means there are 18 core electrons (figure 3). The valence electron does not experience the full force of attraction of the 19 protons that provide the nuclear charge. The 18 core electrons partially cancel this positive charge and the effective nuclear charge is approximately 1:

$$Z_{eff} \approx Z - S = 19 - 18 = 1$$

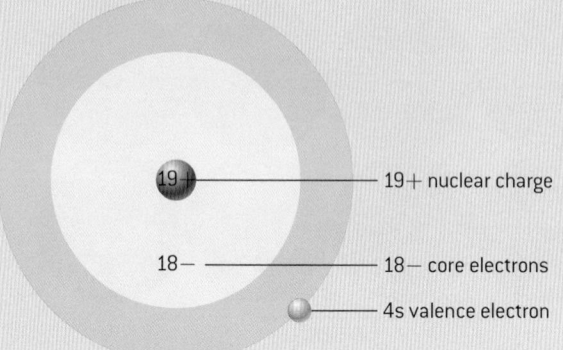

▲ Figure 3 Shielding of the outer valence electron in the potassium atom

Periodic trends in atomic radius

Across a period from left to right, atomic radii *decrease*. This is because of the increasing effective nuclear charge, Z_{eff}, going from left to right across the period. This pulls the valence (outer-shell) electrons closer to the nucleus, reducing the atomic radius.

Down a group from top to bottom, atomic radii *increase*. In each new period the outer-shell electrons enter a new energy level so are located further away from the nucleus. This has a greater effect than the increasing nuclear charge, Z, because of shielding by the core electrons.

These trends are summarized in figures 4 and 5. Figure 5 shows that the atomic radii of the transition elements do not change greatly across a period. The reason for this is that the number of electrons in the outermost energy level of the principal quantum number, n, remains almost constant across the period. As electrons are added they enter the $(n - 1)$ rather than the nth energy level. So the number of valence electrons and hence Z_{eff} remain essentially constant, resulting in little variation in atomic radius.

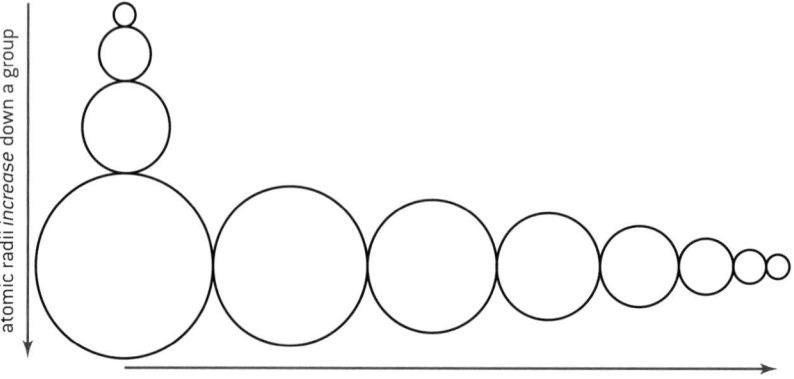

atomic radii *increase* down a group

atomic radii *decrease* across a period

▲ Figure 4 Trends in atomic radii. Some people think of these shapes as snowmen – going down a group the snowman is standing upright, while across a period the snowman is sleeping!

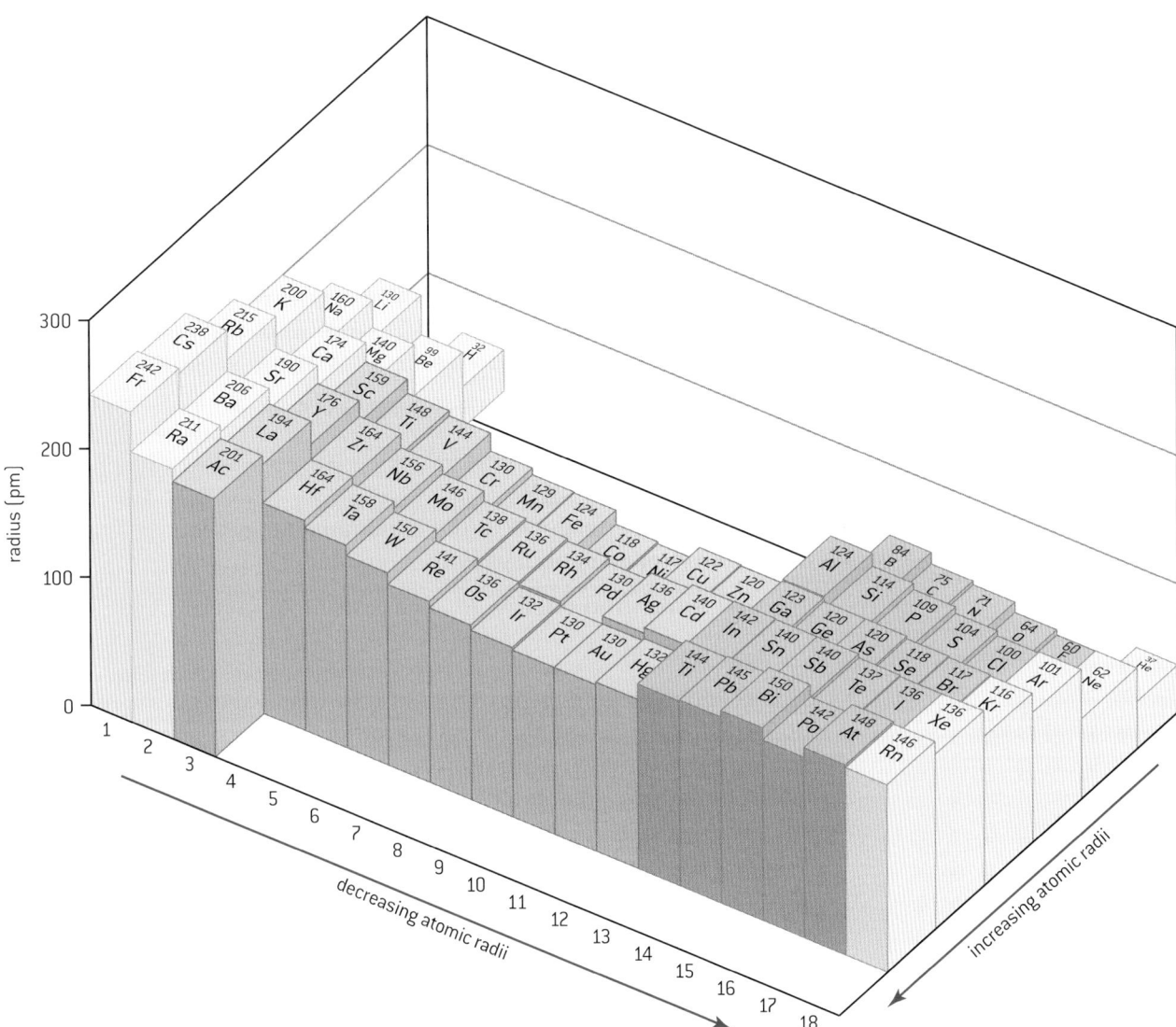

▲ Figure 5 Values of atomic radii of elements in pm. These data can be found in section 9 of the *Data booklet*

Periodic trends in ionic radius

The radii of cations and anions vary from the parent atoms from which they are formed in the following way.

The radii of cations are smaller than those of their parent atoms; for example, the atomic radius of K is 200 pm while the ionic radius of K^+ is 138 pm. The reason for this is that there are more protons than electrons in the cation so the valence electrons are more strongly attracted to the nucleus.

The radii of anions are larger than those of their parent atoms; for example, the atomic radius of F is 60 pm while the ionic radius of F^- is 133 pm. This is because the extra electron in the anion results in greater repulsion between the valence electrons.

Ions

An **ion** is a charged species. Ions are either cations or anions:

- A **cation** is positively charged, such as Na^+, Mg^{2+}.

- An **anion** is negatively charged, such as Cl^-, O^{2-}.

Values for ionic radii are also given in section 9 of the *Data booklet*.

potassium   **fluorine**

K^+ K   F F^-

atomic radius of potassium (K) = 200 pm
ionic radius of K^+ = 138 pm

atomic radius of fluorine (F) = 60 pm
ionic radius of F^- = 133 pm

▲ Figure 6 Atomic and ionic radii for potassium and fluorine

Ionization energy

The **ionization energy**, *IE*, is the minimum energy required to remove an electron from a neutral gaseous atom in its ground-state.

The **first ionization energy**, IE_1, of a gaseous atom relates to the process:

$$X(g) \rightarrow X^+(g) + e^-$$

The second ionization energy relates to the removal of a further electron from the ion $X^+(g)$, and the third ionization energy is associated with the removal of another electron from $X^{2+}(g)$. Values of ionization energy are quoted in kJ mol^{-1} (per mole of atoms).

The values of first ionization energies for the elements are provided in section 8 of the *Data booklet*. Ionization energy values are always positive, as there is an input of energy in order to remove an electron.

Periodic trends in ionization energy

Ionization energies vary across the periodic table. Across a period from left to right ionization energy values *increase* for the following reasons:

1   As the effective nuclear charge, Z_{eff}, increases from left to right across a period the valence electrons are pulled closer to the nucleus, so the attraction between the electrons and the nucleus increases. This makes it more difficult to remove an electron from the atom.

2   Atomic radii decrease across a period – because the distance between the valence electrons and the nucleus decreases, it becomes more difficult to remove an electron from the atom.

Going down a group from top to bottom ionization energy values *decrease* for the following reasons:

1   Atomic radii increase down a group, making it easier to remove an electron from the atom.

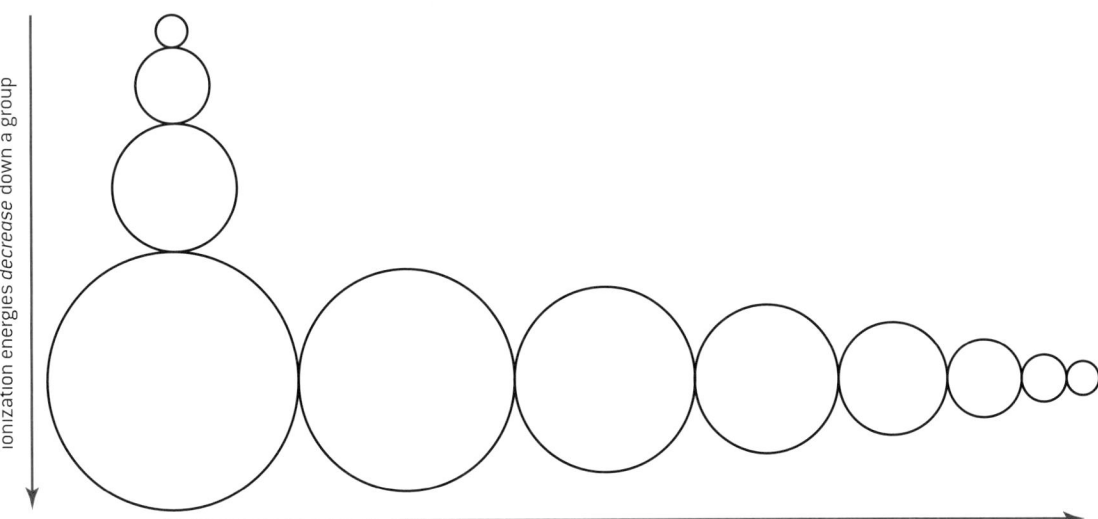

▲ Figure 7 Trends in ionization energy are the opposite of the trends in atomic radius

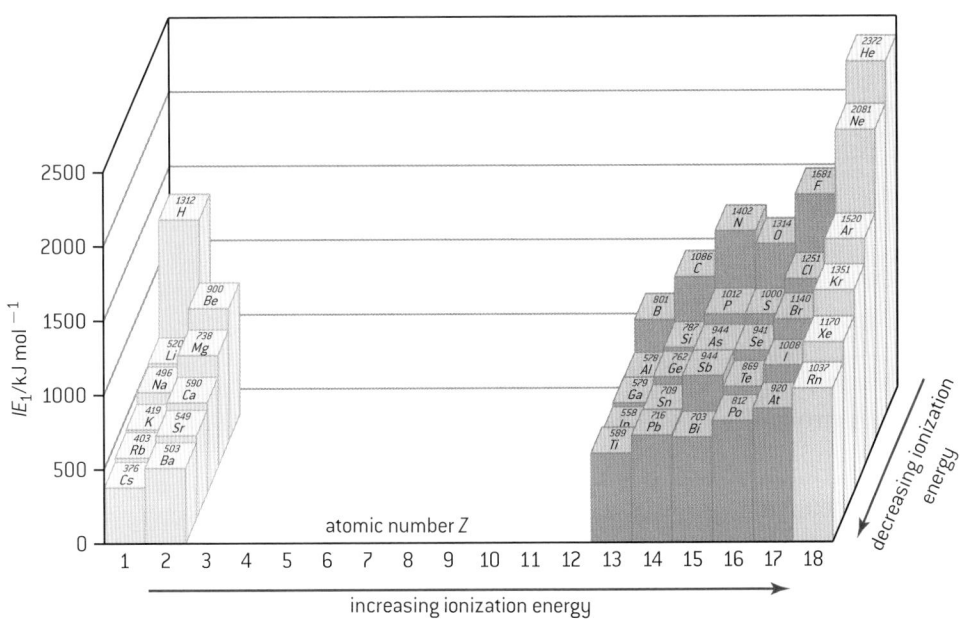

▲ Figure 8 Trends in first ionization energy, IE_1, for groups 1—2 and 13—18 of the periodic table. IE_1 values *increase* across a period and *decrease* down a group

2 The shielding effect of the core electrons increases faster than the nuclear charge, weakening the attractive force between the nucleus and outer electrons in the atom.

If a graph of first ionization energy versus atomic number is plotted, as shown in figure 9, the general trend is that first ionization energy values increase across a period but decrease down a group, though the graph is not smooth across a period. The spikes and dips will be explained in topic 12.

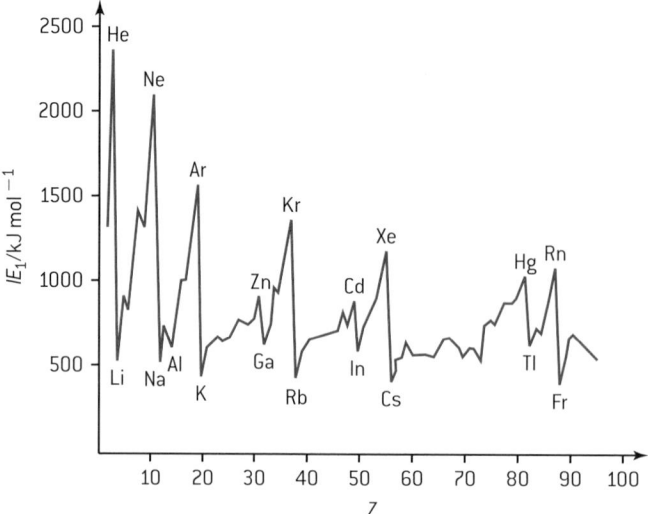

▲ Figure 9 Plot of first ionization energy, IE_1, versus atomic number, Z. Notice the general trend that IE_1 increases across a period but decreases down a group, though the graph is not smooth across a period

Electron affinity

According to IUPAC, the **electron affinity**, E_{ea}, is the energy required to detach an electron from the singly charged negative ion in the gas phase. This is the energy associated with the process:

$$X^-(g) \rightarrow X(g) + e^-$$

A more common and equivalent definition is that the electron affinity is the energy released ($E_{initial} - E_{final}$) when 1 mol of electrons is attached to 1 mol of neutral atoms or molecules in the gas phase:

$$X(g) + e^- \rightarrow X^-(g)$$

Electron affinity values are provided in section 8 of the *Data booklet*. For example, for fluorine:

$$F(g) + e^- \rightarrow F^-(g) \qquad E_{ea} = -328 \text{ kJ mol}^{-1}$$

The negative sign indicates that energy is released during this process: the process is exothermic (in contrast to ionization energies which relate to an endothermic process). The more negative the E_{ea} value, the greater is the attraction of the ion for the electron. However, figure 10 shows that the E_{ea} values for some elements, for example group 18 the noble gases, are positive.

	1	2		13	14	15	16	17	18
1	H -73								He >0
2	Li -60	Be >0		B -27	C -122	N >0	O -141	F -328	Ne >0
3	Na -53	Mg >0		Al -42	Si -134	P -72	S -200	Cl -349	Ar >0
4	K -48	Ca -2		Ga -41	Ge -119	As -78	Se -195	Br -325	Kr >0
5	Rb -47	Sr -5		In -29	Sn -107	Sb -101	Te -190	I 295	Xe >0

▲ Figure 10 Electron affinities E_{ea}, in kJ mol^{-1}, for a selection of main-group elements. Notice that some of the elements have positive E_{ea} values. The group 18 elements have theoretical, calculated values

Periodic trends in electron affinity

Trends in electron affinity across a period

Trends in electron affinity in the periodic table are not as well highlighted as the trends observed for atomic radius and ionization energy. In general, across a period from left to right E_{ea} values become *more negative* (with some exceptions).

The group 17 elements, the halogens, have the most negative E_{ea} values: for example, $E_{ea}(Cl) = -349$ kJ mol^{-1}. This is expected since on gaining an electron these elements attain the stable noble gas configuration. If you look across period 4 ($n = 4$ energy level) in figure 10 you can see that from left to right E_{ea} becomes more negative from -48 kJ mol^{-1} for K to -325 kJ mol^{-1} for Br. However, within each period, as for ionization energies, there are examples of elements that do not follow this trend. For example, arsenic, As, has E_{ea} -78 kJ mol^{-1} while you might expect this to lie between -119 kJ mol^{-1} for Ge and -195 kJ mol^{-1} for Se. The higher E_{ea} value for As can be explained by examining its electron configuration $[Ar]3d^{10}4s^2 4p_x^1 4p_y^1 4p_z^1$: if an electron is added it will enter a 4p orbital that already contains one electron, causing repulsion. A similar argument applies for other members of group 15, in particular for nitrogen where the E_{ea} value is positive.

Trends in electron affinity down a group

In the case of the group 1 alkali metals, values of E_{ea} generally become less negative going down the group (table 1). However, for the last three or four elements there is little difference between E_{ea} values.

Group 1 element	E_{ea} /kJ mol^{-1}
Li	-60
Na	-53
K	-48
Rb	-47
Cs	-46
Fr	-47

▲ Table 1 Electron affinity values for the group 1 elements

The patterns of electron affinity vary by group, so electron affinity values do not show the same clear trends down a group as do atomic radius, ionization energy, and electronegativity (discussed next).

Electronegativity

Electronegativity, symbol χ, is defined as the relative attraction that an atom has for the shared pair of electrons in a covalent bond.

In 1932 the American scientist Linus Pauling proposed the concept of electronegativity and defined it as "the power of an atom in a molecule to attract electrons to itself". There are a number of different electronegativity scales but the one used in section 8 of the *Data booklet* is the **Pauling scale**, which has the symbol χ_p. On this scale fluorine, the most electronegative element in the periodic table, has a value of electronegativity of 4.0.

 Positive E_{ea} values

A positive value for electron affinity suggests that the anion is not stable, so it cannot be formed in the gas phase. For example, E_{ea} for krypton is positive (41 kJ mol^{-1}), so Kr$^-$ does not exist. Interestingly, the N^{3-} anion is well known in the solid state (for example, in sodium nitride, Na$_3$N), despite the fact that E_{ea} for nitrogen is positive (20 kJ mol^{-1}). In crystals, N^{3-} is stabilized by the lattice enthalpy (sub-topic 15.1), which provides sufficient energy to overcome the electron repulsion in the nitride anion.

Periodic trends in electronegativity

As shown in figure 11, electronegativities show periodic trends across a period and down a group that mirror those for ionization energies, for the same reasons (see pages 80–81).

Across a period from left to right electronegativity values *increase* and atomic radii *decrease* because the effective nuclear charge increases.

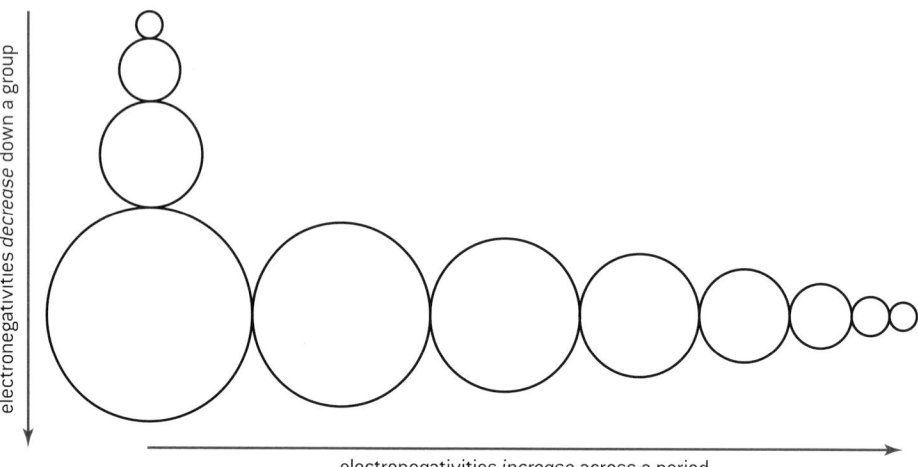

▲ Figure 11 Trends in electronegativity are the same as those in ionization energy and the opposite to the trends in atomic radius

Down a group from top to bottom electronegativity values *decrease* because atomic radii increase and although the nuclear charge, Z, increases, its effect is shielded by the core electrons.

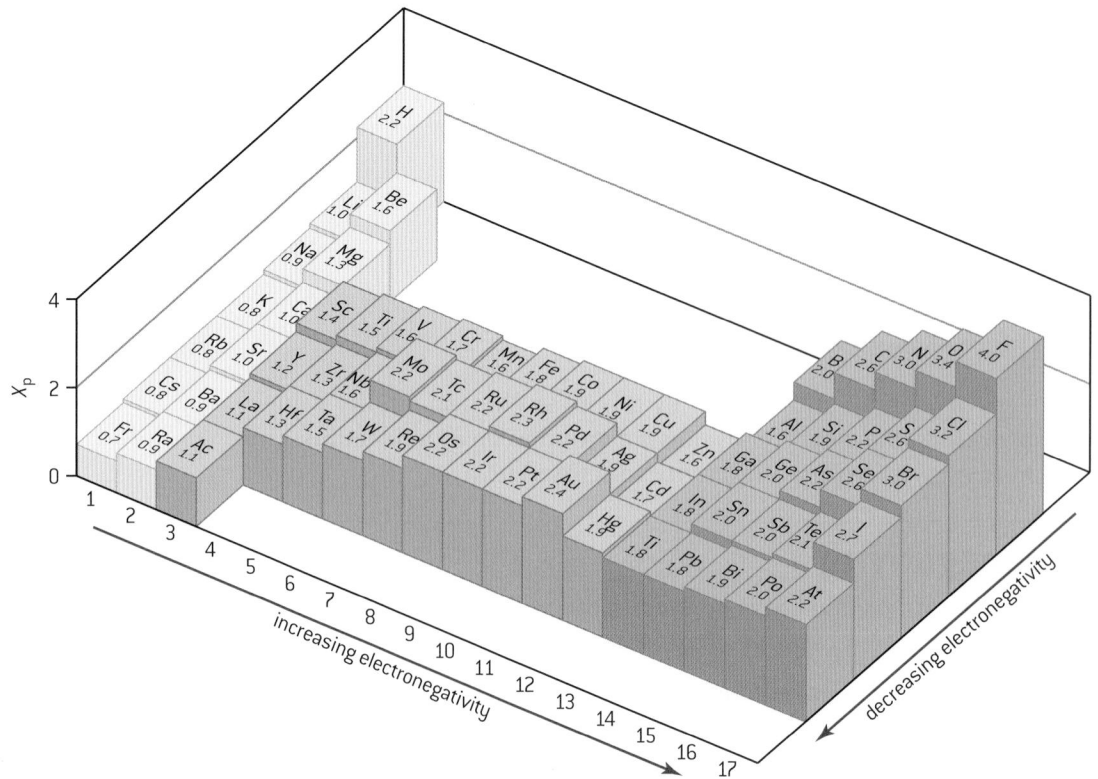

▲ Figure 12 Electronegativity values, χ_p, increase across a period from left to right and decrease down a group from top to bottom. Fluorine is the most electronegative element in the table with a χ_p value of 4.0 on the Pauling scale

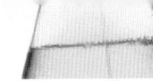

 Science and peace

Pauling was the first person to win two unshared Nobel Prizes, as he also won the Nobel Peace Prize in 1962 for his opposition to weapons of mass destruction.

- Do you know of any other scientists who have promoted peace through their scientific work? What role can scientists play in the promotion of peace in the world today? Discuss this in class.

Pauling also suggested that taking large doses of vitamin C (ascorbic acid) may be effective against the common cold. (The structure of ascorbic acid is given in section 35 of the *Data booklet*.) Was Pauling's suggestion correct? Carry out some research into this aspect, using the library, the scientific literature, and an online search. Discuss your findings in class.

Periodic trends in metallic and non-metallic character

As described in sub-topic 3.1, the elements in the periodic table can be classified into metals, non-metals, and metalloids (see figure 3 in sub-topic 3.1).

Metallic character *decreases* across a period and *increases* down a group, as shown in figure 13.

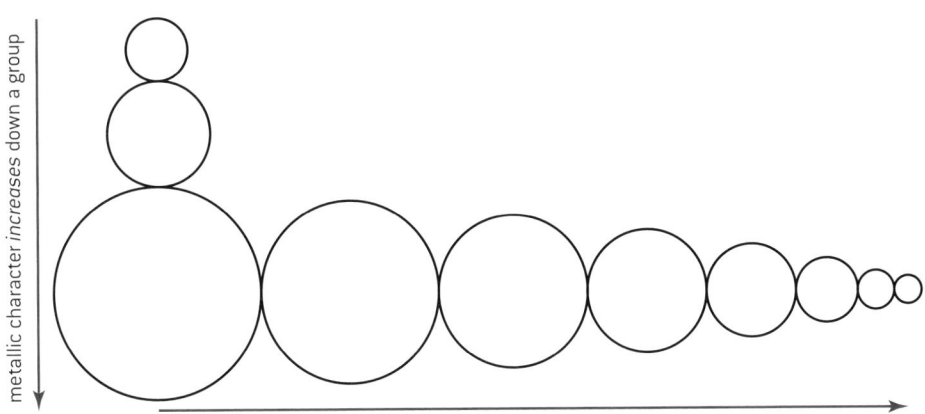

▲ Figure 13 Trends in metallic character in the periodic table

As well as the properties of metals described previously in sub-topic 3.1, metals also have low ionization energy values – they have a tendency to lose electrons during a chemical reaction, that is, they tend to be **oxidized**. We shall explore redox processes further in topic 9.

The properties of non-metals were also described in sub-topic 3.1; in addition, non-metals show highly negative electron affinities – they have a tendency to gain electrons during a chemical reaction, that is, they tend to be **reduced**.

Figure 14 shows the charges of some common ions of metals and non-metals. For the cations of the alkali metals in group 1 the charge is always 1+, and for the alkaline earth metals in group 2 it is always 2+. In topic 13 we shall see that the transition metals form a number of different stable ions.

1							18
1	2	13	14	15	16	17	
2	Li^+				N^{3-}	O^{2-}	F^-
3	Na^+ Mg^{2+}	Al^{3+}		P^{3-}	S^{2-}	Cl^-	
4	K^+ Ca^{2+}				Se^{2-}	Br^-	
5	Rb^+ Sr^{2+}				Te^{2-}	I^-	
6	Cs^+ Ba^{2+}						

▲ Figure 14 The charges of some common ions

Trends in the properties of metal and non-metal oxides

An **oxide** is formed from the combination of an element with oxygen. We make use of the charge on the metal cation as shown in figure 15 to deduce the chemical formula of a metal oxide, taking the charge on the oxide ion to be 2−, for example:

- Na^+ combines with O^{2-} to form Na_2O
- Ca^{2+} combines with O^{2-} to form CaO
- Al^{3+} combines with O^{2-} to form Al_2O_3.

Metal oxides are *basic*: they react with water to form metal hydroxides:

$$CaO(s) + H_2O(l) \rightarrow Ca(OH)_2(aq)$$

$$Na_2O(s) + H_2O(l) \rightarrow 2NaOH(aq)$$

In contrast, oxides of the non-metals are *acidic*: they react with water to form acidic solutions:

$$CO_2(g) + H_2O(l) \rightleftharpoons H_2CO_3(aq) \qquad \text{carbonic acid}$$

$$SO_3(l) + H_2O(l) \rightarrow H_2SO_4(aq) \qquad \text{sulfuric acid}$$

$$SO_2(g) + H_2O(l) \rightleftharpoons H_2SO_3(aq) \qquad \text{sulfurous acid}$$

$$P_4O_{10}(s) + 6H_2O(l) \rightarrow 4H_3PO_4(aq) \qquad \text{phosphoric acid}$$

Naming oxoanions and acids

Students often struggle with the names of the oxoanions and their corresponding oxoacids. Table 2 summarizes some of these names.

Formula of oxoanion	Non-systematic name
CO_3^{2-}	carbonate
$C_2O_4^{2-}$	ethanedioate (oxalate)
NO_2^-	nitrite
NO_3^-	nitrate
SO_3^{2-}	sulfite
SO_4^{2-}	sulfate
PO_3^{3-}	phosphite
PO_4^{3-}	phosphate
ClO^-	hypochlorite
ClO_2^-	chlorite
ClO_3^-	chlorate
ClO_4^-	perchlorate
OH^-	hydroxide

▲ Table 2 The non-systematic names of some oxoanions

In naming oxoanions the following rules are useful:

- If only one oxoanion is formed, the ending is "-ate".
- If two oxoanions are formed, the one with the smaller number of oxygens ends in "-ite" and the one with the greater number of oxygens ends in "-ate".
- If there are four oxoanions, the one with the smallest number of oxygens ends in "-ite" and is prefixed by "hypo"; the next ends in "-ite"; the third ends in "-ate", and the one with the most oxygens is prefixed by "per" and ends in "-ate". The four oxoanions of chlorine, bromine, and iodine follow this system (table 3).

Formula of oxoanion	Non-systematic name	Formula of oxoacid	Non-systematic name
ClO^-	hypochlorite	$HClO$	hypochlorous acid
ClO_2^-	chlorite	$HClO_2$	chlorous acid
ClO_3^-	chlorate	$HClO_3$	chloric acid
ClO_4^-	perchlorate	$HClO_4$	perchloric acid

▲ Table 3 The oxoanions and acids of chlorine

Some interesting oxides

- Silicon dioxide, SiO_2, does not dissolve in water. However, it is classified as an acidic oxide because it can react with sodium hydroxide, NaOH to form sodium silicate, $Na_2SiO_3(aq)$ and water:

$$SiO_2(s) + 2NaOH(aq) \rightarrow Na_2SiO_3(aq) + H_2O(l)$$

- Aluminium oxide, Al_2O_3 is classified as an **amphoteric** oxide. This means it can react both as an acid and as a base. See topic 8 for more information.

Acting as an acid:

$$Al_2O_3(s) + 2NaOH(aq) + 3H_2O(l) \rightarrow 2NaAl(OH)_4(aq)$$
$$\text{sodium aluminate}$$

Acting as a base:

$$Al_2O_3(s) + 6HCl(aq) \rightarrow 2AlCl_3(aq) + 3H_2O(l)$$
$$\text{aluminium}$$
$$\text{chloride}$$

Amphoteric and amphiprotic oxides

The terms amphoteric and amphiprotic are often mixed up. Amphiprotic species are described further in sub-topic 8.1.

- According to the *IUPAC Gold Book*, a chemical species that behaves both as an acid and as a base is termed **amphoteric**. Aluminium oxide is classified as an amphoteric oxide.

- A particular type of amphoteric species is described as **amphiprotic**. These are species that are either proton (H^+) donors or proton acceptors. Examples include self-ionizing solvents (such as water, H_2O and methanol, CH_3OH), amino acids, and proteins.

Table 4 shows how the oxides of some period 3 elements vary. It shows that there is a trend from basic through amphoteric to acidic oxides across the period from left to right.

Formula of oxide	$Na_2O(s)$	$MgO(s)$	$Al_2O_3(s)$	$SiO_2(s)$	$P_4O_{10}(s)$	$SO_3(l)$ and $SO_2(g)$
Nature of oxide	basic	basic	amphoteric	acidic	acidic	acidic

▲ Table 4 Trend in the properties of the oxides of some period 3 elements

 Chemical properties within a group: Group 1, the alkali metals

The group 1 metals are lithium, Li, sodium, Na, potassium, K, rubidium, Rb, caesium, Cs, and francium, Fr (see sub-topic 3.1, figure 3). Note that hydrogen is not a member of the alkali metals – it is a non-metal and a gas.

The group 1 metals are characterized by having one valence electron; they therefore form the ion M^+ in ionic compounds by losing this electron (they are oxidized, topic 9). For example:

$$Na \quad - \quad 1e^- \quad \rightarrow \quad Na^+$$
$$[Ne]3s^1 \qquad\qquad [Ne]$$

On descending group 1 the atomic radius increases and the ionization energy decreases. The reactions of the alkali metals with water therefore become more vigorous further down the group. Less energy is required to remove the valence electron from potassium, K ($IE_1 = 419$ kJ mol^{-1}) than from sodium, Na ($IE_1 = 496$ kJ mol^{-1}), for example.

Reaction with water

The group 1 metals react with water to form a metal hydroxide, MOH(aq), which gives an alkaline solution (table 5). Hydrogen gas is also liberated in this reaction:

$$2M(s) + 2H_2O(l) \rightarrow 2MOH(aq) + H_2(g)$$

Group 1 metal	Reaction with water	Description
Li	$2Li(s) + 2H_2O(l) \rightarrow 2LiOH(aq) + H_2(g)$	Lithium reacts slowly and floats on the water (due to its low density). Bubbling is observed.
Na	$2Na(s) + 2H_2O(l) \rightarrow 2NaOH(aq) + H_2(g)$	Sodium reacts vigorously. Heat is evolved and the sodium melts to form a ball of molten metal which whizzes around on the surface of the water.
K	$2K(s) + 2H_2O(l) \rightarrow 2KOH(aq) + H_2(g)$	Potassium reacts more vigorously than sodium: the reaction is violent. It evolves enough heat to ignite the hydrogen, so bursts into flames instantly. A characteristic lilac-coloured flame is observed.
Rb	$2Rb(s) + 2H_2O(l) \rightarrow 2RbOH(aq) + H_2(g)$	Both rubidium and caesium react explosively with water.
Cs	$2Cs(s) + 2H_2O(l) \rightarrow 2CsOH(aq) + H_2(g)$	

▲ Table 5 Reactions of the alkali metals with water become progressively more violent as you descend the group

Only two elements in the periodic table exist as liquids: bromine, Br_2 and mercury, Hg.

Chemical properties within a group: Group 17, the halogens

The group 17 elements, the halogens, are the non-metals fluorine, F, chlorine, Cl, bromine, Br, iodine, I, and astatine, At (see sub-topic 3.1, figure 3). Their chemistry is characterized by their seven valence electrons, giving them a tendency to gain an electron to attain the noble gas configuration (they are reduced, topic 9). For example:

Cl   + e$^-$ →   Cl$^-$
$[Ne]3s^23p^5$ $[Ne]3s^23p^6$ or simply [Ar]

The group 17 elements exist as diatomic molecules X_2. Fluorine and chlorine are gases, bromine is a liquid, iodine and astatine are solids at room temperature and pressure.

The halogens form ionic compounds with metals, with the X^- anion combining with the metal cation (see topic 4 for details of the structure and bonding of ionic compounds). With non-metals the halogens form covalent compounds. Halogens in general are highly reactive, though the reactivity decreases going down the group with the most reactive halogen being fluorine. The reason for this decrease in reactivity descending the group is that the atomic radius increases down the group making it less easy to gain an electron.

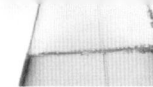

Reaction between halogens and alkali metals

The halogens, X_2, react with the alkali metals, M(s) to form ionic alkali metal halide salts, MX(s). In the ionic compound, MX(s), the cation is M^+ and the anion is X^-:

$$2M(s) + X_2(g) \rightarrow 2MX(s)$$

For example:

$$2Na(s) + Cl_2(g) \rightarrow 2NaCl(s)$$

Reactions between halogens and halides

A solution of a more reactive halogen, X_2(aq), will react with a solution of halide ions, X^-(aq), formed by a less reactive halogen. A summary of these reactions is given in table 6.

In table 6 the reactions are represented as **ionic equations**. A complete balanced equation can also be written. For example, when an aqueous solution of chlorine is added to a colourless solution of potassium bromide, aqueous potassium chloride is formed, which is colourless, and the yellow/orange colour observed is due to the formation of bromine, Br_2(aq) (figure 15):

$$Cl_2(aq) + 2KBr(aq) \rightarrow 2KCl(aq) + Br_2(aq)$$
$$\text{colourless} \qquad\qquad \text{yellow/orange}$$

▲ Figure 15 Gaseous chlorine, Cl_2(g), is bubbled through a solution of potassium bromide, which is initially colourless. On reaction, aqueous bromine is displaced from the potassium bromide solution and the yellow/orange colour of Br_2(aq) is observed

Study tip

You can think of this displacement reaction as being a competition between the chlorine and the bromine for an extra electron. Remember that the atomic radius *increases* down a group (figure 4). The atomic radius of chlorine (100 pm) is smaller than that of bromine (117 pm) so chlorine has a stronger attraction for a valence electron than does bromine. Therefore chlorine forms the chloride anion, Cl^- more readily than bromine forms the bromide anion, Br^-. Going down group 17 the **oxidizing ability**, that is, the ability to gain an electron, decreases.

X_2(aq)	Cl^-(aq)	Br^-(aq)	I^-(aq)
Cl_2(aq)	no reaction	Cl_2(aq) + $2Br^-$(aq) → $2Cl^-$(aq) + Br_2(aq) observation: yellow/orange solution due to formation of Br_2(aq)	Cl_2(aq) + $2I^-$(aq) → $2Cl^-$(aq) + I_2(aq) observation: dark red/brown solution due to formation of I_2(aq)
Br_2(aq)	no reaction	no reaction	Br_2(aq) + $2I^-$(aq) → $2Br^-$(aq) + I_2(aq) observation: dark red/brown solution due to formation of I_2(aq)
I_2(aq)	no reaction	no reaction	no reaction

▲ Table 6 Reactions between halogens X_2(aq) and halides X^-(aq)

Worked example: explaining pH changes

Construct a balanced equation, including state symbols, to explain the pH changes for the reaction of nitrogen dioxide with water (see sub-topic 8.1).

Solution

- Nitrogen is a non-metal and therefore may form an acidic oxide. NO_2 reacts with water to form a 1:1 mixture of nitrous acid, HNO_2, and nitric acid, HNO_3. Nitrous acid is a weak acid and nitric acid is a strong acid.

- We next write the balanced chemical equation:

$$2NO_2 + H_2O \rightarrow HNO_2 + HNO_3$$

- Finally, we include the state symbols:

$$2NO_2(g) + H_2O(l) \rightarrow HNO_2(aq) + HNO_3(aq)$$

- Because a mixture of acids is formed the pH of the solution will be less than 7 (see topic 8).

Questions

1 What is the maximum number of electrons that can occupy a d sublevel?

   A. 2

   B. 5

   C. 6

   D. 10

2 Which of the following elements can be classified as metalloids?

   I.   Al

   II. Si

   III. Te

   A   I and II only

   B   I and III only

   C   II and III only

   D   I, II, and III

3 How many valence electrons does selenium contain?

   A. 2

   B. 6

   C. 16

   D. 34

4 Which of the following elements are alkaline earth metals?

   I.   Rb

   II. Sr

   III. Ba

   A. I and II only

   B. I and III only

   C. II and III only

   D. I, II, and III

5 Which property generally **decreases** across period 3?

   A. Atomic number

   B. Electronegativity

   C. Atomic radius

   D. First ionization energy   [1]

   IB, May 2011

6 Which ion has the largest radius?

   A. Cl^-

   B. K^+

   C. Br^-

   D. F^- [1]

   IB, May 2010

7 What happens when sodium is added to water?

   I.   A gas is evolved.

   II. The temperature of the water increases.

   III. A clear, colourless solution is formed.

   A. I and II only

   B. I and III only

   C. II and III only

   D. I, II, and III   [1]

   IB, November 2009

8 Which oxides produce an acidic solution when added to water?

   I. P_4O_{10}

   II. MgO

   III. SO_3

   A. I and II only

   B. I and III only

   C. II and III only

   D. I, II, and III   [1]

   IB, May 2010

9 Which statement about the elements in group 17 is correct?

   A. Br_2 will oxidize Cl^-.

   B. F_2 has the least tendency to be reduced.

   C. Cl_2 will oxidize I^-.

   D. I_2 is a stronger oxidizing agent than F_2.   [1]

   IB, May 2011

10 How many of the following oxides are amphoteric?

Na_2O, MgO, Al_2O_3, SiO_2

A. None

B. 1

C. 2

D. 4

11 The periodic table shows the relationship between electron arrangement and the properties of elements and is a valuable tool for making predictions in chemistry.

a) Identify the property used to arrange the elements in the periodic table. [1]

b) Outline **two** reasons why electronegativity increases across period 3 in the periodic table and **one** reason why noble gases are not assigned electronegativity values. [3]

IB, May 2010

12 Describe and explain what you will see if chlorine gas is bubbled through a solution of:

a) potassium iodide [2]

b) potassium fluoride. [1]

IB, May 2010

13 The alkali metals are found in group 1 of the periodic table of elements.

a) State the full electron configuration of K and its ion, K^+.

b) Describe what you understand by the term *first ionization energy*.

c) State and explain how the first ionization energies of the alkali metals vary going down group 1.

d) Explain why the ionic radius of K^+ is smaller than the atomic radius of the parent atom, K.

e) Suggest why you should never touch an alkali metal with your fingers when working in the laboratory.

4 CHEMICAL BONDING AND STRUCTURE

Introduction

At the very heart of chemistry lies our understanding of chemical bonding and the structural arrangements in compounds. A chemical bond can be considered as the "glue" that holds atoms together in a molecule, or holds oppositely charged ions (charged species) together in the case of an ionic compound. In this topic we shall explore three different types of bonding – ionic, covalent, and metallic – and look at the differences in structure between ionic and covalent compounds. For covalent compounds we shall see how a simple model, valence shell electron pair repulsion (VSEPR) theory, can be used to determine the shape of a molecule, and we shall also look at some key chemical principles, such as polarity and intermolecular forces.

4.1 Ionic bonding and structure

Understandings

→ Positive ions (cations) form by metals losing valence electrons.

→ Negative ions (anions) form by non-metals gaining electrons.

→ The number of electrons lost or gained is determined by the electron configuration of the atom.

→ The ionic bond is due to electrostatic attraction between oppositely charged ions.

→ Under normal conditions, ionic compounds are usually solids with lattice structures.

Applications and skills

→ Deduction of the formula and name of an ionic compound from its component ions, including polyatomic ions.

→ Explanation of the physical properties of ionic compounds (volatility, electrical conductivity, and solubility) in terms of their structure.

Nature of science

→ Use theories to explain natural phenomena – molten ionic compounds conduct electricity but solid ionic compounds do not. The solubility and melting points of ionic compounds can be used to explain observations.

<table>
<tr><td>

Definition of an ionic bond

An **ionic bond** refers to the electrostatic attraction experienced between the electric charges of a **cation** (positive ion) and an **anion** (negative ion).

</td></tr>
</table>

Ionic bonding

Ions are formed when one or more electrons are transferred from one atom to another. The driving force for this electron transfer is usually the formation of a noble gas electron configuration.

For example, the electron configuration of sodium, Na is:

$[Ne]3s^1$

where [Ne] is the noble gas core. A sodium atom can lose its one valence (outer-shell) electron to form the Na^+ cation, [Ne]. That is:

$Na - e^- \rightarrow Na^+$

We say that sodium is **oxidized** in this process (it loses an electron).

The electron configuration of chlorine, Cl is:

$[Ne]3s^23p^5$

If a chlorine atom gains an electron to form the Cl^- anion it will adopt a noble gas configuration, $[Ne]3s^23p^6$ or [Ar]. That is:

$Cl + e^- \rightarrow Cl^-$

We say that chlorine is **reduced** in this process (it gains an electron).

Hence, the electron that is lost by sodium is gained by chlorine in the formation of the ionic compound sodium chloride, NaCl.

Ionic compounds are generally formed between metals and non-metals, but note that the strict definition involves electrostatic attraction between a cation and an anion (for example, the compound ammonium chloride, NH_4Cl, which consists of the ammonium cation, NH_4^+, and the chloride anion, Cl^-, is ionic, but does not contain a metal).

Let us take another example of an ionic compound, magnesium oxide. Magnesium is a group 2 alkaline earth metal, and so has two valence electrons:

$[Ne]3s^2$

A magnesium atom can lose these two electrons forming Mg^{2+}, which also adopts the [Ne] noble gas core. That is:

$Mg - 2e^- \rightarrow Mg^{2+}$

Magnesium is **oxidized** in this process.

Oxygen is in group 16, the chalcogen group, and so has six valence electrons. The electron configuration of oxygen is:

$[He]2s^22p^4$

An oxygen atom can gain two electrons to form the O^{2-} anion, which adopts a noble gas configuration:

$[Ne]$ or $[He]2s^22p^6$

That is:

$O + 2e^- \rightarrow O^{2-}$

Oxygen is **reduced** in this process.

Hence, the two electrons that are lost by magnesium are gained by oxygen in the formation of the ionic compound magnesium oxide, MgO.

Under normal conditions, ionic compounds are typically solids, and have **lattice-type** structures that consist of three-dimensional repeating units of positive and negative ions (figure 1).

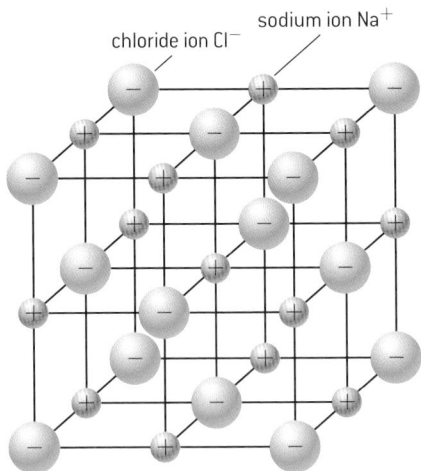

chloride ion Cl^-

sodium ion Na^+

▲ Figure 1 Lattice structure of sodium chloride, which consists of sodium cations, Na^+, and chloride anions, Cl^-. From the ionic radii given in section 9 of the *Data booklet* you can see that Na^+ (102×10^{-12} m = 102 pm) is smaller than Cl^- (181×10^{-12} m = 181 pm)

Ion	Name
NH_4^+	ammonium
OH^-	hydroxide
NO_3^-	nitrate
HCO_3^-	hydrogencarbonate
CO_3^{2-}	carbonate
SO_4^{2-}	sulfate
PO_4^{3-}	phosphate

▲ Table 1 Names of various ions

The octet rule

The **octet rule** has its own place in the discussion of chemical bonding and can be a useful starting point in trying to understand how chemical bonds are formed. The rule states that elements tend to lose electrons (that is, undergo oxidation), gain electrons (reduction), or share electrons in order to acquire a noble gas core electron configuration. The first two processes are the basis of ionic bonding. The third process is the basis of covalent bonding, which we shall discuss in sub-topic 4.2.

Worked example: deduction of the formula and name of an ionic compound

Deduce the formula and name of the ionic compounds formed between the following pairs of elements and/or polyatomic species:

a) magnesium and fluorine

b) aluminium and oxygen

c) sodium and oxygen

d) calcium and nitrate

e) ammonium and phosphate.

Solution

	Combination	Formula	Name
a)	magnesium and fluorine	Mg is in group 2, so forms Mg^{2+}; F is in group 17, so forms F^-; **formula is MgF_2**	magnesium fluoride
b)	aluminium and oxygen	Al is in group 3, so forms Al^{3+}; O is in group 16, so forms O^{2-}; **formula is Al_2O_3**	aluminium oxide
c)	sodium and oxygen	Na is in group 1, so forms Na^+; O is in group 16, so forms O^{2-}; **formula is Na_2O**	sodium oxide
d)	calcium and nitrate	Ca is in group 2, so forms Ca^{2+}; nitrate is NO_3^-; **formula is $Ca(NO_3)_2$**	calcium nitrate
e)	ammonium and phosphate	ammonium is NH_4^+; phosphate is PO_4^{3-}; **formula is $(NH_4)_3(PO_4)$**	ammonium phosphate

▲ Table 2 Formulas and names of some ionic compounds from their component ions. In naming ionic binary compounds, **AB**, consisting of a metal and a non-metal, the ending will be -ide

 # Physical properties of ionic compounds

Melting and boiling points

Ionic compounds have high melting points and high boiling points because of the strong electrostatic forces of attraction between the ions in their lattice structures. For example, the melting point of NaCl is 801 °C and its boiling point is 1413 °C. In order to melt an ionic solid there must be a large input of energy to break apart the electrostatic forces.

The electrostatic force of attraction, F, is directly proportional to the interacting charges, Q_1 and Q_2, and inversely proportional to the square of the distance between them, r^2, as given by **Coulomb's law of electrostatics** from physics:

$$F \propto \frac{Q_1 Q_2}{r^2}$$

Hence, in the case of magnesium oxide, the two charges correspond to 2+ for the magnesium cation, Mg^{2+}, and 2− for the oxide anion, O^{2-}. As these two charges are greater than those of 1+ and 1− in the case of the Na^+ and Cl^- ions, the melting point for MgO is higher, that is 2852 °C.

Volatility

Volatility refers to the tendency of a substance to vaporize. For ionic compounds the electrostatic forces of attraction are strong, and so the volatility of such compounds is very low.

Electrical conductivity

For an ionic compound in the solid state the ions occupy fixed positions in the lattice. Hence the ions are not free to move in the solid state, so solid ionic compounds do not conduct electricity. In contrast, in the molten state, the ions are free to move and conduct electricity.

Solubility

Ionic compounds dissolve in polar solvents such as water, but do not dissolve in non-polar solvents such as hexane. The molecule of water is polar and has partial charges itself, δ^+ on H and δ^{2-} on O. These partial charges are attracted to the ions in the lattice (for example, in the case of sodium chloride, the δ^+ on each H in the water molecule is attracted to the negatively charged chloride anion, Cl^-). As a result individual ions are pulled out of the lattice and become surrounded by water molecules. In the case of a non-polar solvent, there is no attraction between the ions of the ionic compound and the solvent molecules, so the cations and anions remain within the lattice.

Uses of ionic liquids

Ionic liquids are efficient solvents and electrolytes, used in electric power sources and green industrial processes.

4.2 Covalent bonding

Understandings

→ A covalent bond is formed by the electrostatic attraction between a shared pair of electrons and the positively charged nuclei.

→ Single, double, and triple covalent bonds involve one, two, and three shared pairs of electrons, respectively.

→ Bond length decreases and bond strength increases as the number of shared electrons increases.

→ Bond polarity results from the difference in electronegativities of the bonded atoms.

Applications and skills

→ Deduction of the polar nature of a covalent bond from electronegativity values.

 Nature of science

→ Looking for trends and discrepancies – compounds that contain non-metals have different properties from compounds that contain non-metals and metals.

→ Use theories to explain natural phenomena – Lewis introduced a class of compounds which share electrons. Pauling used the idea of electronegativity to explain unequal sharing of electrons.

Definition of a covalent bond

A **covalent bond** is formed by the electrostatic attraction between a shared pair of electrons and the positively charged nuclei. According to IUPAC (the International Union of Pure and Applied Chemistry), a covalent bond is a region of relatively high electron density between nuclei that arises at least partly from the sharing of electrons and gives rise to an attractive force and characteristic internuclear distance.

$\cdot \overset{\displaystyle \cdot}{\underset{\displaystyle \cdot}{\text{B}}}$

▲ Figure 1 Lewis symbols of three elements. Nitrogen has five valence electrons, chlorine has seven valence electrons, and boron has three valence electrons

Study tip

Remember, to deduce the **number of valence electrons** of an element you can use the **group number** from the periodic table of elements. For example, sodium (s-block) is in group 1, so has one valence electron; calcium (also s-block) is in group 2, so has two valence electrons. For the p-block elements you simply drop the '1' in the group number to find the number of valence electrons: silicon (p-block) is in group 14, so has four valence electrons. Fluorine (also p-block) is in group 17, so has seven valence electrons, and so on.

Covalent bonding

In ionic bonding we saw how atoms can either lose or gain electrons in order to attain a **noble gas electron configuration**. A second type of chemical bond exists, however, in which atoms share electrons with each other in order to attain a noble gas electron configuration. This type of bonding is **covalent bonding**, and it usually occurs between non-metals.

In order to look at this type of bonding in detail, it is useful first to introduce the idea of a **Lewis symbol**, which is a simple and convenient method of representing the valence (outer shell) electrons of an element. In sub-topic 4.3 we shall develop this further into what we term the **Lewis (electron dot) structure** of a compound, based on a system devised by the US chemist, Gilbert N. Lewis (1875–1946).

In a Lewis symbol representation, each element is surrounded by a number of dots (or crosses), which represent the valence electrons of the element. Some examples are given in figure 1.

Let us consider the presence of covalent bonding in four different species, F_2, O_2, N_2, and HF.

Fluorine, F_2

- Fluorine is in group 17, so has seven valence electrons. Hence by acquiring one more electron, fluorine would attain a noble gas electron configuration with a complete octet of electrons.

- The Lewis symbol for fluorine is:

$$: \overset{\displaystyle \cdot\cdot}{\underset{\displaystyle \cdot\cdot}{\text{F}}} \cdot$$

- If two fluorine atoms share one electron each with each other, each fluorine atom gains one more electron to attain a complete octet of electrons, which results in the formation of a covalent bond between the two fluorine atoms. This covalent bond is a **single bond** and the shared pair can be represented by a line:

$$: \overset{\displaystyle \cdot\cdot}{\underset{\displaystyle \cdot\cdot}{\text{F}}} \cdot + \cdot \overset{\displaystyle \cdot\cdot}{\underset{\displaystyle \cdot\cdot}{\text{F}}} : \longrightarrow : \overset{\displaystyle \cdot\cdot}{\underset{\displaystyle \cdot\cdot}{\text{F}}} : \overset{\displaystyle \cdot\cdot}{\underset{\displaystyle \cdot\cdot}{\text{F}}} :$$

$$: \overset{\displaystyle \cdot\cdot}{\underset{\displaystyle \cdot\cdot}{\text{F}}} - \overset{\displaystyle \cdot\cdot}{\underset{\displaystyle \cdot\cdot}{\text{F}}} :$$

- Note that in this Lewis structure of F_2 there are a total of **six non-bonding pairs of electrons** (often called **lone pairs**) and **one bonding pair of electrons**.

Oxygen, O_2

- Oxygen is in group 16, so has six valence electrons. Hence by acquiring two more electrons, oxygen would attain a noble gas electron configuration with a complete octet of electrons.

- If two oxygen atoms each share two electrons with each other, this electron configuration can be achieved and results in the formation of a covalent bond between the two oxygen atoms. This covalent bond is a **double bond** and the two shared pairs can be represented by two lines.

$$:\ddot{O}\cdot + \cdot\ddot{O}: \longrightarrow :\ddot{O}::\ddot{O}:$$

$$:\ddot{O} = \ddot{O}:$$

- Note that in this Lewis structure of O_2 there are a total of **four non-bonding pairs of electrons** (the lone pairs) and **two bonding pairs of electrons**.

Nitrogen, N_2

- Nitrogen is in group 15, so has five valence electrons. Hence by acquiring three more electrons nitrogen would achieve a noble gas electron configuration with a complete octet of electrons.

- If two nitrogen atoms each share three electrons with each other, this electron configuration can be achieved and results in the formation of a covalent bond between the two nitrogen atoms. This covalent bond is a **triple bond** and the three shared pairs can be represented by three lines:

$$:\dot{N}\cdot + \cdot\dot{N}: \longrightarrow :N:::N:$$

$$:N \equiv N:$$

- Note that in this Lewis structure of N_2 there are a total of **two non-bonding pairs of electrons** (the lone pairs) and **three bonding pairs of electrons**.

Hydrogen fluoride, HF

- Fluorine is in group 17, so has seven valence electrons. Hence by acquiring one more electron, fluorine would attain a noble gas electron configuration with a complete octet of electrons. Hydrogen is in group 1, so has just one valence electron. Hence by acquiring just one more electron, hydrogen would attain the noble gas configuration of helium.

- Note that hydrogen does not acquire an octet (the octet rule is historical in nature, and the key point to remember here for hydrogen is the formation of a noble gas electron configuration).

- The Lewis symbols for hydrogen and fluorine are:

$$H^{\times}$$

$$:\ddot{F}\cdot$$

For convenience we use different symbols (a cross and a dot) for the electrons in each of the two Lewis symbols to signify different electrons for the two elements.

- To achieve noble gas configurations, fluorine and hydrogen can each share one electron with each other, forming a covalent bond. This covalent bond is a **single bond** and the shared pair can be represented by a line.

$$H^{\times} + \cdot\ddot{F}: \longrightarrow H\overset{\times}{:}\ddot{F}:$$

$$H - \ddot{F}:$$

> In the Lewis structure of a molecule, the electrons involved in the covalent bond are indistinguishable.

> **Activity**
>
> Using a similar approach to that of the examples here, deduce the Lewis structures of the molecules carbon dioxide, CO_2, and water, H_2O, showing the steps involved in the formation of the covalent bonds in each case.

- Note that in this Lewis structure of HF there are a total of **three non-bonding pairs of electrons** (the lone pairs) and **one bonding pair of electrons**.

Bond strength and bond length

The examples above describe molecules with single, double, and triple covalent bonds. These bonds differ in both bond strength and bond length.

Bond strength

The trend in bond strength is:

$$\equiv\ >\ =\ >\ -$$

That is, a triple bond is stronger than a double bond, which in turn is stronger than a single bond.

The bond enthalpies in section 11 of the *Data booklet* show this (table 1). Bond enthalpies will be discussed in sub-topic 5.3.

Bond length

This is the opposite trend to bond strength:

$$-\ >\ =\ >\ \equiv$$

A single bond is longer than a double bond, which in turn is longer than a triple bond. The covalent bond lengths in section 10 of the *Data booklet* illustrate this (table 1).

Bond	Bond enthalpy (at 298 K) / kJ mol^{-1}	Covalent bond length / pm
C≡C	839	120
C=C	614	134
C–C	346	154

▲ Table 1 Bond strengths (enthalpies) and bond lengths

Comparison of covalent bonds and ionic bonds

We now understand the inherent difference between ionic and covalent bonds. Table 2 summarizes some of these differences.

Electronegativity

We saw in the case of both fluorine, F_2, and hydrogen fluoride, HF, that the single covalent bond is made up of a shared pair of electrons for each molecule. In the case of identical atoms, such as the two fluorine atoms in F_2, there is an equal sharing of the electrons in the shared pair between the two atoms. This is not the case, however, in HF, and the shared pair is unequally shared between the hydrogen and fluorine atoms. In fact, you might think of this as a "tug-of-war" between the two atomic partners for the shared pair! In reality, fluorine has a much

Analogy

You can think of **bond strength** in terms of windows – a window that is triple-glazed is stronger than a window that is double-glazed, which in turn is stronger than a window with a single pane.

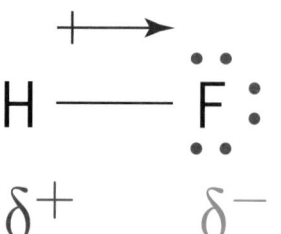

▲ Figure 2 Dipole moment represented by a vector in the polar molecule, HF

Ionic bonding	Covalent bonding
Formed between a cation (usually metal) and an anion (usually non-metal). Some cations (such as NH_4^+) can be comprised of non-metals and some anions (such as MnO_4^-) can contain metals.	Usually formed between non-metals.
Formed from atoms either losing electrons (process of oxidation) or gaining electrons (process of reduction) in order to attain a noble gas electron configuration.	Formed from atoms sharing electrons with each other in order to attain a noble gas electron configuration.
Electrostatic attraction between oppositely charged ions, that is, a cation (positive ion) and an anion (negative ion).	Electrostatic attraction between a shared pair of electrons and the positively charged nuclei.
Ionic compounds have lattice structures.	Covalent compounds consist of molecules.*
Ionic compounds have higher melting points and boiling points.	Covalent compounds have lower melting points and boiling points.
Ionic compounds have low volatilities.	Covalent compounds may be volatile.
Ionic compounds tend to be soluble in water.	Covalent compounds typically are insoluble in water.
Ionic compounds conduct electricity because ions are free to move in the molten state. They do not conduct electricity when solid, however, as the ions are not free to move.	Covalent compounds do not conduct electricity because no ions are present to carry the charge.

▲ Table 2 Differences between ionic and covalent bonding

*We shall discuss covalent network structures that involve lattices later.

greater attraction for the shared pair than hydrogen does and this leads to what we describe as a **polar covalent bond**, with one atom adopting a partial negative charge, δ^-, and one atom adopting a partial positive charge, δ^+. In this case, since fluorine has a greater pulling power for the shared pair of electrons in the covalent bond, it acquires the partial negative charge, δ^-, and hydrogen then adopts the partial positive charge, δ^+. This separation of charge can be represented vectorially by a **dipole moment**, symbol μ (figure 2).

If the two atoms involved in the formation of the covalent bond are identical, the bond is said to be a **pure covalent bond**; that is, the covalent bond is non-polar and has no dipole moment. Hence, the F–F bond in F_2 is a **non-polar covalent bond**.

The US chemist Linus Pauling (1901–1994) introduced the idea of **electronegativity** (χ_P) as the relative attraction that an atom of an element has for the shared pair of electrons in a covalent bond.

Pauling devised a scale of electronegativity values, which can be found in section 8 of the *Data booklet*. On the Pauling scale, fluorine is the most electronegative element in the periodic table with a value of $\chi_P = 4.0$ (sub-topic 3.2).

There are certain trends in electronegativity values that mirror what we have already seen for the ionization energies across a period and down a group.

Trends in electronegativities

- Going from left to right across a period, χ_P values increase.

  Reasons:

  i) decreasing atomic radii

  ii) increasing nuclear charge.

- Going down a group, χ_P values decrease.

  Reasons:

  i) increasing atomic radii

  ii) primary screening (shielding) effect of inner electrons.

4.3 Covalent structures

Understandings

→ Lewis (electron dot) structures show all the valence electrons in a covalently bonded species.

→ The "octet rule" refers to the tendency of atoms to gain a valence shell with a total of eight electrons.

→ Some atoms, like Be and B, might form stable compounds with incomplete octets of electrons.

→ Resonance structures occur when there is more than one possible position for a double bond in a molecule.

→ Shapes of species are determined by the repulsion of electron pairs according to the valence shell electron pair repulsion (VSEPR) theory.

→ Carbon and silicon form covalent network (giant covalent) structures.

Applications and skills

→ Deduction of Lewis (electron dot) structure of molecules and ions showing all valence electrons for up to four electron pairs on each atom.

→ The use of VSEPR theory to predict the electron domain geometry and the molecular geometry for species with two, three, and four electron domains.

→ Prediction of bond angles from molecular geometry and presence of non-bonding pairs of electrons.

→ Prediction of molecular polarity from bond polarity and molecular geometry.

→ Deduction of resonance structures, examples include but are not limited to C_6H_6, CO_3^{2-} and O_3.

→ Explanation of the properties of covalent network (giant covalent) compounds in terms of their structures.

Nature of science

→ Scientists use models as representations of the real world – the development of the model of molecular shape (VSEPR) to explain observable properties.

Nature of science

But a scholar must be content with the knowledge that what is false in what he says will soon be exposed and, as for what is true, he can count on ultimately seeing it accepted, if only he lives long enough.

Ronald Coase (Recipient of the Nobel Prize in Economic Sciences in 1991).

Ronald Coase (1910–2013) was the oldest living Nobel laureate until his death on 2 September 2013.

Scientists use models as representations of the real world – for example, VSEPR theory as a model of molecular shape has been used to explain observable properties. Every model in science is built on certain assumptions – one of the major considerations for a scientist is to appreciate the validity of a model, its limitations, and whether it will withstand the test of time. VSEPR theory is one such model, although not without its limitations.

Lewis (electron dot) structures

Earlier in this topic we introduced the idea of a **Lewis symbol**, which shows the number of valence electrons of an element represented by either dots or crosses. From this we developed the idea of **Lewis (electron dot) structures**, based on the formation of the covalent bond in a molecule. In a Lewis structure, each pair of electrons can be represented in a number of different ways – either by two dots, by two crosses (or a combination of a dot and a cross), or by a line.

For example, some of the ways in which the Lewis structure of phosphine, PH_3, might be represented are shown in figure 1(a).

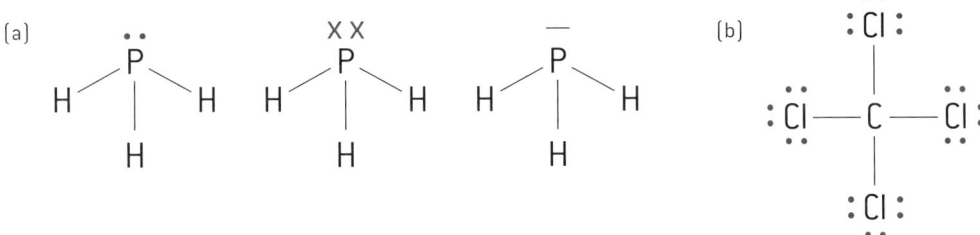

▲ Figure 1 (a) Two dots, two crosses (or a combination of the two), or a line can be used to represent each pair of electrons in a Lewis (electron dot) structure. (b) Lewis (electron dot) structure of CCl_4. Remember the bond angles shown in a Lewis structure do not necessarily represent the actual bond angles in the molecular geometry

In such a representation it is important to distinguish between:

- **bonding pairs of electrons** (showing the covalent bond as single, double, or triple bonds) and

- **non-bonding pairs of electrons**, often called the **lone pairs**, which are pairs of electrons not involved in the bonding.

In the Lewis structure of phosphine there are three bonding pairs of electrons and one lone pair.

Similarly, the Lewis structures of carbon dioxide, CO_2, and carbon monoxide, CO, which contain multiple bonds, can be represented as shown in figure 2.

$$\ddot{O} = C = \ddot{O} \qquad :C \equiv O:$$

▲ Figure 2 Lewis structures of CO

In CO_2, each double bond represents two bonding electron pairs, and in CO, the triple bond represents three bonding electron pairs.

Lewis structures help us understand the different types of covalent bond (single, double, or triple bonds) and the existence of lone pairs in molecules. However, Lewis structures tell us nothing about the **actual shapes** of molecules, and hence the representation of the Lewis structure of a molecule may be drawn with a geometrical arrangement that differs completely from its real shape in space. For example, the Lewis structure of carbon tetrachloride, CCl_4, is typically represented as in figure 1b, which might suggest the existence of 90° Cl–C–Cl bond angles. In fact the shape of the carbon tetrachloride molecule is tetrahedral with 109.5° Cl–C–Cl bond angles. We shall shortly see how to deduce this shape based on a very useful model for predicting molecular geometries, called the **valence shell electron pair repulsion (VSEPR) theory**.

Lewis (electron dot) structures of cations and anions and ionic compounds

Lewis structures can be written not only for neutral molecules but also for cations and anions. In a compound containing both a cation and an anion there is an electrostatic attraction between the oppositely charged ions, which forms the ionic bond. However, the bonding *within* the cation and anion separately may be covalent in nature; for example, in ammonium nitrate, NH_4NO_3 (figure 3(a)) the bonding in $[NH_4]^+$ and in $[NO_3]^-$ is covalent, even though the bonding *between* the cation and the anion is ionic. In the case of ammonium chloride, NH_4Cl (figure 3(b)) the Lewis structure of the chloride anion can be represented with the chlorine surrounded by eight dots to represent the eight valence electrons present in the anion.

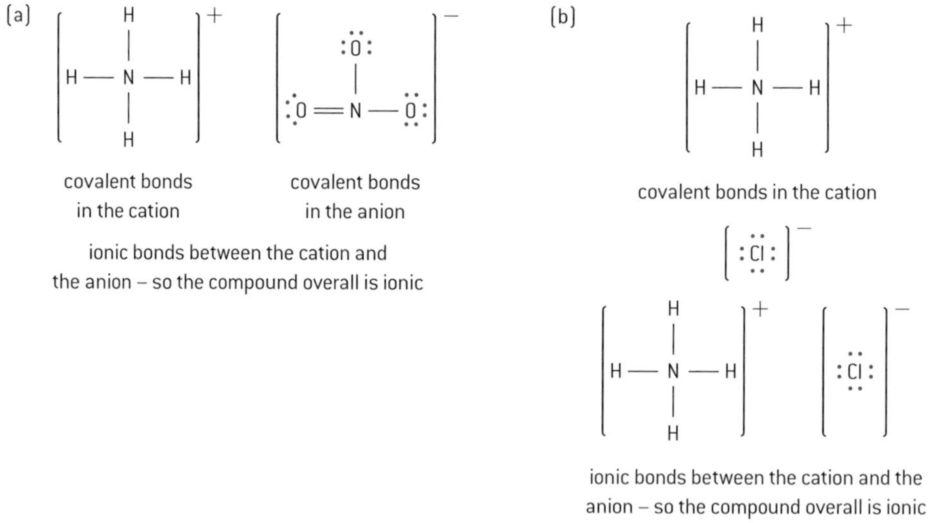

> **Study tip: Use of square brackets in Lewis (electron dot) structures**
>
> The chemical formula of ammonium nitrate is often written as NH_4NO_3, but in reality it is made up of a cation, ammonium, and an oxoanion, nitrate. When you write Lewis structures of cations or anions, including oxoanions, you should always *include square brackets and the charge* in the representation.

(a) covalent bonds in the cation — covalent bonds in the anion

ionic bonds between the cation and the anion – so the compound overall is ionic

(b) covalent bonds in the cation

ionic bonds between the cation and the anion – so the compound overall is ionic

▲ Figure 3 (a) Lewis structure of ammonium nitrate. (b) Lewis structure of ammonium chloride

Valence shell electron pair repulsion (VSEPR) theory

Much of the core understanding of chemistry involves discussions of structure and bonding. Every molecule has a particular shape and as chemists we need to have the ability to always think in three dimensions.

a)

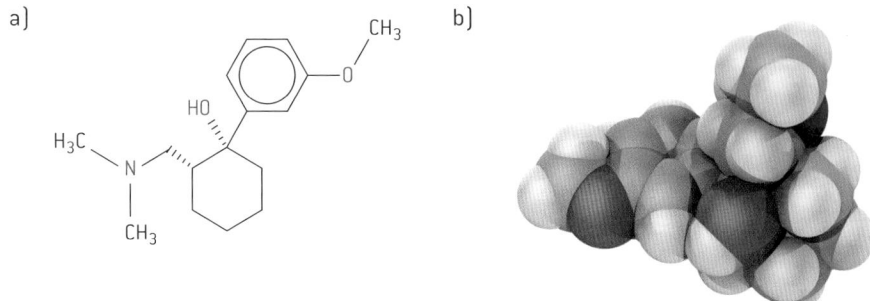

b)

▲ Figure 4 (a) 2D representation of the drug tramadol, whose molecular formula is $C_{16}H_{25}NO_2$. Tramadol is a centrally acting synthetic opioid analgesic used in treating severe pain. (b) Three-dimensional molecular space-filling model of tramadol. The atoms are represented as spheres and are colour coded: carbon (grey), hydrogen (white), nitrogen (blue), and oxygen (red)

As mentioned previously, Lewis structures are two-dimensional representations and ultimately tell us nothing about shape. **Valence shell electron pair repulsion (VSEPR) theory** can be used to deduce the shapes of covalent molecules.

The basis of this theory is as follows – since electrons are negatively charged subatomic particles, *pairs of electrons repel one another to be as far apart as possible in space.*

In order to determine the maximum angle that can be achieved from this electron pair–electron pair repulsion try tying a number of balloons together. Then examine the spatial shape the balloons ultimately adopt. In the case of two balloons a **linear** geometry is obtained, with the two balloons aligning at 180° to each other. Think of dividing a circle up into halves: $\frac{360}{2} = 180°$ (figure 5).

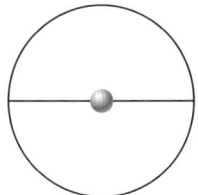 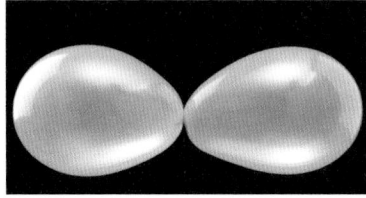

▲ Figure 5 Two balloons tied together showing a linear arrangement in space

In the case of three balloons, a **trigonal planar** arrangement is generated, similar to taking a circle and slicing it into three segments: $\frac{360}{3} = 120°$. Hence the balloons arrange themselves to lie on one plane at 120° to each other (the term planar in chemistry means flat) (figure 6).

Now consider tying four balloons together. Thinking in two dimensions, you might visualize taking a circle and dividing 360° by 4, which would give a bond angle between any two of the balloons of 90°. This is not what happens: in three-dimensional space the balloons maximize their spatial arrangement to be 109.5° apart – *try it!* This shape creates a **tetrahedral** geometry (figure 7).

You might imagine the tetrahedron sitting in the environment of a cube to help you appreciate the three-dimensionality of this geometry based on the repulsion of four electron pairs (figure 8).

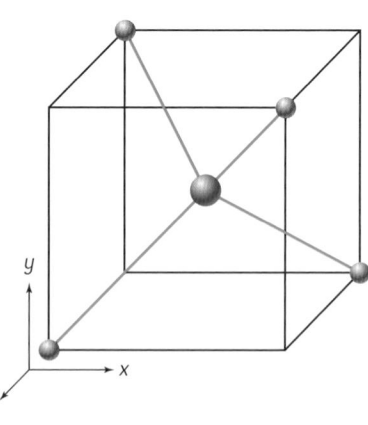

▲ Figure 6 Three balloons tied together showing a trigonal planar arrangement in space

▲ Figure 7 Four balloons tied together showing a tetrahedral arrangement in space

▲ Figure 8 A tetrahedron fits into a cube in three-dimensional space

The basic molecular geometries can therefore be summarized, as shown in table 1, on the basis of two, three, or four pairs of electrons. Each pair of electrons is described as occupying an **electron domain**, which you might like to imagine as being a field of electron density.

▲ Figure 9 The Scottish scientist, engineer, and inventor, Alexander Graham Bell (1847–1922) sitting in his tetrahedral chair. Most famous for his invention of the telephone, Bell was also fascinated by the theory of engineering structures and flight. He championed the cause of **tetrahedral structures**, frameworks based on a series of interlocked tetrahedra. He is seen here watching trials of his kite designs

Number of electron domains	Molecular geometry	Bond angle	Examples of molecules or ions having this shape
two	linear	180°	AB_2 $BeCl_2$, CO_2
three	trigonal planar	120°	AB_3 BF_3, $[NO_3]^-$
four	tetrahedral	109.5°	AB_4 CH_4, $[NH_4]^+$, $[ClO_4]^-$

▲ Table 1 Molecular geometries based on two, three, and four electron domains

The set of three molecular geometries, AB_2 (linear), AB_3 (trigonal planar), and AB_4 (tetrahedral), can also be extended to generate additional shapes for species that have fewer bonding pairs of electrons than the number of electron domains present. In such cases, the electron domains not occupied by the bonding pairs of electrons are filled by non-bonding pairs of electrons (lone pairs). In such cases, three additional molecular geometries are generated: AB_2E (V-shaped or bent), AB_3E (trigonal pyramidal), and AB_2E_2 (V-shaped or bent), where E represents a lone pair of electrons (table 2).

We can therefore distinguish between:

- the **electron domain geometry** (based on the total number of electron domains predicted from VSEPR theory); and

- the **molecular geometry** (which gives the shape of the molecule).

To illustrate this idea let us take the example of the water molecule, H_2O. The number of electron domains predicted from VSEPR theory is four (we shall learn how to deduce this shortly). This means that the electron domain geometry is tetrahedral. However, from the chemical formula we see that there are only two O–H bonds, which suggests the presence of two bonding pairs (not four). The other two domains are occupied by two lone pairs of electrons. This implies that the actual molecular geometry, based on an AB_2E_2 structure, is V-shaped or bent (table 2).

Number of electron domains	Electron domain geometry	Molecular geometry	Bond angle	Examples of molecules or ions having this shape
three	trigonal planar AB_2E	V-shaped (bent)	$<120°$	SO_2, $[NO_2]^-$
four	tetrahedral AB_3E	trigonal pyramidal	$<109.5°$	NH_3, $[SO_3]^{2-}$, $[H_3O]^+$
four	tetrahedral AB_2E_2	V-shaped (bent)	$<109.5°$	H_2O, $[NH_2]^-$

▲ Table 2 Geometries involving lone pairs based on three and four electron domains

Bond angles in molecular geometries:

Lone pairs of electrons affect the bond angles in a molecule. *Lone pairs occupy more space* than bonding pairs, so they decrease the bond angle between bonding pairs. The degree of electron pair–electron pair repulsion follows this order:

LP|LP > LP|BP > BP|BP

where LP represents lone pairs of electrons and BP represents bonding pairs of electrons.

Table 3 illustrates how repulsion between lone pairs of electrons decreases the bond angles.

Balloon analogy for molecular shape

Returning to the balloon analogy, you can see this in action if you again take four balloons and tie them together. This time have two of the balloons blue and two of the balloons yellow, the latter representing lone pairs of electrons. Make the two yellow balloons bigger than the two blue balloons (the text opposite explains why). To emphasize the fact that the lone pairs are non-bonding pairs of electrons take a black marker and mark two dots on each yellow balloon. You still have four electron domains, so the electron domain geometry is designated as tetrahedral, but now it is made up of two bonding electron pairs and two non-bonding electron pairs.

Molecule	Number of electron domains	Molecular geometry	Bond angle
CH_4	four	tetrahedral AB_4	109.5°
NH_3	four	trigonal pyramidal AB_3E	107°
H_2O	four	V-shaped AB_2E_2	104.5°

▲ Table 3 Effect of lone pairs on bond angles

Interpreting the VSEPR model

Using the model of VSEPR theory it is not possible to predict exact bond angles when lone pairs are present. All you can state is that the bond angle will be expected to be less than predicted from the bond angle associated with the basic shape. However, LP|BP and LP|LP repulsions should be taken into account. A common mistake that many students make is to learn the experimentally determined bond angles for ammonia (107°) and water (104.5°) and then assume that all trigonal pyramidal molecular geometries and all V-shaped molecular geometries also have these bond angles. This is a mistaken interpretation of the VSEPR model. For example, phosphine, PH_3, also has an AB_3E structure and is trigonal pyramidal, but its H–P–H bond angle drops to 93.5°. Likewise,

hydrogen sulfide, H_2S, is V-shaped, based on an AB_2E_2 structure, but the H–S–H bond angle is much lower at 92.1°. The bond angles are affected by many factors, so making exact predictions is not feasible. Two other factors that play a role are **electronegativity differences** and **multiple bonds** (the latter also occupy more space, just like lone pairs). For example, in the molecule of ethene, the H–C–H bond angle is 117° and the H–C=C bond angle is 121°, even though both would be predicted to be 120° based on a trigonal planar arrangement about each carbon:

Working method to deduce both Lewis (electron dot) structures and electron domain and molecular geometries

We can combine Lewis structures and VSEPR theory in a simple-to-use working method. The following method can be used to deduce Lewis structures and electron domain and molecular geometries:

1 Draw a ball-and-stick diagram, identifying the central atom. Each stick represents a pair of electrons in the covalent bond. Don't worry about bond angles at this stage – you can draw the sticks in any direction to commence

the process. In the case of oxoanions, localize the negative charges on any terminal oxygen atoms; the remaining bonds should be converted into double bonds. In the case of other anions (not oxoanions) and cations use square brackets and place the charge outside these.

2 For the central atom, deduce from its group number in the periodic table the number of valence electrons.

3 From the number of sticks, count the number of single bonds, which we shall designate as sigma (σ) bonds.

4 Add one electron for each negative charge (*but not for localized charges already assigned to oxygen atoms in oxoanions in step 1*). Delete one electron for a positive charge. Subtract one for each pi (π) bond.

5 Combining steps 2, 3 and 4, divide this number by two to obtain the number of electron pairs, which equals the number of electron domains.

6 Based on the number of electron domains, deduce the electron domain geometry.

7 Determine the number of lone pairs present, if applicable, and deduce the molecular geometry. Then draw an exact representation of the structure, complete with predicted bond angles, taking into account the order of electron-pair repulsion:

LP|LP > LP|BP > BP|BP

8 Finally, draw a Lewis representation by completing the octets on all terminal atoms, *excluding hydrogen* (which will already have attained a noble gas electron configuration of two). Remember to include square brackets for any cation or anion.

9 Draw any resonance structures (explained on page 115) where applicable.

Let us put this working method to the test. There are three types of structure that you are required to work out:

- Basic shapes – AB_2 (linear), AB_3 (trigonal planar), and AB_4 (tetrahedral).

- Species with lone pairs of electrons – AB_2E (V-shaped), AB_3E (trigonal pyramidal), and AB_2E_2 (V-shaped) – all with associated bond-angle considerations.

- Oxoanions.

Worked examples

Example 1: Carbon tetrachloride, CCl_4

- A ball-and-stick diagram for CCl_4:

- C has four valence electrons (it is in group 14);

  four σ bonds;

  so the total number of valence electrons is eight;

  $\frac{8}{2} = 4$ so there are four electron domains.

  Thus the electron domain geometry is tetrahedral (AB_4).

- There are four C–Cl bonds so no lone pairs are present – *the molecular geometry is therefore tetrahedral* and the bond angle will be 109.5°.

- Finally you need to complete the octets on each terminal Cl in order to generate the Lewis structure.

Example 2: Ammonium cation, $[NH_4]^+$

- Ball-and-stick diagram for $[NH_4]^+$:

- N has five valence electrons (as it is in group 15);

  four σ bonds;

  one positive charge;

  so total number of valence electrons = 8;

  $\frac{8}{2} = 4$ so 4 electron domains.

  Electron domain geometry is tetrahedral (AB$_4$).

- There are four N–H bonds so no lone pairs are present – the *molecular geometry is therefore tetrahedral* and the bond angle will be 109.5°.

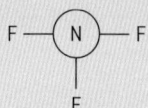

- The above structure is also a valid Lewis structure as hydrogen is surrounded by two electrons, which is the maximum number of electrons permissible.

Example 3: Nitrogen trifluoride, NF$_3$

- Ball-and-stick diagram for NF$_3$:

F —(N)— F
        |
        F

- N has five valence electrons (as it is in group 15); three σ bonds;

  so the total number of valence electrons is eight;

  $\frac{8}{2} = 4$, so there are four electron domains.

  Electron domain geometry is tetrahedral (AB$_3$E).

- There must be one lone pair present, as there are only three N–F covalent bonds. Hence the *molecular geometry is trigonal pyramidal* and the bond angle will be *less than* 109.5° due to the presence of the lone pair (which occupies more space) – the repulsion between LP|BP is greater than that between BP|BP.

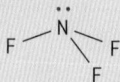

The experimentally determined F–N–F bond angle, which cannot be determined precisely from the model, is 102.2°, suggesting that not only is

the lone pair influencing the bond angle but also the difference in electronegativity is likely to play a role.

- Finally, you need to complete the octets on each terminal F in order to generate the Lewis structure:

Example 4: Sulfur difluoride, SF$_2$

- Ball-and-stick diagram for SF$_2$:

F —(S)— F

- S has six valence electrons (group 16);

  two σ bonds;

  so the total number of valence electrons is eight;

  $\frac{8}{2} = 4$, so four electron domains.

  Electron domain geometry is tetrahedral (AB$_2$E$_2$).

- There must be two lone pairs present, as there are only two S–F covalent bonds. Hence the *molecular geometry is V-shaped* and the bond angle will be *less* than 109.5° due to the presence of the two lone pairs, which occupy much more space. The repulsion between a LP|LP is greater than that between a LP|BP which is greater than that between a BP|BP, so the bond angle is reduced *significantly* from its predicted value of 109.5°.

.S.
 /<109.5°\
F   F

The experimentally determined F–S–F bond angle, which cannot be determined precisely from the model, is 98°, showing the significant role of the LP|LP repulsion in operation (also the electronegativity of fluorine will have an influence).

- Finally you need to complete the octets on each terminal F in order to generate the Lewis structure:

Example 5: Nitrite oxoanion, $[NO_2]^-$

- This is an oxoanion, so in the ball-and-stick diagram we first localize the one negative charge on any one of the two oxygen atoms. Since oxygen has a valency of two, this means the other nitrogen-to-oxygen bond must be a double bond.

Note that in the case of an oxoanion, we first begin with square brackets with the negative charge outside, but we then localize the charge on any one of the two oxygen atoms and remove the square brackets until later in our stepwise working method.

- N has five valence electrons (as it is in group 15);

  two σ bonds;

  one pi (π) bond, which counts as −1 (it is important to remember this; see the box below);

  so the total number of valence electrons is six;

  $\frac{6}{3} = 3$ so there are three electron domains.

  Electron domain geometry is trigonal planar (AB_2E).

How to handle π bonds in VSEPR theory

π bonding involves off-axis bonding, as we will explain in topic 14. Hence, as the shape of a molecule is controlled by the σ bonding framework along the internuclear axis, in counting the valence electrons we subtract 1 for each π bond present. *For SL students, you do not need to go into the reason behind this (the explanation is given at HL) but you do need to know the method involved.*

You can think of it like this – the shape is controlled by the geometrical arrangement along the internuclear axis, where the σ bonding framework lies. A double bond is described as a $(\sigma + \pi)$ bond and a triple bond is described as a $(\sigma + 2\pi)$ bond. To reduce these back to σ bonds, you simply subtract out the π components.

Oxoanions

In the case of oxoanions, we **do not** add an additional electron here for the negative charge on the nitrite oxoanion because this has already been accounted for in the first step. This is a very important point and you need to note this difference for oxoanions.

- There must be one lone pair present, as there are only two nitrogen-to-oxygen covalent σ bonds. Hence the *molecular geometry is V-shaped* and the bond angle will be *less* than 120° due to the presence of the one lone pair, which occupies much more space.

The experimentally determined bond angle, which cannot be determined precisely from the model, is 115°.

- Finally you need to complete the octets on each terminal O in order to generate the Lewis structure:

Note that for the oxygen containing the double bond, completing the octet entails the addition of two lone pairs, whereas for the oxygen with the single bond in this structure, completion of the octet requires the addition of three lone pairs. The actual structure of nitrite is a combination of two contributing Lewis structures. This is **resonance**, which we shall return to shortly. Contributing resonance structures are represented by a double-headed arrow.

The two nitrogen-to-oxygen bond lengths in nitrite are equivalent and intermediate in length between a single and a double bond. The two contributing resonance structures therefore could be combined, each represented with square brackets and the negative charge placed outside.

Example 6: Sulfite oxoanion, $[SO_3]^{2-}$

- This is also an oxoanion, so in the ball-and-stick diagram we first localize the two negative charges on two of the three oxygen atoms. Since oxygen has a valency of two, this means the other sulfur-to-oxygen bond must be a double bond in order to satisfy this valency for oxygen.

- S has six valence electrons (as it is in group 16);

  three σ bonds;

  one π bond (−1);

  so the total number of valence electrons is eight;

  $\frac{8}{2} = 4$, so there are four electron domains.

  Electron domain geometry is tetrahedral (AB_3E).

- There must be one lone pair present, as there are only three sulfur-to-oxygen covalent σ bonds. Hence the *molecular geometry is trigonal pyramidal* and the bond angle will be less than 109.5° because of the presence of the one lone pair, which occupies much more space.

The experimentally determined bond angle, which cannot be determined precisely from the model, is 106°.

- Finally you need to complete the octets on each terminal O in order to generate the Lewis structure.

Note that for the oxygen containing the double bond, completing the octet involves the addition of two lone pairs, whereas for the two terminal oxygen atoms with the single bonds in this structure, completion of the octet requires the addition of three lone pairs. The actual structure of sulfite is a combination of three contributing resonance structures:

The three sulfur-to-oxygen bond lengths in sulfite are equivalent and intermediate in length between a single and a double bond.

Incomplete and expanded octets

In most Lewis structures, the central atom will be surrounded by an octet of electrons. However, in some species, the central atom will have less than an octet of valence electrons: these are **incomplete octets** (for example, the linear molecule beryllium chloride, $BeCl_2$, which has the central beryllium atom surrounded by only four electrons, or the trigonal planar molecule boron trichloride, BCl_3, in which the central boron atom is surrounded by only six electrons):

In other species, an **expanded octet** is possible. This is discussed in topic 14. In such cases alternative Lewis structures involving octets may be used.

Resonance structures

As we saw in the case of the nitrite oxoanion, sometimes it is possible for Lewis structures to have identical arrangements of atoms but different arrangements of the electrons.

The individual Lewis structures that contribute to the overall structure are called **resonance forms**. The actual electronic structure of the species is called a **resonance hybrid** of these resonance forms. In order to represent this idea of **resonance**, the contributing resonance forms are linked via a double-headed arrow.

One of the best known examples of resonance is the molecule benzene, C_6H_6 (figure 10).

(a) 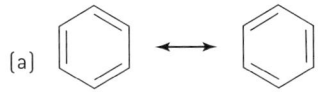 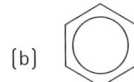 (b)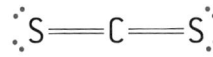

▲ Figure 10 (a) Two Kekulé structures of benzene showing resonance. (b) Representation of benzene showing the delocalized nature of its π electrons

The two resonance forms represented here are termed the **Kekulé structures of benzene**. In benzene, as can be seen from section 10 of the *Data booklet*, each carbon-to-carbon bond length is 140 pm, intermediate between a carbon-to-carbon double bond (134 pm) and a carbon-to-carbon single bond (154 pm). The structure of benzene, therefore, is often drawn as in Figure 10(b), where the circle represents the **delocalization** (which we shall discuss further in topic 10).

In topic 14 we shall discuss resonance in more detail, together with π electrons and bond order.

Molecular polarity

Earlier in this topic we discussed the idea of **bond polarity**. We now focus on **molecular polarity**, that is, whether the molecule itself is polar or non-polar (figure 11). The polarity of molecules is distinct from the polarity of individual bonds; a non-polar molecule may have polar bonds. In order to deduce the molecular polarity we can follow a simple three-step working method described in the box below.

> ## Working method to deduce molecular polarity
>
> 1   Using VSEPR theory, deduce the molecular geometry.
>
> 2   For each bond present, using electronegativity differences, $\Delta\chi_p$, deduce the bond polarity for each bond present and draw the associated dipole moments; these are best represented as vectors.
>
> 3   Using vector addition, sum all the dipole moments present to establish whether there is a net dipole moment, μ, for the molecule. If so, the molecule is polar.

> ## Definitions
>
> - **Resonance** involves using two or more Lewis structures to represent a particular molecule or ion. A resonance structure is one of two or more alternative Lewis structures for a molecule or ion that cannot be described fully with one Lewis structure alone.
>
> - In **delocalization** electrons are shared by more than two atoms in a molecule or ion as opposed to being localized between a pair of atoms.

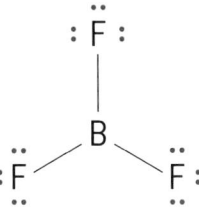

non-polar molecule

non-polar molecule

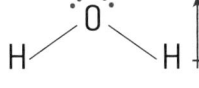

polar molecule

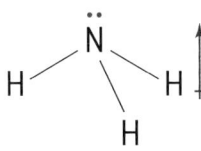

polar molecule

▲ Figure 11 Examples of non-polar and polar molecules. The net dipole moment, μ, of a polar molecule can be represented vectorially. The dipole moment represents the non-symmetrical distribution of charge in a polar molecule (compared with a symmetrical distribution of charge in a non-polar molecule). In the vector the head of the vector represents δ^- and the tail represents δ^+.

Worked example: deducing molecular polarity

Deduce the molecular polarities of the following:

a) SF_2

b) CO_2

Solution

a) **SF_2**

As seen in an earlier worked example on VSEPR theory, the molecular geometry of SF_2 is V-shaped. From section 8 of the *Data booklet*:

$$\chi_P(S) = 2.6 \text{ and } \chi_P(F) = 4.0$$

Hence, fluorine is more electronegative than sulfur and the S–F bond is polar with the following dipole moment:

$$S \xrightarrow{} F$$
$$\delta^+ \qquad \delta^-$$

To deduce the molecular polarity, we need to sum the two S–F vectors. The SF_2 molecule is V-shaped so we add the two vectors using the **parallelogram law** (see Study tip below):

$\vec{\nu}_1 + \vec{\nu}_2 = \vec{\nu}_{net}$. This results in a net dipole moment, μ; the molecule is polar.

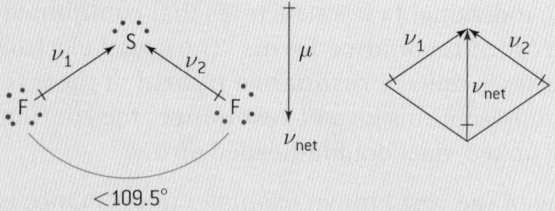

b) **CO_2**

Using VSEPR theory, carbon dioxide is found to be a linear molecule. From section 8 of the *Data booklet*:

$$\chi_P(C) = 2.6 \text{ and } \chi_P(O) = 3.4$$

Hence, each C=O bond ($\Delta\chi_P = 0.8$) is polar. The two vectors are equal in magnitude but opposite in direction and hence cancel each other out, resulting in no net dipole moment, that is, $\mu = 0$. The molecule is non-polar (even though it has two polar bonds).

$$\ddot{O} = C = \ddot{O}$$
$$\mu = 0$$

Study tip

In deducing molecular polarities based on molecular geometries you need to find the vector sum of the individual dipole moments. The parallelogram law is a useful method.

- *The parallelogram law.* If you have two vectors $\vec{\nu}_1$ and $\vec{\nu}_2$, and both vectors start from the same point, the sum of the two vectors, $\vec{\nu}$, can be found by completing the parallelogram. The diagonal will give the resultant (the vector sum).

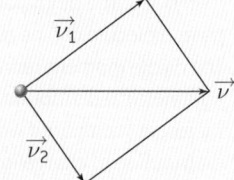

that is:

$$\vec{\nu} = \vec{\nu}_1 + \vec{\nu}_2$$

- A tug-of-war is a model that can be used to consider vectors on the same line (axis).

▲ Figure 12 Students in Montserrat in a tug-of-war. Both teams are pulling along the same axis

More polar bonds (resulting from a greater difference in electronegativity, $\Delta\chi_P$) win the tug-of-war, provided the pull is along the same axis.

Allotropes

Allotropes of the same element can vary in both physical and chemical properties.

Carbon is one of the most fascinating elements in the periodic table, and life forms on Earth are based on carbon. Carbon has a number of allotropes: graphite, diamond, graphene, and C_{60} fullerene.

Covalent network solids

- Graphite, diamond, and graphene are examples of **covalent network solids**. A covalent network solid is one in which the atoms are held together by covalent bonds in a giant three-dimensional lattice structure (in large networks or chains). Another well known example of a covalent network solid is quartz, which is silicon dioxide, SiO_2.

- In contrast, C_{60} fullerene is **molecular**.

Graphite

Graphite is an example of a covalent network solid. In graphite there are layers of hexagonal rings consisting of carbon atoms. These layers are connected by weak intermolecular forces of attraction, which are called **London** forces, leading to the use of graphite as a lubricant and in pencils (the so-called 'lead' in our pencils is not lead but carbon in the form of graphite). Each carbon atom adopts a trigonal planar geometry, and is covalently bonded to three other each carbon atoms at a bond angle of 120°. The coordination number of each carbon is three in the structure. Although the covalent bonds are strong within the sheets, the London forces between the layers are weak, which allows the layers to slide past each other, and thus graphite can be used as a lubricant (figure 13). Unlike other covalent network solids, graphite is a good conductor of electricity as it has delocalized π electrons.

▲ Figure 13 Graphite is a covalent network solid that consists of hexagonal layers of carbon atoms, which can slide past each other. The layers are connected by weak intermolecular forces of attraction (London forces)

Definition of allotropes

As described by IUPAC, **allotropes** are different structural modifications of the same element.

Properties of covalent network solids

- *Melting points.* Covalent network solids have high melting points (typically greater than 1000 °C and much higher than the melting points of molecular substances).

- *Electrical conductivity.* Covalent network solids are poor electrical conductors (though graphite and graphene are clear exceptions – electrical conductivity is one of the characteristics that makes graphene remarkable).

- *Solubility.* They are typically insoluble in common solvents.

- *Hardness.* Generally, covalent network solids are hard, though in graphite the layers can slide past one another.

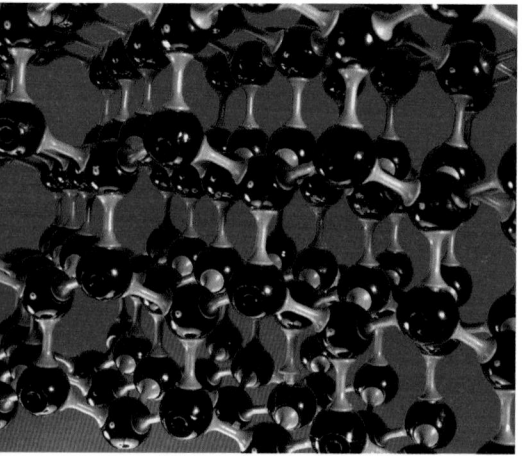

▲ Figure 14 Diamond is an allotrope of carbon with a covalent network lattice structure. Large crystals of diamond are mined for use as gemstones. Small crystals are used as an industrial abrasive. High-quality crystals of diamond are found in South Africa, Russia, Brazil, and Sierra Leone.

International perspective

Throughout history diamonds have often been a potential source of significant global conflict. The term "*blood diamond*" has been coined to describe diamonds mined in regions of conflict and subsequently sold to fund such conflicts. What responsibilities do nations and governments have in the import of products such as gemstones and precious metals?

Useful resource

Look at the history of the discovery of graphene and current research developments in using this material at the University of Manchester, UK, the university where Geim and Novoselov did their research to win the Nobel prize in Physics in 2010, http://www.graphene.manchester.ac.uk/story/

Diamond

Diamond is also a covalent network solid. In the lattice structure of diamond, each carbon atom is covalently bonded to four other carbon atoms in a tetrahedral arrangement, with a C–C–C bond angle of 109.5° (figure 14). The coordination number of each carbon within diamond is four. Diamond is one of the hardest substances known because of this covalently bonded interlocking structural arrangement of tetrahedra. For this reason diamond is often used in heavy-duty cutting tools such as saws, polishing tools, and dental drills.

The melting and boiling points of diamond are very high (3550 and 4827 °C, respectively). Unlike graphite, in diamond the valence electrons are localized in the single σ covalent bonds, and therefore cannot move freely. This means diamond does not conduct electricity.

Strong covalent bonds in diamond make it is insoluble in all common solvents.

 Diamonds are forever?

It has been said that "diamonds are a girl's best friend", but have you ever considered if diamonds last forever?

Unfortunately not! Under ambient conditions, diamond is *thermodynamically* unstable and eventually turns into another allotrope of carbon, graphite. However, at room temperature this process is extremely slow, so diamond is said to be *kinetically* stable. At 1000 °C the conversion of diamond into graphite accelerates and at 1700 °C it completes within seconds. When we talk about stability in chemistry we need to consider both *thermodynamic stability* and *kinetic stability*.

Graphene – the super material!

Graphene is not only one of the thinnest and strongest of known materials, but it is also the first two-dimensional crystal ever discovered. Graphene is a covalent network solid, but differs from graphite in that it consists of a single planar sheet of carbon atoms arranged hexagonally (figure 15), and is only one atom in thickness. As in graphite, each carbon atom is covalently bonded to three other carbon atoms so the coordination number of each carbon in graphene is three. The carbon atoms are densely packed in a honeycomb crystalline lattice, but the lattice is actually planar, which makes it remarkable as a crystalline structure.

The experimental evidence for the existence of graphene was obtained in 2004 by the Russian scientists Andre Geim and Konstantin Novoselov, who won the Nobel Prize in Physics in 2010 for their ground-breaking experiments at the University of Manchester in the UK.

Graphene is an excellent thermal and electrical conductor, 300 times more efficient than copper. A piece of graphite 1 mm thick consists of three million sheets of graphene, with one stacked on top of another. When graphite is prised apart it becomes essentially graphene. If a graphene sheet is rolled up, it forms a **carbon nanotube** (sub-topic A.6). When this, in turn, is folded up into a sphere it becomes a **fullerene**, which looks like a soccer ball (discussed below).

Graphene is a remarkable material, especially because of its superb electrical conductivity, strength, flexibility, and transparency.

Graphene has been described as the "new silicon". Some of the future applications of graphene lie in the following research areas:

- development of graphene–plastic composite materials to replace metals used in the aerospace industry because of their low density and high strength

- liquid-crystal displays (LCD) and flexible touch-screens for mobile devices due to the flexibility, transparency, and electrical conductivity of graphene.

After the discovery of graphene in 2004 a whole class of two-dimensional materials have emerged, which include the single layers of boron nitride, BN and molybdenum disulfide, MoS_2. BN is an excellent lubricant and can be used in a vacuum so it is important in space research and is also used in ceramic materials. MoS_2 is also a very good lubricant.

▲ Figure 15 Graphene

C_{60} fullerene

In 1985 a new form of carbon allotrope called fullerene, with carbon atoms arranged in closed shells, was discovered by Robert F. Curl Jr (working at Rice University in the USA), Sir Harold W. Kroto (working at Sussex University in the UK), and Richard E. Smalley (also working at Rice University). In 1996 these scientists were awarded the Nobel Prize in Chemistry for their discovery of fullerenes. The number of carbon atoms in the shell was found to vary, which led to the discovery of several new carbon structures. Fullerenes were found to form when vaporized carbon condensed in an atmosphere of an inert gas. Clusters of C_{70} and C_{60} were initially synthesized, with more C_{60} than C_{70} clusters being formed. The structure of each C_{60} molecule was found to consist of a truncated icosahedral cage, which has the shape of a soccer ball. The spherically symmetrical C_{60} molecule was unique in nature at its time of discovery.

In the C_{60} polyhedron cage there are 20 hexagonal surfaces and 12 pentagonal surfaces, and each carbon atom is covalently bonded to three others so that the coordination number is three, but the arrangement is not planar (figure 16).

The geodesic dome designed by the US architect R. Buckminster Fuller for the 1967 Montreal World Exhibition in Canada has this shape of a soccer ball and hence C_{60} has been named **buckminsterfullerene**. These spherical fullerenes are sometimes referred to as **buckyballs**.

C_{60} fullerene is not a covalent network solid, and so is different from graphite, diamond, and graphene. C_{60} is composed of individual molecules with strong covalent bonds, but with weak London forces between the molecules.

Activity

Graphene nanoribbons - the future landmark in the field of graphene!

Find out about graphene nanoribbons (GNRs) by accessing the chemical literature or online and why are material scientists so excited about their future development.

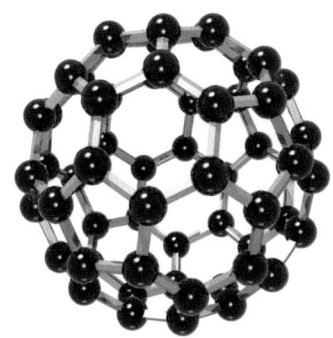

▲ Figure 16 Molecular structure of C_{60} **fullerene (buckminsterfullerene)**, showing 60 carbon atoms arranged in a spherical structure that consists of interlinking hexagonal and pentagonal rings, like a soccer ball

119

Fullerenes are black solids that do not dissolve in water, but can dissolve in some non-polar solvents (for example, benzene). In organic solvents they form coloured solutions; the colour depends on the solvent and varies from red to brown to magenta. C_{60}, unlike both graphite and graphene, does not conduct electricity. Although it does contain delocalized electrons, the electrons do not have the capacity to move from one C_{60} molecule to the next because of the symmetrical nature of C_{60}.

Many new compounds of fullerenes have been synthesized subsequently, with atoms or small molecules enclosed within the fullerene cage. Such **inclusion complexes** can potentially be used as gene and drug carriers; other applications lie in the areas of superconductivity and ferromagnetism because of the unusual electromagnetic properties of inclusion complexes. Fullerenes also have the ability to fit inside the hydrophobic cavity that forms the active site of the human immunodeficiency virus (HIV) protease enzymes and thereby inhibit them.

Carbon nanotubes are tube-shaped molecules, closely related to C_{60}, and have electrical conductivity approximately ten times better than copper and are 100 times stronger than steel. They are also used in many electronic applications, replacing silicon, and in the synthesis of new compounds that allow unstable substances to become stabilized when incorporated within the tubes. The diameter of such tubes is extremely small, in the order of 10^{-9} mm, on the **nano scale**. Buckyballs and carbon nanotubes have become a vibrant and dynamic area of chemical research in materials science, with numerous applications.

 Serendipitous discoveries

The accidental discovery of buckminsterfullerene is a classic example of the importance of *serendipity* in chemistry. Carry out some research in the library and online to find out how buckminsterfullerene was discovered. What other examples of serendipitous discoveries are famous in chemistry?

Silicon dioxide, SiO₂ (quartz)

Silicon dioxide, SiO_2, often called silica, is found in its **amorphous** form (that is, a solid with no ordered structure) as sand. In its most common crystalline form it is called **quartz**. Quartz is another example of a three-dimensional covalent network solid. It consists of arrays of SiO_4 tetrahedra arranged in a lattice (figure 17). Each silicon atom is bonded covalently to four oxygen atoms and each oxygen atom is bonded covalently to two silicon atoms. The Si–O–Si geometrical arrangement is bent because of the presence of two non-bonding pairs of electrons on each oxygen atom. Silicon dioxide has both a high melting point (1710 °C) and a high boiling point (2230 °C) due to the existence of strong covalent bonds.

▲ Figure 17 Structure of quartz, which is a crystalline form of silicon dioxide, SiO_2. Crystals of quartz are used in optical and scientific instruments and in electronics, such in as quartz watches

Both crystalline and amorphous dioxide are insoluble in water and solid crystalline SiO_2 does not conduct electricity (since there are no delocalized electrons present) or heat. Note that molten silicon dioxide can conduct electricity however as electrons are free to move in the molten state.

Coordinate covalent bonding

We have just considered covalent network solids. Another type of covalent bonding is called **coordinate covalent bonding**. In a typical covalent bond, the shared pair of electrons originate from both atoms that form the bond; one atom contributes one electron to the shared pair and the second atom contributes the second electron. In coordinate covalent bonding, the shared pair of electrons comes from only one of the two atoms; this atom donates both electrons to the shared pair.

A number of species have coordinate covalent bonding. Examples include:

* $[NH_4]^+$
* $[H_3O]^+$
* CO
* Al_2Cl_6
* transition metal complexes (discussed in topic 13).

> The term **coordination bond** is often used (based on IUPAC recommendations) to designate a coordinate covalent bond.

Ammonium cation, $[NH_4]^+$

When ammonia, NH_3, reacts with an acid, H^+, the lone pair on the nitrogen in NH_3 combines with the proton, H^+, to form the ammonium cation, $[NH_4]^+$:

$$H_3N: + H^+ \rightarrow [NH_4]^+$$

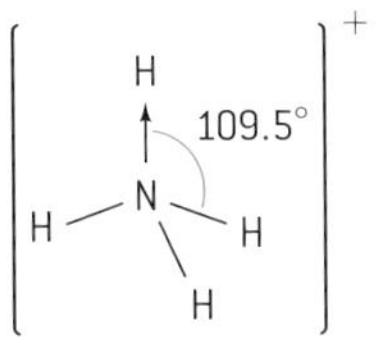

Hydronium cation, $[H_3O]^+$

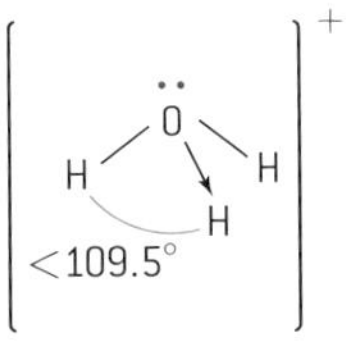

> The coordinate covalent bond is represented by an arrow to signify the origin of the electrons in the bond. Once formed, however, all the bonds are equivalent (whether coordinate covalent or normal covalent). Previously, the term dative covalent bonding was used for this type of bond but, based on IUPAC recommendations, this term is now largely obsolete.

Carbon monoxide, CO

$$:C \overset{\longleftarrow}{\equiv} O:$$

Dimer of aluminium chloride, Al_2Cl_6

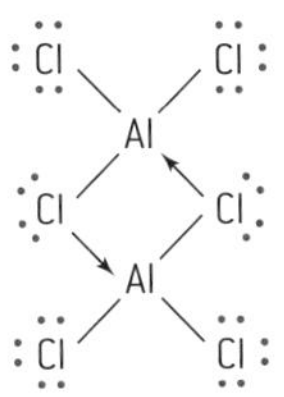

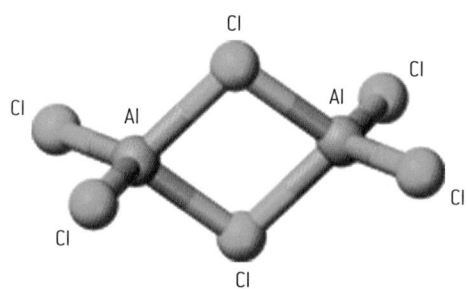

In the solid state aluminium chloride is ionic. $AlCl_3$ is six-coordinate involving an ionic lattice (but with significant covalent characteristics). At atmospheric pressure it sublimes at 180 °C. On increasing the pressure it melts. On melting at 192.4 °C it forms the dimer, Al_2Cl_6, which has coordination bonding. In this structure, aluminium is tetravalent with a coordination number of four. The three-dimensional structure of the dimer is shown in figure 18. The bridging chlorines are on a different plane compared to the terminal chlorines.

Al_2Cl_6 molecules predominate in the gaseous state up to 400 °C. Above this temperature, it dissociates to form molecules of $AlCl_3$ with a trigonal planar geometry (120° bond angles).

▲ Figure 18 The structure of the Al_2Cl_6 dimer in the gaseous phase

Quick questions

1 Apart from aluminium chloride, identify **three** other substances that sublime readily.

2 Deduce whether both solid and molten aluminium chloride conducts electricity.

Representations of structures

For tetrahedral structures it is common to use **wedge-and-dash notation** to show the various planes:

- a wedge indicates that the bond is in front of the defining plane
- a dash indicates that the bond is behind the defining plane
- a solid line indicates that the bond lies on the defining plane.

For example, the tetrahedral structure of methane, CH_4, can be represented as follows using this notation:

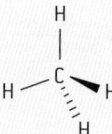

4.4 Intermolecular forces

Understandings

→ Intermolecular forces include London (dispersion) forces, dipole–dipole forces, and hydrogen bonding.

→ The relative strengths of these interactions are London (dispersion) forces < dipole–dipole forces < hydrogen bonds.

Applications and skills

→ Deduction of the types of intermolecular force present in substances, based on their structure and chemical formula.

→ Explanation of the physical properties of covalent compounds (volatility, electrical conductivity, and solubility) in terms of their structure and intermolecular forces.

Nature of science

→ Obtain evidence for scientific theories by making and testing predictions based on them – London (dispersion) forces and hydrogen bonding can be used to explain special interactions. For example, molecular covalent compounds can exist in the liquid and solid states. To explain this, there must be attractive forces between their particles that are significantly greater than those that could be attributed to gravity.

Theories on intermolecular forces

In sub-topics 4.2 and 4.3 we saw that there are **intramolecular forces of attraction** that hold the atoms together within a molecule, resulting in covalent bonding. Such intramolecular forces affect molecular geometries, physical properties, and reactivities of compounds. Intramolecular forces could be described as "**bonding**" forces of attraction.

Another type of attraction, **intermolecular forces**, are interactions between molecules within a compound (figure 1). Intermolecular forces are largely responsible for the **bulk properties of matter**, that is, its physical properties such as melting point and boiling point.

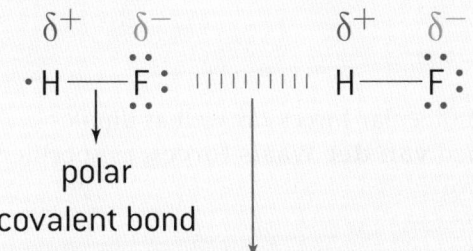

polar
covalent bond

intermolecular
force of attraction

▲ Figure 1 Difference between intramolecular and intermolecular forces of attraction for hydrogen fluoride. The intramolecular forces result in covalent bonding

Intermolecular forces of attraction are much weaker than covalent bonds. For example, the **standard enthalpy change of vaporization of water**, ΔH_{vap} (the enthalpy change associated with the conversion of one mole of pure liquid into a gas at its boiling point at standard pressure, 100 kPa) is 44.02 kJ mol^{-1} (at 298 K) whereas 926 kJ mol^{-1} is required to break the two O–H polar covalent bonds in a molecule of water (see the bond enthalpies for O–H from section 11 of the *Data booklet*: $2 \times 463 = 926$ kJ mol^{-1}). Since the intermolecular forces of attraction are relatively weak, the molecules of a covalent compound are not held strongly together and, for this reason, many covalent compounds are gases (for example, $N_2(g)$, $O_2(g)$, $CO_2(g)$ and $CO(g)$) or liquids (for example, $H_2O(l)$). In contrast ionic compounds have very strong electrostatic forces of attraction between the ions, meaning that ionic compounds are solids at room temperature and have high melting points. For example, the melting point of the ionic compound sodium chloride, NaCl(s), is 801 °C, whereas the melting point of water, $H_2O(s)$, which is a covalent compound, is much lower, at 0 °C.

Chemistry in the kitchen

The next time you are having a drink with cubes of ice in it, take one cube of ice and try to break it with your fingers. As you will discover this is virtually impossible and the reason for this can be associated with the intermolecular forces of attraction in the vast network of water molecules present in the ice.

One of the assumptions made in topic 6 in relation to the kinetic–molecular theory of gases is that collisions between one gaseous particle and another are completely **elastic**. This suggests that gaseous atoms or molecules do not stick or adhere to one another. This is incorrect because every gaseous species can be converted into a liquid at some temperature. It is the existence of intermolecular forces of attraction that enable molecules of a covalent compound to exist in the **condensed phase** (liquid and solid). Figure 2 demonstrates an example of this.

As seen in topic 1, the particles in a solid or liquid are tightly packed together — that is why we use the term **condensed phase**.

▲ Figure 2 Sperm bank shipping containers being filled with liquid nitrogen to keep the sperm frozen. We think of nitrogen as being in the gas phase, but all gases can be converted into liquids at some temperature because of the intermolecular forces of attraction between the molecules. The boiling point of liquid nitrogen is −195.8 °C at atmospheric pressure

In order to understand these bulk physical properties, we need to widen our discussion and take into account intermolecular forces of attraction in all three phases, solid, liquid, and gas. The question, therefore, is what are the various types of intermolecular forces of attraction that can be present?

In science we obtain evidence for scientific theories by making and testing predictions based on the theories. London (dispersion) forces and hydrogen bonding, two types of intermolecular forces of attraction can be used to explain special interactions. For example, how molecular covalent compounds can exist in the liquid and solid states. As stated above, the explanation is related to the existence of attractive forces between their particles that are significantly greater than those that could be attributed to gravity.

> The relative strengths of these interactions in general are:
>
> **London forces < dipole–dipole forces < hydrogen bonds**
>
> However, we shall also consider examples where London forces are stronger than dipole–dipole forces, but the above order is what occurs often!

The main three types of intermolecular forces of attraction that we shall discuss are:

- London forces (also called dispersion forces or instantaneous induced dipole-induced dipole forces)
- dipole–dipole forces
- hydrogen bonding.

Collectively the first two intermolecular forces (as well as dipole-induced dipole forces) are termed **van der Waals forces**, as specified by IUPAC.

> London (dispersion) forces + dipole–dipole forces + dipole-induced dipole = **van der Waals forces**

The strengths of intermolecular forces of attraction between molecules can vary significantly, but it must be emphasized that these forces are considerably weaker than ionic or covalent bonds.

London forces

London (dispersion) forces exist in all molecules. Such forces were first recognized by the German–American theoretical physicist Fritz Wolfgang London (1900–1954), hence the name (which has nothing to do with London, the capital of the United Kingdom!) London forces are also called **dispersion forces** or **instantaneous induced dipole–induced dipole forces**. The origin of the latter term can be understood if we consider what happens to a non-polar molecule such as diatomic hydrogen, H_2, when it is approached by another hydrogen molecule.

Although the hydrogen molecule is non-polar (that is, there is no net dipole moment), each hydrogen molecule consists of positively charged nuclei surrounded by a cloud of negatively charged electrons. The electron clouds constantly change position. The *average* distribution with respect to the time the electrons are located in the cloud throughout the molecule, is spherically symmetrical, as shown in figure 3.

However, if you were to take a random snapshot *at a given instant of time* one part of the molecule might have slightly more electron density than another part. In this case, a temporary dipole moment is generated. This temporary dipole moment is termed an **instantaneous dipole** and has an influence on adjacent hydrogen molecules.

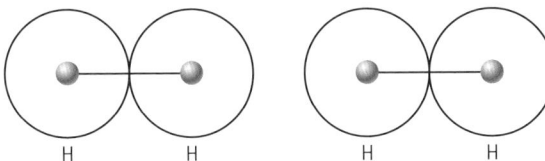

▲ Figure 3 The average distribution over time of the electrons in any H_2 molecule is spherically symmetrical

Therefore, if one hydrogen molecule now approaches a second hydrogen molecule that has acquired a short-lived instantaneous dipole, the nucleus of the first hydrogen molecule will be attracted to the region of higher electron density in the electron cloud of the second hydrogen molecule. At the same time, the two regions of electron density within each molecule will repel each other on approach, as both are negatively charged.

As described by IUPAC, **polarizability** is the ease of distortion of the electron cloud of a molecular entity by an electric field (such as that caused by the proximity of a charged particle). Because the mobile electrons can be dispersed, the repulsion between regions of electron density can be minimized. Therefore, the orbital (region of space where there is a high probability of finding electrons) can effectively change its shape, which results in a non-spherical distribution of the electron cloud; that is, the orbital is pulled out of its symmetrically spherical shape. A temporary dipole is generated that results in electrostatic attractions between the partial positive charge, δ^+, of one hydrogen molecule and the partial negative charge, δ, of the neighbouring hydrogen molecule (figure 4).

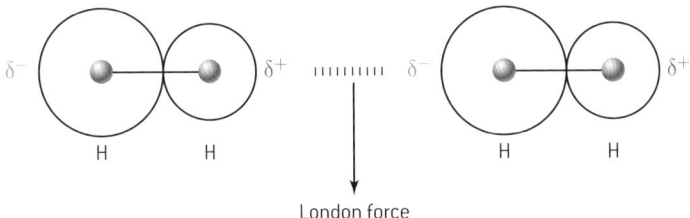

London force

▲ Figure 4 At a given instant in time, a temporary dipole, the instantaneous dipole, is established

This interaction is the basis of a **London force**. Note, however, that such an arrangement is only temporary – in the next instant of time a different pattern of induced dipoles may emerge.

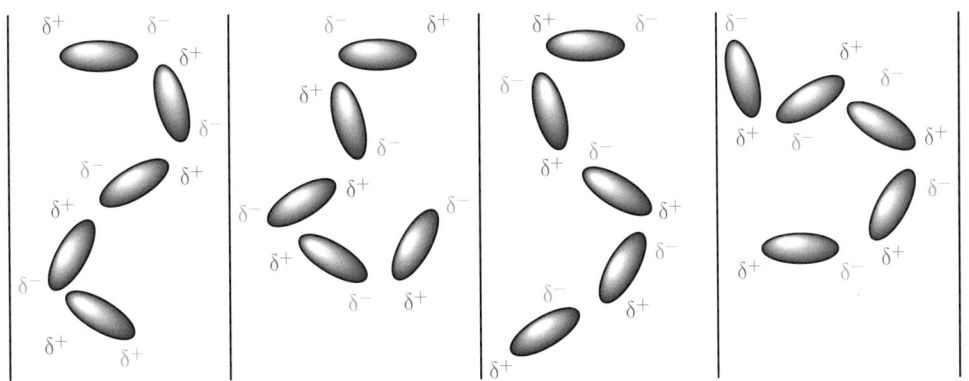

▲ Figure 5 Different arrangements of the interactions of the London forces of attraction between molecules, which result from interactions between an instantaneous dipole on one molecule and an induced dipole on an adjacent molecule

What affects the magnitude of London forces?

There are three factors:

- number of electrons
- size (volume) of the electron cloud
- shapes of molecules.

Number of electrons

The greater the number of electrons, the larger the distance between the valence electrons and the nucleus. The attraction of the valence electrons to the nucleus will be reduced and hence the electron cloud can be polarized more easily.

For example, consider the boiling points of the two noble gases neon, Ne, and krypton, Kr, using the information in section 7 of the *Data booklet* (table 1).

Noble gas	Number of electrons	Boiling point / °C
Ne ($Z = 10$)	10	−246.0
Kr ($Z = 36$)	36	−153.4

▲ Table 1 Boiling points and number of electrons for neon and krypton

London forces decrease rapidly with increasing distance, r^6, based on the relationship:

$$V \propto \frac{1}{r^6}$$

where V is the potential energy associated with the interactions.

Hence, in the case of neon the eight outer electrons are located in the $n = 2$ energy level, but in the case of krypton the eight outer electrons are located much further from the nucleus in the $n = 4$ energy level. This means the attraction of the outer electrons to the nucleus is not as great and in krypton the electron cloud can be polarized more easily. Hence the London forces in krypton are stronger, so the boiling point of krypton is higher than that of neon.

Size (volume) of the electron cloud

The magnitude of the London forces will also depend on the size of the electron cloud, that is its volume in space. In a large electron cloud, the attraction of electrons to the nucleus will not be as great as in a smaller electron cloud, and hence the electrons in a large electron cloud can be polarized more easily.

For example, consider the boiling points of the two alkanes propane, $CH_3CH_2CH_3$ and octane, $CH_3(CH_2)_6CH_3$ (table 2). The number of carbon atoms in octane is greater than in propane, which results in stronger London forces and hence a higher boiling point for octane than for propane.

As the number of dispersed electrons can be linked to the molecular mass the greater the molecular mass, the greater the number of London forces present.

Alkane	Boiling point / °C	Space-filling model
propane (C_3H_8)	−42.0	
octane (C_8H_{18})	125	

▲ Table 2 Boiling points of two alkanes

Shapes of molecules

The molecular shape is the third factor that influences the magnitude of London forces. Let us compare the boiling points of the two isomers of C_5H_{12}, pentane, $CH_3(CH_2)_3CH_3$, and 2,2-dimethylpropane, $(CH_3)_4C$ (table 3).

Both isomers contain the same number of electrons, but the boiling point of 2,2-dimethylpropane is considerably lower than the boiling point of pentane. The reason for this is that the straight-chain nature of pentane's shape allows the molecules to interact with each other across the full length of the molecule; that is, there is a large area of interaction because of the better contact between the molecules of pentane. In contrast, for 2,2-dimethylpropane, the contact area for the molecules is considerably smaller because of the almost soccer-ball shape of the molecule (figure 6).

Isomer	Boiling point / °C
pentane	36.1
2,2-dimethylpropane	9.5

▲ Table 3 Boiling points of two isomers of pentane

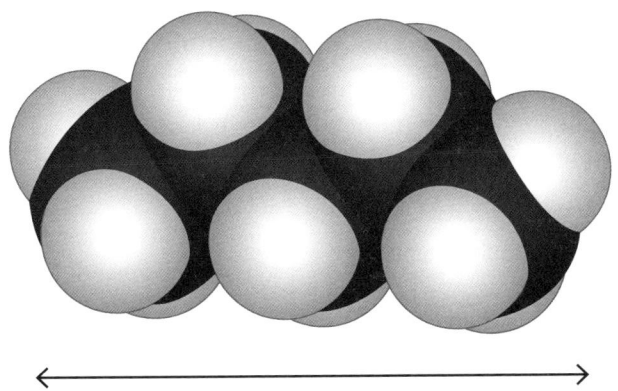

In pentane, there is a large contact area across the entire molecule for adjacent molecules to interact.

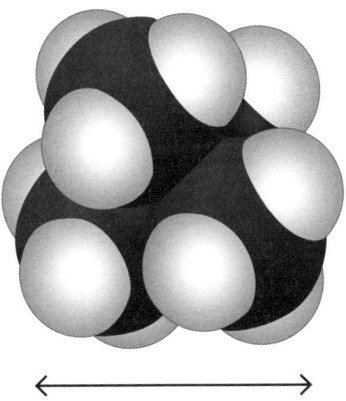

In 2,2-dimethylpropane, there is a much smaller contact area for adjacent molecules to interact.

▲ Figure 6 Space-filling models of pentane and 2,2-dimethylpropane showing areas of contact between adjacent molecules for London forces of attraction

As the contact area is much larger between molecules of pentane, the London forces between the molecules will have a greater magnitude, which results in a higher boiling point for pentane.

 Final points on London forces – a warning on misinterpretation!

- It must be stressed that London forces of attraction between molecules are not attributed to gravitational attraction between molecules. In fact, gravitational attraction between molecules is almost zero, since the masses of individual molecules are very small. A common misinterpretation when explaining London forces is to state *alone* that molecules with greater molecular mass have greater London forces. This is a useful marker, but remember that if a molecule has a greater molecular mass it means a greater number of electrons are able to be polarized, resulting in an increase in the magnitude of London forces.

- London forces are attractive forces between atoms (for example in He) and also occur between non-polar molecules (for example, H_2), but they exist between polar molecules as well (for example, HCl). That is, *every molecule* will experience London forces (whether non-polar or polar).

Dipole–dipole forces

The second type of intermolecular forces are dipole–dipole forces, which exist in all polar molecules with a permanent (*not* instantaneous) dipole moment, μ. Examples of such molecules include HF, ICl, HCl, and CH_3CHO. In this type of intermolecular force, there is an attraction between the positive end of one **permanent dipole** and the negative end of **another permanent dipole on an adjacent molecule**.

Let us compare the boiling points of two molecules that have similar molar masses: the interhalogen, iodine monochloride, ICl, and the halogen, bromine, Br_2 (table 4).

Isomer	Boiling point / °C	Types of intermolecular forces present
ICl ($M = 162.35\ \text{g mol}^{-1}$)	97.4	London forces + dipole–dipole forces
Br_2 ($M = 159.80\ \text{g mol}^{-1}$)	58.8	only London forces

▲ Table 4 Boiling points of ICl and Br_2

Since ICl is highly polar, in addition to London forces it also has dipole–dipole forces of attraction between the ICl molecules (figure 6), which lead to a higher boiling point.

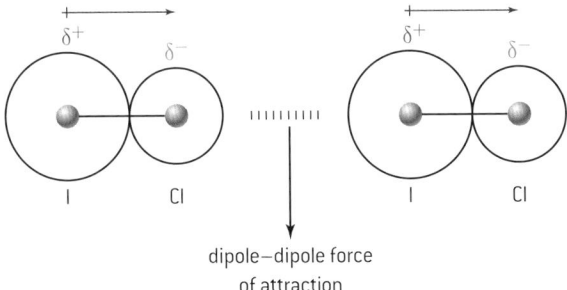

dipole—dipole force
of attraction

▲ Figure 7 Dipole—dipole force of attraction between *permanent* dipoles on adjacent molecules of ICl. Note that iodine has a larger atomic radius compared to chlorine (see section 9 of the *Data booklet*), but chlorine is more electronegative (see section 8 of the *Data booklet*).

Hydrogen bonding

This third type of intermolecular force holds a special place in chemistry and is one of the most important types of intermolecular force.

Hydrogen bonding can occur between molecules when there is a H–F, an O–H, or an N–H bond present.

A typical hydrogen bond may be depicted as:

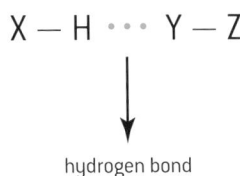

hydrogen bond

where the:

- hydrogen donor is X–H

- acceptor may be an atom or an anion, Y, a fragment or a molecule Y–Z in which Y is bonded to Z

- hydrogen bond is represented by the three dots though dashes are sometimes used.

Hydrogen bonds occur, for example, between

a) water molecules, H_2O

b) ammonia molecules, NH_3

c) hydrogen fluoride molecules, HF

d) water molecules and dimethyl ether molecules, $(CH_3)_2O$

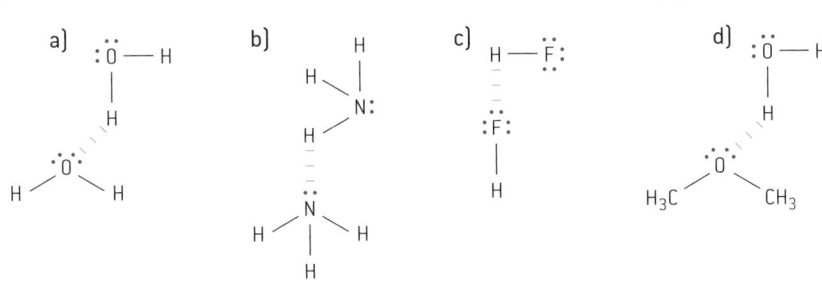

Hydrogen bonding often has a large influence on both the properties and structures of materials.

Definition of the hydrogen bond

As recommended by IUPAC in 2011, a **hydrogen bond** is defined as an attractive interaction between a hydrogen atom from a molecule or a molecular fragment, X–H, in which X is more electronegative than H, and an atom or a group of atoms in the same or a different molecule, in which there is evidence of bond formation.

Pure and Applied Chemistry, 83(8), (2011) pp1637-1641

Key point

The H–F, O–H, and N–H bonds are polar covalent bonds and are not hydrogen bonds.

Representation of hydrogen bonds

In this book we use dashes to represent hydrogen bonds to distinguish them from lone pairs of electrons.

Example of the effect of hydrogen bonding

Let us compare the boiling points of some hydrides of groups 14, 15, 16, and 17.

Figure 8 shows a plot of the boiling points of the series of hydrides versus period number. As you move down a group, the boiling points increase within a particular hydride series, because of an increase in the number of electrons, resulting in a greater number of London forces. However, in the case of the hydrides H_2O, HF, and NH_3, the boiling points are considerably higher. This is because of the existence of hydrogen bonding in these compounds. Methane, CH_4, however, has a lower boiling point as expected, because it does not show hydrogen bonding.

The strength of the hydrogen bond depends on the electrostatic attraction between the lone pair of electrons of the electronegative atom and the nucleus of the proton. Hence the hydrogen bonding in HF is stronger than the hydrogen bonding in H_2O because fluorine is more electronegative than oxygen [$\chi_P(F) = 4.0$, $\chi_P(O) = 3.4$]

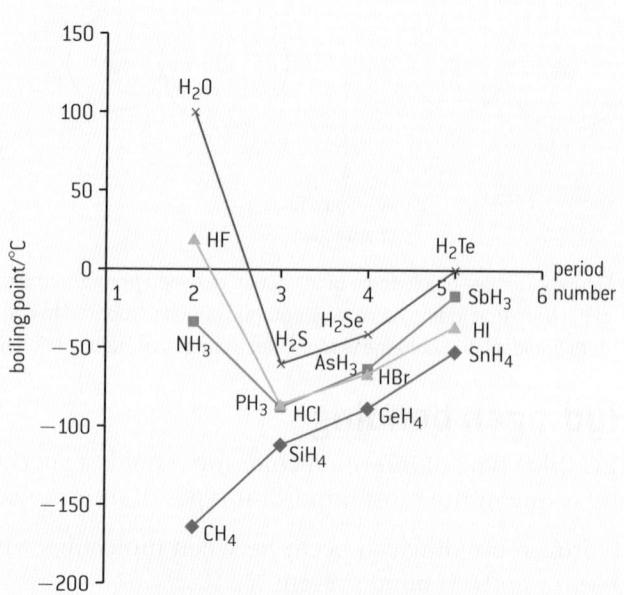

▲ Figure 8 Boiling points for the series of hydrides (HX, H_2X, XH_3, and XH_4) from groups 14, 15, 16, and 17

Hydrogen bonding and water

If water did not show hydrogen-bonding, all the water on Earth would be in the gaseous state. In addition, hydrogen bonding means that the solid phase of water (ice) has a lower density than water in the liquid state. This is why ice floats on water. In ice, each water molecule forms hydrogen bonds with adjacent water molecules, which leads to a regular, very ordered network in the lattice (figure 9). The presence of the hydrogen bonds creates cavities in the lattice. In contrast, in the liquid phase the hydrogen bonding is more random, which results in a higher density compared to ice.

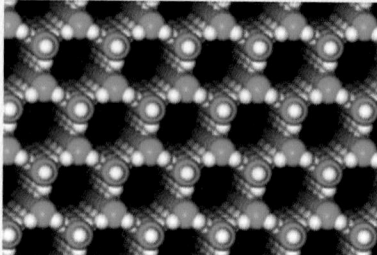

▲ Figure 9 Open cavity structure in the lattice structure of ice

Hydrogen bonding is also present in biomolecules such as in the double helix structure of DNA. (sub-topic B.8)

Type of intermolecular force	Relative strength / kJ mol^{-1}
London forces	weak (1–10) – this can increase with number of electrons, size (volume) of electron cloud, and shape of molecule
dipole–dipole forces	weak to moderate (3–25)
hydrogen bonds	moderate to strong (10–40)

▲ Table 5 Comparison of the various relative strengths of intermolecular forces between molecules

Worked examples

Example 1

Identify the intermolecular forces in the following substances:

- He
- $CH_3(CH_2)_4CH_3$
- NF_3

- $(CH_3)_2O$
- CH_3F
- CH_3CH_2OH

Solution

Substance	Intermolecular forces present	Comment
He	London only	
$CH_3(CH_2)_4CH_3$	Non-polar molecule so London only	
NF_3	Polar molecule so London + dipole–dipole	Since F is more electronegative than N [$\chi_P(F) = 4.0$, $\chi_P(N) = 3.0$], this trigonal-pyramidal molecule has a net dipole moment and therefore is polar: This molecule contains no H atoms, so no hydrogen bonding is possible.
$(CH_3)_2O$	Polar molecule so London + dipole–dipole	Even though the highly electronegative element oxygen is present, there is no O–H bond so therefore no hydrogen bonding is possible.
CH_3F	Polar molecule so London + dipole–dipole	This molecule is tetrahedral; F is more electronegative than H and C [$\chi_P(F) = 4.0$, $\chi_P(C) = 2.6$, $\chi_P(H) = 2.0$] so there is a net dipole moment present making the molecule polar: Even though the highly electronegative element fluorine is present, there is no H–F bond so therefore no hydrogen bonding is possible.
CH_3CH_2OH	The molecule is polar and an O–H bond is present, so: London + dipole–dipole + hydrogen bonding	

Example 2

State and explain which of the following species can form hydrogen bonds with water molecules: ammonia, NH_3, propane, $CH_3CH_2CH_3$, ethanoic acid, CH_3COOH.

Solution

$CH_3CH_2CH_3$ does not contain OH, NH or HF bonds. It cannot form hydrogen bonds with water molecules.

NH_3 and CH_3COOH have N–H and O–H bonds, therefore they have the ability to form hydrogen bonds with water:

Example 3

As a general rule the relative strengths of intermolecular forces follow the order:

London (dispersion) forces < dipole–dipole forces < hydrogen bonds

Comment, basing your answer on intermolecular forces, on the fact that the boiling point of carbon tetrachloride, CCl_4 is 76.72 °C, whereas the boiling point of fluoromethane, CH_3F is −78.2 °C.

Solution

- We first work out the types of intermolecular forces of attraction present in each compound. CCl_4: only London forces, because this is a non-polar molecule with no net dipole moment. CH_3F: London forces and dipole–dipole forces, because this is a polar molecule.

- On the basis of this and following our general rules, we would expect that the boiling point of fluoromethane with its additional intermolecular forces should be much higher. *In fact, the opposite is the case!*

- The reason for this must be associated with the strength of the London forces. In the case of CH_3F, the number of valence electrons is considerably fewer than in CCl_4. In CCl_4 the presence of more valence electrons leads to a greater polarizability of the electron cloud. This results in significantly stronger London forces, which outweigh the dipole–dipole forces. *The key point here is that the above order cited in the syllabus is relative and every example must be challenged based on the data provided!*

Quick question

Suggest a second way in which ethanoic acid CH_3COOH can hydrogen bond with water.

Activity

Researchers have seen hydrogen bonds for the first time!

Researchers in China recently used **atomic force microscopy (AFM)** to produce the first high quality images of hydrogen bonds that exist between molecules of 8-hydroxyquinoline. Find out more about this from the chemical literature or online. What is especially surprising about the atoms involved in this type hydrogen bond (hint – consider the involvement of carbon!)?

4.5 Metallic bonding

Understandings

→ A metallic bond is the electrostatic attraction between a lattice of positive ions and delocalized electrons.

→ The strength of a metallic bond depends on the charge of the ions and the radius of the metal ion.

→ Alloys usually contain more than one metal and have enhanced properties.

Applications and skills

→ Explanation of electrical conductivity and malleability in metals.

→ Explanation of trends in melting points of metals.

→ Explanation of the properties of alloys in terms of non-directional bonding.

Nature of science

→ Use theories to explain natural phenomena – the properties of metals are different from covalent and ionic substances and this is due to the formation of non-directional bonds with a "sea" of delocalized electrons.

Metallic bonding

In topic 3, we saw that metals lie to the left of the stepped line in the periodic table of elements. Metals have low ionization energies (see section 8 of the *Data booklet*), so valence (outer-shell) electrons can be delocalized throughout the metal. The structure of a metal, shown in figure 1, is a regular giant lattice that consists of positive ions (**cations**) surrounded by a "sea" of delocalized electrons.

Definition of a metallic bond

A **metallic bond** is the electrostatic attraction between a lattice of positive ions (cations) and delocalized electrons.

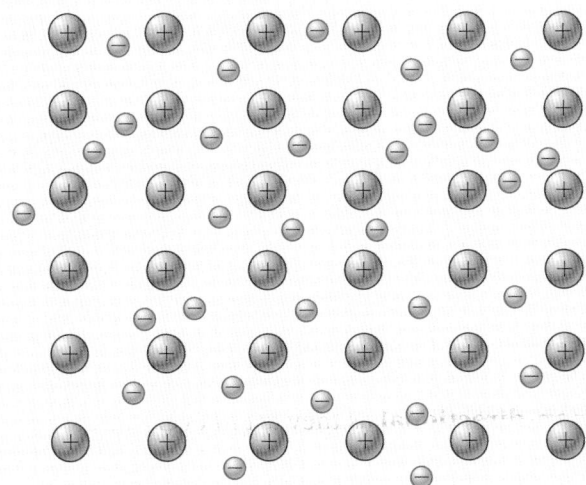

▲ Figure 1 Structure of a metal showing an array of positive ions (cations) surrounded by a "sea" of delocalized electrons

Delocalized electrons are not associated with a particular nucleus of a metal, but instead are free to move throughout the entire crystalline lattice forming a "sea" of mobile electrons.

The strength of a metallic bond depends on three factors:

- the number of valence electrons that can become delocalized
- the charge of the metal ion
- the ionic radius of the metallic positive ion (cation).

Alloys

An alloy is a mixture that consists either of two or more metals, or of a metal (or metals) combined with an alloying element composed of one or more non-metals (for example, cast iron consists of the metal iron and the non-metal carbon). Alloys have enhanced properties, such as strength, hardness, and durability which differ from those of the parent metallic elements (table 1).

Definition of an alloy

An **alloy** can be defined as a metallic material, homogeneous on a macroscopic scale, consisting of two or more elements so combined that they cannot be readily separated by mechanical means. Alloys are to be considered as mixtures for the purpose of classification.

United Nations (2011)
4th Edition.

Composition of alloy mixtures

(major element: a metal**)** + **(alloying element:** can be metal or non-metal**)**

Alloy	Composition	Uses
brass	copper and zinc	door handles, window fittings, screws
steel	Iron, carbon, and other metals such as tungsten	bridges, buildings
dental amalgam	mercury, silver, and tin	used by dentist for teeth fillings

▲ Table 1 Examples of alloys

 Explanation of electrical conductivity and malleability in metals

Electrical conductivity

Metals are good conductors of electricity because of the mobile delocalized electrons. When a potential is applied to the metal, the mobile electrons can move through the metallic structure and hence carry an electric current. The presence of impurities in a metal can restrict the movement of electrons through the metal, resulting in an increase in electrical resistance. Hence copper used in electrical wiring needs a high degree of purity.

Malleability

Metals are malleable. **Malleability** is the ability of a solid to be pounded or hammered into a sheet or other shape without breaking. The reason why metals have this property is that the positive ions (cations) can slide past one another, which leads to a rearrangement of the shape of the solid. The metallic bonds within the lattice do not have any defined direction (they are often described as **non-directional** as they act in every direction about the fixed immobile cations). Thus if pressure is applied by pounding, the cations may slide over one another but there is no disruption to the metallic bonding (figure 2).

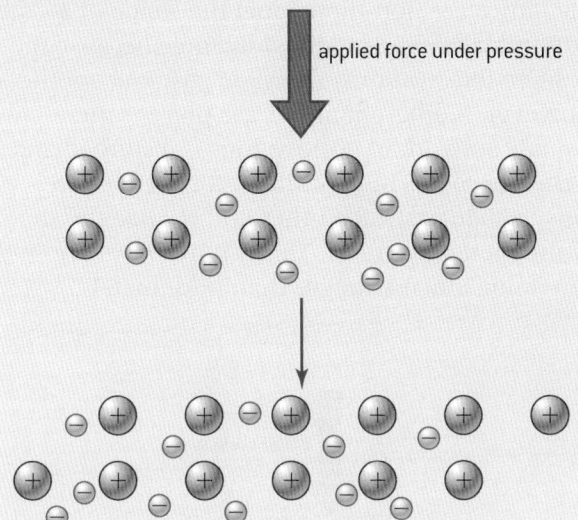

applied force under pressure

structure of metal after being pounded into a sheet

▲ Figure 2 Metallic bonding remains intact even after a metal is hammered into a sheet or other object without breaking. This illustrates the property of malleability

Aluminium foil, often used to wrap food, shows this property of malleability for the metal aluminium.

Explanation of trends in melting points of metals

Metallic bonds are very strong and therefore metals often have high melting points.
For example, the melting point of tungsten is 3414 °C, although some low melting points also occur such as −38.8 °C for mercury.

Table 2 compares the melting points of the alkali metal, potassium and the alkaline earth metal calcium (section 7 of the *Data booklet*).

Metal	Melting point / °C
potassium (K)	63.5
calcium (Ca)	842

▲ Table 2 Melting points of potassium and calcium

The melting point of a metal depends on the strength of the attractive forces that hold the positive ions within the "sea" of delocalized electrons. The melting point of calcium is higher than the melting point of potassium for the following reasons:

- Calcium has two delocalized electrons per atom, whereas potassium has only one delocalized electron per atom. Therefore, the electrostatic attraction between the positive ions and the delocalized electrons will be greater in calcium.

- Calcium forms a 2+ ion whereas potassium forms a 1+ ion. Therefore the electrostatic attraction between the positive ions and the delocalized electrons will also be greater in calcium.

- From section 9 of the *Data booklet* we see that the size of the ionic radius of K^+ is 138 pm, whereas the size of the Ca^{2+} ionic radius is smaller at 100 pm, which implies that the delocalized electrons will be more strongly attracted to the Ca^{2+} ion.

This variation in melting points can also be observed on descending group 1, the alkali metals. As we saw in topic 3, the ionic radius increases (remember the snowman diagram) going down a group, and hence the melting points will decrease with decreasing strength of the attractive forces (table 3).

Metal	Ionic radius of M^+ / pm	Melting point / °C
lithium (Li)	76	180.5
sodium (Na)	102	97.8
potassium (K)	138	63.5
rubidium (Rb)	152	39.3
caesium (Cs)	167	28.5

▲ Table 3 Melting points for the group 1 metals

When comparing the melting points of the alkali metals, the number of delocalized valence electrons per atom (one) and the charge on the cation (1+) does not change within the group, so the only factor that influences the melting point comparison is the size of the ionic radius of M^+.

Explanation of the properties of alloys in terms of non-directional bonding

Alloys can have a number of improved properties compared to the parent metallic element:

- greater strength

- greater resistance to corrosion

- enhanced magnetic properties

- greater **ductility** (a mechanical property that allows a metal to deform under tensile stress, for example being able to stretch the metal into a wire).

Even adding a small amount of an alloying element can dramatically change the properties of an alloy compared with those of the parent metal.

As seen in figure 3, as the metallic bonds within the lattice are non-directional the shape of a pure metal can be modified by force because the positive ions (cations) can slide past one another. However, if different atoms are present the regular network of positive ions is disturbed and it then becomes more difficult for the positive ions to slide past each other and change the shape of the metal (figure 3). This is why alloys are generally, much stronger than pure metals.

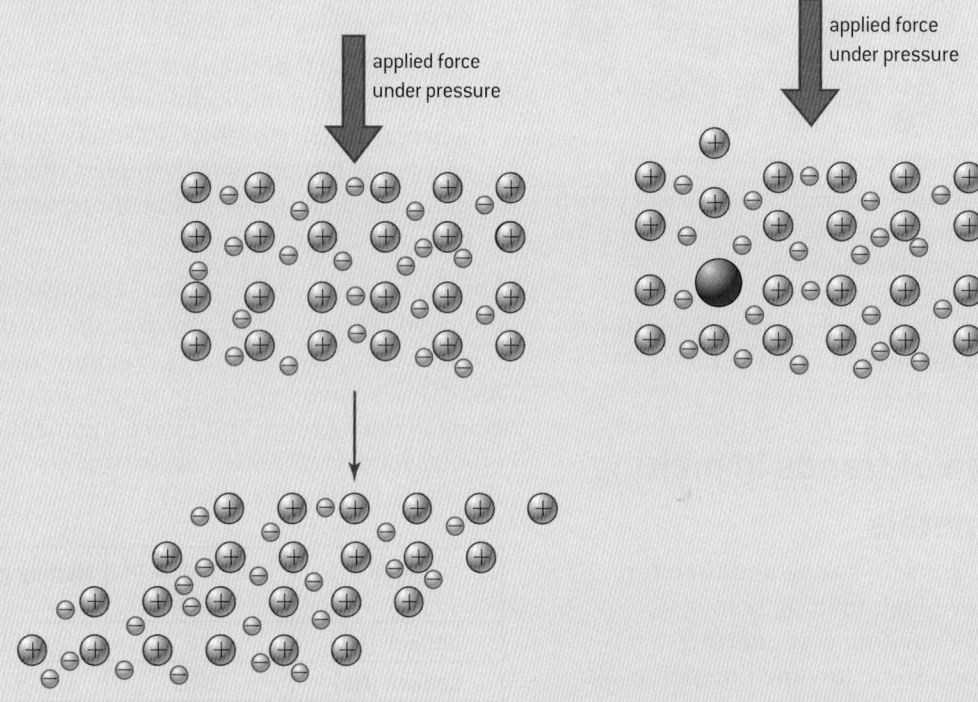

applied force under pressure

applied force under pressure

structure of metal after being pounded into a sheet

▲ Figure 3 The presence of different atoms in an alloy disturbs the regular lattice and hinders the movement of positive ions past one another

Metals – an international dimension

In this sub-topic we explored some key aspects of metals. The availability of metal resources and the means to extract them varies greatly in different countries, and is a factor in determining national wealth. As technologies develop, the demands for different metals change and careful strategies are needed to manage the supply of these finite resources.

Discuss this in class in the context of your own country and the regional economy.

Questions

1 What is the name of K_2SO_4?

A. potassium sulfite

B. calcium sulfite

C. potassium sulfate

D. calcium sulfate

2 What is the formula of magnesium oxide?

A. MgO

B. MgO_2

C. MnO

D. MnO_2

3 What is the formula of sodium nitrate?

A. $NaNO_2$

B. $NaNO_3$

C. Na_3N

D. NaCN

4 Which of the following species is molecular?

A. Na_2O

B. KBr

C. NH_4NO_3

D. N_2O_4

5 Which molecule has the shortest carbon-to-oxygen bond length?

A. CH_3CH_2OH

B. $(CH_3)_2O$

C. $(CH_3)_2CO$

D. CO

6 The electronegativities, χ_P, for four elements are given in table 4.

Element	H	C	O	Cl
χ_P	2.2	2.6	3.4	3.2

▲ Table 4

Which bond is the most polar?

A. C–H

B. O–H

C. H–Cl

D. C–O

7 What is the electron domain geometry, the molecular geometry, and the Cl–P–Cl bond angle for the molecule phosphorus trichloride, PCl_3?

	Electron domain geometry	Molecular geometry	Cl–P–Cl bond angle / °
A.	tetrahedral	tetrahedral	109.5
B.	tetrahedral	trigonal pyramidal	109.5
C.	tetrahedral	trigonal pyramidal	100.3
D.	trigonal pyramidal	trigonal pyramidal	100.3

8 Which of the following allotropes of carbon is molecular?

A. graphite

B. graphene

C. C_{60}

D. diamond

9 What are the intermolecular forces present in the molecule CH_2F_2?

A. London forces

B. London forces and dipole–dipole forces

C. London forces, dipole–dipole forces, and hydrogen bonding

D. only hydrogen bonding

10 Which statement best describes metallic bonding?

A. Electrostatic attractions between oppositely charged ions.

B. Electrostatic attractions between a lattice of positive ions and delocalized electrons.

C. Electrostatic attractions between a lattice of negative ions and delocalized protons.

D. Electrostatic attractions between protons and electrons. [1]

IB, May 2009

11 Consider the following species

BF_2Cl NCl_3 OF_2

For each species

a) deduce:

    (i) its electron domain geometry

    (ii) its molecular geometry

    (iii) its bond angle(s)

    (iv) its molecular polarity

b) draw an appropriate Lewis (electron dot) structure.

12 Consider the following species:

$[NO_3]^-$ $[ClO_3]^-$ $[BF_4]^-$ COF_2

For each species

a) deduce:

    (i) its electron domain geometry

    (ii) its molecular geometry

    (iii) its bond angle(s)

b) draw an appropriate Lewis (electron dot) structure.

13 Deduce the intermolecular forces present in each of the following species:

- Ar
- $CH_3CH_2CH_2OH$
- CH_3Cl
- $CH_3CH_2OCH_2CH_3$

14 Deduce which of the following species may form hydrogen bonds with water molecules:

- $CH_3CH_2OCH_2CH_3$
- NH_3
- C_2H_4
- PH_3

15 Compare and contrast the allotropes of carbon (diamond, graphite, graphene, and C_{60}) in terms of:

- structure
- bonding
- intermolecular forces
- melting points
- electrical conductivity.

16 ⊘ In chemistry both terminology and models can often lead to certain assumptions.

a) Suggest why the term "macromolecular" is incorrect based on IUPAC recommendations for covalent network solids such as graphite.

b) The O–Cl–O bond angle in OCl_2 is 110.9°. Discuss whether this bond angle agrees with predictions of bond angles based on the model of VSEPR theory.

17 ⊘ Suggest why VSEPR theory does not work for the majority of transition metal complexes, such as $[FeCl_4]^{2-}$, but does for a few complexes, such as $[MnO_4]^-$.

5 ENERGETICS AND THERMOCHEMISTRY

Introduction

Chemistry involves the study of reactions of the elements of the periodic table. Conservation of energy is a fundamental principle of science. The use of models, empirical data, mathematics and scientific terminology to explain the energy changes associated with chemical reactions is central to the nature of science. In this topic we examine the relationship that exists between chemistry and energy. We will introduce the state function enthalpy, investigate the applications of Hess's Law and gain a greater understanding of the applications of bond enthalpies.

5.1 Measuring energy changes

Understandings

→ Heat is a form of energy.

→ Temperature is a measure of the average kinetic energy of the particles.

→ Total energy is conserved in chemical reactions.

→ Chemical reactions that involve transfer of heat between the system and the surroundings are described as endothermic or exothermic.

→ The enthalpy change (ΔH) for chemical reactions is indicated in kJ mol^{-1}.

→ ΔH values are usually expressed under standard conditions, known as $\Delta H^{\ominus}$, including standard states.

Applications and skills

→ Calculation of the heat change when the temperature of a pure substance is changed using $q = mc\Delta T$.

→ A calorimetry experiment for an enthalpy of reaction should be covered and the results evaluated.

Nature of science

→ Fundamental principle – conservation of energy is a fundamental principle of science.

→ Making careful observations – measurable energy transfers between systems and surroundings.

139

Calculating enthalpy changes in aqueous solutions

We stated earlier that the change in enthalpy ΔH is defined as the heat transferred by a closed system during a chemical reaction. To calculate ΔH for a reaction we therefore need to find the heat change. When calculating the heat change of a pure substance such as water, we need to have an understanding of the physical quantity, the **specific heat capacity, c**.

> The units for specific heat capacity are kJ kg^{-1} K^{-1}. The specific heat capacity of water is 4.18 kJ kg^{-1} K^{-1} and this can be found in section 2 of the *Data booklet*.

The specific heat capacity of a pure substance is defined as the amount of heat needed to raise the temperature of 1 g of the substance by 1 °C

or 1 K. For example, the specific heat capacity of copper is 0.385 J g^{-1} K^{-1} while that of ethanol is 2.44 J g^{-1} K^{-1}. The lower the specific heat capacity of a given substance, the higher the temperature rise achieved for the same amount of heat transferred to the sample.

Specific heat capacity is an **intensive property** that does not vary in magnitude with the size of the system being described. For example, a 10 cm^3 sample of copper has the same specific heat capacity as a 1 tonne block.

Specific heat capacity is used to calculate the heat q of a system using the relationship:

$$q = mc\Delta T$$

where m is mass in kg and ΔT is the change in temperature.

Worked examples : calculating enthalpy changes

Example 1

When a 1.15 g sample of anhydrous lithium chloride, LiCl was added to 25.0 g of water in a coffee-cup calorimeter, a temperature rise of 3.80 K was recorded. Calculate the enthalpy change of solution for 1 mol of lithium chloride.

Solution

$$q = mc\Delta T$$
$$= 0.025 \text{ kg} \times 4.18 \text{ kJ kg}^{-1}\text{K}^{-1} \times 3.80 \text{ K}$$
$$= 0.397 \text{ kJ}$$

Convert to energy gained for 1 mol of LiCl.

0.397 kJ / 1.15g LiCl $\times$ 42.394 g/mol $= 14.6$ kJ/mol LiCl

$\Delta H = -q = -14.6$ kJ mol^{-1}.

Example 2

180.0 J of heat is transferred to a 100.0 g sample of iron, resulting in a temperature rise from 22.0 °C to 26.0 °C. Calculate the specific heat capacity of iron.

Solution

$\Delta T = (299 - 295)$ K $= 4$ K.

Make c the subject of the equation and solve:

$$c = \frac{q}{m\Delta T}$$
$$= \frac{0.180 \text{ kJ}}{0.100 \text{ g} \times 4 \text{ K}}$$
$$= 0.450 \text{ kJ K}^{-1}$$

Coffee-cup calorimeters

Performing reactions in a polystyrene coffee cup to measure the enthalpy change is a convenient experimental procedure. The methodology introduces systematic errors that can be analysed and the effect of their directionality assessed.

Systematic errors are a consequence of the experimental procedure. Their effect on empirical data is constant and always in the same direction. With the coffee-cup calorimeter, the measured change in enthalpy for a reaction will always be lower than the actual value, as heat will be transferred between the contents and the surroundings in every experiment.

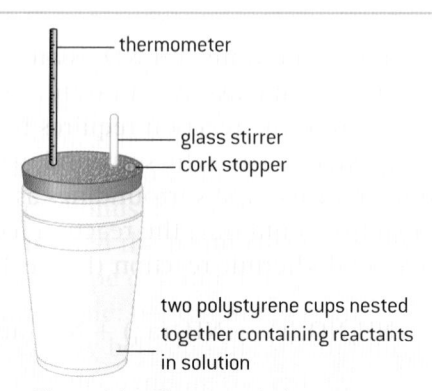

▲ Figure 4 A coffee-cup calorimeter

Investigation to find the molar enthalpy change for a reaction

Earlier we looked at the exothermic metal displacement reaction between zinc and copper(II) sulfate:

$$Zn(s) + CuSO_4(aq) \rightarrow Cu(s) + ZnSO_4(aq)$$

The following method is used to calculate the molar enthalpy change for this reaction from the equation:

$$q = mc\Delta T$$

Experimental method to determine ΔT

1   Using an electronic balance, accurately measure the mass of 25 cm³ of 1.0 mol dm⁻³ $CuSO_4$ solution. Subtract the mass of the cylinder from the mass of the cylinder + solution following the transfer of the solution to the coffee-cup calorimeter.

2   Using a thermometer or a temperature probe and related software, record the temperature of the solution every 30 seconds for up to 3 minutes, or until a constant temperature is achieved.

3   At 3 minutes, introduce powdered zinc (between 1.3 g and 1.4 g, previously weighed) and commence stirring.

4   Continue to take temperature readings for up to 5 minutes after the maximum temperature has been reached.

5   Produce a temperature versus time graph to determine the change in temperature.

Assumptions and errors

A number of assumptions are made when using this method:

•   The heat released from the reaction is completely transferred to the water.

•   The coffee cup acts as an insulator against heat loss to the surroundings. However, the coffee cup also has a heat capacity and heat is transferred to it from the water. It would be difficult to quantify the heat capacity of a polystyrene cup, so it is assumed to be zero.

•   The maximum temperature reached is an accurate representation of the heat evolved during the reaction.

•   The specific heat capacity of an aqueous solution is the same as that of water.

Loss of heat from the system to the surroundings is the main source of error in this experiment and one that is difficult to quantify. The change in temperature ΔT calculated from a graph will include a **systematic** or **directional error**. This loss of heat means that the maximum temperature recorded will be lower than the actual value, making the calculated value of q lower than the actual value. The effect of errors in the procedure on the result of subsequent calculations is important in considering improvements in experimental procedures.

An accepted method of calculating the maximum temperature to compensate for systematic errors in data is to look at the cooling section of the curve after the reaction is complete, and extrapolate this back to the point of introduction of the zinc, as shown in figure 5. A more accurate value for ΔT can then be determined.

Calculation of molar enthalpy change

Mass of copper(II) sulfate solution/g	28.8
Mass of zinc/g	1.37
Change in temperature T(final) $- T$(initial)/°C	39.0

▲ Table 1 Sample results

Taking the results in table 1 we can calculate the molar enthalpy change as follows:

$$q = mc\Delta T$$
$$= 0.0288 \text{ kg} \times 4.18 \text{ kJ kg}^{-1} \text{ K}^{-1} \times 39.0 \text{ K}$$
$$= 4.69 \text{ kJ}$$

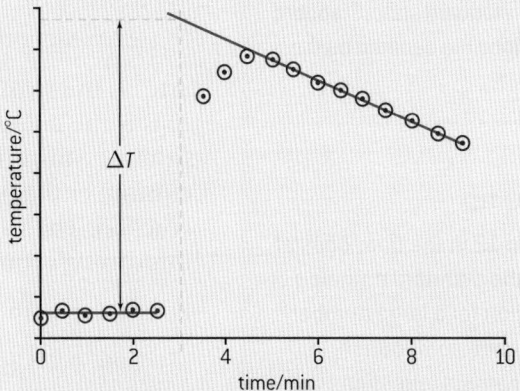

▲ Figure 5 Determination of the change in temperature in calorimetry experiments

$$\text{amount of CuSO}_4 = 1.37 \text{ g} \times \frac{1}{65.38 \text{ g mol}^{-1}}$$
$$= 0.0210 \text{ mol}$$

$$\text{molar enthalpy change} = \frac{4.69 \text{ kJ}}{0.0210 \text{ mol}}$$
$$= 223 \text{ kJ mol}^{-1}$$

The shape of the graph and the change in temperature from a lower to a higher value lead to the conclusion that the reaction is exothermic: $\Delta H = -223 \text{ kJ mol}^{-1}$

TOK

In theory of knowledge there are eight specific ways of knowing. These are: language, sense perception, emotion, reasoning, imagination, faith, intuition, and memory. Scientists perform experiments and process the raw data to enable us to draw conclusions. We compare experimental and theoretical values. What criteria do we use when making these comparisons? Are our judgments subjective or objective? When analysing and appraising experimental limitations and making theoretical assumptions, which of the ways of knowing are we utilizing?

Temperature scales

The SI unit of temperature is the kelvin (K). Note that a *change in temperature* ΔT calculated from experimental data in Celsius will be identical to the value of ΔT calculated in kelvin.

Throughout the world, the majority of countries use the Celsius scale for the everyday description of temperature. As the Celsius and kelvin scales are linked, you will often see both scales being used in an IB question. The USA uses a mixture of metric and imperial units of measurement. For example, the Fahrenheit scale of temperature is used in the USA.

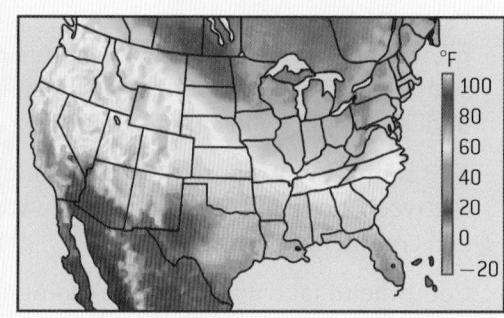

▲ Figure 6 We use SI units in science, but a mixture of imperial and metric systems of measurement is used in different countries

The **standard enthalpy change of a reaction** $\Delta H_{298}^{\ominus}$ is determined at temperature 25 °C/298 K and pressure 100 kPa with all species in their standard state. **Standard conditions** are denoted by the symbol $^{\ominus}$.

Study tip

Section 12 in the *Data booklet* gives the standard enthalpy of formation for a large number of common compounds. In examinations, questions will provide any other values not included in the *Data booklet*.

🌐 Enthalpy change of formation

The change in enthalpy during a reaction can be determined using the following equation:

$$\Delta H^{\ominus}\text{reaction} = \sum(\Delta H_f^{\ominus}\text{products}) - \sum(\Delta H_f^{\ominus}\text{reactants})$$

$\Delta H_f^{\ominus}$ is the **standard enthalpy change of formation** of a substance. This is the energy change upon the formation of 1 mol of a substance from its constituent elements in their standard state. We can use existing enthalpy of formation data to calculate the enthalpy of reaction. The value and sign of the calculated enthalpy of formation informs us about the energetics of the reaction.

For example, the standard enthalpy change of formation for methane is:

$$\text{C(s)} + 2\text{H}_2(\text{g}) \rightarrow \text{CH}_4(\text{g}) \qquad \Delta H_f^{\ominus} = -74.9 \text{ kJ mol}^{-1}$$

It is important to note that the elements carbon and hydrogen are represented in their standard states. Equations for $\Delta H_f^{\ominus}$ must represent the formation of 1 mol of a substance. In some cases, such as the

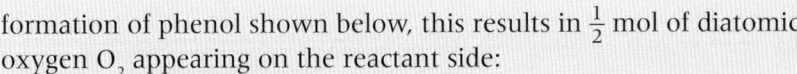

formation of phenol shown below, this results in $\frac{1}{2}$ mol of diatomic oxygen O_2 appearing on the reactant side:

$$6C(s) + 3H_2(g) + \frac{1}{2}O_2(g) \rightarrow C_6H_5OH(s) \qquad \Delta H_f^\ominus = -165.0 \text{ kJ mol}^{-1}$$

Enthalpy change of combustion

The **standard enthalpy change of combustion** $\Delta H_c^\ominus$ is the heat evolved upon the complete combustion of 1 mol of substance.

The enthalpies of combustion found in section 13 of the *Data booklet* are values derived under standard conditions. Butane, one of the gases classified as liquefied petroleum gas (LPG) is highly flammable:

$$C_4H_{10}(g) + \frac{13}{2}O_2(g) \rightarrow 4CO_2(g) + 5H_2O(l)$$

$$\Delta H_c^\ominus = -2878 \text{ kJ mol}^{-1}$$

This **thermochemical equation** can also be written with the enthalpy of combustion value included in the equation. The negative enthalpy change indicates an exothermic reaction so the value would be included on the product side:

$$C_4H_{10}(g) + \frac{13}{2}O_2(g) \rightarrow 4CO_2(g) + 5H_2O(l) + 2878 \text{ kJ mol}^{-1}$$

Working method

Benzene, C_6H_6 is highly flammable, producing a sooty flame:

$$2C_6H_6(l) + 15O_2(g) \rightarrow 12CO_2(g) + 6H_2O(l)$$

$$\Delta H^\ominus \text{reaction} = \sum(\Delta H_f^\ominus \text{products}) - \sum(\Delta H_f^\ominus \text{reactants})$$

$$= [12 \times (-393.5) + 6 \times (-285.8) - 2 \times (+49.0) - 15 \times 0] \text{ kJ}$$

$$= (-4722 - 1714.8 - 98.0) \text{ kJ}$$

$$= -6535 \text{ kJ}$$

> **Quick question**
>
> Write equations to describe the standard enthalpy change of formation for the following compounds and state the enthalpy value by referring to the *Data booklet*.
>
> **a)** propane
>
> **b)** chloromethane
>
> **c)** ethanol
>
> **d)** benzoic acid
>
> **e)** carbon monoxide
>
> **f)** methylamine

Compound	$\Delta H_f^\ominus$/kJ mol^{-1}
$C_6H_6(l)$	+49.0
$CO_2(g)$	−393.5
$H_2O(l)$	−285.8

▲ Table 2 Standard enthalpy changes of formation.

 ## Investigation to find the enthalpy change of combustion

The enthalpy change of combustion of common alcohols can be determined in the laboratory. From a homologous series of alcohols, patterns in enthalpy change of combustion values can be determined and subsequently analysed.

Experimental method

The following procedure utilizes equipment available in a standard school laboratory. Five spirit burners are required, each containing one of the alcohols methanol, ethanol, propan-1-ol, butan-1-ol and pentan-1-ol.

1   Determine the initial mass of the spirit burners using an electronic balance.

2   Accurately determine the mass of 30 cm³ of water contained in a 250 cm³ beaker.

3   Place the beaker or metal calorimeter on a tripod with the spirit burner beneath.

4   Using either a temperature probe or a thermometer, determine and record the initial temperature of the water.

5   A spirit burner is lit under the calorimeter and the alcohol is burnt to heat the water. The period over which it burns can be monitored in one of two different ways:

a)   allow each alcohol to burn until a temperature change of 30 °C is achieved

Write equations to describe the standard enthalpy change of combustion for the following compounds, and state the enthalpy value by referring to section 134 of the *Data booklet*.

a)   octane, C_8H_{18}

b)   chloroethane, C_2H_5Cl (hint: a corrosive strong acid is one of the products)

c)   cyclohexanol, $C_6H_{12}O$

d)   methanoic acid, CH_2O_2

e)   glucose, $C_6H_{12}O_6$

As in all investigations, first determine the dependent and independent variables and the variables that will be controlled.

Data loggers can be used to record temperature changes accurately and the associated software to perform data analysis and graphing. The use of data-logging equipment demonstrates the practical application of technology in the laboratory.

Energetics experiments provide a useful set of raw data and involve experimental procedures that can be evaluated for random and systematic errors (topic 11). The identification of the systematic errors and examination of their directionality is an essential aspect of the analysis of experimental results.

b)   allow each alcohol to burn for a period of 2 minutes.

6   After this time period, extinguish each spirit burner by replacing the cap, re-weigh each one and record the change in mass of the alcohol.

Calculation of enthalpy of combustion

Alcohol	Δm of alcohol/g	$\Delta T/°C$	Mass of water/g
methanol	0.348	30.0	31.2

▲ Table 3 Sample results

$q = mc\Delta T$

$\quad = 0.0312 \text{ kg} \times 4.18 \text{ kJ kg}^{-1} \text{ K}^{-1} \times 30.0 \text{ K}$

$\quad = 3.91 \text{ kJ}$

amount of $CH_3OH = \dfrac{0.348 \text{ g}}{32.05 \text{ g mol}^{-1}} = 0.0109 \text{ mol}$

molar energy change $= \dfrac{3.91 \text{ kJ}}{0.0109 \text{ mol}}$

$\quad\quad\quad\quad\quad\quad\quad = 359 \text{ kJ mol}^{-1}$

$\Delta H = -359 \text{ kJ mol}^{-1}$

Assumptions

- Heat loss to the environment is negligible (in reality, it is significant but cannot be quantified).

- All the alcohols are pure and undergo complete combustion.

Obesity and the energy content of food

The world increase in obesity

Obesity, eating disorders, and unhealthy diet are serious health issues facing many cultures throughout the world, as many societies become more affluent and food is readily available. Obesity is generally defined as an excessive accumulation of fat that can lead to health problems. The **body mass index (BMI)** is found by taking a person's mass (in kilograms) and dividing it by the square of their height (in metres).

An adult with a BMI above 25 is considered overweight while one with a BMI greater than 30 is obese. The World Health Organization (WHO) has been monitoring the effect of changes in diet on different nations for decades. Their research has found the following:

- In 2013 the occurrence of obesity worldwide was more than double the level in 1980.

- In 2008 over 1.4 billion adults worldwide were overweight, with approximately 200 000 000 men and 300 000 000 women being classified as obese.

- 65% of the world's population reside in countries where more people die from obesity-related causes than from being underweight.

- More than 40 000 000 children under the age of 5 years were overweight in 2010.

- Obesity is a preventable disease.

In China, the rapid increase in affluence and the globalization of the economy has seen an unprecedented expansion in the fast-food industry and of nutritional choices. With these has come a significant increase in the number of children who are overweight and obese. Type 2 diabetes is normally associated with adults, but the rise in the prevalence of the disease amongst children in China and in other countries is seen as a significant threat to the wellbeing of future generations.

Food labelling and determination of energy content

Governments throughout the world have a responsibility to their citizens to provide leadership, education, and guidance in health and nutrition. Linked to the globalization of economies and free-trade agreements has been the standardization of labelling of food products to include an analysis of the contents including energy content.

To determine their energy content, foods were traditionally placed in a calorimeter surrounded with water and completely burnt, causing the water to rise in temperature. This temperature change was then used to calculate the energy content (sometimes referred to as "calorific value") of the food. Today the preferred method of calculation is using the Atwater system. This system relies on average energy values for proteins, carbohydrates, fats, and alcohol being applied to foods of a known composition. The National Data Laboratory (NDL) in the USA holds information on the energy content of over 6000 foods.

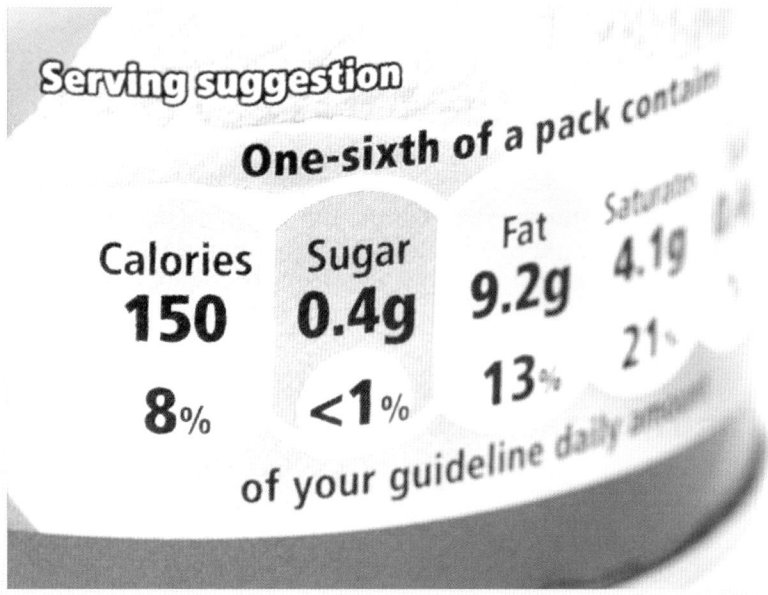

▲ Figure 7 Nutritional information displayed on food packaging

5.2 Hess's law

Understandings

→ The enthalpy change for a reaction that is carried out in a series of steps is equal to the sum of the enthalpy changes for the individual steps.

Applications and skills

→ Application of Hess's law to calculate enthalpy changes.
→ Calculation of ΔH reactions using $\Delta H_f^{\ominus}$ data.
→ Determination of the enthalpy change of a reaction that is the sum of multiple reactions with known enthalpy changes.

Nature of science

→ Hypotheses – based on the conservation of energy and atomic theory, scientists can test the hypothesis that if the same products are formed from the same initial reactants then the energy change should be the same regardless of the number of steps.

TOK

In TOK, a primary focus is on questions about knowledge which are open ended with multiple perspectives and expressed without using subject-specific language. Hess's law can be considered an application of the law of conservation of energy. What are the challenges in applying general principles of a law to something as specific as Hess's law?

Testing hypotheses

Experimental evidence enables scientists to prove or disprove a hypothesis. Based on the principles of conservation of energy and atomic theory, scientists are able to test experimentally the hypothesis that when products are formed from the same set of reactants, the change in enthalpy should be identical regardless of the route taken and the number of chemical reactions involved. Quantitative data can be analysed and used as evidence for this hypothesis.

Overall and net reactions

If you have travelled to New York, Tokyo, London, Hong Kong, Paris, Beijing, Berlin, or Seoul, you will have experienced the subways that criss-cross these enormous cities and transport millions of people every day. In any transport network there is more than one way to travel between point A and B. For the adventurous traveller, half the fun is often working out which is the fastest route.

The same idea can be true in the field of chemistry. A chemical equation usually shows the net reaction – it is a summary of a number of different reactions, which when added together result in an overall reaction.

Hess's law is an application of the conservation of energy law:

Regardless of the route by which a chemical reaction proceeds, the enthalpy change will always be the same providing the initial and final states of the system are the same.

Figure 2 shows that summing all the equations, taking into account the direction and magnitude of each, results in an overall equation. Hess's law can then be applied to find the enthalpy change for the reaction.

Figure 2 Hess's law

▲ Figure 1 You can take many alternative routes on the Paris Metro

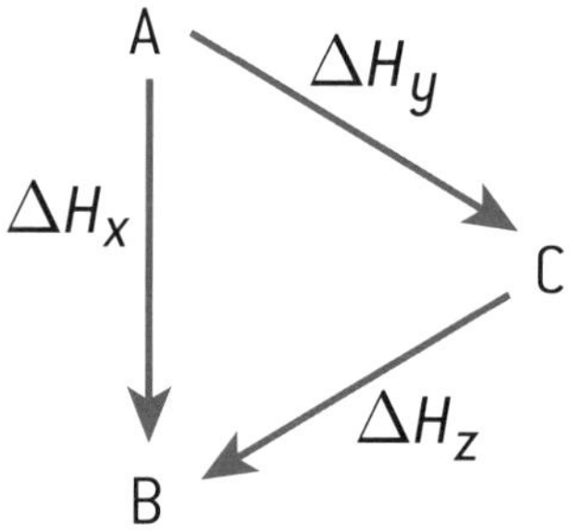

Worked example : calculating enthalpy change

Using Hess's law, and the following information, calculate the enthalpy change ΔH_4 for the reaction:

$$C + 2H_2 + \frac{1}{2}O_2 \rightarrow CH_3OH \quad \Delta H_4 \quad (4)$$

$$CH_3OH + 1\frac{1}{2}O_2 \rightarrow CO_2 + 2H_2O$$
$$\Delta H_1 = -676 \text{ kJ} \quad (1)$$

$$C + O_2 \rightarrow CO_2 \quad\quad \Delta H_2 = -394 \text{ kJ} \quad (2)$$

$$H_2 + \frac{1}{2}O_2 \rightarrow H_2O \quad\quad \Delta H_3 = -242 \text{ kJ} \quad (3)$$

IB, May, 2006

Solution

- Look at the overall equation 4 for the enthalpy of formation of methanol.

- From equations 1–3, the reactant carbon and the product methanol should be the main focus of your methodology.

- For carbon, we require a reaction that uses 1 mol of carbon as a reactant. Carbon is a reactant in equation 2, so this equation can be used as written:

$$C + O_2 \rightarrow CO_2 \quad \Delta H_2 = -394 \text{ kJ}$$

- For methanol we need to use equation 1, but we need to reverse the equation so that methanol is a product not a reactant. When

reversing the chemical equation we must change the sign of the enthalpy value:

$$CO_2 + 2H_2O \rightarrow CH_3OH + 1\frac{1}{2}O_2$$
$$\Delta H_1 = +676 \text{ kJ}$$

- Because oxygen is found in all three equations, the next point of focus should be hydrogen. We require 2 mol of hydrogen as a reactant. Therefore, equation 3 can be used in the direction as written but with double the number of moles. This means that the enthalpy value must be doubled:

$$2H_2 + O_2 \rightarrow 2H_2O \quad \Delta H_3 = -484 \text{ kJ}$$

- Now we can add the three equations together, eliminating those species common to both sides and summing the enthalpy values as shown in table 1.

Reactants		Products	Enthalpy
C + O₂	→	CO₂	$\Delta H_2 = -394$ kJ
CO₂ + 2H₂O	→	CH₃OH + 1½O₂	$\Delta H_1 = +676$ kJ
2H₂ + O₂ ½O₂	→	2H₂O	$\Delta H_3 = -484$ kJ
C + 2H₂ + ½O₂	→	CH₃OH	$\Delta H_4 = -202$ kJ

▲ Table 1

The combination of these reactions can also be represented diagrammatically as shown in figure 3.

You will notice that the enthalpy cycle diagram shows the combustion of methanol equation reversed, in the same way it was during the summation of equations method.

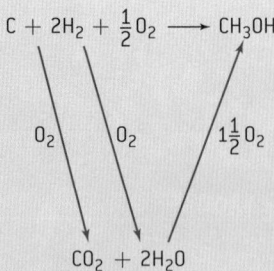

$$C + 2H_2 + \tfrac{1}{2}O_2 \longrightarrow CH_3OH$$

O_2 $\quad$ O_2 $\quad$ $1\tfrac{1}{2}O_2$

$$CO_2 + 2H_2O$$

▲ Figure 3 Enthalpy cycle for the formation of methanol

The energetics of recycling

With the world population exceeding 7 billion people, and increasing international concern over the world's resources, recycling of materials has become mainstream in many countries. Recycling developed from a desire to use raw materials more efficiently, to reduce energy use in the production of goods, to protect the environment from excessive pollution, and to utilize waste materials and thereby reduce landfill waste disposal. However, economic and other pressures to recycle can lead to potentially harmful impacts on the environment. For example, electronic or e-waste is an escalating problem throughout the world. This waste contains heavy metals such as lead and cadmium, highly toxic phosphor dust, and other hazardous substances. Guiyu in China has become a centre for e-waste disposal and the rapid expansion of recycling processes in villages around Guiyu has resulted in heavy metal contamination of the groundwater and soil, and air pollution from the burning of plastics within the waste.

Despite good intentions, the end result of recycling can be extremely negative. What is more, the efficiency of recycling processes in energy terms varies widely. Many countries and environmental organizations are investigating how we can address the long-term effects of recycling programmes on the environment and communities.

▲ Figure 4 Technology waste recycling in Guiyu, China

Worked example

Determine the enthalpy of formation of ethane, C_2H_6 using the enthalpy of combustion data in section 13 of the *Data booklet*.

Solution

- Write a balanced chemical equation for the formation of 1 mol of ethane:

$$2C(s) + 3H_2(g) \xrightarrow{\Delta H_f^\ominus} C_2H_6(g)$$

- Write equations for the combustion of carbon, hydrogen, and ethane and determine their enthalpy values from the *Data booklet*:

  1....$C(s) + O_2(g) \rightarrow CO_2(g)$

  $\Delta H_c^\ominus = -393.5$ kJ

  2....$H_2(g) + \frac{1}{2}O_2(g) \rightarrow H_2O(l)$

  $\Delta H_c^\ominus = -286$ kJ

  3....$C_2H_6(g) + \frac{7}{2}O_2(g) \rightarrow 2CO_2(g) + 3H_2O(l)$

  $\Delta H_c^\ominus = -1561$ kJ

- Multiply or reverse the sign of the enthalpy change for each equation accordingly.

- Now combine the equations to form the net enthalpy of formation equation for ethane:

  1....(×2): $2C(s) + 2O_2(g) \rightarrow 2CO_2(g)$

  $\Delta H_c^\ominus = -787$ kJ

  2....(×3): $3H_2(g) + \frac{3}{2}O_2(g) \rightarrow 3H_2O(l)$

  $\Delta H_c^\ominus = -858$ kJ

  3....(reversed): $2CO_2(g) + 3H_2O(l) \rightarrow C_2H_6(g) + \frac{7}{2}O_2(g)$

  $\Delta H_c^\ominus = +1561$ kJ

In summary:

- The equation for the combustion of carbon is doubled.

- The equation for the combustion of hydrogen is tripled.

- The equation for the combustion of ethane is reversed.

  $2C(s) + 3H_2(g) \rightarrow C_2H_6(g)$

  $\Delta H_c^\ominus = -84$ kJ

Figure 5 shows how the enthalpy of formation can be found using an enthalpy cycle diagram.

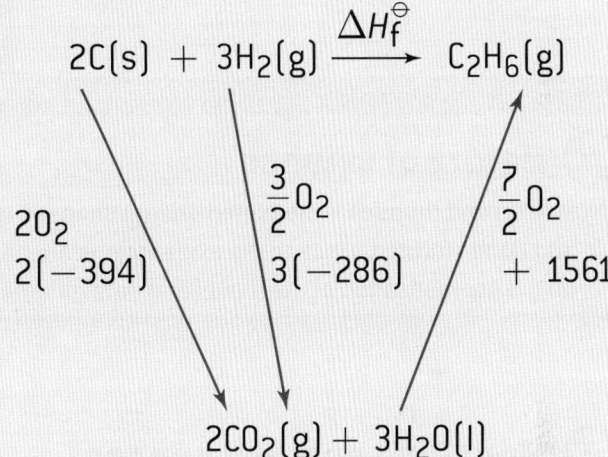

▲ Figure 5 Alternative method: enthalpy cycle to find the enthalpy of formation of ethane

Study tips

You may find the summation of equations method easier when working out the direction of the equations and the mole coefficients. During examinations you may be asked to use the summation of equations method and/or construct an enthalpy cycle.

Often candidates make simple arithmetical errors when calculating the enthalpy of reaction. It is advisable to clearly show your full working rather than simply recording the final answer. This gives the examiner the opportunity to assign part marks where applicable.

5.3 Bond enthalpy

Understandings

→ Bond forming releases energy and bond breaking requires energy.

→ Average bond enthalpy is the energy needed to break 1 mol of a bond in a gaseous molecule averaged over similar compounds.

 Applications and skills

→ Calculation of the enthalpy changes from known bond enthalpy values and comparison of these with experimentally measured values.

→ Sketching and evaluation of potential energy profiles in determining whether reactants or products are more stable and if the reaction is exothermic or endothermic.

→ Discussion of the bond strength in ozone relative to oxygen in its importance to the atmosphere.

 Nature of science

→ Models and theories – measured energy changes can be explained based on the model of bonds broken and bonds formed. Since these explanations are based on a model, agreement with empirical data depends on the sophistication of the model and data obtained can be used to modify theories where appropriate.

 Modelling energy changes

Scientific models are developed to explain certain processes that cannot be observed directly. Based on a theoretical understanding of the processes, such models can produce evidence in support of the theories, or can inform modifications to the theories. Energy changes in reactions can be understood using models of bond breaking and bond making. The degree of agreement between these models and the empirical data obtained in the laboratory is dependent on the validity of the model and the accuracy of the data.

Bond enthalpy

The breaking of the hydrogen molecule into individual hydrogen atoms requires energy. The **bond enthalpy** of a bond (the H–H bond in this example) is defined as the energy required to break 1 mol of bonds in gaseous covalent molecules under standard conditions.

Bond breaking is an endothermic process and has a positive enthalpy value, for example:

$$H_2(g) \rightarrow 2H(g) \quad \Delta H^\circ = +436 \text{ kJ mol}^{-1}$$

Bond enthalpy is also referred to as **bond dissociation enthalpy**, and selected values are provided in section 11 of the *Data booklet* and in table 1. These are **average values** and are therefore only an approximation. They are derived from experimental data involving the breaking of the same bond found in a wide variety of compounds. For example, the C–H bond enthalpy will vary through the alkane series as the chemical environment of the individual bonds changes. If a molecule of methane underwent a series of steps in which one hydrogen atom was removed at a time, the bond dissociation enthalpy would be different each time, as the chemical environment changes upon the removal of successive hydrogen atoms. Additionally, the bond enthalpy for molecules in the gaseous state does not take account of the intermolecular forces that exist. These limitations are not considered to be significant and the average bond enthalpy is an accepted value to use for enthalpy of reaction calculations.

Bond length

Consider the molecules hydrogen fluoride H–F, hydrogen chloride H–Cl, hydrogen bromide H–Br, and hydrogen iodide H–I. In topic 3 we discussed how as you move down group 17 the atomic radius increases with increasing atomic number, Z. The consequence of this is that bond length increases, and bond strength decreases, in the hydrogen halides as you move down group 17 (table 2).

Bond	H–F	H–Cl	H–Br	H–I
Bond length/pm	92	128	141	160

▲ Table 2 Bond lengths of the hydrogen halides

Bond strength

The bond enthalpy reflects the strength of the covalent bond. As we move from single to double to triple bonds the number of electrons in the bond increases resulting in an increase in electrostatic forces and a shortening of the bond length. This trend can be seen in the carbon–carbon bond of the homologous series of alkanes, alkenes, and alkynes.

Bond polarity

The polarity of a bond can be described by the difference in electronegativity of the bonded atoms (tables 3 and 4).

Atom	Electronegativity
H	2.2
F	4.0
Cl	3.2
Br	3.0

▲ Table 3 Electronegativity values

Bond	Δ(Electronegativity)	Bond enthalpy/ kJ mol^{-1}
H–F	1.8	567
H–Cl	1.0	431
H–Br	0.8	366

▲ Table 4 Bond polarity and bond enthalpy at 298 K

Fluorine has the highest electronegativity value of any element. The polarity of the H–F bond results in a partial charge on each atom. The bond is said to have **ionic character**. The partial charges attract one another, increasing the strength of the bond (figure 1).

Bond	Average bond enthalpy/kJ mol^{-1}
H–H	436
O–O	144
O=O	498
O–H	463
C–H	414
C–C	346
C=C	614
C≡C	839
C–O	358
C=O	804
C=N	615
C≡N	890
N–N	158
N=N	470
N≡N	945
Cl–Cl	242
Br–Br	193
I–I	151

▲ Table 1 Average bond enthalpies at 298 K

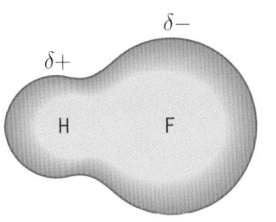

▲ Figure 1 The polar H–F bond

Worked example: using bond enthalpies to find the enthalpy change of reaction

Using the data from section 11 of the *Data booklet* find the enthalpy change for the electrophilic addition of hydrogen bromide to ethene to form bromoethane.

$$C_2H_4 \text{ (g)} + HBr \text{ (g)} \rightarrow C_2H_5Br \text{ (g)}$$

$$\Delta H^\ominus = \sum(\text{BE bonds broken}) - \sum(\text{BE bonds formed})$$

$$\Delta H^\ominus = [4BE_{C-H} + BE_{C=C} + BE_{H-Br}] - [5BE_{C-H} + BE_{C-C} + BE_{C-Br}]$$

$$\Delta H^\ominus = [(4 \times 414) + 614 + 366] - [(5 \times 414) + 346 + 285]$$

$$\Delta H^\ominus = 2636 - 2701 = -65 \text{ kJ mol}^{-1}$$

The bond enthalpy calculated will vary significantly from the calculation using $\Delta H^\ominus$ reaction $= \sum \Delta H_f^\ominus$ (products) $- \sum \Delta H_f^\ominus$ (reactants) (-105.7 kJ) because bond enthalpies are average values. Additionally, when liquids are involved in the reaction, bond enthalpy calculations do not take into account the intermolecular forces within the liquids.

Study tips

A frequent error made by candidates is to confuse the different equations for the calculation of a change in enthalpy. For bond enthalpy, think in terms of bond breaking and formation:

$$\Delta H^\ominus = \sum(\text{BE bonds broken}) - \sum(\text{BE bonds formed})$$

For enthalpy of formation, think in terms of products and reactants:

$$\Delta H^\ominus \text{reaction} = \sum(\Delta H_f^\ominus \text{products}) - \sum(\Delta H_f^\ominus \text{reactants})$$

Bond enthalpy values and enthalpies of combustion

Gasoline or petrol is produced by the fractional distillation of petroleum and used to power various modes of transport. The automotive industry has witnessed significant changes in its markets as the demand for automobiles in developed economies is being overtaken by that of the developing economies.

The enthalpy of combustion of octane, C_8H_{18} can be calculated using bond enthalpy values from section 11 of the *Data booklet*:

$$C_8H_{18}(g) + \frac{25}{2}O_2(g) \rightarrow 8CO_2(g) + 9H_2O(g)$$

$$\Delta H_c^\ominus = \sum(\text{BE bonds broken}) - \sum(\text{BE bonds formed})$$

$$= (18BE_{C-H} + 7BE_{C-C} + \frac{25}{2}BE_{O=O}) - (16BE_{C=O} + 18BE_{O-H})$$

$$= (18(414) + 7(346) + \frac{25}{2}(498)) - (16(804) + 18(463))$$

$$= 16\,099 - 21\,198 = -5099 \text{ kJ mol}^{-1}$$

In comparison, the experimentally determined value for the enthalpy change of combustion $\Delta H_c^{\ominus}$ for octane (section 13 of the *Data booklet*) is $-5470 \text{ kJ mol}^{-1}$.

The reason for the difference is that when calculating the enthalpy change using bond enthalpy values, it is assumed that the reaction takes place in the gaseous state, with no intermolecular forces involved. However, the experimentally derived enthalpy of combustion involves octane and water in their standard states, namely liquid. Additionally, as mentioned above, all bond dissociation enthalpy values are averaged across a wide range of related compounds so they represent only an approximation of the true value.

Ozone

Ozone, O_3 is both created and destroyed in the stratospheric layer of Earth's atmosphere. Ultraviolet (UV) rays from the sun are absorbed by oxygen, O_2, splitting the molecule into single oxygen atoms. These oxygen atoms can then combine with oxygen molecules to form ozone:

$$O_2 \text{ (g)} \xrightarrow{\text{UV}} O^{\bullet} \text{(g)} + O^{\bullet} \text{(g)}$$

$$O_2 \text{ (g)} + O^{\bullet} \text{(g)} \rightarrow O_3 \text{ (g)}$$

Ozone is very effective at absorbing harmful long- and short-wavelength UV radiation. This absorption breaks down the ozone molecule to reform molecular oxygen and a single oxygen atom. Without the presence of ozone in the stratosphere, life on Earth would change forever, as harmful UV radiation would damage cells in both plants and animals.

The bond dissociation enthalpy of an oxygen molecule is 498 kJ mol^{-1}. In comparison, the energy required to break an oxygen–oxygen bond within an ozone molecule is 364 kJ mol^{-1}. The consequence of this is that an ozone molecule is decomposed by UV rays more readily than an oxygen molecule. The ozone photolysis reaction described above is an endothermic reaction with the required energy coming from the UV radiation. The potential energy profile for this reaction is shown in figure 2.

Examination of this energy profile reveals that the oxygen molecule and oxygen atom have a greater combined energy than the reactant ozone molecule. The products of this reaction are said to be less stable, as they exist at a higher energy.

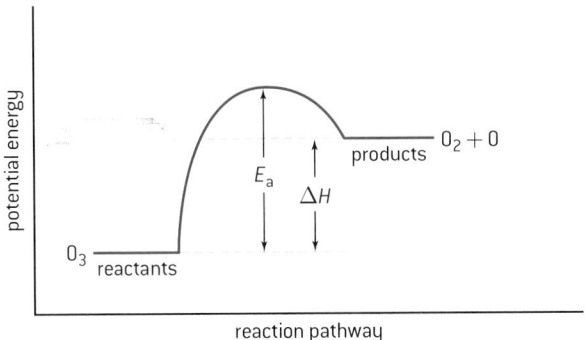

▲ Figure 2 Endothermic energy profile for ozone photolysis

Depletion of the ozone layer

Since the early 1980s scientists have been monitoring ozone depletion globally, particularly the giant holes in the ozone layer which have appeared above the Arctic and Antarctic polar icecaps (figure 3).

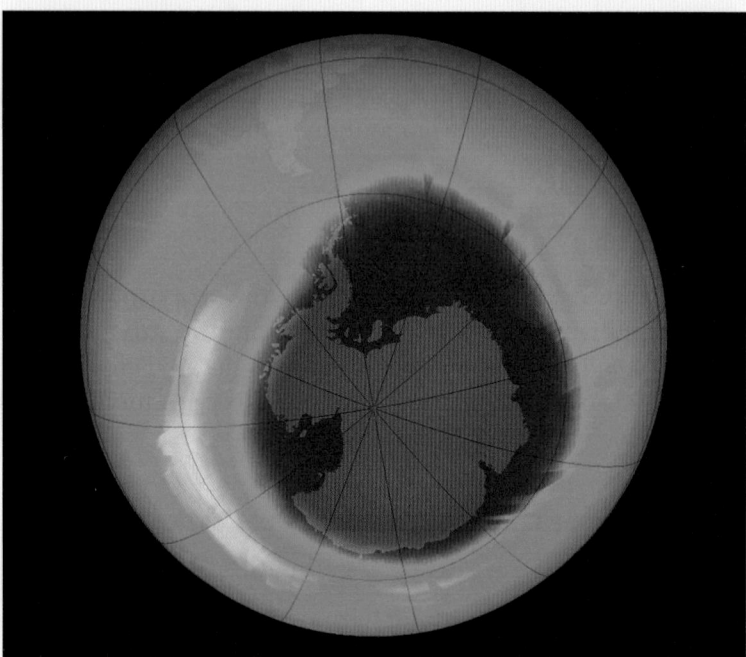

▲ Figure 3 ERS-2 satellite map of Antarctic ozone hole in 2010

Chlorofluorocarbons (CFCs) are a type of hydrocarbon containing carbon, hydrogen, and halogen atoms. The compounds themselves are considered to be non-toxic and have a low level of both flammability and reactivity. First used in industry in the late 1920s, a CFC called Freon gas became an industry standard refrigerant in domestic refrigerators throughout the world. Over the following decades the use of CFCs in commercial air conditioning and the automobile industry, as an aerosol propellant, and as a solvent saw the demand for this class of compounds increase rapidly.

Scientific research in 1973 discovered that while CFCs remained harmless close to the Earth's surface, when exposed to UV radiation in the stratosphere these compounds underwent chemical reactions resulting in the release of chlorine. The chlorine released then had a catalytic effect on the destruction of ozone in this layer of the atmosphere. As the chlorine is not directly consumed in the reaction, small amounts of CFCs were found to be responsible for the destruction of large quantities of ozone.

$$Cl(g) + O_3(g) \rightarrow ClO^{\cdot}(g) + O_2(g)$$
$$ClO^{\cdot}(g) + O_3(g) \rightarrow Cl(g) + 2O_2(g)$$

Science has made many advances that have improved people's daily lives and extended life expectancy and quality. As a consequence the world's population is increasing. The use of CFCs has had a massive economic (positive) and environmental (negative) impact on the world. Whether the blame for the environmental consequences lies with multinational companies who utilize technology or the scientists who invent it is the focus of many discussions.

Questions

1 Consider the specific heat capacity of the following metals (table 4).

Metal	Specific heat capacity/J kg^{-1} K^{-1}
Cu	385
Ag	234
Au	130
Pt	134

▲ Table 4

Which metal will show the greatest temperature increase if 50 J of heat is supplied to a 0.001 kg sample of each metal at the same initial temperature?

A. Cu   **C** Au

B. Ag   **D.** Pt   [1]

IB, May 2007

2 When 40 joules of heat are added to a sample of solid H_2O at $-16.0\ ^\circ C$ the temperature increases to $-8.0\ ^\circ C$. What is the mass of the solid H_2O sample? Specific heat capacity of $H_2O(s) = 2.0$ J g^{-1} K^{-1}

A. 2.5 g   **C.** 10 g

B. 5.0 g   **D.** 160 g   [1]

IB, Nov 2007

3 The temperature of a 2.0 g sample of aluminium increases from 25 °C to 30 °C. How many joules of heat energy were added? (Specific heat capacity of aluminium = 0.90 J g^{-1} K^{-1}.)

A. 0.36   **C.** 9.0

B. 2.3   **D.** 11   [1]

IB, May 2003

4 What is the energy change (in kJ) when the temperature of 20 g of water increases by 10 °C?

A. $20 \times 10 \times 4.18$

B. $20 \times 283 \times 4.18$

C. $\dfrac{20 \times 10 \times 4.18}{1000}$

D. $\dfrac{20 \times 283 \times 4.18}{1000}$ [1]

IB, November 2003

5 Use data from section 11 of the *Data booklet* to calculate the enthalpy change of reaction for each of these chemical reactions:

a) $C_2H_4(g) + H_2(g) \rightarrow C_2H_6(g)$

b) $CH_4(g) + H_2O(g) \rightarrow CO(g) + 3H_2(g)$

6 a) Define the term *standard enthalpy change of formation*, $\Delta H_f^\ominus$.   [2]

b) (i) Use the information in table 5 to calculate the enthalpy change for the complete combustion of but-1-ene according to the following equation.

$C_4H_8(g) + 6O_2(g) \rightarrow 4CO_2(g) + 4H_2O(g)$

Compound	$C_4H_8(g)$	$CO_2(g)$	$H_2O(g)$
$\Delta H_f^\ominus$/kJ mol^{-1}	+1	−394	−242

                                          [3]

▲ Table 5

(ii) Deduce, giving a reason, whether the reactants or the products are more stable.   [2]

(iii) Predict, giving a reason, how the enthalpy change for the complete combustion of but-2-ene would compare with that of but-1-ene based on average bond enthalpies.   [1]

IB, May 2007

7 The $\Delta H^\ominus$ values for the formation of two oxides of nitrogen are given below.

$\frac{1}{2}N_2(g) + O_2(g) \rightarrow NO_2(g)$ $\Delta H^\ominus = -57$ kJ mol^{-1}

$N_2(g) + 2O_2(g) \rightarrow N_2O_4(g)$ $\Delta H^\ominus = +9$ kJ mol^{-1}

Use these values to calculate $\Delta H^\ominus$ for the following reaction (in kJ):

$2NO_2(g) \rightarrow N_2O_4(g)$

a) −105   c) +66

b) −48   d) +123

IB, November 2007

8 The standard enthalpy change of three combustion reactions is given below in kJ.

$$2C_2H_6(g) + 7O_2(g) \rightarrow 4CO_2(g) + 6H_2O(l)$$
$$\Delta H^\circ = -3120$$

$$2H_2(g) + O_2(g) \rightarrow 2H_2O(l) \qquad \Delta H^\circ = -572$$

$$C_2H_4(g) + 3O_2(g) \rightarrow 2CO_2(g) + 2H_2O(l)$$
$$\Delta H^\circ = -1411$$

Based on the above information, calculate the standard change in enthalpy, ΔH°, for the following reaction.

$$C_2H_6(g) \rightarrow C_2H_4(g) + H_2(g) \qquad [4]$$

IB, November 2009

9 Approximate values of the average bond enthalpies, in kJ mol^{-1}, of three substances are shown in table 6.

H–H	430
F–F	155
H–F	565

▲ Table 6

What is the enthalpy change, in kJ, for this reaction?

$$2HF \rightarrow H_2 + F_2$$

A. +545   B. +20   C. –20   D. –545

IB, May 2006

10 The reaction between ethene and hydrogen gas is exothermic.

a) Write an equation for this reaction. [1]

b) Deduce the relative stabilities and energies of the reactants and products. [2]

c) Explain, by referring to the bonds in the molecules, why the reaction is exothermic. [2]

IB, November 2007

11 Two reactions occurring in the manufacture of sulfuric acid are shown below:

Reaction I

$$S(s) + O_2(g) \rightleftharpoons SO_2(g) \qquad \Delta H^\circ = -297 \text{ kJ}$$

Reaction II

$$SO_2(g) + \frac{1}{2}O_2(g) \rightleftharpoons SO_3(g) \qquad \Delta H^\circ = -92 \text{ kJ}$$

a) State the name of the term ΔH°. State, with a reason, whether reaction I would be accompanied by a decrease or increase in temperature. [3]

b) At room temperature sulfur trioxide, SO_3, is a solid. Deduce, with a reason, whether the ΔH° value would be more negative or less negative if $SO_3(s)$ instead of $SO_3(g)$ were formed in reaction II. [2]

c) Deduce the ΔH° value of this reaction: [1]

$$S(s) + 1\frac{1}{2}O_2(g) \rightarrow SO_3(g)$$

IB, November 2005

12 But-1-ene gas burns in oxygen to produce carbon dioxide and water vapour according to the following equation.

$$C_4H_8 + 6O_2 \rightarrow 4CO_2 + 4H_2O$$

a) Use the data in table 7 to calculate the value of ΔH° for the combustion of but-1-ene. [3]

Bond	C–C	C=C	C–H	O=O	C=O	O–H
Average bond enthalpy/ kJ mol^{-1}	348	612	412	496	743	463

▲ Table 7

b) State and explain whether the reaction above is endothermic or exothermic. [1]

IB, May 2006

13 Given the following data:

$$C(s) + 2F_2(g) \rightarrow CF_4(g); \qquad \Delta H_1 = -680 \text{ kJ mol}^{-1}$$

$$F_2(g) \rightarrow 2F(g); \qquad \Delta H_2 = +158 \text{ kJ mol}^{-1}$$

$$C(s) \rightarrow C(g); \qquad \Delta H_3 = +715 \text{ kJ mol}^{-1}$$

calculate the average bond enthalpy (in kJ mol^{-1}) for the C–F bond.

IB, November 2003

14 Methanol is made in large quantities as it is used in the production of polymers and in fuels.

The enthalpy of combustion of methanol can be determined theoretically or experimentally.

$$CH_3OH(l) + 1\tfrac{1}{2}O_2(g) \rightarrow CO_2(g) + 2H_2O(g)$$

a) Using the information from section 11 of the *Data booklet*, determine the theoretical enthalpy of combustion of methanol. [3]

b) The enthalpy of combustion of methanol can also be determined experimentally in a school laboratory. A burner containing methanol was weighed and used to heat water in a test tube as illustrated in figure 4.

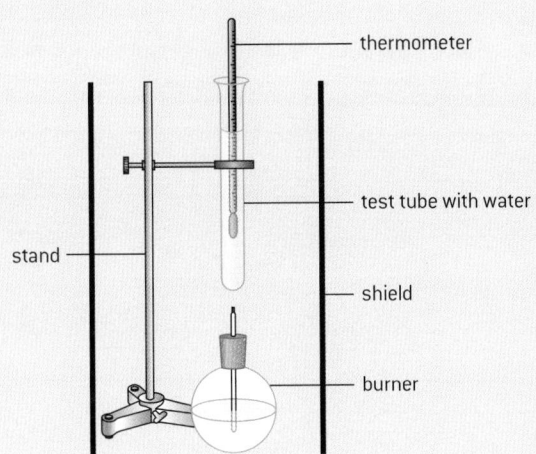

▲ Figure 4

The data shown in table 8 were collected.

Initial mass of burner and methanol/g	80.557
Final mass of burner and methanol/g	80.034
Mass of water in test tube/g	20.000
Initial temperature of water/°C	21.5
Final temperature of water/°C	26.4

▲ Table 8

i) Calculate the amount, in mol, of methanol burned. [2]

ii) Calculate the heat absorbed, in kJ, by the water. [3]

iii) Determine the enthalpy change, in kJ mol⁻¹, for the combustion of 1 mole of methanol. [2]

c) The *Data booklet* value for the enthalpy of combustion of methanol is −726 kJ mol⁻¹. Suggest why this value differs from the values calculated in parts **a)** and **b)**.

(i) Part a) [1]

(ii) Part b) [1]

IB, May 2011

15 One important property of a rocket fuel mixture is the large volume of gaseous products formed which provide thrust. Hydrazine, N_2H_4, is often used as a rocket fuel. The combustion of hydrazine is represented by the equation below.

$$N_2H_4(g) + O_2(g) \rightarrow N_2(g) + 2H_2O(g)$$

$$\Delta H_c^{\ominus} = -585 \text{ kJ mol}^{-1}$$

a) Hydrazine reacts with fluorine to produce nitrogen and hydrogen fluoride, all in the gaseous state. State an equation for the reaction. [2]

b) Draw the Lewis structures for hydrazine and nitrogen. [2]

c) Use the average bond enthalpies given in section 11 of the *Data booklet* to determine the enthalpy change for the reaction in part a) above. [3]

d) Based on your answers to parts **a)** and **c)**, suggest whether a mixture of hydrazine and fluorine is a better rocket fuel than a mixture of hydrazine and oxygen. [2]

IB, May 2010

16 Two students were asked to use information from the *Data booklet* to calculate a value for the enthalpy of hydrogenation of ethene to form ethane.

$$C_2H_4(g) + H_2(g) \rightarrow C_2H_6(g)$$

John used the average bond enthalpies from section 11. Marit used the values of enthalpies of combustion from section 12.

a) Calculate the value for the enthalpy of hydrogenation of ethene obtained using the average bond enthalpies given in section 11. [2]

b) Marit arranged the values she found in section 12 into an energy cycle (figure 5).

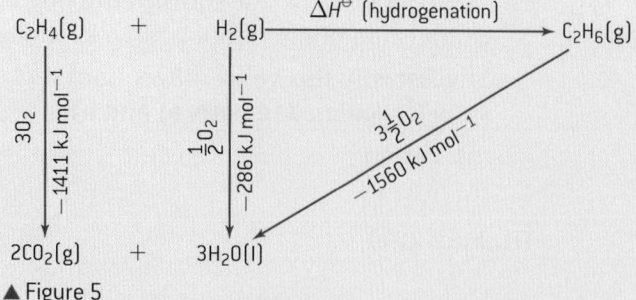

▲ Figure 5

Calculate the value for the enthalpy of hydrogenation of ethene from the energy cycle. [1]

c) Suggest **one** reason why John's answer is slightly less accurate than Marit's answer. [1]

d) John then decided to determine the enthalpy of hydrogenation of cyclohexene to produce cyclohexane.

$$C_6H_{10}(l) + H_2(g) \rightarrow C_6H_{12}(l)$$

(i) Use the average bond enthalpies to deduce a value for the enthalpy of hydrogenation of cyclohexene. [1]

(ii) The percentage difference between these two methods (average bond enthalpies and enthalpies of combustion) is greater for cyclohexene than it was for ethene. John's hypothesis was that it would be the same. Determine why the use of average bond enthalpies is less accurate for the cyclohexene equation shown above, than it was for ethene. Deduce what extra information is needed to provide a more accurate answer. [2]

IB, May 2009

6 CHEMICAL KINETICS

Introduction

In this topic we will discuss a very important aspect of chemical reactions, namely **chemical kinetics**. We begin the chapter by exploring the rate of a chemical reaction. We will then describe the kinetic-molecular theory of gases and collision theory and outline the importance of the principle of Occam's razor in science as a guide to developing theories.

We will also learn that the greater the probability that molecules collide with sufficient energy and proper orientation the higher the rate of reaction. In addition we consider the effect of a catalyst on rate.

Throughout the topic the analysis of graphical and numerical data obtained from rate experiments will form a key part of our treatment of this branch of physical chemistry.

6.1 Collision theory and rates of reaction

Understandings

→ Species react as a result of collisions of sufficient energy and proper orientation.

→ The rate of reaction is expressed as the change in concentration of a particular reactant/product per unit time.

→ Concentration changes in a reaction can be followed indirectly by monitoring changes in mass, volume, and colour.

→ Activation energy (E_a) is the minimum energy that colliding molecules need in order to have successful collisions leading to a reaction.

→ By decreasing E_a, a catalyst increases the rate of a chemical reaction, without itself being permanently chemically changed.

Applications and skills

→ Description of the kinetic theory in terms of the movement of particles whose average kinetic energy is proportional to temperature in kelvin.

→ Analysis of graphical and numerical data from rate experiments.

→ Explanation of the effects of temperature, pressure/concentration, and particle size on rate of reaction.

→ Construction of Maxwell–Boltzmann energy distribution curves to account for the probability of successful collisions and factors affecting these, including the effect of a catalyst.

→ Investigation of rates of reaction experimentally and evaluation of the results.

→ Sketching and explanation of energy profiles with and without catalysts.

Nature of science

→ The principle of Occam's razor is used as a guide to developing a theory — although we cannot directly see reactions taking place at the molecular level, we can theorize based on the current atomic models. Collision theory is a good example of this principle.

Studying reaction rates

When a chemical reaction takes place, four questions are of interest to chemists:

1   What occurs during the reaction?

2   What is the extent of the chemical reaction?

3   Does the reaction happen rapidly or slowly?

4   What energy transfer is involved in the reaction and could a reaction potentially occur given sufficient time?

Answering these four questions involves the following:

1   This information can be obtained from the balanced chemical equation which identifies the reactants, products (and their states), and stoichiometry (sub-topic 1.1).

2   This is discussed in terms of chemical equilibrium sub-topic 7.1).

3   This is the study of the rate of chemical reactions, a branch of chemistry called **chemical kinetics**, which is the focus of this topic.

4   This is the field of thermodynamics (topic 5), which is the study of energy or heat flow in a chemical reaction. Thermodynamics tells us nothing about how quickly or slowly a given reaction takes place.

One of the main considerations when examining a chemical reaction is whether it will take place fast enough for it to be useful. There is not much point in carrying out a reaction if it takes 150 years for the product to be formed! Many chemical reactions occur very quickly, such as the rapid inflation of airbags in cars, while others such as rusting take place over a period of years.

Rate of reaction

The rate of reaction is defined as the change in concentration of reactants or products per unit time.

> Units: $mol\ dm^{-3}\ s^{-1}$, $mol\ dm^{-3}\ min^{-1}$, etc.

Experimental measurements of reaction rates

Δc, the change in concentration, can be measured by monitoring a property that will change when the reactants are converted into products. Examples include:

1   change in pH (for acid–base reactions)

2   change in conductivity (for reactions involving electrolytes)

3   change in mass or volume (for reactions involving solids or gases)

4   change in colour (for reactions involving transition metals or other coloured compounds).

In order to determine the rate of reaction, at time t, a graph of concentration (or the property associated with concentration) versus time is plotted. The rate of reaction is then determined from the slope or gradient of the tangential line at time t.

Let us consider the following reaction between calcium carbonate (limestone), $CaCO_3$ and hydrochloric acid, HCl:

$$CaCO_3(s) + 2HCl(aq) \rightarrow CaCl_2(aq) + CO_2(g) + H_2O(l)$$

In this reaction, gaseous carbon dioxide is one of the products. If the volume, V, of carbon dioxide gas produced is recorded over time t, the rate of reaction can be determined. Figure 1 shows the experimental set-up for this reaction:

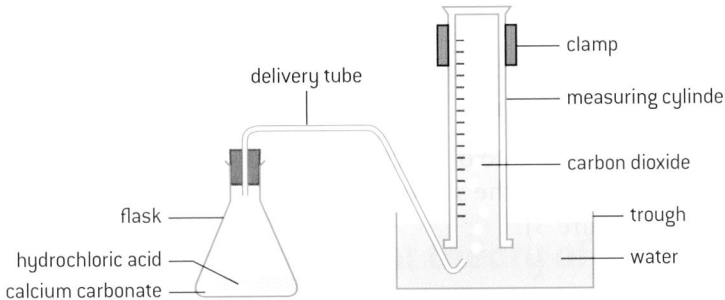

▲ Figure 1 Experimental set-up for measuring the rate at which carbon dioxide is produced in the reaction between calcium carbonate and hydrochloric acid

Table 1 shows the data recorded during this experiment.

t/s	0.0	10.0	20.0	30.0	40.0	50.0	60.0	70.0	80.0	90.0	100.0
$V(CO_2)$/cm³	0.0	19.0	30.0	37.5	45.0	50.0	52.0	53.0	53.0	53.0	53.0

▲ Table 1 Volume, V, of carbon dioxide evolved at time t in the reaction between calcium carbonate and hydrochloric acid

Figure 2 shows a plot of V against t.

The rate of reaction can be expressed in three ways:

- average rate

- instantaneous rate

- initial rate.

Average rate

The **average rate** is a measure of the change in concentration of reactant or product in a given time interval, t.

Mathematically, average rate can be expressed as:

$$\text{average rate} = \frac{\Delta c}{\Delta t}$$

where:

Δc = change in concentration of reactant or product

Δt = time interval over which the change in concentration was measured

For gases, the average reaction rate can be also expressed as the change in volume:

$$\text{average rate} = \frac{\Delta V}{\Delta t}$$

where ΔV is the change in volume of the gas produced or consumed during the reaction.

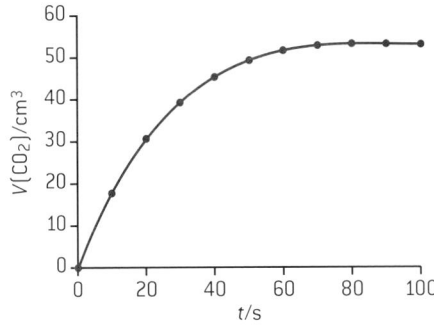

▲ Figure 2 Plot of volume of carbon dioxide gas evolved $V(CO_2)$ against time t in the reaction between calcium carbonate and hydrochloric acid

Collision theory

Occam's razor is a principle attributed to the fourteenth-century English Franciscan friar, theologian, and logician William of Ockham, though many references suggest that the principle was known much earlier than this. The principle states that:

> "Entities should not be multiplied unnecessarily."

Scientists have formulated the principle like this:

> "When two competing theories exist that explain observed facts, both giving essentially the same predictions, then the simpler of the two theories is the optimum one that should be used until more evidence transpires to prove otherwise."

The British theoretical physicist Stephen Hawking of the University of Cambridge and author of *A Brief History of Time* said:

> "We could still imagine that there is a set of laws that determines events completely for some supernatural being, who could observe the present state of the universe without disturbing it. However, such models of the universe are not of much interest to us mortals. It seems better to employ the principle known as Occam's razor and cut out all the features of the theory that cannot be observed."

The principle is often encountered widely in chemistry. For example, it has been used as a justification of the idea of uncertainty in the field of quantum mechanics. The principle of Occam's razor is used as a guide in the development of a theory — although we cannot directly see reactions taking place at the molecular level, we can formulate theories based on current atomic models. Collision theory is a good example of this principle.

For a chemical reaction to occur between two reacting particles, a number of conditions must be fulfilled:

1 The two particles must collide with each other, that is, there must be physical contact.

2 The colliding particles must have the correct mutual orientations.

3 The reacting particles must have sufficient kinetic energy to initiate the reaction.

These three conditions form the basis of the **collision theory**. This theory is a model that helps us understand why rates of chemical reactions depend on temperature. The model is based on the kinetic–molecular theory. For most chemical reactions, only a small fraction of collisions lead to a reaction taking place. In most collisions, the reacting particles bounce off one another and remain unchanged, resulting in no new product formation. In the gaseous reaction between hydrogen, H_2, and iodine, I_2, to form hydrogen iodide, HI, the reaction proceeds very slowly at room temperature, since only a small number of collisions results in a reaction:

$$H_2(g) + I_2(g) \rightleftharpoons 2HI(g)$$

Definition of activation energy, E_a

The activation energy is the *minimum* energy that colliding particles need for a reaction to occur.

Let us consider each of these three conditions in the box above separately:

1 The first, that physical contact is necessary for a reaction to occur, is relatively obvious to appreciate.

2 Figure 5 illustrates the second condition. In (a) the orientation of the colliding particles is not favourable, so no reaction will occur. In (b) the particles have a suitable orientation, so the collision can be effective and result in a chemical reaction, *provided the reacting particles have sufficient kinetic energy.*

(a) ineffective collision

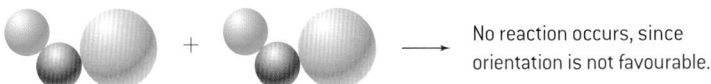

No reaction occurs, since orientation is not favourable.

(b) effective collision

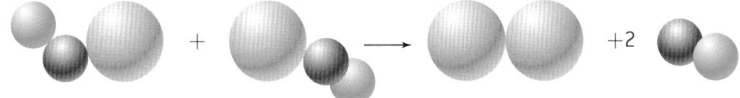

Orientation is favourable, so may result in reaction if there is sufficient kinetic energy.

▲ Figure 5 The possibility of a collision leading to a reaction depends on the orientation of the particles

A useful analogy in understanding activation energy is to imagine somebody trying to push a large rock over a hill (figure 6).

Initially, there must be a minimum input of energy in order for the person to shift the rock over the hill. Once over the hill, the rock can fall to point z.

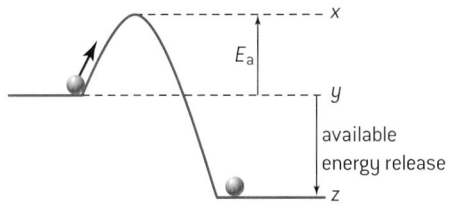

▲ Figure 6 Analogy of activation energy, E_a

Catalysts

The analogy just described can be conveyed in a **potential energy profile** (figure 7).

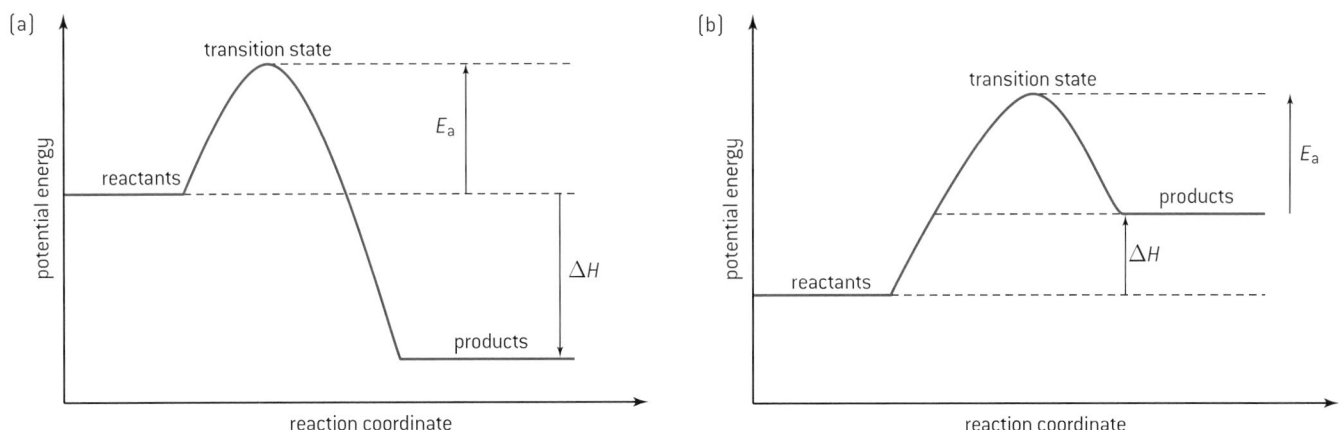

▲ Figure 7 Potential energy profile for (a) an exothermic reaction with $\Delta H < 0$; (b) an endothermic reaction with $\Delta H > 0$

The arrangement of atoms at the crest of the energy profile is termed the **transition state** or **activated complex**.

Catalysts may be either **homogeneous** or **heterogeneous**.

Homogeneous catalysts

A homogeneous catalyst is in the same physical phase or state as the reactants. The destruction of gaseous ozone, O_3, by chlorine atoms is an example of homogeneous catalysis, since chlorine atoms (which act as the catalyst) have the same state as the gaseous reactants.

In the stratosphere (upper atmosphere), ozone in the **ozone layer** absorbs over 95% of the UV radiation reaching Earth from the sun, protecting us from this harmful radiation.

> **Definition of a catalyst**
>
> A catalyst is a substance that increases the rate of a chemical reaction, but is not consumed in the reaction itself. A catalyst provides an alternative pathway for the reaction and lowers the activation energy, E_a.

$$h\nu$$
$$O_3(g) \rightarrow O_2(g) + O\cdot(g)$$
$$O_2(g) + O\cdot(g) \rightarrow O_3(g) + heat$$

Thus, there is a net energy conversion from UV to heat energy.

With the progressive depletion of the ozone layer (figure 8), more UV radiation can reach the Earth's surface. This can be associated with an increased risk of skin cancers (melanomas) and cataracts.

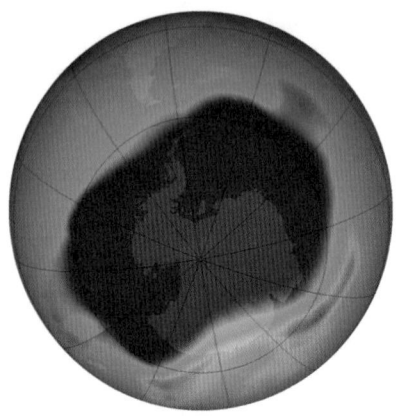

▲ Figure 8 The largest ozone hole to date seen here shown in purple was recorded on September 24, 2006 for the Antarctic Hemisphere

Chlorine atoms are produced in the reaction of a chlorofluorocarbon (CFC) with UV light. CFCs were previously used in air conditioning units, refrigerators, and aerosols. Freon, $CF_2Cl_2(g)$, is one example of a CFC. In the presence of UV light, the weaker C–Cl bond is broken (the C–F bond strength is greater), and radicals are produced. The chlorine radicals then attack ozone.

$$h\nu$$

	$CF_2Cl_2(g)$	$\rightarrow$	$CF_2Cl\cdot(g) + Cl\cdot(g)$
first step:	$Cl\cdot(g) + O_3(g)$	$\rightarrow$	$ClO\cdot(g) + O_2(g)$
second step:	$ClO\cdot(g) + O_3(g)$	$\rightarrow$	$Cl\cdot(g) + 2O_2(g)$
overall reaction:	$2O_3(g)$	$\rightarrow$	$3O_2(g)$

Chlorine acts as a catalyst and is regenerated in the second step. We can represent the potential energy profile for this reaction scheme as shown in figure 9.

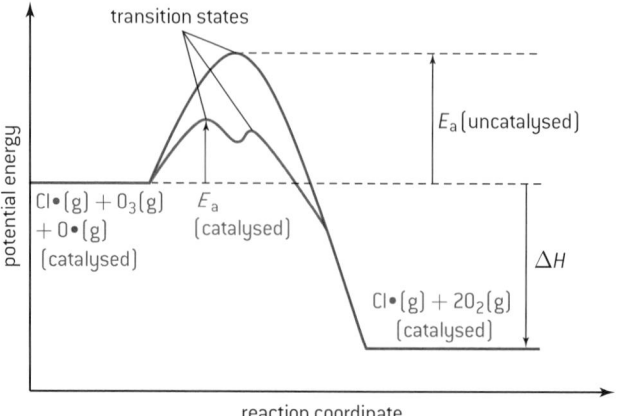

▲ Figure 9 Potential energy profile for the catalytic destruction of ozone, showing both the catalysed and uncatalysed pathways

As can be seen from figure 9, there are actually two transition states for the catalysed pathway. This is specific for this particular reaction and, in general, the more typical representation of a potential energy profile showing the catalysed and uncatalysed pathways shows just one transition state for the catalysed pathway. This is shown in figure 10.

Although ozone in the stratosphere acts as a protective shield from the harmful effects of high-energy UV-a and UV-b radiation, high concentrations of ozone in the troposphere (the lower atmosphere) can lead to respiratory problems such as asthma and emphysema. Ozone may be formed in the reaction of gaseous nitrogen oxides, NO and NO_2, from car exhaust gases with VOCs (volatile organic compounds). Ozone in the troposphere also acts as a **greenhouse gas**.

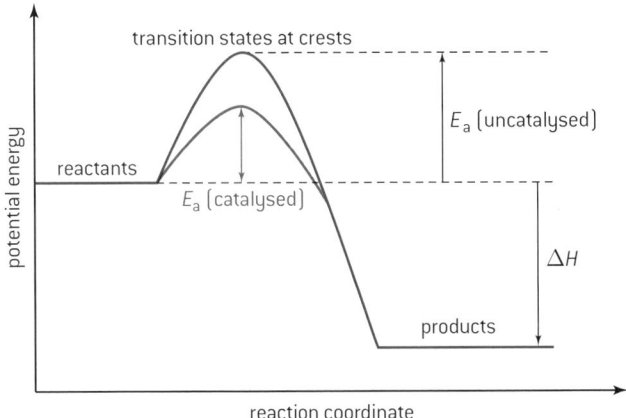

▲ Figure 10 Typical potential energy profile showing catalysed and uncatalysed pathways

Heterogeneous catalysts

A heterogeneous catalyst is in a different phase or state from the reactants. Typically, the heterogeneous catalyst is in the solid phase and the reactants are in the liquid or gaseous states. An example of a heterogeneous catalyst is the catalytic converter used in the exhaust system of a car. In a car engine pollutants such as carbon monoxide, CO, nitrogen monoxide, NO, and unburned hydrocarbons, C_xH_y, are produced. The carbon monoxide comes from the incomplete combustion of hydrocarbon fuels. Nitrogen monoxide is produced from the reaction of atmospheric nitrogen and oxygen at *high temperature* in the engine of the car. The catalytic converter converts these substances into less harmful substances, namely carbon dioxide, CO_2, water, H_2O, and nitrogen, N_2:

$2NO(g) \xrightarrow{\text{catalyst}} N_2(g) + O_2(g)$

$2NO_2(g) \xrightarrow{\text{catalyst}} N_2(g) + 2O_2(g)$

$2CO(g) + O_2(g) \xrightarrow{\text{catalyst}} 2CO_2(g)$

$CH_3CH_2CH_3(g) + 5O_2(g) \xrightarrow{\text{catalyst}} 3CO_2(g) + 4H_2O(g)$
propane fuel

Examples of catalysts used in catalytic converters are platinum, palladium, and rhodium, and transition metal oxides such as vanadium(V) oxide, V_2O_5, copper(II) oxide, CuO, and chromium(III) oxide, Cr_2O_3. Leaded petrol (leaded gasoline) is not used in modern cars – if a car fitted with a catalytic converter used leaded petrol, the lead would poison the catalyst.

Activity

Can you name five countries around the world that still continue to use leaded petrol?

169

Maxwell—Boltzmann energy distribution and temperature

Rates of reaction in the gas phase can be interpreted at the molecular level using different approaches:

1 the collision theory

2 Maxwell—Boltzmann energy distribution curve

3 temperature effects on kinetic energies.

We have already explored the collision theory model. We shall now look at approaches 2 and 3.

Maxwell—Boltzmann energy distribution curve

The kinetic-molecular theory says that particles of gas move randomly in different directions at high velocities. However, these velocities differ and, since the particles are constantly colliding with other particles and with the sides of the container, the velocity of a single gaseous particle changes constantly. It is not realistic to discuss the velocity of an individual gaseous particle in this scenario. What needs to be considered is the **distribution of velocities of the particles**. The distribution of velocities is described by the **Maxwell—Boltzmann energy distribution curve**. This is a plot of the fraction of particles with a given kinetic energy (that is the probability of that value of kinetic energy occurring) versus kinetic energy. As can be seen from figure 11, the representation is asymmetric. The area under the curve represents the total number of gaseous particles in the sample. At a certain temperature, the majority of gaseous particles will have a kinetic energy near the mean value. At a given time, some of the gaseous particles will have either high or low velocities, but the majority will have velocities and hence kinetic energies close to the mean kinetic energy.

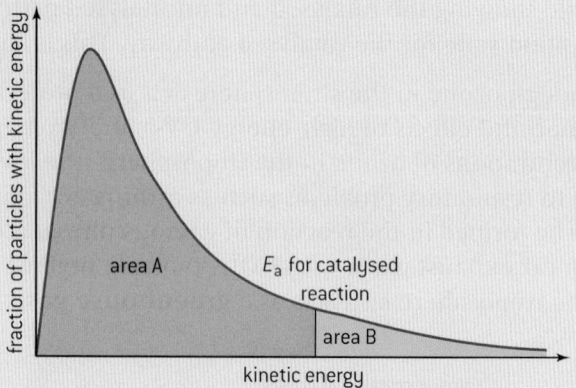

▲ Figure 11 Maxwell—Boltzmann energy distribution curve showing the activation energy for a catalysed reaction. Area A shows the fraction of particles in the sample that do not have sufficient energy to react. Area B shows the fraction of particles in the sample that have the minimum energy required to initiate a reaction with the use of a catalyst

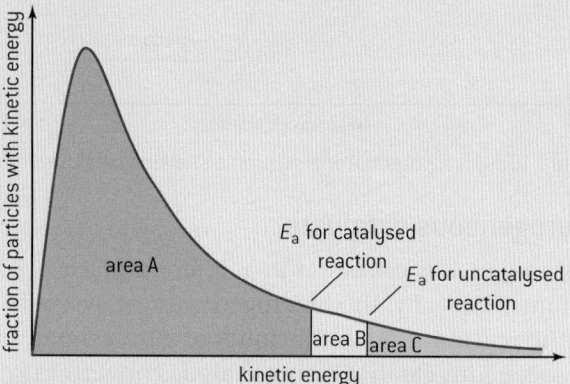

▲ Figure 12 Maxwell—Boltzmann energy distribution curve showing the activation energy for an uncatalysed reaction. Area A shows the fraction of particles in the sample that do not have sufficient energy to react. Area B + area C shows the fraction of particles that have the minimum energy required in order to initiate a reaction using a catalyst. Area C shows the fraction of particles in the sample that have the minimum energy required to initiate a reaction without the use of a catalyst

Temperature effects on kinetic energies

With increasing temperature T, the proportion of particles that have sufficient kinetic energy to overcome the activation energy barrier will increase. As T increases, the mean velocity of the particles increases and so there will be an increase in the proportion of particles having greater kinetic energy. As a result, the Maxwell–Boltzmann energy distribution curve becomes flatter and broader at a higher temperature.

Therefore, with increasing temperature, the frequency of collisions will increase. There will be more successful collisions, since there are now more particles which have sufficient kinetic energy to overcome the activation energy barrier. This results in an increase in the rate of reaction. Typically with an increase in temperature of 10 °C, the reaction rate will double.

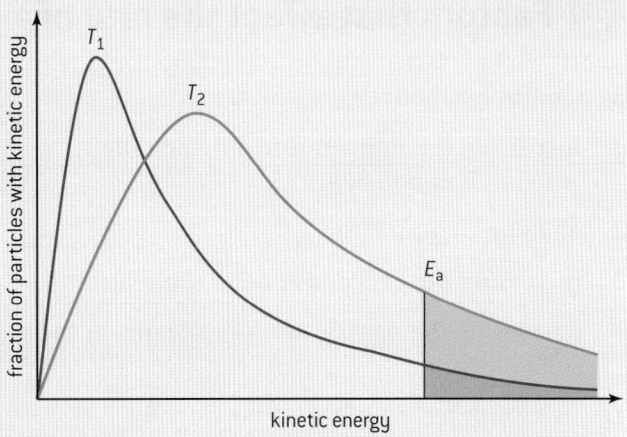

▲ Figure 13 Maxwell–Boltzmann energy distribution curves for two temperatures, $T_2 > T_1$. Notice that, at the higher temperature, the energy distribution is broader and the mean kinetic energy is greater. The proportion of particles that have sufficient thermal energy to overcome the activation energy barrier has increased. The area under both curves is the same as this signifies the total number of gaseous particles in the sample

TOK

Temperature can be considered as a measure of the average amount of kinetic energy of particles, which is the energy due to motion. As the temperature increases, so does the kinetic energy of the particles. The lowest temperature that can be attained *theoretically* is **absolute zero**, which is −273.15 °C. This is taken as the zero point on the Kelvin scale. At 0 K, all thermal energy has been removed from a substance and the motion of particles has effectively ceased. However, if you consider Heisenberg's uncertainty principle, all molecular movement may not cease entirely at 0 K. Consider why this may be the case.

The Kelvin scale gives a natural measure of the kinetic energy of a gas, and is independent of physical properties, whereas the artificial Celsius scale is based on the physical properties of water. The Celsius scale is defined about an arbitrary zero point and hence negative °C values occur. Are physical properties such as temperature invented or discovered? Could Anders Celsius, the Swedish astronomer credited with the Celsius scale, have chosen the melting and boiling points of another substance other than water to devise the artificial Celsius scale?

Study tips

In a potential energy profile, the y-axis is the potential energy and the x-axis is the reaction coordinate, which represents the progress of the reaction. In a Maxwell–Boltzmann energy distribution curve, the y-axis is the fraction of particles with a certain kinetic energy and the x-axis is kinetic energy. Don't mix up the two axis labels in these two representations.

Factors that affect the rate of a chemical reaction

There are four factors that can increase the rate of a chemical reaction:

1 increasing the temperature at which the reaction is conducted

2 addition of a catalyst

3 increasing the concentration of the reactants

4 decreasing the particle size of reactants in the solid phase.

Let us examine each of these individually.

1 Increasing the temperature at which the reaction is conducted

We have just discussed this factor. A good example of the effect of temperature on rate involves the refrigeration of milk. At room temperature, milk can turn sour over a period of time due to bacterial reactions. This process is slowed down if the temperature is brought to just above 0 °C in a refrigerator.

2 Addition of a catalyst

We have also already discussed this in detail.

3 Increasing the concentration of the reactants

If in a fixed volume the concentration of reactant species increases, there will be a corresponding increase in the frequency of collisions. Hence there will be an increase in the number of successful collisions, and therefore the rate of reaction will increase. An example of this involves the destruction of statues made of limestone (calcium carbonate, $CaCO_3$). If the concentration of the pollutant sulfur dioxide, SO_2, in the atmosphere increases, the rate of destruction of the limestone statues increases.

$$2CaCO_3(s) + 2SO_2(g) + O_2(g) \rightarrow 2CaSO_4(s) + 2CO_2(g)$$

The solubility of calcium sulfate is greater than that of calcium carbonate, leading to erosion of the limestone.

4 Decreasing the particle size of reactants in the solid phase

In a heterogeneous reaction involving a gas (or a liquid) and a solid, the rate of reaction will increase if the surface area of the solid is increased by breaking the solid up into smaller pieces. The reason for this is that the reaction takes place only on the surface of the solid reactant. So, for example, in the case of a liquid reacting with a solid, only the surface particles of the solid will have direct contact with the liquid reactant. If a finely divided solid or powder is used, there will be an increase in the surface area and there will be a greater number of solid particles available for reaction. As a result, the rate of reaction will increase. For example, finely divided grain can combust explosively. This has led to a number of major explosions in confined spaces in grain factories.

> **Activity**
>
> Find out some examples from different locations around the world where this has happened.

Measuring the rate of a chemical reaction

We shall look at a number of techniques that can be used to monitor the rate of a chemical reaction as stated previously:

1 Change in pH (for acid–base reactions)

In a reaction where either hydronium cations, H_3O^+ (or simply H^+) or hydroxide ions, OH^-, are present as either the reactant or product species, the change in pH can be monitored using a **pH probe and meter**.

2 Change in conductivity (for reactions involving electrolytes)

Consider the following reaction:

$$IO_3^-(aq) + 5I^-(aq) + 6H^+(aq) \rightarrow 3I_2(aq) + 3H_2O(l)$$

In this reaction of iodate ions with iodide ions in an acidic medium there is a net decrease in the concentration of ions from a total of 12 on

the reactant side to zero on the product side. This decrease in the concentration of ions can be monitored using a **conductivity probe and meter**. If the net number of ions decreases during the reaction, then the total electrical conductivity will also decrease and vice versa.

3 Change in mass or volume (for reactions involving gases)

We have seen an example of this physical method earlier.

4 Change in colour (for reactions involving transition metals or other coloured compounds)

Colorimetry (figure 14) is used to monitor chemical reactions that have a coloured reactant or a coloured product. The change in colour intensity corresponding to the change in concentration of the coloured reactant or product can be monitored using a colorimeter. One example of this involves the ethanedioate–manganate(VII) reaction:

$2MnO_4^-(aq) + 5C_2O_4^{2-}(aq) + 16H^+(aq)$
purple

$\rightarrow 2Mn^{2+}(aq) + 10CO_2(g) + 8H_2O(l)$
pale pink

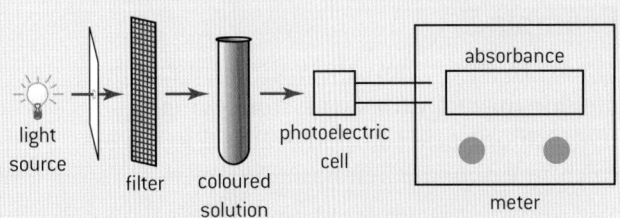

▲ Figure 14 Schematic diagram of a colorimeter that records absorbance. According to Beer's law the absorbance A is directly proportional to the concentration c, that is $A \propto c$ or $A = c\varepsilon l$, where l is the path length and ε is the extinction coefficient.

In colorimetry, light of a certain wavelength is passed through the coloured solution being monitored. The colorimeter or spectrophotometer then measures the intensity of the transmitted light. The absorbance is indicative of the amount of light absorbed by the reaction mixture. A **calibration curve** can be plotted of absorbance versus concentration of the standard coloured solution (figure 15).

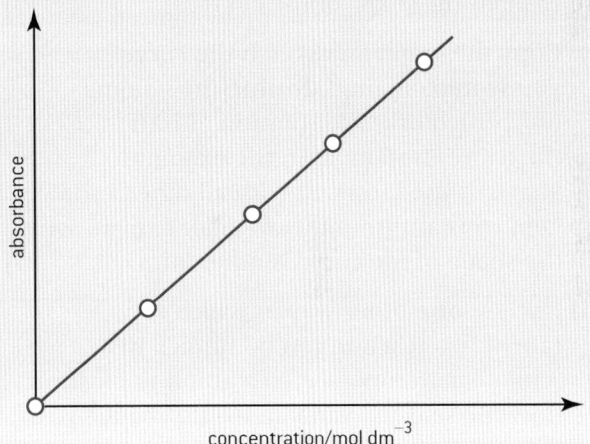

▲ Figure 15 Sketch of a typical calibration curve

🌐 Worked examples: rates of reaction

Example 1

In the chemical reaction of calcium carbonate with hydrochloric acid, 0.25 g of carbon dioxide was generated in 60.0 s.

a) Deduce the balanced chemical equation for the reaction, including state symbols.

b) Calculate the average rate, in mol s^{-1}.

Solution

a) $CaCO_3(s) + 2HCl(aq) \rightarrow CaCl_2(aq) + CO_2(g) + H_2O(l)$

b) $M(CO_2) = 12.01 + 2 \times 16.00 = 44.01$ g mol^{-1}

$n(CO_2) = \dfrac{0.25}{44.01} = 5.7 \times 10^{-3}$ mol

Average rate $= \dfrac{5.7 \times 10^{-3}}{60.0} = 9.5 \times 10^{-5}$ mol s^{-1}

Study tips

Be careful with **significant figures** in a question like this. Since division is involved the answer should be expressed with the smallest number of significant figures from the experimental data, which in this case will be two.

Example 2

Figure 16 shows how the volume of carbon dioxide formed varies with time, when a hydrochloric acid solution is added to **excess** calcium carbonate in a flask.

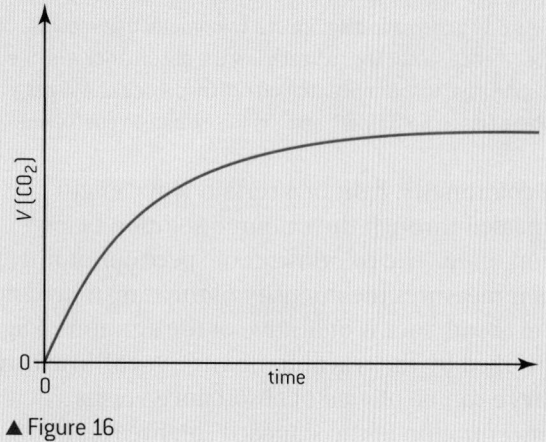

▲ Figure 16

i) Explain the shape of the curve. [3]

ii) Copy the graph and sketch the curve you would obtain if **double** the volume of hydrochloric acid solution of **half** the concentration, as in the example above, is used instead, with all other variables kept constant from the original. Explain why the shape of the curve is different. [4]

iii) Outline **one** other way in which the rate of this reaction can be studied in a school laboratory. Sketch a graph to illustrate how the selected variable would change with time. [2]

iv) Define the term **activation energy** and state **one** reason why the reaction between calcium carbonate and hydrochloric acid takes place at a reasonably fast rate at room temperature. [2]

IB, May 2010

Solution

i) Rate = gradient of graph.

At the beginning of the reaction, carbon dioxide is produced quickest, as the concentration of hydrochloric acid is greatest.

Then as the reaction proceeds with time, the gradient decreases as the amount of carbon dioxide generated slows as the concentration of acid decreases.

The plot eventually flattens as hydrochloric acid is consumed.

ii)

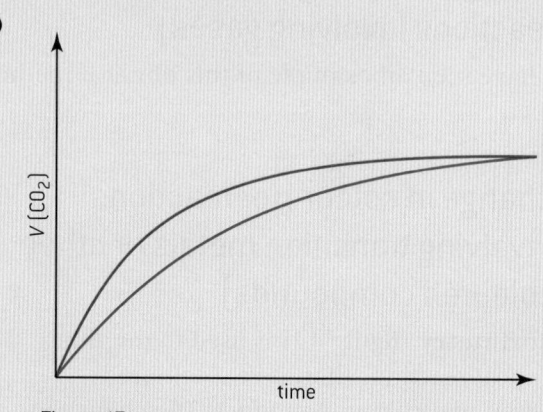

▲ Figure 17

Decreasing the concentration decreases the rate of reaction. In figure 17 the curve (blue) is less steep, but the same maximum volume of gas will be evolved over a longer time. Since the acid concentration is halved, collisions will be less frequent and hence the rate of reaction will be slower. Since there is the same amount of hydrochloric acid, the same volume of carbon dioxide will be produced.

iii) There are a number of possible answers here. One method is to measure the rate at which the mass decreases as the gas is given off. This would simply involve a plot of mass of flask + contents versus time or a plot of mass loss versus time (figure 18).

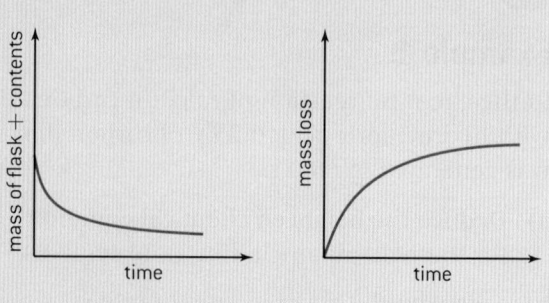

▲ Figure 18

Other possibilities might include monitoring the pH or the pressure.

iv) Activation energy is the *minimum* energy the colliding particles need to have in order for a reaction to occur. The reaction between calcium carbonate and hydrochloric acid takes place at a reasonably fast rate at room temperature because the activation energy barrier is quite low.

Example 3

The data in table 2 were recorded for the decomposition of hydrogen peroxide using a manganese(IV) oxide catalyst. The total volume of oxygen gas collected was measured at different times.

Time/min	Total volume of oxygen gas collected/cm^3
0	0
1	18
2	32
3	42
4	50
5	56
6	61
7	64
8	64
9	64
10	64

▲ Table 2

a) Deduce the balanced chemical equation for the reaction, including state symbols.

b) Draw a graph of total volume of oxygen versus time.

c) Calculate the average rate, in cm^3 min^{-1}, correct to one decimal place.

d) Deduce, in s, how long it took for 40 cm^3 of oxygen to be collected.

e) Determine the initial rate, in cm^3 min^{-1}.

f) Determine the instantaneous rate, in cm^3 min^{-1}, at $t = 4$ min.

g) Explain whether the catalyst used is homogeneous or heterogeneous.

Solution

a) $H_2O_2(aq) \rightarrow H_2O(l) + \frac{1}{2}O_2(g)$

b)

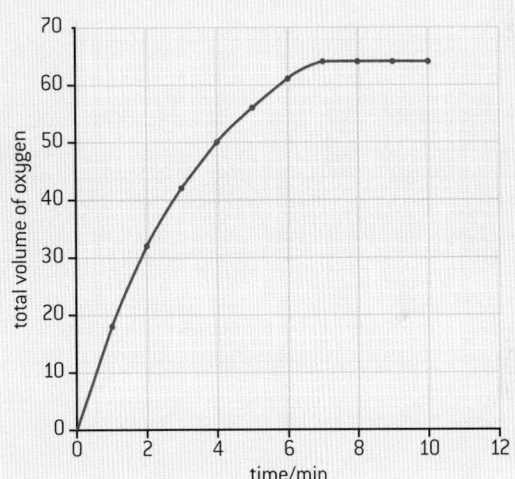

▲ Figure 19 Plot of total volume of oxygen given off versus time

c) Average rate $= \dfrac{64 \text{ cm}^3}{7 \text{ min}} = 9.1 \text{ cm}^3 \text{ min}^{-1}$

d) 2.8 min $= 2.8 \times 60$ s $= 168$ s

e) $(x_1, y_1) = (2,50)$; $(x_2, y_2) = (0,0)$;

Initial rate $= \dfrac{0-50}{0-2} = 25 \text{ cm}^3 \text{ min}^{-1}$

f) $(x_3, y_3) = (6, 64)$; $(x_4, y_4) = (2, 36)$;

Instantaneous rate at $t = 4$ min:

$\dfrac{36-64}{2-6} = 7$ cm^3 min^{-1}

g) Manganese(IV) oxide is solid, so is in a different phase to aqueous hydrogen peroxide. Manganese(IV) oxide is acting as a heterogeneous catalyst in this reaction.

Study tips

There are a number of points that you have to be careful about in graphical questions:

1 Graphs should have a title which involves a plot of y versus x (not the other way round!).

2 Both the x- and y-axes must be labelled and units included.

3 When finding the slope of a tangential line, try to choose two points as far apart as possible.

here

Questions

1 Consider the reaction between gaseous iodine and gaseous hydrogen.

$$I_2(g) + H_2(g) \rightleftharpoons 2HI(g) \qquad \Delta H^\ominus = -9 \text{ kJ}$$

Why do some collisions between iodine and hydrogen not result in the formation of the product?

A. The I_2 and H_2 molecules do not have sufficient energy.

B. The system is in equilibrium.

C. The temperature of the system is too high.

D. The activation energy for this reaction is very low. [1]

IB, May 2011

2 At 25 °C, 200 cm³ of 1.0 mol dm⁻³ nitric acid is added to 5.0 g of magnesium powder. If the experiment is repeated using the same mass of magnesium powder, which conditions will result in the same initial reaction rate?

	Volume of HNO_3/ cm³	Concentration of HNO_3/mol dm⁻³	Temperature / °C
A.	200	2.0	25
B.	200	1.0	50
C.	100	2.0	25
D.	100	1.0	25

[1]

IB, May 2011

3 Which of the following is an appropriate unit for rate of reaction?

A. s

B. min

C. cm³ s

D. mol dm⁻³ min⁻¹

4 Equal masses of powdered calcium carbonate were added to separate solutions of hydrochloric acid. The calcium carbonate was in excess. The volume of carbon dioxide produced was measured at regular intervals. Which curves in figure 20 best represent the evolution of carbon dioxide against time for the acid solutions shown in table 3?

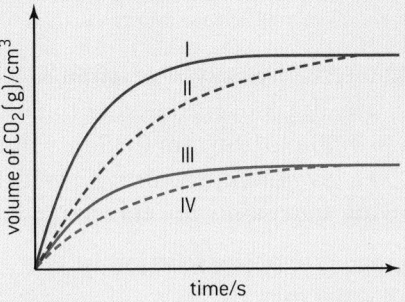

▲ Figure 20

	25 cm³ of 2 mol dm⁻³ HCl	50 cm³ of 1 mol dm⁻³ HCl	25 cm³ of 1 mol dm⁻³ HCl
A.	I	III	IV
B.	I	IV	III
C.	I	II	III
D.	II	I	III

▲ Table 3

[1]

IB, May 2009

5 Hydrochloric acid is reacted with large pieces of calcium carbonate; the reaction is then repeated using calcium carbonate powder. How does this change affect the activation energy and the collision frequency?

	Activation energy	Collision frequency
A.	increases	increases
B.	stays constant	increases
C.	increases	stays constant
D.	stays constant	stays constant

[1]

IB, November 2009

177

6 Which factors can affect the rate of a chemical reaction?

    I.   The concentration of the reactants.

    II. The temperature at which the reaction takes place.

    III. The physical state of the reactants.

    A. I and II only

    B. I and III only

    C. II and III only

    D. I, II, and III

7 In an acid-catalysed hydrolysis reaction of ethyl ethanoate, the concentration of the ester changes from 1.50 mol dm^{-3} to 0.35 mol dm^{-3} in 3.5 min, at a given temperature, T_1. Which of the following statements are correct?

    I.   The average rate of the reaction is 0.33 mol dm^{-3} min^{-1}.

    II. If the reaction is carried out at a higher temperature, T_2, the reaction rate will be greater.

    III. The products of the reaction will be ethanoic acid and ethanol.

    A. I and II only

    B. I and III only

    C. II and III only

    D. I, II, and III

8 Factors that affect the rate of a chemical reaction include particle size, concentration of reactants, and the temperature of the reaction.

    i) Define the term **rate of a chemical reaction**. [1]

    ii) List the **three** characteristic properties of reactant particles that affect the rate of reaction as described by the collision theory. [3]

IB, May 2011

9 a) A solution of hydrogen peroxide, H_2O_2, is added to a solution of sodium iodide, NaI, acidified with hydrochloric acid, HCl. The yellow colour of the iodine, I_2, can be used to determine the rate of reaction.

$$H_2O_2(aq) + 2NaI(aq) + 2HCl(aq)$$
$$\rightarrow 2NaCl(aq) + I_2(aq) + 2H_2O(l)$$

The experiment is repeated with some changes to the reaction conditions. For each of the changes that follow, predict, stating a reason, its effect on the rate of reaction.

    i) The concentration of H_2O_2 is increased at constant temperature. [2]

    ii) The solution of NaI is prepared from a fine powder instead of from large crystals. [2]

  b) Explain why the rate of a reaction increases when the temperature of the system increases. [3]

IB, November 2009

10 Models can prove vital in chemistry. Discuss the principles of the kinetic–molecular theory and the collision theory.

11 Design an appropriate experiment to measure the rate of reaction of a hydrolysis reaction (saponification) of the ester methyl ethanoate in an alkaline medium.

12 Design an appropriate experiment to measure the rate of reaction using the "clock reaction" technique involving the reaction of magnesium with dilute hydrochloric acid solution.

7 EQUILIBRIUM

Introduction

An understanding of reactions that are in equilibrium and how to control the position of the equilibrium is of fundamental importance to science and society. The Haber process, used for the large-scale manufacture of ammonia, is an equilibrium system that has had a profound impact on the history of the World. In this chapter we discuss how reactions can be in a state of equilibrium, examine the equilibrium constant K_c, examine the information it conveys and discuss the effects of changing experimental conditions on the value of K_c applying Le Châtelier's principle. We will also introduce the term reaction quotient, Q, which is a measure of the relative amounts of products and reactants present in a reaction that is not in a state of equilibrium.

7.1 Equilibrium

Understandings

→ A state of equilibrium is reached in a closed system when the rates of the forward and reverse reactions are equal.

→ The equilibrium law describes how the equilibrium constant (K_c) can be determined for a particular chemical equation.

→ The magnitude of the equilibrium constant indicates the extent of a reaction at equilibrium and its temperature dependence.

→ The reaction quotient (Q) measures the relative amount of products and reactants present during a reaction at a particular point in time. Q is the equilibrium constant expression with non-equilibrium concentrations. The position of the equilibrium changes with changes in concentration, pressure, and temperature.

→ A catalyst has no effect on the position of equilibrium or the equilibrium constant.

Applications and skills

→ The characteristics of chemical and physical systems in a state of equilibrium.

→ Deduction of the equilibrium constant expression (K_c) from an equation for a homogeneous reaction.

→ Determination of the relationship between different equilibrium constants (K_c) for the same reaction at the same temperature.

→ Application of Le Châtelier's principle to predict the qualitative effects of changes of temperature, pressure, and concentration on the position of equilibrium and on the value of the equilibrium constant.

Nature of science

→ Obtaining evidence for scientific theories – isotopic labelling and its use in defining equilibrium.

→ Common language across different disciplines – the term dynamic equilibrium is used in other contexts, but not necessarily with the chemistry definition in mind.

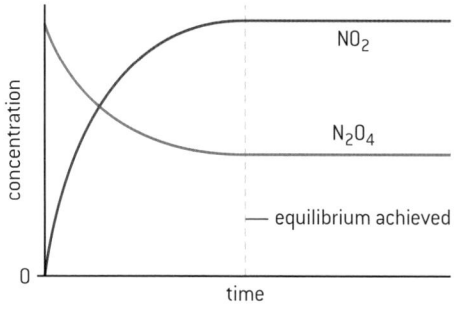

▲ Figure 4 Change in concentration versus time for the reaction $N_2O_4(g) \rightleftharpoons 2NO_2(g)$ approaching and achieving equilibrium

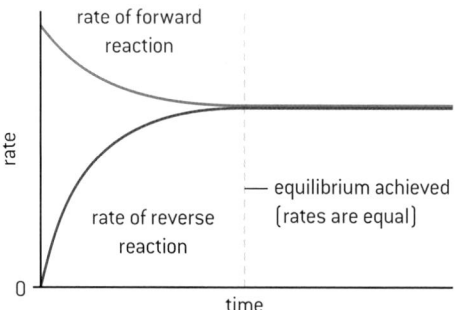

▲ Figure 5 At equilibrium, the rates of the forward and reverse reactions remain constant

Animated computer simulations are available in which the user can decide how to change the conditions of a reaction to illustrate the concept of dynamic equilibrium.

TOK

Scientists have a common terminology and a common reasoning process, which involves using deductive logic and induction through analogies and generalizations. They share mathematics, the language of science, as a powerful tool. Indeed, some scientific explanations exist only in mathematical form.

At equilibrium:

- The forward and reverse reactions are occurring at equal rates.
- There is no change in the concentrations of reactants and products.
- There is no change in macroscopic properties such as colour and density.
- The equilibrium can be approached from either the forward or reverse direction.
- The equilibrium is dynamic. The forward and reverse reactions continue but no overall change in the concentration of reactants and products occurs.
- Any changes in the reaction conditions, such as temperature, pressure, or concentration of reactants or products, can affect the equilibrium, demonstrating its dynamic nature.

Quick question

Which statement is always true for a chemical reaction that has reached equilibrium?

a) The yield of product(s) is greater than 50%.

b) The rate of the forward reaction is greater than the rate of the reverse reaction.

c) The amounts of reactants and products do not change.

d) Both forward and reverse reactions have stopped. [1]

IB, Nov 2005

The equilibrium law

When a reaction system has established equilibrium, the forward and reverse reactions are occurring at equal rates and there is no change in the concentration of reactants and products. To make best use of the reaction we need to understand the *position of the equilibrium*, that is, whether reactants or products are favoured, and how we can manipulate this to maximize the yield of products and the profitability of industry.

The **law of chemical equilibrium** states that at a given temperature the ratio of the concentration of products (raised to the power of their molar coefficients) to the concentration of reactants (raised to the power of their molar coefficients) is a constant. This constant is called the **equilibrium constant, K_c**. The subscript 'c' indicates that concentration values for products and reactants are being used.

For example:

$$O_2(g) + 4HCl(g) \rightleftharpoons 2H_2O(g) + 2Cl_2(g)$$

$$K_c = \frac{[H_2O]^2 \, [Cl_2]^2}{[O_2][HCl]^4}$$

The value of the equilibrium constant is specific for each chemical reaction at a given temperature. Values of K_c have no units.

The magnitude of K_c tells you about the position of the equilibrium. If K_c is a very large number, $K_c \gg 1$, this indicates that at a given temperature, products are favoured over reactants. The larger the value of K_c, the greater the proportion of products that exists compared with reactants at equilibrium. Conversely, a very small value of K_c ($K_c \ll 1$) indicates that the reaction is unfavourable at this given temperature.

Writing equilibrium constant expressions

When constructing the equilibrium constant expression for a homogeneous reaction, the following need to be considered:

1   For an aqueous reaction the concentration of the solvent water does not appear in the equilibrium constant expression, as its concentration does not change during the reaction.

2   If the reaction takes place in a non-aqueous solution (such as the esterification discussed in sub-topic 10.2), water must be included in the K_c expression as any other reactant or product.

Table 1 gives some examples of equilibrium constant expressions for different reactions.

> In a **homogeneous equilibrium** all the reactants and products are present in one phase. The most common example is reversible reactions that occur in the gaseous phase.
>
> In a **heterogeneous equilibrium** the reactants or products exist in more than one phase, such as gaseous and solid, liquid and solid.
>
> The equilibrium constant expressions derived here are for homogeneous equilibria.

Chemical equation	Equilibrium constant expression
$Fe^{3+}(aq) + SCN^-(aq) \rightleftharpoons Fe(SCN)^{2+}(aq)$	$K_c = \dfrac{[Fe(SCN)^{2+}]}{[Fe^{3+}][SCN^{-1}]}$
$CH_3OH(aq) + C_3H_7COOH(aq) \rightleftharpoons C_3H_7COOCH_3(aq) + H_2O(l)$	$K_c = \dfrac{[C_3H_7COOCH_3]}{[CH_3OH][C_3H_7COOH]}$
$N_2(g) + 3H_2(g) \rightleftharpoons 2NH_3(g)$	$K_c = \dfrac{[NH_3]^2}{[N_2][H_2]^3}$

▲ Table 1 Equilibrium constant expressions for some chemical reactions

TOK

Scientists approach their understanding of the universe from varying perspectives. Observations made using our senses examine phenomena at the macroscopic level. The models and theories developed by the scientific community focus on our understanding of the microscopic world. Which of the ways of knowing (WOK) enable us to make the transition from the macroscopic to the microscopic?

Macroscopic properties of substances can be identified using our senses, and can be directly determined by measurements. Examples include colour, texture, and density. **Microscopic** properties exist at an atomic level and can be determined only indirectly.

Quick question

Deduce the equilibrium constant expression for each of the following homogeneous equilibrium reactions.

a) $\frac{1}{2}N_2(g) + \frac{3}{2}H_2(g) \rightleftharpoons NH_3(g)$

b) $ClNO_2(g) + NO(g) \rightleftharpoons NO_2(g) + ClNO(g)$

c) $4NH_3(g) + 5O_2(g) \rightleftharpoons 4NO(g) + 6H_2O(g)$

Worked example: deducing equilibrium constant expressions

For the homogeneous equilibrium:

$$H_2(g) + I_2(g) \rightleftharpoons 2HI(g)$$

the equilibrium concentrations (in mol dm^{-3}) are as follows:

$$[H_2(g)] = 0.300, [I_2(g)] = 0.300, [HI(g)] = 3.00$$

Deduce the equilibrium constant expression, K_c, and determine the value of K_c for the forward and reverse reactions.

Solution

For the forward reaction:

$$K_{c(1)} = \frac{[HI]^2}{[H_2][I_2]}$$

$$= \frac{(3.00)^2}{0.300 \times 0.300}$$

$$= 1.00 \times 10^2$$

For the reverse reaction:

$$2HI(g) \rightleftharpoons H_2(g) + I_2(g)$$

$$K_{c(2)} = \frac{[H_2][I_2]}{[HI]^2}$$

$$= \frac{0.300 \times 0.300}{(3.00)^2}$$

$$= 1.00 \times 10^{-2}$$

Note:

$$K_{c(1)} = \frac{1}{K_{c(2)}}$$

The value of K_c also depends on how the chemical equilibrium reaction is balanced:

$$\tfrac{1}{2}H_2(g) + \tfrac{1}{2}I_2(g) \rightleftharpoons HI(g)$$

$$K_{c(3)} = \frac{[HI]}{[H_2]^{\frac{1}{2}}[I_2]^{\frac{1}{2}}}$$

$$K_{c(3)} = \frac{3.00}{(0.300)^{\frac{1}{2}}(0.300)^{\frac{1}{2}}}$$

$$= 10.0$$

$$K_{c(3)} = \sqrt{K_{c(1)}}$$

This last example shows that dividing the equation by 2 throughout gives a new equilibrium constant which is equal to the square root of the original equilibrium constant.

Quick questions

1 The equilibrium constant for the reaction between hydrogen and chlorine gas to produce hydrogen chloride gas is 2.40×10^{33} at 298 K.

$$H_2(g) + Cl_2(g) \rightleftharpoons 2HCl(g)$$

Calculate the equilibrium constant for the reverse reaction, namely the decomposition of HCl.

2 The equilibrium constant for the equilibrium between N_2O_4 and NO_2 is 7.7×10^{-4} at 273 K:

$$N_2O_4(g) \rightleftharpoons 2NO_2(g)$$

Calculate the equilibrium constant for $\tfrac{1}{2}N_2O_4(g) \rightleftharpoons NO_2(g)$ at 273 K.

Combining equilibrium constants

Table 2 summarizes the ways that the equilibrium constant expressions and equilibrium constants can be combined.

$$K_c \text{ (reverse)} = \frac{1}{K_c}\text{(forward)}$$

Change in reaction equation	Equilibrium constant expression	Equilibrium constant
reverse the reaction	inverse of the expression	$\frac{1}{K_c}$ or K_c^{-1}
halve the coefficients	square root of the expression	$\sqrt{K_c}$
double the coefficients	square the expression	K_c^2
sum equations	product of the expressions	$K_c = K_{c1} \times K_{c2} \times \ldots$

▲ Table 2 The equilibrium constant K_c for the same reaction at the same temperature can be expressed in a number of ways

The effect of changing experimental conditions on the equilibrium constant

When equilibrium is established, the position of the equilibrium remains constant provided that the temperature and pressure do not change. A change in experimental conditions can affect the equilibrium position. However, the value of K_c remains constant unless the temperature changes (table 3).

Change in condition	Equilibrium position	K_c
concentration of product or reactant	changes in response to a change in [reactants] or [products]	no change
pressure	in a reaction with gaseous reactants or products, the pressure can affect the equilibrium position	no change
temperature	usually changes: the direction of change depends on whether the reaction is exothermic or endothermic	changes, unless $\Delta H = 0$
catalyst	no change	no change

▲ Table 3 The effect of changing conditions on the equilibrium position and the value of K_c

Le Châtelier's principle

Le Châtelier's principle is a useful tool for predicting the effect that changing conditions will have on the equilibrium position:

If a change is made to a system that is in equilibrium, the balance between the forward and reverse reactions will shift to offset this change and return the system to equilibrium.

At a given temperature, table 3 shows that changing the concentration of reactants or products does not result in a change in the value of equilibrium constant K_c. The equilibrium position of the reaction will change in response to the change in concentration so as to return K_c to its original value. For example, figure 6 illustrates the chromate–dichromate equilibrium:

$$Cr_2O_7^{2-}(aq) \quad + \quad H_2O(l) \quad \rightleftharpoons \quad 2CrO_4^{2-}(aq) \quad + \quad 2H^+(aq)$$
orange dichromate $\qquad\qquad\qquad$ yellow chromate

If the concentration of a reactant is increased, Le Châtelier's principle tells us that the forward reaction will be favoured to counteract this change. Decreasing the concentration of a product will also favour the forward reaction. The addition of hydroxide ions, OH^-, to the reaction mixture results in a reduction in H^+ concentration as OH^- reacts with H^+ to form water. The equilibrium mixture becomes a paler colour as more orange dichromate reacts to form yellow chromate.

The reverse reaction is favoured if the concentration of a product is increased or the concentration of a reactant is decreased. The addition of H^+ ions in the form of concentrated hydrochloric acid, HCl results in the reverse reaction being favoured and a deeper orange colour being observed. In both cases, the value of equilibrium constant K_c remains unchanged as long as the temperature remains the same.

▲ Figure 6 The chromate–dichromate equilibrium. Aqueous dichromate $Cr_2O_7^{2-}$ (aq) is orange while aqueous chromate CrO_4^{2-} (aq) is yellow. The colour of the solution gives an indication of the position of equilibrium

Le Châtelier's principle allows industrial chemists to manipulate reaction conditions to maximize the amount of the desired product formed in an equilibrium reaction. The Haber process for the manufacture of ammonia uses the following reaction:

$$N_2(g) + 3H_2(g) \rightleftharpoons 2NH_3(g)$$

The yield is increased by using high concentrations of reactant gases nitrogen and hydrogen, and removing the product ammonia from the equilibrium mixture.

 ## Cause and effect

"Cause and effect" or "causation" refers to the situation where a second event (the effect) is a direct consequence of the first event (the cause). Scientists understand this concept and often develop a hypothesis that suggests a relationship or causation between factors that may be true or false. They then test the hypothesis by experimentation and collection of empirical data, which is used to either support, modify, or disprove the hypothesis.

Le Châtelier's principle is a useful tool in helping to predict the qualitative effects of changes in temperature, pressure, and concentration on the position of an equilibrium reaction and the value of the equilibrium constant. However, the principle does not provide an explanation for the effects of these changes on the equilibrium reaction: it does not demonstrate a cause-and-effect relationship between changing these conditions and the resultant change in the equilibrium system.

 ## The Haber process

Over the past century, advances in the science of agriculture have resulted in over one-third of the Earth's land being cultivated with crops and the population of the planet increasing to over 7 billion people. Agriculture is a major employment sector in developing countries and food security remains of great concern to many, including international bodies such as the World Trade Organization (WTO), the Food and Agriculture Organization (FAO), and the World Health Organization (WHO).

Global demand for food increases exponentially every decade and we will not be able to meet demands for food and water without scientific endeavours in the fields of biotechnology and fertilizers. More than 50% of current food production relies on the use of fertilizers.

Fritz Haber (1868–1934) was a German chemist who was awarded the Nobel Prize in Chemistry in 1918 for his work on the synthesis of ammonia, NH_3 from its elements. A shortage of natural fertilizers at the beginning of the twentieth century prompted Haber to research

a process that could manufacture ammonia on a large scale. Ammonia manufacture is now one of the most widespread industrial processes in the world. As well as its major application in fertilizers, ammonia is also used in the manufacture of plastics, fibres, explosives, and pharmaceuticals.

With the onset of the first world war, Haber worked with the German military to advance the production of explosives. Ammonia reacts with nitric acid to form ammonium nitrate, $HNO_3(aq) + NH_3(g) \rightarrow NH_4NO_3(aq)$ which is used to this day as a fertilizer. Ammonium nitrate will also undergo explosive decomposition when detonated:

$$NH_4NO_3(s) \rightarrow N_2O(g) + 2H_2O(g)$$

Haber is sometimes referred to as the "father of chemical warfare", advancing the research into and utilization of many poisonous gases during the first world war. He was not alone in making significant advances in science with military applications during this time. Gustav Hertz and James Franck were physicists who devised

investigations to support Niels Bohr's model of the atom and lay the groundwork for **quantum mechanics** (for more on quantum mechanics, see the Physics course book, topic 12). They were awarded the Nobel Prize in Physics in 1925, while Otto Hahn received the Nobel Prize in Physics in 1944 for his discovery of the fission of heavy atomic nuclei. Later in life he continued to receive an impressive range of awards for his scientific achievements and was nominated on several occasions for the Nobel Peace Prize. He is regarded as the founder of the atomic age.

Hertz, Franck, and Hahn were all members of Haber's research team who worked on developing chemical weapons. Their stories serve to illustrate the Nature of Science (NOS). Their research significantly advanced scientific understanding while raising questions about the ethical considerations of the results of some of their work, the concept of intellectual property and how society goes on to utilize scientific discoveries.

The effect of pressure on reactions in the gas phase

We have seen how Le Châtelier's principle can be used to predict the effect of a change in concentration of a product or reactant on the position of equilibrium. A system that involves substances in the gaseous rather than the aqueous phase will be affected by changes in applied pressure:

$$4HCl(g) + O_2(g) \rightleftharpoons 2H_2O(g) + 2Cl_2(g)$$

In this reaction there are 5 moles of gas on the reactant side and 4 moles on the product side. A change in pressure applied to the system will result in a shift in the equilibrium position. If the pressure is increased, Le Châtelier's principle says that the equilibrium will shift to reverse this change. The forward reaction becomes favoured to reduce the pressure of the system. In the same way, a decrease in pressure will result in the reverse reaction being favoured.

In the Haber process:

$$N_2(g) + 3H_2(g) \rightleftharpoons 2NH_3(g)$$

there are 4 moles of gas on the reactant side and only 2 moles of gas on the product side. A high pressure will favour the forward reaction, to decrease the pressure of the system.

Such a change in the equilibrium position will not affect the value of equilibrium constant, K_c, if the temperature remains constant.

Temperature and the equilibrium constant

An understanding of the thermodynamics of a reaction is required when considering the effect of changing the temperature on the equilibrium constant. For example:

$$N_2(g) + 3H_2(g) \rightleftharpoons 2NH_3(g) \quad \Delta H = -92 \text{ kJ}$$

In this reaction energy can be considered a product and is released to the surroundings.

> **Study tip**
>
> When considering the effect of changes in pressure on a reaction, you must refer to the moles of *gaseous* reactants or products in the reaction. For example:
>
> $$C(s) + H_2O(g) \rightleftharpoons H_2(g) + CO(g)$$
>
> In this heterogeneous reaction, the solid carbon is not included when considering the effect of changes in pressure.

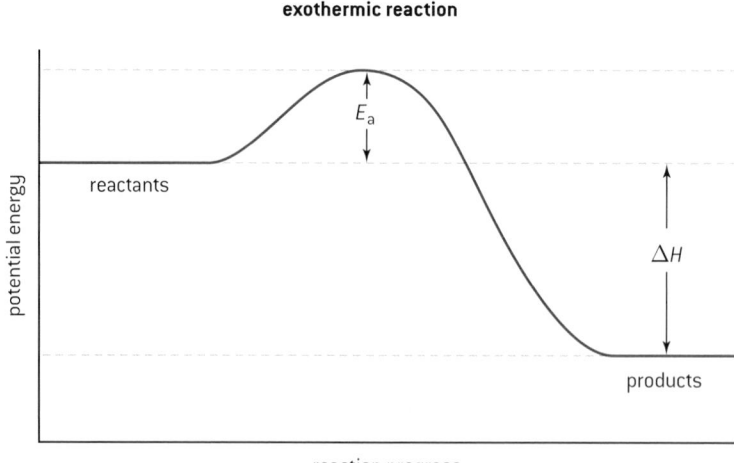

▲ Figure 7 The potential energy profile of an exothermic reaction

The reverse reaction is endothermic – it requires energy from the surroundings.

At equilibrium the forward and reverse reactions occur at equal rates and there is no net change in energy.

For the exothermic reaction, an increase in temperature will shift the equilibrium in the direction that will consume the extra energy. In accordance with Le Châtelier's principle, the equilibrium position will move to the left, favouring the reactants, to minimize the effect of the change; the concentrations of nitrogen and hydrogen will increase. This results in a decrease in the equilibrium constant K_c. Conversely, a decrease in temperature for the exothermic reaction will shift the equilibrium to the right, favouring the forward reaction and increasing the concentration of ammonia, NH_3. This results in an increase in K_c. Table 4 summarizes the effects on the equilibrium system of changing temperature.

Type of reaction	Change in temperature	Equilibrium position	Equilibrium constant K_c
exothermic	increase	moves to the left, favouring reactants	decreases
	decrease	moves to the right, favouring products	increases
endothermic	increase	moves to the right, favouring products	increases
	decrease	moves to the left, favouring reactants	decreases

▲ Table 4 The effects on the equilibrium system of a change in temperature

The effect of a catalyst on equilibrium reactions

The addition of a catalyst provides an alternative pathway for a reaction, lowering the activation energy. In a reaction that goes to completion, a catalyst means that a greater proportion of reactants have sufficient energy to overcome the activation energy barrier and become products. In a reversible reaction, the lowered activation energy has an equal effect on both the forward and reverse reactions. The rates of the forward and reverse reactions increase by an equal amount. The position of the equilibrium will not change and there is no effect on the equilibrium constant K_c.

Quick question

$PCl_5(g) \rightleftharpoons PCl_3(g) + Cl_2(g) \; \Delta H^\ominus = +92.5 \, kJ$

Predict and explain any shift in the equilibrium position when:

a) the temperature of the system is decreased

b) additional chlorine gas is injected into the system

c) the pressure applied to the system is increased

d) a catalyst is added.

Reaction quotient

If a system has not reached equilibrium, the ratio of concentration of product to reactants will not equal K_c. This ratio is called the **reaction quotient Q** and this helps you to determine the progress of the reaction as it moves toward equilibrium and the direction of the reaction that is favoured to establish equilibrium (table 5).

$Q > K_c$	The concentration of products is greater than at equilibrium and the reverse reaction is favoured until equilibrium is reached.
$Q < K_c$	The concentration of reactants is greater than at equilibrium and the forward reaction is favoured until equilibrium is reached.
$Q = K_c$	The system is at equilibrium and the forward and reverse reactions occur at equal rates.

▲ Table 5 The relationship between the reaction quotient Q and the equilibrium constant K_c

Questions

1   The equation for a reversible reaction used in industry to convert methane to hydrogen is shown below.

$$CH_4(g) + H_2O(g) \rightleftharpoons CO(g) + 3H_2(g)$$
$$\Delta H^{\ominus} = +210 \text{ kJ}$$

Which statement is always correct about this reaction when equilibrium has been reached?

A.   The concentrations of methane and carbon monoxide are equal.

B.   The rate of the forward reaction is greater than the rate of the reverse reaction.

C.   The amount of hydrogen is three times the amount of methane.

D.   The value of $\Delta H^{\ominus}$ for the reverse reaction is -210 kJ.   [1]

IB, May 2006

2   Sulfur dioxide and oxygen react to form sulfur trioxide according to the equilibrium:

$$2SO_2(g) + O_2(g) \rightleftharpoons 2SO_3(g)$$

How are the amount of SO_2 and the value of the equilibrium constant for the reaction affected by an increase in pressure?

A.   The amount of SO_3 and the value of the equilibrium constant both increase.

B.   The amount of SO_3 and the value of the equilibrium constant both decrease.

C.   The amount of SO_3 increases but the value of the equilibrium constant decreases.

D.   The amount of SO_3 increases but the value of the equilibrium constant does not change.   [1]

IB, November 2007

3   What will happen to the position of equilibrium and the value of the equilibrium constant when the temperature is increased in the following reaction?   [1]

$$Br_2(g) + Cl_2(g) \rightleftharpoons 2BrCl(g) \qquad \Delta H = +14 \text{ kJ}$$

	Position of equilibrium	Value of equilibrium constant
A.	shifts towards the reactants	decreases
B.	shifts towards the reactants	increases
C.	shifts towards the products	decreases
D.	shifts towards the products	increases

IB, November 2003

4   The equation for one reversible reaction involving oxides of nitrogen is shown below:

$$N_2O_4(g) \rightleftharpoons 2NO_2(g) \qquad \Delta H^{\ominus} = +58 \text{ kJ}$$

Experimental data for this reaction can be represented on the following graph (figure 8).

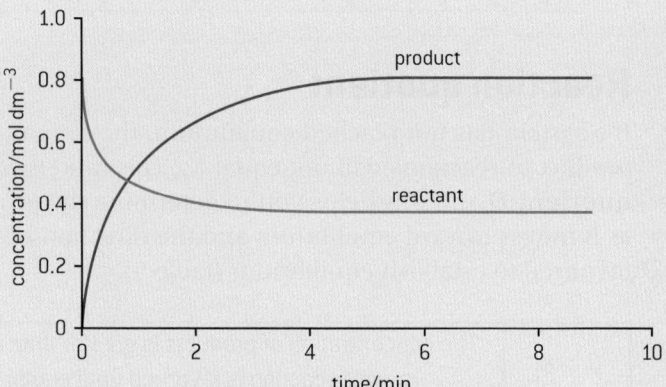

▲ Figure 8

a)   Write an expression for the equilibrium constant, K_c, for the reaction. Explain the significance of the horizontal parts of the lines on the graph. State what can be deduced about the magnitude of K_c for the reaction, giving a reason.   [4]

b)   Use Le Châtelier's principle to predict and explain the effect of increasing the temperature on the position of equilibrium.   [2]

c)   Use Le Châtelier's principle to predict and explain the effect of increasing the pressure on the position of equilibrium.   [2]

d)   State and explain the effects of a catalyst on the forward and reverse reactions, on the position of equilibrium, and on the value of K_c.   [6]

IB, November 2005

8 ACIDS AND BASES

Introduction

The theories of acids and bases and their applications exemplify the work of scientists and the Nature of Science. The early work of scientists laid the foundations for the development of a range of definitions of acids and bases. This chapter examines the Brønsted–Lowry theory that describes acids and bases as proton donors and acceptors, respectively. The general reactions and properties of acids and bases, the pH scale as a measure of the concentration of hydronium ions and the characteristics of strong and weak acids are all explored in detail. The chemistry of acid deposition, a product of the industrial age and a threat to the environment, concludes the topic.

8.1 Theories of acids and bases

Understandings

→ A Brønsted–Lowry acid is a proton/H^+ donor and a Brønsted–Lowry base is a proton/H^+ acceptor.

→ Amphiprotic species can act as both Brønsted–Lowry acids and bases.

→ A pair of species differing by a single proton is called a conjugate acid–base pair.

Applications and skills

→ Deduction of the Brønsted–Lowry acid and base in a chemical reaction.

→ Deduction of the conjugate acid or conjugate base in a chemical reaction.

Nature of science

→ Falsification of theories – HCN altering the theory that oxygen was the element which gave a compound its acidic properties allowed for other acid–base theories to develop.

→ Theories being superseded – one early theory of acidity derived from the sensation of a sour taste, but this has been proven false.

→ Public understanding of science – outside of the arena of chemistry, decisions are sometimes referred to as "acid test" or "litmus test".

The process of **calcination** refers to the heating of materials to very high temperatures in air in order to bring about their thermal decomposition (in the case of limestone), the removal of water from a hydrated compound (for bauxite), or the removal of a volatile matter from minerals and ores.

The role of acids and bases

Acids and bases are familiar in our everyday lives and have a significant role in chemistry. For many hundreds of years scientists have been investigating acid-base reactions and developing a range of definitions and theories that will be discussed in this topic and topic 18. Each of these theories has its strengths and weaknesses and some of the earliest ideas about acids and bases have now been disproved.

It has long been understood that acids and bases behave as opposites. The word **acid** is derived from the Latin *acidus* meaning sour and we still associate a sour taste with many substances that are acidic such as vinegar (ethanoic acid), lemon juice or grapefruit (citric acid), and sour milk (lactic acid). A base that is soluble in water is called an **alkali**. The word alkali is derived from the Arabic word *al-qaly* meaning calcined ashes.

Early theories about acids

Antoine Lavoisier's investigations led him to believe that oxygen, integral to combustion reactions, was present in all acids and was the source of their acidic properties. He gave the name *oxygene* ("acid-forming" in Greek) to the element that had previously been discovered by Joseph Priestly. Lavoisier's work was fundamental in disproving the phlogiston theory (sub-topic 1.1). However, his belief that properties characteristic of acidic compounds were due to the presence of oxygen was subsequently disproved.

Theories developed around the globe are sometimes influenced by differences in the meaning of vocabulary specific to cultures or languages. Early theories about acids and bases, even if later disproved, served an important purpose in generating curiosity and scientific endeavour to explain the phenomenon. Evidence from experimentation and observation is used to develop theories, generalize laws, and propose hypotheses. The testing of these theories over time by the scientific community leads to existing theories being supported, disproved, or even replaced by a new theory. Modern theories on acids and bases focus on trying to explain why they react in the way they do.

▲ Figure 1 The reaction between the vapours of concentrated ammonia and hydrogen chloride solutions produces solid ammonium chloride, visible as a white smoke

Arrhenius's theory of acids and bases

Svante August Arrhenius (1859–1927) was awarded the Nobel Prize in Chemistry in 1903 for his work in the field of acids and bases. He defined an **acid** as a substance that ionizes in water to produce hydrogen ions, H^+. An **alkali**, a soluble base, produces hydroxide ions, OH^-. The combination of an acid and base is well known as a **neutralization** reaction involving the combination of the hydrogen ion and the hydroxide ion.

$$H^+(aq) + OH^-(aq) \rightarrow H_2O(l)$$

An example of this type of neutralization is the reaction of hydrochloric acid in the stomach with aluminium hydroxide contained in an antacid tablet:

$$3HCl(aq) + Al(OH)_3(s) \rightarrow AlCl_3(aq) + 3H_2O(l)$$

Arrhenius's theory had its limitations. The reaction between the weak base ammonia and hydrogen chloride gas (figure 1) could not be explained, as ammonia does not contain hydroxide ions.

$$NH_3(g) + HCl(g) \rightarrow NH_4Cl(s)$$

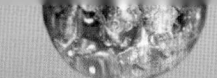

 ## Science in society

Terminology that has an accepted meaning within the scientific community, clearly communicating a specific understanding, can often have a different meaning in everyday life. For example, in general life the term "acid test" or "litmus test" is associated with the testing of the certainty of some event or phenomenon: a teacher might say that "the student's performance in the final examination serves as a litmus test of their ability to study at university." Science is often used to add credibility to a myriad of situations in everyday life, as the scientific process is perceived as providing data that is rigorous, supported by experimentation, and uninfluenced by human bias. Public understanding of science is vital in making informed decisions about scientific findings and issues.

TOK

Scientists employ a wide variety of methodologies to develop knowledge and understanding. Evidence to test hypotheses is obtained in the laboratory through observation and experimentation. One assumption in this process is that scientists are able to recreate conditions in the laboratory that accurately represent what is occurring in the universe outside. How then is this methodology used in chemistry different from the methodologies employed in other areas of knowledge?

Brønsted–Lowry acids and bases

Scientists often work collaboratively, participating in an open exchange of information and ideas that leads to a better understanding of the subject of the research. In other cases scientists work independently, sometimes discovering and subsequently theorizing about the same idea simultaneously. This was the case when Johannes Brønsted and Thomas Lowry developed a definition of acids and bases that broadened Arrhenius's theory.

Referring to a hydrogen ion as a proton, they proposed that an acid could be defined as a proton donor and a base as a proton acceptor.

In an aqueous solution a proton can be represented as either the hydrogen ion, H^+ or as the **hydronium ion**, H_3O^+. The hydronium ion is formed when a water molecule forms a coordinate bond with a proton (figure 2). For example:

$$HCl(aq) + H_2O(l) \rightarrow H_3O^+(aq) + Cl^-(aq)$$

Common acids are often referred to as being monoprotic (such as hydrochloric acid), diprotic (such as sulfuric acid), or triprotic (such as phosphoric acid). Hydrochloric and sulfuric acids are strong acids while phosphoric acid is a weak acid.

$$HCl(aq) \rightarrow H^+(aq) + Cl^-(aq)$$

$$H_2SO_4(aq) \rightarrow 2H^+(aq) + SO_4^{2-}(aq)$$

$$H_3PO_4(aq) \rightleftharpoons 3H^+(aq) + PO_4^{3-}(aq)$$

Ethanoic acid, CH_3COOH is also a weak acid:

$$CH_3COOH(aq) \rightleftharpoons CH_3COO^-(aq) + H^+(aq)$$

or

$$CH_3COOH(aq) + H_2O(l) \rightleftharpoons CH_3COO^-(aq) + H_3O^+(aq)$$

In the last equilibrium ethanoic acid is acting as a Brønsted–Lowry acid and water is acting as a Brønsted–Lowry base. Focusing on the reverse reaction, the ethanoate ion, CH_3COO^- is acting as a proton acceptor and

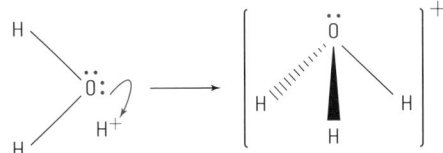

▲ Figure 2 The hydronium ion, H_3O^+

A **strong acid** is assumed to undergo complete dissociation in water (sub-topic 8.4). For example, in hydrogen chloride, HCl the hydrogen ion has almost no affinity for the chloride ion.

A **weak acid** undergoes only partial dissociation in water, establishing an equilibrium, and a solution of a weak acid is only a weak electrolyte.

the hydronium ion as a proton donor. The ethanoate ion is the **conjugate base** of the Brønsted–Lowry acid (ethanoic acid), while the hydronium ion is the **conjugate acid** of another Brønsted–Lowry base, water.

The conjugate acid and base differ from one another by a single proton. This is termed **a conjugate acid–base pair**. Figure 3 shows another example.

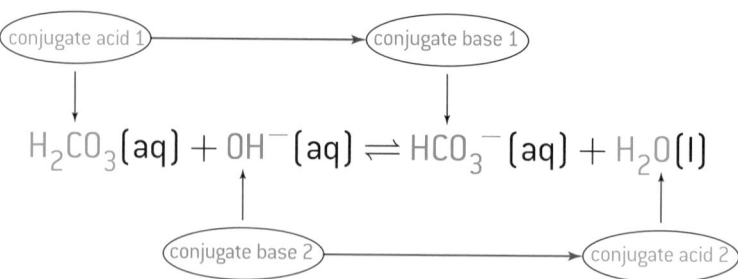

▲ Figure 3 Conjugate acid–base pairs in the neutralization of carbonic acid (Brønsted–Lowry acid) with a hydroxide ion (Brønsted–Lowry base)

Quick questions

1 Identify the conjugate bases of the following acids.

   a) H_2SO_4 d) C_6H_5OH

   b) HNO_3 e) OH^-

   c) C_2H_5OH f) H_2O

2 Identify the conjugate acids of the following bases.

   a) OH^- d) CO_3^{2-}

   b) H_2O e) HNO_3

   c) NH_3 f) $C_2H_5NH_2$

3 Identify the conjugate acid–base pairs in the following equations:

   $$HCO_3^-(aq) + S^{2-}(aq) \rightleftharpoons HS^-(aq) + CO_3^{2-}(aq)$$

   $$CH_3COOH(aq) + HPO_4^{2-}(aq) \rightleftharpoons CH_3COO^-(aq) + H_2PO_4^-(aq)$$

Amphiprotic species

Some substances have the ability to act as either a Brønsted–Lowry acid or a Brønsted–Lowry base depending on the reaction in which they are taking part. These species are said to be **amphiprotic**. For example, the water molecule can donate a proton in a reaction, thus acting as a Brønsted–Lowry acid. It can also accept a proton, acting as a Brønsted–Lowry base.

Polyprotic species are frequently involved in reactions in which they behave amphiprotically for example:

$$H_2PO_4^-(aq) + OH^-(aq) \rightleftharpoons HPO_4^{2-}(aq) + H_2O(l)$$
$$H_2PO_4^-(aq) + H_3O^+(aq) \rightleftharpoons H_3PO_4(aq) + H_2O(l)$$

Amino acids (sub-topic B.2) also act as amphiprotic species. All 2-amino acids contain a weakly acidic carboxyl group and a weakly basic amino group. In the ionized form (a zwitterion, figure 4) the compound acts as an acid in the presence of a strong base, donating a proton. In the presence of a strong acid it acts as a base and accepts a proton.

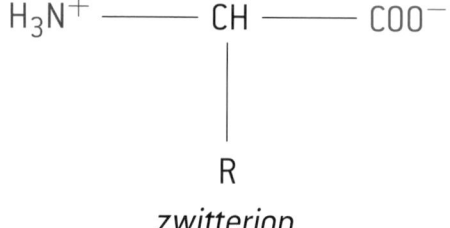

▲ Figure 4 A 2-amino acid is amphiprotic

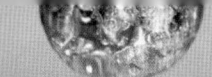

8.2 Properties of acids and bases

Understandings

→ Most acids have observable characteristic chemical reactions with reactive metals, metal oxides, metal hydroxides, hydrogencarbonates, and carbonates.

→ Salt and water are produced in exothermic neutralization reactions.

 ## Applications and skills

→ Balancing chemical equations for the reactions of acids.

→ Identification of the acid and base needed to make different salts.

→ Candidates should have experience of acid–base titrations with different indicators.

 ## Nature of science

→ Obtaining evidence for theories – observable properties of acids and bases have led to the modification of acid–base theories.

 ## Acid–base theories

For scientists, experimentation and observation provide evidence that can either support or refute the theories we have formulated to make sense of our world.

Theories about the reactions of acids and bases have been modified over time as more evidence has come from observed properties of acids and bases. Scientists analyse qualitative and quantitative data, establishing patterns and rationalizing discrepancies with the goal of defining relationships.

Properties of acids and bases

Acids and bases perform many useful functions in daily life. Caustic soda (concentrated sodium hydroxide) dissolves grease and oil deposits that can block domestic and commercial drains. Phosphoric acid is an effective rust remover, changing iron(III) oxide (Fe_2O_3, rust) into iron(III) phosphate, $FePO_4$. Ammonia is used as a general cleaner while mild acids such as vinegar are sometimes put on wasp stings, which are alkaline.

Table 1 shows some properties of acids and bases.

Acids	Bases
taste sour	taste bitter
pH < 7.0	pH > 7.0
litmus is red	litmus is blue
phenolphthalein is colourless	phenolphthalein is pink
methyl orange is red	methyl orange is yellow

▲ Table 1 General properties of acids and bases and their effects on some common indicators

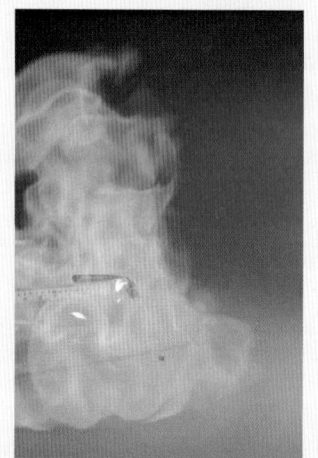

▲ Figure 1 Hydrogen gas explodes upon ignition

▲ Figure 2 Zinc reacting with hydrochloric acid

Many metal oxides act as bases in aqueous solutions whereas most non-metal oxides are acidic (sub-topic 3.2).

The reactions of acids with metals, bases, and carbonates

Most acids react with metals, metal oxides, hydroxides, hydrogencarbonates, and carbonates.

All these reactions produce a **salt**. Sodium chloride is referred to as "common salt" but this is just one example of a salt, which is a compound composed of an anion and cation.

It is important to understand how a salt is derived from an acid and a base. The following reactions illustrate the formation of a wide variety of salts.

Metals that are found above hydrogen in the activity series (sub-topic 9.1) react with acids to form a salt and hydrogen gas:

acid + metal → salt + hydrogen

$$2HCl(aq) + Zn(s) \rightarrow ZnCl_2(aq) + H_2(g)$$

$$H_2SO_4(aq) + Fe(s) \rightarrow FeSO_4(aq) + H_2(g)$$

$$2CH_3COOH(l) + 2Na(s) \rightleftharpoons 2CH_3COONa(l) + H_2(g)$$

These reactions give off hydrogen gas at different rates according to the reactivity of the metal and the strength and concentration of the acid. The salt produced depends on the acid from which it was produced, for example, magnesium chloride is a chloride salt produced in a reaction with hydrochloric acid.

The **standard enthalpy change of neutralization** is the energy change associated with the formation of 1 mol of water from the reaction between a strong acid and a strong base under standard conditions. This enthalpy change has a negative value – neutralization is an exothermic process (sub-topic 5.1):

$$H^+(aq) + OH^-(aq) \rightarrow H_2O(l)$$

The salt produced in neutralization reactions is composed of a cation from the base and an anion from the acid. Common examples of bases include metal hydroxides, metal oxides, and ammonium hydroxide, which is a weak base:

acid + base → salt + water

$$2HCl(aq) + Ca(OH)_2(aq) \rightarrow CaCl_2(aq) + 2H_2O(l)$$

$$H_2SO_4(aq) + CaO(s) \rightarrow CaSO_4(s) + H_2O(l)$$

$$CH_3COOH(aq) + NH_4OH(aq) \rightleftharpoons CH_3COONH_4(aq) + H_2O(l)$$

Calcium oxide does not react directly with aqueous acids. This base dissolves in water to create an alkaline solution of calcium hydroxide, which neutralizes the acid:

$$CaO(s) + H_2O(l) \rightarrow Ca(OH)_2(aq)$$

Calcium hydroxide is slightly soluble in water. A soluble base is called an **alkali**. Many other bases, such as iron(II) hydroxide or aluminium hydroxide, are insoluble in water.

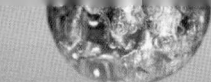

Metal carbonates and hydrogencarbonates react with acids to produce carbon dioxide and water:

acid + metal carbonate/metal hydrogencarbonate → salt + carbon dioxide + water

$$2HCl(aq) + Na_2CO_3(s) \rightarrow 2NaCl(aq) + CO_2(g) + H_2O(l)$$

$$HCl(aq) + NaHCO_3(s) \rightarrow NaCl(aq) + CO_2(g) + H_2O(l)$$

Acid–base titrations

A titration (sub-topic 1.3) is a volumetric analysis technique that involves a reaction between a substance of unknown concentration with a standardized solution (the titrant). The titrant is delivered from a burette into the solution being analysed, in small increments. The progress of the reaction can be monitored using several techniques. Data loggers combined with a pH probe can be used to collect data that can be plotted to produce a pH curve. An acid–base indicator undergoes a colour change as the titration approaches and reaches the equivalence point (sub-topic 18.3).

The colour changes of different indicators can be found in section 22 of the *Data booklet*.

Test for carbon dioxide

Passing carbon dioxide gas through limewater (calcium hydroxide) results in a cloudy (milky) suspension of insoluble calcium carbonate, $CaCO_3$:

$$Ca(OH)_2(aq) + CO_2(g) \rightarrow CaCO_3(s) + H_2O(l)$$

▲ Figure 3 Limewater provides a test for carbon dioxide

8.3 The pH scale

Understandings
→ $pH = -\log[H^+(aq)]$ and $[H^+] = 10^{-pH}$
→ A change of one pH unit represents a 10-fold change in the hydrogen ion concentration $[H^+]$.
→ pH values distinguish between acidic, neutral, and alkaline solutions.
→ The ionic product constant, $K_w = [H^+][OH^-] = 10^{-14}$ at 298 K.

Applications and skills
→ Solving problems involving pH, $[H^+]$, and $[OH^-]$.
→ Students should be familiar with the use of a pH meter and universal indicator.

Nature of science
→ Occam's razor – the pH scale is an attempt to scale the relative acidity over a wide range of H^+ concentrations into a very simple number.

Occam's razor

Scientific theories can be complex and comprehensive models of how the universe works. Derived from experimentation, observations, analysis, and hypothesis, they can be elegant and also imposing.

The principle of Occam's razor is a blueprint for the development of theories in a number of fields of knowledge. Its philosophy is that a theory should remain as simple as is possible while maintaining a high capacity for gaining understanding.

The pH scale is a very effective method of representing a continuous range of hydrogen ion concentration [H^+] as simple numbers that enable ease of interpretation for both students of science and the general population. The pH scale clearly distinguishes between acids, neutral solutions, and basic/alkaline solutions.

The pH scale

The **pH scale** is a simple and effective way of representing the concentration of hydrogen ions, [H^+] in a solution. This concentration is often very low; for example, in water [H^+] = 1.0×10^{-7} mol dm^{-3}. Comparing values of [H^+] for different substances relative to water can result in ratios that are difficult to comprehend:

$$[H^+]_{water} : [H^+]_{oven\ cleaner}$$

$$1 : 0.000\ 001$$

Scientists employ a number of different mathematical approaches to the treatment and presentation of data. The use of a logarithmic scale to display hydrogen ion concentrations results in a simple visual scale that is accessible to non-scientists and valid for scientists (figure 1).

The pH of a solution is defined by the following two expressions:

$$pH = -\log [H^+(aq)] \text{ or } pH = -\log [H_3O^+ (aq)]$$

$$[H^+] = 10^{-pH}$$

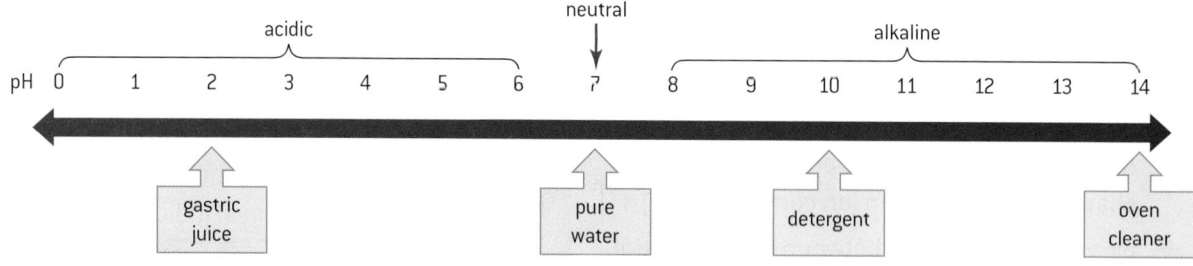

▲ Figure 1 The pH scale

pH	[H^+]
1	1×10^{-1}
2	1×10^{-2}
5	1×10^{-5}
7	1×10^{-7}
10	1×10^{-10}
14	1×10^{-14}

▲ Table 1 pH values and their corresponding hydrogen ion concentrations

As the pH scale is a logarithmic scale to base 10, a change of 1 pH unit is equivalent to a 10-fold change in the hydrogen ion concentration. A small change in the pH of a solution represents a large change in the hydrogen ion concentration (table 1). Note that the pH scale has no units.

Calculating pH

Strong acids and strong bases are assumed to completely ionize in aqueous solutions. Therefore the concentration of a strong monoprotic acid will be the same as the concentration of the hydrogen ion. A 0.1 mol dm^{-3} solution of hydrochloric acid equates to [H$^+$] = 0.1 mol dm^{-3}.

Ionization of water

The pH scale covers both the acidic and alkaline regions of aqueous systems. When considering solutions involving strong and weak bases, we need to examine the relationship between hydrogen ion and hydroxide ion concentrations.

Water can undergo **auto-ionization**, according to the following equilibrium expression (sub-topic 7.1):

$$H_2O(l) \rightleftharpoons H^+(aq) + OH^-(aq)$$

$$K_c = \frac{[H^+][OH^-]}{[H_2O]}$$

as [H$_2$O] is constant, $K_w = [H^+][OH^-] = 1.0 \times 10^{-14}$ at 298 K.

This expression is the **ion product constant** for water. In pure water,

$$[H^+] = [OH^-] = \sqrt{1.0 \times 10^{-14}} = 1.0 \times 10^{-7}$$

Worked examples: calculating pH

Example 1

A solution of fresh milk has a pH of 6.70. Calculate [H$^+$] and [OH$^-$].

Solution

$$[H^+] = 10^{-pH} = 10^{-6.70} = 2.0 \times 10^{-7} \text{ mol dm}^{-3}$$

$$K_w = [H^+][OH^-] = 1.0 \times 10^{-14}$$

$$[OH^-] = \frac{1.0 \times 10^{-14}}{[H^+]} = \frac{1.0 \times 10^{-14}}{2.0 \times 10^{-7}}$$

$$= 5.0 \times 10^{-8} \text{ mol dm}^{-3}$$

Example 2

Calculate the pH of a 1.0×10^{-2} mol dm^{-3} solution of sodium hydroxide.

Solution

Sodium hydroxide is a strong base that completely ionizes in water:

$$NaOH (aq) \rightarrow Na^+(aq) + OH^-(aq)$$

$$[OH^-] = 1.0 \times 10^{-2} \text{ mol dm}^{-3}$$

$$K_w = [H^+][OH^-] = 1.0 \times 10^{-14}$$

$$[H^+] = \frac{1.0 \times 10^{-14}}{[OH^-]}$$

$$[H^+] = \frac{1.0 \times 10^{-14}}{1.0 \times 10^{-2}} = 1.0 \times 10^{-12} \text{ mol dm}^{-3}$$

$$pH = -\log [H^+(aq)]$$

$$= -\log (1.0 \times 10^{-12}) = 12.00$$

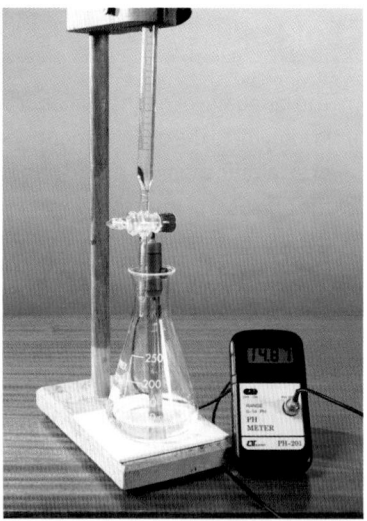

▲ Figure 2 A pH probe can be used to collect data during an acid–base titration

pH and acid–base titrations

The analytical technique of titration (sub-topic 1.3) has been used in the laboratory for the past 200 years. Traditionally a titration is monitored by the addition of an indicator (sub-topic 18.3). Plotting a pH curve illustrates the progress of an acid–base titration and enables analysis of characteristic features of the titration. These curves are generated from data that can be collected using a pH probe and its associated software (figure 2).

8.4 Strong and weak acids and bases

Understandings

→ Strong and weak acids and bases differ in the extent of ionization.

→ Strong acids and bases of equal concentrations have higher conductivities than weak acids and bases.

→ A strong acid is a good proton donor and has a weak conjugate base.

→ A strong base is a good proton acceptor and has a weak conjugate acid.

Applications and skills

→ Distinction between strong and weak acids and bases in terms of the rates of their reactions with metals, metal oxides, metal hydroxides, metal hydrogencarbonates, and metal carbonates, and their electrical conductivities for solutions of equal concentrations.

Nature of science

→ Improved instrumentation – the use of advanced analytical techniques has allowed the relative strengths of different acids and bases to be quantified.

→ Looking for trends and discrepancies – patterns and anomalies in relative strengths of acids and bases can be explained at the molecular level.

→ The outcomes of experiments or models may be used as further evidence for a claim – data for a particular type of reaction supports the idea that weak acids exist in equilibrium.

Predictions, patterns, and anomalies

Advances in computing power, the development of analytical techniques (including sensors such as pH meters), and networking between research institutes provide a wealth of data. Scientists analyse this data looking for trends and discrepancies, patterns and anomalies to gain greater understanding. Quantitative data on the relative strengths of acids and bases allows predictions about their chemical behaviour to be made with high reliability. Patterns and anomalies in the relative strengths of different acids and bases allow explanations of their chemical behaviour to be developed at the molecular level. Empirical evidence also supports the idea that weak acids exist in equilibrium.

Science recognizes that there are limitations in all measurements; however, modern instrumentation provides data that is close to certainty. Science is an inherently human endeavour which is strengthened by advances in technology.

Strengths of acids and bases

The strength of an acid or base depends on the degree to which it ionizes or dissociates in water. A **strong acid** is an effective proton donor that is assumed to completely dissociate in water. Examples include hydrochloric acid, HCl, sulfuric acid, H_2SO_4, and nitric acid, HNO_3:

$$HCl(aq) + H_2O(l) \rightarrow H_3O^+(aq) + Cl^-(aq)$$

$$H_2SO_4(aq) + H_2O(l) \rightarrow H_3O^+(aq) + HSO_4^-(aq)$$

$$HNO_3(aq) + H_2O(l) \rightarrow H_3O^+(aq) + NO_3^-(aq)$$

These reactions are represented by chemical equations that are assumed to go to completion. The conjugate base of a strong acid is a very weak base. For hydrochloric acid, the conjugate base is the chloride ion, Cl^- which has almost no affinity for a proton in aqueous solution.

A **weak acid** dissociates only partially in water; it is a poor proton donor. The dissociation of a weak acid is a reversible reaction that reaches equilibrium (sub-topic 7.1). At equilibrium only a small proportion of the acid molecules have dissociated. The conjugate base of a weak acid has a higher affinity for a proton than does the conjugate base of a strong acid.

$$CH_3COOH(aq) + H_2O(l) \rightleftharpoons H_3O^+(aq) + CH_3COO^-(aq)$$

$$H_2CO_3(aq) + H_2O(l) \rightleftharpoons H_3O^+(aq) + HCO_3^-(aq)$$

In the reactions of both strong and weak acids, water is acting as a base, accepting a proton.

The terms "strong" and "weak" when applied to acids and bases are quite distinct from "concentrated" and "dilute". Table 1 gives some examples.

	Concentrated	Dilute
Strong	6 mol dm^{-3} HCl	0.5 mol dm^{-3} HNO$_3$
Weak	10 mol dm^{-3} CH$_3$COOH	0.1 mol dm^{-3} H$_2$CO$_3$

▲ Table 1 Some examples of strong, weak, dilute, and concentrated acid solutions

Sulfuric acid is a strong acid. However, the first ionization is complete but the second dissociation is incomplete so an equilibrium sign is used:

$$H_2SO_4(aq) + H_2O(l) \rightarrow HSO_4^-(aq) + H_3O^+(aq)$$

$$HSO_4^-(aq) + H_2O(l) \rightleftharpoons SO_4^{2-}(aq) + H_3O^+(aq)$$

Study tip

The terms "ionization" and "dissociation" are interchangeable and are both equally acceptable in examination answers.

Amphoteric: Species that behaves both as an acid and a base eg aluminium hydroxide:

Acting as a base:

$Al(OH)_3(s) + 3HCl(aq) \rightarrow$
$\qquad AlCl_3(aq) + 3H_2O(l)$

Acting as an acid:

$Al(OH)_3(s) + NaOH(aq) \rightarrow$
$\qquad Na[Al(OH)_4](aq)$

Amphiprotic: One type of amphoteric species are amphiprotic molecules. These can act as Brønsted-Lowry acids (proton donors) or Brønsted-Lowry bases (proton acceptors) eg water and amino acids.

▲ Figure 1 Testing the conductivity of a strong acid. The lamp gives a qualitative reading of current; an ammeter would provide quantitative readings

A **strong base** also completely dissociates in water. The group 1 metal hydroxides are all soluble in water and are good examples of strong bases:

$$NaOH(aq) \rightarrow Na^+(aq) + OH^-(aq)$$

$$KOH(aq) \rightarrow K^+(aq) + OH^-(aq)$$

A metal hydroxide does not act as a Brønsted–Lowry base because it does not have the capacity to accept a proton. However, in solution the hydroxide ion acts as a Brønsted–Lowry base, accepting a proton:

$$OH^-(aq) + H_3O^+(aq) \rightarrow 2H_2O(l)$$

Ammonia is an example of a **weak base**. In the reaction with water, ammonia accepts a proton and effectively undergoes ionization.

$$NH_3(aq) + H_2O(l) \rightleftharpoons NH_4^+(aq) + OH^-(aq)$$

In this reaction water displays its amphiprotic nature by acting as a Brønsted–Lowry acid, donating a proton.

Experimental determination of the strength of acids and bases

A number of techniques can be used to compare acids and bases of *equal concentration*, so that they can be assigned an order of strength relative to one another.

Conductivity

All acids and bases dissociate to a degree in water and create ions. The conductivity of an aqueous solution depends on the concentration of ions present. This can be measured in a simple experiment using a power pack and graphite electrodes connected to an ammeter (figure 1). The voltage applied must be identical for each solution so that any difference in current passing through aqueous solutions of different acids or bases reflects the concentration of ions.

Strong acids and bases are strong electrolytes (sub-topic 19.1), so they display higher conductivity than weak acids and bases. For example, a comparison of the conductivity of equimolar solutions of hydrochloric acid and ethanoic acid would demonstrate that the hydrochloric acid gives a higher ammeter reading and so has a higher degree of dissociation than ethanoic acid, which is a weak acid.

Energy changes on neutralization

Neutralization occurs when an acid and a base react together. The reaction is exothermic ($\Delta H < 0$, sub-topic 5.1). The enthalpy change of neutralization for a strong acid is almost identical to that for a weak acid. The neutralization reaction removes ionized species from the dissociation reaction, so driving the reaction to completion.

A strong acid or base is completely dissociated in solution so the only enthalpy consideration in this reaction is the exothermic formation of water from hydrogen and hydroxide ions.

For a neutralization reaction involving a weak acid or base there are other enthalpy considerations. These species exist predominantly in their undissociated forms in aqueous solution. The ionization of a weak acid or base is mildly endothermic. Therefore the enthalpy of neutralization for a strong base–weak acid reaction will be slightly less exothermic than that for a strong base–strong acid reaction. The weaker the acid, the more endothermic the dissociation reaction becomes and thus the lower the enthalpy change of neutralization.

$$HCl(aq) + NaOH(aq) \rightarrow NaCl(aq) + H_2O(l)$$
$$\Delta H^{\ominus}_{neutralization} = -57.1 \text{ kJ mol}^{-1}$$

$$HCl(aq) + NH_3(aq) \rightleftharpoons NH_4Cl(aq)$$
$$\Delta H^{\ominus}_{neutralization} = -53.4 \text{ kJ mol}^{-1}$$

$$CH_3COOH(aq) + NaOH(aq) \rightarrow CH_3COONa(aq) + H_2O(l)$$
$$\Delta H^{\ominus}_{neutralization} = -56.1 \text{ kJ mol}^{-1}$$

$$CH_3COOH(aq) + NH_3(aq) \rightleftharpoons CH_3COONH_4(aq)$$
$$\Delta H^{\ominus}_{neutralization} = -50.4 \text{ kJ mol}^{-1}$$

Monitoring the rate of a reaction

The reactions of strong and weak acids with metals, metal hydrogencarbonates, and metal carbonates all produce a gas. The rate of the reaction (topic 6) can be determined qualitatively through observation (figure 2) followed by analysis and quantitatively by monitoring the rate at which gas is evolved through loss of mass. A series of such experiments enables a strong acid to be distinguished from a weak acid.

The individual reactions of 1 mol dm^{-3} solutions of hydrochloric and ethanoic acids with zinc granules, powdered sodium carbonate, and sodium hydrogencarbonate demonstrate the different rates of reaction shown by strong and weak acids.

Performing the reactions on an electronic balance (figure 3) enables mass data to be collected over time. Graphing these results illustrates the differences in the initial rate of reaction (sub-topic 16.1).

▲ Figure 2 Observation provides qualitative data for a reaction that evolves a gas

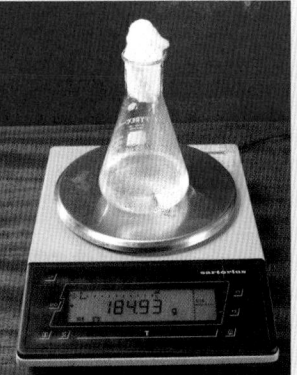
▲ Figure 3 Monitoring the loss of mass provides quantitative data for a reaction that evolves a gas

203

8.5 Acid deposition

Understandings

→ Rain is naturally acidic because of dissolved CO_2. Acid deposition has a pH below 5.0.

→ Acid deposition is formed when nitrogen or sulfur oxides dissolve in water to form HNO_3, HNO_2, H_2SO_4 and H_2SO_3.

→ Sources of the oxides of sulfur and nitrogen and the effects of acid deposition should be covered.

Applications and skills

→ Balancing the equations that describe the combustion of sulfur and nitrogen to their oxides and the subsequent formation of H_2SO_3, H_2SO_4, HNO_2 and HNO_3.

→ Distinction between the pre-combustion and post-combustion methods of reducing sulfur oxides emissions.

→ Deduction of acid deposition equations for acid deposition with reactive metals and carbonates.

Nature of science

→ Risks and problems – oxides of metals and non-metals can be characterized by their acid–base properties.

→ Acid deposition is a topic that can be discussed from different perspectives.

→ Chemistry allows us to understand and reduce the environmental impact of human activities.

Acid deposition

Acid deposition is the process by which acid-forming pollutants are deposited on the Earth's surface. Increased industrialization and economic development in many parts of the world have led to rapidly increasing emissions of the nitrogen and sulfur oxides that cause **acid rain**, the most prevalent form of acid deposition.

Acid deposition affects the environment in many ways. These include: deforestation, the leaching of minerals from soils leading to elevated acid levels in lakes and rivers, the uptake of toxic minerals from soil by plants, reduction in the pH of lake and river systems, increased uptake of toxic metals by shellfish and other marine life which can affect the fishing industry and ultimately people's health, and corrosive effects on marble, limestone, and metal buildings, bridges, and vehicles.

Acid rain

Pure water has a pH of 7.0. Rainwater is naturally acidic due to the presence of dissolved carbon dioxide which forms weak carbonic acid, H_2CO_3. A typical pH value of rainwater is 5.6.

$$CO_2(g) + H_2O(l) \rightleftharpoons H_2CO_3(aq)$$
$$H_2CO_3(aq) \rightleftharpoons H^+(aq) + HCO_3^-(aq)$$
$$HCO_3^-(aq) \rightleftharpoons H^+(aq) + CO_3^{2-}(aq)$$

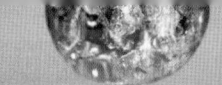

Acid rain has a pH less than 5.6. The major pollutants that cause acid deposition are sulfur dioxide, SO_2 and nitrogen oxides, NO and NO_2. These are products of natural occurrences such as volcanic eruptions and the decomposition of vegetation, as well as man-made primary pollutants from the combustion of fossil fuels containing high levels of sulfur impurities (option C.2). Acid rain results principally from the formation of two strong acids, nitric acid, HNO_3, and sulfuric acid, H_2SO_4 and can be considered as a major global environmental problem.

For example, at high temperature in the internal combustion engine of a car or a jet engine, nitrogen gas reacts with oxygen gas to form the oxide, nitrogen(II) oxide (nitrogen monoxide):

High Temperature

$$N_2(g) + O_2(g) \rightarrow 2NO(g)$$

On reaction with oxygen, the oxide, nitrogen(IV) oxide (nitrogen dioxide) can form:

$$2NO(g) + O_2(g) \rightarrow 2NO_2(g)$$

Nitrogen(IV) oxide causes the brown colour of smog, a common type of air pollution which is often observed in cities such as Los Angeles in the USA and Mexico City.

The reaction between water and nitrogen(IV) oxide produces nitric acid and nitrous acid:

$$2NO_2(g) + H_2O(l) \rightarrow HNO_3(aq) + HNO_2(aq)$$

Another oxide of nitrogen, NO, is easily oxidized to nitrogen(IV) oxide by atmospheric oxygen:

$$2NO(g) + O_2(g) \rightarrow 2NO_2(g)$$

Nitrous acid can be also oxidized by atmospheric oxygen:

$$2HNO_2(aq) + O_2(g) \rightarrow 2HNO_3(g)$$

Therefore, all oxides of nitrogen eventually produce nitric acid, HNO_3.

Sulfur dioxide combines with water to form sulfurous acid:

$$SO_2(g) + H_2O(l) \rightleftharpoons H_2SO_3(aq)$$

$$H_2SO_3(aq) + H_2O(l) \rightleftharpoons HSO_3^-(aq) + H_3O^+(aq)$$

Some coal can contain almost 3% sulfur. On combustion, sulfur dioxide forms:

$$S(s) + O_2(g) \rightarrow SO_2(g)$$

On subsequent reaction with oxygen in the atmosphere, the oxide, sulfur trioxide is generated:

$$2SO_2(g) + O_2(g) \rightleftharpoons 2SO_3(g)$$

Sulfur trioxide can then react with rain in the atmosphere, to form sulfuric acid:

$$SO_3(g) + H_2O(l) \rightarrow H_2SO_4(aq)$$

Acid deposition: A global problem

Acid deposition, a secondary pollutant, can take many different forms including rain, snow, fog and dry dust. The components of acid deposition (the primary pollutants) may be generated in one country and depending on climate patterns may be deposited in neighbouring countries or even different continents. There are no boundaries for acid deposition. For example in Europe industrial conurbations in countries such as Germany and the UK may act as the source of acid rain but due to factors such as prevailing winds, acid deposition may occur in Scandinavian countries further north such as Norway and Sweden. Hence the effects of acid rain may occur away from the actual source leading to widespread deforestation and pollution of lakes and river systems. National and regional environmental protection agencies throughout the world collaborate in an effort to better understand and control acid deposition. The US Environmental Protection Agency and the Acid Deposition Monitoring Network in East Asia (EANET) websites provide data that can be used in the discussion of secondary pollutants and their political implications.

Pre- and post-combustion technologies

Pre-combustion methods to reduce sulfur emissions refer to techniques used on fuels before their combustion. Physical cleaning or mineral beneficiation involves crushing coal, followed by flotation that reduces the amounts of sulfur and other impurities. Combinations of different pre-combustion methods result in the removal of up to 80–90% of inorganic sulfur.

Post-combustion methods focus on several complementary technologies to remove sulfur dioxide, nitrogen oxides, heavy metals and dioxins from the combustion gases before they are released into the atmosphere. For example, calcium oxide or lime will react with sulfur dioxide and remove it from flue gases:

$$CaO(s) + SO_2(g) \rightarrow CaSO_3(s)$$

 ### The effects of acid rain on buildings

Limestone and marble are building materials that are commonly used in monuments and buildings of significant cultural importance throughout the world. Both contain calcium carbonate, differing only in their structural composition.

When calcium carbonate is exposed to acid rain, a neutralization reaction occurs and the building is gradually eroded, causing significant damage.

$$CaCO_3(s) + H_2SO_4(aq) \rightarrow CaSO_4(s) + CO_2(g) + H_2O(l)$$

 ### The role of chemists in studying acid deposition

Science, and chemistry in particular, enables us to understand the ways in which acid deposition occurs and the extent of its impact on the environment. The study of this phenomenon is wide ranging and cross disciplinary (IB Geography option G: Urban environments; HL – Global interactions – environmental change; and Environmental systems and societies sub-topic 5.8: Acid deposition). The interest in this subject reflects the concern of the wider community about acid deposition. Scientific research provides evidence that informs discussions which lead to decisions on how to reduce the impact of acid deposition on the environment. Political and economic cooperation between nations is also needed to achieve success in the control and reduction of acid deposition.

Questions

1　Consider the equilibrium below:

$$CH_3CH_2COOH(aq) + H_2O(l) \rightleftharpoons CH_3CH_2COO^-(aq) + H_3O^+(aq)$$

Which species represent a conjugate acid–base pair?

A.　CH_3CH_2COOH and H_2O

B.　H_2O and $CH_3CH_2COO^-$

C.　H_3O^+ and H_2O

D.　$CH_3CH_2COO^-$ and H_3O^+　　　[1]

IB, May 2011

2　Which is not a conjugate acid–base pair?

A.　HNO_3 and NO_3^-

B.　CH_3COOH and CH_3COO^-

C.　H_3O^+ and OH^-

D.　HSO_4^- and SO_4^{2-}　　　[1]

IB, May 2011

3　Which species behave as Brønsted–Lowry acids in the following reversible reaction?

$$H_2PO_4^-(aq) + CN^-(aq) \rightleftharpoons HCN(aq) + HPO_4^{2-}(aq)$$

A.　HCN and CN^-

B.　HCN and HPO_4^{2-}

C.　$H_2PO_4^-$ and HPO_4^{2-}

D.　HCN and $H_2PO_4^-$　　　[1]

IB, May 2010

4　Explain, using the Brønsted–Lowry theory, how water can act either as an acid or a base. In each case identify the conjugate acid or base formed.

IB, May 2011

5　Which of the following is/are formed when a metal oxide reacts with a dilute acid?

I)　A metal salt

II)　Water

III)　Hydrogen gas

A.　I only

B.　I and II only

C.　II and III only

D.　I, II and III　　　[1]

IB, November 2003

6　An aqueous solution of which of the following reacts with magnesium metal?

A.　Ammonia

B.　Hydrogen chloride

C.　Potassium hydroxide

D.　Sodium hydrogencarbonate　　　[1]

IB, May 2003

7　Which property is characteristic of acids in aqueous solution?

A.　Acids react with ammonia solution to produce hydrogen gas and a salt.

B.　Acids react with metal oxides to produce oxygen gas, a salt, and water.

C.　Acids react with reactive metals to produce hydrogen gas and a salt.

D.　Acids react with metal carbonates to produce hydrogen gas, a salt, and water.　　　[1]

IB, May 2010

8　A solution of acid A has a pH of 1 and a solution of acid B has a pH of 2. Which statement must be correct?

A.　Acid A is stronger than acid B

B.　[A] > [B]

C.　The concentration of H^+ ions in A is higher than in B

D.　The concentration of H^+ ions in B is twice the concentration of H^+ ions in A　　　[1]

IB, November 2010

9　100 cm^3 of a NaOH solution of pH 12 is mixed with 900 cm^3 of water. What is the pH of the resulting solution?

A.　1

B.　3

C.　11

D.　13　　　[1]

IB, May 2009

10 Black coffee has a pH of 5 and toothpaste has a pH of 8. Identify which is more acidic and deduce how many times the $[H^+]$ is greater in the more acidic product. [2]

IB, May 2011

11 Determine the pH of the solution resulting when 100 cm^3 of 0.50 mol dm^{-3} HCl(aq) is mixed with 200 cm^3 of 0.10 mol dm^{-3} NaOH(aq). [5]

IB, May 2011

12 Which 0.10 mol dm^{-3} solution would have the highest conductivity?

A. HCl

B. NH_3

C. CH_3COOH

D. H_2CO_3 [1]

IB, May 2011

13 A student has equal volumes of 1.0 mol dm^{-3} sodium hydroxide and ammonia solutions.

Which statement about the solutions is correct?

A. Sodium hydroxide has a lower electrical conductivity than ammonia.

B. Sodium hydroxide has a higher hydrogen ion concentration than ammonia.

C. Sodium hydroxide has a higher pH than ammonia.

D. Sodium hydroxide has a higher hydroxide ion concentration than ammonia. [1]

IB, May 2010

14 Which list contains only strong acids?

A. CH_3COOH, H_2CO_3, H_3PO_4

B. HCl, HNO_3, H_2CO_3

C. CH_3COOH, HNO_3, H_2SO_4

D. HCl, HNO_3, H_2SO_4 [1]

IB, May 2009

15 Which methods will distinguish between equimolar solutions of a strong base and a strong acid?

  I) Add magnesium to each solution and look for the formation of gas bubbles.

  II) Add aqueous sodium hydroxide to each solution and measure the temperature change.

  III) Use each solution in a circuit with a battery and lamp and see how bright the lamp glows.

A. I and II only

B. I and III only

C. II and III only

D. I, II, and III [1]

IB, Specimen paper

16 Describe **two** different properties that could be used to distinguish between a 1.00 mol dm^{-3} solution of a strong monoprotic acid and a 1.00 mol dm^{-3} solution of a weak monoprotic acid. [2]

IB, May 2011

17 Ethanoic acid, CH_3COOH, is a weak acid.

a) Define the term *weak acid* and state the equation for the reaction of ethanoic acid with water. [2]

b) Vinegar, which contains ethanoic acid, can be used to clean deposits of calcium carbonate from the elements of electric kettles. State the equation for the reaction of ethanoic acid with calcium carbonate. [2]

IB, May 2009

18 The equations of two acid–base reactions are given below.

Reaction A

$NH_3(aq) + H_2O(l) \rightleftharpoons \underline{NH_4^+(aq)} + OH^-(aq)$

The reaction mixture in A consists mainly of reactants because the equilibrium lies to the left.

Reaction B

$NH_2^-(aq) + H_2O(l) \rightleftharpoons \underline{NH_3(aq)} + OH^-(aq)$

The reaction mixture in **B** consists mainly of products because the equilibrium lies to the right.

a) For each of the reactions **A** and **B**, deduce whether water is acting as an acid or a base and explain your answer. [2]

b) In reaction **B**, identify the stronger base, NH_2^- or OH^- and explain your answer. [2]

c) In reactions **A** and **B**, identify the stronger acid, NH_4^+ or NH_3 (underlined) and explain your answer. [2]

IB, November 2009

9 REDOX PROCESSES

Introduction

Redox reactions based on reduction and oxidation lie at the centre of many everyday processes (both chemical and biochemical) and have numerous applications. In this chapter we explore the different ways both oxidation and reduction can be considered and introduce the idea of oxidation state, which can be a useful tool in solving redox titration problems in volumetric chemistry. We will also examine the energy conversion between chemical and electrical energy, which is the basis of electrochemistry. Both voltaic and electrolytic cells will be introduced.

9.1 Oxidation and reduction

Understandings

→ Oxidation and reduction can be considered in terms of oxygen gain/hydrogen loss, electron transfer, or change in oxidation number.

→ An oxidizing agent is reduced and a reducing agent is oxidized.

→ Variable oxidation numbers exist for transition metals and for most main-group non-metals.

→ The activity series ranks metals according to the ease with which they undergo oxidation.

→ The Winkler method can be used to measure biochemical oxygen demand (BOD), used as a measure of the degree of pollution in a water sample.

Applications and skills

→ Deduction of the oxidation states of an atom in an ion or a compound.

→ Deduction of the name of a transition metal compound from a given formula, applying oxidation numbers represented by Roman numerals.

→ Identification of the species oxidized and reduced and the oxidizing and reducing agents in redox reactions.

→ Deduction of redox reactions using half-equations in acidic or neutral solutions.

→ Deduction of the feasibility of a redox reaction from the activity series or reaction data.

→ Solution of a range of redox titration problems.

→ Application of the Winkler method to calculate BOD.

Nature of science

→ How evidence is used – changes in the definition of oxidation and reduction from one involving specific elements (oxygen and hydrogen), to one involving electron transfer, to one invoking oxidation numbers is a good example of the way that scientists broaden similarities to general principles.

▲ Figure 1 The Statue of Liberty, New York, USA. It took many years after restoration in 1986 for the copper to be oxidized and for the statue to reform the green patina

Redox reactions

Three of the main types of reaction that occur in chemistry are:

- acid–base reactions
- precipitation reactions
- redox reactions.

A **redox reaction** involves two processes, **reduction and oxidation**. Both reduction and oxidation can be considered in a number of different ways, and all three descriptions have merit in their own right. The different ways of describing these processes are:

- in terms of specific elements – oxygen and hydrogen
- in terms of electron transfer
- in terms of oxidation number.

Oxidation: Combining with oxygen

At the simplest level oxidation can be considered as a reaction in which a substance combines with oxygen. Examples include:

$$2Mg(s) + O_2(g) \rightarrow 2MgO(s)$$

$$2CH_3OH(l) + 3O_2(g) \rightarrow 2CO_2(g) + 4H_2O(l)$$

$$4Fe(s) + 3O_2(g) \rightarrow 2Fe_2O_3(s)$$

$Fe_2O_3(s)$, iron(III) oxide, is rust. Rusting is an example of the process of **corrosion**.

 Global examples of corrosion

The deterioration of metals caused by electrochemical processes (redox reactions) is described as corrosion.

Oxidation of iron and copper: The Statue of Liberty

The Statue of Liberty in New York, USA was restored in 1986 as it was found that corrosion had occurred between the wrought iron structural support and the outer copper skin. Shellac (a resin secreted by the female lac bug found on trees in Thailand and India) was originally inserted as an insulator between the copper and the iron but over time the insulation had failed and the iron supports rusted. As part of the renovation work a different insulator was

used, the addition polymer polytetrafluoroethene (PTFE):

$$\left[\begin{matrix} F & F \\ | & | \\ C - C \\ | & | \\ F & F \end{matrix}\right]_n$$

PTFE is commonly known by its brand name Teflon®, which is also also used as a non-stick coating for cooking pans.

The copper in the Statue of Liberty oxidized to form an outer green coating called the patina. When the restoration work on the Statue of Liberty was completed the statue was brown in colour. It has taken many years for it to oxidize fully and reform the patina (figure 1).

▲ Figure 2 Howrah bridge in Calcutta, India

Steel and Paan: The Howrah bridge

Another interesting example of corrosion is the Howrah bridge in Calcutta, India (figure 2). In 2010 the bridge was found to be undergoing corrosion caused by an unusual agent, Paan. Paan is a mixture of betel leaf, areca nut, and slaked lime (calcium hydroxide) chewed by millions of people in India. The leaf itself is not harmful, but other substances added to it such as tobacco (shada/zarda), betel nut, and lime can damage the health of a person. Chewing Paan with tobacco can cause mouth cancer. Scientists have found that the lime in Paan is highly corrosive and this is the primary cause of the corrosion of the bridge.

> **Activity**
>
> Can you think of some initiatives that governments and city councils can adopt to inform the general public of the health effects related to chewing Paan in countries where this is particularly prevalent?

In contrast, Paan is also used as a post-meal digestive stimulant, aphrodisiac, and nerve tonic in India and recent research has suggested that many of these properties may be due to the antioxidant nature of Paan. An **antioxidant** is a substance that delays the onset of oxidation or slows down the rate at which oxidation occurs.

Reduction: Removal of oxygen or addition of hydrogen

Reduction may be considered as the removal of oxygen, for example:

$$NiO(s) + C(s) \rightarrow Ni(s) + CO(g)$$

In this reaction nickel(II) oxide is reduced by carbon to give metallic nickel.

Reduction may also be considered as the addition of hydrogen. An example of such a reaction is:

$$WO_3(s) + 3H_2(g) \rightarrow W(s) + 3H_2O(g)$$

This mirrors the previous interpretation of reduction, as oxygen is removed from tungsten(VI) oxide in the process.

Oxidation and reduction in terms of electron transfer

In terms of electron transfer, oxidation and reduction can be defined as follows:

- **Oxidation** involves the loss of electrons and **reduction** involves the gain of electrons.

A useful mnemonic for remembering this is **OILRIG**:

    <u>O</u>xidation <u>I</u>s <u>L</u>oss of electrons

    <u>R</u>eduction <u>I</u>s <u>G</u>ain of electrons

Let us return to the reaction of magnesium metal with oxygen gas to form magnesium oxide.

Magnesium is a member of group 2 (alkaline earth metals), and has the electron configuration $[Ne]3s^2$. It loses its two valence electrons to attain the noble gas core configuration, $[Ne]$:

    $Mg \rightarrow Mg^{2+} + 2e^-$

    $[Ne]3s^2$ $[Ne]$

Oxygen is a member of group 16 (chalcogens) and is a non-metal. It has the electron configuration $[He]2s^22p^4$ and gains two electrons to attain the noble gas configuration $[Ne]$:

    $O + 2e^- \rightarrow O^{2-}$

    $[He]2s^22p^4$ $[Ne]$ or $[He]2s^22p^6$

Hence magnesium is oxidized (loses electrons) and oxygen is reduced (gains electrons).

The overall reaction is:

$$2Mg(s) + O_2(g) \rightarrow 2MgO(s)$$

We can consider the reaction as being two processes:

$$2Mg \rightarrow 2Mg^{2+} + 4e^-$$

$$O_2 + 4e^- \rightarrow 2O^{2-}$$

Considering redox processes in terms of electron transfer is a common approach; however this interpretation must be applied with caution. For example, complete combustion of solid carbon (eg in the form of coal) in oxygen yields carbon dioxide:

$$C(s) + O_2(g) \rightarrow CO_2(g)$$

However, carbon dioxide is molecular, with covalent bonds, so no ionic bonds are formed. We cannot describe this combustion reaction as a redox process in terms of electron transfer as in theory no electrons are lost or gained and carbon dioxide is a neutral species! The original definition of oxidation as the addition of oxygen is more appropriate here as carbon is clearly oxidized in this process.

An application of redox chemistry from optometry

Optometrists often prescribe glasses with photochromic lenses. These lenses darken in the presence of ultraviolet light (from sunlight); this change is based on a redox reaction.

Ordinary glass is composed of silicates while photochromic lenses contain copper(I) chloride, CuCl, and silver chloride, AgCl.

The chloride ions are oxidized to chlorine atoms on exposure to ultraviolet light ($h\nu$).

$$Cl^- \xrightarrow{h\nu} Cl + e^-$$

Electron transfer then takes place causing the silver cation to be reduced to metallic silver atoms.

$$Ag^+ + e^- \rightarrow Ag$$

The silver atoms inhibit the transmittance of light, making the lenses turn dark.

The darkening process is reversed by copper(I) chloride allowing the lenses to become transparent again.

When the lenses are removed from the light, the following reaction takes place:

$$Cu^+ + Cl \rightarrow Cu^{2+} + Cl^-$$

The chlorine atoms formed by the exposure to light are reduced by the Cu^+ ions. The Cu^+ ions are oxidized to Cu^{2+} ions. These Cu^{2+} ions then oxidize silver atoms to Ag^+ ions:

$$Cu^{2+} + Ag \rightarrow Cu^+ + Ag^+$$

The lenses then become transparent again and the silver and chlorine atoms return to the initial Ag^+ and Cl^- species.

▲ Figure 3 Photochromic lenses

Electron book-keeping

Chemists have developed an electron book-keeping model for redox reactions which can be used to track the number of electrons in reactants and products during a chemical process. The development of the definition of oxidation and reduction from a definition involving specific elements (oxygen and hydrogen), to one involving electron transfer, to one invoking oxidation states is a good example of the way that scientists broaden similarities to general principles.

Oxidation and reduction in terms of oxidation states

The **oxidation state** is the apparent charge of an atom in a free element, a molecule, or an ion.

In terms of oxidation state:

- **Oxidation** describes a process in which the oxidation state increases and **reduction** describes a process in which the oxidation state decreases.

Rules for assigning oxidation states

1 The oxidation state of an atom in a free element is 0; for example, S_8, O_2, P_4, and Na all have atoms with an oxidation state of 0.

2 Group 1 metals always have a +1 oxidation state in their ions and compounds.

Group 2 elements always have a +2 oxidation state in their ions and compounds.

Aluminium, which is a member of group 3, has an oxidation state of +3 in the majority of its compounds.

3 The oxidation state of hydrogen is +1 when hydrogen is bonded to a non-metal, such as in HCl and HNO_3. However, when hydrogen is bonded to a metal, for example in a metal hydride such as NaH, the oxidation number of hydrogen is −1.

4 The oxidation state of oxygen is usually −2, such as in H_2O and H_2SO_4. The main exception is in a peroxide (a species with an –O–O– linkage); here the oxidation state of oxygen is −1. A typical example of such a compound is H_2O_2, hydrogen peroxide.

5 The oxidation state of fluorine is −1 in all its compounds, for example HF, OF_2, and LiF. For the other group 17 halogen elements the oxidation state is usually −1 in binary compounds (HI, NaCl, KBr) but in combination with oxygen in oxoanions and oxoacids the oxidation state is positive (for example, in $HClO_4$ chlorine has a +7 oxidation state).

6 In a neutral molecule the sum of the oxidation states of all the atoms is zero. In a polyatomic ion the sum of the oxidation states of all the atoms equals the overall charge of the ion. For example, in NH_3 the oxidation state of nitrogen is −3 and hydrogen is +1; the sum of the oxidation states is $-3 + (3 \times +1) = 0$, which equals the net charge on the ammonia molecule.

In the ammonium cation, NH_4^+, the individual oxidation states of nitrogen and hydrogen are the same as in NH_3 and the sum of the oxidation states now is $-3 + (4 \times +1) = +1$ which equals the net charge on the ammonium cation.

 # An application of redox chemistry at the hair salon

Proteins have a number of functions, one of which involves having a structural role in the body. Proteins are polymers composed of monomeric units called **amino acids** (see sub-topic B.2). The protein molecules in hair contain –SH thiol groups, and hydrogen peroxide can oxidize these to sulfonic acid groups, $-SO_3H$. This oxidation of the thiol groups changes the structure of the proteins and hair can become more brittle.

Activity

Do people with bleached hair use particular conditioners? Find out what a suitable conditioner might be for heavily bleached hair. What might its chemical components be?

International directives

In the European Union (EU) the use of hydrogen peroxide in hair, skin, and oral hygiene products is restricted to maximum concentrations of 12%, 4%, and 0.1%, respectively. As stipulated by the EU Cosmetics Directive such products must be labelled: "*Contains hydrogen peroxide. Avoid contact with eyes. Rinse immediately if product comes in contact with them.*"

The Directive states that when using hair products containing hydrogen peroxide gloves should be worn.

Difference between oxidation number and oxidation state

"Oxidation number" and "oxidation state" are used interchangeably in many textbooks.

Note however IUPAC recommends the use of Roman numerals for oxidation numbers. Interestingly, oxidation states in some compounds may not have integer values; they may be fractional, for example the oxidation state of oxygen in the superoxide anions, O_2^- is $-\frac{1}{2}$.

Variable oxidation states

As mentioned in the rules for assigning oxidation states above, although many elements have fixed oxidation states, in their ions and compounds, such as the group 1 alkali metals (eg +1 for Na) and the group 2 alkaline earth metals (eg +2 for Ca), variable oxidation states exist for many main-group non-metals and in particular for most of the **transition elements** (also called the **transition metals**). Indeed variable oxidation states are a characteristic property of the transition metals. The range of different oxidation states for the d-block elements is shown in figure 4, which is given in section 14 of the *Data booklet*. IUPAC describes transition elements as elements whose atoms have an incomplete d-subshell or which can give rise to cations with an incomplete d-subshell. In the first-row d-block elements, the transition elements are Sc to Cu inclusive (but not Zn, which is explained in topic 13).

Sc	Ti	V	Cr	Mn	Fe	Co	Ni	Cu	Zn
		+1	+1	+1	+1	+1	+1	+1	
+2	+2	+2	+2	+2	+2	+2	+2	+2	+2
+3	+3	+3	+3	+3	+3	+3	+3	+3	
	+4	+4	+4	+4	+4	+4	+4		
		+5	+5	+5					
			+6	+6	+6				
				+7					

type A: Sc, Ti, and V — type B: Cr and Mn — type C: Fe, Co, Ni, Cu, and Zn

▲ Figure 4 Oxidation states of the first-row d-block metals. The most stable oxidation states are marked in green

Oxidizing and reducing agents

An **oxidizing agent** causes another species to be oxidized, and is itself reduced in the process. A **reducing agent** causes another species to be reduced, and is itself oxidized in the process.

Worked example

Deduce the oxidation states of each atom (marked ^{x}) in each of the following species:

a) $K_2Cr^x{}_2O_7$ **b)** $Mn^xO_4^-$

c) $Mg_3N^x{}_2$ **d)** $S^x{}_8$

e) $[NH_4]_2[Fe^x(H_2O)_6][SO_4]_2$

Solution

a) $K_2Cr^x{}_2O_7$

$2(+1) + 2x + 7(-2) = 0$

$x = +6$

b) $Mn^xO_4^-$

$x + 4(-2) = -1$

$x = +7$

c) $Mg_3N^x_2$

$3(+2) + 2x = 0$

$x = -3$

d) S^x_8

$x = 0$, since this is a free element.

e) $[NH_4]_2[Fe^x(H_2O)_6][SO_4]_2$

To answer this question, you should use your knowledge of the charges of ammonium (+1), water (0) and sulfate (−2) species (see sub-topic 4.1).

$2(+1) + x + 6(0) + 2(-2) = 0$

$x = +2$

Quick questions

1 Deduce the oxidation states of each atom (marked with an ^{x}) in each of the following species:

   a) C^xO_2 **b)** HCl^xO_4

   c) $Na_3P^xO_4$ **d)** O^x_3

   e) P^xH_3 **f)** I^xCl

   g) $Fe^x_2(SO_4)_3$ **h)** $H_2C^x_2O_4$

   i) $N^xO_3^-$

2 In the following balanced equation:

   $Cl_2(aq) + 2KI(aq) \rightarrow 2KCl(aq) + I_2(aq)$

   a) Deduce the oxidation states of chlorine and iodine in the reactants and products.

   b) State which element is oxidized and which element is reduced.

   c) Identify the oxidizing agent and the reducing agent.

Oxidation states and the nomenclature of transition metal compounds

As stated previously one of the characteristics of transition elements is that they can have variable oxidation states in their compounds. Traditionally, the Roman numeral system of nomenclature has been used to name such compounds and this system is based on oxidation numbers. The system is called the **Stock nomenclature system**. In the Stock system, Roman numerals (I, II, III etc.) are used to indicate the oxidation number.

 The Stock nomenclature system

In $KMnO_4$, often called by its old name potassium permanganate by many chemists, manganese has an oxidation state of +7. However, from a purely electrostatic perspective the presence of an Mn^{7+} cation is highly improbable. Its name is potassium manganate(VII) using the Stock nomenclature system.

Study tips

1 Remember when writing oxidation states the charge goes before the number and not after it. For example, the oxidation number of hydrogen in HBr is +1 and not 1+.

2 The oxidizing and reducing agents are always the reactants.

1   When deducing the name of a transition metal compound using the Stock system, do not be tempted to use the subscript representing the number of atoms of the other element in the compound to write the oxidation number in Roman numerals. For example, the name of $FeCl_2$ is iron(II) chloride because iron is deduced to have the $+2$ oxidation state $[x + 2(-1) = 0$, so $x = +2]$. It is coincidental that 2 matches the number of chlorine atoms in the formula. In the compound FeO this becomes clearer: the correct name is iron(II) oxide $[x + (-2) = 0$, so $x = +2]$.

2   In working out the names of many transition metal compounds, knowledge of the non-systematic names and charges of the various oxoanions can be useful (table 1).

In naming oxoanions, a good rule of thumb is as follows:

- If there is only one oxoanion, the ending will be -ate.

- If there are two oxoanions, the one with the smaller number of oxygens will end in -ite and the one with the greater number of oxygens will end in -ate.

- If there are four oxanions, the one with the lowest number of oxygens will end in -ite and be prefixed by "hypo", the next will end in -ite, the third will end in -ate, and the one with the greatest number of oxygens will be prefixed by "per" and end in -ate. The four oxoanions of chlorine, bromine, and iodine follow this system.

Formula of oxoanion	Non-systematic name
CO_3^{2-}	carbonate
$C_2O_4^{2-}$	ethanedioate (oxalate)
NO_2^-	nitrite
NO_3^-	nitrate
SO_3^{2-}	sulfite
SO_4^{2-}	sulfate
PO_3^{3-}	phosphite
PO_4^{3-}	phosphate
ClO^-	hypochlorite
ClO_2^-	chlorite
ClO_3^-	chlorate
ClO_4^-	perchlorate
OH^-	hydroxide
SiO_4^{4-}	orthosilicate

▲ Table 1 Formulas and non-systematic names of some oxoanions

TOK

Chemistry has developed a systematic language that has resulted in older names becoming obsolete. What has been lost and gained in this process?

Quick question

Using the Stock nomenclature system, deduce the name of each of the following transition metal compounds:

a) CoF_3

b) V_2O_3

c) $Cu(OH)_2$

d) MnO_2

e) Cu_2O

 Nomenclature

In theory one could include the oxidation state in the names of all inorganic compounds. However, as stated previously many elements have only one oxidation state, such as potassium, $+1$, so the state is not required. In 2005 IUPAC published a new set of guidelines for the systematic naming of oxoanions and the corresponding inorganic oxoacids. Although the new system, based on systematic additive names, has huge merit, IUPAC does recognize that it is unrealistic to completely eliminate old non-systematic names such as carbonate, carbonic acid, nitrate, nitric acid, etc. In this system, carbonate would be called trioxidocarbonate(2–), nitrate would be trioxidonitrate(1–), carbonic acid would be dihydroxidooxidocarbon, and nitric acid would be hydroxidodioxidonitrogen. Can you suggest how these names are formed?

Expressing redox reactions using half-equations in acidic or neutral solutions

Half-equations can be very useful in balancing complex redox reactions. Each half-equation represents the separate oxidation and reduction processes. The following general working method can be used to balance a redox reaction involving oxidation states. In the IB syllabus you are only required to balance an equation in acidic or neutral media.

Working method

Step 1: Assign oxidation states for each atom in the reactant and product species.

Step 2: Deduce which species is oxidized and which species is reduced.

Step 3: State the half-equation for the oxidation process and the corresponding half-equation for the reduction process.

Step 4: Balance these half-equations so that the number of electrons lost equals the number of electrons gained.

Step 5: Add the two half-equations together to write the overall redox reaction.

Step 6: Check the total charge on the reactant and product sides.

Step 7: Balance the charge by adding H^+ and H_2O to the appropriate sides.

Worked example

1 Iron tablets are often prescribed to patients. The iron in the tablets is commonly present as anhydrous iron(II) sulfate, $FeSO_4$. An experiment to determine the percentage by mass of iron in such tablets involves a redox reaction, shown in the following unbalanced equation:

$$Fe^{2+}(aq) + MnO_4^-(aq) \rightarrow Fe^{3+}(aq) + Mn^{2+}(aq)$$

  a) Deduce the balanced redox equation in acid and identify the oxidizing and reducing agents.

  b) ⊘ Consider the oxidation state of manganese in the permanganate anion, MnO_4^-. Comment on the following statement:

    "If oxidation state is considered as the apparent charge that an atom of an element has in an ion, then the oxidation state of manganese here must signify the presence of the corresponding ion!"

Solution

a) **Step 1:**

  Fe^{2+}: Fe, $x = +2$

  MnO_4^-: Mn, $x + 4(-2) = -1$, so $x = +7$;
      O, $x = -2$

  Fe^{3+}: Fe, $x = +3$

  Mn^{2+}: Mn, $x = +2$

Step 2:

The oxidation state of Fe changes from $+2$ in Fe^{2+} to $+3$ in Fe^{3+}, so the oxidation state increases, indicative of oxidation. The oxidation state of Mn changes from $+7$ in MnO_4^- to $+2$ in Mn^{2+}, so the oxidation state decreases, indicative of reduction.

Step 3:

Oxidation (loss of electrons):
$$Fe^{2+}(aq) \rightarrow Fe^{3+}(aq) + e^-$$
Reduction (gain of electrons):
$$MnO_4^-(aq) + 5e^- \rightarrow Mn^{2+}(aq)$$

Step 4:

Oxidation (loss of electrons):
$$5Fe^{2+}(aq) \rightarrow 5Fe^{3+}(aq) + 5e^-$$
Reduction (gain of electrons):
$$MnO_4^-(aq) + 5e^- \rightarrow Mn^{2+}(aq)$$

Step 5:

Oxidation: $5Fe^{2+}(aq) \rightarrow 5Fe^{3+}(aq) + 5e^-$
Reduction: $MnO_4^-(aq) + 5e^- \rightarrow Mn^{2+}(aq)$
Overall: $5Fe^{2+}(aq) + MnO_4^-(aq) \rightarrow$
$$5Fe^{3+}(aq) + Mn^{2+}(aq)$$

Step 6:

  Total charge on reactant side $= 9+$
  Total charge on product side $= 17+$

Step 7:

To balance this equation $8H^+$ must be inserted on the reactant side:

$$5Fe^{2+}(aq) + MnO_4^-(aq) + 8H^+(aq) \rightarrow$$
$$5Fe^{3+}(aq) + Mn^{2+}(aq)$$

Next we need to balance the hydrogens. Water can be included at the very last stage on whichever side of the equation it is required:

$$5Fe^{2+}(aq) + MnO_4^-(aq) + 8H^+(aq) \rightarrow$$
$$5Fe^{3+}(aq) + Mn^{2+}(aq) + 4H_2O(l)$$

The oxidizing agent is $MnO_4^-(aq)$ and the reducing agent is $Fe^{2+}(aq)$.

b) ⊘ Possible response to NOS question: in part a) the oxidation state of manganese in the permanganate anion was found to be +7. However, oxidation states and ionic charges have different meanings, and an oxidation state of +7 does not signify a corresponding ionic charge of 7+. Ionic charges are real properties of ions, whereas oxidation states are theoretical constructs and are not real. Based on electrostatic considerations the presence of a 7+ ionic charge is most unlikely. Oxidation states assume that bonds are ionic. However, in MnO_4^- the manganese–oxygen bonds are covalent in nature.

The activity series

The **activity series** (table 2) ranks metals according to the ease with which they undergo oxidation. Metals higher up in the activity series can displace those lower down from solutions of their respective salts. Although the series is primarily based on metals, hydrogen is often included even though it is a non-metal. The series is given in section 25 of the *Data booklet*. The most reactive metals are found at the top of the series.

Element	Decreasing reactivity	Ease of oxidation increases
lithium		
potassium		
sodium		
magnesium		
aluminium		
manganese	↓	↑
zinc		
iron		
lead		
hydrogen		
copper		
silver		
mercury		
gold		

▲ Table 2 The activity series

Let's consider some examples:

- $Zn(s) + Cu^{2+}(aq) \rightarrow Zn^{2+}(aq) + Cu(s)$

  Zinc metal is above copper metal in the series, so therefore it is more reactive and can displace the Cu^{2+} ions in solution to form copper metal.

- $Zn(s) + 2HCl(aq) \rightarrow ZnCl_2(aq) + H_2(g)$

  Zinc metal is above hydrogen in the series. Therefore, it can displace the hydrogen ions in hydrochloric acid to form hydrogen gas.

- $2Al(s) + Fe_2O_3(s) \rightarrow 2Fe(l) + Al_2O_3(s)$

  Aluminium metal is above iron in the series. Hence molten iron can form according to the reaction above.

- $2Na(s) + 2H_2O(l) \rightarrow 2NaOH(aq) + H_2(g)$

  In this reaction, hydrogen is displaced from water by the very reactive alkali metal, sodium, to liberate hydrogen gas in the process.

A reactivity series can also be written for the group 17 elements, fluorine, chlorine, bromine, and iodine (table 3).

Group 17 element	Atomic radius (pm)	Electronegativity χ_p (Pauling scale)	Increasing reactivity
fluorine	60	4.0	
chlorine	100	3.2	
bromine	117	3.0	↑
iodine	136	2.7	

▲ Table 3 Reactivity series for group 17 elements

Again the more reactive elements are found higher in the series.

$$2KBr(aq) + Cl_2(aq) \rightarrow 2KCl(aq) + Br_2(aq)$$

The atomic radius of chlorine is smaller than that of bromine, so chlorine is more electronegative. The chlorine nucleus therefore has a greater attraction for an electron than does bromine, so chlorine is reduced, gaining an electron to form the chloride ion. Bromine is oxidized by losing an electron to form bromine. Note the change in oxidation states. Chlorine changes from 0 to −1, so there is a decrease in the oxidation state indicative of a reduction process. The oxidation state of bromine changes from −1 to 0, so there is an increase in the oxidation state indicative of an oxidation process.

In the laboratory, when chlorine gas is bubbled through a solution of potassium bromide there is a corresponding colour change from the colourless solution of potassium bromide to yellow/orange, indicating the formation of aqueous bromine. Chlorine is higher up in the reactivity series so can displace bromide ions from potassium bromide to form bromine (see topic 3).

Quick questions

1 Deduce the oxidizing and reducing agents in the reaction of potassium bromide with chlorine.

2 Table 4 shows reactions involving aqueous solutions of halogens with aqueous potassium iodide solution. Copy and complete the table, and in each case:

   a) state whether a reaction will occur or not

   b) identify the colour of the halide solution after reaction

   c) deduce the balanced equation for any reaction that occurs.

Halogen	$Cl_2(aq)$	$Br_2(aq)$
a) Reaction with $KI(aq)$		
b) Colour of halide solution		
c) Balanced equation		

▲ Table 4

Uses of chlorine in everyday life

Chlorine is a powerful oxidizing agent and is widely used as a **disinfectant** and **antiseptic**. Calcium hypochlorite, $Ca(OCl)_2$, is often used in hospitals by healthcare professionals to disinfect their hands (figure 5). Sodium hypochlorite, $NaOCl$, is another disinfectant, often used in our homes as household bleach.

The sharing of needles and syringes among drug users is a contributory factor in the transmission of the human immunodeficiency virus (HIV), which can lead to acquired immune deficiency syndrome (AIDS). The US Centers for Disease Control and Prevention (CDC) has reported that disinfection of syringes and needles with household bleach may go some way to alleviating this risk.

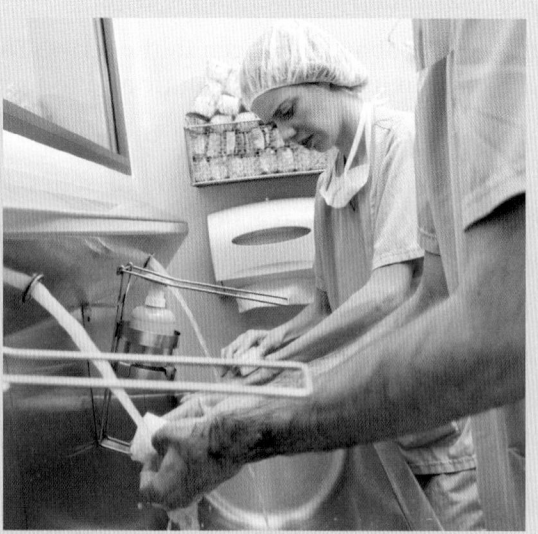

▲ Figure 5 A solution of calcium hypochlorite acts as both a disinfectant and an antiseptic. Can you explain the difference between these two terms?

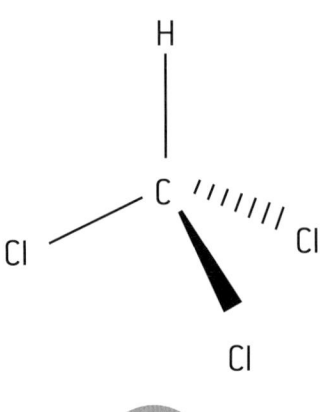
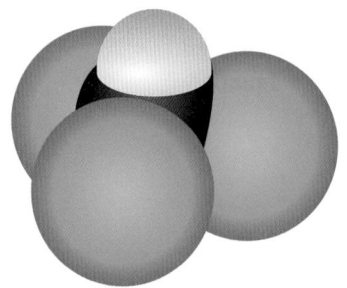

▲ Figure 6 The structure of trichloromethane, $CHCl_3$

Activity

In your country, find out whether chlorine or ozone is used to disinfect municipal water supplies.

Use of chlorine and ozone as disinfectants in drinking water

Access to a supply of clean drinking water has been recognized by the United Nations as a fundamental human right, yet it is estimated that over one billion people worldwide do not have the luxury of such a fundamental resource to mankind. Water supplies are disinfected using strong oxidizing agents such as chlorine, Cl_2 or ozone, O_3 to kill microbial pathogens. In the USA, chlorine is used for this purpose. Chlorine can be added in three forms: chlorine gas, Cl_2; sodium hypochlorite, $NaOCl$; and calcium hypochlorite, $Ca(OCl)_2$. All three of these solutions yield hypochlorous acid, $HOCl$, which is the antibacterial agent.

The use of chlorine can cause problems for the general public. Some people object to the taste and general odour of residual chlorine in water. Residual chlorine can also react with other chemicals to form toxic products such as trichloromethane, $CHCl_3$, commonly known as chloroform (figure 6).

In Europe, France was one of the first countries to use ozone to disinfect water supplies. The first industrial ozonation plant was established in Nice in 1906 for this purpose. Table 5 compares the use of ozone and chlorine for water treatment.

Ozone	Chlorine
can be used to treat viruses	cannot be used to treat viruses
leaves no unpleasant residual taste or odour	leaves a residual taste and unpleasant odour
fewer toxic by-products	can form toxic by-products, often carcinogenic
more expensive	cheaper

▲ Table 5 Advantages and disadvantages of using ozone and chlorine in the treatment of water supplies

Redox titration reactions

In topic 1 titrations were introduced, and these play a pivotal role in the field of volumetric analysis. In addition to reactions involving acid–base titrations, those involving redox reactions are also extremely useful.

In order to solve titration questions involving redox reactions, you need to recall some of the formulae used in volumetric analysis:

1   Amount of substance (in mol) = n

$$n = \frac{m}{M}$$

where m = mass in g; M = molar mass in g mol^{-1}

2   n = volume (in dm^3) × concentration (in mol dm^{-3})

$$= \frac{\text{volume (in cm}^3\text{)} \times \text{concentration (in mol dm}^{-3}\text{)}}{1000}$$

because 1 dm^3 = 1000 cm^3.

3    $\frac{1}{\nu_A}(n_A) = \frac{1}{\nu_B}(n_B)$ and hence

$$\frac{1}{\nu_A}(V_A \times c_A) = \frac{1}{\nu_B}(V_B \times c_B)$$

V_A = volume of reactant A (in dm^3)

c_A = concentration of reactant A (in mol dm^{-3})

V_B = volume of reactant B (in dm^3)

c_B = concentration of reactant B (in mol dm^{-3})

ν_A and ν_B are the stoichiometry coefficients

Working method

Step 1: Deduce the balanced redox equation, using oxidation states.

Step 2: From the information given, state which three pieces of data are given from V_A, c_A, V_B, and c_B and identify the fourth variable that needs to be determined. Identify the **stoichiometry coefficients** ν_A and ν_B from the balanced equation.

Step 3: Set up the following expression and fill in the known data:

$$\frac{1}{\nu_A}(V_A \times c_A) = \frac{1}{\nu_B}(V_B \times c_B)$$

Step 4: Solve for the unknown variable (V_A, c_A, V_B, or c_B).

Step 5: Answer any riders to the question (such as expressing a concentration in particular units).

Worked example

1   Consider the following balanced equation for the reaction of potassium manganate(VII) with ammonium iron(II) sulfate.

$$5Fe^{2+}(aq) + MnO_4^-(aq) + 8H^+(aq) \rightarrow 5Fe^{3+}(aq) + Mn^{2+}(aq) + 4H_2O(l)$$

In a titration to determine the concentration of a potassium manganate(VII) solution, 28.0 cm^3 of the potassium

> **Molarity**
>
> Note that for convenience concentration is often termed the "**molarity**" (the unit is sometimes abbreviated to M), but it is best practice to use the unit mol dm^{-3} in calculations.

> **Study tip**
>
> This type of question frequently appears in Question 1 of Paper 2.

manganate(VII) solution solution reacted completely with 25.0 cm^3 of a $0.0100 \text{ mol dm}^{-3}$ solution of ammonium iron(II) sulfate. Determine the concentration, in g dm^{-3}, of the potassium manganate(VII) solution.

Solution

Step 1: Deduce the balanced redox equation, using oxidation states.

This step is not required here as the equation is given in the question.

Step 2: From the information given, state which three pieces of data are given from V_A, c_A, V_B, and c_B and identify the fourth variable that needs to be determined. Identify the stoichiometry coefficients, ν_A and ν_B, from the balanced equation.

A represents Fe^{2+} and B represents MnO_4^-

V_A = volume of Fe^{2+} = 0.0250 dm^3

c_A = concentration of Fe^{2+} = $0.0100 \text{ mol dm}^{-3}$

V_B = volume of MnO_4^- = 0.0280 dm^3

c_B = concentration of MnO_4^-: this is what must be calculated

$\nu_A = 5$

$\nu_B = 1$

Step 3: Set up the following expression and fill in the known data:

$$\frac{1}{5}(0.0250 \times 0.0100) = \frac{1}{1}(0.0280 \times c_B)$$

Step 4: Solve for the unknown variable (V_A, c_A, V_B, or c_B).

c_B (concentration of MnO_4^-) = $0.00179 \text{ mol dm}^{-3}$

Step 5: Answer any riders to the question (eg expressing a concentration in particular units).

To calculate the concentration in g dm^{-3}, we use dimensional analysis.

1 mol of $KMnO_4 \equiv (39.10) + (54.94) + 4(16.00) \equiv 158.04 \text{ g}$

So:

$$\frac{0.00179 \text{ mol}}{1 \text{ dm}^3} \times \frac{158.04 \text{ g}}{1 \text{ mol}} = 0.283 \text{ g dm}^{-3}$$

An environmental application of redox chemistry: The Winkler method

Aquatic life depends on gases such as carbon dioxide and oxygen dissolved in the water in order to survive. Oxygen, O_2, is a non-polar molecule, but water, H_2O, is polar. Therefore the solubility of oxygen in water will be very low.

The solubility of oxygen in water is temperature dependent. At 273 K (0° C) the solubility is 14.6 mg dm^{-3} (or 14.6 ppm), compared with just 7.6 mg dm^{-3} (7.6 ppm) at 293 K (20 °C). Clearly as the temperature increases, the solubility of the gas decreases.

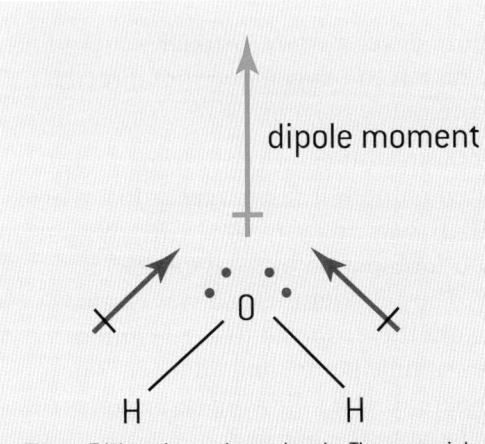

dipole moment

▲ Figure 7 Water is a polar molecule. The vectorial sum of the two individual OH polar bonds results in a net dipole moment for the molecule

Concentrations in parts per million

The concentration of very dilute solutions is often measured in **parts per million, ppm**

concentration in ppm

$$= \frac{\text{mass of component in solution}}{\text{total mass of solution}} \times 10^6$$

$$= \frac{\text{mass of solute in mg}}{\text{volume of solution in dm}^3}$$

The amount of dissolved oxygen is often used as a barometer to indicate the quality of a body of water. The **Winkler method**, based on redox reactions, is one technique that can be used to measure the amount of dissolved oxygen in water. In general, a high concentration of dissolved oxygen indicates a low level of pollution.

▲ Figure 8 The WHO is the directing and coordinating authority for health within the United Nations

The degree of organic pollution in a sample of water can be measured by the **biochemical oxygen demand** or **BOD**. This is defined as the amount of oxygen required to oxidize organic matter in a sample of water at a definite temperature over a period of 5 days. BOD is measured in units of ppm.

In environmental science, ppm is often used as the standard unit of concentration to indicate the maximum allowable upper limit of a potentially toxic or carcinogenic (cancer-causing) substance. For example, according to recommendations from the **World Health Organization** (WHO) the maximum allowed concentration of lead(II) cations, $Pb^{2+}(aq)$, in drinking water is 0.001 mg dm^{-3} or 0.001 ppm.

Activity

Go to the WHO website (http://www.who.int/en/) and try to find data about the maximum allowed concentrations of other metals in drinking water. Compare this data with the limits set by the government of the country where you live or by directives set by a wider union of countries (eg the European Union).

Typical values of BOD

Pure water generally has a BOD less than 1 ppm. Water from a river that has a BOD of 1 ppm would be considered very clean. However, water taken from a river with a BOD of 20 ppm would be considered of poor quality (table 6).

Example source	BOD (ppm)
pure water	less than 1
untreated domestic sewage	350
effluent from a brewery	500
water from an abbatoir	3000

▲ Table 6 Typical biological oxygen demands for water samples

When organic matter is discharged into a body of water, it provides a source of food for any bacteria present. The bacteria break down the organic material into compounds such as carbon dioxide and water in a series of oxidation reactions. The carbon is oxidized to carbon dioxide, the hydrogen is oxidized to water, and any nitrogen present is oxidized to nitrate, NO_3^-. The bacteria multiply and their increased levels mean that more dissolved oxygen is used for these oxidation processes. If the uptake of oxygen by the bacteria is faster than the rate at which dissolved oxygen is replaced from the atmosphere and from photosynthesis, the body of water will eventually become depleted of oxygen. Under such anaerobic conditions the bacteria will produce products such as hydrogen sulfide, H_2S, ammonia, NH_3 (and amines), and phosphine, PH_3. Hydrogen sulfide is the gas commonly associated with the odour from rotten eggs; it is also often liberated from volcanoes. Due to its characteristic unpleasant odour and its potential source H_2S is often referred to as sewer gas.

Element	Substance produced under aerobic conditions	Substance produced under anaerobic conditions
carbon	CO_2	CH_4 (methane, commonly known as marsh gas)
hydrogen	H_2O	CH_4, NH_3, H_2S, and H_2O
nitrogen	NO_3^-	NH_3, amines
sulfur	SO_4^{2-}	H_2S (hydrogen sulfide)
phosphorus	PO_4^{3-}	PH_3 (phosphine)

▲ Table 7 Substances produced by bacteria under aerobic and anaerobic conditions

The reduction in dissolved oxygen can result in the depletion of fish stocks. If the BOD is greater than the dissolved content in the water, aquatic life cannot survive. Typically fish require at least 3 ppm of dissolved oxygen in water, and to sustain a healthy aquatic environment the content of dissolved oxygen in water should not fall below 6 ppm.

Worked example: measuring BOD using the Winkler method

In the Winkler method, an iodine/thiosulfate redox titration is carried out to measure the dissolved oxygen present in a water sample.

The procedure developed by Winkler is an indirect one, as the dissolved oxygen does not directly react with the redox reagent.

A 50.0 cm³ sample of water taken from a location where treated effluent is discharged into a marina in Dubai, UAE, was first saturated with oxygen and then left for a period of 5 days at 293 K in the dark. The Winkler method was carried out to measure the dissolved oxygen content in the water sample before and after the 5-day incubation period. The following is the series of reactions related to the method:

$$Mn^{2+}(aq) + 2OH^-(aq) \rightarrow Mn(OH)_2(s)$$

$$2Mn(OH)_2(s) + O_2(g) \rightarrow 2MnO(OH)_2(s)$$

$$MnO(OH)_2(s) + 4H^+(aq) + 2I^-(aq) \rightarrow Mn^{2+}(aq) + I_2(aq) + 3H_2O(l)$$

$$I_2(aq) + 2S_2O_3^{2-}(aq) \rightarrow 2I^-(aq) + S_4O_6^{2-}(aq)$$

It was found that 5.25 cm³ of a 0.00500 mol dm⁻³ solution of sodium thiosulfate, $Na_2S_2O_3(aq)$ was required to react with the iodine produced.

a) Determine the concentration of dissolved oxygen, in ppm, in the sample of water.

b) Deduce the BOD, in ppm, of the water sample, assuming that the maximum solubility of oxygen in the water is 9.00 ppm at 293 K.

c) Comment on the BOD value obtained.

Solution

Step 1: Deduce the balanced redox equation.

The series of balanced redox equations is given in the question. The important point is to determine the correct stoichiometric ratio between oxygen and thiosulfate; this is needed for the calculation. Careful examination of the three reactions gives the following ratio:

$$1 \text{ mol } O_2(g) \rightarrow 2 \text{ mol } MnO(OH)_2(s) \rightarrow 4 \text{ mol } S_2O_3^{2-}(aq)$$

Step 2: From the information given, state which three pieces of data are given from V_A, c_A, V_B and c_B and identify the fourth variable that needs to be determined. Identify the stoichiometry coefficients, ν_A and ν_B, from the balanced equation.

A represents $S_2O_3^{2-}$ and B represents O_2

V_A = volume of $S_2O_3^{2-}$ = 0.00525 dm^3

c_A = concentration of $S_2O_3^{2-}$
= 0.00500 mol dm^{-3}

V_B = volume of O_2 = 0.0500 dm^3

c_B = concentration of O_2; this must be calculated

ν_A = 4

ν_B = 1

Step 3: Set up the following expression and fill in the known data:

$$\frac{1}{4}(0.00525 \text{ dm}^3 \times 0.00500)$$

$$= \frac{1}{1}(0.0500 \times c_A)$$

Step 4: Solve for the unknown variable (V_A, c_A, V_B or c_B).

c_B (concentration of O_2) = 1.31×10^{-4} mol dm^{-3}

Step 5: Answer any riders to the question.

In order to calculate the concentration in g dm^{-3}, we use dimensional analysis.

1 mol of $O_2 \equiv 2(16.00) \equiv 32.00$ g

So:

$$\frac{1.31 \times 10^{-4} \text{ mol}}{1 \text{ dm}^3} \times \frac{32.00 \text{ g}}{1 \text{ mol}}$$

$$= 4.19 \times 10^{-3} \text{ g dm}^{-3}$$

$$= 4.19 \text{ mg dm}^{-3} = 4.19 \text{ ppm}$$

Hence the oxygen used by the bacteria (BOD) = $9.00 - 4.19 = 4.81$ ppm.

This BOD value shows reasonable water quality for the sample taken at the effluent discharge point in Dubai, suggesting that an effective sewage treatment plan must be in place. Typically untreated domestic sewage has a BOD in the range 100–400 ppm.

 **Activity**

Chemistry is full of abstract concepts, theories, and assumptions. Discuss this statement with reference to the thiosulfate oxoanion, commenting on aspects such as oxidation numbers, formal charge, ionic charge, and negative charge centres. Suggest why electron domain may be a preferable term to negative charge centre in this context.

9.2 Electrochemical cells

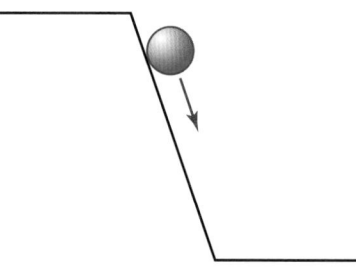

▲ Figure 1 Voltaic cells convert energy from spontaneous exothermic chemical processes to electrical energy

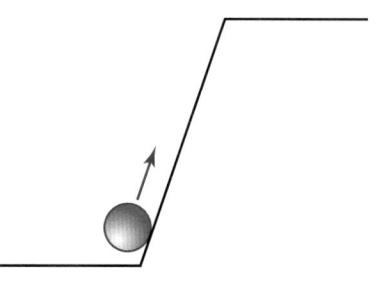

▲ Figure 2 Electrolytic cells convert electrical energy to chemical energy, bringing about a non-spontaneous process

Energy

Energy is the capacity to do work. The SI unit of energy is the joule (J). The **law of conservation of energy** states that energy cannot be created or destroyed but is converted from one form to another.

There are many different forms of energy, such as kinetic energy (energy due to motion), potential energy (stored or positional energy), light energy, heat energy, nuclear energy, sound energy, chemical energy, and electrical energy.

Electrochemical cells

In an electrochemical cell chemical energy–electrical energy conversions take place, which can go in either direction. There are two main types of electrochemical cell:

1 **Voltaic (or galvanic) cells** – these convert chemical energy to electrical energy. Voltaic cells convert energy from spontaneous, exothermic chemical processes to electrical energy.

2 **Electrolytic cells** – these convert electrical energy to chemical energy, bringing about a non-spontaneous process.

Early ideas about electricity

Electrochemistry explores energy conversions between chemical and electrical energy. The Italian physician, physicist, and philosopher Luigi Galvani (1737– 1798) considered electricity essentially biological in its origins, whereas the Italian physicist Alessandro Volta (1745–1827) did not. The initial discovery of electrochemical cells resulted from *serendipitous* observations of scientists, but recent developments have been driven by potential profits from improved technology. The increasing demand for energy has driven innovation in devices and processes.

Electrodes

Electrons are carriers of electric charge in metals. An **electrode** is a conductor of electricity used to make contact with a non-metallic part of a circuit, such as the solution in a cell (the **electrolyte**). An electrochemical cell contains two electrodes, the **anode** and the **cathode**.

In both voltaic and electrolytic cells:

- **oxidation** always takes place at the anode
- **reduction** always takes place at the cathode.

The polarity of the electrodes differs in the different types of cell.

In a voltaic cell:

- the cathode is the positive electrode
- the anode is the negative electrode.

In an electrolytic cell:

- the cathode is the negative electrode
- the anode is the positive electrode.

The voltaic cell

A voltaic cell consists of two half-cells. Oxidation occurs at one half-cell (the anode) and reduction occurs at the other half-cell (the cathode). There are different types of electrode used in voltaic cells, such as metal/metal-ion electrodes, metal ions in two different oxidation states, and the gas-ion electrode. For the IB Chemistry Diploma programme, SL students are required to be familiar only with the metal/metal-ion electrode.

The metal/metal-ion electrode

A metal/metal-ion electrode consists of a bar of metal dipped into a solution containing cations of the same metal. Typical examples of this type of electrode include:

- $Fe(s)|Fe^{2+}(aq)$ (figure 3)
- $Zn(s)|Zn^{2+}(aq)$
- $Cu(s)|Cu^{2+}(aq)$.

In this notation, the vertical line represents **a phase boundary** or **junction**.

In a voltaic cell the two half-cells are separated – if the solutions were allowed to mix in a single container, a spontaneous reaction would occur

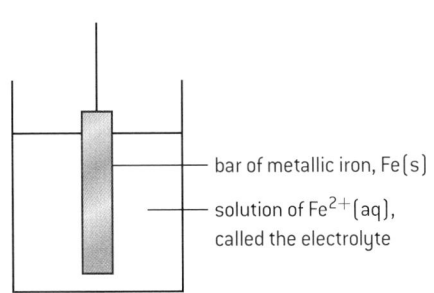

bar of metallic iron, Fe(s)

solution of $Fe^{2+}(aq)$, called the electrolyte

▲ Figure 3 The $Fe(s)|Fe^{2+}(aq)$ electrode

Activity

Discuss the origins of the field of electrochemistry from a Nature of Science perspective, from the serendipitous discoveries of some of the original scientists working in this field to the ethical implications of current research, with our ever-increasing global desire to produce more energy.

TOK

Is energy real, or just an abstract concept used to justify why certain types of changes are always associated with each other?

but there would be no movement of electrons through the external circuit, and hence no current. The two electrodes are in electrical contact via a liquid junction called a **salt bridge**. This has a number of functions:

- It allows physical separation of the cathode and anode and hence the oxidation and reduction processes, preventing mixing of the two solutions.

- It provides electrical continuity – a path for the migration of the positive ions (the cations) and the negative ions (the anions) in the cell.

- It reduces the **liquid-junction potential**. This is the voltage generated when two different solutions come into contact with each other, which occurs due to unequal cation and anion migration across the junction.

A salt bridge contains a concentrated solution of a strong electrolyte. The high concentration of ions in the salt bridge allows ions to diffuse out of it. For example, the **Daniell voltaic cell** consists of the $Cu(s)|Cu^{2+}(aq)$ and $Zn(s)|Zn^{2+}(aq)$ electrodes. Typical compounds used in the salt bridge for this cell could be sodium sulfate, $Na_2SO_4(aq)$ or potassium chloride, $KCl(aq)$. The ions used in the salt bridge must be inert – they should not react with the other ions in the solution.

The Daniell voltaic cell

In the Daniell cell (figure 4), the following half-equations show the redox processes occurring.

- Anode (negative electrode): oxidation.

  $Zn(s) \rightarrow Zn^{2+}(aq) + 2e^-$

- Cathode (positive electrode): reduction.

  $Cu^{2+}(aq) + 2e^- \rightarrow Cu(s)$

- Overall cell reaction:

  $Cu^{2+}(aq) + Zn(s) \rightarrow Zn^{2+}(aq) + Cu(s)$

Once the cell is connected, as the redox processes occur the blue colour of the copper(II) sulfate solution fades. The copper bar increases in size as it becomes coated in more copper, and the zinc bar gets thinner.

When drawing voltaic cells, by convention the cathode is drawn on the right-hand side as in figure 4.

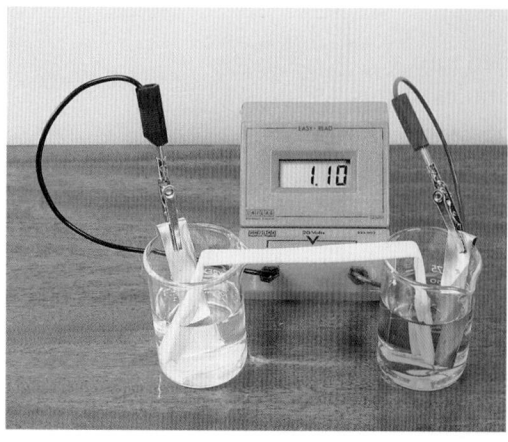

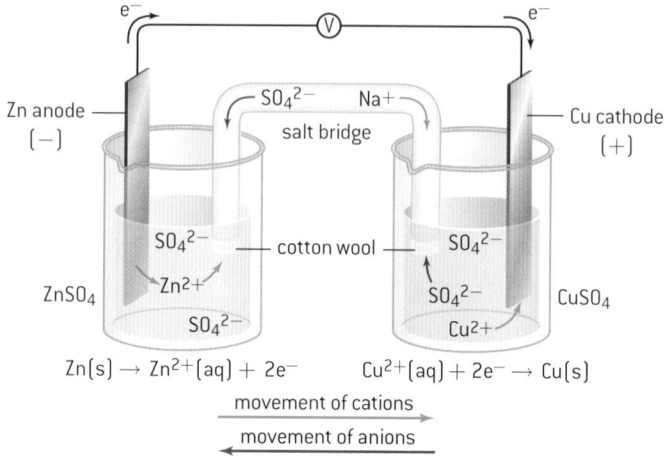

▲ Figure 4 The Daniell cell: a cell consisting of $Zn(s)|Zn^{2+}(aq)$ and $Cu(s)|Cu^{2+}(aq)$ half-cells

How can you determine which metal will be oxidized and which metal will be reduced in a voltaic cell? The answer lies in the activity series. For the Daniell cell, zinc is higher up in the series than copper, so it is more easily oxidized – the zinc half-cell acts as the anode.

Cell diagrams

Cell diagrams are used as a convenient shorthand to represent a voltaic cell. *By convention* the anode is always written on the left and the cathode on the right. The salt bridge is represented by two parallel vertical lines. For the Daniell cell the cell diagram would be written as:

$$Zn(s)|Zn^{2+}(aq) \parallel Cu^{2+}(aq)|Cu(s)$$

Study tip

When answering questions about voltaic cells, make sure you know the direction of flow of the electrons and ions (figure 5).

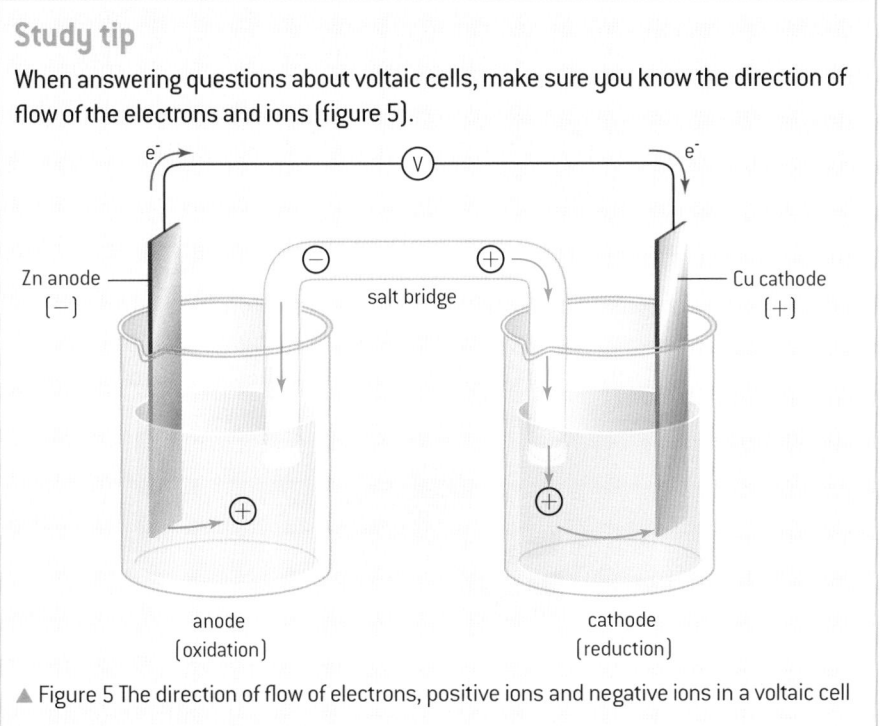

▲ Figure 5 The direction of flow of electrons, positive ions and negative ions in a voltaic cell

Quick question

For a voltaic cell consisting of a $Zn(s)|Zn^{2+}(aq)$ half-cell and an $Fe(s)|Fe^{2+}(aq)$ half-cell:

a) State the cell diagram for the cell.

b) Write half-equations for the reactions occurring at the cathode and the anode.

c) Identify a suitable compound that may be used in the salt bridge.

d) Identify the direction of the movement of electrons and ion flow, both in solution and in the salt bridge.

e) Explain why the cation and anion of the salt used in the salt bridge should have approximately the same size and charge. Identify using section 9 of the *Data booklet* the ionic radii of the cation and anion of the compound given in (c).

f) Cotton wool is often used at the tips of the salt bridge. Suggest the function of this.

Key experimental work

In this topic it is important to have carried out laboratory experiments or seen videos or simulations involving a typical voltaic cell using two metal/metal-ion half-cells.

The global energy perspective

Fuel cells

The combustion of fuels such as oil, coal, or natural gas releases heat energy which is converted into electrical energy in an electric power plant. The energy loss in this process can be 67% or greater. **Fuel cells** however can convert approximately 70% of the energy in a fuel into electrical energy. A fuel cell is a voltaic cell so is based on redox processes. The most common type of fuel cell, the **hydrogen–oxygen fuel cell**, uses the reaction between hydrogen and oxygen:

$$2H_2(g) + O_2(g) \rightarrow 2H_2O(l)$$

Under acidic conditions, the following reactions take place at the anode and cathode:

- anode (negative electrode): oxidation

  $$2H_2(g) \rightarrow 4H^+ + 4e^-$$

- cathode (positive electrode): reduction

  $$O_2(g) + 4H^+ + 4e^- \rightarrow 2H_2O(l)$$

Fuel cells are highly efficient devices. Water is the only product of the hydrogen–oxygen fuel cell so it is non-polluting, (unlike conventional combustion reactions of fossil fuels). Another advantage of a fuel cell is that it does not need recharging. Fuel cells provide a continuous supply of electricity because as the reactants are used up, more reactants are added. Fuel cells have a number of applications (figure 6 and sub-topic C.6). One disadvantage of fuel cells is that they are very expensive to produce. Another disadvantage of fuel cells is that they are prone to poisoning by impurities in the fuel, which reduces their lifetime or requires very complex and expensive purification of the fuel.

One type of fuel cell is the **biological fuel cell**, which uses bacteria to generate electricity from chemical energy in chemicals such as methane or organic waste materials.

Although the hydrogen fuel cell is non-polluting and an efficient alternative to the internal combustion engine, the storage of hydrogen fuel is a major problem. The **methanol fuel cell** uses liquid methanol rather than hydrogen, which is much easier to transport. Methanol can be produced from biomass as a **carbon-neutral fuel** (it does not contribute to the greenhouse effect).

Fuel cells and the International Space Station

The hydrogen–oxygen fuel cell can be used as an energy source in spacecraft. The **International Space Station** (ISS) is a collaborative product of five space agencies representing 15 nations, and has been continuously inhabited by humans since November 2000.

▲ Figure 7 The Japanese pressurized experiment module for the International Space Station, shown here at its manufacturing facility in Nagoya, Japan. The module, called Kibo or "hope" in Japanese, is Japan's first human space facility. Experiments in Kibo focus on space medicine, biology, Earth observations, material science, biotechnology, and communications research

▲ Figure 6 This bus in Reyjavik, Iceland, is powered by a fuel cell that runs on hydrogen

Activity

1 Find out the cathode and anode half-equations and the overall cell reaction for the direct methanol fuel cell. Compare and contrast the methanol fuel cell with the hydrogen–oxygen fuel cell from an environmental perspective.

2 **a)** Discuss some aspects of what is commonly termed the "hydrogen economy". Your answer might address aspects such as the advantages and problems of using hydrogen as a fuel in motor cars, the various methods for generating hydrogen, and the use of renewable energy sources such as wind and solar energy.

   b) Suggest how wind farms may be assisting developing countries in dealing with their energy needs, and so driving the global "hydrogen economy". Explore what problems wind turbines may pose for rural communities. Research and discuss any possible health effects associated with wind turbines. Consider a number of countries worldwide where wind farms are located and compare and contrast any government regulations that may be in place regarding their construction.

▲ Figure 8 Example of an on-shore wind farm in Kilmore, Co. Wexford, Republic of Ireland

Electrolytic cells

Electrolysis is the process by which electrical energy is used to drive a non-spontaneous chemical reaction. An electrolytic cell is used for this purpose, which consists of a single container, two electrodes (the cathode and the anode), a solution (the electrolyte), and a battery which can be considered as an electron pump.

There are many different types of electrolytic cell but at SL you will be only assessed on the electrolysis of a *molten* salt.

Electrolysis of a molten salt such as lead(II) bromide

In the electrolysis of molten lead(II) bromide, $PbBr_2(l)$, inert graphite electrodes are dipped into the $PbBr_2(l)$ electrolyte. The following half-equations show the processes that take place at the electrodes:

- anode (positive electrode): oxidation

  $2Br^- \rightarrow Br_2(g) + 2e^-$

- cathode (negative electrode): reduction

  $Pb^{2+}(l) + 2e^- \rightarrow Pb(l)$

- overall cell reaction:

  $PbBr_2(l) \rightarrow Pb(l) + Br_2(g)$

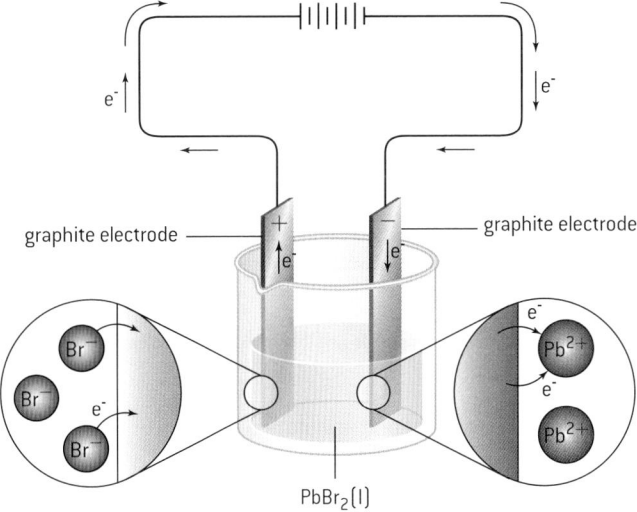

▲ Figure 9 Electrolysis of molten lead bromide, $PbBr_2(l)$

231

Working method for the electrolysis of a molten salt

Step 1: Identify all species present.

Step 2: Identify which species are attracted to the cathode (negative electrode) and which species are attracted to the anode (positive electrode).

Step 3: Deduce the two half-equations taking place at the cathode and anode and the overall cell reaction.

Step 4: Draw and annotate the electrolytic cell and show the direction of the movement of electrons and the direction of ion flow.

Step 5: State what would be observed at each electrode.

Worked example

Describe the electrolysis of molten sodium chloride.

Solution

Step 1:

$$NaCl \rightarrow Na^+ + Cl^-$$

So $Na(l)^+$ and $Cl(l)^-$ ions are present.

Step 2:

Cathode (negative electrode): Na^+

Anode (positive electrode): Cl^-

Step 3:

Anode (positive electrode): oxidation:

$$2Cl(l)^- \rightarrow Cl_2(g) + 2e^-$$

Cathode (negative electrode): reduction:

$$Na(l)^+ + e^- \rightarrow Na(l)$$

This needs to be multiplied by 2 to balance the number of electrons from the anode equation:

$$2Na(l)^+ + 2e^- \rightarrow 2Na(l)$$

Overall cell reaction:

$$2Cl(l)^- + 2Na(l)^+ \rightarrow 2Na(l) + Cl_2(g)$$

Step 4:

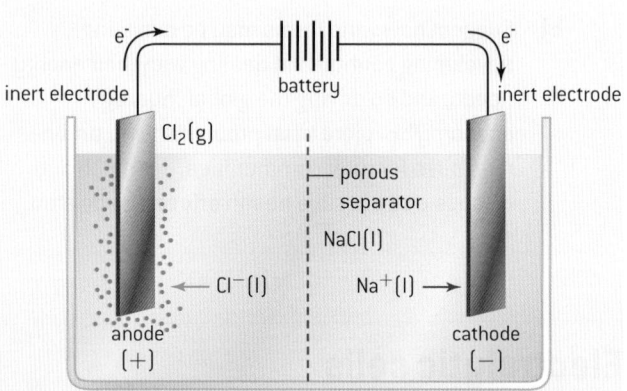

▲ Figure 10 Electrolytic cell for molten sodium chloride, NaCl(l). This experimental set-up is used commercially in the **Downs cell** for the electrolysis of sodium chloride. The liquid sodium metal is less dense than the molten sodium chloride, so it floats on the surface and is collected

Step 5:

At the anode (positive electrode): bubbles of chlorine gas are observed.

At the cathode (negative electrode): a pool of liquid sodium forms.

Quick questions

1 Explain why solid lead(II) bromide does not conduct electricity.

2 **a)** Construct and annotate a diagram of the electrolytic cell for the electrolysis of molten aluminium oxide.

 b) Identify the half-equations occurring at the cathode and at the anode.

 c) State a suitable material for each electrode.

 d) Identify the direction of movement of electrons and ion flow.

 e) State what would be observed at each electrode.

 f) Discuss, with reference to differences in properties, why aluminium is used to replace iron in many applications.

Questions

1 Which species could be reduced to form NO_2?

   A. N_2O

   B. NO_3^-

   C. HNO_2

   D. NO [1]

   IB, May 2011

2 Consider the overall reaction taking place in a voltaic cell.

$$Ag_2O(s) + Zn(s) + H_2O(l) \rightarrow 2Ag(s) + Zn(OH)_2(s)$$

What is the role of zinc in the cell?

   A. The positive electrode and the oxidizing agent.

   B. The positive electrode and the reducing agent.

   C. The negative electrode and the oxidizing agent.

   D. The negative electrode and the reducing agent.   [1]

   IB, May 2011

3 What happens to bromine when bromate ions, BrO_3^-, are converted to bromine molecules, Br_2?

   A. It undergoes reduction and its oxidation state changes from –1 to 0.

   B. It undergoes oxidation and its oxidation state changes from –1 to 0.

   C. It undergoes reduction and its oxidation state changes from +5 to 0.

   D. It undergoes oxidation and its oxidation state changes from +5 to 0.

4 Consider the following reactions of three unknown metals X, Y, and Z.

$$2XNO_3(aq) + Y(s) \rightarrow 2X(s) + Y(NO_3)_2(aq)$$

$$Y(NO_3)_2(aq) + Z(s) \rightarrow \text{no reaction}$$

$$2XNO_3(aq) + Z(s) \rightarrow 2X(s) + Z(NO_3)_2(aq)$$

What is the order of **increasing** reactivity of the metals (least reactive first)?

   A. X < Y < Z

   B. X < Z < Y

   C. Z < Y < X

   D. Y < Z < X   [1]

   IB, May 2011

5 Which statement about the electrolysis of molten sodium chloride is correct?

   A. A yellow-green gas is produced at the negative electrode.

   B. A silvery metal is produced at the positive electrode.

   C. Chloride ions are attracted to the positive electrode and undergo oxidation.

   D. Sodium ions are attracted to the negative electrode and undergo oxidation.   [1]

   IB, May 2011

6 What is the reducing agent in the reaction below?

$$2MnO_4^-(aq) + Br^-(aq) + H_2O(l) \rightarrow 2MnO_2(s) + BrO_3^- + 2OH^-(aq)$$

   A. Br^-

   B. BrO_3^-

   C. MnO_4^-

   D. MnO_2 [1]

   IB, May 2012

7 Which changes could take place at the positive electrode (cathode) in a voltaic cell?

   I. $Zn^{2+}(aq)$ to $Zn(s)$

   II. $Cl_2(g)$ to $Cl^-(aq)$

   III. $Mg(s)$ to $Mg^{2+}(aq)$

   A. I and II only

   B. I and III only

   C. II and III only

   D. I, II, and III   [1]

   IB, May 2010

8 Metal A is more reactive than metal B. A standard voltaic cell is made as shown (figure 11).

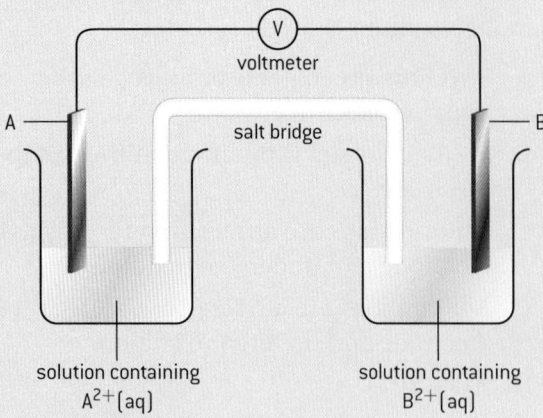

▲ Figure 11

Which statement is correct?

A. Electrons flow in the external circuit from A to B.

B. Positive ions flow through the salt bridge from A to B.

C. Positive ions flow in the external circuit from B to A.

D. Electrons flow through the salt bridge from B to A. [1]

IB, November 2010

9 A 0.1337 g sample of an alkali metal iodate, XIO_3, was dissolved in water, acidified, and an excess of potassium iodide, KI added. The resulting iodine solution required 36.64 cm^3 of the sodium thiosulfate pentahydrate solution, $Na_2S_2O_3.5H_2O$ (25.49 g dm^{-3}) for complete titration using starch solution as an indicator. Calculate the relative atomic mass of X and hence identify the metal. Deduce all relevant half-equations involved.

10 Describe how the dissolved oxygen concentration in a river would decrease if:

a) a car factory releases warm water into the river after using it for cooling [1]

b) a farmer puts large quantities of a fertilizer on a field next to the river. [1]

IB, May 2009

11 Describe how the addition of nitrates or phosphates to water can increase the BOD value of a water sample. [2]

IB, November 2009

12 The Winkler method uses redox reactions to find the concentration of oxygen in water. 100 cm^3 of water was taken from a river and analysed using this method. The reactions taking place are summarized below.

Step 1: $2Mn^{2+}(aq) + 4OH^-(aq) + O_2(aq) \rightarrow$
$$2MnO_2(s) + 2H_2O(l)$$

Step 2: $MnO_2(s) + 2I^-(aq) + 4H^+(aq) \rightarrow$
$$Mn^{2+}(aq) + I_2(aq) + 2H_2O(l)$$

Step 3: $2S_2O_3^{2-}(aq) + I_2(aq) \rightarrow S_4O_6^{2-}(aq) + 2I^-(aq)$

a) State what happened to the O_2 in step 1 in terms of electrons. [1]

b) State the change in oxidation number for manganese in step 2. [1]

c) 0.0002 moles of I$^-$ were formed in step 3. Calculate the amount, in moles, of oxygen, O_2, dissolved in the water. [1]

IB, November 2009

10 ORGANIC CHEMISTRY

Introduction

Organic chemistry is the chemistry of carbon-containing compounds and studies a vast array of compounds and their reactions. From biological systems to biotechnology, from foods to fuels, from paints and dyes to pesticides and fertilizers, organic chemistry is of fundamental importance to the expansion of our understanding of the material world. In this topic, we develop an understanding of the classification system of organic compounds. The application of IUPAC rules of nomenclature will be the main focus, in addition to the identification of important functional groups and their reactions. The chemistry of alkanes, alkenes, alcohols, halogenoalkanes, polymers and benzene will be explored.

10.1 Fundamentals of organic chemistry

Understandings

→ A homologous series is a series of compounds of the same family, with the same general formula, which differ from each other by a common structural unit.

→ Structural formulas can be represented in full and condensed format.

→ Structural isomers are compounds with the same molecular formula but different arrangements of atoms.

→ Functional groups are the reactive parts of molecules.

→ Saturated compounds contain single bonds only and unsaturated compounds contain double or triple bonds.

→ Benzene is an aromatic, unsaturated hydrocarbon.

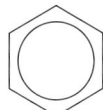

Applications and skills

→ Explanation of the trends in boiling points of members of a homologous series.

→ Distinction between empirical, molecular, and structural formulas.

→ Identification of different classes: alkanes, alkynes, halogenoalkanes, alcohols, ethers, aldehydes, ketones, esters, carboxylic acids, amines, amides, nitriles, and arenes.

→ Identification of typical functional groups in molecules eg phenyl, hydroxyl, carbonyl, carboxamide, aldehyde, ester, ether, amine, nitrile, alkyl, alkenyl and alkynyl.

→ Construction of 3-D models (real or virtual) of organic molecules.

→ Application of IUPAC rules in the nomenclature of straight-chain and branched-chain isomers.

→ Identification of primary, secondary, and tertiary carbon atoms in halogenoalkanes, and alcohols and primary, secondary, and tertiary nitrogen atoms in amines.

→ Discussion of the structure of benzene using physical and chemical evidence.

Nature of science

→ Serendipity and scientific discoveries – PTFE and superglue.

→ Ethical implications – drugs, additives, and pesticides can have harmful effects on both people and the environment.

The theory of "Vitalism" was based on the belief that a vital force was involved in the chemistry of living organisms. Indeed the word "organic" originated from scientists' understanding at that time, that organic compounds could only be synthesized within living organisms. This belief remained until the German chemist Friedrich Wöhler artificially synthesized urea from the inorganic compound ammonium cyanate, NH_4OCN.

Are there other examples in science where vocabulary has developed from a misunderstanding that was the product of the technology of the day? Language plays a vital role in the communication of knowledge and its subsequent understanding. Therefore should language be universal so that misnomers arising from misconceptions may be eliminated?

Introduction to organic chemistry

Organic chemistry is the field of chemistry that studies carbon-based compounds. Carbon atoms have four valence electrons so they can form four bonds to other atoms. Carbon can undergo **catenation**, the process by which many identical atoms are joined together by covalent bonds, producing straight-chain, branched, or cyclic structures. Organic chemistry is therefore a wide and varied field of study.

Understanding natural and synthetic organic compounds requires the study of chemical bonding and nomenclature, chemical structure, stoichiometric relationships, functional groups, and reaction mechanisms. The energetics of reactions and their role in industry, chemical kinetics, and the impact of synthetic medicines and drugs on the health of society are some of the points of focus of organic chemistry.

Homologous series

Classification is a common human activity. Just as biology uses scientific taxonomy to classify organisms on the basis of shared characteristics, chemists utilize a unique system of nomenclature to group and name compounds that share important features and patterns of reactions.

A **homologous series** is a series of compounds that can be grouped together based on similarities in their structure and reactions. A homologous series has the same general formula which varies from one member to another by one CH_2 (methylene) group.

The **alkane** series has the general formula C_nH_{2n+2} (table 1). The alkanes are **hydrocarbons** (they contain carbon and hydrogen only). The **alkenes** and **alkynes** are two more hydrocarbon homologous series that contain carbon–carbon double and triple bonds, respectively.

Homologous series that contain **functional groups** can also be described by a general formula and also show similar physical and chemical properties within the series. The functional groups are the reactive parts of the molecules and commonly contain elements such as oxygen and nitrogen. In the alkene and alkyne homologous series the carbon–carbon double and triple bonds respectively make up the functional groups of the series. Table 2 shows the structures of three homologous series that contain oxygen.

Physical properties of a homologous series

The physical properties of the members of a homologous series change gradually as the length of the carbon chain increases. For example, the boiling points of members of the alkane series can be measured using the apparatus shown in figure 1. Such an experiment shows that the boiling point rises with an increasing number of carbon atoms (or increasing molar mass), as seen in table 1. This can be seen by the state at room temperature within the series: butane is a gas at room temperature, while pentane is a liquid.

Name	Formula	Condensed structural formula	Structural formula	Boiling point /°C
methane	CH_4	CH_4	H—C—H (with H above and below)	−161
ethane	C_2H_6	CH_3CH_3	H—C—C—H (with H's)	−89
propane	C_3H_8	$CH_3CH_2CH_3$	H—C—C—C—H (with H's)	−42
butane	C_4H_{10}	$CH_3CH_2CH_2CH_3$	H—C—C—C—C—H (with H's)	−0.5
pentane	C_5H_{12}	$CH_3CH_2CH_2CH_2CH_3$	H—C—C—C—C—C—H (with H's)	36
hexane	C_6H_{14}	$CH_3CH_2CH_2CH_2CH_2CH_3$	H—C—C—C—C—C—C—H (with H's)	69

▲ Table 1 The homologous series of alkanes

Homologous series	alcohols	aldehydes	ketones
General formula	$C_nH_{2n+1}OH$	$C_nH_{2n}O$	$C_nH_{2n}O$
C_3	H—C—C—C—OH (with H's)	H—C—C—C(=O)—H (with H's)	H—C—C—C—H (with O)
C_4	H—C—C—C—C—OH (with H's)	H—C—C—C—C(=O)—H (with H's)	H—C—C—C—C—H (with O)
C_5	H—C—C—C—C—C—OH (with H's)	H—C—C—C—C—C(=O)—H (with H's)	H—C—C—C—C—C—H (with O)

▲ Table 2 The general formula and structural formulae of the homologous series of alcohols, aldehydes, and ketones

This trend in boiling point results from increasingly strong intermolecular forces (London (dispersion) forces, sub-topic 4.4) as the carbon chain becomes longer. Trends in increasing density and viscosity with carbon chain length are well understood by the petrochemical industry. Crude oil is a mixture of hydrocarbons that vary in the length of their carbon chain. **Fractional distillation** is a physical separation process that uses differences in boiling points to separate the mixture into fractions of similar boiling point. A simple distillation apparatus can effectively separate volatile fractions from long-chain, non-volatile compounds in a school laboratory.

Homologous series have similar chemical properties due to the presence of the same functional group; this is responsible for their overall chemical reactivity and the types of characteristic reactions they undergo.

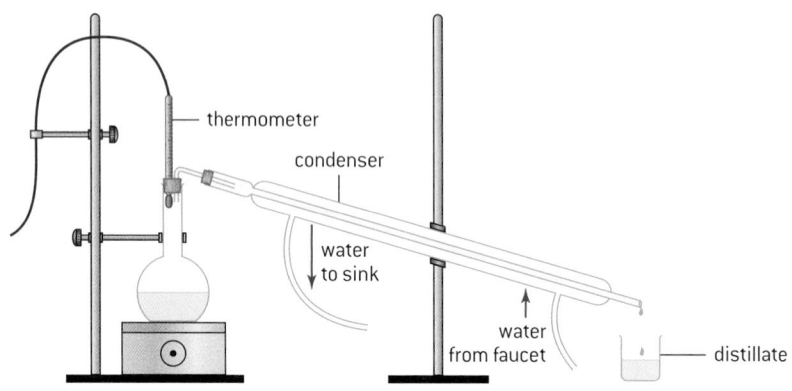

▲ Figure 1 Distillation apparatus incorporating a temperature probe

Quick questions

1 Alkenes are important starting materials for a variety of products.

a) State and explain the trend of the boiling points of the first five members of the alkene homologous series. [3]

b) Describe two features of a homologous series. [2]

IB, May 2011

2 The boiling points of the isomers of pentane, C_5H_{12}, shown in figure 2 are 10 °C, 28 °C, and 36 °C, but not necessarily in that order.

a) Identify the boiling point for each of the isomers **A**, **B**, and **C** in a copy of table 1 and state a reason for your answer. [3]

Isomer	A	B	C
Boiling point			

▲ Table 3

b) Applying IUPAC rules, state the names of isomers **B** and **C**. [2]

IB, Nov 2009

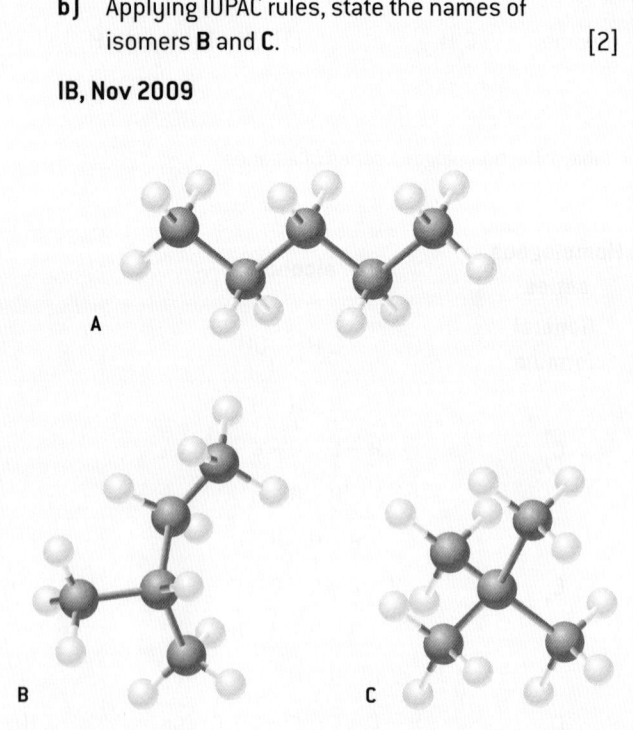

▲ Figure 2

Chemical formulae of organic compounds

The structure of an organic compound may be represented in several different ways providing varying levels of information.

In sub-topic 4.3 we examined the use of **Lewis (electron-dot) structures**. These are useful to visualize the valence electrons present in simple molecular compounds and polyatomic ions.

Empirical formulae (sub-topic 1.2) represent the simplest ratio of atoms present in a molecule. The **molecular formula** describes the actual number of atoms present in the molecule. Both these types of formula offer little or no information about the possible structure of larger, more complex molecules.

Structural formulae take three forms: full, condensed, and skeletal.

- **Full structural formulae** are two-dimensional representations showing all the atoms and bonds, and their positions relative to one another in a compound.

- In a **condensed structural formula** all the atoms and their relative positions are represented but the bonds are omitted.

- A **skeletal formula** is the most basic representation of the structural formula where the carbon and hydrogen atoms are not shown but the end of each line and each vertex represents a carbon atom. The atoms present in functional groups are also included as shown in table 4.

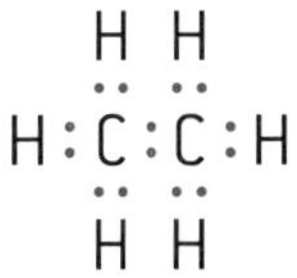

H H
H : C : C : H
H H

▲ Figure 3 Lewis structure of ethane, C_2H_6

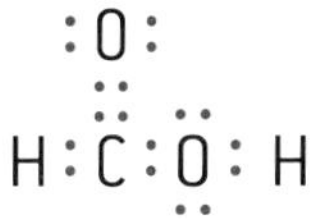

: O :
H : C : O : H

▲ Figure 4 Lewis structure of methanoic acid, HCOOH

> In structural formulae a covalent bond between two atoms is represented by a single line that describes two bonding electrons. For a double bond two lines are used and for a triple bond, three lines are used (sub-topic 4.2).

Name	Full structural formula	Condensed structural formula	Skeletal formula
propane	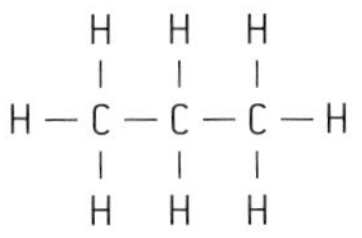 H H H / H—C—C—C—H / H H H	$CH_3CH_2CH_3$	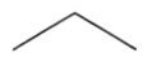 (skeletal drawing)
propan-2-ol	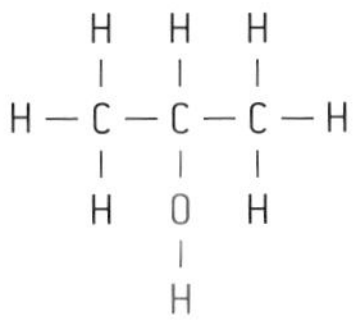 H H H / H—C—C—C—H / H O H / H	$CH_3CH(OH)CH_3$	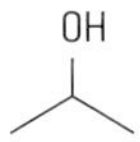 OH (skeletal drawing)
propanal	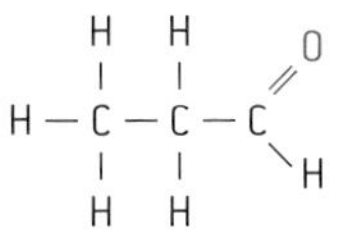 H H O / H—C—C—C / H H H	CH_3CH_2CHO	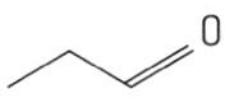 O (skeletal drawing)
propanone	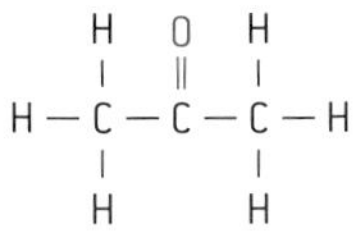 H O H / H—C—C—C—H / H H	$CH_3C(O)CH_3$	 O (skeletal drawing)
propene	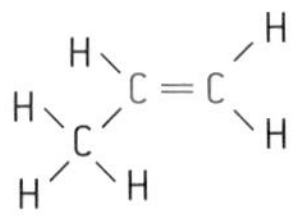 H H / C=C / H H / C / H H	$CH_3CH{=}CH_2$ or CH_3CHCH_2	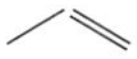 (skeletal drawing)

▲ Table 4 Full, condensed, and skeletal structural formulae can all be used to represent organic compounds

Nomenclature of organic compounds

The International Union of Pure and Applied Chemistry (IUPAC) is the world authority on chemical nomenclature. The name of a chemical substance needs to provide enough information to signpost the class of compound from which the chemical is derived, including any substituents and functional groups present. The name has a number of parts that describe the compound (figure 5).

The alkanes form the backbone of the IUPAC rules for naming organic compounds.

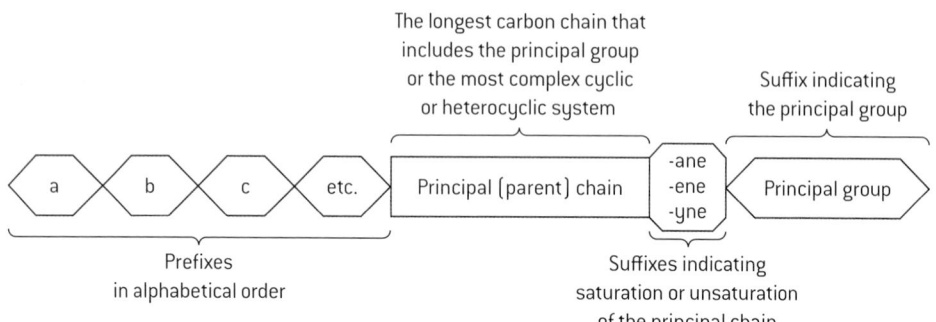

▲ Figure 5 Outline of the nomenclature of organic compounds

Nomenclature of alkanes

1 Examine the structure of the compound and determine the longest continuous carbon chain. This provides the root name for the alkane (table 5).

2 If alkyl substituents are present, creating branched chains, the name for the branch will be determined by the number of carbons (table 5). The *suffix* will change from "-ane" to "-yl".

3 When numbering the longest carbon chain, the position of any substituent must be the lowest numbered carbon. In this example, numbering from left to right results in the methyl substituent being on carbon 2. Numbering from right to left would incorrectly have the substituent on carbon 5.

Length of carbon chain	Name
1	meth-
2	eth-
3	prop-
4	but-
5	pent-
6	hex-

▲ Table 5 The IUPAC root names for the alkane series

Substituent name	Condensed formula
methyl	$-CH_3$
ethyl	$-CH_2CH_3$
propyl	$-CH_2CH_2CH_3$
butyl	$-CH_2CH_2CH_2CH_3$

▲ Table 6 Naming alkyl substituents

4 When there are several different substituents, arrange them in alphabetical order prior to the *root* name.

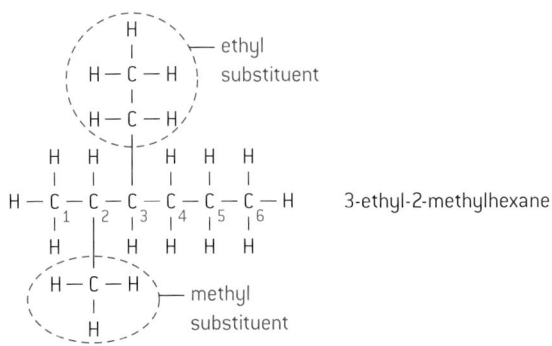

3-ethyl-2-methylhexane

alphabetically, ethyl comes before methyl.

5 Use a comma to separate numbers.

6 Use a hyphen to separate numbers and letters.

7 The number of multiple substituents of the same type is indicated by prefixes shown in table 7.

8 Successive words are merged into one word.

1	mono
2	di
3	tri
4	tetra
5	penta

▲ Table 7 Numerical multipliers in the IUPAC nomenclature system

To demonstrate the application of these rules we shall examine the isomers or **structural isomers** of the hydrocarbon hexane. Structural isomers are compounds that have the same chemical formula but a different structural formula. Isomers have unique physical and chemical properties.

1 Begin by drawing the molecule with the longest straight chain of carbon–carbon atoms.

$$H_3C — CH_2 — CH_2 — CH_2 — CH_2 — CH_3$$

hexane

2 Reduce the longest chain by one carbon and use the removed carbon to act as a methyl substituent. The longest chain is now five carbons so this derivative of hexane is a substituted pentane. The numbering of the chain must result in the methyl group branching off at the lowest numbered carbon.

$$\begin{array}{c} CH_3 \\ | \\ H_3C — CH — CH_2 — CH_2 — CH_3 \\ {}_{1}\quad\ {}_{2}\quad\ {}_{3}\quad\quad {}_{4}\quad\quad {}_{5} \end{array}$$

2-methylpentane

3 Examine the isomer to see if the methyl substituent can be moved to another carbon. If it is moved to C3 another isomer is formed. This is the last of the substituted pentane isomers (4-methylpentane does not exist as it is the same as 2-methylpentane).

$$\begin{array}{c} CH_3 \\ | \\ H_3C — CH_2 — CH — CH_2 — CH_3 \\ {}_{1}\quad\ {}_{2}\quad\quad {}_{3}\quad\ {}_{4}\quad\quad {}_{5} \end{array}$$

3-methylpentane

Isomers

Isomers may differ from one another in their physical properties. The ability of molecules of one isomer to pack closer together will result in increased intermolecular forces and therefore an increase in the boiling point. Hexane molecules (boiling point 69 °C) can approach each other more closely than those of the branched-chain hexane derivative 2,3-dimethylbutane (boiling point 58 °C).

241

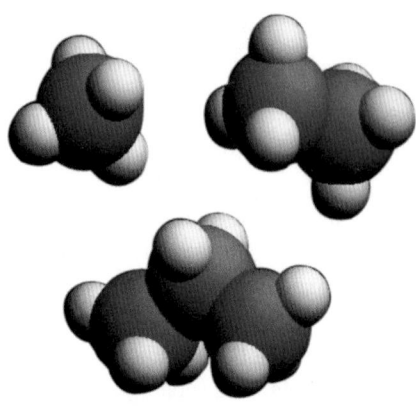

▲ Figure 7 Computer-generated 3D models of methane, ethane, and propane, the first three members of the alkane series

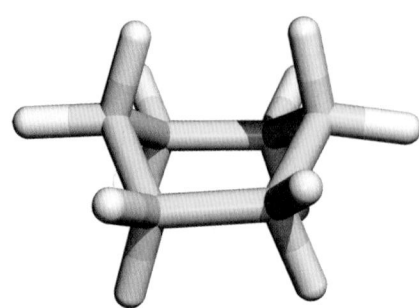

▲ Figure 8 Computer-generated model of cyclohexane, a cycloalkane

4   In a similar way, now remove another carbon atom and create substituted butane compounds from the original hexane:

```
        CH₃                        CH₃  CH₃
         |                          |    |
H₃C — C — CH₂ — CH₃        H₃C — CH — CH — CH₃
    1  2|   3    4              1    2    3    4
        CH₃
```

  2,2-dimethylbutane   2,3-dimethylbutane

Saturated and unsaturated hydrocarbons

Hydrocarbons are organic compounds consisting of carbon and hydrogen atoms only. In a **saturated** compound all the carbon–carbon bonds are single bonds. **Unsaturated** compounds contain double and/or triple carbon–carbon bonds. The simplest example of a saturated hydrocarbon is methane, CH_4, a member of the **alkane** family. Alkanes are **aliphatic** or straight-chain compounds (figure 7).

The majority of naturally occurring hydrocarbons come from crude oil. This mixture is extracted from beneath the Earth's surface, refined, and separated by fractional distillation into useful substances such as petroleum, butane, and kerosene.

The mixture of hydrocarbons that makes up crude oil is a combination of mainly alkanes, **cycloalkanes** and **aromatic hydrocarbons**. Cycloalkanes are ring structures that contain single carbon–carbon bonds (figure 8), whereas aromatic hydrocarbons or **arenes** are ring structures consisting of alternating single and double carbon–carbon bonds.

Functional groups

Tens of millions of organic compounds exist in the world and the number is constantly rising as new compounds are synthesized by pharmaceutical companies and chemical industries. **Synthetic compounds** are the products of reactions involving both natural and man-made compounds. **Natural compounds** found in plants and animals are synthesized by organisms. All these substances are organized into **classes of organic compounds** containing specific functional groups (table 7).

When naming compounds that contain a functional group, the position of the group is identified by giving the number of the carbon atom to which it is attached. When numbering the carbon atoms, functional groups take priority over substituents and carbon–carbon multiple bonds.

Unsaturated hydrocarbons

The primary chain in unsaturated hydrocarbons must include the double or triple carbon–carbon bond. If the molecule below was numbered from left to right, the methyl substituent would branch off from C2 and the double bond would be at C3. However, the double bond takes priority so numbering is from right to left as shown.

```
      H    H    H    H    H
      |    |    |    |    |
  H — C — C — C = C — C — H
      |5   |4   3    2    |1
      H   CH₃             H
```

4-methylpent-2-ene. Note the use of hyphens and the fact that the substituent is named before the functional group

Class	Functional group	Suffix	General formula	Example
alkanes	——	-ane	C_nH_{2n+2}	Propane C_3H_8
alkenes	alkenyl >C=C<	-ene	C_nH_{2n}	but-2-ene $CH_3CH=CHCH_3$
alkynes	—C≡C— alkynyl	-yne	C_nH_{2n-2}	but-2-yne $CH_3C≡CCH_3$
arenes	phenyl ⬡	——	C_nH_{2n-6}	benzene C_6H_6
halogenoalkanes	—X (X = F, Cl, Br, I)	——	$C_nH_{2n+1}X$	2-chlorobutane $CH_3CH(Cl)CH_2CH_3$
alcohols	—OH hydroxyl	-ol	ROH	butan-2-ol $CH_3CH(OH)CH_2CH_3$
aldehydes	aldehyde —C(=O)H	-al	RCHO	ethanal CH_3CHO
ketones	C=O carbonyl	-one	RC(O)R′	propanone $CH_3C(O)CH_3$
carboxylic acids	—C(=O)OH carboxyl	-oic acid	RCOOH	ethanoic acid CH_3COOH
esters	—C(=O)O— ester	-oate	RCOOR′	methyl ethanoate CH_3COOCH_3
ethers	—O— ether	——	ROR′	ethoxyethane $CH_3CH_2OCH_2CH_3$
amines	—N(H)H / —N<(H) / —N< amino	-amine	RNH_2 RNHR′ RN(R′)R″	propan-1-amine $CH_3CH_2CH_2NH_2$ N-methylethanamine $CH_3CH_2NHCH_3$
amides	—C(=O)NH₂ amido	-amide	$RCONH_2$	ethanamide CH_3CONH_2
nitriles	—C≡N cyano	-nitrile	RCN	propanenitrile CH_3CH_2CN

▲ Table 7 A summary of classes of organic compounds showing their functional groups

Constructing 3-D models of organic compounds is an excellent interactive technique, enhancing visualization of the molecule and mutual orientation of individual atoms. Models can enhance understanding of a variety of concepts from the naming of organic molecules, visualizing stereoisomers (including optical isomers) to complex reactions mechanisms (sub- topic 20.1).

Aliphatic compounds

H H H H
| | | |
H — C — C — C — C — H
| | | |
H H OH H

butan-2-ol

Number the carbon chain from right to left so that the functional group is on the lowest numbered carbon atom.

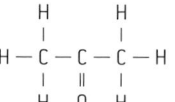

H H H O
| | | ∥
H — C — C — C — C
| | | \
H H H H

butanal

The functional group for an aldehyde is by definition on the terminal carbon atom (C1), so there is no need to state this number.

H H
| |
H — C — C — C — H
| ∥ |
H O H

propanone

The ketone has a general formula RC(O)R′. In this three-carbon compound the functional group can only be on C2.

H Cl H H H
| | | | |
H — C — C — C — C — C — H
| | | | |
H Cl H H H

2,2-dichloropentane

For multi-substituted halogenoalkanes, the carbon number is used along with a prefix to signify the number of halogens present. If different halogens are present, they are listed in alphabetical order.

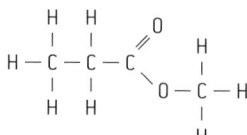

H H H O
| | | ∥
H — C — C — C — C
| | | \
H H H O — H

butanoic acid

Note that the carbon atom in the functional group is counted in the longest carbon chain. As the functional group is in the terminal position, there is no need to include the number C1 in the name.

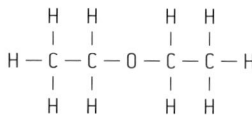

H H O
| | ∥
H — C — C — C
| | \ H
H H O — C — H
|
H

methyl propanoate

Esters are acid derivatives. The alkyl substituent replaces the hydrogen atom on the functional group —COOH. The alkyl substituent is named first, followed by the name of the acid anion.

H H H H
| | | |
H — C — C — O — C — C — H
| | | |
H H H H

ethoxyethane

Ethers are named as substituted alkanes. The —O—R′ group is called the alkoxy group so —OCH$_2$CH$_3$ is the ethoxy group.

By convention, the functional group of the carboxylic acids, aldehydes, esters and amides are positioned at the right-hand end of the structural formulae of organic compounds.

PTFE, a fortunate discovery

PTFE or polytetrafluoroethene is a synthetic polymer that is more commonly known by its commercial name, Teflon. It is a **thermoplastic** polymer which means it can be moulded when heated, retaining its new shape upon cooling.

Its properties include high chemical resistance, a low coefficient of friction, high melting point, electrical and thermal insulation, non-stick surface qualities, and very low solubility in all known solvents. The most common use of PTFE is as a non-stick surface on cooking implements.

This fluoropolymer was accidently discovered in 1938, a product of the iron-catalysed polymerization of tetrafluoroethene gas.

Serendipity describes fortunate accidental discoveries within science. Scientific endeavour stems from flashes of inspiration, imagination, and serendipity. Methyl cyanoacrylate or "superglue" is another example of an accidental discovery. *"Ethical discussions, risk-benefit analyses, risk assessment and the precautionary principle are all parts of the scientific way of addressing the common good."* (IB Chemistry syllabus.)

Classifying molecules: primary, secondary, and tertiary compounds

We shall look at three classes of compound to understand the effect on their chemical reactions of changing the position of the functional group on the carbon chain. We shall look at the carbon attached to the functional group: a **primary** carbon atom is bonded to one other carbon atom, a **secondary** carbon atom to two other carbon atoms, and a **tertiary** carbon atom to three other carbon atoms (figure 9).

Classifying halogenoalkanes

Alkanes undergo free-radical substitution reactions with halogens (sub-topic 10.2) and the resulting mono-substituted alkanes are known as halogenoalkanes. Whether a primary, secondary, or tertiary halogenoalkane is formed depends on the conditions and mechanism of the reaction. This idea will be developed in sub-topic 20.1.

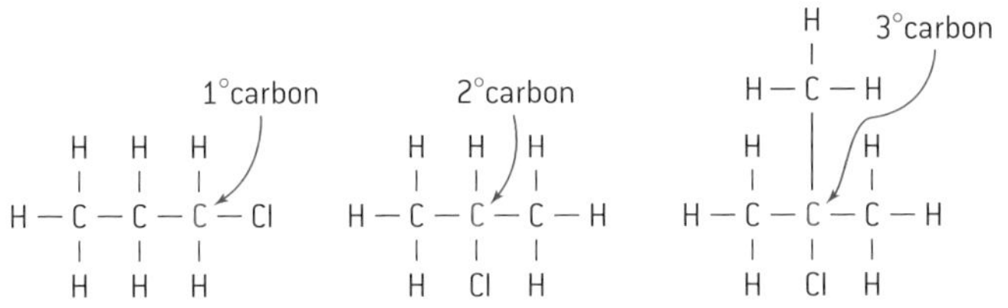

▲ Figure 9 Primary (1°), secondary (2°), and tertiary (3°) halogenoalkanes

Classifying alcohols

Alcohols can be classified in the same way as halogenoalkanes (figure 10). The position of the hydroxyl group determines the products formed when the alcohol undergoes oxidation in the presence of acidified potassium dichromate(VI) or potassium manganate(VII) (sub-topic 10.2).

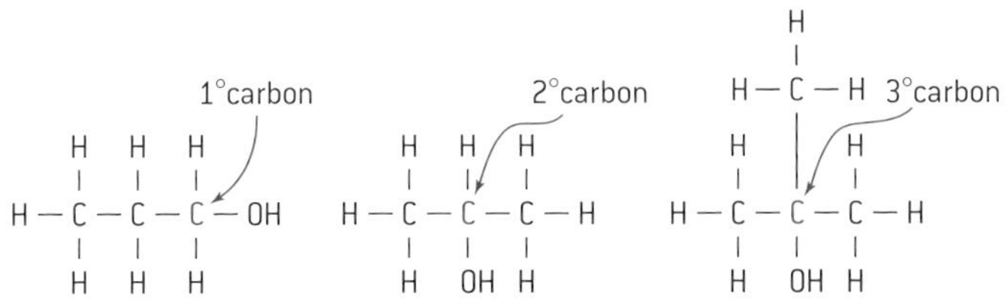

▲ Figure 10 Primary (1°), secondary (2°), and tertiary (3°) alcohols

Biofuels are substances whose energy is derived from carbon fixation in plants. Alcohols and other biofuels are increasingly being used as alternative fuels to petrol (gasoline) and diesel. Brazil has undertaken large-scale production of ethanol from sugar cane for decades, adding it to traditional fossil fuels. Fossil fuels remain the primary source of energy on a global scale, but the complex mixture that makes up crude oil contains components that can be used for the synthesis of various products – from dyes and cosmetics to pesticides and polymers. The ever-increasing combustion of valuable non-renewable fossil fuels could result in the depletion not only of the fuels themselves, but also of valuable raw materials for a vast array of substances that are a part of our daily lives.

Classifying amines

An amine is classified as a primary, secondary or tertiary amine depending on the number of alkyl groups bonded to the *nitrogen* atom of the functional group (unlike halogenoalkanes and alcohols, which consider the *carbon* atom next to the functional group).When naming amines the root loses the -e and is replaced by "amine". "*N*" signifies that the substituent, namely the methyl group in figure 11, is bonded to the nitrogen atom rather than the carbon atom.

propan-1-amine | *N*-methylpropan-1-amine | *N*, *N*-dimethylpropan-1-amine

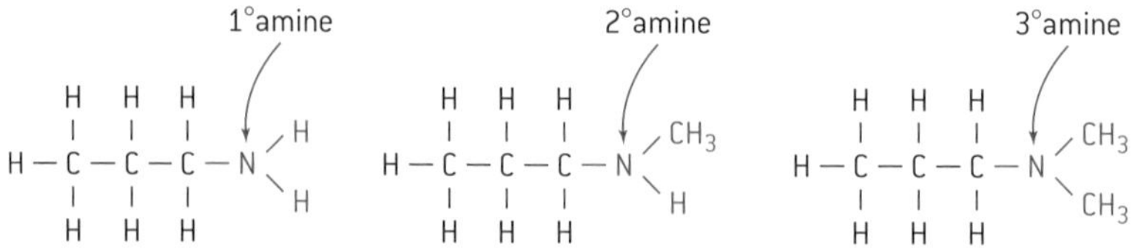

▲ Figure 11 Primary (1°), secondary (2°), and tertiary (3°) amines

Aromatic hydrocarbons

The alkanes, alkenes, and alkynes are differentiated by the presence of single, double, or triple bonds respectively. **Aromatic hydrocarbons** are characterized by the presence of the benzene ring. The German scientist August Kekulé (1829–96) proposed a ring structure for benzene C_6H_6 composed of six carbons bonded together by alternating double and single bonds. This structure would result in an unsymmetrical molecule with carbon–carbon bonds of different bond lengths.

Benzene crystallizes upon cooling and analysis of X-ray diffraction patterns generated from the crystalline substance revealed that all six carbon–carbon bonds in its molecule have identical bond lengths of 140 pm. It is now understood that the carbon–carbon bonds in benzene are intermediate in length between single (154 pm) and double (134 pm) carbon–carbon bonds and thus have a bond order of 1.5. Electrostatic potential mapping of benzene (figure 12) confirms that all the carbon atoms have equal electron density, so the molecule is symmetrical.

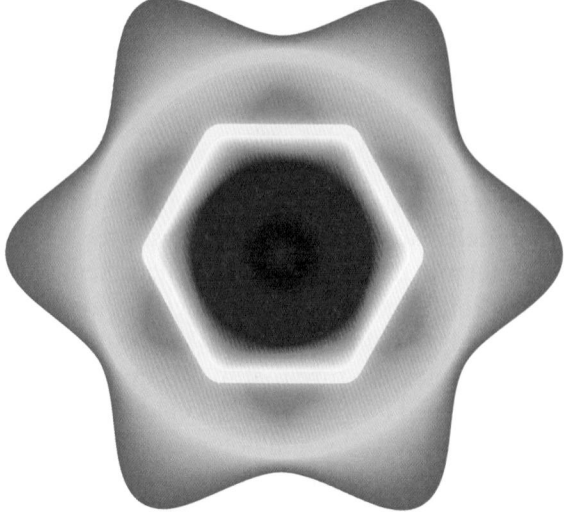

▲ Figure 12 An electrostatic potential map of benzene

Octane numbers of fuels

The fractional distillation of crude oil or petroleum separates this mixture of hydrocarbons into various fuels including natural gas, gasoline or petrol, kerosene used in the aviation industry, diesel fuel used in transport, construction, and agriculture, and fuel oil for heating.

The **octane number** is a standard method of describing the performance of fuels used in cars and aircraft. The octane number is not indicative of the energy content of the fuel but is rather a way of describing its ability to combust in a controlled manner without causing excessive **engine knocking**. In simple terms, engine knocking is the result of uncontrolled detonation of the air–fuel mixture in a combustion engine. This is more common in fuels that have a low octane number.

Automobile manufacturers are multinational companies that produce and sell their products globally. The octane rating system is expressed differently in different countries, making comparisons difficult. The **research octane number** (RON) is the most common method of determining the octane number, though the **motor octane number** (MON) is also used either in isolation or in combination with the RON. Thus methods of reporting fuel octane numbers at the pump can vary from country to country, leading to inconsistency in communication to consumers.

Benzene contains six sp^2 hybridized carbon atoms (sub-topic 14.2) bonded to one another and each carbon is bonded to a single hydrogen atom by sigma bonds. The p orbitals of the six sp^2 hybridized carbon atoms overlap one another, forming a continuous π bond that lies above and below the plane of the six carbon atoms. The delocalization of the π electrons over the six carbon nuclei can be represented by the resonance structures of benzene shown in figure 13.

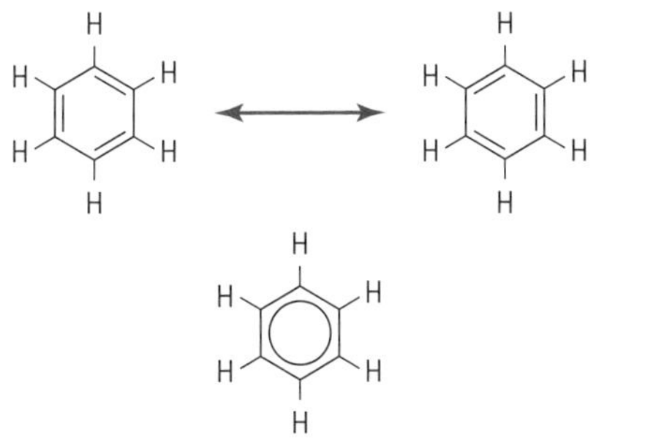

▲ Figure 13 Resonance structures of benzene

The Kekulé structure of benzene was drawn as a series of alternating double and single bonds. The circle within the six-member ring structure represents a system of delocalized pi electrons that are evenly distributed between each of the six carbons.

Hydrogenation is the addition of hydrogen to unsaturated hydrocarbons in the presence of a catalyst (sub-topic 10.2). If benzene contained three carbon–carbon double bonds, the enthalpy change of hydrogenation of benzene would be approximately three times the enthalpy change of hydrogenation of cyclohexene. However, experiments show that it is in fact much less than this, and this difference in energy, known as the **resonance energy** or **delocalization energy**, is evidence of the enhanced stability of the benzene molecule resulting from the delocalization of the π electrons. A consequence of this stability is that benzene readily undergoes electrophilic substitution reactions (sub-topic 20.1) but does not show addition reactions as other unsaturated aliphatic and cyclic compounds do; for example, benzene does not decolorize bromine water. Only one isomer exists for 1,2-disubstituted benzene compounds; there would be two isomers if the benzene ring had alternating single and double bonds as suggested by Kekulé.

TOK

August Kekulé is best known for his discovery of the structure of benzene. While many scientific discoveries are the product of reasoning supported by evidence obtained through observation and experimentation, some discoveries are born from moments of inspiration or flashes of intuition, as well as a healthy imagination with a high degree of creativity. Kekulé is said to have visualized the cyclic structure of benzene in a dream. While the importance of evidence is universally accepted in the scientific community, there is an understanding of the role of less analytical ways of knowledge in the acquisition of scientific knowledge. To what degree do these ways of knowing play a part in the acquisition of new knowledge, and can you think of some recent examples?

▲ Figure 14 August Kekulé

247

10.2 Functional group chemistry

Understandings

Alkanes:

→ Alkanes have low reactivity and undergo free-radical substitution reactions.

Alkenes:

→ Alkenes are more reactive than alkanes and undergo addition reactions. Bromine water can be used to distinguish between alkenes and alkanes.

Alcohols:

→ Alcohols undergo nucleophilic substitution reactions with acids (also called esterification or condensation) and some undergo oxidation reactions.

Halogenoalkanes:

→ Halogenoalkanes are more reactive than alkanes. They can undergo (nucleophilic) substitution reactions. A nucleophile is an electron-rich species containing a lone pair that it donates to an electron-deficient carbon.

Polymers:

→ Addition polymers consist of a wide range of monomers and form the basis of the plastics industry.

Benzene:

→ Benzene does not readily undergo addition reactions but does undergo electrophilic substitution reactions.

Nature of science

→ Use of data – much of the progress that has been made to date in the developments and applications of scientific research can be mapped back to key organic chemical reactions involving functional group interconversions.

Applications and skills

Alkanes:

→ Writing equations for the complete and incomplete combustion of hydrocarbons.

→ Explanation of the reaction of methane and ethane with halogens in terms of a free-radical substitution mechanism involving photochemical homolytic fission.

Alkenes:

→ Writing equations for the reactions of alkenes with hydrogen and halogens and of symmetrical alkenes with hydrogen halides and water.

→ Outline of the addition polymerization of alkenes.

→ Relationship between the structure of the monomer to the polymer and repeating unit.

Alcohols:

→ Writing equations for the complete combustion of alcohols.

→ Writing equations for the oxidation reactions of primary and secondary alcohols (using acidified potassium dichromate(VI) or potassium manganate(VII) as oxidizing agents). Explanation of distillation and reflux in the isolation of the aldehyde and carboxylic acid products.

→ Writing the equation for the condensation reaction of an alcohol with a carboxylic acid, in the presence of a catalyst (eg concentrated sulfuric acid) to form an ester.

Halogenoalkanes:

→ Writing the equation for the substitution reactions of halogenoalkanes with aqueous sodium hydroxide.

 Converting one functional group to another

Chemical reactions involving functional group interconversions form the basis for the synthesis of organic compounds. Scientific research into natural compounds and their production involves the determination of their structure and subsequently the design of pathways to achieve their synthesis so that they can be produced in a laboratory. A wealth of data on the chemistry of organic functional groups has enabled chemists to utilize various reaction pathways to create new organic molecules. Society's desire to advance the health of communities, ensure food supplies for both the developed and developing global populations, and improve quality of life drives research to develop new organic compounds.

Alkanes

Alkanes are the simplest hydrocarbons. With low bond polarity and strong covalent carbon–carbon bonds (bond energy 346 kJ mol^{-1}) and carbon–hydrogen bonds (bond energy 414 kJ mol^{-1}), they are relatively inert. However, alkanes do undergo some important reactions.

The combustion of alkanes

Alkanes are commonly used as fuels, releasing large amounts of energy in combustion reactions. Volatility (the tendency to change state from liquid to gas) decreases as the length of the carbon chain increases. Alkanes used as fuels tend to be short-chain molecules such as butane and octane.

Alkanes undergo **complete combustion** in the presence of excess oxygen. This highly exothermic reaction produces carbon dioxide and water. Carbon dioxide has a significant environmental impact, contributing to global warming. Petrol or gasoline is a mixture of hydrocarbons with octane present in the highest proportion.

$$C_8H_{18}(l) + 12\tfrac{1}{2}O_2(g) \rightarrow 8CO_2(g) + 9H_2O(g) \qquad \Delta H = -5470 \text{ kJ}$$

 Incomplete combustion

When oxygen is in limited supply alkanes undergo **incomplete combustion**. In this reaction carbon monoxide, which is a poisonous gas, is produced. It irreversibly binds to hemoglobin in the blood thus reducing its oxygen-carrying capacity.

$$C_5H_{12}(l) + 5\tfrac{1}{2}O_2(g) \rightarrow 5CO(g) + 6H_2O(g) \quad \Delta H = -1830 \text{ kJ}$$

Quick question

Deduce balanced equations for the complete combustion of:

a) propane

b) pentane

c) hexane.

Writing organic mechanisms

"Curly arrows" are used to illustrate the movement of electrons in organic reaction mechanisms as bonds are broken and made. A fish-hook arrow is used to show **homolytic fission**, breaking a bond to produce two particles that both have a single unpaired electron, a radical. The half arrow represents the movement of a single electron as the bond breaks:

A full arrowhead shows the movement of a *pair* of electrons during **heterolytic fission**, when both electrons move together to form a new bond:

When drawing mechanisms using curly arrows:

- The base of the arrow must originate from the exact location of the electrons being moved.

- The arrowhead must accurately finish at the exact destination of the electrons.

- The arrow commences at an electron-rich region and ends at an electron-poor region of the molecule.

Students are not required to make the distinction between fish-hook arrows and full arrowheads when drawing reaction mechanisms. In this text, only full arrowheads will be used in mechanisms.

Global efforts to reduce greenhouse gas emissions

The work of environmental organizations has long focused on the fight to reduce greenhouse gas emissions and slow the rate of pollution that accompanies economic development throughout the world. The US Environmental Protection Agency estimates that globally 80 million tonnes of methane annually (28% of global methane emissions) can be attributed to ruminant livestock. Countries such as Brazil, Uruguay, Argentina, Australia, and New Zealand contribute disproportionally large amounts of greenhouse gases for their levels of population and economic development, due to the scale of their livestock industries.

Landfill in developed countries contains an increasing amount of organic "green" waste and domestic kitchen waste. In anaerobic conditions common in landfill sites, microbes produce methane in vast quantities. This form of anaerobic respiration is known as **methanogenesis**. Governments and environmental agencies are developing technologies to reduce these emissions, using the gas to generate electricity for domestic power grids through methane capture systems. Governments and local councils in some countries are collecting green waste to compost, avoiding the waste going to landfill and contributing to the production of methane.

The halogenation of alkanes

Alkanes are relatively inert, and chemists often work to *activate* alkanes to increase their reactivity. One way of achieving this is to halogenate the alkane.

Common reactions studied in organic chemistry include **substitution, addition**, and **elimination**. Substitution is the replacement of individual atoms with other single atoms or with a small group of atoms. In an addition reaction two molecules are added together to produce a single molecule, while elimination is the removal of two substituents from the molecule.

Alkanes can undergo free-radical substitution and elimination to form unsaturated alkenes and alkynes. Alkenes and alkynes can undergo all three types of reaction listed above.

Free-radical substitution

An example of free-radical substitution is the reaction between methane and chlorine in the presence of UV light.

The term **free-radical** refers to a species that is formed when a molecule undergoes **homolytic fission**: the two electrons of a covalent bond are split evenly between two atoms resulting in two free-radicals that each have a single electron:

$$A : B \longrightarrow A \cdot + \cdot B$$

Heterolytic fission of a bond creates a cation and an anion, as the electrons involved in the bond are unevenly split between the two atoms:

$$A : B \longrightarrow A^+ + :B^-$$

When methane reacts with chlorine in the presence of UV light the halogenoalkane chloromethane is produced:

$$H - \overset{\overset{\displaystyle H}{|}}{\underset{\underset{\displaystyle H}{|}}{C}} - H + Cl - Cl \xrightarrow{h\nu} H - \overset{\overset{\displaystyle H}{|}}{\underset{\underset{\displaystyle H}{|}}{C}} - Cl + H - Cl$$

methane   chlorine   chloromethane

There are three stages involved in such free-radical substitution reactions: **initiation**, **propagation**, and **termination**, described below.

Initiation

The homolytic fission of the chlorine molecule in the presence of UV light produces two chlorine radicals that have a short lifespan.

$$:\!\ddot{C}l\!:\!\ddot{C}l\!: \xrightarrow{h\nu} 2 \; :\!\ddot{C}l\!\cdot$$

initiation

Propagation

The first propagation stage involves a reaction between methane and a chlorine free-radical.

$$:\!\ddot{C}l\!\cdot + H\!:\!CH_3 \longrightarrow H\!:\!\ddot{C}l\!: + \cdot CH_3$$

propagation 1

The production of the methyl radical allows the reaction to continue as a chain reaction is set up. The methyl radical reacts with a chlorine molecule producing the desired halogenoalkane, chloromethane, along with a chlorine radical that can now take part in the first propagation reaction.

$$\cdot CH_3 + :\!\ddot{C}l\!:\!\ddot{C}l\!: \longrightarrow :\!\ddot{C}l\!: CH_3 + :\!\ddot{C}l\!\cdot$$

propagation 2

Termination

A termination step reduces the concentration of radicals in the reaction mixture. Termination reactions become more prevalent when the concentration of the hydrocarbon begins to decrease. They "mop up" the radicals, slowing the rate of reaction and eventually stopping it completely.

$$:\!\ddot{C}l\!\cdot + \cdot \ddot{C}l\!: \longrightarrow :\!\ddot{C}l\!:\!\ddot{C}l\!:$$

$$:\!\ddot{C}l\!\cdot + \cdot CH_3 \longrightarrow :\!\ddot{C}l\!:CH_3$$

$$H_3C\cdot + \cdot CH_3 \longrightarrow H_3C\!:\!CH_3$$

termination reactions

Alkenes

Alkenes are unsaturated hydrocarbons that contain at least one carbon–carbon double bond. The presence of the double bond makes alkenes more reactive than the corresponding saturated alkanes. Alkenes undergo addition reactions.

Test for unsaturation

The presence of a double bond in a hydrocarbon can be demonstrated using the addition of bromine water, $Br_2(aq)$. A mixture of the alkene and bromine water will undergo a colour change from brown to colourless:

$$C_2H_4(g) + Br_2(aq) \rightarrow C_2H_4Br_2(aq)$$
$$\text{brown} \qquad \text{colourless}$$

The mechanism of this reaction is shown in figure 1. If there is no colour change with bromine water this is a negative result, indicating the absence of the carbon–carbon double bond.

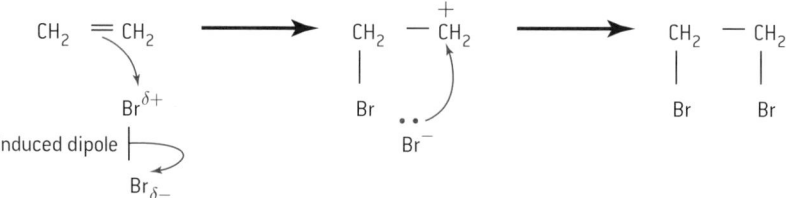

▲ Figure 1 Testing for the presence of a C=C bond by the addition of bromine water

▲ Figure 2 Bromine water is decolorized by gaseous ethene

Addition of hydrogen: hydrogenation

Large quantities of ethene are used in the chemical industry. The product of thermal decomposition or catalytic cracking of long-chain hydrocarbons, ethene is an important raw material in the production of organic polymers.

In the presence of a finely divided nickel catalyst at a temperature of 150 °C, ethene will undergo an addition reaction with hydrogen gas to produce the saturated alkane ethane:

$$C_2H_4(g) + H_2(g) \xrightarrow{Ni, \Delta} C_2H_6(g)$$

This type of reaction is very important in the food industry. The addition of hydrogen to unsaturated fats and oils occurs in the manufacture of margarine. Removing the carbon–carbon double bonds increases the melting point, making a substance that is solid rather than liquid at room temperature.

The partial hydrogenation of polyunsaturated vegetable oils therefore results in an elevation of the melting point, creating margarine which is a solid at room temperature. The food industry uses partially hydrogenated oils as they have a prolonged shelf life, the length of time a product can be stored in a supermarket and remain fit for consumption. However, saturated fats and oils in the diet increase the blood concentration of low-density lipoproteins (LDLs) which are involved in the transport of cholesterol in the blood. Raised levels of LDLs are associated with increased health risks (see sub-topic B.3).

Partial hydrogenation also results in the conversion of *cis*-carbon–carbon double bonds into *trans*-carbon–carbon double bonds. As a result of the work of the scientific community, many governments over the past decade have recognized the dangers of *trans*-fats and legislation has been introduced to reduce the use of these harmful products and make the consumer more aware of the content of processed foods. Many large multinational food producers have also distanced themselves from *trans*-fats to avoid negative publicity and resulting loss of sales and market share. *Cis-* and *trans-* fats are discussed in detail in sub-topics B.3 and B.10.

Halogenation of alkenes

The electrophilic halogenation of symmetrical alkenes involves the addition of elemental halogens such as chlorine, Cl_2 or bromine, Br_2, resulting in a dihalogenated alkane:

$$C_4H_8(g) + Br_2(g) \rightarrow C_4H_8Br_2(l)$$

```
 H  H  H  H                          H  H  H  H
 |  |  |  |                          |  |  |  |
 H—C—C=C—C—H + Br₂    ───────→     H—C—C—C—C—H
 |        |                          |  |  |  |
 H        H                          H  Br Br H

   but-2-ene                      2,3-dibromobutane
```

> **Ethene and ripening fruit**
>
> Ethene is used in the food industry to accelerate the ripening of fruit. Fruits are generally picked while they are still unripe to enable them to be transported from farm to supermarket without becoming damaged and looking unappealing to the consumer. Ethene is a natural part of food ripening, as it is released by ripening fruit. Exposure of the fruit to ethene increases the rate of ripening, while preventing the build-up of ethene around fruit by introducing carbon dioxide to the container holding the fruit slows the rate of ripening.

The addition of a hydrogen halide, HX, to a symmetrical alkene results in a single mono-halogenated alkane. With an unsymmetrical alkene two alternative products are possible; however, only symmetrical alkenes are discussed in this topic (unsymmetrical alkenes are covered at HL in sub-topic 20.1).

$$C_4H_8(g) + HBr(g) \rightarrow C_4H_9Br(l)$$

H H H H
| | | |
H — C — C = C — C — H + HBr   ⟶   H — C — C — C — C — H
| | | | | | |
H H   H H Br H

but-2-ene   2-bromobutane

The large-scale production of ethanol is achieved by reacting ethene with steam in the presence of a catalyst, phosphoric(V) acid, at 300 °C and a pressure of 6–7 MPa:

$$C_2H_4(g) + H_2O(g) \rightarrow C_2H_5OH(g)$$

Ethanol has a variety of uses including as an additive to gasoline or petrol, creating a biofuel (sub-topic 10.1).

Polymerization of alkenes

The plastics industry is one of the largest manufacturing bodies in the world, producing a broad range of addition polymers used widely for a variety of purposes. Polymers improve the quality of our lives although they can have a negative impact on the environment.

Addition polymerization is the reaction of many small **monomers** that contain a carbon–carbon double bond, linking together to form a **polymer**. The ethene monomer, supplied to the plastics industry by the petrochemical industry, undergoes addition polymerization to produce the monomer polyethene:

$$n\,C_2H_4 \rightarrow \{CH_2{-}CH_2\}_n$$

> **Study tip**
>
> When drawing diagrams to represent polymerization, it is important to draw continuation bonds through the brackets.

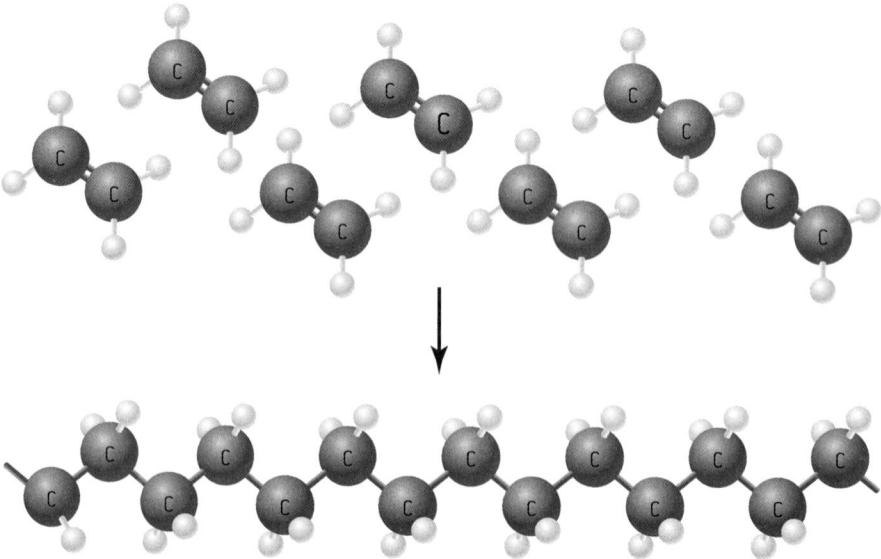

▲ Figure 3 The polymerization of ethene

Any monomer that contains a carbon–carbon double bond can undergo polymerization. The repeating structural unit of the polymer reflects the structure of the monomer that formed it, with the double bond replaced by a single bond, and the electrons released forming new bonds to the adjacent monomers. For example, figure 4 shows the polymerization of propene.

$$n\ CH_2 = CH \longrightarrow \left[CH_2 - CH \right]_n$$
$$\qquad\quad | \qquad\qquad\qquad\quad |$$
$$\qquad\quad CH_3 \qquad\qquad\qquad CH_3$$

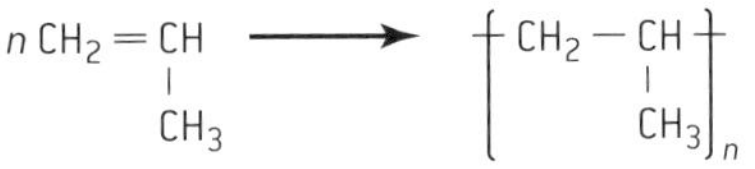

▲ Figure 4 The polymerization of propene

Alcohols

Alcohols form a diverse group of compounds that have a wide range of applications and play a significant part in synthetic reactions.

Alcohols can undergo complete combustion, releasing carbon dioxide and water:

$$C_2H_5OH(l) + 3O_2(g) \rightarrow 2CO_2(g) + 3H_2O(g) \qquad \Delta H = -1367 \text{ kJ}$$

Oxidation of alcohols

Acidified potassium dichromate(VI) (figure 5) or potassium manganate(VII) can be used for the oxidation of alcohols. The half-equations for these oxidizing agents are as follows. (The working method to construct these equations was developed in sub-topic 9.2.)

$$MnO_4^-(aq) + 8H^+(aq) + 5e^- \rightarrow Mn^{2+}(aq) + 4H_2O(l)$$
$$Cr_2O_7^{2-}(aq) + 14H^+(aq) + 6e^- \rightarrow 2Cr^{3+}(aq) + 7H_2O(l)$$

The oxidation products of alcohols depend on the type of alcohols involved.

Primary alcohols

The oxidation of a primary alcohol is a two-stage process that first produces an **aldehyde** followed by a **carboxylic acid**. Potassium dichromate(VI), $K_2Cr_2O_7$ is a milder oxidizing agent than potassium manganate(VII), $KMnO_4$. When the primary alcohol ethanol, C_2H_5OH is heated with acidified $K_2Cr_2O_7$, the aldehyde ethanal, CH_3CHO is produced. This aldehyde can be further oxidized to the carboxylic acid ethanoic acid, CH_3COOH.

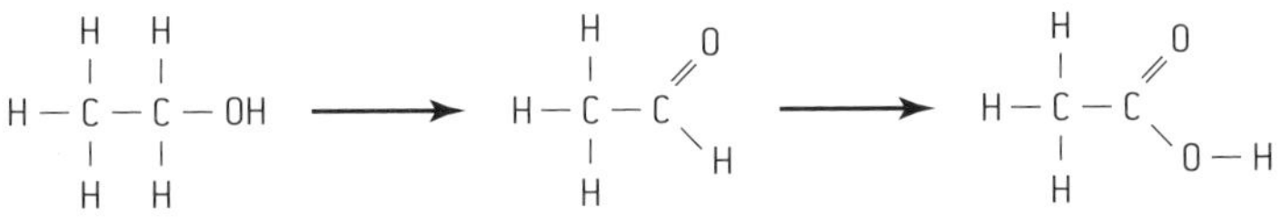

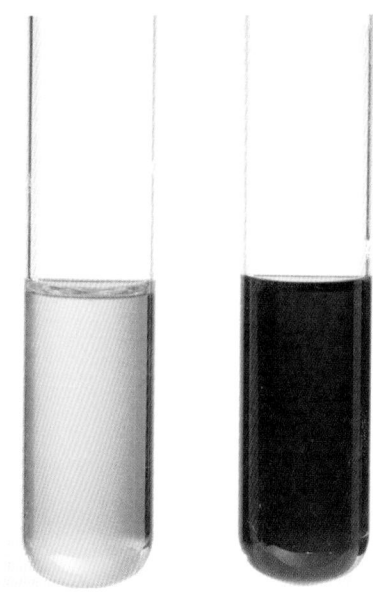
▲ Figure 5 Colour change during the reduction of orange $Cr_2O_7^{2-}$ (aq) ions to green Cr^{3+}(aq) ions

Social implications of alcohol consumption

Excessive alcohol consumption is a growing problem in many countries. **Binge drinking** is sometimes defined as drinking 5 (for a man) or 4 (for a woman) alcohol units over a 2-hour period and increasing the blood alcohol content above 0.08% by volume. An alcohol unit varies from country to country. In general, one alcohol unit is equivalent to approximately 10 cm³ of pure alcohol. Having reached alarming proportions amongst adolescents and young adults in many western societies, binge drinking is having a significant impact on economies, social structure, law and order, and ultimately health systems. Associated health risks include physical and psychological dependence on alcohol, liver and brain damage, elevated risk of cancer of the throat, mouth, and esophagus, depression, anxiety and social problems at work and within the family.

$$\underset{\text{ethanol}}{\underset{\displaystyle H-\overset{\displaystyle |}{\underset{\displaystyle |}{C}}-\overset{\displaystyle |}{\underset{\displaystyle |}{C}}-OH}{H\ H}} \longrightarrow \underset{\text{ethanal}}{H-\overset{H}{\underset{H}{C}}-C\overset{O}{_{\diagdown H}}} \longrightarrow \underset{\text{ethanoic acid}}{H-\overset{H}{\underset{H}{C}}-C\overset{O}{_{\diagdown O-H}}}$$

Carboxylic acids are capable of forming dimers, paired molecules held together by hydrogen bonds (figure 6). The increased size of the molecule leads to stronger van der Waals' forces and a higher boiling point.

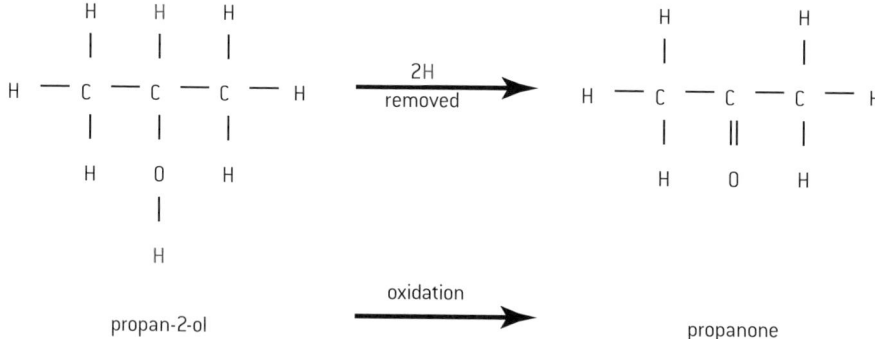

▲ Figure 6 A dimer of ethanoic acid

The aldehyde can be recovered by the process of **distillation**, preventing its further oxidation. The aldehyde has a lower boiling point than the carboxylic acid due to differences in the intermolecular forces: aldehydes have weak dipole–dipole intermolecular forces while carboxylic acids have stronger intermolecular hydrogen bonds and so have higher boiling points.

If the carboxylic acid is the desired product, the aldehyde must remain in the reaction mixture with the oxidizing agent for a longer period of time. Instead of the distillation apparatus a reflux column is used. **Refluxing** is a technique that involves the cyclic evaporation and condensation of a volatile reaction mixture, preserving the solvent as it does not evaporate.

Secondary alcohols

The oxidation of a secondary alcohol such as propan-2-ol results in the formation of a ketone:

propan-2-ol

oxidation

propanone

Upon formation of the ketone, no further oxidation is possible as the carbon atom of the functional group has no hydrogens attached to it.

Condensation reaction of an alcohol and a carboxylic acid

Esters are derived from carboxylic acids and have a variety of applications ranging from flavouring agents and medications to solvents and explosives.

Esterification is a reversible reaction that occurs when a carboxylic acid and an alcohol are heated in the presence of a catalyst, normally concentrated sulfuric acid:

$$CH_3CH_2COOH(l) + CH_3OH(l) \xrightarrow{H_2SO_4 \text{ (conc)}} CH_3CH_2COOCH_3(l) + H_2O(l)$$
propanoic acid   methanol   methyl propanoate

The IUPAC names of esters consist of two words. The first word is derived from the name of the alkyl chain in the alcohol (in this example, "methyl"). The second word is composed of the root name of the carboxylic acid (in this case, "prop"), followed by the suffix "anoate" (figure 7).

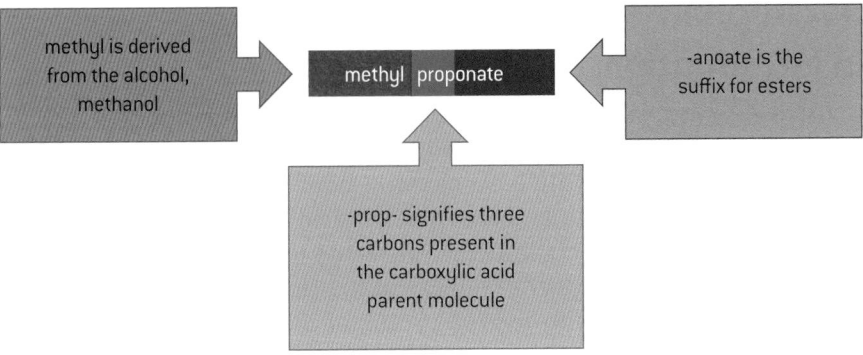

▲ Figure 7 Naming esters

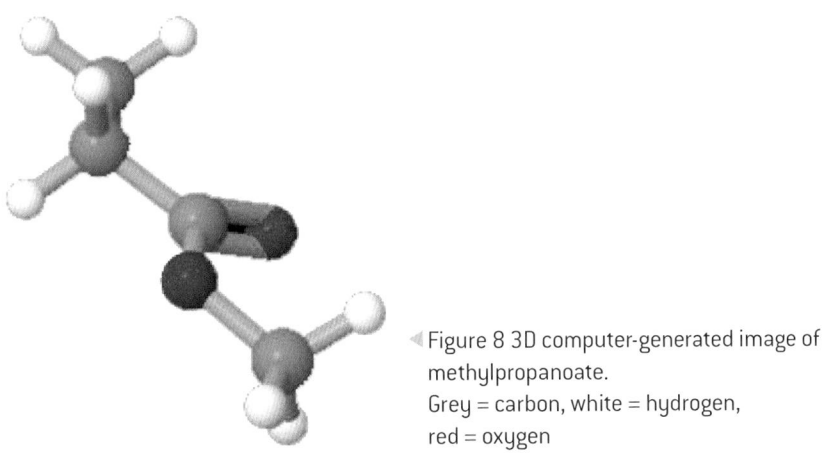

◀ Figure 8 3D computer-generated image of methylpropanoate.
Grey = carbon, white = hydrogen, red = oxygen

Nucleophilic substitution reactions: An introduction

Once an unreactive alkane has undergone a substitution reaction to form a halogenoalkane, the resulting molecule can take part in other types of reaction. Since halogenoalkanes contain a polar carbon–halogen bond, C–X, the electron-deficient carbon is now open to attack by electron-rich species known as **nucleophiles**. Nucleophiles contain a lone pair of electrons and sometimes carry a full negative charge. Nucleophiles act as Lewis bases (sub-topic 18.1).

An aqueous solution of sodium hydroxide, NaOH(aq), contains the nucleophile :OH⁻. The partial positive charge on the carbon atom in the polarized C–X bond makes it susceptible to attack by the hydroxide ion. The substitution of the halogen atom by the nucleophile will result in an alcohol being formed.

There are two distinct reaction mechanisms for this **nucleophilic substitution reaction**. The mechanism that occurs depends on the class of halogenoalkane present, namely primary, secondary, or tertiary. These mechanisms will be discussed in sub-topic 20.1.

An example of nucleophilic substitution is the reaction of chloroethane with aqueous sodium hydroxide. This reaction produces an alcohol (ethanol) and releases a chloride ion, Cl⁻, the leaving group.

$$CH_3CH_2Cl(g) + OH^-(aq) \rightarrow CH_3CH_2OH(aq) + Cl^-(aq)$$

Electrophilic substitution reactions: An introduction

As discussed in sub-topic 10.1, benzene does not readily undergo addition reactions. Instead it will undergo **electrophilic substitution reactions**. An **electrophile** is an electron-poor species capable of accepting an electron pair. It acts as a Lewis acid (sub-topic 18.1). While electron-poor electrophiles are attracted to the π electrons in the aromatic benzene ring, the stability of the ring leads to substitution rather than addition. In sub-topic 20.1 the mechanism of the nitration (electrophilic substitution) reaction of benzene will be developed, illustrating the unique reaction properties of benzene.

An example of electrophilic substitution is the reaction of benzene with elemental bromine. This reaction takes place in an anhydrous environment and requires a Lewis base ($FeBr_3$ or $AlBr_3$) as the catalyst.

$$C_6H_6 + Br_2 \xrightarrow{FeBr_3} C_6H_5Br + HBr$$

One bromine atom from Br_2 replaces a hydrogen atom in benzene; the remaining bromine and hydrogen atoms form the inorganic by-product, hydrogen bromide. Please note that this reaction takes place in an organic (non-aqueous) environment, so the states of reactants and products are omitted.

Questions

1 Which three compounds can be considered to be a homologous series?

A. CH_3OH, CH_3CH_2OH, $CH_3CH_2CH_2OH$

B. CH_3CH_2OH, CH_3CHO, CH_3COOH

C. $CH_3CH_2CH(OH)CH_3$, $CH_3CH_2CH_2CH_2OH$, $(CH_3)_3COH$

D. $CH_3CH_2CH_2CH_2OH$, $CH_3CH_2OCH_2CH_3$, $(CH_3)_2CH_2CHO$ [1]

IB, May 2009

2 What is the IUPAC name for $CH_3CH_2CH(CH_3)CH_3$?

A. 1,1-dimethylpropane

B. 2-ethylpropane

C. 2-methylbutane

D. 3-methylbutane [1]

IB, May 2009

3 Which conditions are required to obtain a good yield of a carboxylic acid when ethanol is oxidized using potassium dichromate(VI), $K_2Cr_2O_7(aq)$?

I. Add sulfuric acid

II. Heat the reaction mixture under reflux

III. Distil the product as the oxidizing agent is added

A. I and II only

B. I and III only

C. II and III only

D. I, II, and III [1]

IB, May 2009

4 Alkenes are an economically and chemically important family of organic compounds.

a) The reaction of alkenes with bromine water provides a test for unsaturation in the laboratory. Describe the colour change when bromine water is added to chloroethene. [1]

b) Deduce the Lewis structure of chloroethene and identify the formula of the repeating unit of the polymer poly(chloroethene). [2]

c) Besides polymerization, state **two** commercial uses of the reactions of alkenes. [2]

IB, May 2010

5 State and explain whether the following molecules (figure 9) are primary, secondary, or tertiary halogenoalkanes. [4]

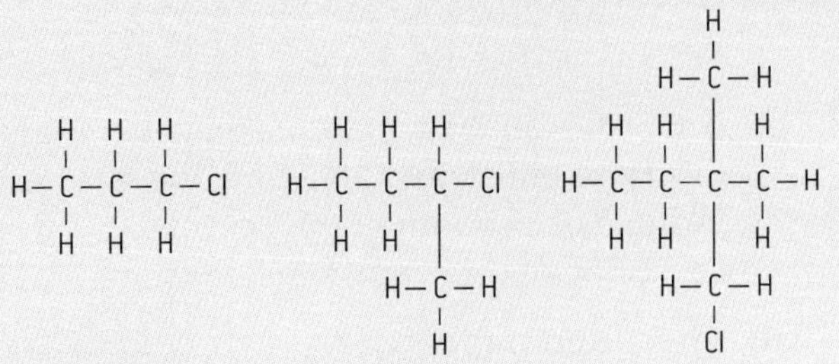

▲ Figure 9

IB, May 2011

6 Consider the following sequence of reactions:

$$RCH_3 \xrightarrow{reaction\ 1} RCH_2Br \xrightarrow{reaction\ 2} RCH_2OH \xrightarrow{reaction\ 3} RCOOH$$

RCH_3 is an unknown alkane in which R represents an alkyl group.

a) The alkane contains 81.7% by mass of carbon. Determine its empirical formula, showing your working. [3]

b) Equal volumes of carbon dioxide and the unknown alkane are found to have the same mass, measured to an accuracy of two significant figures, at the same temperature and pressure. Deduce the molecular formula of the alkane. [1]

c) (i) State the reagent and conditions needed for *reaction 1*. [2]

(ii) State the reagent(s) and conditions needed for *reaction 3*. [2]

d) *Reaction 1* involves a free-radical mechanism. Describe the stepwise mechanism, by giving equations to represent the initiation, propagation and termination steps. [4]

IB, November 2010

7 a) Identify the formulas of the organic products, **A–E**, formed in the reactions, **I–IV**:

   I. $CH_3(CH_2)_8OH + K_2Cr_2O_7 \xrightarrow{H^+} A \xrightarrow{H^+} B$

   II. $(CH_3)_3CBr + NaOH \longrightarrow C$

   III. $(CH_3)_2CHOH + K_2Cr_2O_7 \xrightarrow{H^+} D$

   IV. $H_2C=CH_2 + Br_2 \longrightarrow E$ [5]

   b) $H_2C=CH_2$ can react to form a polymer. Name this **type** of polymer and draw the structural formula of a section of this polymer consisting of three repeating units. [2]

IB, Specimen paper

8 Two compounds, **A** and **D**, each have the formula C_4H_9Cl.

Compound **A** is reacted with dilute aqueous sodium hydroxide to produce compound **B** with a formula of $C_4H_{10}O$. Compound **B** is then oxidized with acidified potassium manganate(VII) to produce compound **C** with a formula of C_4H_8O. Compound **C** resists further oxidation by acidified potassium manganate(VII).

Compound **D** is reacted with dilute aqueous sodium hydroxide to produce compound **E** with a formula of $C_4H_{10}O$. Compound **E** does not react with acidified potassium manganate(VII).

Deduce the structural formulas for compounds **A**, **B**, **C**, **D**, and **E** [5]

IB, May 2011

9 Chloroethene, C_2H_3Cl, is an important organic compound used to manufacture the polymer poly(chloroethene).

   a) Draw the Lewis structure for chloroethene and predict the H – C – Cl bond angle. [2]

   b) Draw a section of poly(chloroethene) containing six carbon atoms. [1]

   c) Outline why the polymerization of alkenes is of economic importance and why the disposal of plastics is a problem. [2]

IB, May 2010

Introduction

Analytical techniques lie at the very core of chemistry. As chemists, not only do we need to appreciate the power of analysis, but in addition we must realize that any measurement has a limit of precision and accuracy, and this must be taken into account when evaluating experimental results. In this topic we will explore the principles that underpin uncertainty in measurement, how we can effectively represent data by graphical means, and examine the spectroscopic identification of organic compounds, looking at the analytical techniques of infrared spectroscopy (IR), mass spectrometry (MS), and proton nuclear magnetic resonance spectroscopy (^{1}H NMR).

11.1 Uncertainties and errors in measurement and results

Understandings

→ Qualitative data includes all non-numerical information obtained from observations not from measurement.

→ Quantitative data are obtained from measurements, and are always associated with random errors/uncertainties, determined by the apparatus, and by human limitations such as reaction times.

→ Propagation of random errors in data processing shows the impact of the uncertainties on the final result.

→ Experimental design and procedure usually lead to systematic errors in measurement, which cause a deviation in a particular direction.

→ Repeat trials and measurements will reduce random errors but not systematic errors.

Applications and skills

→ Distinction between random errors and systematic errors.

→ Record uncertainties in all measurements as a range ($\pm$) to an appropriate precision.

→ Discussion of ways to reduce uncertainties in an experiment.

→ Propagation of uncertainties in processed data, including the use of percentage uncertainties.

→ Discussion of systematic errors in all experimental work, their impact on the results, and how they can be reduced.

→ Estimation of whether a particular source of error is likely to have a major or minor effect on the final result.

→ Calculation of percentage error when the experimental result can be compared with a theoretical or accepted result.

→ Distinction between accuracy and precision in evaluating results.

Nature of science

→ Making quantitative measurements with replicates to ensure reliability – precision, accuracy, systematic, and random errors must be interpreted through replication.

According to IUPAC, qualitative and quantitative analysis can be distinguished as:

Qualitative analysis

Substances are identified or classified on the basis of their chemical or physical properties, such as chemical reactivity, solubility, molar mass, melting point, radiative properties (emission, absorption), mass spectra, nuclear half-life, etc.

Quantitative analysis

The amount or concentration of an analyte may be determined (estimated) and expressed as a numerical value in appropriate units.

'Nomenclature in evaluation of analytical methods including detection and quantification capabilities' *Pure and Applied Chemistry, 67(1699), (1995), p1701*

Qualitative and quantitative analysis

The analytical chemist is often described as the chemical detective. Analysis can be of two types:

- qualitative analysis
- quantitative analysis.

Uncertainty in measurement

In science numerical data can be divided into two types:

- data involving *exact numbers* (that is, the values are known exactly – there is no uncertainty)
- data involving *inexact numbers* (for these types of numbers there is a degree of uncertainty).

As scientists, when we carry out a particular experiment involving measurement, there will always be some uncertainty associated with the measured data, that is the data will involve inexact numbers to some degree. Such uncertainty may be associated with factors such as the instruments used in the laboratory. For example, the mass of a sample of potassium bromide, KBr(s), will depend on the type of balance used. In a typical school laboratory, top-pan balances often read to 0.01 g, but an analytical balance, used in more precise analytical experiments, can read to at least 0.0001 g or better) (figure 1). Uncertainty may also depend on human error.

Data may also be classified as qualitative or quantitative data:

Qualitative data

Qualitative data includes all non-numerical information obtained from observations not from measurement.

Quantitative data

Quantitative data are obtained from measurements, and are always associated with random errors/uncertainties (defined shortly) determined by the apparatus and by human limitations, such as reaction times.

▲ Figure 1 (a) A top-pan balance used to measure mass in a typical school laboratory can read to 0.01 g. (b) An analytical balance used to measure mass to a high degree of precision can often read to 0.0001 g. The shutters on the balance should be closed to reduce both airflow and dust collecting which can both affect the reading

An example of a flawed experiment in science

The OPERA experiment – a case-study involving CERN and LNGS

Neutrinos are extremely small, electrically neutral particles produced in nuclear reactions and are one of the fundamental particles that make up the universe. They might be considered as being similar to the electron but, unlike the electron, they do not carry an electrical charge. In 2011, results from the OPERA experiment (an international research collaboration between Conseil Européen pour la Recherche Nucléaire (CERN), in Geneva, Switzerland, and the Laboratori Nazionali del Gran Sasso (LNGS) in Gran Sasso, Italy) suggested that neutrinos appear to travel at a greater velocity than the velocity of light. The proposed scientific discovery made international headlines all over the world, but many physicists were very concerned about the findings as it was considered that nothing travels faster than light, as postulated by Albert Einstein. OPERA, however, was staunchly defensive of their findings and said that there was no flaw in the experiment. Subsequently, it was reported that there were, in fact, flaws in the set-up of the equipment, which led to possible timing problems in the original set of measurements. In 2012 it was reported in *Nature* that the velocities of neutrinos are, indeed, consistent with the velocity of light. This shows the importance of testing the reliability and validity of experimental results in science and understanding the idea of uncertainty in measurement.

An example of the impact of errors from research space

Crash of the Mars Climate Orbiter Spacecraft

In 1998, NASA launched the Mars Climate Orbiter, a space probe designed to examine the climate on the planet Mars. However in 1999, the spacecraft crashed and completely disintegrated as it approached Mars at an incorrect altitude (http://mars.jpl.nasa.gov/msp98/orbiter/). The reason for the crash related to an error in the transfer of information between the spacecraft team based in Colorado, and the mission navigation team based in California, USA.

Dr. Edward Weiler, NASA's Associate Administrator for Space Science stated the following:

> People sometimes make errors. The problem here was not the error, it was the failure of NASA's systems engineering, and the checks and balances in our processes to detect the error. That's why we lost the spacecraft.

It was subsequently reported that one team used imperial units while the other team used SI units for a pivotal operation for the space probe, which led to the incorrect trajectory required to place the probe on Mars.

Difference between precision and accuracy

In order to understand the idea of uncertainty in measured values, we need to consider the difference between precision and accuracy.

Precision

According to IUPAC, precision is the closeness of agreement between independent test results obtained by applying the experimental procedure under stipulated conditions. The smaller the random part of the experimental errors (defined shortly) which affect the results, the more precise the procedure.

Accuracy

According to IUPAC, accuracy is the closeness of the agreement between the result of a measurement and a true value of the **measurand** (which is the particular quantity to be measured).

'Nomenclature for the presentation of results of chemical analysis' *Pure and Applied Chemistry*, 66(595), (1994), p598

The number of digits in the mantissa indicates the number of significant figures for a logarithmic entity (this covers both logs to the base 10 and logs to the base e, that is natural logs, ln). Calculations that involve base 10 logs are common in pH, whereas in kinetics the natural log, ln, is frequently used, for example in the Arrhenius equation. For example, in:

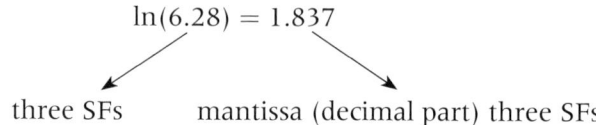

$$\log_{10}(2.7) = 0.43 = 4.3 \times 10^{-1}$$

two SFs   mantissa (decimal part) two SF

the number 2.7 contains two SFs, so the answer should be reported with the mantissa (the decimal part) having two SFs. Here is another example:

$$\ln(6.28) = 1.837$$

three SFs   mantissa (decimal part) three SFs

the number 6.28 contains three SFs, so the answer should be reported with the mantissa (the decimal part) having three SFs.

Experimental errors

As stated already, every single measurement has a degree of uncertainty associated with it. This is termed **experimental error**. There are two types of experimental error:

- systematic error
- random error.

Systematic errors

Systematic errors are associated with a flaw in the actual experimental design or with the instrumentation used. Systematic errors imply that the measured quantity will always be greater or less than the true value.

Systematic errors can be further classified into three types:

- **instrumentation errors**
- **experimental methodology errors**
- **personal errors.**

Examples of systematic errors:

- Faulty gas syringes that have associated leakage (*instrumentation error*).
- Errors in the readings taken from a pH meter due to faulty calibration of the instrument (*instrumentation error*).
- Poorly insulated calorimeter in a thermochemistry experiment (*experimental methodology error*).
- Measuring the volume of a colourless liquid in a graduated cylinder or burette from the top of the meniscus instead of from the bottom.

Systematic error

According to IUPAC, systematic error is the mean that would result from an infinite number of measurements of the same measurand carried out under repeatability conditions minus a true value of the measurand.

Random error

According to IUPAC, random error is the result of a measurement minus the mean that would result from an infinite number of measurements of the same measurand carried out under repeatability conditions.

International Vocabulary of Basic and General Terms in Metrology, Second Edition, ISO, 1993.

Such a systematic error would lead to greater volumes in the manipulation of data (*experimental methodology error*).

- Evaporation of volatile liquids on heating a sample (*experimental methodology error*).

- Occurrence of side-reactions which can interfere with the parameter being measured (*experimental methodology error*).

- The exact colour of a solution at its end point (*personal error*).

- parallax error associated with reading a graduated cylinder incorrectly

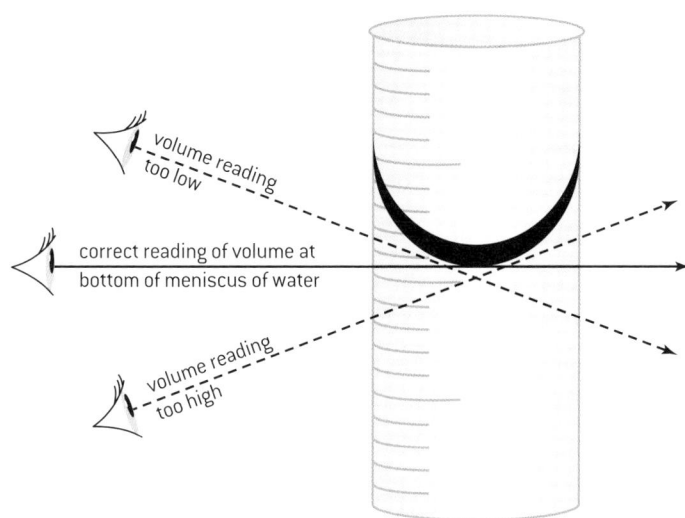

Systematic errors can often be reduced by **adopting greater care to the experimental design**. Such errors are consistent and can be detected and ultimately corrected.

Systematic errors will affect the **accuracy** of the results.

Random errors

Random errors occur because of uncontrolled variables in an experiment and hence **cannot be eliminated**. They can, however, be **reduced** by repeated measurements. Random errors affect the **precision** of the results.

Examples of random errors:

- estimating a quantity which lies between marked graduations of a particular instrument eg with a spectrophotometer) or measuring apparatus (for example, a burette).

- not being able to read an instrument due to fluctuations in readings that occur during measurements due to changes in changes in the surroundings (for example, temperature variations, airflow, changes in pressure)

- reaction time.

Useful resource

The BPIM document *Evaluation of Measured Data – Guide to the Expression of Uncertainty in Measurement* can be accessed at http://www.bipm.org/en/publications/guides/gum.html.

BIPM

The mission of BIPM (*Bureau International des Poids et Mesures*, whose headquarters are based in Paris, France) is to ensure worldwide uniformity of measurements and their traceability to the **International System of Units** (SI).

With the authority of the Convention of the Metre, a diplomatic treaty between 55 nations, BIPM functions through a series of consultative committees, whose members are the national metrology laboratories of the signatory states, and through its own experimental programmes. BIPM carries out measurement-related research. Part of the work of BIPM involves looking at international comparisons of national measurement standards as well as performing calibrations for its member states. As a result of collaboration between seven international organizations, including IUPAC and the International Organization for Standards (ISO), the *Evaluation of Measurement Data – Guide to to the Expression of Uncertainty in Measurement* was first published in 1995. This was revised in 2008 and has been widely adopted in most countries; it has been translated into several languages.

Absolute and relative uncertainty

Suppose in an experiment you use a top-pan balance to measure the mass, m, of a sample of aspirin (2-acetoxybenzoic acid) that you have synthesized in the laboratory. Your recorded mass was $m = 3.56$ g.

As stated, there will be a degree of uncertainty associated with every single measurement. So here there is an uncertainty associated with the instrument used, in this case the top-pan balance. The mass could have been between 3.55 g and 3.57 g. Hence, the uncertainty is 0.01 g and the mass should be reported as follows in your laboratory notebook:

$$m = (3.56 \pm 0.01) \text{ g}$$

Any experimental result should be reported in the form:

$$\text{experimental result} = (A \pm \Delta A) \text{ unit}$$

Where A represents the measured experimental result and ΔA represents the uncertainty in A or, strictly speaking the **absolute uncertainty**.

An uncertainty can be absolute or relative.

- **Absolute uncertainty** is the margin of uncertainty associated with the result from a given measurement. Its symbol is ΔA.

- **Relative uncertainty** is the ratio comparing the size of the absolute uncertainty, ΔA, to the size of the measured experimental result, A.

$$\text{relative uncertainty} = \frac{\Delta A}{A}$$

Example

A calibrated burette has an absolute uncertainty of ± 0.02 cm³. During a titration, the volume of a 0.15 mol dm⁻³ solution of hydrochloric acid at the end point was recorded as 12.25 cm³. Calculate the associated relative uncertainty.

Solution

absolute uncertainty $(\Delta A) = \pm 0.02$ cm³

measured experimental result $(A) = (12.25 \pm 0.02)$ cm³

relative uncertainty $\dfrac{\Delta A}{A} = \dfrac{0.02 \text{ cm}^3}{12.25 \text{ cm}^3} = 2 \times 10^{-3}$

Note that relative uncertainty is dimensionless, since the units cancel each other out!

The relative uncertainty is often expressed as the **percentage relative uncertainty**, so in this example the percentage relative uncertainty would be 0.2%.

Propagation of uncertainty

After identifying the uncertainties associated with experimentally measured quantities, the next step is to figure out how these different uncertainties combine to give the resultant uncertainty. This is what is termed the **propagation of uncertainties**, and in order to do this two rules are applied.

> **Percentage (%) relative uncertainty =**
> relative uncertainty $\left(\frac{\Delta A}{A}\right) \times 100\,\%$

> **Rule 1**
> When adding or subtracting measurements, the absolute uncertainty associated with the net measured parameter is the square root of the sum of the squares of the absolute uncertainties that is $\sqrt{\Sigma \Delta A^2}$.
>
> **Rule 2**
> When multiplying or dividing measurements, the relative uncertainty associated with the net measured parameter is the square root of the sum of the squares of the relative uncertainties.

Percentage error

Percentage relative uncertainty is different to percentage error.

For example, the literature value for the standard enthalpy change of the decomposition reaction of calcium carbonate, $CaCO_3(s)$, was found to be +178.1 kJ:

$$CaCO_3(s) \rightarrow CaO(s) + CO_2(g) \qquad \Delta H^\ominus = +178.1 \text{ kJ}$$

The experimental value was found to be +172.0 kJ. Hence the percentage error is given by:

$$\text{percentage error} = \left| \frac{178.1 - 172.0}{178.1} \right| \times 100\%$$

$$= \left| \frac{6.1}{178.1} \right| \times 100\% = 3.4\%$$

> Percentage error $=$
>
> $$\frac{literature\ value - experimental\ value}{literature\ value} \times 100\%$$

TOK

The vertical lines, $|...|$, used in the expression for percentage error represent the idea of a modulus mathematically, that is any negative value is considered positive. This is also used in describing the modulus of a complex number in mathematics, $|z|$, which is the distance the complex number is, in the form $z = x + iy$, from the origin. $|z|$ is expressed as $\sqrt{(x^2 + y^2)}$. This entity has to be positive.

The same symbols are often used to represent alternative meanings in different scientific disciplines. For example, in chemistry we use vertical lines in cell diagram notations to represent different phase boundaries. In both physics and chemistry, vertical lines of different lengths represent a cell, used in a battery, that is $|||$ where the shorter vertical line represents the negative pole and the longer vertical line represents the positive pole. Equally, even in chemistry we often use the square brackets, [ ], symbol for different purposes, for example concentration, idea of a complex, etc.

TOK

Science has been described as a self-correcting and communal public endeavour. To what extent do these characteristics also apply to the other areas of knowledge?

Worked examples

Example 1

The mass of a sample bottle and a piece of titanium metal is 33.2901 g. The mass of the empty sample bottle is 26.3505 g. If the density of titanium is 4.506 g cm^{-3} at 298 K, calculate the volume of water, in cm³, displaced at this temperature by the metal.

Solution

- d is the density of titanium;
  m is the mass of titanium;

V is the volume of titanium.

$$d = \frac{m}{V}$$

The mass (m) of titanium sample is:

$$m = 33.2901 \text{ g} - 26.3505 \text{ g} = 6.9396 \text{ g}$$

(this operation involves subtraction so you express the reported result based on the smallest number of decimal places, which is four).

$$V = \frac{m}{d} = \frac{6.9396 \text{ g}}{4.506 \text{ g cm}^{-3}} = 1.540 \text{ cm}^3$$

(this operation involves division so you express the reported result based on the smallest number of SF, which is four).

Example 2

State the number of significant figures associated with the following:

a) 0.00390 kg of Cu(s)

b) 136.250 g of NaCl(s)

Solution

First express both masses in scientific notation

3.90×10^{-3} kg of Cu(s)

1.36250×10^2 g of NaCl(s)

a) three SFs

b) six SFs (note the last zero is significant).

Example 3

Calculate the pH of a 0.020 mol dm^{-3} solution of perchloric acid, $HClO_4(aq)$.

Solution

Perchloric acid is a strong acid, so is assumed to be completely dissociated in solution:

$$HClO_4(aq) + H_2O(l) \rightarrow H_3O^+(aq) + ClO_4^-(aq)$$

$$pH = -\log_{10}[H_3O^+] = -\log_{10}(0.020)$$

$$0.020 \text{ mol dm}^{-3} = 2.0 \times 10^{-2} \text{ mol dm}^{-3}$$

which involves two SFs. Therefore:

$$pH = 1.70$$

(since the mantissa must contain two SFs).

Also note also there are **no units for pH** since it is based on a logarithmic expression.

Example 4

The pH of a carton of orange juice was found to be 3.75. Calculate the hydrogen ion concentration, [H$^+$], in mol dm^{-3}.

Solution

$$pH = -\log_{10}[H^+] = 3.75$$

The mantissa here contains two SFs.

$$[H^+] = \text{anti-log}_{10} \text{ of } -3.75 = e^{-3.75}$$
$$= 1.8 \times 10^{-4} \text{ mol dm}^{-3}$$

expressed as two SFs.

Example 5

A calibrated burette has an absolute uncertainty of ± 0.02 cm^3. During a titration, the volume of a 0.10 mol dm^{-3} solution of hydrochloric acid at the end point was recorded as 22.18 cm^3. Calculate its percentage relative uncertainty.

Solution

absolute uncertainty $(\Delta A) = \pm 0.02$ cm^3

measured experimental result $(A) = (22.18 \pm 0.02)$ cm^3

relative uncertainty $\left(\dfrac{\Delta A}{A}\right) = \left(\dfrac{0.02 \text{ cm}^3}{22.18 \text{ cm}^3}\right)$
$= 9 \times 10^{-4}$

percentage (%) relative uncertainty =
$(9 \times 10^{-4}) \times 100\% = 0.09\%$

Example 6

During a titration the following titres were recorded for a 0.10 mol dm^{-3} solution of hydrochloric acid from a burette:

initial titre $= (5.00 \pm 0.02)$ cm^3

final titre $= (21.35 \pm 0.02)$ cm^3

Calculate the volume delivered, in cm^3, and the uncertainty of this volume.

Solution

volume delivered $= 21.35 - 5.00 = 16.35$ cm^3.

In order to obtain the uncertainty in this volume we need to use the expression:

$$\sqrt{\Sigma \Delta A^2}$$

since a subtractive operation is involved.

$$\text{uncertainty} = \sqrt{[(0.02)^2 + (0.02)^2]} = 0.03$$

The volume would be reported as (16.35 ± 0.03) cm^3

Example 7

(13.3 ± 0.1) g of sodium chloride salt, NaCl(s), is dissolved in a flask containing (2.0 ± 0.1) dm^3 of water, H$_2$O(l). Calculate the concentration of sodium chloride, in g dm^{-3}, in the solution and the percentage relative uncertainty of this concentration. Assume that the salt is fully dissolved in the solution.

Solution

$$\text{concentration} = \frac{13.3 \text{ g}}{2.0 \text{ dm}^3} = 6.7 \text{ g dm}^{-3}$$

Notice that this answer is expressed as two SFs because division is involved and rounding up is required on calculation. *Check this using your calculator!*

When multiplying or dividing measurements, the percentage uncertainty is the square root of the sum of the squares of the percentage relative uncertainties:

percentage relative uncertainty in mass =

$$\frac{\Delta A}{A} \times 100\% = \frac{0.1}{13.3} \times 100\%.$$

percentage relative uncertainty in volume =

$$\frac{\Delta A}{A} \times 100\% = \frac{0.1}{2.0} \times 100\%.$$

percentage relative uncertainty in concentration =

$$\sqrt{\left[\left(\frac{10}{13.3}\right)^2 + \left(\frac{10}{2.0}\right)^2\right]} = 5\%$$

$$\text{concentration} = 6.7 \text{ g dm}^{-3} \ (\pm 5\%)$$

The absolute uncertainty can also be found subsequently from this:

$$\Delta A = \frac{5 \times 6.7}{100} = \pm 0.3 \text{ g dm}^{-3}$$

$$\text{concentration} = (6.7 \pm 0.3) \text{ g dm}^{-3}$$

Example 8

The literature value for the standard enthalpy change of combustion of methanol, CH$_3$OH(l), was found to be -726.0 kJ mol^{-1}:

$$CH_3OH(l) + \frac{3}{2}O_2(g) \rightarrow CO_2(g) + 2H_2O(l)$$

$$\Delta H_c^{\ominus} = -726.0 \text{ kJ mol}^{-1}$$

The experimental value was found to be -680.0 kJ mol^{-1}.

Calculate the percentage error, correct to **two** decimal places.

Solution

$$\text{percentage error} = \left|\frac{(-726.0) - (-680.0)}{(-726.0)}\right| \times 100\%$$

$$= \left|\frac{-46.0}{-726.0}\right| \times 100\% = 6.34\%$$

 ## Writing thermochemical equations

Thermochemical equations can be written with non-integer stoichiometric coefficients (for example, $\frac{3}{2}O_2(g)$) indicative of mole ratios. However, when an equation is considered in terms of molecules, only integers are used, as we would not consider two-thirds of a molecule of oxygen being broken in a reaction! Hence, when considering molecules it is better to write the equation for this reaction as:

$$2CH_3OH(l) + 3O_2(g) \rightarrow 2CO_2(g) + 4H_2O(l)$$

11.2 Graphical techniques

Understandings

→ Graphical techniques are an effective means of communicating the effect of an independent variable on a dependent variable, and can lead to determination of physical quantities.

→ Sketched graphs have labelled but unscaled axes, and are used to show qualitative trends, such as variables that are proportional or inversely proportional.

→ Drawn graphs have labelled and scaled axes, and are used in quantitative measurements.

Applications and skills

→ Drawing graphs of experimental results, including the correct choice of axes and scale.

→ Interpretation of graphs in terms of the relationships of dependent and independent variables.

→ Production and interpretation of best-fit lines or curves through data points, including an assessment of when these can and cannot be considered as a linear function.

→ Calculation of quantities from graphs by measuring slope (gradient) and intercept, including appropriate units.

Nature of science

→ The idea of correlation can be tested in experiments, the results of which can be displayed graphically.

Graphs and correlation

In science a graph can be a very useful way of representing data, which can subsequently be interpreted. **Dependence** is considered any statistical relationship between two sets of data or between two random variables. In a graph of Y versus X, the **independent variable** (that is, the **cause**) is plotted on the **x-axis** and the **dependent variable** (that is, the **effect**) is plotted on the **y-axis**.

We have already described the analytical chemist as the chemical detective. Part of the role of an analytical chemist is to explore statistical relationships that involve data and statistics, and the joy of analytical chemistry is all about the discovery of patterns embedded within a set of data (hence the chemical detective analogy).

Correlation can be described as a statistical measure and technique that indicates the degree and direction of the relationship between two sets

of variables; that is, it is a measure of the extent to which the two variables change with one another. A **positive correlation** is where the two variables increase or decrease in parallel to one another. A **negative correlation** is one in which one variable increases while the second variable decreases or vice versa. This idea of correlation can be tested in experiments whose results can be displayed graphically.

Correlations can be deduced from the **correlation coefficient**, represented by the symbol, r. This coefficient is a measure of the strength of the relationship between two variables. Data are often represented by **scatter plots** that show the scatter of various points on a graph. The correlation coefficient is a useful way to quantify the extent of a possible linear relationship between the two variables in the data set. The value of r can range from -1 to $+1$ (figure 1):

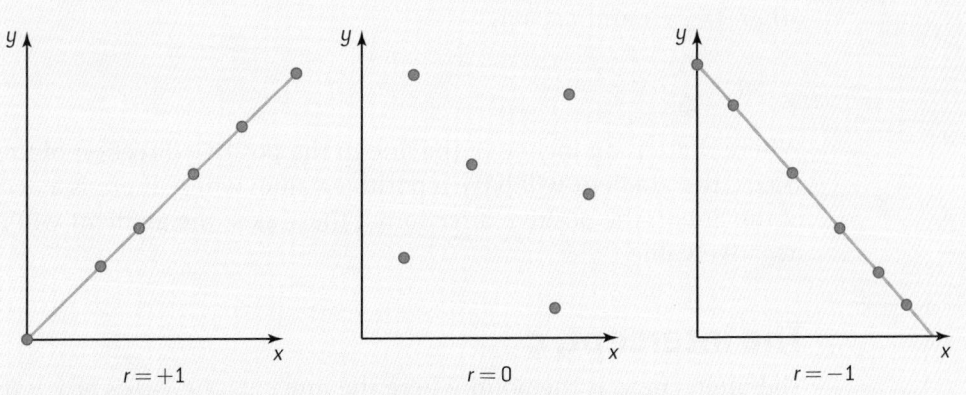

▲ Figure 1 Sketches of various scatter plots showing different correlation coefficients, *r*. The independent variable is the **cause**, represented on the *x*-axis. The dependent variable is the **effect**, represented by the *y*-axis

- $r = +1$, is indicative of a perfect positive linear relationship (all points lie on a straight line)
- $r = 0$, no linear relationship exists (there is complete scatter of points)
- $r = -1$, is indicative of a perfect negative linear relationship (one variable increases, the other decreases – all the data points will lie on a straight line but the gradient will be negative).

Graphical representations of data are widely used in diverse subject fields such as population analysis, finance, and climate modelling. Interpretation of these statistical trends can often lead to predictions, and so underpins the setting of government policies in many areas, such as health and education.

Charts and graphs, which largely transcend language barriers, can facilitate communication between scientists worldwide. At a research conference in Chile, where the language used by presenters is Spanish, a chemist from Oman whose native tongue is Arabic may not understand the presentation given in the Spanish language, but would clearly understand graphical data of the research findings. This shows the benefit of pictorially represented scientific data!

- Graphical techniques are an effective means of communicating the effect of an independent variable on a dependent variable, and can lead to the determination of physical quantities.

- Sketched graphs have labelled but unscaled axes, and are used to show qualitative trends, such as variables that are proportional or inversely proportional. Units generally would not need to be shown on a sketch, only the variables.

- Drawn graphs have labelled and scaled axes, and are based on quantitative measurements. Drawn graphs always display the appropriate units for variables.

There are a number of features that you are required to know for graphs:

- the slope or gradient of a line, *m*
- the intercept, *c*
- the idea of a "best-fit" line.

The slope or gradient of a line, *m*

Mathematically, the slope of a line, *m*, is the tangent of the angle, θ, that the line makes with the positive direction of the *x*-axis. In order to find the slope of a line, you need to

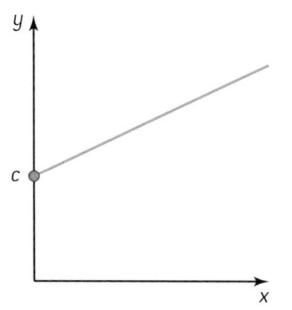

▲ Figure 2 The intercept, c, is the point where the line cuts the y-axis at $x = 0$

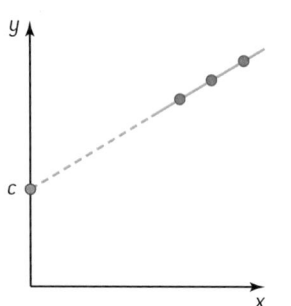

▲ Figure 3 The method of extrapolation involves extending a line back to the y-axis to find c

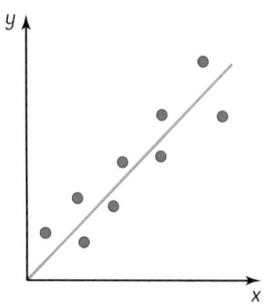

 Figure 4 Line of "best fit"

choose two points on the line, ideally well separated from each other, (x_1, y_1) and (x_2, y_2).

$$m = \frac{\Delta y}{\Delta x} = \frac{y_2 - y_1}{x_2 - x_1}$$

Where there is an incline of the line in the positive direction of the x-axis, the gradient will have a positive value; where there is a decline of the line in the positive direction of the x-axis, the gradient will have a negative value.

The intercept, c

The intercept, c, is the point where the line cuts the y-axis at $x = 0$ (figure 2).

The intercept can be found by two methods:

- using **extrapolation**
- using the **equation of a line**, $y = mx + c$.

Sometimes when you plot a graph, it is more convenient to draw the graph with appropriate scales where the x-axis scale begins at a point greater than zero as the data points may not be located at zero on the x-axis. If this is the case, by **extrapolation** you can simply extend the line back to the y-axis to find c (figure 3).

Alternatively, you could choose some point on the line, (x_c, y_c), and use the equation of the line $y = mx + c$ to find c, as long as you know m:

$$c = y_c - mx_c$$

The idea of a "best-fit" line

When you plot data obtained from an experiment you may find that, although there is a linear relationship, not all the data points lie exactly on the line. For this purpose, it is best to draw a line of best fit (figure 4). *Remember, this line may not necessarily contain all the experimental data points.*

🌐 Worked examples: using graphs

Example 1

Beer's law is based on the relationship:

$$A = \varepsilon cl$$

where A is the absorbance, ε is the extinction coefficient, c is the concentration, and l is the path length.

The following data was recorded using six standard solutions in order to determine the concentration of an unknown sample of copper(II) sulfate using atomic absorption spectroscopy.

Concentration, c/mol dm^{-3}	Absorbance, A
0.1002	0.130
0.2008	0.270
0.2819	0.380
0.4000	0.540
0.5082	0.685
0.6000	0.810
Unknown sample	0.460

a) Explain what you understand by a standard solution.

b) Draw a suitable plot showing the linear relationship at low concentrations that proves Beer's law.

c) Calculate the concentration, in mol dm⁻³, of the unknown copper(II) sulfate solution.

d) Calculate the slope, m, of the line, and state its units.

Solution

a) A standard solution is one whose concentration is known exactly.

b) Based on the data a graph is now drawn on graph paper (never drawn on just white paper!) or using a computer programme such as Microsoft Excel. The graph must have a title and all the axes must be labelled with the appropriate units and appropriately scaled. A "best-fit" line is then plotted. In this case all the data points lie exactly on the line, so proving Beer's law; that is, the absorbance is directly proportion to the concentration.

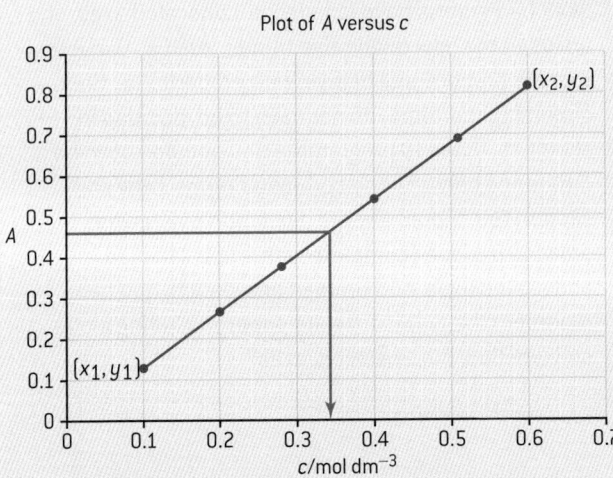

Plot of A versus c

c) When A is 0.460, $C = 0.34$ mol dm⁻³.

d) $(x_1, y_1) = (0.1002, 0.130)$ and
$(x_2, y_2) = (0.6000, 0.810)$

$$m = \frac{\Delta y}{\Delta x} = \frac{0.810 - 0.130}{0.6000 - 0.1002} = \frac{0.680}{0.4998}$$

$$= 1.36 \text{ mol}^{-1} \text{ dm}^3$$

Example 2

The graph below is a plot of concentration of reactant A versus time based on the data given in the table for a decomposition reaction of reagent A → products.

Time t/s	$[A]$/mol dm⁻³
4.00×10^2	2.30×10^{-3}
1.00×10^3	2.00×10^{-3}
2.00×10^3	1.50×10^{-3}
3.00×10^3	1.00×10^{-3}
4.00×10^3	5.0×10^{-4}

The graph can be expressed mathematically as:

$$[A] = -kt + [A]_o$$

where k represents the rate constant and $[A]_o$ the initial concentration.

a) Calculate the slope, m, of the graph and state its units.

b) Determine the intercept, c, of the graph and state its units.

c) Calculate the rate constant, k, and state its units.

d) Deduce the initial concentration, $[A]_o$, and state its units.

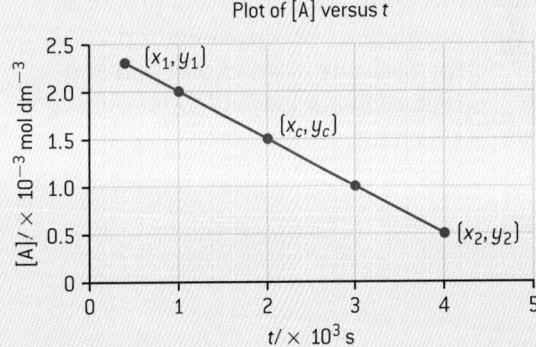

Plot of $[A]$ versus t

Solution

a) $(x_1, y_1) = (0.400, 2.30)$ and
$(x_2, y_2) = (4.00, 0.50)$

$$m = \frac{\Delta y}{\Delta x} = \frac{0.500 - 2.30}{4.00 - 0.400} = \frac{-1.80}{3.60} = -0.500.$$

Next we need to take the units of m into account:

$$10^{-3} \text{ mol dm}^{-3}/10^3 \text{ s} = 10^{-6} \text{ mol dm}^{-3} \text{ s}^{-1}$$

275

Therefore $m = -0.500 \times 10^{-6}$ mol dm^{-3} s^{-1}
$= 5.00 \times 10^{-1} \times 10^{-6}$ mol dm^{-3} s^{-1} =
-5.00×10^{-7} mol dm^{-3} s^{-1}.

b) Let $(x_c, y_c) = (2.00, 1.50)$

$$y_c = mx_c + c$$

$$c = y_c - mx_c = 1.50 - (-0.500 \times 2.00)$$
$$= 2.50 \times 10^{-3} \text{ mol dm}^{-3}$$

> **Study tip**
>
> Note that the units of c will always correspond to the units of the y-axis variable, which in this case is [A].

c) The expression $[A] = -kt + [A]_o$ is in the form of $y = mx + c$.

So $m = -k$, and hence:

$$k = -m = -(-5.00 \times 10^{-7} \text{ mol dm}^{-3} \text{ s}^{-1})$$
$$= 5.00 \times 10^{-7} \text{ mol dm}^{-3} \text{ s}^{-1}.$$

d) $c = [A]_o$, so $[A]_o = 2.50 \times 10^{-3}$ mol dm^{-3}.

> For HL students *only*. This data is based on a zero-order reaction, discussed further in topic 16.

Example 3

Boyle's law is one of the gas laws and states that with the temperature, T, and the amount of gas, n, constant, the pressure is inversely proportional to the volume. State how you would represent this law mathematically as an expression and suggest the type of linear graph you would sketch to illustrate this relationship.

Solution

The key phrase here is inverse proportionality. Hence:

$$p \propto \frac{1}{V}$$

$$p = k\left(\frac{1}{V}\right)$$

where k is a constant of proportionality.

> **Study tip**
>
> Mathematically, any proportionality sign, "$\propto$", in a mathematical expression can be replaced by an equality expression plus a constant, for example "$= k$".

This is in the form $y = mx + c$, so a suitable linear sketch would be to plot p on the y-axis and $\left(\frac{1}{V}\right)$ on the x-axis. m would equate to the constant k. As there is no term to the right of the $k\left(\frac{1}{V}\right)$ part of the expression, mathematically the intercept, c, will be zero.

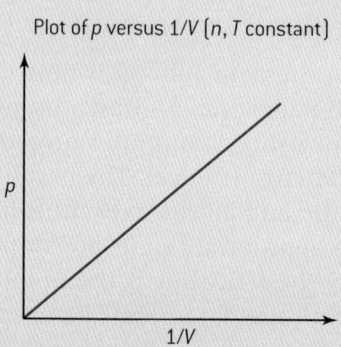

Plot of p versus $1/V$ (n, T constant)

> **Study tip**
>
> *Sketched* graphs have labelled but unscaled axes, and are used to show qualitative trends, such as variables that are proportional or inversely proportional. Units generally, therefore, do not have to be included in sketches. In the IB programme you should know the difference between the command terms *draw* and *sketch*. You should always use graph paper when *drawing a graph* (complete with a title, labelled and scaled axes, and units). For a sketch, you can represent the graph on white paper (including a title and labelled, but unscaled, axes).

11.3 Spectroscopic identification or organic compounds

Understandings

→ The degree of unsaturation or index of hydrogen deficiency (IHD) can be used to determine from a molecular formula the number of rings or multiple bonds in a molecule.

→ Mass spectrometry (MS), proton nuclear magnetic resonance spectroscopy (^{1}H NMR), and infrared spectroscopy (IR) are techniques that can be used to help identify compounds and to determine their structure.

Applications and skills

→ Determination of the IHD from a molecular formula.

→ Deduction of information about the structural features of a compound from percentage composition data, MS, ^{1}H NMR, or IR.

Nature of science

→ Improvements in instrumentation – mass spectrometry, proton nuclear magnetic resonance and infrared spectroscopy have made identification and structural determination of compounds routine.

→ Models are developed to explain certain phenomena that may not be observable – for example, spectra are based on the bond vibration model.

Degree of unsaturation or index of hydrogen deficiency (IHD)

The **degree of unsaturation** or **index of hydrogen deficiency** (IHD) can be used to determine from a molecular formula the number of rings or multiple bonds in a molecule.

The degree of unsaturation is used to calculate the number of rings and π bonds present in a structure, where:

- a double bond is counted as one degree of unsaturation
- a triple bond is counted as two degrees of unsaturation
- a ring is counted as one degree of unsaturation
- an aromatic ring is counted as four degrees of unsaturation.

The IHD can be worked out two ways:

- from the structure
- from the molecular formula.

From the structure

Compound	Structure	Number of rings, double bonds, and triple bonds	IHD
benzene		• one ring • three double bonds (in Kekulé structure)	4
cyclobutane		• one ring	1
cyclohexane (chair conformation)		• one ring	1
cyclopentadiene		• one ring • two double bonds	3
2-acetoxybenzoic acid (aspirin)	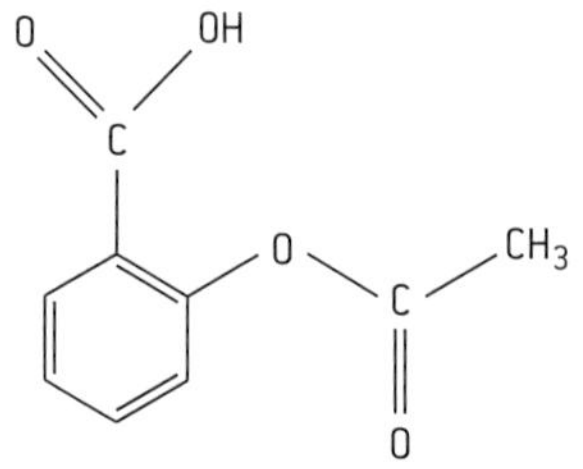	• one ring • five double bonds	6
ethyne	$H - C \equiv C - H$	• one triple bond	2

From the molecular formula

In order to deduce the IHD for the generic molecular formula $C_cH_hN_nO_oX_x$, where X is a halogen atom (F, Cl, Br, or I), we can use the following expression:

$$IHD = (0.5)(2c + 2 - h - x + n)$$

Hence for $C_4H_8O_2$:

$c = 4$

$h = 8$

$n = 0$

$o = 2$

$x = 0$

$IHD = (0.5)(8 + 2 - 8 - 0 + 0) = 1$

Therefore the molecule contains either one double bond or one ring. There are several isomers of $C_4H_8O_2$. Here are just three for illustration:

Isomer of $C_4H_8O_2$	Structure	IHD
methyl propionate	$H_3C - O - \overset{\overset{O}{\|\|}}{C} - CH_2CH_3$	1
ethyl ethanoate	$H_3C - \overset{\overset{O}{\|\|}}{C} - O - CH_2CH_3$	1
tetrahydro-3-furanol		1

Let us take four compounds with different molecular formulas and deduce their IHD using the formula:

Molecular formula	IHD
$C_{17}H_{21}NO_4$ (cocaine)	8
$C_{27}H_{46}O$ (cholesterol)	5
C_6H_7N (aniline)	4
$C_{15}H_{10}ClN_3O_3$ (clonazepam)	12

Activity

Using the ChemSpider RSC database (www.chemspider.com) look at the structures of these molecules and check to see if the above calculations agree with what you expect the IHD to be from the respective structures.

Electromagnetic spectrum (EMS)

The electromagnetic spectrum (figure 1) is given in section 3 of the *Data booklet*.

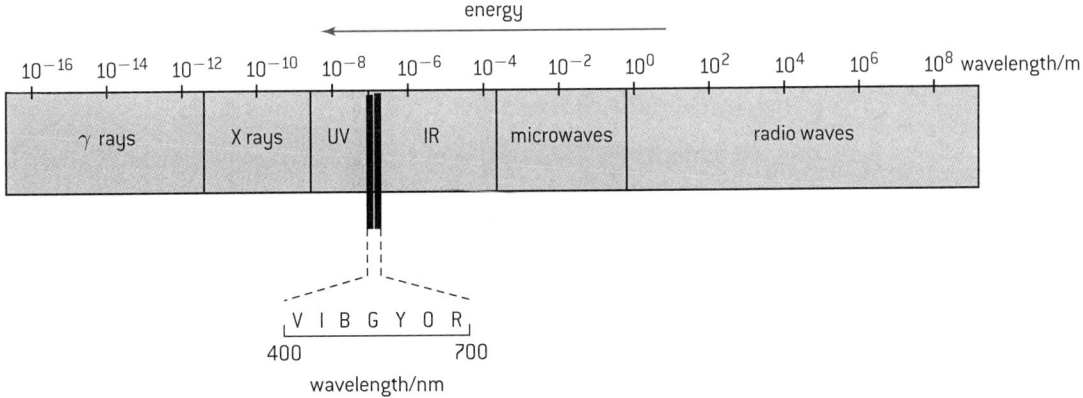

▲ Figure 1 The electromagnetic spectrum

The energy of electromagnetic radiation, E, is related to the frequency, ν, of the radiation by **Planck's equation**:

$$E = h\nu = \frac{hc}{\lambda}$$

where:

h = Planck's constant = 6.63×10^{-34} J s (given in section 2 of the *Data booklet*)

E = energy of radiation (measured in J)

ν = frequency of radiation (measured in Hz)

c = speed of light = 3.00×10^8 m s^{-1} (given in section 2 of *Data booklet*)

λ = wavelength (measured in m).

From this relationship it can be seen that the energy is directly proportional to the frequency and inversely proportional to the wavelength, that is:

$$E \propto \nu \text{ and } E \propto \frac{1}{\lambda}$$

The various regions of the EMS are the basis of different types of **spectroscopy** (which is the study of the way matter interacts with radiation) and various techniques are used to identify the structures of substances:

- **X-rays** – as their energy is high, these cause electrons to be removed from the inner energy levels of atoms. Diffraction patterns can lead to information such as the bond distances and bond angles in a structure and form the basis of **X-ray crystallography**.

- **Visible and ultraviolet (UV) light** give rise to electronic transitions and hence this type of spectroscopy gives information about the electronic energy levels in an atom or molecule. This is the basis of **UV-vis spectroscopy**.

- **Infrared radiation** causes certain bonds in a molecule to vibrate (for example, stretch and bend) and as such provides information on the functional groups present. This is the basis of **IR spectroscopy**.

- **Microwaves** cause molecular rotations and can give information on bond lengths.

- **Radiowaves** can cause nuclear transitions in a strong magnetic field because radiowaves can be absorbed by certain nuclei, which causes their spin states to change. **Nuclear magnetic resonance (NMR) spectroscopy** is based on this and information on different chemical environments of atoms can be deduced, which leads to information on the connectivity of the atoms present in a molecule.

Let's now consider three different types of spectroscopy that form the cornerstone of the spectroscopic identification of organic compounds:

- infrared (IR) spectroscopy

- proton nuclear magnetic resonance (^{1}H NMR) spectroscopy

- mass spectrometry (MS).

Infrared spectroscopy

Unlike UV and visible radiation, IR radiation does not have sufficient energy to result in electronic transitions, but can cause molecular vibrations, which result from the vibration of certain groups of molecules about their bonds. Hence, using this type of spectroscopy various functional groups can be identified in a molecule. The vibrational transitions correspond to definite energy levels. The basis of IR spectroscopy is the **spring model**.

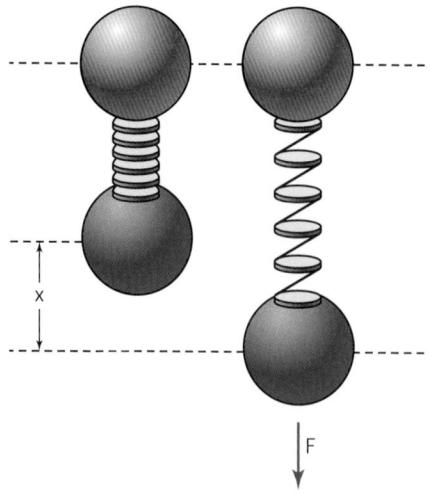

▲ Figure 2 Hooke's law is the basis of the spring model used to understand the vibration of molecules

In the spring model, every covalent bond is considered as a spring. Such a spring can be stretched (both symmetrically and asymmetrically), bent, or twisted, giving rise to a distortion. The force required to cause the vibration is based on a law from physics called **Hooke's law**:

$$F \propto x$$

That is, the expansion of the length of a spring (x) from its equilibrium position will be directly proportional to the force (F) caused by the load applied to the spring (figure 2). By convention the law is typically expressed with a negative sign where F represents the restoring force exerted by the spring:

$$F = -kx$$

where k is a constant of proportionality, called the **spring constant**.

The fundamental frequency of the vibration, ν, based on a system obeying Hooke's law can be related to the mass, m, by the expression:

$$\nu = \frac{1}{2\pi}\sqrt{\frac{k}{m}}$$

Hence, for the vibration of an atom it can be seen that lighter atoms will vibrate at higher frequencies, ν, and heavier atoms will vibrate at lower frequencies, ν. The same applies for multiple bonds (for example double and triple bonds). If you imagine two atoms connected by a spring the stronger the bond connecting the two atoms the tighter the string will be and therefore more energy is required to stretch it.

For a diatomic molecule, such as hydrogen chloride, HCl, only one form of molecular vibration is possible, that is stretching (figure 3).

If we compare the frequencies of HCl, HBr, and HI we find that because HCl has the smallest mass and greatest bond enthalpy (see section 11 of the *Data booklet*) it will have the greatest frequency.

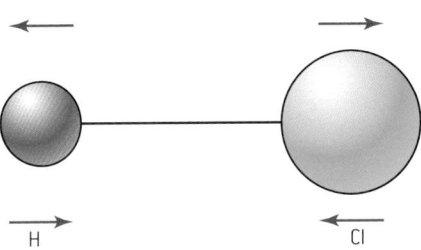

▲ Figure 3 Stretching in HCl molecule

Molecule	Bond enthalpy/kJ mol^{-1}	Wavenumber/cm^{-1}
H—Cl	431	2886
H—Br	366	2559
H—I	298	2230

▲ Table 1 Bond enthalpies and wavenumbers for selected HX molecules

Different molecules absorb at different frequencies because the energy required to execute a vibration will depend on the bond enthalpy. As can be seen from table 1, IR absorptions are typically cited as the reciprocal of the wavelength $\left(\frac{1}{\lambda}\right)$. This is the **wavenumber** and has units of cm^{-1}.

For polyatomic species, there may be several different modes of vibration. For example, the water molecule has three modes of vibration (figure 4):

- a symmetric stretch (3652 cm^{-1})
- an asymmetric stretch (3756 cm^{-1})
- a symmetric bend (1595 cm^{-1}).

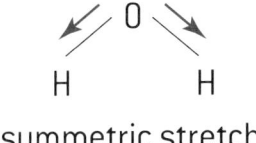

symmetric stretch

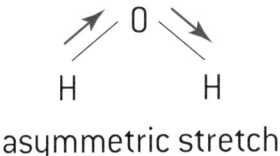

asymmetric stretch

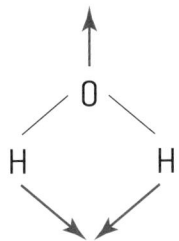

symmetric bend

▲ Figure 4 Modes of vibration of the water molecule. All three modes of vibration are IR active

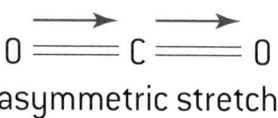

symmetric stretch

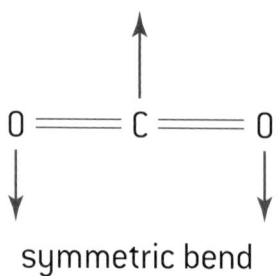

asymmetric stretch

symmetric bend

▲ Figure 5 Modes of vibration of the CO_2 molecule. The asymmetric stretch and the symmetric bend are IR active because of a change in the molecular dipole moment that results from the molecular vibration

However, for a covalent bond to absorb IR radiation there must be a **change** in the **molecular dipole moment** associated with the **vibration mode**. Let us take some examples:

Molecule	Polarity of molecule	Absorption of IR radiation
H_2	non-polar	no (IR inactive)
O_2	non-polar	no (IR inactive)
CO_2 (figure 5)	non-polar	symmetric stretch – no (IR inactive)
		asymmetric stretch – yes (IR active) (2349 cm^{-1})
		symmetric bend – yes (IR active) (667 cm^{-1})

The absorbance, A, of a sample can be related to the transmittance by the expression:

$$A = -\log_{10} T$$

An IR spectrum is a plot of the percentage transmittance, $\%T$, versus the wavenumber (in cm^{-1}), where $\%T$ ranges from 0% to 100%. In an IR spectrum, functional groups can be identified. The characteristic ranges for the IR absorptions of various bonds in different classes of molecules and of different functional groups are given in section 26 of the *Data booklet*; for example, the C=C absorption in alkenes typically occurs in the range 1620–1680 cm^{-1}, etc.

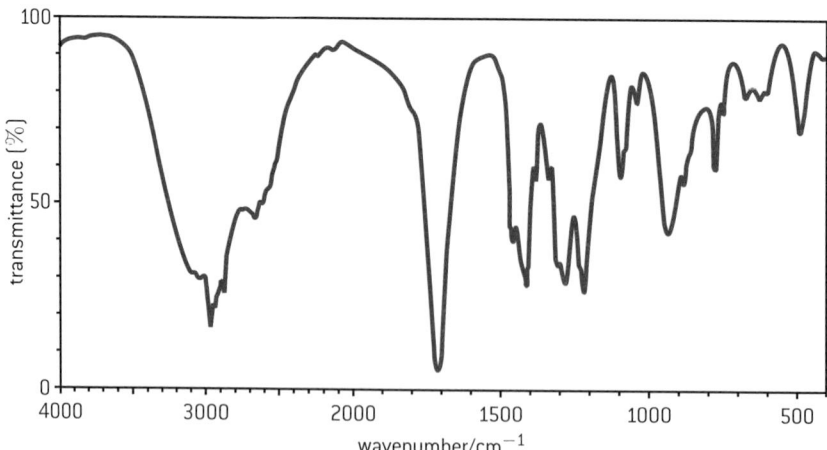

▲ Figure 6 IR spectrum (in liquid film) of butanoic acid

In figure 6, note the following absorptions:

- strong, broad peak in the range 2500–3000 cm^{-1} characteristic of the O–H bond of a carboxylic acid

- strong peak in the range 1700–1750 cm^{-1} characteristic of the C=O group
- peak in the range 2850–3090 cm^{-1} characteristic of the C—H bond.

Typically in the region 300–1400 cm^{-1} more complex vibrations can be identified. This is termed the **fingerprint region** of an IR spectrum.

IR is often termed a supporting analytical technique as the only information it really provides relates to the presence or absence of certain chemical bonds associated with different functional groups in molecules – it provides little other structural information. However, it is a powerful technique in making some key decisions at the beginning of the structural elucidation of an organic (and, sometimes, an inorganic) compound and is often the starting point for the organic chemist on his or her journey into the detective work of analytical chemistry. IR spectroscopy is also used in physics for heat sensors and remote sensing.

Proton nuclear magnetic resonance (^{1}H NMR) spectroscopy

^{1}H NMR spectroscopy gives information on the different chemical environments of hydrogen atoms in a molecule and is possibly the most important structural technique available to the organic chemist. The nuclei of hydrogen atoms can exist in two possible spin states. As such they can behave as tiny magnets. When there is no magnetic field the two spin states have the same energy and random orientation. When the nuclei are placed in a magnetic field, the spin states may align with the magnetic field or against it. This results in two nuclear energy levels. The parallel alignment of the nuclear spin with the external magnetic field is of a lower energy compared to the anti-parallel alignment configuration. This difference in energy, ΔE, between the two levels corresponds to the radiowave region of the EMS. As the applied magnetic field is increased, ΔE will increase. The energy difference depends on the different chemical environments of the hydrogen atoms.

In a ^{1}H NMR spectrum, the position of the NMR signal relative to a standard (tetramethylsilane, TMS) is termed **the chemical shift, δ,** expressed in parts per million (ppm), of the proton. δ for TMS is assigned as 0 ppm. Hydrogen nuclei in the different chemical environments have different chemical shifts (see section 27 of the *Data booklet*).

Therefore the number of signals on a ^{1}H NMR spectrum shows the number of different chemical environments in which the hydrogen atoms are found. For example, in the ^{1}H NMR spectrum of methanoic acid, HCCOH, two signals are found, which shows the two different chemical environments of the two hydrogen atoms, A and B.

Type of proton	Chemical shift (ppm)
— CH$_3$	0.9–1.0
— CH$_2$—R	1.3–1.4
— R$_2$CH	1.5
RO—C(=O)—CH$_2$—	2.0–2.5
R—C(=O)—CH$_2$—	2.2–2.7
phenyl—CH$_3$	2.5–3.5
— C≡C—H	1.8–3.1
— CH$_2$—Hal	3.5–4.4
R—O—CH$_2$—	3.3–3.7
R—C(=O)—O—CH$_2$—	3.7–4.8
R—C(=O)—O—H	9.0–13.0
R—O—H	1.0–6.0
— HC=CH$_2$	4.5–6.0
phenyl—OH	4.0–12.0
phenyl—H	6.9–9.0
R—C(=O)—H	9.4–10.0

▲ Table 2 Typical proton chemical shift values (δ) relative to tetramethylsilane (TMS). R represents an alkyl group, and Hal represents F, Cl, Br, or I. These values may vary in different solvents and conditions

The signals (figure 7) can be assigned as follows:

Type of proton	Predicted δ/ppm (from section 27 of *Data booklet*)	Actual δ/ppm (from ^{1}H NMR spectrum)
A: $-$HCO	9.4–10.0	8.06
B: $-$COOH	9.0–13.0	10.99

Another useful feature of a ^{1}H NMR spectrum is that it contains an **integration trace** that shows the relative number of hydrogen atoms present. In the case of the ^{1}H NMR spectrum for methanoic acid this will be 1:1.

An important application of ^{1}H NMR spectroscopy is associated with the fact that the protons in water molecules within human cells can be detected by **magnetic resonance imaging** (MRI), which gives a three-dimensional view of organs in the human body (discussed further in topic 21).

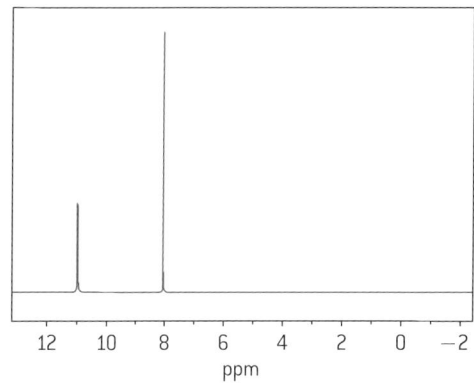

▲ Figure 7 ^{1}H NMR spectrum of methanoic acid, HCOOH.

Mass spectrometry (MS)

In topic 2 we introduced the principles of mass spectrometry (MS). When a gaseous molecule is ionized its molecular ion, M^+, is formed. The **molecular ion peak** in a mass spectrum corresponds to the molecular mass of the compound. Owing to the highly energetic ionization process involved in a mass spectrometer, the molecule can, in fact, break up into smaller fragments, some of which will be ions. The **fragmentation pattern** observed in a mass spectrum provides further information on certain functional groups present in a molecule. Section 28 of the *Data booklet* lists some of these fragments and their masses. Here are some examples:

- $(M_r - 15)^+$ results from the loss of $-CH_3$
- $(M_r - 17)^+$ results from the loss of $-OH$
- $(M_r - 29)^+$ results from the loss of $-CHO$ *or* the loss of $-CH_2CH_3$
- $(M_r - 31)^+$ results from the loss of $-OCH_3$
- $(M_r - 45)^+$ results from the loss of $-COOH$.

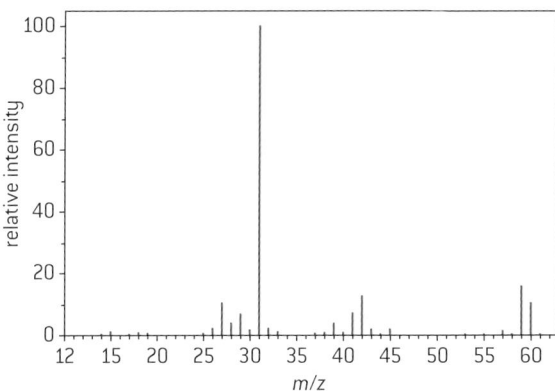

▲ Figure 9 MS of propan-1-ol

▲ Figure 8 A nuclear magnetic resonance (NMR) spectrometer. NMR spectroscopy measures the resonance between an applied magnetic field and the magnetic moment of a molecule's atoms. It allows identification of molecules in a sample. This 400 MHz Agilent Unity Inova NMR Spectrometer is located at the Magnetic Resonance Facility of the National Renewable Energy Laboratory (NREL). The NREL is based in Golden, Colorado, USA

Some of the peaks for propan-1-ol (figure 8) can be assigned as follows:

m/z	Predicted δ / ppm (from section 27 of *Data booklet*)
60	molecular ion peak, $CH_3CH_2CH_2OH^+$
31 (high relative intensity)	$(M_r - 29)^+$ from the loss of $-CH_2CH_3$
29	$(M_r - 31)^+$ from the loss of $-CH_2OH$

🌐 Worked examples

Example 1

Deduce the index of hydrogen deficiency of:

a) Zanamivir (an inhibitor used to treat infections caused by the influenza A and B viruses) using Section 37 of the *Data booklet*.

b) Carbolic acid which has the molecular formula C_6H_6O.

Solution

a) Zanamivir has four double bonds and one ring, so its IHD = 5.

b) For C_6H_6O:

   IHD $= (0.5)(12 + 2 - 6 - 0 + 0) = 4$

> Carbolic acid is, in fact, phenol which has the Kekulé structure:

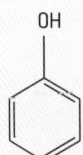
OH

Example 2

For each of the following two compounds:

    $(CH_3)_3COH$

    CH_3COOCH_3

a) State how many signals each compound shows in its ^1H NMR spectrum.

b) State what you expect the integration trace to be for each ^1H NMR.

c) For the MS of $(CH_3)_3COH$, state possible *m/z* values, giving a reason for your answer.

Solution

For $(CH_3)_3COH$:

a) 2

b) 9:1

For CH_3COOCH_3:

a) 2

b) 1:1

c) $m/z = 74$ molecular ion peak $(CH_3)_3COH^+$

    $m/z = 57$ $(M_r - 17)^+$ from loss of $-OH$.

    $m/z = 59$ $(M_r - 15)^+$ from loss of $-CH_3$.

> **Study tip**
>
> In MS, don't forget the + sign for species that remain after a fragment has been lost.

Example 3

An unknown compound, X, of molecular formula C_2H_4O, has the following IR and ^1H NMR spectra.

IR spectrum (in liquid film):

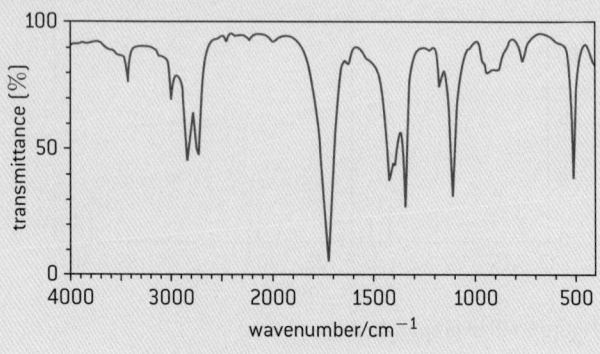

¹H NMR spectrum (in CDCl₃):

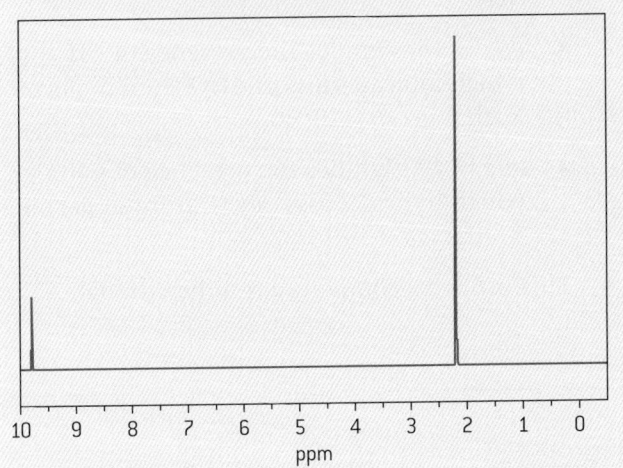

The MS spectrum of X showed peaks at *m/z* values of 15, 29, and 44 (other peaks were also found).

Deduce the structure of X using the information given and any other additional information from the *Data booklet*. For each spectrum assign as much spectroscopic information as possible based on the structure of X.

Solution

- As the molecular formula of X is given, it is worth first finding out the IHD, which indicates the **index of hydrogen deficiency** or **degree of unsaturation**.

For the generic molecular formula $C_cH_hN_nO_oX_x$:

$$IHD = (0.5)(2c + 2 - h - x + n)$$

For C_2H_4O the IHD is:

$$(0.5)(4 + 2 - 4 - 0 + 0) = 1$$

Therefore the molecule contains either one double bond or one ring.

- We note that, based on the molecular formula, X contains just one oxygen atom. The classes for X could be an ether (C−O−C), a ketone (C−CO−C), an aldehyde (C−CHO), or an alcohol (C−OH).

- Based on the above we now examine the IR spectrum and see whether there is a strong IR

absorption for C=O in the wavenumber range 1700–1750 cm⁻¹, based on section 26 of the *Data booklet*. Indeed, there is a strong peak at approximately 1727 cm⁻¹, which suggests the presence of a C=O bond.

- If C=O is present, then X might be either an aldehyde or a ketone. An aldehydic proton is quite characteristic in the ¹H NMR spectrum, with a chemical shift, δ, in the range 9.4–10.0 ppm, as seen from section 27 of the *Data booklet*. In fact, there does appear to be a single signal with $\delta = 9.8$ ppm.

- If X is an aldehyde that means we now have identified a portion of the molecule, that is −CHO. As the remaining number of atoms must contain one carbon and three hydrogens, this indicates a methyl group, −CH₃, which suggests that X is ethanal, CH₃CHO.

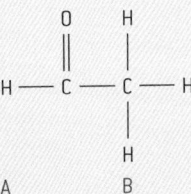

- Let's now test this proposed structure based on the spectroscopic data gained from the ¹H NMR spectrum. Two types of hydrogen atoms are present in different chemical environments, A and B.

Type of proton	Predicted δ/ppm (from Section 27 of *Data booklet*)	Integration trace	Actual δ/ppm (from ¹H NMR spectrum)
A: −CHO	9.4–10.0		9.8
B: −COCH₃	2.2–2.7	1:3	2.2

- Having established the structure of X, it is worth returning to the IR spectrum to confirm the additional characteristic range for the infrared absorption due to the CH bonds in the wavenumber range 2850–3090 cm^{-1}. As can be seen from the IR spectrum, there are, indeed, absorptions within this range.

- Finally, consider the MS. There should be a molecular ion peak at $m/z = 44$, corresponding to the relative molar mass of C_2H_4O, calculated as 44.06. This, indeed, is present. In addition, the other dominant m/z values in the MS can be assigned as follows (using section 28 of the *Data booklet*):

  - $m/z = 15$, signifies the presence of CH_3, which indicates loss of CHO from molecule X, that is $(M_r - 29)^+$

  - $m/z = 29$, signifies the presence of CHO^+, which indicates loss of CH_3 from molecule X, that is $(M_r - 15)^+$.

- This confirms compound X to be ethanal.

Useful resources

- Spectral Database for Organic Compounds, SDBS, hosted by *National Institute of Advanced Industrial Science and Technology* (AIST), Japan. http://sdbs.db.aist.go.jp/sdbs/cgi-bin/cre_index.cgi

- *EURACHEM* – a network of organisations in Europe having the objective of establishing a system for the international traceability of chemical measurements and the promotion of good quality practices. There is an excellent guide on uncertainty in measurement which might be useful for your IA and other laboratory work. http://www.eurachem.org/

- *NIST – National Institute of Standards and Technology, USA*. The chemistry portal is worth accessing for spectroscopic data etc. http://www.nist.gov/chemistry-portal.cfm

Questions

1 How many significant figures are in 0.0200 g?

   A. 1
   B. 2
   C. 3
   D. 5

2 A burette reading is recorded as 27.70 $\pm$ 0.05 cm³. Which of the following could be the actual value?

   I.   27.68 cm³
   II. 27.78 cm³
   III. 27.74 cm³

   A. I and II only
   B. I and III only
   C. II and III only
   D. I, II, and III   [1]

   IB May 2011

3 A piece of metallic aluminium with a mass of 10.044 g was found to have a volume of 3.70 cm³.

   A student carried out the following calculation to determine the density:

   $$\text{density (g cm}^{-3}) = \frac{10.044}{3.70}$$

   What is the best value the student could report for the density of aluminium?

   A. 2.715 g cm⁻³
   B. 2.7 g cm⁻³
   C. 2.71 g cm⁻³
   D. 2.7146 g cm⁻³   [1]

   IB May 2011

4 Which experimental procedure is most likely to lead to a large systematic error?

   A. Determining the concentration of an alkali by titration with a burette
   B. Measuring the volume of a solution using a volumetric pipette
   C. Determining the enthalpy change of neutralization in a beaker
   D. Measuring the volume of a gas produced with a gas syringe   [1]

   IB May 2010

5 Which are likely to be reduced when an experiment is repeated a number of times?

   A. Random errors
   B. Systematic errors
   C. Both random and systematic errors
   D. Neither random nor systematic errors   [1]

   IB November 2009

6 Deduce the IHD for codeine using section 37 of the *Data booklet*.

7 Deduce the IHD for a molecule of molecular formula $C_5H_{10}N_2$.

8 The ¹H NMR spectrum of **X** with molecular formula C_3H_6O is shown below.

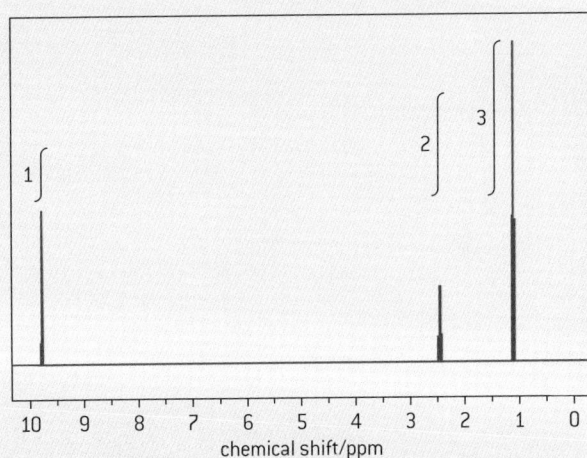

Source: SDBSWeb, http://sdbs.riodb.aist.go.jp (National Institute of Advanced Industrial Science and Technology)

   a) Deduce which of the following compounds is **X** and explain your answer.   [2]

   $CH_3-CO-CH_3$; CH_3-CH_2-CHO;

   $CH_2=CH-CH_2OH$

   b) Deduce which one of the signals in the ¹H NMR spectrum of **X** would also occur in the spectrum of one of the other isomers, giving your reasoning.   [2]

c) The infrared and mass spectra for **X** were also recorded.

   (i) Apart from absorptions due to C−C and C−H bonds, suggest **one** absorption, in wavenumbers, that would be present in the infrared spectrum. [1]

   (ii) Apart from absorptions due to C−C and C−H bonds, suggest **one** absorption, in wavenumbers, absent in this infrared spectrum, but present in one of the other compounds shown in part **a)**. [1]

d) Suggest the formulas and m/z values of **two** species that would be detected in the mass spectrum. [2]

IB May 2011

12 ATOMIC STRUCTURE (AHL)

Introduction

The quantized nature of energy transitions is related to the energy states of electrons in atoms and molecules. In this topic we see how empirical evidence gained from line emission spectra has been used to provide strong evidence for the existence of energy levels.

12.1 Electrons in atoms

Understandings

→ In an emission spectrum, the limit of convergence at higher frequency corresponds to the first ionization energy.

→ Trends in first ionization energy across periods account for the existence of main energy levels and sublevels in atoms.

→ Successive ionization energy data for an element give information that shows relations to electron configurations.

Applications and skills

→ Solving problems using $E = hv$.

→ Calculation of the value of the first ionization energy from spectral data which gives the wavelength or frequency of the convergence limit.

→ Deduction of the group of an element from its successive ionization energy data.

→ Explanation of the trends and discontinuities in data on first ionization energy across a period.

Nature of science

→ Experimental evidence to support theories – emission spectra provide evidence for the existence of energy levels.

CERN

Scientific theories must be supported by experimental evidence to be accepted. **CERN (Conseil Européen pour la Recherche Nucléaire)** is the European Organization for Nuclear Research. The CERN laboratory, located near Geneva on the border between Switzerland and France has some of the largest and most advanced sophisticated scientific instruments in the world, used to study the fundamental particles of matter. The CERN project involves over 20 member states across Europe, and several countries outside Europe. CERN is an excellent example of extensive international collaborative scientific research. The main focus of its research is particle physics. Scientists at CERN are trying to discover the origins of the universe, and what it is made of. When CERN was set up in 1954 after the second world war most research at the time concentrated on exploring inside the atom. Our understanding of matter since then has progressed significantly beyond the discovery of the nucleus by Ernest Rutherford in 1911 (sub-topic 2.1).

CERN was also the birthplace of the World Wide Web (the internet), invented by the British scientist Tim Berners-Lee; it was originally developed so that scientists and universities around the globe could instantly share information and data.

In 2013 preliminary experimental results from the **Large Hadron Collider** (LHC, figure 1) at CERN suggested evidence for the existence of the **Higgs boson** particle. The Higgs boson is an elementary particle whose existence is predicted by a theoretical model in particle physics termed the **Standard Model**, which describes how the universe is constructed. In this model everything in the universe, from people to plants to stars to trees is considered to be composed of just a few building blocks. These are the particles of matter. Such particles themselves are governed by forces. The Higgs boson is one of 17 **fundamental particles** (another is the photon, for example, see sub-topic 2.2). Some fundamental particles are constituents of everyday matter whereas others (including the Higgs particle and the photon) are responsible for all the forces that occur in nature, excluding gravity.

▲ Figure 1 The Large Hadron Collider (LHC) at CERN is the world's largest and most powerful particle accelerator. It consists of a 27 km ring of superconducting magnets

In 1964 a number of physicists suggested a theoretical mechanism outlining the existence of the Higgs boson. However, the main problem for scientists for almost 50 years was that there was no experimental evidence for its existence until 2013. The particle is named after British physicist Professor Peter Higgs, one of six scientists who proposed its existence originally in 1964. Empirical evidence for the existence of the Higgs boson has been described as the greatest scientific discovery in 100 years, linking theory with experimental evidence.

The Nobel Prize in Physics 2013 was awarded jointly to François Englert, Université Libre de Bruxelles, Belgium, and Peter W. Higgs, University of Edinburgh, Scotland, for the theoretical discovery of a mechanism that contributes to our understanding of the origin of mass of subatomic particles, and which was recently confirmed through the discovery of the predicted fundamental particle, by the ATLAS and CMS experiments at CERN's LHC.

TOK

In topic 2 we discussed a number of key discoveries associated with the structure of the atom. In this topic we continue to look at historical developments in this field.

de Broglie: Wave properties of electrons

In 1924 the French scientist Louis de Broglie brought the wave–particle dual theory of the electron to the fore in the **de Broglie equation**:

$$\lambda = \frac{h}{p}$$

where:

λ = wavelength

h = Planck's constant

p = momentum = mv = mass × velocity

In this equation, wavelength is inversely proportional to momentum.

You might find that interpretation of this expression may warrant an exploratory journey of how we consider the movement of particles. Suppose you have a particle of large mass, such as a tennis ball. If the mass is large then so will be the momentum, p. The wavelength, λ, associated with the tennis ball when moving at high velocity becomes negligible. This agrees with what we observe: you don't see a tennis ball moving in a wave pattern across a tennis court!

However, if the mass of the particle is tiny, such as the electron with mass $m_e = 9.109 \times 10^{-31}$ kg, then the wavelength will be large, suggesting that a wave motion can be associated with the electron in an atom. So de Broglie suggested that not only does light have wave properties, but so does matter!

The de Broglie equation shows that macroscopic particles have too short a wavelength for their wave properties to be observed.

Is it meaningful to talk of properties that cannot be observed with our senses?

Heisenberg's uncertainty principle

In 1927 another theoretical physicist, Werner Heisenberg from Germany, published the ground-breaking theory termed **Heisenberg's uncertainty principle**. Professor Heisenberg was one of the pioneers in the field of **quantum mechanics** and the basis of his principle is as follows:

> The more precisely the position is determined, the less precisely the momentum is known in this instant, and vice versa.
>
> *Heisenberg, uncertainty paper, 1927.*

Theoretical scientists often work in fields where the application of their research to real life may be difficult to predict, or difficult to comprehend by non-scientists.

Should governments and funding bodies fund basic theoretical research, or should they concentrate on applied or strategic research where the end application may be more economically tangible? Can you think of examples of scientific discoveries resulting from basic research that have resulted in unforeseen applications?

Heisenberg's uncertainty principle can be expressed mathematically as follows:

$$\Delta p \times \Delta q \geq \frac{h}{4\pi}$$

where:

Δp = uncertainty of momentum measurement

Δq = uncertainty of position measurement

h = Planck's constant

Suppose you wish to measure the location of a moving electron. If the position is measured with high accuracy, then Δq will approach zero. What then happens simultaneously to Δp? To explore this we can rearrange the equation:

$$\Delta p \geq \left(\frac{h}{4\pi}\right)\frac{1}{\Delta q}$$

as $\Delta q \to 0$, then $\frac{1}{\Delta q} \to \infty$

so $\Delta p \to \infty$

that is, as the uncertainty of position measurement approaches zero, the uncertainty of momentum measurement approaches infinity so the momentum becomes effectively undefined.

Heisenberg said: "What we observe is not nature itself, but nature exposed to our method of questioning." Can our senses give us objective knowledge about the world?

The idea of uncertainty lying at the heart of Heisenberg's thinking is an example of a historical journey of discovery that embraces not just physics but the persona of an individual as well.

Find out more about Heisenberg and consider what is meant by this statement.

The Schrödinger wave equation

The dual wave–particle nature of the electron has been one of the great discussions in the history of subatomic particles. The Austrian physicist Erwin Schrödinger (1887–1961) was an advocate of **wave mechanics**, expressed in the **Schrödinger wave equation**. See sub-topic 2.2 for more information on the Schrödinger wave equation.

Schrödinger's wave equation accurately predicted the energy levels of atoms.

The values of some first ionization energies, in kJ mol^{-1}, are given in section 8 of the *Data booklet*.

Useful resource

Chemsoc Timeline – This is a visual exploration of key events in the history of science with particular emphasis on chemistry. It was developed by Murray Roberston in collaboration with *ChemSoc*, the chemical network of the *Royal Society of Chemistry* (RSC). You can even make predictions for inventions or discoveries that you think will be made in years to come!

http://www.rsc.org/chemsoc/timeline/pages/timeline.html

Emission spectra and ionization

In sub-topic 2.2 the line emission spectrum of hydrogen was introduced. Emission spectra provide experimental evidence for the existence of atomic energy levels.

In sub-topic 3.2 ionization energy was defined as the minimum energy required to remove an electron from a neutral gaseous atom or molecule in its ground-state. The **first ionization energy (IE_1)** of a gaseous atom is related to the process:

$$X(g) \rightarrow X^+(g) + e^-$$

Successive ionizations are also possible; for example, the **second ionization energy (IE_2)** is associated with the process:

$$X^+(g) \rightarrow X^{2+}(g) + e^-$$

The nth **ionization energy (IE_n)** relates to the process:

$$X^{(n-1)+}(g) \rightarrow X^{n+}(g) + e^-$$

For a given element the IE increases for successive ionizations, in the order:

$$IE_1 < IE_2 < IE_3 < IE_4 < IE_5 \ldots$$

This is because with each successive ionization an electron is being removed from an increasingly positive species, and hence more energy is required. For example, for magnesium, Mg:

$IE_1 = 737.7$ kJ mol^{-1}

$IE_2 = 1450.7$ kJ mol^{-1}

$IE_3 = 7732.7$ kJ mol^{-1}

$IE_4 = 10542.5$ kJ mol^{-1}

$IE_5 = 13630$ kJ mol^{-1}

In sub-topic 2.2 we saw that in the emission spectrum of the hydrogen atom, the lines converge at higher energies. At the **limit of convergence** the lines merge, forming a **continuum**. Beyond the continuum the electron can have any energy, so is no longer under the influence of the nucleus: the electron is outside the atom (ionization has occurred). The increase in principal quantum number from $n = 1$ to $n = \infty$ shown in figure 2 represents the process of ionization of the atom.

The frequency of the radiation in the emission spectrum at the limit of convergence can be used to determine IE_1. In the Lyman series for the hydrogen atom (UV region), the frequency at the limit of convergence relates to the energy given out when an electron falls from $n = \infty$ and returns to the ground-state, $n = 1$, as shown in figure 6 of sub-topic 2.2.

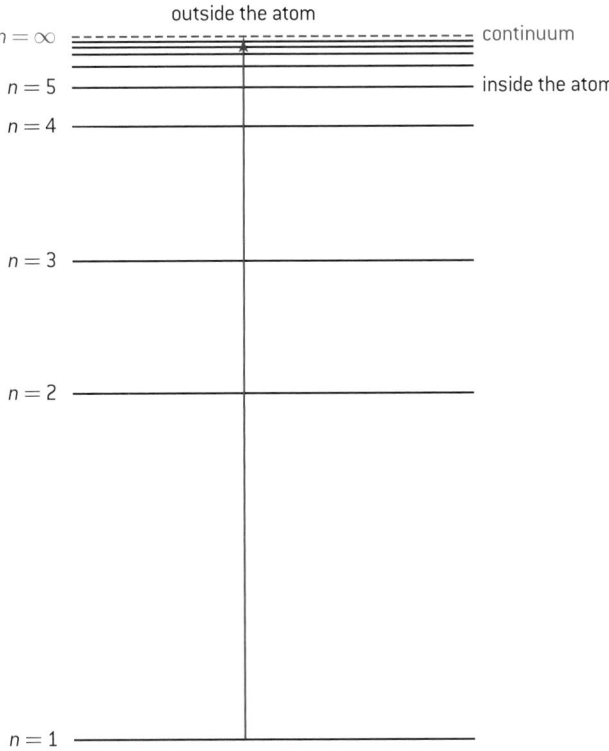

▲ Figure 2 The ionization process for the hydrogen atom

Determining the wavelengths of lines in spectra: The Rydberg equation

The **Rydberg equation** can be used to find the wavelengths of all the spectral lines in the emission spectrum of hydrogen, and is given by the expression:

$$\frac{1}{\lambda} = R_H\left(\frac{1}{n_i^{\,2}} - \frac{1}{n_f^{\,2}}\right)$$

where:

λ = wavelength

R_H = Rydberg constant = 1.097×10^7 m^{-1}

i = initial state

f = final state

n = principal quantum number

Note that n_f is greater than n_i.

The *IE* can then be determined as follows:

$$\Delta E = h\nu$$

where:

h = Planck's constant = 6.626×10^{-34} J s

ν = frequency

$$IE = E_\infty - E_1 = h\nu = \frac{hc}{\lambda}$$

(since $c = \nu\lambda$)

c = speed of light = 2.998×10^8 m s^{-1}

From the Rydberg equation, $\frac{1}{\lambda}$ can be inserted into the expression for *IE* and rearranged to:

$$IE = hcR_H\left(\frac{1}{1^2}\right) - \left(\frac{1}{\infty^2}\right)$$

$$= (6.626 \times 10^{-34} \text{ J s})(2.998 \times 10^8 \text{ m s}^{-1})$$
$$(1.097 \times 10^7 \text{ m}^{-1})(1 - 0)$$

$$= 2.179 \times 10^{-18} \text{ J}$$

The energy in kJ mol^{-1} is found by:

$$IE = (2.179 \times 10^{-18} \text{ J})(6.022 \times 10^{23} \text{ mol}^{-1})$$

$$= 1.312 \times 10^6 \text{ J mol}^{-1}$$

$$= 1312 \text{ kJ mol}^{-1}$$

This value calculated for the first ionization energy (IE_1) for hydrogen is given in section 8 of the *Data booklet*.

Study tips

- In any calculation you should use the data given in the question or otherwise data from the *Data booklet*. In example 1 you should use the values of *h* and *c* provided. Note that the question requires an answer to a given number of significant figures.

- Always read the question carefully. Also, make sure you include the units throughout the various stages of your answer. This will help you obtain the correct units related to the final numerical answer. In this question you must remember to convert nm to m.

Worked examples: determining energy

Example 1

Determine the energy, in J, of a photon of red light, correct to **four** significant figures, given that the wavelength $\lambda = 650.0$ nm. $h = 6.626 \times 10^{-34}$ J s; $c = 2.998 \times 10^8$ m s^{-1}.

Solution

$$\Delta E = h\nu = \frac{hc}{\lambda}$$

$$\Delta E = \frac{6.626 \times 10^{-34} \text{ J s} \times 2.998 \times 10^8 \text{ m s}^{-1}}{650.0 \times 10^{-9} \text{ m}} = \mathbf{3.056 \times 10^{-19} \text{ J}}$$

Example 2

Calculate the first ionization energy, in kJ mol^{-1}, for hydrogen given that its shortest-wavelength line in the Lyman series is 91.16 nm. $h = 6.626 \times 10^{-34}$ J s; $c = 2.998 \times 10^8$ m s^{-1}; $N_A = 6.022 \times 10^{23}$ mol^{-1}.

Solution

The shortest-wavelength line in the Lyman series corresponds to a transition of $n = \infty$ to $n = 1$.

$$IE_1 = h\nu = \frac{hc}{\lambda}$$

$$IE_1 = \frac{6.626 \times 10^{-34} \text{ J s} \times 2.998 \times 10^8 \text{ m s}^{-1}}{91.16 \times 10^{-9} \text{ m}} = 2.179 \times 10^{-18} \text{ J}$$

expressed in kJ mol^{-1}:

$$IE_1 = (2.179 \times 10^{-18} \text{ J}) \times (6.022 \times 10^{23} \text{ mol}^{-1}) = 1.312 \times 10^6 \text{ J mol}^{-1}$$

$$= \mathbf{1312 \text{ kJ mol}^{-1}}$$

Periodic trends in ionization energies

Figure 3 shows ten successive ionization energies, in kJ mol^{-1}, for the group 2 alkaline earth metal calcium, Ca and the group 4 transition metal titanium, Ti. In the case of Ca there is a significant jump going from IE_2 to IE_3; the third ionization energy corresponds to the removal of an electron from the fully occupied 3p sublevel. As a result Ca^{3+} species do not occur. This supports observations that for the group 2 metals there is one stable oxidation state, +2 (forming 2+ ions, eg Ca^{2+}). In contrast, Ti exhibits oxidation states of +2, +3, and +4 (see topic 13 and section 14 of the *Data booklet*). The most stable oxidation state of Ti is +4. In figure 3 there is a large jump in *IE* for Ti going from IE_4 to IE_5, corresponding to the removal of the fifth electron, supporting the observation that species with Ti in the +5 oxidation state do not occur.

The electron configurations for the most stable ions of Ca and Ti are deduced as follows:

Ca: [Ar]4s^2

Ca^{2+}: [Ar]

Ti: [Ar]3d^{2}4s^2

Ti^{4+}: [Ar]

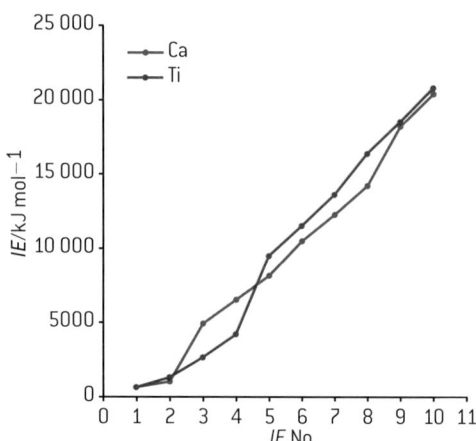

▲ Figure 3 Ten successive ionization energies for calcium and titanium

It is interesting to note that in the case of Ti, the ionization energies increase more gradually than for Ca as electrons are being removed from the 3d and 4s orbitals, which are much closer in energy compared to the 3p and 4s orbitals.

The group 1 alkali metal potassium, K ($Z = 19$) has the electron configuration:

K: $1s^2 2s^2 2p^6 3s^2 3p^6 4s^1$

From this configuration we might expect a large jump going from IE_1 to IE_2, from IE_9 to IE_{10}, and from IE_{17} to IE_{18}. These significant jumps are associated with the removal of electrons from energy levels of different principal quantum number n. Table 1 shows the first 19 IEs for K.

IE number	IE/kJ mol^{-1}
1	418.8
2	3052
3	4420
4	5877
5	7975
6	9590
7	11343
8	14944
9	16963.7
10	48610
11	54490
12	60730
13	68950
14	75900
15	83080
16	93400
17	99710
18	444870
19	476061

▲ Table 1 Values of ionization energy (IE) for the first 19 ionization energies for potassium. Significant jumps are evident between IEs 1 and 2, 9 and 10, and 17 and 18

TOK

In plotting ionization energies, a **logarithmic scale** allows all data points to be plotted on a single graph.

The difference between IE_1 and IE_2 for K is 2633.2 kJ mol^{-1}, while the difference between IE_{18} and IE_{19} is 31191 kJ mol^{-1}. Therefore it is difficult to represent all 19 ionization energies for K on a linear scale. Look at the unreasonably long y-axis when comparing the plot of IE versus IE No. to the plot of $\log_{10} IE$ versus IE No. in figure 4 (a) and (b).

Can you think of examples in chemistry or other sciences that present data in a particular way to support the scientist's postulates, theories, and hypotheses? Where else in chemistry do we use logarithmic plots? Do you know the difference between $\log_{10}$ and $\log_e$ (ln) and can you give examples of where each type of log is used in chemistry?

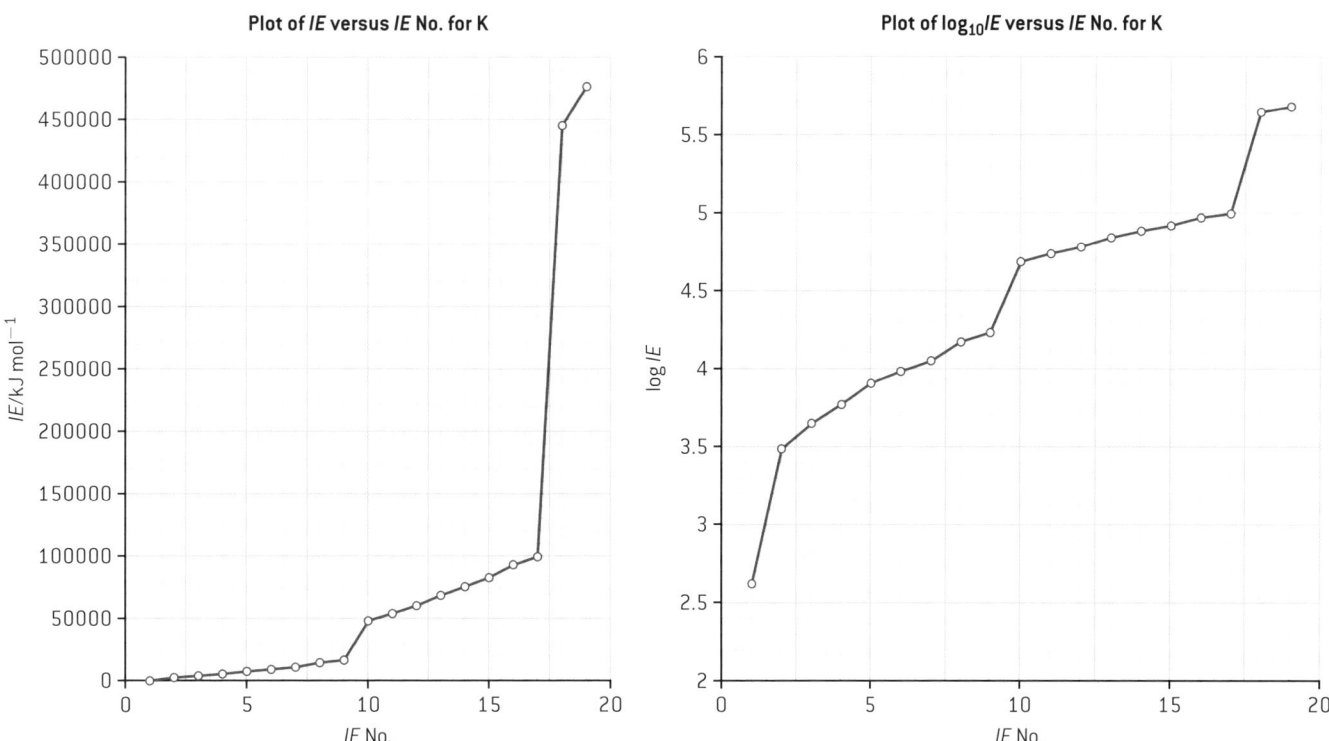

▲ Figure 4 (a) Plot of IE versus IE No. for potassium; (b) Plot of $\log_{10} IE$ versus IE No. for potassium

Worked examples

Example 1

Values for the successive *IE*s for an unknown element X are given in table 2. Deduce in which group of the periodic table of elements you would expect to find X. State the name of this group.

IE number	*IE*/kJ mol⁻¹
IE_1	899
IE_2	1757
IE_3	14850
IE_4	21005

▲ Table 2 Ionization energies (*IE*s) for X.

Solution

The largest jump in *IE* occurs between IE_2 and IE_3, corresponding to $\Delta IE = 13093$ kJ mol⁻¹. This must correspond to a change in energy level; therefore X must be in group 2, the alkaline earth metals.

Example 2

Figure 5 represents the successive ionization energies of sodium. The vertical axis plots log (ionization energy) instead of ionization energy to allow the data to be represented without using an unreasonably long vertical axis.

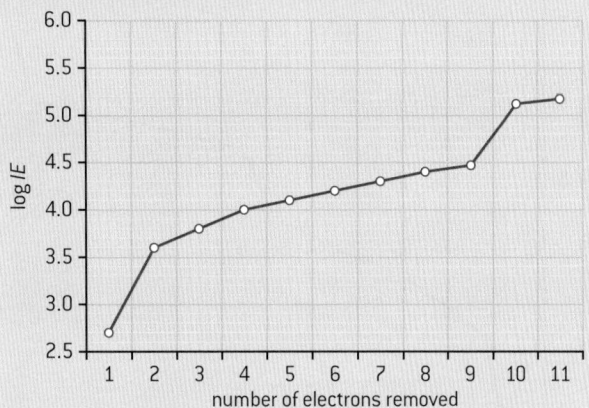

▲ Figure 5 Successive ionization energies of sodium

State the full electron configuration of sodium and explain how the successive ionization energy data for sodium are related to its electron configuration. [4]

IB, May 2010

Study tip

When writing electron configurations, electrons in individual orbitals should be presented as superscript: $1s^2 2s^2 2p^6 3s^1$ rather than 1s2, 2s2, 2p6, 3s1. In addition, always take note of the type of electron configuration requested – this question asks for the full electron configuration so you should not write a condensed electron configuration such as $[Ne]3s^1$.

Solution

Na: $1s^2 2s^2 2p^6 3s^1$

The plot in figure 5 shows that the first electron is the easiest to remove. This is because it is furthest from the nucleus, being the valence electron occupying the outermost $n = 3$ energy level. There is a large increase going from IE_1 to IE_2 because the next electron is removed from the $n = 2$ level. However, for the next seven electrons the small, the gradual increase in *IE* reflects the fact that all eight electrons occupy the same $n = 2$ energy level. There is another large jump in *IE* going from IE_9 to IE_{10}, associated with the removal of an electron from the $n = 1$ energy level. This electron is closest to the nucleus and so will be very difficult to remove. The eleventh electron also comes from the 1s sublevel, so IE_{11} shows only a small increase over IE_{10}.

Example 3

Figure 6 shows the variation in first ionization energies for the second-row elements in the periodic table from Li to Ne.

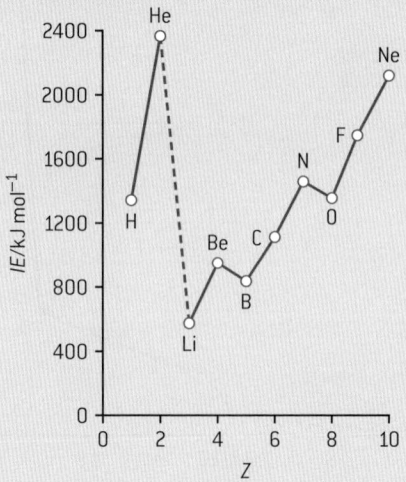

▲ Figure 6 Ionization energies for the first 10 elements

a) Explain why as you go across a period, *IE*s increase.

b) Although there a general increase in IE_1 across the second period as expected, there is evidence of some discontinuity. This is often referred to as a dog-teeth plot. Explain why:

    i) IE_1 for oxygen (1314 kJ mol^{-1}) is lower than IE_1 for nitrogen (1402 kJ mol^{-1})

    ii) IE_1 for boron (801 kJ mol^{-1}) is lower than IE_1 for beryllium (900 kJ mol^{-1}).

Solution

a) *IE*s increase across a period for two reasons:

- decreasing atomic radii across a period from left to right

- increasing nuclear charge, *Z*.

b) i) First consider the orbital diagrams of the elements oxygen and nitrogen:

O: $[He]2s^2 2p_x^2 2p_y^1 2p_z^1$

       $2s^2$ $2p_x^2$ $2p_y^1$ $2p_z^1$

The first electron to be removed from a neutral gaseous atom will come from the highest occupied sublevel of the highest energy level. In this case this is the 2p sublevel which is higher than 2s in energy. This difference in energy may not be obvious from an orbital diagram which is often just represented with the outermost energy levels shown horizontally. Remember that *within an energy level* the order of the energies of the sublevels is s < p < d < f (sub-topic 2.2).

So an electron will be removed from the 2p sublevel. Which is the most loosely bound electron? There will be maximum repulsion in an orbital that contains paired electrons, so the most loosely bound electron will be a 2p$_x$ electron as circled in the orbital diagram for oxygen above. This is the reason why IE_1 is lower for oxygen than for nitrogen, whose orbital diagram is shown on the here:

N: $[He]2s^2 2p_x^1 2p_y^1 2p_z^1$

 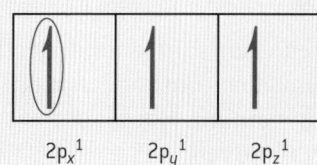

       $2s^2$ $2p_x^1$ $2p_y^1$ $2p_z^1$

For nitrogen the most loosely bound electron is any one of the three 3p electrons which are all are degenerate (have the same energy).

ii) Again start by drawing orbital diagrams for Be and B:

Be: $[He]2s^2$

 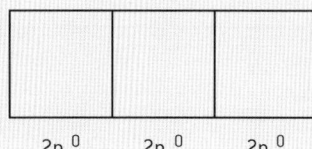

       $2s^2$ $2p_x^0$ $2p_y^0$ $2p_z^0$

B: $[He]2s^2 2p_x^1 2p_y^0 2p_z^0$

 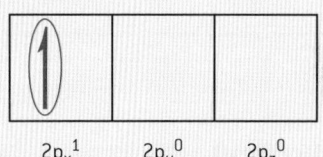

       $2s^2$ $2p_x^1$ $2p_y^0$ $2p_z^0$

The most loosely bound electron for B occupies the 2p$_x$ orbital while for Be it is one of the two electrons in the 2s level. It will be easier to remove the electron from the 2p$_x$ orbital in B since 2p is higher in energy than 2s. This criterion overrides any consideration of paired electrons in an orbital. Hence IE_1 for B (801 kJ mol^{-1}) is lower than IE_1 for Be (900 kJ mol^{-1}).

> **Analogy**
>
> Suppose you have a two-story building and you need to remove one floor in order to meet new height regulations. Which floor would you remove? Obviously the top floor (floor 2) – the building would collapse if you removed the ground floor (floor 1)! It is the same when removing electrons from energy levels and sublevels – electrons are removed from the energy level of highest principal quantum number *n* first, and from the sublevel with the greatest energy, within that energy level.

Questions

1 Figure 7 represents the energy needed to remove nine electrons, one at a time, from an atom of an element. Not all of the electrons have been removed.

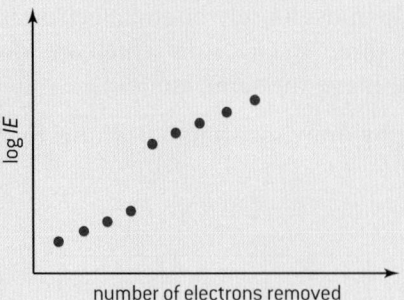

▲ Figure 7

Which element could this be?

A. C

B. Si

C. P

D. S [1]

IB, May 2010

2 Between which ionization energies of boron will there be the greatest difference?

A. Between 1st and 2nd ionization energies

B. Between 2nd and 3rd ionization energies

C. Between 3rd and 4th ionization energies

D. Between 4th and 5th ionization energies [1]

IB, November 2009

3 Which of the following is correct?

A. $IE_3 > IE_4$

B. Molar ionization energies are measured in kJ.

C. The third ionization energy represents the process:

$X^{2+}(g) \rightarrow X^{3+}(g) + e^-$

D. Ionization energies decrease across a period going from left to right. [1]

4 The graph of the first ionization energy plotted against atomic number for the first twenty elements shows periodicity (figure 8).

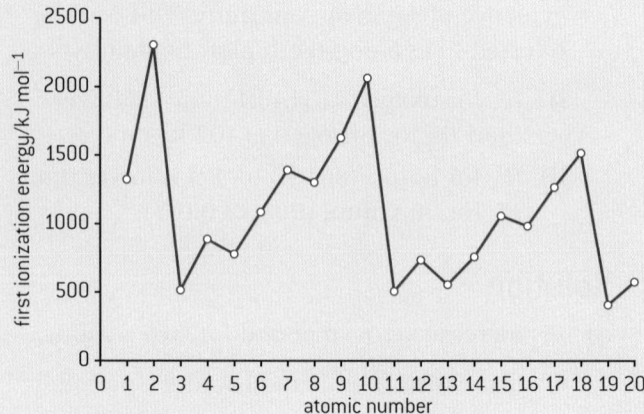

▲ Figure 8

i) Define the term *first ionization energy* and state what is meant by the term *periodicity*. [2]

ii) Explain how information from this graph provides evidence for the existence of main energy levels and sublevels within atoms. [4]

iii) State what is meant by the term *second ionization energy*. [1]

IB, May 2009

Introduction

Transition elements have characteristic properties. These properties can be associated with the incomplete d sublevels of such metals. In this topic we explore these properties and in particular consider one simple theory, crystal field theory which can help us in understanding why the complexes of transition metal are often coloured.

13.1 First-row d-block elements

Understandings

→ Transition elements have variable oxidation states, form complex ions with ligands, have coloured compounds, and display catalytic and magnetic properties.

→ Zn is not considered to be a transition element as it does not form ions with incomplete d orbitals.

→ Transition elements show an oxidation state of +2 when the s- electrons are removed.

Applications and skills

→ Explanation of the ability of transition metals to form variable oxidation states from successive ionization energies.

→ Explanation of the nature of the coordinate bond within a complex ion.

→ Deduction of the total charge given the formula of the ion and ligands present.

→ Explanation of the magnetic properties in transition metals in terms of unpaired electrons.

Nature of science

→ Looking for trends and discrepancies – transition elements follow certain patterns of behaviour. The elements Zn, Cr, and Cu do not follow these patterns and are therefore considered anomalous in the first-row d-block.

IUPAC definition of a transition element

According to IUPAC, a **transition element** is an element that has an atom with an incomplete d-sublevel or that gives rise to cations with an incomplete d-sublevel. These refer to elements in groups 3–11.

At the centre of the periodic table of elements lies a very important family of elements, called the **transition elements**, whose physical and chemical properties often play a key role in many facets of everyday life.

In the periodic table (figure 1) the first-row transition elements are the elements in period 4 from scandium (Sc) to copper (Cu) inclusive. The elements below these elements in periods 5, 6, and 7 are also described as transition elements.

The **lanthanoids** are the elements from $Z = 57$ to $Z = 71$ and the **actinoids** are the elements from $Z = 89$ to $Z = 103$. La ($Z = 57$) and Ac ($Z = 89$) have electron configurations of [Xe]$5d^1 6s^2$ and [Rn]$6d^1 7s^2$, respectively (so do not contain *f*-electrons in their outer energy levels), but all the other lanthanoids and actinoids contain *f*-electrons in their electron configurations. The **f-block** elements are sometimes described as the **inner transition elements**.

1	2	3	4	5	6	7	8	9	10	11	12	13	14	15	16	17	18
1 H hydrogen [1.007; 1.009]																	2 He helium 4.003
3 Li lithium [6.938; 6.997]	4 Be beryllium 9.012											5 B boron [10.80; 10.83]	6 C carbon [12.00; 12.02]	7 N nitrogen [14.00; 14.01]	8 O oxygen [15.99; 16.00]	9 F fluorine 19.00	10 Ne neon 20.18
11 Na sodium 22.99	12 Mg magnesium 24.31											13 Al aluminium 26.98	14 Si silicon [28.08; 28.09]	15 P phosphorus 30.97	16 S sulfur [32.05; 32.08]	17 Cl chlorine [35.44; 35.46]	18 Ar argon 39.95
19 K potassium 39.10	20 Ca calcium 40.08	21 Sc scandium 44.96	22 Ti titanium 47.87	23 V vanadium 50.94	24 Cr chromium 52.00	25 Mn manganese 54.94	26 Fe iron 55.85	27 Co cobalt 58.93	28 Ni nickel 58.69	29 Cu copper 63.55	30 Zn zinc 65.38(2)	31 Ga gallium 69.72	32 Ge germanium 72.63	33 As arsenic 74.92	34 Se selenium 78.96(3)	35 Br bromine 79.90	36 Kr krypton 83.80
37 Rb rubidium 85.47	38 Sr strontium 87.62	39 Y yttrium 88.91	40 Zr zirconium 91.22	41 Nb niobium 92.91	42 Mo molybdenum 95.96(2)	43 Tc technetium	44 Ru ruthenium 101.1	45 Rh rhodium 102.9	46 Pd palladium 106.4	47 Ag silver 107.9	48 Cd cadmium 112.4	49 In indium 114.8	50 Sn tin 118.7	51 Sb antimony 121.8	52 Te tellurium 127.6	53 I iodine 126.9	54 Xe xenon 131.3
55 Cs caesium 132.9	56 Ba barium 137.3	57-71 lanthanoids	72 Hf hafnium 178.5	73 Ta tantalum 180.9	74 W tungsten 183.8	75 Re rhenium 186.2	76 Os osmium 190.2	77 Ir iridium 192.2	78 Pt platinum 195.1	79 Au gold 197.0	80 Hg mercury 200.6	81 Tl thallium [204.3; 204.4]	82 Pb lead 207.2	83 Bi bismuth 209.0	84 Po polonium	85 At astatine	86 Rn radon
87 Fr francium	88 Ra radium	89-103 actinoids	104 Rf rutherfordium	105 Db dubnium	106 Sg seaborgium	107 Bh bohrium	108 Hs hassium	109 Mt meitnerium	110 Ds darmstadtium	111 Rg roentgenium	112 Cn copernicium		114 Fl flerovium		116 Lv livermorium		
		57 La lanthanum 138.9	58 Ce cerium 140.1	59 Pr praseodymium 140.9	60 Nd neodymium 144.2	61 Pm promethium	62 Sm samarium 150.4	63 Eu europium 152.0	64 Gd gadolinium 157.3	65 Tb terbium 158.9	66 Dy dysprosium 162.5	67 Ho holmium 164.9	68 Er erbium 167.3	69 Tm thulium 168.9	70 Yb ytterbium 173.1	71 Lu lutetium 175.0	
		89 Ac actinium	90 Th thorium 232.0	91 Pa protactinium 231.0	92 U uranium 238.0	93 Np neptunium	94 Pu plutonium	95 Am americium	96 Cm curium	97 Bk berkelium	98 Cf californium	99 Es einsteinium	100 Fm fermium	101 Md mendelevium	102 No nobelium	103 Lr lawrencium	

▲ Figure 1 IUPAC periodic table of the elements

The elements of group 12, that is Zn, Cd, Hg, and Cn, are *not* classified as transition elements according to IUPAC as all four elements have full d-sublevels containing ten d-electrons (for example, Zn: [Ar]$3d^{10}4s^2$). Both scandium and yttrium are classified as transition elements as they have an incomplete d-sublevel (Sc, [Ar]$3d^1 4s^2$; Y, [Kr]$4d^1 5s^2$). In 1920, when only Sc^{3+} and Y^{3+} compounds were known, they were widely considered to be non-transition elements because their ions contained no d-electrons (Sc^{3+}, [Ar], Y^{3+}, [Kr]). Since then many lower oxidation state compounds of these elements have been synthesized, most of which involve metal–metal bonding. For example, scandium can exist in the +2 oxidation state, and because its electron configuration is [Ar]$3d^1$, scandium *is* considered a transition element according to the IUPAC definition. An example of a compound in which scandium is in the +2 oxidation state is $CsScCl_3$.

Useful resource

The electron configurations of all the elements in the periodic table can be found by accessing http://www.webelements.com/, compiled by Professor Mark Winter at the University of Sheffield, UK.

Collectively, the elements in groups 3–12 inclusive (including La and Ac) are referred to as the **d-block elements**. The elements which comprise the **f-block** are those in which the 4f and 5f orbitals are filled. These elements are formal members of group 3 but they form a separate f-block in the periodic table (figure 2).

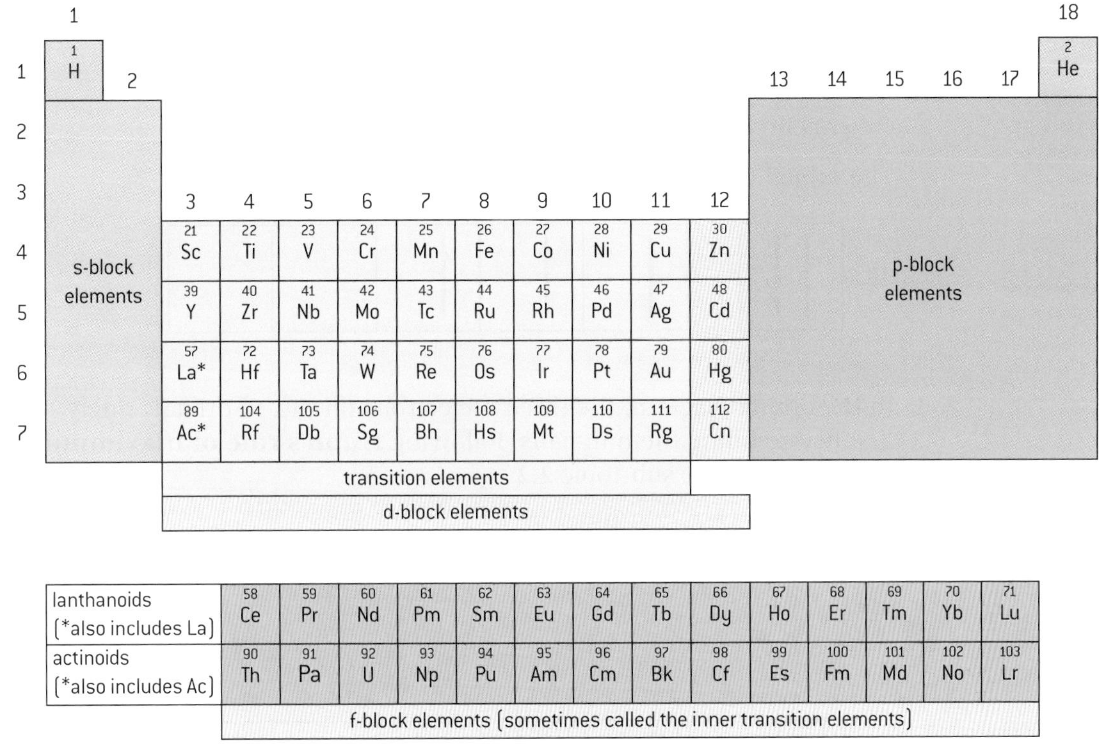

main-group elements	group 1 (excluding H), group 2 and groups 13–18
transition elements	groups 3–11 (the f-block elements are sometimes described as the inner transition elements)
s-block elements	groups 1 and 2 and He
p-block elements	groups 13–18 (excluding He)
d-block elements	groups 3–12 [including $Z = 57$ (La) and $Z = 89$ (Ac), but excluding $Z = 58$ (Ce) to $Z = 71$ (Lu) and $Z = 90$ (Th) to $Z = 103$ (Lr), which are classified as f-block elements]
f-block elements	elements from $Z = 58$ (Ce) to $Z = 71$ (Lu) and from $Z = 90$ (Th) to $Z = 103$ (Lr)
lanthanoids	elements from $Z = 57$ (La) to $Z = 71$ (Lu)
actinoids	elements from $Z = 89$ (Ac) to $Z = 103$ (Lr)

▲ Figure 2 Periodic table of the elements showing the main-group elements, the transition elements, the s-, p-, d-, and f-block elements, the lanthanoids and the actinoids

The metallic nature of the transition elements means they are often described as the **transition metals**.

Electron configurations of first-row d-block elements and their ions

The following are some examples of full and condensed electron configurations of the first-row d-block elements, their ions, and their corresponding orbital diagrams:

- For vanadium, V ($Z = 23$):

    $1s^2 2s^2 2p^6 3s^2 3p^6 3d^3 4s^2$ (full electron configuration)

    $[Ar]3d^3 4s^2$ (condensed electron configuration)

The orbital diagram is:

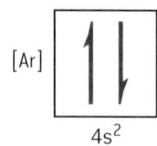 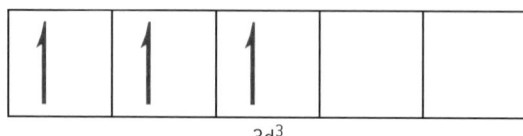

In the orbital diagram, the three d-electrons fill the 3d orbitals singly first before filling them in pairs, following **Hund's rule of maximum multiplicity**, see sub-topic 2.2.

- For nickel, Ni ($Z = 28$):

    $1s^2 2s^2 2p^6 3s^2 3p^6 3d^8 4s^2$ (full electron configuration)

    $[Ar]3d^8 4s^2$ (condensed electron configuration)

The orbital diagram is:

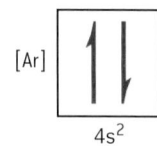 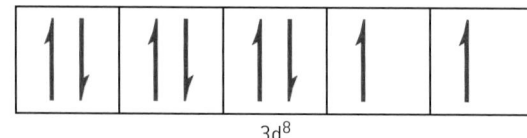

- For Ni^{2+} ($Z = 28$):

    $1s^2 2s^2 2p^6 3s^2 3p^6 3d^8$ (full electron configuration)

    $[Ar]3d^8$ (condensed electron configuration)

The orbital diagram is:

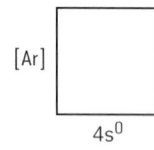 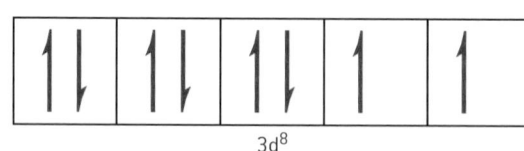

In the case of Ni^{2+} the electrons are removed from the 4s level before the 3d level.

🔬 Transition metals and the Aufbau principle

In an article by L.G. Vanquickenborne, K. Pierloot, and D. Devoghel published in the *Journal of Chemical Education* (71, (1994), p469-471), the relative energies of the 3d and 4s orbitals are discussed and an explanation is given with respect to the filling of these orbitals in both the ground-state and in transition metal ions (that is, why electrons are removed from the 4s level before the 3d for the first-row). The scope of this is beyond the current IB syllabus, but the article is interesting to read in relation to some misconceptions that appear in some sources with regards to what orbital energies need to be considered.

Electron configurations involving exceptions

In the first-row transition metals, there are two exceptions in terms of electron configurations that you have to be careful with: Cr ($Z = 24$) and Cu ($Z = 29$). In the case of Cr, you may be tempted to write the electron configuration as [Ar]$3d^44s^2$. This is incorrect. Chromium has a condensed electron configuration of [Ar]$3d^54s^1$. A similar anomaly occurs for Cu. One would expect the electron configuration to be [Ar]$3d^94s^2$. The correct electron configuration is [Ar]$3d^{10}4s^1$. At a simplistic level attempts are often made to rationalize this in terms of the extra stability of the half-filled (d^5) and fully-filled (d^{10}) d-sublevel. However, this approach is far too simplistic and a much more detailed explanation (in the previously mentioned *J. Chem. Educ.*, paper and references therein) relates to the effect of increasing nuclear charge on the energies of the 4s and 3d levels and interactions between electrons that occupy the same orbital. This explanation involves finding the sum of the energies of all electrons with their respective interactions. A strong correlation has been found between experimental data and theoretical data based on advanced computational calculations.

An earlier explanation proposed by R.L. Rich, based on electron–electron interactions, was more schematic in nature and can be useful at an introductory level in understanding d-block electron configurations.

An orbital is assumed to have one energy level. However, when two electrons occupy an orbital, because of their electrostatic repulsion (both are negatively charged), there is an additional factor to be considered, termed the **pairing energy**, *P*. As the nuclear charge (Z) increases, there is greater attraction of the electrons: d orbitals are not shielded (screened) from the nucleus to the same extent as s orbitals. As a result electrons will occupy the lowest available orbitals, which is what we have been doing previously in earlier topics in writing electron configurations. For example, vanadium ($Z = 23$) has a condensed electron configuration of [Ar]$3d^34s^2$ as already stated, but as can be seen from the energy levels cited from Rich's work, there is a crossover point after vanadium, leading on to chromium (note that after nickel, leading onto copper, there is another a crossover point). This process produces two lines that represent the energies of the individual electrons in each subshell. As Rich points out, the lower line is followed until the subshell is half-filled; thereafter, the upper line is also used. Hence, for every element the outer electrons are simply given the lowest energies available. As can be interpreted from the diagram, this does not lead exactly to half-filled (d^5) or fully-filled (d^{10}) (or empty d^0) subshells. However, these often occur because some additional energy is required to go beyond them.

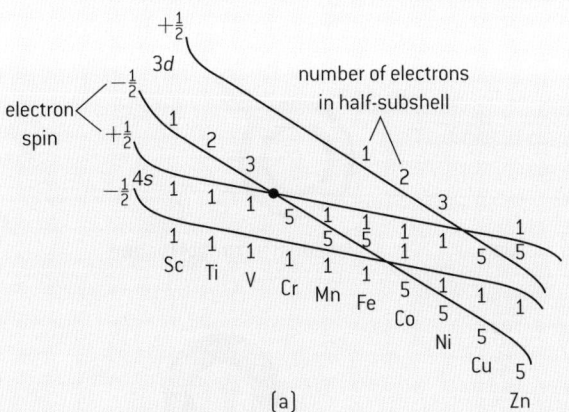

(a)

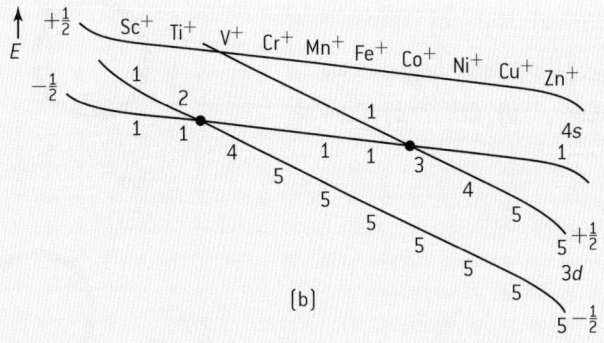

(b)

▲ Figure 3 Schematic representation of Rich's interpretation of electron configurations for transition elements in terms of intra-orbital repulsion and trends in subshell energies. In the first diagram, the order in which the levels are occupied is presented. In both diagrams key crossover points feature and in the second diagram one sees how an electron is removed from the 4s level before one from the 3d level

In this approach, the commonly held rationale based on the model of a perceived extra stability of the half-filled and fully-filled d-sublevels could be considered somewhat invalid. In the case of the formation of the transition metal cation, M^{n+}, when electrons are removed, the overall electron repulsion is decreased and the energy of the d orbitals is lowered to a greater extent compared to that of the s orbitals.

A more detailed account of this discussion can be found in G.L. Miessler, P.J. Fischer and D.A. Tarr., *Inorganic Chemistry* (5th Edition), 2013, Prentice Hall.

Activity

On closer examination of the electron configurations of the entire periodic table, other elements also convey deviations from expected patterns. Look at the webelements website and try to find four other d-block elements with electron configurations that differ from what is expected.

Study tip

For the IB Chemistry Diploma programme, you are only required to know anomalous electron configurations of the elements chromium and copper from the first-row transition metals. These configurations are: Cr, $[Ar]3d^5 4s^1$, and Cu, $[Ar]3d^{10}4s^1$. All other first-row transition metals will have electron configurations as predicted based on their position in the periodic table.

Note the following point, however. Once deduced, *do not* be tempted to modify electron configurations of cations further to follow this $3d^5$ and $3d^{10}$ pattern. For example, Fe has an electron configuration of $[Ar]3d^6 4s^2$. However, the electron configuration of Fe^{2+} is $[Ar]3d^6$. Do not then be inclined to rearrange this electron configuration further to $[Ar]3d^5 4s^1$. This is an incorrect electron configuration for Fe^{2+}. In summary, just note the two exceptions of Cr and Cu for the IB Diploma Chemistry programme.

Quick questions

1   Deduce the full electron configurations of:

   a) Co; **b)** Zn; **c)** Ti^{3+}

   and explain why Zn is not described as a transition element, according to IUPAC recommendations.

2   Deduce the condensed electron configurations of:

   a) V; **b)** Mn; **c)** Mn^{2+}

3   Deduce the orbital diagrams of:

   a) Co^{3+}; **b)** Cr^{3+}; **c)** Cu^+

TOK

The medical symbols for the female and male genders originate from the symbols used for copper and iron by the alchemists. These symbols have been used since Renaissance times (see the *Royal Society of Chemistry* (RSC) Visual Elements Periodic Table, www.rsc.org/periodic-table/alchemy).

▲ Mars symbol – symbolizes a male organism

▲ Alchemist's symbol for iron

▲ Venus symbol – symbolizes a female organism

▲ Alchemist's symbol for copper

Iron and copper are two of the seven metals of alchemy (gold, silver, mercury, copper, lead, iron and tin). Alchemists are often considered as the first chemists. Alchemists developed a unique language to describe not only chemical reactions, but also philosophical doctrines. Some commentators claim that the pseudoscience of alchemy has played a key role in the development of modern medicine and chemistry. Alchemists made a significant contribution to the chemical industries of that period, in areas such as the metallurgical industry, the dye industry and the glass-manufacturing industry. Alchemists extracted metals from their ores and tried to arrange the information known at that time of the various substances. The original idea of a periodic table might therefore in part be attributed to the alchemists. During the early days of alchemy the astronomical signs of the planets were used as alchemical symbols. Alchemical symbols were used to represent some elements and their compounds until the 18th century.

Characteristics of transition elements

As mentioned in topic 3, going from left to right across the periodic table, the nuclear charge, Z, increases and the atomic radii decrease. As a result of these two factors, the first ionization energy (IE) will increase across a period. In the case of transition elements, although there is a gradual increase in the first IE across the period, the rate of increase is much lower compared to that of the corresponding main-group elements. This difference can be attributed to the fact that for transition elements, the electrons enter an inner-shell orbital, whereas for main-group elements, the electrons enter a valence shell orbital. Inner-shell electrons have a greater shielding (screening) effect than valence electrons. This trend is shown in figure 4.

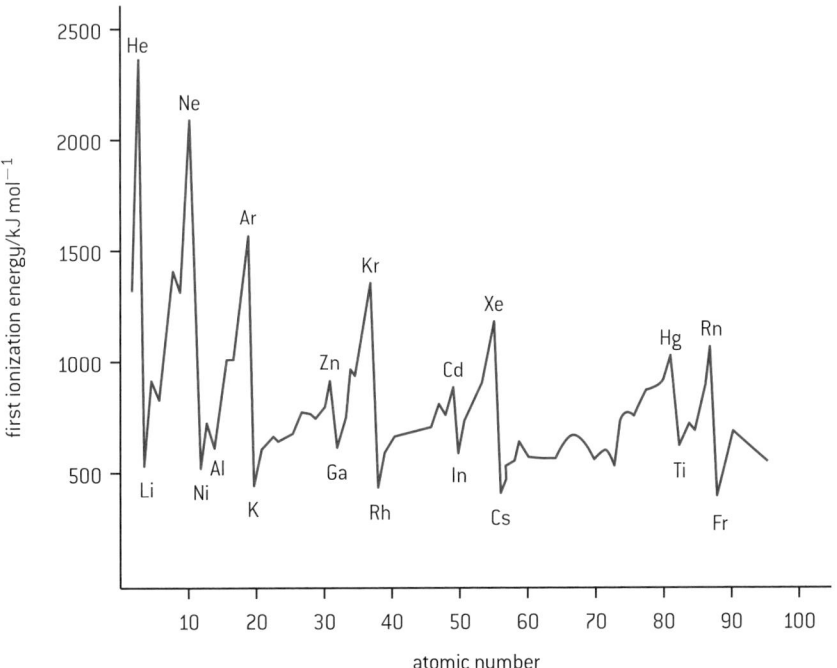

▲ Figure 4 Trends in the first IE for main-group and transition elements. Notice that the rate of increase in the first IE across the period is much more gradual for the transition elements compared to that for the main-group elements

Robert Boyle

Robert Boyle, who was born in Lismore in Ireland in 1627, is often described as "The Father of Chemistry" (see www.robertboyle.ie/).

Boyle was not only a devotee of "natural philosophy", an advocate of the experimental sciences, but also a key founder of the *Royal Society* in England. Boyle proved the inverse relationship between the volume of a gas and its pressure, known as **Boyle's law** (sub-topic 1.3). Although Boyle made the transition to modern science, much of his thinking centred around what is termed **"scholasticism"**, which has its basis as the extension of knowledge by reasoning and inference. All these attributes are still important to the modern day chemical practitioner. Have any of the principles of the earlier alchemists also been carried through the ages to modern day scientific methodology and chemical practice? You might wish to reflect on the importance of hypothesis and observation in chemical experiments you carry out in the laboratory.

TOK

Robert Boyle was a scientist and a philosopher. In many countries such as France the study of philosophy is mandatory at high school. In France the philosophy curriculum aims at producing enlightened citizens capable of intelligent criticism.

Find out what other countries prescribe the teaching of philosophy as mandatory at school and discuss the role and importance of taking a philosophical view in scientific discourse.

Laboratory tips

- When reading the meniscus for potassium permanganate, the top of the meniscus should always be read (B in figure 6), because it is convex upwards. This is because the deep colour makes it very difficult to read the meniscus. This is in contrast to normal practice for most clear solutions (A) where the meniscus is read from the bottom, that is concave upwards.

▼ Figure 6 How to read a meniscus: A, for clear solutions; B, for KMnO₄

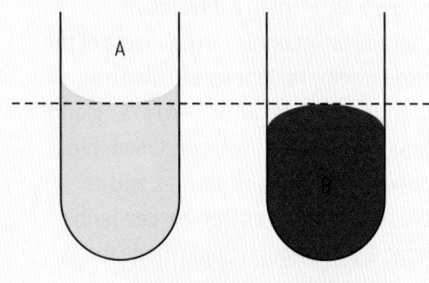

- In carrying out a redox titration involving potassium permanganate the colour change is typically from purple to colourless (with a faint pink tinge, signifying the $+7$ to $+2$ oxidation state change for manganese.). If, however, the colour changes from purple to brown, this would signify the formation of the intermediate ion of manganese, Mn^{4+}, with an associated oxidation state of $+4$, which is also a stable oxidation state. This may occur if there is insufficient acid in the conical flask.

Transition metals have a number of key characteristics:

- they have variable oxidation states
- compounds of transition elements and their ions are often coloured
- transition metals form complexes with ligands
- transition metals are often used as catalysts
- magnetic properties of transition metals depend on their oxidation states and coordination number.

Variable oxidation states

In contrast to an alkali metal such as sodium, where the oxidation state is always $+1$ in its ion and compounds, transition metals are often found with different oxidation states. The range of different oxidation states for the first-row d-block elements (see sections 9 and 14 of the *Data booklet*) can be seen from the diagram shown in figure 5. The d-block elements can be split according to their oxidation states into three types – A, B, and C.

Sc	Ti	V	Cr	Mn	Fe	Co	Ni	Cu	Zn
		+1	+1	+1	+1	+1	+1	+1	
+2	+2	+2	+2	+2	+2	+2	+2	+2	+2
+3	+3	+3	+3	+3	+3	+3	+3	+3	
	+4	+4	+4	+4	+4	+4	+4		
		+5	+5	+5					
			+6	+6	+6				
				+7					

type A: Sc, Ti, and V	type B: Cr and Mn	type C: Fe, Co, Ni, Cu, and Zn

▲ Figure 5 Range of oxidation states of the first-row d-block metals. The most common oxidation states are marked in green

The characteristics of **Type A** are dominated by:

- stable high oxidation states (for example, V is $+5$ in VO_3^-)
- unstable low oxidation states.

The characteristics of **Type B** are dominated by:

- stable high oxidation states (for example, Mn is $+7$ in MnO_4^-, Cr is $+6$ in $Cr_2O_7^{2-}$)

- stable low oxidation states (for example, Mn is +2 in $[Mn(H_2O)_6]^{2+}$, Cr is +3 in $[Cr(H_2O)_6]^{3+}$).

The characteristics of **Type C** are dominated by:

- unstable high oxidation states

- stable low oxidation states (for example, Fe is +2 in $[Fe(H_2O)_6]^{2+}$).

Manganese is characterized by oxidation states that range from +1 to +7. In the chemical laboratory you may often use the reagent potassium permanganate, in redox titrations (more correctly named potassium manganate(VII) though this compound is rarely named this way in practice in the chemical workplace!) This reagent is characterized by a deep burgundy (purple) colour. In redox reactions, manganese with an oxidation state of +7 is reduced to manganese with an oxidation state of +2, which is almost colourless:

$$Mn^{7+} + 5e^- \rightarrow Mn^{2+}$$

oxidation state: $\quad$ +7 $\qquad\qquad$ +2

species: $\qquad\quad$ $[MnO_4]^-$ $\quad$ $[Mn(H_2O)_6]^{2+}$

Another transition metal, chromium, can also exist in various oxidation states. In its highest oxidation state, +6, chromium forms orange and yellow compounds, which can be reduced to green complexes with chromium in a +3 oxidation state.

$$Cr^{6+} + 3e^- \rightarrow Cr^{3+}$$

oxidation state: $\quad$ +6 $\qquad\qquad$ +3

The oxidation of primary alcohols is a two-step process. A primary alcohol is first oxidized into an aldehyde, which in turn is oxidized further into the corresponding carboxylic acid.

Primary alcohols can be oxidized by strong oxidizing agents such as potassium dichromate(VI), $K_2Cr_2O_7$, in sulfuric acid, H_2SO_4, to form the corresponding carboxylic acid, under reflux (as discussed in sub-topic 10.2):

$$CH_3CH_2OH \xrightarrow[H^+]{K_2Cr_2O_7} CH_3CHO \xrightarrow[H^+]{K_2Cr_2O_7} CH_3COOH$$

$\quad$ ethanol $\qquad\qquad$ ethanal $\qquad\qquad$ ethanoic acid
$\;$ (primary alcohol) $\quad\;$ (aldehyde) $\qquad$ (carboxylic acid)

Oxidation of a primary alcohol

Secondary alcohols can also be oxidized by potassium dichromate(VI) in sulfuric acid to form the corresponding ketone:

$$CH_3CH_2CH(CH_3)OH \xrightarrow[H^+]{K_2Cr_2O_7} CH_3CH_2C(O)CH_3$$

$\quad$ butan-2-ol $\qquad\qquad\qquad\qquad$ butan-2-one
$\;$ (secondary alcohol) $\qquad\qquad\qquad$ (ketone)

As outlined in topic 9, each redox process involves two half-reactions, oxidation and reduction. In the case of this reaction with potassium dichromate(VI), the chromium is reduced from an oxidation state of +6 to +3.

Laboratory tips

- In the oxidation of a primary alcohol, the aldehyde can be isolated by **distilling** it off as it forms. Distillation is a technique used to separate liquids that have different boiling points (boiling point of ethanal is 20.2°C; that of ethanoic acid is 118°C).

- Alternatively, if a milder oxidizing agent is used, such as pyridinium chlorochromate (PCC), with an organic solvent such as tetrahydrofuran (THF), the aldehyde forms as the final product of the reaction.

$$CH_3CH_2OH \xrightarrow[THF]{PCC} CH_3CHO$$

$\quad$ ethanol $\qquad\qquad$ ethanal

(primary alcohol) $\;$ (aldehyde)

Oxidation half-reaction:

$$CH_3CH_2OH(aq) + H_2O(l) \rightarrow CH_3COOH(aq) + 4H^+(aq) + 4e^-$$

Reduction half-reaction:

$$Cr_2O_7^{2-}(aq) + 14H^+(aq) + 6e^- \rightarrow 2Cr^{3+}(aq) + 7H_2O(l)$$

Overall equation:

$$3CH_3CH_2OH(aq) + 2Cr_2O_7^{2-}(aq) + 16H^+(aq) \rightarrow 3CH_3COOH(aq) + 4Cr^{3+}(aq) + 11H_2O(l)$$

Breathalyser test

The redox reaction involving potassium dichromate(VI) is the basis of the **breathalyser test** used by police forces worldwide to determine if a driver of a vehicle has consumed alcohol. In this test, crystals of potassium dichromate(VI), which are orange/yellow in colour, change to green, which signifies the formation of the Cr^{3+} species.

Since 2012, it has been required by French law that all vehicles need to be equipped with a breathalyser. As seen in figure 7, the simple version of this on-board vehicle breathalyser test kit involves the colour change from orange/yellow to green. This type of breathalyser does not record the **blood alcohol concentration** (BAC), which is the concentration of ethanol in a person's blood. BAC is the mass, in milligrams, of ethanol per 100 cm³ of blood.

In order to measure the BAC three devices can be used:

- semiconductor oxide-based sensor
- fuel-cell sensor
- intoximeter, which is an IR spectrometer; this type of technology is often used in large, table-top breathalysers found at police stations.

Semiconductor oxide sensors

These are relatively new to the market and have a number of advantages, such as their low cost, low power consumption, and portability. The disadvantage of this type of breathalyser is that their sensors need to be calibrated more frequently than fuel-cell based testers. Incorrect calibration can result in systematic errors.

Fuel-cell sensors

Another type of breathalyser is based on the **fuel cell**.

Ethanol is oxidized initially into ethanoic acid and then into carbon dioxide and water. The fuel cell converts chemical energy generated from the oxidation process into electrical energy. The electric potential is used to determine the concentration of ethanol.

This type of fuel cell can also be quite basic and the results typically determined may not be sufficiently accurate to

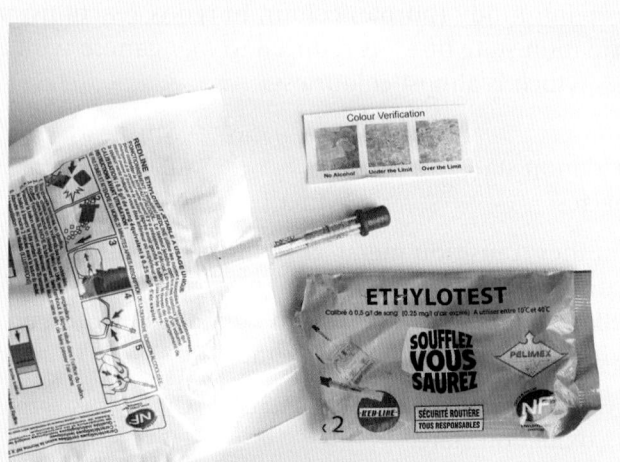

- If the crystals are all yellow/orange, the result is zero – you are clear to go!

- If the crystals are green below the line, according to the tube, you are under the maximum limit. But you *do* have alcohol in your blood and your judgement and reaction times will almost certainly be affected – *you should consider waiting a while and retesting*.

- If the crystals are green *beyond* the line – you are definitely over the limit. **DO NOT DRIVE!**

▲ Figure 7 Example of a simple breathalyser test kit used in France. Notice the orange/yellow to green colour change, which signifies the $Cr^{6+} + 3e^- \rightarrow Cr^{3+}$ reduction caused by ethanol

support a legal case in a court of justice. For this reason, positive tests obtained by preliminary screening need to be confirmed by more advanced analytical techniques, such as gas liquid chromatography (GLC), in which a sample is sent to a forensic science laboratory and the exact concentration of ethanol in the blood is determined. GLC is used to analyse volatile substances.

Intoximeter

A third type of breathalyser is the intoximeter based on IR spectroscopy. This is discussed in detail in option D.

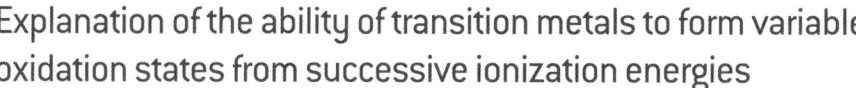

Explanation of the ability of transition metals to form variable oxidation states from successive ionization energies

As stated already one of the key characteristics of transition metals is that they exhibit variable oxidation states. This is in stark contrast to the s-block metals, which have only one fixed oxidation state. For example, calcium is an alkaline earth metal and occurs with a +2 oxidation state in its ion and compounds. In contrast, the transition metal titanium occurs with oxidation states of +2, +3, and +4. The reason for this difference between the two types of metal is related to the patterns in the successive ionization energies.

Coloured compounds of transition metals and their ions

Transition metal compounds and ions are often coloured, for example:

$KMnO_4$	burgundy (purple)
$[Mn(H_2O)_6]^{2+}$	almost colourless, with a faint pink tinge
$K_2Cr_2O_7$	orange
$[Cr(H_2O)_6]^{3+}$	green
$CuSO_4 \cdot 5H_2O$	blue
$[NH_4]_2[Fe(H_2O)_6][SO_4]_2$	pale green

Crystalline hydrated copper(II) sulfate, $CuSO_4 \cdot 5H_2O$, is Mediterranean blue in colour (figure 8). Upon heating the compound loses its water of crystallization, and the solid, anhydrous $CuSO_4$ forms, which is a white powder.

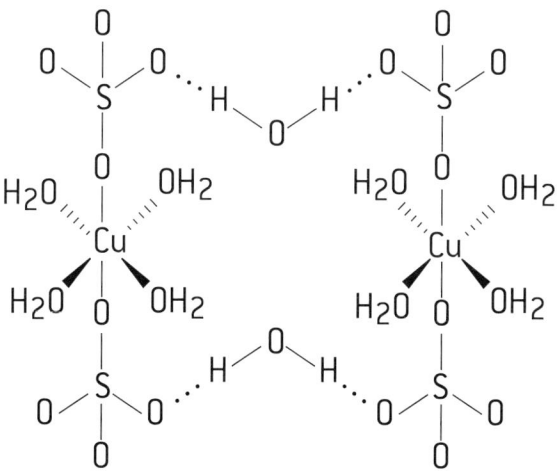

▲ Figure 8 Structure of $CuSO_4 \cdot 5H_2O$. Note the presence of hydrogen bonding and that Cu^{2+} has an octahedral stereochemistry, which may not be obvious from the formula

As stated previously, zinc is not classified as a transition element, as it has a complete d-sublevel, $[Ar]3d^{10}4s^2$. Its ion, Zn^{2+}, has the electron configuration, $[Ar]3d^{10}$.

Compounds of zinc(II) are usually colourless, unless the ligands (explained below) in the complex have a **chromophore** (group of atoms responsible for the absorption of electromagnetic radiation), which can absorb in the visible region of the electromagnetic spectrum.

Coordinate bonding

In coordinate bonding the pair of electrons comes from the same atom, unlike typical covalent bonding where the shared pair consists of electrons that originate from both atoms, A and B, which form the covalent bond, as discussed in topic 4. The older name for coordinate bonding was dative covalent bonding. The use of this older name is no longer recommended by IUPAC.

IUPAC recommends the term coordination bonding but in this text we will use the more widely used term coordinate bonding as applied in the IB Chemistry guide.

Complexes of transition metals

Compounds that contain transition elements and in which the central metal ion, M^{n+}, is bonded, via **coordinate bonding**, to a group of molecules or ions (termed the **ligands**) are termed transition metal complexes. Such compounds are often described as **coordination compounds**, to signify the coordinate bonding present between the ligand(s) and the central metal ion.

Examples of species with coordinate bonding:

- hydronium cation, $[H_3O]^+$
- carbon monoxide, CO

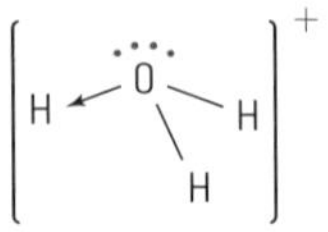

- ammonium cation, $[NH_4]^+$
- a transition metal complex, for example $[Ni(NH_3)_6]^{2+}$

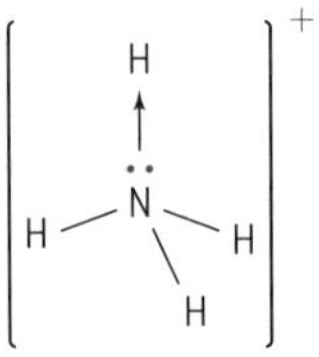

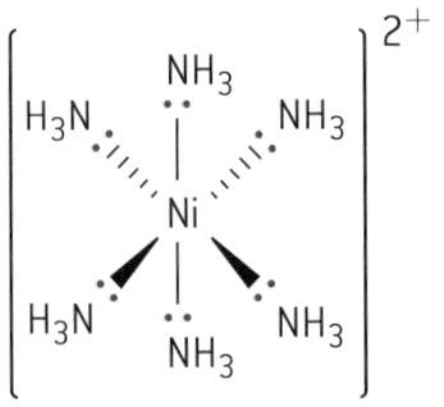

Description of a ligand

A ligand is an atom, molecule, or ion that contains a lone pair of electrons (non-bonding pair) that coordinates, through **coordinate bonding**, to a central transition metal ion to form a **complex**.

The term **proligand** is often used to describe a species that has the ability to act as a ligand in a complex, but is not yet coordinated. Hence, H_2O, because of its two lone pairs, is a neutral proligand, but in the complex $[Cr(H_2O)_6]^{3+}$, water acts as a ligand.

Bonding models of transition metal complexes

Pauling's electroneutrality principle is an approximate method of estimating how charge is distributed in a molecule or complex ion. The basis of this principle is that the charge on any individual atom in the molecule or ion is restricted to a range between 1− to 1+ and ideally the charge should be close to zero.

Figure 9 shows various representations of the cationic complex, $[Fe(H_2O)_6]^{3+}$.

- In figure 9(a) a typical representation of the cationic complex is given. As the lone pair of electrons on each water ligand contributes to the coordinate bond, an arrow is used instead of a straight line.

- If we were to adopt the model proposed in figure 9(b), it would mean a net transfer of charge from each water ligand to the metal centre. The charge distribution that results from this 100% covalent bonding model would confer 3− on Fe and 1+ on each water.

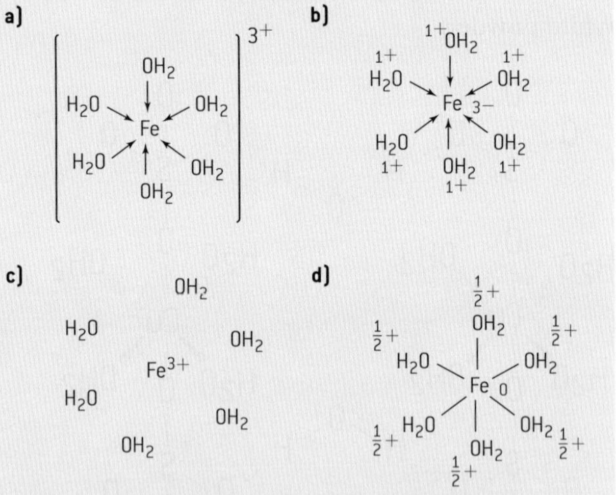

▲ Figure 9 Various representation and bonding models for the cationic complex, $[Fe(H_2O)_6]^{3+}$. (a) Conventional representation of the cationic complex $[Fe(H_2O)_6]^{3+}$. The lone pair on each water ligand forms the coordinate bond with the central Fe^{3+} ion. Square brackets here represent the complex, which has an octahedral stereochemistry. The overall charge on the complex is 3+. (b) Charge distribution in the cationic complex $[Fe(H_2O)_6]^{3+}$ based on a 100% covalent bonding model. (c) Charge distribution in $[Fe(H_2O)_6]^{3+}$ based on a 100% ionic bonding model. (d) Approximate charge distribution in $[Fe(H_2O)_6]^{3+}$ based on Pauling's electroneutrality principle

(b) is not a valid model, however, as negative charges residing on metals is atypical.

- In figure 9(c), a 100% ionic bonding model is shown. The 3+ charge resides on the iron ion and the water molecules stay effectively neutral. This theoretical model is also invalid, as experimental results have shown the existence of the $[Fe(H_2O)_6]^{3+}$ species in aqueous solution; that is, it remains a single unit in solution.

- In figure 9(d), Pauling's electroneutrality principle is applied and the approximate charge distribution means that now the net charge on iron, the central metal, should be zero. As there are a total of six water ligands in the compound, the Fe^{3+} cation needs, effectively, three electrons to confer on it a net zero charge. The charge distribution, then, on each water ligand will be three electrons/six ligands $= \frac{1}{2}+$. Hence, in this model, coordinate bonds in $[Fe(H_2O)_6]^{3+}$ would be 50% covalent and 50% ionic.

This is a good example of evaluating various models to try to understand the nature of a scientific idea.

The Nobel Prize in Chemistry 2013 was awarded jointly to Martin Karplus (Université de Strasbourg, France and Harvard University, USA), Michael Levitt (Stanford University School of Medicine, USA) and Arieh Warshel (University of Southern California, Los Angeles, USA) for the development of multi-scale models for complex chemical systems. Chemists have always used models ranging from spheres and sticks to sophisticated computational programmes to explore further the structures of molecules, complexes and proteins. Their properties help chemists understand chemical processes. What was remarkable about the work of the recipients of the 2013 Noble Prize in Chemistry was that their models combined the two approaches of both traditional **classical mechanics** and **quantum mechanics**. For example in their research of simulating how a drug interacted with a target protein in the body, quantum mechanical calculations were performed on the atoms in the protein which interact with the target drug and classical mechanics was used to simulate the remainder of the protein. As outlined in the press release given by the *The Royal Swedish Academy of Sciences* on the 2013 prize they remark that *"Today the computer is just as important a tool for chemists as the test tube. Simulations are so realistic that they predict the outcome of traditional experiments"*.

Activity

Potassium permanganate, $KMnO_4$, is frequently used as an oxidizing agent. The manganate(VII) ion has the formula $[MnO_4]^-$.

(i)   Comment, giving a reason, whether or not VSEPR theory could be used to deduce the geometry of the manganate(VII) ion. Draw the structure of the ion, and identify the geometry (including the bond angles).

(ii) If the bonding in this anion was considered in terms of a 100% ionic bonding model, deduce what the charge would be on the manganese and oxygen atoms and explain why this model may be invalid for the manganate(VII) ion.

(iii) The American chemist, Linus Pauling, is well known for his development of the scale of electronegativities, but Pauling is perhaps less known for his electroneutrality principle. If Pauling's electroneutrality principle was applied to the manganate(VII) ion, suggest what the approximate charge distribution might be on each oxygen if manganese resulted in a net charge of 1+. Determine on this basis the percentage covalent character and the percentage ionic character.

Classification of ligands

The number of coordinate bonds formed by one ligand with a metal ion depends on the number of donor centres (atoms with lone electron pairs) in the ligand. **Monodentate ligands** are able to form only one coordinate bond with a metal ion while **polydentate ligands** (also known as **chelate ligands**) can form two or more such bonds.

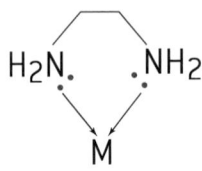

▲ Figure 10 Notice that the water proligand contains two lone pairs of electrons, but only one contributes to the coordinate bond in a transition metal complex

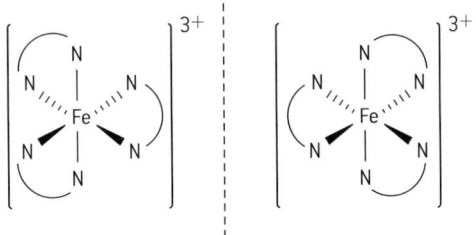

▲ Figure 11 Optical isomers of $[Fe(en)_3]Cl_3$

Monodentate ligands

Monodentate ligands contain a single donor atom and have one lone pair contributing to the coordinate bond in a complex. Typical examples include water, ammonia, and the halides such as Cl^- etc.

$$: \overset{..}{\underset{..}{Cl}} : \longrightarrow$$

Polydentate (chelate) ligands

These are ligands which have two or more donor atoms that form coordinate bonds with a transition metal centre. Some common examples are given below.

1,2-ethanediamine (en), $H_2NCH_2CH_2NH_2$

$$H_2N. \qquad .NH_2$$
$$M$$

The structure of en is given in section 16 of the *Data booklet*. en is a **bidentate** ligand because it has two donor atoms that coordinate to the transition metal centre in a complex. en is still sometimes referred to by its older name, ethylenediamine.

Polydentate ligands are often described as **chelate ligands** (coming from the Greek language, meaning *crab-claw*) as the ligands look like they are grabbing the metal between two or more donor atoms, just like a crab can grab your toes on a beach! The complexes formed from chelate ligands are very stable. An example of a chelate complex is $[Fe(en)_3]Cl_3$. The **coordination number** of the iron is six as each en ligand is bidentate. The complex has optical isomers (two non-superimposable mirror images, figure 11).

Ethanedioate (ox), $(C_2O_4)^{2-}$

Ethanedioate, often referred to by its older name, oxalate, is a bidentate, dianionic ligand.

Ethylenediaminetetraacetate, $(EDTA)^{4-}$

(EDTA)^4 is a polydentate ligand that can form up to six coordinate bonds. It has the ability to wrap itself around a transition metal centre in an octahedral complex. For example, in the anionic complex [Co(EDTA)]$^-$, the EDTA acts as a hexadentate ligand.

EDTA is used in:

- *Removal of heavy metals.* The ligand has a number of applications, such as its use in the treatment of lead poisoning. EDTA can coordinate with other metal ions present in blood. When Na$_2$[Ca(EDTA)] is administered to a patient, lead can displace calcium to form the anionic complex [Pb(EDTA)]$^{2-}$:

$$[Ca(EDTA)]^{2-} + Pb^{2+} \rightarrow [Pb(EDTA)]^{2-} + Ca^{2+}$$

  Once formed, [Pb(EDTA)]$^{2-}$ can be passed by the kidneys into the urine.

- *Chelation therapy.* Another medical application of EDTA is its potential use in heart by-pass surgery. Chelation therapy has been considered as a potential treatment for atherosclerosis ("hardening of the arteries"). The presence of EDTA in the bloodstream reduces the concentration of free calcium ions and effectively removes calcium from the atherosclerotic tissue. This can reduce cholesterol-filled plaque, which potentially reduces the risk of cardiovascular problems. However, to date the use of chelation therapy has shown somewhat limited benefit for heart disease.

- *Water softening.* EDTA is also used in water softening to ensure that no free calcium or magnesium ions remain (which can precipitate with soaps). It is used in shampoos for the same reason.

- *Food preservation.* Ca-EDTA is often added to food products (for example mayonnaise). Metal ions can catalyse reactions leading to rancidity, loss of taste or colour. **Rancidity** occurs in fats and oils. It is perceived by the senses to be when a food has "gone off" because of the development of a bad odour, taste, or appearence. In **hydrolytic rancidity** the lipids are broken down into their components, fatty acids and propan-1,2,3-triol. In **oxidative rancidity**, the fatty acid chains are oxidized and oxygen is added across the carbon-to-carbon double bond in the unsaturated lipid. Volatile aldehydes and carboxylic acids form, which can have noxious odours. The process involves radical reactions catalysed by light or metal ions. EDTA acts as a scavenger for such metal ions.

- *Restorative sculpture.* EDTA can also be used in the restoration of sculptured artwork pieces. Old brass or copper sculptures develop a coating of the insoluble solid complex, brochantite, $CuSO_4 \cdot 3Cu(OH)_2$. Upon the addition of EDTA, [Cu(EDTA)]$^{2-}$ can form, which is soluble and easily removed.

- *Cosmetics.* EDTA is sometimes used as a preservative in cosmetics.

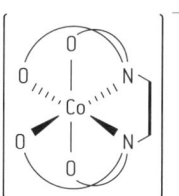

▲ Figure 12 Structure of [Co(EDTA)]$^-$

▲ Figure 13 Use of EDTA as a preservative in cosmetics

Coordination numbers

The majority of transition metal complexes have **coordination numbers** of six (octahedral geometry) or four (tetrahedral or square planar geometries).

Stereochemistry	Bond angles/°	Coordination number	Example
octahedral	90 (and 180)	6	$[Fe(H_2O)_6]Cl_2$
tetrahedral	109.5	4	$K_2[CoCl_4]$
square planar	90 (and 180)	4	$K_2[Ni(CN)_4]$

VSEPR theory cannot be used to deduce the geometry of transition metal complexes because of the incomplete d-sublevels of the transition metal ions. The structures of such complexes can be determined by the structural technique of **X-ray crystallography** if single crystals of the complex are available. Structural features, such as all the bond angles and bond distances present in the structure, can be elucidated using this technique.

Many of the platinum(II) complexes of coordination number four have square planar geometries. Cisplatin, used in the treatment of ovarian, bladder and testicular cancer, is square planar. Its geometrical isomer, transplatin, shows no anticancer activity.

Transition metals as catalysts

Transition metals are often used as catalysts in chemical reactions. Here are some examples of reactions that you may be familiar with from other sections of the programme.

- **Haber process:**

  $$N_2(g) + 3H_2(g) \rightleftharpoons 2NH_3(g)$$

  catalyst: Fe(s)

- **decomposition of hydrogen peroxide:**

  $$2H_2O_2(aq) \rightarrow 2H_2O(l) + O_2(g)$$

  catalyst: $MnO_2(s)$

- **hydrogenation of alkenes:**

  $$H_2C{=}CH_2(g) + H_2(g) \rightarrow CH_3CH_3(g)$$
  ethene   ethane

  catalyst: Ni(s), Pd(s), or Pt(s)

- **hydrogenation of oils**

  $$RCH{=}CHR' + H_2(g) \rightarrow RCH_2CH_2R'$$

  catalyst: Ni(s)

Unsaturated oils can be hydrogenated, to form a semi-solid (or solid) instead of a liquid. This is advantageous for cooking purposes. The product also has greater chemical stability due to a reduced rate of oxidation. The texture (that is, its hardness and plasticity) of the product is controlled. The main disadvantages of hydrogenation are:

- Mono- and polyunsaturated fats are healthier for the heart than saturated fats.

- *Trans* fatty acids can be formed in partial hydrogenation. These metabolize with difficulty and therefore may accumulate in the fatty tissues of the body. *Trans* fatty acids increase the levels of low-density lipoprotein (LDL) cholesterol (colloquially known as "bad cholesterol"), which may result in cardiovascular problems because of the narrowing of the arteries. This will be discussed further in sub-topic B.3.

Catalytic converters in cars

In a running car engine, gaseous nitrogen and oxygen react under high-temperature conditions (1500 °C) to form nitrogen monoxide:

$$N_2(g) + O_2(g) \rightarrow 2NO(g)$$

When NO(g) is released into the atmosphere, it combines with $O_2(g)$ to form nitrogen dioxide $NO_2(g)$:

$$2NO(g) + O_2(g) \rightarrow 2NO_2(g)$$

Nitrogen dioxide is a secondary pollutant that is primarily responsible for the brown colour of photochemical smog. Nitrogen dioxide is toxic and can result in respiratory problems.

Carbon monoxide, CO(g), a highly toxic, odourless, and colourless gas, is also emitted from the exhaust of a car, as well as unburned hydrocarbons. Most modern cars now are equipped with catalytic converters that reduce NO(g) and $NO_2(g)$ to $N_2(g)$ while oxidizing CO(g) and unburned hydrocarbons to $CO_2(g)$ and $H_2O(g)$, which are less harmful substances:

$$2NO(g) + 2CO(g) \rightarrow N_2(g) + 2CO_2(g)$$

$$CH_3CH_2CH_3(g) + 5O_2(g) \rightarrow 3CO_2(g) + 4H_2O(g)$$

Ethane and propane in exhausts can result in ozone formation.

In one chamber of the catalytic converter (figure 14) beads of Pt, Pd, and Rh oxidize CO(g) and unburned hydrocarbons. However, it increases the temperature of the exhaust gases and produces additional amounts of NO(g) so there is a second chamber that contains a different catalyst, often CuO or Cr_2O_3, which operates at a much lower temperature, reducing NO(g) to $N_2(g)$.

Catalysts in green chemistry

Catalysts play an important role in **green chemistry**. According to the *American Chemical Society*, green chemistry is the design, development, and implementation of chemical products and processes to reduce or eliminate the use and generation of substances hazardous to human health and the environment.

Biological catalysts

An **enzyme** is a biological catalyst. In the human body there are many enzyme-catalysed reactions that occur in cells and involve transition metals. One example is heme (figure 15), which is the iron centre of hemoglobin (Hb). Hemoglobin (figure 16) is the protein that transports oxygen in the blood. The vibrant red colour of blood stems from heme. Each subunit of hemoglobin contains an atom of iron, to which oxygen binds.

▲ Figure 14 Catalytic converter on the underside of a car. Three-way catalysts convert oxides of nitrogen, carbon monoxide and hydrocarbons into nitrogen, carbon dioxide and water. However, unleaded fuel has to be used in vehicles fitted with catalytic converters. If leaded fuel is used (that is, fuel containing added lead compounds used as anti-knocking agents) the catalyst can be poisoned

Homogeneous and heterogeneous catalysts

Homogeneous catalyst

A homogeneous catalyst is one that is in the same phase or physical state as the substances involved in the reaction that it is catalysing.

Heterogeneous catalyst

A heterogeneous catalyst is one that is in a different phase to the substances involved in the reaction that it is catalysing. Industrial catalysts that involve transition metals are usually heterogeneous catalysts.

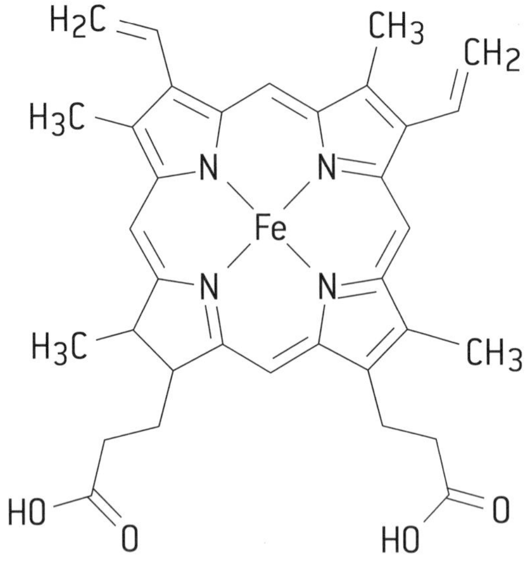

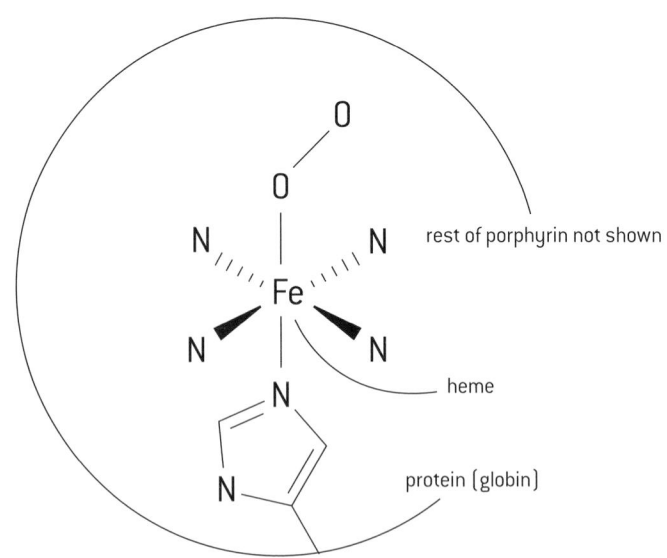

▲ Figure 15 Structure of heme. In heme, iron has a coordination number of four and a square planar geometry. The Fe^{2+} ion is at the centre of a large nitrogenous heterocyclic ring called a **porphyrin**. Each Hb molecule contains four heme groups. The iron can bind to one O_2 molecule and therefore a single Hb molecule can transport up to four O_2 molecules

▲ Figure 16 Structure of human oxyhemoglobin. Oxygen is carried through the blood stream by the formation of a weak bond with heme. The $O_2–Fe^{2+}$ bond is then broken relatively easily. When O_2 bonds to Fe^{2+}, O_2 functions as a monodentate ligand and Fe^{2+} then adopts an octahedral stereochemistry, with a coordination number of six, as heme is linked to the protein (the globin) via an additional Fe—N bond.

Magnetic properties of transition metals

Magnetic properties of transition metals and their complexes depend on many factors, including the oxidation state of the metal, its coordination number, and the geometry of the complex. **Paramagnetic materials** contain unpaired electrons that behave as tiny magnets and are attracted by an external magnetic field. In contrast, **diamagnetic materials** do not contain unpaired electrons and therefore are repelled by external magnetic fields. Para- and diamagnetic properties of metals, ions, and compounds are further discussed in sub-topic A.2.

13.2 Coloured complexes

Understandings

→ The d-sublevel splits into two sets of orbitals of different energy in a complex ion.

→ Complexes of d-block elements are coloured, as light is absorbed when an electron is excited between the d orbitals.

→ The colour absorbed is complementary to the colour observed.

Applications and skills

→ Explanation of the effect of the identity of the metal ion, the oxidation state of the metal and the identity of the ligand on the colour of transition metal ion complexes.

→ Explanation of the effect of different ligands on the splitting of the d-orbitals in transition metal complexes and colour observed using the spectrochemical series.

Nature of science

→ Models and theories – the colour of transition metal complexes can be explained through the use of models and theories based on how electrons are distributed in d-orbitals.

→ Transdisciplinary – colour linked to symmetry can be explored in the sciences, architecture, and the arts.

In an isolated atom, d orbitals have the same energy but in a complex ion, they split into two sublevels. The electronic transitions between these sublevels leads to absorption and emission of photons of visible light, which are responsible for the colour of the complex.

Theories on complexes

A number of different theories have been proposed to explain the bonding of d-block metals in complexes. These theories are listed below in chronological order:

- **Valence bond theory (VBT).** VBT was developed by Linus Pauling in the 1930s, which had hybridization as its basis. This theory is rarely used today.

- **Crystal field theory (CFT).** CFT is based on an electrostatic model. CFT does have its limitations, for example, it cannot explain the order of ligands in the spectrochemical series (this will be considered later).

- **Molecular orbital theory (MOT).** In this theory, covalent interactions between the

transition metal centre and the ligands are considered.

- **Ligand field theory (LFT).** LFT is an extension of CFT, but differs from CFT as it is not based on an electrostatic model. LFT is often considered a combination of the CFT and MOT models. The bonding description associated with LFT is more detailed and can be discussed in terms of electronic energy levels involving **frontier orbitals***.

- **Angular overlap model.** In this model, the relative sizes of orbital energies are estimated in a molecular orbital (MO) calculation.

- These theories and models help us explain many of the characteristics of transition metal complexes, such as colour, electronic spectra, and magnetic properties. The comprehensive details of these models are beyond the scope of the IB Chemistry Diploma programme. In this book, we shall use only the CFT model to explain the colour of transition metal complexes.

*As outlined in the *IUPAC Gold Book* (http://goldbook.iupac.org/), **frontier orbitals** refer to the highest-energy occupied molecular orbital (HOMO) (filled or partly filled) and the lowest-energy unoccupied molecular orbital (LUMO) (completely or partly vacant) of a molecular entity. The *IUPAC Gold Book* is an invaluable source for chemists.

Crystal field theory (CFT)

The d-sublevel consists of five d-orbitals (figure 1) d_{xy}, d_{yz}, d_{xz}, $d_{x^2-y^2}$, and d_{z^2}.

As can be seen from figure 1 three of these orbitals have their lobes of electron density pointing at 45° to the Cartesian axes (d_{xy}, d_{yz}, d_{xz}). In contrast, the remaining two orbitals ($d_{x^2-y^2}$, and d_{z^2}) have their lobes of electron density pointing along the Cartesian axes. However, in the free metal ion, M^{n+}, with ligands (L) at an infinite distance away, these five d-orbitals are degenerate.

CFT is based on an electrostatic model, where the ligands are considered as point charges that surround the metal cation, M^{n+}. If the electrostatic field created by the ligand point charges is isotropic (that is, spherically symmetrical), the energies of the d orbitals will remain degenerate but will increase in energy uniformly. If, however, the electrostatic field created by the ligand point charges is octahedral, then the d orbitals will split into two sets of degenerate energy, the t_{2g} set and the e_g set. Three of the orbitals (the t_{2g} set) will decrease in energy (that is, they are stabilized) and two of the orbitals (the e_g set) will increase in energy (that is, they are destabilized). The stabilized orbitals that comprise the

Useful resource

Look at the Orbitron website to see the shapes of the d orbitals: http://winter.group.shef.ac.uk/orbitron/

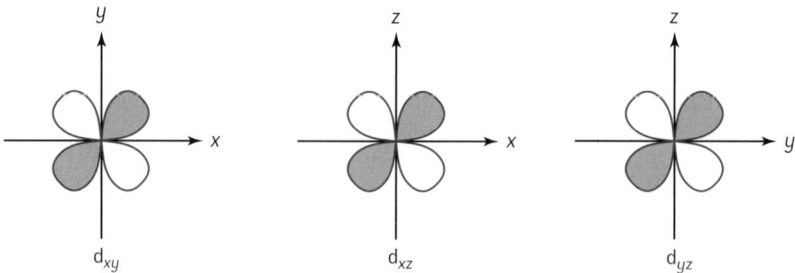

third cartesian axis in each case is orthogonal (90°) to the 2D plane

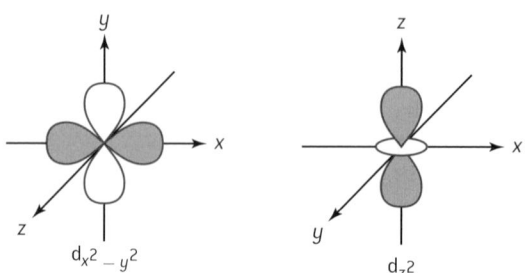

▲ Figure 1 Five d-orbitals

t_{2g} set are the d_{xy}, d_{yz}, and d_{xz} orbitals. The reason for this stabilization is associated with the fact that these three orbitals have their lobes of electron density lying at 45° to the Cartesian axes. In contrast, the $d_{x^2-y^2}$ and d_{z^2} orbitals (e_g) are destabilized because their lobes of electron density are directed along the Cartesian axes. The energy separation between the two split degenerate sets of orbitals is defined as Δ_0, the **crystal field splitting energy**.

For the first three d^n electron configurations, d^1, d^2, and d^3, the electrons will occupy the t_{2g} set of degenerate orbitals in an octahedral crystal field, and will fill the orbitals singly before filling them in pairs, following Hund's rule of maximum multiplicity. However, after d^3, the fourth electron has a choice – it can either occupy the destabilized e_g level or else occupy the stabilized t_{2g} level. Although the electron would enter a stabilized energy level, to do so would require additional energy to pair the electron with another electron in an already filled orbital. This additional energy is termed the **pairing energy, P**.

So what are the factors that affect the crystal field splitting energy parameter, Δ_0? First of all, it is important to stress that Δ_0 is an experimental quantity. The following are the factors that affect the size of Δ_0:

- identity of the metal ion
- oxidation state of the metal ion
- nature of the ligands
- geometry of the complex ion.

 Symmetry

Science is peppered with symbolic representations which form part of the universal language of science. The origin of such symbols can be historically interesting. As chemists we should never just accept symbols at their face value and should always try to grasp the origin of such representations. Part of the IB learner profile is that as IB learners we strive to be inquirers.

The t_{2g} and e_g notations used in an octahedral crystal field energy splitting diagram have their origin in symmetry:

- g comes from the term *gerade*, meaning that the wavefunction does not change sign upon inversion; that is, there is no change in parity of the orbital. u comes from the term *ungerade*, meaning that the wavefunction changes sign upon inversion; that is, there is a change in the parity of the orbital. g and u are only used if a geometrical entity has a centre of inversion. Hence, as there is a centre of inversion in the octahedral stereochemistry (but not in a tetrahedron), g is used. Gerade is the German term for even.

- t refers to a triply degenerate set of orbitals. e refers to a doubly degenerate set of orbitals. The symbols a or b are used if there is only one orbital involved. Ungerade is the German term for odd.

- The number 2 is used if the sign of the wavefunction changes upon rotation about the Cartesian axes (figure 2). For example, let us look at what happens to the sign of the wavefunction with respect to the d_{xy} orbital on rotation about the x-axis:

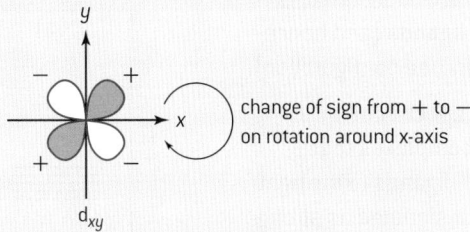

▲ Figure 2 Explanation of the 2 symmetry label

As the sign changes from + to −, the number 2 is used.

Group 9 complex	Δ_o / cm^{-1}
$[Co(NH_3)_6]^{3+}$	22900
$[Rh(NH_3)_6]^{3+}$	34100
$[Ir(NH_3)_6]^{3+}$	41100

Complex	Δ_o / cm^{-1}
$[Co(NH_3)_6]^{2+}$	10200
$[Co(NH_3)_6]^{3+}$	22900

Complex	Δ_o / cm^{-1}
$[Co(H_2O)_6]^{3+}$	18200
$[Co(NH_3)_6]^{3+}$	22900

Identity of the metal ion

The identity of the metal ion can influence the extent of the crystal field splitting. In general, Δ_o increases descending a group with the metal in the same oxidation state.

Oxidation state of the metal ion

For a given metal, Δ_o increases as the oxidation state increases. Since the metal–ligand interaction is partly electrostatic in nature, as the charge on the metal increases, the distances between the metal and ligands decrease resulting in a better overlap between the metal orbitals and the ligand orbitals.

Nature of the ligands

Ligands may have different charge densities. For example, the ammonia ligand, NH_3, has a greater charge density compared to water, H_2O, and hence the crystal field splitting caused by ammonia will be greater.

Spectrochemical series

$$I^- < Br^- < Cl^- < F^- < [C_2O_4]^{2-} \approx H_2O < NH_3 < en < bpy < phen < NO_2^- < CN^- \approx CO$$

weak-field ligands → strong-field ligands

increasing Δ_0

In the case of weak-field ligands, the configuration adopted involves a **spin-free** configuration (figure 3), whereas in the case of strong-field ligands, such as CN^-, the configuration adopted involves a **spin-paired** arrangement (figure 4).

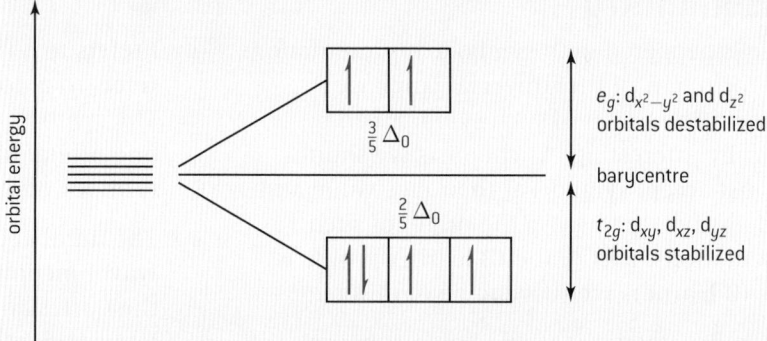

▲ Figure 3 Crystal field splitting for $[Fe(H_2O)_6]^{2+}$, which involves the H_2O weak-field ligand. The $t_{2g}^4 e_g^2$ configuration adopted is a **spin-free** configuration

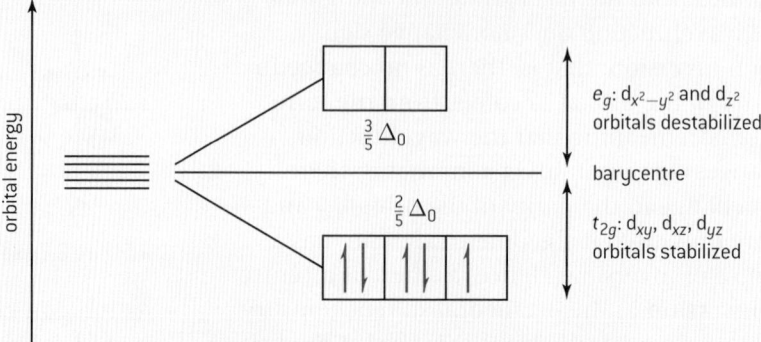

▲ Figure 4 Crystal field splitting for $[Fe(CN)_6]^{3-}$, which involves the CN^- strong-field ligand. The $t_{2g}^5 e_g^0$ configuration adopted is a **spin-paired** configuration

The following is a guideline when considering whether a configuration involves a spin-paired or a spin-free arrangement:

- **M^{2+}:** In the spectrochemical series, for complexes that involve M^{2+}, ligands to the right of NO_2^- are designated as **strong-field ligands** (and hence adopt a spin-paired configuration), whereas complexes with ligands to the left of NO_2^- are designated as **weak-field ligands** (and hence adopt a spin-free configuration).

- **M^{3+}:** In the spectrochemical series, for complexes that involve M^{3+}, ligands to the right of H_2O are designated as strong-field ligands whereas complexes with ligands to the left of H_2O are designated as weak-field ligands.

A Japanese chemist, R. Tsuchida, suggested that ligands could be arranged into a **spectrochemical series**, based on order of increasing Δ_o. The spectrochemical series, which is given in section 15 of the *Data booklet* is based on empirical evidence.

The geometry of the complex ion

The geometry of the complex ion can also influence the crystal field splitting parameter. For example, Δ_t for a tetrahedral complex is $\sim \frac{4}{9}\Delta_o$.

It must be emphasized, however, that because the spectrochemical series is empirical in nature, as a model CFT cannot account for the relative values of the crystal field energy splitting parameters.

Explanation of the colour of transition metal complexes

The colour of transition metal ions is associated with partially filled d orbitals. For example, Cu^{2+} has an $[Ar]3d^9$ condensed electron configuration, so its d sublevel is incomplete, and thus one would expect Cu^{2+} ions to be coloured. Cu^{2+} ions are often blue in colour. For example, crystals of $CuSO_4 \cdot 5H_2O$ have a dominant Mediterranean blue colour. In contrast, ions of the d-block metal zinc, Zn^{2+}, have an $[Ar]3d^{10}$ configuration, and so zinc's d-sublevel is fully filled. As a result, ions of Zn^{2+} are typically colourless.

White light consists of all the colours of the visible spectrum. Transition metal complexes absorb some of these colours, allowing other colours to be transmitted. The **colour wheel** (figure 5) can be used to determine the colour of the light transmitted, that is the **complementary colour** of the absorbed light. For example, $[Ti(H_2O)_6]^{3+}$ absorbs yellow–green light. The complementary colour to yellow-green is red-violet, which lies at the opposite side of the colour wheel. Therefore, $[Ti(H_2O)_6]^{3+}$ ions will transmit the complementary colour and appear red-violet. Let us now examine the nature of light absorption in transition metal complexes.

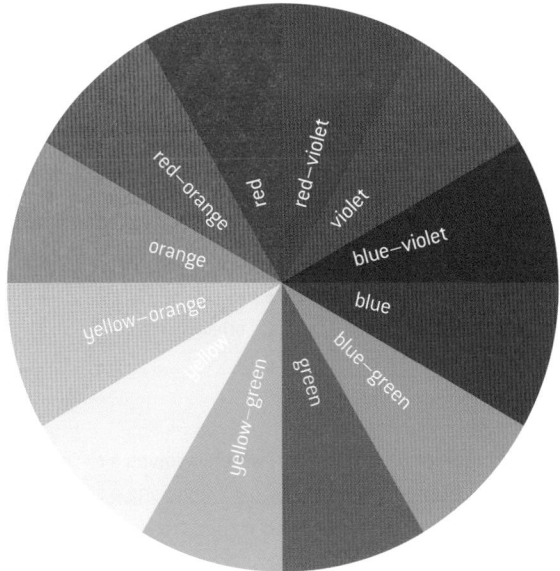

▲ Figure 5 The colour wheel

As outlined previously, the five d-orbitals in an octahedral crystal field are split into two sets of degenerate orbitals – the stabilized t_{2g} set and the destabilized e_g set. If the d orbitals are partially filled, **d-to-d electronic transitions** can occur. In other words, an electron can jump from the lower-energy t_{2g} set of orbitals to the higher-energy e_g set of orbitals. This d-to-d electronic transition is the origin of the colour of transition metal complexes. In the case of $[Ti(H_2O)_6]^{3+}$, such an electronic transition requires a photon of yellow-green light to be absorbed. The frequency of the yellow–green light absorbed is a measure of Δ_o. Since ΔE represents the energy change, the frequency of the light, v (or f), is related to ΔE via the expression:

$$\Delta E = hv = \frac{hc}{\lambda}$$

where:

h = Planck's constant = 6.63×10^{-34} J s
c = speed of light in a vacuum = 3.00×10^8 m s^{-1}
λ = wavelength, in m

Since ΔE is related to Δ_o, the actual colour of any complex will depend on all the factors described previously, including the identity and oxidation state of the metal ion, the nature of the ligands, and the geometry of the complex ion.

a)

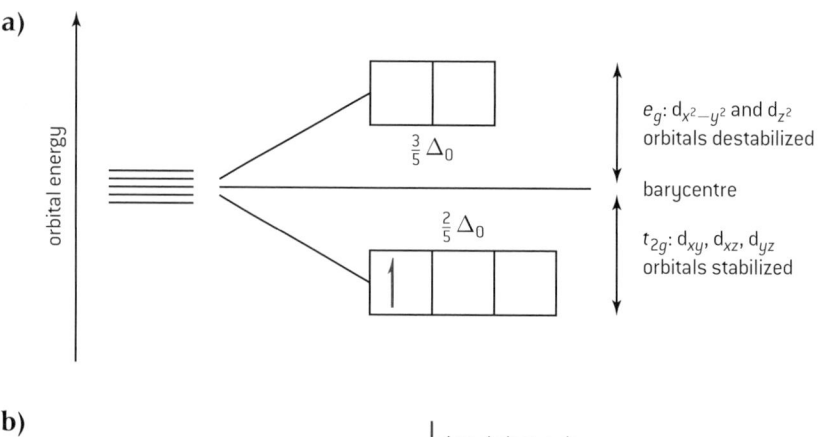

b)

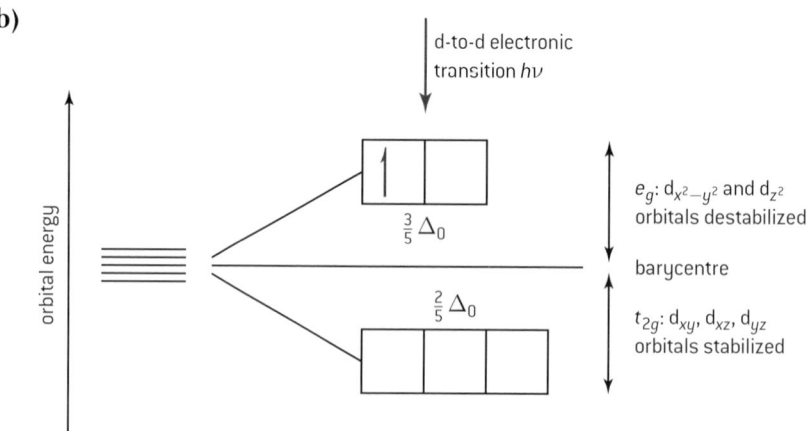

▲ Figure 6 (a) Crystal field splitting for ground-state $[Ti(H_2O)_6]^{3+}$, which involves a $t_{2g}^{1} e_g^{0}$ configuration. (b) Crystal field splitting for excited-state $[Ti(H_2O)_6]^{3+}$ involving a $t_{2g}^{0} e_g^{1}$ configuration

Worked examples

Example 1

For the complex $K_3[Fe(ONO)_6]$, deduce:

a) The oxidation state of the transition metal in the complex.

b) The condensed electron configuration of the transition metal in this oxidation state.

c) The coordination number of the metal in the complex.

d) The stereochemistry (geometry) of the complex.

e) The charge on the complex.

Solution

a) Let x = the oxidation state of iron in the complex.

$3(+1) + x + 6(-1) = 0$, so $x = +3$.

b) Fe is $[Ar]3d^6 4s^2$, so Fe^{3+} is $[Ar]3d^5$.

c) Six, assuming a monodentate nitrito ligand, $(ONO)^-$.

d) Octahedral, assuming a monodentate nitrito ligand, $(ONO)^-$.

e) Each potassium has a charge of $1+$, so the net charge for three K^+ ions will be $3+$, meaning that the charge on the anionic complex part (that is nested in the square brackets) will be $3-$.

> **Study tip**
>
> Oxidation states are written with the sign first and then the number (for example, here iron has an oxidation state of $+3$); ions are written with the number first and then the charge (for example, Fe^{3+} ion).

Example 2

$Ni(ClO_4)_2$ reacts with water to form the complex ion $[Ni(H_2O)_6][ClO_4]_2$. Explain this reaction in terms of an acid–base theory, and outline how the bond is formed between Ni^{2+} and H_2O.

Solution

A Lewis acid is an electron-pair acceptor and a Lewis base is an electron-pair donor. Hence Ni^{2+} acts as the Lewis acid and H_2O acts as the Lewis

base. Each water molecule acts as a monodentate ligand, forming a coordinate bond with Ni^{2+}. As there are six water ligands involved, the geometry of the cationic complex, $[Ni(H_2O)_6]^{2+}$, is octahedral, with Ni^{2+} having a coordination number of 6. The perchlorate ions are in the lattice and do not form part of the cationic complex.

$$\left[\begin{array}{c} OH_2 \\ H_2O \underset{H_2O}{\overset{Ni}{}} \overset{OH_2}{\underset{OH_2}{}} 90° \end{array} \right] [ClO_4]_2$$

Example 3

Consider the complex $[Ni(NH_3)_6]Cl_2$.

a) Deduce the condensed electron configuration of the transition metal in its associated oxidation state in this complex.

b) State the geometry of the transition metal complex and draw a diagram of the complex.

c) Identify the nature of the bonding between the ligand and the transition metal ion in the complex.

d) State the denticity of the ammonia ligand.

e) Draw a diagram showing the splitting of the d-sublevel. Label the orbitals involved and populate each of the orbitals with electrons.

f) Explain whether the complex is paramagnetic or diamagnetic.

Solution

a) Let x = the oxidation state of nickel in the complex.

$x + 6(0) + 2(-1) = 0$, so $x = +2$;

Ni is $[Ar]3d^8 4s^2$, Ni^{2+} is $[Ar]3d^8$.

b) Octahedral, $CN^0 = 6$.

$$\left[\begin{array}{c} NH_3 \\ H_3N \underset{H_3N}{\overset{Ni}{}} \overset{NH_3}{\underset{NH_3}{}} 90° \end{array} \right] Cl_2$$

c) Coordinate bonding.

d) NH_3 has one lone pair involved in the coordinate bond to Ni so it is monodentate.

e) $t_{2g}^{6} e_{g}^{2}$.

f) Paramagnetic since there are two unpaired electrons.

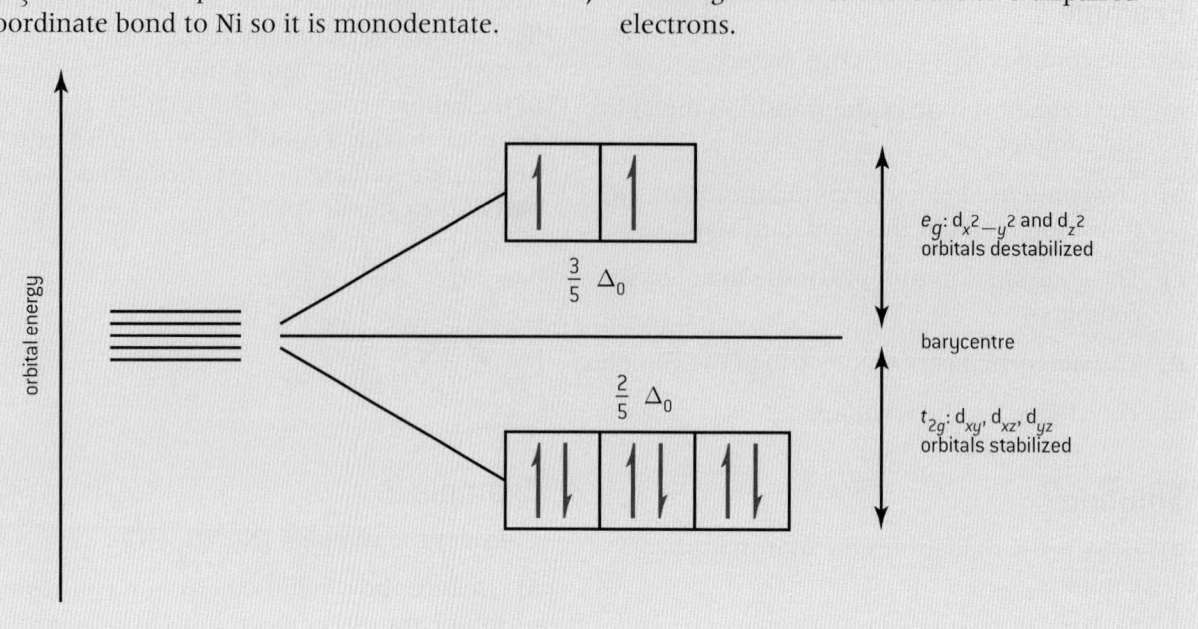

e_g: $d_{x^2-y^2}$ and d_{z^2} orbitals destabilized

barycentre

t_{2g}: d_{xy}, d_{xz}, d_{yz} orbitals stabilized

$\frac{3}{5}\Delta_0$

$\frac{2}{5}\Delta_0$

orbital energy

Questions

1 Which of the following elements is not a transition element?

   A. Fe

   B. Cu

   C. Sc

   D. Zn

2 What is the condensed electron configuration of Co^{2+}?

   A. $[Ar]4s^23d^7$

   B. $[Ar]4s^23d^5$

   C. $[Ar]3d^7$

   D. $[Ar]4d^7$

3 What is the condensed electron configuration of Fe^{2+}?

   A. $[Ar]4s^23d^6$

   B. $[Ar]4s^13d^5$

   C. $[Ar]3d^6$

   D. $[Ar]4s^23d^4$

4 What is the ligand in the complex $[NH_4]_2[Fe(H_2O)_6][SO_4]_2$?

   A. Fe^{2+}

   B. $[SO_4]^{2-}$

   C. H_2O

   D. $[NH_4]^+$

5 What is the oxidation state of iron in the complex $Na[Fe(EDTA)] \cdot 3H_2O$?

   A. $+1$

   B. $+2$

   C. $+3$

   D. $+6$

6 What is the total charge, n, in the following complex of Ni(II), $[Ni(NH_3)_6]^n$?

   A. 0

   B. $1+$

   C. $2+$

   D. $3+$

7 What of the following can act as a ligand?

   I. PH_3

   II. H_2O

   III. NO_2^-

   A. I and II only

   B. I and III only

   C. II and III only

   D. I, II, and III

8 Which electron transitions are responsible for the colours of transition metal compounds?

   A. Between d orbitals and s orbitals

   B. Among the attached ligands

   C. From the metal ion to the attached ligands

   D. Between d orbitals   [1]

   IB, May 2009

9 Which salts form coloured solutions when dissolved in water?

   I. $ZnCl_2$

   II. $FeBr_2$

   III. $Co(NO_3)_3$

   A. I and II only

   B. I and III only

   C. II and III only

   D. I, II and III

10 Which of the following statements is correct for the complex $[Cr(H_2O)_6]Cl_3$?

   A. It is paramagnetic.

   B. It is diamagnetic.

   C. The coordination number of the chromium ion is 3.

   D. H_2O acts as a bidentate ligand in the complex.

11 Explain, by referring to successive ionization energies, why Ti forms variable oxidation states, but Ca only occurs in the $+2$ oxidation state.

12 Explain why $[Ni(H_2O)_6][BF_4]_2$ is coloured.

13 Consider the complex, $K_4[Fe(C_2O_4)_3]$.

a) Deduce the condensed electron configuration of the transition metal in this complex.

b) State the geometry of the transition metal complex and draw a diagram of the complex.

c) Identify the nature of the bonding between the ligand and the transition metal ion in the complex.

d) State the denticity of the ethanedioato ligand.

e) Draw a diagram showing the splitting of the d-sublevel. Label the orbitals involved and populate each of the orbitals with electrons.

f) Explain whether the complex is paramagnetic or diamagnetic.

14 In an article written by W.B. Jensen in the *Journal of Chemical Education* (85, 9, (2008), p1182-3), it was reported that minute quantities of HgF_4 have been detected, using matrix isolation techniques, at 4 K under extreme non-equilibrium conditions. Suggest why, on this basis, in the first instance mercury might now be considered a transition metal. In the publication, however, Jensen challenges this claim. Explore why Jensen's counterargument may have merit, in view of conventional thinking on what IUPAC considers as a transition element.

14 CHEMICAL BONDING AND STRUCTURE (AHL)

Introduction

More in-depth explanations of bonding systems and closer analysis of structural arrangements often require more sophisticated concepts and theories of bonding to be considered. In this topic we expand the principles of VSEPR Theory introduced in topic 4 to explore species involving five and six electron domains and discuss the roles that formal charge and delocalization play in such a treatment of structure and bonding. Hybridization is also introduced as a mathematical model and we see how hybridization schemes can be deduced from an examination of the number of electron domains around a central interior atom in a species.

14.1 Further aspects of covalent bonding and structure

Understandings

→ Covalent bonds result from the overlap of atomic orbitals. A sigma bond (σ) is formed by the direct head-on/end-to-end overlap of atomic orbitals, resulting in electron density concentrated between the nuclei of the bonding atoms. A pi bond (π) is formed by the sideways overlap of atomic orbitals, resulting in electron density above and below the plane of the nuclei of the bonding atoms.

→ Formal charge (FC) can be used to decide which Lewis (electron dot) structure is preferred from several. The FC is the charge an atom would have if all atoms in the molecule had the same electronegativity. FC = (Number of valence electrons) $- \frac{1}{2}$(Number of bonding electrons) $-$ (Number of non-bonding electrons). The Lewis (electron dot) structure with the atoms having FC values closest to zero is preferred.

→ Exceptions to the octet rule include some species having incomplete octets and expanded octets.

→ Delocalization involves electrons that are shared by more than two atoms in a molecule or ion as opposed to being localized between a pair of atoms.

→ Resonance involves using two or more Lewis (electron dot) structures to represent a particular molecule or ion. A resonance structure is one of two or more alternative Lewis (electron dot) structures for a molecule or ion that cannot be described fully with one Lewis (electron dot) structure alone.

Applications and skills

→ Prediction whether sigma (σ) or pi (π) bonds are formed from the linear combination of atomic orbitals.

→ Deduction of the Lewis (electron dot) structures of molecules and ions showing all valence electrons for up to six electron pairs on each atom.

→ Application of FC to ascertain which Lewis (electron dot) structure is preferred from different Lewis (electron dot) structures.

→ Deduction using VSEPR theory of the electron domain geometry and molecular geometry with five and six electron domains and associated bond angles.

→ Explanation of the wavelength of light required to dissociate oxygen and ozone.

→ Description of the mechanism of the catalysis of ozone depletion when catalysed by CFCs and NO_x.

Nature of science

→ Principle of Occam's razor – bonding theories have been modified over time. Newer theories need to remain as simple as possible while maximizing explanatory power, for example the idea of formal charge.

Theories of bonding and structure

To study large structures and consider in-depth explanations of bonding systems requires more sophisticated concepts, models, and theories of bonding than we have met so far. In this topic we look at species based on five and six electron domains, and consider (and challenge) some of the models and theories used in chemical bonding and structure.

The principle of **Occam's razor** is a blueprint for the development of theories in a number of different fields of knowledge. The philosophy underpinning this principle is that a theory should remain as simple as is possible while maintaining a capacity for maximum discovery of understanding and application.

Bonding theories have been modified over time. New theories need to remain as simple as possible while maximizing their explanatory power. In this chapter several models and theories of structure and bonding are presented, each with its pros and associated limitations.

Formal charge

In sub-topic 4.3 we introduced the idea of a Lewis (electron dot) structure as a convenient way of showing how the valence electrons are distributed in a covalent molecular species or a polyatomic ion. Sometimes a number of different Lewis structures can be drawn that all obey the **octet rule**. A useful approach in deciding which Lewis structure is the most appropriate is to determine the **formal charge (FC)** of the atoms present in the molecule or ion. The calculation of FC can be considered as a process involving electronic book-keeping; it is a hypothetical charge worked out as follows:

$$FC = \text{(number of valence electrons)} - \frac{1}{2}\text{(number of bonding electrons)} - \text{(number of non-bonding electrons)}$$

For example, in the molecule tetrachloromethane, CCl_4 (figure 1) the formal charge on each atom is calculated as follows:

$$FC(C) = (4) - \frac{1}{2}(8) - 0 = 0$$

$$FC(Cl) = (7) - \frac{1}{2}(2) - 6 = 0$$

In the case of the carbonate anion, CO_3^{2-} (figure 2), the FCs on the carbon and oxygen atoms are:

$$FC(C) = (4) - \frac{1}{2}(8) - 0 = 0$$

$$FC(O_A) = (6) - \frac{1}{2}(2) - 6 = -1$$

$$FC(O_B) = (6) - \frac{1}{2}(4) - 4 = 0$$

If there are a number of possible Lewis structures that all obey the octet rule, the most reasonable one will be:

- the one with FC difference $\left(\Delta FC = FC_{max} - FC_{min}\right)$ closest to 0, and

- the one that has the negative charges located on the most electronegative atoms.

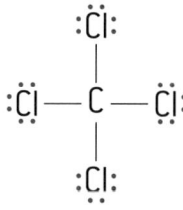

▲ Figure 1 Lewis structure of tetrachloromethane, CCl_4

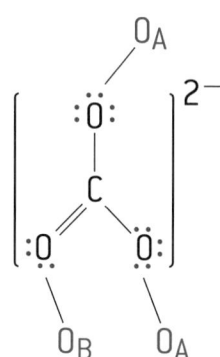

▲ Figure 2 Lewis structure of the carbonate anion, CO_3^{2-}

For example, figure 3 shows two Lewis structures for boron trifluoride, BF_3. For structure (a):

$$FC(B) = (3) - \frac{1}{2}(6) - 0 = 0$$

$$FC(F) = (7) - \frac{1}{2}(2) - 6 = 0$$

$$\Delta FC = 0$$

For structure (b):

$$FC(B) = (3) - \frac{1}{2}(8) - 0 = -1$$

$$FC(F_A) = (7) - \frac{1}{2}(2) - 6 = 0$$

$$FC(F_B) = (7) - \frac{1}{2}(4) - 4 = +1$$

$$\Delta FC = FC_{max} - FC_{min} = (+1) - (-1) = +2$$

Since ΔFC is closest to zero for Lewis structure (a), this is the most reasonable representation of BF_3.

Although Lewis structure (b) obeys the octet rule, structure (a) is preferred based on FC considerations even though boron has an **incomplete octet** of electrons (fewer than 8 valence electrons). Species can also be found with **expanded octets** of electrons (more than 8 valence electrons surrounding the central atom), as we shall see later in this topic.

 ## Different interpretations of "charge"

The idea of charge has many connotations in chemistry (oxidation state, formal charge, ionic charge, partial charge, total charge), and we need to interpret the intended meaning depending on the context.

To distinguish the terms oxidation state, formal charge, partial charge, and ionic charge, consider the hydrogen fluoride molecule, HF (figure 4).

Oxidation states

hydrogen: +1

fluorine: −1

Formal charges

$$FC(H) = (1) - \frac{1}{2}(2) - 0 = 0$$

$$FC(F) = (7) - \frac{1}{2}(2) - (6) = 0$$

$$\Delta FC = FC_{max} - FC_{min} = 0$$

Partial charges

From section 8 of the *Data booklet*, the electronegativity for H = 2.2 and for F = 4.0, so you would expect the partial charge on fluorine to be more negative than the partial charge on hydrogen. One rather approximate but simple way of calculating

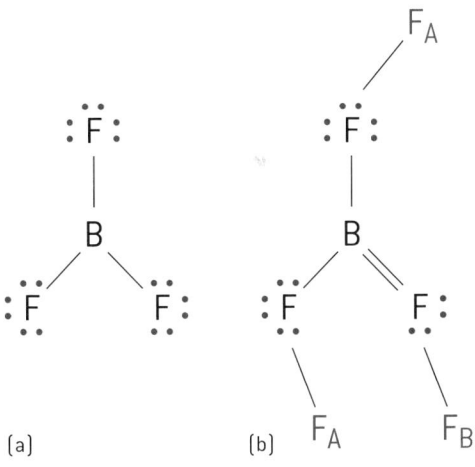

(a)   (b)

▲ Figure 3 Two possible Lewis structures for boron trifluoride, BF_3

$$H - \ddot{\underset{\cdot\cdot}{F}}:$$

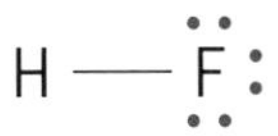

▲ Figure 4 Lewis structure for hydrogen fluoride, HF

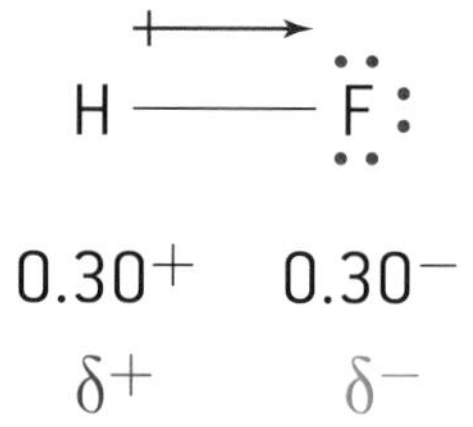

H $\xrightarrow{\hspace{1.5cm}}$ F:

0.30+   0.30−

δ+   δ−

▲ Figure 5 The partial charge in a molecule of hydrogen fluoride, HF, based on simple approximations

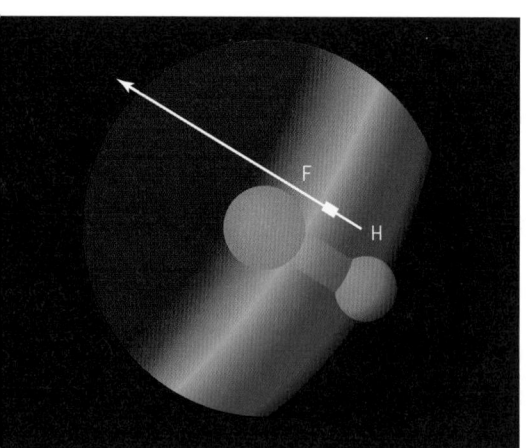

▲ Figure 6 The molecular electrostatic potential (MEP) on the van der Waals surface of the hydrogen fluoride molecule

the partial charge based on electronegativity values is as follows:

H: $\dfrac{(2.2)}{(2.2 + 4.0)} = 0.35$ of the charge of the bonding pair

F: $\dfrac{(4.0)}{(2.2 + 4.0)} = 0.65$ of the charge of the bonding pair

Hence, since the single covalent bond in HF consists of two electrons, fluorine will have $0.65 \times 2e = 1.30e$. Therefore fluorine in HF will have 0.30e more negative charge than a neutral atom of fluorine. This equates to a 0.30− partial charge on fluorine and a 0.30+ partial charge on hydrogen. HF is a polar molecule with a net dipole moment, which is represented by the vector shown in figure 5. The experimentally calculated dipole moment for HF is 1.86 D.

The **molecular electrostatic potential** (MEP) for HF is shown in figure 6. The red region shows the area of greatest electron density and the blue region shows the area of lowest electron density.

Another method of finding partial charges involves taking the H−F bond length of 92 pm (section 10 of the *Data booklet*) and the experimentally calculated dipole moment 1.86 D:

$$Q = \frac{\mu}{r}$$

where:

Q = apparent charge on each end of the molecule

μ = dipole moment

r = H−F bond length

We can use the following conversion:

$$1 \text{ D} = 3.34 \times 10^{-30} \text{ C m}$$

hence:

$$Q = \frac{(1.86 \times 3.34 \times 10^{-30} \text{ C m})}{(92 \times 10^{-12} \text{ m})}$$

$$= 6.75 \times 10^{-20} \text{ C}$$

The charge of an isolated electron $= 1.60 \times 10^{-19}$ C (section 4 of the *Data booklet*), therefore the partial charge δ is given by:

$$\delta = \frac{(6.75 \times 10^{-20} \text{ C})}{(1.60 \times 10^{-19} \text{ C})} = 0.42$$

Computer programs can calculate partial charges much more accurately, and from the MEP model shown in figure 6, H has δ+ = 0.54+ and F has δ− = 0.54−.

Net charge on the HF molecule

The net charge on the HF molecule is 0.

Molecular geometries based on five and six electron domains

In topic 4 we examined molecular geometries based on two, three, and four electron domains. Molecular geometries based on five and six electron domains are summarized in table 1, and can be deduced using the procedure outlined in topic 4.

For molecular geometries based on five electron domains, remember that lone pairs (non-bonding pairs) occupy equatorial positions in the first instance. This is based on the fact that in terms of repulsion, the order of interactions is as follows (topic 4):

$$LP|LP > LP|BP > BP|BP$$

(LP = lone pair; BP = bonding pair)

We shall examine the reason for this in a worked example later in this topic.

> **Study tip**
>
> **Equatorial** and **axial** positions occur only for geometries based on five electron domains, not those based on six electron domains. We usually do not refer to axial or equatorial positions for an octahedral geometry.

Number of electron domains	Electron domain geometry	Molecular geometry	Notes
5	trigonal bipyramidal	trigonal bipyramidal	AB_5 5 BPs example: PF_5
5	trigonal bipyramidal	see-saw	AB_4E 4 BPs and 1 LP example: SF_4
5	trigonal bipyramidal	T-shaped	AB_3E_2 3 BPs and 2 LPs example: ClF_3

Number of electron domains	Electron domain geometry	Molecular geometry	Notes
5	trigonal bipyramidal	**Linear**	AB_2E_3 2 BPs and 3 LPs example: I_3^-
6	octahedral	**octahedral**	AB_6 6 BPs example: SF_6
6	octahedral	**square-based pyramidal**	AB_5E 5 BPs and 1 LP example: BrF_5
6	octahedral	**square planar**	AB_4E_2 4 BPs and 2 LPs example: XeF_4

▲ Table 1 Electron domain geometries and molecular geometries based on five and six electron domains. B_a = axial substituent; B_e = equatorial substituent; E (in formula) = lone pair of electrons; BP = bonding pair of electrons; LP = lone pair of electrons

Overlap of atomic orbitals: Sigma (σ) and pi (π) bonding

In topic 4 we discussed the difference between a single covalent bond and a multiple bond such as a double or triple bond. A **single covalent bond** consists of two electrons shared between two atoms A and B:

The single bond, represented by a stick, is a **sigma bond** (σ).

A **double covalent bond** consists of four electrons, two pairs, shared between two atoms A and B.

The double bond, represented by two sticks, is a sigma plus **pi bond** ($\sigma + \pi$).

A **triple covalent bond** consists of six electrons, or three pairs, shared between two atoms A and B.

The triple bond, represented by three sticks, is a sigma plus 2 two pi bonds ($\sigma + 2\pi$).

A Lewis structure is a simple model showing how the valence (outer-shell) electrons are distributed in a molecule or polyatomic ion. In sub-topic 14.2 we shall look at a more sophisticated theory based on quantum mechanics called **molecular orbital theory** (MOT), which is helpful in visualizing the difference between a sigma bond and a pi bond.

For atomic orbitals to overlap and form molecular orbitals they must be relatively close in energy, and the symmetry of the atomic orbitals must be identical. X atomic orbitals combine to form x new molecular orbitals. There are three possible outcomes:

1 bonding orbital: sigma (σ) or pi (π) orbital

2 anti-bonding orbital: sigma star ($\sigma*$) or pi star ($\pi*$) orbital

3 non-bonding situation.

Table 2 shows a number of combinations of atomic orbitals.

$$A \overset{\sigma}{\rule{2cm}{0.4pt}} B$$

$$A : B$$

$$A \overset{\sigma + \pi}{=\!=\!=} B$$

$$A \overset{\bullet\bullet}{\underset{\bullet\bullet}{:}} B$$

$$A \overset{\sigma + 2\pi}{\equiv} B$$

$$A \overset{\bullet\bullet}{\underset{\bullet\bullet}{:}} B$$

▲ Figure 7 Covalent bonds

In order of bond length:

$A{-}B > A{=}B > A{\equiv}B$

In order of bond strength:

$A{\equiv}B > A{=}B > A{-}B$

Combination of atomic orbitals	Molecular orbitals formed	Type
s + s	(s + s orbital diagram)	σ bonding
	(s + s anti-bonding diagram)	$\sigma*$ anti-bonding
s + p_x	(s + p_x orbital diagram)	σ bonding
	(s + p_x anti-bonding diagram)	$\sigma*$ anti-bonding
s + p_y	(s + p_y diagram) NB	non-bonding
s + p_z	(s + p_z diagram) NB	non-bonding
p_x + p_x	(p_x + p_x orbital diagram)	σ bonding
	(p_x + p_x anti-bonding diagram)	$\sigma*$ anti-bonding

Combination of atomic orbitals	Molecular orbitals formed	Type
$p_y + p_y$		π bonding
		π^* anti-bonding
$p_z + p_z$		π bonding
		π^* anti-bonding
$p_x + p_y$	NB	non-bonding
$p_x + p_z$	NB	non-bonding
$p_y + p_z$	NB	non-bonding

▲ Table 2 Combination of atomic orbitals (LCAOs). Black represents the positive wavefunction, Ψ_+, and white represents the negative wavefunction, Ψ_-

Description of a sigma and pi bond

In the formation of a **sigma bond** there is a direct head-on overlap of the atomic orbitals along the internuclear axis and the electron density is located along this axis.

In the formation of a **pi bond** there is a sideways overlap of the atomic orbitals and the electron density is located above and below the internuclear axis.

Delocalization and resonance

Figure 8 shows two Lewis structures for the nitrite oxoanion, NO_2^- that have identical arrangements of atoms but different arrangements of electrons.

▲ Figure 8 Lewis structures of the nitrite oxoanion, NO_2^-

Each Lewis structure shows one N—O single bond and one N=O double bond, but the position of the double bond is different in the two structures. An ion with one of these structures would have

one shorter N=O bond (114 pm) and one longer N—O bond (136 pm). In fact the experimentally measured bond lengths are both 125 pm, intermediate between a single and double NO bond. Both Lewis structures, called **resonance forms**, contribute to the electronic structure which is called a **resonance hybrid**. This idea of **resonance** is shown by linking the contributing resonance forms by a double-headed arrow (figure 9).

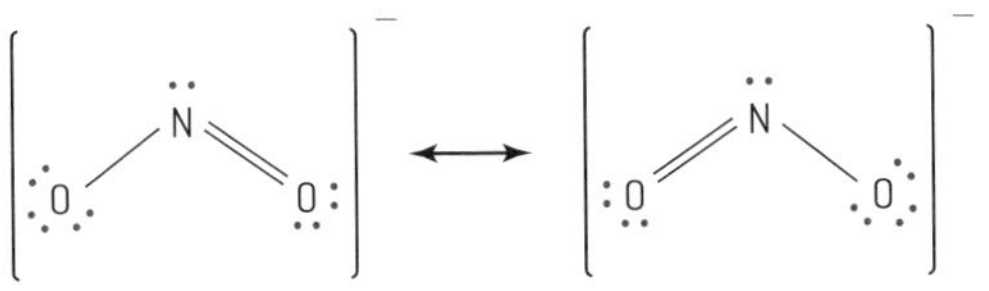

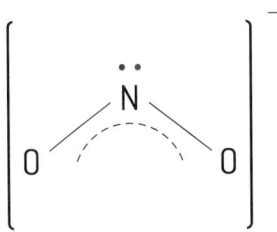

▲ Figure 9 Resonance in the nitrite oxoanion, NO_2^-

The resonance in the NO_2^- ion can also be represented as shown in figure 10. The dashed curve conveys **delocalization**. As specified by IUPAC, delocalization is a quantum mechanical concept used to describe the pi bonding in a conjugated system.

A **conjugated system** is a molecular entity whose structure can be represented as a system of alternating single and multiple bonds. Conjugation is the interaction of one p orbital with another across an intervening sigma bond. A conjugated system may also form the interaction of a double bond and a p orbital containing a lone pair of electrons. The bonding in a conjugated system is not localized between two atoms but instead each link has a "fractional double bond character" or **bond order**. In the case of the NO_2^- oxoanion the bond order of each N—O bond is 1.5, since both bonds are equivalent and intermediate between a single and a double bond:

$$\text{bond order in } NO_2^- = \frac{\text{total number of NO bonding pairs}}{\text{total number of NO positions}} = \frac{3}{2} = 1.5$$

Another oxoanion that has resonance structures is the carbonate oxoanion, CO_3^{2-}. Three resonance structures can be written for CO_3^{2-} (figure 11).

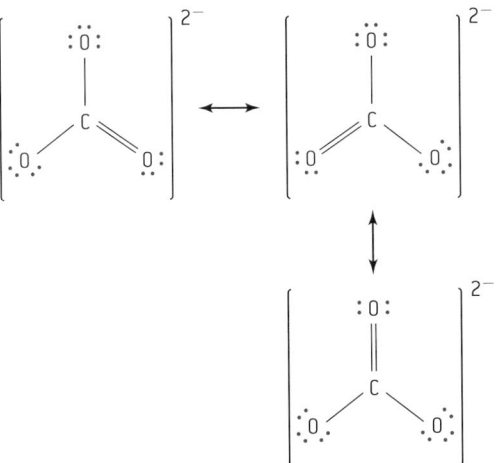

▲ Figure 11 Resonance in the carbonate oxoanion, CO_3^{2-}

In CO_3^{2-} the three C—O bond lengths are equivalent (129 pm), intermediate between a single C—O bond (143 pm) and a double C=O bond length (122 pm). The bond order is worked out as follows:

$$\text{bond order in } CO_3^{2-} = \frac{\text{total number of CO bonding pairs}}{\text{total number of CO positions}} = \frac{4}{3} = 1.33$$

Resonance

In **resonance**, two or more Lewis structures can represent a particular molecule or ion.

A **resonance structure** is one of two or more alternative Lewis structures for a molecule or ion that cannot be described fully with one Lewis structure alone.

Note that resonance structures are purely hypothetical species that do not actually exist.

▲ Figure 10 An alternative representation of resonance in the NO_2^- oxoanion

Delocalization

Delocalization involves electrons that are shared by more than two atoms in a molecule or ion as opposed to being localized between a pair of atoms.

Study tip

Covalent bond lengths (both single and multiple) are provided in section 10 of the *Data booklet*.

Worked examples

Example 1

Consider the following six species:

a) BrF_3 **b)** IF_5 **c)** $[ICl_2]^-$

d) SOF_4 **e)** $[ICl_4]^-$

For each species, deduce:

  (i)   the electron domain geometry

 (ii)   the molecular geometry

(iii)   the approximate bond angle(s)

(iv)   a valid Lewis (electron dot) structure.

Solution

a) Ball-and-stick diagram of the BrF_3 molecule (ignoring the bond angles):

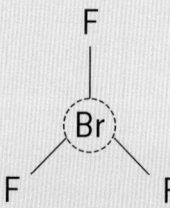

For Br:

number of valence electrons	= 7
number of sigma bonds	= 3
total number of valence electrons	= 10
number of electron domains	= 5

  (i)   From table 1, electron domain geometry: trigonal bipyramidal.

 (ii)   This is an example of an AB_3E_2 system, so the two lone pairs are located in the equatorial positions:

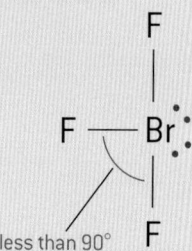

less than 90°

Molecular geometry: T-shaped.

(iii)   Bond angles: less than 90°. (The F—Br—F experimental value is 86.2°, but this cannot be predicted from VSEPR theory.)

(iv)   To draw the Lewis structure, complete the octets on fluorine atoms:

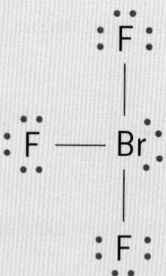

b) Ball-and-stick diagram of the IF_5 molecule (ignoring bond angles):

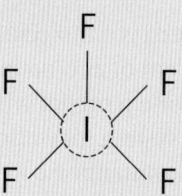

For I:

number of valence electrons	= 7
number of sigma bonds	= 5
total number of valence electrons	= 12
number of electron domains	= 6

  (i)   From table 1, electron domain geometry: octahedral.

 (ii)   This is an example of an AB_5E system, so the one lone pair is located in any of the six equivalent positions.

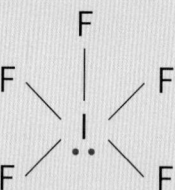

Molecular geometry: square-based pyramidal.

(iii)   Bond angles: six F–I–F bond angles are less than 90°. (Each one of these experimental F–I–F bond angles is 80.9°.)

(iv) To draw the Lewis structure, complete the octets on fluorine atoms:

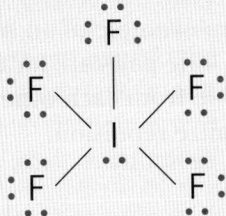

c) Ball-and-stick diagram of the $[ICl_2]^-$ anion (ignoring bond angles):

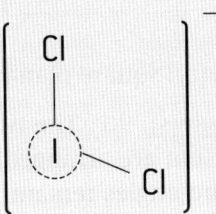

For I:

number of valence electrons	= 7
number of sigma bonds	= 2
−ve charge	= 1
total number of valence electrons	= 10
number of electron domains	= 5

(i) From table 1, electron domain geometry: trigonal bipyramidal.

(ii) This is an example of an AB_2E_3 system, so the three lone pairs are located in equatorial positions of the trigonal bipyramidal electron domain geometry:

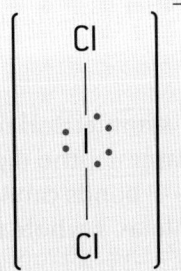

Molecular geometry: linear.

(iii) The bond angle remains 180°, as the lone pairs are arranged symmetrically.

(iv) To draw the Lewis structure, complete the octets on chlorine atoms:

$$\left[\begin{array}{c} :\overset{\displaystyle ..}{\underset{\displaystyle ..}{Cl}}: \\ | \\ :\overset{\displaystyle ..}{I}: \\ | \\ :\overset{\displaystyle ..}{\underset{\displaystyle ..}{Cl}}: \end{array} \right]^{-}$$

> **Study tip**
>
> When drawing the Lewis structure of an anion or cation, include square brackets and the respective charge.
>
> Don't forget to complete the octets on the terminal atoms in a Lewis structure, unless the terminal atom is hydrogen which has only the two electrons in a bonded pair, so attaining the noble gas configuration of helium.

d) Ball-and-stick diagram of the SOF_4 molecule (ignoring bond angles); note that the maximum valency of oxygen is 2 so oxygen cannot be the central atom. Sulfur, as a member of period 3, can expand its octet, so S must be the central atom in SOF_4.

For S:

number of valence electrons	= 6
number of sigma bonds	= 5
1 pi bond	= −1
total number of valence electrons	= 10
number of electron domains	= 5

> **Pi bonds in VSEPR theory**
>
> Pi bonding is off-axis bonding. Because the shape of a molecule is controlled by the sigma-bonding framework along the internuclear axis, when counting the valence electrons we subtract 1 for each pi bond present.

Atom	FC =	$\Delta FC = FC_{max} - FC_{min}$
Lewis structure A		
terminal N	$(5) - \frac{1}{2}(4) - (4) = -1$	
central N	$(5) - \frac{1}{2}(8) - (0) = +1$	$(+1) - (-1) = +2$
O	$(6) - \frac{1}{2}(4) - (4) = 0$	
Lewis structure B		
terminal N	$(5) - \frac{1}{2}(4) - (4) = -1$	$(+2) - (-1) = +3$
O	$(6) - \frac{1}{2}(8) - (0) = +2$	

▲ Table 3 Formal charges (FCs) for the dinitrogen oxide molecule, N_2O: Lewis structures A and B

Since ΔFC for A is less than ΔFC for B, the preferred Lewis structure will be A.

b) Table 4 shows the FCs for Lewis structures A, B, and C for the cyanate anion, OCN^-.

Atom	FC =	$\Delta FC = FC_{max} - FC_{min}$
Lewis structure A		
O	$(6) - \frac{1}{2}(6) - (2) = +1$	
C	$(4) - \frac{1}{2}(8) - (0) = 0$	$= (+1) - (-2) = +3$
N	$(5) - \frac{1}{2}(2) - (6) = -2$	
Lewis structure B		
O	$(6) - \frac{1}{2}(4) - (4) = 0$	
C	$(4) - \frac{1}{2}(8) - (0) = 0$	$(0) - (-1) = +1$
N	$(5) - \frac{1}{2}(4) - (4) = -1$	
Lewis structure C		
O	$(6) - \frac{1}{2}(2) - (6) = -1$	
C	$(4) - \frac{1}{2}(8) - (0) = 0$	$= (0) - (-1) = +1$
N	$(5) - \frac{1}{2}(6) - (2) = 0$	

▲ Table 4 Formal charges (FCs) for the cyanate anion, OCN^-: Lewis structures A, B, and C

Based on ΔFC the two preferred Lewis structures are B and C. However, O is the most electronegative atom (electronegativities: O = 3.4; N = 3.0; C = 2.6). Since both B and C have the same ΔFC, the second criterion is that the negative FC should reside on the more electronegative element, in this case oxygen. The preferred Lewis structure is therefore C.

Example 4

Explain why bromine trifluoride, BrF_3 has its lone pairs of electrons located in equatorial positions.

Solution

In example 1(a) the molecular geometry of BrF_3 was deduced to be T-shaped. The geometrical arrangement is based on an AB_3E_2 system. Below are three Lewis structures for BrF_3 with the lone pairs surrounding Br located in different positions:

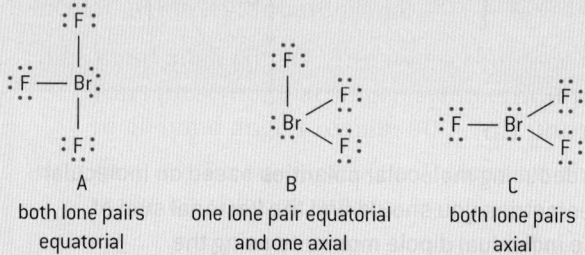

	A	B	C
	both lone pairs equatorial	one lone pair equatorial and one axial	both lone pairs axial

Table 5 shows the interactions for these three possible Lewis structures.

Interactions	Lewis structure A	Lewis structure B	Lewis structure C
Position of lone pairs	both equatorial	equatorial and axial	both axial
Angle between LP|LP	120°	90°	180°
Angle between LP|BP	4 × 90° 2 × 120°	3 × 90° 2 × 120°	6 × 90°
Angle between BP|BP	2 × 90°	1 × 120° 2 × 90°	3 × 120°

▲ Table 5 Interactions for three Lewis structures A, B, and C for bromine trifluoride, BrF_3. (LP = lone pair; BP = bonding pair)

Remember that:

$$LP|LP > LP|BP > BP|BP$$

The greatest bond angle for LP|LP interactions occurs in structures A and C, so we can discard Lewis structure B. Next considering LP|BP interaction: structure C gives 6 × 90° bond angles while structure A gives only 4 × 90° angles, so A will be the preferred structure, with the lone pairs both occupying equatorial positions.

🌐 An environmental perspective: Catalysis of ozone depletion

Ozone, O_3 is a V-shaped (bent) molecule with a bond angle of 116.8° and its two O−O bond lengths are equal (128 pm). Two contributing resonance forms can be written for ozone, as shown in figure 13.

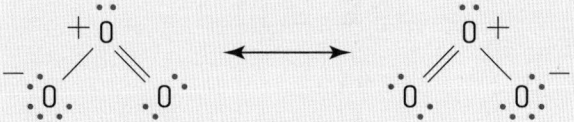

▲ Figure 13 Resonance forms of ozone, O_3

The bond order for the O−O bond in ozone is calculated as follows:

each bond order in O_3

$$= \frac{\text{total number of O–O bonding pairs}}{\text{total number of O–O positions}} = \frac{3}{2} = 1.5$$

The wavelength of light required to dissociate oxygen and ozone

In topic 6 we mentioned that in the stratosphere, the ozone layer absorbs over 95% of harmful UV radiation from the sun:

$$O_3(g) \xrightarrow{h\nu} O_2(g) + O\bullet(g)$$

$$O_2(g) + O\bullet(g) \rightarrow O_3(g) + \text{heat}$$

There is a net energy conversion from UV radiation to heat energy.

The progressive depletion of the ozone layer allows more UV radiation to reach the Earth's surface, resulting in an increased risk of skin cancers (melanomas) and cataracts.

The bonds in ozone can be broken by UV radiation ($h\nu$). The bond order in ozone (1.5) is lower than the bond order in oxygen (2), so the O=O double bond in oxygen is stronger. Radiation of shorter wavelength is required to break the stronger bond in oxygen. The reason is that the energy, E, of a photon of light is inversely proportional to the wavelength λ, so the greater the energy, the shorter the wavelength and vice versa.

$$E = h\nu = \frac{hc}{\lambda}$$

where:

h = Planck's constant = 6.63×10^{-34} J s

ν = frequency of the radiation

c = speed of light = 3.00×10^{-8} m s^{-1}

λ = wavelength of the radiation

Worked example

The average bond enthalpy in ozone is 362 kJ mol^{-1}. Using the relationships given in section 1 and the bond enthalpy data given in section 11 of the *Data booklet*, calculate the maximum wavelength, in nm, of the UV radiation required to break the O=O double bond in oxygen and the O−O bond in ozone.

Solution

Oxygen: The bond enthalpy for O=O is 498 kJ mol^{-1}. Next calculate the energy of photons in J. First convert kJ to J and then use Avogadro's constant, L:

$$E = \frac{498 \times 1000 \text{ J mol}^{-1}}{6.02 \times 10^{23} \text{ mol}^{-1}} = 8.27 \times 10^{-19} \text{ J}$$

$E = \frac{hc}{\lambda}$, so by cross-multiplication $E\lambda = hc$. We then make λ the subject of the equation:

$$\lambda = \frac{hc}{E} = \frac{(6.63 \times 10^{-34} \text{ J s})(3.00 \times 10^8 \text{ m s}^{-1})}{8.27 \times 10^{-19} \text{ J}}$$

$$= 2.41 \times 10^{-7} \text{ m}$$

1 nm = 10^{-9} m, so λ = (2.41×10^{-7} m) × (1 nm/10^{-9} m) = **241 nm**

Ozone: The bond enthalpy for the O−O bond in ozone is 362 kJ mol^{-1}.

$$E = \frac{362 \times 1000 \text{ J mol}^{-1}}{6.02 \times 10^{23} \text{ mol}^{-1}} = 6.01 \times 10^{-19} \text{ J}$$

$$\lambda = \frac{hc}{E} = \frac{(6.63 \times 10^{-34} \text{ J s})(3.00 \times 10^8 \text{ m s}^{-1})}{6.01 \times 10^{-19} \text{ J}}$$

$$= 3.31 \times 10^{-7} \text{ m}$$

λ = (3.31×10^{-7} m) × (1 nm/10^{-9} m) = **331 nm**

The depletion of the ozone layer was discussed in greater detail in topic 6.

Models and theories of structure and bonding

You come to nature with all her theories, and she knocks them all flat.

Pierre-Auguste Renoir (1841–1919), French impressionist artist.

Chemical bonding and structure are riddled with models and theories which can often be interpreted in specific ways to suit our requirements. Models should be used with caution in science. Let us take an example of a Lewis structure of an oxoanion and its associated molecular geometry to see what we mean by this.

It is the theory that decides what we can observe.

Albert Einstein (1879–1955), German theoretical physicist who the Nobel Prize in Physics in 1921 for his services to theoretical physics, especially for his discovery of the law of the photoelectric effect.

A question of approach: Expanded octets, octets, or formal charge considerations

In topic 4 we looked at one method of determining the shape of an oxoanion such as the sulfate oxoanion, SO_4^{2-}.

Using this method the molecular geometry of sulfate is correctly worked out as tetrahedral and can be represented by six contributing resonance structures. *Try to write these!* All S—O bonds are equivalent in length.

The final step involves completing the octets on the four terminal oxygen atoms, in order to write six valid Lewis structures for sulfate. Here are two of these as an example:

This model shows that S has an **expanded octet** of electrons: in each of the six Lewis structures, the central S atom is surrounded by 12 electrons.

Another approach is to simply draw a Lewis structure with the S having a complete octet of electrons:

Here although S obeys the octet rule, the valency of O shown is incorrect. However, as S is surrounded by four electron domains we arrive at the same geometry for the oxoanion, namely tetrahedral.

Both models result in each S—O bond being equivalent, intermediate between a double and a single S—O bond. The question then is which Lewis structure is the most valid.

If we turn now to formal charge considerations we get the results shown in table 6.

Table 6 shows that in both Lewis structures the more negative FC resides on the more electronegative oxygen atom, so this is not a factor here. ΔFC is closer to zero for the Lewis structure in which sulfur has an **expanded octet**, so this supports the suggestion that this is the preferred structure.

Atom	FC =	ΔFC = FC$_{max}$ − FC$_{min}$
Lewis structure: expanded octet		
S	$(6) - \frac{1}{2}(12) - (0) = 0$	
O—	$(6) - \frac{1}{2}(2) - (6) = -1$	$(0) - (-1) = +1$
O=	$(6) - \frac{1}{2}(4) - (4) = 0$	
Lewis structure: octet		
S	$(6) - \frac{1}{2}(8) - (0) = +2$	$(+2) - (-1) = +3$
O—	$(6) - \frac{1}{2}(2) - (6) = -1$	

▲ Table 6 Formal charges (FC) for the Lewis structures of the sulfate oxoanion, SO_4^{2-}

A similar analysis with the phosphate oxoanion, $[PO_4]^{3-}$ reveals that the Lewis structure in which phosphorus has an expanded octet is favourable based on FC calculations. However, theoretical quantum mechanical calculations suggest that the most favoured single Lewis structure is the one with the octet of electrons and many chemists argue that when choosing between alternative Lewis structures for oxoanions, it is better to opt for the Lewis structure in which the octet rule is obeyed rather than the expanded octet alternative.

If we go one step further and examine the molecular geometry of sulfuric acid, experimental evidence points to two shorter S=O bonds, supporting a Lewis structure for the sulfate oxoanion with an expanded octet:

As we said at the outset, approach models and theories with an open scientific mind!

14.2 Hybridization

Understandings

→ A hybrid orbital results from the mixing of different types of atomic orbitals on the same atom.

Applications and skills

→ Explanation of the formation of sp^3, sp^2, and sp hybrid orbitals in methane, ethene, and ethyne.
→ Identification and explanation of the relationships between Lewis (electron dot) structures, electron domains, molecular geometries, and types of hybridization.

 ## Nature of science

→ The need to regard theories as uncertain – hybridization in valence bond theory can help explain molecular geometries, but is limited.

Quantum mechanics involves several theories explaining the same phenomena, depending on specific requirements.

 ## Models, theories, assumptions, and deductions

What we see depends mainly on what we look for.

Sir John Lubbock (1834–1913), FRS, English biologist, banker and politician.

Each model in chemistry has its own merits and each theory has its own limitations. Sometimes the merits of a particular theory may depend on our perspective. We perhaps should treat models and theories in chemistry with the same degree of critical perspective as Samuel Beckett (figure 1). Theories should not be seen in black and white – as scientists we should also look for shades of grey and question every assumption, challenge theories with experimental evidence, and consider using a combination of models and theories to allow us to see the bigger picture. "Real science" becomes possible with this perspective!

In topic 4 we saw the value of a Lewis (electron dot) structure in providing a simple model which shows how valence or outer-shell electrons are distributed in a molecule. Sub-topics 4.3 and 14.1 showed how the number of electron domains can be deduced from a Lewis structure, enabling the

▲ Figure 1 The Irish playwright Samuel Beckett. John Calder, author of *The Theology of Samuel Beckett*, said to Beckett "It is a fine day," to which Beckett replied "So far!"

prediction of the electron-domain geometry and ultimately the molecular geometry using VSEPR theory. Neither of these models (Lewis and VSEPR theory) however offers an explanation of the detailed electronic structure of covalent bonds or even why chemical bonds exist. To fully explain the formation of chemical bonds we need to look to quantum mechanics. Physical phenomena can be explored on the microscopic scale, with models that explain the behaviour of sub-atomic particles. Schrödinger's wave equation forms the launch-pad for the field of **quantum mechanics**. It helps us understand chemical bonding and provides an understanding of the shapes of molecules.

Two theories in quantum mechanics that aid an understanding of molecular geometries are:

- **valence bond theory (VBT)**

- **molecular orbital theory (MOT).**

We shall discuss each one separately.

Valence bond theory (VBT)

The VBT model considers that atoms approach each other to form a molecule. The bond formed results from the overlap of atomic orbitals resulting in a bonding orbital with the electrons localized between the two atoms. In VBT it is assumed that when the atoms interact they retain their own respective atomic orbitals.

Figure 2 shows the interaction of two hydrogen atoms H_A and H_B, each having one electron, to form the diatomic molecule H_2. The two atoms approach each other and subsequently there is an interaction between their electrons and nuclei. If there is a decrease in the energy of the system due to the interaction, a chemical bond is formed.

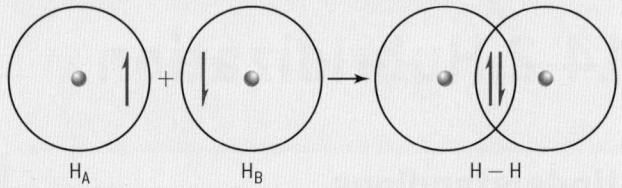

▲ Figure 2 Formation of the H_2 diatomic molecule based on VBT. The two electrons are localized between the two atoms and each atom involved in the bonding retains its own respective atomic orbital

We can represent this bonding model by a sketch of potential energy versus the distance, d, separating the two atomic nuclei (figure 3).

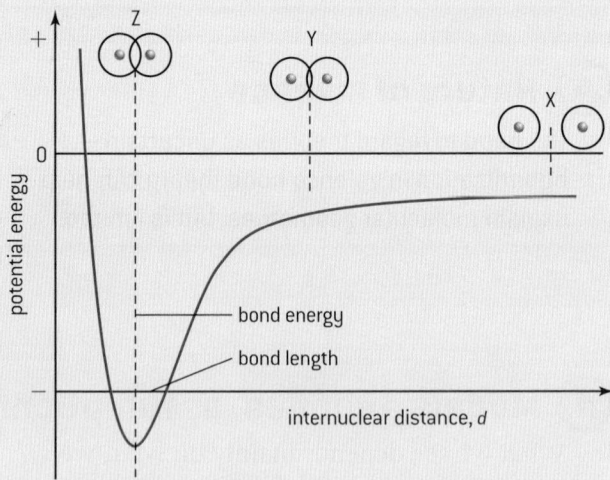

▲ Figure 3 Interaction energy diagram showing potential energy versus internuclear distance, d, for the H_2 molecule

On figure 3:

- At X the potential energy is essentially zero, as the hydrogen atoms are too far apart to interact.

- At Y the hydrogen atoms approach each other. The electron on atom A, e_A, is attracted to the nucleus on atom B, H_B. Simultaneously, e_A and e_B repel each other and nuclei H_A and H_B repel each other. There is a decrease in potential energy going from X to Y as the attraction is greater than the repulsion.

- At Z the minimum potential energy is achieved. This represents the most stable state of H_2. The H–H bond length is 74 pm (see section 10 of the *Data booklet*).

Question

1 Explain why the potential energy rises significantly to the left of point Z on figure 3.

Study tip

Remember from chapter 5 that bond breaking is an endothermic process (ΔH positive) and bond formation is an exothermic process (ΔH negative).

In VBT, the greater the degree of orbital overlap (interaction), the stronger the bond will be.

VBT can be applied to other homonuclear diatomic molecules such as F_2 and to heteronuclear diatomic molecules such as HCl or HF. One limitation of the Lewis structure model is that all covalent bonds are considered the same and so the model does not explain inherent differences between covalent bonds. VBT considers the changes in energy that occur on formation of the chemical bond.

VBT can also be applied to the electronic structure of polyatomic molecules. We shall shortly expand the model to **hybridization**, a concept used in VBT to explain the number of bonds that an atom can form and the spatial orientation of these bonds.

Before discussing hybridization note that VBT also has its limitations and has been superseded as a model by molecular orbital theory.

Molecular orbital theory (MOT)

VBT assumes that when bonds are formed from the overlap of atomic orbitals, the original nature of the atoms is retained. MOT involves atomic orbitals overlapping, but the overlap results in the formation of new orbitals called **molecular orbitals**. The electrons are assigned to these molecular orbitals, and associated with the whole molecule rather than individual atoms.

Figure 4 shows the formation of the molecular orbitals for the H_2 molecule. The two 1s atomic orbitals on the hydrogen atoms combine to form two new molecular orbitals. One combination, which is additive, results in a **bonding molecular orbital** sigma (σ), which is of lower energy, and the other combination, which is a difference, results in an **anti-bonding molecular orbital** sigma star (σ^*), which is of higher energy.

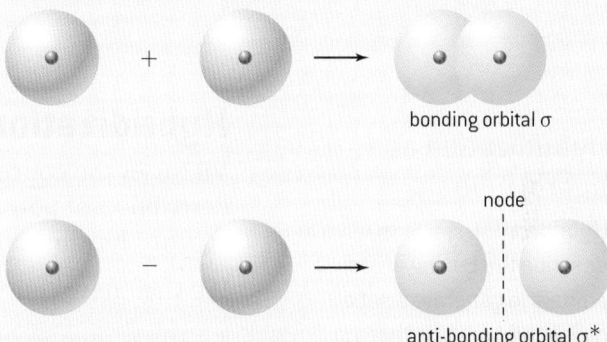

▲ Figure 4 Molecular orbital theory (MOT) model for the H_2 molecule. There is a build-up of electron density in the bonding molecular orbital, σ, between the two nuclei. If we think of a covalent bond as the "glue" that holds atoms in a molecule together, the electron density is the negative "glue" that holds the two positively charged nuclei together. In the anti-bonding molecular orbital, σ^*, there is a nodal plane (node) between the two nuclei, which means that there is zero electron density here

We can represent this model by a **molecular orbital diagram** as shown in figure 5.

For molecular orbitals to be formed two conditions must be met.

1 The atomic orbitals must be relatively close in energy for effective overlap.

2 The symmetry of the atomic orbitals, that is, the sign of the wavefunction Ψ, must be identical. For example, the signs of both Ψ could be positive (both Ψ_+), and sum to form the σ bonding molecular orbital. In contrast the subtractive combination would have one wavefunction positive, Ψ_+, and one wavefunction negative, Ψ_-, resulting in the σ^* anti-bonding molecular orbital.

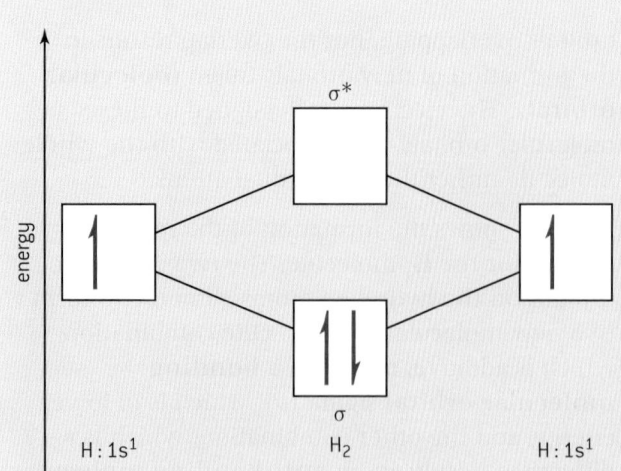

In MOT x atomic orbitals combine to form x new molecular orbitals; for example, two $1s^1$ atomic orbitals combine to form two new molecular orbitals, σ and σ^*.

Many theories have their advantages but equally each theory has its own constraints. Quantum mechanics provides several theories that can explain the same phenomena, depending on specific requirements.

▲ Figure 5 Molecular orbital diagram for the H_2 molecule

Hybridization and hybrid orbitals

Hybridization is a term used to describe the mixing of atomic orbitals to generate a set of new hybrid orbitals that are equivalent. Hybridization is a mathematical procedure.

A **hybrid orbital** results from the mixing of different types of atomic orbital on the same atom.

Hybridization

VBT uses the concept of hybridization to provide an electronic description of polyatomic molecules such as CH_4 and NH_3, but it can also account for the geometries of molecules.

We shall now consider three different types of hybridization:

1 sp^3 as seen in methane

2 sp^2 as seen in ethene

3 sp as seen in ethyne.

The hybridization type can be determined from the number of electron domains around the central atom.

TOK

Hybridization is a mathematical tool that allows us to relate the bonding in a molecule to its symmetry. What is the relationship between the natural sciences, mathematics, and the natural world?

Symmetry is key in several different areas of knowledge. In the field of mathematics, for example, there is great emphasis on the concept of symmetry. Early Greek mathematicians such as Pythagoras and Euclid used the idea of the symmetry of specific shapes and objects to develop their theories. The structure of DNA is an excellent example of the symmetry of the helix, and it was this symmetry that ultimately led Crick and Watson to unravel the structure of DNA.

Supersymmetry (SUSY) is an important theory met in particle physics. What is meant by supersymmetry, why is it discussed so much, and to which elementary particles does it relate? How are architects, musicians, artists, and dancers influenced by symmetry in their creativity?

The formation of sp³ hybrid orbitals in methane

Methane, CH_4 is a hydrocarbon and is one of the major components of natural gas. Let's consider the electron configuration of carbon:

full electron configuration: $\qquad 1s^22s^22p^2$

condensed electron configuration: $\qquad [He]2s^22p^2$

ground-state orbital diagram:

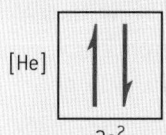
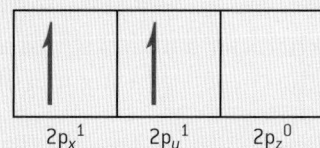

Hydrogen has the electron configuration $1s^1$. You might expect that since carbon has two unpaired electrons in its 2p sublevel, it would form two bonds with hydrogen in its ground-state, forming CH_2 rather than CH_4. Although CH_2 does exist it is extremely unstable.

The tetrahedral nature of methane involves the hybridization of its one 2s and three 2p orbitals on the C atom in an excited-state as follows.

Step 1

One of the electrons in the 2s orbital of the ground-state configuration of carbon is promoted to the vacant $2p_z$ orbital, to form an excited-state:

This arrangement gives four potential C—H bonds. However, we know from topic 4 that methane has a tetrahedral molecular geometry with H—C—H bond angles of 109.5°. So although this model for the excited-state accounts for the four C—H bonds, the geometry is incompatible because the three 2p orbitals are at 90° to each other (topic 2).

Step 2

The four atomic orbitals 2s, $2p_x$, $2p_y$, and $2p_z$ combine to form a set of four new hybrid orbitals. The four new sp³ hybrid orbitals are entirely equivalent and each one consists of 25% s character and 75% p character, since they were formed from one s orbital and three p orbitals.

The shape of each sp³ hybrid orbital will have 75% of the characteristics of a p orbital (dumbbell shaped) combined with 25% of the characteristics of the spherical s orbital. The shape of an sp³ hybrid orbital is shown in figure 6.

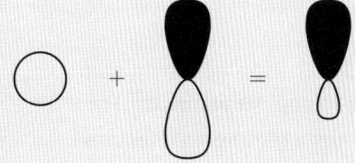

▲ Figure 6 One of four sp³ orbitals which each have 75% p character and 25% s character. This hybrid orbital is aligned in the z— direction

The four sp³ hybrid orbitals point to the corners of a tetrahedron (figure 7).

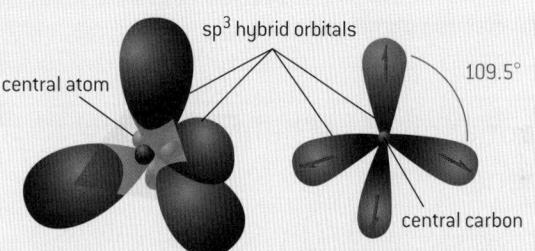

▲ Figure 7 (a) The tetrahedral orientation of the four sp³ hybrid orbitals on carbon. Each of these sp³ hybrid orbitals contains one electron. (b) For simplicity, the smaler of the two lobes is often omitted on diagrams

Step 3

The final step involves overlap of each carbon sp³ orbital with a hydrogen 1s atomic orbital. Hydrogen has a $1s^1$ electron configuration and the s orbital is spherically symmetrical. Each $1s^1$ atomic orbital on hydrogen and each sp³ hybrid orbital on the central carbon contains one electron (figure 8).

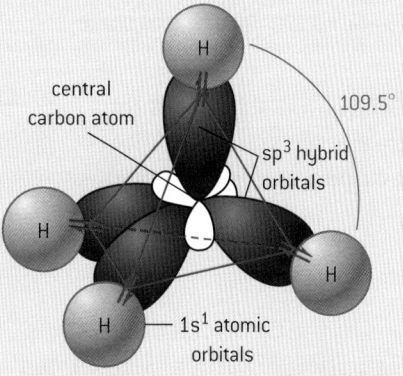

▲ Figure 8 In methane, each carbon sp³ orbital overlaps with a hydrogen 1s orbital to form a tetrahedral geometry

Any molecule with a tetrahedral electron domain geometry on its interior central atom (based on four electron domains) would be predicted to have sp³ hybridization.

Study tip

Hybridization is deduced from the electron domain geometry rather than the molecular geometry. For example, ammonia, NH_3 has a tetrahedral electron domain geometry, corresponding to four electron domains (three bonding pairs and one non-bonding or lone pair). Its molecular geometry is trigonal pyramidal (with a bond angle of 107° due to the repulsion between the lone pair and bonding pairs, which reduces the bond angle from 109.5°). The hybridization of ammonia is sp³ since it is based on the four electron domains.

The formation of sp² hybrid orbitals in ethene

Ethene, C_2H_4, is another hydrocarbon. Ethene is an alkene with one C=C double bond. Around each carbon there are three electron domains so the electron domain geometry (and molecular geometry) is trigonal planar. The H−C=C bond angle is 121.3° and the H−C−H bond angle is 117° (figure 9).

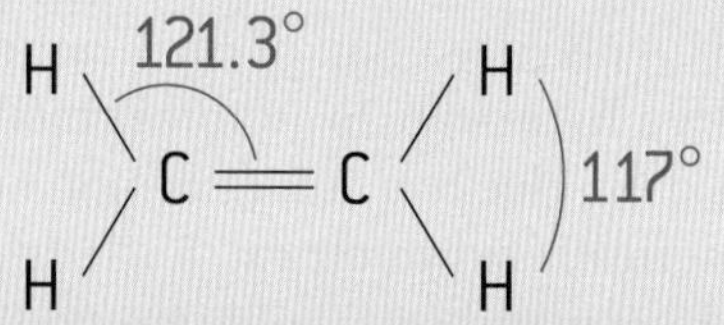

$$H \diagdown \overset{121.3°}{} \diagup H$$
$$C = C 117°$$
$$H \diagup \diagdown H$$

▲ Figure 9 The molecular geometry of ethene. The double bond occupies more space so the H−C−H bond angle is reduced from the predicted 120° for a trigonal planar geometry

To deduce the hybridization scheme of the carbon atom in ethene, we start with the orbital diagram of carbon as shown previously for methane.

Step 1

One of the electrons in the 2s orbital of the ground-state configuration of carbon is promoted to the vacant 2p$_z$ orbital to form an excited-state, as for methane above.

Step 2

To account for the approximate 120° bond angles (based on three electron domains), the next step involves hybridization of three of the atomic orbitals 2s, 2p$_x$ and 2p$_y$. These combine to form a set of three new sp² hybrid orbitals. The 2p$_z$ orbital remains unhybridized.

The three new sp² hybrid orbitals on the carbon atoms are entirely equivalent and each one has 33.3% s character and 66.7% p character. They point to the corners of a trigonal planar system (figure 10).

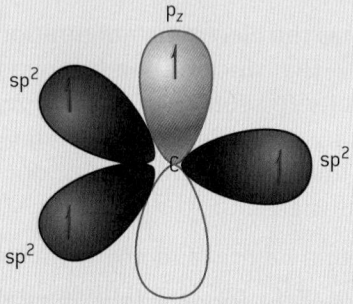

▲ Figure 10 Each of the three sp² hybrid orbitals contains one electron. The 2p$_z$ unhybridized atomic orbital also contains one electron. As before, for simplicity of representation, the smaller of the two lobes is omitted

Step 3

The next step involves the formation of a sigma bond along the internuclear axis by the overlap of two sp² hybrid orbitals, one on each carbon atom. A pi bond is formed from the sideways overlap of the two p$_z$ unhybridized atomic orbitals, with the overlap regions above and below the internuclear axis (figure 11).

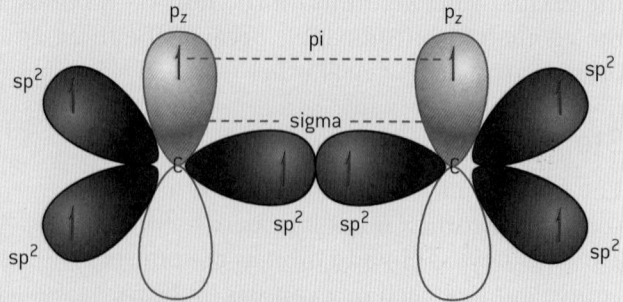

▲ Figure 11 The formation of the C=C double bond: a sigma bond forms from the overlap of two sp² orbitals, and a pi bond from the overlap of two p$_z$ orbitals

Step 4

The final step involves overlap of each remaining sp^2 orbital on carbon with a hydrogen 1s atomic orbital (figure 12).

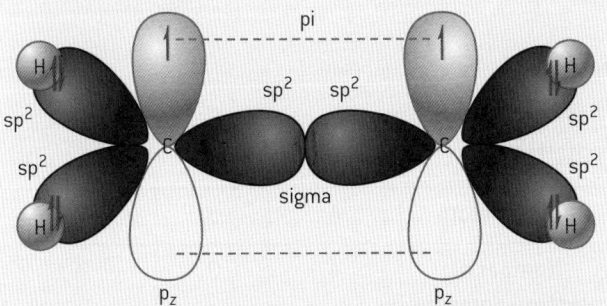

▲ Figure 12 In ethene, four carbon sp^2 orbitals overlap with four hydrogen 1s orbitals to form a trigonal planar geometry on each carbon atom

Any molecule with a trigonal-planar electron domain geometry will be predicted to have sp^2 hybridization.

The formation of sp hybrid orbitals in ethyne

Ethyne, C_2H_2, is an alkyne and has one C≡C triple bond. As seen in sub-topic 14.1, around each carbon atom there are two electron domains so the electron domain geometry (and molecular geometry) is linear, with a 180° bond angle (figure 13).

$$\overset{180°}{H - C \equiv C - H}$$

▲ Figure 13 The molecule of ethyne is linear

To deduce the hybridization scheme of the carbon atom in ethyne, again we start with the orbital diagram of carbon as shown for methane.

Step 1

One of the electrons in the 2s orbital of the ground-state configuration of carbon is promoted to the vacant $2p_z$ orbital to form an excited-state, as for methane.

Step 2

To account for the 180° bond angle (based on two electron domains), the next step involves hybridization of two of the atomic orbitals 2s

and $2p_x$. These combine to form a set of two new sp hybrid orbitals. The remaining $2p_y$ and $2p_z$ orbitals remain unhybridized.

The two new sp hybrid orbitals on the carbon atoms are entirely equivalent and each one consists of 50% s character and 50% p character. They point in opposite directions (figure 14). Each of these sp hybrid orbitals contains one electron. The remaining orbitals are the $2p_y$ and $2p_z$ unhybridized atomic orbitals, also containing one electron each.

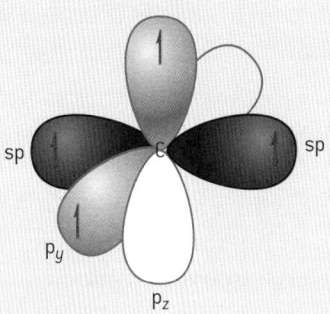

▲ Figure 14 Each of the two sp hybrid orbitals contains one electron. The $2p_y$ and $2p_z$ unhybridized atomic orbitals also contain one electron each

Step 3

The next step involves the formation of a sigma bond along the internuclear axis by the overlap of two sp hybrid orbitals, one on each carbon atom. Two pi bonds are formed from the sideways overlap of the two p_y and two p_z unhybridized atomic orbitals, with the overlap regions above and below the internuclear axis (figure 15).

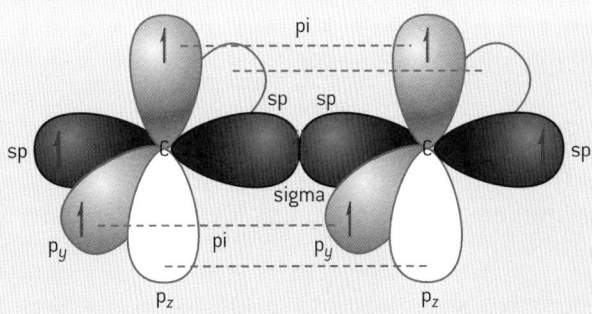

▲ Figure 15 The formation of the C≡C triple bond in ethyne: a sigma bond forms from the overlap of two sp orbitals, a pi bond from the overlap of two p_y orbitals, and another pi bond from the overlap of two p_z orbitals

Step 4

The final step involves overlap of each remaining sp orbital on carbon with a hydrogen 1s atomic orbital (figure 16).

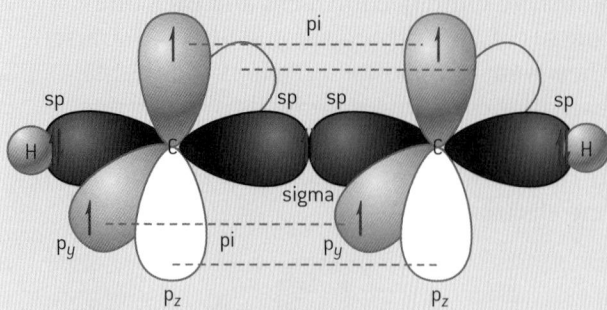

▲ Figure 16 In ethyne, two carbon sp orbitals overlap with two hydrogen 1s orbitals to create a linear geometry on each carbon atom

Any molecule with a linear electron domain geometry on its interior central atom would be predicted to have sp hybridization.

Summary

Number of electron domains	Electron domain geometry	Hybridization
4	tetrahedral	sp^3
3	trigonal planar	sp^2
2	linear	sp

▲ Table 1 Electron-domain geometry and hybridization

Worked examples

Example 1

Deduce the hybridization of the central nitrogen interior atom in ammonia.

Solution

As seen in topic 4, because there are four electron domains in ammonia, the hybridization must be sp^3.

Example 2

For each interior atom A, B, and C in a molecule of methyl propanoate (figure 17), deduce the electron domain geometry, the molecular geometry, the bond angles, and the hybridization state.

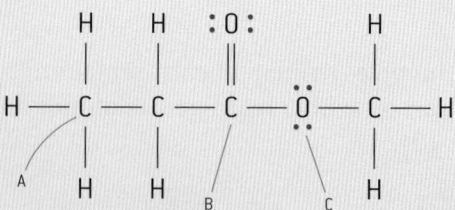

▲ Figure 17 Methyl propanoate

Solution

Interior atom	A	B	C
Number of electron domains	4	3	4
Electron domain geometry	tetrahedral	trigonal planar	tetrahedral
Molecular geometry	tetrahedral	trigonal planar	V-shaped (bent)
Bond angle(s)	H—C—H: 109.5°	C—C—O: 120°	C—O—C: less than 109.5°
Hybridization	sp^3	sp^2	sp^3

▲ Table 2

⊘ Example 3

The condensed structural formula of phenylamine (traditional name aniline) is $C_6H_5NH_2$.

a) Using VSEPR theory, deduce the electron domain and molecular geometries of the carbon and nitrogen atoms in phenylamine.

b) A model of the molecule is shown in figure 18.

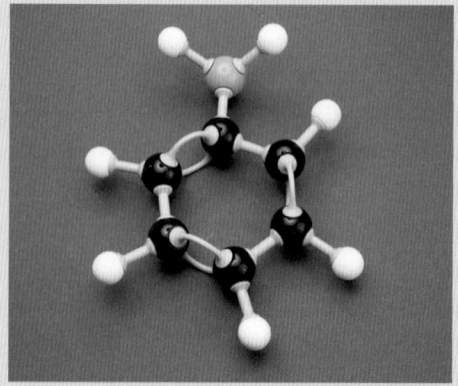

▲ Figure 18 Molecular model of phenylamine

In this model the two hydrogen atoms attached to nitrogen appear to be above the horizontal plane of the molecule.

Figure 19 shows a space-filling model of phenylamine, which is the basis of an electrostatic potential map showing electron charge density.

▲ Figure 19 Space-filling model of phenylamine

Deduce the hybridization of the nitrogen from figure 18 and from figure 19 and comment on the two models.

c) A theoretical study of the electronic structure of phenylamine found the H−N−H bond angle in phenylamine to be 112.79°, which is very close to the experimental value from gas-phase microwave studies. Discuss what you may conclude about the molecular geometry around the nitrogen in the −NH$_2$ group in the structure of phenylamine, and deduce its hybridization state on this basis.

Solution

a) Using VSEPR theory to consider the geometry at carbon, start by drawing the ball-and-stick diagram (ignoring bond angles):

$$C = C - C$$
$$|$$
$$H$$

For C:

number of valence electrons	= 4
number of sigma bonds	= 3
one pi bond	= −1
total number of valence electrons	= 6
number of electron domains	= 3

electron domain geometry: trigonal planar

molecular geometry: trigonal planar

bond angle: 120°

predicted hybridization based on 3 electron domains: sp^2

Ball-and-stick diagram for nitrogen (ignoring bond angles):

$$C = N - H$$
$$|$$
$$H$$

For N:

number of valence electrons	= 5
number of bonds	= 3
total number of valence electrons	= 8
number of electron domains	= 4

electron domain geometry: tetrahedral

predicted molecular geometry: trigonal pyramidal

predicted H−N−H bond angle: less than 109.5°

predicted hybridization based on 4 electron domains: sp^3

See figure 20.

▲ Figure 20 VSEPR theory suggests sp^3 hybridization at nitrogen in phenylamine

b) The model in figure 18 shows a trigonal pyramidal geometry, suggesting sp^3 hybridization at nitrogen as outlined above, while that in figure 19 shows a trigonal planar geometry around nitrogen with a 120° bond angle, suggesting sp^2 hybridization.

Although VSEPR theory is a useful model for predicting structure, in some cases it does not agree with the geometry found experimentally; in such cases it cannot be used to accurately deduce the hybridization.

In phenylamine the $-NH_2$ functional group is slightly flattened resulting in a trigonal planar-type geometry about N, due to resonance interactions with the aromatic ring. The p orbital on N interacts with the pi system on the aromatic ring resulting in delocalization of the non-bonding electron pair (lone pair). The most stable Lewis structure of phenylamine is shown in figure 21 along with three dipolar resonance structures.

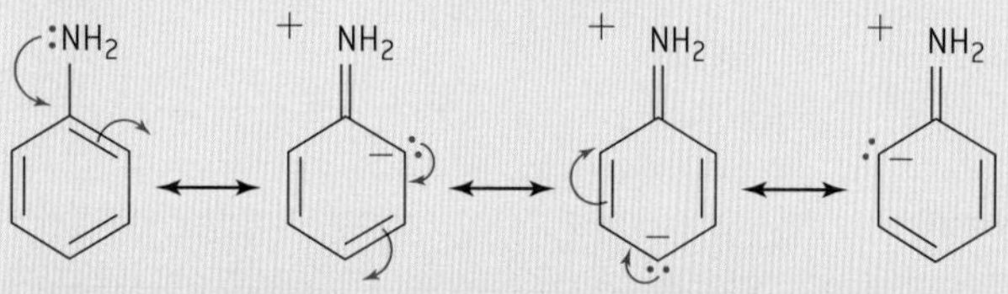

▲ Figure 21 Lewis structures for phenylamine

Delocalization of the non-bonding pair on N reduces the electron density in the p orbital

on N and increases the electron density of the pi system on the aromatic ring. A more planar geometry is therefore adopted around the NH_2 group. We can work out the electron domain geometry around N in one of these three dipolar resonance structures:

number of valence electrons	= 5
number of sigma bonds	= 3
1 pi bond	= -1
1 +ve charge	= -1
total number of valence electrons	= 6
number of electron domains	= 3

electron domain geometry: trigonal planar

molecular geometry: trigonal planar

This is consistent with the planar geometry from the space-filling model, suggesting sp^2 hybridization.

c) The experimental H−N−H bond angle of 113° suggests a geometry somewhere between trigonal pyramidal (109.5° bond angle) and trigonal planar (120°), and a hybridization state somewhere between sp^2 and sp^3.

Useful resource

The "ChemEd DL" website contains numerous digital resources. It is a collaboration with the *Journal of Chemical Education*, the education division of the American Chemical Society, and the ChemCollective Project, initiated through the National Science Foundation in the USA. It has an excellent library of three-dimensional models.

Questions

1 What are the formal charges on P and O in the Lewis (electron dot) structure of the phosphate oxoanion represented in figure 22?

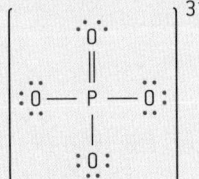

▲ Figure 22

A. P is 1− and O is 0

B. P is +5 and O is −2

C. P is 0 and O is 0

D. Both are −3

2 What are the formal charges on the carbon and oxygen atoms and the formal charge difference, ΔFC, in the Lewis (electron dot) structure of carbon dioxide shown in figure 23?

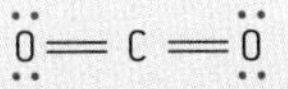

$$\ddot{O} = C = \ddot{O}$$

▲ Figure 23

	FC(C)	FC(O)	ΔFC
A.	+4	−2	+6
B.	+4	+4	0
C.	+4	+2	+2
D.	0	0	0

3 What is the electron domain geometry of the sulfite oxoanion, $[SO_3]^{2-}$?

A. Trigonal planar

B. Trigonal pyramidal

C. Tetrahedral

D. V-shaped (bent)

4 What is the molecular geometry of BrF_5?

A. Octahedral

B. Square planar

C. T-shaped

D. Square pyramidal

5 What is the molecular geometry of $[PF_6]^-$?

A. Trigonal planar

B. Trigonal bipyramidal

C. Square pyramidal

D. Octahedral

6 Which of the following molecules is non-polar?

A. SF_4

B. ClF_3

C. $BrCl_5$

D. SeF_6

7 Which of the following combinations of atomic orbitals shown in figure 24 results in a sigma bond?

I.

II.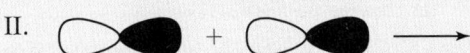

III.

▲ Figure 24

A. I and II only

B. I and III only

C. II and III only

D. I, II, and III

8 What is the hybridization of the oxygen atom in ethanol?

A. sp

B. sp^2

C. sp^3

D. dsp^3

9 What is the hybridization of the carbon atom in methanal?

A. sp

B. sp^2

C. sp^3

D. dsp^3

10 How many sigma and pi bonds are present in a molecule of propyne, H_3CCCH?

	sigma	pi
A.	5	3
B.	6	2
C.	7	1
D.	8	0

[1]

IB, May 2011

11 Consider the following species:

CCl_4 NH_3 CS_2

For each species,

a) Deduce:

   i)   its electron-domain geometry

   ii) its molecular geometry

   iii) bond angle

   iv) the hybridization state of the central atom

   v)   its molecular polarity.

b) Draw an appropriate Lewis (electron dot) structure and calculate the formal charge on each atom.

12 Consider the following species:

$[NO_3]^-$ $[SiF_6]^-$ $[IF_4]^+$ SCl_4

For each species:

a) Deduce:

   i)   its electron domain geometry

   ii) its molecular geometry

   iii) the bond angle(s).

b) Draw an appropriate Lewis (electron dot) structure and calculate the formal charge on each atom.

13 The following reactions take place in the ozone layer by the absorption of ultraviolet light.

   I $O_3 \rightarrow O_2 + O\bullet$

   II $O_2 \rightarrow O\bullet + O\bullet$

State and explain, by reference to the bonding, which of the reactions, **I** or **II**, requires a shorter wavelength. [2]

IB, May 2011

Introduction

In sub-topic 4.1 we examined how ionic compounds form and their three-dimensional lattice structures. This topic begins with energy cycles, derived from Hess's Law (topic 5) for the calculation of energy values that cannot be determined directly from empirical evidence. While some reactions go to completion, others reach equilibrium between reactants and products, as discussed in topic 7. Entropy, associated with molecular randomness or disorder, can be considered as the driving force behind physical and chemical changes and Gibbs free energy enables chemists to assess the spontaneity of the reaction.

15.1 Energy cycles

Understandings

→ Representative equations (eg $M^+(g) \rightarrow M^+(aq)$) can be used for enthalpy/energy of hydration, ionization, atomization, electron affinity, lattice, covalent bond, and solution.

→ Enthalpy of solution hydration enthalpy and lattice enthalpy are related in an energy cycle.

Nature of science

→ Making quantitative measurements with replicates to ensure reliability – energy cycles allow for the calculation of values that cannot be determined directly.

Applications and skills

→ Construction of Born–Haber cycles for group 1 and 2 oxides and chlorides.

→ Construction of energy cycles from hydration, lattice, and solution enthalpy. For example, dissolution of solid NaOH or NH_4Cl in water.

→ Calculation of enthalpy changes from Born–Haber or dissolution energy cycles.

→ Relate size and charge of ions to lattice and hydration enthalpies.

→ Perform laboratory experiments which could include single replacement reactions in aqueous solutions.

Models for finding enthalpy changes of reaction

We summarize many chemical and biological processes in a single chemical equation representing one pathway from reactants to products. These processes in reality often take place involving several reactions with alternative pathways. We can use these alternative pathways to determine changes in enthalpy values that cannot be measured directly.

In doing this we develop models that represent the energy changes taking place within a system. We can assess the extent to which these models are in agreement with empirical data by focusing on bonds in reactants that are broken and bonds that are formed to make products. Using empirical data to confirm or modify proposed models is a central methodology in science.

The Born–Haber cycle and enthalpy of formation

The **standard enthalpy of formation** of an ionic compound can be represented by a single equation:

$$Na(s) + \tfrac{1}{2}Cl_2(g) \rightarrow NaCl(s) \quad \Delta H_f^\ominus = -411 \text{ kJ mol}^{-1}$$

An application of Hess's law (sub-topic 5.2), the **Born–Haber cycle**, is a series of reactions that can be combined to determine the enthalpy of formation of an ionic compound.

Constructing a Born–Haber cycle

The Born–Haber cycle combines the enthalpy changes associated with several steps in the formation of an ionic compound, defined below.

Lattice enthalpy

The **lattice enthalpy** is defined as the standard enthalpy change that occurs on the formation of 1 mol of gaseous ions from the solid lattice:

$$MX(s) \rightarrow M^+(g) + X^-(g) \quad \Delta H_{lat}^\ominus > 0$$

The process is endothermic. Experimental values of lattice enthalpy at 298K can be found in section 18 of the *Data booklet*.

Enthalpy of atomization

The enthalpy of atomization $\Delta H_{at}^\ominus$ is the standard enthalpy change that occurs on the formation of 1 mol of separate gaseous atoms of an element in its standard state:

$$M(s) \rightarrow M(g) \qquad \Delta H_{at}^\ominus > 0$$
$$\tfrac{1}{2}X_2(g) \rightarrow X(g) \qquad \Delta H_{at}^\ominus > 0$$

Ionization energy

As introduced in topic 3, the **ionization energy**, ΔH_{IE}, is the standard enthalpy change that occurs on the removal of 1 mol of electrons from 1 mol of atoms or positively charged ions in the gaseous phase. For metal ions with multiple valence electrons the first, second, and sometimes third ionization energies are defined.

IE_1: $M(g) \rightarrow M^+(g) + e^-$ $\Delta H^{\ominus}_{IE_1} > 0$

IE_2: $M^+(g) \rightarrow M^{2+}(g) + e^-$ $\Delta H^{\ominus}_{IE_2} > 0$

Electron affinity

The **electron affinity**, $\Delta H^{\ominus}_{EA}$, is the standard enthalpy change on the addition of 1 mol of electrons to 1 mol of atoms in the gaseous phase:

$X(g) + e^- \rightarrow X^-(g)$ $\Delta H^{\ominus}_{EA} < 0$

As discussed in topic 3, electron affinity is typically negative, but there are exceptions, such as the electron affinity for helium.

Constructing the Born–Haber cycle

The lattice enthalpy, the enthalpy of atomization, the ionization energy, and the electron affinity are combined to construct the Born–Haber cycle and find the enthalpy of formation of an ionic compound. The standard Born–Haber cycle (figure 1) focuses on the processes involved and the relationships between the individual steps rather than the magnitude of each energy change.

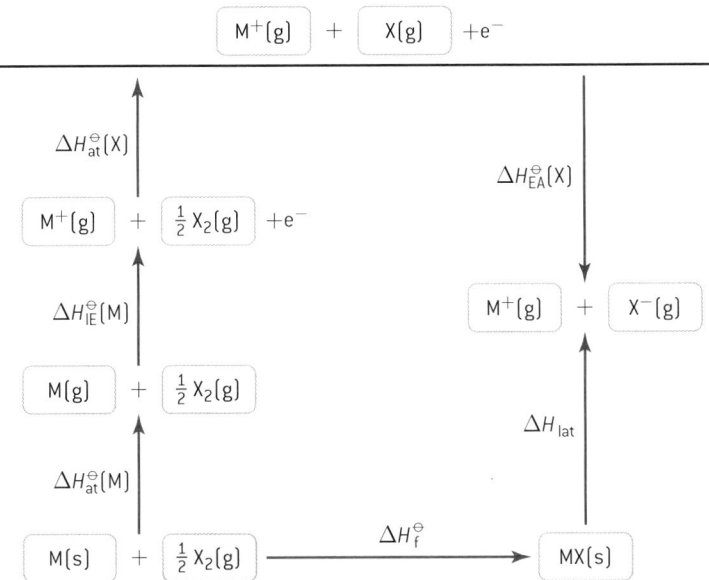

▲ Figure 1 A generalized Born–Haber cycle

Study tip

Values for lattice enthalpies can be found in the *Data booklet* (section 18), along with enthalpies of aqueous solutions (section 19), and enthalpies of hydration (section 20) which will be used later in this topic. The *Data booklet* will be available during the examination, except in Paper 1.

Enthalpy changes in solution

Many reactions studied in chemistry take place in solution. It is useful therefore to consider the **enthalpy of solution** and the relationship between the enthalpy change of solution, the hydration enthalpy, and the lattice enthalpy.

> Enthalpy changes of solution can have either positive or negative values.

The **standard enthalpy change of solution**, $\Delta H^{\ominus}_{sol}$, is the change in enthalpy when 1 mol of a substance is dissolved in a large excess of a pure solvent:

$$NH_4Cl(s) \rightarrow NH_4^+(aq) + Cl^-(aq) \quad \Delta H^{\ominus}_{sol} = +14.78 \text{ kJ mol}^{-1}$$

$$LiBr(s) \rightarrow Li^+(aq) + Br^-(aq) \quad \Delta H^{\ominus}_{sol} = -48.83 \text{ kJ mol}^{-1}$$

It is possible to calculate the enthalpy change of solution empirically, or by using an energy cycle that involves the lattice enthalpy of the ionic solid and the subsequent hydration enthalpy of the gaseous ions produced.

> The enthalpy change of hydration always has a negative value.

The **enthalpy change of hydration**, $\Delta H^{\ominus}_{hyd}$, for an ion is the enthalpy change when 1 mol of the gaseous ion is added to water to form a dilute solution. The term **solvation** is used in place of hydration for solvents other than water.

$$M^+(g) \rightarrow M^+(aq) \quad \Delta H^{\ominus}_{hyd} = - \text{ kJ mol}^{-1}$$

$$X^-(g) \rightarrow X^-(aq) \quad \Delta H^{\ominus}_{hyd} = - \text{ kJ mol}^{-1}$$

Worked example

Find the enthalpy change of solution ΔH_{sol} for sodium hydroxide using the enthalpy cycle in figure 5.

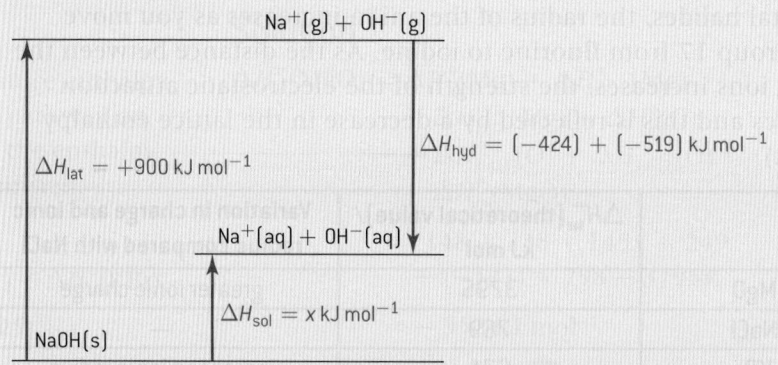

▲ Figure 5 Enthalpy cycle to calculate the enthalpy change of solution for sodium hydroxide

Solution

$$\Delta H_{sol} = \Delta H_{lat}(NaOH) + \Delta H_{hyd}(Na^+) + \Delta H_{hyd}(OH^-)$$

$$= 900 + (-424) + (-519) \text{ kJ mol}^{-1}$$

$$= -43 \text{ kJ mol}^{-1}$$

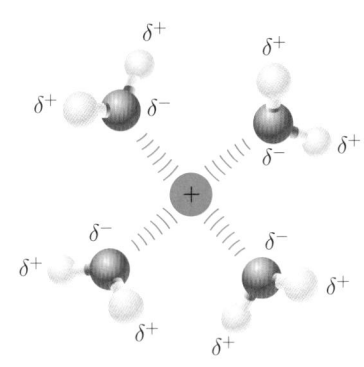

Solvation, dissolution, and hydration

Three terms commonly used when describing the interactions between solvents and solutes and the subsequent solutions formed, are described briefly here:

Solvation is described by the IUPAC *Gold Book* as "any stabilizing interaction of a solute and the solvent or a similar interaction of solvent with groups of an insoluble material. Such interactions generally involve electrostatic forces and van der Waals' forces, as well as chemically more specific effects such as hydrogen bond formation."

Water is a polar solvent. The difference in electronegativity between oxygen and hydrogen, combined with the geometry of the water molecule (bent) due to the repulsive forces between the lone pairs of electrons on the oxygen atom, result in this polar molecule having partial negative charges on the oxygen atom and partial positive charges on the hydrogen atoms. Water molecules orientate themselves so that their partial charges surround cations and anions, forming a **solvation shell**, also known as a **hydration shell** when the solvent is water (figure 6). When solid sodium hydroxide is mixed with liquid water, a new homogeneous phase is formed known as the solution. This is the process of **dissolution**.

The enthalpy of hydration is a way of quantifying the amount of energy released during the process of solvation. The magnitude of the enthalpy of hydration is influenced by the charge and size of the ion (table 3).

▲ Figure 6 Solvation shell: the water (solvent) molecules surround a positively charged sodium ion

Cation	ΔH_{hyd}/kJ mol^{-1}	Anion	ΔH_{hyd}/kJ mol^{-1}
Li$^+$	−538	F$^-$	−504
Na$^+$	−424	Cl$^-$	−359
Mg^{2+}	−1963	Br$^-$	−328
Al^{3+}	−4741	I$^-$	−287

▲ Table 3 Enthalpies of hydration (more data is available in section 20 of the *Data booklet*)

As you move down a group in the periodic table the enthalpy of hydration decreases as the ionic radius increases. Lithium has the greatest hydration enthalpy in group 1 while fluorine has the highest value in group 17. For cations, an increase in charge on the ion combined with a decrease in size results in a significantly larger enthalpy of hydration.

Quick question

Calculate the enthalpy change of solution of barium chloride given the following data:

$BaCl_2(s) \rightarrow Ba^{2+}(g) + 2Cl^-(g)$ $\Delta H_{lat} = +2069$ kJ mol^{-1}

$Ba^{2+}(g) \rightarrow Ba^{2+}(aq)$ $\Delta H_{hyd} = -1346$ kJ mol^{-1}

$Cl^-(g) \rightarrow Cl^-(aq)$ $\Delta H_{hyd} = -359$ kJ mol^{-1}

15.2 Entropy and spontaneity

Understandings

→ Entropy (S) refers to the distribution of available energy among the particles. The more ways the energy can be distributed the higher the entropy.

→ Gibbs free energy (G) relates the energy that can be obtained from a chemical reaction to the change in enthalpy (ΔH), change in entropy (ΔS), and absolute temperature (T).

→ Entropy of gas > liquid > solid under the same conditions.

Applications and skills

→ Prediction of whether a change will result in an increase or decrease in entropy, by considering the states of the reactants and products.

→ Calculation of entropy changes (ΔS) from standard entropy values ($S^{\ominus}$).

→ Application of $\Delta G^{\ominus} = \Delta H^{\ominus} - T\Delta S^{\ominus}$ in predicting spontaneity and calculation of various conditions of enthalpy and temperature that will affect this.

→ Relation of ΔG to position of equilibrium.

Nature of science

→ Theories can be superseded – the idea of entropy has evolved through the years as a result of developments in statistics and probability.

Spontaneous changes

Chemists work to understand the conditions under which chemical reactions will proceed, so that they can modify and control chemical systems to achieve the desired outcomes.

A reaction is said to be **spontaneous** when it moves towards either completion or equilibrium under a given set of conditions without external intervention. Reactions that are spontaneous can occur at different rates and may be either endothermic or exothermic. Reactions that do not take place under a given set of conditions are said to be **non-spontaneous**.

The enthalpy change of a reaction, whether positive or negative, is just one aspect to be considered when examining the spontaneity of a reaction. Exothermic reactions are usually spontaneous but there are many exceptions to this rule.

The **first and second laws of thermodynamics** are of fundamental importance in practical applications of chemistry. The first law, the law of conservation of energy, concerns energy in the physical world. The second law of thermodynamics focuses on entropy and the spontaneity of chemical reactions. **Entropy** (S) is a measure of the distribution of total available energy between the particles. The greater the shift from energy being localized to being widespread amongst the particles, the lower the chance of the particles returning to their original state and the higher the entropy of the system. Spontaneous reactions lead to an increase in the total entropy within the system and surroundings. If we can gain an understanding of this freedom of movement and so quantify the total entropy change for a system, this will allow us to predict the direction of the reaction.

Changes in entropy

Figure 1 shows condensation on the outside of a glass containing iced water. The temperature difference that exists between the system (iced water and the glass) and the surroundings (everything outside the system) results in thermal energy being transferred from the surrounding atmosphere to the glass and its contents, until they reach an equilibrium. With this thermal energy transfer, the entropy of the water/ice mixture will increase while the entropy of the surroundings will decrease as energy is transferred from it. The condensed water on the surface of the glass is lower in entropy than the water vapour in the atmosphere.

▲ Figure 1 Changes in entropy are associated with every chemical and physical process

Predicting changes in entropy

Simple representations of particles in the different states of matter show an increasing entropy as the particles gain more freedom of movement and more ways of distributing the energy as the particles move from solids through liquids to gases.

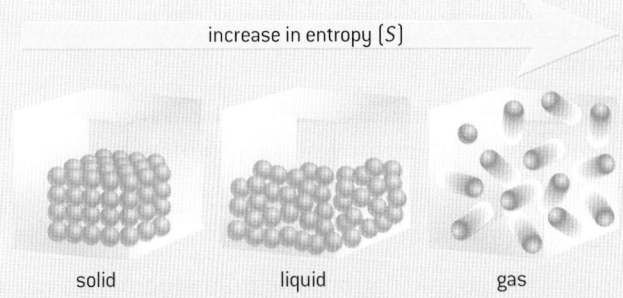

increase in entropy (S)

solid   liquid   gas

▲ Figure 2 Entropy increases from the solid through to the liquid to the gaseous phase

Achieving a change of state from solid to liquid to gas is sometimes described in terms of energy being absorbed which results in the kinetic energy of the particles increasing. In terms of entropy we can say

that under the same conditions, the entropy of a gas is greater than that of a liquid which in turn is greater than that of a solid.

Entropy, S is a state function, so a change in entropy is determined by the difference between its final and initial values:

$$\Delta S^{\ominus}_{298}(\text{reaction}) = \sum \Delta S^{\ominus}_{298}(\text{products}) - \sum \Delta S^{\ominus}_{298}(\text{reactants})$$

The conditions must be specified for a particular entropy change, and the subscript "298" refers to a temperature of 298 K.

The **second law of thermodynamics** says that chemical reactions that result in an overall increase in the entropy of the universe are spontaneous. When the overall entropy of the universe remains unchanged, the system is in equilibrium. If the overall entropy of the universe is found to be negative, rather than describing a reaction that is non-spontaneous, this describes a reaction that is spontaneous in the opposite direction to the way in which it is written (table 1).

$$\Delta S_{\text{total}} = \Delta S_{\text{system}} + \Delta S_{\text{surroundings}}$$

$\Delta S_{\text{total}} > 0$	spontaneous
$\Delta S_{\text{total}} = 0$	equilibrium
$\Delta S_{\text{total}} < 0$	non-spontaneous

▲ Table 1 The second law of thermodynamics allows us to predict the direction of a reaction

An increase in heat energy (enthalpy) within the system will result in increased movement of the particles, leading to greater disorder and an increase in the entropy of the system. Therefore, the changes in both enthalpy and entropy affect the spontaneity of a chemical reaction.

- Exothermic reactions are more likely to be spontaneous, as this leads to a reduction in enthalpy and greater stability of the reaction products.

- An increase in entropy makes reactions more likely to be spontaneous, as greater disorder leads to more uniform distribution of energy within the system.

This will be revisited in greater depth later in this topic.

▲ Figure 3 Jostedalsbreen glacier, Norway.

Calculating entropy changes

The entropy change ΔS of a system can be calculated from thermodynamic data which is provided in section 12 of the *Data booklet*. The standard molar entropy values, $S^\ominus$, relate to standard conditions of temperature and pressure.

To calculate the standard entropy change associated with a reaction we find the difference between the total entropy of the products and the total entropy of the reactants:

$$\Delta S^\ominus_{298}(\text{reaction}) = \sum \Delta S^\ominus_{298}(\text{products}) - \sum \Delta S^\ominus_{298}(\text{reactants})$$

When performing entropy change calculations the following points need to be considered:

- Remember that values for entropy are specific for different states of matter, for example, $S^\ominus(H_2O(g)) = 188.8$ J K^{-1}mol^{-1} while $S^\ominus(H_2O(l)) = 70.0$ J K^{-1}mol^{-1}.

- The coefficients used to balance the equation must be applied to molar entropy values when calculating the overall entropy change.

- Examine the chemical reaction and predict whether you expect the reaction to have positive or negative entropy change based on the degree of disorder in the products and reactants. This prediction can be used to check your final calculation.

 Worked example

Calculate the standard entropy change for the following reactions:

a) $H_2(g) + \frac{1}{2}O_2(g) \rightarrow H_2O(l)$ $S^{\ominus}_{298}(H_2)$ 130.7 J K^{-1} mol^{-1}

b) $NH_4Cl(s) \rightarrow NH_3(g) + HCl(g)$ $S^{\ominus}_{298}(O_2)$ 205.1 J K^{-1} mol^{-1}

Solution

a) $H_2(g) + \frac{1}{2}O_2(g) \rightarrow H_2O(l)$

$\Delta S^{\ominus}_{298} = [\Delta S^{\ominus}_{298}(H_2O)] - [\Delta S^{\ominus}_{298}(H_2) + \frac{1}{2}\Delta S^{\ominus}_{298}(O_2)]$

$= [70.0] - [130.7 + \frac{1}{2} \times 205.1]$

$= -163.3$ J K^{-1}

The negative entropy change associated with this chemical reaction indicates a decrease in disorder (greater order), with $1\frac{1}{2}$ mol of gas changing into 1 mol of a liquid.

b) $NH_4Cl(s) \rightarrow NH_3(g) + HCl(g)$

$\Delta S^{\ominus}_{298} = [\Delta S^{\ominus}_{298}(NH_3) + \Delta S^{\ominus}_{298}(HCl)] - [\Delta S^{\ominus}_{298}(NH_4Cl)]$

$= [192.5 + 186.9] - [94.85]$

$= +284.55$ J K^{-1}

Transforming 1 mol of a solid into 2 mol of a gas results in a large increase in disorder, hence the large positive entropy. Thermodynamic data can be found in section 12 of the *Data Booklet*.

Study tip

Standard molar entropy has the unit J K^{-1} mol^{-1}; compare this with kJ mol^{-1} for standard enthalpy of formation. These values are combined in Gibbs free energy calculations (see later in this topic). When combining these quantities be sure to convert units appropriately.

Quick questions

1   Predict whether the following reactions will have a positive or negative entropy change, $\Delta S^{\ominus}$.

    a) $NH_4NO_3(s) \rightarrow N_2O(g) + 2H_2O(g)$

    b) $N_2(g) + 3H_2(g) \rightarrow 2NH_3(g)$

    c) $N_2O_4(g) \rightarrow 2NO_2(g)$

    d) $CaCO_3(s) \rightarrow CaO(s) + CO_2(g)$

    e) $2C_2H_6(g) + 7O_2(g) \rightarrow 4CO_2(g) + 6H_2O(l)$

2   The equation for the reaction between carbon dioxide and hydrogen is shown below.

    $CO_2(g) + 4H_2(g) \rightarrow CH_4(g) + 2H_2O(g)$

Table 2 shows the standard entropy values of the substances in the reaction above.

Substance	$CO_2(g)$	$H_2(g)$	$CH_4(g)$	$H_2O(g)$
$S^{\ominus}$/J K^{-1} mol^{-1}	214	131	186	189

▲ Table 2

Calculate the standard entropy change for the reaction. [3]

Explain how the sign can be predicted from the equation for the reaction. [2]

IB, Nov 2008

Sustainable energy

"Sustainable energy is a UN initiative with a goal of doubling of global sustainable energy resources by 2030." *Chemistry Syllabus sub-topic 15.1*

"Sustainable energy for all" is a United Nations (UN) initiative that aims to reduce the inequalities that exist in the provision of clean and efficient energy services, improve energy efficiency thereby reducing energy demand, and increase the proportion of energy that comes from renewable resources (http://www.sustainableenergyforall. org/objectives/universal-access). The project has three main objectives:

1   Ensure universal access to modern energy services which focuses on improving the lives and economic conditions of people throughout the world. Approximately one-fifth of the world's population do not have electricity in their home and almost 40 per cent utilize fuel sources such as animal waste, charcoal, and wood to provide heat for cooking. Toxic products from this form of combustion result in the deaths of over 2 million people annually, mainly women and children. "Electricity enables children to study after dark. It enables water to be pumped for crops, and foods and medicines to be refrigerated."

2   Energy efficiency is the part of the project that looks at countries, including the way in which we use power. From industry to households, office and accommodation buildings to transportation, lighting to electrical appliances, a variety of people, agencies, and governments are being encouraged to both educate and legislate, with the aim of decreasing the global electricity demand. Energy-saving light bulbs, energy-efficient televisions, buildings that require less energy to heat and cool, and the use of information technology in industry to better manage power usage are all examples of how the global community is reducing the demand for power. This ultimately saves governments, individuals, and businesses money and lessens the impact of coal-fired power stations on the environment.

3   Renewable energy – the UN has set a target of doubling the share of renewable energy contributed to global energy production by 2030. The cost of development of renewable energy sources has decreased appreciably over the decades and now represents a viable option for governments, businesses, and individuals. Where resources are available, renewable energy can play a major role in power generation. For example, hydroelectric dams in Brazil generate 83% of the country's electricity.

▲ Figure 4 Itaipu dam, built between Brazil and Paraguay, is the second largest hydroelectric power plant in the world

Gibbs free energy

The Gibbs free energy G is a state function, along with enthalpy H, entropy S, and absolute temperature T. Having established the importance of entropy in defining the spontaneity of a reaction, we shall now look at the relationship between total entropy, enthalpy, and the temperature of the system.

For a spontaneous reaction:

$$\Delta S^{\ominus}_{total} = \Delta S^{\ominus}_{sys} + \Delta S^{\ominus}_{surroundings} > 0$$

A chemical reaction may be either exothermic or endothermic: the transfer of heat across the system/surroundings boundary is directionally dependent on the change in enthalpy. For an exothermic reaction in an open system, heat is transferred from the system to the surroundings. This results in an increase in the entropy of the surroundings.

The impact that the enthalpy change of a reaction has on the entropy of the surroundings is dependent on the conditions existing in the system. Imagine transferring heat energy into two separate systems, one at low temperature and one at high temperature, such as a block of ice at $0\,°C$ and a bowl of water at $60\,°C$. The transfer of the same amount of energy into each system will have a different effect. The ice will begin to melt as the kinetic energy of the water molecules increases, resulting in a significant change in the level of entropy. However, the hot water already has significant disorder compared with the ice so the additional energy will have a much less marked effect on the level of entropy.

The combination of enthalpy, entropy, and temperature of system can be used to define a new state function called **Gibbs free energy, G**:

$$G = H - TS$$
$$\Delta G^{\ominus} = \Delta H^{\ominus} - T\Delta S^{\ominus}$$

The Gibbs free energy provides an effective way of focusing on a reaction system at constant temperature and pressure to determine its spontaneity. For a reaction to be spontaneous the Gibbs free energy must have a negative value ($\Delta G < 0$).

> Reactions that are spontaneous and are therefore thermodynamically favourable can sometimes be kinetically improbable, due to the existence of very high activation energies (see sub-topic 16.2).

ΔH	ΔS	ΔG	Spontaneity	Explanation
positive (> 0): *endothermic*	positive (> 0): *more disorder*	negative at high T positive at low T	dependent on temperature	spontaneous only at high temperatures when $T\Delta S > H$
positive (> 0): *endothermic*	negative (< 0): *more order*	always positive > 0	never spontaneous	reverse reaction spontaneous at all temperatures
negative (< 0): *exothermic*	positive (> 0): *more disorder*	always negative < 0	always spontaneous	forward reaction spontaneous at all temperatures
negative (< 0): *exothermic*	negative (< 0): *more order*	negative at low T positive at high T	dependent on temperature	spontaneous only at low temperatures when $T\Delta S < H$

▲ Table 3 Factors affecting ΔG and the spontaneity of a reaction

It is not always possible to predict whether a chemical reaction will be spontaneous or not (table 3). Exothermic reactions that involve increasing disorder will always be spontaneous, with $\Delta G < 0$. Similarly, endothermic reactions of increasing order will always be non-spontaneous, with $\Delta G > 0$. The spontaneity of other reactions depends on the temperature of the system.

Gibbs free energy change of formation

The **Gibbs free energy change of formation**, $\Delta G_f^\ominus$, represents the free energy change when 1 mol of a compound is formed from its elements under standard conditions of 298 K and a pressure of 100 kPa:

$$\Delta G_r^\ominus = \Sigma \Delta G_f^\ominus(\text{products}) - \Sigma \Delta G_f^\ominus(\text{reactants})$$

 ## Worked example: finding $\Delta G_r^\ominus$ from $\Delta G_f^\ominus$ values

Calculate the Gibbs free energy change of reaction, $\Delta G_r^\ominus$ for the combustion of ethanol, C_2H_5OH to give $CO_2(g)$ and $H_2O(g)$.

Substance	$C_2H_5OH(l)$	$H_2O(g)$	$CO_2(g)$
$\Delta G_f^\ominus$ values/ kJ mol^{-1}	-175	-228.6	-394.4

Solution

$$C_2H_5OH(l) + 3O_2(g) \rightarrow 2CO_2(g) + 3H_2O(g)$$

$$\Delta G_r^\ominus = \Sigma \Delta G_f^\ominus(\text{products}) - \Sigma \Delta G_f^\ominus(\text{reactants})$$
$$= [2\Delta G_f^\ominus(CO_2) + 3\Delta G_f^\ominus(H_2O)]$$
$$- [\Delta G_f^\ominus(C_2H_5OH)]$$
$$= [2 \times -394.4 + 3 \times -228.6] - [-175]$$
$$= -1299.6 \text{ kJ mol}^{-1}$$

Substance	$\Delta G_f^\ominus$/kJ mol^{-1}
$SO_3(g)$	-371.1
$H_2SO_4(l)$	-690.0
$NH_4Cl(s)$	-202.9
$CaCO_3(s)$	-1129.1
$CaO(s)$	-604.0
$CaCl_2(s)$	-748.1
$NH_3(g)$	-16.5

▲ Table 4 $\Delta G_f^\ominus$ values not found in section 12 of the *Data booklet*

Quick question

Calculate the Gibbs free energy change for the following reactions. Values for $\Delta G_f^\ominus$ can be found in section 12 of the *Data booklet*; additional data is listed in table 4.

a) $SO_3(g) + H_2O(l) \rightarrow H_2SO_4(l)$

b) $2NH_4Cl(s) + CaO(s) \rightarrow CaCl_2(s) + H_2O(l) + 2NH_3(g)$

c) $C_2H_4(g) + H_2O(l) \rightarrow C_2H_5OH(l)$

Calculating the Gibbs free energy change of a reaction from enthalpy and entropy data

To determine the spontaneity of a reaction from $\Delta G^\ominus = \Delta H^\ominus - T\Delta S^\ominus$, we need to calculate the Gibbs free energy change for the reaction under standard conditions (298 K and 100 kPa). If the Gibbs free energies of formation of reactants and/or products are unknown, we need first to calculate the enthalpy and entropy changes for the reaction.

Worked example: calculating $\triangle G^\ominus$ from $\triangle H^\ominus - T\triangle S^\ominus$

Standard enthalpy change of combustion reactions are given below:

$$2C_2H_6(g) + 7O_2(g) \rightarrow 4CO_2(g) + 6H_2O(l) \qquad \Delta H^\ominus = -3120 \text{ kJ}$$

$$2H_2(g) + O_2(g) \rightarrow 2H_2O(l) \qquad \Delta H^\ominus = -572 \text{ kJ}$$

$$C_2H_4(g) + 3O_2(g) \rightarrow 2CO_2(g) + 2H_2O(l) \qquad \Delta H^\ominus = -1411 \text{ kJ}$$

a) Based on the above information, calculate the standard change in enthalpy, $\Delta H^\ominus$, for the following reaction:

$$C_2H_6(g) \rightarrow C_2H_4(g) + H_2(g)$$

b) Predict, stating a reason, whether the sign of $\Delta S^\ominus$ for the above reaction would be positive or negative.

c) Calculate the standard entropy change for the reaction.

d) Determine the value of $\Delta G^\ominus$ for the reaction at 298 K.

e) Determine the temperature at which this reaction will occur spontaneously.

IB, Nov 2009

Solution

a) Rearrange the three combustion reactions to find the standard change in enthalpy.

The first equation will occur in the same direction but only half of the stoichiometry is needed so halve the enthalpy value:

$$C_2H_6(g) + 3\tfrac{1}{2}O_2(g) \rightarrow 2CO_2(g) + 3H_2O(l)$$
$$\Delta H^\ominus = -1560 \text{ kJ}$$

The second equation needs to be reversed and halved:

$$H_2O(l) \rightarrow H_2(g) + \tfrac{1}{2}O_2(g) \quad \Delta H^\ominus = +286 \text{ kJ}$$

The third equation needs to be reversed:

$$2CO_2(g) + 2H_2O(l) \rightarrow C_2H_4(g) + 3O_2(g)$$
$$\Delta H^\ominus = +1411 \text{ kJ}$$

Summation of these equations determines the standard enthalpy change:

$$\Delta H^\ominus = -1560 + 286 + 1411 = +137 \text{ kJ}$$

$$C_2H_6(g) \rightarrow C_2H_4(g) + H_2(g) \; \Delta H^\ominus = +137 \text{ kJ}$$

b) The sign for the change in entropy is positive: an increase in disorder is evident as the number of moles of gas increases from 1 to 2 in the reaction.

c) $\Delta S^\ominus_{298} = [\Delta S^\ominus_{298}(C_2H_4)(g) + \Delta S^\ominus_{298}(H_2)(g)] - [\Delta S^\ominus_{298}(C_2H_6)(g)]$

$$= [220 + 131] - [230]$$

$$= 120 \text{ J K}-1$$

d) $\Delta G^\ominus = \Delta H^\ominus - T\Delta S^\ominus$

$$= +137 - \left(298 \times \frac{120}{1000}\right)$$

$$= +101 \text{ kJ}$$

In this calculation, the entropy value is converted from joules to kilojoules by dividing by 1000. The positive value for the change in Gibbs free energy indicates that the reaction is non-spontaneous. This can be predicted by examining the positive value for change in enthalpy (endothermic) and the low temperature.

e) To determine the temperature at which this reaction will occur spontaneously, we make the assumption that the value for Gibbs free energy is zero and solve for T.

$$\Delta G^\ominus = \Delta H^\ominus - T\Delta S^\ominus$$

$$0 = \Delta H^\ominus - T\Delta S^\ominus$$

$$T = \frac{\Delta H^\ominus}{\Delta S^\ominus} = \frac{137}{120 \times 10^{-3}} = 1142 \text{ K}$$

The reaction becomes spontaneous at temperatures greater than 1142 K.

Gibbs free energy and chemical equilibrium

We have established that reactions taking place at constant temperature and pressure are spontaneous when $\Delta G < 0$. From the time when a reversible reaction commences to the point where it reaches equilibrium, the Gibbs free energy is changing as the ratio of reactants to products alters. As the amount of products increases and the reaction moves towards completion (for non-reversible reactions) or equilibrium (for reversible reactions) the Gibbs free energy decreases. *At the point of equilibrium the system has reached its minimum Gibbs free energy* (figure 5).

From figure 5 we can see that as the reaction proceeds the Gibbs free energy decreases towards a minimum. In this region (A) the forward reaction is favoured. As the reaction continues, at the point of equilibrium the Gibbs free energy reaches a minimum and then increases, during which time the forward reaction becomes non-spontaneous (B). The reverse reaction is then spontaneous and the Gibbs free energy again reaches a minimum in the same way as during the forward reaction. The relationship between the Gibbs free energy change of a reaction and the equilibrium constant will be examined in detail in topic 17.

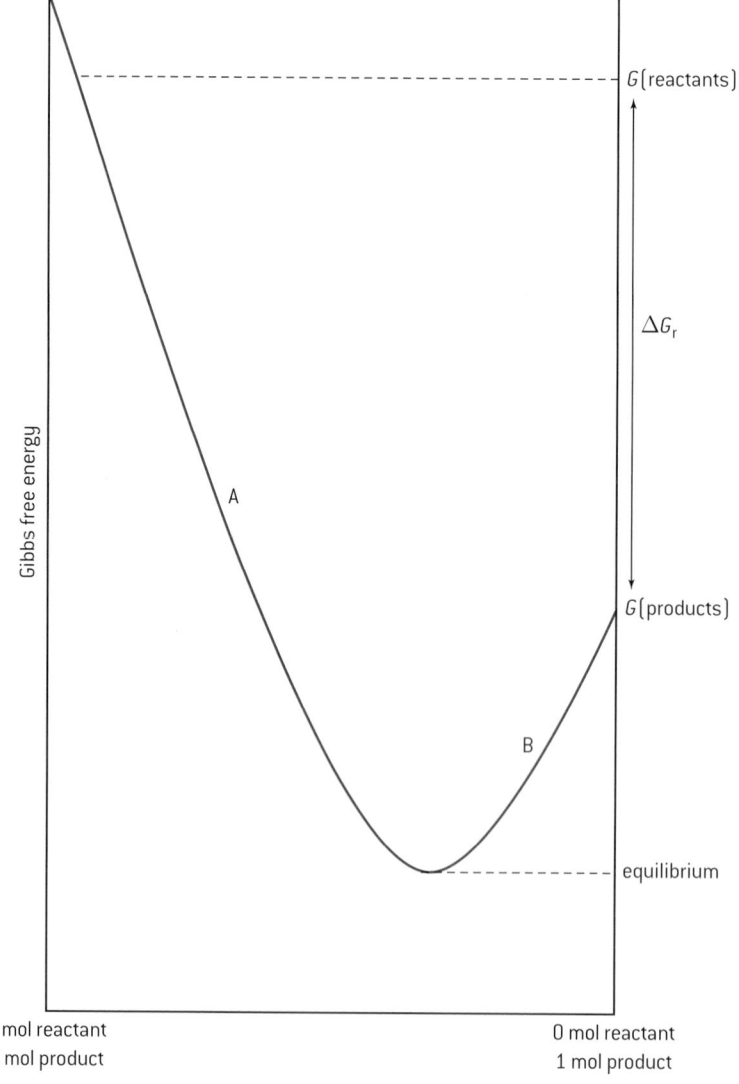

▲ Figure 5 How the Gibbs free energy changes as the reaction proceeds

Questions

1 The lattice enthalpy of magnesium chloride can be calculated from the Born–Haber cycle shown in figure 6.

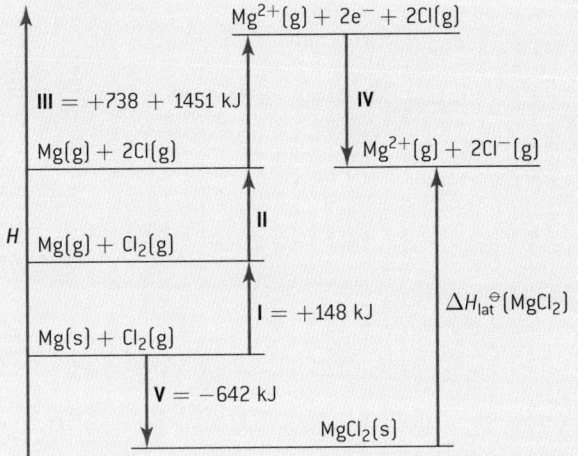

▲ Figure 6

a) Identify the enthalpy changes labelled by **I** and **V** in the cycle. [2]

b) Use the ionization energies given in the cycle above and further data from the *Data booklet* to calculate a value for the lattice enthalpy of magnesium chloride. [4]

c) The theoretically calculated value for the lattice enthalpy of magnesium chloride is +2326 kJ. Explain the difference between the theoretically calculated value and the experimental value. [2]

d) The experimental lattice enthalpy of magnesium oxide is given in section 18 of the *Data booklet*. Explain why magnesium oxide has a higher lattice enthalpy than magnesium chloride. [2]

IB, November 2010

2 The lattice enthalpy values for lithium fluoride and calcium fluoride are shown below.

$LiF(s)$ $\Delta H^{\ominus} = +1022$ kJ mol^{-1}
$CaF_2(s)$ $\Delta H^{\ominus} = +2602$ kJ mol^{-1}

Which of the following statements help(s) to explain why the value for lithium fluoride is less than that for calcium fluoride?

I) The ionic radius of lithium is less than that of calcium.

II) The ionic charge of lithium is less than that of calcium.

A. I only

B. II only

C. I and II

D. Neither I nor II [1]

IB, May 2004

3 Which reaction occurs with the largest increase in entropy?

A. $Pb(NO_3)_2(s) + 2KI(s) \rightarrow PbI_2(s) + 2KNO_3(s)$

B. $CaCO_3(s) \rightarrow CaO(s) + CO_2(g)$

C. $3H_2(g) + N_2(g) \rightarrow 2NH_3(g)$

D. $H_2(g) + I_2(g) \rightarrow 2HI(g)$ [1]

IB, May 2004

4 The $\Delta H^{\ominus}$ and $\Delta S^{\ominus}$ values for a certain reaction are both positive. Which statement is correct about the spontaneity of this reaction at different temperatures?

A. It will be spontaneous at all temperatures.

B. It will be spontaneous at high temperatures but not at low temperatures.

C. It will be spontaneous at low temperatures but not at high temperatures.

D. It will not be spontaneous at any temperature. [1]

IB, May 2004

5 The following reaction is spontaneous only at temperatures above 850° C.

$CaCO_3(s) \rightarrow CaO(s) + CO_2(g)$

Which combination is correct for this reaction at 1000 °C?

	ΔG	ΔH	ΔS
A.	−	−	−
B.	+	+	+
C.	−	+	+
D.	+	−	−

[1]

IB, May 2007

6 Explain in terms of $\Delta G^\ominus$, why a reaction for which both $\Delta H^\ominus$ and $\Delta S^\ominus$ values are positive can sometimes be spontaneous and sometimes not. [4]

IB, May 2004

7 Throughout this question, use relevant information from the *Data booklet*.

a) Define the term *standard enthalpy change of formation* and illustrate your answer with an equation, including state symbols, for the formation of nitric acid. [4]

b) Propyne undergoes complete combustion as follows:

$$C_3H_4(g) + 4O_2(g) \rightarrow 3CO_2(g) + 2H_2O(l)$$

Calculate the enthalpy change of this reaction, given the following additional values: [4]

$\Delta H_f^\ominus$ of $CO_2(g) = -394$ kJ mol^{-1}
$\Delta H_f^\ominus$ of $H_2O(l) = -286$ kJ mol^{-1}

c) Predict and explain whether the value of $\Delta S^\ominus$ for the reaction in part (b) would be negative, close to zero, or positive. [3]

IB, May 2005

16 CHEMICAL KINETICS (AHL)

Introduction

In this topic we explore the various mathematical equations that relate to the rate of a chemical reaction. Rate equations can only be determined empirically and in many cases are limited by the slowest step of the reaction. A detailed understanding of the reaction mechanism allows chemists to control a reaction and optimize the reaction conditions in terms of yield, reaction time, product cost and the environmental impact.

16.1 Rate expression and reaction mechanism

Understandings

→ Reactions may occur by more than one step and the slowest step determines the rate of reaction (rate determining step/RDS).

→ The molecularity of an elementary step is the number of reactant particles taking part in that step.

→ The order of a reaction can be either integer or fractional in nature. The order of a reaction can describe, with respect to a reactant, the number of particles taking part in the rate-determining step.

→ Rate equations can only be determined experimentally.

→ The value of the rate constant (k) is affected by temperature and its units are determined from the overall order of the reaction.

→ Catalysts alter a reaction mechanism, introducing a step with lower activation energy.

Applications and skills

→ Deduction of the rate equation from experimental data and solving problems involving the rate equation.

→ Sketching, identifying, and analysing graphical representations for zero, first, and second order reactions.

→ Evaluation of proposed reaction mechanisms to be consistent with kinetic and stoichiometric data.

Nature of science

→ Principle of Occam's razor – newer theories need to remain as simple as possible while maximizing explanatory power. The low probability of three-molecule collisions means stepwise reaction mechanisms are more likely.

The reaction mechanism representing the sequence of molecular events leading from reactants to products is actually composed of two elementary steps:

step 1: $NO_2(g) + NO_2(g) \rightarrow NO(g) + NO_3(g)$
                    step 1 is bimolecular

step 2: $NO_3(g) + CO(g) \rightarrow NO_2(g) + CO_2(g)$
                    step 2 is also bimolecular

overall reaction: $NO_2(g) + CO(g) \rightarrow NO(g) + CO_2(g)$

In this mechanism, $NO_3(g)$ is described as a **reaction intermediate**, as it is formed in step 1 and then is consumed subsequently in step 2.

Therefore, reactions may occur by more than one step and the **slow step** determines the rate of the reaction. The slow step is termed the **rate-determining step** (**RDS**).

Deduction of a rate equation from a proposed reaction mechanism

In order to deduce the rate equation from a proposed reaction mechanism:

1   Decide on which step is the RDS. The rate of the overall reaction is equal to the rate of this slow step.

2   From (1) deduce the rate equation for the RDS.

For temperatures less than 498 K, the **experimental** rate equation for the reaction just discussed has been found to be:

     rate $= k[NO_2]^2$

- In effect, the reaction mechanism is essentially a **hypothesis** of the sequence of events that has led to the overall reaction converting the reactants into products. There might, therefore, be a number of possible reaction mechanisms that equate with the experimental rate equation.

For the example just discussed, here is a proposed reaction mechanism.

Consider step 1 as the slow step (so is the RDS) and step 2 as the fast step:

step 1: $NO_2(g) + NO_2(g) \xrightarrow{k_1} NO(g) + NO_3(g)$ (slow)

step 2: $NO_3(g) + CO(g) \xrightarrow{k_2} NO_2(g) + CO_2(g)$ (fast)

overall reaction: $NO_2(g) + CO(g) \rightarrow NO(g) + CO_2(g)$

Hence:

     rate of overall reaction = rate of the slow step (in this case step 1)
     $= k[NO_2]^2$

where k represents the rate constant for the overall reaction. This proposed mechanism is consistent with the experimentally determined rate equation.

- In contrast, at temperatures greater than 498 K, the **experimental** rate equation for the reaction just discussed has been found to be:

     rate $= k[NO_2][CO]$

A proposed reaction mechanism here might be a single-step bimolecular process:

single step: $NO_2(g) + CO(g) \xrightarrow{k} NO(g) + CO_2(g)$ (slow)

Hence:

> rate of overall reaction = rate of the slow step (in this case the single step) = $k[NO_2][CO]$

This proposed mechanism is consistent with the experimentally determined rate equation.

TOK

A reaction mechanism can be supported by indirect evidence. What is the role of empirical (experimental) evidence in the formulation of scientific theories? Can we ever be certain in science?

Cancer research, for example, is all about identifying mechanisms for carcinogens as well as for cancer-killing agents and inhibitors.

Worked example: deduction of the rate equation from experimental data and solving problems involving the rate equation

1 Consider the balanced equation, and note the stoichiometry coefficients of the reactants and products. For example,

$$xA + yB \rightarrow qC + pD$$

x, y, q, and p are the **stoichiometry coefficients**.

2 Write down the rate equation, where m and n represent the **orders with respect to each reactant**:

$$\text{rate} = k[A]^m[B]^n$$

3 From the given data for each of the experiments, deduce each of the following ratios (as appropriate):

$$\frac{(\text{rate 1})}{(\text{rate 2})}, \frac{(\text{rate 2})}{(\text{rate 3})}, \frac{(\text{rate 3})}{(\text{rate 4})}, \text{etc.}$$

Look for pairs of rate data where the concentration does not change in one of them going from one experiment to another.

4 From each ratio obtained in step 3, deduce the orders m and n. Use of some fundamental mathematical tools with respect to indices and logs may be helpful here, for example:

$$x^0 = 1$$

$$\log (XY) = \log X + \log Y$$

$$\log \left(\frac{X}{Y}\right) = \log X - \log Y$$

$$\log X^p = p\log X$$

5 Deduce the overall order of the reaction:

$$\text{overall reaction order} = m + n$$

6 Determine the rate constant, k, for each experiment (1, 2, 3, etc.). Find the mean of these values to give the mean value of k and deduce the appropriate units for k.

Example 1

Consider the reaction:

$$A(g) + B(g) \rightarrow C(g) + D(g)$$

Based on the experimental initial rate data below:

- Deduce the orders with respect to each reactant and the overall reaction order.

- Deduce the rate equation.

	[A(g)] / mol dm^{-3}	[B(g)] / mol dm^{-3}	Initial rate / mol dm^{-3} s^{-1}
Experiment 1	1.00×10^{-2}	1.00×10^{-2}	4.20×10^{-3}
Experiment 2	2.00×10^{-2}	1.00×10^{-2}	8.40×10^{-3}
Experiment 3	2.00×10^{-2}	2.00×10^{-2}	3.36×10^{-2}

- Calculate the value of the rate constant, k, for the reaction from experiment 3 and state its units.

- Determine the rate of the reaction when $[A(g)] = 3.00 \times 10^{-2}$ mol dm^{-3} and $[B(g)] = 4.00 \times 10^{-2}$ mol dm^{-3}.

Solution

In order to solve this question we can use the working method to deduce the rate equation from the method of initial rates:

- There are two reactants in the chemical equation so the rate equation is given by:

$$\text{rate} = k[A]^m[B]^n$$

- You next have to choose the appropriate ratios to use. In order to decide this, look for pairs of data in which one of the concentrations does not change – this helps reduce the problem down to just one order. For example:

$$\frac{\text{rate 1}}{\text{rate 2}} = \frac{k(0.010)^m(0.010)^n}{k(0.020)^m(0.010)^n} = \frac{0.00420}{0.00840}$$

Hence:

$$(0.5)^m = 0.5, \text{ so } m = 1$$

Therefore, the reaction is first order with respect to reactant A.

Next, do the same for the other ratio, $\frac{\text{ratio 2}}{\text{ratio 3}}$ which also has pairs of data in which one of the concentrations does not change:

$$\frac{\text{rate 2}}{\text{rate 3}} = \frac{k(0.020)^1(0.010)^n}{k(0.020)^1(0.020)^n} = \frac{0.00840}{0.0336}$$

Hence:

$$(0.5)^n = 0.25, \text{ so } n = 2$$

Therefore, the reaction is second order with respect to reactant B.

- The overall reaction order $= m + n = 1 + 2 = 3$, so the reaction is third order overall.

- The rate equation is therefore:

$$\text{rate} = k[A][B]^2$$

- We next have to rearrange this rate equation to make k the subject of the expression:

$$k = \frac{\text{rate}}{[A][B]^2}$$

Then substituting the data from experiment 3:

$$k = \frac{(3.36 \times 10^{-2} \text{ mol dm}^{-3} \text{ s}^{-1})}{(2.00 \times 10^{-2} \text{ mol dm}^{-3})(2.00 \times 10^{-2} \text{ mol dm}^{-3})^2}$$

$$= 4.20 \times 10^3 \text{ mol}^{-2} \text{ dm}^6 \text{ s}^{-1}$$

The units were worked out as follows:

$$\text{units of } k = \frac{\text{mol dm}^{-3} \text{ s}^{-1}}{\text{mol dm}^{-3} \times \text{mol dm}^{-3} \times \text{mol dm}^{-3}}$$

$$\text{units of } k = \text{mol}^{-2} \text{ dm}^6 \text{ s}^{-1}$$

> The orders may also be deduced by inspection. By keeping [B] constant in experiments 1 and 2 and doubling [A], the initial rate is seen to double. Hence the order with respect to A will be one. Likewise, by keeping [A] constant in experiments 2 and 3, [B] doubles. However, this time the initial rate is seen to increase by a factor of four, meaning that the order with respect to B is two. This is a quick way of deducing the orders, but with more difficult numbers finding the orders by this method might be quite tricky – *following the working method using ratios will always allow you to find the correct answer.*

- In order to determine the rate of the reaction when $[A(g)] = 3.00 \times 10^{-2}$ mol dm^{-3} and $[B(g)] = 4.00 \times 10^{-2}$ mol dm^{-3}, we may use the rate equation:

$$\text{rate} = k[A][B]^2$$
$$= (4.20 \times 10^3 \text{ mol}^{-2} \text{ dm}^6 \text{ s}^{-1})$$
$$(3.00 \times 10^{-2} \text{ mol dm}^{-3})$$
$$(4.00 \times 10^{-2} \text{ mol dm}^{-3})^2$$
$$\text{rate} = 2.02 \times 10^{-1} \text{ mol dm}^{-3} \text{ s}^{-1}$$

Study tip

Always watch out for significant figures in questions.

Graphical representations of zero order, first order and second order reactions

First and second order reactions are found to occur most frequently; in contrast, zero order reactions are not common.

Zero order reactions

For the zero-order reaction:

A → products

the rate equation will be:

rate $= k[A]^0 = k$

Using calculus, the following equation can be derived:

$$[A] = -kt + [A]_o$$

where:

$[A]$ = concentration of reactant A

k = rate constant

t = time

$[A]_o$ = initial concentration.

This equation is of the form:

$$y = mx + c$$

Where:

m = slope = $-k$

c = intercept = $[A]_o$

Hence, a plot of $[A]$ versus t would yield a straight-line plot for a zero order reaction. The gradient of the line would be $-k$ and the graph would cut the y-axis when $x = 0$, at $[A]_o$.

In the plot shown in figure 1 notice that the gradient, corresponding to $-k$, is negative.

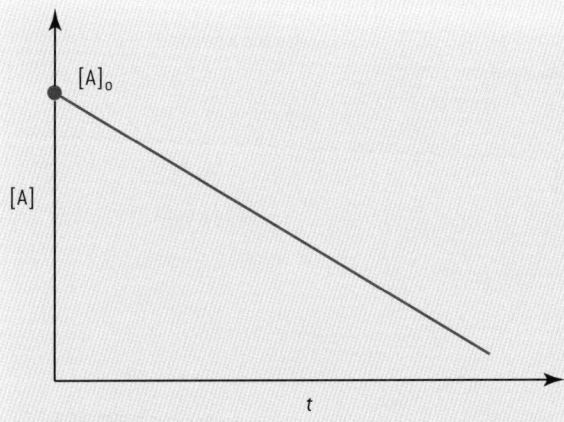

▲ Figure 1 Sketch of a concentration–time plot for a zero order reaction

The plot in figure 2 is of rate versus concentration for a zero order reaction – notice how the rate is constant and independent of the concentration (that is, rate = k).

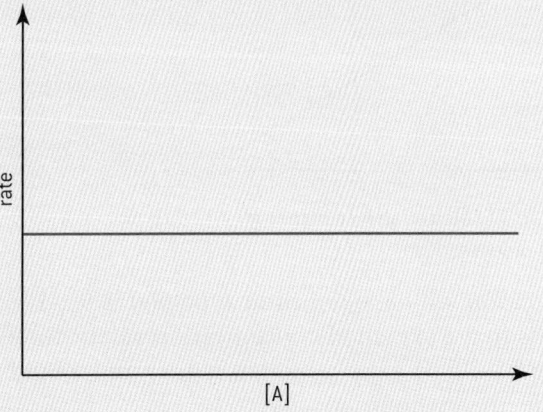

▲ Figure 2 Sketch of a rate–concentration plot for a zero order reaction

First order reactions

For the first-order reaction:

A → products

the rate equation will be:

rate $= k[A]^1 = k[A]$

From calculus, the following equation can be derived:

$$\ln[A] = -kt + \ln[A]_o$$

where:

$[A]$ = concentration of reactant A

k = rate constant

t = time

$[A]_o$ = initial concentration.

'ln' represents the natural log to the base e.

This equation is of the form:

$$y = mx + c$$

where:

m = slope = $-k$

c = intercept = $\ln[A]_o$

Hence, a plot of $\ln[A]$ versus t would yield a straight-line plot for a first order reaction (figure 3). The gradient of the line would be $-k$ and the graph would cut the y-axis when $x = 0$, at $\ln[A]_o$.

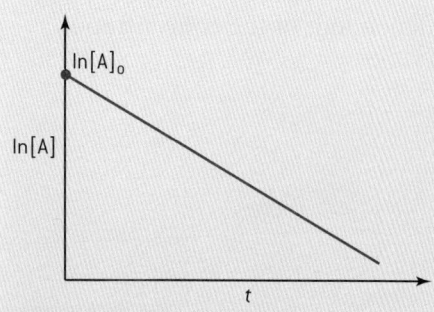

▲ Figure 3 Sketch of an ln(concentration)–time plot for a first order reaction

The reason why a logarithmic type plot is used here is that a sketch of concentration versus time alone would be exponential in nature and would not be linear (figure 4).

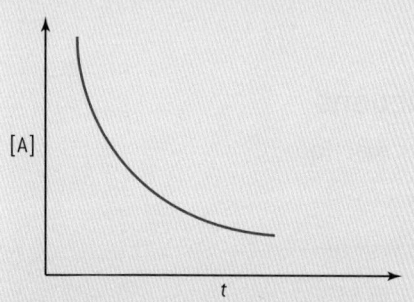

▲ Figure 4 Sketch of a concentration–time plot for a first order reaction

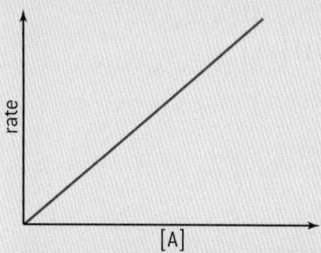

▲ Figure 5 Sketch of a rate–concentration plot for a first order reaction

In a rate–concentration plot for a first order reaction (figure 5), the rate is directly proportional to the concentration, because rate = $k[A]$ for a first order reaction.

Second order reactions

For the second-order reaction:

A → products

the rate equation will be:

rate = $k[A]^2$

From calculus, the following equation can be derived:

$$\frac{1}{[A]} = kt + \frac{1}{[A]_o}$$

where:

[A] = concentration of reactant A

k = rate constant

t = time

$[A]_o$ = initial concentration.

This equation is of the form:

$$y = mx + c$$

where:

m = slope = k

c = intercept = $\frac{1}{[A]_o}$

Hence, a plot of $\frac{1}{[A]}$ versus t would yield a straight-line plot for a second order reaction (figure 6). The gradient of the line would be k and the graph would cut the y-axis when $x = 0$, at $\frac{1}{[A]_o}$. The gradient of the line is positive, corresponding to k.

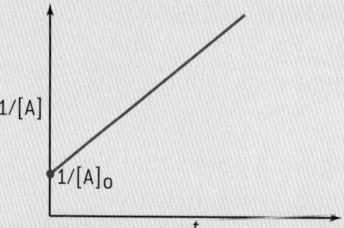

▲ Figure 6 Sketch of $\frac{1}{[A]}$ versus time for a second order reaction

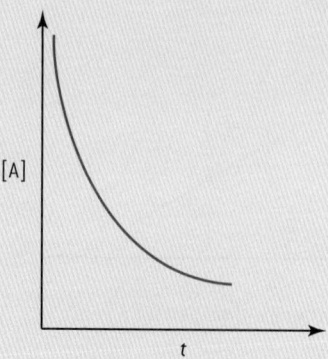

▲ Figure 7 Sketch of a concentration–time plot for a second order reaction

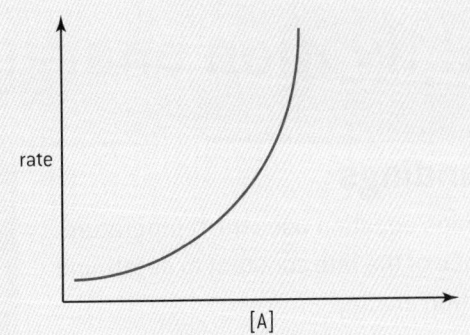

In a concentration–time plot for a second order reaction (figure 7) the curve appears somewhat deeper than the corresponding curve in figure 4 of [A] versus t for a first order reaction.

In a rate–concentration plot for a second order reaction (figure 8), the rate is directly proportional to the square of the concentration, because rate $= k[A]^2$ for a second order reaction. As a square term is involved, a straight line will not be observed, unlike that seen in the corresponding sketch for a first order reaction.

▲ Figure 8 Sketch of a rate–concentration plot for a second order reaction

Study tip

For the IB Chemistry syllabus, you need to know the following for zero order, first order, and second order reactions:

- rate equation for each order

- sketch of rate versus concentration for each order (can be deduced from the rate equation)

- sketch of concentration versus time for each order

Remember, if you are asked to a **sketch** a graph, no units are required for the labels of the x- and y-axes. If you are asked to **draw** a graph using experimental data, you should always include units for each axis label (unless a parameter is logarithmic for which there will be no units involved). For both (a sketch or a drawing of a graph) a title should always be included.

Worked example: evaluation of proposed reaction mechanisms to be consistent with kinetic and stoichiometric data

Consider the following two-step reaction mechanism:

step 1: $N_2O(g) \rightarrow N_2(g) + O(g)$ (slow)

step 2: $N_2O(g) + O(g) \rightarrow N_2(g) + O_2(g)$ (fast)

a) Deduce the overall reaction.

b) Identify the intermediate in the reaction.

c) Identify the molecularity of the rate-determining step.

d) Deduce the rate of the overall reaction and state the order of the reaction.

Solution

a) In order to deduce the overall reaction we simply add the reactants and products from the two steps:

$$2N_2O(g) \rightarrow 2N_2(g) + O_2(g)$$

b) A **reaction intermediate** is formed in one step and then consumed in the subsequent step. The intermediate here will be the oxygen atom, O(g).

c) Step 1 is the slow step, which is the rate-determining step (RDS). This elementary step is unimolecular.

d) Rate of RDS $= k_1[N_2O]$, so the rate of the overall reaction $= k[N_2O]$. The reaction is first order with respect to $N_2O(g)$, so first order overall.

Study tip

Molecularity and order are completely different!

16.2 Activation energy

Understandings

→ The Arrhenius equation uses the temperature dependence of the rate constant to determine the activation energy.

→ A graph of $\ln k$ against $\frac{1}{T}$ is a linear plot with gradient $\frac{-E_a}{R}$ and intercept $\ln A$.

→ The frequency factor (or pre-exponential factor) (A) takes into account the frequency of collisions with proper orientations.

 ## Applications and skills

→ Analysing graphical representation of the Arrhenius equation in its linear form:

$$\ln k = \frac{-E_a}{RT} + \ln A$$

→ Using the Arrhenius equation

$$k = Ae^{\frac{-E_a}{RT}}$$

→ Describing the relationships between temperature and rate constant; frequency factor and complexity of molecules colliding.

→ Determining and evaluating values of activation energy and frequency factors from data.

 ## Nature of science

→ Theories can be supported or falsified and replaced by new theories – changing the temperature of a reaction has a much greater effect on the rate of reaction than can be explained by its effect on collision rates. This resulted in the development of the Arrhenius equation, which proposes a quantitative model to explain the effect of temperature change on reaction rate.

Arrhenius equation

In topic 6 we saw that temperature increases the rate of a chemical reaction and that this temperature effect can be explained in terms of the **kinetic-molecular theory**. We also discussed **collision theory**, which is a model that allows us to understand why rates of reaction depend on temperature. The collision theory itself is based on the kinetic-molecular theory.

For a chemical reaction to occur between two reactant particles, a number of conditions must be fulfilled:

• The two particles must collide with each other, that is there must be physical contact.

• The colliding particles must have correct mutual orientation.

• The colliding particles must have sufficient kinetic energy to initiate the reaction itself.

The reaction rate constant, therefore, can be expressed as follows:

$$k = p \times Z \times e^{\frac{-E_a}{RT}}$$

where:

k = rate constant;

p = steric factor (fraction of collisions where the particles have correct mutual orientation)

Z = collision number (constant related to the frequency of collisions)

e = the base of natural logarithms (2.718....)

E_a = activation energy (in J mol^{-1})

R = universal gas constant = 8.31 J K^{-1} mol^{-1}

T = temperature (in K).

In this expression, $e^{\frac{-E_a}{RT}}$ represents the fraction of molecules that have sufficient energy for a reaction to take place and is termed the **exponential factor**.

As p, the **steric factor**, and Z, the **collision number**, are both almost (not totally, however) independent of the temperature, the expression can be approximated to the following equation, called the **Arrhenius equation**:

$$k = Ae^{\frac{-E_a}{RT}}$$

pre-exponential factor (frequency factor)   exponential factor

In this expression, A, is a constant termed the **pre-exponential factor** (or **frequency factor**), which takes into account the frequency

of collisions with the correct orientations. The frequency factor is essentially the number of times reactants will approach the activation energy barrier in unit time.

The Arrhenius equation can be rearranged by applying natural logarithms to give:

$$\ln k = \frac{-E_a}{RT} + \ln A$$

This form of the expression is very useful as the plot of this function is a straight line, that is:

$$y = mx + c$$

By plotting a graph of $\ln k$ versus $\frac{1}{T}$, the slope of the line, m, is $\frac{-E_a}{R}$ and the intercept, c, is $\ln A$.

Both forms of the Arrhenius equation can be found in section 1 of the *Data booklet*.

Worked examples

Example 1

Consider the plot of $\ln k$ versus $\frac{1}{T}$ for a given decomposition reaction.

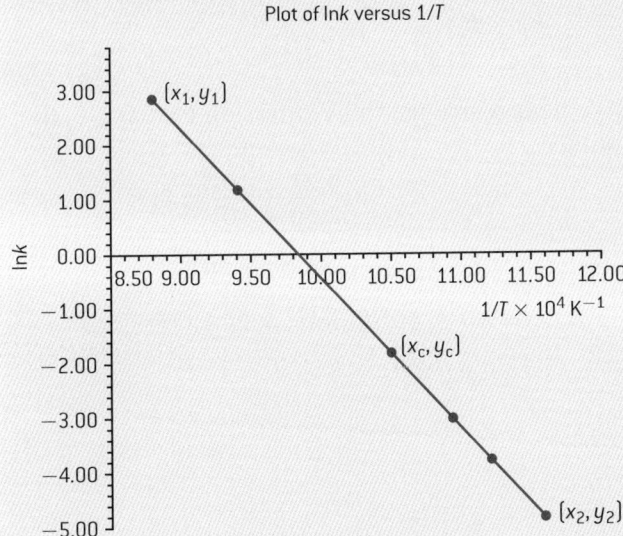

Plot of $\ln k$ versus $1/T$

The units of k are $mol^{-1}\ dm^3\ s^{-1}$

a) Show that the rate constant, k, varies with the temperature, T.

b) Determine the activation energy, E_a, for the reaction, correct to **three** significant figures, and state the units of E_a.

c) Calculate the frequency factor, A, correct to **one** significant figure and state its units.

Solution

a) The rate constant, k, increases with increasing temperature, T. Note, however, that k is *not* directly proportional to T, as seen from the Arrhenius equation.

b) From the plot choose two points on the line as far apart as possible:

$(x_1, y_1) = (8.80, 2.90)$ and
$(x_2, y_2) = (11.60, -4.80)$

Then:

$$m = \frac{\Delta y}{\Delta x} = \frac{y_2 - y_1}{x_2 - x_1} = \frac{-4.80 - 2.90}{11.60 - 8.80}$$

$$= \frac{-7.70}{2.80} = -2.75$$

The units of m are 10^4 K. From the Arrhenius equation:

$$m = \frac{-E_a}{R}$$

Hence:

$$-E_a = m \times R$$

and so:

$$E_a = -m \times R = -(-2.75 \times 10^4 \text{ K})$$

$$(8.31 \text{ J K}^{-1} \text{ mol}^{-1}) = -228525 \text{ J mol}^{-1}$$

Based on three significant figures, $E_a = 2.29 \times 10^2 \text{ kJ mol}^{-1}$.

c) In order to find the intercept, c, choose any one point on the line, for example:

$$(x_c, y_c) = (10.49, -1.80)$$

Then:

$$y_c = mx_c + c$$

So make c the subject of the equation:

$$c = y_c - mx_c$$

$$c = (-1.80) - (-2.75)(10.49)$$

$$= 27.0 = \ln A$$

> The intercept c is the point where the line cuts the y-axis at $x = 0$. As the y-axis is logarithmic in nature, c will have no units. You could also find c by extrapolating back to the y-axis on the plot.

Hence:

$$A = \text{anti-log}_e(27.0) = e^{27.0} = 5 \times 10^{11}$$

The units of A will be the same as the units of k, so $A = 5 \times 10^{11} \text{ mol}^{-1} \text{ dm}^3 \text{ s}^{-1}$.

Example 2

The rate constant, k_1, of a reaction is $5.21 \times 10^3 \text{ s}^{-1}$ at 27 °C and the corresponding rate constant, k_2, is $2.50 \times 10^4 \text{ s}^{-1}$ at 77 °C.

Deduce the activation energy, E_a, in kJ mol^{-1}, correct to **two** significant figures.

Solution

- First, write down all the data and convert all temperatures into kelvin:

$$k_1 = 5.21 \times 10^3 \text{ s}^{-1}$$

$$T_1 = 27 \text{ °C} = (27 + 273) \text{ K} = 300 \text{ K}$$

$$k_2 = 2.50 \times 10^4 \text{ s}^{-1}$$

$$T_2 = 77 \text{ °C} = (77 + 273) \text{ K} = 350 \text{ K}$$

- Next, write the Arrhenius equation for both sets of conditions, and solve the two equations to make E_a the subject:

$$\ln k_1 = -E_a/(RT_1) + \ln A$$

$$\ln k_2 = -E_a/(RT_2) + \ln A$$

$$\ln k_1 - \ln k_2 = \frac{-E_a}{RT_1} + \frac{E_a}{RT_2}$$

From the rules of logs:

$$\log \frac{X}{Y} = \log X - \log Y$$

Hence:

$$\ln \frac{k_1}{k_2} = \frac{E_a}{R}\left(\frac{1}{T_2} - \frac{1}{T_1}\right)$$

We next rearrange this expression to make E_a the subject:

$$E_a = \frac{\ln \frac{k_1}{k_2} \times R}{\frac{1}{T_2} - \frac{1}{T_1}} = \frac{\ln \frac{5.21 \times 10^3}{2.50 \times 10^4} \times 8.31}{\frac{1}{350} - \frac{1}{300}}$$

$$= 2.7 \times 10^4 \text{ J mol}^{-1} = 27 \text{ kJ mol}^{-1}$$

Questions

1 Bromine and nitrogen(II) oxide react according to the following equation.

$$Br_2(g) + 2NO(g) \rightarrow 2NOBr(g)$$

Which rate equation is consistent with the experimental data?

$[Br_2]/$ mol dm^{-3}	$[NO]/$ mol dm^{-3}	Rate/ mol dm^{-3} s^{-1}
0.10	0.10	1.0×10^{-6}
0.20	0.10	4.0×10^{-6}
0.20	0.40	4.0×10^{-6}

A. rate = $k[Br_2]^2[NO]$

B. rate = $k[Br_2][NO]^2$

C. rate = $k[Br_2]^2$

D. rate = $k[NO]^2$ [1]

IB May 2011

2 The rate information below was obtained for the following reaction at a constant temperature.

$$2NO_2(g) + F_2(g) \rightarrow 2NO_2F(g)$$

$[NO_2]/$ mol dm^{-3}	$[F_2]/$ mol dm^{-3}	Rate/ mol dm^{-3} s^{-1}
2.0×10^{-3}	1.0×10^{-2}	4.0×10^{-4}
4.0×10^{-3}	1.0×10^{-2}	8.0×10^{-4}
4.0×10^{-3}	2.0×10^{-2}	1.6×10^{-3}

What are the orders of the reaction with respect to NO_2 and F_2?

A. NO_2 is first order and F_2 is second order.

B. NO_2 is second order and F_2 is first order.

C. NO_2 is first order and F_2 is first order.

D. NO_2 is second order and F_2 is second order. [1]

IB May 2011

3 Which step is the rate-determining step of a reaction?

A. The step with the lowest activation energy.

B. The final step.

C. The step with the highest activation energy.

D. The first step. [1]

IB May 2011

4 A student experimentally determined the rate expression to be:

$$rate = k[S_2O_3^{2-}(aq)]^2$$

Which graph is consistent with this information? [1]

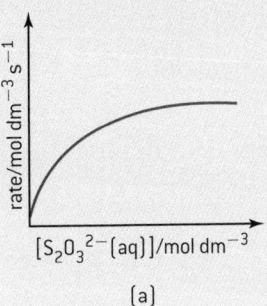

(a)

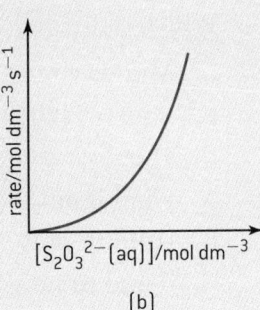

(b)

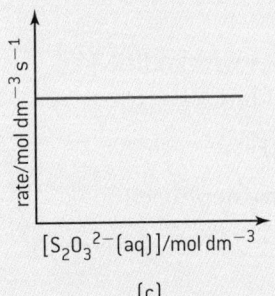

(c)

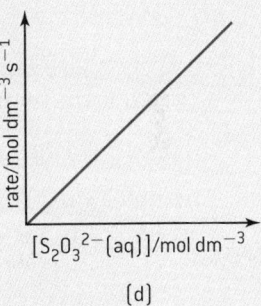

(d)

IB May 2010

5 Consider the following reaction:

$$NO_2(g) + CO(g) \rightarrow NO(g) + CO_2(g)$$

At $T < 227\ °C$ the rate expression is rate = $k[NO_2]^2$. Which of the following mechanisms is consistent with this rate expression?

A. $NO_2 + NO_2 \rightleftharpoons N_2O_4$ *fast*
   $N_2O_4 + 2CO \rightarrow 2NO + 2CO_2$ *slow*

B. $NO_2 + CO \rightarrow NO + CO_2$ *slow*

C. $NO_2 \rightarrow NO + O$ *slow*
   $CO + O \rightarrow CO_2$ *fast*

D. $NO_2 + NO_2 \rightarrow NO_3 + NO$ *slow*
   $NO_3 + CO \rightarrow NO_2 + CO_2$ *fast* [1]

IB May 2010

6 Consider the following reaction.

$$2Q(g) + R(g) \rightarrow X(g) + Y(g)$$

This reaction occurs according to the following mechanism:

$$Q(g) + R(g) \rightarrow X(g) + M(g) \qquad slow$$
$$M(g) + Q(g) \rightarrow Y(g) \qquad fast$$

Which of the following is correct?

I. M(g) is a reaction intermediate.

II. Rate = $k[Q][R]$

III. The slow-step is the rate-determining step.

A. I and II only

B. I and III only

C. II and III only

D. I, II, and III

7 Hydrogen gas, $H_2(g)$, reacts with iodine gas, $I_2(g)$, to form hydrogen iodide, HI(g):

$$H_2(g) + I_2(g) \rightarrow 2HI(g)$$

The mechanism of the two-step reaction is considered to be:

$$\text{step 1:} \quad I_2(g) \underset{k_{-1}}{\overset{k_1}{\rightleftharpoons}} 2I(g) \qquad fast$$

$$\text{step 2:} \quad 2I(g) + H_2(g) \overset{k_2}{\rightarrow} 2HI(g) \qquad slow$$

What is the rate equation for the overall reaction?

A. rate = $k[H_2][I]^2$

B. rate = $k[H_2]$

C. rate = $k[I_2]$

D. rate = $k[H_2][I_2]$

8 What are the units of the frequency factor in the Arrhenius equation?

A. kJ mol^{-1}

B. J mol^{-1}

C. s^{-1}

D. Depends on the units of k.

9 Ozone is considered to decompose according to the following two-step mechanism:

$$\text{step 1:} \quad O_3(g) \underset{k_{-1}}{\overset{k_1}{\rightleftharpoons}} O_2(g) + O(g) \qquad fast$$

$$\text{step 2:} \quad O(g) + O_3(g) \overset{k_2}{\rightarrow} 2O_2(g) \qquad slow$$

Which of the following are correct?

I. The overall reaction is $2O_3(g) \rightarrow 3O_2(g)$.

II. O(g) is a reaction intermediate.

III. The rate equation is:

$$\text{rate} = k[O_3]^2[O_2]^3$$

A. I and II only

B. I and III only

C. II and III only

D. I, II, and III

10 Consider the following reaction:

$$A(g) + B(g) \rightarrow C(g) + D(g)$$

and the following experimental initial rate data:

	[A(g)]/ mol dm^{-3}	[B(g)]/ mol dm^{-3}	Initial rate/ mol dm^{-3}s^{-1}
Experiment 1	1.50×10^{-2}	1.50×10^{-2}	2.32×10^{-3}
Experiment 2	1.50×10^{-2}	3.00×10^{-2}	4.64×10^{-3}
Experiment 3	3.00×10^{-2}	1.50×10^{-2}	4.64×10^{-3}

a) Deduce the orders with respect to each reactant and the overall reaction order.

b) Deduce the rate equation.

c) Calculate the value of the rate constant, k, for the reaction from experiment 2 and state its units.

d) Determine the rate of the reaction when

$$[A(g)] = 2.00 \times 10^{-2} \text{ mol dm}^{-3} \text{ and}$$
$$[B(g)] = 4.00 \times 10^{-2} \text{ mol dm}^{-3}$$

11 The rate constant, k_1, of a first-order reaction is 6.30×10^3 s^{-1} at 32 °C and the corresponding rate constant, k_2, is 2.25×10^5 s^{-1} at 83 °C.

a) Deduce the activation energy, E_a, in kJ mol^{-1}, correct to **two** significant figures.

b) Calculate the rate constant, k_3, in s^{-1}, at 20 °C.

Introduction

This topic examines the equilibrium law and develops methodology for calculations of the equilibrium constant. The role of Gibbs free energy in describing the relationship that exists with temperature of a reaction and its spontaneity is also examined. This topic highlights the significance of mathematics in the study of chemistry.

17.1 The equilibrium law

Understandings

→ Le Châtelier's principle for changes in concentration can be explained by the equilibrium law.

→ The position of equilibrium corresponds to a maximum value of entropy and a minimum in the value of the Gibbs free energy.

→ The Gibbs free energy change of a reaction and the equilibrium constant can both be used to measure the position of an equilibrium reaction and are related by the equation, $\Delta G = -RT \ln K$.

Applications and skills

→ Solution of homogeneous equilibrium problems using the expression for K_c.

→ Relationship between ΔG and the equilibrium constant.

→ Calculations using the equation $\Delta G = -RT \ln K$.

Nature of science

→ Employing quantitative reasoning – experimentally determined rate expressions for forward and backward reactions can be deduced directly from the stoichiometric equations and allow Le Châtelier's principle to be applied.

The position of equilibrium

In topic 7 we discussed the characteristics of a reversible reaction and dynamic equilibrium. We developed an understanding of the equilibrium constant expression K_c for a reversible reaction, how the equilibrium constant is determined, why it is significant, and what the reaction quotient Q conveys. We began a discussion of how Le Châtelier's principle can explain the effect on the equilibrium of changes in concentration, temperature, and pressure.

The **equilibrium law** can be used to quantify the equilibrium position at a given temperature.

⊕ Calculating the equilibrium constant using concentration data

To calculate the equilibrium constant K_c for a reaction at a given temperature, we follow a series of steps using initial concentrations and equilibrium concentrations. This chapter will focus on homogeneous systems: reactions in which all reactants and products are in the same phase, that is, all gases, all miscible liquids, or all aqueous solutions.

The worked example that follows shows a method that can also be applied to calculations involving weak acids and bases (sub-topic 18.2). A complete understanding of this method is essential.

ICE method for determining the equilibrium constant

1   Deduce the balanced chemical equation for the reaction.

2   Arrange the data according to the **ICE** method:

   I: Initial concentration of the reactants. Initially, [products] = 0.

   C: Change in concentration. This is the amount by which [reactants] decrease and [products] increase. These changes must be consistent with the stoichiometric ratios shown by the coefficients in the balanced equation (sub-topic 1.3).

   E: Equilibrium concentration is the concentration of reactants and products when equilibrium is established. E = I +/− C. See the worked examples that follow for more details.

3   Substitute the values into the equilibrium constant expression and determine the equilibrium constant.

⊕ Worked example

The esterification reaction between ethanol and ethanoic acid produces the ester, ethyl ethanoate, and water. 1.0 mol of ethanol and 1.0 mol of ethanoic acid are dissolved in an inert organic solvent to produce 1 dm³ of the solution and heated in the presence of the catalyst sulfuric acid. When equilibrium is reached 0.60 mol of each reactant remains. Calculate the equilibrium constant K_c.

Solution

$$CH_3CH_2OH(l) + CH_3COOH(l) \overset{H_2SO_4}{\rightleftharpoons} CH_3COOCH_2CH_3(aq) + H_2O(l)$$

I	**1.0**	**1.0**	0.0	0.0
C	$-\alpha$	$-\alpha$	$+\alpha$	$+\alpha$
E	$1.0 - \alpha = \mathbf{0.60}$	$1.0 - \alpha = \mathbf{0.60}$	$0.0 + \alpha$	$0.0 + \alpha$

- The coefficients for α, the change in concentration, must reflect the coefficients of the balanced equation. In this reaction the ratios of reactants and products are 1:1:1:1, so the coefficients for α are all 1. We shall develop this in the following examples.

- If 0.6 mol of reactants remains at equilibrium, then $1.0 - \alpha = 0.60$ and so $\alpha = 0.40$. We can therefore complete the calculations above and substitute the values into the equilibrium expression.

- The volume of the reaction mixture is 1 dm³ so the concentrations in mol dm⁻³ are the same as the amounts of reactants and products in mol.

$$K_c = \frac{[CH_3COOCH_2CH_3][H_2O]}{[CH_3CH_2OH][CH_3COOH]}$$

$$= \frac{(0.0 + 0.40)(0.0 + 0.40)}{(1.0 - 0.40)(1.0 - 0.40)} = \frac{(0.40)^2}{(0.60)^2} = 0.44$$

The equilibrium constant has no units.

TOK

Mathematics is an integral part of the universe. From the symmetry in nature to the presence of geometric shapes in structures and organisms and Fibonacci spirals in plants and animals, mathematics is all around us.

Mathematics can be used to create models that explain the equilibrium systems that we investigate in chemistry. Do scientists create mathematical models that mirror what occurs in an equilibrium system or is mathematics a part of the models that we use because reality is intrinsically mathematical?

▲ Figure 1 The spiral structure of the Nautilus sea shell and its relationship to the Fibonacci sequence and the golden ratio has been the subject of much debate.

Many problems in science involve assumptions that simplify the mathematics. In solving quantitative problems of equilibrium systems, assumptions are made. What is the role of intuition in problem solving?

 Worked example

Sulfur dioxide, SO_2 is oxidized in the presence of the catalyst vanadium(V) oxide, V_2O_5. In this reaction 2.0 mol of SO_2 and 1.4 mol of O_2 are mixed in a 3.0 dm³ sealed container and the system is allowed to come to equilibrium. At 700 K a conversion rate of 15% is achieved. Calculate the equilibrium constant for this reaction.

Solution

$$2SO_2(g) + O_2(g) \rightleftharpoons 2SO_3(g)$$

I	**2.0**	**1.4**	0.0
C	-2α	$-\alpha$	$+2\alpha$
E	1.7	1.25	0.30

- The change in concentrations reflects the coefficients of the balanced equation. For example for SO_2 the change is -2α.

- The conversion rate of 15% means that 15% of SO_2 is converted into product. Hence the equilibrium amount for SO_3 will be:

$2.0 \times 15\% = 0.30$

α can then be found as follows:

$0.30 = +2\alpha$

$\alpha = 0.15$

- Substitute this value into the equilibrium constant expression, remembering to convert the amounts into concentrations.

$$K_c = \frac{[SO_3]^2}{[SO_2]^2[O_2]}$$

$$K_c = \frac{\left(\dfrac{0.30}{3.0}\right)^2}{\left(\dfrac{1.7}{3.0}\right)^2\left(\dfrac{1.25}{3.0}\right)} = 7.5 \times 10^{-2}$$

Gibbs free energy and equilibrium

The Gibbs free energy G describes the spontaneity and temperature dependence of a reaction (sub-topic 15.2). The free energy will change as reactants are converted into products. The reaction will be spontaneous in the direction that results in a decrease in free energy (or the direction in which the free energy value becomes more negative). During this discussion we shall explore the relationship between the free energy, entropy, and position of the equilibrium.

When the equilibrium constant K is determined for a given reaction, its value indicates whether products or reactants are favoured at equilibrium. The Gibbs free energy change ΔG for a given reaction is an indication of whether the forward or reverse reaction is favoured. The relationship between Gibbs free energy and the equilibrium constant K is summarized in table 1.

Equilibrium constant	Description	Gibbs free energy change
$K = 1$	at equilibrium, neither reactants nor products favoured	$\Delta G = 0$
$K > 1$	products favoured	$\Delta G < 0$ (negative value)
$K < 1$	reactants favoured	$\Delta G > 0$ (positive value)

▲ Table 1 The relationship between the equilibrium constant and the Gibbs free energy change

At a given temperature, a negative ΔG value for a reaction indicates that the reaction is spontaneous and the equilibrium concentrations of the products are larger than the equilibrium concentrations of the reactants. The equilibrium constant is greater than 1. The more negative the value of ΔG, the more the forward reaction is favoured and the larger the value of K.

The quantitative relationship between standard Gibbs free energy change, temperature, and the equilibrium constant is described in the equation:

$$\Delta G^{\ominus} = -RT \ln K$$

By rearranging this equation it is possible to calculate the equilibrium constant, and hence deduce the position of equilibrium for the reaction.

$$\ln K = -\frac{\Delta G^{\ominus}}{RT}$$

The standard Gibbs free energy change can be calculated using the methods described in sub-topic 15.2.

> **Study tip**
>
> The expression $\Delta G = -RT \ln K$ is provided in section 1 of the *Data booklet*.
>
> The gas constant (R) has the value and units of 8.31 J K^{-1} mol^{-1}. This is provided in section 2 of the *Data booklet*. The standard Gibbs free energy, $\Delta G°$, has units of kJ mol^{-1}. When using this expression, as shown in the worked example below, either R has to be changed to 0.00831 kJ K^{-1} mol^{-1} or $\Delta G°$ converted to J mol^{-1}.

 Worked example

Calculate the equilibrium constant at 300 K for the oxidation of iron:

$$2Fe(s) + \tfrac{3}{2}O_2(g) \rightleftharpoons Fe_2O_3(s)$$

$$\Delta H^{\ominus} = -824.2 \text{ kJ mol}^{-1}$$

$$\Delta S^{\ominus} = -270.5 \text{ J K}^{-1} \text{ mol}^{-1}$$

Solution

First find $\Delta G^{\ominus}$:

$$\Delta G^{\ominus} = \Delta H^{\ominus} - T\Delta S^{\ominus}$$

$$= -824.2 - (300 \times -0.2705) \text{ kJ mol}^{-1}$$

$$= -743.1 \text{ kJ mol}^{-1}$$

Rearranging the equation to solve for K,

$$\Delta G^{\ominus} = -RT \ln K$$

$$\ln K = -\frac{\Delta G^{\ominus}}{RT}$$

$$\ln K = -\frac{-743.1 \times 10^3 \text{ J mol}^{-1}}{8.31 \text{ J K}^{-1} \text{ mol}^{-1} \times 300 \text{ K}} = 298$$

$$K = e^{298} = 2.6 \times 10^{129}$$

The very large value of K demonstrates that the oxidation of iron at room temperature is highly favoured. Reactions of this nature are said to be irreversible.

17 EQUILIBRIUM (AHL)

Questions

1 Consider the following equilibrium reaction.

$$Cl_2(g) + SO_2(g) \rightleftharpoons SO_2Cl_2(g) \quad \Delta H^\ominus = -84.5 \text{ kJ}$$

In a 1.00 dm³ closed container, at 375 °C, 8.60×10^{-3} mol of SO_2 and 8.60×10^{-3} mol of Cl_2 were introduced. At equilibrium, 7.65×10^{-4} mol of SO_2Cl_2 was formed.

a) Deduce the equilibrium constant expression K_c for the reaction. [1]

b) Determine the value of the equilibrium constant K_c. [3]

c) If the temperature of the reaction is changed to 300 °C, predict, stating a reason in each case, whether the equilibrium concentration of SO_2Cl_2 and the value of K_c will increase or decrease. [3]

d) If the volume of the container is changed to 1.50 dm³, predict, stating a reason in each case, how this will affect the equilibrium concentration of SO_2Cl_2 and the value of K_c. [3]

e) Suggest, stating a reason, how the addition of a catalyst at constant pressure and temperature will affect the equilibrium concentration of SO_2Cl_2. [2]

IB, November 2009

2 When a mixture of 0.100 mol NO, 0.051 mol H_2 and 0.100 mol H_2O were placed in a 1.0 dm³ flask at 300 K, the following equilibrium was established.

$$2NO(g) + 2H_2(g) \rightleftharpoons N_2(g) + H_2O(g)$$

At equilibrium, the concentration of NO was found to be 0.062 mol dm⁻³. Determine the equilibrium constant, K_c, of the reaction at this temperature.

IB, May 2009

3. 0.50 mol of $I_2(g)$ and 0.50 mol of $Br_2(g)$ are placed in a closed flask. The following equilibrium is established.

$$I_2(g) + Br_2(g) \rightleftharpoons 2IBr(g)$$

The equilibrium mixture contains 0.80 mol of IBr(g). What is the value of K_c?

A. 0.64

B. 1.3

C. 2.6

D. 64 [1]

IB, May 2010

4. a) The production of ammonia is an important industrial process.

$$N_2(g) + 3H_2(g) \rightleftharpoons 2NH_3(g)$$

i) Using the average bond enthalpy values in Table 10 of the *Data Booklet*, determine the standard enthalpy change for this reaction. [3]

ii) The standard entropy values, S, at 298 K for $N_2(g)$, $H_2(g)$ and $NH_3(g)$ are 193, 131 and 192 $JK^{-1} mol^{-1}$ respectively. Calculate $\Delta S^\ominus$ for the reaction and with reference to the equation above, explain the sign of $\Delta S^\ominus$. [4]

iii) Calculate $\Delta G^\ominus$ for the reaction at 298 K. [1]

iv) Describe and explain the effect of increasing temperature on the spontaneity of the reaction. [2]

b) The reaction used in the production of ammonia is an equilibrium reaction. Outline the characteristics of a system at equilibrium. [2]

c) Deduce the equilibrium constant expression, K_c, for the production of ammonia. [1]

d) i) 0.20 mol of $N_2(g)$ and 0.20 mol of $H_2(g)$ were allowed to reach equilibrium in a 1 dm³ closed container. At equilibrium the concentration of $NH_3(g)$ was 0.060 mol dm⁻³. Determine the equilibrium concentrations of $N_2(g)$ and $H_2(g)$ and calculate the value of K_c. [3]

ii) Predict and explain how increasing the temperature will affect the value of K_c. [2]

IB, May 2010

18 ACIDS AND BASES (AHL)

Introduction

As our understanding of the reactions of acids and bases has increased, theories have evolved and the range of reactions considered as acid and base reactions has broadened. In this topic, we define Lewis acids and bases and examine their reactions. In topic 7, the equilibrium law described how the equilibrium constant can be determined for a specific chemical reaction at equilibrium. Weak acids and bases partially ionize in water with the reactants and products being in a state of equilibrium. We extend our understanding of K_c and develop our understanding of the acid and base dissociation constants, K_a and K_b respectively. As a quantitative analytical technique, the acid-base titration has wide-ranging applications in scientific research and industry. Increased power of instrumentation has improved the reliability of this technique. This chapter concludes with an in-depth analysis of the pH curve, its features and the chemistry of buffer solutions; a product of specific types of acid-base reactions.

18.1 Lewis acids and bases

Understandings
→ A Lewis acid is a lone pair acceptor and a Lewis base is a lone pair donor.
→ When a Lewis base reacts with a Lewis acid a coordinate bond is formed.
→ A nucleophile is a Lewis base and an electrophile is a Lewis acid.

Applications and skills
→ Application of Lewis acid–base theory to inorganic and organic chemistry to identify the role of the reacting species.

Nature of science
→ Theories can be supported, falsified, or replaced by new theories – acid–base theories can be extended to a wider field of applications by considering lone pairs of electrons. Lewis theory doesn't falsify Brønsted–Lowry but extends it.

Acid–base theories have resulted from collaboration and competition within the global scientific community. Brønsted (Danish), Lowry (British), and Lewis (American) were chemists who lived and worked during the late nineteenth and early twentieth centuries, before computers, the internet, or high-speed communication and transportation. Their endeavours built on the work of scientists before them in moving our understanding of acid–base theory forward.

A straight line between bonded atoms represents a covalent bond in which each atom contributes an equal number of electrons. An arrow between bonded atoms represents a **coordinate bond** in which one atom contributes both electrons involved in forming the covalent bond (see sub-topic 4.2).

Extending our understanding

In developing acid–base theories chemists collected evidence through observation and experimentation and used it to support, refute, or replace existing theories. Rather than falsifying the Brønsted–Lowry theory, the Lewis theory of acids and bases extends our understanding of acid–base reactions, enabling further applications in this field.

Defining Lewis acids and bases

In sub-topic 8.1 a Brønsted–Lowry base was defined as a substance that can accept a proton. It is the presence of at least one lone pair of electrons that allows a Brønsted–Lowry base to form a coordinate bond with a proton. The hydroxide ion and ammonia are good examples of Brønsted–Lowry bases:

G.N. Lewis defined a **Lewis acid** as "an electron pair acceptor" and a **Lewis base** as "an electron pair donor". Lewis focused on a more general definition of acids and bases than Arrhenius and Brønsted–Lowry, enabling a wider range of substances to be included. Ammonia and the hydroxide ion are acting as Lewis bases, donating a pair of electrons to the hydrogen ion. The hydrogen ion is a Lewis acid, as it accepts the electron pair.

Forming coordinate bonds

In the reaction between boron trifluoride and ammonia, no proton is involved. Neither compound in this reaction acts as a Brønsted–Lowry acid or base:

Ammonia donates a lone pair of electrons to form a coordinate bond. Boron, $1s^2 2s^2 2p^1$, forms three sp^2 hybrid orbitals, resulting in a vacant unhybridized $2p_z$ orbital (figure 1). The lone pair on the nitrogen atom forms a coordinate bond with the empty $2p_z$ orbital of the boron atom.

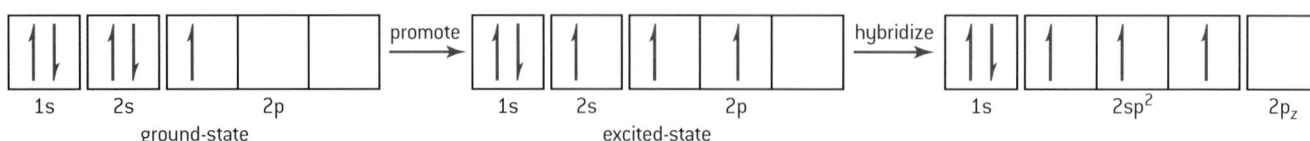

▲ Figure 1 Hybridization of boron in boron trifluoride

Transition elements have a partially occupied d subshell (sub-topic 13.2) so they can form complex ions with ligands that possess a lone pair of electrons. The metal atom or ion is acting as a Lewis acid and the ligand as a Lewis base:

$$Co^{2+}(aq) + 6H_2O(l) \rightarrow [Co(H_2O)_6]^{2+}(aq)$$

$$Ni^{2+}(aq) + 6NH_3(aq) \rightarrow [Ni(NH_3)_6]^{2+}(aq)$$

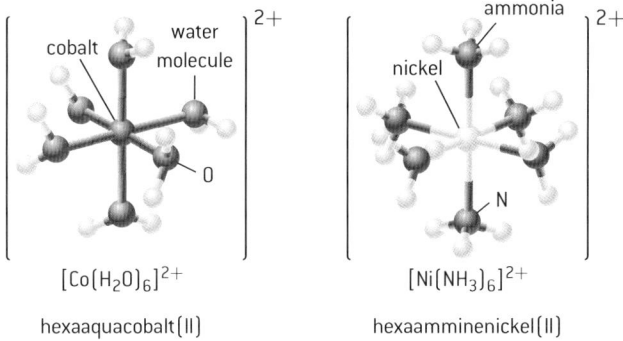

hexaaquacobalt(II)   hexaamminenickel(II)

▲ Figure 2 The ligands in these complex ions are acting as Lewis bases

Water, H_2O and ammonia, NH_3 act as Lewis bases in forming complexes (figure 2). The cyanide ion, CN^-, chloride ion, Cl^-, and hydroxide ion, OH^- can also act as Lewis bases. As well as acting as ligands these species can also act as **nucleophiles** in nucleophilic substitution reactions (sub-topic 20.1). They are electron rich with at least one lone pair of electrons.

An **electrophile** is an electron-deficient species that can accept a lone pair from a nucleophile, in the same way that the nickel(II) ion accepts electron pairs from ammonia.

> A **nucleophile** is a Lewis base and an **electrophile** is a Lewis acid.

18.2 Calculations involving acids and bases

Understandings

→ The expression for the dissociation constant of a weak acid (K_a) and a weak base (K_b).

→ For a conjugate acid base pair, $K_a \times K_b = K_w$.

→ The relationship between K_a and pK_a is $(pK_a = -\log K_a)$, and between K_b and pK_b is $(pK_b = -\log K_b)$.

Applications and skills

→ Solution of problems involving $[H^+(aq)]$, $[OH^-(aq)]$, pH, pOH, K_a, pK_a, K_b, and pK_b.

→ Discussion of the relative strengths of acids and bases using values of K_a, pK_a, K_b, and pK_b.

Nature of science

→ Obtaining evidence for scientific theories – application of the equilibrium law allows strengths of acids and bases to be determined and related to their molecular structure.

The strengths of acids and the acid dissociation constant

Strong acids are assumed to be completely ionized in water, the reaction effectively going to completion. The conjugate base of a strong acid has almost no affinity for a proton. The consequence is that to determine the concentration of the hydrogen ion and subsequently the pH is a simple calculation. For a monoprotic acid, the concentrations of each of the two ions produced is the same as the initial concentration of the strong acid.

$$HCl(aq) + H_2O(l) \rightarrow H_3O^+(aq) + Cl^-(aq)$$

concentration/mol dm^{-3} 0.5   0.5   0.5

Weak acids, such as ethanoic acid, only partially ionize in water. At equilibrium a majority of the ethanoic acid molecules remains unreacted.

$$CH_3COOH(aq) + H_2O(l) \rightleftharpoons CH_3COO^-(aq) + H_3O^+(aq)$$

We can determine the concentration of the dissociated weak acid using the relationship between concentrations of reactants and products and the equilibrium position.

The following general equilibrium constant expression, K_c, can be written for the reaction of a weak acid HA with water:

$$HA(aq) + H_2O(l) \rightleftharpoons A^-(aq) + H_3O^+(aq)$$

$$K_c = \frac{[A^-][H_3O^+]}{[HA][H_2O]}$$

In this reaction [H$_2$O] is considered a constant, and can be removed from the expression. The resulting expression represents the **acid dissociation constant K_a**:

$$K_a = \frac{[A^-][H_3O^+]}{[HA]}$$

A weak base B will also ionize in water. The following expression represents the **base dissociation constant K_b**:

$$B(aq) + H_2O(l) \rightleftharpoons BH^+(aq) + OH^-(aq)$$

$$K_b = \frac{[BH^+][OH^-]}{[B]}$$

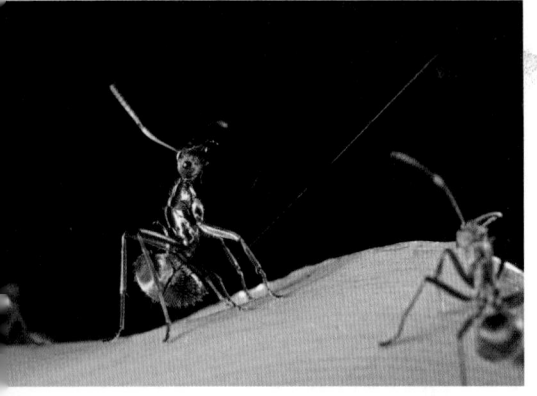
▲ Figure 1 Arboreal ants spray intruders of their nests with methanoic acid (commonly called formic acid)

Calculating K_a and K_b

There are many weak acids and bases that undergo partial ionization in water. The strength of a weak acid or weak base can be expressed quantitatively by determining the dissociation constant at a given temperature.

The stronger the acid, the greater the concentration of hydrogen ions in solution at equilibrium. This corresponds to a larger K_a value. Similarly, the stronger the base, the larger the value of K_b.

Worked examples: dissociation constant

Example 1

Calculate the acid dissociation constant K_a at 298 K for a
0.010 mol dm^{-3} solution of propanoic acid, CH_3CH_2COOH.
The pH of this solution is 3.43.

Solution

Use the pH to calculate [H$^+$] at equilibrium (sub-topics 8.3 and 17.1):

$$[H^+] = 10^{-pH} = 10^{-3.43} = 3.7 \times 10^{-4}$$

$$CH_3CH_2COOH(aq) \rightleftharpoons CH_3CH_2COO^-(aq) + H^+(aq)$$

	CH_3CH_2COOH	$CH_3CH_2COO^-$	H^+
I/mol dm^{-3}	**0.010**	**0.000**	**0.000**
C/mol dm^{-3}	-3.7×10^{-4}	$+3.7 \times 10^{-4}$	$+3.7 \times 10^{-4}$
E/mol dm^{-3}	$0.010 - 3.7 \times 10^{-4}$	3.7×10^{-4}	3.7×10^{-4}

$$K_a = \frac{[CH_3CH_2COO^-][H^+]}{[CH_3CH_2COOH]}$$

$$K_a = \frac{(3.7 \times 10^{-4})(3.7 \times 10^{-4})}{0.010} = 1.4 \times 10^{-5} \text{ mol dm}^{-3}$$

In this calculation, because 3.7×10^{-4} is a very small value, the term
$0.01 - (3.7 \times 10^{-4})$ is rounded to 0.01, which is a valid approximation
within the boundaries of experimental error.

Example 2

Calculate the base dissociation constant K_b at 298 K for a
0.001 00 mol dm^{-3} solution of the base 1-phenylmethanamine,
$C_6H_5CH_2NH_2$. The pH of this solution is 10.17.

Solution

$$pH + pOH = 14$$

$$pOH = 14 - pH = 14 - 10.17 = 3.83$$

$$[OH^-] = 10^{-3.83} = 1.5 \times 10^{-4}$$

$$C_6H_5CH_2NH_2(aq) + H_2O(l) \rightleftharpoons C_6H_5CH_2NH_3^+(aq) + OH^-(aq)$$

	$C_6H_5CH_2NH_2$	$C_6H_5CH_2NH_3^+$	OH^-
I/mol dm^{-3}	**0.001 00**	0.00	0.00
C/mol dm^{-3}	-1.5×10^{-4}	$+1.5 \times 10^{-4}$	$+1.5 \times 10^{-4}$
E/mol dm^{-3}	$0.001 00 - 1.5 \times 10^{-4}$	1.5×10^{-4}	1.5×10^{-4}

$$K_b = \frac{(1.5 \times 10^{-4})^2}{0.001 00 - 1.5 \times 10^{-4}} = 2.6 \times 10^{-5}$$

Study tip

When performing equilibrium
calculations, always state
any approximations and then
explain why they are valid.

In Example 1, the
approximation is valid since
the expression:

$$\frac{3.7 \times 10^{-4}}{0.010} \times 100\% = 3.7\%,$$

which is less than 5%.

Use of approximations

In Example 2, the
approximation is invalid since
the expression:

$$\frac{1.5 \times 10^{-4}}{0.00100} \times 100\% = 15\%,$$

which is greater than 5%.

K_a and K_b for a conjugate acid–base pair

The relationship between the acid dissociation constant for a weak acid and the base dissociation constant of its conjugate base can be useful in calculations. For example, ethanoic acid partially dissociates in water:

$$CH_3COOH(aq) \rightleftharpoons CH_3COO^-(aq) + H^+(aq)$$

$$K_a = \frac{[CH_3COO^-][H^+]}{[CH_3COOH]}$$

The conjugate base of ethanoic acid is the ethanoate ion, CH_3COO^-. It reacts with water according to the following equation.

$$CH_3COO^-(aq) + H_2O(l) \rightleftharpoons CH_3COOH(aq) + OH^-(aq)$$

$$K_b = \frac{[CH_3COOH][OH^-]}{[CH_3COO^-]}$$

Combining these expressions:

$$K_a K_b = \frac{[CH_3COO^-][H^+]}{[CH_3COOH]} \times \frac{[CH_3COOH][OH^-]}{[CH_3COO^-]} = [H^+][OH^-] = K_w$$

In summary:

$$K_a K_b = K_w$$

Other forms of this equation are useful in applying the relationship:

$$K_a = \frac{K_w}{K_b} \quad \text{and} \quad K_b = \frac{K_w}{K_a}$$

Analysing these relationships reinforces the following conclusions:

The stronger the acid:

- the larger the K_a
- the weaker the conjugate base
- the smaller the K_b of the conjugate base.

The stronger the base:

- the larger the K_b
- the weaker the conjugate acid
- the smaller the K_a of the conjugate acid.

The temperature dependence of K_w

In sub-topic 8.3 the ionic product constant K_w was defined:

$$H_2O(l) \rightleftharpoons H^+(aq) + OH^-(aq)$$

$$K_w = [H^+][OH^-] = 1.0 \times 10^{-14} \text{ at 298 K}$$

The ionization of water is an endothermic process. In accordance with Le Châtelier's principle, a change in the temperature of the system will result in a change in the position of equilibrium.

A rise in temperature will result in the forward reaction being favoured, increasing the concentration of the hydrogen and hydroxide ions. This represents an increase in the magnitude of K_w and a decrease in the pH (table 1). Here we make the distinction between the neutrality and the pH of the solution. The pH of the solution decreases with an increase

The value of K_w at different temperatures can be found in section 23 of the *Data booklet*.

Temperature/°C	K_w	pH
15	0.453×10^{-14}	7.17
20	0.684×10^{-14}	7.08
25	1.00×10^{-14}	7.00
30	1.47×10^{-14}	6.92
35	2.09×10^{-14}	6.84

▲ Table 1 The temperature dependence of K_w and pH

in the concentration of hydrogen ions. However, as the concentration of hydroxide ions increases by an equal amount, the solution remains neutral.

pK_a and pK_b

The pH scale is a model that represents very small concentrations of hydrogen ions in a way that is easy to interpret, eliminating the use of negative exponents.

In a similar way, while acid and base dissociation constants are good descriptors of the strengths of weak acids and bases their values can be very small and so difficult to compare; for example, the K_a of ethanoic acid is 1.74×10^{-5}. Therefore K_a and K_b values are represented as pK_a and pK_b respectively:

$$-\log_{10} K_a = pK_a \qquad K_a = 10^{-pK_a}$$
$$-\log_{10} K_b = pK_b \qquad K_b = 10^{-pK_b}$$

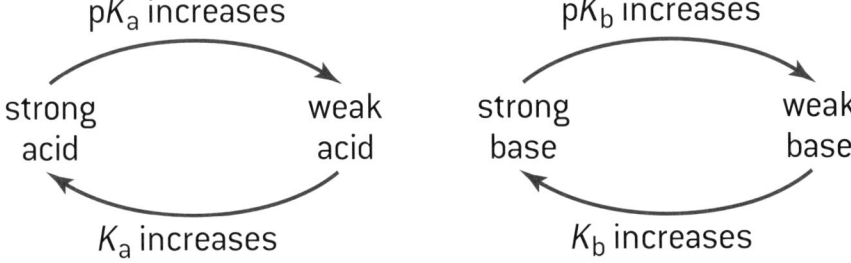

▲ Figure 2 pK_a and pK_b values give a model of strength of acids and bases that is easy to interpret

🌐 Worked example

Calculate the pH of a solution of 0.080 mol dm^{-3} methanoic acid, HCOOH, for which pK_a = 3.75 at 298 K.

Solution

$$K_a = 10^{-pK_a} = 10^{-3.75} = 1.8 \times 10^{-4}$$

$$\text{HCOOH (aq)} \rightleftharpoons \text{HCOO}^-\text{(aq)} + \text{H}^+\text{(aq)}$$

	HCOOH	HCOO⁻	H⁺
I/mol dm^{-3}	0.080	0.000	0.000
C/mol dm^{-3}	$-\alpha$	$+\alpha$	$+\alpha$
E/mol dm^{-3}	$0.080 - \alpha$	$0.000 + \alpha$	$0.000 + \alpha$

$$K_a = \frac{[\text{H}^+][\text{HCOO}^-]}{[\text{HCOOH}]}$$

$$K_a = 1.8 \times 10^{-4} = \frac{\alpha^2}{0.080 - \alpha}$$

$$K_a = 1.8 \times 10^{-4} = \frac{\alpha^2}{0.080}$$

$$\alpha = \sqrt{1.8 \times 10^{-4} \times 0.080}$$

$$\alpha = 3.8 \times 10^{-3}$$

$$[\text{H}^+] = 3.8 \times 10^{-3}$$

$$\text{pH} = 2.42$$

Finding α would require solving a quadratic expression. The small K_a value for this weak acid means that very little dissociation occurs so the value of α is very small. It is acceptable to use the approximation that the initial concentration of the weak acid or base is equal to its equilibrium concentration.

Worked example

Calculate the pH of a 0.25 mol dm^{-3} solution of triethylamine, $(C_2H_5)_3N$. The pK_b of triethylamine at 298 K is 3.25.

Solution

$$(C_2H_5)_3N \text{ (aq)} + H_2O\text{(l)} \rightleftharpoons (C_2H_5)_3NH^+\text{(aq)} + OH^-\text{(aq)}$$

I/mol dm^{-3}	0.25	0.00	0.00
C/mol dm^{-3}	$-\alpha$	$+\alpha$	$+\alpha$
E/mol dm^{-3}	$0.25 - \alpha$	$0.00 + \alpha$	$0.00 + \alpha$

$$K_b = \frac{[(C_2H_5)_3NH^+][OH^-]}{[(C_2H_5)_3N]}$$

$$K_b = 10^{-pK_b} = 10^{-3.25} = 5.6 \times 10^{-4}$$

$$= 5.6 \times 10^{-4} = \frac{\alpha^2}{0.25 - \alpha}$$

$$\alpha = \sqrt{5.6 \times 10^{-4} \times 0.25}$$

$$= 1.2 \times 10^{-3}$$

$$[OH^-] = 1.2 \times 10^{-3}$$

$$pOH = -\log_{10}(1.2 \times 10^{-3}) = 1.9$$

$$pH = 14 - pOH$$

$$= 14 - 1.9 = 12.1$$

18.3 pH curves

Understandings

→ The characteristics of the pH curves produced by the different combinations of strong and weak acid and bases.

→ An acid–base indicator is a weak acid or a weak base where the components of the conjugate acid–base pair have different colours.

→ The relationship between the pH range of an acid–base indicator, which is a weak acid, and its pK_a value.

→ The buffer region on the pH curve represents the region where small additions of acid or base result in little or no change in pH.

→ The composition and action of a buffer solution.

Applications and skills

→ The general shapes of graphs of pH against volume for titrations involving strong and weak acids and bases with an explanation of their important features.

→ Selection of an appropriate indicator for a titration, given the equivalence point of the titration and the end point of the indicator.

→ While the nature of the acid–base buffer always remains the same, buffer solutions can be prepared either by mixing a weak acid/base with a solution of a salt containing its conjugate, or by partial neutralization of a weak acid/base with a strong acid/base.

→ Prediction of the relative pH of aqueous salt solutions formed by the different combinations of strong and weak acid and base.

Nature of science

→ Increased power of instrumentation and advances in available techniques – development in pH meter technology has allowed for more reliable and ready measurement of pH.

Titration

As described in sub-topic 1.3, titration is a quantitative analytical technique used to determine the concentration of a reactant from a reaction of known stoichiometry. Different types of titration have been utilized in a range of industry and research settings for over 150 years. Quantitative data such as that resulting from titrations is subject to mathematical analysis and can help chemists identify patterns and formulate relationships. The many features shown in titration or pH curves unlock a wealth of chemical knowledge.

Buffer solutions

The addition of a single drop of a strong acid or base to water can result in a significant change in pH. A **buffer** is a solution that resists a change in pH upon the addition of small amounts of a strong base or strong acid, or upon the dilution of the buffer through the addition of water.

A buffer may be composed of a weak acid and its conjugate base, or a weak base and its conjugate acid.

The aqueous solution resulting from the reaction between equal amounts of a weak acid and a strong base is alkaline.

The sodium ion will not undergo hydrolysis but the ethanoate ion is the conjugate base of a weak acid and so has a strong affinity for hydrogen ions. The ethanoate ions are hydrolysed with water, producing hydroxide ions:

$$CH_3COO^-(aq) + H_2O(l) \rightleftharpoons CH_3COOH(aq) + OH^-(aq)$$

Figure 3 shows the pH curve for the titration of ethanoic acid with sodium hydroxide.

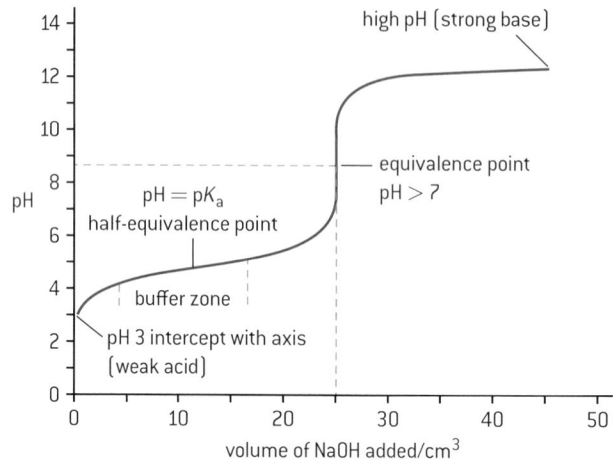

▲ Figure 3 pH curve for the titration of a weak acid with a strong base

- The weak acid gives an initial pH reading ≈ 3.0.

- The initial rise is steep, as a strong base is being added to a weak acid and neutralization is rapid.

- As the weak acid begins to be neutralized the strong conjugate base sodium ethanoate is formed, creating a buffer that resists change in pH. Ethanoic acid is in equilibrium with the ethanoate ion:

$$CH_3COOH(aq) + H_2O(l) \rightleftharpoons CH_3COO^-(aq) + H_3O^+(aq)$$

- The continued addition of base to the solution uses up hydrogen ions, hence the forward reaction is favoured. This results in a very gradual change in pH in this region of the curve.

- The **half-equivalence point** is the stage of the titration at which half of the amount of weak acid has been neutralized:

$$[CH_3COOH(aq)] = [CH_3COO^-(aq)]$$

$$K_a = \frac{[CH_3COO^-][H_3O]^+}{[CH_3COOH]}$$

$$K_a = [H_3O]^+$$

$$\mathbf{pK_a = pH}$$

- There is a sharp rise in pH at the equivalence point (pH > 7). The equivalence point is the result of salt hydrolysis.

- With no remaining acid to be neutralized, the curve flattens and finishes at a high pH due to the presence of excess strong base. This section of the curve is identical to that in figure 2.

The titration of a weak base with a strong acid

The reaction between hydrochloric acid and ammonia is shown in the following equation:

$$HCl(aq) + NH_3(aq) \rightarrow NH_4Cl(aq)$$

or, in ionic form:

$$H^+(aq) + NH_3(aq) \rightarrow NH_4^+(aq)$$

$$NH_4Cl(aq) \rightarrow NH_4^+(aq) + Cl^-(aq)$$

The chloride ion, Cl^- is the conjugate base of the strong acid hydrochloric acid, HCl and has almost no affinity for hydrogen ions.

The ammonium ion, NH_4^+ is the conjugate acid of the weak base ammonia, NH_3. It will donate a proton in the reaction with water, forming the hydronium ion:

$$NH_4^+(aq) + H_2O(l) \rightleftharpoons NH_3(aq) + H_3O^+(aq)$$

Figure 4 shows the pH curve for the titration of ammonia with hydrochloric acid. In this titration ammonia is put into the conical flask and the burette is filled with hydrochloric acid.

> The aqueous solution resulting from the reaction between equal amounts of a strong acid and a weak base is acidic.

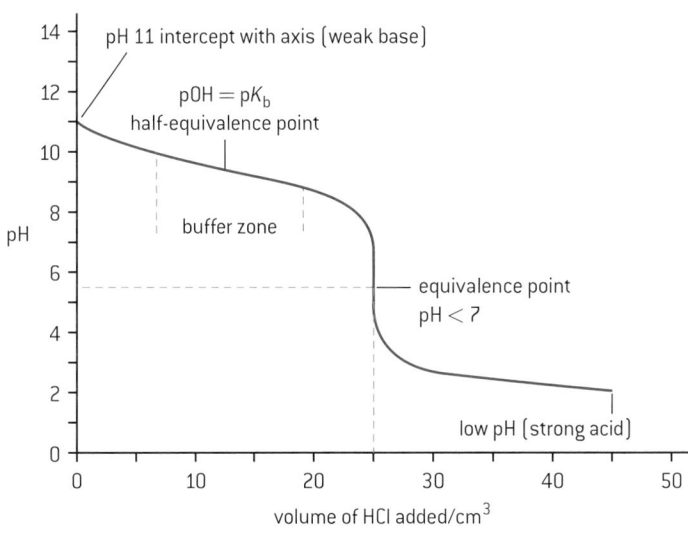

▲ Figure 4 pH curve for the titration of a weak base with a strong acid

- The weak base gives an initial pH reading ≈ 11.0.

- As the weak base begins to be neutralized, the ammonium ion NH_4^+, the conjugate acid, is created resulting in a buffer that resists change in pH. Ammonia is in equilibrium with the ammonium ion:

$$NH_3(aq) + H_2O(l) \rightleftharpoons NH_4^+(aq) + OH^-(aq)$$

- At the half-equivalence point half of the amount of weak base has been neutralized. At this point, $pOH = pK_b$.

- There is a gradual fall in the pH due to the buffering effect as the titration approaches the equivalence point.

- The pH falls sharply at the equivalence point (pH < 7). The equivalence point is the result of salt hydrolysis.

- With no remaining base to be neutralized, the curve flattens and ends at a low pH due to the presence of excess strong acid.

The titration of a weak base with a weak acid

Salts derived from a weak acid and a weak base will undergo hydrolysis in water and the resulting pH of the aqueous solution depends on the relative strengths of the acid (K_a) and base (K_b). Ammonium ethanoate, CH_3COONH_4 forms a neutral aqueous solution:

$$NH_3(aq) + CH_3COOH(aq) \rightarrow CH_3COONH_4(aq)$$

or, in ionic form:

$$NH_3(aq) + CH_3COOH(aq) \rightarrow CH_3COO^-(aq) + NH_4^+(aq)$$

Figure 5 shows the pH curve for the titration of ammonia with ethanoic acid.

● The weak base gives an initial pH reading ≈ 11.0.

● The change in pH throughout the titration is very gradual.

● The point of inflection in the pH curve is not as steep as in the previous pH curves. The point of equivalence is difficult to determine, so this kind of titration has little or no practical use.

● With no remaining base to be neutralized, the curve flattens and ends at a pH that indicates the presence of a weak acid.

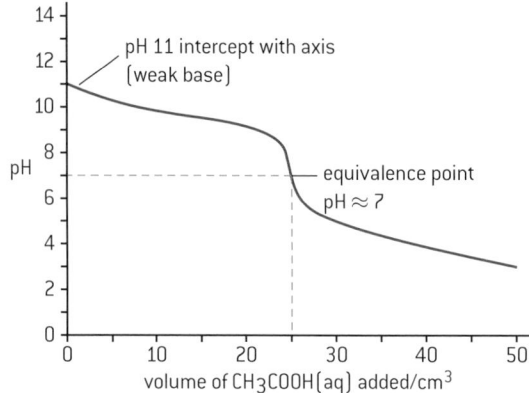

▲ Figure 5 pH curve for the titration of a weak base with a weak acid

Indicators

An indicator is typically a weak acid or a weak base that displays a different colour in acidic or alkaline environments. Many indicators in aqueous solutions behave as weak acids:

$$HIn(aq) \rightleftharpoons H^+(aq) + In^-(aq)$$
colour A   colour B

$$K_a = \frac{[H^+][In^-]}{[HIn]}$$

The above formula can be rearranged as follows:

$$\frac{K_a}{[H^+]} = \frac{[In^-]}{[HIn]}$$

The midpoint of the colour change is observed when $[HIn] = [In^-]$.

At this point:

$$[H^+] = K_a$$
$$pH = pK_a$$

The colour change for most indicators takes place over a range of pH = pK_a ± 1.

The colour of a given indicator depends on the pH of the solution. In acidic solutions the indicator exists in protonated form HIn(aq), so colour A is observed. In basic solutions hydrogen ions in the equilibrium are consumed and the forward reaction is favoured. In this case, the indicator exists as In⁻(aq) and colour B becomes dominant.

An indicator can also be a weak base.

$$BOH(aq) \rightleftharpoons B^+(aq) + OH^-(aq)$$
colour A   colour B

For such indicators, colour A is observed in alkaline solutions while colour B appears in the presence of acids.

Selection of an indicator

The choice of indicator for an acid–base titration depends on the relative strengths of the acid and base and therefore on the pH of the equivalence point. The midpoint of an indicator's colour change must correspond to the equivalence point of the titration.

The titration of a strong acid with a strong base such as hydrochloric acid with sodium hydroxide has an equivalence point of pH 7.0. Phenol red has a pK_a of 7.9 and a pH range of 6.8–8.4. However, the titration curve of a strong acid with a strong base shows a very steep rise near the equivalence point. This rise covers the pH range of most acid–base indicators, so all common indicators, such as phenolphthalein or methyl orange, can be used in such titrations.

The titration of hydrochloric acid with the weak base ammonia has an equivalence point at pH < 7.0. Methyl orange (pK_a = 3.7) is an effective indicator for this titration.

The titration of the weak acid ethanoic acid with sodium hydroxide has an equivalence point at pH > 7.0. Phenolphthalein (pK_a = 9.6) is an effective indicator for this titration (figure 7):

▲ Figure 6 Methyl orange indicator is red in acidic solutions and yellow in alkaline solutions. Most indicators are weak acids but methyl orange is in fact a weak base

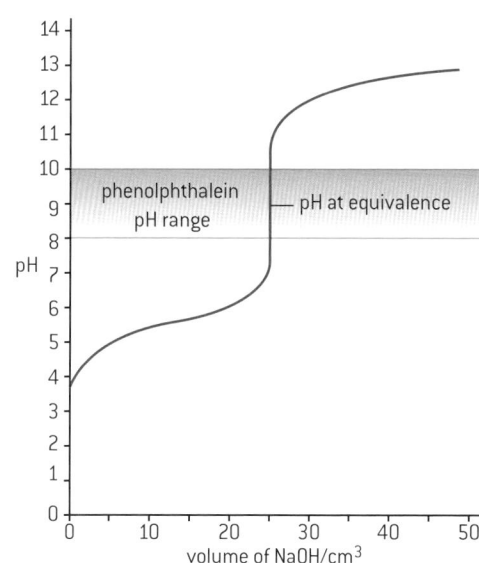

▲ Figure 7 Phenolphthalein indicator is suitable for the titration of ethanoic acid with sodium hydroxide

Titration	Indicator	pK_a	pH range	Acidic colour	Alkaline colour
strong acid–strong base	phenol red	7.9	6.8–8.4	yellow	red
strong acid–weak base	methyl orange	3.7	3.1–4.4	red	yellow
weak acid–strong base	phenolphthalein	9.6	8.3–10.0	colourless	pink

▲ Table 1 Some acid–base indicators commonly used in titrations

Study tip

Examples of acid–base indicators, their pK_a values, and their colour changes are listed in section 22 of the *Data booklet*, which will be available during the examination.

Questions

1 Which of the following could be added to a solution of ethanoic acid to prepare a buffer?

   A. Sodium hydroxide

   B. Hydrochloric acid

   C. Sodium chloride

   D. More ethanoic acid   [1]

   IB, May 2010

2 Which mixture of acid and alkali would produce a buffer solution?

	Acid	Alkali
A.	40 cm³ 0.1 mol dm⁻³ HCl	60 cm³ 0.1 mol dm⁻³ NaOH
B.	60 cm³ 0.1 mol dm⁻³ HCl	40 cm³ 0.1 mol dm⁻³ NaOH
C.	40 cm³ 0.1 mol dm⁻³ HCl	60 cm³ 0.1 mol dm⁻³ NH₃
D.	60 cm³ 0.1 mol dm⁻³ HCl	40 cm³ 0.1 mol dm⁻³ NH₃

[1]

   IB, November 2009

3 When the following 1.0 mol dm⁻³ aqueous solutions are arranged in order of **increasing** pH, which is the correct order?

   I   Ammonium chloride

   II Ammonium ethanoate

   III Sodium ethanoate

   A. I, II, III

   B. II, I, III

   C. III, I, II

   D. III, II, I   [1]

   IB, November 2003

4 Predict and explain, using equations where appropriate, whether the following solutions are acidic, alkaline, or neutral.

   a) 0.1 mol dm⁻³ $FeCl_3$(aq)   [1]

   b) 0.1 mol dm⁻³ $NaNO_3$(aq)   [1]

   c) 0.1 mol dm⁻³ Na_2CO_3(aq)   [1]

   IB, Specimen paper

5 Equal volumes and concentrations of hydrochloric acid and ethanoic acid are titrated with sodium hydroxide solutions of the same concentration. Which statement is correct?

   a) The initial pH values of both acids are equal.

   b) At the equivalence points, the solutions of both titrations have pH values of 7.

   c) The same volume of sodium hydroxide is needed to reach the equivalence point.

   d) The pH values of both acids increase equally until the equivalence points are reached.   [1]

   IB, November 2010

6 A 0.10 mol dm⁻³ ammonia solution is placed in a flask and titrated with a 0.10 mol dm⁻³ hydrochloric acid solution.

   a) Explain why the pH of the ammonia solution is less than 13.   [2]

   b) Estimate the pH at the equivalence point for the titration of hydrochloric acid with ammonia and explain your reasoning.   [2]

   c) State the equation for the reaction of ammonia with water and write the K_b expression for NH_3(aq).   [2]

   d) When half the ammonia has been neutralized (the half-equivalence point), the pH of the solution is 9.25. Deduce the relationship between [NH_3] and [NH_4^+] at the half-equivalence point.   [1]

   e) Determine pK_b and K_b for ammonia based on the pH at the half-equivalence point.   [3]

   f) Describe the significance of the half-equivalence point in terms of its effectiveness as a buffer.   [1]

   IB, May 2010

7 Which species can act as a Lewis acid?

   A. BF_3 C. H_2O

   B. OH^- D. NH_3 [1]

   IB, Specimen paper

8   Which statement explains why ammonia can act as a Lewis base?

A. Ammonia can donate a lone pair of electrons.

B. Ammonia can accept a lone pair of electrons.

C. Ammonia can donate a proton.

D. Ammonia can accept a proton.   [1]

IB, May 2011

9   Which equation represents an acid–base reaction according to the Lewis theory **but not** the Brønsted–Lowry theory?

A. $NH_3 + HCl \rightleftharpoons NH_4Cl$

B. $2H_2O \rightleftharpoons H_3O^+ + OH^-$

C. $NaOH + HCl \rightleftharpoons NaCl + H_2O$

D. $CrCl_3 + 6NH_3 \rightleftharpoons [Cr(NH_3)_6]^{3+} + 3Cl^-$ [1]

IB, November 2003

10 The equilibrium reached when ethanoic acid is added to water can be represented by the following equation:

$$CH_3COOH(l) + H_2O(l) \rightleftharpoons CH_3COO^-(aq) + H_3O^+(aq)$$

Define the terms Brønsted–Lowry acid and Lewis base, and identify two examples of each of these species in the equation.   [4]

IB, November 2005

11 a) Define a Brønsted–Lowry acid.   [1]

b) Deduce the two acids and their conjugate bases in the following reaction: [2]

$H_2O(l) + NH_3(aq) \rightleftharpoons OH^-(aq) + NH_4^+(aq)$

c) Explain why the following reaction can also be described as an acid–base reaction.   [2]

$F^-(g) + BF_3(g) \rightleftharpoons BF_4^-(s)$

IB, May 2009

12 a) Define a Lewis acid and state an example that is not a Brønsted–Lowry acid.   [2]

b) Draw structural formulas to represent the reaction between the Lewis acid named in (a) and a Lewis base and identify the nature of the bond formed in the product.   [4]

IB, November 2009

13 When these 1.0 mol dm^{-3} acidic solutions are arranged in order of increasing strength (weakest first), what is the correct order?

acid in solution X
$\quad K_a = 1.74 \times 10^{-5}$ mol dm^{-3} at 298 K
acid in solution Y
$\quad K_a = 1.38 \times 10^{-3}$ mol dm^{-3} at 298 K
acid in solution Z
$\quad K_a = 1.78 \times 10^{-5}$ mol dm^{-3} at 298 K

A. X < Z < Y   C. Z < X < Y

B. X < Y < Z   D. Y < X < Z   [1]

IB, May 2010

14 pK_w for water at 10 °C = 14.54. What is the pH of pure water at this temperature?

A. 6.73   C. 7.27

B. 7.00   D. 7.54   [1]

IB, May 2010

15 What is K_b for the aqueous fluoride ion given that K_w is 1.0×10^{-14} and K_a for HF is 6.8×10^{-4} at 298 K?

A. $\dfrac{1}{6.8 \times 10^{-4}}$

B. $(6.8 \times 10^{-4})(1.0 \times 10^{-14})$

C. $\dfrac{1.0 \times 10^{-14}}{6.8 \times 10^{-4}}$

D. 6.8×10^{-4} [1]

IB, May 2010

16 Ammonia acts as a weak base when it reacts with water. What is the K_b expression for this reaction?

A. $\dfrac{[NH_4^+][OH^-]}{[NH_3][H_2O]}$ C. $\dfrac{[NH_3]}{[NH_4^+][OH^-]}$

B. $\dfrac{[NH_3][H_2O]}{[NH_4^+][OH^-]}$ D. $\dfrac{[NH_4^+][OH^-]}{[NH_3]}$ [1]

IB, May 2009

17 Ammonia, NH_3, is a weak base. It has a pK_b value of 4.75.

Calculate the pH of a 1.00×10^{-2} mol dm^{-3} aqueous solution of ammonia at 298 K.   [4]

IB, May 2011

18 Ammonia can be converted into nitric acid, $HNO_3(aq)$, and hydrocyanic acid, $HCN(aq)$. The pK_a of hydrocyanic acid is 9.21.

   a) Distinguish between the terms *strong* and *weak acid* and state the equations used to show the dissociation of each acid in aqueous solution. [3]

   b) Deduce the expression for the ionization constant, K_a, of hydrocyanic acid and calculate its value from the pK_a value given. [2]

   c) Use your answer from part (b) to calculate the $[H^+]$ and the pH of an aqueous solution of hydrocyanic acid of concentration 0.108 mol dm^{-3}. State **one** assumption made in arriving at your answer. [4]

   IB, November 2010

19 0.100 mol of ammonia, NH_3, was dissolved in water to make 1.00 dm^3 of solution. This solution has a hydroxide ion concentration of 1.28×10^{-3} mol dm^{-3}.

   a) Determine the pH of the solution. [2]

   b) Calculate the base dissociation constant, K_b, for ammonia. [3]

   IB, November 2009

20 Consider an acid–base indicator solution.

   $$HIn(aq) \rightleftharpoons H^+(aq) + In^-(aq)$$
   colour A   colour B

   What is the effect on this acid–base indicator when sodium hydroxide solution is added to it?

   A. Equilibrium shifts to the right and more of colour B is seen.

   B. Equilibrium shifts to the left and more of colour B is seen.

   C. Equilibrium shifts to the right and more of colour A is seen.

   D. Equilibrium shifts to the left and more of colour A is seen. [1]

   IB, May 2010

21 Which indicator would be the most appropriate for titrating aqueous ethylamine, $CH_3CH_2NH_2$, with nitric acid, HNO_3?

   A. Bromophenol blue ($pK_a = 4.1$)

   B. Bromothymol blue ($pK_a = 7.3$)

   C. Phenol red ($pK_a = 8.0$)

   D. Thymolphthalein ($pK_a = 10.0$) [1]

   IB, November 2009

22 The graph below (figure 8) shows the titration curve of 25 cm^3 of 0.100 mol dm^{-3} of hydrochloric acid with sodium hydroxide, of 0.100 mol dm^{-3} concentration. The indicator methyl orange was used to determine the equivalence point. Methyl orange has a pH range of 3.1–4.4.

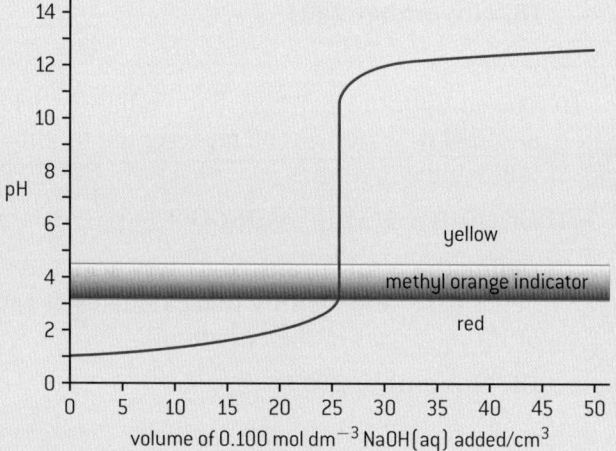

▲ Figure 8

If the hydrochloric acid was replaced by ethanoic acid of the same volume and concentration, which property of the titration would remain the same?

A. The initial pH.

B. The pH at the equivalence point.

C. The volume of strong base, NaOH, needed to reach the equivalence point.

D. The colour of the titration mixture just before the equivalence point is reached. [1]

IB, May 2011

Introduction

In this topic we will explore further voltaic and electrolytic cells which we first met in topic 9. In particular we will see the role that standard electrode potentials play in both types of electrochemical cell. The electrolysis of aqueous solutions will be discussed in this topic and the quantitative aspects of electrolysis.

19.1 Electrochemical cells

Understandings

→ A voltaic cell generates a voltage resulting in the movement of electrons from the anode (negative electrode) to the cathode (positive electrode) via the external circuit. The voltage is termed the cell potential (E).

→ The standard hydrogen electrode (SHE) consists of an inert platinum electrode in contact with 1 mol dm^{-3} hydrogen ions and hydrogen gas at 100 kPa and 298 K. The standard electrode potential ($E^{\ominus}$) is the potential (voltage) of the reduction half-equation under standard conditions measured relative to the SHE. Solute concentration is 1 mol dm^{-3} or 100 kPa for gases. $E^{\ominus}$ of the SHE is 0 V.

→ When aqueous solutions are electrolysed, water can be oxidized to oxygen at the anode and reduced to hydrogen at the cathode.

→ $\Delta G^{\ominus} = -nFE^{\ominus}$. When $E^{\ominus}$ is positive, $\Delta G^{\ominus}$ is negative, indicative of a spontaneous process. When $E^{\ominus}$ is negative, $\Delta G^{\ominus}$ is positive, indicative of a non-spontaneous process. When $E^{\ominus}$ is 0, then $\Delta G^{\ominus}$ is 0.

→ Current, duration of electrolysis, and charge on the ion affect the amount of product formed at the electrodes during electrolysis.

→ Electroplating involves the electrolytic coating of an object with a metallic thin layer.

Applications and skills

→ Calculation of cell potentials using standard electrode potentials.

→ Prediction of whether a reaction is spontaneous or not using $E^{\ominus}$ values.

→ Determination of standard free-energy changes ($\Delta G^{\ominus}$) using standard electrode potentials.

→ Explanation of the products formed during the electrolysis of aqueous solutions.

→ Perform laboratory experiments that could include single replacement reactions in aqueous solutions.

→ Determination of the relative amounts of products formed during electrolytic processes.

→ Explanation of the process of electroplating.

Nature of science

→ Employing quantitative reasoning – electrode potentials and the standard hydrogen electrode.

→ Collaboration and ethical implications – scientists have collaborated to work on electrochemical cell technologies and have to consider the environmental and ethical implications of using fuel cells and microbial fuel cells.

413

Voltaic (galvanic) cells
EMF and the standard cell potential

In topic 9 we examined voltaic (galvanic) cells, which convert chemical to electrical energy. As described by IUPAC an **electromotive force (EMF)** is the energy supplied by a source divided by the electric charge transported through the source. In a voltaic cell the EMF is equal to the electric potential difference **for zero current** through the cell. The EMF is the maximum voltage that can be delivered by the cell.

A helpful analogy

The idea of potential difference can often be difficult to understand. However, here is one useful analogy. Imagine you have two water barrels, A and B, with different volumes of water at two distinct levels in each barrel and with the two barrels connected by a pipe. When the connecting pipe is opened the water will flow from the barrel where the water is at a higher level (that is barrel A) through the open pipe to barrel B, where the water level is less.

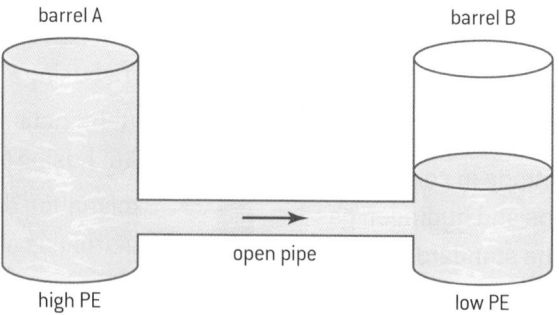

▲ Figure 1 (a) Analogy for the idea of potential difference, of water spontaneously flowing from one barrel, barrel A, where the water is at a higher level, through the pipe when open to barrel B, where the water level is lower. Barrel A could be described as having a greater potential energy (PE) than barrel B

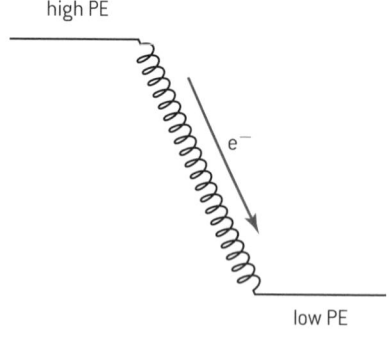

▲ Figure 1(b) The photograph to the left shows the Rio Latus waterfall in Ecuador. The same principle applies at the top of a waterfall with water flowing down into a lake when a suitable pathway is available for this to happen. At the top of the waterfall, the PE is greatest, whereas at the bottom of the waterfall the PE is lowest. In a similar way electrons flow downhill from high PE to low PE, once a suitable pathway is present, eg a conducting wire. The difference in PE between the two electrodes is the EMF

In a voltaic cell **a cell potential** is generated, resulting in the movement of electrons from the anode (negative electrode) to the cathode (positive electrode) via the external circuit. The **cell potential** is then defined

as the potential difference between the cathode and the anode when the cell is operating and is always less than the maximum voltage that can be delivered by the cell. The cell potential also depends on the concentrations of the species involved (that is reactants and products) and the operating temperature (taken in general as 298 K or 25 °C). Under standard conditions (1 mol dm^{-3} concentration for reactants in solution and 100 kPa for gaseous reactants), the cell potential is termed the **standard cell potential** ($E^{\ominus}_{cell}$).

In order to calculate the overall standard cell potential, we use the expression:

$$E^{\ominus}_{cell} = E^{\ominus}_{rhe} - E^{\ominus}_{lhe}$$

where:

$E^{\ominus}_{rhe}$ represents the standard electrode potential at the cathode, which by convention is taken as the right-hand side electrode in a voltaic cell;

$E^{\ominus}_{lhe}$ represents the standard electrode potential at the anode, which by convention is taken as the left-hand side electrode in a voltaic cell.

Section 24 of the *Data booklet* contains a number of standard electrode potentials, given in units of volts. By international agreement with the scientific community standard electrode potentials are always expressed as reductive processes. In order to calculate $E^{\ominus}_{cell}$ for a **spontaneous** cell, the cathode (and hence $E^{\ominus}_{rhe}$) is taken as the more positive value chosen from two standard electrode potentials and the anode (and hence $E^{\ominus}_{lhe}$) is taken as the least positive value.

Let us return to the Daniell cell, first introduced in topic 9 for redox processes. The two electrodes are Zn(s)|Zn^{2+}(aq) and Cu(s)|Cu^{2+}(aq). From section 24 of the *Data booklet*, the two standard electrode potentials are as follows:

$$Zn^{2+}(aq) + 2e^- \rightleftharpoons Zn(s) \qquad E^{\ominus} = -0.76 \text{ V}$$
$$Cu^{2+}(aq) + 2e^- \rightleftharpoons Cu(s) \qquad E^{\ominus} = +0.34 \text{ V}$$

Notice that both are written as reduction half-equations with an equilibrium sign, and signs are always included (+ or −) in this tabular format. Based on the two $E^{\ominus}$ values, we see that, since +0.34 V is more positive, the Cu^{2+}(aq)|Cu(s) electrode is the cathode (positive electrode), and therefore reduction will take place at this electrode. The half-equations corresponding to the processes occurring at the cathode and anode can then be written as follows:

Cathode (positive electrode): Reduction takes place here.

$$Cu^{2+}(aq) + 2e^- \rightarrow Cu(s) \qquad E^{\ominus}_{rhe} = +0.34 \text{ V}$$

Anode (negative electrode): Oxidation takes place here.

$$Zn(s) \rightarrow Zn^{2+}(aq) + 2e^- \qquad E^{\ominus}_{lhe} = -0.76 \text{ V}$$

Overall cell reaction:

$$Cu^{2+}(aq) + Zn(s) \rightarrow Zn^{2+}(aq) + Cu(s)$$

The overall standard cell potential for the Daniell cell, $E^{\ominus}_{cell}$, can be calculated as follows:

$$E^{\ominus}_{cell} = E^{\ominus}_{rhe} - E^{\ominus}_{lhe} = (+0.34 \text{ V}) - (-0.76 \text{ V}) = +1.10 \text{ V}$$

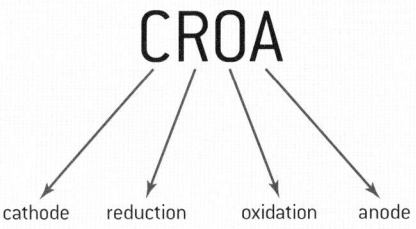

Study tip

Consider the Daniell cell:

$$Zn(s)|Zn^{2+}(aq) \parallel Cu^{2+}(aq)|Cu(s)$$

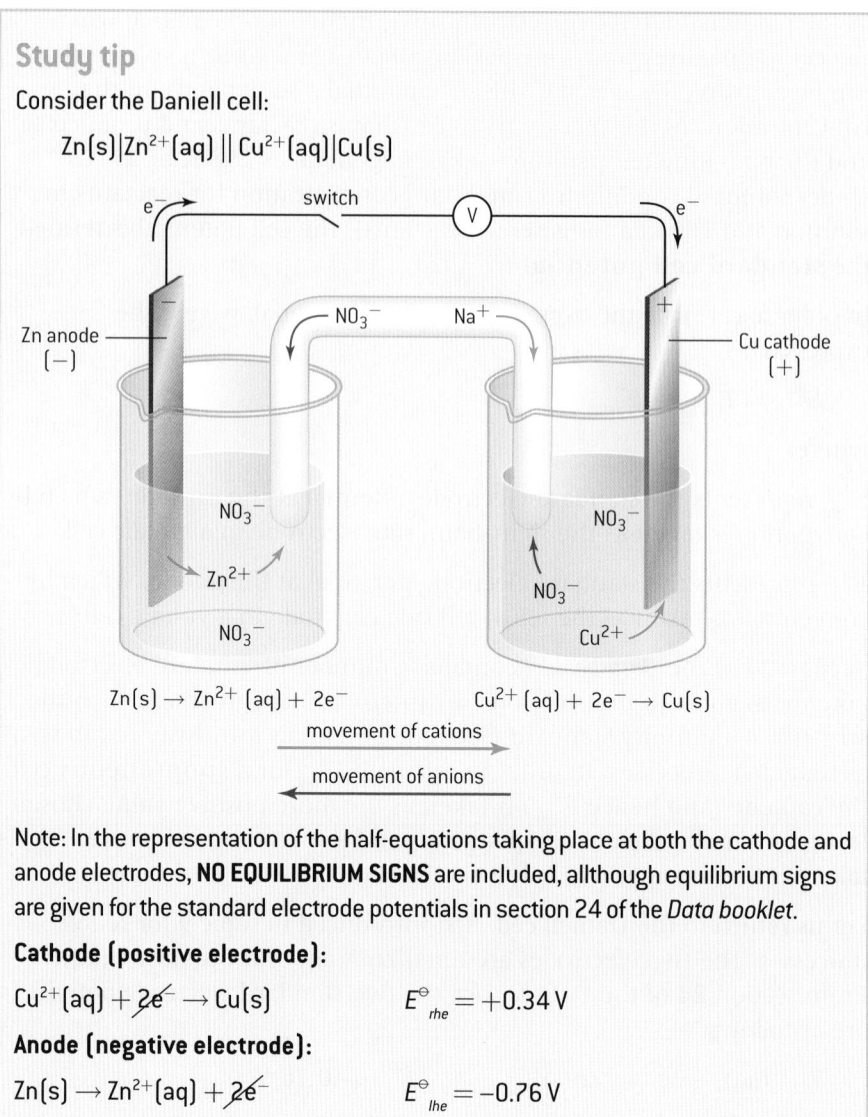

Note: In the representation of the half-equations taking place at both the cathode and anode electrodes, **NO EQUILIBRIUM SIGNS** are included, allthough equilibrium signs are given for the standard electrode potentials in section 24 of the *Data booklet*.

Cathode (positive electrode):

$$Cu^{2+}(aq) + 2e^- \rightarrow Cu(s) \qquad E^{\ominus}_{rhe} = +0.34\,V$$

Anode (negative electrode):

$$Zn(s) \rightarrow Zn^{2+}(aq) + 2e^- \qquad E^{\ominus}_{lhe} = -0.76\,V$$

The standard hydrogen electrode

It is not possible to measure the electrode potential of a single half-cell, as in order to measure the potential we require a potential energy difference for the electrons, which must be in two chemically different set-ups. For this reason, electrode potentials are measured relative to an internationally agreed standard, which has been chosen as the **standard hydrogen electrode** (SHE). The SHE is the universal reference electrode and is a **gas electrode**. The standard hydrogen electrode (SHE) consists of an inert platinum electrode in contact with 1 mol dm⁻³ hydrogen ions and hydrogen gas at 100 kPa.

Therefore, the **standard electrode potential** ($E^{\ominus}$) is the potential (voltage) of the reduction half-equation under standard conditions measured relative to the SHE. The standard concentration for a solute is 1 mol dm⁻³ and under standard conditions the pressure is 100 kPa for gases. $E^{\ominus}$ of the SHE is taken as 0 V at all temperatures. The potentials of other electrodes are then compared to the SHE reference at the same temperature. The reduction half-equation corresponding to the SHE half-cell is:

$$2H^+(aq) + 2e^- \rightleftharpoons H_2(g)$$

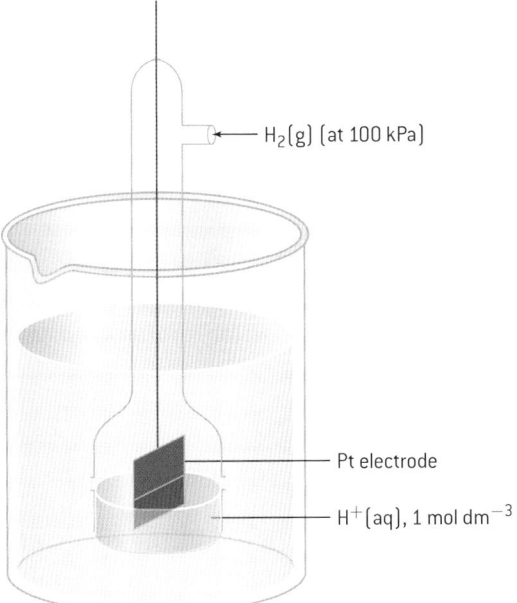

▲ Figure 2 The standard hydrogen electrode (SHE)

The standard electrode potential of another half-cell is then determined simply by connecting the half-cell, under standard conditions to the SHE, using a connecting wire with a voltmeter attached and a salt bridge. The cell potential can then be determined.

Let us consider the following cell consisting of the $Cu^{2+}(aq)|Cu(s)$ half-cell connected to the SHE.

Cathode (positive electrode):

$$Cu^{2+}(aq) + 2e^- \rightarrow Cu(s) \qquad E^\ominus_{rhe} = +0.34 \text{ V}$$

Anode (negative electrode):

$$H_2(g) \rightarrow 2H^+(aq) + 2e^- \qquad E^\ominus_{lhe} = 0.00 \text{ V}$$

This cell can be represented by the following cell diagram:

$$Pt(s)|H_2(g)|H^+(aq) \parallel Cu^{2+}(aq)|Cu(s)$$

$$E^\ominus_{cell} = E^\ominus_{rhe} - E^\ominus_{lhe} = (+0.34 \text{ V}) - (0.00 \text{ V}) = +0.34 \text{ V}$$

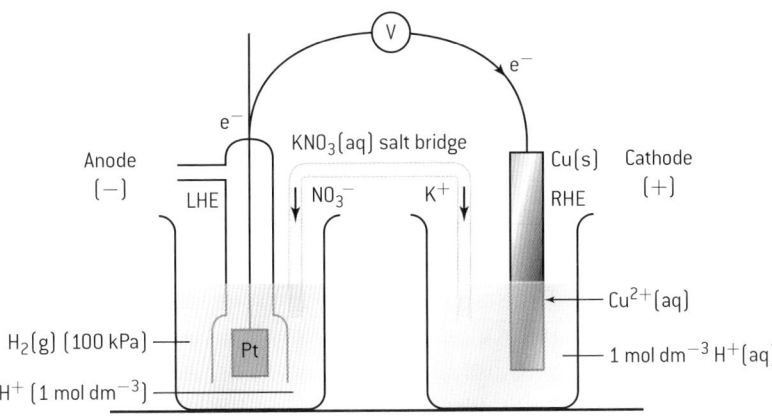

▲ Figure 3 A voltaic cell with a $Cu^{2+}(aq)|Cu(s)$ half-cell connected to the SHE

TOK

The absolute electrode potential of the standard hydrogen electrode under standard conditions has the estimated value: $E^\ominus_{abs} = (+4.44 \pm 0.02)$ V at 298 K. However, for comparison purposes with all other electrode reactions, hydrogen's standard electrode potential is assigned as 0 V **at all temperatures**. The SHE is a universal reference electrode and is an example of an arbitrary reference.

Since the SHE is an example of an arbitrary reference, consider whether or not our scientific knowledge would be the same if we chose different references?

Can you think of other examples of arbitrary references in chemistry, which we met already in an earlier topic? See Example 2 on page 419.

Study tip

From physics, energy = potential × charge. The SI unit of potential (voltage) is the volt and the SI unit of charge is the coulomb. Since the joule is the SI unit of energy:

$$1\,J = 1\,V \times 1\,C$$

Cell potential and Gibbs free energy

If the standard cell potential, $E^{\ominus}_{cell}$, is positive, a redox reaction will be spontaneous. If the standard cell potential, $E^{\ominus}_{cell}$, is negative, the redox reaction is non-spontaneous. In topic 15 we saw that $\Delta G^{\ominus}$, the standard change in Gibbs free energy, is negative for a spontaneous reaction and positive for a non-spontaneous reaction. $\Delta G^{\ominus}$ is related to $E^{\ominus}_{cell}$ by the following expression:

$$\Delta G^{\ominus} = -nFE^{\ominus}_{cell}$$

where:

n = amount, in mol, of electrons transferred in the balanced equation;

F = Faraday's constant = 96500 C mol^{-1} (given in section 2 of the *Data booklet*);

$E^{\ominus}_{cell}$ = standard cell potential (calculated from $E^{\ominus}_{cell} = E^{\ominus}_{rhe} - E^{\ominus}_{lhe}$);

Faraday's constant (F) is the charge in coulombs of 1 mol of electrons.

Hence in the Daniell cell:

$$Cu^{2+}(aq) + Zn(s) \rightarrow Zn^{2+}(aq) + Cu(s)$$

$\Delta G^{\ominus} = -nFE^{\ominus}_{cell} = -(2\ mol\ e^-)(96500\ C\ mol^{-1}\ e^-)(+1.10\ V)$
$= -212300\ VC = -2.12 \times 10^5$ J. Since $\Delta G^{\ominus}$ is negative, this reaction is spontaneous under standard conditions.

 Worked examples

Example 1

Consider the following table of standard electrode potentials.

	$E^{\ominus}/V$
$Al^{3+}(aq) + 3e^- \rightleftharpoons Al(s)$	−1.66
$Cr^{3+}(aq) + 3e^- \rightleftharpoons Cr(s)$	−0.74
$Co^{2+}(aq) + 2e^- \rightleftharpoons Co(s)$	−0.28
$Sn^{4+}(aq) + 2e^- \rightleftharpoons Sn^{2+}(aq)$	+0.15
$\frac{1}{2}Cl_2(g) + e^- \rightleftharpoons Cl^-(aq)$	+1.36

a) Deduce the species which is the strongest oxidizing agent.

b) Deduce the species which can reduce $Cr^{3+}(aq)$ to $Cr(s)$ under standard conditions.

c) Deduce the species which can reduce $Sn^{4+}(aq)$ to $Sn^{2+}(aq)$ but not $Cr^{3+}(aq)$ to $Cr(s)$ under standard conditions.

d) The standard electrode potential for the half-cell made from cobalt metal, Co(s), in a solution of cobalt(II) ions, $Co^{2+}(aq)$ has the value of −0.28 V. Explain the significance of the negative sign in −0.28 V.

Solution

a) The higher the standard electrode potential, the greater the ability of the species to gain electrons, so the strongest oxidizing agent is $Cl_2(g)$ with $E^\ominus = +1.36$ V

b) To reduce $Cr^{3+}(aq)$ to $Cr(s)$, we need a species with an $E^\ominus$ of less than -0.74 V. The only species with such potential ($E^\ominus = -1.66$ V) involves $Al(s)$ as the reducing species.

c) The species that can reduce $Sn^{4+}(aq)$ to $Sn^{2+}(aq)$ but not $Cr^{3+}(aq)$ to $Cr(s)$ must have $E^\ominus$ lower than $+0.15$ V but greater than -0.74 V. The only such species in the table is $Co(s)$ with $E^\ominus = -0.28$ V.

d) When a cell with a negative $E^\ominus$ is connected to the SHE ($E^\ominus = 0.00$ V), the SHE will act as the cathode (positive electrode), and reduction will take place here, whereas the $Co(s)|Co^{2+}(aq)$ half-cell will act as the anode (negative electrode), where oxidation takes place. This means that, at the anode, there will be a loss of electrons and hence electrons will flow from the $Co(s)|Co^{2+}(aq)$ half-cell, the anode, to the SHE, the cathode.

Example 2

The standard hydrogen electrode (SHE) is an example of an arbitrary reference.

a) Describe the SHE, using an annotated diagram.

b) Describe the functions of the platinum electrode in the SHE.

c) State **one** other example of an arbitrary reference in chemistry.

Solution

a)

- The SHE consists of a platinum electrode, $Pt(s)$, with hydrogen gas, $H_2(g)$, at a pressure of 100 kPa bubbled into a 1 mol dm^{-3} H^+ solution (eg HCl). The conditions involved are standard-state conditions. Hydrogen is bubbled through the tube and into the solution, where the following reaction takes place:

$$2H^+(aq) + 2e^- \rightleftharpoons H_2(g)$$

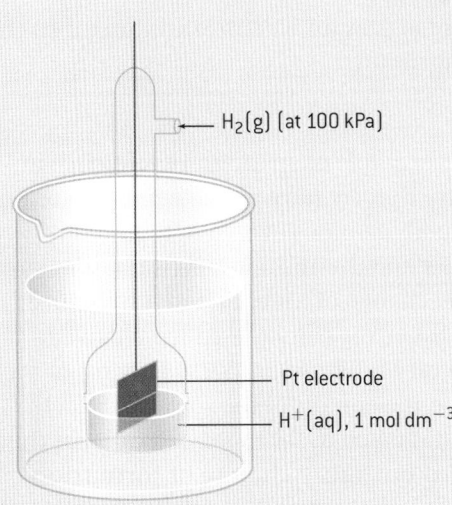

H$_2$(g) (at 100 kPa)

Pt electrode

H$^+$(aq), 1 mol dm^{-3}

The temperature is often quoted as 298 K (that is 25 °C) possibly due to the fact that the absolute electrode potential of the hydrogen electrode under standard conditions has the estimated value: $E^\ominus_{abs} = (4.44 \pm 0.02)$ V at 298 K. However, under such standard-state conditions the potential for the reduction of $H^+(aq)$ to $H_2(g)$, $E^\ominus$, is taken to be 0 V **at all temperatures.**

b)

Note: the platinum electrode, $Pt(s)$, a is actually **platinized platinum**, that is the platinum metallic surface is coated with finely divided platinum, thereby increasing its surface area.

The functions of the $Pt(s)$ electrode are:

- Platinum is an inert metal and does not corrode or ionize. It will not act as an electrode in the system.

- Platinum can act as a heterogeneous catalyst. It provides a surface to allow the dissociation of the molecules of hydrogen. Hydrogen absorbs on its surface (if the SHE acts as the anode):

$$H_2(g) \rightarrow 2H^+(aq) + 2e^-$$

The platinum provides the surface where transfer of electrons can occur.

Note that the reverse reaction would be the case if the SHE was acting as the cathode.

- An equilibrium is established between the adsorbed molecules of hydrogen, $H_2(g)$, on the Pt surface and the hydrogen ions, $H^+(aq)$:

$$2H^+(aq) + 2e^- \rightleftharpoons H_2(g)$$

419

- Platinum acts as an electrical conductor to the external circuit.

c)

- Another example of an arbitrary reference in chemistry that we have already encountered in thermodynamics, covered in topic 5 on energetics and thermchemistry, is the standard enthalpy change of formation of a substance, ΔH_f^o(substance). This is described relative to the arbitrary reference of ΔH_f^o(element) of the component elements of the substance, each of which is taken as 0 kJ mol^{-1}. Recall that the **standard enthalpy change of formation** of a substance is defined as the enthalpy change when 1 mol of a compound is formed from its elements in their standard states (that is at 100 kPa). The standard enthalpy of formation, ΔH_f^o could be described as a "thermodynamic sea-level" reference just like the SHE could be described as an "electrochemical sea-level" reference.

Note: **standard state** itself is also an example of an arbitrary reference. Recall that standard state is the most stable state of a substance under standard conditions. It is the state of a system chosen as standard for reference by convention. As defined by IUPAC, for a gas phase it is the hypothetical state of the pure substance in that state at 100 kPa. For a pure phase, or a mixture, or a solvent in the liquid or solid state, it is the state of the pure substance in the liquid or solid phase at 100 kPa. For a solute in solution, it is the hypothetical state of the solute at 1 mol dm^{-3} concentration at 100 kPa and showing infinitely dilute solution behaviour.

Carbon 12, as discussed in topic 1, was chosen as the arbitrary reference standard for the SI unit of amount of substance (the mole).

Example 3

Consider the following electrochemical reaction, which takes place in a voltaic cell at 298 K:

$Sn^{2+}(aq) + 2Fe^{3+}(aq) \rightarrow Sn^{4+}(aq) + 2Fe^{2+}(aq)$

Given: $Sn^{4+}(aq) + 2e^- \rightleftharpoons Sn^{2+}(aq)$ $E^\ominus = +0.15$ V

a) Identify the half-equations occurring at the cathode and anode electrodes.

b) Identify the oxidizing and reducing agents.

c) State the cell diagram convention for the cell.

d) Calculate the standard potential, in V, for this cell.

e) **(i)** Determine $\Delta G^\ominus$, the standard change in Gibbs free energy at 298 K, in kJ, for the electrochemical reaction.

(ii) Comment on the spontaneity of the reaction.

Solution

a)

- Using the information given for the $Sn^{2+}(aq)|\, Sn^{4+}(aq)$ electrode and section 24 of the *Data booklet*:

$Sn^{4+}(aq) + 2e^- \rightleftharpoons Sn^{2+}(aq)$ $E^\ominus = +0.15$ V

$Fe^{3+}(aq) + e^- \rightleftharpoons Fe^{2+}(aq)$ $E^\ominus = +0.77$ V

- Since $E^\ominus$ for the $Fe^{2+}(aq)|Fe^{3+}(aq)$ half-cell is more positive, this is deemed the RHE, that is the cathode where reduction takes place. Since $E^\ominus$ for the $Sn^{2+}(aq)|\, Sn^{4+}(aq)$ half-cell is less positive, this is the LHE, that is the anode where oxidation takes place. Hence the two half-equations taking place at the two electrodes are as follows:

Cathode (positive electrode): Reduction takes place here.

$Fe^{3+}(aq) + e^- \rightarrow Fe^{2+}(aq)$ $E^\ominus_{rhe} = +0.77$ V

Anode (negative electrode): Oxidation takes place here.

$Sn^{2+}(aq) \rightarrow Sn^{4+}(aq) + 2e^-$ $E^\ominus_{lhe} = +0.15$ V

The electrons are not balanced, so the cathode half-equation must be multiplied by two to generate the overall reaction:

Cathode (positive electrode):

$2Fe^{3+}(aq) + 2e^- \rightarrow 2Fe^{2+}(aq)$ $E^\ominus_{rhe} = +0.77$ V

b)

Oxidizing agent:
$Fe^{3+}(aq)$.

Reducing agent:
$Sn^{2+}(aq)$.

c)

Since all reacting species are present in the aqueous phases, inert electrodes, Pt(s), must be used:

$Pt(s)|Sn^{2+}(aq), Sn^{4+}(aq) \,||\, Fe^{3+}(aq), Fe^{2+}(aq)|Pt(s)$

LHE	RHE
anode	cathode

Study tip

Note that the standard cell potential $E^\ominus$ values are **not** affected by coefficients and you should **never** multiply the $E^\ominus$ values by the integer mole ratios. The reason for this is that $E^\ominus$ is an example of an **intensive property**, that is it is independent of quantity of sample (other examples of intensive properties include density, temperature, and melting point). This differs from **extensive properties** (such as volume and mass) which depend on amount of substance. In contrast, the standard Gibbs free energy change, $\Delta G^\ominus$, is an example of an extensive property and if an equation is multiplied by a factor, n will change and hence also $\Delta G^\ominus$, from the expression $\Delta G^\ominus = -nFE^\ominus_{cell}$, as seen in (e) in Example 3.

If we return to the analogy of the waterfall conveying the idea of PE difference, it does not matter whether 15000 dm^3 of water or 30000 dm^3 of water falls from the top to the bottom of the waterfall – as long as there exists a pathway for the water to gush down the waterfall, the difference in height between the top and bottom will stay constant. Hence, if an equation is multiplied by a factor, the stoichiometry coefficients will change and hence the number of electrons will change but **not** the potential difference through which electron transfer occurs.

Question

Construct and annotate an example of a voltaic cell consisting of a Ni(s)|Ni^{2+}(aq) half-cell and a Cu(s)|Cu^{2+}(aq) half-cell.

a) Identify the half-equations occurring at the cathode and anode electrodes.

b) Deduce the equation for the **spontaneous** reaction occurring in this cell.

c) State the cell diagram convention for the cell.

d) Identify the direction of the movement of electrons and ion flow, both in solution and in the salt bridge.

e) Calculate the standard potential, in V, for this cell.

f) Determine $\Delta G^\ominus$, the standard change in Gibbs free energy, in J, for the electrochemical reaction.

Note that the spontaneity of the reaction can also be deduced from the sign of $E^\ominus_{cell}$. A positive $E^\ominus_{cell}$ is indicative of a spontaneous redox reaction whereas a negative $E^\ominus_{cell}$ is indicative of a non-spontaneous redox reaction.

Study tip

Relationships between $\Delta G^\ominus$ and $E^\ominus_{cell}$ $(\Delta G^\ominus = -nFE^\ominus_{cell})$

$\Delta G^\ominus$	$E^\ominus_{cell}$	Reaction under standard-state conditions
negative	positive	spontaneous, so will favour formation of products
positive	negative	non-spontaneous, so will favour formation of reactants
zero	zero	both products and reactants will be favoured equally

d)

$$E^\ominus_{cell} = E^\ominus_{rhe} - E^\ominus_{lhe} = (+0.77\text{ V}) - (+0.15\text{ V})$$
$$= +0.62\text{ V}$$

e)

(i) $\Delta G^\ominus = -nFE^\ominus_{cell} = -(2\text{ mol e}^-)(96500\text{ C mol}^{-1}\text{ e}^-)$
$(+0.62\text{ V}) = -119660\text{ V C}$
$= -1.2 \times 10^5\text{ J} = -1.2 \times 10^2\text{ kJ}$.

(ii) Since $\Delta G^\ominus$ is negative this reaction is spontaneous under standard conditions.

Study tip

You should remember **standard-state conditions** which can be summarized simply as follows:

(s), (l), (g) as pure substances at a pressure of 100 kPa

Solutes at 1 mol dm^{-3} concentration

Note that temperature is **not** a formal requirement in the description of standard state, but 298 K (25 °C) is often quoted in thermodynamic tables as the specified temperature!

Electrolytic cells

In topic 9 we also examined a second type of electrochemical cell, the electrolytic cell, which converts electrical to chemical energy. In topic 9, we looked at one type of electrolytic cell, the electrolysis of a molten salt. We shall now examine another type of electrolysis, namely the electrolysis of aqueous solutions.

We will consider the following examples of electrolysis of aqueous solutions:

a) Electrolysis of aqueous sodium chloride

    (i) Concentrated solution

    (ii) Dilute solution

b) Electrolysis of aqueous copper(II) sulfate

    (i) Using inert graphite (carbon) electrodes

    (ii) Using active copper electrodes

c) Electrolysis of water

(a) Electrolysis of *concentrated* aqueous sodium chloride

Unlike the electrolysis of molten sodium chloride discussed in topic 9, in the electrolysis of **concentrated** aqueous sodium chloride there is an additional species to be considered, namely water! Let us consider the species present at each electrode:

Cathode (negative electrode):

    $Na^+(aq)$, $H_2O(l)$

Anode (positive electrode):

    $Cl^-(aq)$, $H_2O(l)$

- In order to determine the most relevant half-equation corresponding to each electrode process, you should first write down the two half-equations taking place at each electrode and the corresponding $E^{\ominus}$ values using section 24 of the *Data booklet*. (*Remember that in the Data booklet the standard electrode potential values relate to reductive processes. Hence, since reduction takes place at the cathode, the sign of $E^{\ominus}$ taken from section 24 in the Data booklet will be correct for the reductive process. However, when you write any half-equation for an oxidation reaction taking place at the anode, the sign of $E^{\ominus}$ will have to be switched if using the section 24 Data booklet values.*)

- In addition, if you examine section 24, you will pick out two different equations involving water as a single species on either side of the equilibrium sign (*you can ignore any other equations involving water where it is not written on its own on either side of the equilibrium sign*):

$$H_2O(l) + e^- \rightleftharpoons \frac{1}{2}H_2(g) + OH^-(aq) \qquad E^{\ominus} = -0.83 \text{ V}$$

Water on its own as a single species on one side of the equilibrium sign

$$\frac{1}{2}O_2(g) + 2H^+(aq) + 2e^- \rightleftharpoons H_2O(l) \qquad E^{\ominus} = +1.23 \text{ V}$$

Study tip

Remember the mnemonic **CNAP** for electrolytic cells:

Cathode – **Negative**
Anode – **Positive**

This differs from voltaic cells, where the cathode is the positive electrode and the anode is the negative electrode. However, for both electrochemical cells, reduction takes place at the cathode and oxidation takes place always at the anode.

Let us look first at the possible half-equations that take place at the cathode, as there will be no change in the sign of $E^{\ominus}$ here, since it will be a reductive process.

Cathode (negative electrode):

$$Na^+(aq) + e^- \rightarrow Na(s) \qquad\qquad E^{\ominus} = -2.71 \text{ V}$$

$$H_2O(l) + e^- \rightarrow \frac{1}{2}H_2(g) + OH^-(aq) \qquad E^{\ominus} = -0.83 \text{ V}$$

The half-equation for water is chosen as written, since $H_2O(l)$ will be the species present at the cathode (not $O_2(g)$, as seen in the other half-equation showing the reduction process).

Anode (positive electrode):

$$2Cl^-(aq) \rightarrow Cl_2(g) + 2e^- \qquad\qquad E^{\ominus} = -1.36 \text{ V}$$

$$H_2O(l) \rightarrow \frac{1}{2}O_2(g) + 2H^+(aq) + 2e^- \qquad E^{\ominus} = -1.23 \text{ V}$$

At the anode, oxidation takes place, so any sign of $E^{\ominus}$ for a reductive process taken from section 24 of the *Data booklet* will have to be inverted (as will the half-equation) to indicate the oxidative process. This is the reason why the signs have been changed above. In addition, since we have used the first of the two half-equations for water given in the table to describe the cathode half-equation, we now use the second one here to describe a possible anode half-equation.

We now have to decide which of these two half-equations is preferred. *In general, the more positive $E^{\ominus}$ value would indicate the preferred reaction, suggesting that oxidation of water would be preferential.* **HOWEVER, in the case of concentrated aqueous sodium chloride this is not as simple as the rule suggests.** The reason for this is the phenomenon termed **overvoltage**. In an electrolytic experiment the applied potential needed to carry out the electrolysis is always greater than the potential calculated from the standard redox potentials. This extra difference in potential or voltage is the overvoltage. Many reactions taking place at electrodes are extremely slow. Therefore, this additional voltage effectively is the extra voltage required for a reaction with a slow rate to proceed at a reasonable rate in an electrochemical cell. As a result the oxidation reaction taking place at the anode would actually require a potential greater than 1.23 V to occur (the overvoltage for oxygen gas formation is quite high compared to chlorine gas formation), suggesting that, at the anode in a concentrated solution of sodium chloride, the chloride ions, $Cl^-(aq)$, are actually reduced to chlorine gas, $Cl_2(g)$, which is what is observed experimentally. Note that this can only be confirmed from experimental evidence.

In the case of a **dilute** aqueous solution of sodium chloride, overvoltage does not play such a role and the half-equation taking place at the anode can simply be worked out using $E^{\ominus}$ values.

The half-equation taking place at the anode in the electrolysis of **concentrated** aqueous sodium chloride will be:

$$2Cl^-(aq) \rightarrow Cl_2(g) + 2e^- \qquad\qquad E^{\ominus} = -1.36 \text{ V}$$

Hence, let's combine the two electrode half-equations to generate the overall cell reaction:

$$2H_2O(l) + 2Cl^-(aq) \rightarrow Cl_2(g) + H_2(g) + 2OH^-(aq) \quad E^{\ominus}_{cell} = -2.19 \text{ V}$$

> **Study Tip**
>
> Always try to use your chemical intuition in working out chemical reactions, products and processes. It would be very unlikely here to have sodium metal forming, since sodium reacts vigorously with water generating hydrogen gas and sodium hydroxide solution! Thinking as a real chemist makes IB chemistry much more accessible and fun as you discover the power of chemical prediction!

> *If you look at the two $E^{\ominus}$ values, the one with the more positive $E^{\ominus}$ will correspond to the favoured reduction. Hence, the cathode half-equation will be:*
>
> $H_2O(l) + e^- \rightarrow \frac{1}{2}H_2(g) + OH^-(aq)$
>
> $E^{\ominus} = -0.83 \text{ V}$

Sodium is not involved in the electrode reactions and simply acts as a spectator cation. In essence, as the electrolytic process progresses, the solution of sodium chloride is converted to a solution of sodium hydroxide. The **electrolysis of concentrated sodium chloride solution (brine)** is a very important industrial process, the basis of **the chlor-alkali industry**. Three very important industrial products are produced by this process: chlorine gas, hydrogen gas and sodium hydroxide.

Uses:

- **Uses of chlorine:** Chlorine can be used to make the polymer polyvinylchloride (PVC), which is used in pipes, floor tiles, transparent film for the packaging of meats and rain jackets (sub-topic A.5). Chlorine can also be used as a bleaching agent (used in the textile and paper industries) and as a disinfectant. It can be used in the purification of water.

- **Uses of hydrogen:** Hydrogen is a valuable fuel and can be used in the Haber process to produce ammonia: $N_2(g) + 3H_2(g) \rightleftharpoons 2NH_3(g)$. The ammonia produced is important in the manufacturing of fertilizers such as ammonium nitrate, NH_4NO_3 (see topic 7).

- **Uses of sodium hydroxide:** Sodium hydroxide is used in the manufacturing of soap and paper.

Observations at each electrode

Cathode (negative electrode): $2H_2O(l) + 2e^- \rightarrow H_2(g) + 2OH^-(aq)$	Bubbles of colourless hydrogen gas are observed. You could test the gas by taking a sample in a closed test tube and lighting a match in the gas. The gas will ignite with a small popping sound heard. The sample will be mainly **pure** hydrogen, whereas typically a much louder pop is obtained if a mixture of hydrogen and air is present, which you might have carried out in the laboratory, in a separate experiment.
Anode (positive electrode): $2Cl^-(aq) \rightarrow Cl_2(g) + 2e^-$	Bubbles of chlorine gas are observed (pale yellow colour may be seen perhaps). Pungent odour of chlorine gas can be experienced (similar to odour found in bleach). Note that the chlorine formed at the anode can combine with sodium chloride to form bleach which is sodium hypochlorite, NaOCl. This can be tested by using some moist blue litmus paper which can be effectively bleached. At higher temperatures, sodium chlorate, $NaClO_3$, may form instead of NaOCl. **Safety note:** Chlorine gas is a toxic gas so, when working with it even in small amounts, a fumehood should be used in the chemical laboratory.
Electrolyte: $2H_2O(l) + 2Cl^-(aq) \rightarrow$ $Cl_2(g) + H_2(g) + 2OH^-(aq)$	pH of the electrolyte will increase due to the formation of $OH^-(aq)$, producing a more basic solution. This can be observed by testing the solution with indicator paper.

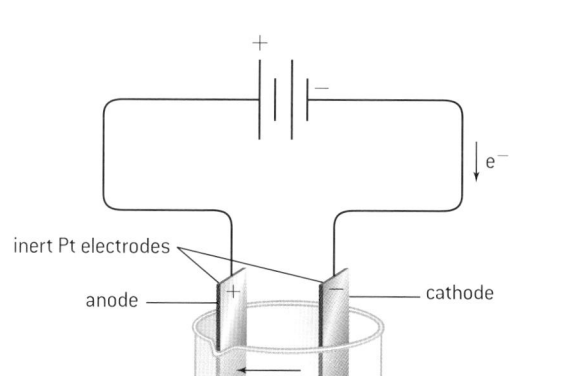

▲ Figure 4 Electrolysis of **concentrated** aqueous sodium chloride, NaCl(aq).

(b) Electrolysis of aqueous copper(II) sulfate

(i)Using inert graphite (carbon) electrodes

The formula of aqueous copper(II) sulfate is $CuSO_4(aq)$.

Let us consider the species present at each electrode.

Cathode (negative electrode):

$Cu^{2+}(aq)$, $H_2O(l)$

Anode (positive electrode):

SO_4^{2-} (aq), $H_2O(l)$

Let us look next at the possible half-equations that may take place at the cathode,

Cathode (negative electrode): Reduction takes place here.

$$Cu^{2+}(aq) + 2e^- \rightarrow Cu(s) \qquad E^{\ominus} = +0.34 \text{ V}$$
$$H_2O(l) + e^- \rightarrow \frac{1}{2}H_2(g) + OH^-(aq) \qquad E^{\ominus} = -0.83 \text{ V}$$

The reduction with more positive $E^{\ominus}$ value will be favoured. Hence, the cathode half-equation will be:

$$Cu^{2+}(aq) + 2e^- \rightarrow Cu(s) \qquad E^{\ominus} = +0.34 \text{ V}$$

Anode (positive electrode): Oxidation takes place here.

Sulfates do not tend to oxidize. In sulfate, the oxidation state of sulfur is +6, corresponding to the stable noble gas core of [Ne] which it will not want to give up.

At the anode the following reaction therefore takes place:

$$H_2O(l) \rightarrow \frac{1}{2}O_2(g) + 2H^+(aq) + 2e^- \qquad E^{\ominus} = -1.23 \text{ V}$$

Hence, let's combine the two electrode half-equations to generate the overall cell reaction:

Cathode (negative electrode):

$$Cu^{2+}(aq) + 2e^- \rightarrow Cu(s) \qquad\qquad E^\ominus = +0.34 \text{ V}$$

Anode (positive electrode):

$$H_2O(l) \rightarrow \frac{1}{2}O_2(g) + 2H^+(aq) + 2e^- \qquad\qquad E^\ominus = -1.23 \text{ V}$$

Overall cell reaction:

$$Cu^{2+}(aq) + H_2O(l) \rightarrow Cu(s) + \frac{1}{2}O_2(g) + 2H^+(aq) \quad E^\ominus_{cell} = -0.89 \text{ V}$$

Observations at each electrode

Cathode (negative electrode): $Cu^{2+}(aq) + 2e^- \rightarrow Cu(s)$	Layer of pink–brown colour of solid copper seen deposited on cathode.
Anode (positive electrode): $H_2O(l) \rightarrow \frac{1}{2}O_2(g) + 2H^+(aq) + 2e^-$	Bubbles of colourless oxygen gas observed at anode.
Electrolyte: $Cu^{2+}(aq) + H_2O(l) \rightarrow Cu(s) + \frac{1}{2}O_2(g) + 2H^+(aq)$	pH of the electrolyte will decrease due to an increase in the concentration of $H^+(aq)$. This can be observed by testing with an indicator. Mediterranean blue colour of $Cu^{2+}(aq)$ ions fades in colour due to the discharge of $Cu^{2+}(aq)$.

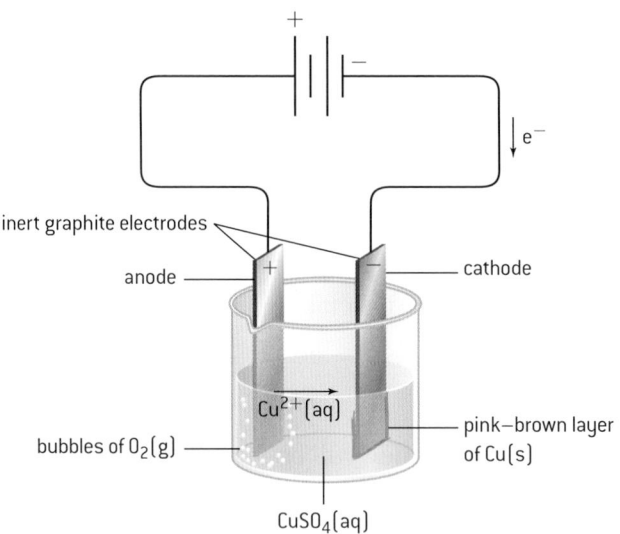

▲ Figure 5 Electrolysis of aqueous copper(II) sulfate, $CuSO_4(aq)$ using **inert graphite** (carbon) electrodes

(ii) Using active copper electrodes

Using copper instead of graphite electrodes means that the copper electrodes now participate in the electrolysis process (they are termed **active electrodes** as opposed to inert electrodes) and the following will be the corresponding half-equations taking place at each electrode:

Cathode (negative electrode):

$$Cu^{2+}(aq) + 2e^- \rightarrow Cu(s)$$

Anode (positive electrode):

$$Cu(s) \rightarrow Cu^{2+}(aq) + 2e^-$$

Observations at each electrode

Cathode (negative electrode): $Cu^{2+}(aq) + 2e^- \rightarrow Cu(s)$	Layer of pink–brown colour of solid copper deposited on cathode (this copper will be pure). Mass of cathode increases.
Anode (positive electrode): $Cu(s) \rightarrow Cu^{2+}(aq) + 2e^-$	Copper anode seen to disintegrate since the mass of the anode decreases. At the bottom a sludge of impurities is seen to form.
Electrolyte:	Mediterranean blue colour of solution does not change, since the concentration of $Cu^{2+}(aq)$ ions remains constant.

One use of this type of electrolysis is in the **electrorefining** of copper, as the purification of copper takes place. In electrical wires the purity of copper needs to be very high. If impure copper wiring is used, the electrical resistance increases.

In this electrolysis the anode consists of impure copper metal and the cathode consists of pure copper metal. The impure copper at the anode is converted into pure copper at the cathode and the residue of impurities (typically platinum, gold and silver) forms a sludge below the anode as seen in Figure 6 below. More easily oxidized impurities, such as iron and zinc, remain in solution as Fe^{2+} and Zn^{2+} species.

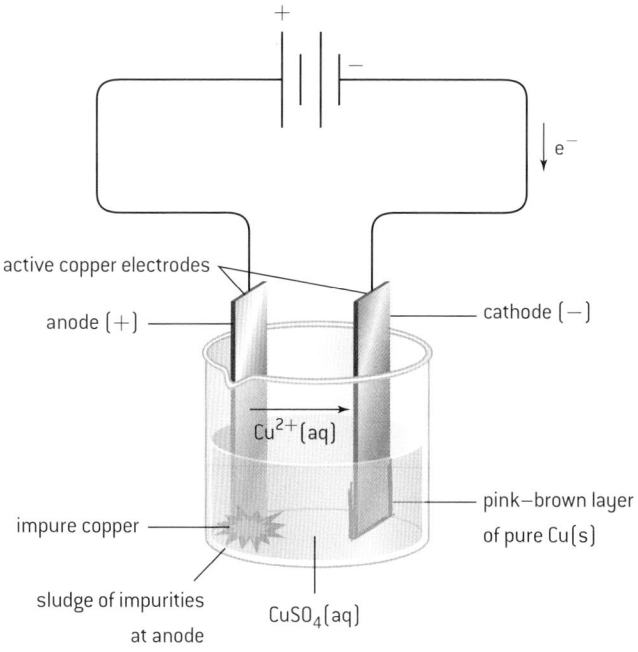

▲ Figure 6 Electrolysis of aqueous copper(II) sulfate, $CuSO_4(aq)$ using **active** copper electrodes

Quick question

Suggest why the sludge might be reprocessed.

427

This idea of using active electrodes is also the basis of the electrochemical process of **electroplating**, which involves using electrolysis to deposit a thin layer (typically 10^{-3} to 10^{-4} mm thick) of one metal onto the cathode of another. This is usually done either to prevent corrosion or for decorative purposes, thereby enhancing the appearance of a particular object. Electroplating is widely used in jewellery and in the plating of steel bumpers in cars with chromium.

Let's consider the electroplating of an object with silver using a solution of $Na[Ag(CN)_2]$ as the electrolyte. Jewellery is commonly electroplated with gold whereas cutlery is typically electroplated with silver (hence the name "silver service" in the restaurant business). You might be inclined to think that silver nitrate, $AgNO_3$, would be an appropriate solution to use for this purpose, but it has been found that the rate at which the silver deposits is too quick and hence has been found not to adhere effectively to the object being plated. For this reason, a more appropriate solution involves the complex, sodium dicyanoargentate(I), $Na[Ag(CN)_2]$. The anode consists of a bar of silver which disintegrates. The cathode is the metal object to be plated (for example, a spoon). The following are the half-equations corresponding to the cathode and anode processes:

Anode (positive electrode):

$$Ag(s) + 2CN^-(aq) \rightarrow [Ag(CN)_2]^-(aq) + e^-$$

Cathode (negative electrode):

$$[Ag(CN)_2]^-(aq) + e^- \rightarrow Ag(s) + 2CN^-(aq)$$

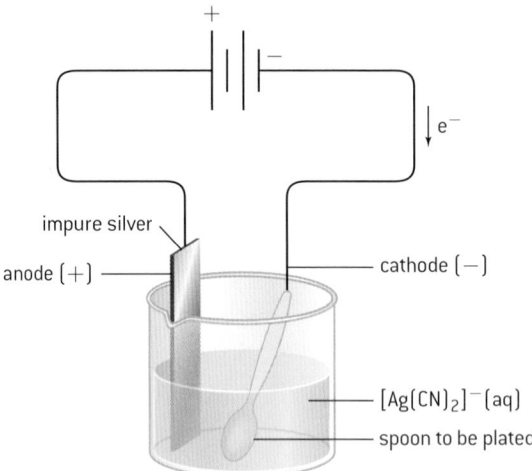

▲ Figure 7 Electroplating of a spoon with silver

(c) Electrolysis of water

Pure water is a poor conductor of electricity, but when even a tiny amount of ions are present the electrical conductivity of water increases. The electrolysis of water can be carried out using a **dilute** solution of sulfuric acid, $H_2SO_4(aq)$ or a **dilute** solution of sodium hydroxide, $NaOH(aq)$, using inert Pt electrodes.

Let us consider electrolysis of water using dilute sulfuric acid:

Consider the species present at each electrode.

Cathode (negative electrode):

$H^+(aq)$

Anode (positive electrode):

$SO_4^{2-}(aq)$, $H_2O(l)$

Cathode (negative electrode):

$H^+(aq) + e^- \rightarrow \frac{1}{2}H_2(g)$ $\qquad E^\ominus = 0.00$ V

Anode (positive electrode):

As stated previously sulfates do not tend to oxidize, so at the anode the following reaction therefore takes place:

$H_2O(l) \rightarrow \frac{1}{2}O_2(g) + 2H^+(aq) + 2e^-$ $\qquad E^\ominus = -1.23$ V

Hence, let's combine the two electrode half-equations to generate the overall cell reaction:

Cathode (negative electrode):

$H^+(aq) + e^- \rightarrow \frac{1}{2}H_2(g)$ $\qquad E^\ominus = 0.00$ V

This needs to be multiplied by two to balance the number of electrons.

$2H^+(aq) + 2e^- \rightarrow H_2(g)$ $\qquad E^\ominus = 0.00$ V

Anode (positive electrode):

$H_2O(l) \rightarrow \frac{1}{2}O_2(g) + 2H^+(aq) + 2e^-$ $\qquad E^\ominus = -1.23$ V

Overall cell reaction:

$H_2O(l) \rightarrow \frac{1}{2}O_2(g) + H_2(g)$ $\qquad E^\ominus = -1.23$ V

Observations at each electrode

Cathode (negative electrode): $2H^+(aq) + 2e^- \rightarrow H_2(g)$	Bubbles of colourless hydrogen gas observed. pH at cathode increases with the discharge of $H^+(aq)$.
Anode (positive electrode): $H_2O(l) \rightarrow \frac{1}{2}O_2(g) + 2H^+(aq) + 2e^-$	Bubbles of colourless oxygen gas observed. pH at anode decreases because $H^+(aq)$ ions are produced
Electrolyte: $H_2O(l) \rightarrow H_2(g) + \frac{1}{2}O_2(g)$	The ratio of the two gases by volume is: $1H_2(g) : \frac{1}{2}O_2(g)$ or $2H_2(g) : O_2(g)$

Quick question

What is the purpose of the dilute $H_2SO_4(aq)$ (eg 0.1 mol dm^{-3})?

▲ Figure 8 Electrolysis of water using a Hoffman apparatus. Reactions at the two electrodes (black hooks dipped in the beaker of water) are powered by the electric current from the battery (lower left). Oxygen and hydrogen gas bubbles are evolved at the anode (left electrode) and cathode (right electrode) respectively. As water molecules consist of two hydrogen atoms and one oxygen atom, twice as much hydrogen as oxygen is trapped in the test tubes (upper right). Use of a burning splint will ignite the hydrogen gas, while the oxygen will relight a glowing splint.

Hydrogen cells and research

The electrolysis of water is an important source of hydrogen gas (sub-topic C.6). The term "hydrogen economy" was coined by Professor John Bokris, an electrochemist born in South Africa during a presentation given to General Motors in 1970, who proposed generating energy using hydrogen. Many countries are trying to move away from the generation of energy using fossil fuels. There are extensive energy demands associated with the production of hydrogen. The availability of hydrogen is limited for this important use in fuel cells. Remember that hydrogen is an energy carrier and is not a resource *per se*.

The most common type of fuel cell, the **hydrogen–oxygen fuel cell** involves the reaction between hydrogen and oxygen to yield water as the product:

$$H_2(g) + \frac{1}{2}O_2(g) \rightarrow H_2O(l)$$

In order to produce electricity in the fuel cell, hydrogen is required. Hydrogen can be produced by the electrolysis of water in a solar-powered electrolytic cell. In the development of this technology extensive collaborative research between scientists and engineers from different fields (chemists, biologists, material scientists, etc.) is necessary. In the design of such technologies scientists often have to consider environmental, socio-economic, safety and ethical aspects of energy production. Although vehicles powered by hydrogen–oxygen fuel cells are environmentally preferred (greenhouse gases such as carbon dioxide, $CO_2(g)$ are not produced,) there are other considerations which need to be considered such as the safe storage of the hydrogen fuel used, which is a highly flammable substance. Other fuel cells such as the DMFC, the **direct methanol fuel cell**, do generate carbon dioxide. These fuel cells do have some advantages over conventional batteries. They are lighter and hence are often used in smart phone technologies. Market cost, environmental issues, and ethical aspects play a role in the development of any new technology. If liquid methanol is used in fuel cells it can be produced from biomass as a **carbon neutral fuel** (one which does not contribute to the greenhouse effect).

Many electrochemical cells can act as energy sources alleviating the world's energy problems but some cells such as super-efficient **microbial fuel cells** (MFCs also termed biological fuel cells sub-topic C.6) can also reduce the environmental impact of human activities.

- *How do national governments and the international community decide on research priorities for funding purposes?* Although the intended outcomes may be clear to a large extent in the development of strategic and applied research, do governments and funding agencies ignore the ongoing development of basic research at their peril? Can you think of any examples from electrochemistry (or the broader fields of physical chemistry, inorganic chemistry or materials science) where this might be the case?

Worked example

Construct and annotate the electrolytic cell for the electrolysis of **dilute** sodium chloride.

a) Identify the half-equations occurring at the cathode and anode electrodes and the equation for the overall cell reaction.

b) State a suitable material that can be used for each electrode.

c) Identify the direction of the movement of electrons and ion flow.

d) State what would be observed at each electrode.

Solution

a) First, consider the ions present, noting that this is a **dilute** solution of sodium chloride (not concentrated). The ions present are generated from:

$NaCl(aq) \rightarrow Na^+(aq) + Cl^-(aq)$ and

$H_2O(l) \rightleftharpoons H^+(aq) + OH^-(aq)$

Water slightly dissociates into hydrogen and hydroxide ions.

Cathode (negative electrode):

$Na^+(aq)$, $H_2O(l)$, $H^+(aq)$

Anode (positive electrode):

$Cl^-(aq)$, $H_2O(l)$

Possible processes at the cathode (negative electrode): Reduction takes place here

$$Na^+(aq) + e^- \rightarrow Na(s) \qquad E_1^{\ominus} = -2.71 \text{ V}$$

$$H_2O(l) + e^- \rightarrow \frac{1}{2}H_2(g) + OH^-(aq) \qquad E_2^{\ominus} = -0.83 \text{ V}$$

$$2H^+(aq) + 2e^- \rightarrow H_2(g) \qquad E_3^{\ominus} = 0.00 \text{ V}$$

The third process takes place since $E_3^{\ominus}$ is the most positive.

Possible processes at the anode (positive electrode): Oxidation takes place here

$$2Cl^-(aq) \rightarrow Cl_2(g) + 2e^- \qquad E_1^{\ominus} = -1.36 \text{ V}$$

$$H_2O(l) \rightarrow \frac{1}{2}O_2(g) + 2H^+(aq) + 2e^- \quad E_2^{\ominus} = -1.23 \text{ V}$$

Since $E_2^{\ominus} > E_1^{\ominus}$, the second process takes place.

Note: There is no overvoltage as the solution is dilute which differs from the electrolysis of concentrated sodium chloride where chlorine gas is evolved at the anode.

Overall cell reaction:

$$H_2O(l) \rightarrow H_2(g) + \frac{1}{2}O_2(g)$$

b) **Material for electrodes:** inert metal, such as Pt(s), or graphite, C(s).

c) **Electrolytic cell:**

Electron flow: from anode to cathode (through external circuit).

Ion flow: $H^+(aq)$ flow from anode to cathode (in solution)

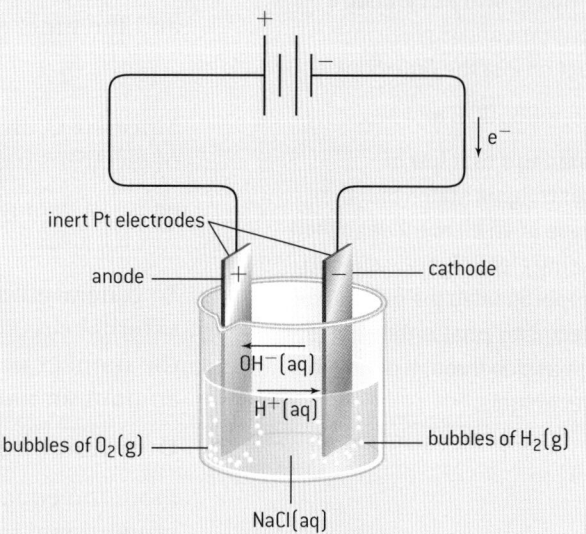

▲ Figure 9 Electrolysis of **dilute** aqueous sodium chloride, NaCl(aq)

d) **Observation at cathode:** bubbles of colourless gas (hydrogen).

Observation at anode: bubbles of colourless gas (oxygen), two times fewer bubbles than those of hydrogen.

Question

Consider the electrolysis of water in **dilute** sodium hydroxide.

a) Identify the half-equations occurring at the cathode and anode electrodes and the equation for the overall cell reaction.

b) (i) State what would be observed at each electrode.

(ii) Deduce the ratio by volumes of any species produced.

Quantitative aspects of electrolysis

The following factors affect the amount of product formed at the electrodes during electrolysis:

1 Current (I)

2 Duration of electrolysis (t)

3 Charge on the ion (z)

431

1 mol of e⁻ carries an approximate charge of 96500 C. This known as **Faraday's constant** and is given in section 2 of the *Data booklet*.

$$1 F = 96500 \text{ C mol}^{-1}$$

Faraday's first law of electrolysis states that the mass of an element deposited during electrolysis is directly proportional to the quantity of electricity (that is the charge, Q) passing through during the electrolysis.

Let us consider each of these separately:

1 Current (I)

From physics, the charge, Q (in C) is related to the current, I (in A), and the time, t (in s), as follows:

$$Q = It$$

Since $I \propto Q$, this means that I will also be proportional to the number of electrons passing through the external circuit.

eg $Al^{3+}(aq)$ + $3e^- \rightarrow$ $Al(s)$
                                              3 mol of e⁻   1 mol of Al(s)

To generate 1 mol of Al(s), 3 mol of e⁻ must pass through the external circuit. If I is doubled at time t, Q will be doubled and hence the number of electrons will be doubled. This then means that the amount, in mol, of Al(s) will also increase.

2 Duration of electrolysis (t)

From the equation above $t \propto Q$. This means that t will also be directly proportional to the number of electrons passing through the external circuit.

3 Charge on the ion (z)

The amount, in mol, of e⁻ needed to discharge 1 mol of an ion at an electrode is equal to the charge on the ion, z. This is termed **Faraday's second law**.

 Worked examples

Example 1

Calculate the mass, in g, of copper produced at the cathode when a current of 1.50 A is passed through a solution of aqueous copper(II) sulfate for 3.25 hours.

Solution

- $Q = It$

  $Q = 1.50 \times 3.25 \times 60.0 \times 60.00 = 17550$ C

- $1 F = 96500$ C mol⁻¹

  So 96500 C equates to 1 mol of e⁻

  1 C equates to (1/96500) mol of e⁻

  17550 C equates to (17550/96500) mol of e⁻ = 0.182 mol of e⁻ (correct to three significant figures)

- At the cathode the following reduction half-reaction takes place:

  $Cu^{2+}(aq) + 2e^- \rightarrow Cu(s)$

  2 mol e⁻ ≡ 1 mol Cu

  1 mol e⁻ ≡ $\frac{1}{2}$ mol Cu

  0.182 mol e⁻ ≡ (0.182 × 0.5) mol
  Cu = 0.0910 mol Cu

- $n = m/M$, so $m = n \times M$

  $m = (0.0910 \text{ mol})(63.55 \text{ g mol}^{-1}) = 5.78$ g

Example 2

Two electrolytic cells are connected in series, so that the same current passes through each individual cell. The first cell, cell A, contains silver electrodes in a silver nitrate solution. The second cell, cell B, contains copper electrodes in copper(II) sulfate solution. In an experiment, 0.658 g of silver was found to deposit in cell A. Calculate the mass, in g, of copper deposited in cell B, showing your working.

Electrolytic cells connected **in series** means that they are connected one after another in a circuit so that the same current passes through each one.

Solution

- **In cell A:**

$$Ag^+(aq) + e^- \rightarrow Ag(s)$$

$$1 \text{ mol } e^- \equiv 1 \text{ mol Ag(s)}$$

$$107.87 \text{ g Ag(s)} \equiv 1 \text{ mol } e^-$$

$$0.658 \text{ g Ag(s)} \equiv \left(\frac{0.658}{107.87}\right) \text{ mol } e^-$$

$$= 6.10 \times 10^{-3} \text{ mol } e^-$$

- **In cell B:**

$$Cu^{2+}(aq) + 2e^- \rightarrow Cu(s)$$

$$6.10 \times 10^{-3} \text{ mol } e^- \equiv (0.5 \times 6.10 \times 10^{-3}) \text{ mol } Cu(s)$$

$$\equiv (0.5 \times 6.10 \times 10^{-3} \text{ mol})$$
$$(63.55 \text{ g mol}^{-1}) \text{ Cu(s)}$$

$$= 0.194 \text{ g}$$

Example 3

A current of 2.55 A is passed through a **concentrated** aqueous solution of sodium chloride for 3.00 h. Calculate the volume, in dm³, of chlorine gas produced at the anode at a pressure of 100 kPa and a temperature of 273 K.

Solution

- $Q = It$

$$Q = (2.55) \times (3.00 \times 60.0 \times 60.0) = 27540 \text{ C}$$

- $1\ F = 96500 \text{ C mol}^{-1}$

So 96 500 C equates to 1 mol of e^-

1 C equates to $\left(\frac{1}{96500}\right)$ mol of e^-

27540 C equates to $\left(\frac{27540}{96500}\right)$ mol of e^- = 0.285 mol of e^- (correct to three significant figures)

- At the anode the following oxidation half-reaction takes place:

$$2Cl^-(aq) \rightarrow Cl_2(g) + 2e^-$$

$$0.285 \text{ mol } e^- \equiv (0.285 \times 0.5) \text{ mol}$$
$$= 0.143 \text{ mol } Cl_2(g)$$

- Using section 2 of the *Data booklet*:

Molar volume of an ideal gas at 100 kPa and 273 K = 22.7 dm³ mol⁻¹

$$1 \text{ mol} \equiv 22.7 \text{ dm}^3 \ Cl_2(g)$$

$$0.143 \text{ mol} \equiv 22.7 \times 0.143 \text{ dm}^3 \text{ volume}$$
$$\text{of } Cl_2(g) = 3.25 \text{ dm}^3$$

Questions

1 What conditions are necessary to directly measure a standard electrode potential ($E^\ominus$)?

I. A half-cell with an electrode in a 1.0 mol dm^{-3} solution of its ions.

II. Connection to a standard hydrogen electrode.

III. A voltmeter between half-cells to measure potential difference.

A. I and II only

B. I and III only

C. II and III only

D. I, II and III [1]

IB, May 2010

2 Consider the following standard electrode potentials.

$Zn^{2+}(aq) + 2e^- \rightleftharpoons Zn(s)$ $E^\ominus = -0.76$ V

$Cl_2(g) + 2e^- \rightleftharpoons 2Cl^-(aq)$ $E^\ominus = +1.36$ V

$Mg^{2+}(aq) + 2e^- \rightleftharpoons Mg(s)$ $E^\ominus = -2.37$ V

What will happen when zinc powder is added to an aqueous solution of magnesium chloride?

A. No reaction will take place.

B. Chlorine gas will be produced.

C. Magnesium metal will form.

D. Zinc chloride will form. [1]

IB, May 2010

3 The standard electrode potentials for two metals are given below.

$Al^{3+}(aq) + 3e^- \rightleftharpoons Al(s)$ $E^\ominus = -1.66$ V

$Ni^{2+}(aq) + 2e^- \rightleftharpoons Ni(s)$ $E^\ominus = -0.23$ V

What is the equation and cell potential for the spontaneous reaction that occurs?

A. $2Al^{3+}(aq) + 3Ni(s) \rightarrow 2Al(s) + 3Ni^{2+}(aq)$
   $E^\ominus = 1.89$ V

B. $2Al(s) + 3Ni^{2+}(aq) \rightarrow 2Al^{3+}(aq) + 3Ni(s)$
   $E^\ominus = 1.89$ V

C. $2Al^{3+}(aq) + 3Ni(s) \rightarrow 2Al(s) + 3Ni^{2+}(aq)$
   $E^\ominus = 1.43$ V

D. $2Al(s) + 3Ni^{2+}(aq) \rightarrow 2Al^{3+}(aq) + 3Ni(s)$
   $E^\ominus = 1.43$ V [1]

IB, May 2011

4 How do the products compare at each electrode when aqueous 1 mol dm^{-3} magnesium bromide and molten magnesium bromide are electrolysed?

	$E^\ominus$/V
$Mg^{2+}(aq) + 2e^- \rightleftharpoons Mg(s)$	-2.37
$\frac{1}{2}Br_2(l) + e^- \rightleftharpoons Br^-(aq)$	$+1.07$
$\frac{1}{2}O_2(g) + 2H^+(aq) + 2e^- \rightleftharpoons H_2O(l)$	$+1.23$

	Positive electrode (anode)	Negative electrode (cathode)
A.	same	same
B.	same	different
C.	different	same
D.	different	different

IB, November 2009 [1]

5 What condition is necessary for the electroplating of silver, Ag, onto a steel spoon?

A. The spoon must be the positive electrode.

B. The silver electrode must be the negative electrode.

C. The spoon must be the negative electrode.

D. The electrolyte must be acidified. [1]

IB, May 2010

6 The same quantity of electricity was passed through separate molten samples of sodium bromide, NaBr, and magnesium chloride, MgCl$_2$. Which statement is true about the amounts, in mol, that are formed?

A. The amount of Mg formed is equal to the amount of Na formed.

B. The amount of Mg formed is equal to the amount of Cl$_2$ formed.

C. The amount of Mg formed is twice the amount of Cl$_2$ formed.

D. The amount of Mg formed is twice the amount of Na formed. [1]

IB, May 2011

7 What is the mass, in g, of copper produced at the cathode when a current of 1.00 A is passed through a solution of aqueous copper(II) sulfate for 60 minutes?

($1 F = 96500$ C mol^{-1})

A. $\dfrac{63.55 \times 3600}{2 \times 96500}$

B. $\dfrac{63.55 \times 3600}{96500}$

C. $\dfrac{63.55 \times 60}{2 \times 96500}$

D. $\dfrac{63.55 \times 60}{1}$

8 a) Construct and annotate an example of a voltaic cell consisting of a Ag(s)|Ag$^+$(aq) half-cell and a Co(s)|Co^{2+}(aq) half-cell.

Given: $Co^{2+}(aq) + 2e^- \rightleftharpoons Co(s) \quad E^\ominus = -0.28$ V

i) Identify the half-equations occurring at the cathode and anode electrodes.

ii) Deduce the equation for the **spontaneous** reaction occurring in this cell.

iii) State the cell diagram convention for the cell.

iv) Identify the direction of the movement of electrons and ion flow, both in solution and in the salt bridge.

v) Calculate the standard potential, in V, for this cell.

vi) Determine $\Delta G^\ominus$, the standard change in Gibbs free energy, in J, for the electrochemical reaction.

9 a) Construct and annotate the electrolytic cell for the electrolysis of **concentrated** potassium iodide.

(i) Identify the half-equations occurring at the cathode and anode electrodes and the equation for the overall cell reaction.

(ii) State a suitable material that can be used for each electrode.

(iii) Identify the direction of the movement of electrons and ion flow.

(iv) State what would be observed at each electrode.

10 A current of 2.35 A is passed through an electrolytic cell for the electrolysis of water, using a dilute sulfuric acid solution, for a duration of 5.00 h.

a) Construct and annotate a suitable electrolytic cell for this experiment.

b) Identify the half-equations occurring at the cathode and anode electrodes and the equation for the overall cell reaction.

c) dentify the direction of the movement of electrons and ion flow.

d) Determine the volume, in cm^3, of the two gases generated in the process at SATP conditions, using information from section 2 of the *Data booklet*.

11 Electroplating is an important application of electrolytic cells with commercial implications. Copper may be plated using an electrolytic cell with an aqueous acidified copper(II) sulfate electrolyte. For the copper plating of tin to make jewellery, state the half-equation at **each** electrode.

Assume the tin electrode is inert. Suggest **two** observations that you would be able to make as the electroplating progresses. [4]

IB, May 2010

12 Two electrolytic cells are connected in series as shown in the diagram below. In one there is molten magnesium chloride and in the other, dilute sodium hydroxide solution. Both cells have inert electrodes. If 12.16 g of magnesium is produced in the first cell, deduce the identity and mass of products produced at the positive and negative electrodes in the second cell. [4]

IB, May 2010

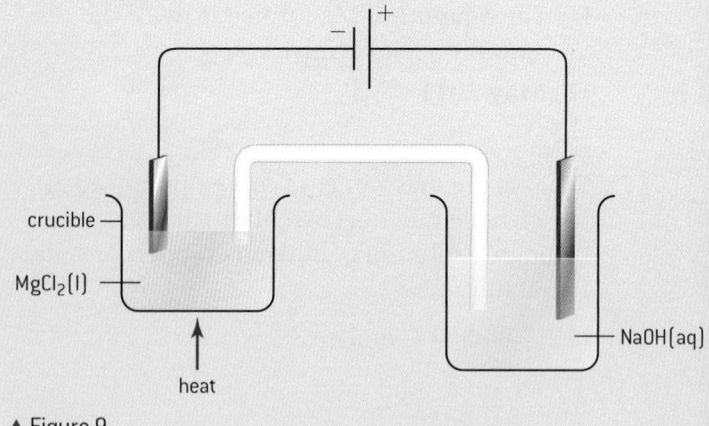

▲ Figure 9

20 ORGANIC CHEMISTRY (AHL)

Introduction

Synthesis and associated reaction mechanisms lie at the core of organic chemistry. An understanding of the properties of organic compounds, their reactions and the mechanisms by which they react is fundamental to the advancement of research in such important fields of science as medicine, biotechnology, food production and the energy industry. In this topic, we examine nucleophilic substitution, electrophilic substitution, addition and reduction reactions. Synthetic routes and the methodologies involved in retro-synthesis are examined. Stereoisomerism concludes the topic with a broad examination of a range of different types of isomers, their nature, properties, nomenclature and importance to the field of chemistry. In both topics 10 and 20, students will learn how to represent organic compounds as both two- and three-dimensional structures, describe key organic chemistry reactions using balanced equations and draw reaction mechanisms using curly arrow notation to represent the synthetic reactions they undergo.

20.1 Types of organic reaction

Understandings

Nucleophilic substitution reactions:

→ S_N1 represents a nucleophilic unimolecular substitution reaction and S_N2 represents a nucleophilic bimolecular substitution reaction. S_N1 involves a carbocation intermediate. S_N2 involves a concerted reaction with a transition state.

→ For tertiary halogenoalkanes the predominant mechanism is S_N1 and for primary halogenoalkanes it is S_N2. Both mechanisms occur for secondary halogenoalkanes.

→ The rate-determining step (slow step) in an S_N1 reaction depends only on the concentration of the halogenoalkane, rate $= k$[halogenoalkane]. For S_N2, rate $= k$[halogenoalkane][nucleophile]. S_N2 is stereospecific with an inversion of configuration at the carbon.

→ S_N2 reactions are best conducted using aprotic, polar solvents and S_N1 reactions are best conducted using protic, polar solvents.

Electrophilic addition reactions:

→ An electrophile is an electron-deficient species that can accept electron pairs from a nucleophile. Electrophiles are Lewis acids.

→ Markovnikov's rule can be applied to predict the major product in electrophilic addition reactions of unsymmetrical alkenes with hydrogen halides and interhalogens. The formation of the major product can be explained in terms of the relative stability of possible carbocations in the reaction mechanism.

Electrophilic substitution reactions:

→ Benzene is the simplest aromatic hydrocarbon compound (or arene) and has a delocalized structure of π bonds around its ring. Each carbon to carbon bond has a bond order of 1.5. Benzene is susceptible to attack by electrophiles.

Reduction reactions:

→ Carboxylic acids can be reduced to primary alcohols (via the aldehyde). Ketones can be reduced to secondary alcohols. Typical reducing agents are lithium aluminium hydride (used to reduce carboxylic acids) and sodium borohydride.

Applications and skills

Nucleophilic substitution reactions:

→ Explanation of why hydroxide is a better nucleophile than water.

→ Deduction of the mechanism of the nucleophilic substitution reactions of halogenoalkanes with aqueous sodium hydroxide in terms of S_N1 and S_N2 mechanisms. Explanation of how the rate depends on the identity of the halogen (i.e. the leaving group), whether the halogenoalkane is primary, secondary, or tertiary and the choice of solvent.

→ Outline of the difference between protic and aprotic solvents.

Electrophilic addition reactions:

→ Deduction of the mechanism of the electrophilic addition reactions of alkenes with halogens/ interhalogens and hydrogen halides.

Electrophilic substitution reactions:

→ Deduction of the mechanism of the nitration (electrophilic substitution) reaction of benzene (using a mixture of concentrated nitric acid and sulfuric acid).

Reduction reactions:

→ Writing reduction reactions of carbonyl-containing compounds: aldehydes and ketones to primary and secondary alcohols and carboxylic acids to aldehydes, using suitable reducing agents.

→ Conversion of nitrobenzene to aniline via a two-stage reaction.

 Nature of science

→ Looking for trends and discrepancies – by understanding different types of organic reaction and their mechanisms, it is possible to synthesize new compounds with novel properties which can then be used in several applications. Organic reaction types fall into a number of different categories.

→ Collaboration and ethical implications – scientists have collaborated to work on investigating the synthesis of new pathways and have considered the ethical and environmental implications of adopting green chemistry.

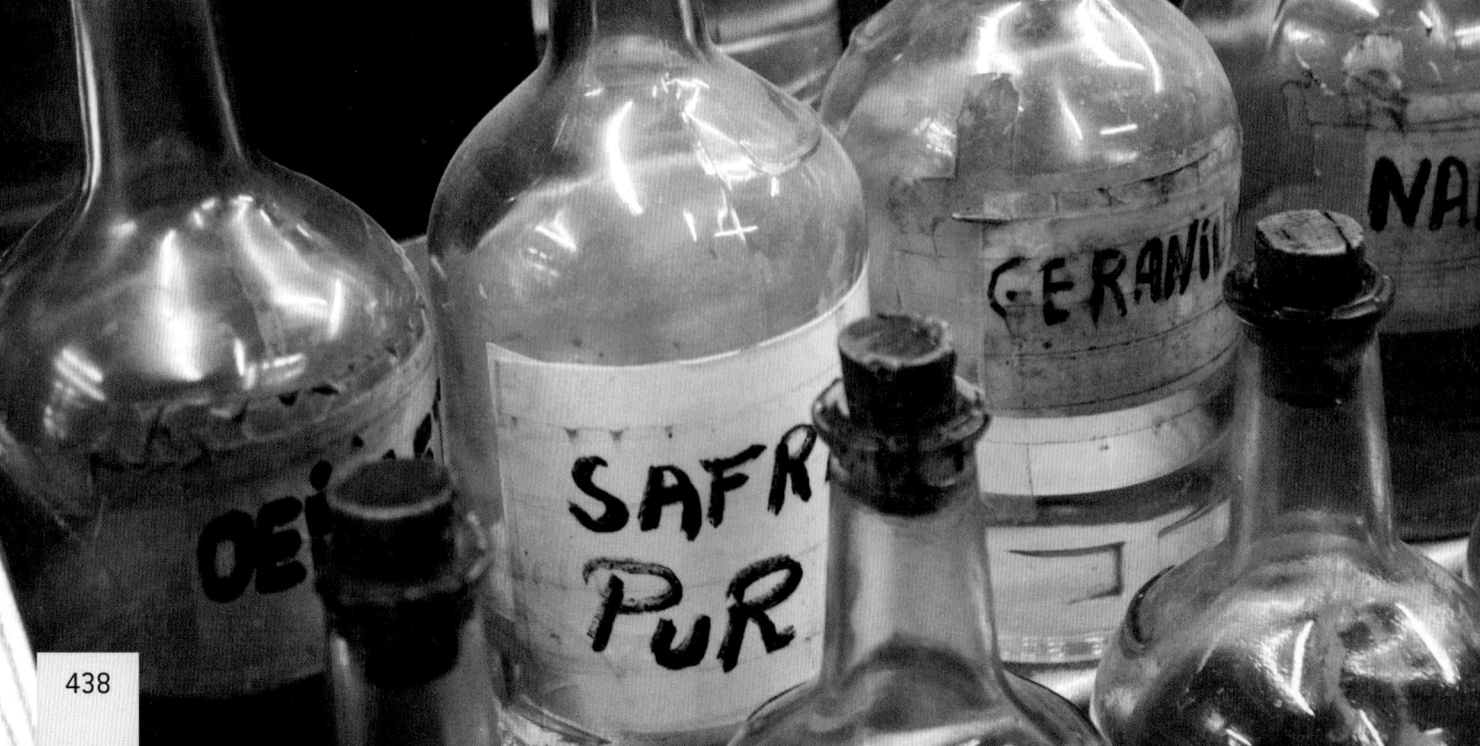

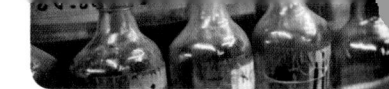

 Organic synthesis

Organic synthesis reactions are of fundamental importance in the field of drug design, leading to the production of new synthetic drugs for the treatment of many different conditions. Knowledge and understanding of the properties of organic compounds and the mechanisms of their reactions is central to the development of new useful compounds.

Associated with the development of new synthetic compounds are ethical considerations concerning the implications of their use. A risk–benefit analysis is stringently applied to all new drugs under consideration and scientists have the responsibility to place first the wellbeing of the public who put their trust in the scientific community and its methodologies. Many new synthetic compounds are the product of collaboration, ranging from small-scale projects in the laboratories of research institutes to international ventures on a historic scale.

As well as the development of new drugs, the fields of food science and nutrition, biotechnology, plastics and textiles, fuels and explosives, paints and dyes, and pesticides and fertilizers are among the applications of synthetic organic chemistry.

Nucleophilic substitution reactions

The substitution reactions of saturated alkanes (sub-topic 10.2) involve homolytic fission, creating free-radicals that possess unpaired electrons. The formation of chloroethane from the reaction between ethane and chlorine is an example:

$$C_2H_6(g) + Cl_2(g) \xrightarrow{\text{UV light}} C_2H_5Cl(g) + HCl(g)$$

The chloroalkane that is produced has very different properties from those of the alkane, and can undergo substitution reactions, producing a wide variety of compounds with different functional groups. The reason for this increased reactivity is the highly electronegative chlorine atom and the polar halogen–carbon bond (table 1 and figure 1).

The partial positive charge makes the carbon atom electron deficient and therefore susceptible to attack by nucleophiles, electron-rich species that are capable of donating a pair of electrons to form a covalent bond (sub-topic 10.2).

There are two types of nucleophilic substitution (S_N). The mechanism of an S_N reaction depends on whether the halogenoalkane is primary, secondary, or tertiary.

S_N2 reactions and primary halogenoalkanes

Nucleophilic substitution in primary halogenoalkanes proceed in one step. The rate-determining step (slow-step) (sub-topic 16.1) involves both the halogenoalkane and the nucleophile so the rate of reaction is dependent on the concentrations of both reactants. It is described as a second-order reaction.

rate = k[halogenoalkane][nucleophile]

As there are two reactant molecular entities involved in the 'microscopic chemical event' termed the elementary reaction, the molecularity is described as bimolecular.

Understand that molecularity is not the same as the order of the reaction (sub-topic 16.1).

Element	Electronegativity X_p
C	2.6
F	4.0
Cl	3.2
Br	3.0
I	2.7

▲ Table 1 Halogen atoms have high electronegativity and form polar bonds with carbon

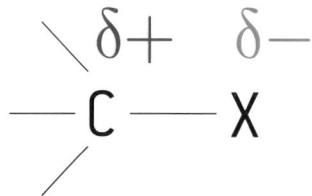

▲ Figure 1 Representation of the partial charges within the polar carbon–halogen bond

Concerted reactions

The S_N2 reaction is an example of a concerted reaction. It is a single-step reaction through which reactants are converted directly into products. The mechanism does not involve an intermediate.

The Walden inversion

When a chemical species with a sp^3 carbon center and tetrahedral geometry undergoes a backside attack by a nucleophile in a S_N2 reaction, a configurational change occurs. Imagine the shape of an umbrella. The handle of the umbrella points towards the attacking nucleophile and the ribs that support the fabric layer are representative of the shape formed by the three atoms bonded to the carbon. The leaving group is represented by the tip of the umbrella. Can you picture this? Just as the strong wind of a storm can blow an umbrella inside out, the approaching nucleophile results in an inversion of configuration. This inversion provides room for the entering nucleophile while the product has the same relative configuration as the reactant.

▲ Figure 3 The inversion of an umbrella is visually analogous to the inversion of configuration that occurs in S_N2 reactions.

For example, the reaction between bromoethane and aqueous hydroxide ion yields ethanol and the leaving group, the bromide ion (figure 2).

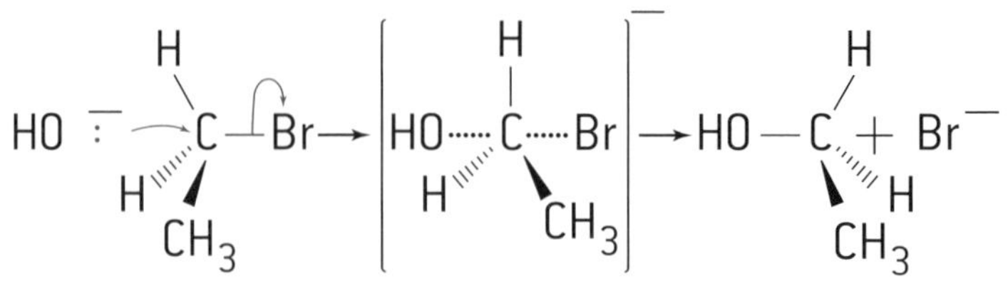

$$HO \overset{\cdot\cdot}{} \cdots C - Br \rightarrow \left[HO \cdots C \cdots Br \right]^- \rightarrow HO - C + Br^-$$

transition state

▲ Figure 2 S_N2 mechanism for the reaction between the primary halogenoalkane bromoethane and hydroxide ion

The hydroxide nucleophile attacks the partially positively charged carbon atom, forming a transition state that involves both the halogenoalkane and the nucleophile. This has a partially formed covalent bond between the nucleophile and the carbon atom, and a weakened carbon–bromine bond that has not completely broken.

In this "backside attack" the nucleophile attacks the electrophilic centre at 180° to the position of the bromine leaving group, the large halogen atom creating **steric hindrance**. Such hindrance by bulky substituents prevents "frontal attack" by a nucleophile. As the reaction proceeds, the entering nucleophile causes an **inversion of configuration**, in the same way that an umbrella blows inside out in a storm. This is known as the **Walden inversion**. Hence, the S_N2 reaction is said to be **stereospecific**. This is a reaction where starting reagents differing only in their configuration are converted into stereoisomeric products.

Drawing mechanisms for S_N2 reactions

- The curly arrow from the nucleophile originates from its lone pair or negative charge, terminating at the carbon atom.

- The curly arrow representing the bromine leaving group originates at the bond between the carbon and bromine atoms. This can be shown on bromoethane or on the transition state.

- Partial bonds, $HO-C-Br$, are represented by dotted lines.

- The transition state is enclosed in square brackets with a single negative charge.

- The formation of the product *and* the leaving group must be shown.

S_N1 reactions and tertiary halogenoalkanes

Tertiary halogenoalkanes undergo nucleophilic substitution reactions that involve two steps. The rate-determining step involves only the halogenoalkane: the bond to the leaving group breaks, forming a carbocation. The reaction is a first-order reaction.

$$rate = k[\text{halogenoalkane}]$$

For example, the reaction between 2-chloro-2-methyl propane and aqueous hydroxide ion yields 2-methylpropan-2-ol and the chloride ion (figure 4). The reaction has a molecularity of one, as there is only

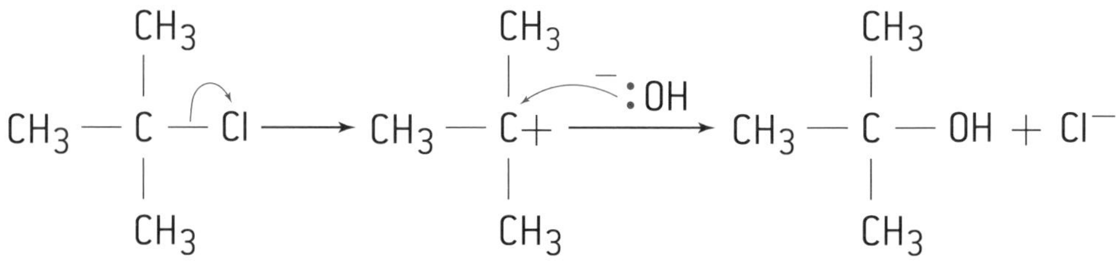

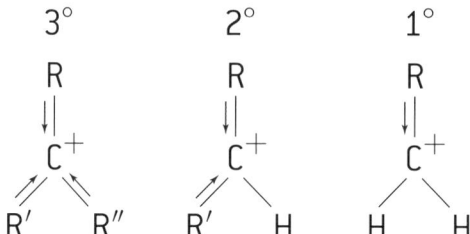

▲ Figure 4 S_N1 mechanism for the reaction between a tertiary halogenoalkane and aqueous hydroxide ion

one molecular entity involved in the elementary reaction and hence is termed unimolecular.

The atoms present in a molecule influence the stability of possible intermediates (S_N2) and carbocations (S_N1) and help determine the likely reaction mechanism.

Inductive effects in organic compounds have a significant effect on which nucleophilic substitution mechanism occurs between a nucleophile and a halogenoalkane. The most important factor is differences in electronegativity between atoms present in the molecule. In the C−H bond the carbon atom has slightly greater electronegativity than hydrogen, creating a weak dipole and a shift in position of the bonding electrons closer to the carbon atom. Other atoms, such as more electronegative halogens, have far greater polarizing effects on the sigma bond.

Alkyl groups bonded to a carbocation have a positive inductive effect, stabilizing the charged carbocation by donating electron density and reducing the positive charge on the carbon atom. In a primary carbocation just one alkyl group contributes to the inductive effect so it receives the least stabilization. A tertiary carbocation is bonded to three alkyl groups so will be more stable (figure 6). This is one reason why tertiary halogenoalkanes have a tendency to undergo reactions via the S_N1 mechanism.

Drawing mechanisms for S_N1 reactions

- The curly arrow representing the halogen leaving group originates at the bond between the carbon and the halogen.

- The representation of the carbocation clearly shows a positive charge centred on the carbon atom.

- The curly arrow from the nucleophile originates from its lone pair or negative charge, terminating at the carbon atom.

- The formation of the product *and* the leaving group must be shown.

Factors affecting the rate of nucleophilic substitution

The rate of a nucleophilic substitution reaction depends on three main factors.

[1] The identity of the halogen

The presence of a good leaving group in a reactant undergoing nucleophilic substitution increases the rate of reaction of both S_N1 and S_N2 mechanisms. In both cases the rate-determining step involves the heterolytic fission of the carbon–halogen bond, in which the two bonding electrons move to the more electronegative atom. The quicker this rate-determining step is completed, the higher the rate of reaction and a better leaving group will help achieve this.

A special arrow is used in organic chemistry to represent the polarization of a bond and the movement of electron density within the σ bond (figure 5).

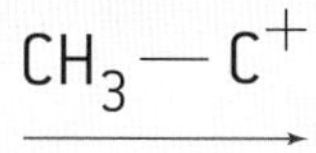

$$CH_3 — C^+$$

▲ Figure 5 Representing the movement of electron density in the inductive effect

3°	2°	1°
R	R	R
C^+	C^+	C^+
R′ R″	R′ H	H H

▲ Figure 6 Decreasing stability due to diminishing inductive effects moving from tertiary to primary carbocations

Carbon—halogen bond	Bond enthalpy/ kJ mol⁻¹
C−F	492
C−Cl	324
C−Br	285
C−I	228

▲ Table 2 Bond dissociation energies of carbon—halogen bonds

Halogenoalkane	Mechanism
primary	S_N2
secondary	S_N2/S_N1
tertiary	S_N1

▲ Table 3 Prevalence of the S_N2/S_N1 mechanisms in different classes of halogenoalkanes

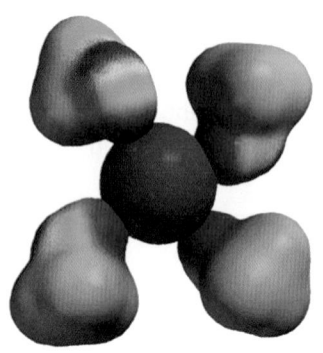

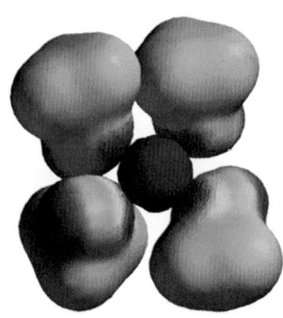

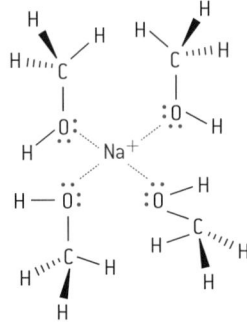

▲ Figure 7 Solvation by the polar, protic solvent methanol

Bond strength as well as electronegativity is important in choosing a leaving group (table 2). Fluoroalkanes are virtually inert due to the short length (138 pm) and the high strength of the C−F bond (492 kJ mol⁻¹). As you move down the halogen group the strength of the carbon–halogen bond decreases as the size of the halogen atom increases. A larger halogen results in longer, weaker bonds. Additionally, the stability of the halogen anion formed during these reactions is directly related to its effectiveness as the leaving group. The larger iodine atom can better dissipate the negative charge compared with the chlorine atom and so iodine is a better leaving group than fluorine or chlorine.

(2) The classes of halogenoalkane: Primary, secondary, or tertiary

As we have seen, tertiary halogenoalkanes predominantly undergo nucleophilic substitution via the S_N1 mechanism while primary halogenoalkanes favour the S_N2 mechanism. For secondary halogenoalkanes, both mechanisms are possible (table 3).

The class of halogenoalkane has a direct effect on the rate of formation of the carbocation (the rate-determining step) and hence the overall rate of reaction. A tertiary carbocation has greater stability than a primary carbocation as a consequence of the inductive effects of the alkyl groups bonded to the carbon atom. Formed following cleavage of the carbon-halogen bond, the more stable tertiary carbocation rapidly forms and immediately reacts with the nucleophile.

(3) The choice of solvent

S_N2 reactions are best performed in **aprotic**, polar solvents while S_N1 reactions are carried out in **protic**, polar solvents.

Aprotic, polar solvents are suitable for S_N2 reactions because they:

- possess no O−H or N−H groups so they cannot form a hydrogen bond to the nucleophile
- cannot solvate the nucleophile so leaving it "naked" and maintaining its effectiveness as a nucleophile in forming the transition state.

Examples of aprotic solvents include ethyl ethanoate, $CH_3COOCH_2CH_3$, and propanone, $CH_3C(O)CH_3$.

Protic, polar solvents are suitable for S_N1 reactions because they:

- are polar in nature due to the presence of polar bonds
- possess either an O−H or N−H group so can form hydrogen bonds with the nucleophile
- solvate the nucleophile, thus inhibiting its ability to attack electrophiles such as the δ^+ carbon atom.

Solvation is the process by which solvent molecules surround the dissolved ions. The smaller the nucleophile, the more effective the solvation. Because the nucleophile is encapsulated by the solvation shell, it is less effective as a nucleophile in forming an S_N2 intermediate; therefore S_N1 reactions are favoured.

Examples of such solvents include methanol, CH_3OH (figure 7), water, H_2O, ammonia, NH_3, methanoic acid, COOH, and hydrogen fluoride, HF.

What makes a good nucleophile?

A nucleophile is an electron-rich species capable of donating a pair of electrons to an electrophile to create a covalent bond. The strength of a nucleophile depends on the ease with which it can make these electrons available.

It is understood that the hydroxide ion is a better nucleophile than the water molecule. These two nucleophiles can be used to demonstrate the factors that influence the strength of a nucleophile.

In summary:

- both nucleophiles possess at least one pair of electrons, so by definition, they can act as Lewis bases (sub-topic 18.1)

- the negatively charged hydroxide ion is a stronger nucleophile than the water molecule which is its conjugate acid (sub-topic 8.1). A negatively charged ion has a far greater attraction for an electrophile than does a neutral molecule.

Since the early 1990s governments and scientific organizations have been supporting the development of green chemistry, recognizing the need to reduce the impact of rapid global development on the environment. *The American Chemical Society* (ACS) formulated the 12 principles of green chemistry. Examples of their incorporation into synthetic organic chemistry include the pharmaceutical industry reducing the need for toxic organic solvents in manufacturing processes and the food industry's development of biodegradable food packaging from corn starch, as an alternative to plastics. Adhesives derived from soya proteins are being used in the building industry in place of adhesives and resins that contain the carcinogen methanal (formaldehyde), a product of the petrochemical industry.

Supercritical carbon dioxide is of increasing importance as an industrial and commercial solvent; see sub-topic D.6 for more information. With a minimal environmental impact, its list of applications continues to grow. The decaffeination of coffee has traditionally been achieved by solvent extraction using dichloromethane, a known carcinogen. Increasingly, the coffee industry is using supercritical carbon dioxide as a non-toxic, green alternative.

Electrophilic addition reactions

In contrast to a nucleophile, an **electrophile** is an electron-deficient species that will accept a pair of electrons, acting as a Lewis acid. Electrophiles include the nitronium ion, NO_2^+ and the methyl cation, CH_3^+.

Electrophiles have either a formal positive charge (a cation) or a partial positive charge ($\delta+$) generated by the presence of a highly electronegative species resulting in the polarization of the bond.

Alkenes are unsaturated compounds that contain electron-rich carbon–carbon double bonds. They undergo addition reactions in which the double bond breaks and two additional atoms bond with the molecule, creating a saturated compound. An electrophile can act as the source of the new additional atoms.

Green chemistry

Green chemistry is an approach that focuses on designing synthetic processes so that they are sustainable, and do not have a negative impact on the environment and society through the production of toxic substances.

The 12 principles of green chemistry

1 Prevent waste

2 Use of renewable feedstock

3 Atom economy

4 Reduce derivatives

5 Less hazardous waste

6 Catalysts

7 Design benign chemicals

8 Design for degradation

9 Benign solvents and auxiliaries

10 Real-time analysis for pollution prevention

11 Design for energy efficiency

12 Inherently benign chemistry for accident prevention

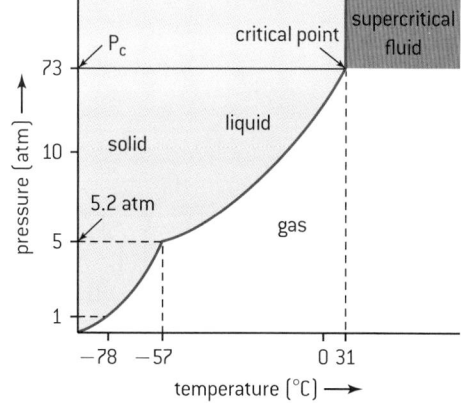

▲ Figure 8 Carbon dioxide phase diagram. In its supercritical state, carbon dioxide exhibits both gas and liquid properties. The temperature, pressure, and additives that control the polarity of the liquid can be varied, resulting in an increasing range of its applications as a solvent

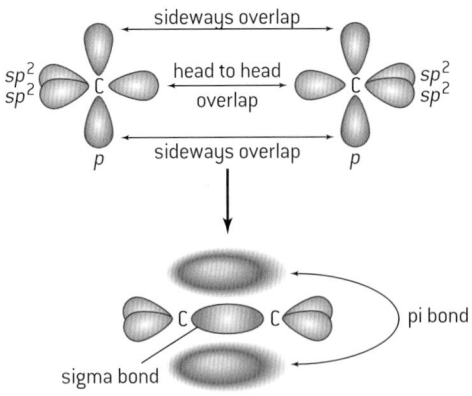

▲ Figure 9 Carbon–carbon double bonds contain both sigma and pi bonds

A carbon–carbon double bond contains both a sigma (σ) bond and a pi (π) bond (figure 9). A sigma bond is formed by the end-to-end overlap of atomic orbitals and electron density is centred between the nuclei of the bonding atoms, along the inter-nuclear axis (sub-topic 14.1). A pi bond is formed by the sideways overlap of atomic orbitals and electron density found above and below the plane of the nuclei of the bonding atoms.

The VSEPR theory (sub-topic 14.1) regards a double bond as a region of high electron density. With sp^2 hybridization in the carbon atoms and a bond angle of approximately 120°, the carbon–carbon double bond provides a reduced level of steric hindrance to the attacking electrophile.

Distinction must be made between the strength of a bond and its reactivity. In terms of bond dissociation energy, a double bond is stronger than a single bond. However, the high density of electrons in a double bond means that the bond is more susceptible to electrophilic attack.

Markovnikov's rule

The major products of the electrophilic addition of hydrogen halides to unsymmetrical alkenes (see below) can be predicted using **Markovnikov's rule**. The hydrogen atom will preferentially bond to the carbon atom of the alkene that is already bonded to the largest number of hydrogen substituents.

This comes about because the carbocation formed when the pi bond is broken has its positive charge centred on the most substituted carbon. A tertiary carbocation has greater stability than a primary carbocation due to the reduction in density of the positive charge through the inductive effects of the three alkyl substituents (table 4).

Type of carbocation	3°	2°	1°
Level of stability	most stable		least stable
Structure	CH₃ | C⁺ / \ CH₃ CH₃	H | C⁺ / \ CH₃ CH₃	H | C⁺ / \ CH₃ H

▲ Table 4 The relative stabilities of primary, secondary and tertiary carbocations form the basis of Markovnikov's rule

Electrophilic addition of hydrogen halides to alkenes

In the electrophilic addition reaction between but-1-ene and hydrogen iodide, the major product is 2-iodobutane as the 2° carbocation is formed preferentially. Hydrogen iodide is split heterolytically, creating the hydrogen cation, H^+ and the iodide anion, I^-. The initial attack on the pi electrons of the C=C bond comes from the cation, followed by rapid reaction between the unstable carbocation and the halogen ion (figure 10).

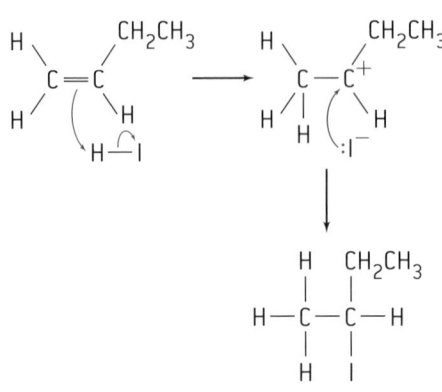

▲ Figure 10 The mechanism of the electrophilic addition of hydrogen iodide to but-1-ene forming 2-iodobutane

Drawing mechanisms for electrophilic addition reactions

- A curly arrow originates from the carbon–carbon double bond to the hydrogen atom of hydrogen iodide.

- The curly arrow representing the iodine leaving originates at the bond between the hydrogen and iodine atoms.

- A curly arrow goes from the lone pair or the negative charge on I^- to C^+ (the carbocation).

- The structural formula of the product 2-iodobutane is shown.

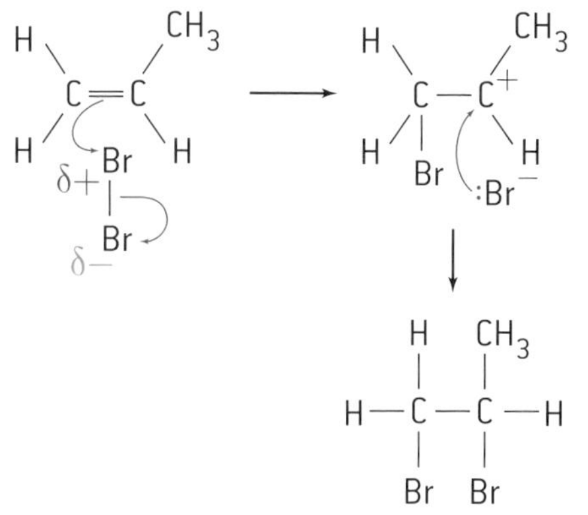

Electrophilic addition of halogens to alkenes

Often used as a test for unsaturation in organic molecules, the electrophilic addition of a halogen (specifically bromine) to an alkene follows the same mechanism as shown in figure 11. The halogen is a non-polar molecule with a net dipole of zero. It is polarized as it approaches the electron-rich C=C of the alkene: electrons within the halogen molecule are repelled, resulting in a temporary dipole.

Electrophilic addition of interhalogens to alkenes

Interhalogens are compounds in which two or more halogens are combined in a molecule. Differences in electronegativity between the halogens will result in an electrophilic region of the molecule and this determines which halogen will attack the pi bond (figure 12).

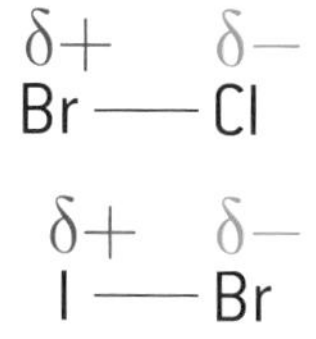

The addition reactions of halogens, interhalogens, and hydrogen halides to symmetrical alkenes all undergo the same mechanism as with unsymmetrical alkenes (figure 13). The difference is that Markovnikov's rule does not apply.

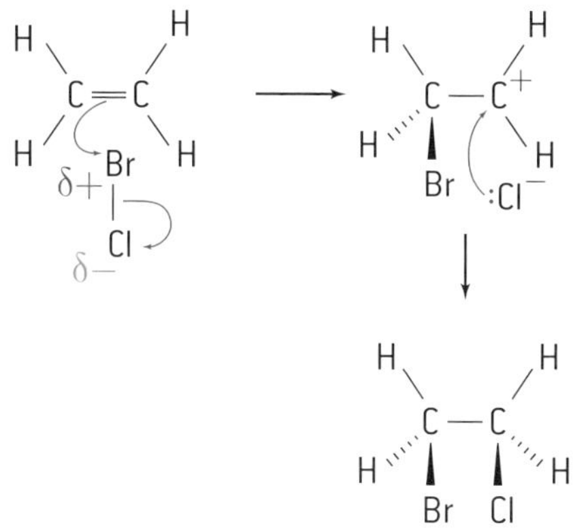

Electrophilic substitution reactions

In sub-topic 10.2 it was stated that benzene does not readily undergo addition reactions, preferring substitution reactions. The electrophilic substitution mechanism of these reactions can be illustrated by the nitration of benzene.

The first step requires the nitronium ion electrophile, NO_2^+ to be generated. Pure nitric acid contains only a small concentration of this electrophile, but a **nitrating mixture** of sulfuric acid and nitric acid at 50 °C generates a higher concentration of nitronium ions, allowing the reaction to proceed at an acceptable rate. Sulfuric acid protonates nitric acid, which subsequently releases a water molecule to generate the electrophile:

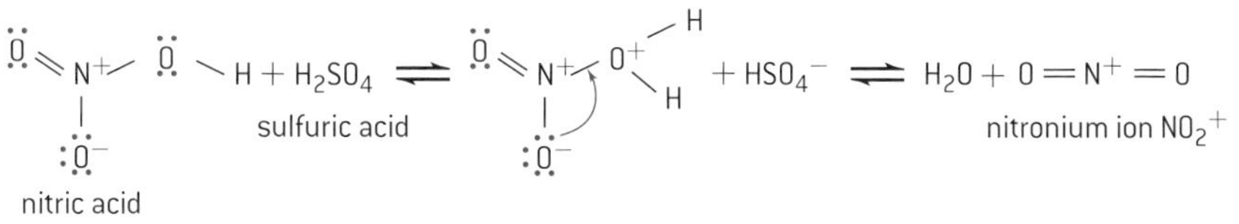

nitric acid   sulfuric acid   nitronium ion NO_2^+

The nitronium ion is a strong electrophile. As the electrophile approaches the delocalized pi electrons of the benzene ring, the nitronium ion is attracted to the ring. Two electrons from the ring are donated and form a new C−N bond. Additionally, a pi electron from the N=O bond of the nitronium ion moves onto the oxygen atom:

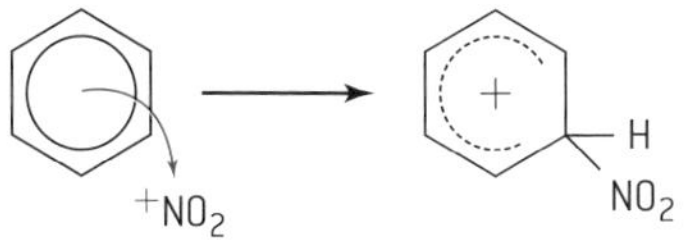

▲ Figure 11 The mechanism of the electrophilic addition of bromine to propene forming 1,2-dibromopropane

$$\overset{\delta+}{Br} \!-\! \overset{\delta-}{Cl}$$

$$\overset{\delta+}{I} \!-\! \overset{\delta-}{Br}$$

▲ Figure 12 Polarity in interhalogen molecules

▲ Figure 13 The addition of the interhalogen BrCl to ethene forms 1-bromo-2-chloroethane

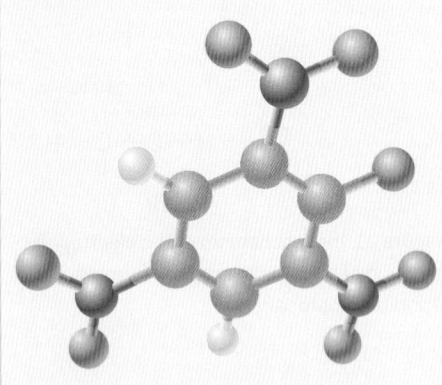

Molecular models are useful aids for visualization of the shapes of molecules and can help with working out mechanisms and nomenclature.

▲ Figure 14 A molecular model of 2,4,6-trinitrophenol (picric acid)

This is the rate-determining step of the mechanism. The addition of the nitronium ion to the C=C bond eliminates the aromaticity of the arene.

Water then acts as a base, deprotonating the carbocation intermediate and restoring the aromaticity of the system.

$$C_6H_6 + HNO_3 \xrightarrow[50\ °C]{H_2SO_4} C_6H_5NO_2 + H_2O$$

The product nitrobenzene is a yellow oil that can be isolated from the reaction mixture.

Drawing mechanisms for electrophilic substitution reactions

- A curly arrow originates from delocalized electrons in benzene and terminates at the $^+NO_2$ electrophile.

- Structural representation of the carbocation shows a partial, broken-line circle and a positive charge on the ring.

- A curly arrow representing the hydrogen ion leaving originates at the bond between the carbon and hydrogen atoms and terminates at the benzene ring cation.

- The structural formula of the organic product nitrobenzene is shown along with the released hydrogen ion, H^+.

Reduction of carboxylic acids

Carboxylic acids are reduced to aldehydes and eventually to primary alcohols while ketones are reduced to secondary alcohols, in reactions that are the reverse of the oxidation of alcohols (sub-topic 10.2).

Two commonly used reducing agents are lithium aluminium hydride $LiAlH_4$ and sodium borohydride $NaBH_4$. Lithium aluminium hydride is regarded as a nucleophilic reducing agent that will reduce polar C=O bonds present in carboxylic acids, aldehydes, and ketones.

$LiAlH_4$ is the stronger reducing agent; it can reduce carboxylic acids to primary alcohols while $NaBH_4$ can reduce only aldehydes and ketones to alcohols.

The reduction equation is often represented in a simplified manner using the symbol [H] to represent the reducing agent. The reduction of an aldehyde to a primary alcohol can be represented by the general equation:

$$R-CHO + 2[H] \rightarrow R-CH_2OH$$

and the reduction of ketones to secondary alcohols by this general equation:

$$R-CO-R' + 2[H] \rightarrow R-CH(OH)-R'$$

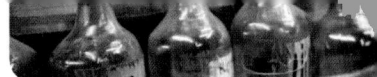

 Conversion of nitrobenzene to phenylamine (aniline)

The nitration of benzene by electrophilic addition occurs when benzene is heated at 50 °C with a mixture of sulfuric acid and nitric acid. This initial step is described earlier in this topic:

$$C_6H_6 + HNO_3 \xrightarrow[\text{50 °C}]{\text{H}_2\text{SO}_4} C_6H_5NO_2 + H_2O$$

The subsequent conversion of nitrobenzene to phenylamine (aniline) (systematic name phenylamine) is described in two stages.

Stage 1: Reduction of nitrobenzene

Nitrobenzene is heated in a water bath under reflux with a mixture of zinc and concentrated hydrochloric acid. The phenylammonium ion is formed and zinc is oxidized to zinc(II).

$$C_6H_5NO_2(l) + 3Zn(s) + 7H^+(aq) \rightarrow C_6H_5NH_3^+(aq) + 3Zn^{2+}(aq) + 2H_2O(l)$$

Stage 2: Formation of aniline

Aniline is formed by the deprotonation of the ammonium salt through the addition of sodium hydroxide:

$$C_6H_5NH_3^+(aq) + OH^-(aq) \rightarrow C_6H_5NH_2(l) + H_2O(l)$$

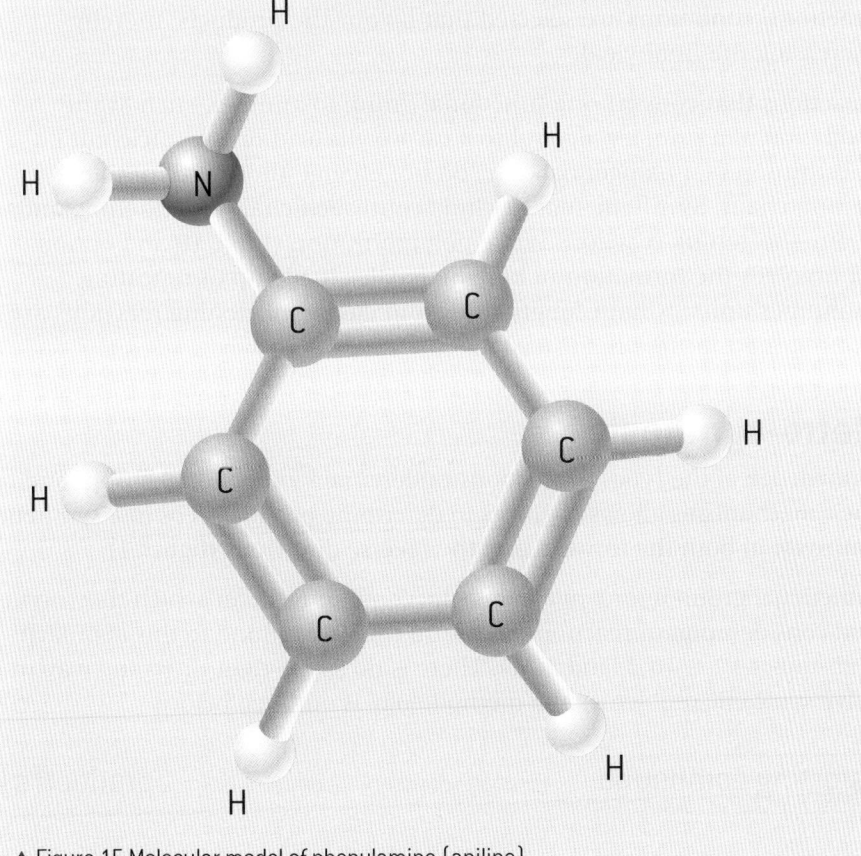

▲ Figure 15 Molecular model of phenylamine (aniline)

Reflux

Reflux is a process in which a reaction mixture is heated under controlled conditions for a period of time. A condenser is used to cool the vapours from volatile solvents and condense them back into the reaction mixture. The process ensures that the temperature remains constant over time and optimal conditions for the reaction are achieved.

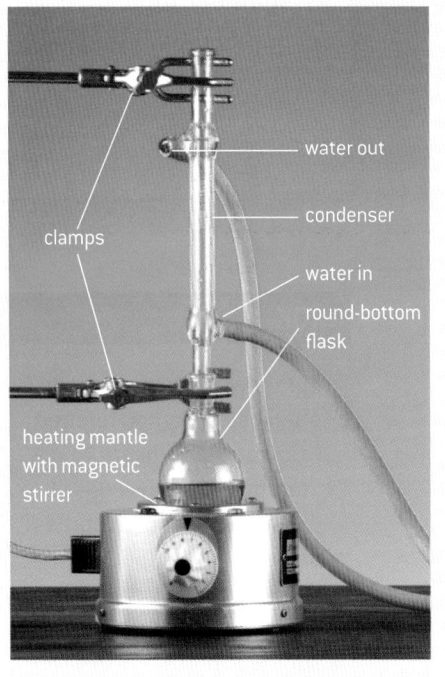

▲ Figure 16 Reflux apparatus

20.2 Synthetic routes

Understandings

→ The synthesis of an organic compound stems from a readily available starting material via a series of discrete steps. Functional group interconversions are the basis of such synthetic routes.

→ Retro-synthesis of organic compounds.

 Applications and skills

→ Deduction of multi-step synthetic routes given starting reagents and the product(s).

 Nature of science

→ Scientific method – in synthetic design, the thinking process of the organic chemist is one which invokes retro-synthesis and the ability to think in a reverse-like manner.

Background to designing a synthetic route

Organic synthesis takes a starting material and converts it via a series of reactions into the desired product. Each step produces an intermediate in quantities less than the theoretical yield, so an efficient synthetic pathway will involve the smallest number of steps. For equilibrium reactions, conditions are selected that favour the products, thereby increasing the final yield.

Reactions that convert one functional group to another, such as the oxidation of a primary alcohol to a carboxylic acid or the nucleophilic substitution of a halogenoalkane, do not change the length of the carbon chain. Synthetic tools include controlled chain-lengthening and chain-shortening reactions while polymerization (topic 10 and option A) involves the formation of long molecules made up of repeating monomer units. (Chain-lengthening and chain-shortening reactions are not required for IB Chemistry.)

Retro-synthesis

Knowledge of the types of reactions undergone by functional groups and their mechanisms allows chemists to determine possible steps in a synthetic pathway, in both the forward and the reverse directions (figure 1).

Functional group interconversions, the conditions under which they occur, and consideration of reaction rates form the background to this approach (sub-topics 10.1, 10.2, and 20.1). There is no single right or wrong way to solve synthetic problems; one methodology is outlined below.

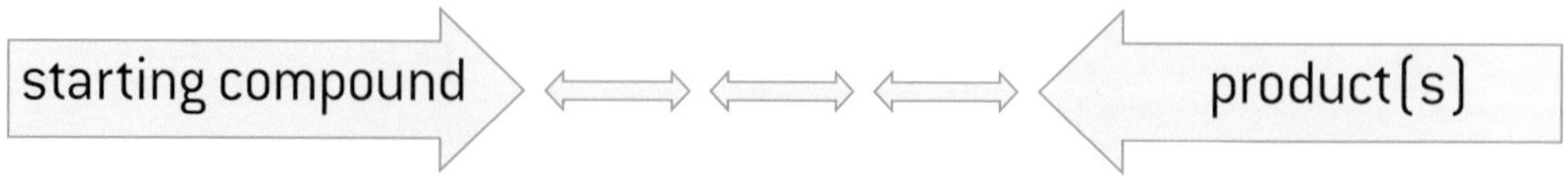

starting compound ⟺ ⟺ ⟺ product(s)

▲ Figure 1 Designing a synthetic pathway. The reversible arrows do not indicate equilibrium but rather a problem-solving approach that involves working both forwards from the reactant and backwards from the product (retro-synthesis).

Step 1

- Draw the structural formulae of both the starting compound and the desired product(s).

- Identify the functional group(s) present in the product.

Step 2

- List possible reactions that would produce the desired functional group(s).

Step 3

- Identify the functional group(s) present in the starting material and identify any relationship between the starting reagent and any intermediate compounds you have listed in step 2.

Step 4

- Design a reaction pathway that has the minimum number of steps. Include all the reaction conditions and reagents required.

 Worked example

Design a synthetic route to produce ethyl methanoate from chloromethane.

Solution

Step 1

The starting material, chloromethane, is a halogenoalkane while the product, ethyl methanoate, is an ester (figure 2).

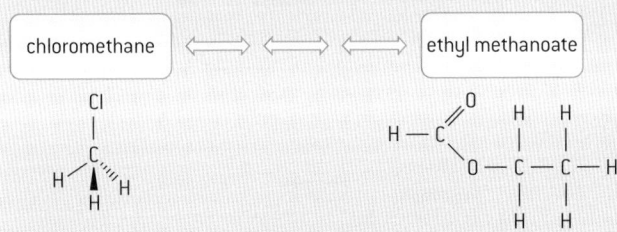

▲ Figure 2 Structural formula of the starting compound, chloromethane and the ester product, ethyl methanoate

Step 2

Ethyl methanoate is the product of a condensation (esterification) reaction between a carboxylic acid (methanoic acid, HCOOH) and an alcohol (ethanol, CH_3CH_2OH):

methanoic acid + ethanol

The condensation reaction is achieved by heating the carboxylic acid and alcohol in the presence of concentrated sulfuric acid, creating the sweet-smelling ester (sub-topic 10.2). Water is also formed in this reaction.

Step 3

The starting material, chloromethane, needs to be converted to methanoic acid. Methanoic acid is the product of the oxidation of a primary alcohol (sub-topic 10.2). Chloromethane, a halogenoalkane, is a reactive compound that contains a polarized C–Cl bond which is susceptible to attack by nucleophiles such as OH^- (sub-topic 20.1).

Step 4

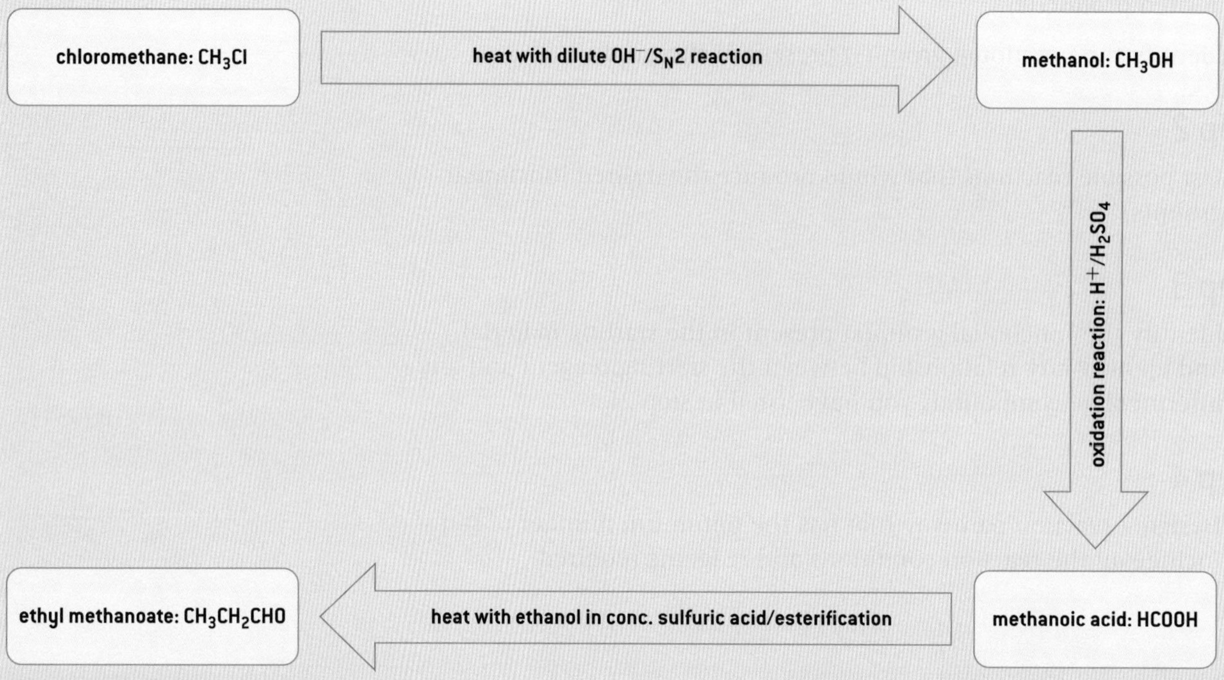

chloromethane: CH_3Cl
heat with dilute OH^-/S_N2 reaction
methanol: CH_3OH
oxidation reaction: H^+/H_2SO_4
methanoic acid: HCOOH
heat with ethanol in conc. sulfuric acid/esterification
ethyl methanoate: CH_3CH_2CHO

▲ Figure 3 A reaction pathway showing how ethyl methanoate is produced from chloromethane, including reaction conditions and reagents

Summary reaction pathways

Figures 4 to 6 show reaction pathways required for the SL/HL organic chemistry topics. When a mechanism is required, this is signified by "**M**".

aldehyde
carboxylic acid
ketone
alcohol
ester
M
halogenoalkane
M
alkane
dihalogenoalkane
M
alkene

▲ Figure 4 Reaction pathways for aliphatic compounds

Examination questions may require you to deduce synthetic routes of up to four steps.

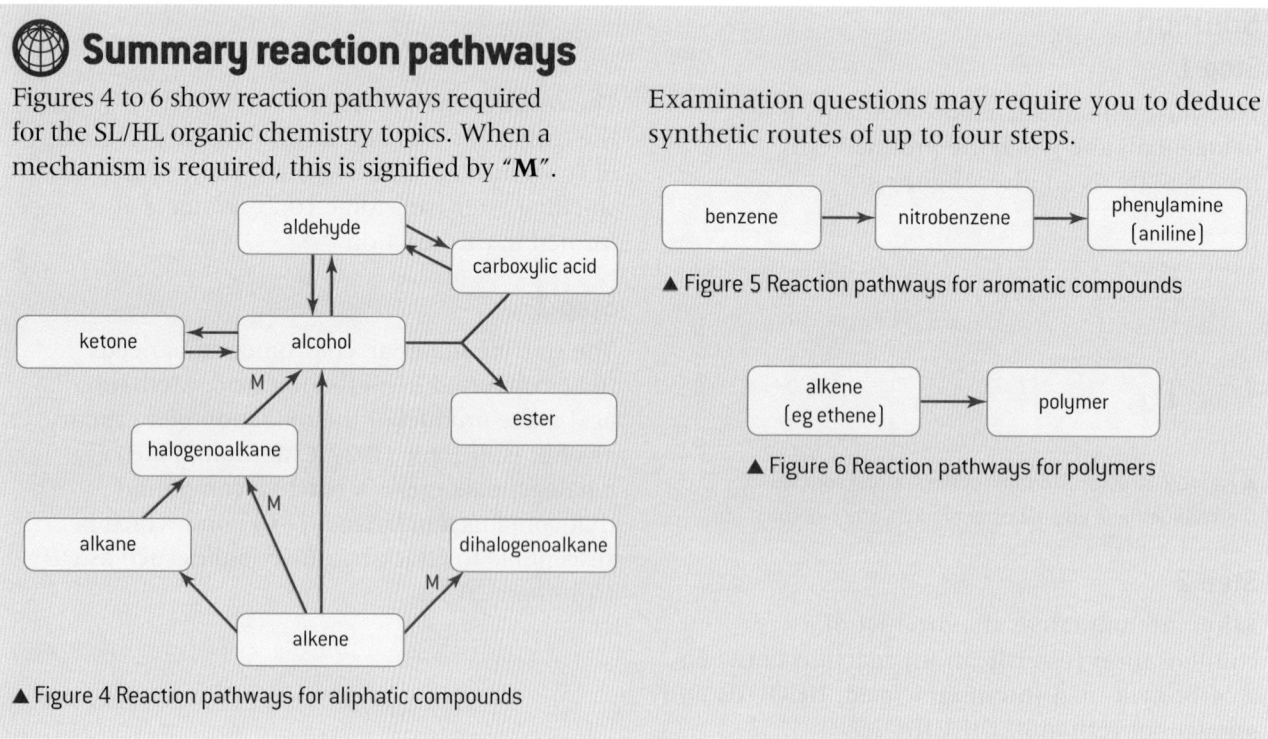

benzene
nitrobenzene
phenylamine
(aniline)

▲ Figure 5 Reaction pathways for aromatic compounds

alkene
(eg ethene)
polymer

▲ Figure 6 Reaction pathways for polymers

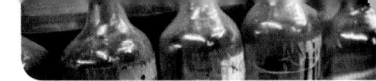

20.3 Stereoisomerism

Understandings

→ Stereoisomers are subdivided into two classes – conformational isomers, which interconvert by rotation about a σ bond, and configurational isomers that interconvert only by breaking and reforming a bond.

→ Configurational isomers are further subdivided into *cis–trans* and *E/Z* isomers, and optical isomers.

→ *Cis–trans* isomers can occur in alkenes or cycloalkanes (or heteroanalogues) and differ in the positions of atoms (or groups) relative to a reference plane. According to IUPAC, E/Z isomers refer to alkenes of the form $R_1R_2C{=}CR_3R_4$ ($R_1 \neq R_2$, $R_3 \neq R_4$) where neither R_1 nor R_2 need be different from R_3 or R_4.

→ A chiral carbon is a carbon joined to four different atoms or groups.

→ An optically active compound can rotate the plane of polarized light as it passes through a solution of the compound. Optical isomers are enantiomers. Enantiomers are non-superimposable mirror images of each other. Diastereomers are not mirror images of each other.

→ A racemic mixture (or racemate) is a mixture of two enantiomers in equal amounts and is optically inactive.

Applications and skills

→ Construction of 3-D models (real or virtual) of a wide range of stereoisomers.

→ Explanation of stereoisomerism in non-cyclic alkenes and C_3 and C_4 cycloalkanes.

→ Comparison between the physical and chemical properties of enantiomers.

→ Description and explanation of optical isomers in simple organic molecules.

→ Distinction between optical isomers using a polarimeter.

Nature of science

→ Transdisciplinary – the three-dimensional shape of an organic molecule is the foundation pillar of its structure and often its properties. Much of the human body is chiral.

Types of isomerism

Stereoisomers have an identical molecular formula and bond multiplicity but show different spatial arrangements of the atoms. Stereoisomers can be subdivided into two major classes, **conformational isomers** and **configurational isomers** (figure 1).

Conformational isomers can be interconverted by rotation about the σ bond, without breaking any bonds. **Configurational isomers** can

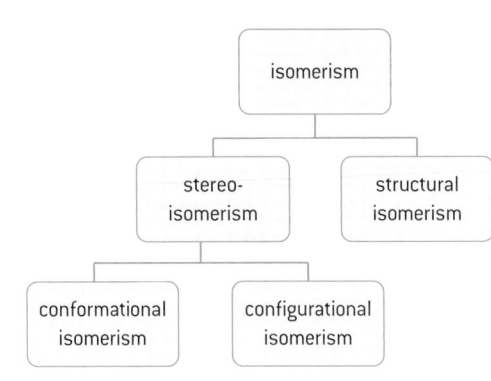

▲ Figure 1 Classes of isomerism

only be interconverted by the breaking of the σ or π bond or through rearrangement of the stereocentres.

Conformational isomers

Substituents and functional groups joined together by single σ-bonds can rotate freely, changing the three-dimensional arrangement of the atoms relative to one another. In contrast, a carbon–carbon double bond is composed of both a σ and a π bond and the arrangement of electron density above and below the internuclear axis means that no rotation is possible without breaking the π bond.

Conformational isomers therefore differ from one another in the arrangement of atoms around a single bond. The rapid interconversion from one conformer to the other means that the separation of the individual isomers is virtually impossible.

🌐 Ethane

In ethane, rotation about the carbon–carbon bond results in two different conformations: **eclipsed** and **staggered**. In the eclipsed conformation, the substituents (hydrogen atoms) on adjacent carbons are as close to one another as is possible. When one half of the molecule rotates about the carbon–carbon bond, the relative positions of the substituents change until the three hydrogen atoms on each carbon are as far apart as possible – this is the staggered conformation.

A **Newman projection** is a representation of the three-dimensional structure which shows the conformation of the molecule by looking along the carbon–carbon bond (figures 2 and 3). The front carbon and its substituents are represented by lines projecting out of the centre. The circle represents the carbon at the rear and bonds coming out of the circle show its substituents.

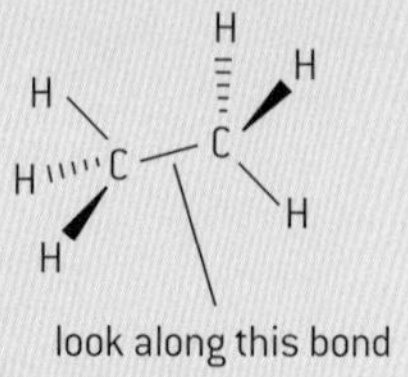

▲ Figure 2 Look along the carbon–carbon bond to construct the Newman projection of ethane

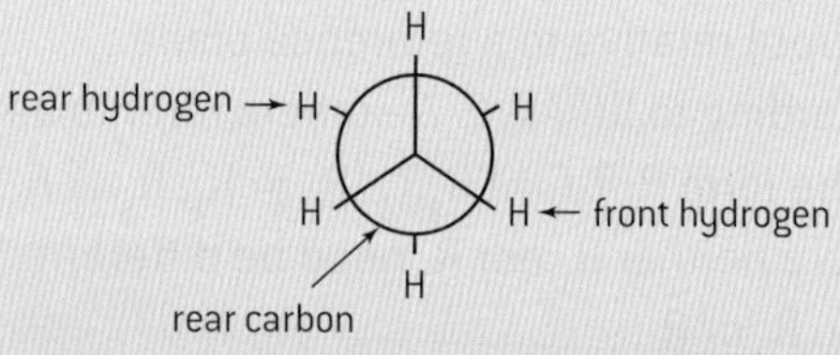

▲ Figure 3 The Newman projection of ethane shows the relative positions of the substituents

The Newman representation of the eclipsed conformation shows the hydrogen atoms slightly askew for clarity (figure 4), but the actual angle between hydrogens on the adjacent carbon atoms is 0°.

The staggered conformation, with the hydrogens on the adjacent carbon atoms positoned at 60° to each other, is more stable. The eclipsed conformation is of higher energy (less stable) due to repulsive interactions between the electrons of C–H bonds.

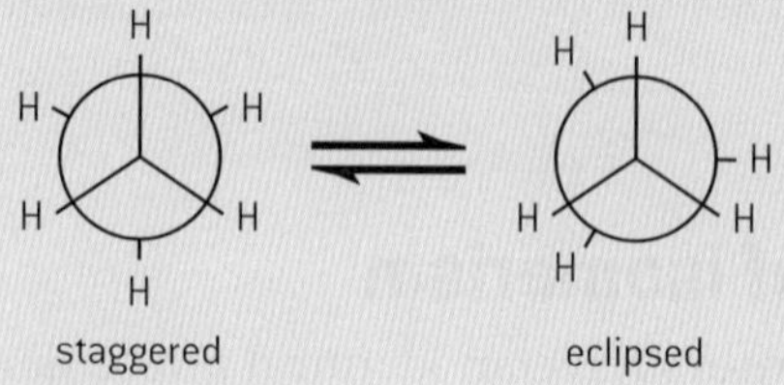

staggered   eclipsed

▲ Figure 4 Newman projections of staggered and eclipsed ethane conformers

🌐 Conformational isomerism in cyclic hydrocarbons

Cycloalkanes also show conformational isomerism. The structural consequences arising from the bond angles in C_3, C_4, and C_5 cycloalkanes have been the subject of extensive research. **Torsional strain** or **torsional energy** is the energy difference between the staggered and eclipsed conformations. It is the result of the repulsion between bonding electrons. In an eclipsed conformation, pairs of bonding electrons in the C–H bonds will repel one another.

Cyclopropane

The ring structure of cyclopropane lacks stability as the molecule experiences **ring strain** for two reasons. It exhibits **torsional strain** from repulsion of adjacent bonding electrons in C–H bonds due to the ring rigidity (figure 5). It also exhibits **angle strain**: the sp³ orbital angle is 109.5° (tetrahedral) but the internuclear bond angle in cyclopropane is only 60° resulting in a misalignment of the orbitals when they overlap end on to create the σ bond. The result is a

bent C–C bond, confirmed by electron-density mapping: the bonding electron density is greatest outside the carbon–carbon internuclear axis.

Cyclobutane

The carbon ring bond angle increases from 60° in cyclopropane to 90° in cyclobutane; the molecule of cyclobutane still experiences angle strain (90° < 109.5°) and torsional strain from the eclipsed arrangement of adjacent C–H bonds. The strained four-membered ring in penicillin shows similar angle strain; this is discussed in sub-topic D.2.

One way of minimizing the strain placed on this conformer is to **pucker** the ring. One of the four carbon atoms moves out of the plane of the ring, slightly increasing angle strain but significantly decreasing torsional strain (figure 6).

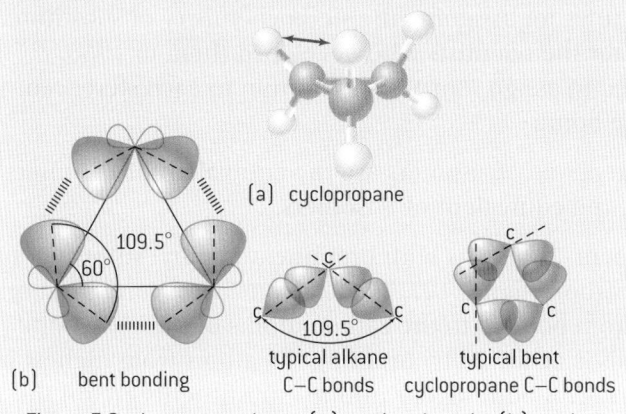

(a) cyclopropane

(b) bent bonding

typical alkane C–C bonds

typical bent cyclopropane C–C bonds

▲ Figure 5 Cyclopropane shows (a) torsional strain; (b) angle strain: the bond angle of 60° is much less than the ideal 109.5° for sp³ hybridized bonds

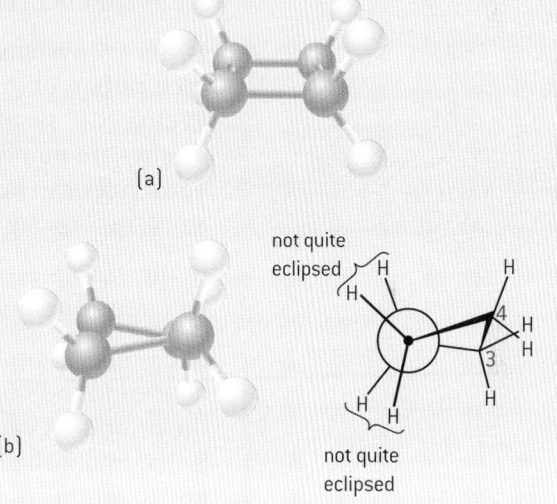

(a)

(b)

not quite eclipsed

not quite eclipsed

▲ Figure 6 (a) The planar ring structure of cyclobutane, with eclipsed C–H bonds, exhibits torsional and angle strain. (b) Puckering of the cyclobutane ring reduces torsional strain

Configurational isomers

Configurational isomers can be interconverted only by the breaking of bonds or through the rearrangement of the stereocentres. Configurational isomers are subdivided into *cis–trans* and *E/Z* isomers on the one hand and optical isomers on the other (figure 7).

As mentioned above, configurational isomers exist due to the lack of rotation around the carbon–carbon double bond such as that present in aliphatic alkenes.

Cis–trans isomers are determined by the positions of substituents relative to a reference plane. For alkenes this reference plane is the

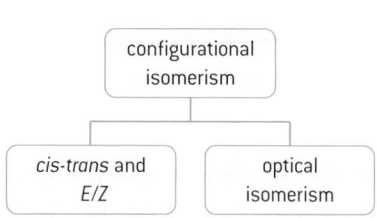

▲ Figure 7 Classes of configurational isomerism

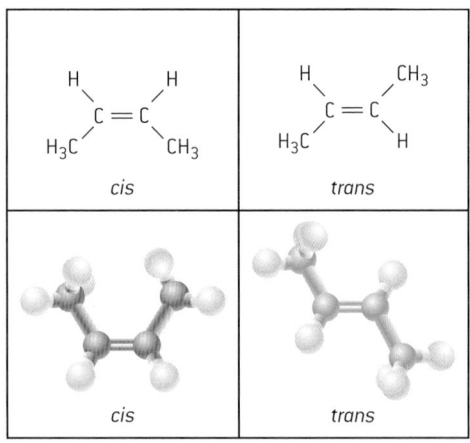

▲ Figure 8 *Cis–trans* configurational isomers of but-2-ene

carbon–carbon double bond. *Cis*-isomers have substituents on the same side of the reference plane while in *trans*-isomers the substituents are on opposite sides (figure 8).

Disubstituted cycloalkanes also exhibit *cis–trans* isomerism with the plane of symmetry being the ring (figure 9).

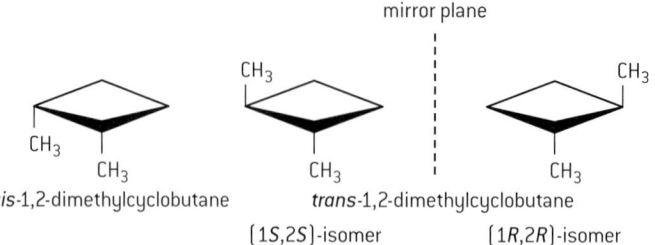

▲ Figure 9 *Cis–trans* isomerism in 1,2-dimethylcyclobutane

It is a relatively simple process to identify *cis–trans* isomers of disubstituted alkenes. However, a different naming convention is adopted for the naming of tri- and tetra-substituted alkenes.

According to IUPAC, for isomers of the form $R_1R_2C = CR_3R_4$, where $R_1 \neq R_2$, $R_3 \neq R_4$, and neither R_1 nor R_2 need be different from R_3 or R_4, the *E/Z* nomenclature rules outlined below can be applied.

In the *E/Z* system relative priorities are assigned to the substituents on each carbon of the carbon–carbon double bond. The **Cahn–Ingold–Prelog** (CIP system) rules for assigning the stereochemistry of substituted alkenes are named after the scientists who developed them in 1966. Their rules can be used to assign *R* or *S* configuration to each stereocentre and *E* or *Z* to a double bond.

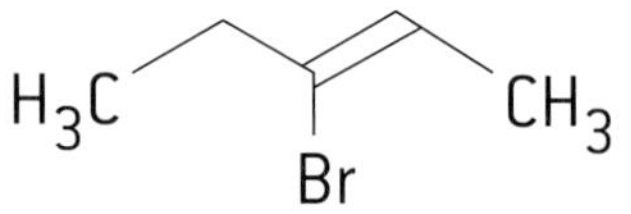

In the examples given above, *cis*-isomers are the equivalent of *Z* isomers and *trans*-isomers are the equivalent of *E* isomers.

The isomer 3-bromopent-2-ene has both bromine and ethyl substituents bonded to the same carbon (figure 10). To assign an *E* or *Z* configuration, the priority of each atom bonded to the carbon atoms of the C=C bond is first established. This is achieved by ordering the atoms from highest to lowest atomic number, *Z*. If both higher priority substituents are on the same side of the double bond, the isomer is designated *Z* (comparable to *cis*). If they are on opposite sides it is designated *E* (comparable to *trans*).

In 3-bromopent-2-ene, one carbon of the double bond has methyl and hydrogen substituents and the other carbon has bromine and ethyl substituents. Table 1 lists substituents in order of priority and the two possible isomers are shown below.

▲ Figure 10 3-bromopent-2-ene, $CH_3CH=C(Br)CH_2CH_3$

Element	Z	
I	53	
Br	35	
Cl	17	highest priority
F	9	↓
O	8	
N	7	lowest priority
C	6	
H	1	

▲ Table 1 To assign an *E* or *Z* configuration, substituents are prioritized according to their atomic number, *Z*

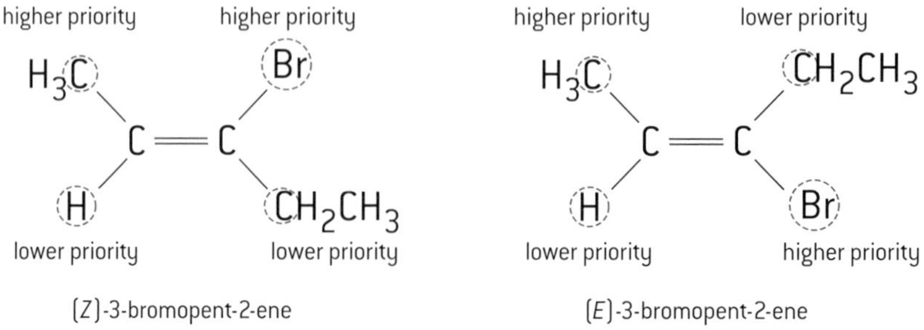

higher priority   higher priority

H_3C (Br)

C = C

(H)   (CH₂CH₃)

lower priority   lower priority

(*Z*)-3-bromopent-2-ene

higher priority   lower priority

H_3C (CH₂CH₃)

C = C

(H)   (Br)

lower priority   higher priority

(*E*)-3-bromopent-2-ene

▲ Figure 11 If both higher priority substituents are the same side of the double bond, the isomer is designated *Z*. If they are on opposite sides it is designated *E*

Stereoisomerism in carotenoids

Carotenoids are a large group of organic pigments (see sub-topic B.2) that display a wide range of stereochemical properties. Research into the effects carotenoids have on visual and motor integration within the human body focuses on individual stereoisomers in isolation from their isomeric partners. Visual motor integration measures a child's ability to make sense of visual information and then use it appropriately for a motor task such as writing, playing sports, or using tools and utensils.

Optical isomerism

Optical isomerism is a type of configurational isomerism determined by the presence of **chiral** carbon atoms. Also known as a **stereocentre** or **asymmetric centre**, a chiral carbon is bonded to four different atoms or groups of atoms. Optical isomers have the ability to rotate plane-polarized light and exist in pairs that are called **enantiomers** or **diastereomers**.

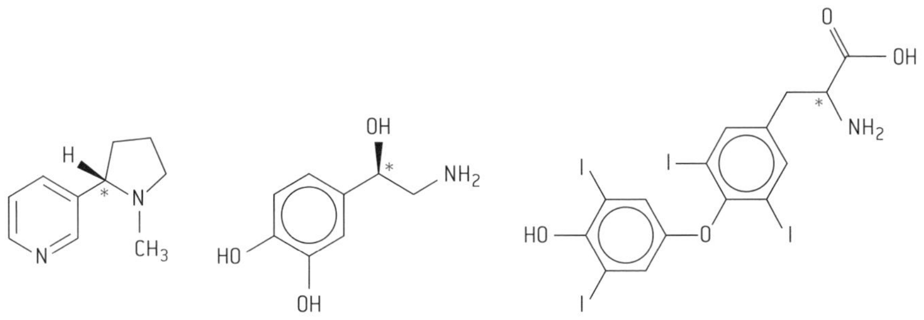

nicotine   norepinephrine   thyroxine

▲ Figure 12 Nicotine is naturally synthesized by the tobacco plant, norepinephrine is a neurotransmitter, and thyroxine is a hormone from the thyroid gland. All of these compounds are chiral: they each contain a stereocentre identified by *

Enantiomers are non-superimposable mirror images of each other. They have no plane of symmetry and their optical activity is most readily assigned when the molecules are represented as three-dimensional images (figure 13).

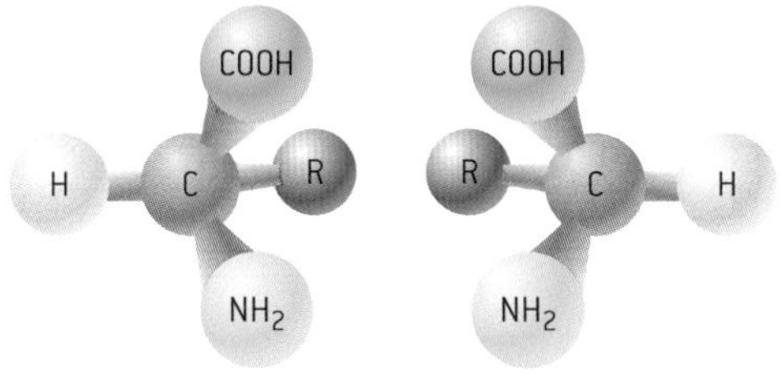

▲ Figure 13 Non-superimposable mirror images of the general formula of a 2-amino acid

Chemists have established that the enantiomers of many chemical compounds have varying effects on the human olfactory system (our sense of smell). Different enantiomers of the same compound have different odours. Recognition of this is of vital importance to the perfume industry that has an annual income of billions of US dollars. Natural products may exist as individual enantiomers or their mixtures. Precise analysis is required to establish their olfactory properties in the development of new perfumes.

▲ Figure 14 The art of making perfumes can be traced back to ancient times in Egypt. Other ancient civilizations such as those of the Arabs, Romans and Persians made perfume an important aspect of their daily lives, promoting the development of scientific techniques used in the extraction of scents from plants. Today, the markets in Fez, Morocco, offer a large variety of scents that can be mixed by artisans

Optical isomers and plane-polarized light

Under the same conditions, two optical isomers with the same general formula rotate the plane of polarized light by the same angle but in opposite directions (figure 15). One enantiomer rotates the plane of polarization in a clockwise direction; this is designated the (+) enantiomer. The other enantiomer rotates the plane of polarization in an anticlockwise direction and is designated the (−) enantiomer. A 50:50 mixture of the two enantiomers is called a **racemic mixture** (or **racemate**) and does not rotate plane-polarized light.

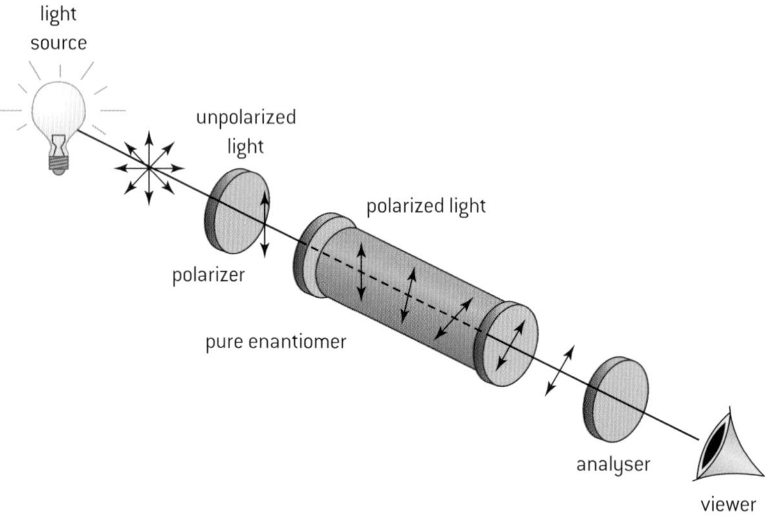

▲ Figure 15 Rotation of plane-polarized light by a pure enantiomer

A polarimeter can be used to determine the optical purity of the products of synthetic reactions (see sub-topic D.7). This technique is commonplace in industry producing optically active products, examples of which can be found throughout the pharmaceutical, fragrance food, and chemical industries. The product's effect on plane-polarized light can be compared with literature values to determine the purity of the desired enantiomer.

TOK

TOK identifies eight specific ways of knowing (WOK). Individual learners undergo experiences which leads to the recognition of new information. The combination of this new information, coupled with prior learning and understanding, enables the learner to determine what knowledge they will construct and how this will integrate into their understanding of the world.

Optical isomers and their structure give indirect evidence of the existence of a tetrahedrally bonded carbon atom. How do we use indirect evidence to construct understanding? What ways of knowing help us to establish relationships between established theories and indirect evidence?

The symbols *d* (for dextro) and *l* (for levo) are now obsolete as stipulated by IUPAC and have been replaced by (+) and (−).

Stereoisomerism in medicine

Many therapeutic drugs are chiral molecules with only one enantiomer having the desired pharmacokinetic and pharmacodynamic properties. **Pharmacokinetics** studies the body's response to foreign compounds and changes caused by the administered drug. It is associated with the absorption, distribution, metabolism and excretion of the drug by the body. **Pharmacodynamics** studies the action of the drug on the systems of the body and how a drug binds to its target site.

The separation of enantiomers can be a very expensive process and such drugs are often administered as racemic mixtures rather than as the pure active enantiomer. For example, synthetic compounds used as anaesthetics may be administered as a racemic mixture with one of the enantiomers having the intended therapeutic effects while the other may have undesired effects that can

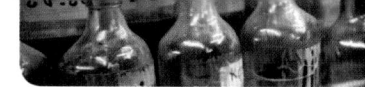

interfere with the actions of the active component or, more seriously, have damaging effects on the body.

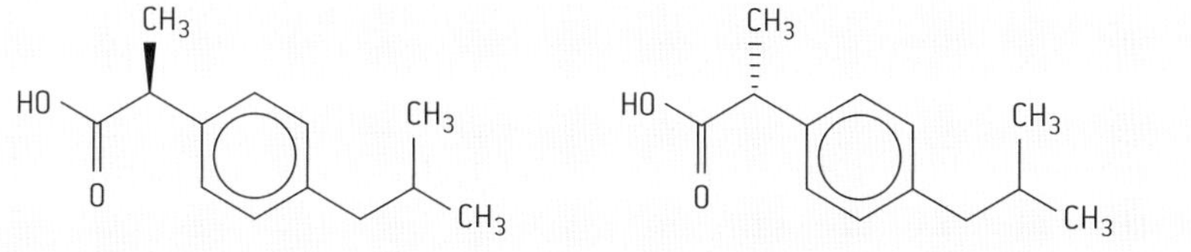

(+)-ibuprofen (active)   (−)-ibuprofen (inactive)

▲ Figure 16 Ibuprofen is a chiral molecule; one enantiomer is active in the body while the other is not

Physical and chemical properties of optical isomers

The two enantiomers of a particular substance have identical physical properties such as boiling and freezing points, viscosity, density, and solubility. Many of their chemical properties are also identical except for their reactions with other optical isomers, often in biological systems. Enzymes within the body are chiral and they can distinguish between the enantiomers of their substrate (see sub-topic B.7).

For example, limonene is a chiral molecule and the body can distinguish between the two enantiomers in both taste and odour. Lemons and oranges contain the same isomer, (+)-limonene. In contrast, (−)-limonene is found in pine needles, star anise, peppermint, and spearmint and has no similarity in smell or taste to the (+)-isomer (figure 17).

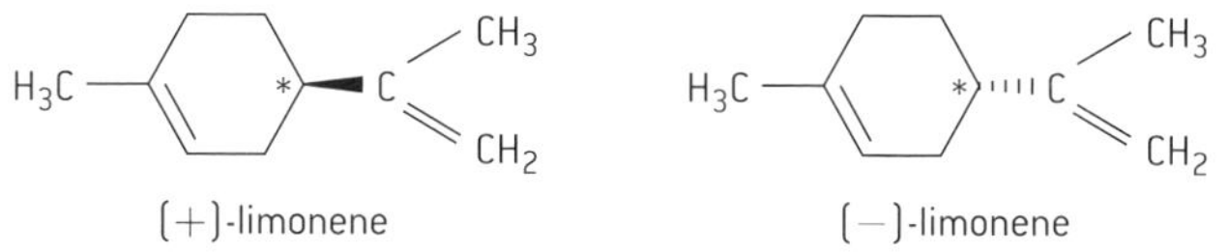

(+)-limonene   (−)-limonene

▲ Figure 17 The (+)- and (−)-enantiomers of limonene

Another example of a chiral molecule is the drug thalidomide (figure 18). It became available in the late 1950s and was prescribed by medical practitioners to pregnant women for the treatment of morning sickness, which is nausea associated with pregnancy. One enantiomer of thalidomide is the effective drug, free of clinical side effects, with the intended therapeutic effect. The other enantiomer is a **teratogen**: it caused birth defects in babies born to some mothers who took the drug.

In the human body the thalidomide enantiomers rapidly interconvert due to the relatively high acidity of the proton at the stereocentre. Regardless of the enantiomer administered, a racemic mixture is soon produced in the body. The drug is now used for the treatment of leprosy and cancer in males and those female patients who use contraception and undergo regular pregnancy tests.

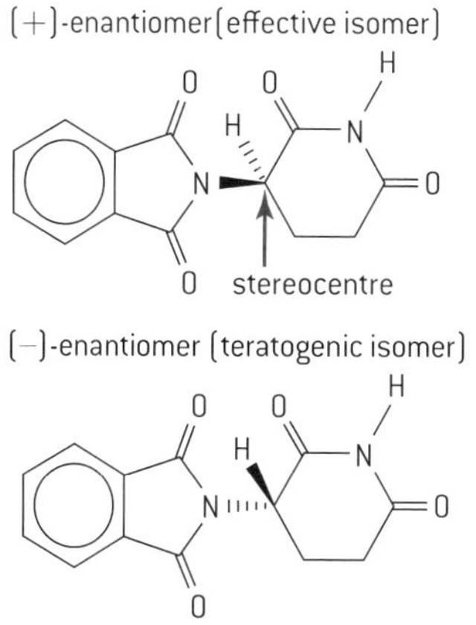

(+)-enantiomer (effective isomer)

stereocentre

(−)-enantiomer (teratogenic isomer)

▲ Figure 18 The enantiomers of thalidomide

TOK

Chemists represent complex molecular structures as three-dimensional models. Modern-day models have become sophisticated and detailed through advances in computing power.

How do the scientists who elucidate complex structures in this way accurately represent them in two dimensions? What are the similarities and differences in the two approaches and what is the role of the different ways of knowing?

Diastereomers

Diastereomers are different from the enantiomers of optical isomers. Like enantiomers they are non-superimposable but they do not form mirror images. They have two or more stereocentres and differ in the configuration of at least one centre. In contrast to enantiomers, diastereomers with the same general formula have different physical and chemical properties. In option B.10 D-xylose and L-ribose are identified as diastereomers due to the configuration at the $C-2$ and $C-4$ positions (figure 19).

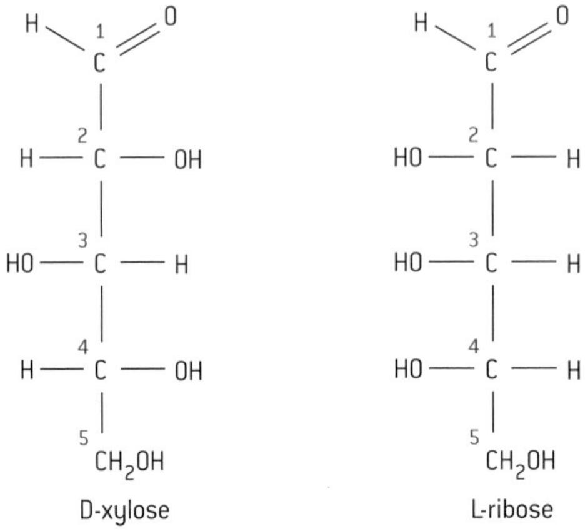

▲ Figure 19 D-xylose and L-ribose are diastereomers due to the configuration at the $C-2$ and $C-4$ positions. They are non-superimposable but they are not mirror images

Questions

1 What is the correct order of reaction types in the following sequence?

$$C_3H_7Br \xrightarrow{\text{I}} C_3H_7OH \xrightarrow{\text{II}} C_2H_5COOH \xrightarrow{\text{III}} C_2H_5COOC_2H_5$$

	I	II	III
A.	substitution	oxidation	condensation
B.	addition	substitution	condensation
C.	oxidation	substitution	condensation
D.	substitution	oxidation	substitution

IB, May 2011

2 a) Identify the reagents used in the nitration of benzene. [2]

   b) Write an equation or equations to show the formation of the species NO_2^+ from these reagents. [1]

   c) Give the mechanism for the nitration of benzene. Use curly arrows to represent the movement of electron pairs. [2]

IB, May 2006

3 Which process can produce a polyester?

   A. Addition polymerization of a dicarboxylic acid

   B. Condensation polymerization of a diol and a dicarboxylic acid

   C. Addition polymerization of a diol and dicarboxylic acid

   D. Condensation polymerization of a dicarboxylic acid

IB, November 2010

4 Which statements about substitution reactions are correct?

   I. The reaction between sodium hydroxide and 1-chloropentane predominantly follows an S_N2 mechanism.

   II. The reaction between sodium hydroxide and 2-chloro-2-methylbutane predominantly follows an S_N2 mechanism.

   III. The reaction of sodium hydroxide with 1-chloropentane occurs at a slower rate than with 1-bromopentane.

   A. I and II only

   B. I and III only

   C. II and III only

   D. I, II and III [1]

IB, May 2009

5 Which statement is correct about the enantiomers of a chiral compound?

   A. Their physical properties are different.

   B. All their chemical reactions are identical.

   C. A racemic mixture will rotate the plane of polarized light.

   D. They will rotate the plane of polarized light in opposite directions. [1]

IB, May 2009

6 Which molecule has a chiral centre?

   A. $CH_3CH=CHCHO$

   B. $(CH_3)_2C=CHCH_2OH$

   C. $CH_3OCH_2CH_3$

   D. $CH_3CHOHCH_2CH_3$ [1]

IB, May 2011

7 Which two molecules in figure 20 are *cis–trans* isomers of each other?

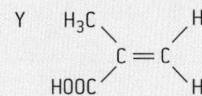

▲ Figure 20

   A. X and Z

   B. X and Y

   C. W and Y

   D. W and Z [1]

IB, May 2011

8 Halogenoalkanes can undergo substitution reactions with potassium hydroxide solution.

a) State an equation for the reaction of C_4H_9Cl with KOH. [1]

b) Substitution reactions may occur by either of two mechanisms namely S_N1 or S_N2.

Outline the meaning of the term S_N1. [2]

c) Predict the mechanism (S_N1 or S_N2) expected for the reaction of the following halogenoalkanes with aqueous KOH.

1-chlorobutane to form butan-1-ol

2-chloro-2-methylpropane to form 2-methylpropan-2-ol. [2]

d) Explain the mechanism of each reaction in part (c) using curly arrows to represent the movement of electron pairs. [6]

IB, November 2009

9 There are several structural isomers with the molecular formula $C_5H_{11}Br$.

a) Deduce the name of **one** of the isomers which can exist as enantiomers and draw three-dimensional representations of its **two** enantiomers. [3]

b) All the isomers react when warmed with a dilute aqueous solution of sodium hydroxide according to the equation below.

$C_5H_{11}Br + NaOH \rightarrow C_5H_{11}OH + NaBr$

i) The reaction with 1-bromopentane proceeds by an S_N2 mechanism. Describe this mechanism using structural formulas and curly arrows to represent the movement of electron pairs. [3]

ii) The reaction with 2-bromo-2-methylbutane proceeds by an S_N1 mechanism. Describe this mechanism using structural formulas and curly arrows to represent the movement of electron pairs. [3]

iii) Explain why 1-bromopentane reacts by an S_N2 mechanism whereas 2-bromo-2-methylbutane reacts by an S_N1 mechanism. [3]

iv) Explain whether the boiling point of 1-bromopentane will be higher, lower, or the same as that of 2-bromo-2-methylbutane. [3]

v) The product $C_5H_{11}OH$ formed from the reaction with 1-bromopentane is warmed with ethanoic acid in the presence of a few drops of concentrated sulfuric acid. State the name of the type of reaction taking place and the structural formula of the organic product. [2]

IB, May 2011

10 Deduce a multi-step synthesis for each of the following conversions. For each step state the structural formulae of the reactants and products and the conditions used for the reactions.

(i) 2-chlorobutane to butan-2-one (2 steps)

(ii) propene to propyl ethanoate (2 steps)

(iii) benzene to aniline (phenylamine) (3 steps)

Introduction

Although spectroscopic characterization techniques form the backbone of structural identification of compounds, typically no one technique results in a full structural identification of a molecule. This combined approach forms the basis of this topic.

21.1 Spectroscopic identification of organic compounds

Understandings

→ Structural identification of compounds involves several different analytical techniques including IR, ^{1}H NMR and MS.

→ In a high resolution ^{1}H NMR spectrum, single peaks present in low resolution can split into further clusters of peaks.

→ The structural technique of single crystal X-ray crystallography can be used to identify the bond lengths and bond angles of crystalline compounds.

Applications and skills

→ Explanation of the use of tetramethylsilane (TMS) as the reference standard.

→ Deduction of the structure of a compound given information from a range of analytical characterization techniques (X-ray crystallography, IR, ^{1}H NMR, and MS).

Nature of science

→ Improvements in modern instrumentation – advances in spectroscopic techniques (IR, ^{1}H NMR and MS) have resulted in detailed knowledge of the structure of compounds.

461

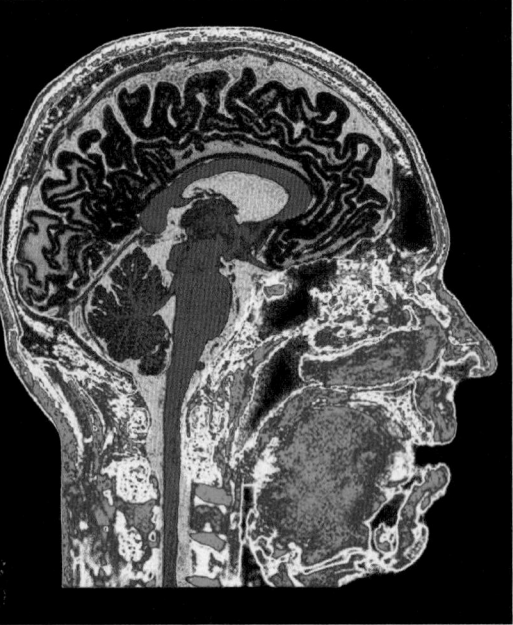

▲ Figure 1 MRI scan of a healthy human brain. With the advances made in the instrumentation used in MRI (due largely to the development of more powerful magnets), MRI instruments can now even detect chemical changes in the brain stemming from external stimuli such as a flash of light. This has allowed neuroscientists to pinpoint specific regions of the brain itself where brain activity is taking place and gain an understanding of the chemical principles underpinning our actual thought processes. Advances in MRI technology now allow advanced research to be carried out into the dysfunctionality of the brain which is important in deepening our understanding of neurological disorders and diseases such as schizophrenia and Alzheimer's disease

Advances in analytical techniques

Improvements in modern instrumentation have led to advances in spectroscopic techniques (IR, ^1H NMR, and MS) resulting in detailed knowledge of the structure of compounds. Analytical techniques have a wide variety of applications such as:

● Testing for drug abuse by high-performance athletes.

● MS (in combination with other chromatographic techniques such as gas-chromatography (GC-MS) etc) can be used in forensic investigations for crimes.

● Protons in water molecules within human cells can be detected by **magnetic resonance imaging** (MRI), giving a three-dimensional view of organs in the human body (figure 1).

In sub-topic 11.3, we discussed a number of the key analytical techniques used to identify the structure of an organic compound. The structural identification of compounds typically involves a combination of several different analytical techniques including IR, ^1H NMR, and MS.

^1H NMR spectroscopy

In topic 11 we introduced the principles of proton nuclear magnetic resonance spectroscopy (^1H NMR). In this chapter we will now revisit this technique and look at some of the features of high-resolution ^1H NMR.

High-resolution ^1H NMR spectroscopy

In practice, most ^1H NMR spectra do not consist of sets of single peaks, which may appear to be the case at low resolution.

A **high-resolution ^1H NMR spectrum** can show further splitting of some absorptions. Splitting patterns result from **spin–spin coupling**. To understand spin–spin coupling, let us take the example of the ^1H NMR spectrum of 1,1,2-trichloroethane, whose structure is shown in figure 2(a). The molecule contains two types of hydrogen in different chemical environments. Let's call these two different types of hydrogen H_a and H_b, respectively.

Since protons have nuclear spin, they hence have a magnetic field associated with them. Every proton can act as a tiny magnet. H_b can adopt two alignments with respect to the applied magnetic field, of

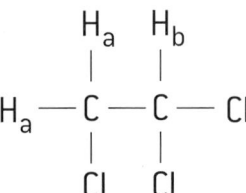

▲ Figure 2(a) Structure of 1,1,2-trichloroethane, which has two types of hydrogen (H_a and H_b) in different chemical environments

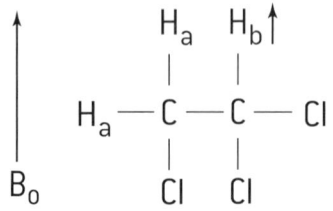

H_b spin aligned with B_o, so hence deshields H_a

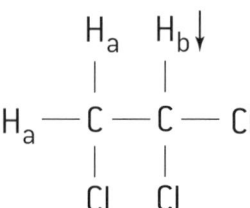

H_b spin opposed to B_o, so hence shields H_a

▲ Figure 2(b) Spin–spin coupling observed in 1,1,2-trichloroethane. Two combinations are seen for the –CH (H_b) proton

magnetic flux density, B_o. The magnetic moment of H_b aligns with B_o for approximately 50% of the molecules in the sample. The other 50% of the molecules will have the magnetic moment of H_b opposing B_o.

Therefore, the signal that will be observed for the methylene protons, $-CH_2$ (H_a) will appear as a doublet, (d), in the high-resolution ^{1}H NMR spectrum. This doublet consists of two lines of the same relative intensity. One of the two lines is located slightly **upfield** from the original single peak observed in the low-resolution spectrum – this is due to 50% of the molecules having their H_as **shielded** by H_b (H_b spin opposing B_o) and the peak will appear at a **lower chemical shift**, δ.

In the other case where the signal has moved slightly **downfield** from the original single peak in the low-resolution spectrum, 50% of the molecules will have their H_as **deshielded** by H_b (spin of H_b aligning with B_o) and the peak will appear at a **higher chemical shift**, δ. The ratio of the intensities of the two lines of the doublet, (d), can be deduced using Pascal's triangle and will be 1:1.

In the case of the methylene $-CH_2$ (H_a) protons, four combinations are possible:

Combination 1: H_{a1} and H_{a2} magnetic moments aligned with B_o (deshields H_b, so signal is shifted downfield to a higher δ).

Combination 2: H_{a1} magnetic moment aligned with B_o and H_{a2} magnetic moment aligned against B_o.

Combination 3: H_{a1} magnetic moment aligned against B_o and H_{a2} magnetic moment aligned with B_o.

Combination 4: H_{a1} and H_{a2} magnetic moments aligned against B_o (shields H_b, so signal is shifted upfield to a lower δ).

In the case of combinations 2 and 3, the shielding effect of one cancels the deshielding effect of the other. Hence, the net effect is that there is no change in the chemical shift of the single peak seen in the original low-resolution ^{1}H NMR spectrum.

Therefore, the signal observed for the neighbouring $-CH$ (H_b) proton will split into three lines, which we call a triplet, (t). The ratio of the intensities of these lines again deduced from Pascal's triangle will be 1:2:1.

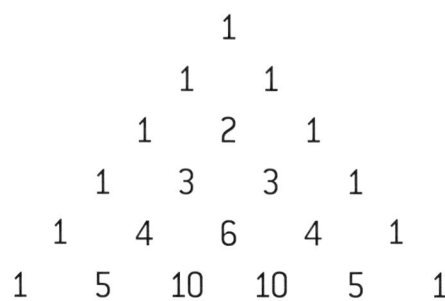

▲ Figure 2(c) Pascal's triangle. This can be used to deduce the splitting patterns in high-resolution ^{1}H NMR spectra

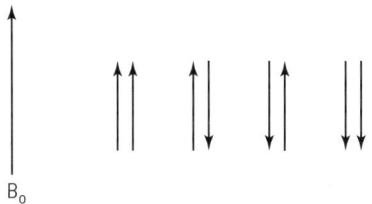

▲ Figure 2(d) Spin–spin coupling observed in 1,1,2-trichloroethane. Four combinations are seen for the methylene $-CH_2$ (H_a) protons

Type of proton	Splitting	Actual δ/ppm from ^{1}H NMR spectrum
A: $-CH_2$	(d)	3.960
B: $-CH$	(t)	5.762

Spin–spin coupling is actually transmitted through the electrons in the individual bonds. Therefore, spin–spin coupling depends on the way the hydrogens are related to each other in the bonding arrangements within the molecule.

It is not necessary, however, to go through the above detailed treatment each time in order to determine the individual splitting patterns and associated intensities of lines resulting from spin–spin coupling. Deductions in fact can be made quite simply using two very simple rules.

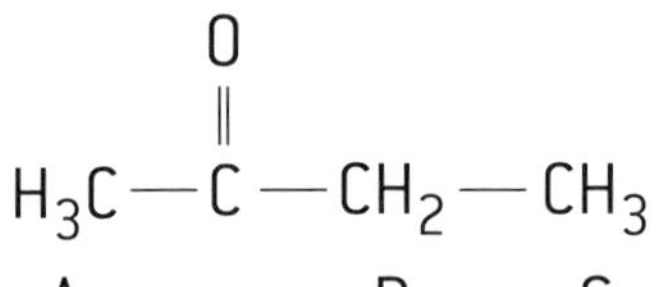

$$H_3C - \overset{\overset{\displaystyle O}{\|}}{\underset{A}{C}} - \underset{B}{CH_2} - \underset{C}{CH_3}$$

▲ Figure 3 Structure of butan-2-one

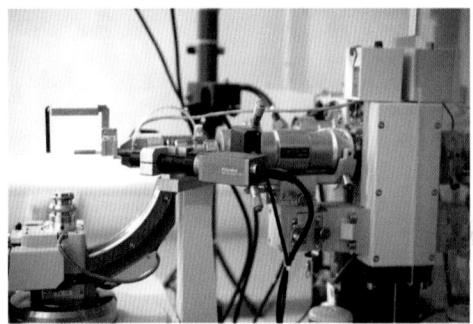

▲ Figure 5 An X-ray diffractometer at a university research unit. X-ray crystallography is a powerful structural technique used to determine the arrangement of atoms within a crystal using X-rays. The X-rays strike the crystal and are diffracted into many specific directions depending on the location of electrons within the sample. From this, a 3D model of the electron density can be created and the mean position of the atoms calculated

Rule 1: If a proton, H_a, has n protons as its nearest neighbours, that is $n \times H_b$, then the peak of H_a will be split into $(n + 1)$ peaks.

Rule 2: The ratio of the intensities of the lines of the split peak can be deduced from Pascal's triangle.

Let's consider this in the case of the high-resolution ^1H NMR spectrum of butan-2-one, whose structure is given in figure 3.

As can be seen, butan-2-one consists of hydrogens present in three different chemical environments. Let's call these hydrogens, A-type, B-type and C-type. The integration trace therefore is in the ratio 3:2:3. Using section 27 of the *Data booklet* and Pascal's triangle, the peaks can be assigned as follows:

Type of proton	Predicted δ/ppm from section 27 of *Data booklet*	Splitting	Actual δ/ppm from ^1H NMR spectrum
A: $-CH_3$	–	(s)	2.139
B: $-CH_2$	2.2–2.7	(q)	2.449
C: $-CH_3$	0.9–1.0	(t)	1.058

The following is the actual high-resolution ^1H NMR spectrum for butan-2-one:

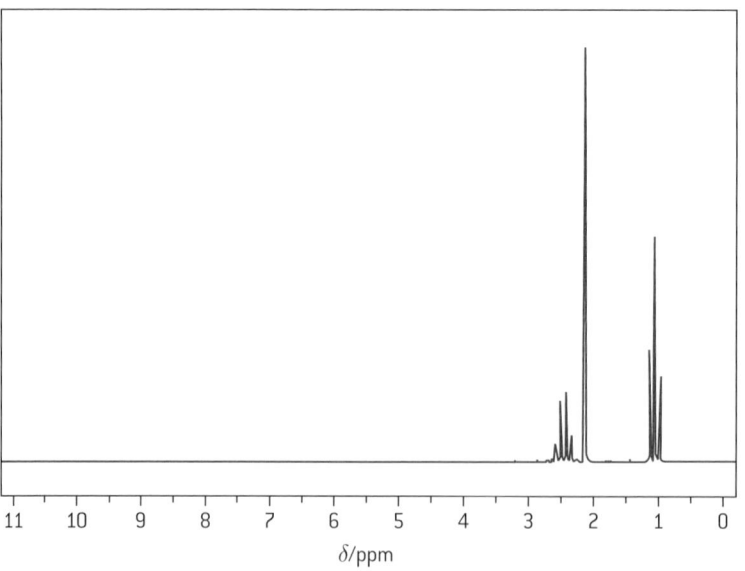

▲ Figure 4 High-resolution ^1H NMR spectrum for butan-2-one (90 MHz in $CDCl_3$), consisting of a quartet ($-CH_2 -B$), a singlet ($-CH_3 -A$) and a triplet ($-CH_3 -C$)

Single-crystal X-ray crystallography

The structural technique of **single crystal X-ray crystallography** can be used to identify the bond lengths and bond angles of crystalline compounds.

The chemical community often shares chemical structural information on the international stage. The *Cambridge Crystallographic Database*, *ChemSpider* developed by the *Royal Society of Chemistry (RSC)* and the *Worldwide Protein Data Bank (wwPDB)* are examples that highlight the international dimension of the global scientific community.

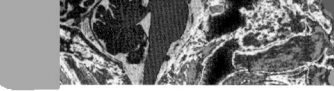

Useful resources

1 *ChemSpider*, developed by the Royal Society of Chemistry (RSC) is a free chemical structure database which provides rapid text and structure search access to over 29 million structures from hundreds of data sources (http://www.chemspider.com/).

2 The *Worldwide Protein Data Bank (wwPDB)* consists of organizations that act as deposition, data processing and distribution centres for PDB data. The RCSB Protein Data Bank is an information portal to biological macromolecular structures (http://www.rcsb.org/pdb/).

Explanation of the use of tetramethylsilane (TMS) as the reference standard in ^{1}H NMR spectroscopy

In IR spectroscopy (and also UV–Vis spectroscopy), the position of the absorption peaks can be associated with a wavelength, λ, or a frequency, ν. In ^{1}H NMR this is not possible, as the position of the ^{1}H NMR signals depends on the strength of the external magnetic field.

Therefore, the frequencies can be variable, as no two magnets will be identical. In order to address this problem, a universal reference standard has been agreed and hence the exact position of the ^{1}H NMR signal can be found relative to the reference signal from the standard. This standard is **TMS, tetramethylsilane**, whose structure is shown in figure 6.

TMS is used as a standard as it has the following advantages:

1 The 12 protons are in the same chemical environment, so there will be just one single peak, which will be strong. The chemical shift of this signal for the TMS reference standard is assigned $\delta = 0$ ppm. (All other chemical shifts are measured relative to this.)

2 TMS is inert (that is, it is fairly unreactive, so will not interfere with the sample being analysed).

3 It will absorb upfield ($\delta = 0$ ppm), well removed from most other protons involved in organic compounds which typically absorb downfield (the terms upfield and downfield have been discussed earlier in this topic).

4 It can be easily removed from the sample after measurement, as it is volatile, having a low boiling point of 26–27 °C.

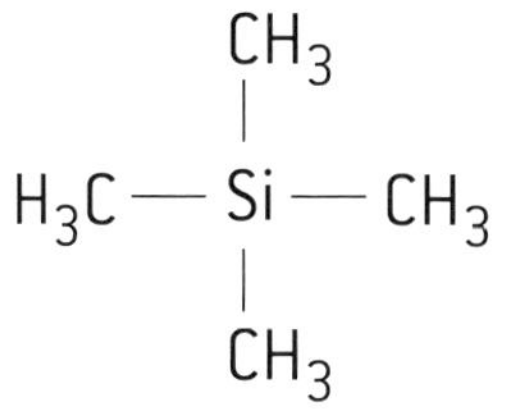

Figure 6 TMS is the universally agreed reference standard used in ^{1}H NMR spectroscopy

Worked example 1

a) Deduce the full structural formula of ethyl ethanoate.

b) Using section 27 of the *Data booklet*, predict the high-resolution ^{1}H NMR spectrum of ethyl ethanoate. Your answer should refer to the integration trace on the spectrum, the approximate chemical shifts of the various protons, in ppm, any possible splitting patterns and the relative intensities of the lines of the splitting patterns.

Solution

a)

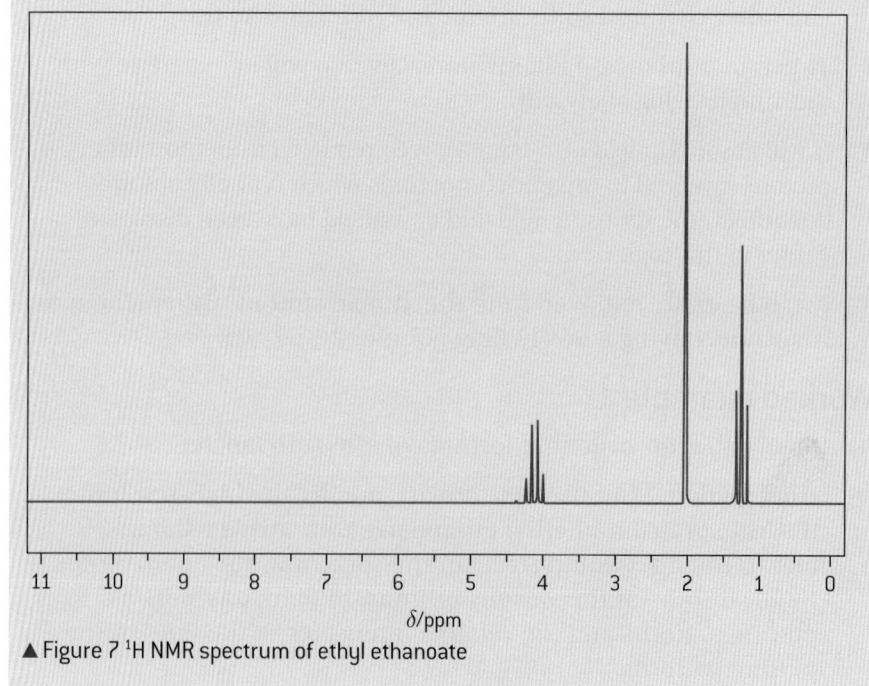

```
    H   O          H   H
    |   ||         |   |
H — C — C — O — C — C — H
    |              |   |
    H              H   H

    A              B   C
```

b)

- Three types of hydrogen atoms are present in different chemical environments.

- Integration trace showing ratio of hydrogen atoms: 3:2:3.

Type of proton	Predicted δ/ppm from section 27 of *Data booklet*	Splitting	Relative intensities of lines of the splitting patterns	Actual δ/ppm from ^{1}H NMR spectrum
A: $-CH_3$	2.0–2.5	(s)	–	2.038
B: $-CH_2$	3.7–4.8	(q)	1:3:3:1	4.119
C: $-CH_3$	0.9–1.0	(t)	1:2:1	1.260

The actual high-resolution of ^{1}H NMR spectrum of ethyl ethanoate recorded at 90 MHz in $CDCl_3$ is shown in figure 7.

▲ Figure 7 ^{1}H NMR spectrum of ethyl ethanoate

Study tip

Sometimes the chemical shifts that you predict from section 27 of the *Data booklet* differ slightly from what you may observe in the actual ^{1}H NMR spectrum. In fact, chemical shifts may vary in different solvents and conditions. For example, some solvents have π-electron capabilities and/or can be involved in hydrogen bonding networks leading to variations in the chemical shifts depending on the solvents used. Notice this is the case with respect to the C type of protons in figure 7.

Worked example 2

An unknown compound, X, of molecular formula, C_3H_6O, with a characteristic fruity odour, has the following IR and ^1H NMR spectra.

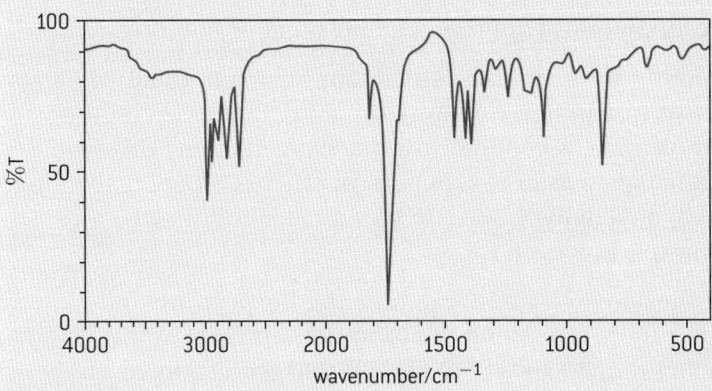

▲ Figure 8 IR spectrum (in CCl_4 solution) of X

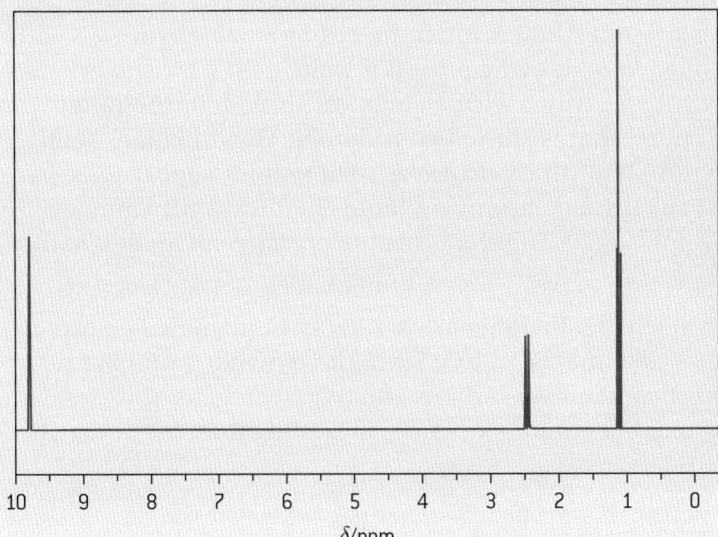

▲ Figure 9 ^1H NMR spectrum (300 MHz in $CDCl_3$) of X

The MS of X showed peaks at m/z values = 58 and 29 (other peaks were also found).

Deduce the structure of X using the information given and any other additional information from the *Data booklet*. For each spectrum assign as much spectroscopic information as possible based on the structure of X.

Solution

- As the molecular formula of X is provided, it is worth deducing first the IHD, which indicates the **index of hydrogen deficiency** or **degree of unsaturation**.

  For the generic molecular formula $C_cH_hN_nO_oX_x$:

  $$IHD = (0.5)(2c + 2 - h - x + n)$$

467

Hence for C_3H_6O:

$c = 3, h = 6, n = 0, o = 1, x = 0$

$IHD = (0.5)(6 + 2 - 6 - 0 + 0) = 1$

Therefore, X contains either one double bond or one ring.

- The compound has a typical fruity odour, which normally is characteristic of an ester. However, we note that, based on the molecular formula, X contains just one oxygen atom. This would rule out X being an ester $(-COO-)$. An ether $(C-O-C)$, a ketone $(C-CO-C)$, an aldehyde $(C-CHO)$ or an alcohol $(C-OH)$ are possible classes, however, for X.

- Based on the above, we now examine the IR spectrum and see whether there is a strong IR absorption for $C=O$ in the wavenumber range 1700–1750 cm^{-1}, based on section 26 of the *Data booklet*. Indeed, there is a strong peak at approximately 1740 cm^{-1}, suggesting the presence of $C=O$.

- If $C=O$ is present, then X might be either an aldehyde or a ketone. An aldehydic proton is quite characteristic in the ^{1}H NMR spectrum, with a chemical shift, δ, in the range 9.4–10.0 ppm, that is shifted considerably downfield, as seen from section 27 of the *Data booklet*. There does appear to be a single peak, in fact, at quite a large chemical shift value of $\delta = 9.8$ ppm, suggesting the probable presence of an aldehydic proton.

- If X is an aldehyde, that means we now have identified a portion of the molecule, that is $-CHO$. Since the remaining number of atoms must contain two carbons and five hydrogens, this could only be the ethyl group, $-CH_2CH_3$, suggesting that X is propanal, CH_3CH_2CHO.

- Let's now test this proposed structure, based on the spectroscopic data obtained from the ^{1}H NMR spectrum. Three types of hydrogen atoms are present in different chemical environments, A, B, and C.

Type of proton	Predicted δ/ppm from section 27 of *Data booklet*	Splitting	Relative intensities of lines of the splitting patterns	Actual δ/ppm from ^{1}H NMR spectrum
A: $-CHO$	9.4–10.0	(s)	–	9.8
B: $-CH_2$	2.2–2.7	(q)	1:3:3:1	2.5
C: $-CH_3$	0.9–1.0	(t)	1:2:1	1.1

Notice that, for the C protons, the chemical shift observed on the spectrum is slightly outside the range predicted from section 27 of the *Data booklet*, but this is often the case as chemical shifts may vary in different solvents and conditions.

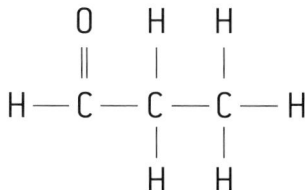

- Having established all the spectroscopically assigned peaks in the ^{1}H NMR spectrum for the proposed structure of X, it is worth returning to the IR spectrum at this stage to confirm the additional characteristic range for the infrared absorption due to the CH bonds in the wavenumber range 2850–3090 cm^{-1}. As can be seen from the IR spectrum, there are indeed absorptions occurring within this range.

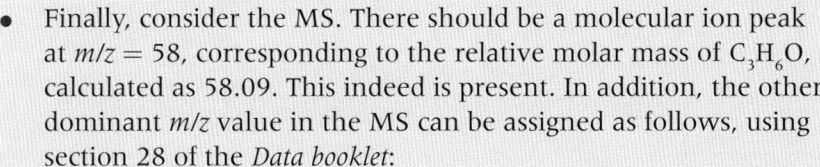

▲ Figure 10 Propanal

- Finally, consider the MS. There should be a molecular ion peak at $m/z = 58$, corresponding to the relative molar mass of C_3H_6O, calculated as 58.09. This indeed is present. In addition, the other dominant m/z value in the MS can be assigned as follows, using section 28 of the *Data booklet*:

  $m/z = 29$ ... signifies presence of $CH_3CH_2^+$, indicating loss of CHO from molecule X, that is $(M_r - 29)$.

- This confirms compound X to be propanal.

Note: The actual MS of propanal is as follows:

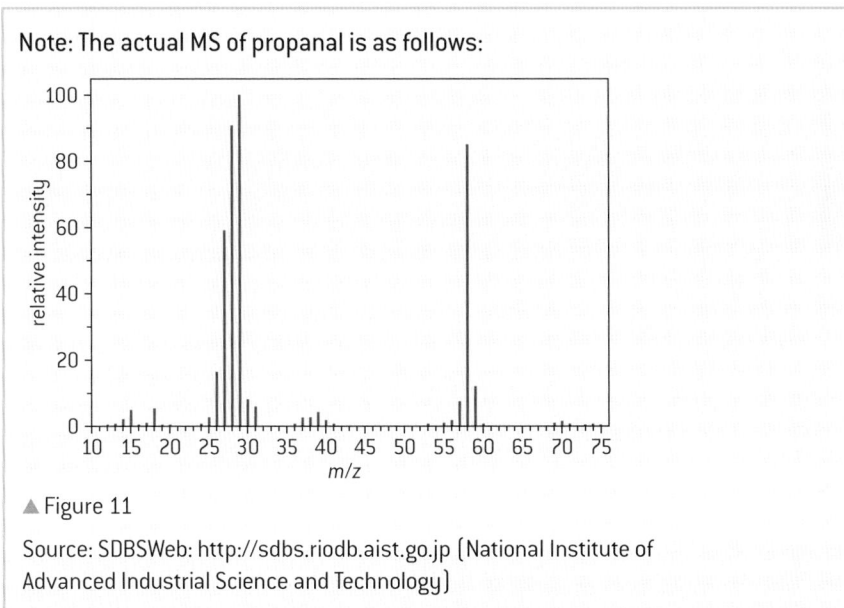

▲ Figure 11

Source: SDBSWeb: http://sdbs.riodb.aist.go.jp (National Institute of Advanced Industrial Science and Technology)

Quick question

Can you explain the peaks on the MS of X greater than $m/z = 58$?

469

Questions

1 An unknown compound, X, of molecular formula, $C_3H_6O_2$, has the following IR and ¹H NMR spectra.

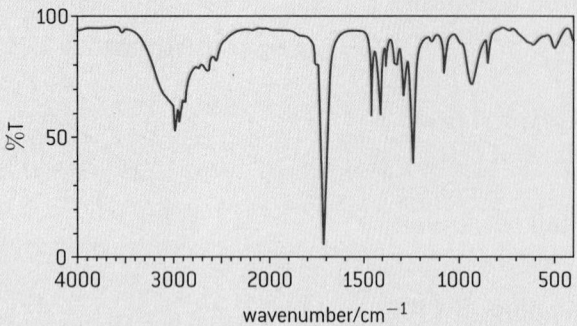

▲ Figure 12 IR spectrum of X (in CCl_4) solution

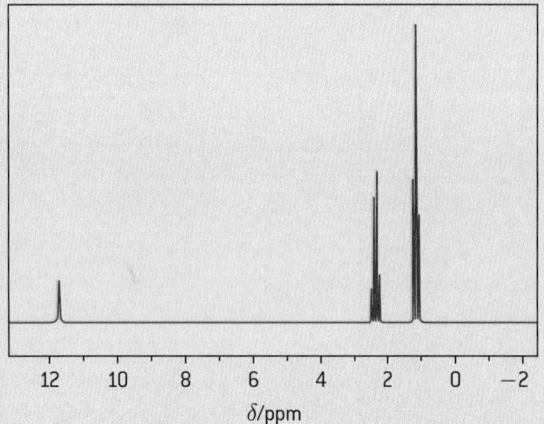

▲ Figure 13 ¹H NMR spectrum (90 MHz in $CDCl_3$) of X

The MS of X showed peaks at *m/z* values = 74, 45, and 29 (other peaks were also found).

Deduce the structure of X using the information given and any other additional information from the *Data booklet*. For each spectrum assign as much spectroscopic information as possible, based on the structure of X.

2 An unknown compound, Y, of molecular formula, $C_5H_{10}O_2$, has the following IR and ¹H NMR spectra.

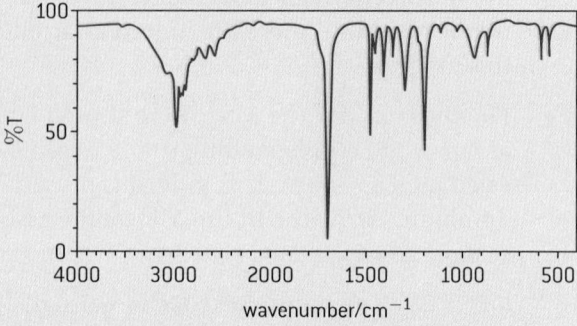

▲ Figure 14 IR spectrum of Y (in CCl_4) solution

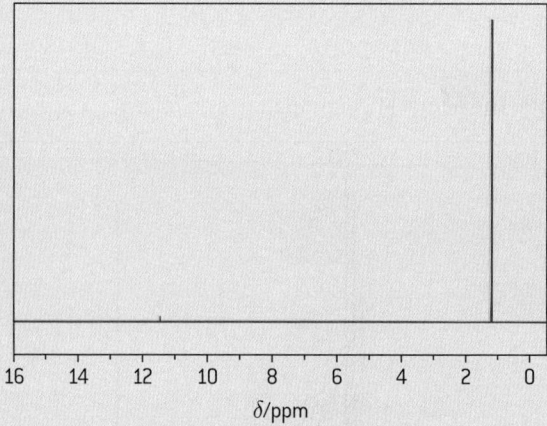

▲ Figure 15 ¹H NMR spectrum (300 MHz in $CDCl_3$) of Y

The MS of, Y, showed peaks at *m/z* values = 102 and 57 (other peaks were also found).

Deduce the structure of Y using the information given and any other additional information from the *Data booklet*. For each spectrum assign as much spectroscopic information as possible, based on the structure of Y.

A MATERIALS

Introduction

History has characterized civilizations by the materials they use: Bronze Age, Stone Age, and Iron Age. Uses of materials were developed based on observations of their properties before an explanation of those properties had been proposed. Using the ideas of bonding and structure, materials are now classified and their properties manipulated to be put to many extraordinary uses. From metals to nanotechnology, research into the properties and uses of materials is sometimes serendipitous. Research often begins with an end use in mind, in advance of specific knowledge about the composition and construction of the material.

A.1 Materials science introduction

Understandings

→ Materials are classified based on their uses, properties, or bonding and structure.

→ The properties of a material based on the degree of covalent, ionic, or metallic character in a compound can be deduced from its position on a bonding triangle.

→ Composites are mixtures in which materials are composed of two distinct phases, a reinforcing phase that is embedded in a matrix phase.

Applications and skills

→ Use of bond triangle diagrams for binary compounds from electronegativity data.

→ Evaluation of various ways of classifying materials.

→ Relating physical characteristics (melting point, permeability, conductivity, elasticity, brittleness) of a material to its bonding and structures (packing arrangements, electron mobility, ability of atoms to slide relative to one another).

Nature of science

→ Improvements in technology – different materials were used for different purposes before the development of a scientific understanding of their properties.

→ Patterns in science – history has characterized civilizations by the materials they used: Stone Age, Bronze Age, and Iron Age. There are various ways of classifying materials according to desired patterns.

Classifying materials

Materials science involves understanding the structure and properties of a material and matching these to suitable applications. Type of bonding is one classification system employed by materials science, and four common types of material are metallics, ceramics, polymers, and composites. Each type is suitable for different end uses.

Metallic substances exhibit **metallic bonding** (sub-topic 4.5). This makes them strong, malleable, and good conductors of heat and electricity. Approximately two-thirds of all elements are metals. The development of alloys has designed metallic materials suitable for many applications. At the atomic level, the freely moving electrons in metallic bonding confer ductility and strength, as well as conductivity.

Ceramics are traditionally inorganic non-metallic solids formed between metals and non-metals. They have a crystalline structure and their **ionic bonding** (sub-topic 4.1) means they are brittle and usually show insulating properties, such as the ceramics familiar in plates and cups. However, a wide variety of ways of combining metals and non-metals leads to many ceramics with various uses. A compound of thallium, barium, calcium, copper, and oxygen forms a superconductor whereas glass is an amorphous insulating ceramic material. Bricks, tiles, electric capacitors, abrasives, and cement are other types of ceramics. Because of their numerous applications they are sometimes classified based on their uses.

Polymers (also known as plastics) form a third classification of materials based on bonding. Plastics are **covalently bonded** long-chain molecules. (The formation of addition polymers is covered in sub-topic 10.2.) There are many uses and types of plastics and the industry is growing rapidly.

In general, because of strong covalent bonds that exist throughout polymer molecules these materials tend to be resistant to chemicals and do not corrode. With no free electrons or metal atoms they are generally good insulators of both heat and electricity and lighter than ceramics. Polymers can be engineered into many forms including thin flexible fibres, soft flexible films such as plastic wrap, or hardened plastics such as polyvinyl chloride (PVC) pipe.

▲ Figure 1 Strain gauge and thermocouple attached to a carbon–carbon composite sample for a stress test. Carbon–carbon is a composite material that consists of carbon fibres in a matrix of graphite. It has been used in high temperature applications such as for the thermal tiles on NASA's Space Shuttles

Pure and applied sciences

Pure science aims to establish a common understanding of the universe; applied science and engineering develop technologies that result in new processes and products. However, the boundaries between pure and applied sciences are not clear cut. Naturally occurring polymers such as rubber and silk have laid the foundations for the plastics industry leading to many products from engineering synthetic polymers.

Composites are mixtures composed of two distinct phases: a reinforcing phase embedded in a matrix. Each substance retains its own properties (as in any mixture, sub-topic 1.1); however the composite has specific properties not shown by either part of the mixture individually. Straw and clay formed an early composite used to build huts. In a composite the reinforcing phase is made up of fibres, particles, or a mesh which is embedded in a tough or ductile matrix, depending on the use. Each phase can be a metal, a ceramic, or a polymer. Examples of composites include fibreglass, carbon fibre, and concrete. Aircraft wings, for example, can be made lighter and stronger by the use of composites.

Identifying the desired properties of a material for a specific application, such as whether it needs to be fire resistant, strong, porous or non-porous, a conductor or not, and then engineering materials to suit these properties forms the basis of the field of materials science.

Designer materials

Understanding bonding and structure allows materials scientists to design and manufacture new materials to desired specifications. For example, waterproof breathable fabrics such as Gore-Tex® allow perspiration to evaporate while protecting the wearer from rain. Liquid water has extensive hydrogen bonding which produces large grouped particles, whereas water vapour exists as individual water molecules without hydrogen bonds. In Gore-Tex the material is layered so that the packing arrangement provides pores that are large enough for individual water molecules to escape through but are too small for the passage of grouped hydrogen-bonded molecules in liquid water. This type of materials development needs to take into account any intermolecular forces between the material and water, which might affect the movement of water vapour or liquid water. Bonding and structure are intrinsically linked with properties.

Quick questions

1   For the statement below identify the structural feature or property from each pair that best serves the application.

    A crucible for heating substances over a Bunsen burner: high melting point/ low melting point; permeable to moisture / not permeable to moisture; loose packing arrangement / highly structured crystalline structure; metallic / ceramic.

2   Classifying materials according to the type of bonding is a useful system, but other classifications have their place. For example, a metallurgist may be more interested in the grade of stainless steel in the alloy than in the type of bonding. Can you suggest when the following classifications may be used? Nanomaterials, biomaterials, textiles, alloys, semiconductors.

Bond triangle diagrams

"The nature of the chemical bond is the problem at the heart of all chemistry."

Bryce Crawford Jr, 1953

Bonds between metals and non-metals vary from ionic to covalent in relation to the electronegativity difference between the two types of atom. The greater the electronegativity difference, the higher the ionic character. Strongly ionic compounds are crystalline, non-conductors of electricity but moderate conductors of heat and have high melting points, whereas strongly covalent compounds have low melting and boiling points, are soft, and are poor conductors of both heat and electricity. A material with polar covalent bonds exhibits some ionic and some covalent character. Figure 2 shows how this can be illustrated simply in a triangle of bonding, while figure 3 gives a more comprehensive version.

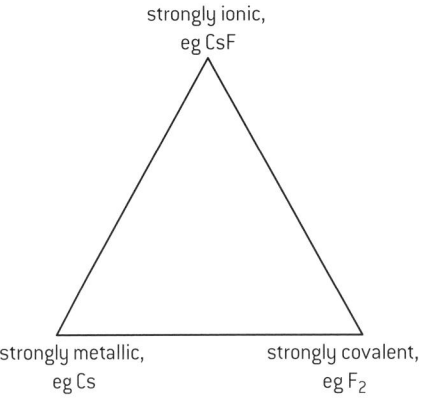

▲ Figure 2 A simple triangle of bonding. The most metallic (least electronegative) element is caesium, while the most electronegative is fluorine

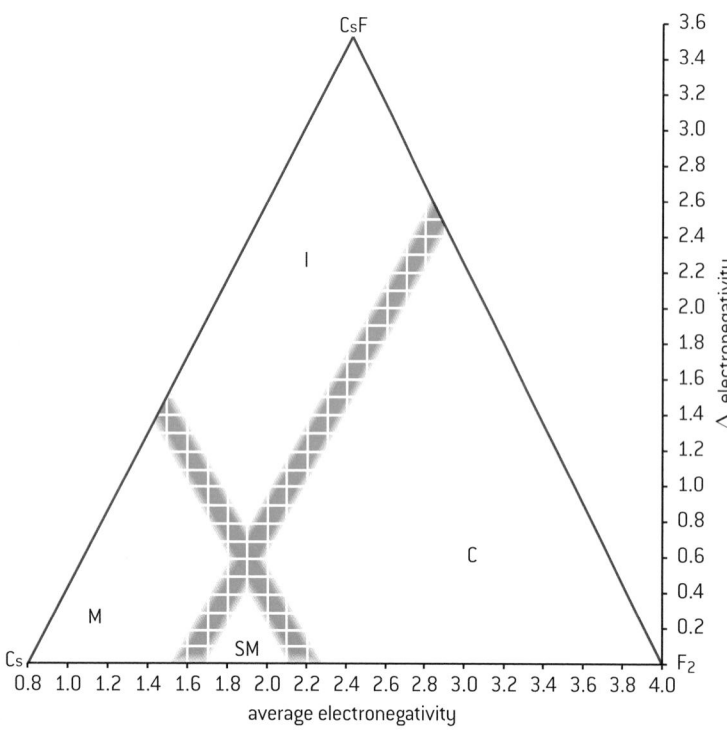

◀ Figure 3 The diagram shows how the bond types ionic (I), metallic (M), covalent (C), and semi-metallic (SM) depend on the difference in electronegativity between the atoms involved. A triangular bonding diagram is provided in section 29 of the *Data booklet*

🌐 Worked example

Tin(II) chloride has a melting point of 247 °C while lead(II) chloride has a melting point of 500 °C. Both are used in the production of aurene glass, an iridescent artwork glassware. One of these two substances exists as discrete molecules in the vapour phase. Using electronegativity tables and the triangular bonding diagram in figure 3 identify which one, and explain your reasoning.

Solution

Electronegativity values: Sn 2.0, Pb 1.8, Cl 3.2

Average electronegativity: $SnCl = \frac{(2.0 + 3.2)}{2} = 2.6$; $PbCl_2 = \frac{(1.8 + 3.2)}{2} = 2.5$

Electronegativity difference: $SnCl = 3.2 - 2.0 = 1.2$; $PbCl_2 = 3.2 - 1.8 = 1.4$

Plotting (x, y) coordinates of (2.6, 1.2) on the bonding triangle diagram classifies $SnCl_2$ as mostly covalent (60–70% covalent character).

Plotting (x, y) coordinates of (2.5, 1.4) on the bonding triangle diagram classifies $PbCl_2$ as more ionic (approximately 60% ionic, 40% covalent).

Because $SnCl_2$ shows more covalent property than does $PbCl_2$ it is more likely to exist as discrete molecules when vaporized. It will also have a lower melting point as covalent substances have lower melting and boiling points than do ionic substances.

Questions

1   Magnesium oxide and manganese(II) oxide are both used in ceramics, but for different purposes.

   a) Using sections 8 and 29 in the *Data booklet*:

   (i) state the *average* electronegativity for magnesium and oxygen and for manganese and oxygen

   (ii) estimate the percentage covalent character for each compound.

   b) Deduce which of the two compounds is more likely to be used to make kilns and crucibles because of its ability to retain its strength at high temperatures. Explain your answer in terms of bonding and structure, mentioning specifically how the degree of ionic or covalent character influences packing arrangements and the ability of atoms to slide relative to one another.

2   One method of sorting materials for recycling is to classify them as plastics, glass, paper, cardboard, and metals. Outline the advantages of using this classification rather than the metallic, ceramic, polymers, and composites system.

A.2 Metals and inductively coupled plasma (ICP) spectroscopy

Understandings

→ Reduction by coke (carbon), a more reactive metal, or electrolysis are means of obtaining some metals from their ores.

→ The relationship between charge and the number of moles of electrons is given by Faraday's constant, F.

→ Alloys are homogeneous mixtures of metals with other metals or non-metals.

→ Diamagnetic and paramagnetic compounds differ in electron spin pairing and their behaviour in magnetic fields.

→ Trace amounts of metals can be identified and quantified by ionizing them with argon gas plasma in Inductively Coupled Plasma (ICP) Spectroscopy using Mass Spectroscopy ICP-MS and Optical Emission Spectroscopy ICP-OES.

Applications and skills

→ Deduction of redox equations for the reduction of metals.

→ Relating the method of extraction to the position of a metal on the activity series.

→ Explanation of the production of aluminium by the electrolysis of alumina in molten cryolite.

→ Explanation of how alloying alters properties of metals.

→ Solving stoichiometric problems using Faraday's constant based on mass deposits in electrolysis.

→ Discussion of paramagnetism and diamagnetism in relation to electron structure of metals.

→ Explanation of the plasma state and its production in ICP-MS/OES.

→ Identification of metals and abundances from simple data and calibration curves provided from ICP-MS and ICP-OES.

→ Explanation of the separation and quantification of metallic ions by MS and OES.

→ Uses of ICP-MS and ICP-OES.

Nature of science

→ Development of new instruments and techniques – ICP spectroscopy, developed from an understanding of scientific principles, can be used to identify and quantify trace amounts of metals.

→ Details of data – with the discovery that trace amounts of certain materials can greatly enhance a metal's performance, alloying was initially more of an art than a science.

Reduction of metals

Some metals such as gold can be mined directly as the element. However, most metals exist in nature in their oxidized states in compounds; for example, aluminium is found in bauxite as aluminium oxide, Al_2O_3. These metals can be extracted from their ores and are then often alloyed to give them useful properties.

▲ Figure 1 Prospectors could pan for elemental gold, Au, because this metal is at the bottom of the activity series (section 25 in the *Data booklet*) and therefore commonly found in its reduced form (zero oxidation state). It would be impossible to find elemental lithium as Li sits at the top of the activity series and can only be obtained by reduction

Because metals in ores are in an oxidized state, they need to be reduced to a zero oxidation state in the elemental form. Reduction by coke (carbon), a more reactive metal, or electrolysis are methods used to obtain metals from their ores.

Reduction of iron ore in the blast furnace

Reduction is carried out on a large scale industrially to obtain iron from **iron ore**. Most of the iron extracted is then processed further to produce steel. Iron ore is mainly the oxides Fe_2O_3 and Fe_3O_4 which are reduced (sub-topic 9.1) by carbon in the form of coke in a blast furnace.

Coke is heated to form carbon dioxide, which reacts with more coke to form carbon monoxide in the reducing furnace:

$$C(s) + O_2(g) \rightarrow CO_2(g)$$

$$CO_2(g) + C(s) \rightarrow 2CO(g)$$

Carbon monoxide is a good reducing agent (it is easily oxidized) and reacts with the iron ore to produce molten iron, which is collected from the furnace:

$$Fe_2O_3(s) + 3CO(g) \rightarrow 2Fe(l) + 3CO_2(g)$$

At the very high temperatures in the blast furnace the coke can react directly with the iron ore and also act as a reducing agent itself:

$$Fe_2O_3(s) + 3C(s) \rightarrow 2Fe(l) + 3CO(g)$$

The carbon monoxide produced in this reaction can reduce more ore.

Reduction by a more reactive metal

A second means of obtaining elemental metals is reduction by a more active metal (sub-topic 9.1). Pure copper can be obtained from aqueous copper(II) sulfate by a single replacement reaction with solid zinc, for example:

$$Zn(s) + CuSO_4(aq) \rightarrow Cu(s) + ZnSO_4(aq)$$

Other redox reactions can be used to reduce the oxidized metal. For example, passing hydrogen gas over heated copper(II) oxide reduces the copper(II) oxide to elemental copper, while the hydrogen is oxidized to the +1 state:

$$CuO(s) + H_2(g) \rightarrow Cu(s) + H_2O(g)$$

Reduction by a more active metal or by carbon cannot be used to extract metals near the top of the activity series such as lithium, rubidium, or potassium. In this case **electrolysis** (sub-topic 9.2) allows us to obtain the pure metals. Once obtained the elemental metals must not be exposed to air or they will become oxidized again.

Lithium is used in lithium batteries and obtained by electrolysis of molten lithium chloride to produce lithium metal and chlorine gas:

$$2LiCl(l) \xrightarrow{\text{electrolysis}} 2Li(l) + Cl_2(g)$$

The quantity of metal reduced at the cathode during electrolysis can be calculated using the current passed in the electrolysis, the time it is passed for, and the Faraday constant. This is the charge in coulombs (C) on 1 mol of electrons and has the value 96500 C mol^{-1}. For example, in the reduction of lithium from its ions:

$$Li^+ + e^- \rightarrow Li$$

the equation shows that 1 mol of electrons are required to reduce 1 mol of lithium ions. Providing 1 mol of electrons requires 96500 C of charge from the electrolysis equipment. The amount of charge, Q, transferred can be calculated from the current I (in amperes, A) and time t (in s):

$$Q = It$$

The SI unit, the ampere, is one coulomb per second; 1 A $= 1$ C s^{-1}.

To reduce 1 mol of copper by electrolysis would take 2 mol of electrons:

$$Cu^{2+} + 2e^- \rightarrow Cu$$

 ## Worked example

Two electrolytic cells are connected in series so that the same current flows through both cells for the same length of time. One contains an aqueous solution of tin(II) sulfate and the other an aqueous solution of silver nitrate (figure 2). Calculate the mass of each metal produced at their respective cathodes if a current of 2.5 A is allowed to run for 20 minutes.

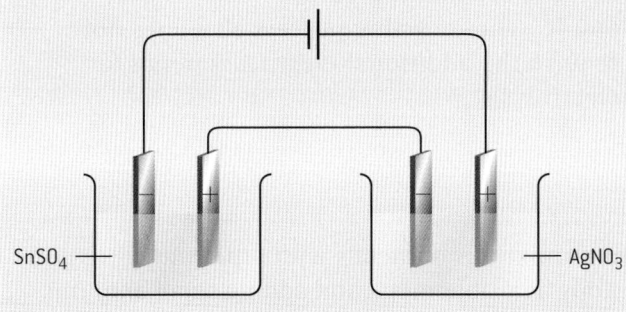

▲ Figure 2 Two electrolytic cells in series

Solution

$$Q = It$$

$$= 2.5 \text{ C s}^{-1} \times 1200 \text{ s} = 3000 \text{ C}$$

$$\frac{3000 \text{ C}}{96500 \text{ C mol}^{-1}} = 0.031 \text{ mol e}^-$$

Reduction equations:

$$Sn^{2+} + 2e^- \rightarrow Sn$$

2 mol e$^-$ produce 1 mol Sn, therefore 0.031 mol e$^-$ produces $0.031/2 = 0.016$ mol Sn

$$0.016 \text{ mol Sn} \times 118.71 \text{ g mol}^{-1} = 1.9 \text{ g Sn}$$

$$Ag^+ + e^- \rightarrow Ag$$

1 mol e$^-$ produces 1 mol Ag, therefore 0.031 mol e$^-$ produces 0.031 mol Ag

$$0.031 \text{ mol Ag} \times 107.87 \text{ g mol}^{-1} = 3.3 \text{ g Ag}$$

The production of aluminium

Aluminium is one of society's most useful metals. It is obtained from "alumina" or aluminium(III) oxide, Al_2O_3. Aluminium is a relatively active metal so needs to be obtained from its ore by electrolysis which must take place in the molten state. The exceedingly high melting point of Al_2O_3 (over 2000 °C) makes electrolysis of the native ore economically unfeasible.

The Hall–Héroult process was developed to overcome this problem. Molten cryolite, Na_3AlF_6 is used as a solvent in the electrolysis allowing the process to be carried out at lower temperatures. A large density difference between cryolite and molten aluminium also makes extraction of the pure metal easier. The process is outlined on the next page and in figure 3.

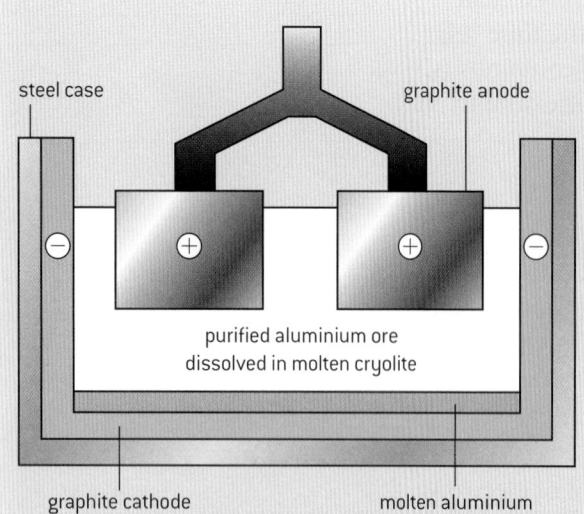

steel case

graphite anode

⊖ ⊕ ⊕ ⊖

purified aluminium ore
dissolved in molten cryolite

graphite cathode

molten aluminium

▲ Figure 3 The Hall–Héroult process for the extraction of aluminium. Molten aluminium is more dense than the cryolite–alumina mixture so the product sinks below the reaction mixture and can be run off

1 Alumina is dissolved in molten cryolite. This has a melting point under 1000 °C so reduces the energy needed to create the molten ore for electrolysis.

2 The steel case surrounding the molten substance is lined with graphite which serves as the cathode. Graphite anodes are inserted into the electrolyte.

3 An electric current is passed which reduces the aluminium ions. The oxide ions react with the anodes and oxidize them. The anodes therefore need to be replaced periodically.

The overall net equation is:

$$2Al_2O_3(l) + 3C(s) \rightarrow 4Al(l) + 3CO_2(g)$$

If anodic oxidation can be avoided, then oxygen gas is produced. In this instance the reactions are:

cathode: $Al^{3+} + 3e^- \rightarrow Al(l)$

anode: $2O^{2-} \rightarrow O_2(g) + 4e^-$

Molten aluminum has a density of 2.35 g cm^{-3}, which is approximately the same as that of molten cryolite. However, a molten mixture of cryolite saturated with alumina has a much lower density (approximately 1.97 g cm^{-3}). This allows the molten aluminium metal to sink to the bottom of the reaction vessel and be tapped off. If it remained in the reaction mixture it could short-circuit the electrolysis apparatus. It is also important to keep the cryolite saturated with alumina. For this reason the ore is continually fed into the vessel as the aluminium metal is drained off.

The production of aluminium uses much more energy than recycling the metal because the melting point of Al(s) is only 660 °C. The recycling process also does not require the additional electrical energy for electrolysis.

Alloys

Alloys are homogeneous mixtures of metals with other metals or non-metals. By taking a readily available metal and adding small amounts of another material to it, certain desired properties can be greatly enhanced. For example, steel is stronger than iron, and stainless steel is produced in many grades of different composition depending on the purpose. Copper alloys such as bronze and brass have increased resistance to corrosion compared with pure copper. Trace amounts of titanium or scandium added to aluminium can greatly increase its strength and durability without compromising its low density for lightweight applications. Alloying aluminium allowed the development of technologies as diverse as compact hard disk drives and strong but light aircraft wings.

The lanthanoids and actinoids (rare earth metals) are finding many alloying applications including in superconductors and lasers. Because the lanthanoids contain f-level electrons they have sharp 4f–4f electron transitions which makes them useful in optics applications such as for

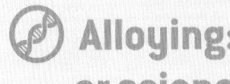

amplifying signals in optical fibres and in phosphors in computer screens and TVs. Materials such as ceramics can also be alloyed into metals, for example for use in dentistry.

Paramagnetic and diamagnetic materials

One property of interest in a metal or alloy is its response to a magnetic field. **Paramagnetic materials** are attracted to a magnetic field whereas **diamagnetic materials** create a magnetic field opposed to the applied field; and are therefore weakly repelled by an external magnetic field. In the atoms of a diamagnetic material the electrons are spin paired; for example, neon is diamagnetic with electron configuration $1s^2 2s^2 2p^6$ (valence electron configuration shown below:)

$\uparrow\downarrow$	$\uparrow\downarrow$	$\uparrow\downarrow$	$\uparrow\downarrow$
2s	$2p_x$	$2p_y$	$2p_z$

so all 10 electrons exist in a paired state. Aluminium atoms, electron configuration $1s^2\ 2s^2\ 2p^6\ 3s^2\ 3p^1$ have one unpaired p electron that is capable of being attracted to an external electric field. Aluminium is paramagnetic:

$\uparrow\downarrow$	$\uparrow$		
3s	$3p_x$	$3p_y$	$3p_z$

A spinning electron creates a magnetic dipole. The spins of unpaired electrons can be temporarily aligned in an external field, causing the material to be attracted to the applied magnetism. This is what happens in paramagnetic materials.

In a ferromagnetic material the electron alignment induced by the magnetic field can be retained, making a permanent magnet. For example iron, electron configuration $[Ar]4s^2 3d^6$, can be heated and cooled in a magnetic field and as the metal cools the unpaired electrons align themselves such that the magnetic field created by their spin is aligned with the applied field. Banging or heating a permanent magnet can disrupt this alignment and weaken the magnet. Paramagnetic materials do not form permanent magnets in this way; their electrons are only temporarily aligned by the external field.

As we have seen, in diamagnetic materials all the electrons are paired. In an external magnetic field the paired electrons orientate themselves such that the field created by their spin opposes the applied field (Lenz's law, which is studied in IB Physics) and so the material will weakly repel the external field. A superconductor exhibits perfect diamagnetism (sub-topic A.8).

Spectroscopic methods

Trace concentrations of elements such as heavy metals in water are difficult to determine by chemical tests but can be detected by spectrophotometry techniques. Qualitative analysis showing which metals are present can be carried out by exciting electrons to higher

Alloying: Art or science?

With the discovery that trace amounts of certain materials can greatly enhance a metal's performance, alloying was initially more of an art than a science. Science has developed many ways to investigate matter indirectly, based upon established scientific foundations.

Quick questions

1   Explain which ion Mn^{2+} or Zn^{2+} will be attracted by a magnetic field and which will be repelled.

2   Deduce how much copper can be electroplated from an aqueous solution of copper(II) sulfate by a current of 2 A running for 20 minutes.

energy levels and detecting the characteristic wavelength of light emitted as these electrons return to lower energy levels; this is the process employed in **atomic emission spectroscopy (AES)** (figure 4). Concentrations (quantitative information) can be detected by the level of absorbance of this radiation in **optical emission spectroscopy (OES)**.

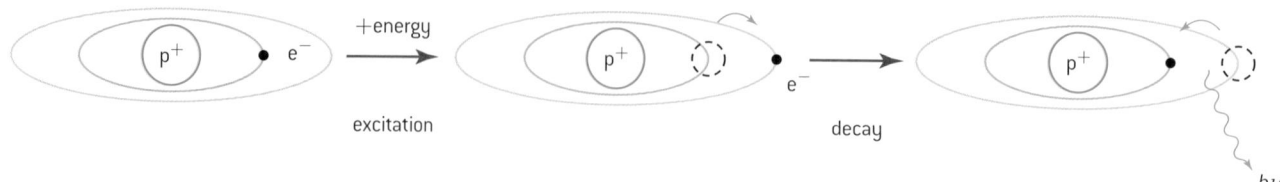

▲ Figure 4 As excited electrons return from an excited-state to a lower energy level they emit a characteristic wavelength of light

In **mass spectrometry (MS)** ions are introduced into a mass spectrometer and separated according to their mass-to-charge ratio. The detector receives a signal proportional to the concentration of the ion reaching it thus allowing both identification (qualitative analysis) and quantification. (Mass spectrometry is explained in more detail in topics 11 and 21.)

These spectroscopic techniques require that atoms are in the excited-state, or that they are fully ionized. Substances must also be atomized for spectroscopic analysis and this is usually accomplished by heating and/or electrical discharge, which bombards atoms with high-energy electrons to excite or ionize them.

Plasma can also be used for the atomization and/or excitation of samples for spectroscopy. Plasma is one of the four states of matter and it consists of free electrons, positive ions, and neutral atoms or molecules. In plasma 1% or more of the electrons are dissociated from their atoms so plasma can conduct electricity and be influenced by magnetic fields. Lightning, electric sparks, and the coloured neon lights used in advertising are all examples of matter in the plasma state.

Argon is the gas that is ionized into plasma in **inductively coupled plasma** (ICP) discharges. Plasma can exist at temperatures much hotter than those reached in furnaces or other discharges (around 10 000 K), and can atomize and ionize any type of material. The plasma discharge ionizes or excites the substance being analysed by MS or OES. The argon is swirled through three concentric tubes of quartz in the torch. The tubes and the swirling action allow the high-temperature plasma to be contained in the centre so as to not melt the torch. A high frequency oscillating current is supplied to a coil surrounding the torch; this creates electromagnetic fields oscillating in resonance at a high frequency, approximately 30 MHz or more. An electric spark is passed to initiate the plasma by ionizing argon atoms, knocking off electrons. These charged particles (Ar^+ and e^-) accelerate back and forth in the electromagnetic fields, occasionally colliding with other argon atoms and creating more Ar^+ and e^-. The process continues until Ar^+ ions are being created at the same rate as electrons are recombining with the ions to re-form argon atoms. The resulting "fireball" of plasma reaches temperatures over 10000 K. This process

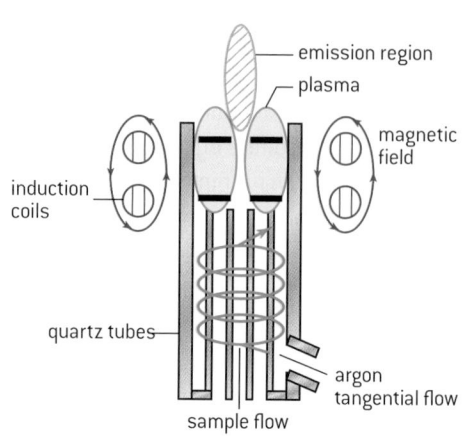

▲ Figure 5 Schematic diagram of an ICP torch. The emission region is further analysed by MS or OES

of heating by magnetic induction is somewhat similar to the process used in induction hobs for cooking.

The plasma is held in the centre of the torch until discharge by the swirling argon gas and strong magnetic field surrounding it. The process of plasma formation is outlined in figure 6.

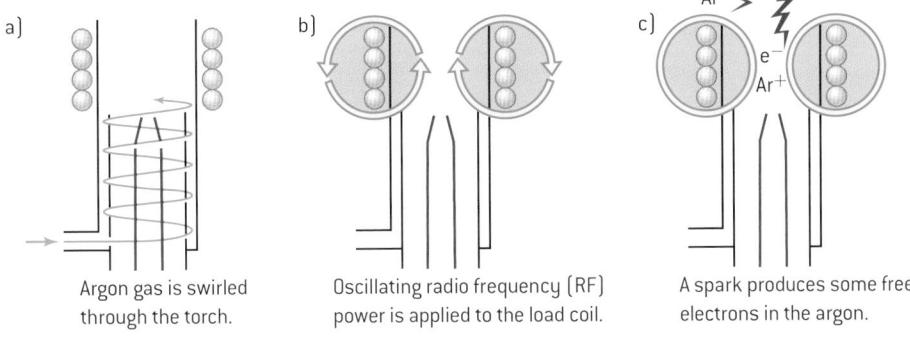

a) Argon gas is swirled through the torch.

b) Oscillating radio frequency (RF) power is applied to the load coil.

c) A spark produces some free electrons in the argon.

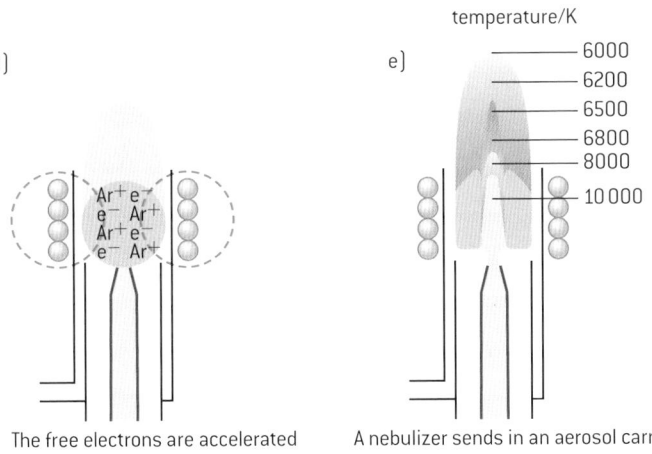

d) The free electrons are accelerated by the RF fields causing further ionization and forming a plasma.

e) A nebulizer sends in an aerosol carrying the sample and this punches a hole in the plasma, creating an ICP discharge.

temperature/K
— 6000
— 6200
— 6500
— 6800
— 8000
— 10 000

▲ Figure 6 Cross-section of an ICP torch and load coil depicting an ignition sequence

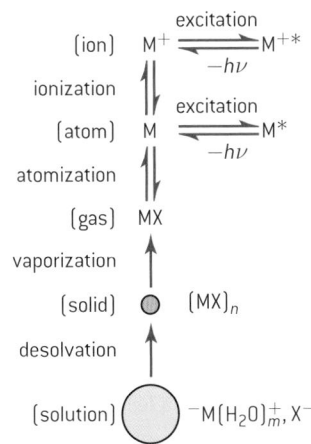

▲ Figure 7 Metals and metallic compounds can be vaporized and ionized in an ICP discharge

A nebulizer sends a spray containing the sample that is to be analysed into the plasma. The sample is ionized in the plasma ready for spectroscopic analysis by OES or MS (figure 8).

Figure 9 gives an overview of ICP-OES. The technique allows the accurate detection of very small traces of many elements. The chart in figure 10 shows some detection limits which are constantly being improved.

The limiting factor in ICP-MS or ICP-OES is not the plasma but rather the quality of the samples and the accuracy of the calibration curve. Known standards are used for calibration which have uncertainties that must be allowed for. Once plotted the calibration curves can then be used to provide values for unknown concentrations. Creating a calibration curve that is accurate at very low concentrations requires a solution of known concentration to be prepared very accurately, and successive dilutions are then made to create the lower concentrations used in calibration.

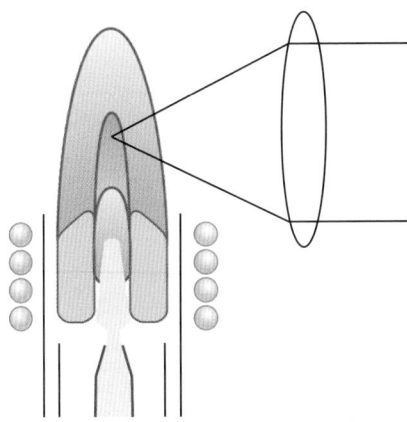

▲ Figure 8 The sample is discharged in the high-temperature plasma and viewed side-on by instrumentation for wavelength detection and absorbance in optical (atomic) emission spectroscopy (OES/AES)

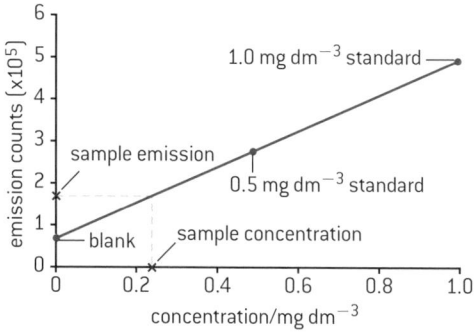

▲ Figure 11 An example calibration curve

▲ Figure 12 ICP torch in an ICP Atomic Emission Spectrometer, photos courtesy of Brian Young, Virginia Tech Chemistry Dept.

▲ Figure 13 ICP torch emission

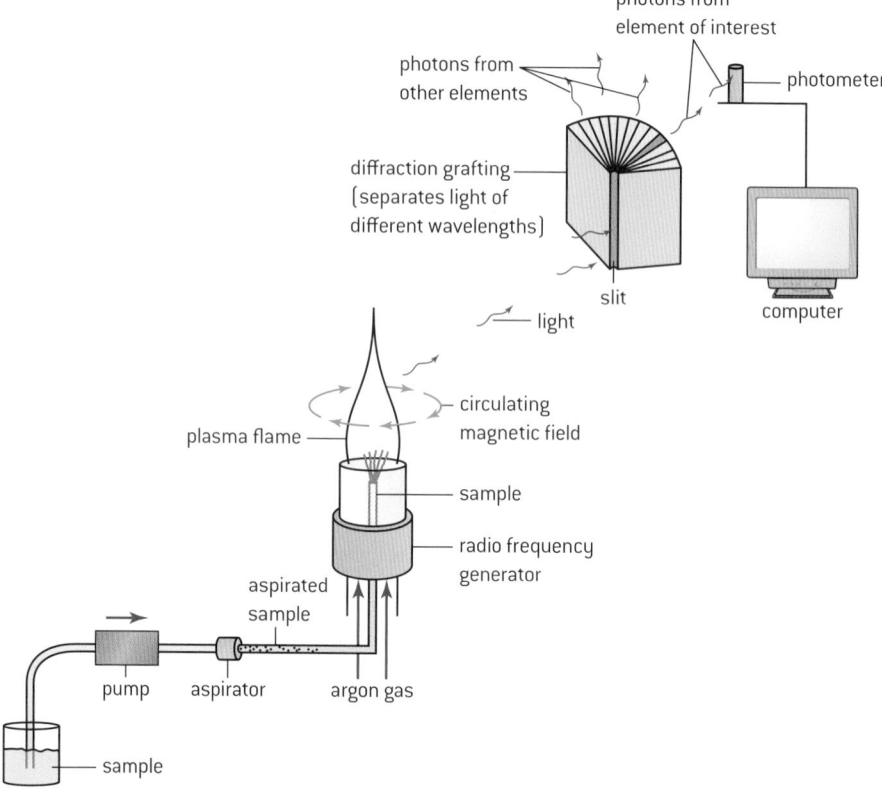

▲ Figure 9 Conceptual diagram of ICP-OES

ICP-OES detection limits/$\mu g\ dm^{-3}$

Li	Be											B	C	N			
0.3	0.1											1	40	na			
Na	Mg											Al	Si	P	S	Cl	
3	0.1											3	4	30	30	ns	
K	Ca	Sc	V	Ti	Cr	Mn	Fe	Co	Ni	Cu	Zn	Ga	Ge	As	Se	Br	
20	0.02	0.3	0.5	0.5	2	0.4	2	1	5	0.4	1	4	20	20	50	na	
Rb	Sr	Y	Nb	Zr	Mo		Ru	Th	Pd	Ag	Cd	In	Sn	Sb	Te	I	
30	0.06	0.3	5	0.8	3		6	5	3	1	1	9	30	10	10	na	
Cs	Ba	La	Hf	Ta	W	Re	Os	Ir	Pt	Au	Hg	Ti	Pb	Bi			
10	0.1	1	4	15	8	5	0.4	5	10	4	1	30	10	20			

	Ce	Pr	Nd		Sm	Eu	Gd	Tb	Dy	Ho	Er	Tm	Yb	Lu
	5	1	1		2	0.1	1	2	2	0.4	1	0.5	0.3	0.2
	Th		U											
	70		15											

▲ Figure 10 Detection limits for different elements by ICP-OES

Some obvious limits apply to ICP-MS and ICP-OES. Argon cannot be analysed, and neither can carbon dioxide as any argon supply usually contains this. Using water or organic solvents prevents H, O, or C atoms from being analysed because of the quantity of these elements in the solvent. At the high temperatures used in ICP-MS and ICP-OES, any solvent sprayed from the nebulizer not only vaporizes but also atomizes as any covalent bonds are broken.

Advantages of ICP-MS and ICP-OES over other analytical techniques include a larger linear calibration, and the ability to detect multiple elements at low concentrations.

Questions

1 Magnesium is an essential component of chlorophyll and traces of it can be found in various fluids extracted from plants. Its concentration may be estimated using inductively coupled plasma optical emission spectroscopy (ICP-OES).

a) Describe the plasma state as used in ICP.

b) The calibration curve shown in figure 14 was set up using ICP-OES.

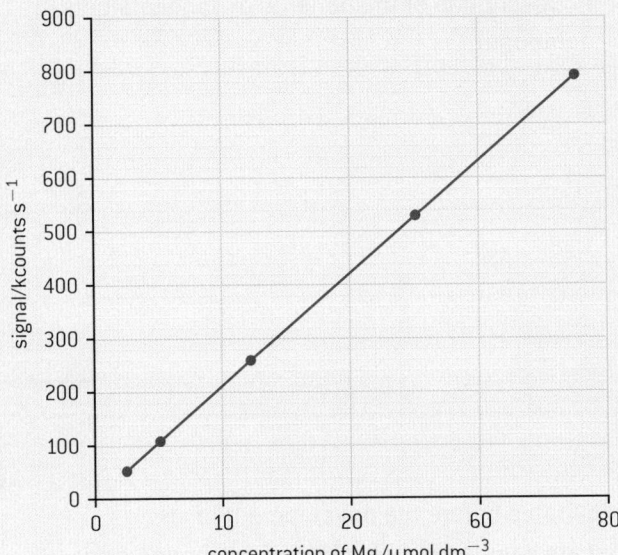

▲ Figure 14

   i) Outline how such a calibration curve might have been produced.

   ii) Comment on the use of this calibration curve for detecting magnesium in plants which has an approximate concentration of 1000 mg dm^{-3} in solution.

   iii) Explain how ICP-MS or ICP-OES/AES is used in determining whether the metal taken up by the plant was magnesium or manganese, based on the separation techniques of MS and OES/AES used.

2 A sample of germanium is analysed in a mass spectrometer. The first and last processes in mass spectrometry are vaporization and detection.

a) State the names of the other three processes in the order in which they occur in a mass spectrometer. [2]

b) For each of the processes named in a), outline how the process occurs. [3]

IB, November 2005

3 Some vaporized magnesium is introduced into a mass spectrometer. One of the ions that reaches the detector is $^{25}Mg^+$.

a) Identify the number of protons, neutrons, and electrons in the $^{25}Mg^+$ ion. [1]

b) State how this ion is accelerated in the mass spectrometer. [1]

c) The $^{25}Mg^{2+}$ ion is also detected in this mass spectrometer by changing the magnetic field. Deduce and explain, by reference to the m/z values of these two ions of magnesium, which of the ions $^{25}Mg^{2+}$ and $^{25}Mg^+$ is detected using a stronger magnetic field. [2]

IB, November 2006

4 Which ion would undergo the greatest deflection in a mass spectrometer?

   A. $^{16}O^+$

   B. $^{16}O^{2+}$

   C. $^{18}O^{2+}$

   D. $(^{16}O^{18}O)^+$ [1]

IB, November 2004

5 **a)** Traditionally, the raw materials for the production of iron are iron ore, coke, limestone, and preheated air. Iron oxides are reduced in a blast furnace by both carbon and carbon monoxide to form iron. Give the equation for the reduction of iron(III) oxide by carbon monoxide. [1]

b) In many modern blast furnaces, hydrocarbons (such as methane) are also added to the preheated air. This produces carbon monoxide and hydrogen. The hydrogen formed can also act as a reducing agent. Give the equation for the reduction of magnetite, Fe_3O_4, by hydrogen. [1]

IB, November 2003

A.3 Catalysts

Understandings

→ Reactants adsorb onto heterogeneous catalysts at active sites and the products desorb.

→ Homogeneous catalysts chemically combine with the reactants to form a temporary activated complex or a reaction intermediate.

→ Transition metal catalytic properties depend on the adsorption/absorption properties of the metal and the variable oxidation states.

→ Zeolites act as selective catalysts because of their cage structure.

→ Catalytic particles are nearly always nanoparticles that have large surface areas per unit mass.

Applications and skills

→ Explanation of factors involved in choosing a catalyst for a process.

→ Description of how metals work as heterogeneous catalysts.

→ Description of the benefits of nanocatalysts in industry.

Nature of science

→ Use of models – catalysts were used to increase reaction rates before the development of an understanding of how they work. This led to models that are constantly being tested and improved.

Models of catalysis

Models of how catalysts work have been developed based on observations and theories, and these are constantly being tested and reworked. The use of catalysts has had tremendous benefits, but is not without risk as many catalysts are toxic and their disposal can be difficult. Can new theories or advances in areas such as nanotechnology find catalysts that are even more effective and environmentally sound?

Homogeneous and heterogeneous catalysts

A **catalyst** increases the rate of a reaction and is left unchanged at the end of the reaction. A **homogeneous catalyst** is in the same phase as the reactants, takes the part of a reactant, and is reformed as a product at the end of the reaction. A **heterogeneous catalyst** is in a different phase than that of the reactants. Catalysts work by providing an alternative reaction pathway for the reaction that lowers the activation energy, as shown in a potential energy profile diagram (figure 1; see also sub-topic 6.1). A higher proportion of reactant particles therefore achieve the required activation energy as a result (see the Maxwell–Boltzmann distribution curve in sub-topic 6.1).

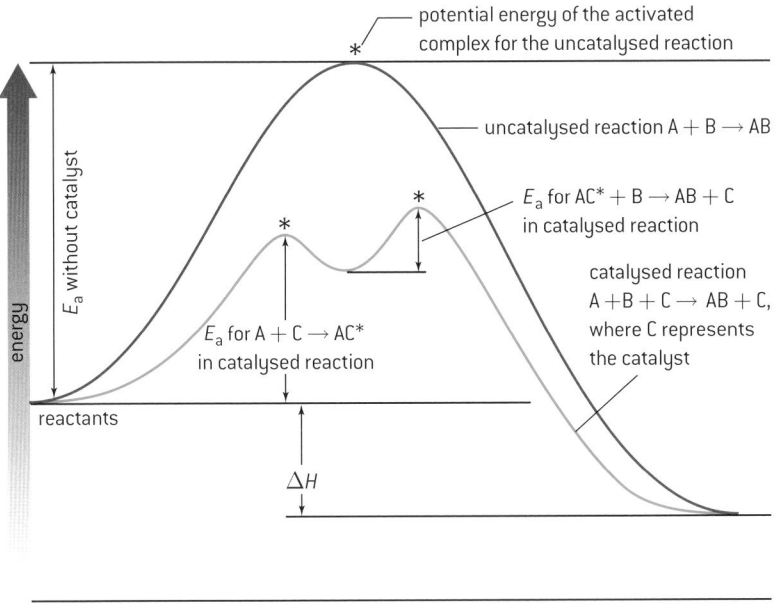

Figure 1 The activation energy for the catalysed reaction is lower than that for the uncatalysed reaction

Mechanisms of catalysis

The mechanism by which the activation energy is lowered varies between homogeneous and heterogeneous catalysts. A homogeneous catalyst forms bonds with one or more of the reactants resulting in either a **reaction intermediate**, which then further reacts, or an **activated complex**, which is a temporary transition state. In either case the energy needed for the reactant molecules to complete the reaction is reduced as the reaction occurs between reactant and catalyst rather than between one reactant and another.

For example, in the general reaction:

A + B →AB

A and B need to collide and overcome the activation energy for the reaction. In the catalysed reaction:

A + C → AC*

AC* + B → AB + C

the collision between A and the catalyst C to form the intermediate AC* requires less energy than does the collision between A and B to form AB. The intermediate (or activated complex) then reacts with B to form the product AB and regenerate the catalyst. In this way the homogeneous catalyst enters the reaction, but is left unchanged at the end.

You will recall from topic 10 that esterification is a reversible reaction between a carboxylic acid and an alcohol. The reactants are heated in the presence of a catalyst, usually concentrated sulfuric acid:

$$CH_3CH_2COOH + CH_3OH \underset{\Delta}{\overset{H^+}{\rightleftharpoons}} CH_3CH_2COOCH_3 + H_2O$$

In an **activated complex** (marked * in figure 1) bonds are both forming and breaking and the reaction could fall either side of the hill. The "valley" on the blue line shows a stable intermediate. This intermediate then reacts in step 2 forming the products and regenerating the catalyst. A one-step reaction has just an activated complex without an intermediate.

The H$^+$ ion from sulfuric acid forms an intermediate with the reactants which allows the water molecule to leave and the ester to form at a much lower activation energy than is the case without the acid catalyst. The catalyst is regenerated in the reaction.

A heterogeneous catalyst is in a different phase from the reactants, usually a solid catalyst for a gaseous reaction or a reaction in solution. Transition metals are common heterogeneous catalysts. These solids bring reactant molecules together in an orientation that enables them to react readily, thus reducing factors that inhibit the reaction.

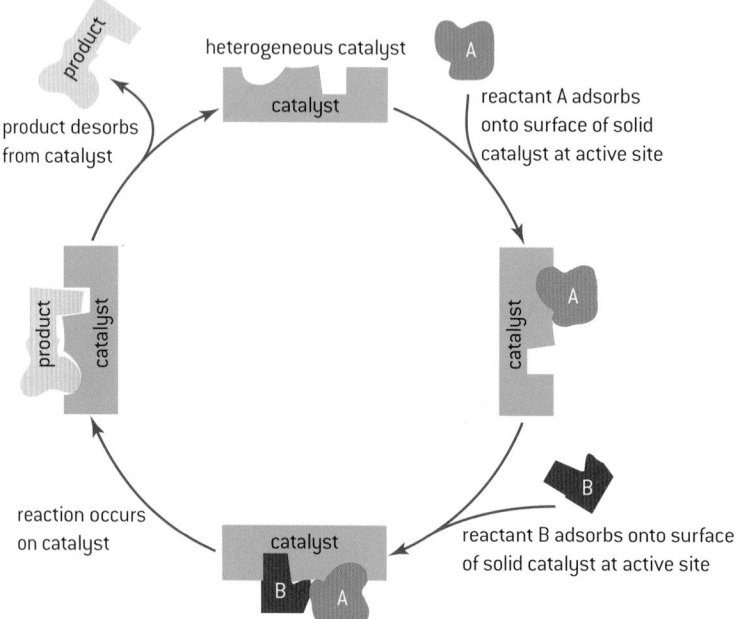

heterogeneous catalyst

product desorbs from catalyst

reactant A adsorbs onto surface of solid catalyst at active site

reactant B adsorbs onto surface of solid catalyst at active site

reaction occurs on catalyst

▲ Figure 2 The action of a heterogeneous catalyst

The most common primary catalysts are platinum, palladium, and rhodium. Copper is sometimes used, but is less common in industrial catalytic processes.

Reactant molecules adsorb onto the heterogeneous catalyst at active sites. The process of adsorbing onto the surface of a solid catalyst affects the bonds in the reactants so that they are stretched, weakened, and sometimes broken. The reaction occurs on the surface of the catalyst, in one or several steps, and the products desorb from the surface of the catalyst.

Homogeneous and heterogeneous catalysts compared

Homogeneous catalysts are in the same phase as the reactants resulting in close contact between reactant and catalyst molecules. They work under mild conditions and have good selectivity for the desired products. A disadvantage of being in the same phase is that the catalyst needs to be removed after the reaction. This is usually accomplished by distillation, which might destroy the catalyst if a high temperature is required to distil off either the product or the catalyst. In industrial processes it is generally easier to separate large quantities of product from a heterogeneous catalyst than from a homogeneous catalyst.

With a heterogeneous catalyst there is a lower effective concentration of catalyst because the reaction can only occur on the surface of the solid.

Forming the catalyst into a mesh is one way of increasing the effective surface area.

A distinct disadvantage of heterogeneous catalysis is that catalytic poisoning can occur when other compounds react with the surface of the catalyst. These might be products that remain, or foreign substances in the reaction mixture. In either case accumulation on the catalyst surface will reduce its effectiveness. For example, the majority of cars have a catalytic converter (figure 3) to reduce the emission of harmful pollutants such as nitrogen oxides and incomplete combustion products including carbon monoxide. The catalyst converts carbon monoxide to carbon dioxide and nitrogen oxides to nitrogen. "Leaded" petrol cannot be used in a car with a catalytic converter because the lead strongly adsorbs onto the surface of the catalyst and prevents the adsorption of carbon monoxide.

Nanocatalysts

The use of nanoparticles has bridged the gap between homogeneous and heterogeneous catalysts. Due to their small size nanocatalysts are sometimes referred to as catalytic particles. Most heterogeneous nanocatalysts are metal nanoparticles which supply catalytically active sites. Catalytic nanoparticles have a large surface area per unit mass. They can provide a large contact area and can be introduced to a reaction mixture in the same way as homogeneous catalysts while providing adsorption/desorption sites as a heterogeneous catalyst.

Common properties that are considered when selecting a catalyst include: selectivity for only the desired product; conversion efficiency; the ability to work in the conditions necessary for the reaction; environmental impact; cost; lifetime; and susceptibility to poisoning. Nanocatalysts generally have a high conversion efficiency because of their small size and large surface area. They can be engineered for maximum selectivity which reduces catalytic poisoning by unwanted substances. Enzymes, for example, can achieve greater than 99.99% selectivity, meaning less than 1 in 10000 conversions gives an unwanted by-product. This level of selectivity is rarely found in synthetic catalysts. Nanocatalysts can also provide low energy consumption and a long lifetime.

Many nanocatalysts contain various forms of carbon, including graphite, carbon nanotubes, fullerenes, and graphene.

Transition metal catalysts

Ceramics provide some useful catalysts, but the most widely used inorganic catalysts are transition metals due to their variable oxidation states and high adsorption capacity. The variable oxidation states of transition metals allow them to enter many reactions as homogeneous catalysts. They can form complexes (topic 13), allowing them to form bonds with many substances. This frequently involves a change in oxidation state, which is returned to the original state when the reaction is over.

▲ Figure 3 Molecular graphic showing how a catalytic converter converts pollutants into harmless gases. The yellow framework is a lattice of copper atoms – the catalyst. Atoms of the pollutant gases carbon monoxide, CO and nitrogen monoxide, NO are shown adsorbed onto the copper surface (oxygen = red; carbon = green; nitrogen = blue). The reaction produces the harmless gases nitrogen, N_2 (right) and carbon dioxide, CO_2 (upper centre). The blue and white zones show molecular orbitals involved in breaking apart an NO molecule over a cluster of 10 copper atoms (brown)

Nanoparticles are particles that have dimensions less than 100 nm and exhibit properties that differ from those of the bulk material. Individual molecules are usually not considered to be nanoparticles but small clusters of them may be.

▲ Figure 4 Pumice, a porous rock with a density less than water, is a naturally occurring zeolite

▲ Figure 5 Molecular graphic of a zeolite structure. Zeolites have a cage-like structure composed of channels, cavities, and various frameworks. This structure acts as a molecular sieve as well as providing a large surface area for catalysis

Transition metals also make good heterogeneous catalysts: many gases will adsorb to their surface. Weak bonds form between the reactant gases and the catalyst surface, locally increasing the reactant concentration at adsorption sites. Reactant bonds are weakened as described earlier, resulting in an increased rate of reaction.

Some transition metals are toxic and should be used only if a suitable alternative is not available.

Zeolites

Zeolites are microporous substances made of alumina silicate which has a cage-like structure providing a large surface area. Zeolites are cheap, plentiful, and occur naturally in many forms including pumice (figure 4). There are over 100 different structures of cages, cavities, channels, and other types of framework. A zeolite can act as a microscopic sieve, allowing only certain molecules through depending on their size and structure. Zeolites work by both adsorption and cation exchange.

Zeolites are used to remove heavy metal ions from water supplies. Water molecules pass through the "molecular sieve" while larger complex ions of the metals are trapped by the sieve. Zeolites are also used as catalysts for cracking in the petroleum industry. Their cation exchange properties are used to remove the "hard water" ions of calcium and magnesium: these ions stay on the zeolite while potassium and sodium ions exchange out. Many washing powders contain zeolites to make washing in hard-water locations more effective.

Questions

1 Catalysts may be homogeneous or heterogeneous.

   a) Distinguish between *homogeneous* and *heterogeneous* catalysts. [1]

   b) Discuss **two** factors which need to be considered when selecting a catalyst for a particular chemical process. [2]

   c) Identify the catalyst used in the catalytic cracking of long chain hydrocarbons and state **one** other condition needed. [2]

   IB, May 2011

2 a) State **one** advantage and **one** disadvantage that homogeneous catalysts have over heterogeneous catalysts. [2]

   b) Apart from their selectivity to form the required product and their cost, discuss **two**

   other factors which should be considered when choosing a suitable catalyst for an industrial process. [2]

   IB, May 2010

3 Compare the modes of action of homogeneous and heterogeneous catalysts. State **one** example of each type of catalysis using a chemical equation **and** include state symbols. [4]

   IB, May 2009

4 Carbon nanotubes can be used as catalysts.

   a) Suggest **two** reasons why they are effective heterogeneous catalysts.

   b) State **one** potential concern associated with the use of carbon nanotubes.

A.4 Liquid crystals

Understandings

→ Liquid crystals are fluids that have physical properties (electrical, optical, and elasticity) that are dependent on molecular orientation to some fixed axis in the material.

→ Thermotropic liquid-crystal materials are pure substances that show liquid-crystal behaviour over a temperature range.

→ Lyotropic liquid crystals are solutions that show the liquid-crystal state over a (certain) range of concentrations.

→ Nematic liquid crystal phase is characterized by rod-shaped molecules which are randomly distributed but on average align in the same direction.

Applications and skills

→ Discussion of the properties needed for a substance to be used in liquid-crystal displays (LCD).

→ Explanation of liquid-crystal behaviour on a molecular level.

Nature of science

→ Serendipity and scientific discoveries – Friedrich Reinitzer accidently discovered flowing liquid crystals in 1888 while experimenting on cholesterol.

The discovery of liquid crystals

The observation of two separate melting points for the substance cholesteryl benzoate (figure 1) led to the serendipitous discovery of liquid crystals by Friedrich Reinitzer in 1888. Continued experimentation and improvements in instrumentation over the years have developed this field into an industry of ultra-high-definition liquid-crystal displays.

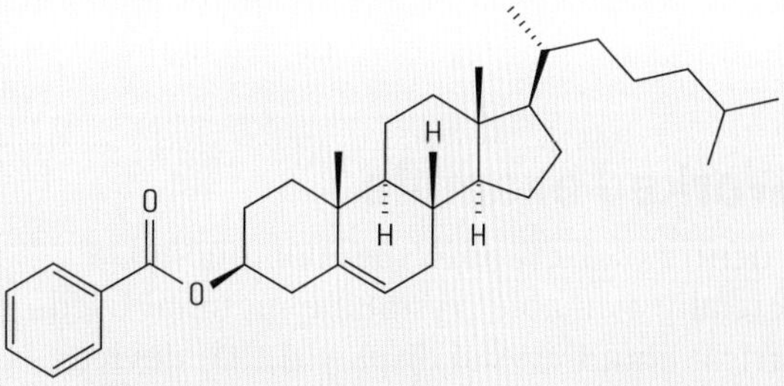

▲ Figure 1 The molecular structure of cholesteryl benzoate. Friedrich Reinitzer, an Austrian botanist, noticed two melting points for this compound in 1888. It turned cloudy at one temperature and clear at another, which led to further research into liquid-crystal behaviour

The properties of liquid crystals

Liquid crystals are a state of matter intermediate between crystalline and liquid. They are fluids whose physical properties are dependent on molecular orientation relative to some fixed axis in the material: a liquid crystal has molecules that can flow like a liquid but line themselves up in a crystalline order. Liquid crystal molecules maintain this orientational order, aligning mostly the same way, but not their positional order; they can slide over each other as in a liquid.

The molecular shape of many liquid crystals is linear or flat with very little branching. They frequently contain long-chain alkyl groups which form long, thin, rigid, rod-shaped molecules, or linear chains of aromatic rings that form flat disc shapes, or a combination of both. The ability of these chains to align when a weak electric field is applied is what forms the liquid-crystal state. Liquid crystals are often polar molecules so they change orientation when an electric field is applied. They normally show a fairly rapid switching speed, changing in orientation when the field is reversed. Molecular lengths of 1.3 nm are typically needed for visual displays.

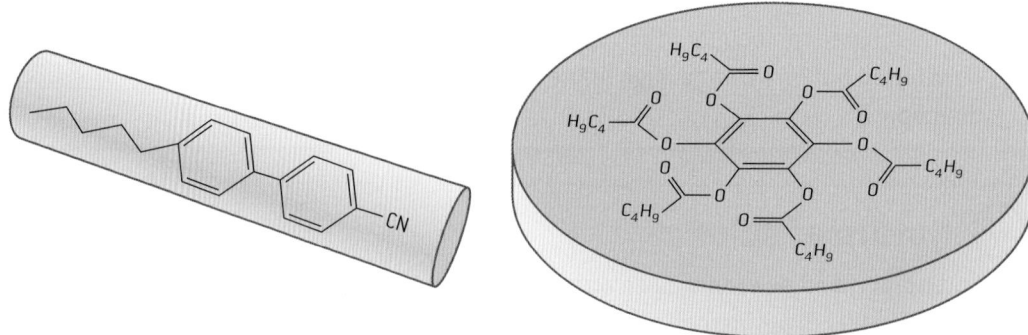

▲ Figure 2 Liquid crystal molecules have an alkyl group and a polar end, as in 4-cyano-4'-pentylbiphenyl, shown on the left. A series of aromatic benzene rings can form flat disk shapes as in the benzene alkanoate derivative pictured on the right

To form useful liquid crystals a substance needs to be chemically stable and have a liquid-crystal phase that is stable over a suitable temperature range.

Worked example

The structure of 4-cyano-4'-pentylbiphenyl, a commercially available crystalline material used in electrical liquid-crystal display (LCD) devices, is shown in figure 3.

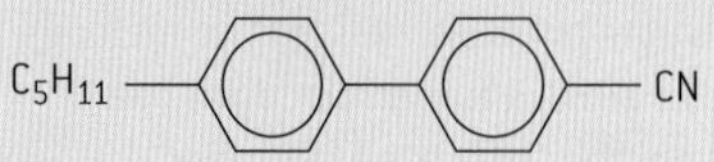

▲ Figure 3

Explain how the three different parts of the molecule – CN, C_5H_{11}, and the biphenyl group – contribute to the properties of the compound used in LCD electrical display devices.

Solution

It is essential for liquid crystals to be polar so that they can be influenced by a weak electric field. The nitrile group ($C\equiv N$) confers a degree of polarity so that the molecules align in a common direction when a weak electric field is applied.

The biphenyl group makes the molecules more rigid and rod-shaped. The benzene ring is also chemically stable and will not decompose under stress such as UV radiation, pressure, or at slightly elevated temperatures.

The long alkyl chain C_5H_{11} ensures that the molecules cannot pack together too closely and so helps maintain the liquid-crystal state. Its length gives a rod-like shape and as it is an alkyl chain it is chemically stable.

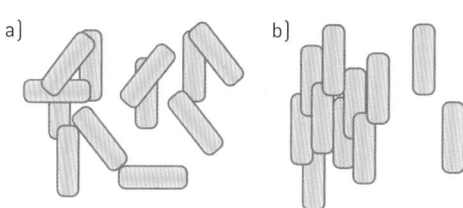

Transmitting light: Forming LCD displays

Liquid-crystal displays are used in many lightweight applications including digital watches, calculators, and laptops because of their low energy consumption. The ability of liquid crystal molecules to transmit light depends on the orientation of the molecules. This can be controlled by the application of a small voltage across a thin film of the material, forming light and dark areas of the display.

Thermotropic liquid-crystal materials are pure substances that show liquid-crystal behaviour over a temperature range between the solid and liquid states. **Lyotropic** liquid crystals are solutions that show the liquid-crystal state at certain concentrations. The **nematic** liquid-crystal phase is characterized by rod-shaped molecules that are randomly distributed but on average align in the same direction.

In an electric field the molecules of a nematic liquid crystal become orientated as shown in figure 4(b). The molecules can still slide over each other but in general they maintain their alignment.

In LCD displays nematic liquid crystals are often placed in layers at right angles to each other with each pixel containing liquid crystal sandwiched between two polarized glass plates. These plates each have a set of grooves and the two sets are at 90° to each other. The liquid crystal molecules in contact with the glass line up with the grooves and the molecules in between form a twisted arrangement between the plates that is held by intermolecular bonds. Light passing through the first filter becomes polarized. When the polar nematic liquid crystal molecules are aligned with the grooves they allow the polarized light to pass through the film and the pixel appears bright. As a voltage is applied across the film the polar molecules all align with the field rather than with the grooves. The plane-polarized light is no longer aligned with the orientation of the liquid crystal molecules and so the pixel appears dark.

Thermotropic liquid crystals change behaviour over a range of temperatures. Biphenyl nitriles are thermotropic liquid crystals that naturally exist in the nematic phase as shown in figure 4(b), their rod-shaped molecules distributed randomly but on average pointing in the same direction. Increased thermal agitation disrupts this directional order until it is lost, as in figure 4(a), when the normal liquid phase is formed.

Lyotropic liquid crystals have a hydrophilic end that is polar and easily attracted to polar molecules such as water, and a hydrophobic end that is non-polar and repelled by polar molecules. These substances take on liquid-crystal arrangements as the concentration increases. Soap is an example of a lyotropic liquid crystal (figure 6).

Soap forms spherical **micelles** (figure 7). The polar ends on the outside of the sphere are surrounded by water molecules with non-polar oil or grease encapsulated in the centre. At high concentrations rod-like micelles are formed that have liquid-crystal properties and can form bilayer sheets.

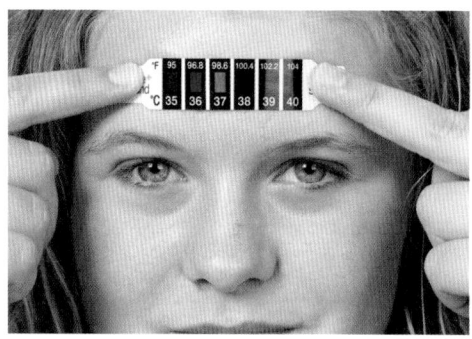

▲ Figure 4 Nematic liquid crystal molecules align in an electric field

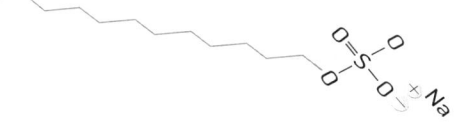

▲ Figure 5 A liquid-crystal strip thermometer being used to measure body temperature. The different areas of the strip contain thermotropic liquid crystals that are designed to respond at different temperatures

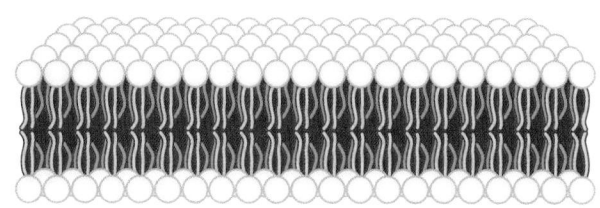

▲ Figure 6 Soap contains a polar end that is soluble in water and a non-polar end capable of dissolving oils and fats. This enables soapy water to wash greasy dishes

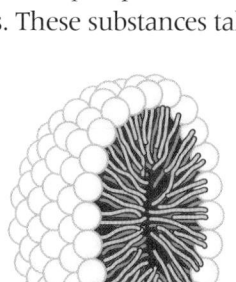
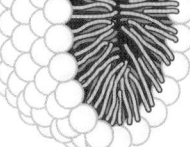

micelle   bilayer sheet

▲ Figure 7 Spherical and bilayer sheets are formed by soap at increasing concentrations

491

In lyotropic liquid crystals rigid structures occur at higher concentrations. Micelles group together into hexagonal layers and then rod-shaped liquid-crystal structures (figures 8 and 9).

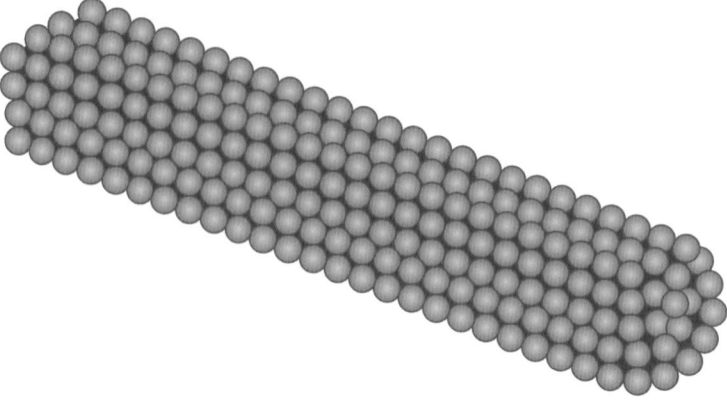

▲ Figure 8 Rod-like micelle structure of a lyotropic liquid crystal. The entire micelle will flow like a liquid but will retain its orientation

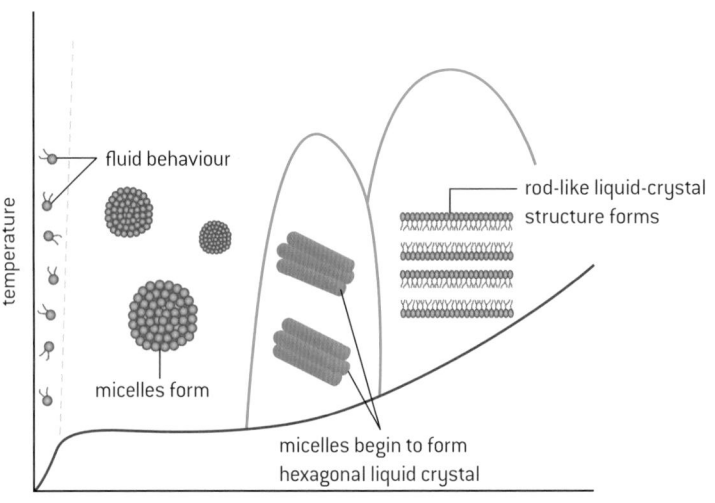

▲ Figure 9 Lyotropic liquid crystals restructure according to the concentration. At certain concentrations they exhibit liquid-crystal properties; at other concentrations they do not

Kevlar (figure 10) is a lyotropic liquid crystal. The linked benzene rings make the rod-shaped molecules rigid. The alignment of these molecules depends on the concentration of the solution. There are strong intermolecular hydrogen bonds between the chains giving a very ordered and strong structure. These bonds can be broken with concentrated sulfuric acid, as oxygen and nitrogen atoms become protonated, breaking the hydrogen bonds. Kevlar is discussed in sub-topic A.9.

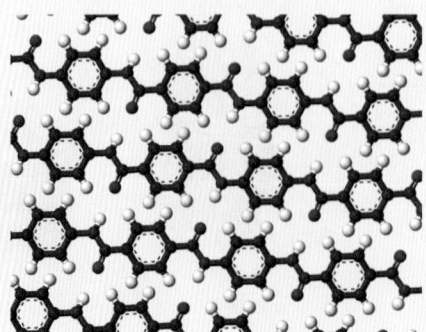

◀ Figure 10 Molecules of Kevlar. Strong hydrogen bonding occurs between the chains. Can you identify where the hydrogen bonding occurs?

Questions

1 Liquid-crystal displays are used in digital watches, calculators, and laptops.

   Describe the liquid-crystal state, in terms of molecular arrangement, and explain what happens as temperature increases. [3]

   IB, May 2011

2 Discuss **three** properties a substance should have if it is to be used in liquid-crystal displays. [3]

   IB, May 2011

3 Kevlar is an example of a lyotropic liquid crystal. Outline what is meant by *lyotropic liquid crystal*. [2]

   IB, May 2010

4 Detergents are one example of lyotropic liquid crystals.

   State **one** other example of a lyotropic liquid crystal and describe the difference between lyotropic and thermotropic liquid crystals. [3]

   IB, May 2010

5 a) Name a thermotropic liquid crystal. [1]

   b) Explain the liquid-crystal behaviour of the thermotropic liquid crystal named in part (a), on the molecular level. [4]

   IB, May 2010

6 Describe the meaning of the term *liquid crystals*. State and explain which diagram in figure 11, a or b, represents molecules that are in a liquid crystalline phase. [2]

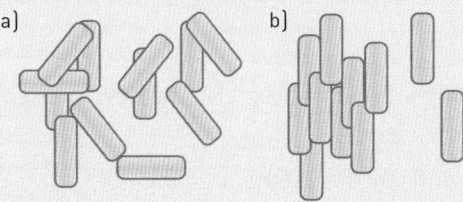

▲ Figure 11

IB, May 2009

7 Distinguish between *thermotropic* and *lyotropic* liquid crystals and state **one** example of each type. [4]

   IB, May 2009

8 The structure of 4-pentyl-4-cyanobiphenyl, a commercially available nematic crystalline material used in electrical display devices, is shown in figure 12.

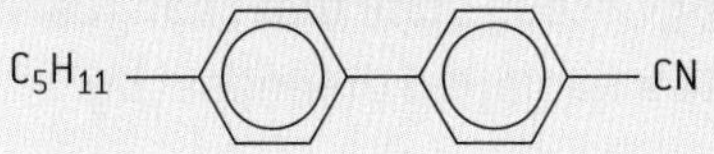

$$C_5H_{11} \longrightarrow \bigcirc\!\!\!\!\bigcirc \longrightarrow CN$$

▲ Figure 12

   Describe and explain in molecular terms the workings of a twisted nematic liquid crystal. [4]

   IB, November 2009 (part)

9 a) Compare the positional and directional order in a crystalline solid, a nematic phase liquid crystal, and a pure liquid. Show your answer by stating **yes** or **no** in a copy of table 1. [2]

	Crystalline solid	Nematic phase liquid crystal	Pure liquid
Positional order			
Directional order			

▲ Table 1

   b) Outline any **two** principles of a liquid-crystal display device. [2]

   IB, May 2009

A.5 Polymers

Understandings

→ Thermoplastics soften when heated and harden when cooled.

→ A thermosetting polymer is a prepolymer in a soft solid or viscous state that changes irreversibly into a hardened thermoset by curing.

→ Elastomers are flexible and can be deformed under force but will return to nearly their original shape once the stress is released.

→ High density polyethene (HDPE) has no branching allowing chains to be packed together.

→ Low density polyethene (LDPE) has some branching and is more flexible.

→ Plasticizers added to a polymer increase the flexibility by weakening the intermolecular forces between the polymer chains.

→ Atom economy is a measure of efficiency applied in green chemistry.

→ Isotactic addition polymers have substituents on the same side.

→ Atactic addition polymers have the substituents randomly placed.

Applications and skills

→ Description of the use of plasticizers in polyvinyl chloride and volatile hydrocarbons in the formation of expanded polystyrene.

→ Solving problems and evaluating atom economy in synthesis reactions.

→ Description of how the properties of polymers depend on their structural features.

→ Deduction of structures of polymers formed from polymerizing 2-methylpropene.

Nature of science

→ As a result of advances in technology (X-ray diffraction, scanning tunnelling electron microscopes, etc.), scientists have been able to understand what occurs on the molecular level and manipulate matter in new ways. This allows new polymers to be developed.

→ Theories can be superseded – Staudinger's proposal of macromolecules made of many repeating units was integral in the development of polymer science.

→ Ethics and risk assessment – polymer development and use has grown quicker than an understanding of the risks involved, such as recycling or possible carcinogenic properties.

High density and low density polyethene

The word "polymer" means "many parts". Polymers (also called plastics) are made up of repeating **monomer** units whose structures can be manipulated in various ways to give materials with desired properties. Polyethene (sub-topic 10.2) is an addition polymer made of ethene monomer units. The same monomer can be linked together to form **high density polyethene** (HDPE) or **low density polyethene** (LDPE), depending on the degree of branching in the polymer chain (figure 1).

The branching in LDPE molecules makes the polymer more flexible. HDPE can have M_r values of 200000 upwards. The linear structure of the molecules allows for very close packing, improving the material's strength which increases with weight. Ultra-high-molecular-weight polyethene (UHMWPE) can have M_r of 2–6 million; this is linear HDPE of very high strength which shows resistance to cutting and abrasion and has been used in synthetic ice-skating rinks and to replace Kevlar in bullet-proof vests.

HDPE and LDPE are produced from the same monomer using different methods and catalysts. LDPE is produced by free-radical polymerization involving an initiator, whereas a Ziegler–Natta catalyst is used to produce HDPE. (Knowledge of the mechanisms of these processes is not necessary for IB Chemistry.)

a molecule of high density polyethene (HDPE)

a molecule of low density polyethene (LDPE)

▲ Figure 1 Little or no branching in the polymer chain produces HDPE, which is stronger than LDPE whose molecules are highly branched

▲ Figure 2 HDPE is used for making bottles, like the one pictured on the left. The water bottle on the right is made from another polymer, polyethylene terephthalate (PETE)

Thermoplastics and thermosets

Polymers can be classified as **thermoplastics** or **thermoset plastics** based on their behaviour when heated. **Thermoplastics** generally do not have straight molecules but rather are formed of a massive weave of polymers bound together by intermolecular (van der Waals') forces that give them their shape. As a result they can be melted and then cooled in moulds to produce different shapes. The melting breaks down the intermolecular forces and on cooling new intermolecular forces form.

Thermoset plastics are made by heating the raw materials (monomers) and forming them into a single large network instead of many molecules. This results in a much stronger plastic because its shape is held by covalent bonds rather than intermolecular forces. The molecules may contain rings, linear chains, and side branches all bonded into one giant molecule. The structure cannot be melted and reformed into a different shape because melting would require sufficient heat to break the covalent bonds, hence decomposing the molecule rather than melting it. Thermoset plastic products are moulded when hot and they set as hardened plastic with the desired shape. They are harder, more rigid, and have higher strength than thermoplastics.

Polyethene, polystyrene, polyvinyl chloride (PVC), and polypropene are some recyclable thermoplastics, whereas resins, epoxies, polyurethanes, Bakelite and polyesters are formed from thermosetting prepolymers into hardened thermosets. Figure 3 outlines the difference between thermoplastics and thermosets.

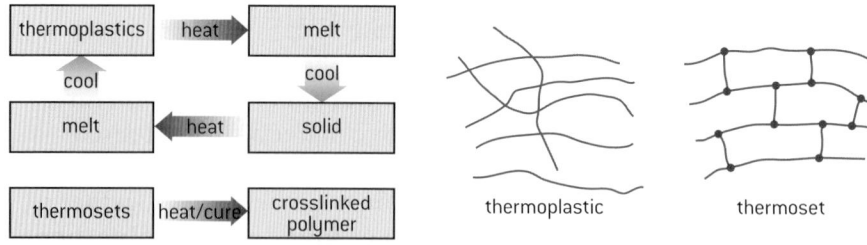

▲ Figure 3 Thermoplastics have cross-links held by intermolecular forces and can be melted and reformed. Thermoset plastics are chemically bonded during formation and cannot be reformed

Elastomers

Elastomers are flexible polymers that return to their original shape after being deformed. They can be manufactured from either thermoplastics or thermoset polymers but thermosets are usually chosen because of their higher strength. When the material is not under stress the polymer chain is tangled, loose, and flexible. Under stress the molecules assume a more linear form but retain their shape afterwards due to the covalently bonded cross-links (figure 4).

PVC and the use of plasticizers

Polychloroethene or polyvinyl chloride (PVC) was discovered in 1835. It is formed from the monomer chloroethene, also called vinyl chloride:

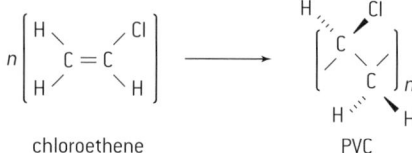

chloroethene   PVC

This addition polymer (sub-topic 10.2) was hard and brittle until Waldo Semon developed the technique of adding a **plasticizer** to the polymer to keep its strands somewhat separated. This reduces the intermolecular forces, softening the polymer and making it flexible and durable. This more flexible material had the added advantages of being water repellent and fire resistant.

Plasticizers work by embedding themselves between polymer chains, thus reducing the intermolecular forces between these chains (figure 5). This increases the volume, thereby lowering the density. The addition of plasticizers also lowers the melting point and makes the material more flexible and fluid. One of the first uses of plasticizers was to make PVC shower curtains. Plasticizing molecules such as bis(2-ethylhexyl)phthalate (figure 5) contain both polar and non-polar groups. The polar group locks the plasticizer in the polymer and the non-polar group weakens some of the attractive forces in the polymer chain, thus enhancing flexibility.

Higher concentrations of plasticizer produce softer and more flexible polymers. The plasticizer tends to evaporate over time so if flexible PVC is left in a hot dry place for a long period the material will become brittle. The distinctive smell in a new car is associated with plasticizer evaporation. Plasticizers can be used to expand other materials such as concrete, but more than 90% of plasticizer use is for polymers.

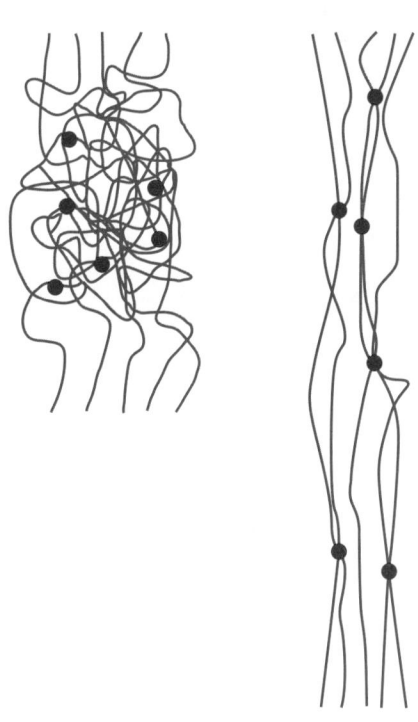

▲ Figure 4 An unstressed elastomer has tangled long-chain strands (left) which straighten out when the elastomer is stretched (right). The covalent cross-links between polymer strands provide strength and the elastomer will return to the unstretched state once the stress is removed. Rubber is an elastomer

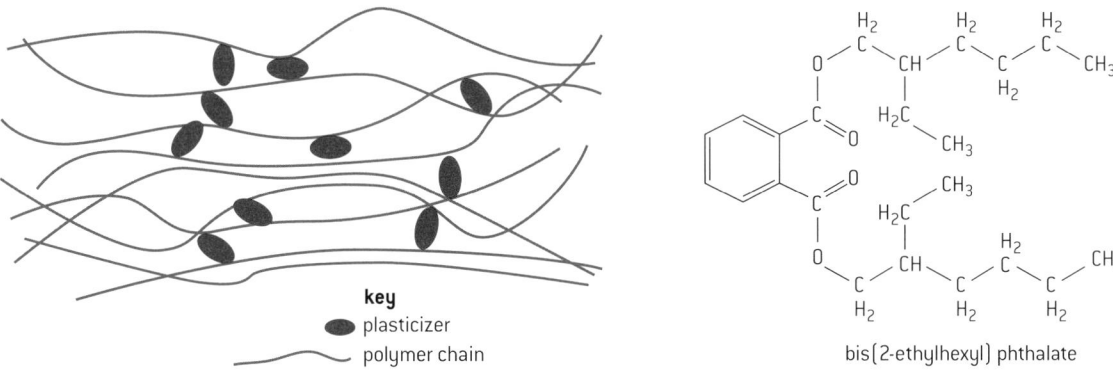

bis(2-ethylhexyl) phthalate

▲ Figure 5 Plasticizers such as bis(2-ethylhexyl) phthalate contain polar and non-polar groups which allow them to embed between polymer chains, keeping them apart and reducing the intermolecular forces

Polystyrene

Polystyrene (polyphenylethene) is a thermoplastic polymer made from the monomer styrene (figure 6), a liquid hydrocarbon that is commercially manufactured from petroleum by the chemical industry. Polystyrene can also be expanded, forming the lightweight and insulating expanded polystyrene familiar in food containers and packaging (figure 7). It is produced from a mixture of polystyrene and a gaseous blowing agent (usually carbon dioxide, pentane, or another volatile hydrocarbon). The solid plastic expands into a foam when heated by steam.

styrene

▲ Figure 6 Phenylethene (styrene)

Isotactic, atactic, and syndiotactic addition polymers

Phenylethene undergoes addition polymerization similar to that shown by ethene in forming polyethene. Chloroethene (vinyl chloride), which polymerizes to polychloroethene (also known as polyvinylchloride, PVC), is another example of an addition polymer.

Phenylethene and chloroethene can undergo several types of polymerization. These monomers are ethene with one hydrogen substituted by the phenyl group in phenylethene and by chlorine in chloroethene. **Isotactic** addition polymers have these substituents on the same side of the molecule, while **atactic** addition polymers have them randomly placed. **Syndiotactic** polymers have the substituents alternating one side to the next. Figure 8 shows these three forms in the case of polystyrene.

Another important polymer is polypropene. This again has isotactic, syndiotactic, and atactic forms depending on the placement of the methyl ($-CH_3$) group. The isotactic form, with all the methyl groups on the same side, is the most common commercial form.

Isotactic propene is more compact than the atactic form, having a regular repeating pattern that allows the molecules to come closer together, increasing the van der Waals' forces between them. Isotactic polypropene is harder, more rigid, and has a higher melting point than the atactic form.

▲ Figure 7 Polystyrene has three common forms. Solid/extruded polystyrene has many applications including models (pictured), disposable cutlery and CD cases. Expanded polystyrene foam (shown below the model) is used in packing materials and disposable cups. Extruded polystyrene foam has good insulating properties making it important as a non-structural construction material

497

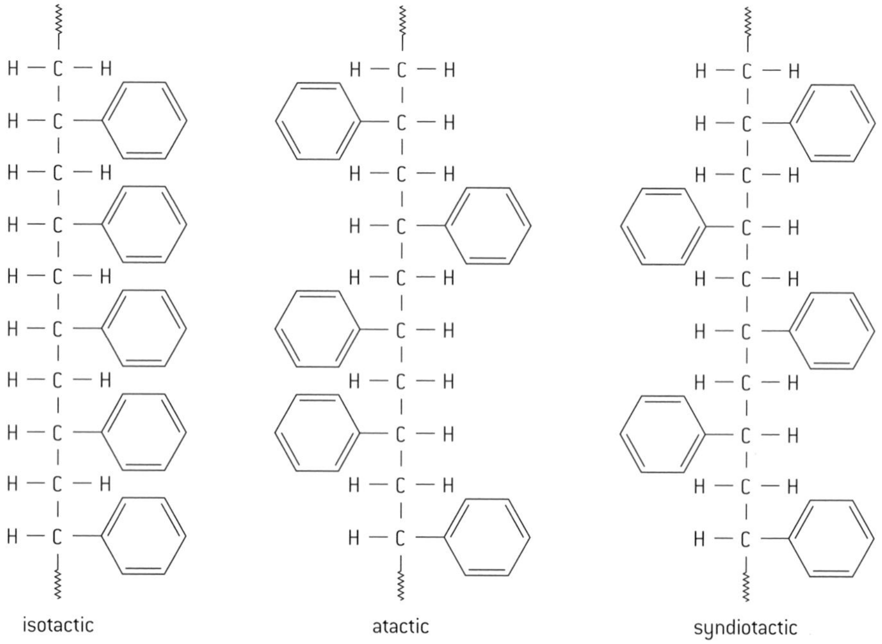

Figure 8 Isotactic, atactic, and syndiotactic polystyrene (polyphenylethene)

Syndiotactic polypropene has some stereoregularity: the regular, alternating placement of the –CH$_3$ groups allows closer packing and stronger intermolecular forces than is the case for the atactic form, which is more amorphous and much softer than the isotactic or syndiotactic polymer. Atactic polypropene has weaker intermolecular forces; this hinders crystallization.

🌐 Identifying monomers

You need to be able to identify up to three repeating units in a polymer. For example, a monomer of 2-methylpropene can undergo addition polymerization (figure 9) by cleavage of the double bond. This polymerization produces butyl rubber, a synthetic rubber used during the second world war. The bond breaking is initiated by an acid (H$^+$) catalyst.

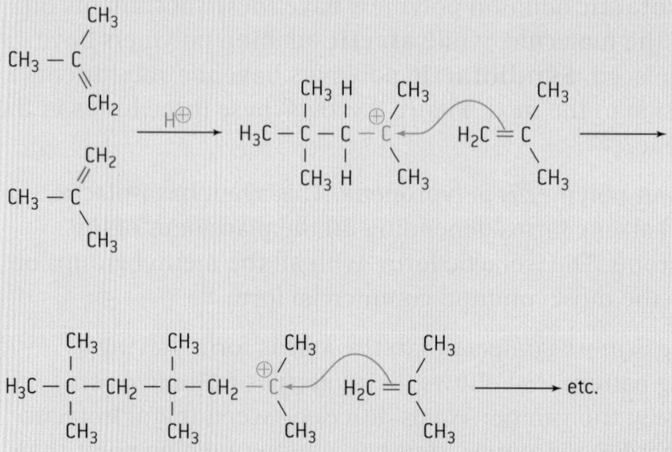

▲ Figure 9 Addition polymerization of 2-methylpropene initiated by an acid catalyst

The repeating unit with alternating methyl and hydrogen substituents is favoured over having two methyl substituents next to each other (figure 10). A lower energy path results if the bulky substituent groups are not on neighbouring atoms in the polymer.

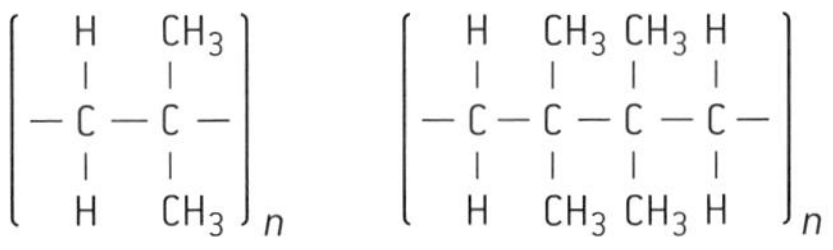

▲ Figure 10 The polymer configuration with alternating substituents (left) is favoured in the addition polymerization of 2-methylpropene

Atom economy of polymerization reactions

In addition polymerization, all the reactant molecules (monomers) end up in the product which provides good atom economy (sub-topic 1.1). The atom economy is distinct from the percentage yield in that it is a measure of the mass of reactant molecules that end up in the desired product:

$$\% \text{ atom economy} = \frac{\text{molar mass of desired product}}{\text{molar mass of all reactants}} \times 100\%$$

For example, producing hydrogen by passing steam over coke may be a highly efficient process if all the reactants are converted to product which can be recovered, but this process does not represent good atom economy:

$$C(s) + H_2O(g) \rightarrow CO(g) + H_2(g)$$

The desired product, hydrogen, has M_r 2.02 while the total M_r of the reactants is $12.01 + 18.02 = 30.03$. The atom economy is therefore $\frac{2.02}{30.03} \times 100\% = 6.7\%$, meaning that 93.3% of the mass of the reactants does not end up being in the desired product.

Atom economy is a measure used in green chemistry, which takes into account not only the efficiency but also the degree of waste produced. Efficient processes with high atom economy are important in sustainable development as they create less waste and use fewer resources. For example, ibuprofen was initially produced in a six-step process with an atom economy of about 40%. Research developed a three-step method which improved the atom economy to 77%. The production of addition polymers represents 100% atom economy as all of the reactant monomer molecules end up in the product.

> **Study tip**
>
> The equation for atom economy is provided in the *Data booklet*, which is available during examinations.

Worked example

Calculate the percentage yield (sub-topic 1.3) and percentage atom economy if 1000 kg of iron(III) oxide, Fe_2O_3 is reduced to 600 kg of iron by carbon monoxide in a blast furnace:

$$Fe_2O_3(s) + 3CO(g) \rightarrow 3CO_2(g) + 2Fe(l)$$

Solution

Theoretical yield = percentage of Fe in Fe_2O_3

$$= \frac{2 \times 55.85}{2 \times 55.85 + 3 \times 16.00} \times 100\% = 69.9\%$$

$$\text{Percentage yield} = \frac{\text{theoretical yield}}{\text{actual yield}}$$

$$= \frac{600 \text{ kg}}{\frac{69.9}{100} \times 1000 \text{ kg}} \times 100\%$$

$$= 85.8\%$$

Atom economy

$$= \frac{2 \times 55.85}{2 \times 55.85 + 3 \times 16.00 + 3 \times 28.01} \times 100\%$$

$$= 45.8\%$$

Questions

1. a) Many of the compounds produced by cracking are used in the manufacture of addition polymers. State the essential structural feature of these compounds and explain its importance. [2]

   b) The polymers often have other substances added to modify their properties. One group of additives are plasticizers. State how plasticizers modify the physical properties of polyvinyl chloride and explain at the molecular level how this is achieved. [2]

   IB, May 2011

2. During the formation of poly(styrene), a volatile hydrocarbon such as pentane is often added. Describe how this affects the properties of the polymer and give **one** use for this product. [2]

   IB, May 2010

3. Addition polymers are extensively used in society. The properties of addition polymers may be modified by the introduction of certain substances.

   a) For two different addition polymers, describe and explain **one** way in which the properties of addition polymers may be modified. [4]

   b) Use high-density poly(ethene) and low density poly(ethene) as examples to explain the difference that branching can make to the properties of a polymer. [3]

   c) Discuss **two** advantages and **two** disadvantages of using poly(ethene). [2]

   IB, May 2010

4. Polyvinyl chloride (PVC) and polyethene are both polymers made from crude oil.

   a) Explain why PVC is less flexible than polyethene. [2]

   b) State how PVC can be made more flexible during its manufacture and explain the increase in flexibility on a molecular level. [2]

   c) PVC can exist in isotactic and atactic forms. Draw the structure of the isotactic form showing a chain of at least six carbon atoms. [1]

   IB, November 2009

5. The manufacture of low density poly(ethene) is carried out at very high pressures and at a temperature of about 500 K. A catalyst (either an organic peroxide or a trace of oxygen) is added to the ethene. Explain how the catalyst reacts and write equations to show the mechanism of the polymerization. [3]

   IB specimen paper, 2008

6. Plastics, such as PVC and melamine resin, are essential to modern society.

   a) PVC is *thermoplastic* whereas melamine resin is *thermosetting*. Explain how differences at a molecular level affect the physical properties of these two types of polymer. [2]

   b) State one other way in which scientists have tried to classify plastics and outline why the classification you have chosen is useful. [2]

   c) After its discovery it took almost a century for PVC to be turned into a useful plastic, when Waldo Semon discovered the effect of adding plasticisers. Explain how these affect the properties of PVC and how they produce this effect. [2]

   d) Justify why, in terms of atom economy, the production of PVC could be considered "green chemistry"? [1]

   e) In spite of the conclusion in D, many consider that the production of PVC is not very environmentally friendly because its decomposition and combustion can lead to pollution. Identify one specific toxic chemical released by the combustion of PVC. [1]

   IB specimen paper, 2013

A.6 Nanotechnology

Understandings

→ Molecular self-assembly is the bottom-up assembly of nanoparticles and can occur by selectively attaching molecules to specific surfaces. Self-assembly can also occur spontaneously in solution.

→ Possible methods of producing nanotubes are arc discharge, chemical vapour deposition (CVD), and high pressure carbon monoxide (HiPCO).

→ Arc discharge involves either vaporizing the surface of one of the carbon electrodes, or discharging an arc through metal electrodes submersed in a hydrocarbon solvent, which forms a small rod-shaped deposit on the anode.

Applications and skills

→ Distinguishing between physical and chemical techniques in manipulating atoms to form molecules.

→ Description of the structure and properties of carbon nanotubes.

→ Explanation of why an inert gas, and not oxygen, is necessary for CVD preparation of carbon nanotubes.

→ Explanation of the production of carbon from hydrocarbon solvents in arc discharge by oxidation at the anode.

→ Deduction of equations for the production of carbon atoms from HiPCO.

→ Discussion of some implications and applications of nanotechnology.

→ Explanation of why nanotubes are strong and good conductors of electricity.

Nature of science

→ Improvements in apparatus – high-power electron microscopes have allowed for the study of positioning of atoms.

→ The need to regard theories as uncertain – the role of trial and error in the development of nanotubes and their associated theories.

→ *"The principles of physics, as far as I can see, do not speak against the possibility of manoeuvering things atom by atom. It is not an attempt to violate any laws; it is something, in principle, that can be done; but in practice, it has not been done because we are too big."*

— Richard Feynman, Nobel Prize winner in Physics.

What is nanotechnology?

"There's plenty of room at the bottom" was the title of a 1959 talk by Richard Feynman proposing the feasibility of **nanotechnology**. Nanotechnology deals with the manipulation and control of atoms, molecules, and objects with dimensions of less than 100 nm (about 1000 atoms or less across). Chemical techniques place atoms in molecules using chemical reactions, whilst physical techniques allow atoms and molecules to be manipulated and positioned to specific requirements.

There are two approaches to nanomanufacturing: top-down or bottom up. The **top-down approach** reduces large pieces of material down to the nanoscale. Optical lithography, for example, uses short wavelengths

$$1\,\text{nm} = 10^{-9}\,\text{m}$$

of light (under 100 nm) in etching, such as in the design of integrated circuits. There is always some waste with the top-down approach as not all the material is used.

The **bottom-up approach** uses **molecular self-assembly** of nanoparticles, in which molecules are selectively attached to specific surfaces. The principles of **bimolecular recognition** and **self-ordering** are used to build up particles in perfect order without any external driving forces. Examples of bimolecular recognition and self-ordering principles in molecular self-assembly include building up DNA strands via complementary base pairing, or other non covalently bonding principles like hydrogen bonding or metal coordination. The two key elements in molecular self-assembly are chemical complements and structural compatibility. Weak non-covalent interactions bind the substances together during the building process. Bottom-up molecular self-assembly produces more homogeneous nanostructures with less defects than results from the top-down process, largely due to the bimolecular recognition involved.

 ## Nanotubes

One form of self-assembled nanoparticles is **nanotubes**, which are a type of fullerene molecule. Fullerene is an allotrope of carbon with atoms arranged into interlinking hexagonal and pentagonal rings. Each carbon is bonded to three rather than four other carbons, resulting in sp^2 hybridized carbons, which confers good electrical conductivity. Also, because all the carbon atoms are covalently bonded rather than held together by intermolecular forces, nanotubes are very strong.

The main cylinder is made only from carbon hexagons, with pentagons needed to close the structure at the ends. Theoretically, a wide range of shapes can be engineered at the molecular level using fullerenes. Single-, double-, or multiple-walled nanotubes made from concentric nanotubes can be formed. Bundles of tubes have high tensile strength as strong covalent bonding extends along the nanotube. The behaviour of electrons depends on the length of the tube; some forms are conductors and others are semiconductors. Such structures have a wide range of technological and medical uses.

Constructing nanotubes

Methods of producing nanotubes include **arc discharge**, **chemical vapour deposition** (CVD), and **high pressure carbon monoxide disproportionation** (HiPCO). Arc discharge was initially used to produce fullerenes, C_{60}, and involves either vaporizing the surface of a carbon electrode or discharging an arc through metal electrodes submersed in a hydrocarbon solvent, forming a small rod-shaped deposit on the anode.

Arc discharge using carbon electrodes

Two carbon rods are placed about 1 mm apart in a container of inert gas (helium or argon) at low pressure. A direct current produces a high-temperature discharge between the two electrodes, vaporizing parts of one carbon anode and forming a small rod-shaped deposit on the other. The anode may be doped with small quantities of a catalytic metal such as cobalt, nickel, yttrium, or molybdenum; in this case single-walled nanotubes are formed. If pure graphite is used, multi-walled nanotubes tend to be formed. Single-walled nanotubes have a diameter of 0.5–7 nm whereas multi-walled nanotubes have concentric tubes with an inner wall diameter of 1.5–15 nm and outer wall diameter of up to 30 nm (figure 1).

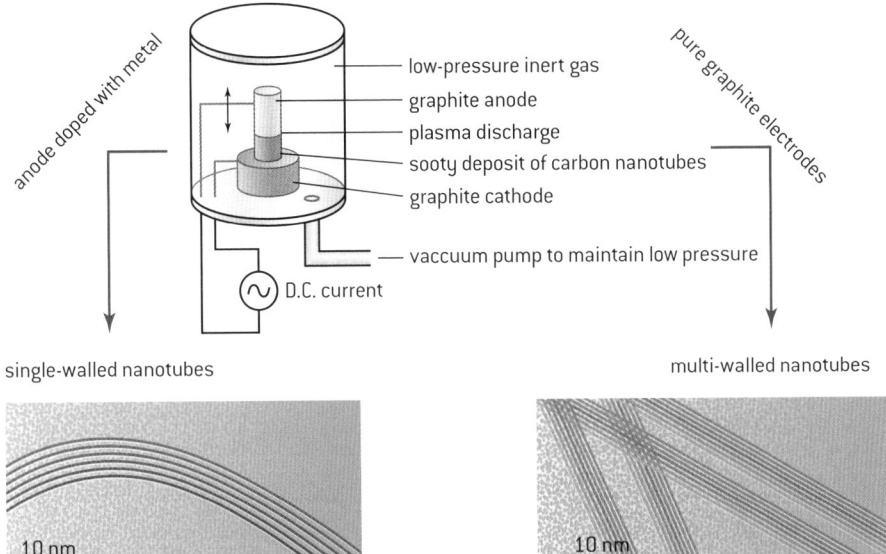

▲ Figure 1 Arc discharge using carbon electrodes produces either single-walled or multi-walled nanotubes

International collaboration in space exploration is growing. Would a carbon nanotube space elevator be feasible? What are the implications of such an advance, and would it be desirable?

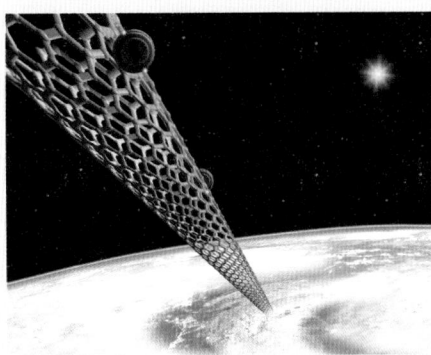

▲ Figure 2 Computer image of a cylindrical fullerene rising from the ground to Earth's orbit, acting as a space elevator. Such an elevator would allow people and materials to ascend and descend to and from space

How do nanotubes grow?

There are several theories about the exact mechanism of growth of nanotubes. For example, the diameter of nanotubes can vary depending on the helium/argon concentrations. The same catalyst can give different results using the arc-discharge method from those obtained using the chemical vapour deposition method. Because theories must accommodate the assumptions and premise of other theories, no universally accepted theory has yet been formulated and trial and error plays a large part in this field of research.

Arc discharge using metal electrodes

Electrodes of a metal such as nickel can be used for discharge in a hydrocarbon solvent (figure 3), for example toluene (methylbenzene, C_7H_8) or cyclohexane (C_6H_{12}). The solvent is the source of carbon atoms for the nanotubes as the hydrocarbon is decomposed by the arc and soot is produced either at the anode (as occurs with toluene) or dispersed throughout the solvent.

Chemical vapour deposition

In CVD gaseous carbon atoms are deposited onto a substrate. This is achieved by the decomposition of a hydrocarbon gas such as methane or ethyne, or carbon monoxide over a transition metal catalyst. The

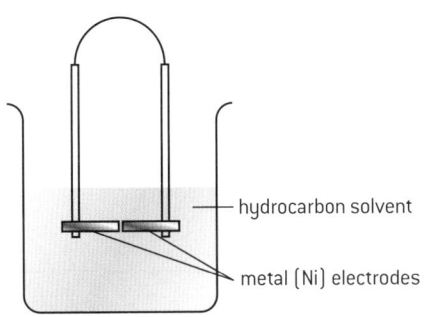

▲ Figure 3 Experimental apparatus for arc discharge using metal electrodes and a hydrocarbon solvent

covalent bonds in the gas are broken by either plasma discharge or heat, cracking the molecule, and the carbon atoms diffuse towards a substrate which is coated with a catalyst. The catalyst is usually iron, nickel, or cobalt and is attached to the substrate by heating or etching.

Once prepared the substrate is heated in an oven to over 600 °C and the hydrocarbon gas is slowly introduced. The gas decomposes and the carbon atoms reform into nanotubes on the substrate. The container must be free of oxygen or any other reactive substances to prevent the formation of carbon dioxide or any other impurities. The carbon atoms move to the substrate by diffusion and form either single-walled or multi-walled nanotubes depending on conditions. Either methane or carbon monoxide is heated to over 900°C to form single-walled nanotubes while ethyne is heated to 600–700 °C for multi-walled nanotubes. Single-walled nanotubes have a higher enthalpy of formation than multi-walled nanotubes (sub-topic 5.1).

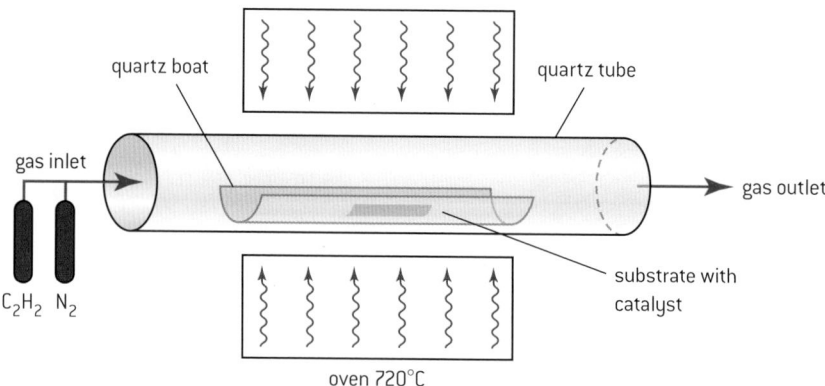

▲ Figure 4 Chemical vapour deposition (CVD)

One method of CVD is high pressure carbon monoxide **disproportionation** (HiPCO). In a disproportionation reaction the same substance is both oxidized and reduced. In HiPCO hot carbon monoxide is continuously supplied at high pressure into the reaction mixture. The catalyst iron pentacarbonyl, $Fe(CO)_5$ is also fed in. The iron pentacarbonyl produces iron nanoparticles that provide a nucleation surface for the reaction. No substrate is needed and the reaction can take place with a continuous feed making it suitable for industrial-scale production. In HiPCO carbon monoxide is reduced to carbon, which forms nanotubes, and is also oxidized to carbon dioxide:

$$2CO(g) \xrightarrow{Fe(CO)_5} C(s) + CO_2(g)$$

As mentioned above, high temperature plasma rather than heating can be used to bring about CVD. A technique known as **laser ablation** uses a laser instead of an arc discharge to vaporize graphite. Either a continuous laser or pulses can be used and again single-, double-, or multi-walled nanotubes can be generated depending on conditions and the catalysts used.

Table 1 summarizes methods used to produce nanotubes.

Technique	Arc discharge	Laser ablation	Chemical vapour depositon (CVD)	High pressure carbon monoxide disproportionation (HiPCO)
Method	Electric plasma discharge vaporizes a graphite electrode, depositing it on the other electrode as single- or multi-walled nanotubes	Laser pulse strikes and vaporizes graphite	Uses heat to crack a gaseous hydrocarbon into carbon atoms which are deposited on a substrate containing an etched-on transition metal catalyst	Carbon atoms produced in the disproportionation reaction from carbon monoxide react with vaporized $Fe(CO)_5$ catalyst to produce nanotubes
Alternative version of the method	In a hydrocarbon solvent using metal electrodes	Continuous wave instead of pulse laser	Plasma discharge instead of heat	Co–Mo catalyst instead of $Fe(CO)_5$
Specific conditions	Inert gas low-pressure atmosphere; ultra-pure graphite rods; $T >$ 3,000 °C; gap between rods 1 mm or less	Inert gas low-pressure atmosphere; gaseous flow; $T \sim$ 1,200 °C; graphite powder or block/rods	Catalyst etched and deposited on substrate; CH_4, C_2H_2 or CO; $T >$ 1,000 °C to crack hydrocarbon or carbon monoxide	High pressure; $T >$ 1,000 °C to crack hydrocarbon (lower with Co–Mo catalyst); temperature affects size of single-walled nanotubes
Yield	About 50% per batch; electrodes replaced each time; about 10 g per day	About 70% per batch before replacing electrode; graphite powder; less than 1 g in a day	About 50%; can produce large quantities (over 1,000 kg per day) due to continuous flow and substrate size	Higher than 95% yield, can run continuously with gas flow, producing about 1 kg per day
Advantages	Mostly defect-free nanotubes	Very high quality single-walled nanotubes engineered to desired specifications with fine control of diameter size	Easiest to scale up to industrial production	Very high yields
Disadvantages	Small tubes with random sizes and directions, difficult to purify	Very expensive	Produces mostly multi-walled nanotubes with many defects; difficult to separate single- from multi-walled nanotubes	Some defects and random production

▲ Table 1 Summary of methods of nanotube production

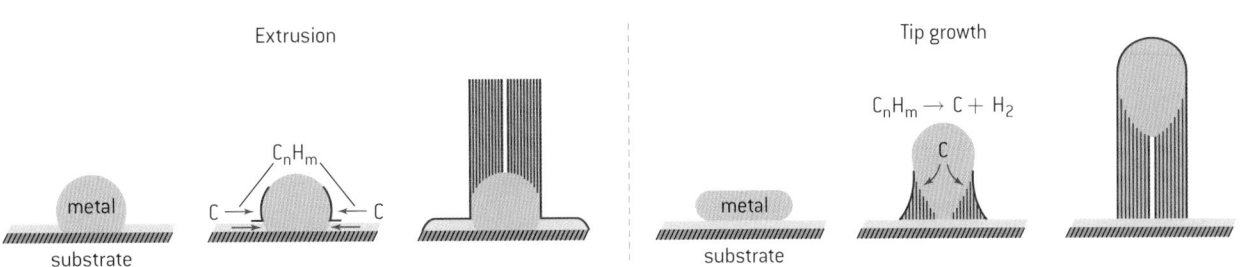

Extrusion   Tip growth

$C_nH_m \rightarrow C + H_2$

▲ Figure 5 Nanotubes form on metal catalyst nanoparticles. Two possible mechanisms are extrusion from the substrate and catalyst or tip growth from the catalyst to substrate

Physical techniques in nanotechnology

Scanning probe techniques

tip is scanned relative to the sample
(or sometimes the sample is scanned)

a feedback mechanism
is used to maintain the
tip at a constant height
above the sample

tip measures some
property of the surface

▲ Figure 6 Scanning probe techniques use various feedback
mechanisms to probe a nanosurface

In the chemical processes of nanotechnology
production just described, carbon is vaporized
or obtained by cracking gaseous hydrocarbons
and fullerenes are allowed to reform into
nanotubes. Physical techniques that manipulate
materials at the molecular level include
scanning probe microscopy techniques which
can be used to probe and manipulate a
molecular surface (figure 6).

Atomic force microscopy

Atomic force microscopy (AFM) is a scanning
probe technique that uses a cantilever with a
crystal tip of radius less than 10 nm made of
microfabricated silicon or silicon nitride, Si_3N_4.
The cantilever is attracted or repelled either by
contact or by interatomic van der Waals' forces. A
laser is reflected off the tip and the reflected beam
gives information about the surface.

AFM is used to measure friction between surfaces.
Hair product manufacturers, for example, use
AFM to study the effect of additives on hair at the
molecular level. AFM can also be used to measure
weak electrical forces on the surface of conductive
or semiconductive nanotubes.

A distinct advantage of AFM is that it can be used
in non-contact mode. The tip oscillates at a regular
harmonic frequency and is brought to within a
few nanometres of the surface. Intermolecular
forces interfere with the oscillations and the
resulting change in oscillations gives a picture of
the surface without contact.

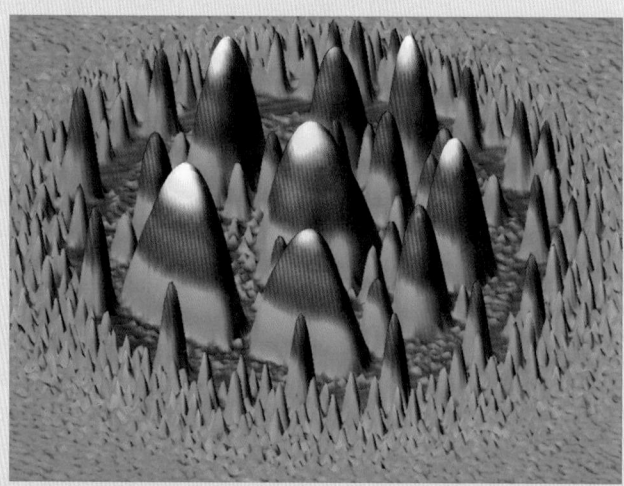

▲ Figure 8 Coloured atomic force micrograph (AFM) of
molecules of yttrium oxide, Y_2O_3 on a thin film of yttrium.
Yttrium compounds are used in superconductors and lasers

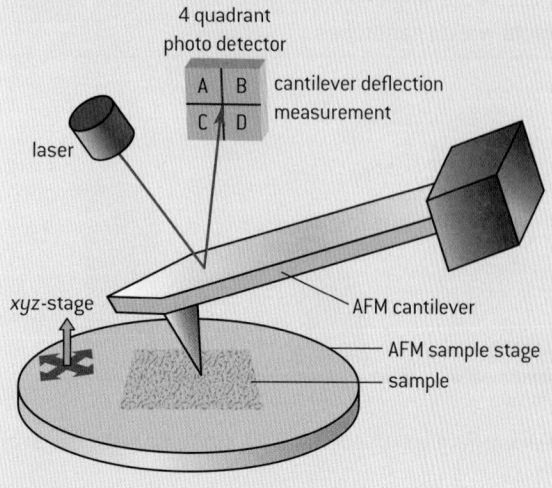

4 quadrant
photo detector

A	B
C	D

cantilever deflection
measurement

laser

xyz-stage

AFM cantilever

AFM sample stage

sample

▲ Figure 7 Atomic force microscopy

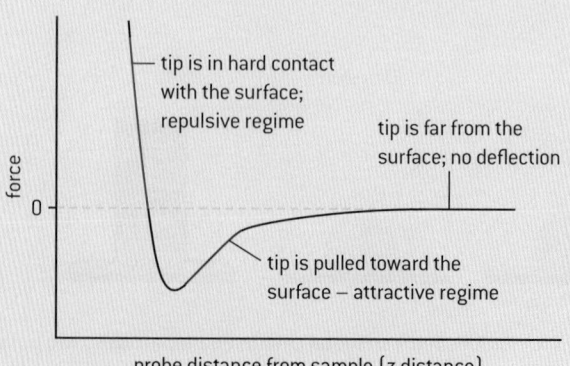

force

0

tip is in hard contact
with the surface;
repulsive regime

tip is far from the
surface; no deflection

tip is pulled toward the
surface – attractive regime

probe distance from sample (z distance)

Scanning tunnelling microscopy

There are many other forms of scanning probe microscopy. **Scanning tunnelling microscopy** (STM) uses a metal tip with a small voltage applied to it to study electrical forces at the surface. An STM image of a sample surface allows surface atoms to be identified. The image is formed by moving a fine point just above the sample surface and electronically recording the height of the point as it scans. The electron clouds surrounding surface nuclei in the point of the STM and the electron clouds in the sample surface overlap as they approach each other. An electric tunnelling current develops which can lead to an exchange of electrons. This rapid change of tunnelling current can be used to produce an image at the atomic level. This technique has applications in data storage and logic gates.

Electron-beam-induced deposition

The technique of **electron-beam-induced deposition** is analogous to a 3-D printer on the nanoscale. An electron beam from a scanning electron microscope is used to direct the synthesis of nanostructures, for example by HiPCO (figure 9).

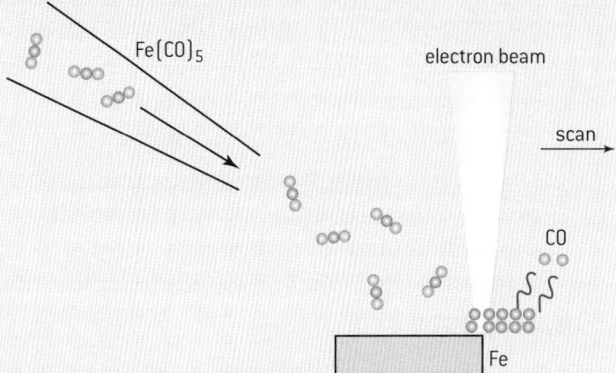

▲ Figure 9 Electron-beam-induced deposition

Virtually any nanostructure shape can be produced quite accurately using this method. Small magnets, superconducting nanowires, and nanogears are all theoretically possible.

Implications and applications of nanotechnology

Nanotechnology not only produces miniaturized products but also uses revolutionary molecular manufacturing processes to make large products from small machines. Nanomanufacturing has the potential to produce life-saving medical applications and new products but also untraceable weapons of mass destruction. It enables the production of cheap, efficient, light, strong structures including electrical and power storage equipment.

Molecular self-assembly does not require assembly lines and factories so once initiated it can become an almost self-sustained process. The technology has the potential to produce exponentially smaller computers which are faster and require less power. New stronger materials at a fraction of the mass are being developed. Because molecular self-assembly works on bimolecular recognition it has many possibilities for advances in medical applications. All these uses have implications to some of today's most pressing problems, such as food shortages, climate change, pollution, clean water, and life-saving applications.

Nanotechnology also brings new problems. There are health risks associated with nanoparticles and their toxicity can vary depending on the size of the particles. How will the human immune system cope with particles on the nanoscale? How can the world control nano-weapons, which are easier and cheaper to build and less detectable than conventional weapons? As new materials and techniques are developed, regulations for their control need to be developed. Materials that are safe on the macro scale may not be safe on a nanoscale and nanoparticle waste products need new disposal methods. How should decisions regarding funding be made, and by whom?

Questions

1 Many recent developments in chemistry have involved making use of devices that operate on a nanoscale.

   a) (i) State the scale at which nanotechnology takes place and outline the importance of working at this scale. [2]

   (ii) State **one** public concern regarding the development of nanotechnology. [1]

   b) One development has been the production of nanotubes. Describe the way in which the arrangement of carbon atoms in the wall and sealed end of a nanotube differ. [2]

   IB, May 2011

2 Exciting developments have taken place in recent years in the area of nanotechnology.

   a) Define the term *nanotechnology*, and state why it is of interest to chemists. [2]

   b) Carbon nanotubes can be used to make *designer catalysts*.

   (i) Describe the structure of carbon nanotubes. [2]

   (ii) State **one** physical property of carbon nanotubes. [1]

   c) Suggest **two** concerns about the use of nanotechnology. [2]

   IB, November 2010

3 Nanotechnology could provide new solutions for developing countries where basic services such as good health care, education, safe drinking water and reliable energy are often lacking. Discuss some of the potential risks associated with developing nanotechnology. [4]

   IB, May 2009

4 Nano-sized *"test-tubes"* with one open end can be formed from carbon structures.

   a) Describe these *"test-tubes"* with reference to the structures of carbon allotropes. [2]

   b) These tubes are believed to be stronger than steel. Explain the strength of these *"test-tubes"* on a molecular level. [1]

   c) Carbon nanotubes can be used as catalysts.

   (i) Suggest **two** reasons why they are effective heterogeneous catalysts. [2]

   (ii) State **one** potential concern associated with the use of carbon nanotubes. [1]

   IB, May 2011

5 Describe the chemical vapour deposition (CVD) method for the production of carbon nanotubes. [2]

   IB, Specimen paper 2013

6 Outline what is meant by *bimolecular recognition*. Explain why it is essential for molecular self-assembly.

7 Explain why allotropes of carbon, graphene and fullerene (used in producing nanotubes) are conductive but diamond is not.

A.7 Environmental impact – plastics

Understandings

→ Plastics do not degrade easily because of their strong covalent bonds.

→ Burning of polyvinyl chloride releases dioxins, HCl gas, and incomplete hydrocarbon combustion products.

→ Dioxins contain unsaturated six-member heterocyclic rings with two oxygen atoms, usually in positions 1 and 4.

→ Chlorinated dioxins are hormone disrupting, leading to cellular and genetic damage.

→ Plastics require more processing to be recycled than other materials.

→ Plastics are recycled based on different resin types.

Applications and skills

→ Deduction of the equation for any given combustion reaction.

→ Discussion of why the recycling of polymers is an energy-intensive process.

→ Discussion of the environmental impact of the use of plastics.

→ Comparison of the structures of polychlorinated biphenyls (PCBs) and dioxins.

→ Discussion of the health concerns of using volatile plasticizers in polymer production.

→ Distinguish possible Resin Identification Codes (RICs) of plastics from an IR spectrum.

 ## Nature of science

→ Risks and problems – scientific research often proceeds with perceived benefits in mind, but the risks and implications also need to be considered.

Challenges of materials science

Green chemistry, also known as sustainable chemistry, is the design of chemical products and processes that reduce or eliminate the use or generation of hazardous substances. Green chemistry applies across the life cycle of a chemical product, including its design, manufacture, and use.

US Environmental Protection Agency

Although materials science has developed countless useful products, it raises challenges associated with the recycling and toxicity of some new materials.

Plastics are polymers composed mainly of carbon and hydrogen. These have strong covalent bonds which are not easily broken so plastics do not decompose readily. Some polymers such as polyvinylchloride (PVC, polychloroethene) also contain chlorine and can release hydrogen chloride, HCl, or dioxins upon combustion. Other environmental concerns associated with plastics include the presence of volatile plasticizers.

The oceans have rotating currents or **gyres**, each with a calm spot at the centre. Here floating plastic garbage collects on such a scale that the raft of plastic waste in the north Pacific gyre is estimated to be the size of Texas. How should nations deal with the international problem of garbage in the oceans which affects the whole ecosystem?

The effect of plastic waste and POPs on wildlife

Large plastic bottles and bags break down to much smaller pieces in the ocean due to the action of the sun and abrasion by the waves. These smaller pieces can be mistaken for prey by marine animals. Over a million sea birds, marine mammals, and turtles are killed each year from ingesting plastic.

Persistent organic pollutants (POPs) such as highly toxic dioxins can enter the food chain, having long-term effects on the health of animals throughout the food chain.

As POPs are passed along the food chain their concentrations increase and can reach very high levels in top predators (sub-topic B.6). This process, known as **biomagnification**, has been largely responsible for the extinction or significant population reduction of many birds of prey and large marine animals across the globe, including in regions far distant from the places where the POPs were released to the environment (figure 1).

Burning is not a viable means of waste disposal for plastics because polymers frequently undergo

incomplete combustion, producing carbon monoxide and fine carbon soot particles. As mentioned above, chlorinated compounds such as PVC can release HCl gas and dioxins on combustion. The combustion reaction of the monomer chlorothene, for example, is given here:

$$CH_2=CHCl + 2\frac{1}{2}O_2 \rightarrow 2CO_2 + H_2O + HCl$$

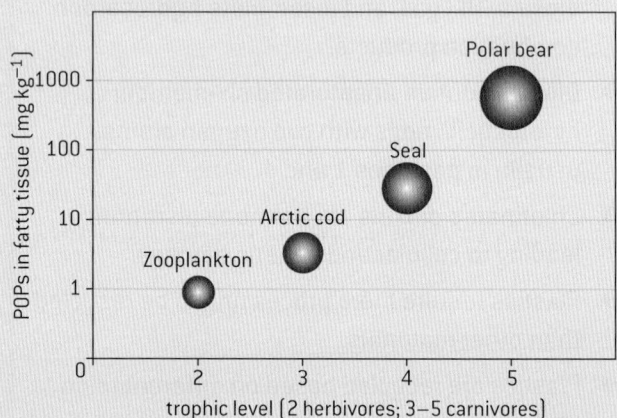

▲ Figure 1 Biomagnification of persistent organic pollutants (POPs) in a food chain

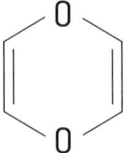

▲ Figure 2 Dioxins contain unsaturated six-membered heterocyclic rings with two oxygen atoms. This is 1,4-dioxin

Dioxins and PCBs

The name "**dioxins**" refers to a class of environmental pollutants that are POPs. Dioxin molecules contain unsaturated six-membered heterocyclic rings with two oxygen atoms, usually in positions 1 and 4 (figure 2). The most toxic member of this class is 2,3,7,8- tetrachlorodibenzodioxin (TCDD).

Certain dioxin-like **polychlorinated biphenyls** (**PCBs**) with similar toxic properties are sometimes included in the term "dioxins". Over 400 types of dioxin-related compounds have been identified though only about 30 of these are considered to have significant toxicity, with TCDD being the most toxic. Section 31 of the *Data booklet* gives the formulas of some representative dioxins (figure 3).

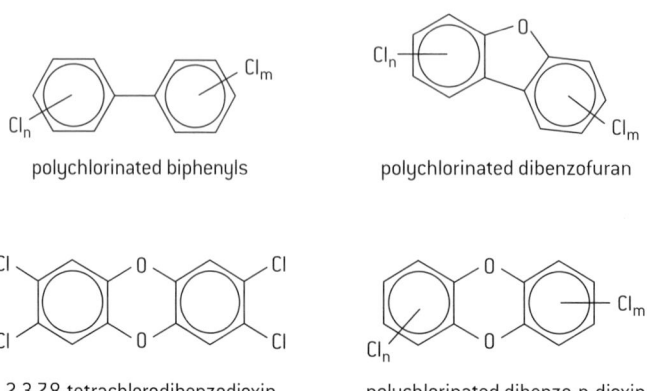

▲ Figure 3 Some examples of dioxin-related compounds

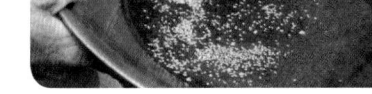

Polychlorinated biphenyls (PCBs) are synthetic organic molecules containing two benzene rings with some or all hydrogen atoms replaced by chlorine. Figure 3 shows that PCBs do not have a dioxin centre ring in their structure, but they have the same toxic effects as dioxins so are considered to be dioxin like. Figure 4 shows an example of a PCB.

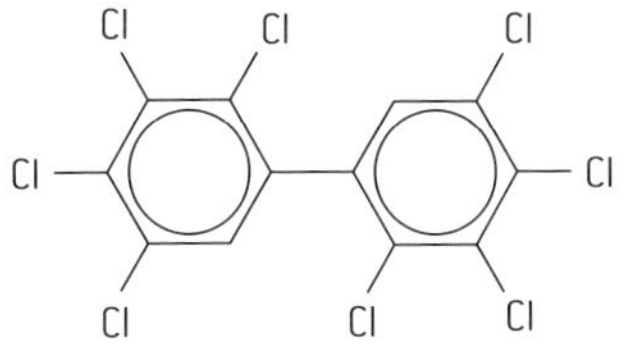

▲ Figure 4 An example of a PCB

Dioxins are highly carcinogenic (cause cancer) and they accumulate in fat tissue so their concentration increases up the food chain. According to the World Health Organization more than 90% of all dioxins found in humans come from food, mostly meat and dairy products or fish and shellfish. The combustion of chlorinated plastics can also lead to the production of dioxins. Dioxin-like substances act on a receptor present in all cells and can cause reproductive and developmental problems. They damage the immune system and interfere with hormone action.

> **Study tip**
>
> The combustion of hydrocarbons releases carbon dioxide and water on complete combustion. Chlorine-containing plastics can release hydrogen chloride gas and dioxins while sulfur-containing compounds can release sulfur dioxide. Given the reactants and products you should be able to balance the equation for any combustion reaction.

🌐 Reducing the environmental effect: Plasticizers and chlorine-free plastics

Plasticizers such as phthalates are readily released into the environment because they are embedded in the plastic only by intermolecular forces rather than by covalent bonds. As plastics age they release plasticizer molecules which can find their way into biological systems by inhalation or ingestion. Whilst not as toxic as dioxins there is some evidence that they disrupt the endocrine system, affecting the release of hormones which leads to cellular and genetic damage. Phthalates are now being replaced with less environmentally harmful plasticizers.

Chlorine-free plastics are also being used as substitutes for PVC. In the event of a house fire such halogen-free plastics are less likely to release dioxins, HCl, or other toxic combustion products.

Recycling of plastics

Recycling rather than disposal is one of the most obvious ways of reducing environmental damage from any material. The **atom economy** increases while the need for the manufacture of new materials is reduced. Recycling of plastics, however, offers significant challenges. Thermosets cannot be melted down and recycled. Heating chlorine-containing polymers carries the risk of releasing dioxins so the method of remoulding needs to take account of this.

> The international symbol for "Recycle, Reuse, Reduce" is a Mobius strip designed in the late 1960s (figure 5). Recycling of plastics can be energy intensive. Should the "Reuse, Reduce" components of the symbol take on a greater emphasis? Has the use of this symbol increased environmental awareness? What factors influence the recognition of symbols?

Section 30 in the *Data booklet* provides a list of RICs.

Plastics are recycled based on their polymer type, identified by a **resin identification code** (**RIC**). This coding system was developed by the Society of the Plastics Industry (SPI) in 1988. Its primary purpose was the efficient identification of plastic polymer types, but it was soon applied to the classification of plastics for recycling. The number on the code gives information about the polymer type rather than its hardness, how frequently it can be recycled, difficulty in recycling, or colour.

Recycling is an energy-intensive process. Plastic bottles for recycling need to be collected and separated from other material. The labels and any other debris are removed and the plastic is washed. It is automatically sorted using near-infrared scanning techniques and then manually checked again as incomplete sorting can lead to difficulties with the process. The separated plastics are then ground into flakes and any remaining water or debris is removed from the flakes by centrifugation. The flakes are then washed and dried again and any further foreign substances such as metals are removed. The recycled end product is not used in food containers as a safety precaution.

Some plastics cannot be recycled into new products. For example, the plastic cases of some cell phones contain bromine which is a fire retardant and these plastics cannot be put through a recycling process. The products and problems associated with recycling are summarized in table 1.

▲ Figure 5 The international symbol for "Recycle, Reuse, Reduce"

Resin Identification Code (RIC)	Properties	Applications	Recycling
1 PETE	**Polyethylene terephthalate** (PET or PETE), also referred to as polyester, has high resistance to chemical solvents and makes a good barrier to gases and liquids. It is clear and the resin can be spun into threads or can make good optical surfaces.	Bottles for water and other drinks including carbonated drinks, dishwashing liquids, and food jars such as for peanut butter. Also used in carpet fibres and microwave trays.	PET bottles can be rinsed and reused, especially as they do not contain phthalate plasticizers. PET softens at about 80 °C so it is mechanically washed and crushed for recycling. Different coloured bottles are separated.
2 HDPE	**High-density polyethylene** (high-density polyethene, HDPE) has high tensile strength and is stiffer than most plastics. It is usually opaque due to its high density and can withstand high temperatures. HDPE is resistant to most solvents and relatively impermeable to gas and moisture. HDPE is widely used.	Bottle caps, bottles for milk, cosmetics, and toiletries such as shampoo, grocery and trash bags, shipping containers, hard hats, buckets, recycling bins. Injection moulding plastics for conduits, wire cable covering, and 3D printing.	HDPE is cleaned, shredded and ground. It can be melted and recycled for non-food plastic applications such as plastic lumber (timber-like mouldings) for decking and garden furniture, mouldings, and bins.

3 PVC	**Polyvinyl chloride** (polychloroethene, PVC) is resistant to grease and chemicals and can be used to produce a variety of shapes and strengths due to the addition of plasticizers. Very stable to corrosion and can be made flexible or stiff.	Both rigid and flexible applications including gaskets, gloves, pipes, window frames, construction materials, credit cards, clothing, and sporting equipment. Early uses included "vinyl" records and plastic shower curtains.	PVC contains chlorine and plasticizers so should not be melted or softened by heat. It can be reused for a similar application, or formed into smaller items such as plastic ties and binders. Very difficult plastic to recycle: can contaminate batches of recycled PET or HDPE.
4 LDPE	**Low-density polyethylene** (low-density polyethene, LDPE) is tough, flexible, and transparent. Good barrier to moisture.	Cling wrap and stretch films, coatings inside milk cartons and hot and cold beverage cups, flexible container lids. Injection mouldings, adhesives, and sealants.	LDPE is melted and turned into plastic sheets which are then manufactured into other goods such as envelopes, bin liners, tiles, films and sheets, carpets, and clothing. LDPE is not recycled into food containers.
5 PP	**Polypropylene** (polypropene, PP) is a thermoplastic polymer. It is strong, inert, and resistant to acids and bases. PP provides a good barrier to moisture and oils. Its properties vary depending on whether its structure is isotactic or atactic.	Containers for yoghurt, medicines, take-away meals, microwave containers, bottle tops and closures for condiments. Also flexible chairs, hinges, coat hangers, toilet seats, and fishing nets. Can form fibres as well as having some electrical applications.	PP is reused but not frequently recycled because of the need for accurate sorting to be successful. Cleaned PP can be melted and remanufactured into various products. Recycled PP products are often made by mixing virgin and recycled PP. Recycling alters the structure of PP so it can only be recycled a limited number of times.
6 PS	**Polystyrene** can be rigid or foamed, both forms showing a fair degree of rigidity.	Styrofoam containers, protective foam packaging, egg cartons, plastic cutlery. Expanded and extruded polystyrene have different uses.	Polystyrene is resistant to decay and is a major contributor to plastic waste in the north Pacific gyre. It can be recycled as it is easily compressed and reblown; it can then be used in packaging. Polystyrene can be converted back to the styrene monomer in a continuous process, rather than melting and remoulding, but this process is energy intensive so it is more often compressed and reformed.
7 OTHER	This code is used if the resin is not one of the six types above or if it is a mixture of the resin types. Polycarbonate is one such polymer under this code.	Depending on type; eg large (20 litre) water bottles.	Dependent on resin type and not usually commercially recycled.

▲ Table 1 Plastic resin identification codes

Sorting plastics

While most plastics can be recycled the main challenge in this process is sorting. One bottle of PVC material can pollute up to 100000 HDPE bottles if not separated, resulting in being melted with them. Sorting by hand is cost prohibitive in many cases. Scanning plastic bottles using infrared (IR) or near infrared spectroscopy can identify the bonds in the molecules quickly. PVC, for example, will show the characteristic C−Cl bond (wavenumber 600-800 cm^{-1}) while the aromatic C−C, C=C bond in polystyrene will give a different absorption wavenumber (1500cm^{-1}) from an alkene C=C bond (1650 cm^{-1}). Reflection can be used to distinguish HDPE from LDPE.

> Section 26 of the *Data booklet* gives information on IR wavelength absorption.

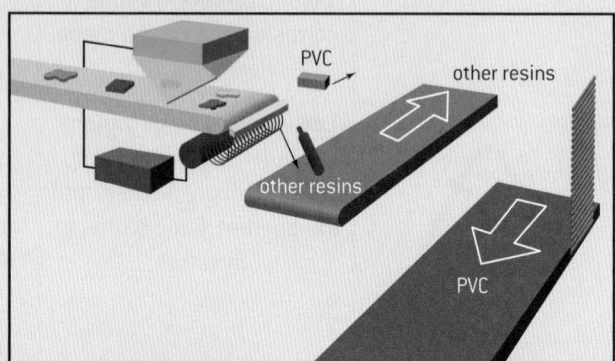

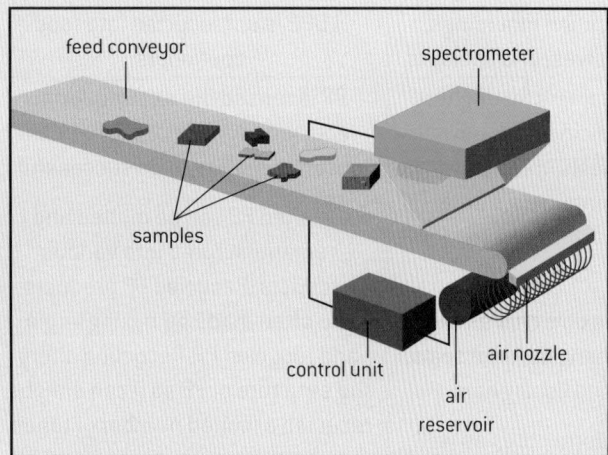

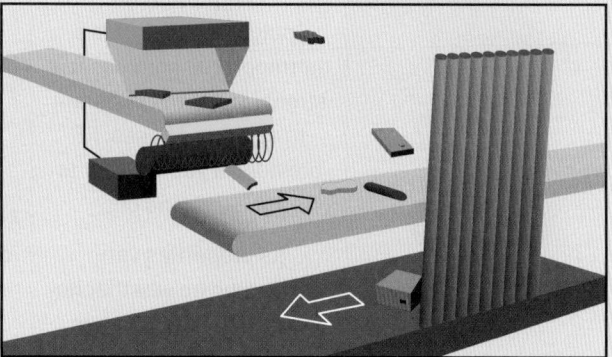

▲ Figure 6 Detecting the C−Cl bond in PVC by IR spectroscopy allows this plastic to be separated from other types

Questions

1 Scientific research often proceeds with perceived benefits in mind, such as the many uses of PVC, but the risks and implications also need to be considered.

   a) Discuss, in terms of atom economy, bond strength, and combustion products whether the production of PVC could be considered "green chemistry"?

   b) Using sections 26 and 30 from the *Data booklet*, identify the structural features of peaks A and B in figure 7 and give the resin identification code (RIC) of the plastic in question. Explain your choice.

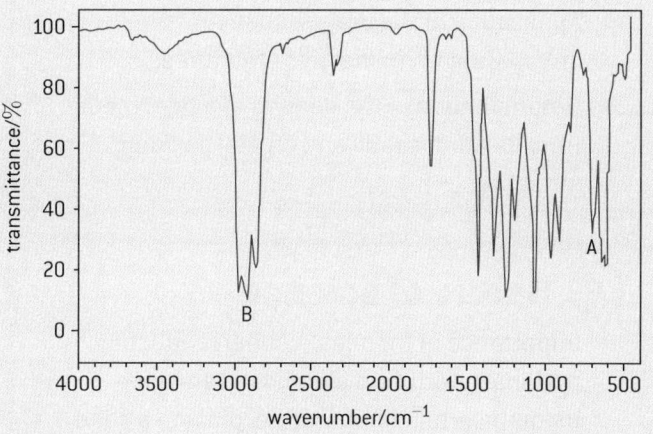

▲ Figure 7

2 Outline why is PCB considered "dioxin like", but not a dioxin.

3 Atom economy is one of the key aspects of green chemistry.

   a) Define the meaning of "atom economy".

   b) Calculate the percentage atom economy of the following reaction if the target product is *N*-methylphenylamine, $C_6H_5NHCH_3$:

   $$C_6H_5NH_2 + (CH_3O)_2CO \rightarrow C_6H_5NHCH_3 + CH_3OH + CO_2$$

   c) Dimethyl carbonate can be synthesized as follows:

   $$4CH_3OH + 2CO + O_2 \rightarrow 2(CH_3O)_2CO + 2H_2O$$

   Suggest how the amounts of waste produced in the synthesis of *N*-methylphenylamine can be further reduced.

4 LDPE has a specific gravity of 0.92 g cm⁻³ and HDPE has a density of 0.95 g cm⁻³. Suggest a reason why flotation is not a good method of separating these two plastics.

5 Many plastic materials are disposed of by combustion. State **two** disadvantages of disposing of polyvinyl chloride in this way.

6 Identify **two** difficulties and **two** advantages in recycling plastics.

515

A.8 Superconducting metals and X-ray crystallography (AHL)

Understandings

→ Superconductors are materials that offer no resistance to electric currents below a critical temperature.

→ The Meissner effect is the ability of a superconductor to create a mirror image magnetic field of an external field, thus expelling it.

→ Resistance in metallic conductors is caused by collisions between electrons and positive ions of the lattice.

→ The Bardeen–Cooper–Schrieffer (BCS) theory explains that below the critical temperature electrons in superconductors form Cooper pairs which move freely through the superconductor.

→ Type 1 superconductors have sharp transitions to superconductivity whereas type 2 superconductors have more gradual transitions.

→ X-ray diffraction can be used to analyse structures of metallic and ionic compounds.

→ Crystal lattices contain simple repeating unit cells.

→ Atoms on faces and edges of unit cells are shared.

→ The number of nearest neighbours of an atom/ion is its coordination number.

Applications and skills

→ Analysis of resistance versus temperature data for type 1 and type 2 superconductors.

→ Explanation of superconductivity in terms of Cooper pairs moving through a positive ion lattice.

→ Deduction or construction of unit cell structures from crystal structure information.

→ Application of the Bragg equation, $n\lambda = 2d\sin\theta$, in metallic structures.

→ Determination of the density of a pure metal from its atomic radii and crystal packing structure

Nature of science

→ Importance of theories – superconducting materials, with zero electrical resistance below a certain temperature, provide a good example of theories needing to be modified to fit new data. It is important to understand the basic scientific principles behind modern instruments.

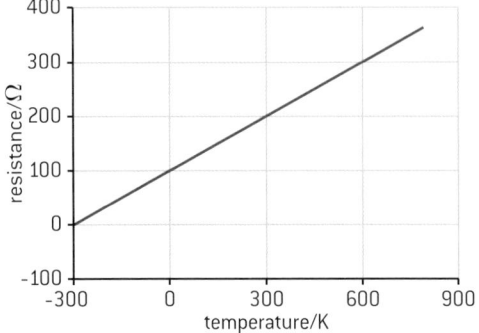

▲ Figure 1 Resistance increases linearly with temperature for many conducting materials

Superconducting materials

Metals are good conductors of electricity because the metallic structure contains electrons that are free to move. As the thermal energy in metals increases, atoms in the lattice vibrate more and there are more collisions between electrons and ions. Some kinetic energy is converted to heat with each collision. It is these collisions that are the cause of electrical resistance in metals, the resistance increasing with temperature (figure 1). By decreasing the temperature there are fewer collisions, the electrons move in a more direct path, and the resistance is reduced: the conductance of the material increases.

Superconductors are materials that offer no resistance to electric current below a critical temperature. At very low temperatures many materials

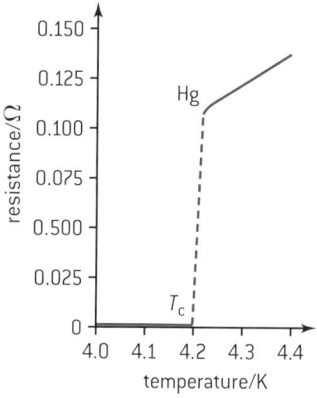

can exhibit this property. For some materials, at low temperatures energy transfer becomes quantized rather than continuous – energy is exchanged in discrete bundles that have a minimum size. If that minimum size is not achieved then transfer does not occur, there is no loss of kinetic energy, and the material becomes a superconductor at this critical temperature (figure 2).

In 1933 Walther Meissner and Robert Ochenfeld found that superconducting materials will repel a magnetic field. This is similar to perfect diamagnetism (sub-topic A.2) in which all external fields are repelled; this is what occurs in a superconductor. The **Meissner effect** (figure 3) is the ability of a superconductor to create a mirror image of an external magnetic field, thus excluding it. When a magnet is brought near the surface of a superconductor, the superconductor responds by creating a magnetic field that is the exact mirror image of the magnet's field. The superconductor behaves as an identical copy of the magnet with like poles facing each other. When the magnet is removed from the superconductor the magnetic field disappears.

▲ Figure 2 Superconductivity was first observed in mercury, Hg in 1911. The mercury had to be near 4 K, the critical temperature for this substance, before the quantized effects of energy transfer were observed

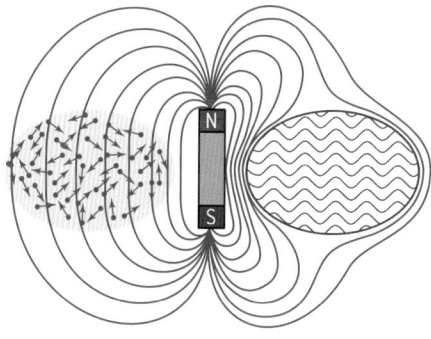

▲ Figure 3 The Meissner effect. An ordinary conductor (left) shows random electron movement and allows magnetic field penetration, whereas a superconductor (right) excludes any magnetic field penetration, creating a mirror-image magnetic field of any magnet brought near

▲ Figure 4 Demonstration of magnetic levitation as a result of the Meissner effect. A high-temperature superconductor, yttrium–barium– copper oxide creates a mirror-image magnetic field of a small, cylindrical magnet. The magnet is floating freely above a nitrogen-cooled, cylindrical specimen of a superconducting ceramic. The glowing vapour is liquid nitrogen, which maintains the ceramic within its superconducting temperature range

 A landmark discovery

Understanding how matter behaves at low temperatures was a landmark in scientific research. Superconductivity at room temperature may need different explanations from cold-temperature superconductivity and theories explaining this phenomenon are constantly evolving.

Type 1 and type 2 superconductors

We have seen that superconductivity is limited by the critical temperature, T_c, for the material. Above T_c the superconducting properties and Meissner effect are no longer exhibited.

Research into superconductors has shown that superconducting properties can also be disrupted by sufficiently high magnetic fields even if the temperature is below T_c. The magnetic field, B, also has a critical value B_c. Any field strength larger than B_c will cause the material to revert from superconducting to a normal conduction band for that material. The value of B_c increases slightly as the temperature is lowered below T_c.

In the search for room-temperature superconductors, the superconducting properties of various alloys and ceramics as well as elements were examined. It was observed that as the magnetic field strength is increased, materials behave in one of two ways.

- A **type 1 superconductor** demonstrates a sharp transition from superconducting (showing the Meissner effect of expelling magnetic fields) to normal behaviour (magnetic fields again penetrate the material and resistance returns to normal).

- A **type 2 superconductor** displays a range of properties with a gradual transition. It shows a superconducting band when the temperature is below T_c and when any external magnetic field B, is also at a minimum. Above B_c but below T_c the material exhibits zero resistance but not perfect diamagnetism (it does not show the Meissner effect) – some of the external magnetic field can penetrate the material in a type of vortex. As long as the vortices remain in one location the material still superconducts

(shows zero resistance); however if these magnetic vortices move then losses in conductivity begin to occur.

In both types of material the current must remain small because moving electrons create a magnetic field. Most metallic elements that can superconduct below T_c are type 1 superconductors whereas alloys and metal oxide ceramics are largely type 2 superconductors. Type 2 superconductors have a higher critical temperature and can therefore act as superconductors at higher temperatures.

Bardeen–Cooper–Schrieffer (BCS) theory

One of the first theories to explain how superconductivity works on the molecular level was developed by John Bardeen, Leon Cooper,

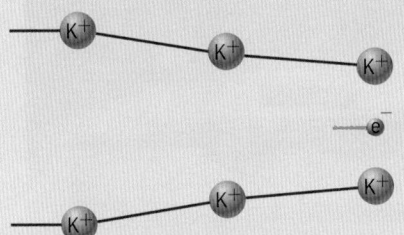

superconducting state

As a negatively charged electron passes between the metal's positively charged atoms in the lattice, the atoms are attracted inward. This distortion of the lattice creates a region of enhanced positive charge which attracts another electron to the area

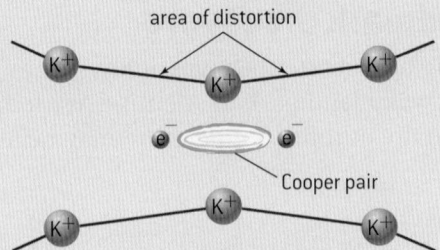

superconducting state

The two electrons, called Cooper pairs, become locked together and will travel through the lattice.

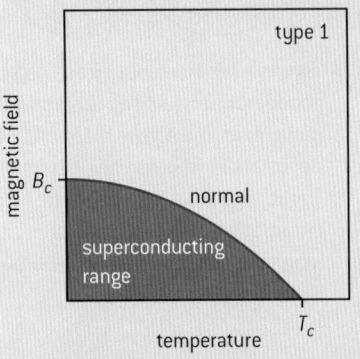

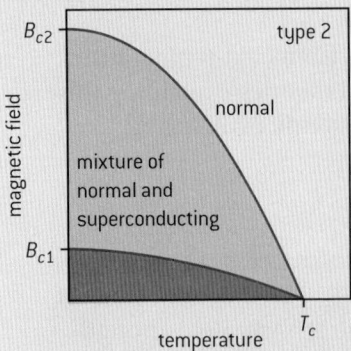

▲ Figure 5 Type 1 superconductors exhibit a sharp transition from superconducting to normal behaviour above a critical temperature T_c and applied external magnetic field B_c, whereas type 2 superconductors show a range of properties below T_c and above B_c

▲ Figure 6 Lattice distortion occurs in a wave-like manner as a negative electron distorts the lattice. This enhanced region of positive charge attracts a second electron and the two pair up, forming a Cooper pair which will travel through the lattice together. Cooper pairs form and reform; they are responsible for superconductivity in type 1 superconductors

and Robert Schrieffer. The **Bardeen–Cooper–Schrieffer** (**BCS**) theory explains that below the critical temperature electrons in superconductors form Cooper pairs which move freely through the superconductor.

At low temperatures the positive ions in the lattice are attracted to a passing electron, distorting the lattice slightly. A second electron is attracted to this slight positive deformation and a coupling of these two electrons occurs. Such electron pairs, called **Cooper pairs**, are not paired by the Pauli exclusion principle (sub-topic 2.2) and they behave differently from single electrons (figure 6). In a Cooper pair any momentum that might be dissipated in a collision between a single electron and the lattice is gained by the second electron. Any energy gained by the lattice from the first electron propagates along the lattice in a wave-like motion called a **phonon**. The phonon is transferred to the second electron, and because phonons are quantized it transfers its entire bundle of energy. Because there is no loss of energy there is no resistance. It is as if the atoms of the lattice oscillate, creating slight positive and negative regions which push and pull the Cooper pair along. If the material is not cold enough the vibrational energy of the lattice is too great for phonon energy transfer, which is why superconductivity can only occur below a critical temperature.

▲ Figure 7 A funicular, like this one in Lisbon, has two cars operating as a pair. As one car goes down it gives some of its energy to the other car pulling it up. Less work needs to be done by the motor as gravitational potential energy is transferred between them and the two cars form a pair with zero total momentum. In early funiculars water was placed in the top car and emptied in the bottom one so that even less work needed to be done by the motor. Cooper pairs operate at zero total momentum with the lattice absorbing and re-emitting phonons to the Cooper pair, allowing the pair to travel through the lattice unimpeded

Applications of superconductors

Ceramic materials, which are normally insulators, have become some of the best high-temperature superconductors ($T > 138$ K). Cuprate superconductors have blocks of alternating planes of atoms; for example, $TlBa_2Ca_2Cu_3O_9$ has conducting CuO_2 layers sandwiched between heavier atom layers of BaO, TlO, and Ca. Experimenting with these layers in type 2 superconductors has raised the critical temperature, with $(Tl_5Pb_2)Ba_2Mg_2Cu_9O_{17}$ demonstrating superconductivity properties above room temperature.

Superconductors have been used to detect small magnetic fields. Superconducting quantum interference devices (SQUIDs) can detect small changes in the tiny electromagnetic fields created by brain activity and are used in neural studies. SQUIDs are also used in submarines detecting undersea mines. Superconducting magnets are used in instruments such as magnetic resonance imaging (MRI) and nuclear magnetic resonance (NMR) machines as well as particle accelerators.

Superconductors could theoretically aid electricity production, but a major challenge is that type 2 superconductors that can operate at higher temperatures are ceramic, so not suitable for making into wires and electrical components. Their material properties (sub-topic A.1) as well as their superconducting properties need to be considered.

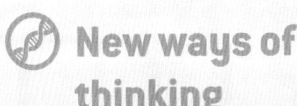

New ways of thinking

BCS theory can explain type 1 superconductivity but cannot explain the transition state of type 2 superconductors or certain high temperature superconductors termed "**strange metals**". New ways of thinking, and perhaps even a paradigm shift in perceiving how matter behaves, may result from research into these materials.

Unit cells

X-ray crystallography (sub-topic 21.1) enables analysis of the structures of crystalline substances. Type 2 superconductors have large complex crystal structures and knowledge of the arrangement of the atoms (ions) in their crystals can help explain their behaviour.

Crystal lattices can be viewed as simple repeating **unit cells**, with atoms at the corners, faces, and edges of each cell shared with neighbouring cells. For example, sodium chloride, NaCl has a cubic crystalline structure in which each unit cell is a simple cube and each corner atom is surrounded by six others: each Na^+ ion is attracted to six neighbouring Cl^- ions and each Cl^- ion is attracted to six neighbouring Na^+ ions. The number of nearest neighbours for an atom in a lattice is its **coordination number**, and for a simple cubic structure such as NaCl the coordination number is 6.

The unit cell is the simplest repeating pattern in a crystal. Different crystals form unit cells of many different shapes, including orthorhombic, hexagonal, and rhombohedral to name a few. However, in this topic we will consider only pure metals forming simple cubic, body centred cubic (BCC), and face centred cubic (FCC) unit cells (figure 8).

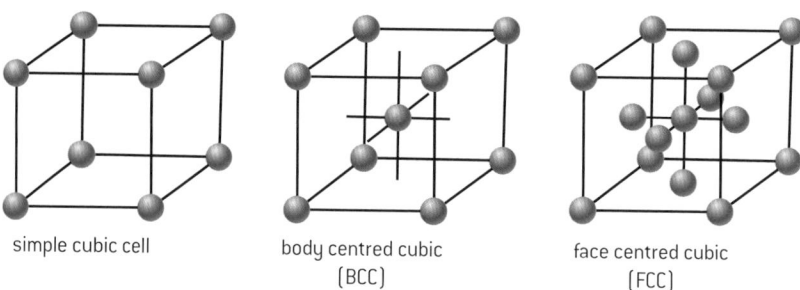

simple cubic cell   body centred cubic   face centred cubic
                                 (BCC)   (FCC)

▲ Figure 8 Cubic structures have all sides equal; length = width = height

A simple cubic cell has eight atoms, one at each corner of the cell. A BCC cell has an additional atom in the centre of the cell making nine in total, whereas a FCC cell has the eight corner atoms plus an additional atom at the centre of each face of the cube, making 14 in total.

In a simple cubic cell structure each atom forms the corner of not just one but eight cells. As mentioned above this structure has a coordination number 6, as each atom is in close contact with six others, but the unit cell itself is only equivalent to one atom (figure 9):

$$8 \text{ corners} \times \frac{1}{8} \text{ atom per corner} = 1 \text{ atom per unit cell}$$

Like the simple cubic cell, the BCC cell has eight corner atoms each shared between eight unit cells, but the centre atom is not shared with any neighbouring cells. This cell is equivalent to 2 atoms:

$$\left(8 \text{ corners} \times \frac{1}{8} \text{ atom per corner}\right) + 1 \text{ central atom} = 2 \text{ atoms per unit cell}$$

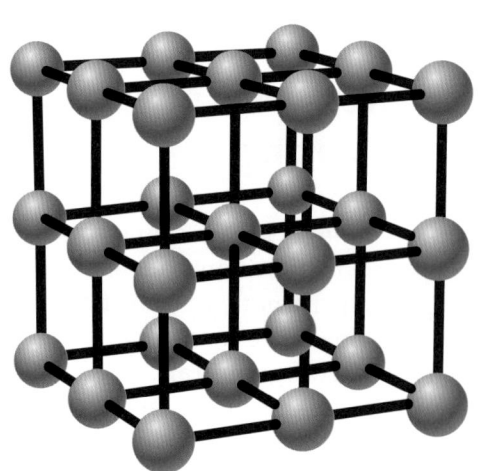

▲ Figure 9 In this model of the simple cubic crystal structure the atom in the centre forms a corner atom for eight different unit cells and has a co-ordination number of 6 as it is equally attracted to its 6 nearest neighbours

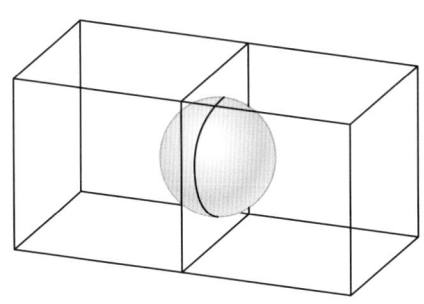

An FCC cell has eight corner atoms each shared between eight unit cells plus an atom at the centre of each cube face. Each face of a cell is shared with the neighbouring cell so each of the six face atoms represents half an atom per cell (figure 10).

The 14 atoms contributing to an FCC cell make up a cell representing 4 atoms:

$$\left(8 \text{ corners} \times \frac{1}{8} \text{ atom per corner}\right) + \left(6 \text{ faces} \times \frac{1}{2} \text{ atom per face}\right) =$$
4 atoms per unit cell

▲ Figure 10 The atom at the face of a cube is shared between two unit cells, so represents half an atom per cell

In addition to the three structures discussed above with atoms on a face or a corner of a cell, atoms can also lie on the edge of a cell. An edge atom is shared by four cells, as shown in figure 11.

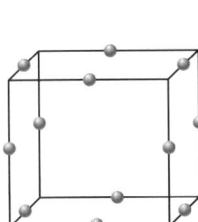 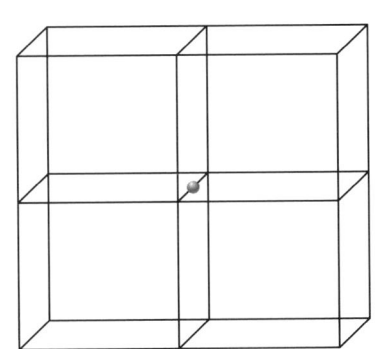

▲ Figure 11 An edge atom is shared by four unit cells. The representative number of atoms per unit cell in the diagram on the left is 3: 12 edge atoms $\times \frac{1}{4}$ atom per edge for each

Figure 12 summarizes the atom contributions in simple cubic, BCC, and FCC unit cells.

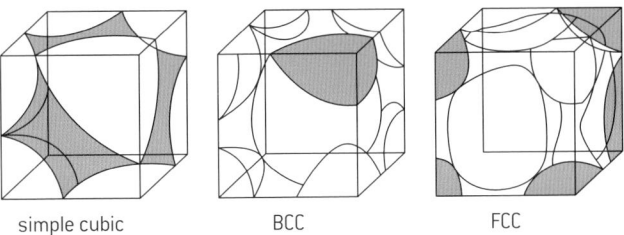

simple cubic   BCC   FCC

▲ Figure 12 Simple cubic structures represent 1 atom per unit cell, BCC represents 2 atoms per unit cell and FCC 4 per unit cell

The coordination number for a simple cubic unit cell is 6, as explained previously; for a BCC unit cell the coordination number is 8 as the centre atom is in contact with eight other atoms (see figure 12).

An FCC unit cell has a coordination number of 12: each "face" atom is in contact with four corner atoms. It is also in contact with four face atoms from each of the two cells it is shared between.

Some unit cell structures have a closer packing structure than others. The simple cubic structure with only 1/8 atom at each corner contains more open space than the BCC cell in which the space is filled with an additional atom. The FCC structure has the closest packing and metals with this type of unit cell are more dense than the other two. Table 1 summarizes the simple cubic, BCC, and FCC crystal structures.

Unit cell	Number of atoms involved	Number of atoms per unit cell	Coordination number	Percentage of cell volume occupied by atoms
simple cubic	8; 1 at each corner	1	6	52%
body centred cubic (BCC)	9; 1 at each corner and a central atom	2	8	68%
face centred cubic (FCC)	14; 1 at each corner and 1 on each face	4	12	74%

▲ Table 1 Summary of the structure and properties of different metallic crystal structures

 ## Worked example

The length of a cubic crystal edge can be determined by X-ray diffraction, and this can be used to determine the type of packing structure. In lithium, for example, the side length of the unit cell is 0.351 nm. The density of lithium is 0.535 g cm^{-3}. Determine the packing structure for lithium.

Solution

- volume of cube = length × width × height = $(0.351 \times 10^{-9})^3$ m^3 = 4.32×10^{-29} m^3 = 4.32×10^{-23} cm^3

- mass of lithium atom = 6.94 g·mol^{-1}/6.02 × 10^{23} mol^{-1} = 1.15×10^{-23} g

- density assuming simple cubic structure (1 atom per unit cell) = 1.15×10^{-23} g/4.32×10^{-23}cm^3 = 0.266 g cm^{-3}

- density assuming BCC structure (2 atoms per unit cell) = $2(1.15 \times 10^{-23}$ g)/4.32×10^{-23} cm^3 = 0.532 g cm^{-3}

- density assuming FCC structure (4 atoms per unit cell) = $4(1.15 \times 10^{-23}$ g)/4.32×10^{-23} cm^3 = 1.06 g cm^{-3}

- The density of lithium at 0.535 g cm^{-3} shows that lithium crystallizes in a BCC structure.

Quick question

Nickel has a density of 8.91 g cm^{-3} and X-ray diffraction shows that the unit cell edge length is 0.3524 nm. Determine the packing structure for Ni.

X-ray crystallography

About 95% of all solids are crystalline. Metals form mainly cubic, BCC, FCC, and hexagonal close packing structures with a regularly repeating pattern. By reflecting X-rays of known wavelength off different layers of the crystal the distance between the layers and hence the unit cell edge length can be determined.

X-rays incident on a crystal are scattered in all directions. Either the wavelength of X-rays used or the angle of incidence of the X-rays can be adjusted until constructive interference occurs, when two light waves one layer apart bounce off the crystal in the same phase. This can then be used to determine the distance between these two layers.

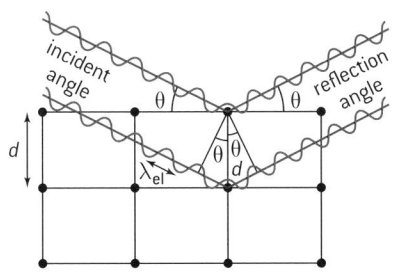
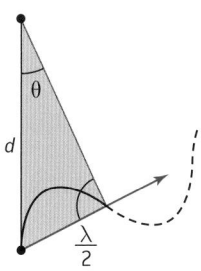

▲ Figure 13 When light waves are reflected from two layers in the same phase, constructive interference occurs and a bright line appears as the two reflected waves reinforce each other. The two light waves are an integral number of wavelengths, $n\lambda$, apart when constructive interference occurs

If the waves reflected from the upper and lower layer are in phase, the wave travelling to the lower layer must have travelled an integer number of one-half wavelengths further to reach the lower layer and the same integer number of half-wavelengths to return to the top layer. Because crystals are arranged in a regular repeating pattern, repeated constructive interference between the layers reinforces the beam to a level where it is detectable.

The distance between the layers in the crystal can be determined by the **Bragg equation:**

$$n\lambda = 2d\sin\theta$$

n is an integer, representing the number of wavelengths difference between the two reflected X-rays; in a first order diffraction pattern $n = 1$ wave, in a second order diffraction pattern $n = 2$ waves, etc.

λ is the wavelength of the X-rays that gave the diffraction pattern

d is the distance between two layers in the crystal

θ is the angle of incident radiation to the crystal.

> **Study tip**
>
> The Bragg equation is provided in section 1 of the *Data booklet*, which will be available in the examination.
>
> Note that d and λ must have the same units, eg nm, pm.

 Worked example

Tantalum, Ta is a type 1 superconductor. When X-rays of wavelength 154 pm are directed at a crystal of Ta the first order diffraction pattern is observed at 13.49°. Calculate the separation of the layers of atoms in the crystal and the density of Ta in g cm^{-3} given that it forms a BCC crystal.

Solution

$n\lambda = 2d\sin\theta$

$d = n\lambda/2\sin\theta$

$= 154$ pm/2sin (13.49°) $= 330$ pm

length of unit cell $= 330$ pm $= 330 \times 10^{-12}$ m $= 330 \times 10^{-10}$ cm

volume of unit cell $= (330 \times 10^{-10}$ cm$)^3 = 3.59 \times 10^{-23}$ cm^3

mass of one Ta atom $= 180.95$ g mol$^{-1}/6.02 \times 10^{23}$ mol$^{-1} = 3.01 \times 10^{-22}$ g

BCC structure contains 2 atoms per unit cell so:

density $= 2(3.01 \times 10^{-22}$ g$)/3.59 \times 10^{-23}$ cm$^3 = 16.8$ g cm^{-3}

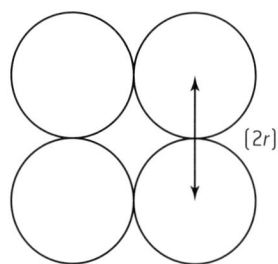

▲ Figure 14 For a simple cubic structure the atomic radius is given by 2r

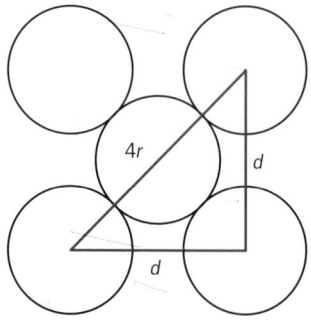

▲ Figure 15 For the FCC structure the atomic radius is given by $(4r)^2 = d^2 + d^2$

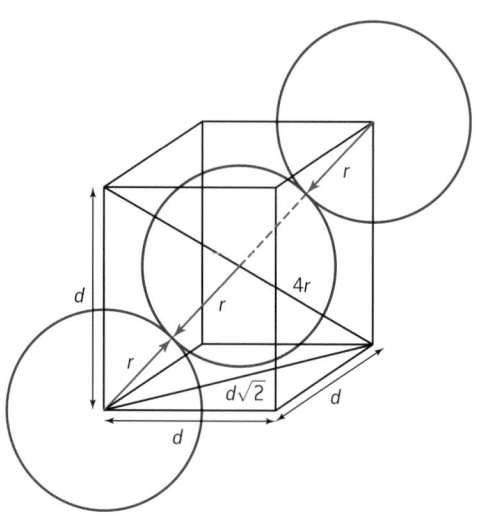

 Figure 16 For the BCC structure the atomic radius is given by $(4r)^2 = 3d^2$

Finding the atomic radius from X-ray crystallography data

Atomic radii can be determined from the packing structure and the distance between atom layers as determined by X-ray crystallography. A simple cubic cell has atoms touching as shown in figure 14. The length of a side of the cube, d, as determined by X-ray crystallography is therefore equal to the diameter of the atom, hence $r = \frac{d}{2}$.

For an FCC structure the atoms touch along a diagonal, but not along the edge (figure 15). The diagonal represents 4 atomic radii, so using Pythagorean theorem:

$$(4r)^2 = d^2 + d^2 \text{ or } (4r)^2 = 2d^2$$

$$\text{so } r = \frac{\sqrt{2}d}{4}$$

For a BCC structure the atoms touch along the diagonal of the cube rather than the diagonal of a face (figure 16). Pythagorean theorem says that:

$$(\text{diagonal of cube body})^2 = \text{length}^2 + \text{width}^2 + \text{height}^2$$

The cube diagonal is $4r$ (the diameter of the central atom plus the atomic radii of each corner atom), so since length = width = height = d:

$$(4r)^2 = 3d^2$$

$$r = \frac{\sqrt{3}d}{4}$$

Table 2 summarizes these results.

Unit cell	Atomic radius
simple cubic	$\dfrac{d}{2}$
body centred cubic (BCC)	$\dfrac{\sqrt{3}d}{4}$
face centred cubic (FCC)	$\dfrac{\sqrt{2}d}{4}$

▲ Table 2 Atomic radius in terms of the length of the unit cell for different metallic crystal structures

Covalent and atomic radii

The atomic radius calculated from X-ray crystallography data assumes atoms are touching; this is equivalent to a covalent radius which may differ from the atomic radius given in section 9 of the *Data booklet*. The idea of atoms as spheres with a fixed volume is no longer an accepted model and the functional use of the concept of atomic radius depends upon the context. The boundary of the outer electrons is not clearly defined and depends on other interactions.

X-ray crystallography data take into account the strong forces between atoms in a crystal.

Worked example

Determine the density of gold, in g cm^{-3}, if it has a FCC structure and an atomic radius of 144 pm.

Solution

For a FCC structure:

radius of atom r = unit cell length $d \times \dfrac{\sqrt{2}}{4}$

$$d = \frac{4r}{\sqrt{2}}$$

$$= \frac{4(144)}{1.414} = 407.4 \text{ pm}$$

volume of unit cell = $(407.4 \times 10^{-10} \text{ cm})^3 = 6.762 \times 10^{-23} \text{ cm}^3$

FCC unit cell has 4 atoms so:

$$\text{density} = 4 \times \frac{196.97 \text{ g mol}^{-1}}{6.02 \times 10^{23} \text{ mol}^{-1} \times 6.762 \times 10^{-23} \text{ cm}^3} = 19.4 \text{ g cm}^{-3}$$

Questions

1   Superconductors are now widely employed in devices such as MRI scanners and MagLev trains. Many superconductors involve niobium.

a) Niobium is most commonly found in a crystalline form having the cubic unit cell shown in figure 17.

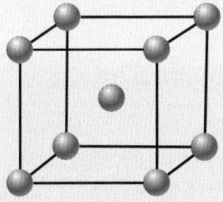

▲ Figure 17

Classify the crystal structure, the coordination number of the atoms and the number of atoms to which the unit cell is equivalent.

b) X-rays of wavelength $\lambda = 154$ pm are diffracted from this crystal at an angle of 14.17 degrees.

  (i) Assuming $n = 1$, calculate the distance, in pm, between layers of the crystal.

  (ii) Use your answer from b)(i) to find the volume of a niobium unit cell in cm^3.

(iii) Use sections 6 and 9 of the *Data booklet* to calculate the density of niobium in g cm^{-3}.

(iv) Determine the atomic radius of niobium and explain why this value may differ from the one in the *Data booklet*.

c) The ground-state electron configuration for niobium is [Kr]4d^{4}5s^1.

  (i) Compare and contrast paramagnetic and diamagnetic materials and explain whether niobium is more likely to be paramagnetic or diamagnetic.

  (ii) Niobium exhibits type 1 superconducting properties at low temperatures **when doped with other materials**. Sketch a graph that illustrates type 1 superconductivity and explain how it is different from type 2 superconductivity.

(iii) According to Bardeen–Cooper–Schrieffer (BCS) theory, Cooper pairs confer superconductivity. Outline how Cooper pairs are formed and the role of the positive ion lattice in their formation at low temperatures.

(iv) ⊘ Electrical resistance has been viewed as a collision between conducting electrons and localized electrons in the lattice causing some loss of energy. There is a gradual decrease of resistance as materials get colder and electrons' interactions lose strength. State what a paradigm shift is and justify why type 1 superconductivity and high temperature superconductivity could possibly require a paradigm shift.

2 The unit cells are shown for two ionic compounds, Q and R.

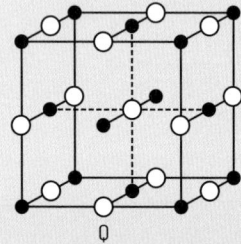

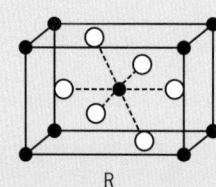

The solid circles (●) represent the metal ion (M) and the open circles (O) represent negative ions (X).

a) What is a unit cell? [1]

b) Which analytical technique would distinguish between the two compounds Q and R? [1]

c) Explain how the technique distinguishes the two compounds. [3]

d) Deduce the simplest formula of R and Q from the unit cell. [2]

IB, May 1998

3 When *monochromatic* X-rays are directed towards the surface of a crystal, some undergo diffraction.

a) What is meant by the term *monochromatic* and why is this important in X-ray crystallography? [2]

b) When X-rays with a wavelength of 154 pm are directed at a crystal of chromium the first order diffraction is found at 15.5°. Calculate the separation of the layers of atoms in the crystal. (1 pm = 1.0×10^{-12} m) [1]

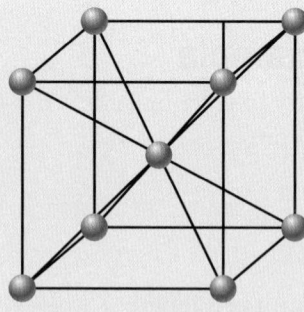

▲ Figure 18

c) (i) Figure 18 shows a representative unit cell of chromium. How many chromium atom equivalents does the unit cell contain? [3]

(ii) Use appropriate data from the *Data booklet* and the information about the dimensions of the unit cell to calculate the density and atomic radius of chromium. (If you could not calculate an answer for part b), use a value of 250 pm, although this is not the correct value.) [2]

d)

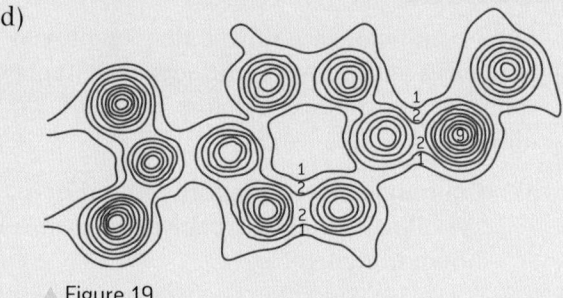

▲ Figure 19

Figure 19 is an electron density map of 4-methylbenzoic acid obtained by X-ray diffraction.

(i) What must have been the physical state of the compound to obtain this map? [1]

(ii) Which atoms in the molecule do not appear on this map? Why is this? [2]

(iii) Comment on the electron density between atoms with reference to the type of bonding present. [1]

IB, November 2000

4 Draw structures representing a face centred cubic and body centred cubic unit cell.

5 Platinum has a lattice edge length of 392.42 pm and crystallizes in a cubic rather than hexagonal form.

a) Determine the expected diffraction angle for a first-order reflection when monochromatic radiation of 0.1542 nm is used.

b) The density of platinum is 21.09 g cm^{-3}. Determine the packing structure of a unit cell.

6 Sketch a graph of resistance versus temperature for a conductor and a superconductor.

7 Deduce which part of figure 20 represents:

a) a normal conductor or a superconductor above its critical temperature

b) a type 1 superconductor exhibiting the Meissner effect below the critical temperature

c) a type 2 superconductor showing mixed transition state?

8 Copy and complete table 3.

Element	Density /g cm^{-3}	Length of cubic crystal edge/ pm	Radius of atom from X-ray diffraction data/pm	Crystal type
iron			125	BCC
sodium	0.968	429		
platinum	21.09			FCC

▲ Table 3

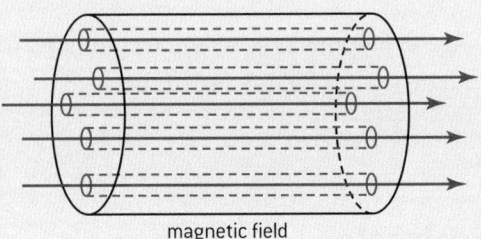

magnetic field

(a)

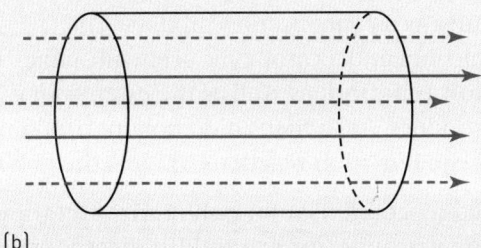

(b)

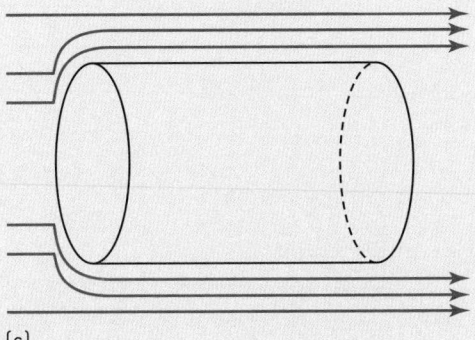

(c)

▲ Figure 20

A.9 Condensation polymers (AHL)

Understandings

→ Condensation polymers require two functional groups on each monomer.

→ NH_3, HCl, and H_2O are possible products of condensation reactions.

→ Kevlar® is a polyamide with a strong and ordered structure. The hydrogen bonds between O and N can be broken with the use of concentrated sulfuric acid.

 Applications and skills

→ Distinguishing between addition and condensation polymers.

→ Completion and descriptions of equations to show how condensation polymers are formed.

→ Deduction of the structures of polyamides and polyesters from their respective monomers.

→ Explanation of Kevlar's strength and its solubility in concentrated sulfuric acid.

Nature of science

→ Speculation – we have had the Stone Age, Bronze Age, and Iron Age. Is it possible that today's age is the Age of Polymers, as science continues to manipulate matter for desired purposes?

Condensation polymerization

Condensation polymers are formed by a reaction that joins monomers and also produces small molecules as a condensation product. The formation of an ester from an alcohol and a carboxylic acid (sub-topic 10.2) is an example of a **condensation reaction**: as well as the ester, water is formed as the condensation product. In condensation polymerization, many monomers are joined by condensation reactions to form the polymer.

For two monomers to be joined by condensation polymerization they must each contain two functional groups, for example, a dicarboxylic acid and a diol:

$$
\underset{\text{a dicarboxylic acid}}{HO - \overset{\overset{\displaystyle O}{\|}}{C} - R - \overset{\overset{\displaystyle O}{\|}}{C} - OH} \ + \ \underset{\text{a diol}}{HO - R' - OH}
$$

$$\downarrow$$

$$
\underset{\text{a polyester}}{\left[\overset{\overset{\displaystyle O}{\|}}{C} - R - \overset{\overset{\displaystyle O}{\|}}{C} - O - R' - O \right]_n}
$$

In the polyester product shown the carboxyl group on the left can react with a further alcohol molecule and the hydroxyl group on the right can react with a further carboxylic acid molecule, and so the polymer chain can continue to grow.

Instead of a dicarboxylic acid and a diol, the reaction may proceed with only one monomer that contains two functional groups: for example, 3-hydroxypentanoic acid contains both an OH group and a COOH group so can polymerize with itself:

3-hydroxypentanoic acid monomer

The esterification reaction

Although you will not be examined on the mechanism for esterification, you need to be aware that the reaction is acid catalysed and that it is the OH group from the acid and the H atom from the alcohol that join to form the condensation product water:

carboxylic acid   tetrahedral   ester
                         intermediate

Acyl chlorides are a class of organic compound in which the OH group of a carboxylic acid is replaced by a chlorine atom: $R(C=O)Cl$ rather than $R(C=O)OH$. Acyl chlorides react with alcohols to form esters even more readily than do carboxylic acids. The condensation product is hydrogen chloride, HCl rather than water. The mechanism is the same and it is the chlorine atom that leaves the intermediate. As before, two functional groups are needed in the monomer; an example is hexanedioyl dichloride which reacts with a diol to form a polyester:

hexanedioyl dichloride

Acyl chlorides react with amines rather than alcohols in a condensation reaction that forms an amide. For example, ethanoyl chloride, CH_3COOH and methylamine, CH_3NH_2 react to form N-methylethanamide, $CH_3NHCOCH_3$. A hydrogen from the amine and the OH group from the acid condense to form hydrogen chloride:

ethanoic acid   methylamine   methylethanamide   hydrogen chloride

Atom economy

Addition polymerization has 100% atom economy because all the monomer ends up in the desired product. This is not the case for condensation polymerization as the second condensation product is lost from the polymer.

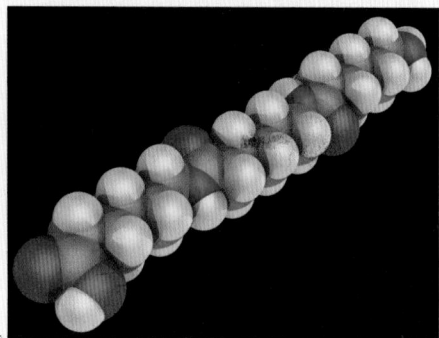
The mechanism is the same as for the esterification reaction except that there is an N rather than an O next to the carbonyl group forming an amide linkage. Polymerization again requires two functional groups per molecule.

The building up of proteins from amino acid monomers is a type of condensation polymerization. The amino acids contain two functional groups: an amino group, NH_2 and a carboxyl, COOH. The type of protein formed depends on the number, type, and sequence of the amino acid monomers (sub-topic B.2).

Phenol–methanal plastics

Phenol–methanal plastics are another example of condensation polymers. The first step in the reaction involves electrophilic substitution (see topic 20) of a hydrogen atom at the benzene ring with methanal:

phenol   methanal

The OH group in phenol is an ortho–para director, meaning that substitution of the hydrogen will occur on the number 2 (ortho) or number 4 (para) carbon atom in the benzene ring.

The second part of the reaction is the condensation step:

The reaction can continue with substitution occurring in either the 2- and/or the 4- position depending on the ratio of methanal to phenol.

Phenol–methanal polymers are **thermoset plastics**. They form resins and are used in laminates and adhesives. Because of their ability to withstand high temperatures and electric fields they are used as electrical insulators in construction and brake linings in vehicles.

Polyurethanes

Polyurethanes are another type of condensation polymer, a polyamide, with a wide variety of uses:

They form foams such as those used in padded chairs, elastomers used in paint, and fibres to produce spandex (elastane), a synthetic fabric with elastic properties. Monomers used to form polyurethanes are often a diol or diamine and a dicyanate (cyanates have the N=C=O functional group).

Modifying polymers

PVC is modified by adding plasticizers to soften the material (sub-topic A.5). Another example of polymer modification is blowing air through plastics to manufacture foams such as expanded polystyrene or padded polyurethane used in seat cushions. Polymers can also be doped with a substance to add a desired property; for example, polyethene may be doped with iodine to increase its conductivity. Fibres are also blended for comfort.

The same chemical backbone, polyurethane for example, can be modified to form elastomers and adhesives, high-density material such as rubber soles for shoes, or padded cushions by air injection.

Other ways of modifying the properties of polymers include changing the polymer chain length; for example, having more CH_2 units in the molecule increases the melting point as larger molecules have stronger intermolecular forces. The orientation of substituent groups also has an influence. The *trans* orientation of functional groups such as that seen in Kevlar (see sub-topic A.4, figure 11) allows close approach of the polymer chains and increases the degree of hydrogen bonding between the chains, conferring strength to the polymer. Isotactic and atactic orientation in addition polymers were explained in sub-topic A.5.

Ion interaction can also alter polymer properties. **Ion implantation** involves bombarding the polymer with large numbers of ions. This process can selectively modify the surface without changing the material's bulk properties, for example to increase or reduce friction. The ions can interact with polar ends of polymers and increase intermolecular forces. If negative ions are added this allows metal complexes to form. **Cross-linking** between polymer chains can add strength to elastomers. Rubber, for example, is a natural polymer which can be **vulcanized** in a process that adds sulfur to the polymer which creates strong covalent bonds between polymer chains. Natural rubber is soft and temperature sensitive, being brittle when cold and deforming easily when warm. Vulcanized rubber products

include rubber tyres which are not so temperature sensitive and are elastomers.

Covalent bonds between polymer chains prevent the chains from moving independently and strengthen the elastomer. For example, **Bakelite** is a phenol–methanal polymer that has cross-linking between the 2- and 4-positions in the benzene ring. This cross-linking makes Bakelite strong, rigid, and resistant to heat:

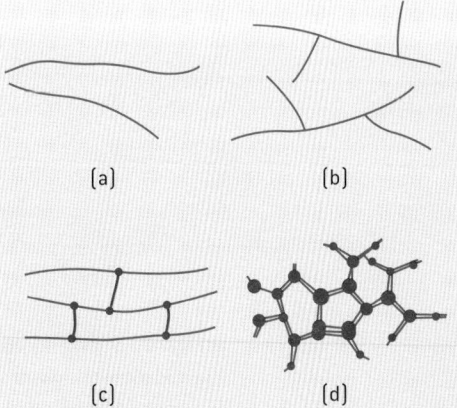

The degree of branching of the chain also influences polymer properties. HDPE, for example, has linear chains with little branching while LDPE has highly branched chains (sub-topic A.5).

▲ Figure 2 (a) Linear molecules can pack close together, eg HDPE. (b) Branched polymers are less dense and are held together by weak intermolecular forces, eg LDPE and natural rubber. (c) Cross-linking in polymers involves covalent bonds joining polymer chains to each other and greatly increases strength, eg vulcanized rubber. (d) Networked polymers, such as Bakelite and epoxy resins, are particularly strong and rigid

Structural property	Physical property	Example
Chain length	The longer the chain, the stronger the polymer.	Longer polymer chains have higher melting point, increased strength, and increased impact resistance due to increased van der Waals' forces.
Branching and packing structures	Straight unbranched chains can pack more closely. A higher degree of branching keeps strands apart and weakens intermolecular forces.	HDPE with no branching is more rigid than the more branched LDPE. Use of plasticizers in PVC to soften the polymer.
Side groups on monomers	Hydrogen bonding can increase strength, eg Kevlar. Atactic and isotactic placement can influence strength, eg polystyrene.	Polystyrene
Cross-linking	Extensive covalently bonded cross-linkage increases polymer strength.	Vulcanized rubber, Bakelite

▲ Table 1 Summary of polymer properties

Breaking down condensation polymers

Condensation polymers are formed from two monomers, releasing a small molecule in the process. These polymers can be broken down by the reverse reaction. Proteins, for example, are hydrolysed (a reaction that adds water) to amino acids during digestion. Polyamides with strong hydrogen bonding such as Kevlar can dissolve in sulfuric acid: the acid donates a proton to the oxygen and nitrogen atoms involved in hydrogen bonding. This breaks hydrogen bonds between chains of Kevlar fibres and the substance dissolves.

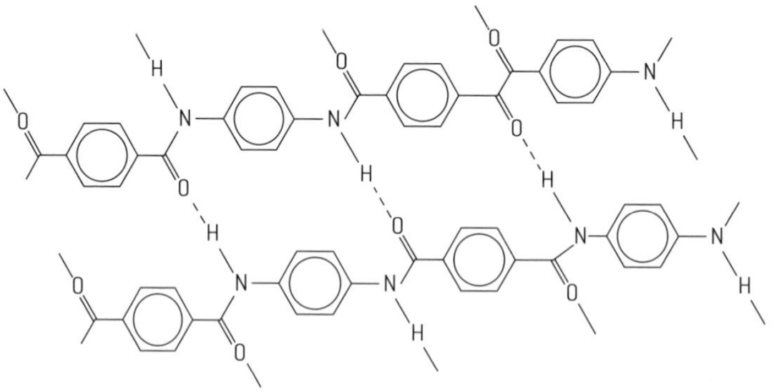

▲ Figure 3 Strong hydrogen bonds between polymer chains in Kevlar. Care must be taken to avoid interfering with hydrogen bond formation during production; for example the solvents must be free of ion impurities

Nylon, another polyamide, reacts readily with dilute acids in a hydrolysis reaction. The amide linkages in Kevlar are somewhat more resistant to acid attack than is nylon, but acids break the hydrogen bonds reducing the strength of the polymer. In breaking down amides to amines and carboxylic acids the condensation product, water, must be added and the reaction proceeds faster at high temperatures. Steam at a pH much greater or less than 7 can be used to break down a polyamide as an H^+ or OH^- ion will initiate the hydrolysis reaction.

Questions

1 Which pair of compounds can be used to prepare CH_3COOCH_3?

   A. Ethanol and methanoic acid

   B. Methanol and ethanoic acid

   C. Ethanol and ethanoic acid

   D. Methanol and methanoic acid [1]

   IB, November 2006

2 Nylon is a condensation polymer made up of hexanedioic acid and 1,6-diaminohexane.

   Which type of linkage is present in nylon?

   A. Amide

   B. Ester

   C. Amine

   D. Carboxyl [1]

   IB, May 2007

3 Kevlar is a condensation polymer that is often used in liquid-crystal displays. A section of the polymer is shown in figure 4.

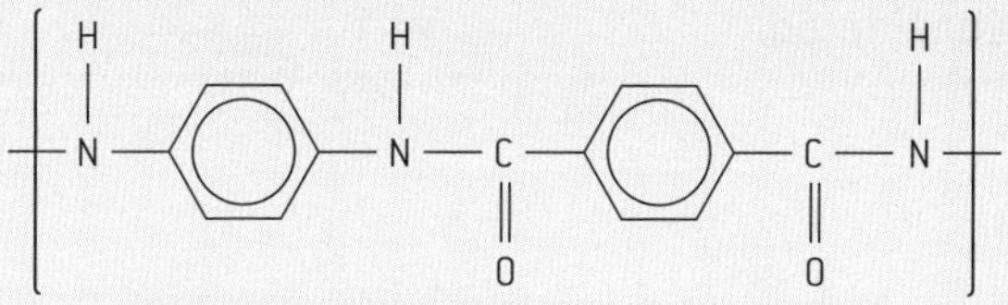

▲ Figure 4

   a) Explain the strength of Kevlar in terms of its structure and bonding. [2]

   b) Explain why a bullet-proof vest made of Kevlar should be stored away from acids. [2]

   IB, May 2011

4 Polymers, used extensively worldwide, are large molecular mass substances consisting of repeating monomer units.

   a) Distinguish between *addition* and *condensation* polymers in terms of how the monomers react together. [2]

   b) Describe and explain how the properties of condensation polymers depend on three structural features. [3]

   IB, May 2009

5 a) Kevlar can be made by reacting 1,4-diaminobenzene, $H_2NC_6H_4NH_2$, with 1,4-benzenedicarbonyl chloride, $ClOCC_6H_4COCl$. Write the equation for the reaction of n molecules of 1,4-diaminobenzene reacting with n molecules of 1,4-benzenedicarbonyl chloride. [2]

   IB, May 2010

A.10 Environmental impact – heavy metals (AHL)

Applications of heavy metals

"Heavy metals" is a term that refers to toxic metals such as lead, mercury, and cadmium which have cumulative effects on health. Such metals have many uses: lead, nickel, and cadmium are used in batteries; arsenic, bismuth, and antimony are often found in semiconductors; and mercury has many uses including in instruments such as thermometers, barometers, and diffusion pumps and has been used in mining, amalgams, and manufacturing. Heavy metals are commonly used as catalysts and have historical uses such as lead for pipes, lead paint, and petrol additives.

Heavy metals accumulate in biological systems over time. They are stored in living organisms and passed on in the food chain (see biomagnification in sub-topic B.6). The toxicity and carcinogenic properties of heavy metals are the result of their ability to form coordinated compounds, exist in various oxidation states, and act as catalysts in the human body.

Toxic metals can react with enzyme binding sites and inhibit or over-stimulate these enzymes. For example, cadmium belongs to the same group as zinc, and competes with zinc during absorption into the body. Lead can compete with and replace calcium in much the same way. Even when we take in more zinc and calcium in the diet, the toxic metals are not eliminated and tend to accumulate.

Toxic doses of transition metals can disturb the normal oxidation–reduction balance in cells through various mechanisms. They can disrupt the endocrine system because they compete for active sites of enzymes and cellular receptors. They exhibit multiple oxidation states so can participate in redox reactions, and they can initiate (free) radical reactions in electron transfer. Their ability to form complex ions enables them to bind with enzymes: iron, for example, forms a complex with hemoglobin which is essential for oxygen transport. Finally, transition metals are very good catalysts (topic 13).

Haber–Weiss and Fenton reactions

Free-radicals (sub-topic 10.2) can be generated naturally in biological systems; for example, the superoxide free-radical ion, $\cdot O_2^-$ is a product of cell metabolism. The Haber–Weiss reaction offers an explanation of how a more toxic hydroxyl radical, $\cdot OH$, could be formed. It was recognized that transition metals can catalyse this reaction, with the iron-catalysed (Fenton) reaction providing a mechanism for generating these reactive hydroxyl radicals.

The **Haber–Weiss reaction** is a slow process that generates hydroxyl radicals, $\cdot OH$, from hydrogen peroxide and the superoxide free-radical ion, $\cdot O_2^-$:

$$\cdot O_2^- + H_2O_2 \rightarrow O_2 + OH^- + \cdot OH$$

The products include a hydroxide ion as well as a hydroxyl radical. The peroxide reactant is formed by certain enzymes acting on the superoxide free-radical to catalyse a disproportionation reaction:

$$2 \cdot O_2^- + 2H^+ \rightarrow O_2 + H_2O_2$$

The peroxide–superoxide reaction is much quicker when catalysed in a two-step reaction, the **Fenton reaction**:

$$Fe^{3+} + \cdot O_2^- \rightarrow Fe^{2+} + O_2$$
$$Fe^{2+} + H_2O_2 \rightarrow Fe^{3+} + \cdot OH + OH^-$$

Notice that in accordance with Hess's law, the two steps of the Fenton reaction result in the Haber–Weiss reaction.

The highly reactive $\cdot OH$ radical is one of the most damaging free-radicals in the body. It reacts with almost any molecule it encounters including macromolecules such as DNA, membrane lipids, and enzymes. Because it is so reactive it can be used to break down pollutants such as pesticides and phenols and the Fenton reaction is carried out in waste-water treatment plants. For example, benzene derivatives, which are not very reactive, can be oxidized to less toxic phenols:

$$2 \cdot OH + C_6H_6 \rightarrow C_6H_5OH + H_2O$$

The $\cdot OH$ radical created by the Fenton reaction is a first step in many industrial processes. It can be used to eliminate some greenhouse gases such as methane from plant emissions, and to reduce odour from waste-water treatment sites. The highly reactive radical can break $C=C$ double bonds, open up aromatic rings, degrade hydrocarbons, and even initiate polymerization.

Global implications of scientific research

Fritz Haber is best known for fixing nitrogen (synthesizing ammonia), and he received the Nobel Prize in Chemistry in 1918 for this work. His final paper in 1934 proposed that the reactive hydroxyl radical could be generated from the superoxide ion and hydrogen peroxide. This greatly enhanced understanding of the role of radicals in biochemistry.

Haber's synthesis of ammonia for fertilizers enabled mass food production, alleviating hunger. It is ironic that his research in chemical warfare went side by side with this. The ethics of scientific research have global implications.

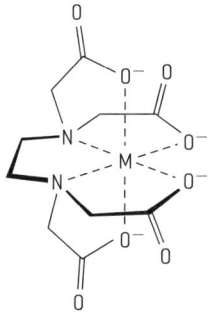

▲ Figure 1 EDTA is a polydentate ligand that can form up to six coordinate bonds to a central metal ion. It is used in chelation therapy to treat lead poisoning and remove excess iron from the blood of patients with thalassemia

Chelating effects

Apart from the Fenton reaction, other methods of removing heavy metals include precipitation, adsorption, and **chelation**. Chelation takes advantage of a metal's ability to form complex ions. The word "chelate" is derived from the Greek for "claw" and refers to polydentate ligands (sub-topic 13.2; some common polydentate ligands are given in section 16 of the *Data booklet*). Chelating agents are used to remove heavy metals such as lead, arsenic, and mercury from the body. Once chelated the complex ion is too large to enter cells but being an ion is water soluble so can be excreted from the body.

▲ Figure 2 Workers at a heavy metal recycling factory. 'Heavy metals' refers to toxic metals such as lead, mercury, and cadmium which have accumulative effects. Heavy metals not recycled must be carefully disposed of in toxic landfill sites

▲ Figure 4 An industrial waste treatment plant in Argentina. Toxins are removed and the water purified before being put back into the environment. Water treatment is needed in many places as many of the world's major rivers show high pollution

To act as ligands, chelating agents must have lone pairs of electrons that can form coordinate covalent bonds to a central atom. "Polydentate" refers to their ability to form more than one such coordinate bond. Figure 1 shows that EDTA can form two, four, or up to six coordinate covalent bonds with a central atom.

Ethylenediamine (ethane-1,2-diamine) is a bidentate chelating agent:

$$H_2\ddot{N}-CH_2-CH_2-\ddot{N}H_2$$

Heme in hemoglobin forms four coordinate covalent bonds to iron.

Dimercaptol and mustard gas

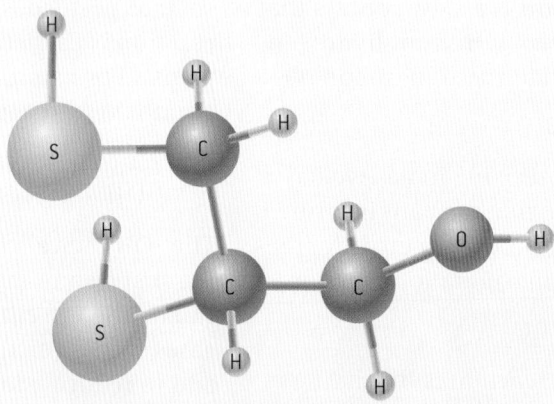

▲ Figure 3 Dimercaptol is a bidentate chelating agent that uses the lone pairs of electrons on its two sulfur atoms to form coordinate bonds with mercury, arsenic, antimony, and gold. Dimercaptol was used to treat arsenic-containing mustard gas during the first world war. Chelated metals cannot enter cells and can be excreted from the body

Polydentate ligands such as EDTA are usually more effective than monodentate ones and will replace them in reactions. Competition in ligands was discussed in topic 13, and one factor influencing this is the increase in entropy involved. Nickel, for example, can form a complex ion with six molecules of water $[Ni(H_2O)_6]^{2+}$. EDTA will replace the six water molecules in this reaction forming a larger complex and releasing the six smaller molecules thus increasing the overall entropy:

$$EDTA^{4-}(aq) + [Ni(H_2O)_6]^{2+}(aq) \rightarrow [Ni(EDTA)]^{2-}(aq) + 6H_2O(l)$$

The existence of a greater number of smaller molecules rather than one larger one yields more ways of distributing the effective energy, and hence represents an increase in entropy. This is one reason why chelation is effective at removing metals, as the polydentate ligand will replace larger numbers of existing ligands, usually water.

🌐 Solubility product constant, K_{sp}

Metal ions from group 1, including K^+, Li^+, and Na^+, form highly soluble compounds whereas the heavy metal ions generally form compounds of low solubility. Their salts precipitate easily and this means heavy metal ions can be removed during waste-water treatment. Many heavy metal hydroxides are only slightly soluble so hydroxide ions are often added to precipitate the metal ions as the level of hydroxide ions can be monitored by measuring the pH. Lime, $Ca(OH)_2$, is commonly used as it is a relatively cheap and abundant material.

The solubility of metal ions can be expressed as the position of equilibrium of the solid salt with its aqueous ions. For example, the equilibrium expression can be written for the highly insoluble lead(II) hydroxide, $Pb(OH)_2$:

$$Pb(OH)_2(s) \rightleftharpoons Pb^{2+}(aq) + 2OH^-(aq)$$

The position of equilibrium at standard conditions (at 298) can be expressed as a constant (topic 7). Solids are not included in equilibrium expressions as they have zero concentration in the solution. This particular equilibrium constant is referred to as the **solubility product constant**, K_{sp} and the value of K_{sp} for $Pb(OH)_2$ at 298 K is 1.43×10^{-20}. A table of solubility product constants can be found in the *Data booklet*.

Worked examples

Example 1

Calculate the solubility in mol dm^{-3} of lead(II) hydroxide.

Solution

Using the ICE method as explained in topic 17:

$$Pb(OH)_2(s) \rightleftharpoons Pb^{2+}(aq) + 2OH^-(aq)$$

$$K_{sp} = [Pb^{2+}][OH^-]^2$$

I	0	0
C	$+x$	$+2x$
E	x	$2x$

$$1.43 \times 10^{-20} = (x)(2x)^2 = 4x^3$$

$$x = 1.53 \times 10^{-7}$$

The concentration of Pb^{2+} is the same as the molarity of $Pb(OH)_2$, so the molar solubility of lead(II) hydroxide is 1.53×10^{-7} mol dm^{-3}.

Example 2

Cadmium is a heavy metal frequently removed from waste water by precipitation. The water is adjusted to pH 11 by adding lime (calcium hydroxide). Calculate the molar solubility of the Cd^{2+} ion at this pH.

Solution

$$K_{sp}[Cd(OH)_2] = 7.2 \times 10^{-15} \text{ (from section 32 of the Data booklet)}$$

At pH 11, $[OH^-] = 10^{-3}$

$$Cd(OH)_2(s) \rightarrow Cd^{2+}(aq) + 2OH^-(aq)$$

	$[Cd^{2+}]$	$[OH^-]$
I	0	10^{-3}
C	$+x$	$10^{-3} + 2x$
E	x	$10^{-3} + 2x$

$7.2 \times 10^{-15} = x(10^{-3} + 2x)^2$. Because the degree of dissociation of $Cd(OH)_2$ is small compared with the 0.001 mol dm^{-3} concentration of the hydroxide ion, the $+ 2x$ in the $[OH^-]$ term can be ignored.

$$7.2 \times 10^{-15} = x(10^{-3})^2 = x(10^{-6})$$

$$x = 7.2 \times 10^{-9} \text{ mol dm}^{-3}$$

Notice the low solubility of Cd^{2+} ions at this pH. See if you can confirm for yourself that the solubility of Cd^{2+} without adjusting the pH is 1.2×10^{-5}, or about 10000 times higher.

Adsorption of heavy metals

Another method of removing heavy metals is by adsorption onto a solid surface. There are many methods including activated carbon, charcoal filters and clays. Biomass such as brewer's yeast has also been found to be effective. Ion-exchange mechanisms which exchange heavy metal ions for calcium or sodium ions can also remove heavy metal contaminants. The treated water then undergoes further purification processes such as ultraviolet treatment to kill bacteria.

Activated charcoal is an expensive adsorbent. Cheaper agricultural methods are proving useful in many developing countries. Coconut shells, rice husks, and sugar cane have adsorbent properties which might be effective in removing heavy metals.

Questions

1 Hydroxyl free-radicals can be generated naturally in the body. This process is catalysed by iron in the following two steps:

reaction 1: $Fe^{3+} + \cdot O_2^- \rightarrow Fe^{2+} + O_2$

reaction 2: $Fe^{2+} + H_2O_2 \rightarrow Fe^{3+} + \cdot OH + OH^-$

a) Use the above information to write the uncatalysed reaction.

b) Deduce whether iron is acting as a heterogeneous or homogeneous catalyst and justify your answer.

c) Show that reactions 1 and 2 are redox equations by writing the oxidation half-equation and the reduction half-equation for each.

2 Heavy metals are often removed from solutions by precipitation.

a) Use section 32 of the *Data booklet* to calculate the concentration of sulfuric acid necessary to precipitate mercury(I) ions at a concentration of 3 μmol dm^{-3}.

b) Evaluate the effectiveness of this method for mercury removal and suggest improvements.

c) Magnesium ion concentrations can be determined by precipitation as magnesium hydroxide. Given that the solubility product, K_{sp}, of magnesium hydroxide is 1.20×10^{-11} calculate the concentration, in mol dm^{-3}, of magnesium ions required to form a precipitate in a solution where the final hydroxide ion concentration is 2.00 mol dm^{-3}.

3 Chromium(III) ions form a hexa-aqua complex ion, $[Cr(H_2O)_6]^{3+}$(aq). Write a balanced equation for the reaction of this complex with EDTA^{4-}(aq) and explain why EDTA will replace the water in the complex ion.

4 Explain the difference between precipitation, chelation, and adsorption as methods of removing heavy metal contamination.

5 Use section 32 of the *Data booklet* to calculate the molar solubility of zinc at pH 11 and explain why zinc ions are more soluble in acidic solutions.

B BIOCHEMISTRY

Introduction

Biochemistry studies chemical processes in living organisms at the molecular level. Despite the diversity of life forms and complexity of biological structures, life functions can be interpreted in chemical terms, because the constitution and properties of biomolecules are governed by the same principles as the constitution and properties of any other form of matter. The processes in the living cells resemble the reactions of traditional chemistry and therefore can be studied and replicated in the laboratory or utilized in industry, agriculture, and medicine. Biochemical studies enhance our understanding of the phenomenon of life and our own place in the natural world.

B.1 Introduction to biochemistry

Understandings

→ Shapes and structures of biomolecules define their functions.

→ Metabolic processes take place in aqueous solutions in a narrow range of pH and temperature.

→ Anabolism is the biosynthesis of complex molecules from simpler units that requires energy.

→ Catabolism is the biological breakdown of complex molecules that provides energy for living organisms.

→ Condensation reactions produce biopolymers that can be hydrolysed into monomers.

→ Photosynthesis transforms light energy into chemical energy of organic molecules synthesized from carbon dioxide and water.

→ Respiration is a set of catabolic processes that produce carbon dioxide and water from organic molecules.

Applications and skills

→ Deduce condensation and hydrolysis reactions and explain the difference between these processes.

→ Describe the balancing of carbon and oxygen in the atmosphere by summary equations of photosynthesis and respiration.

Nature of science

→ Biochemical systems are complex and involve a large number of simultaneous chemical reactions. The development of analytical techniques allows us to collect enough experimental data to reveal certain patterns in biochemical processes and eventually understand metabolic processes.

- **Metabolism** is all the chemical processes that take place within a living organism to maintain life.

- **Anabolism** is the biosynthesis of complex molecules from simpler units that usually requires energy.

- **Catabolism** is the breakdown of complex molecules in living organisms into simpler units that is usually accompanied by the release of energy.

- A **metabolic pathway** is a biochemical transformation of a molecule through a series of intermediates (metabolites) into the final product.

What is biochemistry?

Biochemistry studies chemical processes in living cells at the molecular level. Biochemical processes, collectively known as **metabolism**, are very complex and involve many chemical reactions occurring in the same place and at the same time. Some of these reactions (**anabolic** reactions) produce large organic molecules from simpler organic or inorganic substances while in other reactions (**catabolic** reactions), complex molecules are broken down into smaller fragments.

 A historical perspective

In the nineteenth century the main goals of biochemical studies were the isolation and identification of chemical substances present in living organisms. Progress in analytical techniques allowed more data to be collected, which eventually led to the discovery of certain patterns in distribution of these substances in organisms and their possible roles in biochemical processes. These findings in turn stimulated more focused research and the utilization of a wide range of physicochemical methods that became available to scientists in the twentieth century. As more complex molecules and reactions became known, the focus of biochemistry gradually shifted towards the study of metabolic pathways and eventually to better understanding of the basic functions of living organisms and the phenomenon of life.

What drives metabolism?

Anabolic reactions increase the complexity and order of biochemical systems and thus reduce their entropy (sub-topic 15.2). Such processes cannot be spontaneous; they require energy, which is supplied by catabolic reactions or in **photosynthesis** is received in the form of light from the sun. Photosynthesis is the major source of energy for green plants and some bacteria. Other organisms, including humans, rely entirely on the chemical energy obtained from food by a complex set of metabolic processes known as **respiration**. Photosynthesis and respiration will be discussed later in this topic.

The life functions of all organisms depend on a sophisticated balance between anabolic and catabolic processes in their cells, intake of nutrients, excretion of waste products, and exchange of energy with the environment. The variety of metabolic pathways allows living organisms to adapt to the constantly changing natural world. Life in turn affects the environment on both the local and global scale. Therefore a detailed understanding of metabolism is essential for all life sciences, from pharmacology and nutrition to ecology and agriculture, so biochemistry is increasingly becoming their common language.

Molecules of life

The primary chemical element in all biologically important molecules is carbon. Its relatively small size, moderate electronegativity, and the electronic configuration of the outer shell ($2s^2 2p^2$ in the ground state and $2s^1 2p^3$ in the excited state) allow carbon to form up to four single or multiple covalent bonds with many elements, including itself. The energies of these bonds are high enough to produce stable molecules and at the same time low enough to allow such molecules to undergo various transformations. This combination of stability and reactivity makes organic molecules the chemical basis of life.

The unique ability of carbon to form single and multiple bonds with itself allows for the formation of molecules of any size and complexity – from simple inorganic compounds such as carbon dioxide to giant biopolymers like proteins and nucleic acids. However, from a virtually unlimited number of possible combinations of carbon atoms with other elements, only a small set of relatively simple organic molecules is particularly important for living organisms. These molecules, composed of carbon, hydrogen, oxygen, nitrogen, and some other bioelements (table 1), are used as building blocks for biopolymers of hierarchically increasing complexity (figure 1).

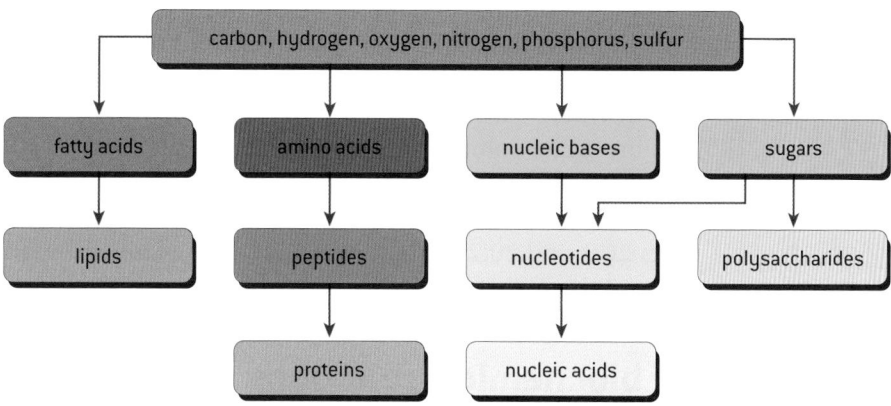

▲ Figure 1 The hierarchy of biomolecules

Macrobioelement	Percentage by mass in the body	Microbioelement	Percentage by mass in the body
oxygen	65	iron	0.006
carbon	19	fluorine	0.004
hydrogen	9.5	zinc	0.003
nitrogen	2.8	silicon	0.002
calcium	1.5	copper	1×10^{-4}
phosphorus	1.1	boron	7×10^{-5}
sulfur	0.25	iodine	2×10^{-5}
potassium	0.30	selenium	2×10^{-5}
sodium	0.15	manganese	2×10^{-5}
chlorine	0.15	nickel	1×10^{-5}
magnesium	0.05	molybdenum	1×10^{-5}
total	99.8	other bioelements	1×10^{-5}

▲ Table 1 Macro- and microbioelements in the human body

TOK

In the study of the intermediate processes of metabolism we have to deal not with complex substances which elude ordinary chemical methods, but with the simple substances undergoing comprehensible reactions.

Sir Frederick Gowland Hopkins. 1914. "The dynamic side of biochemistry". In Report on the 83rd Meeting of the British Association for the Advancement of Science. P653.

Water: Solvent, reactant, and product

The most common types of biochemical reaction are condensation, hydrolysis, oxidation, and reduction, in which water plays the role of both the solvent and, at the same time, the reactant or product. Nearly all biopolymers form by **condensation reactions** that release water as a by-product. For example, amylose (a component of starch, topic B.10) is produced in green plants by polycondensation of glucose:

$$n\mathrm{C_6H_{12}O_6} \rightarrow \mathrm{H{-}(C_6H_{10}O_5)}_n{-}\mathrm{OH} + (n-1)\mathrm{H_2O}$$

$$\text{glucose} \qquad \text{amylose} \qquad\qquad \text{water}$$

This reaction is reversible – in the human body, amylose is **hydrolysed** into glucose:

$$\mathrm{H{-}(C_6H_{10}O_5)}_n{-}\mathrm{OH} + (n-1)\mathrm{H_2O} \rightarrow n\mathrm{C_6H_{12}O_6}$$

$$\text{amylose} \qquad\qquad \text{water} \quad \text{glucose}$$

Worked example

Cyclodextrins are structurally similar to amylose but the fragments of glucose in their molecules form a large ring instead of a chain. Deduce an equation for complete hydrolysis of the cyclodextrin containing six glucose residues. How many molecules of water will be required to balance this equation?

Solution

Cyclic polymers do not have terminals, so the formula of the cyclodextrin is $(\mathrm{C_6H_{10}O_5})_6$. Because each glucose residue needs one oxygen and two hydrogen atoms to produce glucose, the number of water molecules in the balanced equation will be also six:

$$(\mathrm{C_6H_{10}O_5})_6 + 6\mathrm{H_2O} \rightarrow 6\mathrm{C_6H_{12}O_6}$$

$$\text{cyclodextrin} \qquad\qquad \text{glucose}$$

Up to 65% of the human body mass is composed of water, with intracellular fluids and blood plasma containing 70–80 and 90–93% water, respectively. Biochemical reactions proceed in a highly controlled aqueous environment where most of the reactants, products, and catalysts (enzymes) are water soluble or form soluble complexes with other molecules. This fact makes the chemical transformations in living organisms very different from those of traditional organic chemistry (topic 10, sub-topics 20.1 and 20.2), where the reactions usually proceed in organic solvents and the presence of water is carefully avoided.

The nature of biochemical reactions

The reactions responsible for the synthesis and hydrolysis of peptides and proteins (sub-topics B.2 and B.7), fats and phospholipids (sub-topic B.3), nucleotides and nucleic acids (sub-topic B.8), and many other biologically important molecules are very similar.

In contrast to traditional organic reactions, which often require high temperatures and long reaction times, and almost never give products with 100% yield, biochemical reactions usually proceed very fast and with near quantitative yields at body temperature (around 310 K in humans). The reason for this striking difference is the action of **enzymes** – highly specific and efficient biological catalysts. Enzymes and enzymatic catalysis will be discussed in sub-topics B.2 and B.7.

Life and energy

Oxidation and **reduction** of organic substances in living organisms proceed stepwise and involve a series of metabolites that transfer and store energy in chemical bonds of their molecules. As has already been explained (sub-topic 9.1), redox processes can be described in terms of oxidation numbers, transfer of electrons, or combination with certain elements (oxygen and hydrogen). In the aqueous environment of the organism reactions involving protons or water are prevalent and so biochemists rely on the third method, occasionally referring to electron transfer when half-equations are discussed.

- **Oxidation** is the loss of two hydrogen atoms or the gain of an oxygen atom.
- **Reduction** is the gain of two hydrogen atoms or the loss of an oxygen atom.

or

- **Oxidation** is the loss of electrons.
- **Reduction** is the gain of electrons.

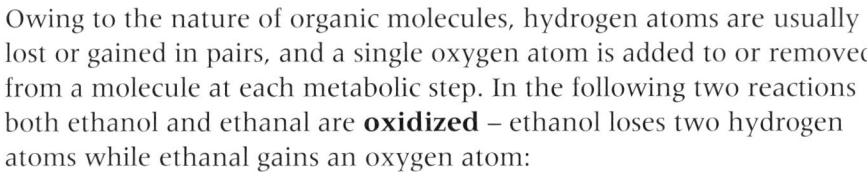

Owing to the nature of organic molecules, hydrogen atoms are usually lost or gained in pairs, and a single oxygen atom is added to or removed from a molecule at each metabolic step. In the following two reactions both ethanol and ethanal are **oxidized** – ethanol loses two hydrogen atoms while ethanal gains an oxygen atom:

$$CH_3CH_2OH + [O] \rightarrow CH_3CHO + H_2O$$
$$\text{ethanol} \qquad\qquad \text{ethanal}$$

$$CH_3CHO + [O] \rightarrow CH_3COOH$$
$$\text{ethanal} \qquad\qquad \text{ethanoic acid}$$

The oxidation of ethanol and ethanal can be also described in terms of electron transfer using half-equations, in which both molecules lose electrons:

$$CH_3CH_2OH \rightarrow CH_3CHO + 2H^+ + 2e$$

$$CH_3CHO + H_2O \rightarrow CH_3COOH + 2H^+ + 2e^-$$

The next examples show two **reduction** processes – in the first ethanal gains two hydrogen atoms and in the second hydrogen peroxide loses an oxygen atom:

$$CH_3CHO + [2H] \rightarrow CH_3CH_2OH$$

$$H_2O_2 \rightarrow H_2O + [O]$$

The reduction of ethanal and hydrogen peroxide can be presented as half-equations in which both molecules gain electrons:

$$CH_3CHO + 2H^+ + 2e^- \rightarrow CH_3CH_2OH$$

$$H_2O_2 + 2H^+ + 2e^- \rightarrow 2H_2O$$

Worked example

In the human body, a series of metabolic processes can lead to the following summary equation:

$$CH_3C(O)COOH + CH_3CH(OH)CH_3 \rightarrow CH_3CH(OH)COOH + CH_3C(O)CH_3$$
2-oxopropanoic   propan-2-ol   2-hydroxypropanoic   propanone
acid (pyruvic acid)   acid (lactic acid)

Which of the two reactants, 2-oxopropanoic acid or propan-2-ol, is oxidized and which one is reduced? Deduce redox half-equations for both processes using protons or water where necessary.

Solution

Propan-2-ol (C_3H_8O) has two more hydrogen atoms than propanone (C_3H_6O), so propan-2-ol is oxidized. If one reactant undergoes oxidation, another reactant (in our case, 2-oxopropanoic acid) must undergo reduction. Indeed, 2-oxopropanoic acid gains two hydrogen atoms and forms 2-hydroxypropanoic acid. Because all the reactants and products are neutral molecules, the number of lost or gained electrons in each half-equation must be equal to the number of protons:

$$CH_3CH(OH)CH_3 \rightarrow CH_3C(O)CH_3 + 2H^+ + 2e^- \qquad \text{(reduction)}$$
propan-2-ol   propanone

$$CH_3C(O)COOH + 2H^+ + 2e^- \rightarrow CH_3CH(OH)COOH \qquad \text{(oxidation)}$$
2-oxopropanoic   2-hydroxypropanoic acid
acid

The stepwise nature of biochemical reactions

The energy liberated when substrates undergo air oxidation is not liberated in one large burst, as was once thought, but is released in a stepwise fashion. The process is not unlike that of locks in a canal. As each lock is passed in the ascent from a lower to a higher level a certain amount of energy is expended.

Eric Glendinning Ball. 1942. "Oxidative mechanisms in animal tissues". In A symposium on respiratory enzymes. P22.

- **Photosynthesis** is the biosynthesis of organic molecules from carbon dioxide and water using the energy of light.

- **Respiration** is the metabolic processes that release energy from nutrients consumed by living organisms.

- **Aerobic respiration** is the reverse process of photosynthesis, in which carbon dioxide and water are formed from organic molecules and oxygen.

- **Anaerobic respiration** is the catabolism of organic compounds that does not involve molecular oxygen as an electron acceptor.

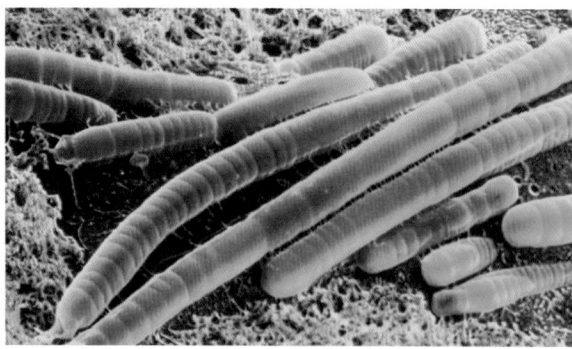 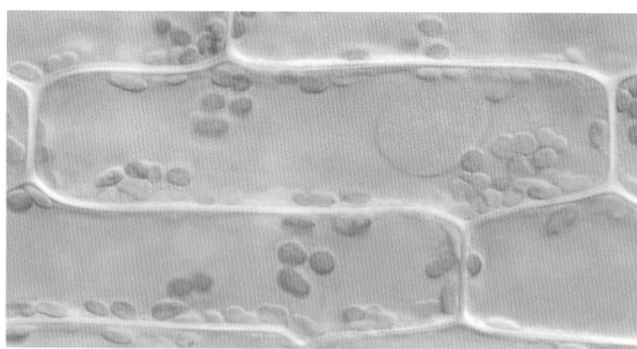

▲ Figure 2 Photosynthesizing blue-green algae (left) and green leaf cells containing chloroplasts (right)

🌐 Photosynthesis

The process of photosynthesis begins when light energy is absorbed by chlorophylls (sub-topic B.9). In plants chlorophylls are held inside organelles called chloroplasts. The absorbed light energy is used in a series of anabolic reactions that ultimately leads to the reduction of carbon dioxide into energy-rich organic molecules such as glucose, and the release of oxygen:

$$6CO_2 + 6H_2O \xrightarrow{\text{light}} \underset{\text{glucose}}{C_6H_{12}O_6} + 6O_2$$

When sunlight is not available this reduction can be reversed, and the energy needed for life functions can be produced by the oxidation of glucose by oxygen in a process called **aerobic respiration**:

$$\underset{\text{glucose}}{C_6H_{12}O_6} + 6O_2 \rightarrow 6CO_2 + 6H_2O + \text{energy}$$

Aerobic respiration also takes place in the cells of humans and other animals, who cannot utilize sunlight and are completely dependent on the chemical energy of nutrients supplied by photosynthesizing green plants.

Photosynthesis, respiration, and the atmosphere

Photosynthesis and respiration are responsible for the global balance of oxygen and carbon dioxide. Nearly all the oxygen in the Earth's atmosphere and oceans is a by-product of photosynthesis, the process

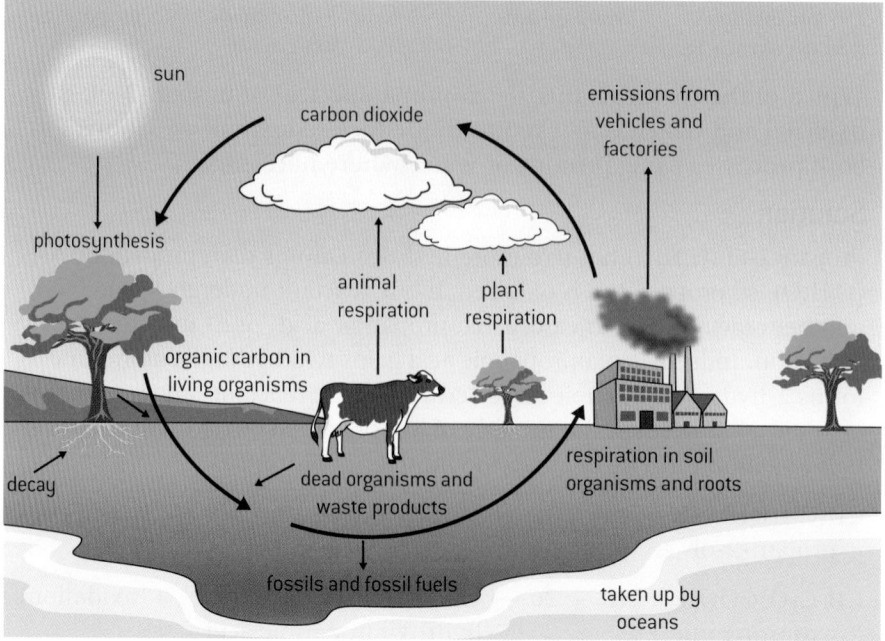

▲ Figure 3 The carbon cycle

that started in blue-green algae (cyanobacteria) over two billion years ago and dramatically changed our planet. Along with the production of oxygen, photosynthesizing bacteria consumed most of the atmospheric carbon dioxide and made the Earth habitable for higher life forms, such as plants, humans, and other animals.

Over the past half a million years, the level of carbon dioxide in the atmosphere has remained almost constant, at about 0.02–0.03%. However, this fragile balance is being increasingly threatened by human activities – the biosphere and oceans are capable of removing less than two-thirds of recent anthropogenic carbon dioxide emissions (those produced by human activities). As a result, the level of carbon dioxide in the atmosphere reached 0.04% in 2013 and continues to rise by about 2 ppm per year, leading to global warming and other climate changes. Biochemical studies allow us to predict the impact of these changes on metabolism, life cycles, and ultimately on the survival of various organisms, including our own species.

Worked example

The human brain receives almost all its energy from glucose, which is completely oxidized to carbon dioxide and water in aerobic respiration. Determine the mass of carbon dioxide produced in the brain per day if its daily consumption of glucose is 135 g.

Solution

The molecular mass of glucose is 180.16 g mol^{-1}, so the amount of glucose is 135 g/180.16 g mol^{-1} = 0.749 mol. During aerobic respiration 1 mol of glucose releases 6 mol of carbon dioxide, so the daily amount of carbon dioxide produced in the brain is 6×0.749 mol = 4.49 mol, and its mass is 4.49 mol $\times$ 44.01 g mol^{-1} = 198 g.

TOK

There is evidence that certain species of fungi exposed to high levels of gamma-radiation after the explosion of the Chernobyl nuclear power plant in 1986 developed an ability to convert this radiation into chemical energy and use it for growth. How does this affect our view of life-supporting environments, both on our planet and beyond?

Questions

1 State the difference between anabolism and catabolism.

2 Nucleic acids are a class of biopolymer. List two other classes of biopolymer.

3 State three differences between metabolic processes in living organisms and the reactions of traditional organic synthesis in the laboratory.

4 Define oxidation in terms of oxidation numbers. [1]

IB, May 2009

5 Define reduction in terms of:

a) hydrogen atoms

b) oxygen atoms

lost or gained by the substrate.

6 In living organisms, two molecules of the 2-amino acid cysteine, $HSCH_2CH(NH_2)COOH$, can combine together to form cystine, $HOOCCH(NH_2)CH_2S–SCH_2CH(NH_2)COOH$. State and explain, in terms of loss or gain of hydrogen atoms, whether the formation of cystine from cysteine is an oxidation or a reduction reaction.

7 Describe aerobic respiration of glucose in the human body, with reference to oxidation and reduction. [4]

IB, November 2007

8 Explain how photosynthesis represents a conversion of energy from one form to another.

9 State the sources of:

a) carbon

b) hydrogen

c) oxygen

in photosynthesis.

10 Determine the mass (in g) of carbon dioxide required to produce 3.15 g of glucose, $C_6H_{12}O_6$.

11 Some bacteria can synthesize all the components of their cells from inorganic materials and sunlight, while humans and other animals are unable to do this and must obtain certain essential organic compounds from their diet. Discuss whether metabolic processes in these bacteria are more complex and sophisticated than the metabolism in our own bodies.

12 The following redox reactions represent bacterial decomposition of organic waste under different conditions.

$$C_6H_{12}O_6 + 6O_2 \rightarrow 6CO_2 + 6H_2O$$

$$CH_3COO^- + H_2O \rightarrow CH_4 + HCO_3^-$$

$$2CH_2O + SO_4^{2-} \rightarrow 2CO_2 + H_2S + 2OH^-$$

$$2CH_2O + O_2 + 2OH^- \rightarrow 2HCOO^- + 2H_2O$$

Identify the most likely environment (aerobic or anaerobic) for each reaction. [2]

IB, November 2012

13 Explain the difference between hydrolysis and condensation reactions.

14 Complete and balance the following equation, and identify its reaction type (hydrolysis or condensation).

$$C_6H_{12}O_6 \rightarrow H–(C_6H_{10}O_5)_6–OH + …$$

15 Determine the mass (in g) of the inorganic product formed in the above reaction if the mass of the biopolymer produced was 4.95 g.

B.2 Proteins and enzymes

Understandings

→ Proteins are polymers of 2-amino acids, joined by amide links (also known as peptide bonds).

→ Amino acids are amphoteric and can exist as zwitterions, cations, and anions.

→ Protein structures are diverse and can be described at the primary, secondary, tertiary, and quaternary levels.

→ Three-dimensional shapes of proteins determine their roles in metabolic processes or as structural components.

→ Most enzymes are proteins that act as catalysts by binding specifically to a substrate at the active site.

→ As enzyme activity depends on the conformation, it is sensitive to changes in temperature, pH, and the presence of heavy metal ions.

→ Chromatography separation is based on different physical and chemical principles.

Applications and skills

→ Deduction of the structural formulae of reactants and products in condensation reactions of amino acids, and hydrolysis reactions of peptides.

→ Explanation of the solubilities and melting points of amino acids in terms of zwitterions.

→ Application of the relationships between charge, pH, and isoelectric point for amino acids and proteins.

→ Description of the four levels of protein structure, including the origin and types of bonds and interactions involved.

→ Deduction and interpretation of graphs of enzyme activity involving changes in substrate concentration, pH, and temperature.

→ Explanation of the processes of paper chromatography and gel electrophoresis in amino acid and protein separation and identification.

Nature of science

→ Collaboration and peer review – several different experiments on several continents led to the conclusion that DNA, and not proteins as originally thought, carried the information for inheritance.

The central role of proteins in biochemistry

Proteins are the most diverse and abundant class of biopolymers, responsible for over 50% of the dry mass of cells. This fact reflects the central role of proteins in metabolic processes, transport and sensory functions, structural integrity, and virtually all other molecular aspects of life.

 ## Proteins and heredity

At the end of the nineteenth century scientists believed that genetic information was stored in certain proteins, which varied across the species and between individuals. However, a series of biochemical experiments in different scientific groups on several continents revealed that nucleic acids, not proteins, were the true carriers of genetic information (sub-topic B.8). This discovery underlines the role of international collaboration and peer review of scientific publications in the development of our understanding of the natural world.

Simple proteins are linear polymers of **2-amino acids**. The structural units of proteins are joined together by **amide linkages** (also known as **peptide bonds**) in strict order and orientation. Most proteins contain several hundred to several thousand structural units. Shorter polymers composed of less than 20 residues of 2-amino acids are called **peptides**. The term "**polypeptides**" refers to longer peptides or small proteins with 20–50 structural units, although its meaning varies in literature. In particular, some biochemists differentiate polypeptides and proteins by their ability to fold and adopt specific conformations in aqueous solutions, which will be discussed later in this topic.

2-Amino acids and peptides

From more than 500 naturally occurring amino acids, only 20 are **proteinogenic**, that is, used by living organisms as building blocks of proteins. The molecules of these amino acids share several structural features. In particular, they all have an amino group and a carboxyl group attached to the same carbon atom. According to substitutive IUPAC nomenclature, this carbon atom is numbered as C-2, so all proteinogenic amino acids are called "2-amino acids". In the past, the same carbon atom was labelled as the "α-carbon", so the term "α-amino acids" is still commonly used in literature as an alternative name for 2-amino acids.

The substituent R, often referred to as a **side-chain**, may be a hydrocarbon fragment or contain various functional groups (table 1). In glycine (2-aminoethanoic acid), R = H while in proline, the side-chain forms a five-membered heterocyclic ring with the 2-amino group.

Common name	Abbreviation	Structural formula	Isoelectric point
alanine	Ala	$H_2N-CH-COOH$ $\quad\quad\vert$ $\quad\quad CH_3$	6.0
arginine	Arg	$H_2N-CH-COOH$ $\quad\quad\vert$ $\quad\quad CH_2(CH_2)_2NH-C-NH_2$ $\quad\quad\quad\quad\quad\quad\quad\vert\vert$ $\quad\quad\quad\quad\quad\quad\quad NH$	10.8
asparagine	Asn	$H_2N-CH-COOH$ $\quad\quad\vert$ $\quad\quad CH_2-C-NH_2$ $\quad\quad\quad\quad\vert\vert$ $\quad\quad\quad\quad O$	5.4
aspartic acid	Asp	$H_2N-CH-COOH$ $\quad\quad\vert$ $\quad\quad CH_2COOH$	2.8
cysteine	Cys	$H_2N-CH-COOH$ $\quad\quad\vert$ $\quad\quad CH_2SH$	4.1
glutamic acid	Glu	$H_2N-CH-COOH$ $\quad\quad\vert$ $\quad\quad CH_2CH_2COOH$	3.2
glutamine	Gln	$H_2N-CH-COOH$ $\quad\quad\vert$ $\quad\quad CH_2CH_2-C-NH_2$ $\quad\quad\quad\quad\quad\quad\vert\vert$ $\quad\quad\quad\quad\quad\quad O$	5.7

glycine	Gly	$H_2N—CH_2—COOH$	6.0
histidine	His	$H_2N—CH—COOH$ (imidazole side chain)	7.6
isoleucine	Ile	$H_2N—CH—COOH$ $CH—C_2H_5$ CH_3	6.0
leucine	Leu	$H_2N—CH—COOH$ $CH_2CH(CH_3)_2$	6.0
lysine	Lys	$H_2N—CH—COOH$ $CH_2(CH_2)_3NH_2$	9.7
methionine	Met	$H_2N—CH—COOH$ $CH_2CH_2SCH_3$	5.7
phenylalanine	Phe	$H_2N—CH—COOH$ CH_2 (benzene ring)	5.5
proline	Pro	$HN—CH—COOH$ (pyrrolidine ring)	6.3
serine	Ser	$H_2N—CH—COOH$ CH_2OH	5.7
threonine	Thr	$H_2N—CH—COOH$ $CH—CH_3$ OH	5.6
tryptophan	Trp	$H_2N—CH—COOH$ CH_2 (indole ring)	5.9
tyrosine	Tyr	$H_2N—CH—COOH$ CH_2 (phenol ring, OH)	5.7
valine	Val	$H_2N—CH—COOH$ $CH(CH_3)_2$	6.0

▲ Table 1 Proteinogenic 2-amino acids

Essential 2-amino acids

While certain 2-amino acids can be synthesized in the human body from simple molecules, other proteinogenic amino acids must be supplied in the diet, usually in the form of proteins. The latter amino acids are termed **essential** and include histidine, isoleucine, leucine, lysine, methionine, phenylalanine, threonine, tryptophan, and valine. However, many non-essential amino acids can become essential under various conditions. For example, arginine, cysteine, and tyrosine must be present in the balanced diet of infants and growing children; the latter amino acid is also required for people with phenylketonuria (sub-topic B.8).

2-Amino acids as zwitterions

From the chemical point of view, 2-amino acids are amphoteric species that contain a weakly acidic group (–COOH) and a weakly basic group (–NH$_2$) in the same molecule. In neutral aqueous solutions, both the carboxyl group and the amino group are almost completely ionized and exist as –COO$^-$ and –NH$_3^+$, respectively. This ionization can be represented as an intramolecular neutralization reaction or a migration of a proton (H$^+$) from the –COOH group to the –NH$_2$ group:

$$H_2\overset{}{N} - CH - COOH \longrightarrow H_3\overset{+}{N} - CH - COO^-$$
$$\qquad\qquad | \qquad\qquad\qquad\qquad\qquad |$$
$$\qquad\qquad R \qquad\qquad\qquad\qquad\qquad R$$

molecular form   zwitterion

The resulting species with two ionized groups has net zero charge and is called **zwitterion** (from the German *Zwitter*, which means "hybrid"). The –NH$_3^+$ group in the zwitterion is the acidic centre that can lose a proton in strongly alkaline solutions and produce the **anionic form** of the amino acid:

$$H_3\overset{+}{N} - CH - COO^- + OH^- \rightleftharpoons H_2N - CH - COO^- + H_2O$$
$$\qquad | \qquad\qquad\qquad\qquad\qquad\qquad\qquad |$$
$$\qquad R \qquad\qquad\qquad\qquad\qquad\qquad\qquad R$$

zwitterion   strong base   anionic form

The –COO$^-$ group in the zwitterion is the basic centre that can be protonated in strongly acidic solutions and produce the **cationic form** of the amino acid:

$$H_3\overset{+}{N} - CH - COO^- + H^+ \rightleftharpoons H_3\overset{+}{N} - CH - COOH$$
$$\qquad | \qquad\qquad\qquad\qquad\qquad\qquad\qquad |$$
$$\qquad R \qquad\qquad\qquad\qquad\qquad\qquad\qquad R$$

zwitterion   strong acid   cationic form

The exact ratios of the cationic, zwitterionic and anionic forms of an amino acid depend on the pH of the solution and the nature of the side-chain (R). At pH ≈ 6, amino acids with neutral side-chains (R = CH$_3$, CH$_2$OH, etc.) exist almost exclusively as zwitterions while the concentrations of cationic and anionic forms are negligible. In this case the sum of the positive and negative charges of all forms of the amino acid is zero, so this pH is called the **isoelectric point (pI)** of the amino acid. Each amino acid has a specific pI value, which typically falls in the range from 5.5 to 6.3 (table 1). The presence of an additional carboxyl group in the side-chain lowers the pI to 2.8–3.2 while extra amino groups increase the pI to 7.6–10.8.

At pH < pI, the net electric charge of the amino acid species becomes positive, the concentration of the cationic form increases, and the zwitterion concentration decreases. At pH > pI, the amino acid has a negative net electric charge, with more anionic and fewer zwitterionic species present in the solution. However, zwitterions remain the most abundant species in the solution over a broad pH range (usually pI ± 3) while cationic and anionic forms become dominant only in strongly acidic and strongly alkaline solutions, respectively (figure 1).

The ability of amino acids and their derivatives (peptides and proteins) to exist in various forms and neutralize both strong acids and strong bases is important in maintaining the acid–base balance in living organisms (sub-topic B.7).

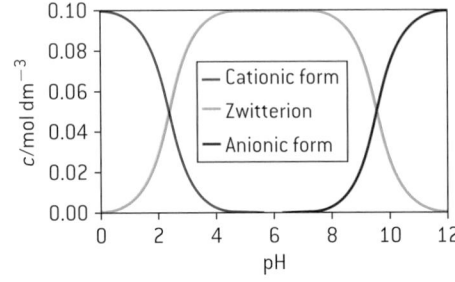

▲ Figure 1 Acid–base equilibria in 0.1 mol dm^{-3} aqueous solution of alanine (pI = 6.0)

Gel electrophoresis

Amino acids, peptides, proteins, and other ionizable compounds can be separated and identified by **gel electrophoresis**. In a typical experiment, a mixture of amino acids is placed in the centre of a square plate covered with agarose or a polyacrylamide gel. The gel is saturated with a buffer solution (sub-topic B.7) to maintain a constant pH during the experiment. Depending on the pH, the amino acids in the mixture will have various net charges – the greater the difference between the pH of the buffer and the pI of the amino acid, the greater the charge. For example, at pH = 6.0, glutamic acid (pI = 3.2) will be charged negatively, alanine (pI = 6.0) will exist as a zwitterion with zero net charge, while both histidine (pI = 7.6) and arginine (pI = 10.8) will be charged positively.

When two electrodes are connected to the opposite sides of the gel and an electric current is applied, negatively charged glutamic acid will move to the positively charged electrode (anode), non-charged alanine will not move, while positively charged histidine and arginine will move to the negatively charged electrode (cathode). Since the pI of arginine is much further from the buffer pH (10.8 – 6.0 = 4.8) than the pI of histidine (7.6 – 6.0 = 1.6), arginine will move faster than histidine and travel further from the centre of the plate.

When the separation is complete, the gel is developed with a locating agent, ninhydrin, that forms coloured compounds with amino acids. The composition of the mixture can be determined by comparing the distances of the coloured spots from the centre of the plate with those of known amino acids (figure 2).

If the separation is incomplete, the plate can be rotated 90 degrees and the electrophoresis repeated at a different pH. The amino acids will move perpendicular to their original direction, separating overlapping spots and producing a 2D map of the mixture. This 2D technique is particularly useful in protein and DNA studies, when complex mixtures containing hundreds or thousands of compounds are analysed.

Gel electrophoresis is widely used in biochemistry and medical diagnostics, in particular, for the analysis of unusual protein content in blood serum or urine.

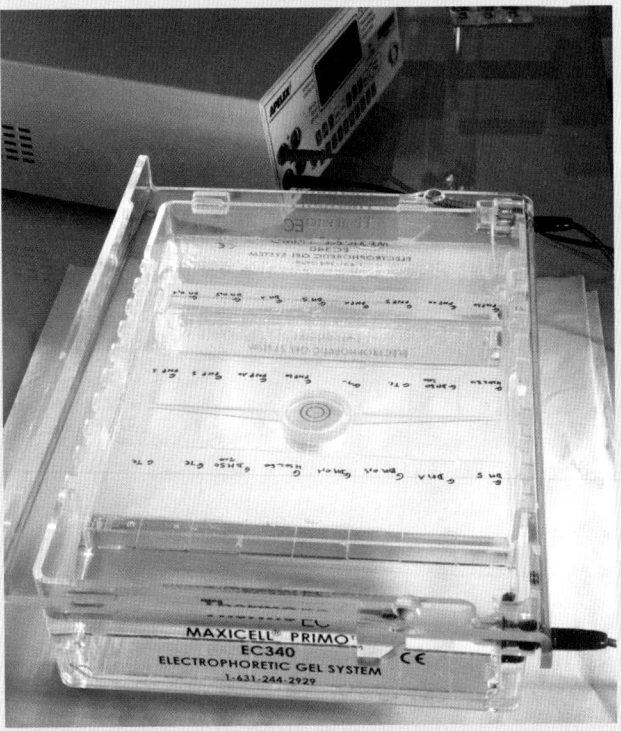
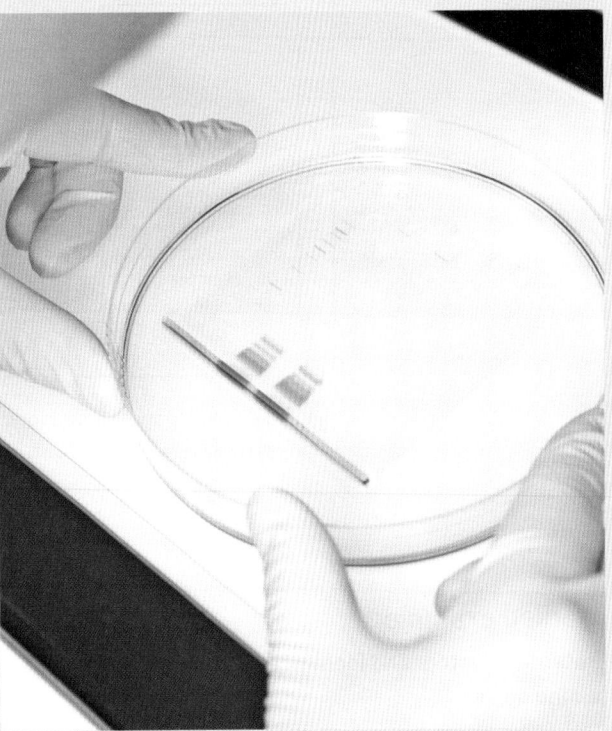

▲ Figure 2 A gel electrophoresis unit (left) and a developed map of a protein mixture (right)

Paper chromatography

Paper chromatography is another common technique used for the identification of amino acids and other organic compounds. A spot of a liquid sample containing the amino acids is placed on the start line near the bottom of a rectangular piece of chromatographic paper (which forms the **stationary phase**). Separate spots of solutions containing known amino acids are placed on the same start line, and the paper is put into a beaker containing a suitable solvent (the **mobile phase** or **eluent**). Due to capillary action, the solvent rises up the paper and eventually reaches the spots of amino acids. As the solvent moves further up the paper, the amino acids partition between the mobile and stationary phases according to their affinities for the solvent and the chromatographic paper. The compounds with higher solubility spend more time in the mobile phase and move up faster than less soluble compounds with a greater tendency to adsorb on the stationary phase. When the solvent front reaches almost the top of the paper, the chromatogram is removed from the beaker,

dried, and developed using a locating agent (ninhydrin) to make the spots visible.

Figure 3 shows a chromatogram of a sample containing a mixture of amino acids. A single spot of the sample has been separated into three isolated spots (A, B, and C) at certain distances (L_A, L_B, and L_C) from the start line. Although these distances can vary from experiment to experiment, the *ratio* of the distances travelled by each spot to the distance travelled by the solvent front (L_0) remains constant. This ratio is known as the **retention factor (R_f)**:

$$R_f(A) = \frac{L_A}{L_0} \qquad R_f(B) = \frac{L_B}{L_0} \qquad R_f(C) = \frac{L_C}{L_0}$$

Each amino acid (or any other compound) has a specific R_f value that is independent of L_0 but depends on the experimental conditions (solvent, paper type, temperature, pH, etc.). Retention factors of all common amino acids determined under standard experimental conditions can be found in reference books and used for the identification of individual components in mixtures.

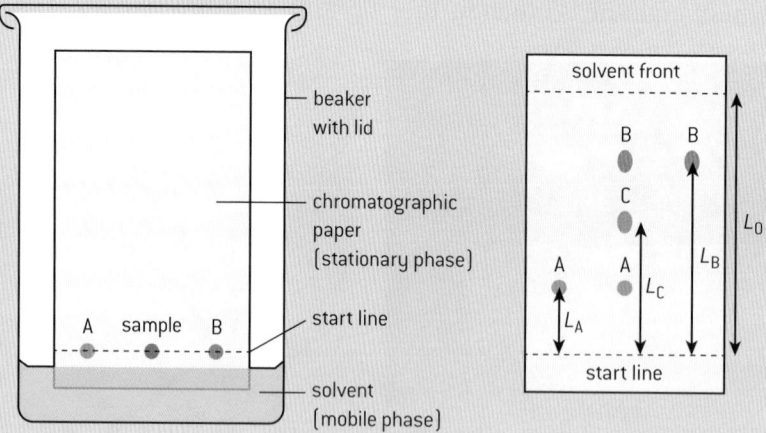

▲ Figure 3 A typical paper chromatography experiment (left) and the chromatogram obtained (right)

Worked example

Retention factors and identification of unknown amino acids

Under certain conditions, proteinogenic 2-amino acids have the retention factors shown in table 2.

In the experiment shown in figure 3 distances L_A, L_B, L_C, and L_0 are 14, 39, 27 and 54 mm, respectively. Identify the unknown amino acid C (figure 3) if A is glycine and B is leucine.

Amino acid	R_f	Amino acid	R_f	Amino acid	R_f	Amino acid	R_f	Amino acid	R_f
histidine	0.11	glutamine	0.13	lysine	0.14	arginine	0.20	aspartic acid	0.24
glycine	0.26	serine	0.27	glutamic acid	0.30	threonine	0.35	alanine	0.38
cysteine	0.40	proline	0.43	tyrosine	0.45	asparagine	0.50	methionine	0.55
valine	0.61	tryptophan	0.66	phenylalanine	0.68	leucine	0.72	isoleucine	0.73

▲ Table 2 The R_f values for amino acids under certain conditions

Solution

First we must confirm that our experimental conditions are the same as those used in the reference experiment. Indeed, $R_f(A) = 14/54 \approx$ 0.26 (glycine) and $R_f(B) = 39/54 \approx 0.72$ (leucine) match the values given in the table, so the retention factor of the unknown amino acid C can be used for its identification. Thus $R_f(C) = 27/54 =$ 0.50 (asparagine).

Experimental conditions for paper chromatography

Depending on the type of compounds present in the mixture, the stationary and mobile phases must be chosen carefully. Standard chromatographic paper consists of the polysaccharide cellulose (sub-topic B.10) that readily adsorbs polar compounds. If a non-polar solvent (for example, a hydrocarbon) is used, highly polar amino acids will remain at the start line ($R_f = 0$) and no separation will be achieved. At the same time, in a highly polar solvent (such as water), amino acids will stay in the mobile phase and travel with the solvent front ($R_f = 1$). Therefore the most common solvents used for amino acid separation are moderately polar alcohols, esters, or chlorinated hydrocarbons. In modern laboratories the use of chlorinated solvents is avoided due to environmental concerns (sub-topic B.6).

If two or more components have similar R_f values, the experiment can be repeated by rotating the paper through 90 degrees and using a different solvent, pH, or even separation method, such as gel electrophoresis. The latter approach was successfully employed in 1951 by Frederick Sanger for the identification of the amino acid composition of insulin.

Modern chromatographic techniques

As well as paper chromatography, many other chromatographic methods have been developed. In **thin-layer chromatography (TLC)** the adsorbent (silica, alumina, or cellulose) is fixed on a flat, inert plate, usually made of aluminium foil or glass. TLC plates offer a wide choice of stationary phases and usually allow faster and more efficient separation than chromatographic paper.

In **column chromatography** the stationary phase (usually silica or alumina) is packed into a long tube with a tap at the bottom. The sample is placed at the top, followed by the solvent (mobile phase). When the tap is opened the mobile phase moves down by gravity and carries the components of the sample, which travel at various speeds and leave the column at different times. While TLC and paper chromatography are primarily used for the identification of organic compounds, column chromatography allows chemists to isolate individual compounds and determine the quantitative composition of the mixture.

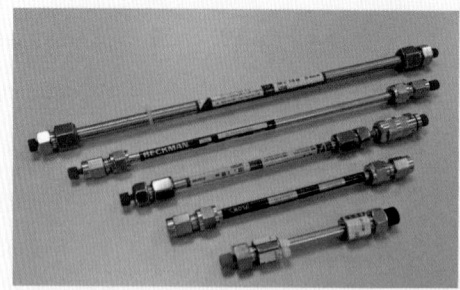

▲ Figure 4 Left: high-performance liquid chromatography (HPLC) columns. Right: a modern gas chromatography (GC) instrument

Various modifications of column chromatography include **high-performance liquid chromatography (HPLC)** that uses solid or liquid stationary phases and a liquid mobile phase pushed through the column at high pressure, and **gas chromatography (GC)** with a gaseous mobile phase and a liquid or solid stationary phase. In HPLC with a liquid stationary phase, the components of the mixture are partitioned between two liquids according to their relative solubilities. GC is primarily used for the identification of volatile compounds in environmental, medical, and forensic studies.

Intermolecular forces in amino acids

In the solid state amino acids exist as zwitterions held together by strong ionic forces between oppositely charged $-NH_3^+$ and $-COO^-$ groups. As a result all proteinogenic amino acids are crystalline solids with high melting points, readily soluble in water, and almost insoluble in non-polar organic solvents. In aqueous solutions the ionic forces are replaced by ion-dipole interactions and hydrogen bonds (sub-topic 4.4) between zwitterions and polar water molecules. In contrast, the molecules of non-polar solvents can form only van der Waals' interactions, which are too weak for overcoming the lattice energy of ionic solids (sub-topic 15.1).

Peptide bonds

Despite the fact that molecular (non-ionized) forms of 2-amino acids do not exist, they are convenient theoretical abstractions that allow us to simplify reaction schemes and the nomenclature of large organic molecules. In this book molecular formulae of amino acids will be used in all cases except when acid–base equilibria are discussed and the exact structures of reacting species must be known.

As mentioned earlier, 2-amino acids may undergo condensation reactions and produce peptides. When the –COOH group of one amino acid reacts with the –NH₂ group of another amino acid, a molecule of water is released and a **peptide linkage** (also known as an **amide linkage** or an **amide bond**) is formed:

peptide linkage

Dipeptides

The product of the above reaction contains the residues of two amino acids and is called a **dipeptide**. If the side-chains of participating amino acids are different, more than one dipeptide can be formed. For example, four different dipeptides can be produced from a mixture of alanine (Ala) and serine (Ser):

$$H_2N-CH-C-N-CH-COOH$$
(with O double bond to C)
CH_3, H, CH_2OH

alanyl-serine (Ala–Ser)

$$H_2N-CH-C-N-CH-COOH$$
CH_2OH, H, CH_3

seryl-alanine (Ser–Ala)

$$H_2N-CH-C-N-CH-COOH$$
CH_3, H, CH_3

alanyl-alanine (Ala–Ala)

$$H_2N-CH-C-N-CH-COOH$$
CH_2OH, H, CH_2OH

seryl-serine (Ser–Ser)

Naming peptides

The names of peptides are formed by changing the suffixes of all but the last amino acid residue from "ine" or "ic acid" to "yl" (i.e., alan*ine* + serine = alan*yl*-serine). Alternatively, abbreviated names of amino acids (table 1) can be joined together by dashes (for example, Ala + Ser = Ala–Ser).

The order of amino acid residues in peptides is very important – for example, the dipeptides Ala–Ser and Ser–Ala are two different compounds that might have very different physiological properties. The first amino acid in a peptide has a free $-NH_2$ group, described as **N-terminal**, while the last amino acid has an unreacted –COOH group (**C-terminal**). Both N- and C-terminals can participate in further condensation reactions that produce larger peptides and proteins. In living organisms the synthesis of peptides usually begins from their N-terminals, so the sequence of amino acids is traditionally recorded in the same way.

An example peptide

The structural formula of a tetrapeptide, Gly–Asp–Pro–Lys, is drawn below. Note that the amino group of proline makes unusual peptide linkages (CO–N instead of CO–NH), and that the side-chains of amino acids remain unchanged when peptides are formed.

N-terminal   peptide linkages   C-terminal

$$H_2N-CH_2-C-N-CH-C-N-CH-C-N-CH-COOH$$
H, CH_2
COOH
(Pro ring)
H, (CH_2)_4
NH_2

Gly   Asp   Pro   Lys

How many peptides can we make?

Amino acids can be joined together in any combinations and produce a virtually limitless number of peptides. Twenty proteinogenic amino acids can form $20 \times 20 = 400$ dipeptides, $20 \times 20 \times 20 = 8000$ tripeptides, etc. For a polypeptide chain of 50 amino acid residues the number of possible combinations reaches 20^{50}, or approximately 10^{65}. If a single molecule of each of these polypeptides could be made, their combined mass would be 2×10^{43} g, which is greater than the entire mass of the Earth (6×10^{27} g), solar system (2×10^{33} g), and even our galaxy (10^{43} g)!

Quick question

Draw the structural formulae of tripeptides Ala–Ser–Pro and Pro–Ser–Ala. Label the peptide linkages, N-terminals, and C-terminals. How many water molecules are released when one molecule of a tripeptide is formed?

From the chemical point of view, both the formation and the hydrolysis of peptide linkages are nucleophilic substitution (S_N) reactions (sub-topic 20.1). However, this term is rarely used in biochemistry while the names "condensation" and "hydrolysis" are much more common.

Peptides in the human body

In the human body, peptides perform various regulatory and signalling functions. Some peptides act as growth hormones that regulate cell reproduction and tissue regeneration. Another group of peptides, endorphins, mimics the effects of opiates (sub-topic D.3), inhibiting the transmission of pain signals and inducing a feeling of well-being. Glutathione, a tripeptide containing a residue of cysteine, is an efficient natural antioxidant (sub-topic B.3). Finally, peptides are easily digestible and can be used as a source of 2-amino acids for the biosynthesis of proteins.

Protein sequencing

Primary structures of proteins can be determined by various techniques, collectively known as **protein sequencing**, including mass spectrometry (sub-topics 2.1 and 11.3), NMR (sub-topics 11.3 and 21.1), and sequential hydrolysis followed by gel electrophoresis or chromatography (see above). The primary structure of the first sequenced protein, bovine insulin, was determined by Frederick Sanger in 1951, in a study that took over ten years and was later recognized with the Nobel Prize in Chemistry. Today protein sequencing is a routine, fast, and highly efficient process that is widely used in **proteomics** for large-scale analysis of proteins.

The hydrolysis of peptides

In the presence of strong acids, strong bases, or enzymes, peptides can be hydrolysed into individual amino acids, for example:

$$H_2N-CH(CH_3)-C(=O)-N(H)-CH(CH_2OH)-COOH + H_2O \longrightarrow H_2N-CH(CH_3)-COOH + H_2N-CH(CH_2OH)-COOH$$

Ala–Ser → Ala + Ser

The hydrolysis of each peptide linkage requires one molecule of water. In a peptide with n amino acid residues, the number of peptide linkages will be $n-1$ and therefore $n-1$ water molecules will be needed to balance the equation.

Properties of peptides

The acid–base properties of peptides are similar to those of 2-amino acids. Terminal $-NH_2$ and $-COOH$ groups, together with the functional groups of the peptide side-chains, can be ionized to various extents and, depending on the pH of the solution, produce polyions with multiple positive and negative charges. Each peptide has a characteristic isoelectric point (pI), which can be used to separate and analyse peptide mixtures by gel electrophoresis. Together with proteins and individual amino acids, peptides act as acid–base buffers and maintain a constant pH of biological fluids (sub-topic B.7).

Proteins: Primary structure

Proteins are the most diverse biopolymers that vary greatly in size, shape, and composition. Simple proteins consist of a single chain of 2-amino acid residues connected to one another in strict order and orientation. The exact sequence of amino acid residues joined together by peptide linkages is known as the **primary structure** of a protein. Similar to peptides, proteins have N- and C-terminals, and the primary structure is traditionally written from left to right starting from the N-amino acid. A fragment of the primary structure of a relatively simple protein, bovine insulin, is shown below:

Gly–Ile–Val–Glu–Gln–Cys–Cys–Ala–Ser–Val–Cys–Ser–Leu–Tyr–Gln–...

Proteins: Secondary structure

Long chains of amino acid residues in proteins tend to adopt certain highly ordered conformations, such as α-**helix** and β-**pleated sheet**. These local and regularly repeating conformations are stabilized by intramolecular hydrogen bonds between carbonyl and amino fragments of peptide linkages and are collectively known as the **secondary structure** of a protein.

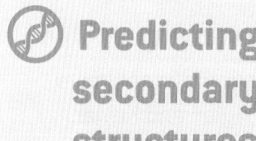

The α-**helix** is a rod-like arrangement of amino acid residues with the side-chains (R) extending outward from a tightly coiled backbone of repeating –NH–CH–CO– units. In the α-helix, the C=O group of each amino acid residue forms a hydrogen bond with the NH group of the amino acid residue that is situated four units ahead in the sequence. In diagrams and models of proteins α-helices are commonly represented as twisted ribbons or rods. Certain proteins such as *tropomyosins* (responsible for the regulation of muscle contraction) consist of nearly 100% α-helix while in other proteins α-helical fragments might be completely absent.

The β-**pleated sheet**, or simply β-**sheet**, contains two or more chains of amino acid residues (known as β-**strands**) which are almost completely extended. The adjacent β-strands can run in the same or opposite directions, producing parallel or antiparallel β-sheets, respectively. If only two β-strands are present they are linked by hydrogen bonds in a ladder-like fashion; hydrogen bonds between three or more β-strands form a regular two-dimensional network. In diagrams and models of proteins β-pleated sheets are usually represented as broad ribbons, often with an arrow pointing toward the C-terminal. Similar to α-helices, the occurrence of β-sheets in proteins can vary from almost zero to nearly 100%. For example, many fatty acid-binding proteins (responsible for lipid metabolism – see sub-topic B.3) consist almost entirely of β-pleated sheets.

Predicting secondary structures

The structures of both the α-helix and the β-pleated sheet were proposed by Linus Pauling and Robert Corey in 1951, six years before the first experimental evidence of the protein conformations could be obtained. This was one of the major achievements in biochemistry because it clearly demonstrated that the conformation of a complex molecule can be predicted if the properties of its constituent parts are known.

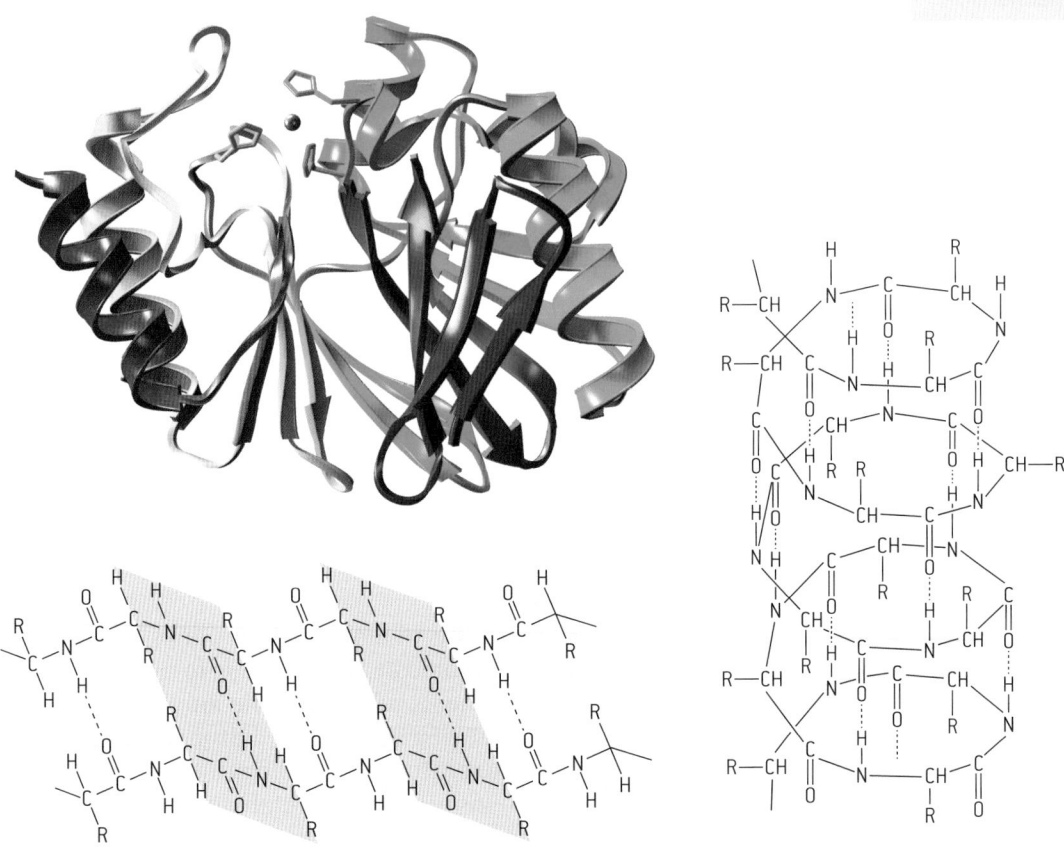

▲ Figure 5 Secondary structures of proteins: α-helix (right) and β-pleated sheet (bottom). A computer model of a protein-based antibiotic resistance enzyme (top) shows several α-helices and multiple β-pleated sheets; the arrows point toward the C-terminal

Certain synthetic polymers, such as nylon and Kevlar, belong to the class of polyamides and closely resemble proteins. Like proteins, synthetic polyamides have a primary structure of repeating units joined together by amide (peptide) linkages. In addition, most polyamides have a highly regular secondary structure stabilized by hydrogen bonds between amide linkages of adjacent polymeric chains. Multiple hydrogen bonds in Kevlar are largely responsible for the exceptional mechanical strength of this polymer, which is five times stronger than steel of the same mass and therefore used for making personal armour and sports equipment. The structures and properties of synthetic polyamides are discussed in sub-topic A.9

Interactions between side-chains: Tertiary structure

While the secondary structure of proteins is stabilized exclusively by the hydrogen bonds between peptide linkages, the side-chains of amino acid residues can also participate in various types of intra- and intermolecular interactions. For example, two non-polar or slightly polar side-chains (such as $-CH_2CH(CH_3)_2$ in leucine or $-CH_2C_6H_5$ in phenylalanine) can interact via weak van der Waals' forces (sub-topic 4.4) while oppositely charged ionized groups (such as $-CH_2-COO^-$ in aspartic acid and $-(CH_2)_4NH_3^+$ in lysine) can experience electrostatic attraction and form ionic bonds. Hydrogen bonds are often formed between non-ionized hydroxyl and/or amino groups (such as $-CH_2-C_6H_4-OH$ in tyrosine and the heterocyclic fragment $-C_3H_3N_2$ in histidine). Finally, covalent bonds can also be formed between certain functional groups of the side-chains. This includes additional peptide linkages between carboxyl and amino groups, ester bonds between carboxyl and hydroxyl groups, and disulfide bridges between two –SH groups of cysteine residues.

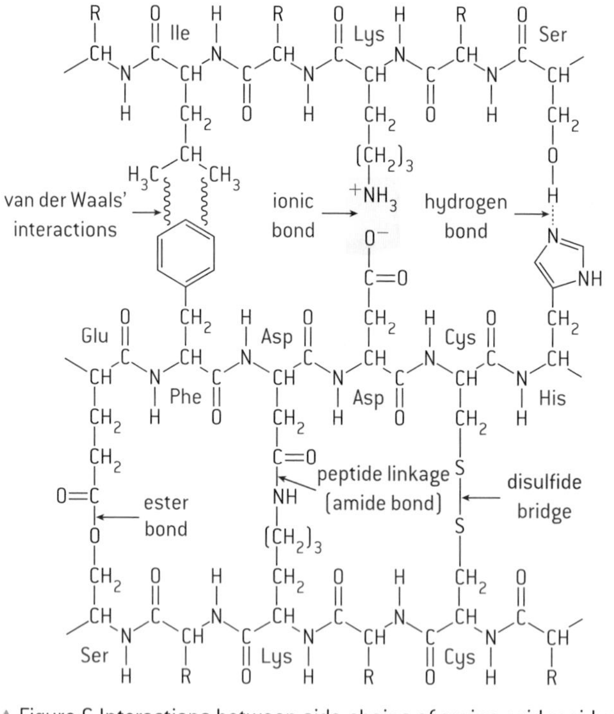

▲ Figure 6 Interactions between side-chains of amino acid residues in proteins

The interactions between side-chains of amino acid residues can cause additional folding of the protein molecule, which leads to a specific arrangement of α-helices and β-sheets relative to one another. The resulting three-dimensional shape of a single folded protein molecule is known as its **tertiary structure**. Under physiological conditions, tertiary structures of most proteins are compact globules with non-polar (hydrophobic) side-chains buried inside and polar groups facing outwards. Such **globular proteins** are readily soluble in water and easily transported by biological fluids. Globular proteins often act as biological catalysts (enzymes), chemical messengers (hormones), or carriers of physiologically active molecules. In contrast, **fibrous**

proteins (also known as scleroproteins) tend to adopt rigid, rod-like conformations, are insoluble in water, and usually perform structural or storage functions in living organisms.

Three-dimensional arrangement: Quaternary structure

Folded protein molecules often interact with one another and form larger assemblies containing multiple polypeptide chains (**protein subunits**) and sometimes non-protein components (**prosthetic groups**), such as heme in hemoglobin (sub-topic B.9) or lipids in lipoproteins (sub-topic B.3). The three-dimensional arrangement of protein and non-protein components in such assemblies is known as their **quaternary structure**. The individual subunits in a quaternary structure are held together by van der Waals' forces (often referred to as "hydrophobic interactions") although hydrogen bonding or ionic interactions between adjacent polypeptide chains can also contribute to the overall stability of resulting multicomponent assemblies.

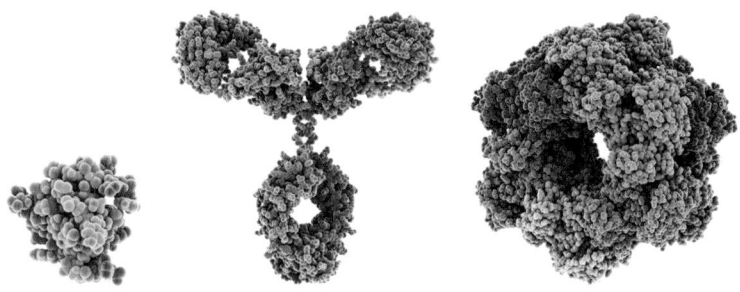

▲ Figure 7 Tertiary and quaternary structures of globular proteins: insulin (left), immunoglobulin G (centre), and glutamine synthetase (right). Separate polypeptide chains are shown in different colours

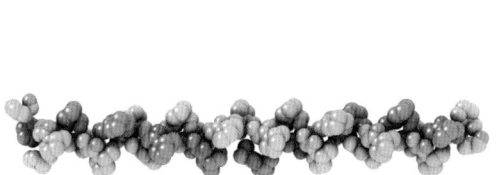

▲ Figure 8 Structures of fibrous proteins. Left: the tertiary structure of collagen is composed of three α-helices wrapped around one another ("coiled coil"). Right: a scanning electron microscopy image of collagen fibres in the human tendon

Secondary, tertiary, and quaternary structures of proteins and other biomolecules under physiological conditions are collectively known as their **native states** or **native structures**. A protein in its native state is properly folded and contains all the subunits required for performing its functions in the living organism. In contrast, **denatured** proteins do not possess their native three-dimensional structures and are unable to perform their physiological functions. Denaturation of proteins is caused by organic solvents, heavy metal ions, high concentrations of inorganic salts, or changes in pH or temperature (figure 9).

Perms and disulfide bridges

A permanent wave, or "perm", is a hairstyling technique based on chemical modification of the tertiary structures of keratin, the main structural component of human hair. Keratin is a fibrous protein that contains multiple disulfide bridges between adjacent polypeptide chains. When these disulfide bridges are temporarily broken by a reducing reagent, the hair loses its elasticity and can be curled or folded easily. After that, the disulfide bridges are re-formed by applying an oxidizing reagent, and the new hair shape is fixed for a period of up to several months.

▲ Figure 9 Denaturation of proteins. Albumins in egg white lose their native structure when exposed to high temperature

Proteomics and international collaboration

Large-scale analysis of proteins, known as **proteomics**, allows biochemists to predict their functions in living organisms, study metabolic pathways, and develop new drugs. The Universal Protein Resource (UniProt) is a consortium of bioinformatics institutes that provides the international scientific community with comprehensive, reliable, and freely accessible data on protein sequences, structures, and functions. UniProt is funded by governmental organizations across the world and offers a range of free services at www.uniprot.org that allow you to search, download, and analyse various proteomics data.

Acid-base properties of proteins

Similar to individual amino acids, proteins are amphoteric species with multiple acidic and basic functional groups in the side-chains of their constituent amino acid residues. Depending on the pH of the solution these functional groups can be ionized to various extents, producing protein polyions with different charges (figure 10). Each protein has a specific isoelectric point (pI) where the numbers of positive and negative charges are equal and the net charge of the polyion is zero. Therefore, proteins with different isoelectric points can be separated by gel electrophoresis in the same way as individual amino acids.

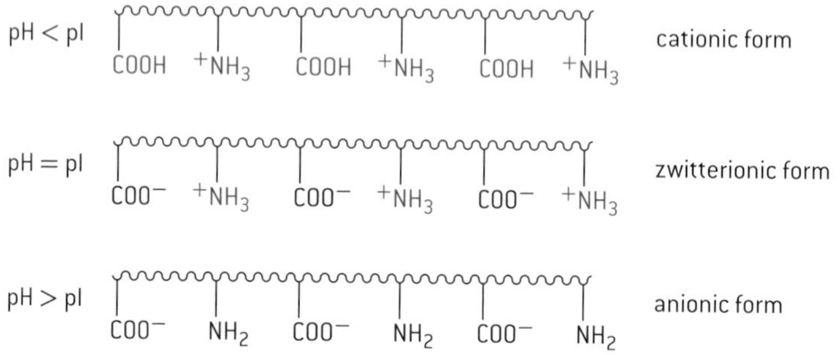

▲ Figure 10 Cationic, zwitterionic, and anionic forms of a protein. Wavy lines represent polypeptide backbones

A modification of the gel electrophoresis technique known as **isoelectric focusing** allows biochemists to concentrate proteins in certain areas of the polyacrylamide gel. This is achieved by using two different buffer solutions at the opposite sides of the gel, which creates a pH gradient. Each protein moves in the electric field until it reaches the area of the gel with pH = pI. At this point the protein acquires net zero charge and becomes immobile, so eventually all proteins spread across the gel according to their individual pI values. The gel is then developed with a locating agent such as silver nitrate or Coomassie® Brilliant Blue dye. Alternatively, the gel material can be cut into narrow strips containing individual proteins for further analysis.

The presence and approximate concentration of proteins and peptides in solutions can be determined by the **biuret test**, which will be discussed in sub-topic B.7.

Enzymes

Most proteins in the human body act as **enzymes** – highly specific and efficient biological catalysts that control virtually all biochemical processes, from the digestion of food to the interpretation of genetic information. Enzymes are classified by the nature of the reaction they catalyse, and their names usually end with the suffix "-ase". For example, *oxidoreductases* catalyse redox reactions (such as the oxidation of ethanol to ethanal catalysed by *alcohol dehydrogenase*) while *transferases* are responsible for the transfer of functional groups (such as the transfer of a phosphate group by *phosphotransferases*). Other enzymes are known by trivial or semi-trivial names such as *catalase* (*hydrogen-peroxide oxidoreductase*) or *lactase* (the enzyme responsible for the hydrolysis of the disaccharide lactose). The absence or insufficient activity of the latter enzyme in adults is a common medical condition known as *lactose intolerance* (sub-topic B.4).

The efficiency of enzymes greatly exceeds the catalytic power of synthetic catalysts. Some enzymes can accelerate reactions as much as 10^{16} times, so chemical transformations that would normally take millions of years proceed in milliseconds in living organisms. At the same time, every enzyme is very specific and catalyses only one or few chemical reactions. This allows enzymes to operate with high precision and distinguish between very similar reactants such as the amino acids valine, leucine, and isoleucine.

Molecules that are modified by enzymes are called **substrates**. Enzymes are large molecules, and the substrate interacts with a relatively small region of the enzyme known as the **active site**. The catalytic process begins when the substrate comes into close proximity with the active site. If the substrate and the active site have complementary structures and correct orientations, a chemical "recognition" occurs and an **enzyme–substrate complex** is formed. Multiple intermolecular interactions in this complex distort and weaken existing chemical bonds in the substrate, making it more susceptible to certain chemical transformations within the active site. The catalytic cycle completes when the reaction product detaches from the enzyme, leaving the active site available for the next substrate molecule.

The above description is a variation of the "**lock-and-key model**" (figure 11) developed in 1894 by the Nobel laureate Emil Fisher. According to modern views, the active site and the substrate molecule do not fit exactly and change their shapes slightly during the catalytic processes. This theory, known as the "**induced fit model**" (sub-topic B.7), suggests that the initial enzyme–substrate interactions are relatively weak but sufficient to induce the conformational changes in the active site that strengthen the binding.

Protein deficiency

Proteins are the main source of amino acids and so they must be present in a healthy diet in sufficient quantities (sub-topic B.4). Protein deficiency causes various diseases that are widespread in many developing countries. One of these diseases, **kwashiorkor**, is characterized by a swollen stomach, skin discoloration, irritability, and retarded growth.

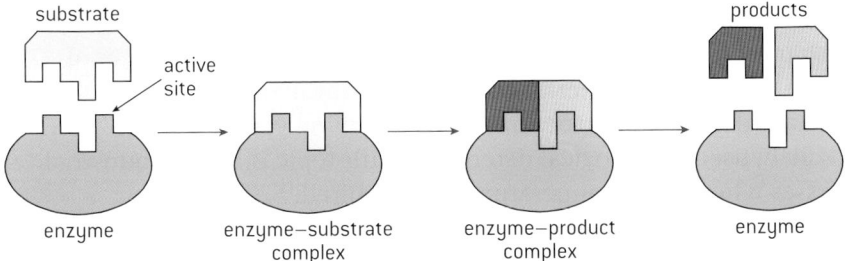

▲ Figure 11 The "lock-and-key" model of enzyme catalysis

Like all catalysts, enzymes cannot change the equilibrium position of the chemical reactions they catalyse. However, by providing alternative reaction pathways with low activation energies (sub-topic 16.2), enzymes facilitate the transfer of energy between different biochemical processes and thus allow the equilibrium of one reaction to be affected by another. In the human body, the energy required for anabolic processes is usually supplied by the hydrolysis of ATP (sub-topic B.8).

The efficiency of an enzyme as a biological catalyst depends on the configuration and charge of its active site, which are very sensitive to pH and temperature. The amino acid residues of both the enzyme backbone and the active site contain ionizable side-chains that undergo reversible protonation or deprotonation. Any change in pH affects the charges of these side-chains and their ability to form ionic and hydrogen bonds with one another. The weakening and breaking of these bonds alter the three-dimensional structure of the enzyme and the shape of its active site, which can no longer accommodate the substrate molecule. In addition, the substrate itself often contains ionizable functional groups that must have specific charges in order to interact with the active site. These charges are also affected by pH, making the enzyme–substrate complex stable over a limited pH range. As a result, most enzymes work best at physiological pH (7.4) or within a narrow pH interval, typically between 6 and 8 (figure 12). Outside this range the enzymes become denatured and rapidly lose their activity. However, certain enzymes can perform their functions under strongly acidic or basic conditions. For example, *pepsin*, a component of the gastric juice, has an optimum pH between 1.5 and 2.0 while *arginase*, an enzyme responsible for the hydrolysis of arginine, shows its maximum activity at pH = 9.5–10.

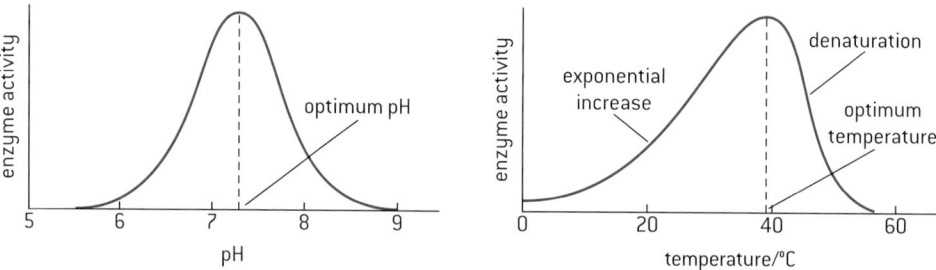

▲ Figure 12 Effects of pH and temperature on the activity of a typical enzyme

Like most chemical reactions, the rates of enzymatic processes generally obey the Arrhenius equation (sub-topic 16.1) and increase exponentially when the temperature rises from 0 to approximately 30 °C. After that point the enzyme activity increases more gradually, reaches its maximum at or slightly above the body temperature (37 °C), and then falls sharply due to thermal denaturation. However, the enzymes of certain thermophilic bacteria reach their optimum activity at 80–90 °C and retain their native structures even in boiling water. Such enzymes are widely used in biological detergents (sub-topic B.6) and industrial processes where high temperatures are required.

Side-chains and polypeptide backbones of enzymes contain many oxygen, nitrogen, and sulfur atoms that can act as ligands and form chelate complexes with various metals (sub-topics 13.1 and 13.2). Heavy metal ions such as lead(II), mercury(II), and cadmium(II) preferentially bind to the –SH groups in the side-chains of cysteine residues, disrupting the formation of disulfide bridges or replacing them with sulfur–metal–sulfur fragments. As a result enzymes become denatured and lose their activity, which is the primary cause of heavy metal toxicity. At the same time, certain heavy metals are essential components of prosthetic groups in some enzymes and metalloproteins (sub-topic B.9).

For many enzymatic processes the reaction rate (υ) varies with the substrate concentration ([S]) as shown in figure 13. When the substrate concentration is low υ is proportional to [S], so the process is a first order reaction (sub-topic 16.1) with respect to S. At higher substrate concentrations υ is nearly independent of [S], and the process becomes a zero order reaction with respect to S.

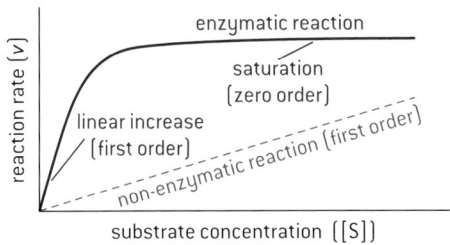

Such unusual dependence of the reaction rate on the substrate concentration is caused by the enzyme–substrate complex ES that forms when the substrate binds to the active site of the enzyme (figure 11). In order to complete its chemical transformation a substrate molecule must remain at the active site for a certain period of time, making the enzyme unavailable for other substrate molecules. When [S] is very low most active sites are vacant, so every substrate molecule can bind to the nearest enzyme without delay. However, at very high [S] nearly all active sites are occupied, and the enzyme works at its maximum capacity. New substrate molecules must wait until active sites become available again, so any further increase of [S] does not affect the reaction rate. The rates of enzymatic processes are quantitatively described by the Michaelis–Menten equation, which will be discussed in sub-topic B.7.

▲ Figure 13 Kinetics of a typical enzyme-catalysed reaction

Questions

1 Individual 2-amino acids have different structures depending on the pH of the solution they are dissolved in. The molecular formula of serine is given in the *Data booklet*.

a) Deduce the structure of serine in a solution with a pH of: i) 2 ii) 12. [2]

b) Deduce the structure of serine at the isoelectric point. [1]

IB, May 2010

2 Explain why 2-amino acids are soluble in water and have high melting points.

3 The primary structure of proteins describes how the different 2-amino acids are linked to each other in a linear chain. Draw the structures of the two different dipeptides that can be formed when glycine reacts with serine. [2]

IB, May 2012

4 Deduce a balanced equation for a condensation reaction that produces the tripeptide Ser–Lys–Phe. Label the peptide linkages, N-terminal, and C-terminal in the resulting peptide.

5 Proteins are products of polycondensation of 2-amino acids. Explain the differences between the primary and secondary structures of proteins and state the bond types responsible for maintaining these structures. [2]

IB, November 2012

6 The tertiary structures of proteins made up of 2-amino acid residues such as serine and cysteine are the result of interactions between amino acids to give a three-dimensional shape. State five different types of interaction that

can occur; in each case identify the atoms or groups joined together. [5]

IB, May 2010

7 The tertiary structure of proteins describes the overall folding of the chains to give the protein its three-dimensional shape. This is caused by interactions between the side-chains of distant amino acid residues. Consider the two segments of a polypeptide chain shown in figure 14.

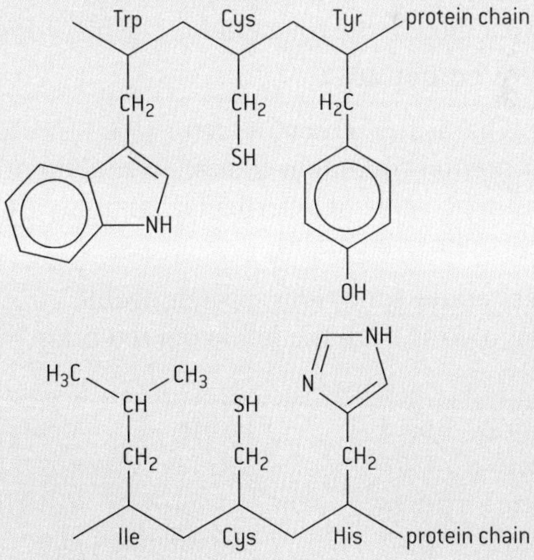

▲ Figure 14

a) Deduce the type of interaction that can occur between the side-chains of Trp and Ile, Cys and Cys, and Tyr and His. [3]

b) State the name of one other type of interaction that can occur between the side-chains of amino acid residues. [1]

IB, May 2012

8 Proteins are natural polymers. List four major functions of proteins in the human body. [2]

IB, May 2010

9 Describe the *quaternary* structure of proteins. [1]

IB, May 2012

10 Proteins are macromolecules formed from 2-amino acids. Once a protein has been hydrolysed, chromatography and electrophoresis can be used to identify the amino acids present.

  a) State the name of the linkage that is broken during the hydrolysis of a protein and draw its structure. [2]

  b) Explain how electrophoresis is used to analyse a protein. [4]

IB, November 2011

11 Chromatography is one of the most universal analytical techniques.

  a) State one qualitative and one quantitative use of chromatography. [2]

  b) Using column chromatography as an example, explain how components of a mixture interact with the stationary and mobile phases, and explain how the separation of the components is achieved. [4]

IB, November 2012

12 State what is the retention factor (R_f). List the experimental conditions that affect and do not affect the R_f value of a particular 2-amino acid in paper chromatography.

13 Under certain conditions, proteinogenic 2-amino acids have the following retention factors:

histidine	0.11	lysine	0.14
glycine	0.26	serine	0.27
alanine	0.38	cysteine	0.40
tyrosine	0.45	asparagine	0.50
valine	0.61	leucine	0.72

A paper chromatogram of a mixture of unknown 2-amino acids showed three spots at distances 10, 28, and 35 mm from the start line. Identify the amino acids if the distance between the start line and the solvent front was 70 mm.

14 Figure 15 represents a thin layer chromatogram of an amino acid.

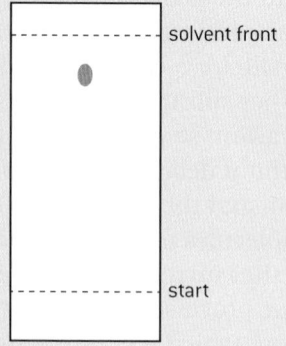

▲ Figure 15

  a) Outline the principle of thin layer chromatography. Refer in your answer to the nature of the mobile and stationary phases and the reason why a mixture of amino acids can be separated using this method. [2]

  b) State one advantage of thin layer chromatography over paper chromatography. [1]

  c) Calculate the R_f of the amino acid. [1]

IB, May 2009

15 Describe how locating agents are used in paper chromatography and gel electrophoresis.

16 Discuss the differences between a traditional catalyst and an enzyme.

17 At a very low concentration of a certain substrate, the rate of the enzyme-catalysed reaction doubles when the substrate concentration increases two times. Explain whether the same effect would be observed at a very high substrate concentration.

18 Enzymes are protein molecules that catalyse specific biochemical reactions. The phosphorylation of glucose is the first step of glycolysis (the oxidation of glucose) and is catalysed by the enzyme hexokinase.

  a) Explain how enzymes such as hexokinase are able to catalyse reactions. [2]

  b) State and explain the effect of increasing the temperature from 20 °C to 60 °C on an enzyme-catalysed reaction. [4]

IB, November 2011

B.3 Lipids

Lipids in living organisms

Lipids are a broad group of naturally occurring substances that are largely non-polar and therefore insoluble in water. Unlike other classes of biomolecules, lipids are defined in terms of their properties rather than structure or chemical behaviour.

In living organisms lipids perform various functions, including energy storage, chemical messaging and transport, thermal insulation of the body, and physical separation of the cell content from biological fluids. Most lipids are relatively small and predominantly hydrophobic molecules that tend to form large assemblies with regular structures. However, in contrast to covalently bonded subunits of biopolymers, individual molecules of lipids in such assemblies are held together by weak van der Waals' forces (sub-topic 4.4).

Fatty acids and triglycerides

Fatty acids is a common name for long-chain unbranched carboxylic acids (table 1). While free fatty acids are not normally classified as lipids themselves, their residues are important components of triglycerides and phospholipids that will be discussed later in this topic.

Chemical formula	Common name	IUPAC name
$CH_3CH_2CH_2COOH$	butyric acid	butanoic acid
$CH_3(CH_2)_6COOH$	caprylic acid	octanoic acid
$CH_3(CH_2)_{10}COOH$	lauric acid	dodecanoic acid
$CH_3(CH_2)_{12}COOH$	myristic acid	tetradecanoic acid
$CH_3(CH_2)_{14}COOH$	palmitic acid	hexadecanoic acid
$CH_3(CH_2)_{16}COOH$	stearic acid	octadecanoic acid
$CH_3(CH_2)_7CH{=}CH(CH_2)_7COOH$	oleic acid	octadec-9-enoic acid
$CH_3(CH_2)_4(CH{=}CHCH_2)_2(CH_2)_6COOH$	linoleic acid (ω−6)	octadeca-9, 12-dienoic acid
$CH_3CH_2(CH{=}CHCH_2)_3(CH_2)_6COOH$	linolenic acid (ω−3)	octadeca-9,12, 15-trienoic acid

▲ Table 1 Common fatty acids

Most fatty acids in the human body contain an even number of carbon atoms, typically from 4 to 18, although some plants and animals produce fatty acids with up to 28 carbon atoms. **Saturated** fatty acids contain only single carbon–carbon bonds and have the general formula $C_nH_{2n+1}COOH$. **Unsaturated** fatty acids with one or more –CH=CH– groups in their molecules are described as **monounsaturated** and **polyunsaturated**, respectively. Naturally occurring unsaturated fatty acids have *cis*-configurations of double carbon–carbon bonds (sub-topic 20.3) while *trans*-fatty acids are often formed as unwanted by-products in food processing (sub-topic B.10).

Physical properties of fatty acids

Melting points of fatty acids generally increase with their molecular masses, from −8 °C for butanoic acid to +70 °C for stearic acid. Saturated fatty acids with 10 and more carbon atoms in their molecules are solid at room temperature as a result of close packing and multiple van der Waals' bonds between rod-shaped carbon chains. The presence of double carbon–carbon bonds distorts carbon chains (figure 1) and prevents them from packing closely, which reduces the intermolecular forces and lowers melting points. As a result, all unsaturated fatty acids are liquid at room temperature. Double carbon–carbon bonds in triglycerides have a similar effect on the molecular packing; this explains why unsaturated fats (oils) have lower melting points than their saturated analogues.

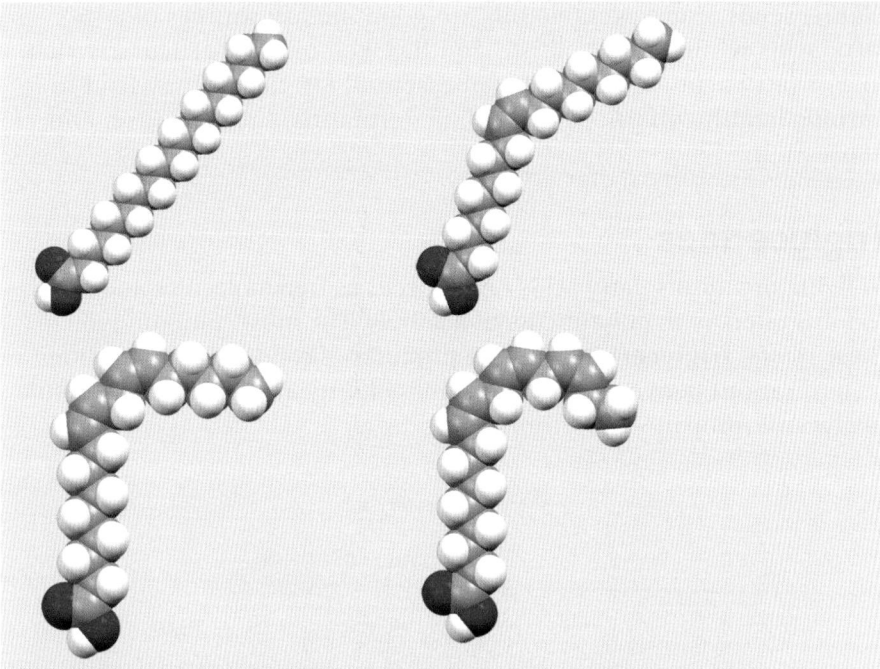

▲ Figure 1 Molecular structures of saturated and unsaturated fatty acids (clockwise, from top left): stearic, oleic, linolenic and linoleic

Essential fatty acids

Certain polyunsaturated fatty acids cannot be synthesized in the human body and therefore must be obtained in sufficient quantities from food. Two **essential fatty acids**, linoleic and linolenic, contain double carbon–carbon bonds at the sixth (ω–6, "omega six") and third (ω–3) carbon atoms from the end of the hydrocarbon chain (when the primary chain of the molecule is numbered from the furthest atom from the carboxylic group). According to this classification, oleic acid (which can be made in the human body and so is a non-essential fatty acid) is an ω–9 fatty acid (figure 2).

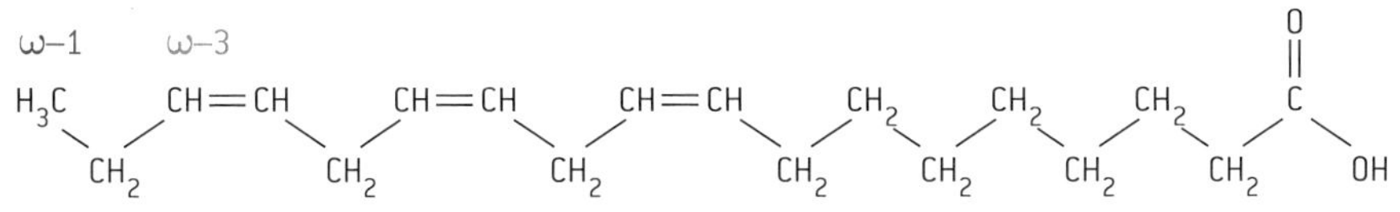

linolenic acid (ω–3)

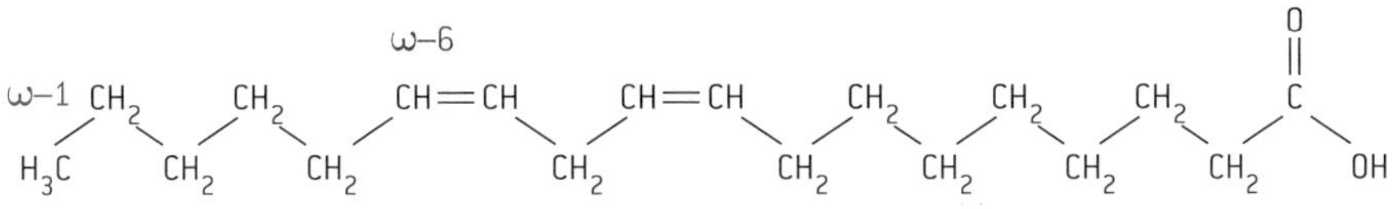

linoleic acid (ω–6)

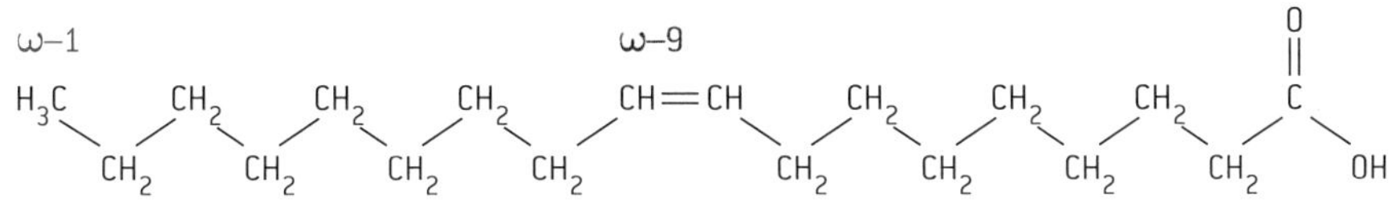

oleic acid (ω–9)

▲ Figure 2 The numbering of carbon chains in the fatty acid molecules

Plants, seeds, and vegetable oils are good dietary sources of ω–6 fatty acids while fish, shellfish, and flaxseed oil are particularly rich in ω–3 fatty acids. A deficiency of essential fatty acids may lead to various health conditions, including dermatitis, heart disease, and depression.

Triglycerides

In living organisms, fatty acids rarely occur as free molecules and tend to form esters with polyfunctional alcohols. The most common type of these esters, **triglycerides**, are the products of condensation reactions (esterification) between three molecules of fatty acids and one molecule of glycerol (propane-1,2,3-triol):

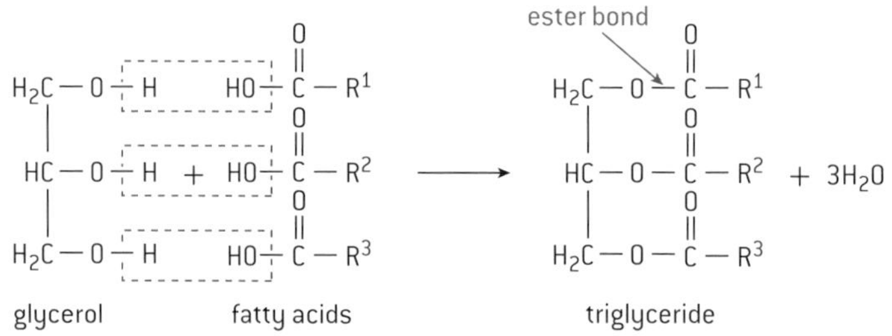

In simple triglycerides all three fatty acid residues are identical ($R^1 = R^2 = R^3$) while mixed triglycerides contain residues of two or three different fatty acids. For example, a molecule of trilauroylglycerol contains three residues of lauric acid while dioleoylstearoylglycerol has two residues of oleic acid and one residue of stearic acid. The latter triglyceride can exist as two structural isomers:

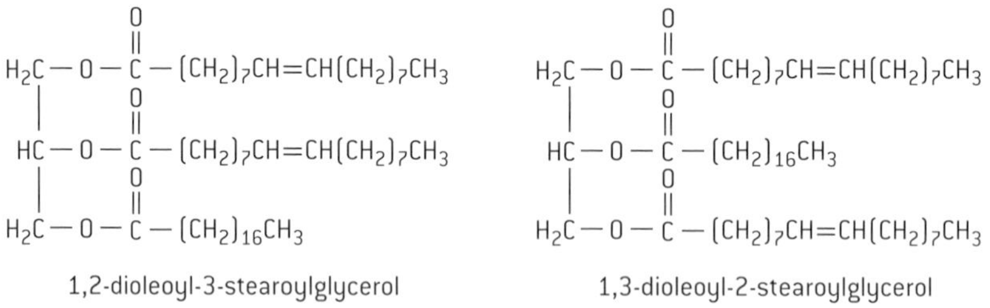

1,2-dioleoyl-3-stearoylglycerol   1,3-dioleoyl-2-stearoylglycerol

The physical properties of triglycerides depend on the nature of the fatty acid residues in their molecules. Similar to free fatty acids, saturated triglycerides (**fats**) are solid at room temperature because their rod-shaped hydrocarbon chains can pack together closely and form multiple van der Waals' interactions with one another. Liquid triglycerides (**oils**) contain residues of unsaturated fatty acids that prevent close packing and weaken intermolecular forces. However, most animal fats contain significant proportions of unsaturated fatty acid residues while certain plant oils such as coconut oil are composed almost exclusively of saturated triglycerides (table 2). Therefore the words "fats" and "oils" usually refer to aggregate states or natural sources of triglycerides rather than their chemical structures.

Triglycerides in chocolate

Chocolate has a relatively low melting point, which accounts for its "melt-in-the-mouth" property. At the same time chocolate must remain solid at room temperature, so the melting point of chocolate produced in hot countries is typically higher than that made in countries with colder climates. Since most types of chocolate contain 33–37% of triglycerides, the melting point can be raised by using fatty acid residues with longer chains or fewer carbon–carbon double bonds. The main chocolate ingredient, cocoa butter, can be partially hydrogenated (sub-topic B.10) to decrease its unsaturation and convert some *cis*-fatty acid residues into their *trans*-isomers. The resulting *trans fats* have higher melting points but their consumption increases the risk of coronary heart disease by raising the levels of LDL cholesterol, which will be discussed later in this sub-topic.

Fat or oil	Saturated fatty acids/%				Unsaturated fatty acids/%			
	Lauric	Myristic	Palmitic	Stearic	Oleic	Linoleic	Linolenic	Other
butter	3	11	29	9	26	4	–	18
lard	–	1	28	12	48	6	–	5
human fat	2	6	26	8	48	4	–	6
fish oil	–	8	15	6	12	–	–	59
olive oil	–	–	7	2	84	5	–	2
sunflower oil	–	–	6	3	25	66	–	–
linseed oil	–	–	6	3	19	24	47	1
coconut oil	45	18	11	2	8	–	–	16

▲ Table 2 Average percentage composition of common fats and oils

The iodine number

Naturally occurring fats and oils are complex mixtures of triglycerides containing the residues of various fatty acids in all possible combinations. Since the exact amount of each triglyceride in a mixture is unknown, the **degree of unsaturation** (sub-topic 11.3) of fats and oils is often expressed as the *average* number of double carbon–carbon bonds per unit mass of the fat or oil. This number can be determined by the reaction of a triglyceride mixture with elemental iodine or another reagent that quantitatively combines with C=C bonds via electrophilic addition reactions (sub-topic 20.1). For example, each residue of monounsaturated oleic acid in 1,2-dioleoyl-3-stearoylglycerol will react with one molecule of I_2 while the residue of saturated stearic acid remains unchanged:

$$H_2C-O-\overset{O}{\overset{\|}{C}}-(CH_2)_7CH=CH(CH_2)_7CH_3$$
$$HC-O-\overset{O}{\overset{\|}{C}}-(CH_2)_7CH=CH(CH_2)_7CH_3 + 2I_2 \longrightarrow$$
$$H_2C-O-\overset{O}{\overset{\|}{C}}-(CH_2)_{16}CH_3$$

$$H_2C-O-\overset{O}{\overset{\|}{C}}-(CH_2)_7\underset{|}{CH}-\underset{|}{CH}(CH_2)_7CH_3$$
$$HC-O-\overset{O}{\overset{\|}{C}}-(CH_2)_7\underset{|}{CH}-\underset{|}{CH}(CH_2)_7CH_3$$
$$H_2C-O-\overset{O}{\overset{\|}{C}}-(CH_2)_{16}CH_3$$

If a solution of iodine is added in small portions to an unsaturated oil or fat, the reaction mixture will stay colourless as long as all the added iodine is consumed by the triglyceride. At the point where the reaction mixture starts to turn yellow or brown the reaction is complete and all double carbon–carbon bonds in the sample have reacted with iodine. The maximum mass of iodine in grams that can be consumed by 100 g of a triglyceride or other unsaturated substance is known as its **iodine number**. Animal fats contain relatively few double carbon–carbon bonds and thus have low iodine numbers, typically between 40 and 70. Vegetable and fish oils have a greater degree of unsaturation so their iodine numbers usually vary from 80 to 140, but can be as low as 10 for coconut oil or as high as 200 for linseed and fish oils.

Worked example

A sample of vegetable oil (5.0 g) has reacted completely with 38 cm^3 of a 0.50 mol dm^{-3} iodine solution. What is the iodine number of the oil? Estimate the average number of double carbon–carbon bonds per molecule of this oil if its average molecular mass is 865 g mol^{-1}.

Solution

Since 38 cm^3 = 0.038 dm^3, the amount of iodine, I_2 in the solution is 0.038 dm^3 × 0.50 mol dm^{-3} = 0.019 mol. The molecular mass of I_2 is 126.9 × 2 = 253.8 g mol^{-1}, so the mass of iodine is 253.8 g mol^{-1} × 0.019 mol ≈ 4.8 g. Therefore the iodine number of the oil is 4.8 g × 100/5.0 g = 96.

If 100 g of the oil react with 96 g of I_2, then 1 mol (865 g) would react with 96 × 865/100 ≈ 830 g or 830 g/253.8 g mol^{-1} ≈ 3.3 mol I_2. Each molecule of iodine reacts with one double bond, so the oil contains approximately 3.3 double carbon–carbon bonds per triglyceride molecule.

Please note that 3.3 is only an average value and the oil might contain triglycerides with any number (typically from 0 to 9) of carbon–carbon double bonds.

Saponification

The alkaline hydrolysis of fats is used in the process of soap-making, known as **saponification**. The fat or oil is treated with a hot solution of sodium hydroxide until the hydrolysis is complete. The sodium salts of fatty acids are separated by precipitation and cooled in moulds to produce soap bars of the desired size and shape. The reaction by-product, glycerol, is often added to the soap as a softening and moisturizing agent.

The saponification of triglycerides with potassium hydroxide produces potassium soaps, which have low melting points and are used as components of liquid detergents. The **saponification number** is the mass of potassium hydroxide in milligrams required for the complete hydrolysis of 1 g of a fat. This value can be used to determine the average molecular mass of triglycerides in the fat and, together with the iodine number, its approximate chemical composition.

Hydrolysis of triglycerides

In the human body the ester bonds in triglycerides are cleaved by a group of enzymes (**lipases**) produced in the pancreas and small intestine. In the laboratory triglycerides can be hydrolysed by hot aqueous solutions of strong acids or bases. Acid hydrolysis gives a molecule of glycerol and three molecules of fatty acids:

Strong bases form salts with fatty acids, so the base acts as a reactant:

Rancidity of fats

The chemical or biological decomposition of fats and oils in dietary products is largely responsible for the unpleasant odours and flavours that are commonly associated with "spoiled" or **rancid** food. **Hydrolytic rancidity** is caused by the hydrolysis of ester bonds in triglycerides and occurs when the food is exposed to moisture or has a naturally high water content. The hydrolysis is accelerated by enzymes (lipases), organic acids (such as ethanoic acid in vinegar or citric acid in lemon juice), and elevated temperatures, especially when the food is acidified and cooked for a prolonged period of time. Butyric and other short-chain fatty acids produced as a result of hydrolytic rancidity have particularly unpleasant smells and further increase the rate of hydrolysis, so the process becomes autocatalytic. Hydrolytic rancidity can be prevented by storing the foods at low temperatures, reducing their water content, and adding any acidic components of the recipe at the latest stage of cooking. Enzyme-catalysed hydrolytic rancidity caused by microorganisms (**microbial rancidity**) can be minimized by sterilization or food processing that reduces the activity of lipases.

Carbon–carbon double bonds in unsaturated fatty acids and triglycerides can be cleaved by free-radical reactions (sub-topic 10.2) with molecular oxygen. This process, known as **oxidative rancidity**, is accelerated by sunlight and is typical for polyunsaturated vegetable and fish oils. Free-radical oxidation of such oils produces volatile aldehydes and ketones with unpleasant odours. Oxidative rancidity can be prevented by light-proof packaging, a protective (oxygen-free) atmosphere, and food additives – natural or synthetic **antioxidants** such as sodium hydrogensulfite, substituted phenols, thiols, and vitamins A, C, and E (sub-topic B.5). Many antioxidants are reducing agents that are readily oxidized by molecular oxygen or reactive free-radical intermediates, effectively terminating chain reactions and inhibiting other oxidation processes (figure 3).

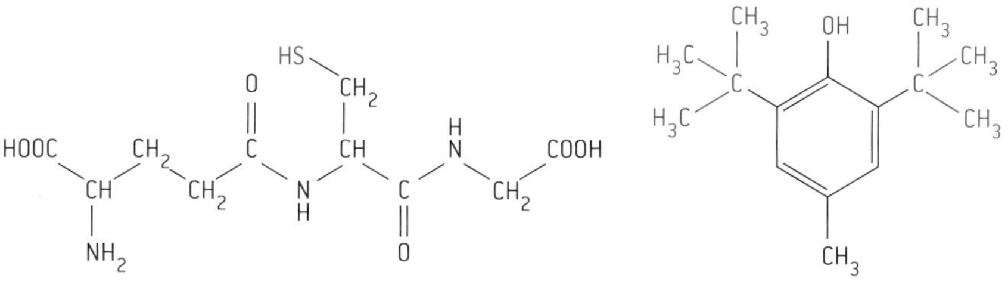

▲ Figure 3 The natural antioxidant glutathione (left) and the artificial antioxidant butylated hydroxytoluene (BHT, right). The functional groups responsible for antioxidative properties are shown in red

🌐 Energy values of fats

Long hydrocarbon chains of triacylglycerides contain many reduced carbon atoms and thus are rich in energy. The complete oxidation of fats produces more than twice as much heat as the oxidation of carbohydrates or proteins of the same mass (table 3). In addition, the hydrophobic nature of triglycerides allows them to form compact aggregates with a low water content. These properties make fats efficient stores of chemical energy. At the same time, the energy accumulated in the fatty tissues of animals and humans is not readily accessible because triglycerides are insoluble in water and take a long time to transport around the body and metabolize. In contrast, hydrophilic molecules of carbohydrates (sub-topic B.4) are already partly oxidized and store less energy but can release it quickly when and where it is needed. Therefore carbohydrates are used as a short-term energy supply while fats serve as long-term energy storage.

Nutrient	Energy/kJ g^{-1}	Energy/kcal g^{-1}
fats	38	9.0
carbohydrates	17	4.0
proteins	17	4.0
ethanol	30	7.1
dietary fibre	0–8	0–2

▲ Table 3 Energy values of food components (1 cal = 4.184 J). Dietary fibre is indigestible by humans but may be metabolized by bacteria in the digestive tract

Alcohol and energy

Ethanol contains almost twice as much energy as carbohydrates and only 20% less than fats. In addition to other negative health effects, excessive consumption of alcoholic drinks may contribute to body weight gain and the development of obesity.

Lipids and health

Fats and oils, along with other nutrients, are important components of any diet. However, excessive consumption of foods that are rich in triglycerides may lead to various health conditions, including obesity, heart disease, and diabetes. In addition, the composition of dietary fats and oils must be balanced in terms of saturation and the level of essential fatty acids. Although dietary sources and amounts of consumed triglycerides vary greatly in different countries and cultures, fats and oils should provide 30–40% of the daily energy intake (60–90 g of fats per day for a healthy adult on a 2000 kcal diet), with at least two-thirds of this amount supplied as unsaturates. Another 50–60% of energy should be obtained from carbohydrates (250–300 g per day) and the remaining 10–15% from proteins (50–75 g per day). The approximate composition and energy values of various foods shown in table 4 can be used as a guideline for creating a balanced diet.

Food	Mass/g per 100 g of food				Energy per 100 g of food	
	Fats	Carbo-hydrates	Proteins	Dietary fibre	kJ	kcal
bacon, grilled	12	0	30	0	970	228
chicken, broiled	11	0	27	0	880	207
cod fillet, baked	2	0	21	0	430	102
fish fingers, fried	13	17	14	0	1020	241
bread, white	2	46	7	3	980	230
cheese, cheddar	34	0	26	0	1730	410
potatoes, boiled	0	20	1	1	360	84
potato crisps	36	48	7	2	2300	544
rice, white boiled	0	28	3	2	530	124
egg, boiled	23	0	10	0	1040	247
yoghurt, natural	2	6	6	0	280	66
milk, whole	4	4	3	0	270	64
orange juice	0	9	1	1	170	40
soft drink, sweetened	0	12	0	0	200	48
chocolate, plain	29	65	5	2	2290	541

▲ Table 4 The composition and energy values of selected foods

Calculating energy content

The energy value of a food can be either calculated from the percentages of its main ingredients or determined experimentally using a calorimeter (sub-topic 5.1). An outline of a simple **bomb calorimeter** is shown in figure 4. If a food sample of a known mass is mixed with oxygen and combusted in the reaction chamber of the bomb calorimeter, the released energy is absorbed by the water, and can be calculated from the temperature change, the mass of water and the heat capacity of the calorimeter itself. Alternatively, a series of samples with known combustion enthalpies can be used for plotting a calibration curve, from which the combustion enthalpies of food samples can be determined.

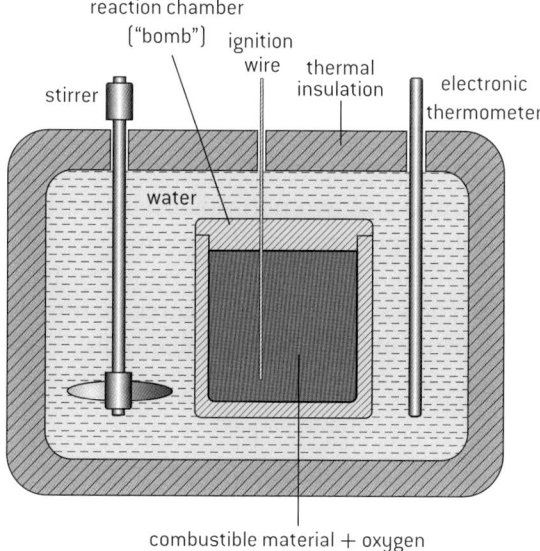

▲ Figure 4 The bomb calorimeter

Phospholipids

Glycerophospholipids, or simply **phospholipids**, are structurally similar to triglycerides except that one residue of a fatty acid in a phospholipid is replaced with a phosphate group:

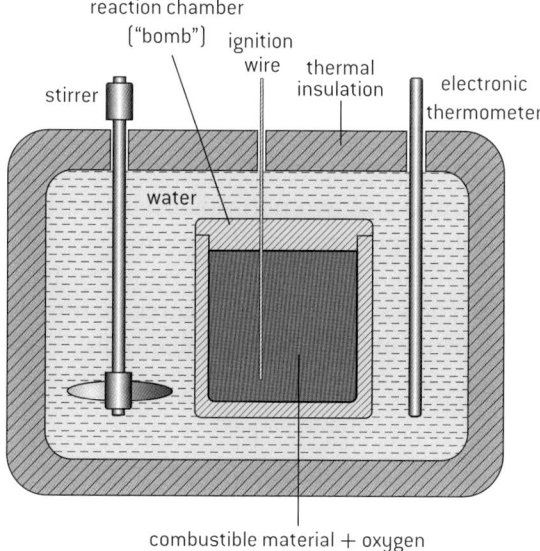

One of the two remaining hydroxyl groups in the phosphoric acid residue may be further esterified with an aminoalcohol or serine. However, in this book we shall discuss only the simplest phospholipids with a single phosphate ester bond.

Worked example

A sample of cheesecake (2.00 g) contains 6.40% of proteins, 44.5% of carbohydrates and unknown amounts of fats and water. Complete combustion of the sample produced 37.4 kJ of heat. Calculate the percentages of fats and water in the cheesecake.

Solution

The sample contains 2.00 g × 6.4/100 = 0.128 g of proteins and 2.00 g × 44.5/100 = 0.890 g of carbohydrates. According to table 3, the combustion of these proteins and carbohydrates will produce (0.128 + 0.890) g × 17 kJ g^{-1} = 17.3 kJ of heat. Because water is not combustible, the remaining 37.4 − 17.3 kJ = 20.1 kJ of energy was produced by fats. Therefore, the mass of fats is 20.1 kJ/38 kJ g^{-1} = 0.529 g. The total mass of combustible nutrients is 0.128 + 0.890 + 0.529 g = 1.55 g, so the mass of water is 2.00 − 1.55 g = 0.45 g. The percentages of fats and water in the cheesecake are 0.529 × 100/2.00 = 26.5% and 0.45 × 100/2.00 = 22.5%, respectively.

In the presence of acids, bases, or enzymes, phospholipids can be hydrolysed into glycerol, phosphoric acid, and fatty acids or their salts, for example:

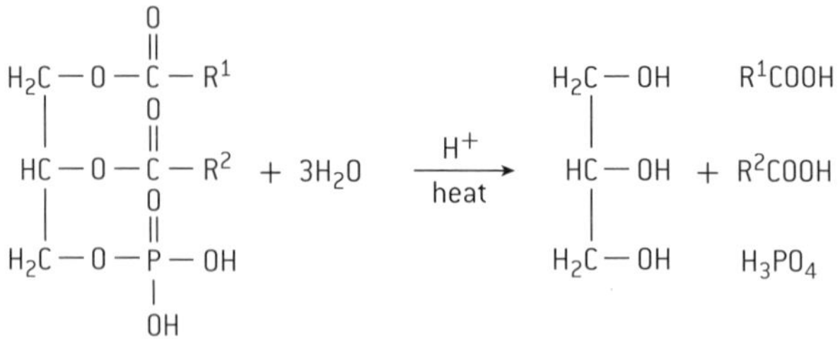

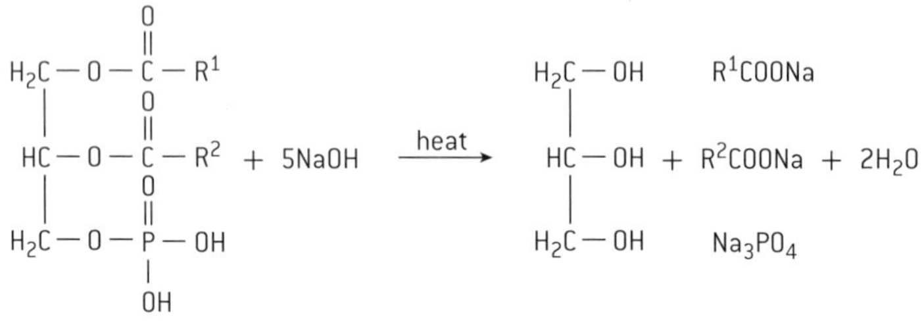

$$H_2C-O-\overset{\overset{O}{\|}}{C}-R^1$$
$$HC-O-\overset{\overset{O}{\|}}{C}-R^2 \;+\; 3H_2O \;\xrightarrow[\text{heat}]{H^+}\; $$
$$H_2C-O-\overset{O}{\underset{\overset{|}{OH}}{\overset{\|}{P}}}-OH$$

$$H_2C-OH \qquad R^1COOH$$
$$HC-OH \;+\; R^2COOH$$
$$H_2C-OH \qquad H_3PO_4$$

$$H_2C-O-\overset{\overset{O}{\|}}{C}-R^1$$
$$HC-O-\overset{\overset{O}{\|}}{C}-R^2 \;+\; 5NaOH \;\xrightarrow{\text{heat}}\; $$
$$H_2C-O-\overset{O}{\underset{\overset{|}{OH}}{\overset{\|}{P}}}-OH$$

$$H_2C-OH \qquad R^1COONa$$
$$HC-OH \;+\; R^2COONa \;+\; 2H_2O$$
$$H_2C-OH \qquad Na_3PO_4$$

The presence of a polar phosphate group and two non-polar hydrocarbon chains makes the molecules of phospholipids **amphiphilic** (they demonstrate both hydrophilic and hydrophobic properties, figure 5). In aqueous solutions amphiphilic molecules spontaneously aggregate into bilayers with hydrophilic "heads" facing out and hydrophobic "tails" facing inwards (figure 6).

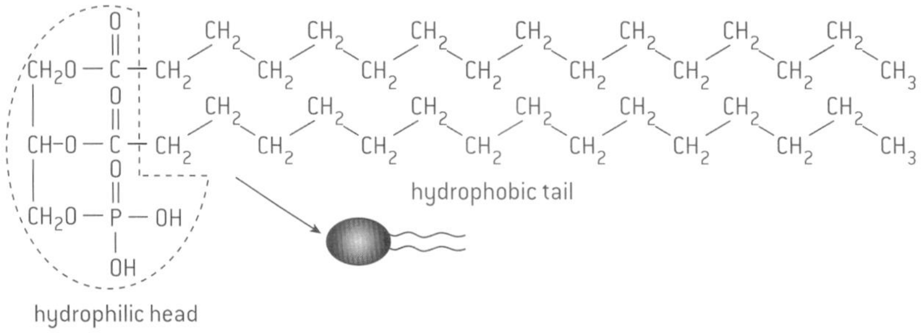

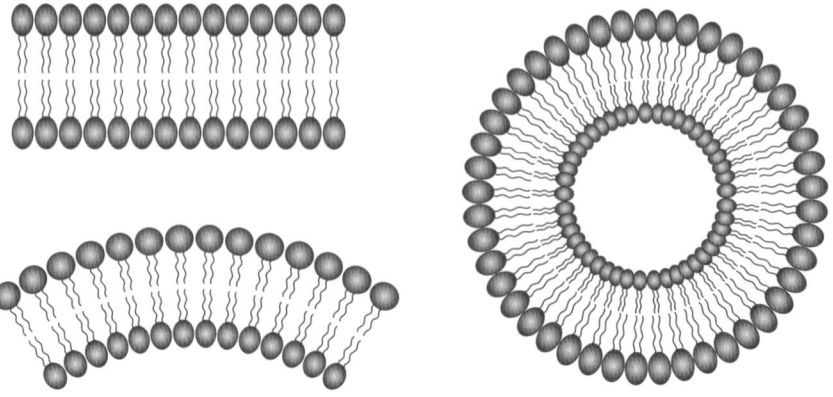

hydrophobic tail

hydrophilic head

▲ Figure 5 Phospholopid molecules are amphiphilic

▲ Figure 6 Phospholopid molecules form bilayers in aqueous solutions

This arrangement maximizes the van der Waals' interactions between the hydrocarbon tails within the bilayer and at the same time allows the

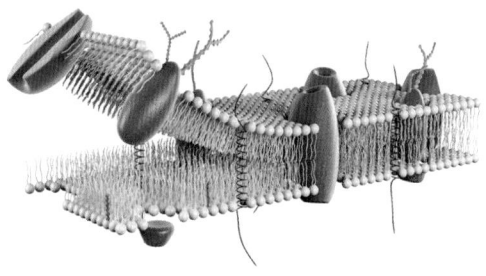

hydrophilic heads to form multiple hydrogen bonds and dipole–dipole interactions with water and one another. These intermolecular forces increase the stability of bilayers and allow them to automatically repair themselves if they are damaged. These properties make phospholipids ideal building blocks for cell membranes, which separate the internal contents of living cells from the surroundings (figure 7).

The hydrophobic nature of fatty acid residues makes phospholipid bilayers impermeable to ions and polar molecules. However, proteins and steroids embedded in cell membranes allow controlled transport of ions, nutrients, and metabolites between the cell and the environment.

▲ Figure 7 The structure of a cell membrane. The membrane is composed of a lipid bilayer (yellow) with embedded proteins (red), protein-based ion channels (blue), and carbohydrate chains (green)

Steroids

Steroids are a class of lipids with a characteristic arrangement of three six-membered and one five-membered hydrocarbon rings fused together in a specific order. The carbon atoms in this four-ring structure, known as the **steroidal backbone**, are traditionally numbered as shown in figure 8.

Almost all steroids contain two methyl groups attached to the steroidal backbone at positions 10 and 13, as well as other functional groups, usually at positions 3 and 17. In addition, many steroids have one or more double carbon–carbon bonds at positions 4, 5, and 6. For example, **cholesterol** contains two methyl groups at positions 10 and 13, a hydroxyl group at position 3 of the first six-membered ring, a double carbon–carbon bond between atoms 5 and 6 of the second six-membered ring, and a long-chain hydrocarbon substituent at position 17 of the five-membered ring (figure 9).

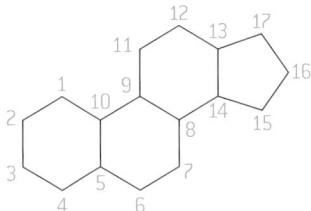

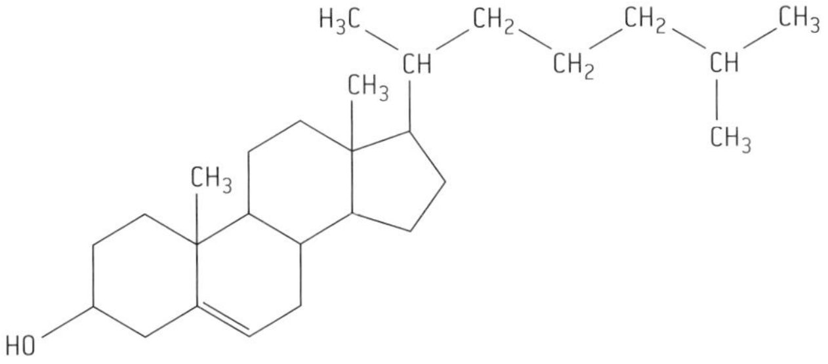

▲ Figure 8 The steroidal backbone

> **Study tip**
>
> The structure of cholesterol is provided in section 34 of the *Data booklet*.

▲ Figure 9 The structure of cholesterol

Cholesterol is an essential component of cell membranes and the main precursor of all steroidal hormones produced in the human body. In cell membranes, the hydroxyl groups of cholesterol molecules hydrogen bond to phosphate groups of phospholipids while the non-polar hydrocarbon backbone and the substituent at the five-membered ring form van der Waals' interactions with the fatty acid residues. As a result, embedded cholesterol molecules increase the rigidity of cell membranes and regulate their permeability to metabolites.

Since cholesterol is largely hydrophobic, its solubility in blood plasma is extremely low. In the human body cholesterol is transported as

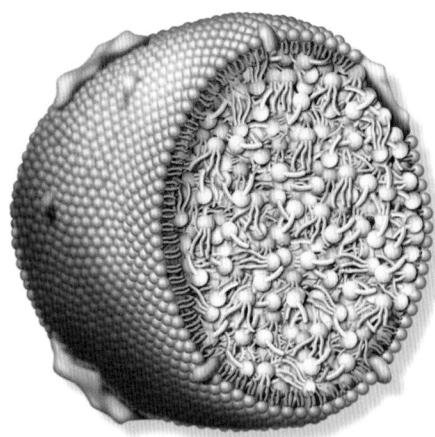

▲ Figure 10 An LDL-C complex containing cholesterol and triglycerides (yellow), a phospholipid membrane (blue), and LDLs (beige)

a component of lipid–protein complexes known as **lipoproteins**. Depending on their composition and density these complexes are classified as **low-density lipoproteins (LDL)** or **high-density lipoproteins (HDL)**. Generally, the density and solubility of lipoproteins in water decrease with increasing lipid content, so the amount of cholesterol carried by LDLs is significantly higher than that by HDLs. LDLs are primarily responsible for the transport of cholesterol from the liver where it is synthesized to various body tissues (figure 10), while HDLs are capable of transporting excess cholesterol back to the liver where it can be metabolized and excreted into the digestive tract.

 ## Lipoproteins and health

Excessive consumption of cholesterol-rich foods or saturated fats increases the levels of cholesterol complexes with low-density lipoproteins (**LDL–C**), which are commonly referred as "bad cholesterol". High levels of LDL–C in the bloodstream may result in cholesterol deposition in the artery walls and eventually lead to cardiovascular disease. In contrast, HDLs form more stable complexes with cholesterol (**HDL–C** or "good cholesterol") and can reduce its deposition in the blood vessels. Therefore the correct balance between LDL–C and HDL–C levels in the human body is very important for preventing heart problems and other health conditions.

 ## Dietary choices

The total cholesterol level and the ratio of HDL to LDL levels in the human body are affected by many factors including genetics, body mass index, dietary intake, food additives, and medications. High levels of LDL–C in the blood can be reduced by a low-cholesterol diet and certain drugs, **statins**, that inhibit the enzymes responsible for the biosynthesis of cholesterol in the liver. Extensive scientific evidence about the negative effects of diets rich in cholesterol, saturated fats, and trans-unsaturated fats have influenced dietary choices and led to the development of new food products.

Steroid hormones

Besides cholesterol, several hundred other steroids with various biological functions are known. In the human body all steroids are synthesized from cholesterol, which loses its side-chain at carbon 17 and undergoes a series of enzymatic transformations. Most steroids are **hormones** – the chemical messengers that regulate metabolism and immune functions (**corticosteroids**), sexual characteristics and reproductive functions (**sex hormones**), or the synthesis of muscle and bone tissues (**anabolic steroids**).

The male sex hormones are produced in the testes and include testosterone and androsterone:

testosterone   androsterone

In addition to androgenic functions (the development of male sex characteristics), male sex hormones act as natural anabolic steroids.

The **female sex hormones** are produced in the ovaries and include progesterone and estradiol:

progesterone   estradiol

Estradiol and progesterone are responsible for controlling sexual development and menstrual and reproductive cycles in women. Estradiol is one of the few steroids that contains an aromatic ring in the steroidal backbone.

The term **anabolic steroids** usually refers to synthetic drugs that mimic the effects of testosterone and other hormones that accelerate protein synthesis and cellular growth, especially in the muscle and bone tissues. Anabolic hormones were initially developed for medical purposes but soon became substances of abuse in sports and bodybuilding. The structures of anabolic steroids such as dianabol or nandrolone are very similar to those of male sex hormones, often with a single substituent added to or removed from the molecules of their natural analogues:

dianabol   nandrolone

Anabolic steroids and health

Aside from giving unfair advantages to athletes, the non-medical use of anabolic steroids presents significant health risks ranging from acne to high blood pressure and liver damage. In addition, many anabolic steroids suppress the production of natural sex hormones and increase the LDL cholesterol level in the body.

Anabolic steroids are banned by most sports organizations including the International Olympic Committee. Athletes are regularly required to provide urine and blood samples for laboratory analyses in which steroids and their metabolites can be detected by a combination of gas chromatography, high-performance liquid chromatography (sub-topic B.2), and mass spectrometry (sub-topics 2.1 and 11.3).

▲ Figure 11 Lance Armstrong, a professional cyclist and winner of seven Tour de France races, has been banned from cycling competitions for life after being found guilty of doping offences by the United States Anti-Doping Agency (USADA) in 2012

Questions

1   Unsaturated fats contain C=C double bonds. The amount of unsaturation in a fat or oil can be determined by titrating with iodine solution.

    a)   Define the term iodine number.   [1]

    b)   Linoleic acid ($M_r = 281$) has the following formula:

    $$CH_3(CH_2)_4CH{=}CHCH_2CH{=}CH(CH_2)_7COOH$$

    Calculate the volume of 1.00 mol dm^{-3} iodine solution required to react exactly with 1.00 g of linoleic acid.   [3]

    IB, May 2010

2   Examples of straight-chain fatty acids include $C_{19}H_{39}COOH$, $C_{19}H_{31}COOH$, and $C_{19}H_{29}COOH$.

    a)   Deduce the number of C=C bonds present in one molecule of each fatty acid.   [2]

    b)   Deduce the least stable of the three fatty acids and explain your reasoning.   [2]

    IB, November 2011

3   Deduce the structural formula of a triester formed from three long-chain carboxylic acid molecules, RCOOH, and one propane-1,2,3-triol molecule, HOCH$_2$–CH(OH)–CH$_2$OH. Identify one of the ester linkages in the structure by drawing a rectangle around it.   [2]

    IB, November 2011

4   There are several types of lipids in the human body. One of these types, triglycerides, might be made of fatty acids with different degrees of saturation.

    a)   State one example of each of the following types of fatty acids: saturated, mono-unsaturated, and poly-unsaturated.   [3]

    b)   Describe, by copying and completing the equation below, the condensation of glycerol and the three fatty acids named in (a) to make a triglyceride.   [2]

    CH$_2$ — OH
    |
    CH — OH   +
    |
    CH$_2$ — OH

    c)   State the names of two other types of lipids present in the human body.   [1]

    d)   Compare their composition with that of triglycerides.   [2]

    IB, November 2012

5   Calculate the mass of sodium hydroxide required for the complete saponification of 5.0 moles of a triglyceride.

6   a)   Fats, such as butter, are solid triglycerides. Explain why fats have a higher energy value than carbohydrates.   [1]

    b)   Explain why linoleic acid has a lower melting point compared to stearic acid.   [2]

    IB, November 2010

7   Predict and explain which fatty acid in each group has the highest melting point:

    a)   butanoic, palmitic, and stearic acids;

    b)   oleic, linoleic, and linolenic acids.

8   Chocolate is a luxury food made from cocoa, sugars, unsaturated vegetable fats, milk whey, and emulsifiers. Bars of chocolate sold in hot climates are made with a different blend of vegetable fats from bars sold in cold climates.

    a)   Explain why fats with different physical properties are used for making chocolate sold in different climates.

    b)   Suggest how the structure of fat molecules used in a hot climate might differ from those used in a cold climate.

    IB, November 2012

9   Food shelf life is the time it takes for a particular foodstuff to become unsuitable for eating because it no longer meets customer or regulatory expectations. As a result, in many parts of the world, packaged foods have a date before which they should be consumed.

    a)   State the meaning of the term *rancidity* as it applies to fats.   [2]

    b)   Rancidity in lipids occurs by hydrolytic and oxidative processes. Compare the two rancidity processes.   [2]

    IB, November 2011

10 Some foods contain natural antioxidants which help to prolong their shelf life. The shelf life of oily fish decreases upon exposure to light.

    a)   Identify the chemical feature in the oil in fish that is susceptible to photo-oxidation.   [1]

    b)   State the specific term given to food that is unsuitable for eating as a result of photo-oxidation.   [1]

c) Suggest how light initiates this process. [1]

d) Some foods contain a yellow spice called turmeric. The active ingredient in turmeric is curcumin, shown below.

Suggest which structural feature of curcumin is responsible for extending the shelf life of such a food. [1]

IB, May 2012

11 A student carried out an experiment to determine the energy value of 100.00 g of a food product by burning some of it. A 5.00 g sample was burned and the heat produced was used to heat water in a glass beaker. She recorded the following data:

Mass of water heated = 100.00 g
Initial temperature of water = 19.2 °C
Highest temperature of water = 28.6 °C
Heat capacity of the glass beaker = 90.2 J K^{-1}
Specific heat capacity of water = 4.18 J g^{-1} K^{-1}

Calculate the energy value for 100.00 g of the food product, in kJ, showing your working. [3]

IB, November 2011

12 Countries have different laws about the use of synthetic colourants in food. Explain why this can be dangerous for the consumer. [1]

IB, May 2011

13 Discuss the responsibilities of governments, industry, and individuals in making healthy choices about diet and maintaining a balance between the protection of public and individual freedom.

14 a) Draw the formula of a glycerophospholipid containing the residues of palmitic and linoleic acids.

b) Deduce the equation for the complete saponification of this glycerophospholipid.

15 Cholesterol belongs to a class of substances named lipids.

a) Identify the characteristic structural feature of cholesterol. [1]

b) Identify two other types of lipids found in the human body. [2]

c) State what the terms HDL and LDL represent. [1]

d) Outline one chemical difference between HDL and LDL. [1]

e) Describe one negative effect of a high concentration of LDL cholesterol in blood. [1]

IB, May 2009

16 Steroidal-based hormones such as estradiol, progesterone, and testosterone all contain a common structure.

a) State what is meant by the term hormone. [1]

b) Deduce the number of hydrogen atoms joined directly to the carbon atoms as part of the steroidal backbone in progesterone. [1]

IB, November 2010

17 Some athletes have abused steroids in order to increase muscular strength and body mass. One such substance is dianabol, which has a structure similar to testosterone.

a) Describe how the structure of dianabol differs from the structure of testosterone. [1]

b) Outline the general function of hormones in the human body. [1]

c) Suggest a reason why male bodybuilders who take dianabol may develop some female characteristics. [1]

IB, May 2012

B.4 Carbohydrates

Introduction to carbohydrates

Carbohydrates are a family of oxygen-rich biomolecules that play a central role in the metabolic reactions of energy transfer (sub-topics B.1 and B.3). Most carbohydrates have the general formula $C_n(H_2O)_m$ ("hydrates of carbon") although this term is also used for deoxyribose ($C_5H_{10}O_4$, see next page) and other structurally similar compounds.

Traditionally carbohydrates are classified as *monosaccharides*, *disaccharides*, and *polysaccharides*, according to the number of carbon chains in their molecules. **Monosaccharides** consist of a single carbon chain, typically five or six atoms long, with a carbonyl group and two or more hydroxyl groups (sub-topic 10.1), for example:

glucose   fructose   ribose   deoxyribose

Monosaccharides with five and six carbon atoms in their molecules are known as **pentoses** and **hexoses**, respectively. For example, glucose and fructose are hexoses while ribose and deoxyribose are pentoses. If the carbonyl group is connected to the terminal carbon atom the monosaccharide belongs to the class of aldehydes and is called an **aldose** ("aldehyde sugar"). Similarly, monosaccharides with a carbonyl group at the second carbon atom are known as **ketoses** ("ketone sugar"). According to this classification glucose, ribose, and deoxyribose are aldoses while fructose is a ketose. Sometimes the number of carbon atoms and the functional group type are combined in a single word. For example, ribose is an aldopentose ("aldose" + "pentose") while fructose is a ketohexose ("ketose" + "hexose").

Due to the presence of a carbonyl group and several hydroxyl groups in the same molecule, straight-chain forms of monosaccharides are unstable and undergo intramolecular nucleophilic addition (A_N) reactions (sub-topic 20.1). The products of these reactions, five- or six-membered **cyclic forms** of monosaccharides, are predominant species in solutions and in the solid state. For aldohexoses such as glucose the most stable form is a six-membered ring of five carbon atoms and one oxygen atom:

straight-chain form of glucose   cyclic form (α-glucose)

Aldopentoses such as ribose and deoxyribose predominantly exist as five-membered cyclic forms:

straight-chain form of ribose   cyclic form (α-ribose)

Deoxysugars contain one oxygen atom less than a "normal" carbohydrate with the same carbon chain length. For example, deoxyribose ($C_5H_{10}O_4$) has four oxygen atoms instead of the five in ribose ($C_5H_{10}O_5$). Ribose and deoxyribose are components of RNA and DNA, respectively (sub-topic B.8).

Similarly, five-membered rings are the most stable forms of ketohexoses such as fructose:

straight-chain form of fructose   cyclic form (α-fructose)

Each cyclic form of a monosaccharide can exist as two stereoisomers (sub-topic 20.3), known as α- and β-forms. Stereoisomerism of monosaccharides is covered by HL only and will be discussed in sub-topic B.10.

Three-dimensional formulae of cyclic carbohydrates are usually represented by **Haworth projections**, in which the carbon atoms in the ring together with their attached hydrogen atoms are omitted (figure 1).

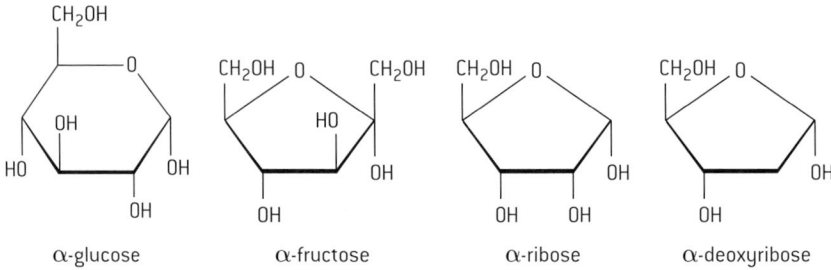

α-glucose   α-fructose   α-ribose   α-deoxyribose

▲ Figure 1 Haworth projections for some monosaccharides

Haworth projections emphasize the nature and positions of the functional groups attached to the ring. The cyclic forms of monosaccharides shown in figure 1 produce space-efficient structures, which is particularly important for polysaccharides and other complex molecules.

Simplified formulae in biochemistry

The omission of certain carbon and hydrogen atoms in Haworth projections simplify formulae and allow biochemists to represent the stereochemistry and three-dimensional arrangement of monosaccharide units in biopolymers by the easily recognizable pentagonal and hexagonal shapes of their backbones. A similar approach was used in sub-topic B.2, where tertiary and quaternary structures of proteins were represented by helices and sheets, allowing us to concentrate on the overall shape and therefore possible properties and biological functions of the whole molecule.

The importance of glucose

Glucose is the most common monosaccharide that occurs in all living organisms. It is the main product of photosynthesis and the primary source of energy for cellular respiration (sub-topic B.1). Glucose is an important intermediate in various metabolic processes including the synthesis of mono-, di-, and polysaccharides (see below), amino acids (sub-topic B.2), vitamins (sub-topic B.5), and many simple biomolecules such as 2-hydroxypropanoic (lactic) acid or ethanol. The latter compound is produced from glucose in an enzymatic process known as **alcoholic fermentation**:

$$C_6H_{12}O_6 \rightarrow 2CH_3CH_2OH + 2CO_2$$
$$\text{glucose} \qquad \text{ethanol}$$

In addition to its use in alcoholic beverages, ethanol is increasingly used as a component of biofuels, reducing consumption of fossil fuels and the net emission of greenhouse gases.

Reducing sugars

The redox properties of monosaccharides depend on the position of the carbonyl group in their molecules (sub-topic 20.1). Glucose and other aldoses are known as **reducing sugars** because their terminal carbonyl (aldehyde) groups are readily oxidized under mild conditions:

In the laboratory reducing sugars can be detected by **Fehling's solution**, which is prepared from aqueous solutions of copper(II) sulfate, sodium potassium tartrate ($NaKC_4H_4O_6$), and sodium hydroxide. In the presence of an aldose the copper(II) ions are reduced to copper(I), the deep blue colour of the original solution disappears, and a red precipitate of copper(I) oxide is formed. Fructose and some other ketoses also give positive tests with Fehling's solution because they quickly isomerize into aldoses under alkaline conditions:

fructose   glucose

The "food versus fuel" problem

The large-scale production of biofuels in many countries has various economical, political and environmental implications. The industry of biofuels can create jobs, stimulate local economies, reduce demand for and therefore the price of oil, and provide a sustainable energy source. However, the diversion of agricultural crops into biofuel production takes up the land, water and other resources that could be used for food production. The ever-increasing demand for biofuels leads to the expansion of cultivated land and results in deforestation, reduction of biodiversity and rising food prices on the global scale.

Control of glucose metabolism

In the human body the glucose concentration in the blood is regulated by the hormone insulin (sub-topic B.2). Insufficient production of insulin or failure of insulin receptors to respond properly to the hormone level lead to a chronic health condition known as **diabetes**. Patients with diabetes must follow a strict dietary regime, regularly check their blood glucose levels and, in some cases, receive insulin injections. About 3% of the global population is currently affected by this disease, with the majority of cases occurring in developed countries. According to the World Health Organization the number of deaths related to diabetes will double between 2005 and 2030.

Instead of Fehling's solution, a mixture of aqueous copper(II) sulfate, sodium citrate ($Na_3C_6H_5O_7$) and sodium carbonate can be used. The resulting solution, known as **Benedict's reagent**, also produces a precipitate of copper(I) oxide in the presence of aldoses and some ketoses. However, the colour of the precipitate varies from green to red depending on the monosaccharide concentration, which can be used for quantitative determination of reducing sugars in solutions.

Disaccharides

In the presence of certain enzymes, monosaccharides or their derivatives undergo condensation reactions and form **disaccharides**. For example, the condensation of two molecules of glucose produces the disaccharide *maltose* and a molecule of water:

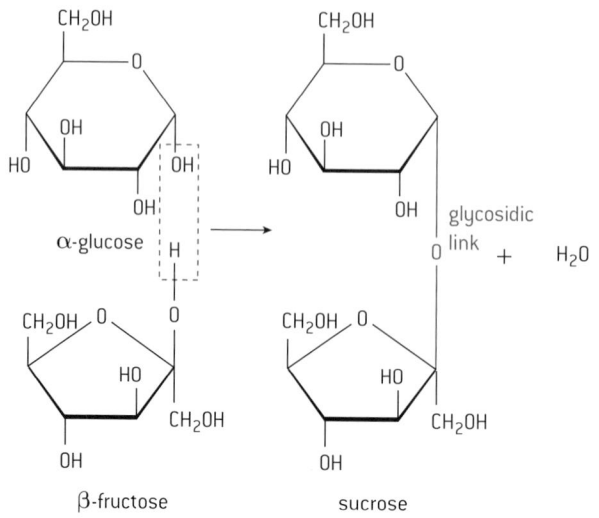

The oxygen bridge between two monosaccharide residues is known as a **glycosidic link**. In the case of maltose, the oxygen atom connects the C–1 atom of the first glucose residue with the C–4 atom of the second glucose unit, so it is called a 1,4-glycosidic link. The stereochemistry of glycosidic links will be discussed in sub-topic B.10.

The most common disaccharide, *sucrose*, is formed by the condensation of α-glucose with β-fructose:

Sucrose, commonly known as table sugar, is an important food ingredient and a major international commodity. Over half of the world's sugar is produced in Brazil and India from sugar cane, which is also cultivated in over 100 other countries with tropical and subtropical climates. In Europe and North America sucrose is extracted from sugar beet, which contributes about a quarter of global sugar production.

Another important disaccharide, *lactose*, contains a residue of the monosaccharide *galactose*. Galactose differs from glucose by the orientation of the hydroxyl group at the C–4 atom:

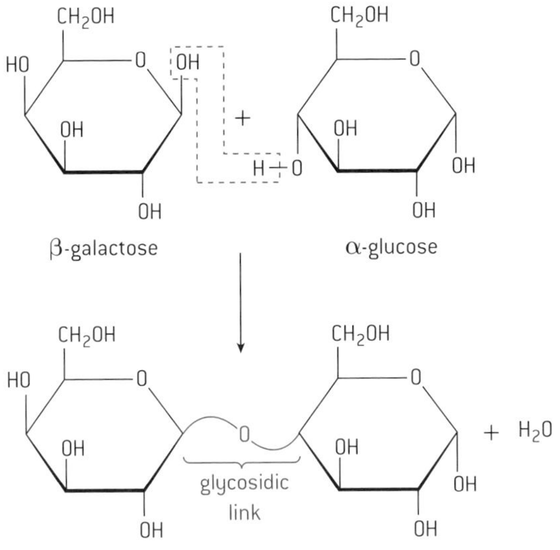

β-galactose   α-glucose

α-lactose

Like most monosaccharides lactose and maltose produce red precipitates of copper(I) oxide when heated with Fehling's or Benedict's solutions, which indicates the presence of aldehyde groups in their molecules. These groups are formed temporarily when the cyclic forms of lactose and maltose undergo reversible ring–chain tautomerism, for example:

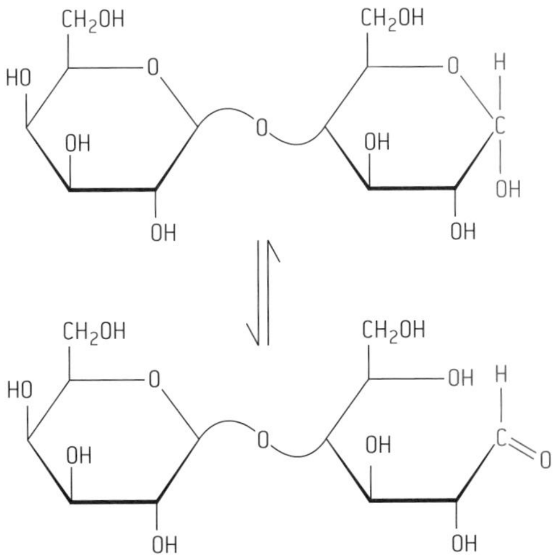

Although the cyclic form is more stable in solution, the equilibrium of the above reaction gradually shifts towards the open-chain form as this is oxidized by copper(II) ions. The process continues until all the molecules of lactose (or other reducing disaccharide) are oxidized.

In contrast, sucrose does not undergo ring–chain tautomerism because both the C–1 atom in glucose and the C–2 atom in fructose are involved in the glycosidic link. As a result, sucrose gives a negative reaction with Fehling's and Benedict's solutions and thus can be distinguished from reducing monosaccharides and disaccharides.

According to traditional classification, both the formation and the hydrolysis of glycosidic links in disaccharides and polysaccharides are nucleophilic substitution (S_N) reactions.

Lactose intolerance

Lactose is the primary carbohydrate of human and cow's milk, providing approximately 40% of their total energy values. In the human body lactose is hydrolysed into glucose and galactose by the enzyme **lactase**, the production of which gradually decreases with maturity. The low level of lactase in adults causes **lactose intolerance**, which is particularly common in certain regions of Africa and eastern Asia. People with this medical condition may experience diarrhoea, flatulence, and other unpleasant symptoms after consuming milk or other lactose-rich dietary products.

The formation of disaccharides is a reversible process. In the presence of acids or enzymes, disaccharides can be hydrolysed into monosaccharides, for example:

$$C_{12}H_{22}O_{11} + H_2O \xrightarrow{H^+} C_6H_{12}O_6 + C_6H_{12}O_6$$
$$\text{sucrose} \qquad\qquad\qquad \text{glucose} \quad \text{fructose}$$

Polysaccharides

Polycondensation reactions of monosaccharides produce long-chain carbohydrates known as **polysaccharides**. One of the most common polysaccharides, *starch*, is a mixture of two polycondensation polymers of glucose. In the first polymer, *amylose*, the glucose residues are connected predominantly by 1,4-glycosidic links and form long unbranched chains:

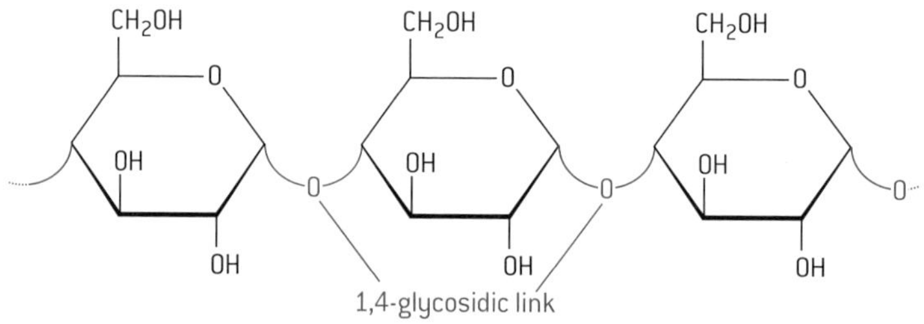

The second component of starch, *amylopectin*, is a branched polymer in which the glucose units are connected by both 1,4- and 1,6-glycosidic links:

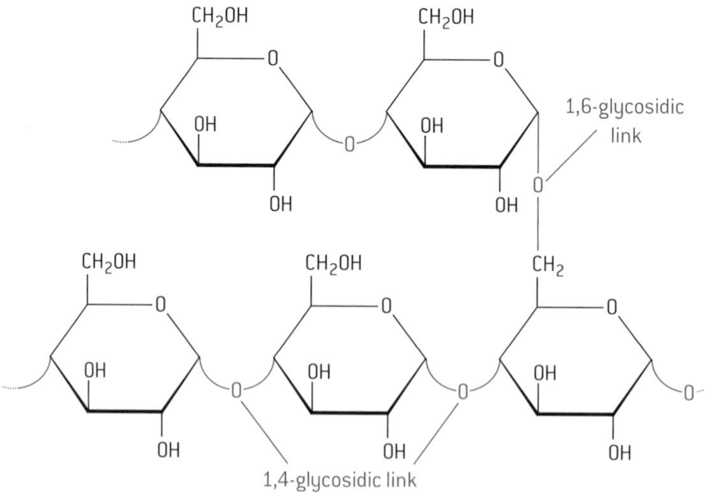

Starch is produced in all green plants, where it is used as the primary energy storage molecule (figure 2). Starch constitutes up to 80% of the dry mass of staple foods such as wheat, corn, rice, and potato, which makes it the most common carbohydrate in the human diet. In the presence of enzymes (such as *amylase*, produced in salivary glands, pancreas and small intestine) or strong inorganic acids, starch can be hydrolysed into glucose, for example:

$$H-(C_6H_{10}O_5)_n-OH + (n-1)H_2O \xrightarrow{H^+} nC_6H_{12}O_6$$
$$\text{amylose} \qquad\qquad\qquad\qquad\qquad \text{glucose}$$

Since the molecular masses of amylose and amylopectin are very large ($n = 300$–$20\ 000$), both starch components are often represented as indefinite chains of glucose residues, $-(C_6H_{10}O_5)_n-$. In this case the number of water molecules needed for complete hydrolysis of these polysaccharides will be approximately the same as the number of monosaccharide units:

$$(C_6H_{10}O_5)_n + nH_2O \xrightarrow{\ H^+\ } nC_6H_{12}O_6$$

amylose   glucose

However, regardless of the way the polysaccharide chains are drawn, the equations for their formation and hydrolysis must be always balanced (sub-topic 1.1).

Worked example

Starch is an important dietary product with high energy content. Although the average energy value of carbohydrates is 17 kJ g^{-1}, the exact energies of combustion of individual mono-, di-, and polysaccharides can vary to some extent.

a) When 2.63 g of starch was completely combusted in a calorimeter, the temperature of 1150 g of water increased from 22.53 to 32.10 °C. Calculate the energy value of starch in kJ g^{-1}.

b) Suggest whether the energy values of sucrose and glucose will be greater than, equal to, or lower than the energy value of starch. Explain your answer.

Solution

a) The energy (Q) absorbed by water in the calorimeter can be calculated as $Q = C \times m \times \Delta T$ (sub-topic 5.1). The temperature of water in the calorimeter increased by $32.10 - 22.53 = 9.57$ K. Since the heat capacity (C) of water is 4.18 J g^{-1} K^{-1} (this value is given in the *Data booklet*), the

amount of heat released was 4.18 J g^{-1} K^{-1} $\times$ 1150 g $\times$ 9.57 K $\approx 46.0 \times 10^3$ J $= 46.0$ kJ (sub-topic 5.1). This amount of heat was produced by 2.63 g of starch, so the energy value of starch is 46.0 kJ/2.63 g ≈ 17.5 kJ g^{-1}.

b) Most carbohydrates can be represented by the general formula $C_x(H_2O)_y$. For glucose $x = y = 6$, or one molecule of water per carbon atom. For sucrose ($x = 12$ and $y = 11$), the carbon-to-water ratio is $12/11 \approx 1.09$. If we draw the formula of starch as $[C_6(H_2O)_5]_n$, the carbon-to-water ratio will be $6n/5n = 1.2$. Of the three carbohydrates starch has the greatest percentage of carbon (which is combustible) and therefore the lowest percentage of water (which is not combustible), so the energy value of starch will be the highest.

You can verify this conclusion by using the enthalpies of combustion for glucose (-2803 kJ mol^{-1}) and sucrose (-5640 kJ mol^{-1}), which are given in the *Data booklet*.

The iodine test for starch

The presence of starch in biological materials can be detected by the **iodine test**. In aqueous solutions of potassium iodide, elemental iodine forms orange coloured tri- and polyiodide ions:

$$KI(s) \rightarrow K^+(aq) + I^-(aq)$$

$$I^-(aq) + I_2(s) \rightarrow I_3^-(aq)$$

colourless   orange

When the resulting orange solution is added to starch, tri- and polyiodide ions react with amylose and produce blue-black complexes:

$$I_3^-(aq) + (C_6H_{10}O_5)_n(aq) \rightarrow [(C_6H_{10}O_5)_n \cdot I_3^-](s)$$

| triiodide (orange) | amylose (colourless) | amylose complex (blue-black) |

Conversely, starch or its individual component amylose can be used for visual detection of iodine and iodide ions in aqueous solutions at concentrations as low as 2×10^{-5} mol dm^{-3}. The complex of amylose with polyiodide ions is also used as an indicator in redox titrations (topic 9.1).

Glycogen and cellulose

In the human body the short-term energy store is in the form of **glycogen**, which is structurally similar to amylopectin but is more densely branched and contains up to a million glucose residues. Glycogen is concentrated in liver and muscle tissue where it is hydrolysed into glucose when the energy is needed.

Another condensation polymer of glucose, **cellulose**, is the major structural polysaccharide in plants and an important component of a healthy diet (dietary fibre). The structure of cellulose and the physiological properties of dietary fibre will be discussed in sub-topic B.10.

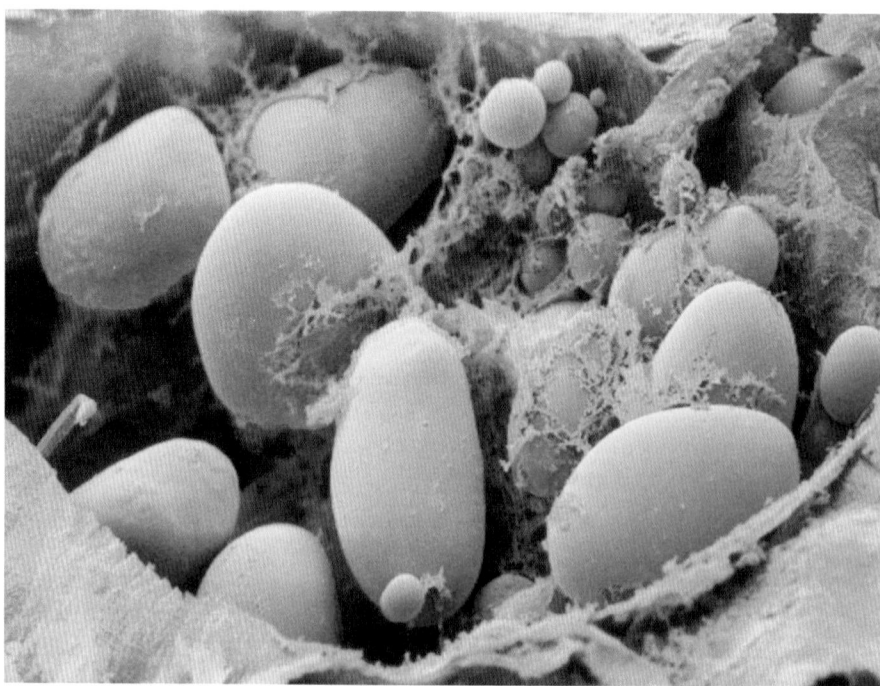

▲ Figure 2 Starch grains in potato cells

Questions

1 Foods such as rice, bread, and potatoes are rich in carbohydrates. There are three main types of carbohydrate – monosaccharides, disaccharides, and polysaccharides.

   a) Glucose, $C_6H_{12}O_6$, is a monosaccharide. When 0.395 g of glucose was completely combusted in a calorimeter, the temperature of 200.10 g of water increased from 20.20 °C to 27.55 °C. Calculate the energy value of glucose in J g^{-1}. [3]

   b) Two α-glucose molecules condense to form the disaccharide maltose. Deduce the structure of maltose. [1]

   c) One of the major functions of carbohydrates in the human body is as an energy source. State one other function of a carbohydrate. [1]

IB, November 2010

2 State three characteristic features of all monosaccharide molecules. [3]

IB, May 2010

3 Glucose is a common monosaccharide.

   a) State the difference in structure between an aldose and a ketose.

   b) State one similarity and one difference between an aldopentose and a ketopentose.

   c) Identify the type of the monosaccharide glucose using the terms "aldose"/"ketose" and "pentose"/"hexose".

4 Explain, in terms of functional group names and types of intermolecular bonds, why all monosaccharides and disaccharides are soluble in water.

5 Fructose is an isomer of glucose, but they differ with regard to one functional group and hence in their redox properties.

   a) Identify the functional group present in glucose, but not fructose. [1]

   b) Identify the functional group present in fructose, but not glucose. [1]

   c) Identify the sugar that acts as a reducing agent. [1]

IB, May 2012

6 Reducing carbohydrates such as glucose exist in solutions predominantly in their cyclic forms, which do not readily undergo oxidation.

   a) Draw the form of glucose that can be oxidized by copper(II) ions.

   b) State the name of the functional group that undergoes oxidation in (a).

   c) Outline the two-step process that leads to the oxidation of the cyclic form of glucose.

   d) State the name of one non-reducing sugar.

7 Lactulose is a synthetic, non-digestible disaccharide that is used in the treatment of chronic constipation and liver disease. This disaccharide contains the residues of galactose and fructose. The formula of α-lactulose is given below.

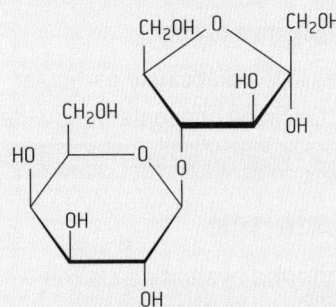

   a) Copy the structure and identify the glycosidic link in lactulose by drawing a circle around it.

   b) Suggest whether lactulose is a reducing or non-reducing sugar. Explain your answer.

8 In making candy or sugar syrup, sucrose is boiled in water with a small amount of organic acid, such as citric acid from lemon juice. Explain why the product mixture tastes sweeter than the initial sucrose solution.

9 The compound olestra has similar properties to saturated fats. It is used in margarine and related products, but it is not digested in the human gut. It is made from a disaccharide with up to eight fatty acid groups attached to it.

   a) Explain what feature of the structure of glycerol (propane-1,2,3-triol) allows fatty acid molecules to become attached to it to make fats, and state the name of the reaction by which this occurs. [2]

   b) Lactose is a typical disaccharide. Suggest a reason why fatty acids can be attached to it. [1]

   c) The fatty acids in olestra are smaller than those in cooking fats. Suggest a reason for this. [1]

IB, November 2010

10 State the name of the two polymeric forms of starch. [1]

IB, May 2009

B.5 Vitamins

Understandings

→ Vitamins are organic micro-nutrients which (mostly) cannot be synthesized by the body but must be obtained from suitable food sources.

→ The solubility (water or fat) of a vitamin can be predicted from its structure.

→ Most vitamins are sensitive to heat.

→ Vitamin deficiencies in the diet cause particular diseases and affect millions of people worldwide.

Applications and skills

→ Comparison of the structures of vitamins A, C, and D.

→ Discussion of the causes and effects of vitamin deficiencies in different countries and suggestion of solutions.

Nature of science

→ Making observations and evaluating claims – the discovery of vitamins ("vital amines") is an example of scientists seeking a cause for specific observations. This resulted in the explanation of deficiency diseases (eg scurvy and beriberi).

Origins of the name

The name "vitamins" reflects a misconception in biochemistry that essential organic micronutrients were amines. In fact, the original spelling of this name, "vitamines", was derived from words "vital" and "amines", that is, "amines of life". However, it soon became obvious that vitamins belong to different classes of organic compounds and some even do not contain nitrogen, so the name was shortened to "vitamins" in order to break the link between these micronutrients and amines.

Do you know other terms that have been developed from misconceptions and still remain in the language, even when their original meaning is proven to be wrong? Do you think that such words can and should be removed from the language?

Introduction to vitamins

Vitamins are organic micronutrients that cannot be synthesized by the organism in sufficient amounts and must either be obtained from suitable foods or taken as food supplements. A lack (deficiency) of vitamins leads to various health conditions and in some cases can be fatal, even if all other food constituents (proteins, fats, carbohydrates, minerals, and water) are present in the diet.

Classification of vitamins

Vitamins are classified according to their biological functions rather than their chemical structures. Many vitamins bind to enzymes as prosthetic groups or cofactors (sub-topics B.2 and B.7) while others act as hormones or antioxidants (sub-topic B.3) or facilitate the transfer of functional groups and electrons (sub-topic B.9). In some cases, a series of structurally similar compounds show the same type of biological activity and therefore are known under the same collective name. For example, the name "vitamin A" refers to a group of organic compounds that includes an alcohol (*retinol*), an aldehyde (*retinal*), and several polyunsaturated hydrocarbons (*carotenes*). Another group of diverse compounds with molecular masses from 123 to 1580 is known as "vitamins B" and includes open-chain and heterocyclic molecules as well as metal–organic complexes. At the same time, the name "vitamin C" refers to a single compound, *ascorbic acid*. Finally, the group of "vitamins D" consists of four structurally similar compounds produced by different metabolic pathways from the same precursor, cholesterol (sub-topic B.3).

Deficiency diseases

The importance of certain foods for maintaining good health was known long before vitamins were discovered. The ancient Egyptians knew that the symptoms of *night blindness* (as we now know, caused by a vitamin A deficiency) would disappear if the affected person consumed liver for a short period of time. Another deficiency disease, *scurvy* (caused by a deficiency of vitamin C), was known from prehistoric times and could be cured by consuming fresh herbs, fruit, and vegetables. These observations were confirmed later by specially designed experiments and eventually convinced scientists that minute amounts of certain organic compounds were essential for the human body and had to be regularly obtained from the diet. The first of these compounds, vitamin B_1 *(thiamine)*, was identified in the beginning of the twentieth century and successfully used for treating *beriberi*, a potentially fatal illness that was common among sailors during long ocean voyages. Other vitamins were soon discovered and linked to specific deficiency diseases, many of which were almost eliminated in developed countries within the next few decades.

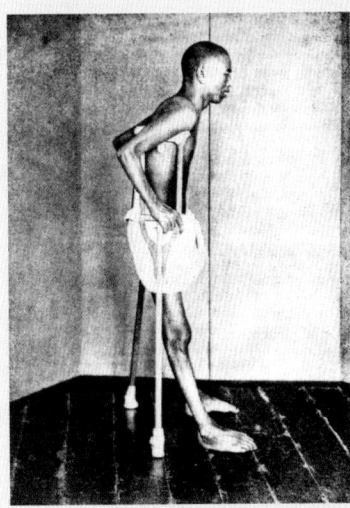

▲ Figure 1 Beriberi is caused by a vitamin B_1 deficiency and leads to weight loss, weakness, limb pains, and nervous system disorders

Preventing deficiencies

To prevent the adverse health conditions associated with vitamin deficiencies, humans must receive vitamins on a regular basis. The optimal frequency of the intake of different vitamins depends on their chemical structures and the way they are distributed and stored in the body. **Water-soluble vitamins** such as vitamin C and some group B vitamins concentrate in blood plasma and intracellular fluids. These vitamins have relatively short half-elimination times, from 30 minutes to several weeks, so they should be supplied to the body on a daily basis. In contrast, **fat-soluble vitamins** such as vitamins A and D are accumulated in the liver and fat tissue, where they can be stored for prolonged periods of time (up to several months). These vitamins can be consumed less frequently without any detrimental health effects.

While **primary vitamin deficiencies** can be prevented by regular intake of vitamins, **secondary deficiencies** may develop as a result of certain health disorders, pregnancy, or risk factors including smoking, excessive alcohol consumption or the use of medical drugs. These and other factors may reduce the absorption or inhibit biological functions of vitamins so that an increase in dose and frequency becomes necessary. At the same time, excessive consumption, of vitamins, especially fat-soluble vitamins, may increase their concentrations in the body tissues to dangerous levels and eventually lead to **vitamin poisoning** or hypervitaminosis. In 2010, about 200 000 cases of vitamin poisoning were registered worldwide, including nearly 100 life-threatening conditions and several fatal incidents.

International support

While scurvy, beriberi, rickets, and other vitamin-related diseases are almost unknown in developed countries, millions of people worldwide still suffer from a lack of vitamins in their diet. This problem can be addressed by providing international support to affected countries, in the form of both vitamin supplements and technologies for their local production and distribution. Some vitamins and minerals can be added to water, salt, and staple foods consumed by the majority of the population of these countries. Finally, people must be educated about the benefits of diverse diet and vitamin supplements, which can be a long and difficult process involving significant changes in the traditional culture.

Study tip

The names and structural formulae of vitamins A, C, and D are given in the *Data booklet*, which will be available during the examination.

Three important vitamins

At present, thirteen vitamins and vitamin groups are known. In this book we shall discuss only three types of vitamin (A, C, and D) that have relatively simple structures and are particularly important for preventing common deficiency diseases and health conditions.

Vitamin A: Retinoids and carotenes

As noted earlier, the collective name **"vitamin A"** refers to several organic compounds, *retinoids* and *carotenes*, that perform similar functions in the human body. The structure of one of these compounds, *retinol*, is shown in figure 2. Another retinoid, *retinal*, will be discussed in sub-topic B.10.

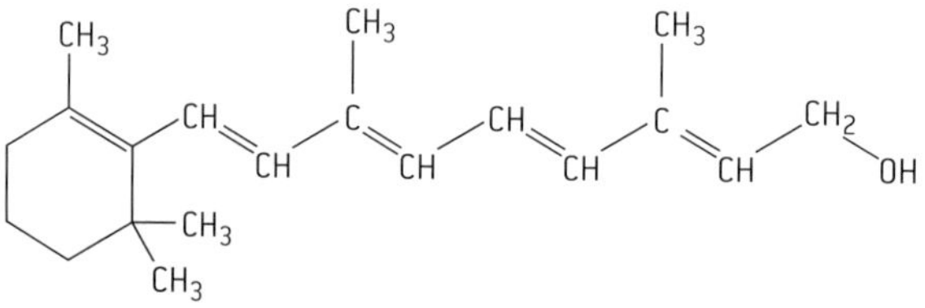

▲ Figure 2 Retinol (vitamin A)

Retinol is a long-chain alcohol with an extensive system of alternating single and double carbon–carbon bonds. Because all carbon atoms involved in such systems have sp^2 hybridization (sub-topic 14.2), the π-electron clouds of adjacent double bonds partly overlap with one another and form a large cloud of delocalized electrons (figure 3).

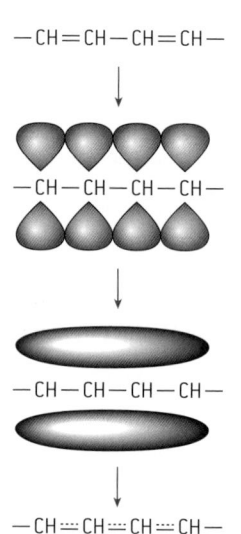

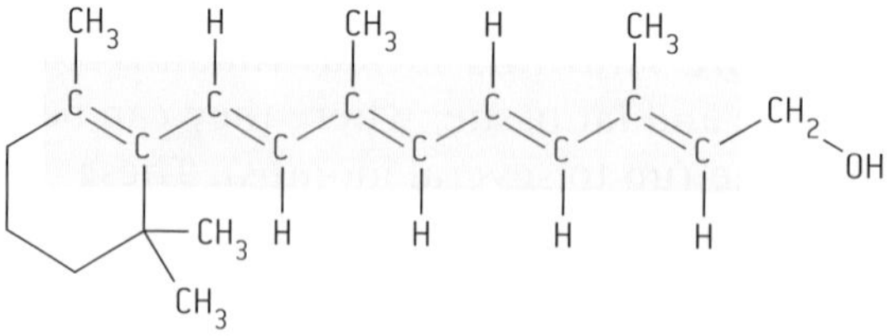

▲ Figure 3 The formation of delocalized electron clouds in retinol

This type of multi-centre chemical bonding, known as **electron conjugation**, is similar to electron delocalization in benzene (sub-topic 20.1) and produces a chain of carbon–carbon bonds with a bond order of 1.5. In retinol the electron conjugation involves 10 carbon atoms, including two carbon atoms in the six-membered ring (figure 4).

▲ Figure 4 Electron conjugation involves 10 carbon atoms in retinol

Carotenes, another group of vitamin A compounds, have even longer conjugation systems that involve up to 22 carbon atoms. Electron conjugation makes retinoids and carotenes efficient antioxidants that readily react with molecular oxygen and free radicals (sub-topic B.3). Also owing to their long conjugation systems, all compounds of the vitamin A group absorb visible light and therefore have bright colours. The optical properties of retinoids will be discussed in sub-topic B.9.

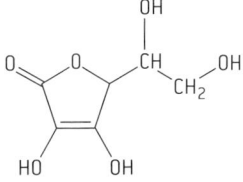

OH

O   O
         CH   OH
            CH₂

HO   OH

▲ Figure 6 Vitamin C (ascorbic acid)

▲ Figure 5 The bright orange colour of carrots is caused by β-carotene, a compound of group A vitamins.

The molecules of retinoids and carotenes contain long hydrocarbon chains with very few or no polar functional groups, which makes these substances predominantly hydrophobic and insoluble in water. However, like all hydrophobic compounds group A vitamins are fat soluble, so their absorption in the intestinal tract and their biological transport depend on certain lipids and lipoproteins (sub-topic B.3). As a result, low-fat diets may lead to secondary vitamin A deficiencies that cannot be corrected by increased intake of retinoids and often require a change in dietary habits.

Vitamin C: Ascorbic acid

Vitamin C or ascorbic acid is a relatively simple oxygen-rich organic molecule containing multiple polar functional groups (figure 6). Several hydroxyl groups and an ester fragment in the molecule can form multiple hydrogen bonds with water, making it a water-soluble vitamin. The same polar functional groups make ascorbic acid insoluble in fats, so it cannot be stored in the body for a long time and requires regular intake.

In the human body vitamin C participates in a broad range of metabolic processes, including the biosynthesis of collagen (sub-topic B.2). This fibrous protein is the main component of connective tissue in the body, which is primarily affected by vitamin C deficiency and shows the most prominent symptoms of scurvy.

Ascorbic acid is a powerful antioxidant and reducing agent capable of donating one or two electrons in biochemical redox reactions, for example:

OH   OH

O   O   O   O
         CH   OH   CH   OH
            CH₂   CH₂ $+ 2H^+ + 2e^-$
C=C   C—C
HO   OH   O   O

ascorbic acid   dehydroascorbic acid
(reduced form)   (oxidized form)

If molecular formulae for ascorbic and dehydroascorbic acids are used, the above equation looks like this:

$$C_6H_8O_6 \rightarrow C_6H_6O_6 + 2H^+ + 2e^-$$

Most animals can synthesize vitamin C in their bodies from galactose, glucose, or other monosaccharides (sub-topic B.4). Humans lack this ability and must obtain ascorbic acid or its derivatives from the diet.

Vitamin C and the common cold

A winner of two Nobel Prizes, Linus Pauling, suggested that vitamin C could reduce the incidence of the common cold and the severity of its symptoms. Although this claim could not be confirmed by double-blind clinical trials (sub-topic D.1), many people still believe in the efficiency of ascorbic acid against infectious diseases and consume it regularly in large doses, typically 10–100 times higher than the recommended daily amount for this vitamin. This example shows the role of authority in communicating scientific knowledge to the public and the importance of experiments in verifying scientific theories.

The oxidized form of vitamin C, dehydroascorbic acid, can be reduced to ascorbic acid by certain enzymes or glutathione (sub-topics B.2 and B.3):

$$C_6H_6O_6 + 2H^+ + 2e^- \rightarrow C_6H_8O_6$$

The concentration of vitamin C in solution can be determined by redox titration (sub-topic 9.1) using DCPIP (2,6-dichlorophenolindophenol, $C_{12}H_7NCl_2O_2$) as an indicator. In the presence of ascorbic acid the pink solution of the protonated indicator, $C_{12}H_8NCl_2O_2^+$, becomes colourless as DCPIP is reduced:

$$\underset{\text{pink}}{C_{12}H_8NCl_2O_2^+} + C_6H_8O_6 \rightarrow \underset{\text{colourless}}{C_{12}H_9NCl_2O_2} + C_6H_6O_6 + H^+$$

During the titration ascorbic acid reacts with the titrant (oxidizing reagent) and the solution remains colourless. When the titration is complete, all the ascorbic acid is oxidized to dehydroascorbic acid and the pink colour of protonated DCPIP reappears:

$$\underset{\text{colourless}}{C_{12}H_9NCl_2O_2} \rightarrow \underset{\text{pink}}{C_{12}H_8NCl_2O_2^+} + H^+ + 2e^-$$

Along with other antioxidants, ascorbic acid is commonly used as food additive E300 for preventing oxidative rancidity (sub-topic B.3).

Vitamin D: Cholecalciferol

The collective name "vitamin D" refers to *cholecalciferol* and three other structurally similar organic compounds with a partly broken steroidal backbone (sub-topic B.3). In the human body small amounts of cholecalciferol can be synthesized from its precursor, *7-dehydrocholesterol* (figure 7).

The biosynthesis of cholecalciferol takes place in the skin and requires ultraviolet (UV) light (which is present in the sunlight spectrum) to open the second six-membered ring of 7-dehydrocholesterol. The human body is normally able to produce enough vitamin D to meet its own metabolic requirements; however when exposure to sunlight is limited (especially at high latitudes during the winter), vitamin D becomes an essential micronutrient that must be obtained from the diet.

The cholecalciferol molecule's large hydrocarbon backbone with only a single hydroxyl group makes it hydrophobic and insoluble in water.

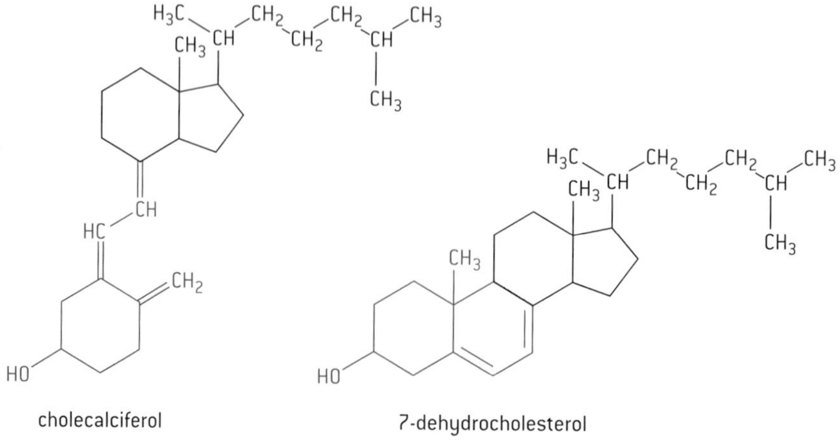

The sunshine vitamin

Urban lifestyles and the widespread use of sunscreen lotions significantly decrease the exposure of skin to sunlight and may lead to a vitamin D deficiency. Even a sunscreen with a minimal *sun protection factor* (SPF) of 15 blocks a significant proportion of UV radiation and reduces the production of cholecalciferol in the skin by 98%. Higher SPF screens can effectively prevent the body from synthesizing vitamin D and make it totally dependent on dietary supplements. A possible solution to this problem involves brief sun exposures without sunscreen, ideally before 10:00 and after 16:00, when the UV radiation is not strong enough to damage the skin but sufficient for vitamin D biosynthesis.

▲ Figure 7 Cholecalciferol (vitamin D) and its precursor 7-dehydrocholesterol

Therefore cholecalciferol and other group D vitamins are fat soluble and can be transported by biological fluids in the form of lipoprotein complexes (sub-topic B.3). In contrast to vitamin A, natural sources and food supplements of vitamin D are usually rich in lipids and do not require additional fat intake for the vitamin to be absorbed in the intestinal tract. Common dietary sources of vitamin D include fish oil, liver (both fish and mammal), eggs, and mushrooms.

Decomposition of vitamins

Vitamins are complex organic compounds and therefore may undergo various chemical transformations when exposed to heat, light, and atmospheric oxygen. The hydrocarbon backbones of fat-soluble vitamins such as A and D are relatively stable to heat and do not decompose significantly when the food is boiled or steamed. In contrast, water-soluble vitamin C is unstable at high temperatures and can be lost by leaching from foods into cooking water. Overcooked or fried foods can lose more than 50% of their fat-soluble vitamins and nearly all their vitamin C.

Extended systems of electron conjugation in vitamins A and C favour free-radical reactions (sub-topic 10.2), so these vitamins are more sensitive to light and air than vitamin D, which contains only three conjugated carbon–carbon double bonds. Transition metals also catalyse free-radical reactions (sub-topic A.10), which are responsible for the accelerated loss of vitamins in canned foods. Finally, ascorbic acid is a strong reducing agent, so foods rich in vitamin C should be protected from atmospheric oxygen during their storage and cooking.

Food fortification

Because many traditional diets do not provide adequate amounts of vitamin D, it is often added artificially to common foods such as vegetable oils, margarine, milk, and breakfast cereals. This practice, known as **food fortification**, increases the nutritional values of dietary products and avoids widespread deficiencies caused by geographical or cultural factors. In particular, rickets, the most common childhood disease of the past, was nearly eradicated in developed countries after the introduction of foods enriched with vitamin D and other micronutrients. While food fortification is beneficial for the majority of the population, it limits the freedom of people to choose their diet and, in rare cases, can lead to vitamin poisoning and allergic reactions. Therefore, similar to other medical or commercial practices, food fortification raises a question about the balance between the interests of society and the rights of individual people.

Questions

1 Describe, in terms of polarity and solubility, the most common properties of vitamins A and D.

2 The formulae of vitamin B$_3$ (niacin) and vitamin E (α-tocopherol) are given below.

vitamin B$_3$ vitamin E

a) Identify two functional groups in vitamin B$_3$ and two functional groups in vitamin E.

b) In the human body, vitamin E acts as antioxidant. Identify the functional group or groups that are responsible for antioxidative properties of this vitamin.

c) Predict, with reference to functional groups and polarity, whether each of these vitamins is water soluble or fat soluble.

d) Suggest which vitamin (B$_3$ or E) must be ingested regularly in small quantities and which one can be taken at much longer intervals but in larger amounts without any detrimental health effects.

3 The American chemist Linus Pauling, who won two Nobel prizes, promoted the taking of vitamin C as a way of preventing the common cold. One of the functions of vitamin C in the body is as an antioxidant. During the process ascorbic acid, C$_6$H$_8$O$_6$, is converted into dehydroascorbic acid, C$_6$H$_6$O$_6$.

Deduce the half-equation to show how vitamin C acts as an antioxidant. [2]

IB, May 2012

4 The structure of vitamin C (ascorbic acid) has some similarities to the structure of carbohydrates.

a) State the name of one functional group that is present both in vitamin C and in all carbohydrates.

b) Predict whether the dietary energy value (in J g^{-1}) of vitamin C will be greater than, equal to, or lower than the energy value of glucose.

5 Vitamins are essential micronutrients that must be obtained from suitable food sources. However, one vitamin can be synthesized in the human body in sufficient quantities even if it is not present in the diet.

a) State the name of this vitamin.

b) Discuss whether a non-essential micronutrient can be classified as vitamin.

6 Explain why vitamin D deficiency in northern countries is more common during the winter.

7 Discuss two solutions for the prevention of nutrient deficiencies. [2]

IB, November 2007

8 Food fortification is a common practice in many countries. Discuss two advantages and two disadvantages of food fortification.

B.6 Biochemistry and the environment

Understandings

→ Xenobiotics refers to chemicals that are found in an organism that are not normally present there.

→ Biodegradable/compostable plastics can be consumed or broken down by bacteria or other living organisms.

→ Host–guest chemistry involves the creation of synthetic host molecules that mimic some of the actions performed by enzymes in cells, by selectively binding to specific guest species such as toxic materials in the environment.

→ Enzymes have been developed to help in the breakdown of oil spills and other industrial wastes.

→ Enzymes in biological detergents can improve energy efficiency by enabling effective cleaning at lower temperatures.

→ Biomagnification is the increase in concentration of a substance in a food chain.

→ Green chemistry, also called sustainable chemistry, is an approach to chemical research and engineering that seeks to minimize the production and release of hazardous chemicals to the environment.

Applications and skills

→ Discussion of the increasing problem of xenobiotics such as antibiotics in sewage treatment plants.

→ Description of the role of starch in biodegradable plastics.

→ Application of host–guest chemistry to the removal of a specific pollutant in the environment.

→ Description of an example of biomagnification, including the chemical source of the substance. Examples could include heavy metals or pesticides.

→ Discussion of the challenges and criteria in assessing the "greenness" of a substance used in biochemical research, including the atom economy.

Nature of science

→ Risk assessment, collaboration, ethical considerations – it is the responsibility of scientists to consider the ways in which products of their research and findings negatively impact the environment, and to find ways to counter this. For example, the use of enzymes in biological detergents, to break up oil spills, and green chemistry in general.

The nature of biochemistry

Biochemistry is a multidisciplinary science that studies the chemical changes associated with living organisms and their interactions with the environment. Our increasing understanding of biochemical processes has greatly enhanced our ability to control biological systems but at the same time created serious ecological problems and raised our awareness of the environmental and ethical implications of science and technology. In this topic we shall discuss the use of biochemical techniques in industrial, agricultural, and household applications, their effects on global and local ecosystems, and the role of biochemistry in reducing the environmental impact of human activities.

Risk assessment

Before carrying out an experiment any scientist must estimate the individual, environmental, and ethical implications of the proposed work. This work, known as **risk assessment**, is particularly important when potentially hazardous chemical or biological materials can be released to the environment, cause unnecessary suffering to laboratory animals, or present a significant risk to human health. In each case, the experimenter is responsible for minimizing the negative impact of his or her work and providing a comprehensive list of emergency procedures to counter any accidental damage to the environment or individuals involved in the research.

Xenobiotics

The rapid development of organic chemistry in the twentieth century led to the industrial production of pesticides, medicinal drugs, and other chemical compounds that had no natural sources and therefore were foreign to living organisms. These compounds, known as **xenobiotics**, are generally toxic to various life forms and are more resistant to biodegradation than naturally occurring organic molecules. Certain xenobiotics (**persistent organic pollutants**, **POPs**) can remain in the soil and in animal fatty tissues for many decades after their release into the environment.

DDT

The abbreviated name of the most notorious insecticide, DDT, is derived from its semi-systematic name, **d**ichloro**d**iphenyl**t**richloroethane (figure 1).

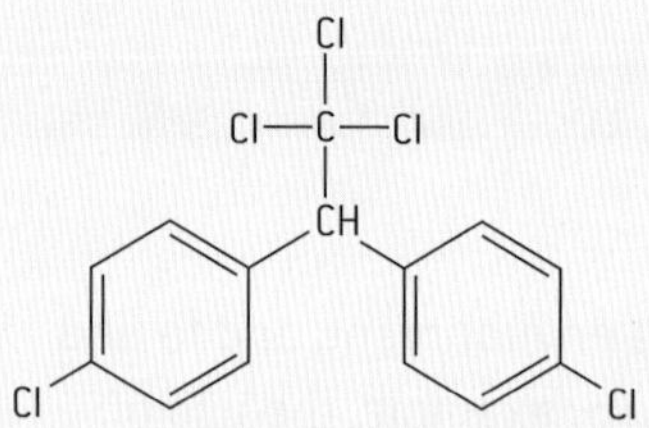

▲ Figure 1 DDT, dichlorodiphenyltrichloroethane

From 1950 to 1980, about 2 million tonnes of DDT were produced and released to the environment worldwide, enabling significant increases in the yields of agricultural crops and nearly eradicating certain diseases such as malaria and dengue fever. However, very soon the widespread use of DDT created resistant insect populations, reducing the effectiveness of this compound and, in many cases, reversing the initial gains in agricultural production and disease control. In addition, it was discovered that DDT was particularly stable in the environment and could accumulate in animals, poisoning wildlife and creating a significant risk to human health. In the 1970s and 1980s this insecticide was banned in most countries, although its limited use is still allowed in regions affected by malaria and other insect-transmitted diseases.

▲ Figure 2 The bald eagle was brought close to extinction by the widespread use of DDT in agriculture. The biomagnification of DDT in these birds of prey led to the thinning of their eggshells, which became too brittle so their chicks could not hatch. Since the ban on DDT introduced in the USA in 1972, the population of bald eagles has increased from several hundred to over 150 000 individuals

The metabolism of xenobiotics

Depending on their chemical structure, some xenobiotics can be completely digested by microorganisms, plants, and animals. However, many synthetic chemicals produce toxic metabolites, alter the metabolic pathways of other compounds, or affect the reproduction, development, and growth of living organisms. Certain xenobiotics cannot be metabolized by existing enzymes (sub-topic B.2) and either remain within the organism or are excreted unchanged.

The nature of its functional groups and the overall polarity of a xenobiotic molecule strongly affects its rate of decomposition in the environment. Polar synthetic chemicals are often soluble in water and are quickly metabolized by living organisms or undergo photochemical oxidation. In contrast, non-polar, hydrophobic xenobiotics easily pass through biological phospholipid membranes (sub-topic B.3) and tend to accumulate within the cells of microorganisms or in fatty tissues of animals. When such compounds are passed along the food chain, their concentrations may increase exponentially and reach very high levels in top predators (figure 3). This process, known as **biomagnification**, has been largely responsible for the extinction or significant population reduction of many birds of prey and large marine animals across the globe, often in regions far distant from the places where the xenobiotics were released to the environment.

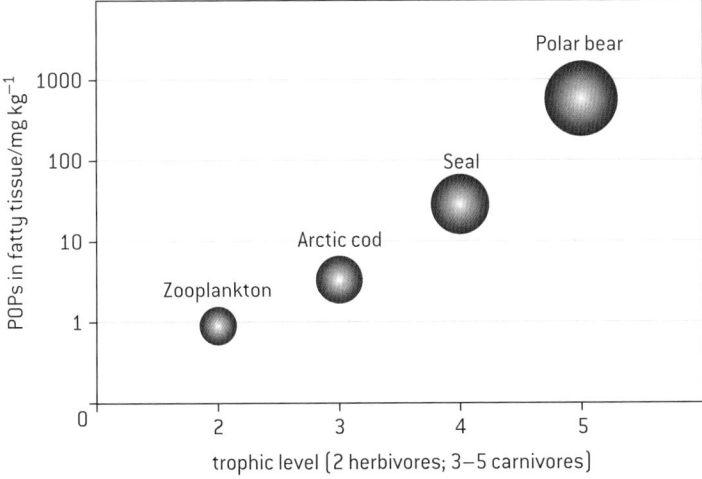

▲ Figure 3 Biomagnification of persistent organic pollutants (POPs) in a food chain

Heavy metal toxicity

Heavy metals, such as mercury, cadmium, and lead, have numerous industrial applications and may be released to the environment at all stages of their production and utilization. These elements cause denaturation of proteins (sub-topic B.2), inhibit the action of enzymes (sub-topic B.7), and affect the redox balance in cells. Although heavy metals are toxic to nearly all living organisms, they often undergo biomagnification and thus are particularly dangerous to predators at the tops of food chains. The environmental impact of heavy metals and common methods of their removal are discussed in sub-topic A.10.

PCBs

Polychlorinated biphenyls (PCBs) are synthetic organic molecules containing two benzene rings with some or all hydrogen atoms replaced by chlorine; an example is shown in figure 4.

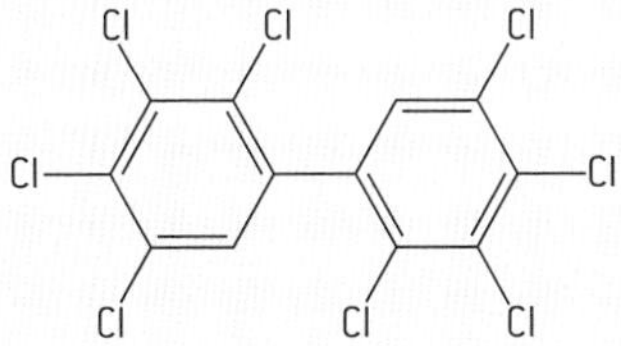

▲ Figure 4 The structure of a polychlorinated biphenyl (PCB)

These compounds were widely used in the twentieth century as coolants, lubricants, plasticizers, and insulating liquids. PCBs were found to cause cancer and liver damage in animals and humans, so their production in most countries was banned in the 1970s. However, PCBs are still present in the environment in significant quantities. In 1996, 20 years after the ban was introduced, the body of a Beluga whale discovered in the St Lawrence River in Canada contained PCBs in excess of 50 mg kg^{-1}. According to local regulations, the whale was classified as hazardous to the environment and had to be disposed of as toxic chemical waste.

Worked example

Atlantic mackerel is a common prey of porbeagle shark, which consumes about 100 mackerel fish per month. Mercury and other heavy metals from consumed mackerel remain in the shark's body for approximately 2 years. Calculate the concentration of mercury in porbeagle shark if mackerel contains 0.05 ppm of mercury (1 ppm = 10^{-4}%), the mass of an average mackerel is 1 kg, and the mass of a porbeagle shark is 120 kg.

Solution

In 2 years (24 months), the shark consumes 2400 mackerel with a total mass of 2400 kg. The mercury level in mackerel is 0.05×10^{-4}% = 5×10^{-6}%, so the mass of mercury in consumed mackerel is 2400 kg $\times 5 \times 10^{-6} / 100 = 1.2 \times 10^{-4}$ kg. Therefore, the concentration of mercury in the shark's body is $(1.2 \times 10^{-4}$ kg/120 kg$) \times 100$% = 1×10^{-4}%, or 1 ppm. As a result of biomagnification, this concentration is 20 times higher than the level of mercury in mackerel.

▲ Figure 5 Left: household batteries contain heavy metals and must be recycled to protect the environment. Right: alternating layers of nickel and cadmium in a rechargeable battery

Pharmaceutically active compounds and detergents

Antibiotics and other **pharmaceutically active compounds** (**PACs**) are a diverse group of xenobiotics commonly found in soil and aquatic ecosystems. At present very little is known about the occurrence, effects, and risks of the release of PACs into the environment. One of the major concerns is the development of resistant bacteria (sub-topics D.3 and D.6), which evolve to survive in the presence of antibiotics and pass their resistance to future generations. Such bacteria may cause serious diseases that cannot be treated effectively by existing medications. In addition, certain PACs affect immune and endocrine systems of aquatic animals, increasing the risk of infectious diseases and inhibiting their reproductive functions.

Another type of common environmental pollutant is household and industrial **detergents** containing amphiphilic molecules (sub-topic B.3) that reduce the surface tension of water and facilitate the cleaning of fabrics and solid surfaces. Many detergents such as *branched alkylbenzenesulfonates (ABSs)* have very poor biodegradability and accumulate in sewage treatment plants, producing persistent foam and altering the bacterial composition of recycled water. In developed countries ABSs have been phased out and replaced by biodegradable *linear alkylbenzenesulfonates (LASs)*, which reduced the levels of surfactants in water and helped to restore the biodiversity of aquatic ecosystems (figure 7, see next page).

Biological detergents contain a variety of enzymes extracted from thermophilic microorganisms. These enzymes facilitate the biological breakdown of fats, proteins, starch, and other organic molecules, providing fast and effective cleaning even in cold water. At the same time, they are more resistant to thermal denaturation (sub-topic B.3) and can be used at temperatures up to 50 °C. Most enzymes used in biological detergents are easily biodegradable and do not have any lasting impact on the environment. In addition, their use saves energy and reduces the amount of non-biological detergents used for cleaning, which is particularly important in densely populated areas with limited capacity of sewage treatment

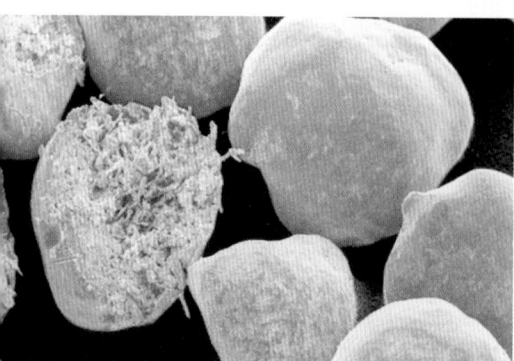

▲ Figure 6 Biological washing powders contain granules of encapsulated enzymes

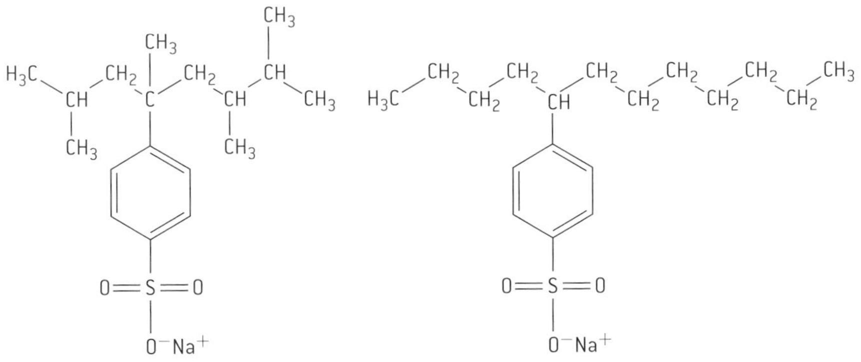

▲ Figure 7 A non-biodegradable branched alkylbenzenesulfonate (ABS) contrasted with a biodegradable linear alkylbenzenesulfonate (LAS)

▲ Figure 8 The Deepwater Horizon disaster in the Gulf of Mexico, 2010

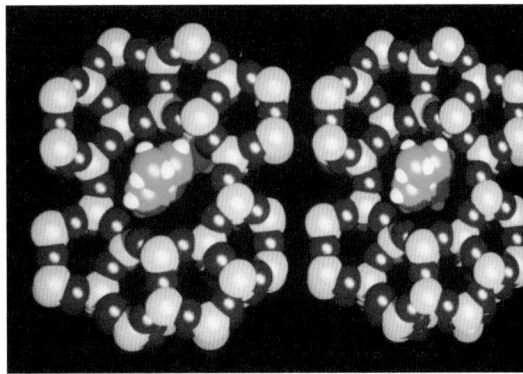

▲ Figure 9 Host–guest complexes of xylene (green and white) with zeolite (yellow and red)

plants. The only known side effect of biological detergents is the possibility of allergic reactions in certain individuals with increased skin sensitivity.

Enzymes and microorganisms are also used to clean up oil spills and industrial wastes. The exact clean-up procedure depends on many factors including the chemical nature and volatility of the waste, location of the spill, temperature, and so on. Generally a mixture of enzymes, surfactants, and other chemicals is used for the initial breakdown of the oil or waste components into biodegradable products, which are further metabolized by common microorganisms. Several strains of oil-degrading bacteria have been discovered near the sites of major oil spills, including the Deepwater Horizon in the Gulf of Mexico, and have been successfully used to break down hydrocarbon-based industrial wastes.

Host–guest complexes

Although enzymatic processes are highly selective and efficient, many enzymes are unstable in the environment and show their optimal activity in narrow ranges of pH and temperature (sub-topic B.2). Certain synthetic molecules are free from these limitations and can selectively bind to environmental pollutants. The resulting **supramolecules**, or **host–guest complexes**, mimic the structures of enzyme–substrate complexes (sub-topic B.2), where the synthetic analogue of the enzyme (**host**) and the environmental pollutant (**guest**) are held together by multiple non-covalent interactions including van der Waals' forces, ionic bonds, and hydrogen bonds (sub-topic 4.4).

To form a stable complex the host and guest molecules must have complementary chemical structures and three-dimensional configurations. In the simplest case the host molecule contains a cavity of a certain size and interacts with a substrate (guest) only by van der Waals' forces. Such host molecules can bind to a broad range of environmental pollutants but have low selectivity and interact with any substances that fit into the cavity. The presence of functional groups that form specific hydrogen or ionic bonds with the substrate increases the

Various host–guest systems have been successfully used for the immobilization and removal of inorganic ions (including heavy metals and radioactive elements such as caesium-137), polychlorinated compounds (PCBs and dioxins), and carcinogenic aromatic amines from water and industrial waste. In addition to environmental applications, host–guest complexes are used in medicine for targeted drug delivery, which is particularly important in cancer treatment.

selectivity of host–guest interactions but often makes the host molecule more sensitive to pH and temperature.

In certain cases the function of the host can be performed by microporous solid materials such as zeolites (aluminosilicate minerals) or branched organic polymers. The pollutants immobilized on the surface of the host material can be mechanically separated from the environment for further processing or incineration.

Plastics and polymers

Non-biodegradable materials such as plastics and other synthetic polymers are the most abundant and persistent environmental pollutants produced by humans. The accumulation of plastic waste is not only unsightly but presents a serious danger to living organisms, especially birds and marine animals. Entanglement and ingestion of non-biodegradable materials reduce the mobility and interfere with the digestive functions of affected species, which often leads to starvation and death. In a recent study over 95% of sea birds were found to have plastic objects in their stomachs, which in some cases prevented the birds from flying due to additional weight and chronic malnutrition.

While many traditional plastics are biologically inert and can remain in the environment for hundreds of years virtually unchanged, **biodegradable plastics** can be digested by microorganisms within a relatively short time. These materials either are composed of renewable biological materials such as starch (sub-topic B.4) and cellulose (sub-topic B.10), or contain additives that alter the structure of traditional plastics and allow microorganisms to digest hydrocarbon-based polymers. In addition, certain non-biodegradable plastics such as aromatic polyesters can be replaced with aliphatic polyesters (sub-topic A.9) that are very similar in structure and properties, but are less resilient to enzymatic hydrolysis. Another important component of biodegradable plastics, *polylactic acid (PLA)*, is a condensation polymer of 2-hydroxypropanoic (lactic) acid:

2-hydroxypropanoic (lactic) acid   polylactic acid (PLA)

Starch-based polymers constitute over 50% of biodegradable plastics. By combining starch with natural plasticizers such as glycerol (sub-topic B.3) and certain carbohydrates, the characteristics of the resulting material can be varied significantly without compromising its biodegradability. Starch plastics are used for making a broad range of products from disposable bags and food packaging to mobile phones and car interiors. In some cases starch is blended with other polymers to create materials with desirable properties and reduce the use of fossil fuels as a hydrocarbon source.

Green chemistry

In traditional chemistry, the efficiency of a synthetic procedure is measured in terms of the product yield and the cost of raw materials while many other factors such as the toxicity of reagents and solvents, energy consumption, and the amount of waste produced are often ignored. A completely different approach, known as **green chemistry**, takes into account the environmental impact of the entire technological process and encourages the synthetic design that minimizes the use and generation of hazardous chemicals. Common practices of green chemistry include aqueous or solvent-free reactions, renewable starting materials, mild reaction conditions, regio- and stereoselective catalysis (sub-topics 20.1 and B.10), and the utilization of any by-products formed during the synthesis.

Atom economy

Another key concept of green chemistry, **atom economy**, expresses the efficiency of a synthetic procedure as the ratio of the mass of the isolated target product to the combined masses of all starting materials, catalysts, and solvents used in the reaction. For example, the atom efficiency of a solvent-free reaction $A + B \rightarrow C$ is equal to the practical reaction yield (sub-topic 1.3) and can potentially reach almost 100%. However, in a reaction $A + B \rightarrow C + D$ with the target product C, the atom efficiency will always be significantly lower than 100% because some of the atoms from reactants A and B form the unwanted by-product D. Solvents and catalysts further reduce the atom efficiency because their constituent atoms do not form the target product and must be disposed of or recycled.

The costs of green chemistry

Green technologies vary in efficiency and in many cases involve expensive equipment, raw materials, and recycling facilities. However, these initial investments reduce the costs associated with environmental remediation, waste management, and energy consumption, so in the long run green chemistry is a commercially attractive and sustainable alternative to traditional organic chemistry.

Increasing adoption of green industrial processes in developed countries has significantly reduced the emissions of many hazardous chemicals such as chlorinated solvents or greenhouse gases, and brought new products to the market. Many of these products including PLA and starch-based plastics are not only biodegradable but also can be produced by "green" technologies, which further decreases their overall environmental impact. At the same time, some non-hazardous substances branded as "green" or "environmentally friendly" still require toxic chemicals or large amounts of energy for their production. In addition, the industrial use of natural products such as plant oils and starch takes up agricultural resources and leads to various ecological and social issues (sub-topic B.4). Therefore the criteria used in assessing the "greenness" of a substance must include all direct and indirect environmental implications of its entire life cycle, which remains one of the most controversial problems in green chemistry.

The term "green chemistry" was coined in 1991 by Paul Anastas and John Warner, who formulated 12 principles of their approach to chemical technology. These principles emphasize the benefits of non-hazardous chemicals and solvents, efficient use of energy and reactants, reduction of waste ("the best form of waste disposal is not to create it in the first place"), choice of renewable materials, and prevention of accidents. The philosophy of green chemistry has been adopted by many educational and commercial organizations and eventually passed into national and international laws, which restricted the use of certain chemical substances and encouraged the use of environmentally friendly technologies.

Examples of atom-efficient reactions are the hydrogenation of alkenes (sub-topic 10.1) and unsaturated fats (sub-topic B.10), which proceed with high yields, require no solvents, and form almost no by-products under appropriate conditions. At the same time many traditional organic reactions such as the oxidation of alcohols (sub-topic 10.2) or electrophilic substitution in aromatic compounds (sub-topic 20.1) are very inefficient because they often require large volumes of solvents, have low yields, and, in some cases, produce mixtures of regio- and stereoisomers instead of individual target products.

Worked example

The alkylation of phenylamine can be carried out using traditional or green chemistry.

a) Dimethyl sulfate, $(CH_3O)_2SO_2$, is a traditional alkylating reagent that has many disadvantages including high toxicity and the possibility of side reactions. Calculate the percentage atom economy of the following reaction if the target product is *N*-methylphenylamine, $C_6H_5NHCH_3$:

$$2C_6H_5NH_2 + (CH_3O)_2SO_2 + 2NaOH \rightarrow 2C_6H_5NHCH_3 + Na_2SO_4 + 2H_2O$$

b) Dimethyl carbonate is a non-toxic and highly efficient alternative to dimethyl sulfate. Calculate the percentage atom economy of the following reaction:

$$C_6H_5NH_2 + (CH_3O)_2CO \rightarrow C_6H_5NHCH_3 + CH_3OH + CO_2$$

c) Dimethyl carbonate can be synthesized as follows:

$$4CH_3OH + 2CO + O_2 \rightarrow 2(CH_3O)_2CO + 2H_2O$$

Suggest how the amounts of waste produced in the synthesis of *N*-methylphenylamine can be further reduced.

Solution

a) The total mass of the products is equal to the total mass of the reactants, so it is sufficient to calculate the molecular masses of the products only: $M_r(C_6H_5NHCH_3) = 107.15$, $M_r(Na_2SO_4) = 142.04$, $M_r(H_2O) = 18.02$. The atom economy is $(2 \times 107.15)/(2 \times 107.15 + 142.04 + 2 \times 18.02) \approx 0.546$ or 54.6%.

b) $M_r(C_6H_5NHCH_3) = 107.15$, $M_r(CH_3OH) = 32.04$, $M_r(CO_2) = 44.01$. The atom economy is $107.15/(107.15 + 32.04 + 44.01) \approx 0.585$ or 58.5%.

c) Methanol formed in reaction (b) can be recycled and converted into dimethyl carbonate using reaction (c). In addition, carbon dioxide from reaction (b) can be recycled by the reaction with elemental carbon at high temperature: $CO_2 + C \rightarrow 2CO$. If both waste products are converted back into reactants, the atom economy of the entire technological process can reach almost 100%.

Questions

1 In environmental research the concentration of pollutants in the air is often reported in molecules per cubic cm. The air in the Ruhr area of Germany contains 3.3 ng m^{-3} of polychlorinated biphenyls (PCBs). Determine the concentration of PCBs over the Ruhr area in molecules per cm^3 if the average molecular mass of PCBs is 320 g mol^{-1}.

2 Explain the meaning of the term "biomagnification".

3 Biomagnification of pollutants is a major environmental concern. An average Far Eastern brown bear has a body mass of 600 kg and consumes 10 kg of fish per day. Calculate the concentration of chlorinated organic pollutants in the bear's body if their concentration in the fish is 2×10^{-6}% and the pollutants remain the bear's body for 5 years.

4 The use of DDT reduces the occurrence of malaria and saves the lives of people in developing countries but at the same time has a serious environmental impact worldwide. Discuss how this conflict between the rights of individuals to protect their health and the right of the global society to protect the environment can be resolved.

5 DDT is a non-biodegradable insecticide that was extensively used worldwide in the twentieth century. When an agricultural field was treated with this insecticide in May 1970, the concentrations of DDT in the soil were measured (1 ppm = 10^{-4}%) (table 1).

a) Plot the concentration of DDT (in ppm) in the soil as a function of time (in months).

b) Determine the half-life (in months) of DDT in the soil.

c) The lowest level of DDT in the soil that can be detected by modern analytical techniques is 0.01 ppb (1 ppb = 10^{-7}%). Estimate the period of time (in years) after the initial application of DDT when its concentration in the soil falls below the detectable level.

6 The extraction and processing of crude oil is essential for the global economy but can have a serious environmental impact. Discuss the role of biochemistry in hydrocarbon waste management and the remediation of accidental oil spills.

7 Describe the bonding between the components of host–guest supramolecules.

8 Atom economy is one of the key aspects of green chemistry.

a) Define "atom economy".

b) Calculate the atom economy of the following reaction:

$$4CH_3OH + 2CO + O_2 \rightarrow 2(CH_3O)_2CO + 2H_2O$$

9 In green chemistry, the use of dangerous materials is generally avoided. Discuss the advantages and disadvantages of using ethanoic acid instead of sulfuric acid as a neutralizing agent for treating alkaline waste.

Year	1970					1971		
Month	May	June	July	September	December	March	July	December
DDT level (ppm)	48.0	46.1	44.2	40.8	36.1	31.9	27.7	22.1

▲ Table 1

B.7 Proteins and enzymes (AHL)

Understandings

→ Inhibitors play an important role in regulating the activities of enzymes.

→ Amino acids and proteins can act as buffers in solution.

→ Protein assays commonly use UV-vis spectroscopy and a calibration curve based on known standards.

Applications and skills

→ Determination of V_{max} and the value of the Michaelis constant K_m for an enzyme by graphical means, and explanation of its significance.

→ Comparison of competitive and non-competitive inhibition of enzymes with reference to protein structure, the active site, and allosteric regulation.

→ Explanation of the concept of product inhibition in metabolic pathways.

→ Calculation of the pH of buffer solutions, such as those used in protein analysis and in reactions involving amino acids in solution.

→ Determination of the concentration of a protein in solution from a calibration curve using the Beer–Lambert law.

Nature of science

→ Theories can be superseded – "lock and key" hypothesis to "induced fit" model for enzymes.

→ Collaboration and ethical considerations – scientists collaborate to synthesize new enzymes and to control desired reactions (i.e. waste control).

Molecular (non-ionized) forms of 2-amino acids do not exist in aqueous solutions and should never be used in acid–base equations (sub-topic B.2).

Introduction to proteins and enzymes

The chemical composition, structural features, and biological functions of amino acids, proteins, and enzymes were discussed in sub-topic B.2. The study of the activity and distribution of these compounds in living organisms is the key area of modern biochemistry. In this sub-topic we shall discuss the acid–base properties of amino acids and proteins, the role of inhibitors in the regulation of enzymatic processes, quantitative interpretation of biochemical data, and spectroscopic techniques used in protein analysis.

Acid–base properties of 2-amino acids

Acid–base equilibria in aqueous solutions of 2-amino acids and proteins were described in sub-topic B.2. Depending on the solution pH (sub-topic 8.3), the carboxyl and amino groups in these amphoteric compounds can be ionized to various extents producing ionic species with different charges. In strongly acidic solutions, amino acids and proteins are protonated and exist as **cations** while in strongly alkaline solutions, deprotonation occurs and **anions** are formed. At a certain pH known as the **isoelectric point** (**pI**) and specific to each amino acid or protein, the positive and negative charges of ionizable groups cancel one another, producing **zwitterions** with net zero charges:

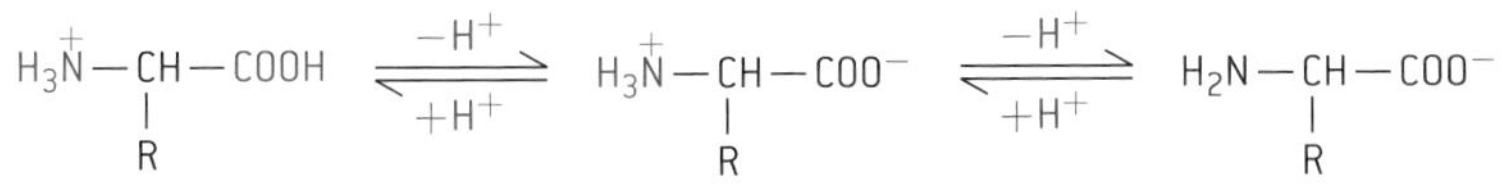

cation (pH < pI) zwitterion (pH = pI) anion (pH > pI)

Each of the two equilibria (cation/zwitterion and zwitterion/anion) in the above scheme involves a pair of species differing by a single proton (H^+). Such pairs are known as **conjugate acid–base pairs**, where the more protonated species is the **conjugate acid** and the less protonated species is the **conjugate base** (sub-topic 8.1). An equilibrium between the components of a conjugate acid–base pair is characterized by the **dissociation constant** (K_a) or, more commonly, its negative logarithm (pK_a, see sub-topic 18.2):

$$K_a = \frac{[\text{conjugate base}][H^+]}{[\text{conjugate acid}]} \qquad pK_a = -\log K_a$$

$$pK_a = -\log \frac{[\text{conjugate base}][H^+]}{[\text{conjugate acid}]}$$

Cationic forms of 2-amino acids with non-ionizable side-chains have two acidic centres, –COOH and $-NH_3^+$, and therefore two dissociation constants, pK_{a1} and pK_{a2} (table 1). The carboxyl group has relatively high acidity and dissociates more easily than the protonated amino group, so the pK_{a1} value characterizes the equilibrium between the cation and the zwitterion:

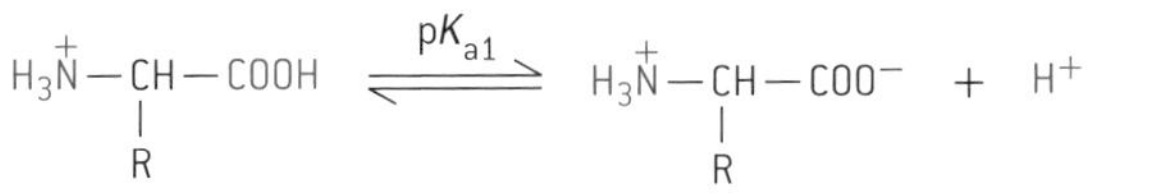

cation (conjugate acid) zwitterion (conjugate base)

Common name	Abbreviation	pK_{a1}	pK_{a2}	Isoelectric point
alanine	Ala	2.3	9.7	6.0
asparagine	Asn	2.1	8.7	5.4
glutamine	Gln	2.2	9.1	5.7
glycine	Gly	2.3	9.6	6.0
isoleucine	Ile	2.4	9.6	6.0
leucine	Leu	2.3	9.6	6.0
methionine	Met	2.2	9.1	5.7
phenylalanine	Phe	1.8	9.1	5.5
proline	Pro	2.0	10.5	6.3
serine	Ser	2.2	9.1	5.7
threonine	Thr	2.2	9.0	5.6
tryptophan	Trp	2.4	9.4	5.9
valine	Val	2.3	9.6	6.0

▲ Table 1 Acid–base properties of selected 2-amino acids

Some proteinogenic 2-amino acids with additional acidic or basic centres in their side-chains (see sub-topic B.2, table 1) have more than two dissociation constants and are able to form several different anionic or cationic species. The acid–base properties of such amino acids will not be discussed in this book.

The pK_{a2} value refers to the equilibrium between the zwitterion and the anion of the amino acid:

$$H_3\overset{+}{N}-CH-COO^- \underset{}{\overset{pK_{a2}}{\rightleftharpoons}} H_2N-CH-COO^- + H^+$$
$$\quad\quad\quad | \quad\quad\quad\quad\quad\quad\quad\quad\quad |$$
$$\quad\quad\quad R \quad\quad\quad\quad\quad\quad\quad\quad\quad R$$

zwitterion (conjugate acid)   anion (conjugate base)

Note that the same zwitterion is the conjugate base in the first acid–base equilibrium but the conjugate acid in the second equilibrium.

In any aqueous solution, only two of the three possible forms of an amino acid can be present at the same time. One of these forms is always the zwitterion while another form can be either the cation or the anion. Both the cation and the anion of the same amino acid cannot exist in the same solution because they will immediately react with one another to produce zwitterions:

$$H_3\overset{+}{N}-CH-COOH + H_2N-CH-COO^- \longrightarrow 2H_3\overset{+}{N}-CH-COO^-$$
$$\quad\quad | \quad\quad\quad\quad\quad\quad\quad | \quad\quad\quad\quad\quad\quad\quad\quad\quad |$$
$$\quad\quad R \quad\quad\quad\quad\quad\quad\quad R \quad\quad\quad\quad\quad\quad\quad\quad\quad R$$

Therefore, acidic solutions (pH < pI) contain mixtures of cations and zwitterions while alkaline solutions (pH > pI) contain zwitterions and anions. The exact ratio between these forms depends on the solution pH and the pK_a of the conjugate acid that is present in the solution. Since pH = −log [H⁺], the pK_a expression can be transformed into the **Henderson–Hasselbalch equation**:

$$pH = pK_a + \log \frac{[\text{conjugate base}]}{[\text{conjugate acid}]}$$

At pH < pI the conjugate acid is the cationic form of the amino acid, the conjugate base is the zwitterion, and $pK_a = pK_{a1}$. At pH > pI the conjugate acid is the zwitterion, the conjugate base is the anion, and $pK_a = pK_{a2}$.

The Henderson–Hasselbalch equation allows calculation of the pH of an amino acid solution with known acid–base composition or the concentration of conjugate acid and base in a solution with known pH. For example, if pH = pK_{a1}, log ([zwitterion]/[cation]) = 0 and thus [zwitterion] = [cation]. Similarly, at pH = pK_{a2} the concentrations of the zwitterion and the anion are equal to each other (figure 1).

Worked example

Calculate the pH of an aqueous solution that contains 0.8 mol dm⁻³ zwitterionic and 0.2 mol dm⁻³ anionic forms of serine.

Solution

The zwitterion contains an extra proton, so it is the conjugate acid while the anion is the conjugate base. The acid–base equilibrium in this solution is characterized by pK_{a2}(serine) = 9.1 (table 1). According to the Henderson–Hasselbalch equation,
pH = 9.1 + log (0.2/0.8)
≈ 9.1 + (−0.6) = 8.5.

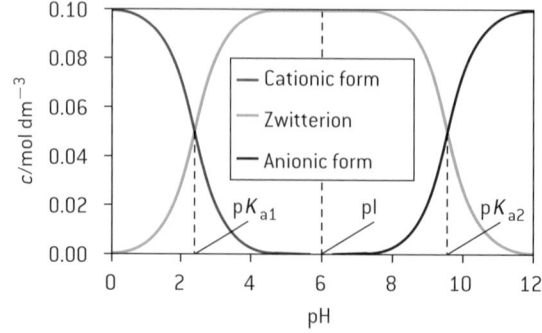

▲ Figure 1 Acid–base equilibria in a 0.1 mol dm⁻³ aqueous solution of alanine (pI = 6.0)

Acid–base buffers

An acid–base **buffer solution** (or **buffer**) containing a weak conjugate acid–base pair can neutralize small amounts of strong acids and bases without significantly changing its pH (sub-topic 18.3). In an amino acid buffer, a strong acid is neutralized by the conjugate base of the buffer while a strong base reacts with the conjugate acid (table 2).

Solution pH	pH < pI (acidic)	pH > pI (alkaline)
pK_a used	pK_{a1}	pK_{a2}
cation	conjugate acid	*does not exist*
zwitterion	conjugate base	conjugate acid
anion	*does not exist*	conjugate base
reaction with a strong acid	zwitterion + H$^+$ → cation	anion + H$^+$ → zwitterion
reaction with a strong base	cation + OH$^-$ → zwitterion + H$_2$O	zwitterion + OH$^-$ → anion + H$_2$O

▲ Table 2 Amino acid buffers

Worked example

Calculate the pH changes after the addition of 1.0 g of solid NaOH to:

a) 1.00 dm^3 of pure water

b) 1.00 dm^3 of a buffer solution containing 0.40 mol of zwitterionic and 0.16 mol of cationic forms of glycine.

Assume that the densities of all solutions are 1.0 kg dm^{-3} and the solution volumes do not change after the addition of NaOH.

Solution

a) The amount of NaOH is 1.0 g/40 g mol^{-1} = 0.025 mol and the concentration of NaOH in the final solution will be 0.025 mol/ 1.00 dm^3 = 0.025 mol dm^{-3}. Since NaOH is a strong base, it will dissociate completely and produce 0.025 mol dm^{-3} hydroxide anions. Therefore, pH = 14 − pOH = 14 + log (0.025) = 14 − 1.6 = 12.4 (sub-topic 8.3). The pH of pure water at 20 °C is 7.0, so ΔpH = 12.4 − 7.0 = 5.4.

b) Cations of amino acids exist in solutions with pH < pI, so the pK_{a1} value of glycine (2.3, table 1) will be used for this buffer. According to the Henderson–Hasselbalch equation, the pH of the original buffer solution is 2.3 + log (0.40/0.16) ≈ 2.3 + 0.4 = 2.7. Sodium hydroxide will react with the conjugate acid (cation) and produce an additional amount of the conjugate base (zwitterion) as follows:

$$H_3N^+CH_2COOH + NaOH → H_3N^+CH_2COONa + H_2O$$

or, in ionic form,

$$H_3N^+CH_2COOH + OH^- → H_3N^+CH_2COO^- + H_2O$$

initial concentration	0.16	0.025	0.40
concentration change	−0.025	−0.025	+0.025
final concentration	0.135	—	0.425

(all concentrations are given in mol dm^{-3}):

Therefore, the pH of the final solution will be $2.3 + \log (0.425/0.135) \approx 2.3 + 0.5 = 2.8$, and $\Delta\text{pH} = 2.8 - 2.7 = 0.1$.

As you can see, the addition of a strong base to a buffer solution causes a much smaller pH change than the pH change in pure water (0.1 versus 5.4 units, respectively).

Buffer pH range

Amino acids can act as acid–base buffers only within certain pH ranges, where both components of a conjugate acid–base pair are present in the solution at sufficient concentrations. At pH = pK_{a1} and pH = pK_{a2}, the amino acid reaches its maximum buffer capacity and can neutralize the greatest amount of strong acid or base before any significant pH change occurs. According to the Henderson–Hasselbalch equation, the ratio between the components of a conjugate acid–base pair increases or decreases 10 times when the pH of the solution changes by one unit, so an amino acid can act as a buffer approximately from pH = $pK_{a1} - 1$ to $pK_{a1} + 1$ and from pH = $pK_{a2} - 1$ to $pK_{a2} + 1$. Outside these ranges the amino acid exists predominantly as a single ionic species (figure 1) and loses its ability to maintain a constant pH of the solution.

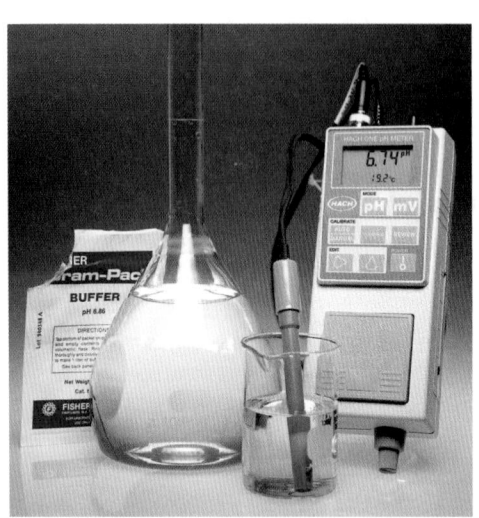

▲ Figure 2 In modern laboratories, buffer solutions are often prepared from commercially available mixtures of dry components. The pH of the solution can be verified using a digital pH meter

Worked example

Identify the conjugate acid and the conjugate base in a 0.500 mol dm^{-3} solution of glycine (pI = 6.0) at pH = 5.0. Calculate the concentrations of both glycine species.

Solution

Since pH < pI, glycine will exist in the solution as a mixture of the zwitterion (conjugate base) and the cation (conjugate acid) with $pK_{a1} = 2.3$ (table 1). According to the Henderson–Hasselbalch equation, $5.0 = 2.3 + \log ([\text{zwitterion}]/[\text{cation}])$ and thus [zwitterion]/[cation] ≈ 501. If [cation] $= x$ mol dm^{-3}, then [zwitterion] $= 501x$ mol dm^{-3}. The total concentration of glycine species is 0.500 mol dm^{-3}, so $501x + x = 0.500$ and $x \approx 9.96 \times 10^{-4}$ mol dm^{-3}. Therefore, [zwitterion] ≈ 0.499 mol dm^{-3} and [cation] ≈ 0.001 mol dm^{-3}. Since the concentration of the cation in this solution is negligible, glycine cannot act as an efficient acid–base buffer at pH = 5.0, i.e., outside the range of $pK_{a1} \pm 1$.

Proteins as biological buffers

Similar to amino acids, proteins can exist in cationic, zwitterionic, and anionic forms due to the presence of ionizable side-chains in their constituent amino acid residues. These side-chains form various polyions that act as biological acid–base buffers. The exact amino acid composition of a protein usually correlates with the pH of the biological fluid where this protein occurs. For example, acidic proteins

Gel electrophoresis and isoelectric focusing are two common techniques that use the differences in acid–base properties of 2-amino acids and proteins for the analysis and separation of these compounds. Both methods were discussed in detail in sub-topic B.2.

containing many residues of aspartic and glutamic acids are more common in the gastric juice while the proteins of the blood plasma and intestinal mucus have a greater proportion of neutral and alkaline amino acid residues, such as lysine or arginine (table 1). In each case, the protein buffers play an important role in maintaining a constant pH of biological fluids, which is essential for the integrity of body tissues and enzyme functions (sub-topic B.2).

Enzyme action and kinetics

The basic concepts of enzymatic reactions were discussed in sub-topic B.2. One of the original theories, the "**lock-and-key**" model, described the process of chemical recognition between the enzyme ("lock") and the substrate ("key") as an exact fit of their complementary structures (figure 11, sub-topic B.2). Although this model could account for the specificity of enzyme catalysis, it was unable to explain certain experimental data, in particular, the enhanced stability of transition states in enzyme–substrate complexes. The development of X-ray crystallography and computer modelling allowed the three-dimensional shapes of active sites in enzymes to be determined, which in many cases did not match the shapes of their substrates.

Enzymatic processes have been known from prehistoric times, when leather processing and milk fermentation were discovered. Brewing and cheese-making are often associated with particular places such as Bordeaux and Camembert in France or Cheddar in England, where these techniques were originally employed to create popular products.

 ## The induced fit theory

The "lock-and-key" model was superseded by the "**induced fit**" **theory**, proposed in 1958 by Daniel Koshland. According to his theory enzymes have flexible structures and continually change their shapes as a result of interactions with the substrate. Therefore the substrate does not simply fit into a rigid active site; instead the active site is dynamically created around the substrate until the most stable configuration of the enzyme–substrate complex is achieved (figure 3). At the same time the substrate also changes its shape slightly, which weakens some chemical bonds, lowers the activation energy of the transition state, and eventually allows the chemical reaction to take place.

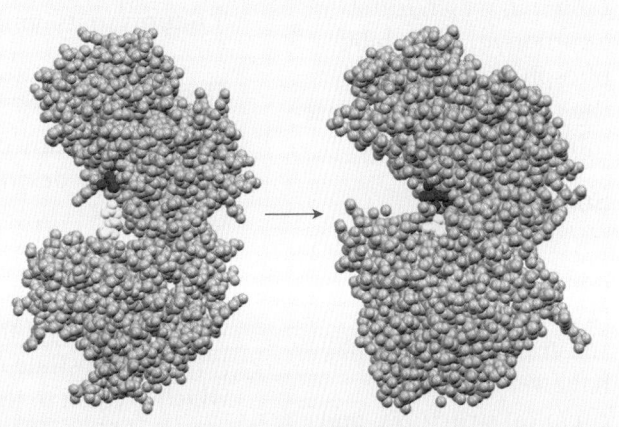

▲ Figure 3 The induced fit of 1,3-bisphosphoglycerate (substrate, yellow) and phosphoglycerate kinase (enzyme, blue). The enzyme wraps around the substrate to create additional intermolecular contacts. The cofactor, ADP, is shown in red

Non-competitive enzyme inhibition

The "induced fit" theory also explains the mechanisms of enzyme inhibition and activation that regulate the metabolic processes of living organisms. Apart from their main active sites, many enzymes have additional **allosteric sites** that can temporarily bind to specific molecules via weak non-covalent interactions. When an allosteric site is occupied the shape of the enzyme molecule changes, which alters the configuration of the main active site. This in turn affects the stability of

611

Worked example

The hydrolysis of glycogen is catalysed by the enzyme phosphorylase. Caffeine, which is not a carbohydrate, inhibits phosphorylase. Identify the type of phosphorylase inhibition by caffeine.

Solution

Glycogen is a carbohydrate (sub-topic B.4) while caffeine is not, so the substrate and the inhibitor have very different chemical structures, and cannot bind to the same active site. Therefore caffeine must bind to an allosteric site, which is a case of non-competitive inhibition.

the enzyme–substrate complex and the ability of the enzyme to act as a catalyst. In most cases allosteric interactions reduce the enzyme's activity, which is known as **allosteric** or **non-competitive inhibition**. The term "non-competitive" refers to the fact that the substrate and the inhibitor have different chemical structures, bind to different sites of the enzyme, and therefore do not compete with one another for the main active site.

Competitive inhibition

Another mechanism of enzyme inhibition, **competitive inhibition**, takes place when the substrate and the inhibitor have similar chemical structures. In this case the inhibitor may occupy the main active site and prevent the substrate from binding to the enzyme. The most common type of competitive inhibition is **product inhibition**, where the active site of the enzyme is blocked by a product of the enzymatic reaction. Product inhibition may also occur via a non-competitive mechanism, where the reaction product binds to an allosteric site and reduces the enzyme activity.

Enzyme inhibition and negative feedback

Competitive and non-competitive product inhibition provide negative feedback to metabolic processes, which is a biochemical equivalent of Le Chatelier's principle (sub-topic 7.1). When the substrate concentration is high, the rate of the forward reaction increases and excess substrate is metabolized. In contrast a high concentration of the product inhibits the enzyme and prevents any further increase of product concentration until it returns to its optimal physiological level.

The Michaelis–Menten equation

The rates of many enzymatic reactions, as was briefly mentioned in sub-topic B.2, are described by the **Michaelis–Menten equation:**

$$v = \frac{V_{max}\,[S]}{K_m + [S]}$$

where v and V_{max} are the actual and maximum reaction rates, respectively, [S] is the substrate concentration, and K_m is the **Michaelis constant**, which is equal to the substrate concentration when $v = 0.5V_{max}$. The values of V_{max} and K_m depend on the enzyme concentration, [E], so the Michaelis–Menten equation can be applied only when [E] = constant.

When the substrate concentration is low, $K_m >> [S]$, so $K_m + [S] \approx K_m$ and therefore $v \approx (V_{max}/K_m)[S]$, which corresponds to a first-order reaction (sub-topic 16.1). At low [S] almost all active sites of the enzyme are available for substrate molecules, so the reaction rate is proportional to the substrate concentration (figure 4). However, as [S] increases more and more enzyme molecules bind to substrate and form enzyme–substrate complexes, ES, reducing the number of available active sites. When all active sites are occupied by the substrate, the enzyme works at its maximum capacity and is said to be **saturated**. Any further increase of [S] will not affect the reaction rate because additional substrate molecules will have to wait until active

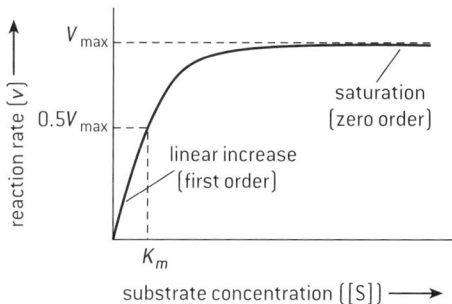

sites become available again. In the Michaelis–Menten equation, this situation corresponds to zero-order kinetics (sub-topic 16.1), where $K_m \ll [S]$ and $\upsilon \approx V_{max}$.

Since V_{max} is limited by the number of available active sites, is must be proportional to the enzyme concentration [E]:

$$V_{max} = k_{cat}[E]$$

where k_{cat}, known as the **turnover number**, is the maximum number of substrate molecules that one molecule of enzyme can convert to product per second.

The Michaelis constant K_m is an inverse measure of the substrate affinity for the enzyme. A small K_m indicates high affinity, which means that enzyme–substrate complex ES is particularly stable and the rate will approach V_{max} even at relatively low substrate concentrations. In contrast, high K_m values are typical for less stable ES complexes where higher substrate concentrations are needed for enzyme saturation.

At $\upsilon = 0.5V_{max}$, half of the enzyme active sites are bound to substrate and the other half remain unoccupied, so [E] = [ES]. At the same time $[S] = K_m$, so the dissociation of the enzyme–substrate complex can be described as follows:

$$ES \rightleftharpoons E + S$$

$$K_c = \frac{[E][S]}{[ES]} = \frac{[ES]K_m}{[ES]} = K_m$$

where K_c is the equilibrium constant (sub-topic 7.1) of the ES dissociation. In other words, K_m is equal to the dissociation constant of the enzyme–substrate complex.

The kinetic constants K_m and V_{max} provide important information about the enzyme activity and metabolic processes in living organisms. In particular they allow us to distinguish between competitive and non-competitive mechanisms of enzyme inhibition. Competitive inhibition can be overcome by increasing the substrate concentration and preventing the inhibitor from binding to the active site. As a result, the V_{max} value in competitive inhibition remains the same while the K_m value increases, as it takes more substrate to reach the $0.5V_{max}$ reaction rate (figure 5, left). In contrast, the binding of non-competitive inhibitors to allosteric sites is not affected by the substrate, so V_{max} will decrease due to less effective binding of the substrate to the main active site. At the same time, the $0.5V_{max}$ value will decrease proportionally to V_{max}, so K_m will not be affected by non-competitive inhibition (figure 5, right).

Figure 4 Michaelis–Menten kinetics

Cofactors

Many enzymes are pure proteins that perform their functions exclusively via the side-chains of amino acid residues. Other enzymes show their full activity only as complexes with non-protein species known as **cofactors**. These species can be either inorganic, such as metal ions, or organic, such as heme (sub-topic B.9) or vitamins (sub-topic B.5). Organic cofactors can either be permanently bound to the enzyme as **prosthetic groups** (sub-topic B.2) or act as **coenzymes**, temporarily altering the structure of the active site and leaving the enzyme after the reaction is complete. Heme is an example of a prosthetic group while vitamins and certain nucleotides (sub-topic B.8) are coenzymes.

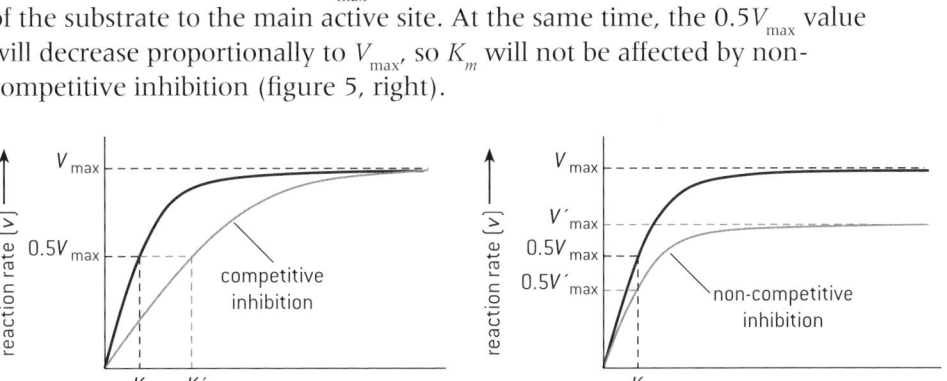

▲ Figure 5 Competitive and non-competitive inhibition

Worked example

A common food ingredient known as "invert sugar" is produced by the hydrolysis of sucrose into glucose and fructose. The reaction is catalysed by the enzyme invertase, which can be inhibited by urea. Using the data in table 3, deduce whether the inhibition of invertase by urea is competitive or non-competitive.

Sucrose concentration/ mmol dm^{-3}	Reaction rate/arbitrary units	
	No urea	2.0 mol dm^{-3} urea
0.029	0.181	0.095
0.058	0.266	0.140
0.088	0.311	0.165
0.117	0.338	0.180
0.175	0.369	0.197
0.320	0.392	0.207
0.485	0.398	0.209

 Table 3

Solution

First plot two kinetic curves of the enzymatic reaction (figure 6).

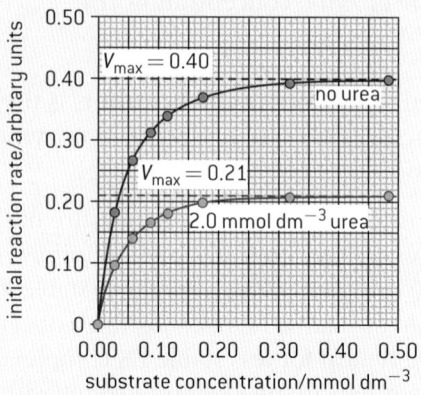

▲ Figure 6 Initial rate versus sucrose concentration in the presence and absence of urea

The inhibitor reduces V_{max} (from 0.40 to 0.21), therefore the inhibition is non-competitive. It can be also shown that the K_m value is the same in both cases (approximately 0.033 mmol dm^{-3}).

🧬 Sharing knowledge

International collaboration is particularly important for biochemistry and other multidisciplinary sciences. The collective efforts of scientists from various research groups allow biological products to be developed for industrial and domestic applications. Advances in protein engineering have produced enzymes that can be used under various conditions including elevated temperatures and extended pH range. Biological detergents (sub-topic B.3), biodegradable plastics (sub-topic B.6), textiles, foods, and beverages are just a few examples of enzyme-based products. New enzymes and microorganisms reduce the amount of waste and mitigate adverse environmental effects of industrial chemicals (sub-topic B.6).

Protein assay

UV-vis spectroscopy

The detection of proteins and the determination of their concentrations in solutions, known as **protein assay**, are the most common analytical procedures in biochemical experiments. In modern laboratories protein assays often involve absorption spectroscopy in the ultraviolet and visible regions of the electromagnetic spectrum (sub-topic 11.3). This technique, often referred to as **UV-vis spectrometry**, measures the absorption of UV and/or visible light by proteins or their complexes with organic dyes and transition metal ions (sub-topic 13.2).

Almost all proteins absorb UV light with a wavelength of 280 nm due to the presence of aromatic rings in phenylalanine, tyrosine, and tryptophan residues (sub-topic B.2, table 1). Certain organic dyes such as Coomassie® Brilliant Blue bind to arginine and aromatic residues and form highly conjugated systems of delocalized electrons (sub-topic B.9) with maximum absorption at 595 nm in the orange region of the visible spectrum. The complexes of proteins with transition metal ions also absorb visible light due to d-orbital electron transitions (sub-topic 13.2).

A typical UV-vis spectrophotometer consists of a light source that produces UV and visible light, a monochromator that allows only a narrow bandwidth of light to pass through, a cuvette that holds the studied sample, a detector and amplifier that convert the light into an electric current and a digital output device or computer that allows analysis of the experimental results (figure 7).

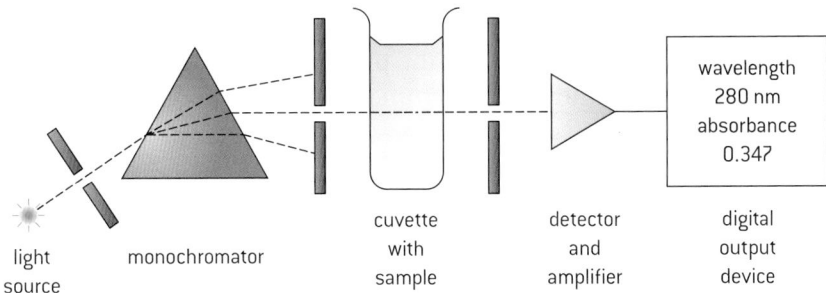

▲ Figure 7 A single-beam UV-vis spectrophotometer. The wavelength and absorbance are shown as examples.

The sample solution containing a protein is put into a transparent cuvette and placed inside the spectrophotometer. Depending on the protein concentration and experimental conditions, the intensity of UV or visible light passed through the sample will be reduced to some degree. The logarithmic ratio between the intensity of light emitted by the monochromator (I_0) and the intensity of light passed through the sample (I) is known as the **absorbance** (A) of the sample:

$$A = \log \frac{I_0}{I}$$

The concentration (c) of the protein in the sample solution can be determined from its absorbance (A) using the **Beer–Lambert law**:

$$A = \varepsilon c L$$

where L is the cuvette length and ε is a constant (known as the **molar absorptivity** or **extinction coefficient**) that depends on the solvent nature and the temperature of the solution. If the same cuvette and experimental conditions are used, the product of ε and L also becomes a constant, so the protein concentration can be determined from a calibration curve plotted as concentration versus absorbance.

The biuret test

The **biuret test** is used for detecting the presence of peptide linkages and estimating the concentration of peptides and proteins in a sample. In a typical experiment an aqueous sample is treated with 2–5 volumes of the **biuret reagent**, which can be prepared from diluted solutions of copper(II) sulfate, sodium potassium tartrate ($NaKC_4H_4O_6$), and sodium hydroxide. Copper(II) ions form coloured complexes (sub-topic 13.2) with peptide linkages and tartrate anions, so in the presence of proteins the solution turns violet, while short-chain peptides may produce a pink colour. According to the Beer–Lambert law the intensity of the colour is proportional to the concentration of peptide linkages, which is in turn proportional to the protein content in the sample. Therefore the concentration of proteins can be determined by measuring the absorption of the solution at 540 nm using a UV-vis spectrometer.

Surprisingly, the biuret reagent does not contain biuret, $[H_2NC(O)]_2NH$. The latter compound contains peptide-like bonds and gives a positive reaction with the biuret reagent, hence the name of the test.

Worked example

A 5.00 cm³ sample of an aqueous protein solution was diluted with a buffer solution to a volume of 0.100 dm³ and analysed by UV-vis spectroscopy. The absorbance of the analysed solution was 0.285. Using the calibration curve in figure 8 determine the concentration of the protein in the original sample.

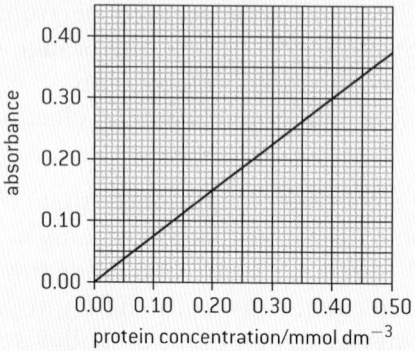

▲ Figure 8 UV-vis spectroscopy calibration curve

Solution

According to the calibration curve, the protein concentration in the analysed (diluted) solution is 0.380 mmol dm⁻³ (figure 9).

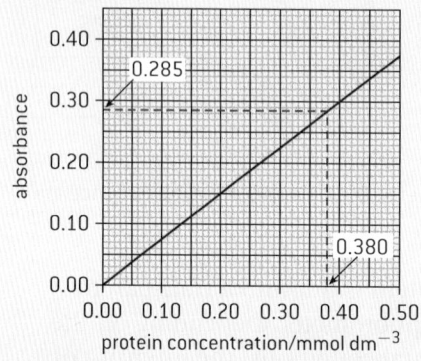

▲ Figure 9 Reading off the concentration from the calibration curve

The amount of protein in the analysed solution is $0.100 \times 0.380 = 0.0380$ mmol. All the protein in the analysed solution came from the sample, so the amount of protein in the sample was the same, 0.0380 mmol. Therefore the concentration of the protein in the sample solution was 0.0380 mmol/5.00 cm³ = 0.007 60 mmol cm⁻³ = 7.60 mmol dm⁻³.

Other analytical techniques

Although UV-vis spectra provide some information about the structure of organic compounds, the identification of proteins on the basis of their UV-vis spectra alone is problematic because the spectra of different proteins are similar to one another and highly sensitive to experimental conditions. However, such identification becomes possible when a UV-vis spectrophotometer is used as a detector in high-performance liquid chromatography (HPLC, topic B.2). In this case, the proteins in the sample are first separated chromatographically and then the UV-vis spectrum of each protein is matched to a large library of known compounds. Unidentified components of the mixture can be further analysed by various techniques including gel electrophoresis (sub-topic B.2), high resolution NMR, and mass spectrometry (sub-topics 11.3 and 21.1).

Questions

1 Proteins are products of polycondensation of 2-amino acids. In addition to their biochemical functions, proteins and individual 2-amino acids may act as acid–base buffers.   [2]

   a) At pH 7, a solution of alanine contains both the zwitterion and negatively charged (anionic) forms of alanine. Deduce the structural formula of each of these forms.   [2]

   b) State equations which show the buffer action of the solution from (a) when a small amount of strong acid is added and a small amount of strong base is added.   [5]

   IB, November 2012

2 An amino acid buffer has been prepared by mixing 0.60 dm³ of 0.20 mol dm⁻³ HCl and 0.40 dm³ of 0.50 mol dm⁻³ glycine solutions. Calculate:

   a) the pH of the original buffer solution

   b) the pH of the solution after the addition of 1.0 cm³ of 1.0 mol dm⁻³ HCl

   c) the pH of the solution after the addition of 0.40 g of solid NaOH.

   Assume that the densities of all solutions are 1.0 kg dm⁻³ and the volume of the buffer solution does not change when small amounts of strong acid or base are added.

3 Compare the behaviour of enzymes and inorganic catalysts, including reference to the mechanism of enzyme action and the ways in which this can be inhibited.

   IB, May 2012

4 The term "lock and key" is a simple and effective metaphor but the "induced fit" model provides a more comprehensive explanation of enzyme catalysis. Discuss how metaphors and models are used in the construction of our knowledge of the natural world.

5 Pepsin is an enzyme found in the stomach that speeds up the breakdown of proteins. Iron is used to speed up the production of ammonia in the Haber process.

   a) Describe the characteristics of an enzyme such as pepsin, and compare its catalytic behaviour to an inorganic catalyst such as iron.   [4]

   b) Enzymes are affected by inhibitors. Lead ions are a non-competitive inhibitor; they have been linked to impaired mental functioning. Ritonavir® is a drug used to treat HIV and acts as a competitive inhibitor. Compare the action of lead ions and Ritonavir® on enzymes, and how they affect the initial rate of reaction of the enzyme with its substrate and the values of K_m and V_{max}.   [5]

   IB, May 2009

6 a) State and explain how the rate of an enzyme-catalysed reaction is related to the substrate concentration.   [3]

   b) When an inhibitor is added, it decreases the rate of an enzyme-catalysed reaction. State the effect that competitive and non-competitive inhibitors have on the value of V_{max}. Explain this in terms of where the inhibitor binds to the enzyme.   [4]

   c) Sketch a graph to show the effect that a change in pH will have on the rate of an enzyme-catalysed reaction.   [1]

   d) Explain why changing the pH affects the catalytic ability of enzymes.   [2]

   IB, May 2012

7 Enzymes are proteins which play an important role in the biochemical processes occurring in the body.

   a) State the major function of enzymes in the human body.   [1]

b) Describe the mechanism of enzyme action in terms of structure. [3]

c) Figure 10 shows how the rate of an enzyme-catalysed reaction changes as the substrate concentration is increased. Use the graph to determine V_{max} and the Michaelis constant, K_m. [2]

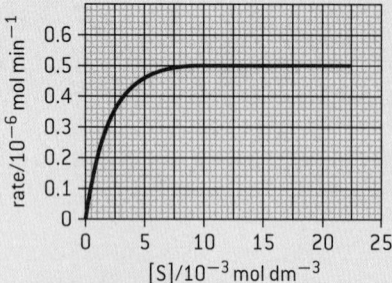

▲ Figure 10

d) Draw a line on a sketch of the graph to represent the effect of adding a competitive inhibitor. [1]

e) State and explain the effects of heavy-metal ions and temperature increases on enzyme activity. [5]

IB, November 2009

8 The kinetics of an enzyme-catalysed reaction are studied in the absence and presence of an inhibitor. Figure 11 represents the initial rate as a function of substrate concentration.

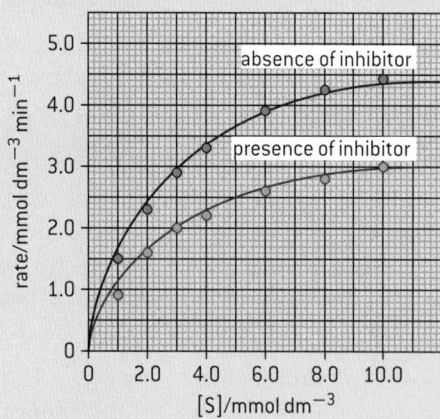

▲ Figure 11

a) Identify the type of inhibition shown in the graph. [1]

b) Determine V_{max} and K_m in the absence of the inhibitor and in the presence of the inhibitor. [3]

c) Outline the relationship between K_m and enzyme activity. [1]

IB, May 2011

9 Describe the operating principles of a UV-vis spectrometer.

10 A 2.00 cm³ sample of an aqueous protein solution was diluted with a buffer solution to a volume of 25.0 cm³ and analysed by UV-vis spectroscopy. The absorbance of the analysed solution was 0.310.

a) State the structural features of proteins that can be detected by UV-vis spectroscopy.

b) Draw the calibration curve using the concentrations (c) and absorbances (A) of standard protein solutions shown in table 4.

c/mmol dm^{-3}	A/arbitrary units
0.100	0.067
0.200	0.135
0.300	0.202
0.400	0.270
0.500	0.337
0.600	0.404

▲ Table 4 Concentrations (c) and UV-vis absorbances (A) of standard protein solutions

c) Determine the concentration (in mmol dm^{-3}) of the protein in the original sample.

B.8 Nucleic acids (AHL)

Understandings

→ Nucleotides are the condensation products of a pentose sugar, phosphoric acid, and a nitrogenous base: adenine (A), guanine (G), cytosine (C), thymine (T), or uracil (U).

→ Polynucleotides form by condensation reactions.

→ DNA is a double helix of two polynucleotide strands held together by hydrogen bonds.

→ RNA is usually a single polynucleotide chain that contains uracil in place of thymine and the sugar ribose in place of deoxyribose.

→ The sequence of bases in DNA determines the primary structure of proteins synthesized by the cell using a triplet code, known as the genetic code, which is universal.

→ Genetically modified organisms have genetic material that has been altered by genetic engineering techniques, involving transferring DNA between species.

Applications and skills

→ Explanation of the stability of DNA in terms of the interactions between its hydrophilic and hydrophobic components.

→ Explanation of the origin of the negative charge on DNA and its association with basic proteins (histones) in chromosomes.

→ Deduction of the nucleotide sequence in a complementary strand of DNA or a molecule of RNA from a given polynucleotide sequence.

→ Explanation of how the complementary pairing between bases enables DNA to replicate itself exactly.

→ Discussion of the benefits and concerns of using genetically modified foods.

Nature of science

→ Scientific method – the discovery of the structure of DNA is a good example of different approaches to solving the same problem. Scientists used models and diffraction experiments to develop the structure of DNA.

→ Developments in scientific research follow improvements in apparatus – double helix from X-ray diffraction provides explanation for known functions of DNA.

Heredity and the storage of biological information

Every living organism contains many thousands of proteins with strictly defined structures and functions (sub-topics B.2 and B.7). The amino acid sequences of specific proteins in all the cells of a particular organism are identical and differ only slightly between individuals of the same species. This fact suggests that there must be a certain mechanism that allows cells to store and interpret biological information, as well as transfer it to other cells and organisms.

The fact that hereditary information resides in the nucleus of the cell has been known since the end of the nineteenth century. Biologists suspected that external and internal characteristics of individuals, such as hair colour or hereditary diseases, were somehow encoded in genes located inside the chromosomes. When nuclear proteins (histones) were discovered, they seemed to be the most obvious candidates for storing genetic information. However, by 1940 the work of Oswald Avery demonstrated that deoxyribonucleic acid (DNA), and not proteins, was the only carrier of hereditary information. Later studies by George Beadle and Edward Tatum showed that each gene in DNA controls the synthesis of one protein and therefore is responsible for a certain internal or external characteristic of the individual. Now we know that not all genes can be related to specific proteins but each gene is responsible for the production of a ribonucleic acid (RNA).

▲ Figure 1 The jaguar and green plants on this picture use identical molecular mechanisms for storing and processing genetic information

It is widely understood that individuals obtain some information from their parents through **heredity**, which allows the passing of anatomical and biochemical characteristics of the species from generation to generation. The transmission of hereditary information takes place in the nucleus of the cell. Certain structures within the nucleus, **chromosomes**, contain intermolecular complexes of basic proteins (**histones**) with acidic biopolymers called **nucleic acids**.

Nucleic acids

Nucleic acids are condensation polymers of **nucleotides**, which in turn are the products of condensation of a **nitrogenous base**, a pentose sugar (ribose or deoxyribose, see sub-topic B.4), and phosphoric acid. In order to understand the functions of nucleic acids in living organisms we need to discuss first the structures and properties of their components, nitrogenous bases and nucleotides.

Nitrogenous bases and nucleotides

Nitrogenous bases are heterocyclic aromatic amines (sub-topic 10.1) that contain several nitrogen atoms and act as proton acceptors in aqueous solutions (sub-topic 8.1). All common nitrogenous bases are derived from two parent amines, **pyrimidine** and **purine** (figure 2).

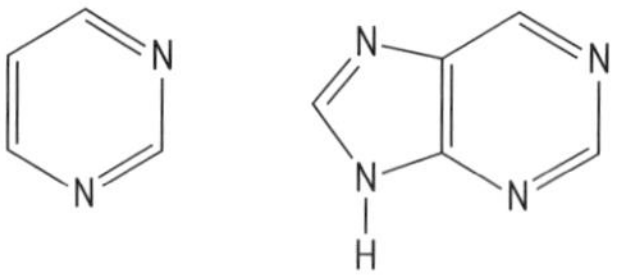

▲ Figure 2 The structures of pyrimidine (left) and purine (right)

Pyrimidine nitrogenous bases, or simply **pyrimidines**, include **cytosine, thymine**, and **uracil** (figure 3).

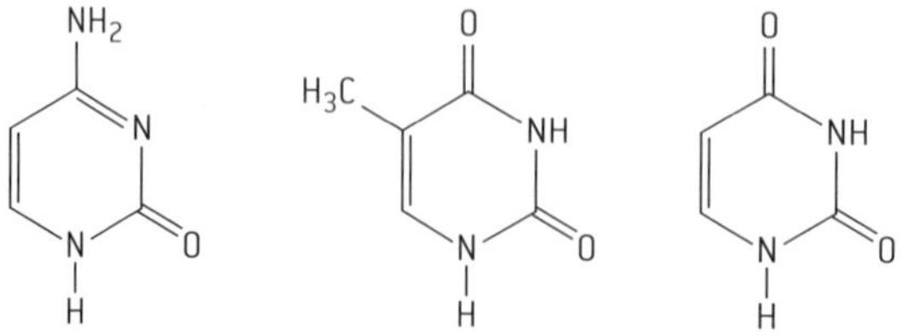

cytosine (C)   thymine (T)   uracil (U)

▲ Figure 3 The three pyrimidines

Purine nitrogenous bases, commonly called **purines**, include **adenine** and **guanine** (figure 4).

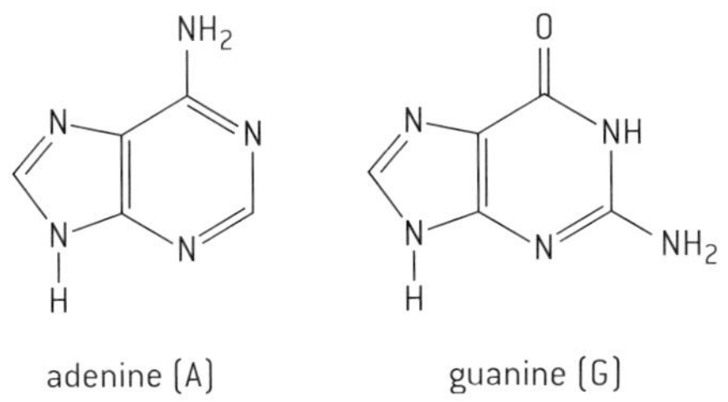

adenine (A)   guanine (G)

▲ Figure 4 The two purines

The names of pyrimidines and purines are often abbreviated to their first letters, such as A for adenine or C for cytosine. Both purines (A and G) and one pyrimidine (C) are found in all nucleic acids. Thymine (T) is normally associated with deoxyribose sugar and is found in deoxyribonucleic acids (DNA). Uracil (U) forms nucleotides with ribose and is found in ribonucleic acids (RNA). Therefore both DNA and RNA contain four nitrogenous bases each, including two purines and two pyrimidines. For DNA these bases are A, G, C, and T while RNA contains A, G, C, and U.

Owing to the presence of multiple polar groups, nitrogenous bases are crystalline substances with high melting points. However, in contrast to amino acids, nitrogenous bases are almost insoluble in water because their molecules are held together by strong hydrogen bonds (sub-topic 4.4). Thymine or uracil can form two hydrogen bonds with adenine while cytosine and guanine bind to each other by three hydrogen bonds (figure 5).

The pairs adenine/thymine (A=T), adenine/uracil (A=U), and guanine/ cytosine (G≡C) are known as **complementary base pairs**. This ability of certain nitrogenous bases to form hydrogen bonds with one another in a specific order and orientation plays an important role in the storing and processing of genetic information, which will be discussed later in this topic.

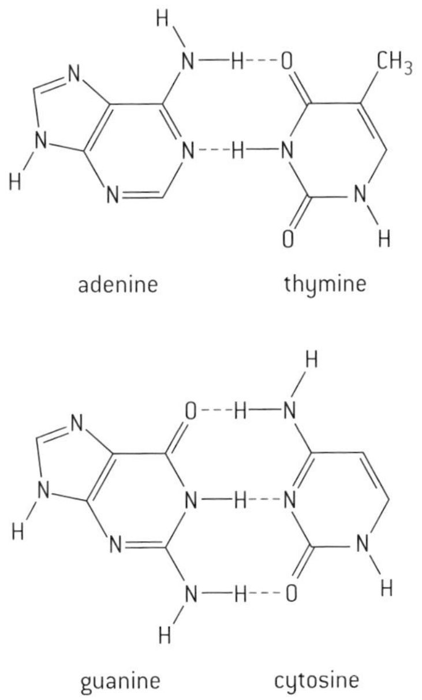

The monomeric units of nucleic acids, **nucleotides**, are composed of a nitrogenous base, a pentose sugar, and phosphoric acid. For example, a condensation reaction between cytosine, deoxyribose, and H_3PO_4 produces deoxycytidine monophosphate:

adenine   thymine

guanine   cytosine

▲ Figure 5 Hydrogen bonds hold complementary base pairs together

deoxycytidine monophosphate

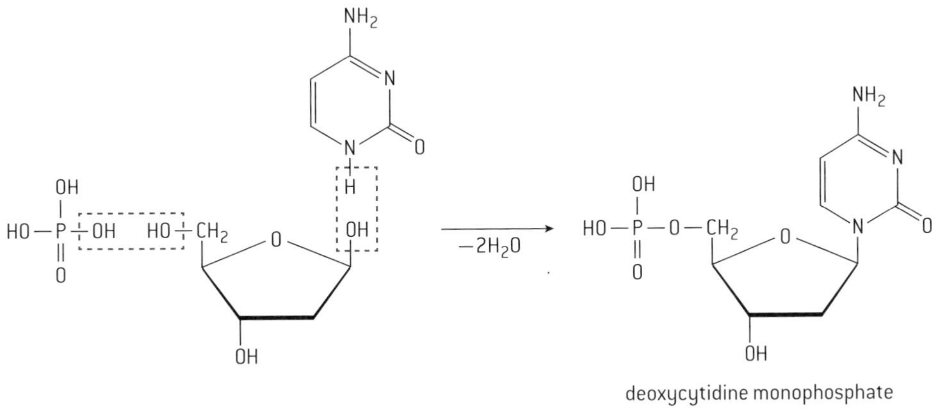

Other common nucleotides are listed in table 1. Note that the names of purine bases in nucleotides change their suffixes from "-ine" to "-osine" while the names of pyrimidines end with the suffix "-idine".

Nitrogenous base	Ribonucleotide (contains ribose)	Deoxyribonucleotide (contains deoxyribose)
adenine (A)	adenosine monophosphate (AMP)	deoxyadenosine monophosphate (dAMP)
guanine (G)	guanosine monophosphate (GMP)	deoxyguanosine monophosphate (dGMP)
cytosine (C)	cytidine monophosphate (CMP)	deoxycytidine monophosphate (dCMP)
thymine (T)	–*	thymidine monophosphate (dTMP)†
uracil (U)	uridine monophosphate (UMP)	–*

▲ Table 1 Common nucleotides containing one phosphate group

* These nucleotides are uncommon and will not be discussed in this book.

† Because thymine normally forms nucleotides with deoxyribose, the prefix "deoxy" is traditionally omitted from their names.

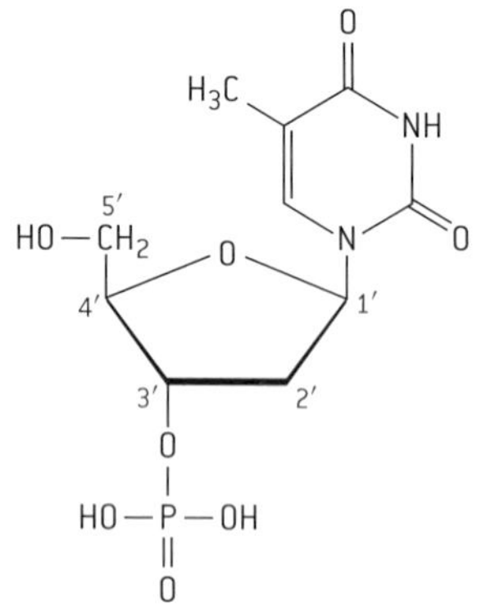

thymidine 3'-monophosphate

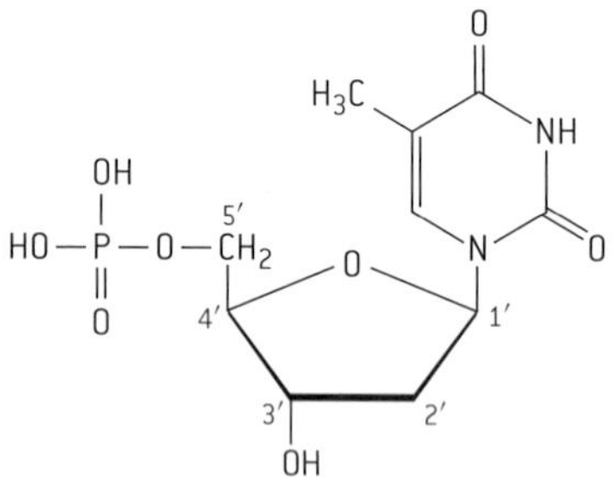

thymidine 5'-monophosphate

▲ Figure 6 Primed numbers show the position of the phosphate group in nucleotides

Phosphoric acid can react with any hydroxyl group in ribose or deoxyribose, producing several isomeric nucleotides in each case. To distinguish between these isomers, the positions of phosphate groups are denoted by primed numbers (numbers without primes are used for nitrogenous bases) (figure 6).

Nucleotides with phosphate groups at 5' positions are much more common, so the number 5' is often omitted.

Adenosine triphosphate

Some nucleotides, such as adenosine 5'-triphosphate (ATP), contain more than one phosphate group in their molecules (figure 7).

Adenosine triphosphate (ATP) is often called the "molecular currency" of energy transfer. The human body contains approximately 250 g of ATP, which is constantly hydrolysed and synthesized again. Depending on the level of physical activity, the mass of ATP converted into energy each day can exceed the mass of the entire body.

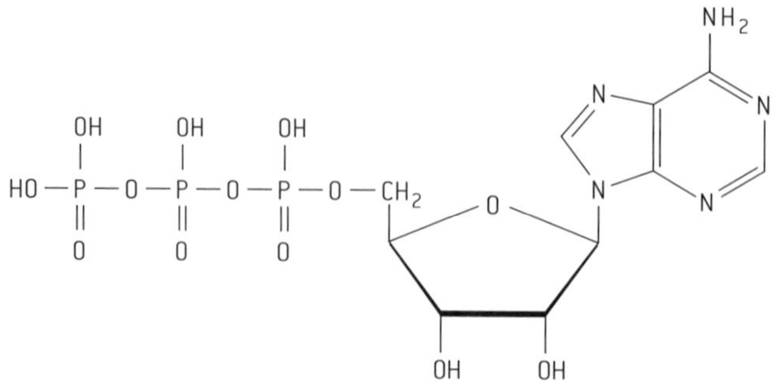

▲ Figure 7 Adenosine 5'-triphosphate (ATP)

Hydrolysis of the terminal phosphate group in ATP releases energy that can be used by other metabolic processes or transformed into mechanical work (see worked example, next page). In addition, ATP molecules act as coenzymes in many biochemical reactions (sub-topic B.7).

Nucleic acids

Living cells contain two types of nucleic acid: ribonucleic acids (RNA) and deoxyribonucleic acids (DNA). As follows from their names, ribonucleic acids are condensation polymers of ribonucleotides (they contain ribose residues) while deoxyribonucleic acids are composed of deoxynucleotides and contain residues of deoxyribose. When nucleotides combine with one another, the phosphate groups form diester bridges between 3' and 5' carbon atoms of adjacent pentose residues, for example:

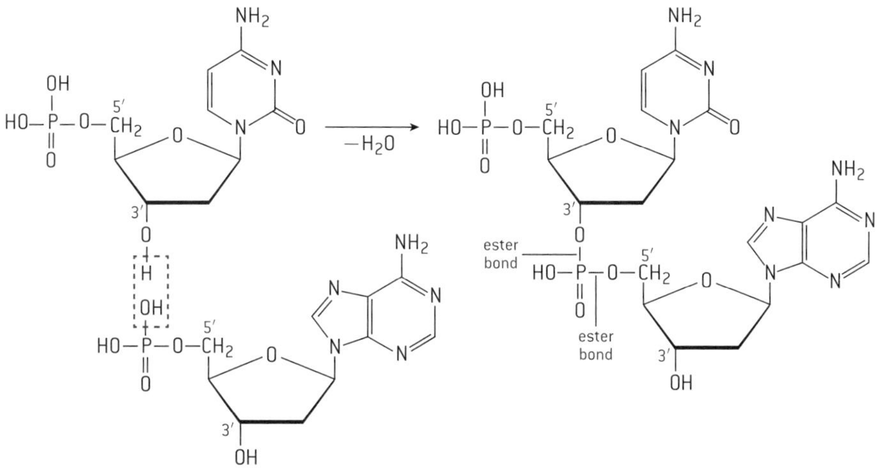

Further condensation reactions produce long polynucleotide chains, known as **strands**, in which monomeric units are joined together in strict order and orientation. Similar to proteins (sub-topic B.2), each DNA or RNA strand has two terminals, which are called 3' and 5' ends. In living cells the synthesis of nucleic acids begins from their 5' ends, so the sequence of nucleotides is traditionally recorded in the same way. For example, the sequence –T–C–A–G– denotes the polynucleotide fragment shown in figure 8.

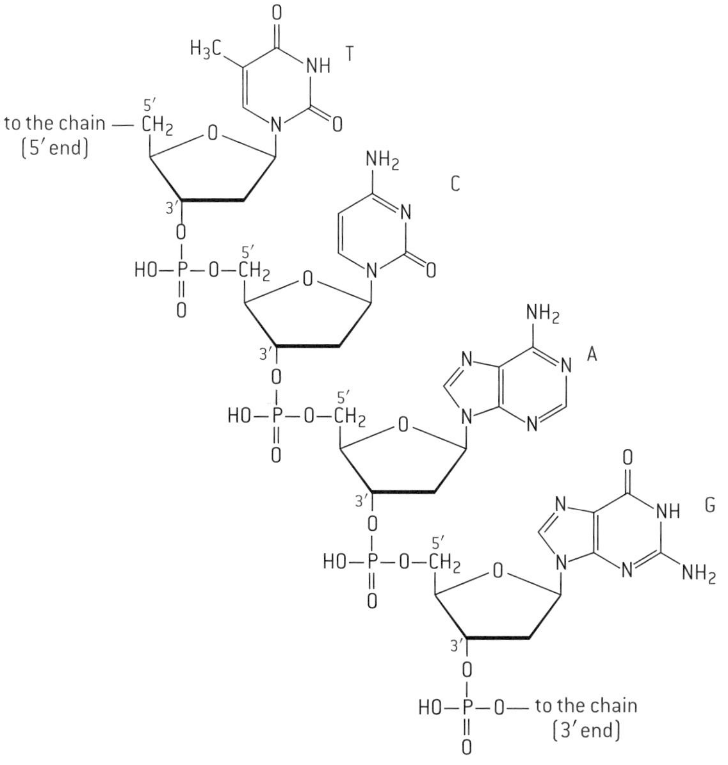

Worked example

Aerobic oxidation of one molecule of glucose in the human body produces 32 molecules of ATP while the hydrolysis of ATP releases 30.5 kJ mol^{-1} of energy. Calculate the efficiency of the energy transfer from glucose to ATP if the enthalpy of glucose combustion is -2803 kJ mol^{-1}.

Solution

The energy released by the hydrolysis of 32 mol ATP is 30.5 kJ mol^{-1} × 32 mol = 976 kJ. Therefore the efficiency of glucose oxidation as the energy source is 976/2803 ≈ 0.348, or 34.8%. The remaining energy is transformed into heat and eventually released to the environment.

▲ Figure 8 The polynucleotide fragment –T–C–A–G–

In DNA and RNA the sequence of nucleotides linked together by phosphodiester covalent bonds is known as the **primary structure** of the nucleic acid.

Discovery of the DNA structure

In 1953 James Watson and Francis Crick established the three-dimensional structure of DNA. They suggested that DNA was composed of two polynucleotide strands wrapped around each other in a double helix (figure 10). This conclusion was partly based on unpublished experimental results obtained by Rosalind Franklin and Maurice Wilkins, but Watson and Crick exhibited a lack of effective collaboration and communication in failing to acknowledge the contribution from their colleagues. Despite this controversy, the work of Watson and Crick was one of the most important achievements in the history of biochemistry and a good example of the scientific method, where experimental evidence from various sources was used to reach the final conclusion. The discovery of the DNA three-dimensional structure became possible only when the development of laboratory techniques and instrumentation allowed the scientists to obtain certain experimental data, including the chemical composition of nucleic acids and X-ray diffraction images of their molecules.

▲ Figure 9 Watson and Crick's DNA molecular model, 1953

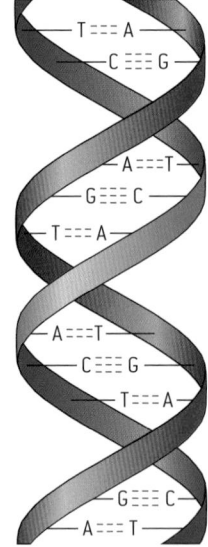

▲ Figure 10 The three-dimensional structure of the DNA double helix.

The structure of DNA

DNA molecules consist of two polynucleotide strands in which each nitrogenous base from one strand forms a complementary pair with a nitrogenous base from another strand. Each pair contains one purine base (A or G) and one pyrimidine base (T or C, respectively). Two hydrogen bonds in A=T base pairs and three hydrogen bonds in G≡C base pairs are shown in figure 10 by dashed lines. The double-helix shape of the DNA molecule stabilized by hydrogen bonds between complementary nitrogenous bases is known as its **secondary structure**.

DNA profiling

DNA profiling, or DNA fingerprinting, is used to identify a person by his or her DNA base sequence. A DNA sample is extracted from a small amount of cellular material or biological fluid such as hair or blood. It is then treated with restriction enzymes that cut the DNA chain into small polynucleotide fragments. Some DNA fragments do not contain genes that code for proteins. These non-coding fragments are unique to each individual (except identical twins) and show characteristic patterns when separated by gel electrophoresis (sub-topic B.2). DNA profiling is often used in court cases to identify criminals and to prove paternity. Paleontologists also use this technique for mapping the evolutionary trees of extinct species.

Intermolecular bonding stabilizes nucleic acids

At physiological pH (7.4), phosphate groups in nucleotides and nucleic acids are almost completely ionized, so the whole molecule of DNA or RNA becomes a negatively charged polyion. In contrast to nitrogenous bases, which are predominantly hydrophobic, the ionized phosphate groups are hydrophilic and form multiple hydrogen bonds with water molecules. In addition, negatively charged DNA interacts with basic chromosomal proteins, histones, which are charged positively at physiological pH. Intermolecular bonds formed by hydrophobic and hydrophilic parts of polynucleotide chains stabilize the double-helical shape of DNA and make it highly resistant to chemical cleavage.

The Human Genome Project (HGP)

The human genome (all the genes in human DNA) contains over 3 billion (3×10^9) complementary base pairs stored in 46 chromosomes. The HGP project, officially started in 1990 and completed in 2000, was a successful international research programme aimed at mapping and sequencing all the genes in the human DNA. Since the genome of any individual is unique, the "human genome" was determined by the analysis of multiple variations of each gene in many individuals. The complete sequence of human DNA is now available free of charge to anyone with internet access.

The sequencing of the human genome can help us in treating various diseases, designing new forms of medication, and understanding human ancestry, migration, evolution, and adaptation to environmental changes. The development in DNA sequencing has completely transformed certain aspects of legal enquiry such as forensic studies and paternity law. At the same time, the success of the HGP raises many ethical, social, and legal issues, including the rights to access our genetic information and possible discrimination of "genetically disadvantaged people". Another concern about the HGP is the possibility of gene modification, which can be used not only for treating genetic diseases but also for "designing" prospective children or creating biogenetic weapons.

DNA replication

The human body contains more than one trillion (10^{12}) cells, most of which have very limited life spans and need to be replaced regularly. Because all cells of an individual organism contain identical DNA, there must be a mechanism by which exact copies of DNA molecules are created. This mechanism, known as **DNA replication**, is facilitated by several families of enzymes and includes three steps: initiation, elongation, and termination.

The first group of enzymes, **initiator proteins**, separate the two DNA strands and create short polynucleotide fragments (**primers**) paired with the separated strands by complementary nitrogenous bases. Another group of enzymes, known as **DNA polymerases**, add more nucleotides to the primers using the existing DNA strands as templates. The resulting new polynucleotide chains are

Worked example

A fragment of a DNA strand has the following nucleotide sequence: –ACGGTATGCA–. Deduce the nucleotide sequence of the complementary strand.

Solution

In DNA the complementary pairs are adenine–thymine (A=T) and guanine–cytosine (G≡C). Each occurrence of A in the first strand will require T in the second strand. Similarly, T will require A, G will require C, and C will require G. Therefore, the nucleotide sequence of the second strand will be –TGCCATACGT–.

Errors in replication: Mutations

DNA replication is a highly accurate and efficient process. At body temperature (37 °C), a DNA polymerase can produce a strand of up to several hundred nucleotides per second with an average error rate of less than 1 in 10^7 base pairs. Many DNA polymerases use various proofreading mechanisms to replace mismatched nucleotides and produce exact copies of the original DNA molecule. However, some errors may still remain uncorrected, which results in the production of altered DNA. Such errors, known as **mutations**, can cause the development of cancer or certain genetic diseases but can be also neutral or even beneficial for the organism. Mutations are essential for the development of the immune system and the evolution of biological species.

complementary to existing DNA strands and therefore produce two identical copies of the original DNA molecule (figure 11). Finally, the replication process is terminated either by a certain sequence in the DNA or by the action of proteins that bind to specific DNA regions.

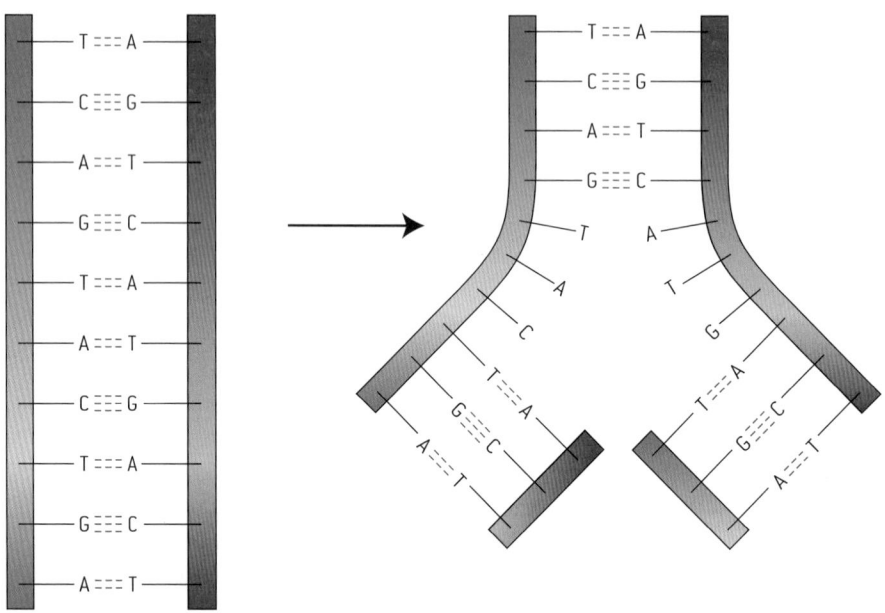

▲ Figure 11 DNA replication

Transcription

A mechanism similar to replication is used when an RNA molecule is created from a DNA template in a process called **transcription**. During transcription a DNA sequence is read by an **RNA polymerase**, which produces an RNA molecule complementary to an existing DNA strand. In contrast to the original DNA, the resulting RNA molecule contains ribose sugar (instead of deoxyribose in DNA) and uracil nitrogenous base (instead of thymine in DNA). In addition, RNA molecules usually exist as single polynucleotide strands with various three-dimensional configurations. The exact shape of an individual RNA molecule, known as its **secondary structure**, is determined by hydrogen bonds between complementary nitrogenous bases from different regions of the same strand.

Each type of nucleic acid plays its own role in heredity. DNA resides in chromosomes, stores genetic information, and acts as a template from which this information is copied to RNA. The resulting RNA molecules transfer the genetic information from chromosomes to other regions of the cell and in turn are used as templates for protein synthesis. The latter process is known as **translation** and occurs in **ribosomes**, which are the largest and most complex molecular machines in cells. All living organisms use the same **genetic code** (table 2) that allows ribosomes to translate three-nucleotide sequences (**triplets**, or **codons**) into sequences of amino acid residues in polypeptide chains.

First base	Second base								Third base
	U		C		A		G		
U	UUU	Phe	UCU	Ser	UAU	Tyr	UGU	Cys	U
	UUC	Phe	UCC	Ser	UAC	Tyr	UGC	Cys	C
	UUA	Leu	UCA	Ser	UAA	Stop	UGA	Stop	A
	UUG	Leu	UCG	Ser	UAG	Stop	UGG	Trp	G
C	CUU	Leu	CCU	Pro	CAU	His	CGU	Arg	U
	CUC	Leu	CCC	Pro	CAC	His	CGC	Arg	C
	CUA	Leu	CCA	Pro	CAA	Gln	CGA	Arg	A
	CUG	Leu	CCG	Pro	CAG	Gln	CGG	Arg	G
A	AUU	Ile	ACU	Thr	AAU	Asn	AGU	Ser	U
	AUC	Ile	ACC	Thr	AAC	Asn	AGC	Ser	C
	AUA	Ile	ACA	Thr	AAA	Lys	AGA	Arg	A
	AUG	Met*	ACG	Thr	AAG	Lys	AGG	Arg	G
G	GUU	Val	GCU	Ala	GAU	Asp	GGU	Gly	U
	GUC	Val	GCC	Ala	GAC	Asp	GGC	Gly	C
	GUA	Val	GCA	Ala	GAA	Glu	GGA	Gly	A
	GUG	Val	GCG	Ala	GAG	Glu	GGG	Gly	G

* The AUG codon also serves as the initiation site and is sometimes called the "Start" codon.

▲ Table 2 The genetic code

Worked example

Hereditary information is stored in DNA and used for protein synthesis. Deduce, using information from table 2, the primary structure of the polypeptide synthesized from the following RNA template: AUG-AUU-UAC-CGC-ACA-GGG-GGU-CAA-UAA.

Solution

According to table 2 AUG is the initiation codon, so the polypeptide synthesis begins with methionine (Met). The second triplet, AUU, encodes isoleucine (Ile), etc., so the polypeptide will have the following primary structure: Met-Ile-Tyr-Arg-Thr-Gly-Gly-Gln. The last triplet, UAA, is a stop codon that does not encode any amino acid but instructs the ribosome to release the polypeptide.

Genetic engineering

Detailed understanding of DNA structure and function led to the development of laboratory techniques for DNA manipulation. These techniques, known as **genetic engineering**, allow scientists to alter DNA sequences in the genes of living organisms, including the transfer of genetic material between different species. The resulting **genetically modified organisms** (**GMOs**) are used in scientific research, biotechnology, and agriculture. Various proteins, medicinal drugs, and other organic compounds are produced by genetically modified bacteria on an industrial scale. The most common GMOs, transgenic plants, possess many unique properties such as resistance to pests, viruses, and herbicides, tolerance to harsh environmental conditions, higher crop yields, and increased nutritional value. For example, golden rice is a species of Asian rice that was genetically modified to produce beta-carotene, a precursor of vitamin A (sub-topic B.5).

Along with these advantages the creation and use of GMOs, especially in genetically modified (GM) food, raises many ethical, health, and environmental issues. Although there is no scientific evidence that GM food is harmful to humans, the long-term effects of its consumption remain unknown. Another major concern is the potential impact of GM crops on the farming industry, especially in developing countries, due to increasing control of the food supply by the companies that make and sell GMOs. In addition, the policies on GM food labelling vary greatly from country to country, which can prevent customers from making informed food choices.

Questions

1 Draw the structure of a complementary base pair formed by uracil and adenine. State the nature of the intermolecular bonds between these nitrogenous bases.

2 Thymine is one of four nitrogen-containing bases present in DNA.

   a) Explain how thymine forms part of a nucleotide in DNA.   [2]

   b) The four nitrogen-containing bases are responsible for the double helix structure of DNA. Using the structure of thymine and the structure of one of the other bases in the *Data booklet*, draw a diagram to explain how thymine is able to play a role in forming a double helix. Identify the type of interactions between the two bases.   [3]

   c) Describe how the order in which the four nitrogen-containing bases occur in DNA provides the information necessary to synthesize proteins in a cell.   [2]

   d) It is now possible to purchase a work of art made from your own DNA profile. Outline the role that restriction enzymes play in making a DNA profile.   [2]

   IB, May 2010

3 a) Draw the scheme for a condensation reaction that produces uridine 5'-monophosphate from a nitrogenous base, pentose sugar, and phosphoric acid.

   b) Identify three different functional groups in this molecule by drawing circles around these groups and stating their names.

4 State two differences in composition and one difference in structure between RNA and DNA.   [3]

   IB, November 2012

5 DNA is the genetic material that individuals inherit from their parents. Genetic information is stored in chromosomes which are very long strands of DNA.

   a) Describe the structure of a nucleotide of DNA.   [1]

   b) Outline how nucleotides are linked together to form polynucleotides.   [1]

   c) Outline the steps involved in the DNA profiling of a blood sample.   [3]

   IB, November 2011

6 James Watson, Francis Crick, and Maurice Wilkins were awarded the 1962 Nobel Prize in Physiology or Medicine "for their discoveries concerning the molecular structure of nucleic acids and its significance for information transfer in living material".

   a) Explain how the two helices are linked in the structure of DNA.   [2]

   b) Describe the role of DNA in the storage of genetic information. The details of protein synthesis are not required.   [3]

   IB, November 2010

7 DNA stores information but not knowledge. Discuss the differences between information and knowledge.

8 The genetic information stored in DNA is expressed in the form of proteins synthesized in the cell. A fragment of a DNA strand has the following nucleotide sequence: TAC-GGG-TCA-CGC-CGA-TCC-GTG-GCA-...

   a) Deduce the RNA sequence complementary to this fragment of a DNA strand.

   b) Deduce, using information from table 2, the primary structure of a protein fragment synthesized from the RNA template produced in (a).

9 The existence of DNA databases raises the issue of individual privacy. Discuss who has the right, and to what extent, to access information about an individual's DNA.

10 Genetically modified (GM) foods are now widely available, although in some countries environmental groups are campaigning against them. Define the term *genetically modified food* and discuss the benefits and concerns of using GM foods.   [5]

   IB, May 2009

B.9 Biological pigments (AHL)

Understandings

→ Biological pigments are coloured compounds produced by metabolism.

→ The colour of pigments is due to highly conjugated systems with delocalized electrons, which have intense absorption bands in the visible region.

→ Porphyrins, such as hemoglobin, myoglobin, chlorophyll, and cytochromes, are chelates of metals with large nitrogen-containing macrocyclic ligands.

→ Hemoglobin and myoglobin contain heme groups with the porphyrin group bound to an iron(II) ion.

→ Cytochromes contain heme groups in which the iron ion interconverts between iron(II) and iron(III) during redox reactions.

→ Anthocyanins are aromatic, water-soluble pigments widely distributed in plants. Their specific colour depends on metal ions and pH.

→ Carotenes are lipid-soluble pigments, and are involved in harvesting light in photosynthesis. They are susceptible to oxidation, catalysed by light.

Applications and skills

→ Explanation of the sigmoidal shape of hemoglobin's oxygen dissociation curve in terms of the cooperative binding of hemoglobin to oxygen.

→ Discussion of the factors that influence oxygen saturation of hemoglobin, including temperature, pH, and carbon dioxide.

→ Description of the greater affinity of oxygen for fetal hemoglobin.

→ Explanation of the action of carbon monoxide as competitive inhibition with oxygen binding.

→ Outline of the factors that affect the stabilities of anthocyanins, carotenes, and chlorophyll in relation to their structures.

→ Explanation of the ability of anthocyanins to act as indicators based on their sensitivity to pH.

→ Description of the function of photosynthetic pigments in trapping light energy during photosynthesis.

→ Investigation of pigments through paper and thin layer chromatography.

Nature of science

→ Use of data – quantitative measurements of absorbance are a reliable means of communicating data based on colour, which was previously subjective and difficult to replicate.

Coloured compounds

Most organic compounds are colourless because they do not absorb electromagnetic radiation in the visible range of the spectrum (sub-topic 2.2). Electron transitions in such compounds require relatively high energy, which corresponds to ultraviolet (UV) light and cannot be detected by the human eye. However, the presence of multi-centre chemical bonds and electron conjugation (sub-topic B.5) lowers the energy of electron transitions and therefore increases the wavelength of absorbed radiation (sub-topic 13.2). As a result, molecules with many delocalized electrons absorb visible light and appear coloured.

Carotenes

Biological pigments are coloured compounds produced in living organisms. Their molecules usually have extensive systems of alternate single and double carbon–carbon bonds. The π-electron clouds of adjacent double bonds overlap and produce long chains of carbon–carbon bonds with delocalized electrons and an average bond order of 1.5. For example, in two members of group A vitamins (sub-topic B.5), retinol and β-carotene, the electron conjugation involves 10 and 22 carbon atoms, respectively (figure 1).

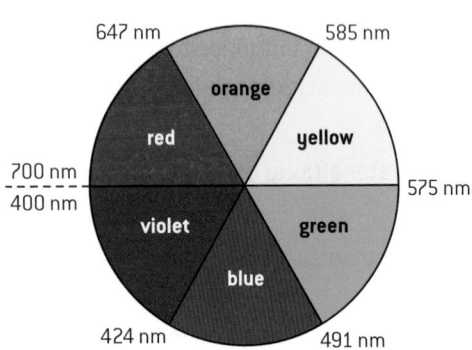

▲ Figure 2 The colour wheel

▲ Figure 1 Electron conjugation in retinol and β-carotene

The colour of a biological pigment depends on its molecular structure and on the number of delocalized electrons. Larger conjugation systems typically absorb light of lower energy, which corresponds to lower frequency and longer wavelength (sub-topic 2.2). If the wavelength at which the maximum of absorption occurs is known, the colour of the pigment can be predicted using the colour wheel (figure 2).

Retinol strongly absorbs violet light at 400–420 nm and appears yellow, as yellow is the complementary colour to violet and lies at the opposite side of the colour wheel (sub-topic 13.2). β-Carotene has a larger system of electron conjugation and therefore a maximum of absorption at longer wavelengths (430–480 nm, blue region), so its colour is orange (complementary to blue). Carotenes and other group A vitamins are fat soluble, so they accumulate in lipid tissues and are largely responsible for the yellowish colour of animal fat. The orange colours of various fruits, vegetables, dry foliage, and the feathers of some birds are also caused by carotenes (figure 3).

▲ Figure 3 The pink colour of flamingos is caused by carotenes they absorb from algae in their diet

Quantitative measurements of colour

The intensity and wavelength of the light absorbed by a biological pigment can be measured quantitatively using a UV-vis spectrophotometer (sub-topic B.7). This gives measurements of colour, which was previously subjective and difficult to describe or replicate. Electronic laboratory equipment improves the precision and accessibility of experimental data, making scientific research efficient and focused on the nature of the problem rather than the experimental skills and techniques.

Carotenes as antioxidants

The ability of carotenes to absorb visible light makes these compounds very sensitive to photo-oxidation. In living organisms carotenes and other group A vitamins act as antioxidants (sub-topic B.3), protecting the cells from UV light, peroxides, and free radicals, including a highly reactive "singlet oxygen" produced by photosynthesis (sub-topic B.1). In addition, carotenes absorb some light energy that cannot be utilized by chlorophyll (see below) and increase the efficiency of the photosynthetic reactions in green plants.

Study tip

The structures of biological pigments and the colour wheel are provided in the *Data booklet*, which will be available during the examination.

Porphyrins

Another important class of biological pigments, **porphyrins**, are complexes of metal ions (sub-topic 13.2) with large cyclic ligands. The organic backbone of porphyrins, known as **porphin**, contains four nitrogen atoms in a highly conjugated aromatic heterocycle (figure 4).

▲ Figure 4 Porphin

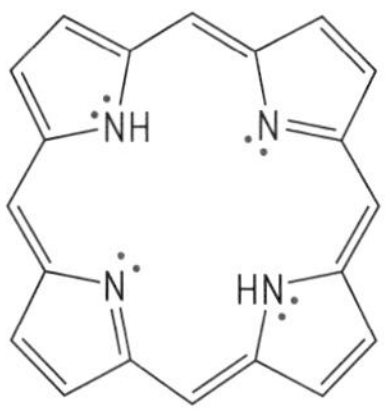

The nitrogen atoms in porphin and porphyrins can bind to metal ions, producing very stable chelate complexes (sub-topic A.10). Iron complexes of porphyrins known as hemes act as prosthetic groups in various metalloproteins, including myoglobin, hemoglobin, and cytochromes (figure 5).

The iron(II) ion in heme can form two additional coordination bonds: one with a histidine residue of the protein and one with an inorganic molecule such as oxygen or water (figure 6).

Myoglobin and hemoglobin are responsible for the transport and release of molecular oxygen to cells so that it can be used for respiration (sub-topic B.1) and other metabolic processes.

▲ Figure 5 Heme B, a porphyrin complex with iron(II)

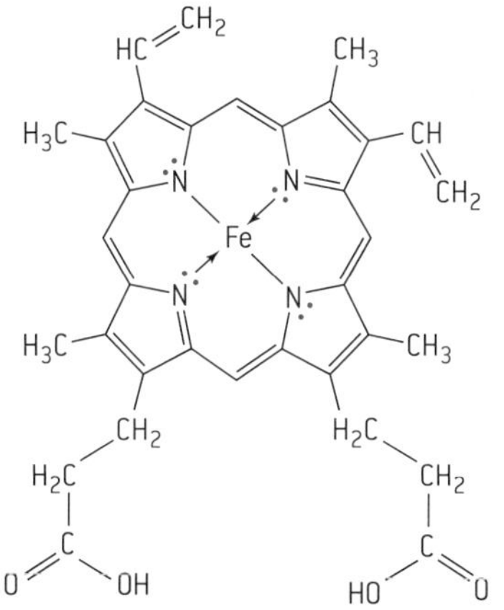

Complexes of porphyrins with d-block elements absorb visible light due to electron transitions in the conjugate macrocyclic system and between d-orbitals of the metal ion. As a result, all proteins with prosthetic heme groups are brightly coloured. Myoglobin, the primary oxygen-binding protein in muscle tissue, contains an oxygen molecule bound to an iron(II) ion in heme, which is responsible for the characteristic red colour of raw meat (figure 7). When meat is cooked the oxygen molecule in myoglobin is replaced with water and the oxidation state of iron changes from +2 to +3. As a result, cooked meat loses its original colour and becomes brown.

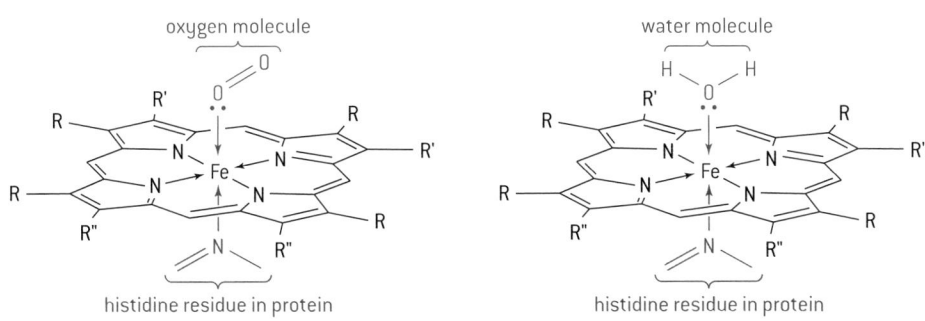

▲ Figure 6 Additional coordination bonds formed by heme B

Hemoglobin is the main oxygen-transporting protein in higher animals. It is composed of four protein subunits that are structurally similar to myoglobin. Each subunit contains a heme prosthetic group that can bind one molecule of oxygen (figure 7). Hemoglobin can therefore carry up to four oxygen molecules from the lungs to other organs and tissues. Red blood cells (erythrocytes) contain over 35% hemoglobin and can absorb 0.5 cm³ of oxygen per 1 g of their mass, which is approximately 70 times more than the solubility of oxygen in blood plasma.

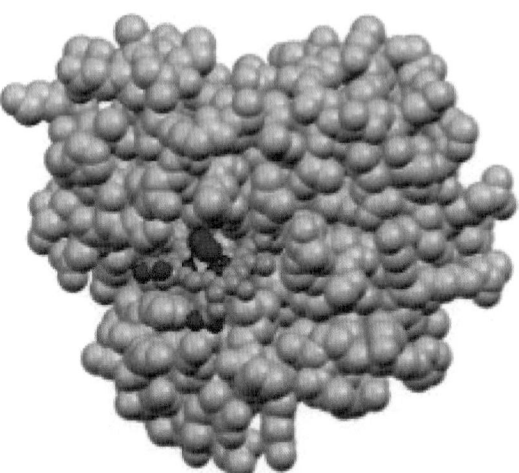

▲ Figure 7 Oxygenated myoglobin. The heme prosthetic group is shown in colour, with an oxygen molecule (two large red spheres) bound to the iron ion (orange–yellow sphere)

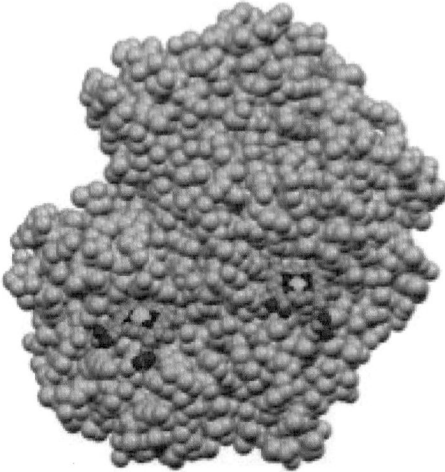

▲ Figure 8 Deoxygenated hemoglobin. Two of the four heme prosthetic groups are shown in colour, with the other two on the opposite side of the protein assembly

Cooperative binding in hemoglobin

As already discussed in sub-topic B.7, the interaction between hemoglobin and molecular oxygen is an example of **cooperative binding**. According to the induced fit model, the binding of a substrate (oxygen molecule) to a free active site (deoxygenated heme) alters the shape of the entire hemoglobin molecule, including the shapes of active sites in all four protein subunits. These changes increase the affinity of partly oxygenated hemoglobin to molecular oxygen. As a result, the kinetic curve of hemoglobin–oxygen interaction does not obey the Michaelis–Menten model (sub-topic B.7) and adopts a characteristic sigmoidal shape (figure 9).

Worked example

Red blood cells (erythrocytes) comprise approximately a quarter of the total cell number in the human body. Calculate the volume of oxygen that can be absorbed by the erythrocytes of an adult human with a blood volume of 5.0 dm³ if the concentration of erythrocytes in the blood is 4.5×10^{12} dm⁻³ and each erythrocyte absorbs 4.8×10^{-11} cm³ of oxygen.

Solution

The total number of erythrocytes in the blood is $5.0 \times 4.5 \times 10^{12} \approx 2.3 \times 10^{13}$,
so the volume of absorbed oxygen will be $2.3 \times 10^{13} \times 4.8 \times 10^{-11} \approx 1100$ cm³ $= 1.1$ dm³.

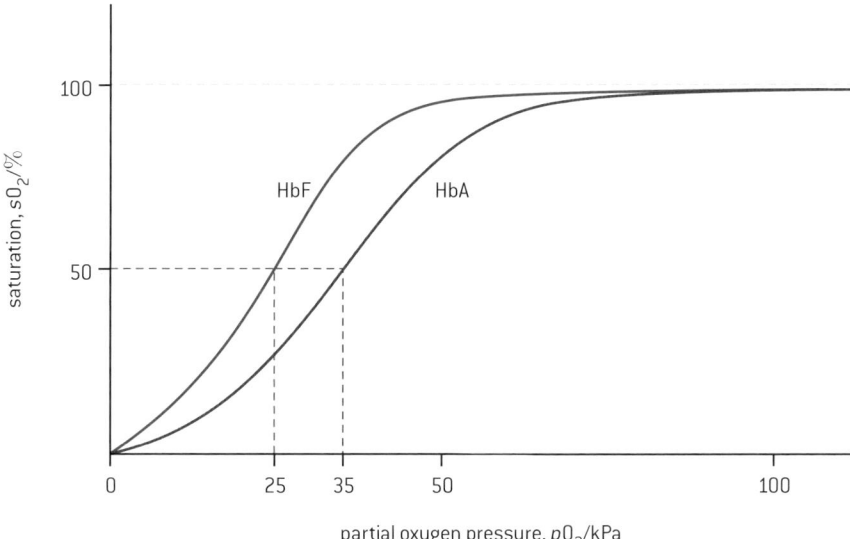

▲ Figure 9 Oxygen saturation curves for adult (HbA) and fetal (HbF) hemoglobin

Cooperative binding increases the efficiency of oxygen transport in the human body. In arterial blood, where the partial pressure of oxygen (pO_2) is high, most of the binding sites in hemoglobin become occupied by oxygen molecules. This further increases its affinity for oxygen, allowing hemoglobin to reach saturation point quickly and carry as much oxygen as possible from the lungs to other tissues. As pO_2 decreases some oxygen molecules are released, which reduces the affinity of hemoglobin for oxygen and accelerates the loss of remaining oxygen molecules. In venous blood with low pO_2, the affinity of hemoglobin for oxygen is minimal, so the last oxygen molecules are released and hemoglobin becomes ready for the next cycle of oxygen transport.

Other factors affecting the affinity of hemoglobin for oxygen

Other factors such as temperature, pH, and concentration of carbon dioxide can also affect the affinity of hemoglobin for oxygen. At abnormally high body temperature (fever), the ability of hemoglobin to carry oxygen decreases due to unfavourable conformational changes of the active sites and the positive entropy of dissociation (sub-topic 15.2) of hemoglobin–oxygen complexes. In contrast, hypothermia (low body temperature) increases the affinity of hemoglobin for oxygen.

In venous blood, which has a slightly lower pH and higher carbon dioxide concentration than arterial blood, protons and carbon dioxide bind to side-chains of amino acids in hemoglobin and act as non-competitive inhibitors (sub-topic B.7). According to some studies, carbon dioxide may also act as a competitive inhibitor that binds directly to heme prosthetic groups. Competitive and non-competitive inhibition facilitate the release of oxygen to venous blood, where it is particularly needed for oxygen-deprived tissues.

Fetal hemoglobin

Fetal hemoglobin (HbF) is structurally different from adult hemoglobin (HbA) and can bind to oxygen more efficiently. A developing fetus receives its oxygen from the partially deoxygenated blood of its mother, so HbF must have greater affinity for oxygen in order to decrease its concentration in fetal blood plasma and allow oxygen to diffuse from adult to fetal blood across the placenta. In addition, fetal hemoglobin is less sensitive to certain inhibitors such as 2,3-bisphosphoglycerate (2,3-BPG), which are present in high concentrations at the placenta and cause HbA to release oxygen. This process further increases the rate of oxygen diffusion through the placenta and the oxygen uptake by HbF. At the same time, the steeper saturation curve of fetal hemoglobin (figure 9) allows HbF to release a greater proportion of oxygen to developing tissues and operate efficiently even at very low partial pressures of oxygen.

After birth, the production of HbF decreases rapidly and the synthesis of HbA is activated. However, this process can be reversed by medication, which is used in the treatment of sickle-cell disease and other hemoglobin-related health conditions.

Worked example

Strenuous physical exercise stimulates respiration and production of 2-hydroxypropanoic (lactic) acid in muscle tissues. Outline, using chemical equations, the effects of these metabolic processes on the affinity of hemoglobin and myoglobin to oxygen.

Solution

Respiration (sub-topic B.1) reduces the concentration of oxygen and produces carbon dioxide:

$$C_6H_{12}O_6 + 6O_2 \rightarrow 6CO_2 + 6H_2O$$

Carbon dioxide reacts reversibly with water to produce the weak acid carbonic acid:

$$CO_2 + H_2O \rightleftharpoons H_2CO_3$$

The dissociation of carbonic and lactic acids produces protons (or hydronium ions), reducing the pH of the blood plasma and cellular tissues:

$$H_2CO_3 \rightleftharpoons H^+ + HCO_3^-$$

$$CH_3CH(OH)COOH \rightleftharpoons CH_3CH(OH)COO^- + H^+$$

The affinity of hemoglobin for oxygen is therefore reduced by three different factors:

- a low concentration of oxygen reduces the degree of its cooperative binding

- carbon dioxide acts as both a non-competitive and a competitive inhibitor

- at low pH, protons (hydronium ions) act as non-competitive inhibitors.

Since the structures of myoglobin and hemoglobin are similar, the affinity of myoglobin for oxygen is also reduced due to inhibition by carbon dioxide and protons. However, myoglobin has only one protein unit with a single heme group, so the concentration of oxygen will have no effect on myoglobin due to the lack of cooperative binding.

Finally, physical exercise can temporarily increase the body temperature, which will further reduce the affinity of hemoglobin and myoglobin for oxygen.

Carbon monoxide poisoning

Carbon monoxide, CO, is a colourless and odourless gas that is highly toxic even at low concentrations. In the human body, carbon monoxide readily combines with heme prosthetic groups and acts as a competitive inhibitor, preventing the delivery of oxygen to body tissues. The complex of carbon monoxide with hemoglobin, carboxyhemoglobin, is very stable and can accumulate in the blood until most active sites in hemoglobin are occupied by CO molecules. Myoglobin reacts with carbon monoxide in the same way as hemoglobin, which further reduces the oxygen supply to cells. The presence of 0.2% of carbon monoxide in the air causes serious health problems while concentrations above 0.5% can be fatal.

As with any competitive inhibitor, carbon monoxide can be displaced from carboxyhemoglobin by inhaling pure oxygen or, in mild cases of CO poisoning, simply by removing the patient from toxic atmosphere to fresh air.

Heavy smokers, truck drivers, and traffic police are regularly exposed to low concentrations of carbon monoxide and often show the symptoms of chronic hypoxia. The decreased oxygen supply leads to higher hemoglobin levels in their blood and increased production of 2,3-bisphosphoglycerate (2,3-BPG, see above) in erythrocytes, which reduces the affinity of hemoglobin to oxygen at low concentrations.

Cytochromes

Molecular oxygen, which is supplied to cellular tissues by hemoglobin, is reduced to water during the final step of aerobic respiration (sub-topic B.1). This step takes place in mitochondria and involves a group of enzymes collectively known as **cytochromes**. One of these enzymes, **cytochrome c oxidase**, is a large metalloprotein assembly containing four heme prosthetic groups and several ions of other metals including copper, magnesium, and zinc. The electrons required for the reduction of molecular oxygen are provided by transition metal ions according to the following simplified scheme:

$$4Fe^{2+}_{(cytochrome\ c)} + 4H^+ + O_2 \rightarrow 4Fe^{3+}_{(cytochrome\ c)} + 2H_2O$$

The above reaction involves four protons, which are pumped through the mitochondrial membrane and used for the synthesis of ATP (sub-topic B.8). Another transition metal in cytochrome c, copper, changes its oxidation state from +1 to +2. When the reduction of oxygen is complete, all transition metal ions (Fe^{3+} and Cu^{2+}) are also reduced (to Fe^{2+} and Cu^+, respectively) using the electrons extracted from glucose metabolites and transferred to cytochrome c by other enzymes. Therefore the net equation for aerobic respiration (sub-topic B.1) includes only glucose and oxygen while all enzymes act as catalysts and return to their original states after the reaction is complete.

Chlorophyll

Chlorophyll is a green pigment found in cyanobacteria and the chloroplasts of green plants. The ability of chlorophyll to absorb energy from visible light is utilized by living organisms in the process of photosynthesis (sub-topic B.1), which is the primary source of nearly all organic compounds on our planet. Chlorophyll is structurally similar to heme but contains a different metal ion (magnesium instead of iron) and different substituents at several positions on its porphin backbone (figure 10).

Photosynthesis is a complex process that involves many pigments and proteins collectively known as **photosystems**. The light energy absorbed by some chlorophyll molecules is passed to other pigments in the reaction centre of the photosystem, where it is used to create a series of energy-rich molecular intermediates. These intermediates, known as the **electron transport chain**, undergo various redox reactions which ultimately lead to the oxidation of water to molecular oxygen and protons:

$$2H_2O \rightarrow O_2 + 4H^+ + 4e^-$$

In green plants, the protons produced by the above reaction are pumped through the chloroplast membrane and used for the synthesis of ATP in the same way as in mitochondria. The molecular oxygen produced by photosynthesis is released to the atmosphere as a part of the carbon–oxygen cycle (sub-topic B.1).

The absorption spectrum of chlorophyll

Chlorophyll absorbs electromagnetic radiation in the blue and red regions of the visible spectrum. At the same time, green and near-green portions of the spectrum are reflected or transmitted, which is responsible for the green colour of plant leaves and other chlorophyll-containing tissues. Carotenes and some other pigments extend the absorption spectrum of chlorophyll and increase the efficiency of photosynthesis (figure 11).

▲ Figure 10 Chlorophyll. $R = CH_3$ (chlorophyll *a*) or CHO (chlorophyll *b*)

The name "chlorophyll" is derived from the Greek words χλωρός (chloros, "green") and φύλλον (phyllon, "leaf"). Chlorophyll is one of the most important biomolecules and occurs in all photosynthesizing organisms. Its structural backbone, chlorin, is very similar to porphin (figure 4) but is more reduced and contains an additional five-membered hydrocarbon ring. Although chlorophyll was isolated from green leaves in 1817, the presence of a magnesium ion in its structure remained unknown until 1900, when this element was detected for the first time in living organisms.

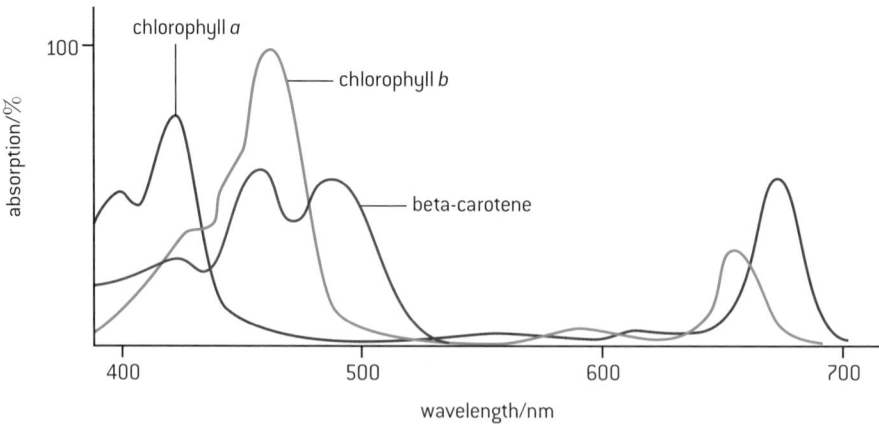

absorption/%

wavelength/nm

▲ Figure 11 Absorption spectra of photosynthetic pigments

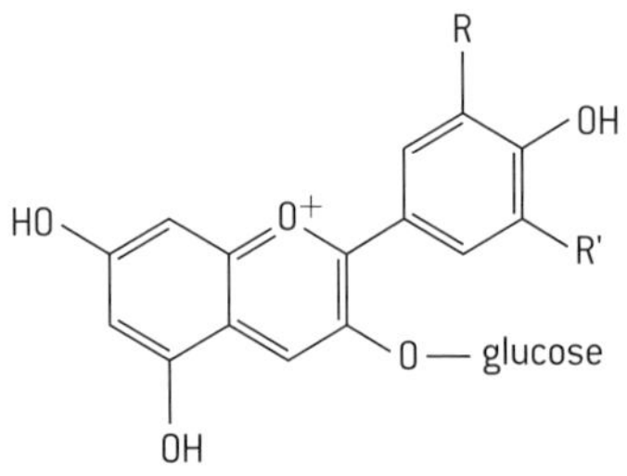

▲ Figure 12 Anthocyanin structure

Anthocyanins

The bright colours of flowers, ripe fruits, berries, and vegetables are largely caused by a group of biological pigments called **anthocyanins**. All anthocyanins are water soluble and concentrate in the vacuoles of plant cells, producing characteristic red, purple, and blue colours of chlorophyll-free plant tissues.

From the chemical point of view, anthocyanins are tricyclic polyphenols with an aromatic backbone (**flavylium ion**) and several substituents including a residue of α-glucose (sub-topic B.4) (figure 12). The composition and occurrence of some common anthocyanins are summarized in table 1.

Paper chromatography

Anthocyanins, chlorophylls, and other biological pigments are often analysed by paper or thin-layer chromatography (sub-topic B.2). The progress of the experiment can be monitored visually because most of these compounds have different colours. In some cases the chromatogram can be developed by dilute solutions of acids or ammonia, allowing detection of pH-sensitive pigments.

Instead of biological pigments, coloured inks from marker pens can be used. Such chromatograms show that most inks are mixtures of several chemicals. For example, black ink may contain red, purple, blue, and brown pigments.

Anthocyanin	R	R'	Common sources
cyanidin	—OH	—H	blackberry, blueberry, cranberry, raspberry, grapes, apples, cherry, plums, red cabbage, red onion
delphinidin	—OH	—OH	flowers (delphiniums and violas), cranberry, pomegranates, some grapes (Cabernet Sauvignon)
pelargonidin	—H	—H	flowers (geraniums), ripe raspberries and strawberries, blueberries, blackberries, cranberries, plums, pomegranates
malvidin	—OCH$_3$	—OCH$_3$	flowers (primulas), grapes (primary pigment of red wine)
petunidin	—OH	—OCH$_3$	flowers (petunias), some berries, some grapes

▲ Table 1 Substituents and natural sources of selected anthocyanins

Colour changes in anthocyanins

The exact colours of anthocyanins in solution depend on the presence of metal ions and the solution pH. Metal ions such as Mg^{2+} and Fe^{3+} form stable complexes with anthocyanins, which are responsible for the colours of flower petals. In the absence of metals the colours of most anthocyanins change from red in acidic solutions to purple in neutral and blue in slightly alkaline solutions. This colour change is the result of acid–base reactions that involve the aromatic backbone of anthocyanins and affect the degree of electron conjugation in their molecules or ions (figure 13).

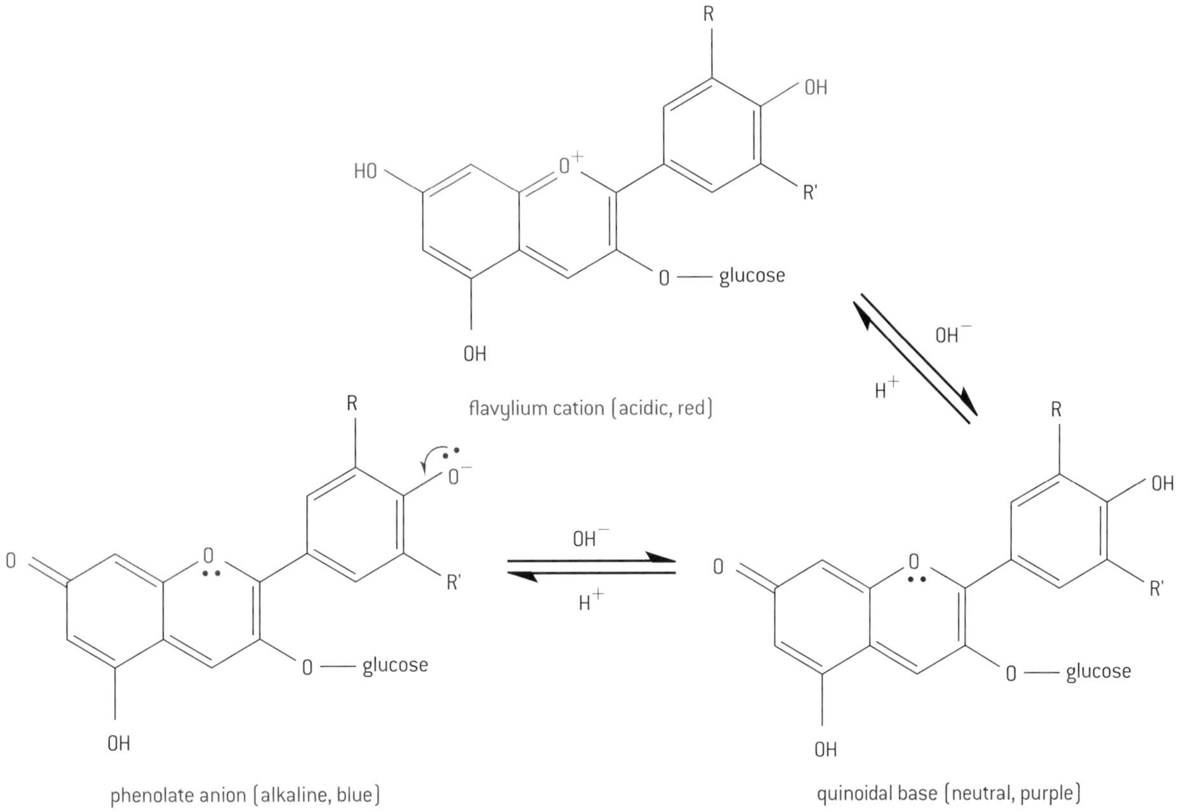

▲ Figure 13 The colours of anthocyanins vary depending on the pH of the solution

In acidic solutions (at low pH), anthocyanins exist as protonated flavylium cations. As the acidity decreases these cations lose one or two protons, producing neutral quinoidal bases or phenolate anions. The loss of protons increases the electron density in the aromatic backbone and lowers the energy of electron transitions. As a result, the wavelength of the absorbed light increases and the maximum of absorption shifts from the green region of the visible spectrum (in flavylium cation) to yellow (in quinoidal base) and finally to orange (in phenolate anion). The actual colours of anthocyanins at different pH (red, purple, and blue) are complementary to their absorption maxima (figure 2).

Because anthocyanins are sensitive to pH and absorb visible light, they can be used as natural acid–base indicators (sub-topic 18.3) and organic components of dye-sensitized solar cells (sub-topic A.8). Anthocyanins are approved as artificial food colours in the EU (E163) and USA although food regulations vary greatly from country to country.

In plants, anthocyanins perform several functions. Bright colours of flowers and fruits attract insects and animals that provide pollination and seed dispersal. In green leaves, anthocyanins absorb certain wavelengths of visible and UV light, protecting photosynthesizing pigments and plant tissues from excessive exposure to solar radiation. The presence of highly conjugated electron systems in anthocyanins makes them efficient antioxidants (sub-topics B.3 and B.5). At the same time, the ability of biological pigments to absorb light and act as free-radical scavengers reduces their own stability and makes them particularly sensitive to photo-oxidation.

Melanin

Melanin is a collective name for black, brown, and red pigments found in most living organisms. These pigments are very complex polyaromatic compounds with extensive systems of electron conjugation. High levels of black and brown melanin (**eumelanin**) are responsible for the darker tones of skin, hair, and eyes of people with African, South American, South Asian, and Australian origins. Red or red-brown **pheomelanin** is more abundant in Europeans and North-Americans, where it is largely responsible for red hair, lighter skin tones, and freckles in some individuals. A lack of melanin, known as albinism, is a common genetic disorder that affects approximately one in 20 000 Europeans and up to one in 3000 people of some African countries.

Skin pigmentation in humans is an important regulatory mechanism that protects the body from harmful UV radiation. Increased exposure to sunlight during summer stimulates melanin production, making the skin darker (tanned) and more resistant to sunburn. However, melanin does not provide total protection against skin cancer and cellular damage so the use of sunscreen lotions is recommended for anyone who is exposed to direct sunlight for prolonged periods of time.

▲ Figure 14 A woman from Tanzania with her two children, one of whom is albino

Questions

1 Carotenes, porphyrins, and anthocyanins are examples of biological pigments.

   a) State what is meant by the term "biological pigment".

   b) State one common structural feature of all biological pigments.

   c) Explain, in terms of their interaction with light, why biological pigments are coloured.

   d) Outline the functions of biological pigments in living organisms.

2 Metal complexes are involved in respiration. State the names of two such complexes and the metals they contain. Explain the role of these complexes in respiration. [4]

   IB, May 2012

3 The oxygen-binding capacity of hemoglobin is affected by pH. The saturation curve of hemoglobin at normal physiological pH (7.4) is given in figure 15.

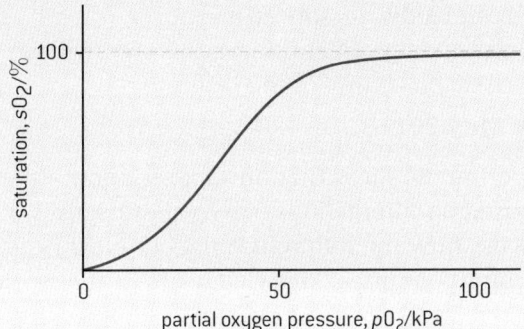

▲ Figure 15

   a) Explain the sigmoidal shape of the saturation curve at pH = 7.4.

   b) Sketch the saturation curve of hemoglobin at pH = 7.2.

   c) Explain, in terms of enzyme inhibition, the effects of protons and carbon monoxide on the saturation curve of hemoglobin.

4 One of the organic compounds shown in figure 16 is colourless while the other is orange.

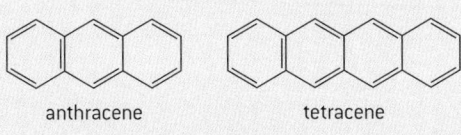

anthracene   tetracene

▲ Figure 16

Predict, with reference to conjugation of double bonds, which compound (anthracene or tetracene) will absorb visible light and, therefore, be coloured. [1]

   IB, November 2012

5 The pigment in blueberries is an anthocyanin.

   a) With reference to the colour wheel in figure 2, explain how the pigment in blueberries causes them to be blue. [2]

   b) State the combination of pH and temperature that produces the strongest colour in anthocyanins. [1]

   IB, November 2010

6 The absorption spectrum of β-carotene is shown in figure 17.

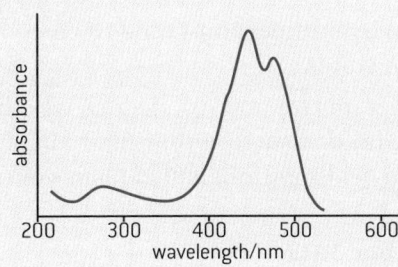

▲ Figure 17

In terms of this spectrum, explain why carotenes have their typical colour. [2]

   IB, May 2012

7 The wavelength of visible light lies between 400 and 750 nm. The absorption spectrum of a particular anthocyanin is shown in figure 18.

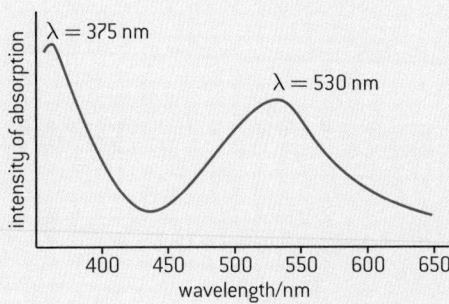

▲ Figure 18

   a) Explain what effect, if any, the absorption at 375 nm will have on the colour of the anthocyanin. [1]

b) Explain what effect, if any, the absorption at 530 nm will have on the colour of the anthocyanin. [1]

IB, May 2010

8 Anthocyanins, the pigments which occur naturally in many flowers and fruits, are water soluble and often change colour as the temperature or pH changes. The diagrams in figure 19 show two structures of the same anthocyanin under different conditions.

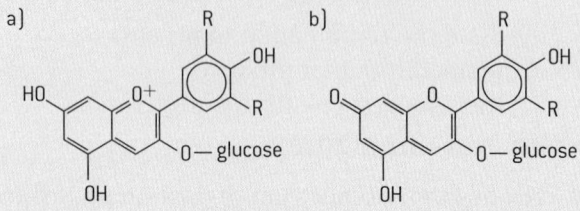

▲ Figure 19

a) Explain why anthocyanins tend to be soluble in water. [2]

b) Using diagrams (a) and (b), deduce whether structure (a) or structure (b) is more likely to exist in acid solution, and explain your answer. [2]

IB, November 2012

9 A sample of food colouring was analysed using thin-layer chromatography to check whether it contained a banned substance. The R_f value of the banned substance is 0.25 under the same conditions.

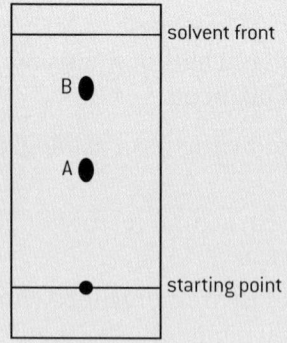

▲ Figure 20

a) State the number of components used to produce the food colouring. [1]

b) Identify a stationary phase commonly used in thin-layer chromatography. [1]

c) Identify the component in this chromatogram that has the greatest attraction for the stationary phase. [1]

d) Explain what is meant by the term R_f value. [1]

e) Predict where you would expect the banned dye to appear on the chromatogram and mark this spot with a circle on a copy of the diagram. [1]

IB, May 2011

10 Experiments show that our appreciation of food is based on interactions between our senses. Discuss how the different senses interact in giving us empirical knowledge about the world.

B.10 Stereochemistry in biomolecules (AHL)

Understandings

→ With one exception, amino acids are chiral, and only the L-configuration is found in proteins.

→ Naturally occurring unsaturated fat is mostly in the *cis* form, but food processing can convert it into the *trans* form.

→ D- and L-stereoisomers of sugars refer to the configuration of the chiral carbon atom furthest from the aldehyde or ketone group, and D-forms occur most frequently in nature.

→ Ring forms of sugars have isomers, known as α and β, depending on whether the position of the hydroxyl group at carbon 1 (glucose) or carbon 2 (fructose) lies below the plane of the ring (α) or above the plane of the ring (β).

→ Vision chemistry involves the light activated interconversion of *cis*- and *trans*-isomers of retinal.

Applications and skills

→ Description of the hydrogenation and partial hydrogenation of unsaturated fats, including the production of *trans*-fats, and a discussion of the advantages and disadvantages of these processes.

→ Explanation of the structure and properties of cellulose, and comparison with starch.

→ Discussion of the importance of cellulose as a structural material and in the diet.

→ Outline of the role of vitamin A in vision, including the roles of opsin, rhodopsin, and *cis*- and *trans*-retinal.

Nature of science

→ Theories used to explain natural phenomena/evaluate claims – biochemistry involves many chiral molecules with biological activity specific to one enantiomer. Chemical reactions in a chiral environment act as a guiding distinction between living and non-living matter.

Stereoisomerism

Most biochemical processes are stereospecific: they involve only molecules with certain three-dimensional configurations. Molecules that have the same sequence of atoms and chemical bonds but different arrangements of atoms in space are known as **stereoisomers** (sub-topic 20.3). Stereoisomers that cannot be transformed into one another without breaking a chemical bond are called **configurational isomers** and include two classes: *cis-/trans-isomers* and **optical isomers**.

Both types of stereoisomerism play important roles in metabolic reactions, which are catalysed by enzymes with specific three-dimensional structures (sub-topics B.2 and B.7). Most enzymes can bind only to those stereoisomers that fit into their active sites; other stereoisomers usually do not participate in normal metabolic processes. However, "wrong" stereoisomers can sometimes be recognized as substrates by different enzymes, act as non-competitive inhibitors (sub-topic B.7), or accumulate in fatty tissues as xenobiotics (sub-topic B.6). Such unwanted stereoisomers are often responsible for side effects of medical drugs (sub-topic D.7) and negative health effects of processed foods.

In this final sub-topic we shall discuss the stereochemistry of biologically important organic compounds, including 2-amino acids, carbohydrates, fatty acids, and retinoids.

The origin of chirality in living organisms

A strictly controlled chiral environment is a distinctive feature of living organisms. However, there is still no satisfactory scientific theory that offers a reasonable explanation of this natural phenomenon. Although many claims have been made about the possible roles of UV light, magnetic fields, mineral templates, and other factors in the spontaneous resolution of racemic mixtures (sub-topic 20.3), all such claims either lack scientific evidence or require some kind of a chiral "seed" that has to be somehow produced in the first place.

Most scientists think that the chirality in the first life forms appeared spontaneously and then became "fixed" by evolution. The first enzymes were probably chiral, because otherwise they would not be able to adopt specific conformations and show any selectivity towards their substrates. Still, no details of early biochemical processes are known, and the origin of chirality remains one of the most challenging puzzles of evolutionary theory.

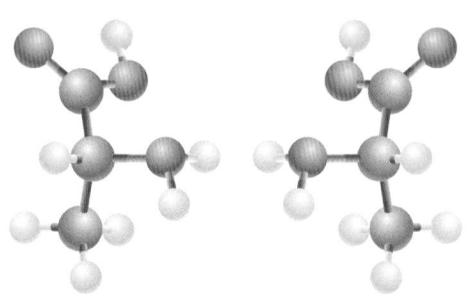

▲ Figure 1 Top: these two pairs of scissors are chiral macroscopic objects that cannot be superimposed on each other. Bottom: molecular models of D- and L-isomers of the 2-amino acid alanine

2-amino acids

Proteinogenic **2-amino acids** (sub-topic B.2) are relatively simple biomolecules of general formula $H_2NCH(R)COOH$, in which the side-chain R can contain additional functional groups (sub-topic B.2, table 1). With a single exception, all 2-amino acids have four different substituents (NH_2, COOH, R, and H) attached to the same carbon atom, C-2. These molecules are **chiral**: they can exist as two **enantiomers** (sub-topic 20.3), which are mirror images of one another (figure 1).

In glycine (2-aminoethanoic acid) R = H, so the C-2 atom has two hydrogens and thus is not chiral. All other proteinogenic amino acids are exclusively L-isomers, although several D-amino acids occur in bacterial cell walls and some antibiotics.

Three-dimensional structures of chiral molecules can be represented on paper using wedge–dash notation (figure 2) or Fischer projections (figure 3). In wedge–dash notation, any two of the four chemical bonds formed by a tetrahedral carbon atom are positioned parallel to the plane of the paper and drawn as plain lines. The third bond coming out of the page toward you is represented as a solid wedge while the fourth bond pointing away from you is drawn as a dashed wedge (sometimes an empty wedge or a dashed line are used instead of the dashed wedge) (figure 2).

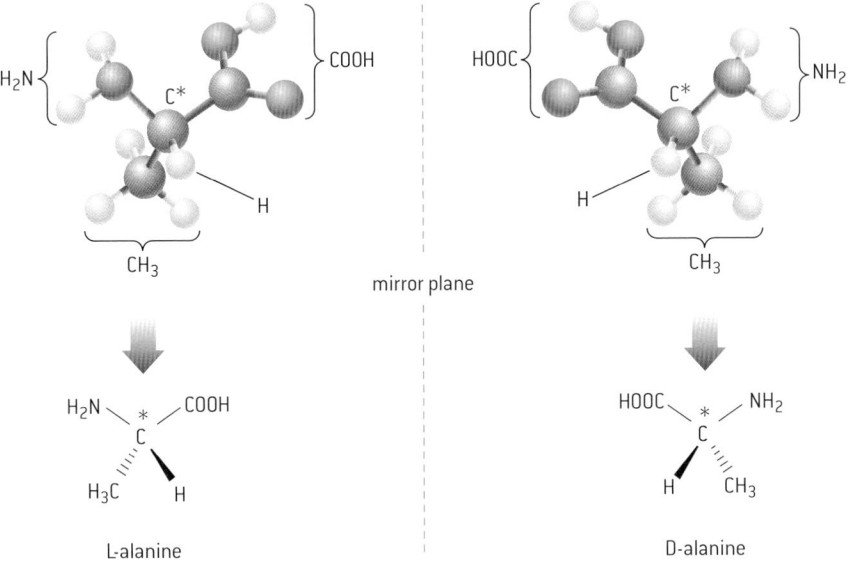

▲ Figure 2 Wedge–dash notation for the two enantiomers of alanine. The resulting stereochemical formulae are mirror images of each other. Chiral carbon atoms are marked with asterisks (*)

Fischer projections are designed to represent stereochemical formulae with plain lines, which is particularly useful for complex molecules with many stereochemical centres. In Fischer projections the carbon chain of the molecule is drawn vertically, with the senior substituent (COOH in amino acids, CHO or C=O in carbohydrates) at the top. Chiral carbon atoms are shown as crosses, with horizontal lines representing chemical bonds coming out of the page (toward you) and vertical lines representing the bonds pointing behind the page (away from you) (figure 3).

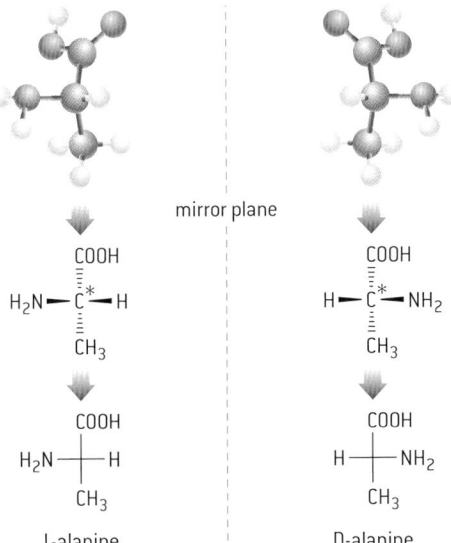

▲ Figure 3 Fischer projections for the two enantiomers of alanine

In Fischer projections, L-enantiomers of 2-amino acids have amino groups on the left of the chiral carbon atom while D-enantiomers have the opposite orientation. In order to identify a particular enantiomer, a wedge–dash formula or three-dimensional model of the chiral molecule can be transformed into its Fischer projection or analysed by the CORN rule, which is described in the box on the next page.

The CORN rule

A simple rule can be used for identifying enantiomers of 2-amino acids:

- remember that the first letters of **COOH**, **R**, and **NH_2** groups form the word "**CO-R-N**"
- position the molecule so that the H atom at the C-2 carbon faces away from you
- "**CO-R-N**" is spelled clockwise for a D-enantiomer and anti-clockwise for an L-enantiomer (figure 4).

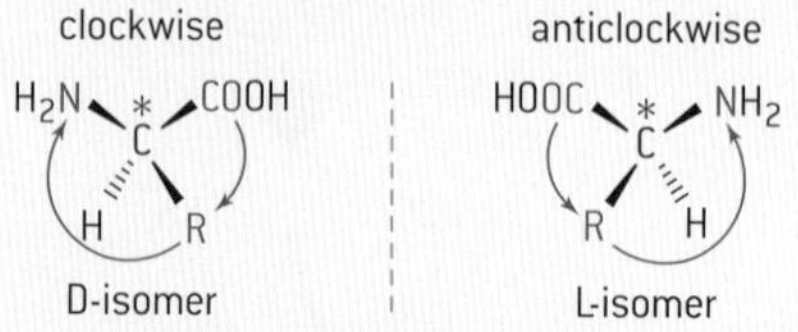

▲ Figure 4 The CORN rule

Carbohydrates

Fischer projections of common monosaccharides were used in sub-topic B.4. Since most monosaccharides have more than one stereogenic centre, their enantiomeric configuration refers to the chiral carbon atom furthest from the senior functional group (CHO or C=O). For example, all the formulae in figure 5 represent the open-chain form of D-glucose, which is the most common monosaccharide in nature.

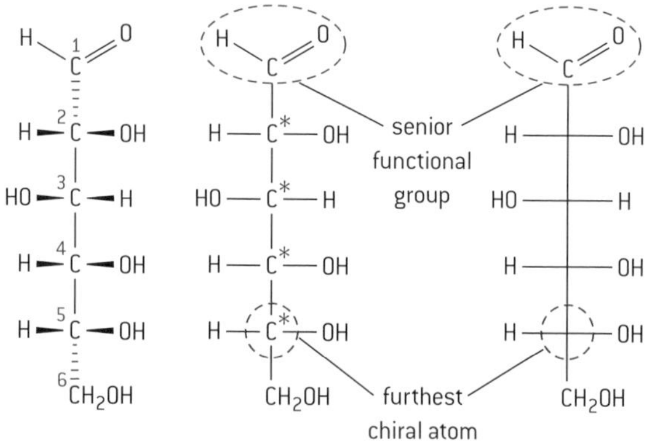

▲ Figure 5 Wedge–dash (left) and Fischer projections (middle and right) of D-glucose

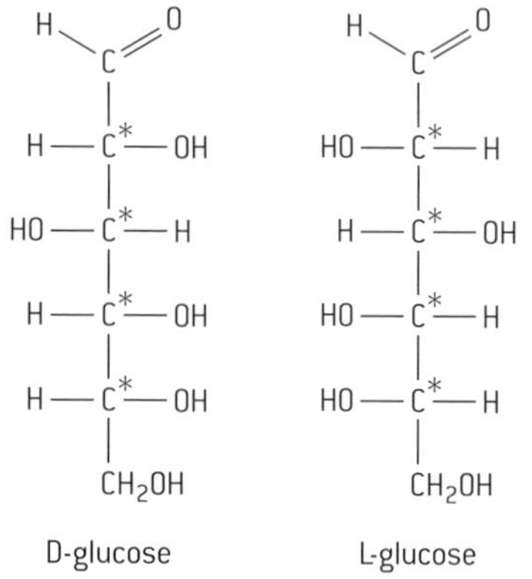

▲ Figure 6 The enantiomers of glucose

The enantiomer of D-glucose, L-glucose, does not occur naturally but can be synthesized in the laboratory. In the L-glucose molecule all chiral carbon atoms have the opposite configuration to the corresponding carbon atoms in D-glucose (figure 6).

Nearly all naturally occurring mono-, di-, and polysaccharides have D-configurations of their molecules or structural units. The majority of living organisms (except some bacteria) lack specific enzymes to metabolize L-sugars. Certain L-monosaccharides, such as L-glucose, are sometimes used as low-calorie sweeteners or inert binders in the pharmaceutical industry (sub-topic B.4).

Study tip

Strictly speaking, the right-hand formula in figure 5 (with the chiral carbon atoms omitted) is the only correct Fischer projection of D-glucose. However, many textbooks and the *Data booklet* show the chemical symbols of all atoms in non-cyclic structures. In examination papers you are expected to draw Fischer projections of monosaccharides in the same way as they are represented in the *Data booklet*, which will be available during the examination.

Another monosaccharide, D-galactose, is a common component of disaccharides such as lactose (sub-topic B.4). The D-galactose molecule differs from that of D-glucose in the configuration of a single chiral carbon atom, C-4 (figure 7).

In contrast to enantiomers, which have identical physical properties and differ only by their ability to rotate plane-polarized light in opposite directions (sub-topic 20.3), D-glucose and D-galactose are **diastereomers** and therefore have different physical properties such as melting point (150 °C and 167 °C, respectively). Diastereomers that differ in the configuration of only one stereogenic centre are known as **epimers**.

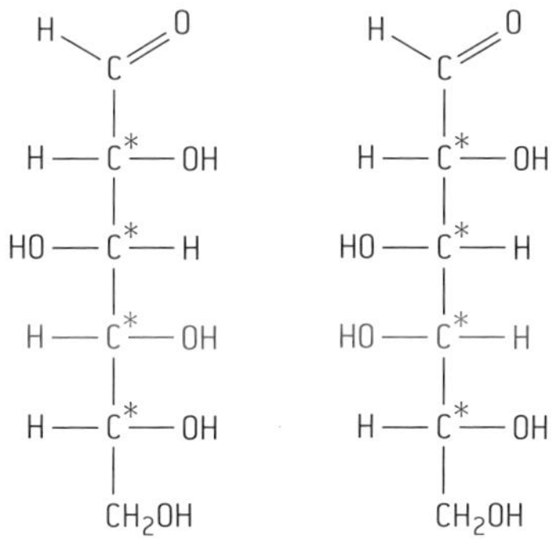

▲ Figure 7 D-glucose (left) and D-galactose (right)

Worked example

Identify enantiomers, epimers, and diastereomers among the carbohydrates in figure 8.

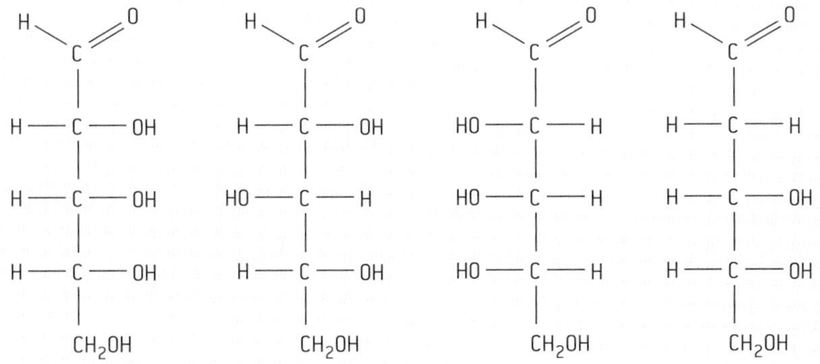

▲ Figure 8

Solution

D- and L-ribose differ in the configurations of all chiral carbon atoms (C-2, C-3, and C-4), so they are enantiomers. D-ribose and D-xylose differ in the configuration of one atom (C-3), so they are both epimers and diastereomers. D-xylose and L-ribose differ in the configurations of two atoms (C-2 and C-4), which is more than one but less than all three, so they are diastereomers. Finally, deoxyribose has a different molecular formula ($C_5H_{10}O_4$) from ribose and xylose ($C_5H_{10}O_5$), so it is not an isomer of the three other sugars.

Cyclic forms of monosaccharides

In the solid state and in aqueous solutions, monosaccharides exist predominantly as cyclic forms which are commonly represented by Haworth projections (sub-topic B.4). Each monosaccharide can produce two cyclic forms that differ in the orientation of the –OH group at the C-1 atom in aldoses or C-2 atom in ketoses (figure 9).

In α-glucose the –OH group at C-1 lies below the plane of the ring while in β-glucose this group lies above the plane of the ring. Similarly, α- and β-isomers of fructose differ in the position of the –OH group directly attached to the C-2 atom. In aqueous solutions, α- and β-isomers of

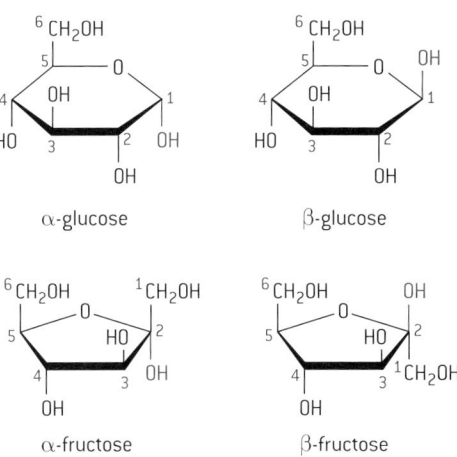

▲ Figure 9 Haworth projections for some cyclic monosaccharides

Nomenclature

In biochemical literature, α- and β-forms of glucose are usually referred to as α-D-glucopyranose and β-D-glucopyranose, respectively. The letter "D" specifies the stereochemical configuration of the fifth carbon atom while the word "pyranose" is derived from the name of a six-membered heterocycle, pyran, which resembles the cyclic forms of glucose. In this book we shall use shortened names of monosaccharides such as α-glucose and β-glucose. These names are also used in the *Data booklet* and examination papers.

monosaccharides can transform into each other through the open-chain form, which exists in equilibrium with both cyclic forms (figure 10).

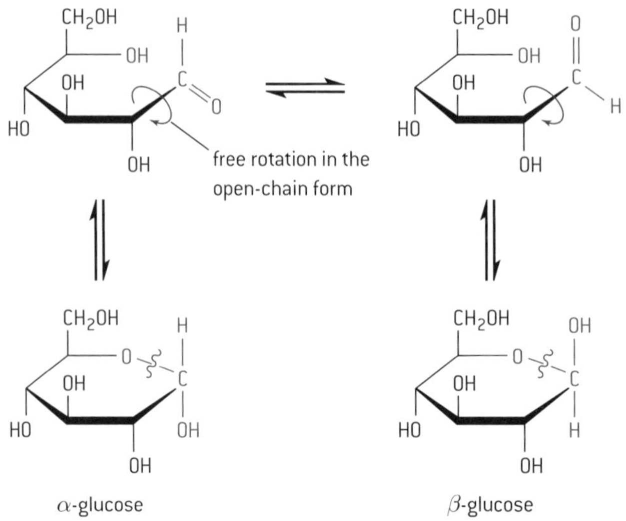

▲ Figure 10 The interconversion of α-glucose and β-glucose

For most monosaccharides, β-isomers are slightly more stable than α-forms. For example, an aqueous solution of glucose at equilibrium contains approximately 64% α-glucose, 36% β-glucose, and very small amount (less than 0.3%) of the open-chain isomer.

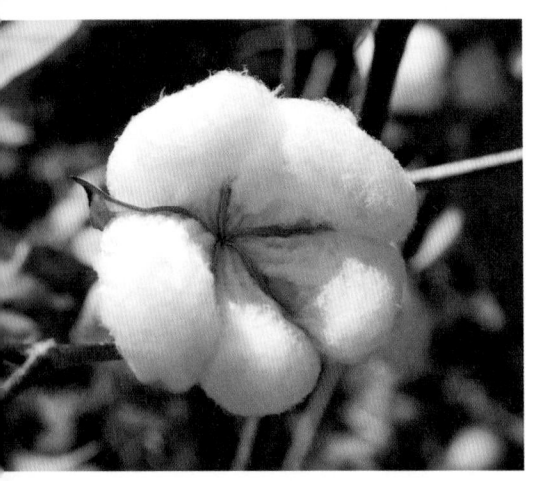

▲ Figure 11 Cotton fibres contain nearly 100% cellulose

Cellulose

A polymer of β-glucose, **cellulose**, is the most abundant structural polysaccharide in plants, comprising up to 50% of the cell wall material of wood and nearly 100% of dry cotton (figure 11). The glucose residues in cellulose are joined together by β-1,4-glycosidic links (figure 12).

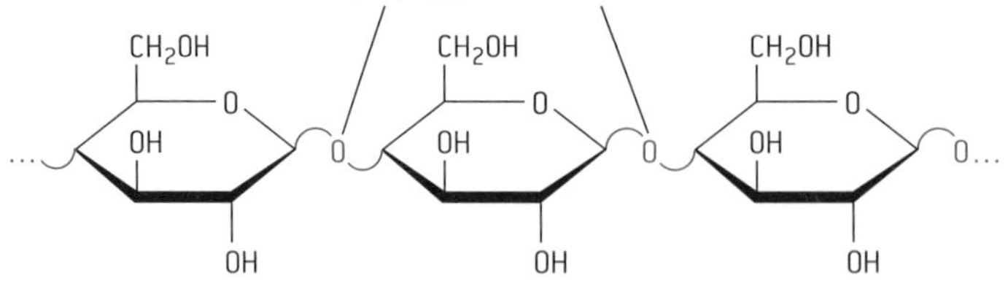

▲ Figure 12 β-1,4-glycosidic links in cellulose

In contrast to starch and glycogen, which are polymers of α-glucose (sub-topic B.4), the polymeric chains of cellulose are unbranched and tend to adopt more extended, rod-like conformations. Linear macromolecules of cellulose can come very close to one another and form multiple intermolecular hydrogen bonds (sub-topic 4.4) between adjacent hydroxyl groups. As a result, cellulose fibres are insoluble in water, have high mechanical strength, and have lower dietary value than other carbohydrates.

Humans and other animals cannot digest cellulose because their bodies produce only α-glucosidases, enzymes that catalyse the hydrolysis of α-glycosidic links in starch and glycogen but not β-glycosidic links in cellulose. In contrast, many microorganisms produce cellulase and other β-glucosidases, which allows them to use cellulose as their principal source of food. Ruminants (such as cattle or sheep), horses, and some insects (such as termites) can extract energy and nutrients from plants and wood using cellulase-producing bacteria in their digestive systems.

Dietary fibre is a common name for cellulose and other indigestible plant materials, which are important components of a healthy diet in humans (sub-topic B.4). Although it cannot be metabolized by humans directly, dietary fibre affects the mechanical properties of food, cleans the intestine, facilitates the passage of food through the digestive system, and prevents constipation. By providing bulk to the diet, dietary fibre reduces appetite and helps to prevent obesity. In addition, dietary fibre regulates the absorption of sugars and bile acids, reducing the risk of diabetes and cholesterol-related heart disease (sub-topic B.4).

A certain amount of dietary fibre can be fermented by cellulase-producing bacteria in the large intestine of humans. These bacteria produce short-chain fatty acids and other metabolites, which help to prevent the development of various health conditions including hemorrhoids, diverticulosis, Crohn's disease, irritable bowel syndrome, and bowel cancer. In many countries dietary fibre is now considered an important macronutrient and is recommended by regulatory authorities for daily consumption.

Fatty acids and triglycerides

Naturally occurring unsaturated fatty acids have a *cis*-configuration of their double carbon–carbon bonds (sub-topic B.3). The hydrocarbon chains in such molecules cannot adopt linear conformations, which prevents them from coming close to one another and reduces intermolecular forces (sub-topic B.3, figure 1 and sub-topic 4.4). As a result *cis*-unsaturated fatty acids and their triglycerides, often referred to as ***cis*-fats**, are usually liquid at room temperature.

Hydrogenation of vegetable oils is commonly used in the food industry to produce saturated fats with high melting points. The reaction of unsaturated triglycerides with hydrogen is similar to the hydrogenation of alkenes (sub-topic 10.2) and takes place at high temperatures in the presence of a nickel or palladium catalyst (figure 13).

Hydrogenated vegetable oils are cheap and offer many benefits, including the absence of cholesterol, a controlled texture, spreadability, increased resistance to heat, and extended shelf life. However, the high temperature of the hydrogenation process leads to partial conversion of *cis*-fatty acid residues into their *trans*-isomers. If the hydrogenation is incomplete, the final product will contain *trans*-unsaturated triglycerides, known as ***trans*-fats**.

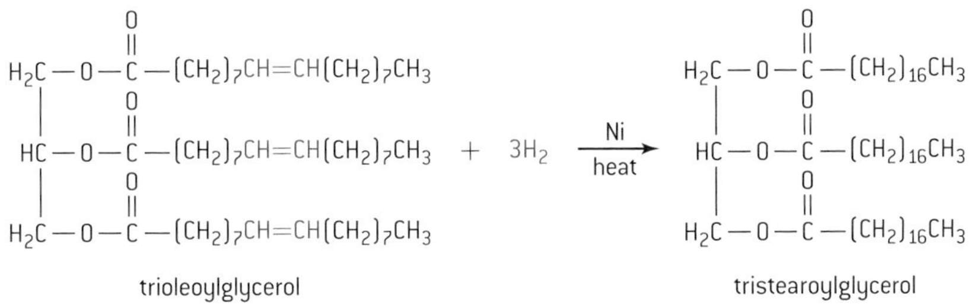

▲ Figure 13 The reaction of an unsaturated triglyceride with hydrogen

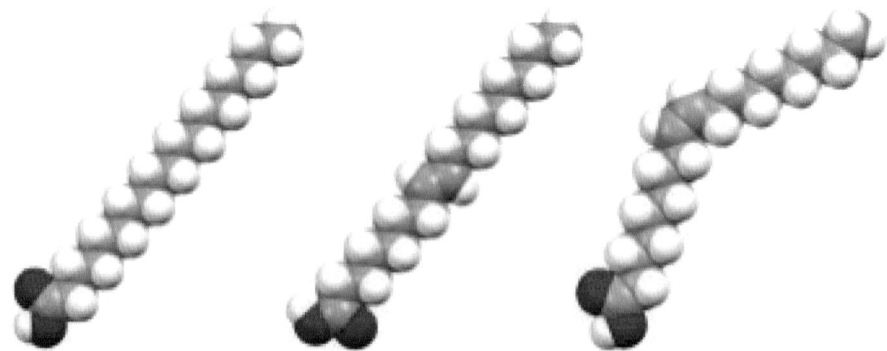

▲ Figure 14 Left to right: molecular models of saturated (stearic), *trans*-monounsaturated (elaidic), and *cis*-monounsaturated (oleic) acids

In contrast to their *cis*-isomers, the hydrocarbon chains of *trans*-fatty acids and their triglycerides are less distorted (figure 14) and can adopt rod-like conformations with stronger intermolecular forces. As a result, the melting points of *trans*-fatty acids are generally higher than those of *cis*-fatty acids but lower than the melting points of saturated fatty acids with the same number of carbon atoms (table 1).

Structural formula	Common name	IUPAC name	Melting point/°C
$\overset{\displaystyle H \qquad\qquad H}{\underset{\displaystyle CH_3(CH_2)_7 \qquad (CH_2)_7COOH}{C=C}}$	oleic acid	*cis*-octadec-9-enoic acid	14
$CH_3(CH_2)_7 \quad H$ $C=C$ $H \quad (CH_2)_7COOH$	elaidic acid	*trans*-octadec-9-enoic acid	43
$CH_3(CH_2)_5 \quad H$ $C=C$ $H \quad (CH_2)_9COOH$	vaccenic acid	*trans*-octadec-11-enoic acid	44
$CH_3(CH_2)_{16}COOH$	stearic acid	octadecanoic acid	70

▲ Table 1 Selected saturated and monounsaturated fatty acids. (The structures and names of other fatty acids are given in sub-topic B.3, table 1.)

Vaccenic acid (table 1) is one of the few naturally occurring *trans*-unsaturated fatty acids. It belongs to a rare family of ω−7 acids (sub-topic B.3) and comprises up to 5% of the total fat in cow and human milk. In contrast to other *trans*-fatty acids, a moderate intake of vaccenic acid and its derivatives is thought to be beneficial for human health, although the exact mechanism of its physiological action is still unknown.

Although trace amounts of *trans*-unsaturated fatty acids are present in all natural products, the levels of *trans*-fats in partly hydrogenated oils are particularly high (up to 15% in margarine and 20–30% in baking shortenings). Due to their increased heat resistance, partly hydrogenated oils are widely used in the fast food industry, resulting in additional intake of *trans*-fats from fried foods. Excessive consumption of *trans*-fats is linked to a high LDL cholesterol level (sub-topic B.3), which increases the risk of coronary heart disease. In addition, *trans*-unsaturated fatty acids take longer to metabolize than their saturated and *cis*-unsaturated analogues, so they accumulate in fatty tissues and increase the risk of obesity. According to some studies, a high level of *trans*-fats in the diet may be responsible for other health conditions, including diabetes and Alzheimer's disease.

The production and labelling of dietary products containing *trans*-fats is regulated by national laws, which vary greatly around the world. While in some countries, such as Austria and Iceland, artificial *trans*-fats are completely banned in the food industry, other countries do not restrict the use of *trans*-fats in any way and do not require producers to specify the levels of these compounds in foods. Such differences affect the international food trade and limit the ability of people to make informed dietary choices.

Retinal and vision chemistry

The collective name "vitamin A" refers to a group of polyunsaturated compounds with diverse biological functions (sub-topics B.5 and B.9). One of these compounds, **retinal**, is a long-chain aldehyde involved in vision chemistry. In the photoreceptor cells retinal exists as two stereoisomers, *cis*-retinal and *trans*-retinal, which can be converted into one another by the action of visible light or enzymes (figure 15).

The aldehyde group of *cis*-retinal can reversibly bind to a lysine residue of the protein **opsin**, producing a light-sensitive pigment **rhodopsin**:

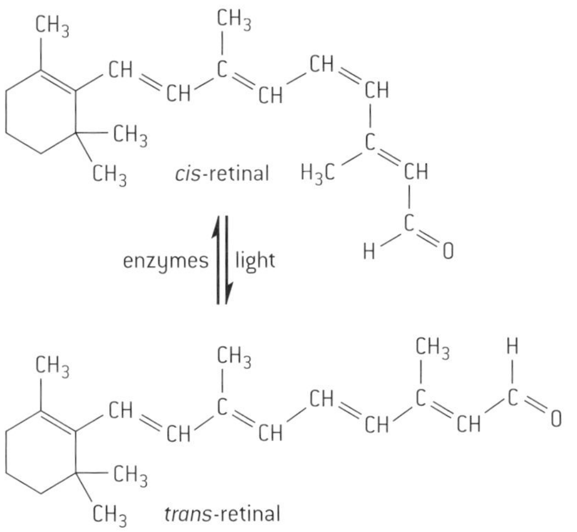

▲ Figure 15 The interconversion of *cis*- and *trans*-retinal

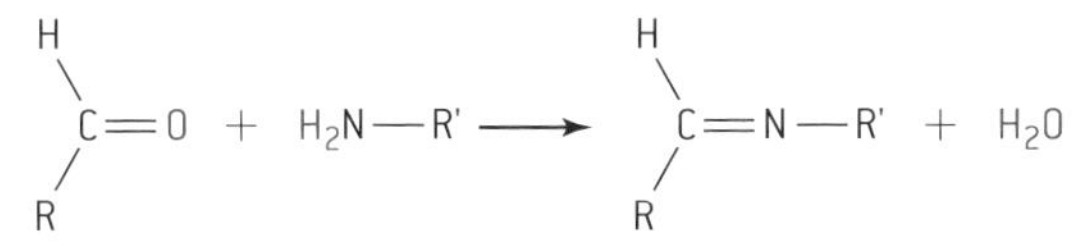

cis-retinal   opsin   rhodopsin

The C=N bond in rhodopsin extends the system of electron conjugation (sub-topic B.5) in *cis*-retinal and allows it to absorb visible light in the blue and green regions of the spectrum. As a result, pure rhodopsin has reddish purple colour and is often called "visual purple". Other proteins of the opsin family produce *cis*-retinal complexes with different absorption spectra, which are responsible for the colour vision of animals and humans.

When rhodopsin absorbs a photon of visible light, the residue of *cis*-retinal isomerizes into *trans*-retinal and the protein conformation changes, triggering a cell response that eventually sends an electrical

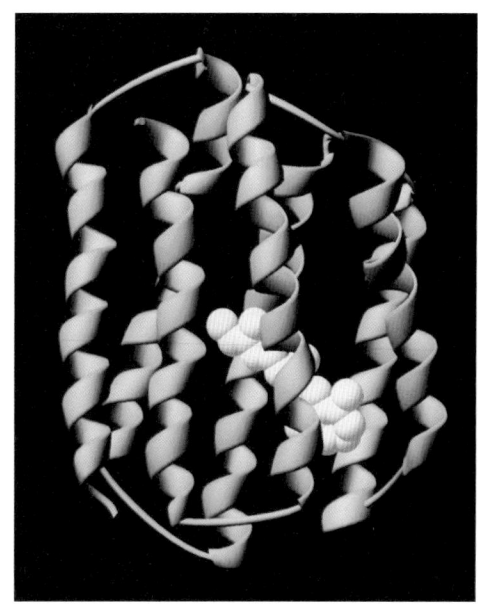

▲ Figure 16 Quaternary structure of rhodopsin, the complex of opsin (blue) with trans-retinal (yellow)

signal to the nervous system. At the same time, *trans*-retinal detaches from opsin and undergoes a series of enzymatic transformations known as the **visual cycle**. At the end of the visual cycle, *trans*-retinal is converted back into *cis*-retinal, which reattaches to opsin and produces a new functional rhodopsin complex.

In the human body, retinal can be synthesized only from other group A vitamins such as retinol or β-carotene (sub-topic B.5). Insufficient production of retinal caused by a vitamin A deficiency leads to night blindness, which is a common medical condition in many developing countries.

Questions

1. a) Amino acids can exist in D and L forms. Describe how the D form of alanine, $H_2NCH(CH_3)COOH$, differs in its physical properties from the L form. [1]

   b) Explain the D and L convention for describing amino acids and draw the D form of alanine to show clearly its three-dimensional structure. [3]

   IB, May 2012

2. Explain how the "CORN" rule can be used to identify an enantiomer of alanine as either D- or L-alanine. [3]

   IB, November 2012

3. Identify, by marking with asterisk (*) symbols, all chiral carbon atoms in the molecules of leucine, isoleucine, valine, proline, and threonine (sub-topic B.2, table 1). Remember that some amino acids contain more than one chiral centre.

4. The reason why only L-enantiomers of 2-amino acids are found in proteins remains a mystery. It is possible that this selection was made arbitrarily and then fixed firmly in evolutionary history.

   a) Discuss the role of random events in evolution.

   b) A photosynthesizing organism based on D-amino acids could potentially be created artificially. Discuss possible environmental implications of such an experiment.

5. The structures of vitamin C and the preservatives 2-BHA and BHT are shown in figure 17.

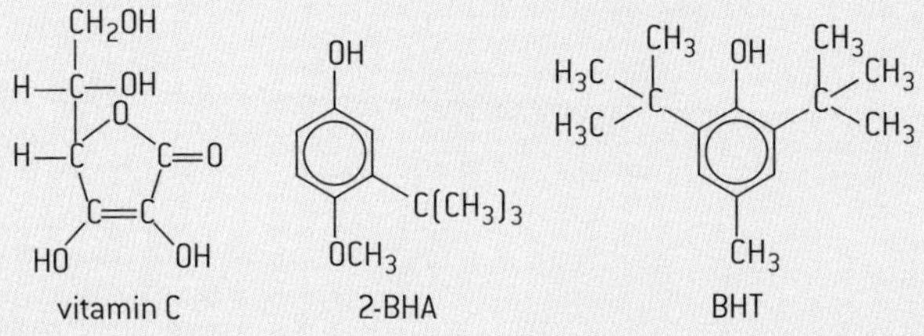

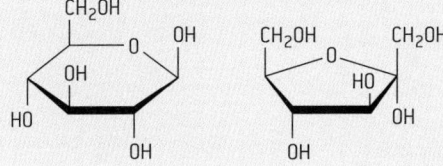

   ▲ Figure 17

   Determine any chiral carbon atoms in these three compounds by placing an asterisk, *, beside them on a copy of the figure. [2]

   IB, May 2012

6. Two stereoisomers with the same molecular formula often have very different biological activities. Explain, with reference to enzymatic catalysis, why only certain stereoisomers can be metabolized by living organisms.

7. Distinguish between the structures of α- and β-glucose. [1]

   IB, November 2010

8. The monosaccharide D-fructose is sweeter than D-glucose, so it can be used in smaller amounts in foods.

   a) Identify the stereoisomer (α- or β-) of the cyclic forms of glucose and fructose in figure 18.

   ▲ Figure 18

   b) Draw the Fischer projection of L-fructose.

   c) Galactose is a C-4 epimer of glucose. Draw the Fischer projection of D-galactose and the Haworth projection of β-galactose.

9. Starch and cellulose are polysaccharides found in many plants.

   a) Compare the structures of starch and cellulose. [3]

   b) Explain why humans cannot digest cellulose. [1]

   IB, November 2011

10 a) State what is meant by the term dietary fibre. [1]

b) Describe the importance of dietary fibre for a balanced diet and the prevention of various health conditions. [3]

IB, November 2012

11 State the names and structural formulae of fatty acids that can be present **only** in:

(a) saturated fats;

(b) *cis*-fats;

(c) *trans*-fats.

12 Fats and vegetable oils are triesters of glycerol and fatty acids.

a) State the conditions required for the hydrogenation of unsaturated oils. [2]

b) Hydrogenation can result in the formation of *trans* fatty acids. Outline the meaning of the term *trans* fatty acids and explain why their formation is undesirable. [2]

IB, May 2011

13 Retinal can reversibly bind to a protein, opsin, to produce a biological pigment "visual purple".

a) State another name for the "visual purple" pigment.

b) Outline the role of this pigment in vision.

C ENERGY

Introduction

All societies depend on energy resources. We extract energy from sunlight, plants, petrochemicals, wind, water, and other sources and convert it to forms that are useful to us; however with each conversion the quality is degraded as some of the available energy is dispersed or converted to heat. Converting energy from one form to another in the world around us results from potential and kinetic energy changes at the molecular level. Exothermic reactions can release potential energy and raise the kinetic energy of the surrounding molecules. The usefulness or quality of the energy becomes lessened the more it is dispersed.

C.1 Energy sources

Understandings

→ A useful energy source releases energy at a reasonable rate and produces minimal pollution.

→ The quality of energy is degraded as heat is transferred to the surroundings. Energy and materials go from a concentrated into a dispersed form. The quantity of the energy available for doing work decreases.

→ Renewable energy sources are naturally replenished. Non-renewable energy sources are finite.

→ Energy density is energy released from fuel/ volume of fuel consumed.

→ The efficiency of an energy transfer is expressed as useful energy output/ total energy input × 100%.

Applications and skills

→ Discuss the use of different sources of renewable and non-renewable energy.

→ Determine the energy density and specific energy of a fuel from the enthalpies of combustion, densities and the molar mass of fuel.

→ Discuss how the choice of fuel is influenced by its energy density or specific energy.

→ Determine the efficiency of an energy transfer process from appropriate data.

Nature of science

→ Use theories to explain natural phenomena— energy changes in the world around us result from potential and kinetic energy changes at the molecular level.

Energy sources: quality and efficiency

What makes a good energy source? It needs not only to contain a large quantity of potential energy but also for this potential energy to be released or converted, at a reasonable rate, to a useful form with minimal pollution and unwanted products. If the conversion is too fast a large quantity of the energy is dispersed, while if it is too slow it is not useful.

The combustion of glucose is an exothermic reaction:

$$C_6H_{12}O_6(s) + 6O_2(g) \rightarrow 6CO_2(g) + 6H_2O(l) \quad \Delta H = -2803 \text{ kJ}$$

The same amount of energy is released when glucose is burnt in a bomb calorimeter as is released by its oxidation in the human body. The slower rate of oxidation in the body allows the energy to be converted to a useful form whereas the rapid oxidation of combustion disperses the energy too quickly, lowering its quality.

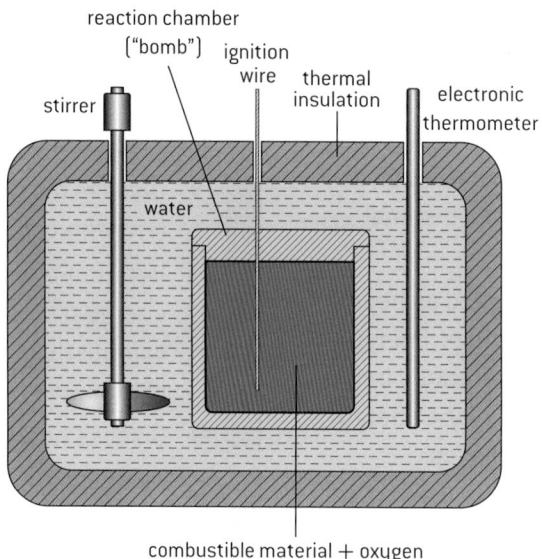

▲ Figure 1 Combustion in a bomb calorimeter is rapid, resulting in the potential energy being dispersed

The term "quality of energy" can have different meanings. Energy companies, for example, may consider the cost per unit energy more important than the efficiency of its conversion. The efficiency of producing electricity from burning coal averages approximately 30% worldwide. This means that 30% of the available thermal energy produced from burning coal becomes electricity. There are also by-products including greenhouse gases and pollutants. Nevertheless, according to the **International Energy Agency** the cost of obtaining electricity from coal is 7% less than from gas and 19% less than from nuclear sources.

All energy conversions undergo some form of quality degradation as some of the energy is dispersed as heat. The energy and materials in the original source change from a concentrated to a dispersed form and the energy available to do useful work diminishes.

The more the quality of energy is degraded, the less **efficient** the fuel is:

$$\text{efficiency of energy transfer} = \frac{\text{useful output energy}}{\text{total input energy}} \times 100\%$$

The **International Energy Agency** is an autonomous organization that works to ensure reliable, affordable, and clean energy for its member countries and beyond.

The distribution of available energy among the particles of a material is known as **entropy**. The more different ways the energy can be distributed, the higher the entropy, and the less energy is available to do useful work.

Worked example

a) Compare the efficiency of coal, oil, or gas for use in home heating. Table 1 gives some typical efficiencies for the conversions in the process.

Conversion	Coal	Oil	Gas
extraction of raw material	0.67	0.35	0.72
processing to a usable form	0.92	0.88	0.97
transporting the fuel to a power station	0.98	0.95	0.95
chemical potential energy to electricity in a power plant	0.35	0.35	0.35
transmission of electricity and conversion to heat in the house	0.90	0.90	0.90

▲ Table 1 The efficiency of some energy conversions in the generation, distribution, and use of electrical energy

b) Suggest a reason why the efficiencies of the last two conversions are the same for each energy source.

Solution

a) We need to combine (multiply) the efficiencies of all the processes from extracting the fuel to converting electricity to heat in the home.

The efficiency of coal as a fuel is:

efficiency = $0.67 \times 0.92 \times 0.98 \times 0.35 \times 0.90$
= 0.19 or 19% efficient

81% of the chemical potential energy available in coal is dispersed and is not used in heating the house.

You should be able to verify for yourself that oil is only 9% and natural gas 21% efficient.

b) Electricity is generated in the same way in each power plant. The fuel boils water to produce steam; this turns turbines which generate electricity. The energy losses are approximately 65% in each case: only 35% of the initial chemical potential energy is converted to useful electrical energy. This electrical energy is transported to the house and converted to heat in a heater in the same way irrespective of the initial fuel source. In this example, it is assumed that gas needs to be used to generate electricity, and then electricity is transmitted to homes. However, gas can be transported to homes and burned in gas furnaces (table 2), which increases the efficiency from 21% to about 56%.

Device	Energy transformation	Efficiency
electric heater	electrical → thermal	nearly 100%
battery	chemical → electrical	~90%
home gas furnace (boiler)	chemical → thermal	~85%
home oil furnace (boiler)	chemical → thermal	~65%
home coal furnace (boiler)	chemical → thermal	~55%
solar cell	light → electrical	~15%
incandescent light bulb	electrical → light	~5%

▲ Table 2 The relative efficiencies of some energy conversions

Energy density and specific energy

The energy density is a useful measure of the quality of a fuel, that compares the energy released per unit volume of fuel:

$$\text{energy density} = \frac{\text{energy released from fuel}}{\text{volume of fuel consumed}}$$

> Note that because the definitions of energy density and specific energy are energy *released* per unit mass/volume, these quantities do not have a negative value.

Study tip

You should be able to convert energy densities to any units required, such as kJ cm^{-3}. See if you can verify that 7.44×10^7 kJ m^{-3} is 74.4 kJ cm^{-3}.

The **International Renewable Energy Agency** (IRENA), based in Abu Dhabi, UAE, was founded in 2009 to promote increased adoption and sustainable use of renewable energy sources (bioenergy, geothermal energy, hydropower, ocean, solar, and wind energy).

In a similar way the specific energy is the energy contained per unit mass of a fuel:

$$\text{specific energy} = \frac{\text{energy released from fuel}}{\text{mass of fuel consumed}}$$

Worked example

The standard enthalpy of combustion of carbon is -394 kJ mol^{-1}. The density of anthracite, one of the purest coals, is 2267 kg m^{-3}. Use this information along with the relative atomic mass of carbon to calculate the energy density and specific energy of this form of coal, assuming it to be 100% carbon.

Solution

specific energy = -394 kJ mol^{-1}/12.01 g mol^{-1} = 32.8 kJ g^{-1}

convert to kJ kg^{-1}: 32.8 kJ g^{-1} × 1000 g kg^{-1} = 32 800 kJ kg^{-1}

energy density = 32 800 kJ kg^{-1} × 2267 kg m^{-3} = 7 435 7600 kJ m^{-3}

Expressed in scientific notation to 3 SF (as the enthalpy of combustion was given to 3 SF) this is 7.44×10^7 kJ m^{-3}.

Renewable energy resources

Some renewable or "green" energy resources include solar energy, wind energy, biomass, water (such as tides, currents, and waves), geothermal energy, and fuel cells.

Geothermal energy is one of the more widely used commercial forms of renewable energy resources. Although it has an efficiency of only about 23%, as with all energy resources it is important to consider not only the efficiency of conversion but also the cost per kilowatt-hour.

▲ Figure 2 A thermal energy production plant in Iceland. Iceland generates 100% of its energy from renewable resources

Questions

1 Decide whether each of the following is true or false.

    a) The energy conversion in an automobile is to convert heat to kinetic energy.

    b) The conversion of heat to electricity is usually more efficient than that of electricity to heat.

    c) The final conversion step in most commercial power plants is work (kinetic energy) to electricity.

    d) Nuclear energy is a renewable resource.

    e) Green energy resources are sustainable, renewable, and produce low pollution.

2 Ethanol is a fuel produced from plant products by fermentation. It has a density of 789 g dm^{-3} and its enthalpy of combustion is -1367 kJ mol^{-1}.

    a) Calculate the energy density for ethanol.

    b) Calculate the specific energy for ethanol.

    c) Write a balanced equation for the combustion of ethanol and state the amount, in mol, of carbon dioxide produced per mole of ethanol burned.

    d) Explain why this method is still considered "green" chemistry even though it produces carbon dioxide in the combustion reaction.

C.2 Fossil fuels

Understandings

→ Fossil fuels were formed by the reduction of biological compounds that contain carbon, hydrogen, nitrogen, sulfur and oxygen.

→ Petroleum is a complex mixture of hydrocarbons that can be split into different component parts called fractions by fractional distillation.

→ Crude oil needs to be refined before use. The different fractions are separated by a physical process in fractional distillation.

→ The tendency of a fuel to auto-ignite, which leads to "knocking" in a car engine, is related to molecular structure and measured by the octane number. The performance of hydrocarbons as fuels is improved by the cracking and catalytic reforming reactions.

→ Coal gasification and liquefaction are chemical processes that convert coal to gaseous and liquid hydrocarbons.

→ A carbon footprint is the total amount of greenhouse gases produced during human activities. It is generally expressed in equivalent tons of carbon dioxide.

 ## Applications and skills

→ Explain the effect of chain length and chain branching on the octane number.

→ Write equations for cracking and reforming reactions, coal gasification and liquefaction.

→ Identify various fractions of petroleum based on volatility and uses, their relative volatility and their uses.

→ Discuss advantages and disadvantages of different fossil fuels.

→ Calculate carbon dioxide production, when different fuels burn and determine carbon footprints for different activities.

 ## Nature of science

→ Scientific community and collaboration – the use of fossil fuels has had a key role in the development of science and technology.

▲ Figure 1 Photosynthesis is the main source of building strong hydrocarbon bonds which form the basis of today's fossil-fuel energy production

Storing energy from photosynthesis

The harnessing of energy from the sun by photosynthesis enabled the emergence of large organisms. As these organisms died out the strong C–C and C–H bonds in them remained intact and these are the source of our main energy supply today.

Energy drives development

The drive for energy has meant that much collaboration and technical development is needed to extract the oil, coal, or gas from often difficult locations. This impetus has led to many innovations in our society that would not have otherwise occurred. International collaboration is necessary for ocean drilling, pipeline construction, and dealing with oil spills.

The discovery of the origins of crude oil gives a fascinating look into the nature of science. The idea that deep carbon deposits existed in the origins of the Earth rather than being of biological origin is still shared by some people today. It was accepted by Dmitri Mendeleev, although the biogenic hypothesis put forth by Mikhail Lomonosov in 1757 is the most widely accepted theory. Could the origin of these fuels influence where and how we look for them?

▲ Figure 2 Mikhail Lomonosov first proposed the idea that oil and gas are 'fossil' fuels

Worked example

Calculate the oxidation states of carbon in methane and methanol and show that carbon in methane is in a more reduced form.

Solution

The oxidation states are deduced as follows:

CH_4: $1C + 4H = 0$;
$C + 4(+1) = 0$; $C = -4$

CH_3OH: $1C + 4H + 1O = 0$;
$C + 4(+1) + (-2) = 0$;
$C = -2$

The oxidation state for carbon is -4 in methane and -2 in methanol, showing that carbon is in a more reduced state in methane.

Fossil fuels store reduced carbon

The formation of fossil fuels from decaying organisms is an example of reduction. You will recall that oxidation can be considered as oxygen gain/hydrogen loss (topic 9) while reduction is hydrogen gain/oxygen loss. Many fossil fuels contain saturated alkanes. During fossil fuel formation carbon atoms become more and more saturated with hydrogen and have fewer bonds to nitrogen, sulfur, and/or oxygen than existed in the living form. The carbon–hydrogen bond is relatively stable and stronger than single bonds between carbon and oxygen, sulfur, or nitrogen (see section 11 of the *Data booklet*).

Crude oil: Fractionating and cracking

There are three main fossil fuels: coal, gas, and crude oil. Crude oil or petroleum is by far the most important yet this "black gold" is difficult to use in its natural form. It contains a vast mixture of hydrocarbons of varying chain lengths. Long-chain hydrocarbons have stronger van der Waals' intermolecular forces between them than do the shorter chains, so their differing boiling points can be used to separate crude oil into "fractions" of various chain lengths. At oil refineries the various fractions are separated by distillation (figure 3).

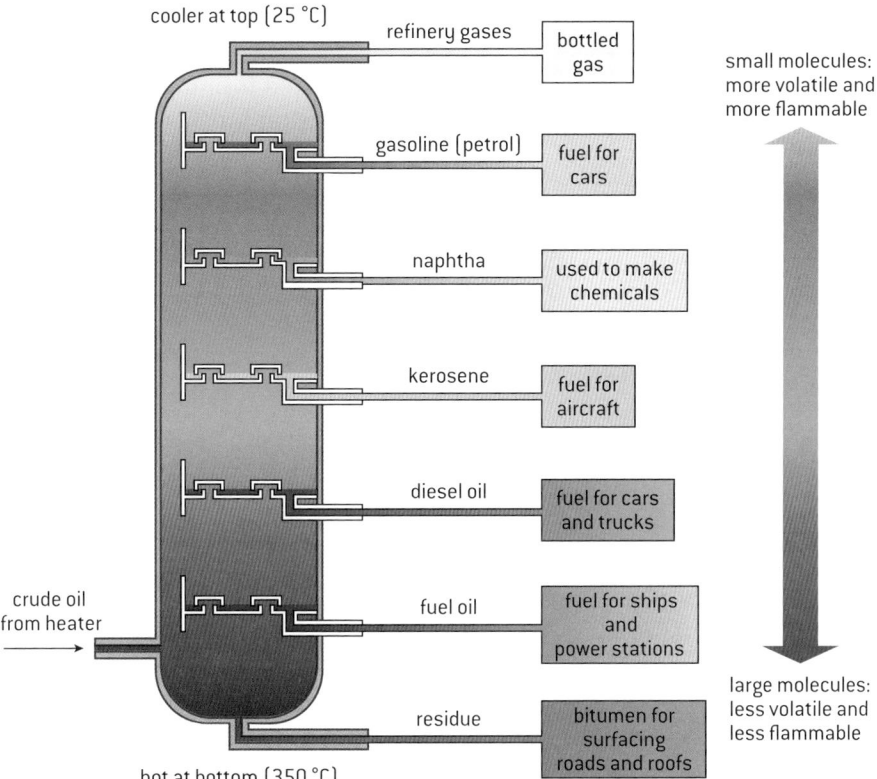

cooler at top (25 °C)

refinery gases — bottled gas

gasoline (petrol) — fuel for cars

naphtha — used to make chemicals

kerosene — fuel for aircraft

diesel oil — fuel for cars and trucks

crude oil from heater

fuel oil — fuel for ships and power stations

residue — bitumen for surfacing roads and roofs

hot at bottom (350 °C)

small molecules: more volatile and more flammable

large molecules: less volatile and less flammable

▲ Figure 3 A fractionating column used to separate crude oil into commercially useful fractions

The crude oil is first heated to make it less viscous, and fed into the bottom of the fractionating column. Temperatures are lower at the top, so low boiling point substances leave the column there whereas the fractions with higher boiling points condense at higher temperatures near the bottom. These longer-chain hydrocarbons are more viscous, darker in colour, and because they are less volatile they have lower flammability.

The more volatile shorter-chain hydrocarbons make better fuels and they burn with a cleaner flame. However there is a much larger percentage of long-chain hydrocarbons in crude oil than short-chain ones. In order to obtain more of the desired short-chain fuels a process called **cracking** is employed. Fractions such as naphtha that contain longer-chain hydrocarbons are heated over a catalyst where they are "cracked" into smaller hydrocarbons including alkenes such as ethene and the more usable alkanes such as the octanes used in petrol (figure 4). Cracking was initially carried out by steam alone; alumina and silica catalysts were then employed. Today zeolites are used as they are more selective in producing the higher octane C5–C10 range of hydrocarbons with more branched hydrocarbons (see sub-topic A.3 for more on zeolites).

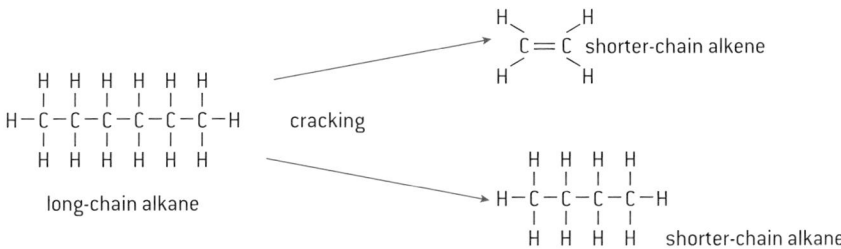

▲ Figure 4 Cracking converts longer-chain hydrocarbons into more useful shorter-chain alkenes and alkanes

Worked example

A C_{15} alkane is heated over a catalyst and cracked forming ethene, propene, and octane. Deduce a balanced equation for this cracking reaction.

Solution

Ethene has 2 carbons, propene has 3, and octane has 8, adding up to 13 carbons. Therefore 2 molecules of ethene must be formed per molecule of $C_{15}H_{32}$ cracked (figure 5).

$$C_{15}H_{32} \rightarrow 2C_2H_4 + C_3H_6 + C_8H_{18}$$

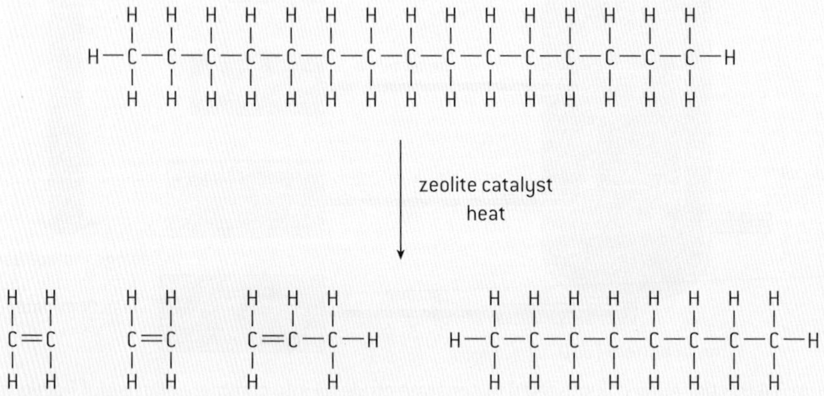

zeolite catalyst
heat

▲ Figure 5

Different octane ratings can be applied in different countries. The octane rating described here is the **research octane number** (RON) which is used in the, Europe, South Africa, and Australia. The **motor octane number** (MON) is typically used in motor sports applications where engines operate under more stressful conditions. The **pump octane number** (PON) is the average of the RON and MON, used in Canada and the USA.

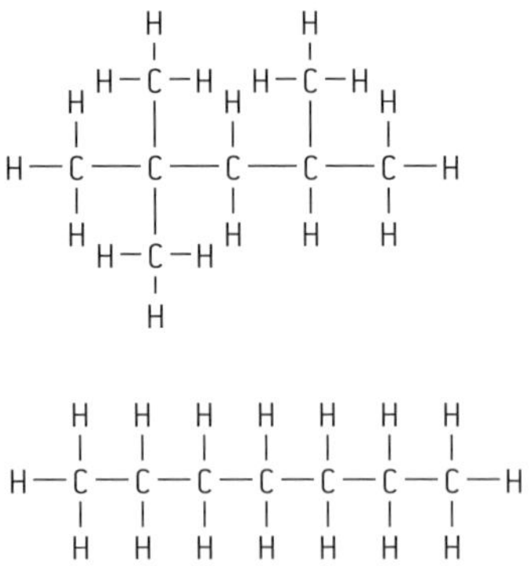

▲ Figure 6 The highly branched 2,2,4-trimethylpentane (top) has an octane rating of 100 whereas heptane (bottom) has an octane rating of 0

Fuels and octane rating

When fuels are burned in automobile engines they are first compressed and then ignited with a spark. Some hydrocarbons have a higher tendency than others to "auto-ignite" during this compression stage. This produces an effect known as "knocking" which can severely damage engines. A measure of the fuel's ability to resist auto-ignition is its **octane rating** (figure 6). A fuel with an octane rating of 87 would have the same "knocking" effect as a mixture of 87% 2,2,4-trimethylpentane and 13% heptane.

Higher-octane fuels can therefore be compressed more and give better performance than fuels with lower octane ratings. Commercial octane boosters added to fuels may contain toluene (methylbenzene) with an octane rating of about 114. You will recall that toluene is an aromatic compound so its use in fuels is limited for environmental reasons.

Petrol or gasoline is a mixture of many different straight- and branched-chain alkanes (aliphatics), cyclic alkanes, and aromatics, but contains no alkenes. It is composed of about 50% aliphatics and 20–30% each of cyclic alkanes and aromatics. The length and degree of branching of the hydrocarbon chain have the following effect on the octane rating:

- Octane rating increases with branching. 2,2,4-trimethylpentane has a higher octane rating than octane. They both contain 8 carbon atoms but the more highly branched 2,2,4-trimethylpentane is more resistant to auto-ignition.

- Octane rating decreases with length of carbon chain. Hexane has a higher octane rating than heptane.

- The octane rating of aromatics is higher than that of straight-chain or branched-chain alkanes with the same number of carbons. Benzene with 6 carbons has a higher octane rating than either hexane or 2-methylpentane.

Catalytic reforming

Catalytic reforming is used to convert low-octane numbered alkanes such as heptane or octane into higher-octane numbered isomers such as methylbenzene or 2,2,4-trimethylpentane. The straight-chain alkanes are isomerized by heating with a platinum catalyst. Their chains break apart and reform, increasing the proportion of branched alkanes. The products are passed over zeolite, which serves as a molecular sieve type catalyst, separating the branched and unbranched alkanes:

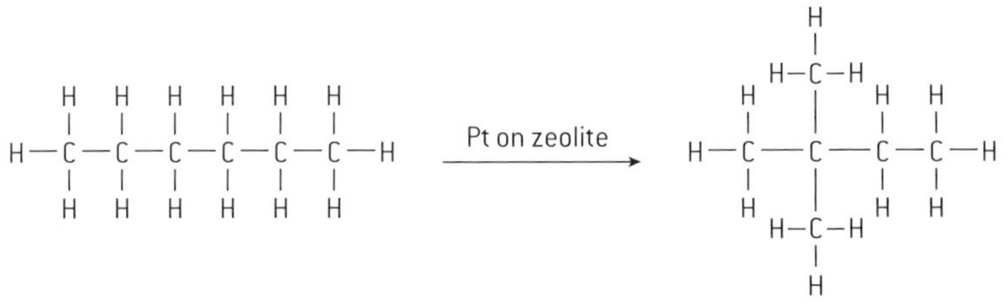

Using a platinum catalyst with aluminium oxide, or other metal catalyst, reforms and dehydrogenates the alkane into an aromatic compound. For example, heptane can be converted into methylbenzene and hydrogen gas:

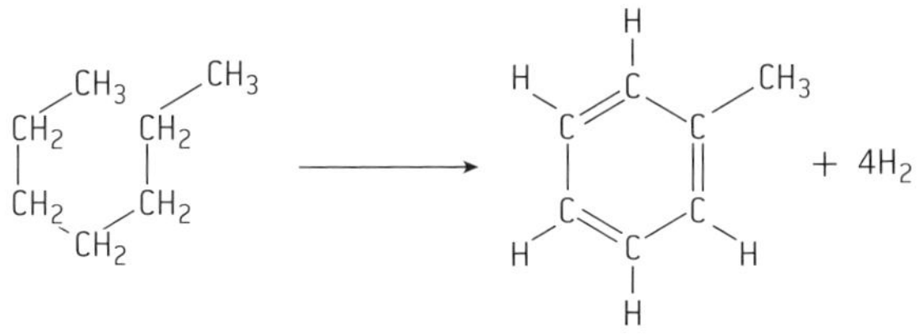

heptane   methylbenzene

Reforming is the summative effect of several reactions such as cracking, unifying, polymerizing, and isomerizing occurring simultaneously. It is used to produce high-octane alkanes or other useful aromatics such as methylbenzene. At an oil refinery crude oil is treated by a combination of distillation, cracking, and reforming to produce the valuable products that drive our society today.

Greener energy

Many developments in fuel technology have emerged in response to the need to limit pollution and greenhouse emissions. Table 1 gives some examples.

Worked example

Calculate the octane rating of a fuel with 80% 2,2,4-trimethylpentane, 10% heptane, and 10% toluene.

Solution

Calculate the weighted averages of the three components:

$$\frac{80}{100} \times 100 + \frac{10}{100} \times 0 +$$
$$\frac{10}{100} \times 114 = 91.4$$

The fuel has an octane rating of about 91.

▲ Figure 7 A mixture of straight-chain C_5–C_{12} hydrocarbons is referred to as naphtha. The naphtha is processed to form branched or aromatic hydrocarbons with the same number of carbon atoms during reforming

Aim	Examples	Advantages
remove sulfur from fossil fuels such as coal	scrubbing, filters and engineered polymers with receptor sites for sulfur compounds	reduces sulfur (SO_2) emissions which could cause acid rain; the sulfur extracted can be used in sulfuric acid production
produce fuels with lower environmental impact	remove lead, benzene and sulfur from petrol; use of catalytic converters in cars	reduces emissions of NO_x, CO, SO_2, lead oxides and carcinogenic benzene
produce alternative or blended petrochemical fuels	mix ethanol with petrol, develop engines that run on LPG (liquefied petroleum gas) or methane	reduces CO_2 emissions, lowers carbon footprint, reduces emissions of NO_x and CO
develop renewable and alternative resources and technologies	bioethanol, biodiesel, electric cars, hybrid cars, fuel cells	reduce dependence on oil, move towards carbon-neutral fuels which absorb CO_2 as they grow (corn, etc.), so are more renewable and sustainable

▲ Table 1 Developments to make the use of fuels "greener"

Coal gasification

Coal is a more abundant fossil fuel than crude oil, and can be converted to other more useful forms that are cheaper than crude oil. One method is **coal gasification** in which **synthesis gas**, also called coal gas or syngas, is produced by reacting coal with oxygen and steam in a gasifier to create hydrocarbons. Inside the gasifier the oxygen reaching the coal is limited so that combustion will not occur.

Coal gasification may occur in a cavity underground, giving low plant costs as no gasifier needs to be built, the coal does not have to be lifted to the surface, and the carbon dioxide formed can be stored underground rather than being released to the atmosphere. This is an example of **carbon capture and storage (CCS)** which involves capturing carbon dioxide from large industrial processes, compressing it, and transporting it to be injected deep into rock formations at selected safe sites. This reduces the amount of carbon dioxide entering the atmosphere.

Pollutants are "washed out" of the synthesis gas leaving a relatively clean efficient fuel. The process of coal gasification is summarized in figure 8 and table 2.

Gasification produces other products including slag which is used in roofing materials or for road construction, methanol, and nitrogen-based compounds for fertilizers.

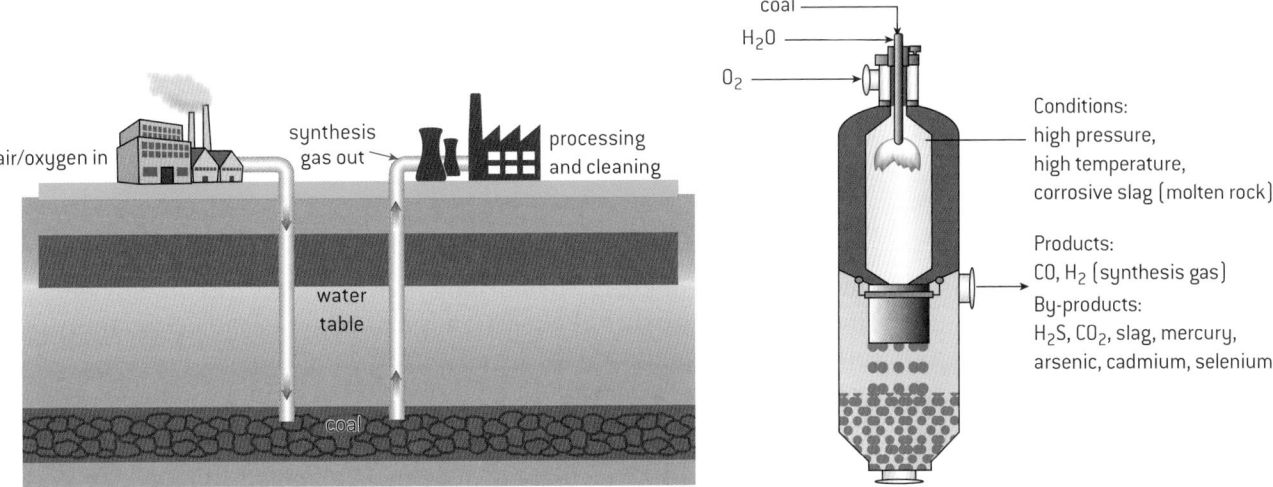

▲ Figure 8 The process of coal gasification which may occur underground or in a gasifier

Reaction	Conditions	Comments
coal → $CH_4 + H_2O$ $C + H_2$ $CO + CO_2$ various hydrocarbons	some oxygen, temperature that will not allow combustion	The first step is **pyrolysis**. Coal is dried and degraded into several gases and char, a charcoal-like substance. This is partial oxidation and generates oxidized compounds including CO and CO_2.
$C + H_2O → CO + H_2$ $C + CO_2 → 2CO$ $CO + 3H_2 → CH_4 + H_2O$ $CO + H_2O → CO_2 + H_2$	increased temperature, decreased oxygen, steam	This is **reduction**. Here synthesis gas (mainly CO and H_2) is produced which can be burnt to generate electricity.
$C + O_2 → CO_2$	Synthesis gas and other desired materials are run through a cooling chamber and removed. Remaining char is burnt off and CO_2 and impurities can be removed and stored underground.	The last stage is the gas **clean-up** in which the desired products are purified and removed. Cleaning can produce other useful materials too.

▲ Table 2 The main reactions that occur during coal gasification

Gasification is not limited to coal; it can be carried out with wood or other biomass materials and has been in use since the late 1790s. The earliest forms of town lighting used gas lights fuelled by gas obtained from coal.

Coal liquefaction

The process of coal liquefaction takes filtered and cleaned synthesis gas and adds water or carbon dioxide over a catalyst. This process is known as **indirect coal liquefaction** (ICL). In **direct coal liquefaction** (DCL) hydrogen, H_2 is added to heated coal in the presence of a catalyst. Both methods adjust the carbon-to-hydrogen ratio and produce synthetic liquid fuels via a process known as the Fischer–Tropsch process, shown by the general equation:

$$n\text{CO} + (2n + 1)\text{H}_2 \xrightarrow{\text{catalyst}} \text{C}_n\text{H}_{(2n + 2)} + n\text{H}_2\text{O}$$

These methods do not necessarily need coal as a feedstock: biofuels can be used to produce synthetic fuels in the same way. For example, methane gas can be converted to synthesis gas by the addition of water. This synthesis gas can then be used to manufacture desirable fuels.

$$CH_4 + H_2O \xrightarrow{\text{catalyst}} CO + 3H_2$$

"Green" fuels and the carbon footprint

The production of energy by burning fuels produces carbon dioxide. The **carbon footprint** of a reaction is a measure of the net quantity of carbon dioxide produced by the process. Even though biofuels may cost more to produce, their carbon footprint is less because carbon dioxide is absorbed by photosynthesis while the fuel is growing.

▲ Figure 9 An algae-growing system used to make ethanol and biodiesel. Algae and other "green" systems also aid carbon capture and storage (CCS)

Worked example

Calculate the carbon footprint, in tonnes of carbon dioxide, of burning 1000 kg (1 tonne) of octane. For simplicity, use integer values of molar masses for this calculation.

Solution

$$C_8H_{18} + 12\tfrac{1}{2}O_2 \rightarrow 8CO_2 + 9H_2O$$

$n(C_8H_{18}) = 1000\,000 \text{ g} / 114 \text{ g mol}^{-1} = 8772 \text{ mol}$

$8772 \text{ mol} \times 8 = 70\,175 \text{ mol } CO_2$

$70\,175 \text{ mol } CO_2 \times 44 \text{ g mol}^{-1} = 3\,087\,700 \text{ g}$

So approximately 3 tonnes of carbon dioxide are introduced to the atmosphere, released from carbon that was previously locked in the Earth in the form of oil.

Questions

1   Write balanced chemical equations, and predict products where necessary, for the following processes:

a) the cracking of $C_{12}H_{26}$ into two ethene molecules and an alkane

b) the reaction of char, C, with steam to produce carbon monoxide and hydrogen gas

c) the production of methane from synthesis gas, CO and H_2

d) catalytic reforming of heptane into methylbenzene

e) liquefaction of synthesis gas, CO and H_2 to produce liquid heptane.

2   Calculate the mass of carbon dioxide produced per gram of the following fuels burned:

a) ethanol

b) methane

c) 2,2,4-trimethylpentane.

3   The enthalpies of combustion of the three fuels given in question 2 are:

ethanol −1367 kJ mol⁻¹; methane −891 J mol⁻¹; 2,2,4-trimethylpentane −5460 kJ mol⁻¹

calculate the energy released in burning 1 g (the specific energy) of each fuel.

4   Use your answers to questions 2 and 3 to discuss the advantages and disadvantages of each fuel and the problems faced by society in using "green" fuels.

C.3 Nuclear fusion and fission

Understandings

→ Light nuclei can undergo fusion reactions as this increases the binding energy per nucleon.

→ Fusion reactions are a promising energy source as the fuel is inexpensive and abundant, and no radioactive waste is produced.

→ Absorption spectra are used to analyse the composition of stars.

→ Heavy nuclei can undergo fission reactions as this increases the binding energy per nucleon.

→ The iron group has the highest binding energy per nucleon. The further away (lighter or heavier) the more energy can be released by fusing the lighter nuclei or splitting the heavier ones.

→ U-235 undergoes a fission chain reaction:

$$^{235}_{92}U + ^{1}_{0}n \rightarrow ^{236}_{92}U \rightarrow X + Y + \text{neutrons.}$$

→ The critical mass is the mass of fuel needed for the reaction to be self-sustaining.

→ Pu-239, used as a fuel in "breeder reactions", is produced from U-235 by neutron capture.

→ Radioactive waste may contain isotopes with long and short half-lives.

→ Half-life is the time it take for one half of the radioactive substance to undergo decay.

Applications and skills

→ Construct nuclear equations for fusion and fission reactions.

→ Explain fusion and fission reactions in terms of binding energy per nucleon.

→ Explain the atomic absorption spectra of hydrogen and helium, including the relationships between the lines and electron transitions.

→ Discuss the storage and disposal of nuclear waste.

→ Solve radioactive decay problems involving integral numbers of half-lives.

Nature of science

→ Assessing the ethics of scientific research – widespread use of nuclear fission for energy production would lead to a reduction in greenhouse gas emissions. Nuclear fission is the process taking place in the atomic bomb and nuclear fusion that in the hydrogen bomb.

The discovery of nuclear fusion

Helium was discovered in the sun by observing the sun's spectra during a solar eclipse in 1868. This was made possible because of developments in spectroscopy in 1859. After Einstein's revelation that mass can be converted directly into energy ($E = mc^2$), observations of radiation from the sun drew the conclusion that nuclear fusion reactions fuel the sun.

Scientific advances often have important ethical and political implications. It was the race for nuclear weapons that helped us to understand nuclear transformations and use controlled fission in nuclear power plants. Nuclear fusion could provide the world with clean, greenhouse gas-free energy, but at what other costs?

▲ Figure 1 The mushroom cloud from the first test of a hydrogen fusion bomb, 1952. The energy released was more than the total of all the explosives detonated in the entire duration of the Second World War

▲ Figure 2 A high-powered laser employed in experiments aimed at producing controlled nuclear fusion

Hydrogen fusion

The fusion of hydrogen nuclei is the source of the sun's energy. This **fusion reaction** (figure 1) releases much more energy than the fission of U-235 or Pu-239, the fuels used in nuclear reactors.

The fusion of hydrogen nuclei to form helium releases tremendous heat and almost no nuclear waste; however it takes a vast amount of energy to initiate the reaction. Hydrogen bombs use a nuclear fission reaction – a small atomic bomb – to provide this energy. The heat released comes from nuclear fusion, but with associated nuclear fallout from the fission reaction.

There exists an abundance of fuel for nuclear fusion, and the lack of waste products makes it an attractive prospect for energy generation. However, there are huge technological issues involved – fusion takes place at such a high temperature that no material can contain it. Nevertheless research into hydrogen fusion continues (figure 2).

It was initially believed that the sun's energy came from some sort of combustion reaction, or from gravitational potential energy due to its massive size being converted to thermal energy. According to these theories the sun would last a few thousand to a few million years. It wasn't until after Einstein's theory of relativity that scientists came to understand nuclear fusion. So, where does the sun's energy come from?

In the sun hydrogen nuclei or protons combine to form the isotope deuterium ^{2}H, which then further combines to form helium nuclei. You will recall that a helium nucleus is composed of 2 protons and 2 neutrons. However, the mass of a helium nucleus is less than the sum of the masses of 2 protons and 2 neutrons. This is known as the **mass defect** (figure 3).

Mass has not been conserved. The missing mass (mass defect) has been converted directly into energy, the amount of which can be predicted using $E = mc^2$. The energy released E is a product of the mass that is lost m times the square of the speed of light c, which is a constant at $3.00 \times 10^8 \, \text{m s}^{-1}$.

Worked example 1

A proton has a rest mass of 1.672622×10^{-27} kg and a neutron has a rest mass of 1.674927×10^{-27} kg. A helium nucleus has a rest mass of $6.644\,77 \times 10^{-27}$ kg. Calculate the sum of the masses of 2 protons + 2 neutrons and use this to calculate the mass defect of the helium nucleus.

Solution

mass defect $= (2 \times 1.672622 + 2 \times 1.674927 - 6.64477 \times 10^{-27}$ kg

$$= 5.0328 \times 10^{-29} \, \text{kg}$$

Worked example 2

Use $E = mc^2$ to calculate the energy released in forming a helium nucleus from the previous worked example.

Solution

The mass defect, 5.0328×10^{-29} kg, has been converted directly to energy.

$$E = (5.0328 \times 10^{-29}\,\text{kg}) \times (3.00 \times 10^8\,\text{m/s})^2$$
$$= 4.52952 \times 10^{-12}\,\text{J}$$

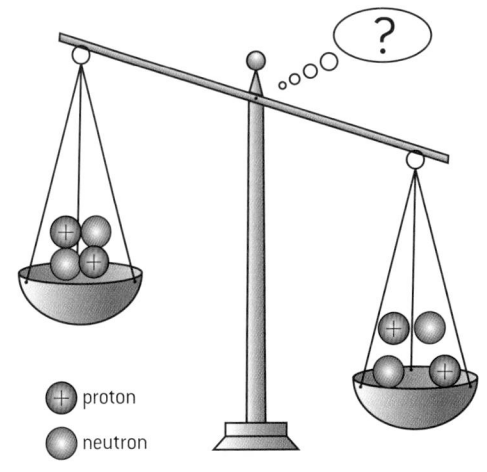

● proton

● neutron

▲ Figure 3 Mass defect: the mass of a helium nucleus is less than the masses of its constituent particles

While the amount of energy calculated in this example may seem small, it represents the energy released per atom. At the atomic scale energy is often referred to in **electronvolts** (eV). The electronvolt is a measure of the energy required to move one electron through a predefined electric field. In terms of joules, $1\,\text{eV} = 1.6022 \times 10^{-19}\,\text{J}$. So, expressed in eV the energy released when the helium nucleus is formed is 28 MeV (28 megaelectronvolts or 28×10^6 eV). You should be able to verify this for yourself:

$$4.52952 \times 10^{-12}\,\text{J}/1.6022 \times 10^{-19}\,\text{J/eV} \sim 28\,000\,000\,\text{eV}$$

So the helium nucleus has a lower potential energy than the sum of 4 unbound protons and neutrons. The mass defect has been converted to a **binding energy**, which for a helium nucleus is 28 MeV. When comparing the binding energy of different elements we calculate the binding energy per nucleon. Because helium has 4 nucleons (2 protons and 2 neutrons) the binding energy per nucleon is 7 MeV. Figure 4 is available in the *Data booklet, section 36*.

Binding Energy

Nuclear binding energy is the energy required to separate a nucleus into its constituent parts, namely protons and neutrons. Rest mass of fundamental particles is in Section 4 of the *Data booklet*.

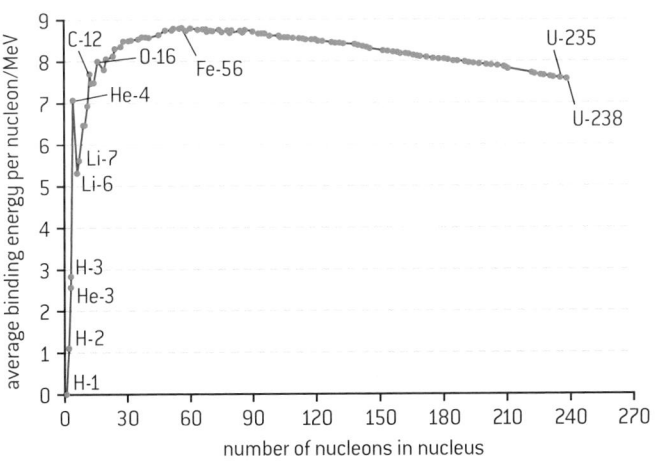

▲ Figure 4 Graph of atomic nuclei binding energies per nucleon plotted against the number of nucleons for the first 94 chemical elements. Lighter elements undergo fusion to become more stable whereas heavier elements can undergo nuclear fission

Nuclear processes: Fusion and fission

Many different chain reaction mechanisms can occur to produce the helium nucleus, and most of them occur in the sun and stars. One of the proposed mechanisms for producing energy by controlled nuclear fusion

A **chain reaction** is self sustaining – a product of the reaction allows further reactions so that the reaction will continue or escalate. An example is the following fission reaction of uranium:

$$^{235}_{92}U + ^{1}_{0}n \rightarrow ^{141}_{56}Ba + ^{92}_{36}Kr + 3^{1}_{0}n$$

The three neutrons that are produced feed in to initiate further atoms of uranium to react.

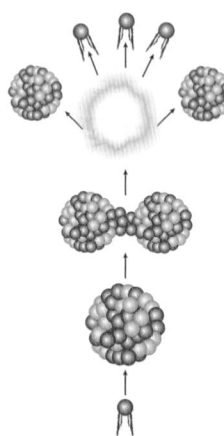

▲ Figure 6 Fission of U-235 produces three neutrons and two daughter nuclei, which are usually radioactive and undergo further decay

▲ Figure 7 Control rods from the reactor at the Chernobyl nuclear power station, Ukraine. The rods are inserted into the reactor to absorb neutrons and so slow down or stop the nuclear chain reaction that generates power. In 1986 power surges and a series of mistakes resulted in the control rods igniting and being rendered useless, emitting radioactive smoke in the world's worst nuclear disaster

here on Earth is the fusion of deuterium (a hydrogen isotope with 1 proton and 1 neutron) with tritium (a hydrogen isotope with 2 neutrons):

$$^{2}_{1}H + ^{3}_{1}H \rightarrow ^{4}_{2}He + ^{1}_{0}n$$

○ proton   ○ neutron

You can see that a neutron is also emitted in this particular fusion reaction. The important point is that there is a significant difference in binding energy per nucleon between helium and the two isotopes of hydrogen. That means the nucleons are bound much more tightly in a stable helium nucleus; there is a mass defect, and that mass is converted directly to energy. Indeed so much energy is released that if the reaction is carried out in quantities any larger than atom by atom experimentally, the heat produced makes the reaction difficult to retain in a vessel.

Iron has the most stable nuclear configuration. By fusing lighter elements to form larger ones the binding energy increases and the mass defect is converted to energy. On the other hand, the heavier transuranium elements (those with atomic number greater than 92) can undergo splitting or **nuclear fission** to form two lighter nuclei. As the sum of the binding energies of the two lighter elements is greater than the binding energy of a uranium-235 isotope, there is a mass defect which is converted directly to energy. Controlled nuclear fission is the process that powers nuclear generating plants today. One such fission reaction is:

$$^{235}_{92}U + ^{1}_{0}n \rightarrow ^{141}_{56}Ba + ^{92}_{36}Kr + 3^{1}_{0}n$$

Three neutrons are created per fission reaction in this example. These neutrons could be used to split other U-235 nuclei, with each split creating 3 more neutrons and initiating a chain reaction. In a nuclear power plant some neutrons are absorbed by control rods in order to prevent a chain reaction from spiralling out of control. The number of control rods and the distance they are inserted into the core can be adjusted as necessary to control the rate of reaction (figures 5, 6).

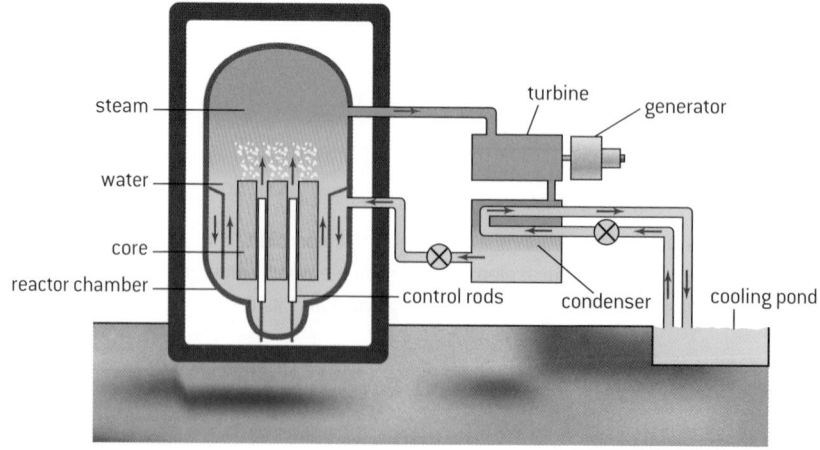

▲ Figure 5 Diagram of the workings of a boiling water reactor (BWR), a type of nuclear reactor. The core is suspended in water. The heat produced by the nuclear reactions boils the water into steam; this turns a turbine which drives a generator. Control rods can be raised or lowered to control the reaction

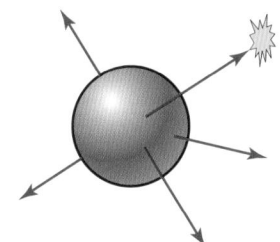

The chain reaction is sustainable provided one neutron from the fission of U-235 strikes another U-235 atom, causing further fission to occur. The amount of material needed for the reaction to remain sustainable is the **critical mass**. At the point where the number of neutrons produced in one generation is equal to the number of neutrons produced in the next generation the reactor is referred to as critical. If the number of neutrons produced becomes greater in successive generations then the reactor is supercritical. The power output increases and the control rods must be used to absorb the extra neutrons to avoid meltdown. On the other hand, if there are fewer neutrons in each successive generation the power generation falls and the reactor becomes subcritical – it is no longer self sustaining (figure 8).

Types of subatomic particle

Fission or fusion reactions involve the capture or emission of subatomic particles. While the number of different subatomic particles and radioactive emissions is large, you should be familiar with those shown in table 1 and be able to use them in balancing nuclear equations.

The conversion of one element to another by capture or emission of a particle is referred to as **transmutation**.

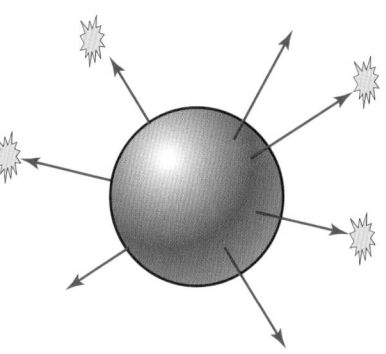

▲ Figure 8 If the amount of fissile material is too small, there are not enough neutrons produced to cause further reaction and the reaction is not sustainable. The critical mass is the amount of material needed to keep the reaction sustainable such that sufficient neutrons can continue to sustain the chain reaction

Particle	Symbol	Description and hazard
alpha particle	α or $^4_2$He	A helium nucleus consisting of 2 protons and 2 neutrons. It is the most massive particle involved in radioactive reactions and can travel only a few centimetres in air. Limited hazard unless inhaled or ingested.
beta particle	β or $^0_{-1}$e	A high speed electron with negligible mass and a charge of -1. Beta particles are a product of nuclear decay. They have a range of a few metres and have enough energy to cause burns to the skin.
gamma ray	γ	High frequency, short wavelength electromagnetic waves. Due to their short wavelength they have a high penetrating ability. They can cause cancer but under controlled conditions are used in medicine for treatment, imaging, and sterilization.
neutron	$^1_0$n	Uncharged nuclear particle with a mass of 1 atomic mass unit. May be emitted in fission and fusion reactions. They have a high penetrating ability and can be damaging to biological material.
positron	$^0_{+1}\beta^+$	The antiparticle of an electron; a positively charged beta particle.
proton	$^1_1$p or $^1_1$H	Nuclear particle that has a mass of 1 atomic mass unit and a charge of $+1$ atomic mass unit.

▲ Table 1 Subatomic particles involved in fusion and fission reactions

Worked example

Write an equation for the transmutation by proton capture followed by alpha decay of Pa-237.

Solution

Write the symbol equation including the proton for the first proton capture reaction. Balance the charge and mass to predict the first product ($Z = 92$: uranium):

$$^{237}_{91}\text{Pa} + ^1_1\text{p} \rightarrow ^{238}_{92}\text{U}$$

Continue the process for the alpha decay reaction:

$$^{237}_{91}\text{Pa} + ^1_1\text{p} \rightarrow ^{238}_{92}\text{U} \rightarrow ^4_2\text{He} + ^{234}_{90}\text{Th}$$

Quick questions

1 Copy and complete the following nuclear equations. For each one, choose an appropriate description: alpha / beta; capture / decay.

   a) $^{131}_{53}\text{I} \rightarrow \; ^{\square}_{\square}\text{—} + \; ^{0}_{-1}\text{e}$

   b) $^{118}_{54}\text{Xe} + \; ^{\square}_{\square}\text{—} \rightarrow \; ^{118}_{53}\text{I}$

   c) $^{226}_{88}\text{Ra} \rightarrow \; ^{\square}_{\square}\text{—} + \; ^{4}_{2}\text{He}$

   d) $^{\square}_{\square}\text{—} \rightarrow \; ^{4}_{2}\text{He} + \; ^{208}_{81}\text{Tl}$

2 Copy and complete these equations to give you more practice at balancing charges and relative atomic masses:

   a) $^{1}_{1}\text{H} + \; ^{3}_{1}\text{H} \rightarrow \underline{\quad}$

   b) $^{235}_{92}\text{U} + \; ^{1}_{0}\text{n} \rightarrow \; ^{139}_{56}\text{Ba} + \; ^{94}_{36}\text{Kr} + \underline{\quad} \; ^{1}_{0}\text{n}$

   c) $^{6}_{3}\text{Li} + \; ^{1}_{0}\text{n} \rightarrow \; ^{0}_{-1}\text{e} + \; ^{4}_{2}\text{He} + \underline{\quad}$

3 Write an equation for:

   a) the beta-decay of sodium-24

   b) positron emission by fluorine-17

   c) alpha decay of americium-241.

4 a) Explain, in terms of binding energy, why energy is released by the fusion of lighter elements but by the fission of heavier elements.

   b) Explain why the fusion of hydrogen nuclei to helium nuclei releases more energy than the fission of uranium-235.

The half-life of a nuclear process

As we have seen, some heavier atoms are radioactive – they undergo spontaneous decay to produce daughter products, releasing alpha, beta, and/or gamma radiation in the process. Radioactive decay is a **first order reaction**, meaning that it has a constant half-life. The **half-life** ($t_{1/2}$) refers to the time it takes for one half of the number of atoms in a sample to decay.

For example, if a radioactive substance has a half-life of 10 years, then in 10 years from now 50% of the present number of atoms will be unchanged and the other 50% will have decayed to daughter products. In another 10 years (20 years from now) half of the remaining 50% will have decayed, leaving only 25% of the original sample. Table 2 shows how this continues for 5 half-lives.

The amount remaining can be expressed as $\frac{1}{2^n}$, where n = the number of half-lives. So after 4 half-lives, $\frac{1}{2^4}$ or $\frac{1}{16}$ of the original amount remains.

Strontium-90 has a half-life of 28.8 years. Figure 9 plots the number of atoms of an original sample of 1000 atoms of Sr-90 that remain against time. The horizontal black lines show that the half-life – the time taken for the number of remaining atoms to halve – is constant.

Number of half-lives passed	Amount remaining	Fraction remaining
0	100%	1
1	50%	$\frac{1}{2}$
2	25%	$\frac{1}{4}$
3	12.5%	$\frac{1}{8}$
4	6.25%	$\frac{1}{16}$
5	3.125%	$\frac{1}{32}$

▲ Table 2 The amount of material remaining after the first 5 half-lives for a decay process

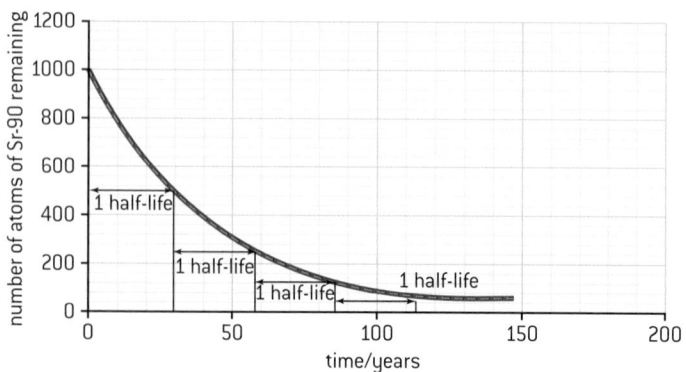

▲ Figure 9 Radioactive decay curve for strontium-90

Half-life calculations

We can find the half-life by plotting a graph as just described. Alternatively, the following equation allows us to calculate the half-life if we know how much material we started with (N_0), how much remains (N), and the time interval (t):

$$t_{1/2} = t \frac{\ln 2}{\ln \left(\frac{N_0}{N} \right)}$$

A rearranged form of this equation allows us to calculate N_0, N or t if we have the values for the other three variables:

$$N_0 = N \times 2^{\text{number of half-lives past}}$$

Worked examples

1   The mass of a radioactive substance was measured, and then re-measured 120 days later. It was found that 56% of the original sample remained. Deduce the half-life of this substance?

Solution

$$t_{1/2} = t \frac{\ln 2}{\ln \left(\frac{N_0}{N} \right)} = \frac{120 \times \ln 2}{\ln \left(\frac{100}{56} \right)} = 143 \text{ days}$$

2   A substance with a half-life of 8 hours has an activity of 450 units after 48 hours. Determine the original radioactivity.

Solution

$$N_0 = N \times 2^{\text{number of half-lives past}}$$
$$= 450 \times 2^6 \text{ (6 half-lives have passed in 48 hours)}$$
$$= 28\,800 \text{ units}$$

3   If Sr-90 has a half-life of 28 years, calculate how much of the original substance remains after 280 years.

Solution

$$N_0 = N \times 2^{\text{number of half-lives past}}$$
$$\frac{N}{N_0} = \frac{100}{2^{10}} \text{ (10 half-lives have passed)}$$
$$= 0.098\% \text{ of the original sample}$$

4   The isotope carbon-14 is taken in by plants during photosynthesis. Carbon-14 has a half-life of 5280 years. If a living redwood tree has a count of 15 counts per minute (cpm), calculate the age of a piece of petrified redwood with a count of 6 cpm.

Solution

$$t = t_{\frac{1}{2}} \times \frac{\ln \left(\frac{N_0}{N} \right)}{\ln 2}$$
$$= 5280 \times \frac{\ln \frac{15}{6}}{\ln 2} = 6980 \text{ years old}$$

Quick questions

1   ^{32}P has a half-life of 14 days. If a sample is registering 10 000 cpm, deduce what it would register after 42 days.

2   Tritium (^{3}H) has a half-life of 12.5 years. Calculate how much of a 20 g sample remains after 25 years.

3   Cobalt-60 is used in radiotherapy. ^{60}Co has a half-life of 5.3 years and undergoes beta-decay. Write an equation for the transmutation of ^{60}Co and identify how much of the daughter product would be formed from a 2.00 mg sample of ^{60}Co after 2.65 years.

4   Use the data in table 3 to plot a graph. Use the graph to determine the half-life of this radioactive substance.

Time passed/min	cpm	Time passed/min	Decay count/cpm
0	7526	21	3784
3	6996	24	3344
6	6512	27	3316
9	5880	30	2788
12	4844	33	2584
15	4508	36	2408
18	4132	39	2148

▲ Table 3

Radioactive waste

▲ Figure 10 Sealing radioactive waste into concrete containers at a French waste storage facility. France is one of the world leaders in electricity generated from nuclear power

As well as generating energy, the process of nuclear fission results in excess neutrons. These are absorbed by control rods in a nuclear reactor. Radioisotopes used in medicine and research can also be made this way by placing them in the reactor as target material.

However, nuclear fission generates a large amount of dangerous radioactive waste which has to be disposed of safely, as well as the possibility of producing materials which could be used to make nuclear weapons. Many of the products of fission reactions have long half-lives and are harmful to living organisms. Used fuel and contaminated control rods can be stored underwater at the nuclear power plant. For long-term storage spent fuel is encased in steel surrounded by an inert gas and covered in concrete for burial.

🌐 Spectroscopy

The race for nuclear weapons during the 1940s and 1950s brought about the development of nuclear power for society. Nuclear fusion in the sun was not understood until after Einstein, and even the composition of the sun and stars was unknown until the development of spectroscopy. As the products of fusion reactions cool and leave the sun's atmosphere, electrons in their atoms undergo transitions to lower-energy states and emit electromagnetic radiation of specific wavelengths. By observing the spectra from the sun and stars scientists are able to deduce their composition.

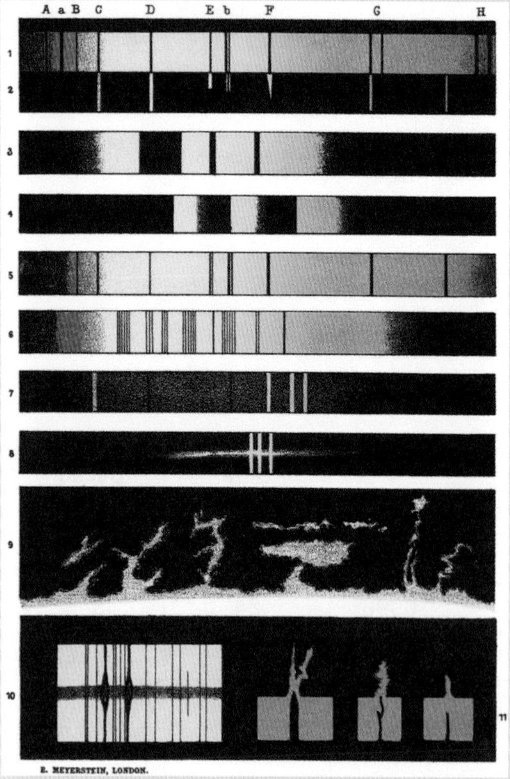

▲ Figure 11 One of the earliest illustrations of solar spectra (from an 1878 article *Chemistry of heavenly bodies* by Dr J. Gladstone). Spectroscopy has shown the composition of stars and comets and led astronomy into astrophysics

▲ Figure 12 The corona of the sun is clearly visible during a solar eclipse. Spectra first observed from these gases led to our understanding of the sun's composition

Questions

1 State what nucleons are.

2 The sun is approximately 91% hydrogren, 8% helium, with trace amounts of carbon, oxygen, nitrogen, silicon, iron, sulfur, and a few other elements. Describe the evidence for how we know this.

3 Explain what is meant by the term half-life.

4 Figure 13 shows the rate of decay versus time for a sample of a radioactive material. Find the half-life for this substance.

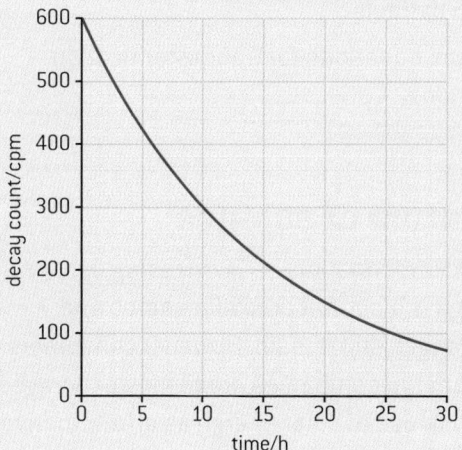

▲ Figure 13

5 The isotope of technetium $^{99m}_{43}$Tc is used in medicine as a source of gamma rays. The $^{99m}_{43}$Tc nucleus is in an excited-state and decays to the radioactive nucleus $^{99}_{43}$Tc by giving off a gamma ray. A $^{99m}_{43}$Tc nucleus is created by causing a molybdenum nucleus $^{98}_{42}$Mo to absorb a neutron and undergo beta decay.

a) Write an equation for the transmutation of $^{98}_{42}$Mo to $^{99m}_{43}$Tc.

b) $^{99m}_{43}$Tc nuclei have a half-life of 6.0 h. Explain the meaning of this statement.

c) A hospital requires 1.0×10^{-9} g of $^{99m}_{43}$Tc. Calculate how many grams must be created if it takes 24 hours to transport it from the reactor to the hospital.

6 Enrico Fermi carried out early experiments on artificial transmutation in the 1930s by bombarding matter with neutrons. He bombarded uranium and suggested the following reactions:

$$^{238}_{92}U + {}^{1}_{0}n \rightarrow {}^{239}_{92}U \qquad\qquad {}^{239}_{92}U \rightarrow {}^{a}_{b}X + {}^{0}_{-1}e$$

$$^{a}_{b}X \rightarrow {}^{c}_{d}Y + {}^{0}_{-1}e$$

a) Determine the values of *a*, *b*, *c*, and *d*.

b) Uranium was named in 1791 after the planet Uranus, which had only been discovered shortly before and was believed to be the furthest planet in the solar system at the time. In 1940 researchers in California isolated X and Y. These were the first transuranium elements to have been produced synthetically. Identify the elements denoted by the letters X and Y in the equations above.

7 Describe the processes of nuclear fission and nuclear fusion.

8 Explain each of these terms:

a) mass defect

b) binding energy

c) binding energy per nucleon.

9 The equation for the fusion of deuterium and tritium is:

$$^{2}_{1}H + {}^{3}_{1}H \rightarrow {}^{4}_{2}He + {}^{1}_{0}n$$

and the atomic masses (in amu, where 1 amu = 1.6605×10^{-27} kg) are:

$^{2}_{1}$H: 2.014 amu; $^{3}_{1}$H: 3.016 amu;
$^{4}_{2}$He: 4.0026 amu; $^{1}_{0}$n: 1.009 amu.

a) Calculate the mass defect, in amu, and the energy released, in MeV, from the fusion of a deuterium nucleus and a tritium nucleus.

b) Given that less energy is released in the fission reaction of a U-235 nucleus than in the fusion reaction of one tritium nucleus with a deuterium nucleus, why would nuclear fusion be preferred for generating energy?

C.4 Solar energy

Understandings

→ Light can be absorbed by chlorophyll and other pigments with a conjugated electronic structure.

→ Photosynthesis converts light energy into chemical energy:

$$6CO_2 + 6H_2O \rightarrow C_6H_{12}O_6 + 6O_2$$

→ Fermentation of glucose produces ethanol which can be used as a biofuel:

$$C_6H_{12}O_6 \rightarrow 2C_2H_5OH + 2CO_2$$

→ Energy content of vegetable oils is similar to that of diesel fuel but they are not used in internal combustion engines as they are too viscous.

→ Transesterification between an ester and an alcohol with a strong acid or base catalyst produces a different ester:

$$RCOOR^1 + R^2OH \rightarrow RCOOR^2 + R^1OH$$

→ In the transesterification process, involving a reaction with an alcohol in the presence of a strong acid or base, the triglyceride vegetable oils are converted to a mixture of mainly alkyl esters and glycerol, but with some fatty acids.

→ Transesterification with ethanol or methanol produces oils with lower viscosity that can be used in diesel engines.

Applications and skills

→ Identify the features of molecules that allow them to absorb visible light.

→ Explain the reduced viscosity of esters produced with methanol and ethanol.

→ Evaluate the advantages and disadvantages of using biofuels.

→ Deduce equations for transesterification reactions.

Nature of science

→ Public understanding – harnessing the sun's energy is a current area of research and challenges still remain. However consumers and energy companies are being encouraged to make use of solar energy as an alternative energy source.

Reproducibility of results

As experts in their particular fields, scientists are well placed to explain to the public their issues and findings. Outside their specializations they may be no more qualified than ordinary citizens to advise others on scientific issues, although their understanding of the processes of science can help them to make personal decisions and to educate the public as to whether claims are scientifically credible.

IB Chemistry syllabus, Nature of Science statement 5.2.

Scientists continue to look for alternative energy sources to reduce our dependence on fossil fuels. In 1989 Stanley Pons and Martin Fleischmann made headlines with claims that they had carried out a nuclear fusion reaction at room temperature – "cold fusion". This "discovery" was missing one key ingredient: good scientific method. The results were not reproducible and the use of fusion as an alternative energy source remains not yet viable. Harnessing of the sun's energy is one of the most researched and trialled alternative energy sources, driven by energy companies and consumers alike.

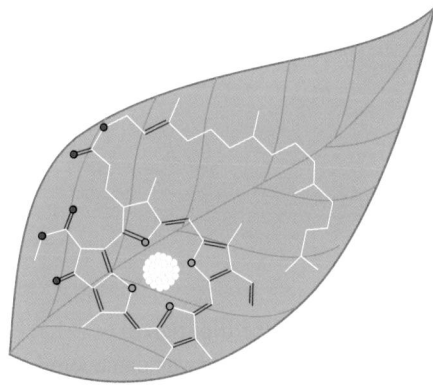

▲ Figure 1 It is the alternating double bonds (conjugated π bonds) that absorb the energy for photosynthesis

Photosynthesis: Harnessing solar energy by chlorophyll

You know that the sun is the source of energy on Earth. We shall look at photovoltaic cells in sub-topic C.8, but here we focus on harnessing solar energy in the process of photosynthesis.

Sunlight is absorbed in chloroplasts by the chemical chlorophyll (figure 1).

Visible light can be absorbed by molecules that have a **conjugated** structure with an extended system of alternating single and multiple bonds. These alternating bonds in chlorophyll can absorb light energy. You will recall from sub-topic 2.2 that absorbing a photon of light excites electrons. In a system of conjugated bonds the excitation of these electrons occurs in the visible wavelength of light rather than requiring higher-energy ultraviolet (UV) radiation. Once excited, electrons normally return to the ground-state emitting a photon of light. During photosynthesis the return of the electron to the ground-state takes place during a complex series of chemical reactions, the net result of which is the transformation of carbon dioxide and water reactants into glucose and oxygen products. The net equation for photosynthesis is:

$$6CO_2 + 6H_2O \rightarrow C_6H_{12}O_6 + 6O_2$$

Pigments in plants are coloured due to conjugated double bond systems. If a certain pigment absorbs red and green, or yellow, light as a result of its extended conjugation, then blue or purple light will be reflected. Violets are blue (or violet) because of anthocyanin pigment in the flower.

Purpurin (1,2,4-trihydroxyanthraquinone) is a pigment found in the rose madder plant. It is often used to dye cotton. Its colour changes in acid and base conditions due to a difference in conjugation in each system (figure 3).

▲ Figure 2 In photosynthesis the excited electrons in the chlorophyll molecule fall through a cascade system, releasing their energy to break bonds in CO_2 and H_2O molecules and reform these atoms to glucose and oxygen

▲ Figure 3 Purpurin: different conjugated double bonds (resonance structures) lead to different colours in acidic and basic conditions

Resonance structures occur when there is more than one possible position for a double bond in a molecule.

Biofuels

The conversion of carbon dioxide to carbohydrates using solar energy by photosynthesis produces our food and fuels. **Biofuels** such as ethanol are obtained from corn sugar or glucose by fermentation:

$$C_6H_{12}O_6 \rightarrow 2C_2H_5OH + 2CO_2$$

The ethanol produced this way can be added to or blended with gasoline (petrol). Many cars have been designed or converted to run on higher blends of ethanol: E10, for example, is a blend of 10% ethanol and 90% petrol. The carbon dioxide produced in the fermentation process is balanced by carbon dioxide taken in for photosynthesis while the plant is growing, so the fuel can be considered **carbon neutral**; its use instead of petrol also conserves fossil fuels.

Biodiesel is another sustainable fuel that can be grown and used as a substitute for diesel. It is produced from vegetable oils, which can release similar amounts of energy to diesel when burnt. However, because they are highly viscous they are unable to flow easily and can clog fuel injectors. A high viscosity implies large intermolecular forces; these oils do not readily vaporize and often undergo incomplete combustion which further damages engines.

These problems are overcome by converting the vegetable oils to a less viscous esters with fewer intermolecular forces. For example in a **transesterification** process a triglyceride is converted to esters and glycerol:

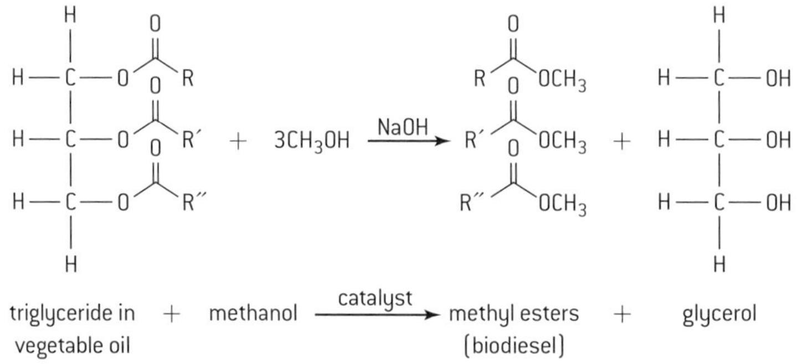

triglyceride in vegetable oil + methanol $\xrightarrow{catalyst}$ methyl esters (biodiesel) + glycerol

A similar transesterification process between a long-chain ester in the vegetable oil and a shorter-chain alcohol using a strong acid or base catalyst produces a different ester:

$$RCOOR' + R''OH \rightarrow RCOOR'' + R'OH$$

The base catalyst is used to deprotonate the alcohol (figure 4). The smaller alkyl group on the alcohol replaces the larger alkyl group producing a less viscous and more volatile ester (figure 6).

In transesterification to form biodiesel the vegetable oil is typically heated with a sodium or potassium hydroxide catalyst along with methanol to produce the methyl ester, or ethanol to produce the ethyl ester of the tryglyceride.

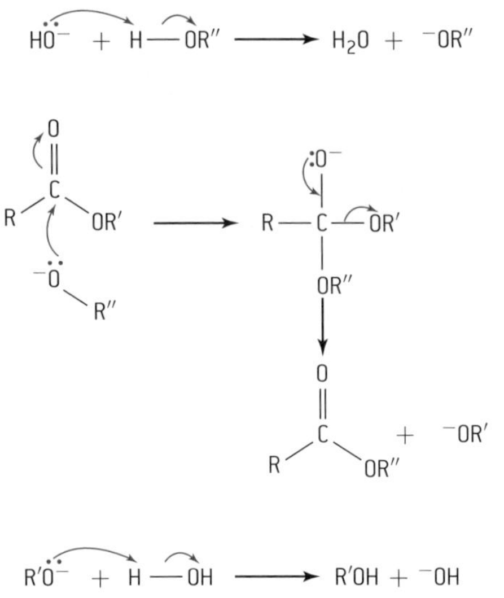

Transesterification of vegetable oils was discovered before the diesel engine was invented. In 1912 the diesel engine's inventor Rudolph Diesel said, "The use of vegetable oils for engine fuels may seem insignificant today but such oils may become, in the course of time, as important as petroleum and the coal-tar products of the present time." The use of biodiesel increased during the Second World War as a result of petroleum shortages.

▲ Figure 4 The mechanism of a transesterification reaction using a strong base catalyst

The source of biodiesel may vary depending on what raw materials are available – it can be produced from fish oil and animal fats as well as from vegetable oils. For example, in Alaska there may be more fish oil than vegetable oil waste available at certain times of year.

Worked example

Deduce the equation for the reaction of pentyloctanoate with methanol in the presence of an alkali catalyst.

Solution

This is a transesterification reaction. The pentyl group of the ester is replaced by the methyl group, lowering its viscosity:

$$C_7H_{15}COOC_5H_{11} + CH_3OH \rightarrow C_7H_{15}COOCH_3 + C_5H_{11}OH$$

Some advantages and disadvantages of biodiesel are summarized in table 1.

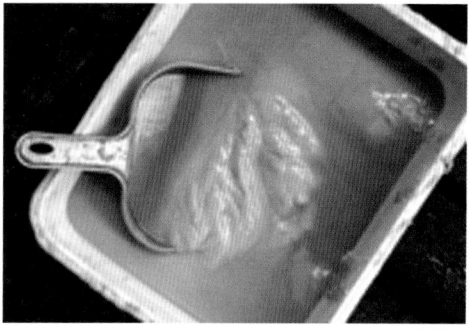

▲ Figure 5 Biodiesel fuel can be produced from vegetable oil wastes from restaurants and caterers by a transesterification process. Using these waste materials rather than virgin oil feedstock lowers the cost of producing biodiesel

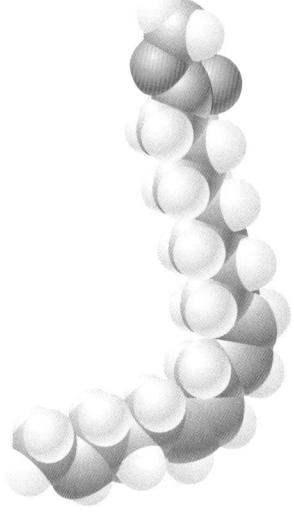

▲ Figure 6 Space-filling model of the ester methyl linolenate or biodiesel, produced by transesterification of soybean and canola triglyceride oils with methanol

Advantages	Disadvantages
High flash point (less flammable than normal diesel)	More viscous than diesel, even when converted to methyl esters – requires pre-warming.
Lower carbon footprint – amount of CO_2 produced is the same, but CO_2 was consumed in growing the plants. For petroleum cars CO_2 is introduced into the atmosphere that wasn't there before.	Slightly lower energy content than petroleum-based diesel.
	Uses agricultural resources resulting in increased food prices on a global scale
More easily biodegradable in the event of an oil spill. Sulfur free so produces no SO_2 emissions.	The production of biodiesel from raw materials is more costly than the production of diesel from fossil fuels.
Sustainable – the raw materials can be grown using solar energy as the source.	Biofuels may contain more nitrogen than fossil fuels and thus release more nitrogen oxides, NO and NO_2, when burned.
A good solvent – cleans engines.	Dirt cleaned from engines tends to clog fuel filters and cause cars to stall. It can also dissolve paint and protective coatings.

▲ Table 1 Some advantages and disadvantages of biodiesel compared with diesel

▲ Figure 7 A diesel power generation plant run on 100% biodiesel produced from fish oil – UniSea's Dutch Harbor seafood processing facility, Alaska

Questions

1 Write the equation for photosynthesis.

2 Explain why ethanol-based fuels are said to have a lower carbon footprint than petroleum-based fuels, even though they both release similar amounts of carbon dioxide on combustion.

3 Outline the reagents and conditions necessary to convert a vegetable oil to a usable fuel for a vehicle such as a car.

4 Explain why the transesterification process is necessary in producing biodiesel. Describe the disadvantages of using vegetable oils as fuels without processing them.

5 Outline what is meant by a system of conjugated double bonds.

6 Identify from section 35 of the *Data booklet* which of vitamins A, C, or D is most likely to appear as a coloured compound. Explain your answer.

7 Write an equation for the fermentation of glucose.

8 Write the general equation for transesterification.

9 Deduce the number of molecules of ester and glycerol produced per molecule of a triglyceride undergoing transesterification.

10 Discuss the advantages and disadvantages of the use of biofuels commercially.

C.5 Environmental impact – global warming

Understandings

→ Greenhouse gases allow the passage of incoming solar short wavelength radiation but absorb the longer wavelength radiation from the Earth. Some of the absorbed radiation is re-radiated back to Earth.

→ There is a heterogeneous equilibrium between concentration of atmospheric carbon dioxide and aqueous carbon dioxide in the oceans.

→ Greenhouse gases absorb IR radiation as there is a change in dipole moment as the bonds in the molecule stretch and bend.

→ Particulates such as smoke and dust cause global dimming as they reflect sunlight, as do clouds.

Applications and skills

→ Explain the molecular mechanisms by which greenhouse gases absorb infrared radiation.

→ Discuss the evidence for the relationship between the increased concentration of gases and global warming.

→ Discuss the sources, relative abundance and effects of different greenhouse gases.

→ Discuss the different approaches for control of carbon dioxide emissions.

→ Examine and evaluate the pH changes in the ocean due to increased concentration of carbon dioxide in the atmosphere.

Nature of science

→ Transdisciplinary – the study of global warming encompasses a broad range of concepts and ideas and is transdisciplinary.

→ Collaboration and significance of science explanations to the public – reports of the Intergovernmental Panel on Climate Change (IPCC).

→ Correlation and cause and understanding of science – CO_2 levels and Earth average temperature show clear correlation but wide variations in the surface temperature of the Earth have occurred frequently in the past.

Global collaboration and climate change

Science is highly collaborative and the scientific community is composed of people working in science, engineering, and technology. It is common to work in teams from many disciplines so that different areas of expertise and specializations can contribute to a common goal that is beyond one scientific field. It is also the case that how a problem is framed in the paradigm of one discipline might limit possible solutions, so framing problems using a variety of perspectives, in which new solutions are possible, can be extremely useful.

IB Chemistry syllabus, Nature of Science statement 4.1

The study of greenhouse gases exemplifies the above paragraph. Findings from the **Intergovernmental Panel on Climate Change** (IPCC) continue to increase our knowledge of the scientific, economical, technical, and social aspects of climate change. Although there is a clear correlation between rising carbon dioxide levels and the Earth's average temperature, extrapolation is difficult because wide variations of Earth's average surface temperature have occurred frequently in the past.

Human influences and climate change

Evidence exists that increased levels of greenhouse gases in the atmosphere produced by human activities are changing the climate. The raised levels of these gases are upsetting the balance between radiation entering and leaving the atmosphere, causing an overall warming of the atmosphere that leads to climate change.

The natural greenhouse effect

The radiation in sunlight has a range of wavelengths (figure 1). The highest frequencies are absorbed by the upper atmosphere, allowing some UV, visible, and longer wavelengths to reach the surface where they are absorbed. The waves re-emitted from the surface are longer-wavelength infrared (IR). These waves interact with carbon dioxide, methane, and water vapour, the main **greenhouse gases**, which capture this energy so that it remains trapped in the Earth's atmosphere. This natural effect of the atmosphere is similar to a greenhouse, hence the term '**greenhouse effect**' (figure 2).

Aside from the greenhouse effect and climate change, human activity has also affected the ozone layer in the stratosphere. Short-wave UV radiation in sunlight is absorbed by the ozone layer. The destruction of the ozone layer by chemicals such as CFCs results in more high-energy UV radiation reaching the Earth, increasing our risk of skin cancer and having harmful effects on plants and other organisms.

As well as destroying the ozone layer, CFCs are also greenhouse gases.

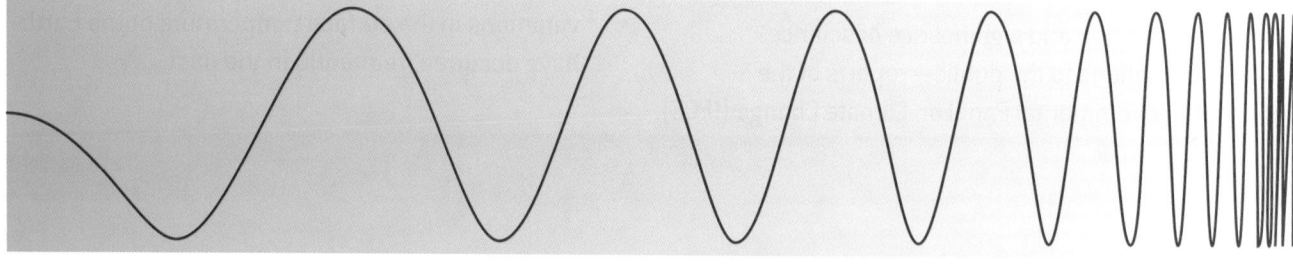

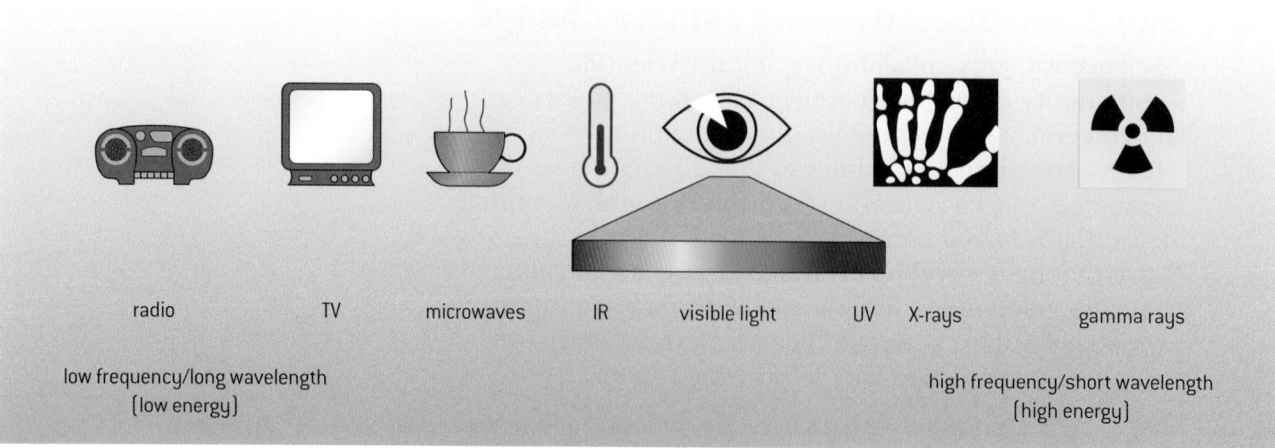

radio   TV   microwaves   IR   visible light   UV   X-rays   gamma rays

low frequency/long wavelength (low energy)

high frequency/short wavelength (high energy)

▲ Figure 1 The electromagnetic spectrum of solar radiation: highest-energy waves have the shortest wavelength and the highest frequency

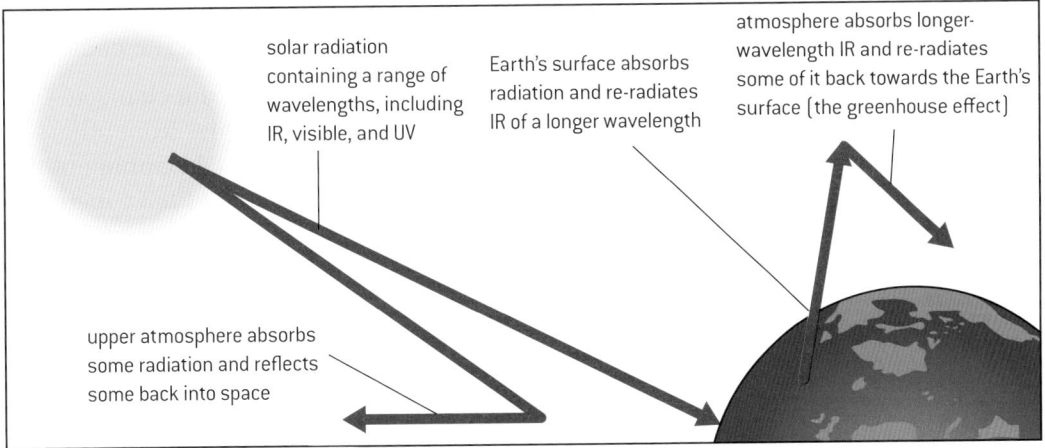

▲ Figure 2 The greenhouse effect

The IR radiation interacts with the covalent bonds of greenhouse gas molecules, causing them to bend and stretch. The natural bending and stretching frequencies of the bonds in these molecules coincides with the frequency of the IR radiation, causing increased vibration at a particular resonant frequency. Certain types of stretching and bending change the dipole moment of the molecule. The polar nature of the molecule is more accentuated, making one end more charged than the other and this can be detected by IR spectroscopy (figure 3).

The C–H, C=O, and O–H bonds in greenhouse gases have resonance frequencies of vibration in the IR region. Figure 4 shows the characteristic absorptions of different types of bonds in an IR spectrum.

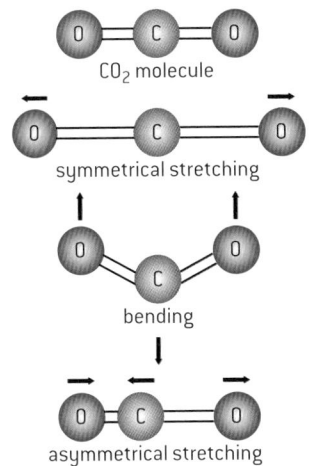

▲ Figure 3 Three modes of vibration in the CO_2 molecule. They each have a particular resonance frequency in the IR range

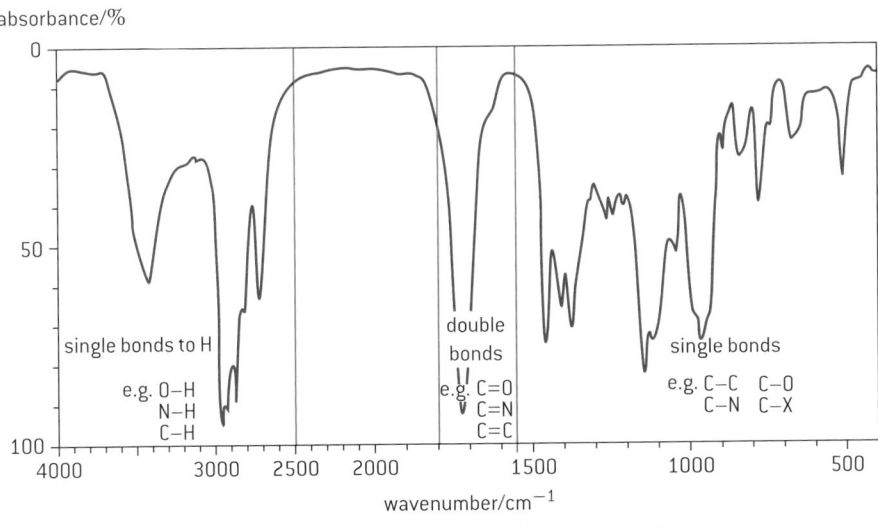

▲ Figure 4 IR absorbance frequencies due to bond bending and stretching

Natural sources of greenhouse gases

The vast majority of atmospheric water vapour is of natural origin and accounts for 95% of all greenhouse gases (figure 5). There is a natural balance between liquid water on the Earth's surface and vapour in the atmosphere. As the Earth warms up, more surface water evaporates and this increases the atmospheric water vapour concentration. The atmosphere then absorbs more IR radiation and causes increased

Clouds also reflect radiation – in this case rather than the bonds absorbing energy and increasing vibration, the solar radiation is physically reflected. Smog, smoke, and other particulate matter in the atmosphere also reflect radiation.

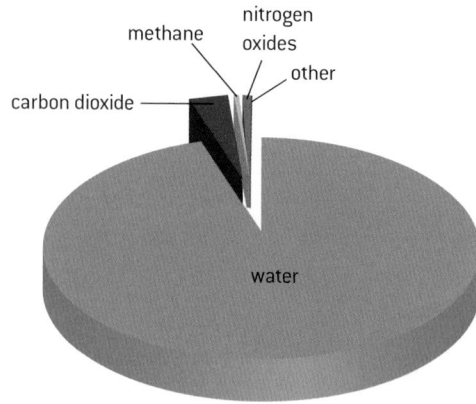

▲ Figure 5 The proportions of different greenhouse gases in the atmosphere

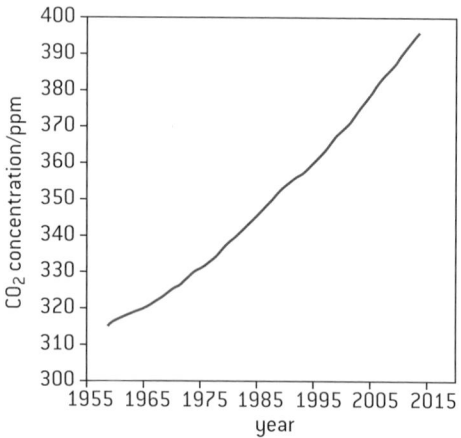

▲ Figure 6 Average carbon dioxide concentrations in the atmosphere during February measured at the NOAA, Mauna Loa, Hawaii

warming. However, much of the water vapour condenses into clouds which block sunlight, causing global dimming and cooling the planet.

While water vapour quantities in the atmosphere have not changed much and appear to be self regulating, the problem comes from other greenhouse gases, particularly carbon dioxide. Since the beginning of the industrial revolution CO_2 emissions from human activities have increased dramatically. Figure 6 shows the average CO_2 concentration as measured by the **National Oceanic and Atmospheric Administration** (NOAA) observatory station at Mauna Loa in Hawaii, while figure 7 shows the change in global temperatures. While there are a lot of variations it does show over a 1-degree increase in average global temperature for the period 1910–2010 when carbon dioxide emissions have been rising.

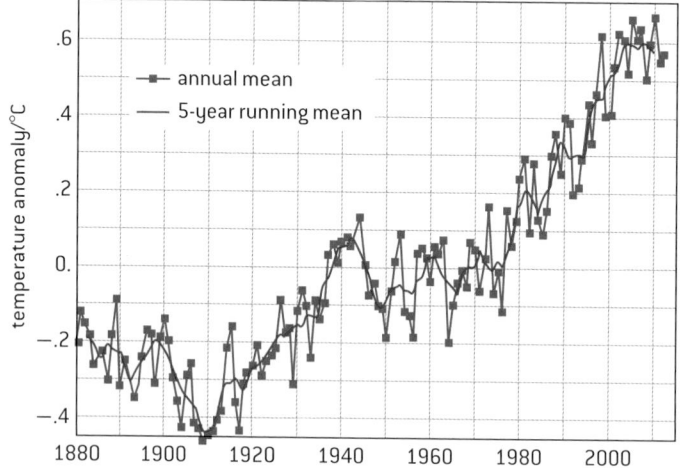

▲ Figure 7 Global land–ocean temperature index. This graph from NASA Goddard Institute for Space Studies uses the period of 1950–1980 as a baseline 0 temperature anomaly. Data courtesy of NASA/GISS/GISTEMP

Greenhouse gas emissions form human activities

The main sources of anthropogenic greenhouse gases (those arising from human activity) are listed below.

- Burning coal, oil, and natural gas for energy production accounts for nearly 50% of anthropogenic greenhouse gases. The carbon dioxide entering the atmosphere as a combustion product comes from hydrocarbons that were previously stored underground, so this increases absolute levels of the gas in the atmosphere. Water vapour is also a combustion product but the increase in water vapour is small compared with the increase in carbon dioxide levels.

- Industrial gases from factories introduce not only carbon dioxide but also new greenhouse gases such as nitrogen oxides (NO_x) accounting for approximately 25% of human greenhouse gas production. Some of these gases, such as chlorofluorocarbons (CFCs), do not occur naturally.

- Agriculture and deforestation account for the remaining 25%, with each contributing nearly equally. Agriculture increases methane concentrations from ruminant animals such as sheep and cows who generate methane in their digestive systems. Deforestation increases carbon dioxide because with fewer trees, less carbon dioxide is absorbed from the atmosphere and used in photosynthesis.

Carbon sinks: The role of the oceans

Of all the carbon dioxide gas released to the atmosphere by human activity, approximately half has remained in the atmosphere. The rest is removed to **carbon sinks** such as the oceans, resulting in CO_2 concentrations rising by about 1% per year for the period 1990 to 2010 (figure 8).

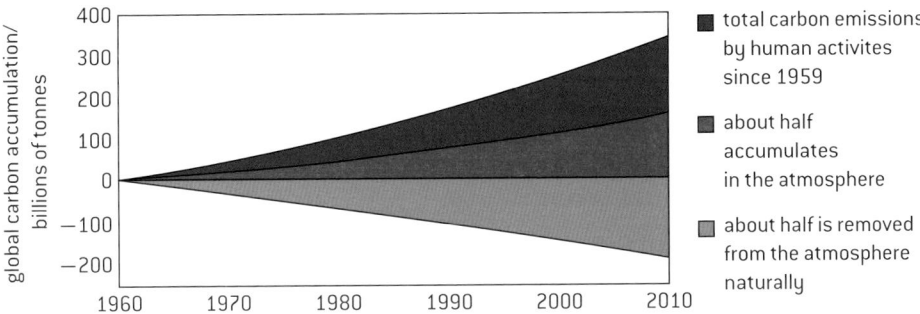

▲ Figure 8 Only half the carbon dioxide emitted remains in the atmosphere. The rest is taken up by carbon sinks

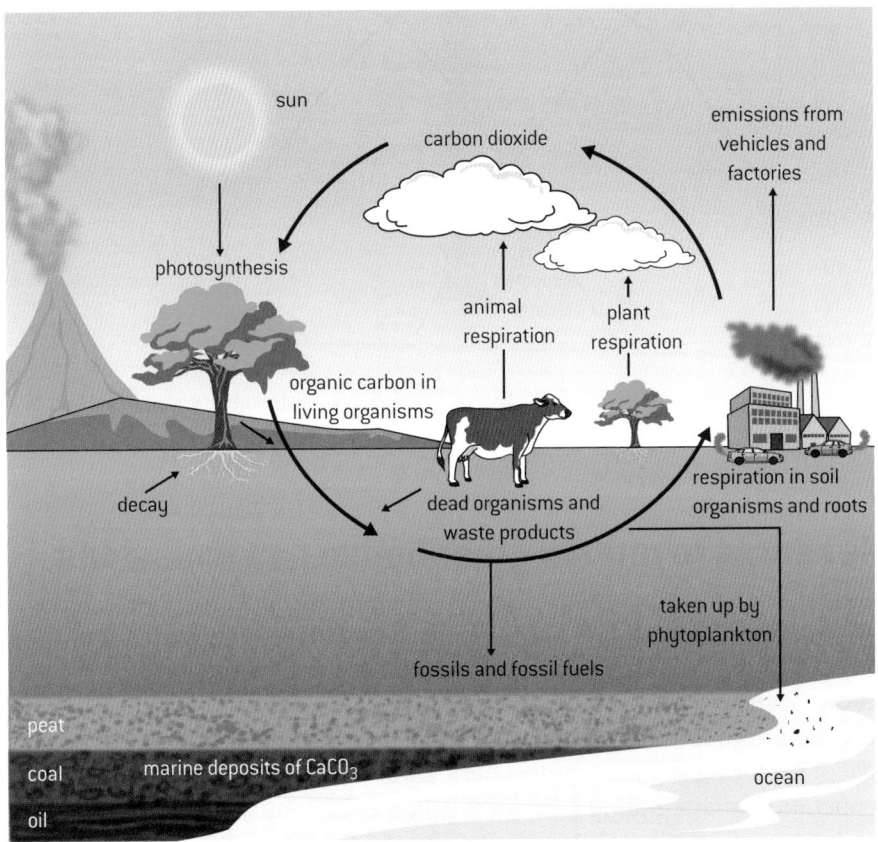

▲ Figure 9 Carbon sinks play a part in the carbon cycle, capturing and storing carbon dioxide. Sinks include the biosphere (animals, plants, soil, fresh water), the geosphere (coal, carbonates, and other minerals), the hydrosphere (oceans), and the atmosphere. The largest carbon sink is the ocean

About 30% of anthropogenic CO_2 is absorbed by the oceans (figure 9). Carbon dioxide itself is not very soluble, with the heterogeneous exchange between carbon dioxide gas and aqueous carbon dioxide occurring at the ocean's surface.

683

$$CO_2(g) \rightleftharpoons CO_2(aq)$$

However, once dissolved an equilibrium between dissolved carbon dioxide and carbonic acid is quickly established.

$$CO_2(aq) + H_2O(l) \rightleftharpoons H_2CO_3(aq)$$

This overall process has a small positive ΔH. An increase in temperature therefore shifts the equilibrium to the left, lowering the ability of carbon dioxide to dissolve in water. Because temperatures are lower near the bottom of the oceans, CO_2 is more soluble in deep water.

The dissolved aqueous carbonic acid releases a proton in water, being a Brønsted–Lowry acid. It is a diprotic weak acid and the following equilibrium reactions occur:

$$H_2CO_3(aq) + H_2O(l) \rightleftharpoons H_3O^+(aq) + HCO_3^-\ (aq)$$

$$HCO_3^-(aq) + H_2O(l) \rightleftharpoons H_3O^+(aq) + CO_3^{2-}(aq)$$

The acidity of water therefore reflects the extent of reaction (figure).

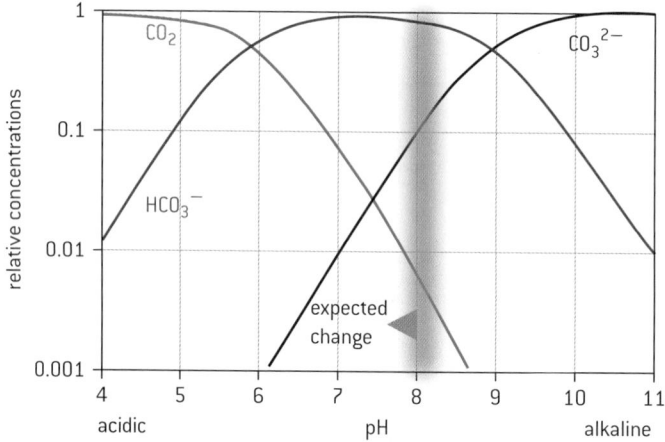

▲ Figure 10 Since the beginning of the Industrial Revolution the pH of the oceans has fallen about 0.1 pH units

Measures to reduce greenhouse gas emissions

International government agencies have begun to cooperate both to reduce the emission of greenhouse gases and to stop deforestation so that more CO_2 can be removed from the atmosphere for photosynthesis. The 1997 **Kyoto Protocol** was an international agreement, which introduced a scheme of carbon trading – countries that signed up agreed to reach the goal of capturing as much atmospheric carbon as they created. International cooperation in attempting to reduce carbon emissions was continued with the **Intergovernmental Panel on Climate Change** (IPCC) and the extension of the Kyoto Protocol in Qatar in 2012.

Industry and energy production

Carbon capture and storage (CCS) is the process of capturing waste carbon dioxide from where it is produced, such as fossil fuel power plants, transporting it to a storage site, and storing it where it will not enter the atmosphere, such as in an underground geological formation (figure 11).

▲ Figure 11 Carbon capture and storage (CCS)

Some approaches to reducing emissions of anthropogenic greenhouse gases are detailed below.

- Many coal power plants use **scrubbers** to remove sulfur dioxide as well as some greenhouse gases from emissions. In a scrubber water and limestone react with SO_2 to produce gypsum, calcium sulfate hydrate $CaSO_4 \cdot 2H_2O$.

- In **sequestration** carbon dioxide is converted to a carbonate in a process that uses silicate (silicon is abundant in the Earth):

$$Mg_2SiO_4(s) + 2CO_2(g) \rightarrow 2MgCO_3(s) + SiO_2(s)$$

- Combustion of fossil fuels liberates carbon dioxide that was previously stored underground, so changing to carbon-neutral alternatives such as synthesis gas (sub-topic C.2) is desirable.

- In **carbon recycling** the aim is to use carbon dioxide as a feedstock for synthetic fuels.

Agriculture and deforestation

Methane, CH_4 and nitrous oxide, N_2O are the main greenhouse gases produced in agriculture. Although these two gases are produced in smaller quantities than carbon dioxide they still have a pronounced effect. Methane is 25 times as powerful a greenhouse gas as carbon dioxide while nitrous oxide has over 300 times the impact. Taking this into consideration rather than simply the quantities of gases produced, the livestock (dairy and beef) industry produces a large percentage of agricultural greenhouse gases by enteric fermentation, anaerobic decomposition of organic matter, and fertilizer use (figure 13).

Careful land use and recycling can reduce the carbon footprint from agriculture. Changing from nitrogen-based fertilizers to crop rotation methods could increase the level of CCS and reduce emissions. Deforestation to create agricultural land should be carbon neutral as crops rather than trees are being grown, but this is not the case if use of fertilizers is increased. The use of urban space to grow crops could subsidize local communities and reduce transport costs.

Global dimming

Smoke, dust particles, and clouds reflect sunlight back to space, causing global dimming which cools the Earth's surface. Particulate matter such as soot and ash can further change the properties of clouds. Small droplets of water start to collect (nucleate) on tiny particulates and intermolecular forces between pollutant particles and water droplets result in the droplets collecting to form clouds.

These polluted clouds reflect more light than non-polluted ones. This was first reported by Atsumu Ohmura who in 1985 claimed that there was a 20% reduction in solar radiation reaching the then Soviet Union between 1960 and 1987. On average across the planet it has been estimated that 2–3% less radiation has reached Earth's surface over the past two decades.

▲ Figure 12 The green base trapping agent used by New Sky Energy, the world's first carbon-negative energy and manufacturing company. New Sky uses a capture process to scrub carbon dioxide from the air or flue gases and convert it into safe, stable solids. These solids can be incorporated into building materials, fertilizers, and other useful products

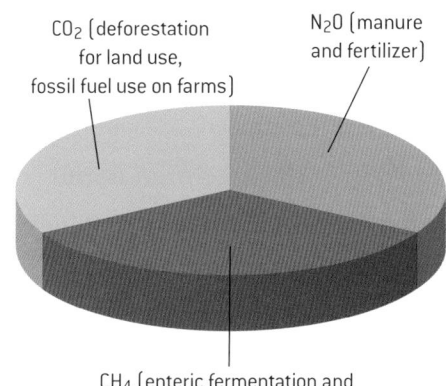

CO_2 (deforestation for land use, fossil fuel use on farms)

N_2O (manure and fertilizer)

CH_4 (enteric fermentation and manure storage and processing)

▲ Figure 13 Agricultural greenhouse gases

▲ Figure 14 Liquid fertilizer being spread onto a farm field in Luxembourg. The fertilizer is a by-product from a nearby biogas factory which processes manure into carbon dioxide and methane gases, providing electricity and heating for the local community

So by the process of global dimming, fossil fuel pollutants reduce as well as increase global warming. However, global dimming has harmful effects such as:

- Certain types of pollutant can cause acid rain.

- Global dimming decreases the rate of evaporation of water, which can reduce monsoon rains and lead to a reduction in crop yields in areas of the world where they are most needed.

- Pollution causes local health problems such as asthma.

The effects of global warming on climate change

Observed measures of climate change include melting permafrosts, less radiation reaching the Earth's surface, more devastating storms occurring, temperatures becoming more extreme (both hotter and colder), and record levels of rainfall and droughts. There appears ample evidence that anthropogenic greenhouse gas emission is raising global temperatures and is linked to global dimming. Radical changes in climate could put pressure on food and water resources for the growing worldwide population.

Questions

1 Even though water vapour is the most common greenhouse gas, carbon dioxide is more frequently discussed. Explain the reason for this. [1]

IB, specimen paper

2 State **three** greenhouse gases and their sources.

3 Discuss the molecular changes that are responsible for the effect of greenhouse gases including what must occur in order for them to absorb infrared light. [2]

IB, specimen paper

4 Explain the mechanism by which greenhouse gases affect the temperature of the Earth's surface.

5 The term *greenhouse effect* is used to describe a natural process for keeping the average temperature of the Earth's surface nearly constant.

a) Describe the greenhouse effect in terms of radiations of different wavelengths. [4]

b) Water vapour acts as a greenhouse gas. State the main natural and man-made sources of water vapour in the atmosphere. [2]

c) Two students disagreed about whether carbon dioxide or methane was more important as a greenhouse gas.

   i) State **one** reason why carbon dioxide could be considered more important than methane as a greenhouse gas. [1]

   ii) State **one** reason why methane could be considered more important than carbon dioxide as a greenhouse gas. [1]

d) Discuss the effects of global warming on Earth. [4]

IB, May 2004

6 After the September 11 2001 terrorist attacks in the USA, all air traffic and much industry was closed down for three days. It was noted that the sky was clearer and that the temperature difference between the hottest part of the day and the coldest part of the day was 1 degree greater than previously, meaning the days were warmer and nights colder. Explain this in terms of the link between global warming and global dimming.

7 The pH of the oceans has dropped slightly over the past century.

a) Explain this, using balanced equilibrium equations and mentioning greenhouse gases.

b) Decaying reefs result in increased CO_3^{2-}(aq) concentrations. Explain how this might affect the equilibrium in (a).

c) Explain why CO_2(g) less soluble in warm than in cold water.

8 State what causes global dimming and outline its effects.

C.6 Electrochemistry, rechargeable batteries and fuel cells (AHL)

Understandings

→ An electrochemical cell has internal resistance due to the finite time it takes for ions to diffuse. The maximum current of a cell is limited by its internal resistance.

→ The voltage of a battery depends primarily on the nature of the materials used while the total work that can be obtained from it depends on their quantity. These variables are related in that work done = voltage × current × time.

→ In a primary cell the electrochemical reaction is not reversible. Rechargeable cells involve redox reactions that can be reversed using electricity.

→ A fuel cell can be used to convert chemical energy, contained in a fuel that is consumed, directly to electrical energy.

→ Microbial fuel cells (MFCs) are a possible sustainable energy source using different carbohydrates or substrates present in waste waters as the fuel.

→ The Nernst equation, $E = E^{\ominus} - \frac{RT}{nF} \ln Q$ can be used to calculate the potential of a half-cell in an electrochemical cell, under non-standard conditions.

→ The electrodes in a concentration cell are the same but the concentration of the electrolyte solutions at the cathode and anode are different.

Applications and skills

→ Distinguish between fuel cells and primary cells.

→ Deduce half equations for the electrode reactions in a fuel cell.

→ Compare and contrast fuel cells and rechargeable batteries.

→ Discuss the advantages of different types of cells in terms of size, mass, and voltage.

→ Solve problems using the Nernst equation.

→ Calculate the thermodynamic efficiency $\left(\frac{\Delta G}{\Delta H}\right)$ of a fuel cell.

→ Explain the workings of rechargeable cells and fuel cells including diagrams and relevant half-equations.

Nature of science

→ Environmental problems – redox reactions can be used as a source of electricity but disposal of batteries has environmental consequences.

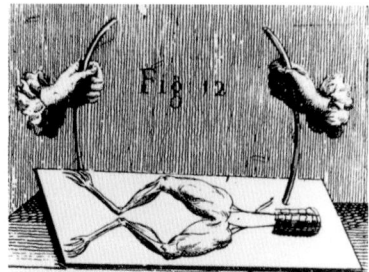

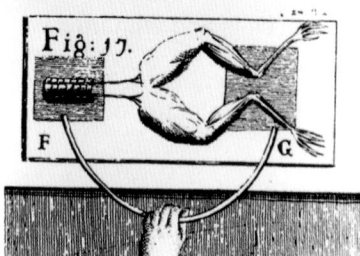

▲ Figure 1 Chemical reactions that produce electrical effects were discovered accidentally

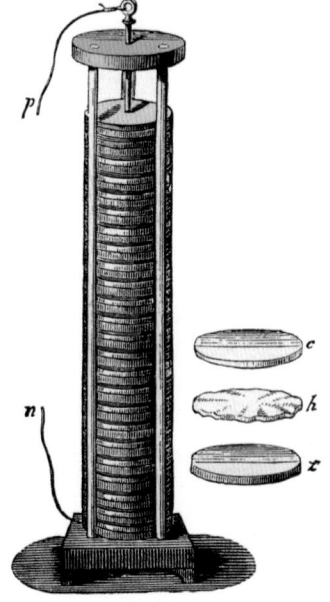

▲ Figure 2 A voltaic pile, the first modern type of electric battery, invented in 1800 by the Italian physicist Alessandro Volta (1745–1827). A voltaic pile consists of alternating plates of two different metals and a piece of wet cardboard or cloth. Wires at the top and bottom carry the electric current, produced by a chemical reaction, to power an electrical device

Challenges in battery technology

Science has been used to solve problems and improve life for humans in many ways. However, scientific advances can inadvertently cause problems. For example, hydrogen is a clean non-polluting fuel used in fuel cells, but it is very difficult to transport and store safely. The heavy metals cadmium and lead used in rechargeable batteries are toxic and can lead to health and environmental problems.

Background to battery technology

You will recall from topics 9 and 19 that in **redox reactions** electrons are transferred from the substance being oxidized to the substance being reduced. Spontaneous redox reactions are exothermic and the energy released in these chemical changes can be used as a portable source of electrical energy in batteries. The push behind moving these electrons, or the voltage of the battery, depends on the nature of the materials.

The mass of the reactive material in the battery or cell is also important. The number of electrons moved is a measure of how much work can be done before the chemical energy is consumed. Let us assume that 1 mol of electrons is moved per mole of atoms in a process. 1 mol of silver, Ag has a mass of 108 g while 1 mol of lithium, Li has a mass of only 7 g. Materials of low molecular mass have a weight advantage, but there are also other factors to consider.

An electric current passing between two dissimilar metals connected by a moist substance was discovered accidently in the 1790s by Luigi Galvani, an Italian anatomy professor. He noticed that he could cause an amputated frog's leg to twitch by touching it with two dissimilar metals. Alessandro Volta, however, doubted that there was electricity that was intrinsic to animal legs. He showed that chemical reactions can produce electricity and made the first 'battery'.

Primary and secondary cells

A **battery** is a series of portable electrochemical cells. In a **primary electrochemical cell** the materials are consumed and the reaction is not reversible. Either the anode, electrolyte, or both need to be replaced or the battery is thrown away, which is usually cheaper. Typically the anode (negative electrode) is oxidized and can no longer be used. Furthermore, the ions travelling through the cell can polarize the cell, which causes the chemical reaction to stop. Polarization can also cause a build-up of hydrogen bubbles on the surface of the anode. These can increase the internal resistance of the cell and reduce its output.

Primary cells do not operate well under high current demands such as electric cars, but are suitable for low-current, long-storage devices such as smoke detectors, wall clocks, or flashlights.

In a **secondary cell** or **rechargeable battery** the chemical reactions that generate electricity can be reversed by applying an electric current to them. Secondary cells can deliver stronger current demands than

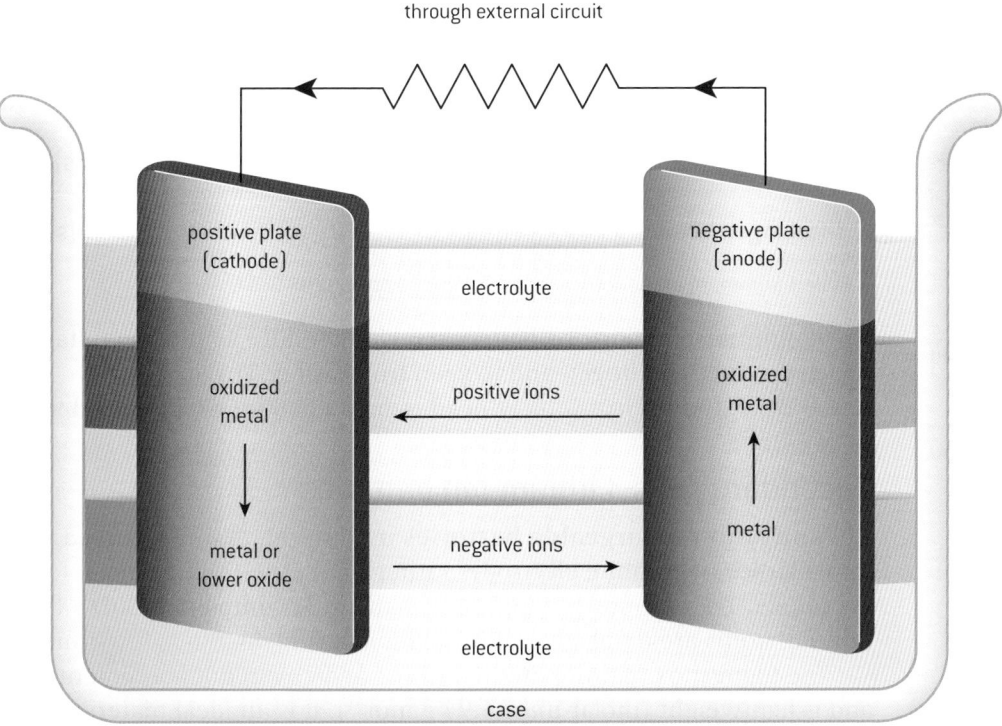

electron flow (current)
through external circuit

positive plate
(cathode)

negative plate
(anode)

electrolyte

oxidized
metal

positive ions

oxidized
metal

metal or
lower oxide

negative ions

metal

electrolyte

case

▲ Figure 3 Structure of an electrochemical cell. In a primary cell the negative anode is
oxidized and the flow of ions causes polarization. This process cannot be reversed in
a primary cell, but can be reversed in a secondary cell or rechargeable battery

primary cells. Secondary cells have a higher rate of self discharge than
do primary cells. When you purchase a replacement battery for a phone,
for example, you would need to charge it before use as it will have self
discharged and so will be only partially charged.

Secondary cells: Lead–acid batteries

Rechargeable batteries are used in cars, for energy storage in the electric
grid (such as to store energy generated from solar cells), in motorized
electric vehicles (hybrid cars, golf carts, etc.), as emergency back-up, and
for many other uses. The typical **lead–acid battery** in a car is recharged
while driving. Electrical energy is used to create ignition and then some
of the energy from combustion is used to reverse the chemical reaction
in the battery, keeping it charged ready for next time. If a car is idle for
a long time the battery could become flat due to self discharge of the
battery.

In the lead–acid battery the electrolyte is sulfuric acid, H_2SO_4. This strong
acid exists in solution as $H^+(aq) + HSO_4^-(aq)$.

The following reactions occur during discharge:

anode: $Pb(s) + HSO_4^-(aq) \rightarrow PbSO_4(s) + H^+(aq) + 2e^-$

cathode: $PbO_2(s) + 3H^+(aq) + HSO_4^-(aq) + 2e^- \rightarrow PbSO_4(s) + 2H_2O(l)$

cell reaction: $Pb(s) + PbO_2(s) + 2H^+(aq) + 2HSO_4^-(aq) \rightarrow 2PbSO_4(s) + 2H_2O(l)$

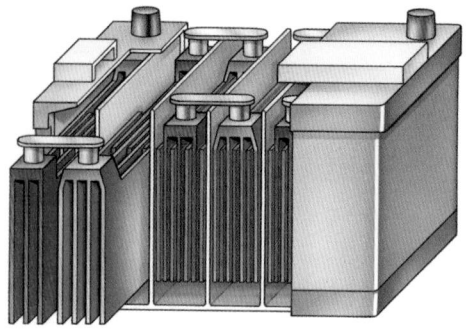

▲ Figure 4 A lead–acid battery consists of a series of cells with lead(IV) oxide plates, lead plates, and sulfuric acid

During charging the above reactions are reversed. Table 1 shows the components of a charged and discharged battery.

	Anode	Electrolyte	Cathode
Fully charged battery	Pb(s)	H_2SO_4(aq)	PbO_2(s)
Discharged battery	$PbSO_4$(s)	H_2SO_4(aq) dilute	$PbSO_4$(s)

▲ Table 1 Summary of the components of a lead–acid battery

The continual charging of a battery tends to produce some overvoltage which produces hydrogen and oxygen from water. This is why non-sealed car batteries occasionally need to be topped up with distilled water.

Secondary cells: Lithium-ion batteries

Lithium-ion rechargeable batteries use lithium atoms absorbed into a lattice of graphite electrodes rather than pure lithium metal for the anode. The cathode is a lithium cobalt oxide complex, $LiCoO_2$. The lithium atoms are oxidized to lithium ions during discharge. As lithium has the highest oxidation potential (most negative reduction potential) and is lightweight (molar mass 6.94 g mol^{-1}), it is an ideal material for lightweight batteries.

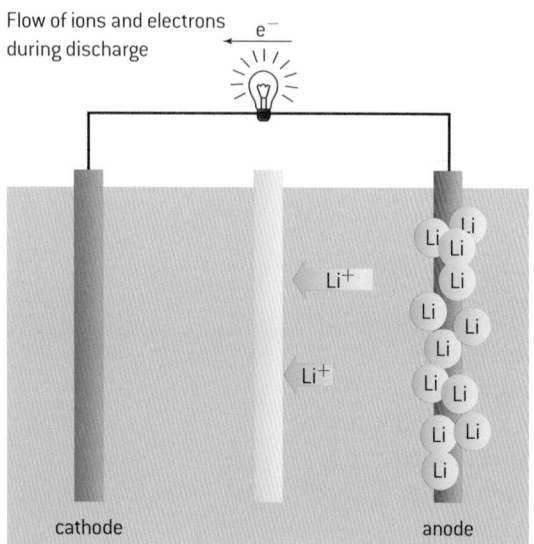

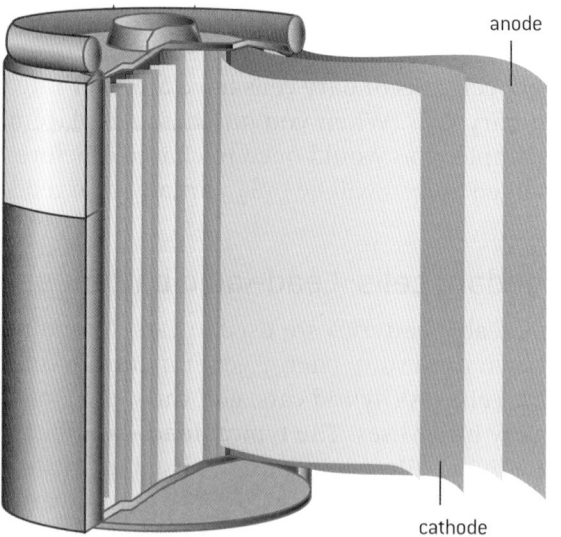

▲ Figure 5 Structure of a typical lithium-ion rechargeable battery. The battery consists of a series of cells composed of cathodes and anodes with a layer (yellow) separating them. When in use, electrons flow from the anode to the cathode in the external circuit and lithium ions from the anode to the cathode inside the cell. When no more lithium ions are left on the anode then the battery is flat. To recharge it the process is reversed, transferring lithium ions back to the anode

During charging the lithium ions in the complex migrate through the electrolyte to the anode where they accept electrons and are reduced to lithium atoms. These atoms become embedded in the graphite lattice, where they can later be oxidized again when the battery is put to use. The electrolyte must be completely non-aqueous, usually a gel polymer, as lithium is an active metal that reacts with water. Table 2 summarizes the reactions during charging and discharge.

Electrode	Charging reaction	Discharging reaction
negative	$Li^+ + e^- \rightarrow Li(s)$ electrons accepted at graphite electrode and Li atoms become embedded in it	$Li(s) \rightarrow Li^+ + e^-$ embedded atoms lose an electron to the external circuit and Li^+ ions migrate to the cathode
positive	$LiCoO_2(s) \rightarrow Li^+ + e^- + CoO_2(s)$	$Li^+ + e^- + CoO_2(s) \rightarrow LiCoO_2(s)$

▲ Table 2 The reactions in the lithium-ion battery

The lithium-ion battery has a very high charge specific density compared with other rechargeable batteries such as lead–acid or nickel–cadmium batteries. Lithium-ion batteries store and deliver 6 times as much energy per kilogram as a lead–acid battery. Some other advantages are:

- they hold charge better than either nickel–cadmium or lead–acid batteries

- they can withstand many recharge cycles

- they contain no heavy metals so used batteries are considered safe for disposal in normal landfill sites.

Disadvantages include the facts that lithium-ion batteries are sensitive to high temperatures, are damaged if allowed to completely run flat, last only a few years, and could possibly explode if overheated or if the separator punctures.

Laptops with lithium-ion batteries left in hot places have been known to explode.

Secondary cells: Nickel–cadmium batteries

The **nickel–cadmium (NiCd) rechargeable cell** was a popular early choice but is losing favour to nickel metal hydride and lithium-ion batteries which both have higher charge specific densities and contain fewer heavy metals, making disposal easier. NiCd batteries have a nickel (III) oxide hydroxide cathode, which becomes reduced to nickel (II) hydroxide during discharge, and a cadmium metal anode, which is oxidized to cadmium hydroxide (table 3).

Electrode	Charging reaction	Discharging reaction
negative	$Cd(OH)_2(s) + 2e^- \rightarrow Cd(s) + 2OH^-(aq)$	$Cd(s) + 2OH^-(aq) \rightarrow Cd(OH)_2(s) + 2e^-$
positive	$2Ni(OH)_2(s) + 2OH^-(aq) \rightarrow 2NiO(OH)(s) + 2H_2O(l) + 2e^-$	$2NiO(OH)(s) + 2H_2O(l) + 2e^- \rightarrow 2Ni(OH)_2(s) + 2OH^-(aq)$

▲ Table 3 The reactions in the nickel–cadmium battery

The solid hydroxides are deposited on the electrodes. Because only hydroxide ions are moving in solution the internal resistance of these cells is low.

Some advantages of NiCad batteries are:

- Their low internal resistance allows for a quick recharge time.

- They can undergo full discharge without damage which allows for high-drain applications.

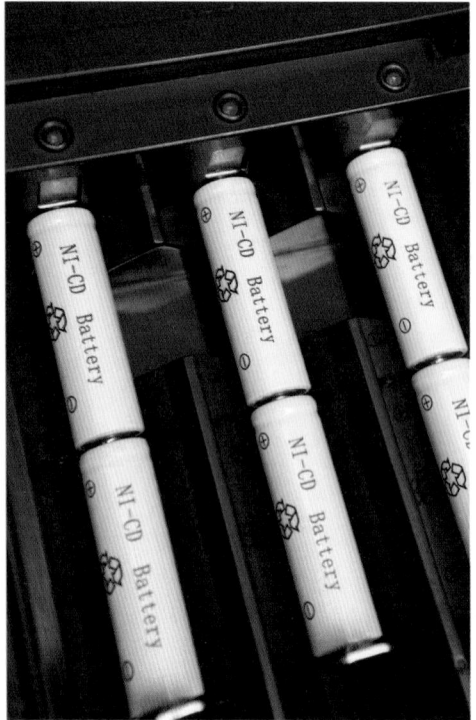

▲ Figure 6 Nickel–cadmium batteries being recharged. NiCd batteries have a quick recharge time

They also have the following disadvantages:

- Their high cost and the use of the heavy metal cadmium makes both production and disposal an environmental concern.

- They quickly lose charge at elevated temperatures.

Nickel metal hydride batteries, nickel–zinc batteries, and fuel cells are proving better substitutes for nickel–cadmium batteries.

The voltage of a cell

The voltage of a battery, whether primary or secondary, depends on the nature of the anode and cathode. The further apart the standard electrode potentials of the oxidizing and reducing materials, the more voltage per cell is available. Placing cells in series provides an increased voltage. Lead–acid car batteries use many such cells and usually provide 12 V.

The total number of electrons moving along with the energy given to them by the cell give a measure of how much work can be done by the current. This in turn depends on the nature and quantity of the materials (the mass and surface area of the electrodes) as well as the specific energy density.

It is the electrons moving in the external circuit that provide us with useful energy but each electrochemical cell also has to move cations and anions inside the cell. A battery's **internal resistance** depends on the ion mobility, the electrolyte conductivity and the electrode surface area.

Reactions occur faster at higher temperatures. At lower temperatures reactions slow down. Ion mobility is reduced, and the battery's internal resistance is increased (figure 7). While batteries have lower resistance at higher temperatures, they also have an increased rate of self discharge, so storing batteries at higher temperatures is not advisable.

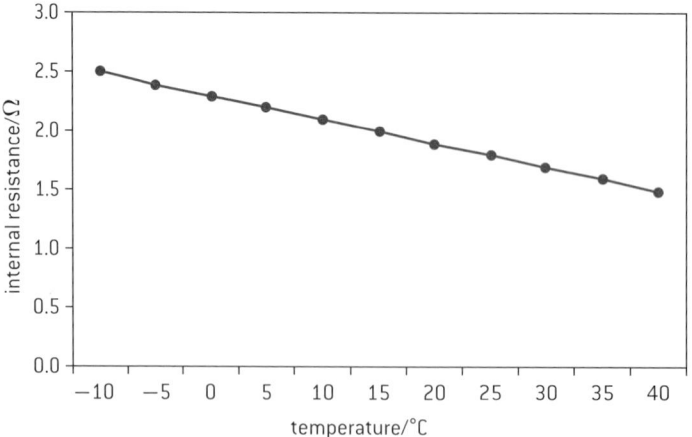

▲ Figure 7 Internal resistance versus temperature for a lead–acid battery

As mentioned above, electrodes with a large surface area allow a higher conductivity. The large plates in a lead–acid battery can produce the high current needed to start a car (figure 4). The maximum current a battery can provide is limited by the internal resistance of the battery.

Hydrogen fuel cells

The PEM fuel cell

A **fuel cell** is an electrochemical device that converts the chemical potential energy in a fuel into electrical energy. In the hydrogen fuel cell the fuel is hydrogen, which is oxidized by oxygen and produces water. There is therefore no pollution and fuel cells are very efficient. The key components of a fuel cell are:

- the **electrolyte** or **separator** which prevents components from mixing – the **proton exchange membrane** (PEM) is a polymer which allows H^+ ions to diffuse through but not electrons or molecules (acts as a salt bridge)

- the oxidizing and reducing electrodes which are catalysts that allow the chemical reactions to occur

- the bipolar plate which collects the current and builds up the voltage in the cell.

Hydrogen is oxidized at the anode and oxygen reduced at the cathode:

anode: $\quad H_2 \rightarrow 2H^+ + 2e^-$

cathode: $\quad O_2 + 4e^- \rightarrow 2O^{2-}$

cell reaction: $2H_2 + O_2 \rightarrow 2H_2O$

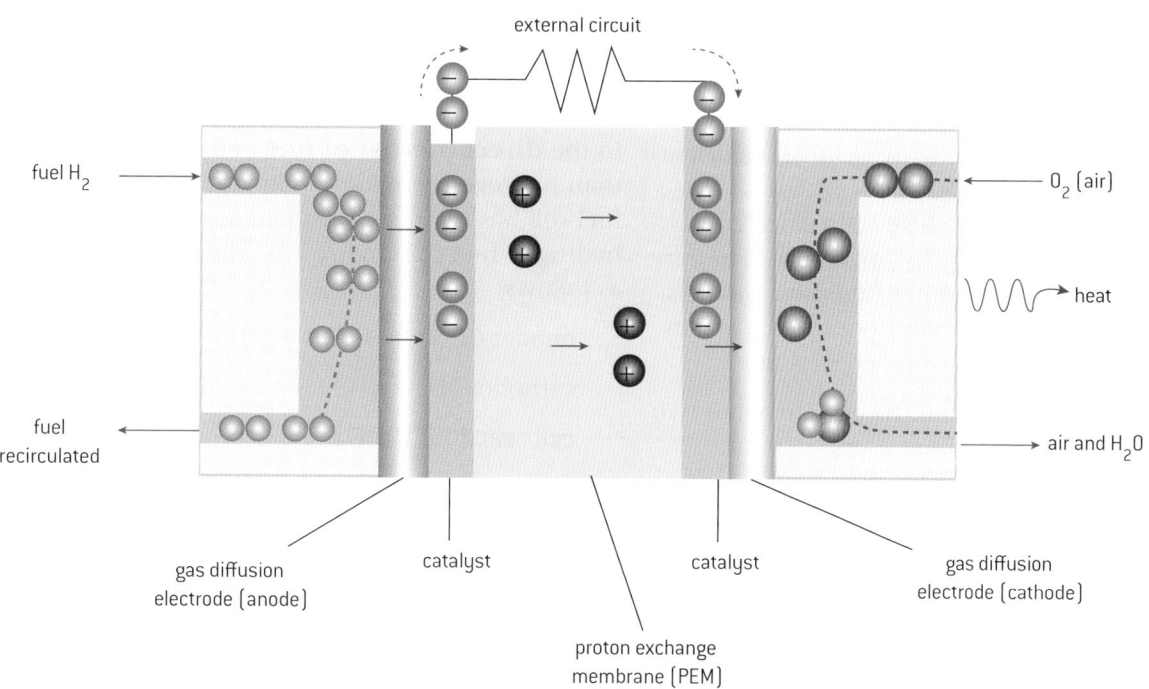

▲ Figure 8 The proton exchange membrane (PEM) in a hydrogen fuel cell. The products are water and heat

Alkali fuel cells

Alkali fuel cells were used as early as the *Apollo* missions to provide electricity and drinking water. The electrolyte in these cells was a solution of potassium hydroxide, providing a source of hydroxide ions. As the OH^- ions migrated towards the anode they reacted with H^+ ions producing water (figure 9).

If an acidic electrolyte such as phosphoric acid is used, then positive H^+ ions in the electrolyte migrate towards the cathode (figure 10).

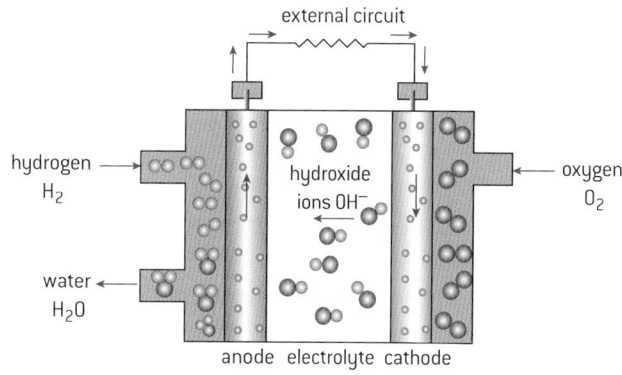

▲ Figure 9 The alkali fuel cell

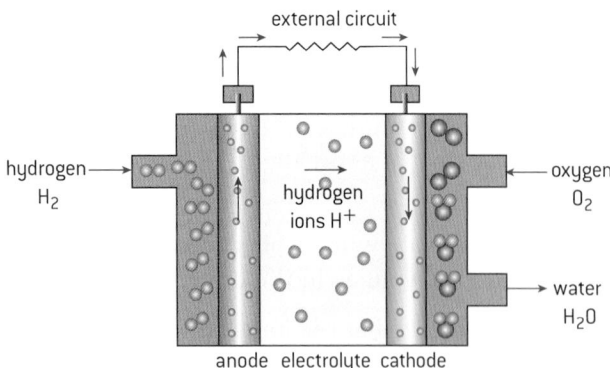
▲ Figure 10 A fuel cell with an acidic electrolyte

In a PEM hydrogen cell, water is formed at the cathode. In an alkali fuel cell it is formed at the anode.

Hydrogen fuel sources

As we have seen, hydrogen fuel cells use hydrogen and oxygen as fuel. These cells are clean and efficient – the heat formed ($H_2(g) + \frac{1}{2}O_2(g) \rightarrow H_2O(l)$ is exothermic) can be used as a heat source, increasing their efficiency. However, the hydrogen has to be very pure and often platinum or other expensive catalysts are impregnated on graphite electrodes which makes them expensive to run on a commercial scale.

Oxygen can be obtained from the air. There are two main sources of hydrogen:

1   Clean hydrogen can be produced by the electrolysis of water. Solar cells or wind generators provide the cleanest form of energy for powering the electrolysis.

2   Hydrogen is made from reforming hydrocarbons or biofuels. Coal gasification or the conversion of methane to synthesis gas (sub-topic C.2) are two such methods. The hydrocarbons are reacted with steam to produce carbon monoxide and hydrogen:

$$C_xH_y + xH_2O \rightarrow xCO + (\tfrac{y}{2} + x)H_2$$

Some carbon dioxide may also be produced. The hydrogen must be separated and purified before it can be used in a fuel cell, adding to the expense of this method. The process is endothermic so again energy needs to be supplied. However, there is an ample source of renewable fuel for this process and it is about 70% efficient. Approximately 85% of hydrogen used in fuel cells is made by this method.

▲ Figure 11 A hydrogen fuel cell bus in Reykjavik, Iceland

The direct methanol fuel cell

In the **direct methanol fuel cell** methanol rather than hydrogen provides H^+ ions at the anode. The fuel cell has the same components as the PEM hydrogen fuel cell (figure 12). The cell reactions are as follows:

anode: $CH_3OH + H_2O \rightarrow 6H^+ + 6e^- + CO_2$

cathode: $\frac{3}{2}O_2 + 6H^+ + 6e^- \rightarrow 3H_2O$

cell reaction: $CH_3OH + \frac{3}{2}O_2 \rightarrow CO_2 + 2H_2O$

The anode reaction requires water, so a dilute solution of approximately 1 mol dm⁻³ methanol is used. Even though this lowers the energy

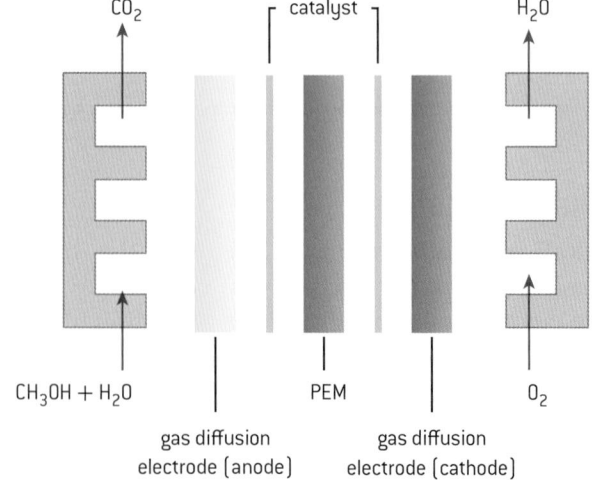
▲ Figure 12 In a direct methanol fuel cell the gas diffusion layer disperses methanol + water and oxygen to their respective catalysts where they react. Carbon dioxide is produced at the anode while steam (H_2O) is produced at the cathode

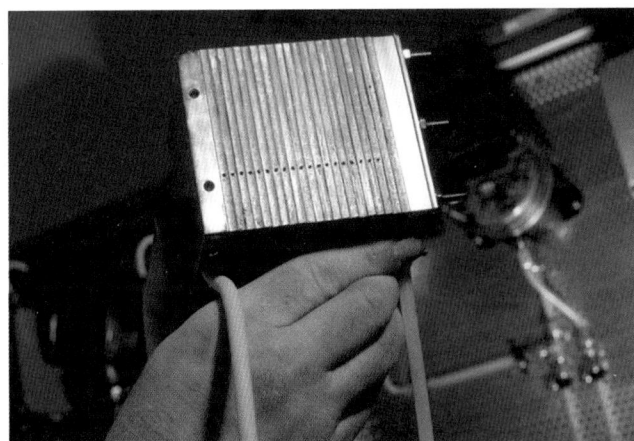

▲ Figure 13 A portable direct methanol fuel cell. It can be used to power demanding items such as laptop computers and video cameras

density from pure methanol, it is still higher than hydrogen as a source in fuel cells. In most cells, pure methanol is continuously fed into the system while water is recirculated, so the concentration of methanol remains constant. Of course this is not as clean as a hydrogen fuel cell as carbon dioxide is a product of the cell reaction.

The direct methanol fuel cell operates at a temperature of 120 °C compared to the lower temperature of 80 °C in the PEM hydrogen fuel cell. The amount of platinum catalyst required in the direct methanol fuel cell is greater than the PEM hydrogen fuel cell however.

Comparing fuels

A distinct advantage of direct methanol fuel cells is their high energy density (figure 14). The slope of the graph gives the energy per unit volume,

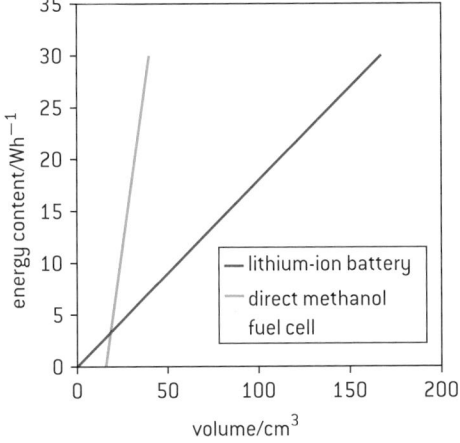

▲ Figure 14 Comparison of the energy density per unit volume for the lithium-ion battery and the direct methanol fuel cell

showing that the direct methanol fuel cell has a much higher energy density than the lithium-ion battery.

When comparing fuels, the energy density (energy per unit volume) and specific energy (energy per unit mass) can give quite different pictures (table 4).

Fuel source	Energy density/ MJ dm^{-3}	Specific energy/ MJ kg^{-1}
compressed hydrogen	1.9	120
methanol	16	20
liquefied natural gas	21	50
liquid propane	27	46
gasoline	32	46

▲ Table 4 Comparing fuels in terms of energy density and specific energy

The specific energy (energy to mass ratio) of hydrogen is more than double that for any other fuel. Because just 2.02 g of hydrogen H_2 contains 1 mol of fuel, compared with 32.05 g for 1 mol of methanol CH_3OH or 114.26 g for 1 mol of octane C_8H_{18}, it would be easy to imagine that hydrogen would be the primary fuel choice.

However, fuels need to be stored and delivered. The molar volume of a gas at room temperature and 1 atm pressure is approximately 24 dm^3. One mol of gaseous hydrogen under these conditions would require a 24 dm^3 storage tank, which adds to the weight if the device is to be portable, such as in a car. One mol of methanol would occupy 40.4 cm^3, and the same 24 dm^3 storage tank could hold over 545 mol of methanol fuel. Even when compressed the hydrogen gas occupies a much larger volume, and regulators and compressors add to the weight.

Octane (gasoline) offers the highest energy density but has associated environmental problems. Polymer electrolyte fuel cells and nanocatalysts are being researched which offer five or more times higher energy densities than methanol (figure 15).

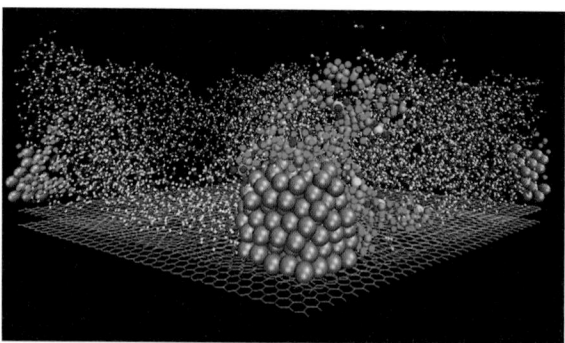

▲ Figure 15 Fuel cell nanocatalysts. Platinum nanoparticles (yellow) on a carbon substrate (green grid). The red and white molecules are water, while the coloured chains are Nafion fragments. Nafion is a perfluorinated polymer with sulfonic acid groups attracted along its backbone used in the proton exchange membranes of some fuel cells. This simulation was produced at the National Energy Research Scientific Computing Center (NERSC), based at the Lawrence Berkeley National Laboratory, California, USA

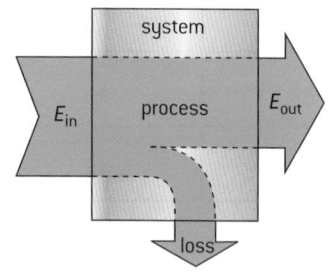

▲ Figure 16 A higher energy output E_{out} corresponds to a lower heat loss to entropy and a higher thermodynamic efficiency

Calculations for electrochemical cells

Thermodynamic efficiency

While octane has a high energy density, fuel cells tend to have a higher thermodynamic efficiency. **Thermodynamic efficiency** is the ratio of the Gibbs energy change to the enthalpy change.

$$\text{thermodynamic efficiency} = \frac{\Delta G}{\Delta H}$$

Recall that $\Delta G^{\ominus} = -nFE^{\ominus}$, and is a measure of the electrical energy output (sub-topic 19.1). ΔH is the total chemical energy that would be released during combustion, and could be considered the energy input. In fuel cells some of the total available energy is lost as heat to entropy (figure 16).

For example, the enthalpy of combustion of hydrogen is -286 kJ mol^{-1}.

$$H_2(g) + \tfrac{1}{2}O_2(g) \rightarrow H_2O(l) \quad \Delta H = -286 \text{ kJ}$$

However, in a fuel cell steam rather than liquid water is produced, so a better equation is:

$$H_2(g) + \tfrac{1}{2}O_2(g) \rightarrow H_2O(g) \quad \Delta H = -242 \text{ kJ}$$

ΔG of the reaction above is -229 kJ.

The thermodynamic efficiency is therefore given by:

$$\text{thermodynamic efficiency} = \frac{\Delta G}{\Delta H}$$

$$= \frac{-229 \text{ kJ}}{-242 \text{ kJ}} = \sim0.95 \text{ or } 95\%$$

The hydrogen fuel cell can theoretically convert 95% of the available chemical energy to electricity. For the methanol fuel cell:

$$CH_3OH(l) + \tfrac{3}{2}O_2(g) \rightarrow CO_2(g) + 2H_2O(g)$$

$$\Delta H^{\ominus} \text{ (reaction)} = \sum \Delta H_f^{\ominus} \text{ (products)} - \sum \Delta H_f^{\ominus} \text{ (reactants)}$$

and

$$\Delta G^{\ominus} \text{ (reaction)} = \sum \Delta G_f^{\ominus} \text{ (products)} - \sum \Delta G_f^{\ominus} \text{ (reactants)}$$

This is a combustion reaction; from section 13 of the *Data booklet*, $\Delta H = -726$ kJ.

$$\Delta G = \Delta G_f^{\ominus}[\theta \; CO_2(g)] + 2\Delta G_f^{\ominus}[H_2O(g)] - (\Delta G_f^{\ominus}[CH_3OH(l)]$$
$$+ \tfrac{3}{2}\Delta G_f^{\ominus}[O_2(g)])$$

Substituting values from section 12 of the *Data booklet*:

$$= -394 + 2(-229) - (-167 + 0) = -685 \text{ kJ}$$

$$\text{thermodynamic efficiency} = \frac{\Delta G}{\Delta H} = -685 \text{ kJ}/-726 \text{ kJ}$$

$$= \sim 94\% \text{ efficient}$$

Fuel cells may not operate at their theoretical maximum efficiency. For example, for a hydrogen fuel cell that typically outputs 0.7 V, we can calculate its thermal efficiency as follows:

$$\Delta G^{\ominus} = -nFE^{\ominus} \text{ (from topic 19)}$$

$$\Delta G^{\ominus} = -(2 \times 96\,500 \times 0.7) = -135\,100 \text{ J or } -135.1 \text{ kJ}$$

$$\text{thermodynamic efficiency} = \frac{\Delta G}{\Delta H} = \frac{-135.1 \text{ kJ}}{242 \text{ kJ}}$$

$$= \sim 56\% \text{ efficient}$$

Internal resistance caused by poor ion mobility or reduced electrolyte conductivity could be a factor contributing to this lowered efficiency.

The potential of a cell under non-standard conditions: The Nernst equation

It is possible to alter the EMF of a cell by changing the concentrations of the mobile ions in the cell. Recall that in any standard cell, the standard conditions are 1 mol dm^{-3} concentrations, 100 kPa pressure for gases, and a temperature of 298 K. When these conditions exist it is possible to predict the EMF of a voltaic cell by adding their half-cell potentials under standard conditions.

The **Nernst equation** can be used to calculate the potential of an electrochemical cell under non-standard conditions:

$$E = E^{\ominus} - \frac{RT}{nF} \ln Q$$

- $E^{\ominus}$ is the EMF of the cell under standard conditions.

- R is the universal gas constant, 8.31 J K^{-1} mol^{-1}.

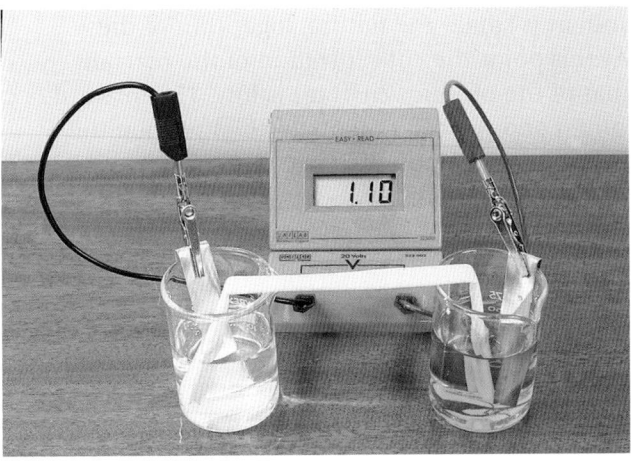
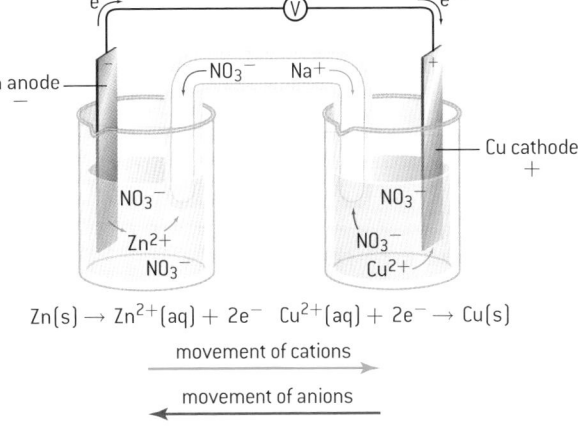

▲ Figure 17 The Daniell cell has an EMF of 1.10 V under standard conditions. Changing the aqueous concentrations of $[Zn^{2+}]$ and $[Cu^{2+}]$ will affect the EMF

- T is the temperature in kelvin, usually 298 K.

- n is the number of moles of electons transferred in the balanced equation. For the Daniell cell (figure 17), $n = 2$.

- F is the Faraday constant, the electric charge on 1 mol of electrons: $F = 96\ 500$ C mol^{-1}.

- Q is the reaction quotient, the ratio of the concentration of ions undergoing oxidation to the concentration of ions undergoing reduction:

$$Q = \frac{[\text{ions being oxidized}]}{[\text{ions being reduced}]}$$

For the Daniell cell:

$$Q = \frac{[Zn^{2+}(aq)]}{[Cu^{2+}(aq)]}$$

For a stoichiometric equilibrium, Q can be expressed as follows:

$$wW + xX \rightleftharpoons yY + zZ$$

$$Q = \frac{[Y]^y[Z]^z}{[W]^w[X]^x}$$

The net equation or overall cell reaction for the Daniell cell is:

$$Cu^{2+}(aq) + Zn(s) \rightarrow Zn^{2+}(aq) + Cu(s)$$

If $[Cu^{2+}]$ is increased to 1.5 mol dm^{-3} and $[Zn^{2+}]$ is decreased to 0.50 mol dm^{-3}, the forward equilibrium is favoured and the cell potential increases. The quantity of that increase is:

$$E = E^\ominus - \frac{RT}{nF} \ln Q$$

$$E = 1.10 \text{ V} - \frac{8.31 \times 298}{2 \times 96\ 500} \times \ln \frac{0.5}{1.5}$$

$$= 1.10 \text{ V} - \frac{8.31 \times 298}{2 \times 96\ 500} \times (-1.10)$$

$$= 1.10 \text{ V} - (-0.0141 \text{ V})$$

$$= 1.11 \text{ V}$$

Figure 18 shows a plot of the potential of the Daniell cell as a function of the natural logarithm of the reaction quotient: $\ln \dfrac{[Zn^{2+}(aq)]}{[Cu^{2+}(aq)]}$

When the reaction quotient is small, the natural logarithm of the quotient is negative. At this point the concentration of the reactant $Cu^{2+}(aq)$ is high and that of the product $Zn^{2+}(aq)$ is low. The forward reaction is greatly favoured and the cell potential is high. When the reaction quotient gets very large the potential becomes negative. This means the reverse reaction is favoured.

> The Nernst equation is provided in section 1 of the *Data booklet*.

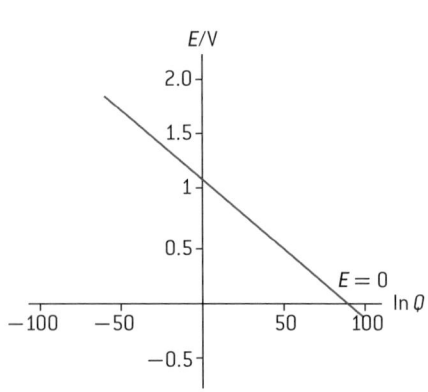

▲ Figure 18 The potential of the Daniell cell against the natural logarithm of the reaction quotient Q, $\ln \dfrac{[Zn^{2+}(aq)]}{[Cu^{2+}(aq)]}$

Study tip

Remember that $\ln\left(\dfrac{a^x}{b^y}\right)$ expands to $(x \times \ln a) - (y \times \ln b)$.

You can use this to find the concentrations of cells required to obtain a certain voltage.

Worked example

Given the standard cell notation Al(s) | Al^{3+}(aq) (0.01 M) || Fe^{2+}(aq) (0.1 M) | Fe(s), calculate the EMF of the cell.

Solution

From section 24 of the *Data booklet*:

$$E^\ominus$$

$$Al^{3+}(aq) + 3e^- \rightleftharpoons \quad Al(s) \quad\quad -1.66$$

$$Fe^{2+}(aq) + 2e^- \rightleftharpoons \quad Fe(s) \quad\quad -0.45$$

The net equation becomes:

$$2Al(s) + 3Fe^{2+}(aq) \rightleftharpoons 2Al^{3+}(aq) + 3Fe(s) \quad\quad E^\ominus = 1.21 \text{ V}$$

The aluminium is oxidized at the anode, as the notation shows. Six electrons are transferred.

$$E = E^\ominus - \left(\frac{RT}{nF}\right) \ln Q$$

$$Q = [Al^{3+}]^2 \_ [Fe^{2+}]_3$$

Assuming the reaction happens at 298 K,

$$E = 1.21 \text{ V} - \frac{8.31 \times 298}{6 \times 96\,500} \ln \frac{(0.01)^2}{(0.1)^3} = 1.22 \text{ V}.$$

A concentration cell

A **concentration cell** has the same electrodes in each half-cell, but the concentration of the ions in each half-cell is different. For example, $Fe(s) \mid Fe^{2+}(aq) \, (0.01 \text{ mol dm}^{-3}) \parallel Fe^{2+}(aq) \, (0.1 \text{ mol dm}^{-3}) \mid Fe(s)$ represents a concentration cell – both the anode and cathode are the same material, solid iron. The oxidation cell has a lower concentration of ions than the reduction cell. As both the oxidation and reduction half-cells are the same, the standard condition $E^\ominus$ is zero.

$$Fe^{2+}(aq) + 2e^- \rightarrow Fe(s) \quad\quad E^\ominus = -0.45 \text{ V}$$

$$Fe(s) \rightarrow Fe^{2+}(aq) + 2e^- \quad\quad E^\ominus = +0.45 \text{ V}$$

However, application of the Nernst equation reveals that a small potential is generated.

$$E = E^\ominus - \frac{RT}{nF} \ln Q \quad\quad Q = \frac{0.01}{0.1}$$

$$E = 0 - \frac{0.02566}{2} \times (-2.3)$$

$$= 0 + 0.029\,54 = \sim 0.03 \text{ V}$$

The most common concentration cells are oxygen concentration cells. The difference in the amount of dissolved oxygen generates a small potential difference between the half-cells. This is often a leading cause of corrosion as the metal may have different concentrations of oxygen around it, especially if scratched or exposed.

Another use of concentration cells is in the combined pH electrode (figure 20). A pH meter has a fixed reference electrode and a temperature sensor, as T is a factor in the Nernst equation. Each pH meter needs to be calibrated to give a correct potential difference between two known concentrations.

▲ Figure 19 The German chemist Walther Hermann Nernst (1864–1941). Nernst was appointed to a professorship in Berlin in 1905. That year he proposed the third law of thermodynamics: entropy change approaches zero at a temperature of absolute zero. This work earned Nernst the 1920 Nobel Prize in Chemistry.

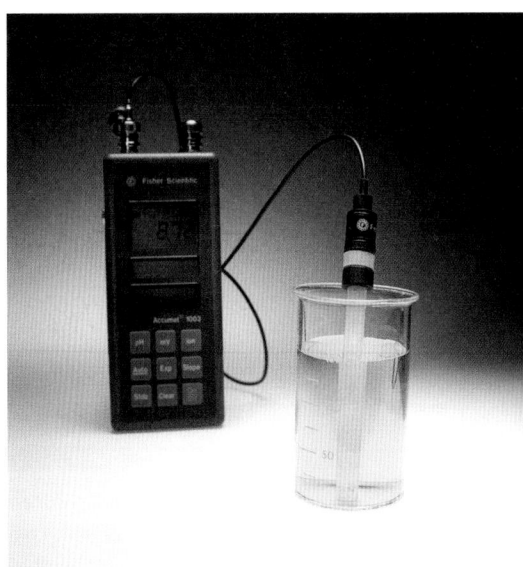

▲ Figure 20 A pH meter that uses a concentration cell

Microbial fuel cells

A **microbial fuel cell** converts chemical energy available from a substrate into electricity by anaerobic oxidation carried out by microorganisms. The aerobic oxidation of glucose produces carbon dioxide and water. However, anaerobic oxidation produces H^+ ions and electrons. These electrons can be harnessed at an anode and the H^+ ions permitted to diffuse through a PEM where they reduce oxygen, forming water. The aerobic and anaerobic oxidations of glucose are given by the equations:

aerobic oxidation:
$$C_6H_{12}O_6 + 6O_2 \rightarrow 6CO_2 + 6H_2O$$

anaerobic oxidation:
$$C_6H_{12}O_6 + 6H_2O \rightarrow 6CO_2 + 24H^+ + 24e^-$$

The bacteria that carry out this oxidation live in the anode half-cell and will work on many substrates such as ethanoate ion, CH_3COO^-, carbohydrates, and waste water.

Bacteria of the *Geobacter* species (figure 22) are proving to be useful in microbial fuel cells because of their ability to transfer electrons to the surfaces of electrodes and their ability to destroy petroleum contaminants and utilize waste water in anaerobic oxidation.

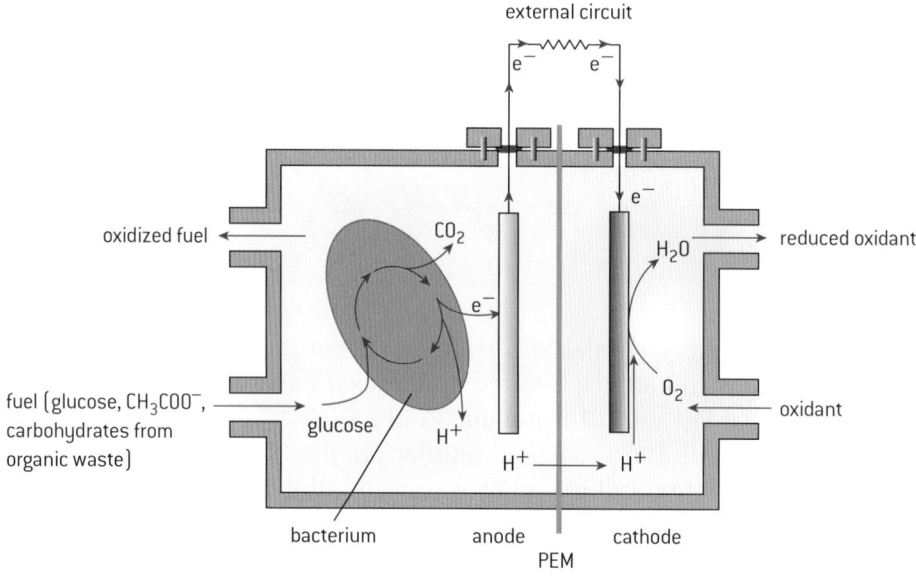

▲ Figure 21 In a microbial fuel cell bacteria oxidize substrates by an electron transfer mechanism

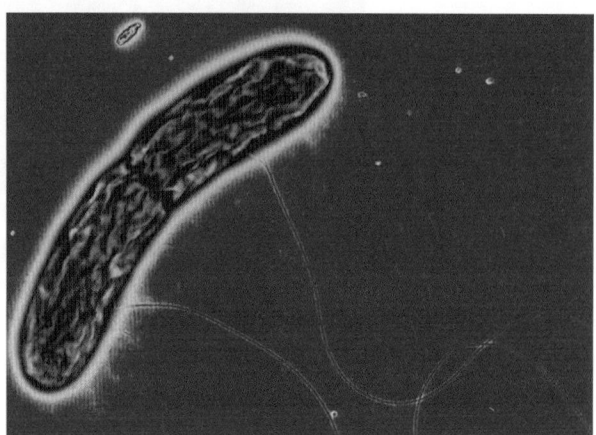

▲ Figure 22 *Geobacter metallireducens* is an anaerobic bacterium that oxidizes organic compounds to form carbon dioxide, using iron (II) oxide or other metals as an electron acceptor. In a microbial fuel cell *Geobacter* can oxidize waste organic matter and transfer surplus electrons directly to an electrode. *Geobacter* grow long filaments known as pili that are electrically conductive

▲ Figure 23 A toy car powered by a microbial fuel cell. The fuel cell is comprised of several beakers of river sediment topped with water. Each beaker contains a graphite anode buried in the sediment, and a graphite cathode suspended in the water. Anaerobic bacteria in the sediment colonize the anode and oxidize the organic matter in the sediment

Geobacter metallireducens gains energy by using iron oxide as we use oxygen. Oxygen oxidizes our food and is reduced in the process. *Geobacter* does the same with iron (II) oxide but because it has electrically conducting filaments (pili) the electron transfer can be captured and utilized.

Microbial fuel cells can be very compact and could possibly be developed to produce electricity from human waste. This would make this form of energy invaluable on a long space mission, such as a 2 year mission to Mars. The possibility of using waste to produce energy makes microbial fuel cells an ideal sustainable energy source worthy of investigation.

Questions

1 The International Baccalaureate Nature of Science statement 4.7 reads: "All science has to be funded and the source of the funding is crucial in decisions regarding the type of research to be conducted." In light of what you have read in this topic suggest which form of alternative energy deserves the most funding, using specific examples in rationalizing your answer.

2 a) State what factors determine the voltage output of a battery.

b) Outline what determines the total energy, or work, a battery can do.

c) Lead–acid batteries employ a large surface area on the anode and cathode plates. Explain the effect that large thick plates on a battery have on the voltage and work a battery can output.

3 Explain what factors influence a battery's internal resistance.

4 Compare and contrast nickel–cadmium batteries with lithium-ion batteries, discussing energy density and internal resistance factors.

5 a) $Li^+ + e^- \rightarrow Li(s)$ is a reaction occurring at one of the electrodes in a lithium-ion battery. State at which electrode (anode/cathode) this reaction occurs and whether this is the charging or discharging reaction.

b) Identify the reaction at the opposite electrode.

c) Explain why lithium-ion batteries must be sealed.

6 Outline the function of the proton exchange membrane (PEM) in fuel cells. Explain why this membrane is important in microbial fuel cells?

7 A direct ethanol fuel cell has the following reactions:

anode reaction:
$$CH_3CH_2OH + 3H_2O \rightarrow 2CO_2 + 12H^+ + 12e^-$$
$$E^\ominus = 0.085 \text{ V}$$

cathode reaction:
$$3O_2 + 12H^+ + 12e^- \rightarrow 6H_2O$$
$$E^\ominus = 1.23 \text{ V}$$

overall reaction:
$$CH_3CH_2OH + 3O_2 \rightarrow 3H_2O + 2CO_2$$
$$E^\ominus = 1.145 \text{ V}$$

a) Calculate the theoretical thermal efficiency of this cell.

b) In practice ethanol fuel cells are less efficient than the direct methanol fuel cell, largely due to polarization and internal resistance. Ethanol has a larger energy density than methanol. Explain why this is an advantage and state two other advantages that a direct ethanol fuel cell might have over a direct methanol fuel cell.

8 Calculate the EMF of a concentration cell that has silver electrodes and $[Ag^+] = 0.10 \text{ mol dm}^{-3}$ in one cell and 2.0 mol dm^{-3} in the other. Which cell is the anode and which is the cathode?

9 a) Sketch an electrochemical cell that has $Zn \mid Zn^{2+}$ in one half-cell and $Al \mid Al^{3+}$ in the other, identifying the anode and cathode.

b) Use section 24 from the *Data booklet* and the Nernst equation to calculate the EMF of this cell if $[Zn^{2+}] = 2.0 \text{ mol dm}^{-3}$ and $[Al^{3+}] = 0.50 \text{ mol dm}^{-3}$.

10 List some advantages and disadvantages of hydrogen, direct methanol, and microbial fuel cells.

C.7 Nuclear fusion and nuclear fission (AHL)

Understandings

Nuclear fusion:

→ The mass defect (Δm) is the difference between the mass of the nucleus and the sum of the masses of its individual nucleons.

→ The nuclear binding energy (ΔE) is the energy required to separate a nucleus into protons and neutrons.

Nuclear fission:

→ The energy produced in a fission reaction can be calculated from the mass difference between the products and reactants using the Einstein mass–energy equivalence relationship $E = mc^2$.

→ The different isotopes of uranium in uranium hexafluoride can be separated using diffusion or centrifugation, causing fuel enrichment.

→ The effusion rate of a gas is inversely proportional to the square root of the molar mass (Graham's law).

→ Radioactive decay is kinetically a first order process with the half-life related to the decay constant by the equation $\lambda = \frac{\ln 2}{t_{1/2}}$.

→ The dangers of nuclear energy are due to the ionizing nature of the radiation it produces which leads to the production of oxygen free-radicals such as superoxide (O_2^-) and hydroxyl ($HO\cdot$). These free-radicals can initiate chain reactions that can damage DNA and enzymes in living cells.

Applications and skills

Nuclear fusion:

→ Calculate the mass defect and binding energy of a nucleus.

→ Apply the Einstein mass–energy equivalence relationship, $E = mc^2$, to determine the energy produced in a fusion reaction.

Nuclear fission:

→ Apply the Einstein mass–energy equivalence relationship to determine the energy produced in a fission reaction.

→ Discuss the different properties of UO_2 and UF_6 in terms of bonding and structure.

→ Solve problems involving radioactive half-life.

→ Explain the relationship between Graham's law of effusion and the kinetic theory.

→ Solve problems on the relative rate of effusion using Graham's law.

Nature of science

→ Trends and discrepancies – our understanding of nuclear processes came from both theoretical and experimental advances. Intermolecular forces in UF_6 are anomalous and do not follow the normal trends.

Background to nuclear technology

Our understanding of nuclear processes comes from both theoretical and experimental advances. Practical difficulties remain for the economic production of energy from fusion reactions.

The ability to **enrich uranium** is a crucial step in the generation of nuclear energy from fission reactions. Uranium hexafluoride is used in uranium processing because its unique physical properties make it very convenient. It can exist as a gas, liquid, or solid at temperatures and pressures commonly used in industrial processes.

Funding for nuclear research has been made available because of the ability to obtain large amounts of energy from small quantities of matter. However, there can be lack of clarity over whether this funding is targeted at nuclear research for peaceful or military purposes.

Nuclear energy

Nuclear energy allows us to obtain large quantities of energy from small quantities of matter, making it a very important industry.

The energy produced in a nuclear reaction can be calculated from the mass difference between the products and reactants using the Einstein mass–energy equivalence relationship. The **mass defect** is the difference between the mass of the nucleus and the sum of the masses of its nucleons (protons and neutrons), and its relationship to the nuclear binding energy was explained in sub-topic C.3. The **nuclear binding energy** (ΔE) is the energy required to separate a nucleus into protons and neutrons.

 Worked example

Example 1

Controlled nuclear fission is the process used in nuclear power plants today. One such fission reaction is:

$$^{235}_{92}U + ^{1}_{0}n \rightarrow ^{89}_{36}Kr + ^{144}_{56}Ba + 3^{1}_{0}n$$

Calculate the energy released if 1 g of $^{235}_{92}U$ undergoes fission in a nuclear reactor. Use the mass data from table 1 (1 amu = 1.66×10^{-27} kg).

Particle	Mass /amu (μ)
neutron	1.008 665
proton	1.007 825
$^{235}_{92}U$	235.043 95
$^{89}_{36}Kr$	89.919 59
$^{144}_{56}Ba$	143.922 953

▲ Table 1 Mass data for particles involved in the fission reaction for U-235

Solution

Step 1: Calculate the mass defect in amu (μ) for 1 atom of U-235.

$$\text{mass defect} = \sum(\text{mass of products}) - \sum(\text{mass of reactants})$$

$$= 3(1.008\ 665) + 89.919\ 59 + 143.922\ 953 - (235.043\ 95 + 1.008\ 665)$$

$$= 0.815\ 92\ \mu$$

Step 2: Convert this mass defect to kg atom^{-1}.

$$0.81592\ \mu \times 1.66 \times 10^{-27}\ \text{kg}\ \mu^{-1}$$
$$= 1.354\ 43 \times 10^{-27}\ \text{kg atom}^{-1}$$

Step 3: Find the number of atoms undergoing fission in this chain reaction mechanism:

$$\frac{1\ \text{g}}{235.043\ 95\ \text{g·mol}^{-1}} = 0.004\ 254\ \text{mol U-235}$$

$$0.004\ 254\ \text{mol} \times 6.022 \times 10^{23}\ \text{atoms mol}^{-1}$$
$$= 2.562 \times 10^{21}\ \text{atoms}$$

Step 4: Find the mass defect in kg for the reaction, and the energy released.

$$2.562 \times 10^{21} \text{ atoms} \times 1.354\,43 \times 10^{-27}\,\text{kg atom}^{-1} = 3.4701 \times 10^{-6}\,\text{kg}$$

$$E = mc^2$$
$$= (3.4701 \times 10^{-6}\,\text{kg})(3.00 \times 10^8\,\text{m s}^{-1})^2$$
$$= 3.123 \times 10^{11}\,\text{J: approx. 310 GJ of energy}$$

Example 2

The nuclear binding energy ΔE is the energy required to separate a nucleus into protons and neutrons.

Calculate ΔE for the U-235 nucleus in MeV, given that $1\,\text{eV} = 1.6022 \times 10^{-19}\,\text{J}$, and compare this with the nuclear binding energy for iron from section 36 of the *Data booklet*.

Solution

$^{235}_{92}\text{U}$ contains 92 protons and $235 - 92 = 143$ neutrons.

total mass of separated nucleons
$$= 92(1.007\,825\,\mu) + 143(1.008\,665\,\mu)$$
$$= 236.958\,995\,\mu$$

$$\text{mass defect} = 236.958\,995\,\mu - 235.043\,95\,\mu$$
$$= 1.915\,045\,\mu$$

convert to kg: $1.915\,045\,\mu \times 1.66 \times 10^{-27}\,\text{kg }\mu^{-1}$
$$= 3.178\,97 \times 10^{-27}\,\text{kg}$$

$$E = mc^2$$
$$= (3.178\,97 \times 10^{-27}\,\text{kg})(3.00 \times 10^8\,\text{m s}^{-1})^2$$
$$= 2.861 \times 10^{-10}\,\text{J}$$
$$= 1786\,\text{MeV (since } 1\,\text{eV} = 1.6022 \times 10^{-19}\,\text{J)}$$

In order to compare this with the nuclear binding energy for iron we calculate the binding energy per nucleon:

$$\frac{1786\,\text{MeV}}{235\,\text{nucleons}} = \text{approx. 7.6 MeV nucleon}^{-1}$$

Iron is the most stable nucleus with a nuclear binding energy of 8.8 MeV/nucleon.

▲ Figure 1 The open-cut Ranger uranium mine in Kakadu National Park, Northern Territory, Australia. Kakadu National Park is a world heritage site. There has been controversy around the mine involving not only environmental issues but also the rights and interests of Indigenous Australians; under 'native title' their traditional laws and customs continue to be observed and a share of the profits from the mine goes to the Indigenous Australian landowners. Should scientists be held morally responsible for how their discoveries are exploited?

Uranium enrichment

The worked examples above show that converting 1 g of enriched U-235 to energy via nuclear fission releases approximately 310 GJ of energy. This energy from 1 g of uranium is equivalent to burning 140 000 kg of coal or about 93 000 litres of gasoline, and no carbon dioxide is produced.

However, not all naturally occurring uranium is fissionable: only U-235 atoms can undergo this type of fission. About 99.28% of naturally occurring uranium is U-238; only 0.72% is U-235. In order to obtain fissile material naturally occurring uranium must be **enriched** so that the percentage of U-235 is large enough. This involves separating the U-235 isotope from the U-238.

Uranium is mined as an ore and contains a mixture of various forms of uranium oxide. The ore is crushed, processed, and purified to uranium(IV) oxide, UO_2, also referred to as uranium dioxide. UO_2 is a dense solid with a melting point of over 2800 °C. It would be convenient to separate U-235 from U-238 in the gaseous state using diffusion as the lighter isotope would diffuse more quickly. However, the melting point and boiling point of the purified ionic UO_2 are too high for this to be a practical option.

To achieve the enrichment process uranium(IV) oxide is converted to gaseous uranium hexafluoride by these reactions:

$$UO_2(s) + 4HF(g) \rightarrow UF_4(s) + 2H_2O(g)$$
$$UF_4(s) + F_2(g) \rightarrow UF_6(g)$$

The uranium hexafluoride, UF$_6$ complex has an octahedral shape (figure 2) and is non-polar. The compound is highly volatile with a boiling point of 56 °C. Below this temperature it has a very high vapour pressure due to its very weak intermolecular forces when compared with uranium tetrafluoride, UF$_4$ or uranium(IV) oxide, UO$_2$.

Uranium hexafluoride has a relatively low boiling point for a compound of its molecular mass. This allows the U-235 isotope to be separated from the U-238. Solid UF$_6$ is vaporized and forced through a porous membrane at high pressure. Because the U-235 isotope is lighter it diffuses through the membrane more easily. The gas with an increased concentration of U-235 is collected and cooled (figure 3). This process increases the concentration by only a small amount so the process is repeated many times.

An alternative enrichment process uses centrifugation instead of diffusion. Gaseous UF$_6$ is introduced into a gas centrifuge in a stream flowing in the opposite direction to the direction of spin of the centrifuge. The heavier U-238 remains closer to the outside wall of the centrifuge due to the centripetal force and the UF$_6$ enriched with U-235 is then withdrawn from the centre (figure 4).

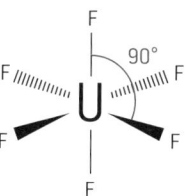

▲ Figure 2 Uranium hexafluoride has weak intermolecular forces. It has an octahedral structure and is non-polar

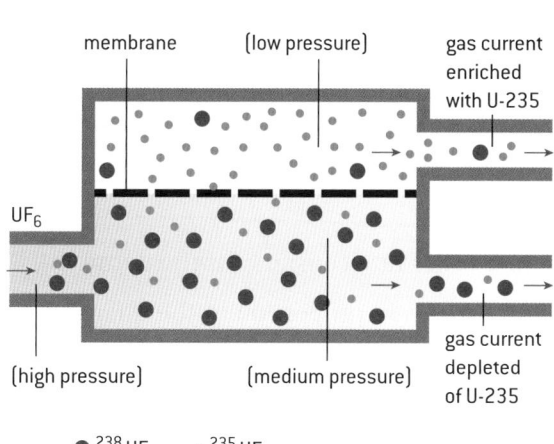

▲ Figure 3 Uranium enrichment by diffusion of UF$_6$ gas through a porous membrane

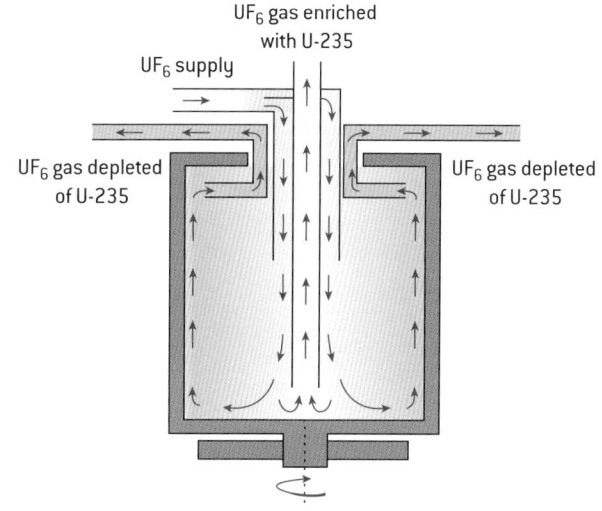

▲ Figure 4 Uranium enrichment by centrifugation of UF$_6$ gas

Following enrichment by either method, the UF$_6$ gas enriched with U-235 is reduced back to uranium metal before being used as a fuel.

Graham's law of effusion

The relative rates of diffusion of the UF$_6$ containing two isotopes of uranium can be calculated using **Graham's law of effusion**. Because all the UF$_6$ is at the same temperature, both isotopes have the same average kinetic energy:

$$KE(^{235}UF_6) = KE(^{238}UF_6)$$

$$\text{or: } \tfrac{1}{2}mv^2(^{235}UF_6) = \tfrac{1}{2}mv^2(^{238}UF_6)$$

Quick question

What is the molar mass of a gas that diffuses 4 times faster than U^{238}F$_6$?

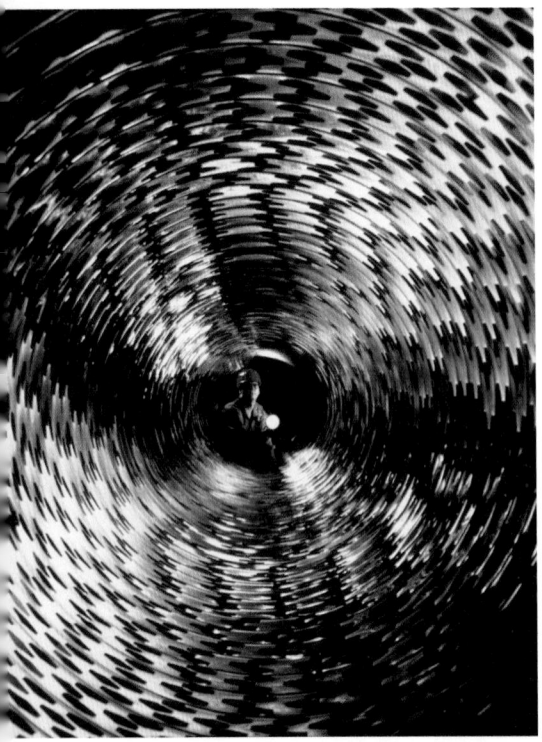

▲ Figure 5 Interior view of a uranium enrichment centrifuge. Mined uranium is converted into uranium hexafluoride gas which is spun in the centrifuge. Molecules containing heavier U-238 tend to collect on the outside and are led off; the remaining gas that is richer in U-235 is passed to further stages of purification before conversion into usable fuel

By rearranging the equation, the ratio of the average velocities of the molecules can be found:

$$\frac{v(^{235}UF_6)}{v(^{238}UF_6)} = \sqrt{\frac{M(^{238}UF_6)}{M(^{235}UF_6)}}$$

In other words, Graham's law states that *the rate of effusion of two gases is inversely proportional to the square root of their molar masses* at the same temperature and pressure:

$$\frac{\text{rate of effusion of gas 1}}{\text{rate of effusion of gas 2}} = \sqrt{\frac{\text{molar mass of gas 2}}{\text{molar mass of gas 1}}}$$

$$M(^{235}UF_6) = 235 + (19 \times 6) = 349\ \mu$$

$$M(^{238}UF_6) = 238 + (19 \times 6) = 352\ \mu$$

therefore:

$$\frac{\text{rate of effusion of } ^{235}U}{\text{rate of effusion of } ^{238}U} = \sqrt{\frac{352}{349}} = 1.004$$

The ratio is very close to 1 so this is why the enrichment process takes a long time and requires many steps to obtain sufficient quantity of U-235.

Radioactive decay

Radioactive decay is kinetically a first order process (sub-topic 16.1). The time it takes for half of the sample to decay is the half-life $t_{1/2}$ (sub-topic C.3). A quantity called the **decay constant**, λ, is related to the half-life by the following equation:

$$\lambda = \frac{\ln 2}{t_{1/2}}$$

The decay constant, λ, is the first order rate constant for the decay.

The level of radioactive decay decreases in proportion to the quantity of material remaining and the rate expression above can also be expressed in terms of the original quantity of material and the quantity remaining after time t has passed:

$$N = N_0\, e^{-\lambda t}$$

where N_0 is the original amount of material and N the amount remaining (*not* the amount decayed) after time t has passed.

 Worked examples: calculating half-life

Example 1

One possible fission product of uranium, ^{144}Ba, has a half-life of 11.5 s. Write the decay equation if ^{144}Ba undergoes beta-decay and calculate the time it takes for its radioactivity to fall to 10% of its original value.

Solution

beta-decay emits a beta-particle:

$$^{144}_{56}Ba \rightarrow\ ^{0}_{-1}\beta +\ ^{144}_{57}La$$

Calculate the rate constant:

$$\lambda = \frac{\ln 2}{t_{1/2}} = \frac{\ln 2}{11.5} = 0.060\ 27 \text{ s}^{-1}$$

Use the rate constant to calculate the time t for its radioactivity to fall to 10% of its original value:

$$t = -\frac{1}{\lambda} \ln \frac{N}{N_0} = -\frac{\ln 0.1}{0.060\ 27} = 38.2 \text{ s}$$

Example 2

The mass of a radioactive substance falls from 100 µg to 0.821 µg in 78 s as it decays. Calculate the half-life of the substance.

Solution

$$N = N_0\ e^{-\lambda t}$$

$$\lambda = -\frac{\ln\left(\frac{N}{N_0}\right)}{t}$$

$$= -\frac{\ln\left(\frac{0.821}{100}\right)}{78} = 0.061\ 57$$

$$t_{1/2} = \frac{\ln 2}{\lambda} = 11.26 \text{ s}$$

Example 3

As-81 has a half-life of 33 s. Calculate the percentage of material which remains after 50.0 s.

Solution

$$\lambda = \frac{\ln 2}{t_{1/2}} = 0.021\ 00$$

$$N = N_0\ e^{-\lambda t}$$

$$\frac{N}{N_0} = 0.35 \text{ or } 35\%$$

The risks associated with nuclear energy

There are serious safety issues associated with nuclear energy, most importantly risks to health and storage problems associated with nuclear waste (sub-topic C.3), as well as the possibility that nuclear fuels may be used in nuclear weapons. While this threat of nuclear weapon development is undoubtedly a cause of worldwide concern, the enrichment process for weapons-grade U-235 is much more involved: generating electric power by steam turbines can use fuel enriched to under 20% U-235, whereas nuclear weapons often require 85% or more enrichment.

One of the biggest dangers of nuclear energy comes from the ionizing radiation emitted by the daughter products. Radiation occurs when unstable nuclei decay and release subatomic particles (sub-topic C.3), which can damage living cells.

The SI unit of ionizing radiation dose is the **sievert**, Sv. It measures the effect that ionizing radiation has on tissue, in J kg^{-1}. The annual worldwide average background radiation is 2.4 mSv year^{-1}. A level of 250 mSv can be detected by blood tests while a radiation dose of 1 Sv gives initial signs of radiation poisoning such as nausea, headaches, and vomiting.

 Terminology

IUPAC states that the term free-radical is obsolete in its Gold Book and that these are now referred to as radicals. Having appropriate terminology is very important. Why do you think the term radicals was adopted?

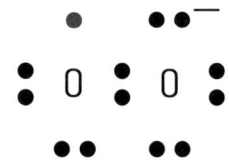

▲ Figure 6 The superoxide free-radical contains an unpaired electron

In biological tissues ionizing radiation can remove electrons from molecules creating radicals such as superoxide, O_2^- (figure 6) and hydroxyl, HO·. These radicals can initiate chain reactions (sub-topic 10.2) that can damage DNA and enzymes in living cells.

The superoxide ion has strong oxidative properties because of the tendency for oxygen to gain electrons (to become reduced, causing oxidation) and the fact that it has an unpaired electron on one oxygen atom, increasing its oxidative properties. The superoxide radical is sometimes created naturally and used by the immune system to kill foreign microorganisms.

Hydroxyl radicals, HO· can also be created in cells either by ionizing radiation or naturally from the superoxide radical via the Haber–Weiss reaction:

$$O_2^- + H_2O_2 \rightarrow O_2 + OH^- + \cdot OH$$

Questions

1 Nuclear power is one potential energy source that does not involve fossil fuels. Current nuclear technology is dependent on fission reactions.

a) Nuclear technology developed very rapidly between 1940 and 1970. Outline why this occurred. [1]

b) The equation for a possible nuclear fission reaction is:

$$^{235}_{92}U + ^{1}_{0}n \rightarrow ^{90}_{38}Sr + ^{136}_{54}Xe + 10^{1}_{0}n$$

The masses of the particles involved in the fission reaction are shown below:

mass of neutron = 1.008 67 μ

mass of U-235 nucleus = 234.993 33 μ

mass of Xe-136 nucleus = 135.907 22 μ

mass of Sr-90 nucleus = 89.907 74 μ

Determine the energy released when one uranium nucleus undergoes fission according to the reaction above. [3]

c) The half-life of strontium-90 is 28.8 years. Using information from section 1 of the *Data booklet*, calculate the number of years required for its radioactivity to fall to 10% of its initial value. [2]

d) Nuclear fuels require the enrichment of natural uranium. Explain how this process is carried out including the underlying physical principle. [3]

IB, Specimen paper

2 Nuclide X has a half-life of 1 day and nuclide Y has a half-life of 5 days. In a particular sample, the activities of X and Y are found to be equal.

When the activity is tested again after 10 days, the activity will be

A. entirely due to nuclide X

B. due equally to nuclides X and Y

C. mostly due to nuclide X

D. mostly due to nuclide Y.

3 Which **one** of the following diagrams (figure 7) best illustrates the first two stages of an uncontrolled fission chain reaction?

a)
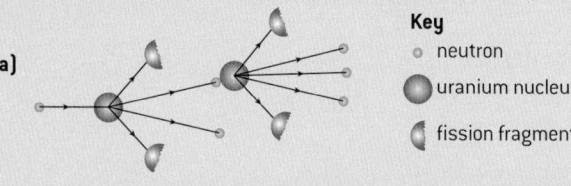

Key
- ○ neutron
- ● uranium nucleus
- ◖ fission fragment

b)
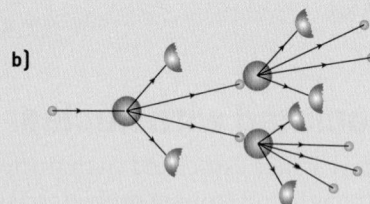

c)
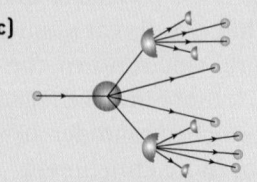

d)
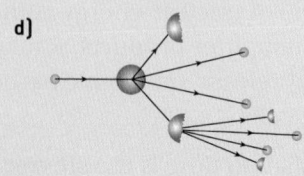

▲ Figure 7

4 This question is about nuclear binding energy.

a) (i) Define *nucleon*.

(ii) Define *nuclear binding energy of a nucleus*.

The axes in figure 8 show values of nucleon number A (horizontal axis) and average binding energy per nucleon E (vertical axis). (Binding energy is taken to be a positive quantity.)

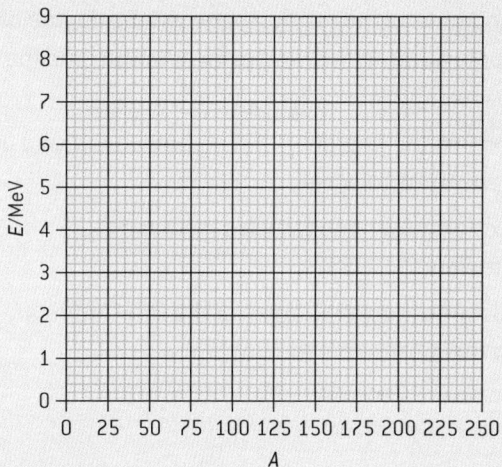

▲ Figure 8

b) On a copy of figure 8, mark on the E axis the approximate position of:

(i) the isotope $^{56}_{26}$Fe (label this F)

(ii) the isotope $^{2}_{1}$H (label this H)

(iii) the isotope $^{238}_{92}$U (label this U).

c) Using the grid in figure 8, draw a graph to show the variation with nucleon number A of the average binding energy per nucleon E.

d) Use the following data to deduce that the binding energy per nucleon of the isotope $^{3}_{2}$He is 2.2 MeV.

nuclear mass of $^{3}_{2}$He = 3.016 03 μ

mass of proton   = 1.007 28 μ

mass of neutron   = 1.008 67 μ

In the nuclear reaction $^{2}_{1}$H + $^{2}_{1}$H → $^{3}_{2}$He + $^{1}_{0}$n energy is released.

e) (i) State the name of this type of reaction.

(ii) Use your graph in (c) to explain why energy is released in this reaction.

5 Consider nuclear power production:

a) With reference to the concept of fuel enrichment explain:

(i) the advantage of enriching the uranium used in a nuclear reactor

(ii) from an international point of view, a possible risk to which fuel enrichment could lead

(iii) the relationship between Graham's law of effusion and the kinetic theory that is involved in fuel enrichment.

b) Uranium enrichment increases the proportion of U-235 isotope in its mixture with more abundant U-238. Before separation, both isotopes must be converted to uranium(VI) fluoride.

(i) Explain the properties of the UF$_6$ complex that make it more suitable for isotope separation rather than using UO$_2$.

(ii) Compare and contrast U-235 isotope separation using diffusion and gas centrifugation of UF$_6$.

6 When ammonia gas, NH$_3$ reacts with hydrogen chloride gas, HCl the white solid ammonium chloride, NH$_4$Cl is formed:

$$NH_3(g) + HCl(g) \rightarrow NH_4Cl(s)$$

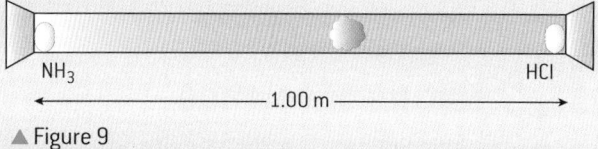

▲ Figure 9

The apparatus shown in figure 9 was set up to test Graham's law. A metre stick is placed beside the tube. Cotton wool soaked in NH$_3$ is placed in the left-hand end and cotton wool soaked in HCl is placed in the right-hand end. A white cloud of solid appears where the HCl and NH$_3$ meet. Assuming the NH$_3$ end to be 0.00 m and the HCl end to be 1.00 m on the metre stick, calculate the position at which you would expect the solid NH$_4$Cl to appear.

7 Draw the Lewis (electron dot) structure of the superoxide and hydroxyl radicals. Explain what a radical is and the steps involved in a radical chain reaction. Explain how radioactivity can cause the initiation step.

C.8 Photovoltaic cells and dye-sensitized solar cells (AHL)

Understandings

→ Molecules with longer conjugated systems absorb light of longer wavelength.

→ The electrical conductivity of a semiconductor increases with an increase in temperature whereas the conductivity of metals decreases.

→ The conductivity of silicon can be increased by doping to produce n-type and p-type semiconductors.

→ Solar energy can be converted to electricity in a photovoltaic cell.

→ DSSCs imitate the way in which plants harness solar energy. Electrons are "injected" from an excited molecule directly into the TiO_2 semiconductor.

→ The use of nanoparticles coated with light-absorbing dye increases the effective surface area and allows more light over a wider range of the visible spectrum to be absorbed.

Applications and skills

→ Understand the relation between the degree of conjugation in the molecular structure and the wavelength of the light absorbed.

→ Explain the operation of the photovoltaic and dye-sensitized solar cell.

→ Explain how nanoparticles increase the efficiency of DSSCs.

→ Discuss the advantages of the DSSC compared to the silicon-based photovoltaic cell.

Nature of science

→ Transdisciplinary – a dye-sensitized solar cell, whose operation mimics photosynthesis and makes use of TiO_2 nanoparticles, illustrates the transdisciplinary nature of science and the link between chemistry and biology.

→ Funding – the level of funding and the source of funding is crucial in decisions regarding the type of research to be conducted. The first voltaic cells were produced by NASA for space probes and were only later used on Earth.

Conjugated systems

Conjugation is the interaction of alternating double bonds, for example in organic molecules, to produce a delocalized array of pi electrons over all the atoms. Molecules with conjugated bonds can absorb visible light, with longer conjugated systems absorbing light of longer wavelength. All the carbon atoms involved in such systems have sp^2 hybridization (sub-topic 14.2): the π-electron clouds of adjacent double bonds partly overlap with one another and form a large cloud of delocalized electrons (figure 1).

This type of multi-centre chemical bonding known as **electron conjugation** is similar to the electron delocalization seen in benzene (sub-topic 20.1) and produces a chain of carbon–carbon bonds with a bond order of 1.5.

Molecules (a) to (c) all have some degree of conjugation. Conjugation occurs across the entire molecule in molecules (a), and (b) but not in molecule (c).

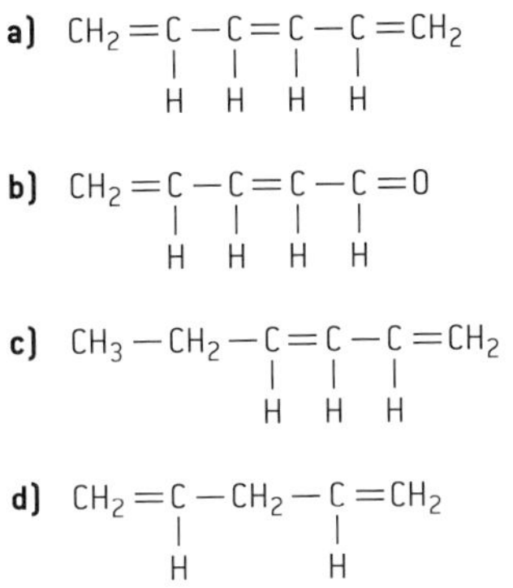

a) $CH_2=C-C=C-C=CH_2$
      | | | |
      H H H H

b) $CH_2=C-C=C-C=O$
      | | | |
      H H H H

c) $CH_3-CH_2-C=C-C=CH_2$
            | | |
            H H H

d) $CH_2=C-CH_2-C=CH_2$
      | |
      H   H

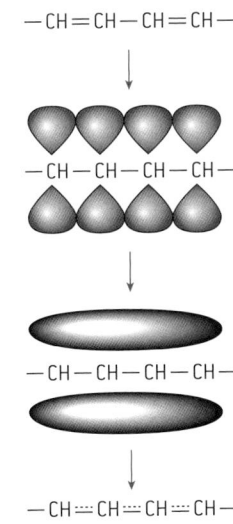

▲ Figure 1 Overlapping orbitals combine to form a cloud of delocalized electrons in a conjugated system

Molecule (d), penta-1,4-diene, does not contain alternating single and double bonds and hence does not show conjugation.

For conjugated alkenes, the higher the degree of conjugation, the longer the wavelength of light can be absorbed. For example, in the group A vitamin, retinol and the carotenoid, beta-carotene, the electron conjugation involves 10 and 22 carbon atoms, respectively (figure 2).

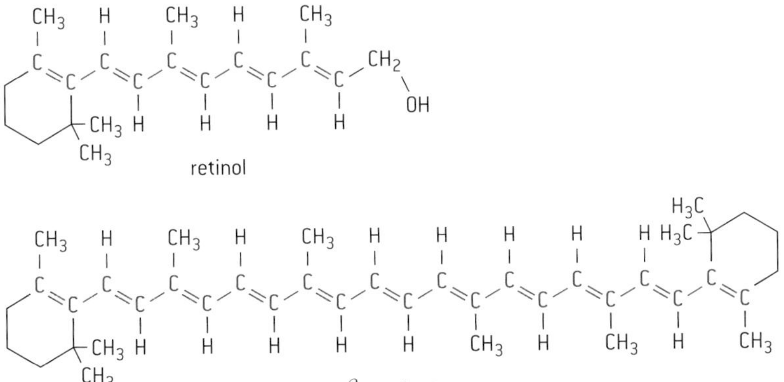

▲ Figure 2 Electron conjugation in retinol and β-carotene

Therefore β-carotene absorbs light of lower energy (longer wavelength/lower frequency) than retinol. The colour absorbed by β-carotene is towards the lower energy red side of visible light (figure 3).

Retinol strongly absorbs violet light at 400–420 nm and appears yellow, as yellow is the complementary colour to violet and lies at the opposite side of the colour wheel (sub-topic 13.2). β-carotene has a larger system of electron conjugation and therefore the maximum of absorption is at longer wavelengths (430–480 nm, blue region), so its colour is orange (complementary to blue).

Table 1 shows the relationship between degree of conjugation in alkenes, wavelength of maximum absorbance, and electron structure.

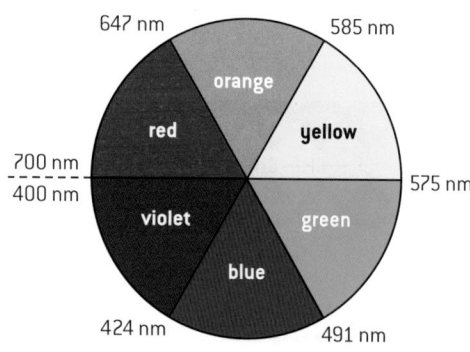

▲ Figure 3 The colour wheel

Molecule	Structure	Conjugation	Wavelength of maximum absorbance /nm
penta-1, 4-diene		no conjugation as this molecule does not have alternating single and double bonds	178
penta-1, 3-diene		delocalization occurs between carbons 1, 2, 3, and 4, but not carbon 5	223
hexa-1,3, 5-triene		delocalization occurs across the entire molecule	274
deca-1,3,5,7, 9-pentaene		all 10 carbon atoms are involved in delocalization	334

▲ Table 1

Worked example

An indicator has a red form and a yellow form. Deduce which of these two colours is due to a molecule with a higher degree of conjugation.

Solution

The red form is due to a molecule that has absorbed its complement on the colour wheel, that is, green light of wavelength around 540 nm. Similarly the yellow form is due to a molecule that has absorbed its complement, namely violet light of wavelength around 410 nm. The longer the wavelength absorbed, the higher the degree of delocalization, hence the red form has a more conjugated system.

Silicon semiconductor photovoltaic cells

Semiconductors have electrical conductivity midway between that of conductors and insulators. The conductivity of a semiconductor increases with temperature, in contrast to that of conductors. Conductors are typically metals with low ionization energies and therefore freely moving electrons. When heated, lattice movement increases which interferes with conduction. However, semiconductors are relatively poor conductors of electricity due to their higher ionization energies. When heated, the extra energy can move an electron into a conduction zone and the electrical conductivity of the material therefore increases.

Photovoltaic cells made of semiconductors can absorb photons of light (sub-topic 2.2) resulting in electrons being knocked free from atoms and creating a potential difference. Semiconductor materials for such cells are often pure group 14 elements such as silicon or germanium. Pure binary compounds of group 13 and 15 elements such as gallium arsenide can also be used. The conductivity of the semiconductor can be increased by "doping" it with small impurities of group 15 elements, such as phosphorus to create an **n-type semiconductor**, or a group 13 element such as boron to create a **p-type semiconductor**.

Silicon has four valence electrons. Doping with an n-type material provides an extra electron which can become mobile with a small potential difference, while doping with a p-type material creates a "hole" that can be used to "hold" an electron (figure 4). This ability to switch between conducting and insulating properties is what makes the material a semiconductor. The **band gap** between valence and mobile electrons (figure 6) is the basic property of a semiconductor, controlling the flow of electrons. This "on-off" property is the foundation of the binary language of 0s and 1s used in computers.

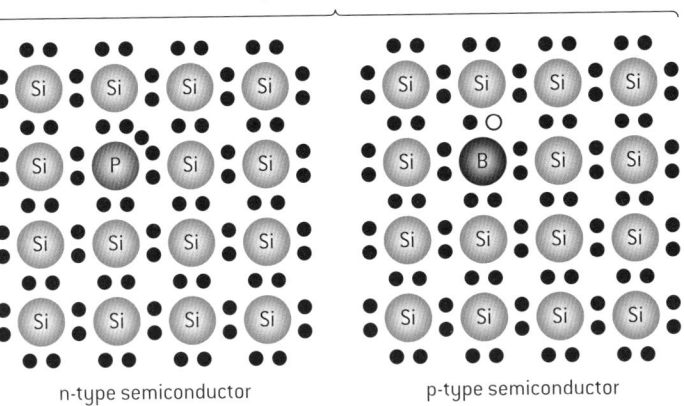

▲ Figure 4 The effect of doping on a semiconductor material

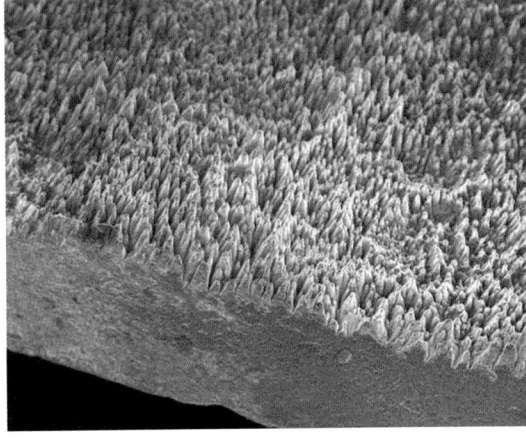

▲ Figure 5 Coloured scanning electron micrograph (×50) of black silicon. Its surface area is increased by monocrystalline silicon needles, thought to be responsible for the material's very high light-absorbing properties

A silicon-based solar cell absorbs a photon of light which excites an electron in the n-type side of the semiconductor from the valence band to the conduction band. This electron is free to move about the semiconductor leaving behind an empty space. An electron from a neighboring atom can move into this empty space and when this electron moves, it leaves behind another space. The continual movement of the space for an electron, called a "hole", is similar to the movement of a positive charge through the semiconductor. So this excitation results in not only an electron in the conduction band but also a "hole" in the valence band. At a p–n junction, an electron may be excited in the n-region while the "hole" is located in the p-region. The electron returns to the p-region through the external circuit and recombines with the hole.

In summary:

- The photovoltaic cell absorbs photons in a semiconducting material, which causes some valence electrons to be removed, resulting in some ionization in the cell.

- A charge separation occurs in the semiconductor which allows for a one-way flow of electrons.

- The cell can be linked to an external circuit where the flow of electrons provides electrical power.

In this way, solar energy is converted to electrical energy.

Dye-sensitized solar cells (DSSC)

In a Grätzel DSSC, photons are absorbed by a dye in a way similar to the absorption of photons by chlorophyll in photosynthesis (sub-topics B.9 and C.4). Electrons in the dye are then injected into a titanium(IV) oxide (TiO2) nanoparticle layer, which conducts the electrons to the anode.

Once a dye molecule has emitted its excited electron it needs to gain another electron. To achieve this, dye-coated TiO_2 nanoparticles are immersed in a solution of iodide ions, I^-. The iodide ions release

> ### Energy conversion in solar cells
>
> When solar energy is converted to electrical energy, light is absorbed resulting in charges being separated. In a photovoltaic cell both of these processes occur in the silicon semiconductor, whereas they happen in separate locations in a dye-sensitized solar cell.

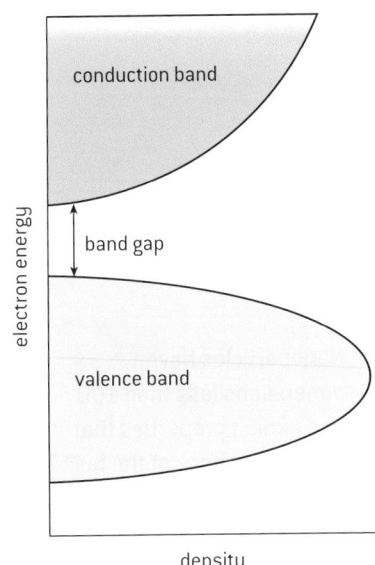

▲ Figure 6 The band gap is the difference in energy between the valence band and the conduction band

A Grätzel cell is named after its Swiss inventor Michael Grätzel, and uses titanium (IV) oxide, TiO_2 instead of silicon. The TiO_2 is coated with light-absorbing dye. The cell generates electricity when the energy captured by the dye makes electrons in the dye molecules jump from one orbital to another. The electrons then are transferred to the TiO_2 particles and diffuse towards one electrode, while the iodide ions carry electrons from the other electrode to regenerate the dye. Grätzel cells are much cheaper to produce than silicon-based ones and could have applications in less economically developed countries.

electrons to the dye on the TiO_2 layer, becoming oxidized to tri-iodide I_3^-. This can accept electrons at the cathode, being reduced back to I^-. An outline of the process is shown in figure 7.

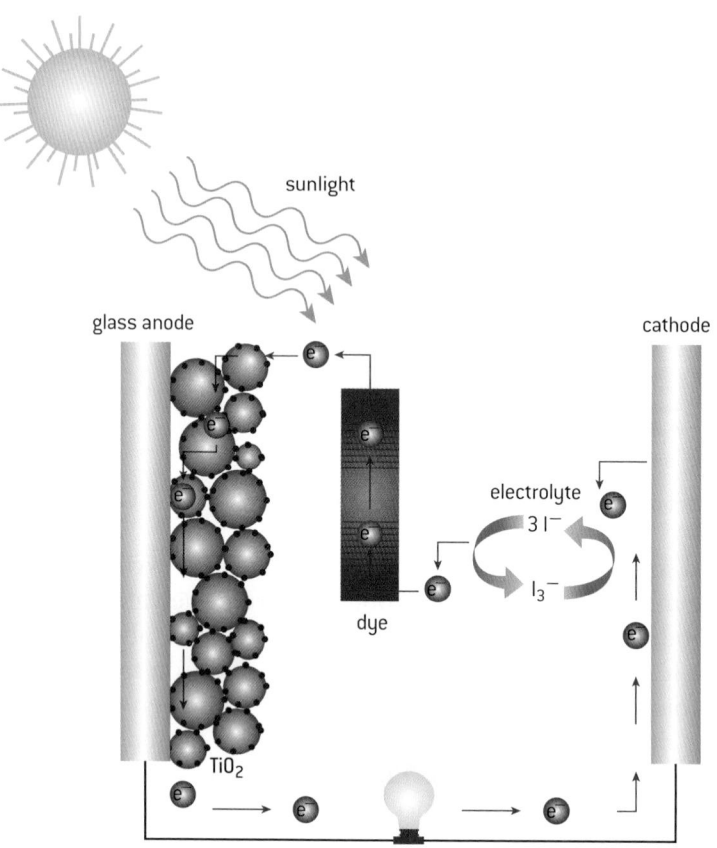

▲ Figure 7 Schematic diagram of the energy flow in the DSSC. Photons excite electrons from the dye coating the conductive TiO_2. These electrons lost from the dye are replaced by oxidation (loss of electrons) of $3I^-$ to I_3^-. The electrons from the TiO_2 conductive layer leave the anode and return to the cathode, where they reduce the I_3^- ions back to $3I^-$

The anode is transparent, allowing sunlight to reach the dye-coated TiO_2. The TiO_2 is laid down as an array of **nanoparticles**, providing a large surface area for the dye. An electrolyte containing iodide ions fills the spaces between the TiO_2 nanoparticles and helps transfer electrons from the cathode back to the dye molecules. TiO_2 nanoparticles are transparent and form a mesh-like structure in which touching nanoparticles act like a wire for the electrons to travel through. The dye coats the TiO_2 particles, except where they are in contact with each other. This mesh arrangement allows electrical conductivity where the nanoparticles touch and also provides a large surface area for exposure of the dye to sunlight where they do not touch.

In summary, in a DSSC:

Nanoparticles have dimensions less than 100 nm and exhibit properties that differ from those of the bulk material. Individual molecules are usually not considered to be nanoparticles but small clusters of them may be classed as nanoparticles.

● Photons excite an electron from the dye, which enters the TiO_2 nanoparticle. This prevents the electron returning to the dye.

● The oxidized dye receives an electron from an iodide ion, which reduces the dye back to its original form. In the process, the iodide ions undergo oxidation:

$$3I^- \rightarrow I_3^- + 2e^-$$

- The electron travels through the nanoparticle mesh and exits at the anode, entering an external circuit.

- The electron returns to the DSSC through the cathode and is used to reduce I_3^- back to iodide ions:

$$I_3^- + 2e^- \rightarrow 3I^-$$

Different dyes on the TiO_2 mesh will absorb different wavelengths of light depending on their colour, which relates to the degree of conjugation in organic molecules or to transition metal properties in metal-based dyes.

> In May 2013 the Mauna Loa observatory claimed that the amount of carbon dioxide in the atmosphere was at the highest level for 3 million years. Could solar cells reduce our dependence on fossil fuels? How might this technology change the economic fortunes of countries with plenty of sunlight and unused land?

Developing the technology

Controlled experimentation on each component of the DSSC, including the electrolyte, the pigment or dye, the electron carrier (TiO_2), and the electrodes, will allow for further development with the aim of producing cheaper, more efficient and more powerful solar cells.

Advantages and disadvantages of DSSCs

Some advantages of DSSCs over silicon-based solar cells are that they are cheaper and use light of lower energy (lower frequency or longer wavelength). The resources to build them are plentiful and renewable. The use of nanoparticles provides a larger surface area exposed to sunlight so the DSSC can absorb more light under cloudy conditions than can silicon-based cells. In addition, because the electrons from the dye are injected into a TiO_2 mesh the conductivity is higher so there is less chance of a promoted electron falling back or finding a "hole" in a semiconductor, as could happen in a silicon-based cell. The DSSC has a thin-layer structure, making it more flexible and durable in low-density solar collectors such as those used on rooftops. Being thinner DSSCs can also radiate heat away better than silicon-based solar cells.

Some disadvantages are that DSSCs are currently not very suitable for large-scale applications such as generating a megawatt or more of power. The liquid electrolyte can freeze at low temperatures or expand and crack the cell at high temperatures, making current versions unsuitable for extreme weather conditions.

Questions

1 A modern solution to the provision of power for remote places is the dye-sensitized solar cell (DSSC). A Gratzel DSSC contains an organic dye molecule on the surface of a titanium dioxide, TiO_2, semiconductor and an electrolyte containing iodide ions.

Explain its operation, including the importance of nanotechnology in its construction and its advantage compared with silicon-based photovoltaic devices.

2 Discuss the use of silicon in photovoltaic cells, with reference to the following:

- why pure silicon is a better electrical conductor than non-metals such as phosphorus and sulfur

- how a p-type semiconductor made from silicon is different from pure silicon

- how sunlight can produce an electric current in a photovoltaic cell. [5]

IB, May 2007

3 Discuss the **two** types of doping of silicon when small amounts of indium and arsenic are added. Name the type of semiconductors produced in each case. [4]

IB, November 2005

4 Explain whether 1,3-hexadiene or 1,5-hexadiene would absorb the longer wavelength of light.

5 Bromothymol blue is an acid–base indicator that is blue in base and yellow in acid:

$$HB \rightleftharpoons H^+ + B^-$$
yellow   blue

Deduce which of the two forms of the indicator has the higher degree of conjugation. Explain it in terms of the wavelength of light absorbed and molecular structure of an organic indicator.

Introduction

Medicinal chemistry is the study of bioactive compounds that can be used in diagnostics and therapy. The discovery, design, and development of such compounds, known as pharmaceutical drugs, is a complex process that requires the combined efforts of scientists from various disciplines, including synthetic organic chemistry, biochemistry, biology, pharmacology, medicine, mathematics, and computer technology. A medicinal chemist must take into account not only the immediate benefits and risks of new drugs but also their long-term effects on individuals, society, and the environment.

D.1 Pharmaceutical products and drug action

Understandings

→ In animal studies, the therapeutic index is the lethal dose of a drug for 50% of the population (LD_{50}) divided by the minimum effective dose for 50% of the population (ED_{50}).

→ In humans, the therapeutic index is the toxic dose of a drug for 50% of the population (TD_{50}) divided by the minimum effective dose for 50% of the population (ED_{50}).

→ The therapeutic window is the range of dosages between the minimum amounts of the drug that produce the desired effect and a medically unacceptable adverse effect.

→ Dosage, tolerance, and addiction are considerations of drug administration.

→ Bioavailability is the fraction of the administered dosage that reaches the target part of the body.

→ The main steps in the development of synthetic drugs include identifying the need and structure, synthesis, yield and extraction

→ Drug–receptor interactions are based on the structure of the drug and the site of activity.

Applications and skills

→ Discussion of experimental foundations for therapeutic index and therapeutic window through both animal and human studies.

→ Discussion of drug administration methods.

→ Comparison of how functional groups, polarity, and medicinal administration can affect bioavailability.

Nature of science

→ Risks and benefits – medicines and drugs go through a variety of tests to determine their effectiveness and safety before they are made commercially available. Pharmaceutical products are classified for their use and abuse potential.

Introduction to medicinal chemistry

Medicinal chemistry is a cross-disciplinary science that links together organic chemistry, pharmacology, biochemistry, biology, and medicine. The primary objective of medicinal chemistry is the discovery, design, and development of new bioactive compounds suitable for therapeutic use. These compounds, known as **pharmaceutical drugs**, have a variety of effects on the body's functioning and may be used to prevent or cure diseases, alleviate the symptoms of health conditions, or assist in medical diagnostics.

Pharmaceutical drugs can be classified according to their physical and chemical properties, routes of administration, and therapeutic effects. Because most drugs are organic compounds, their properties depend on the functional groups present in their molecules (sub-topic 10.2). Drugs with many polar groups are generally water soluble and can be administered **orally** (ingested by mouth). However, some chemical compounds are unstable in the highly acidic gastric juice (sub-topic D.4), so they must be administered **rectally** (in the form of suppositories or enemas) or **parenterally**, that is, injected under the skin (**subcutaneous injection**), into muscle tissue (**intramuscular injection**), or directly into the bloodstream (**intravenous injection**). This last method of injection produces the fastest therapeutic effect as the drug is distributed around the body with the flow of the blood. Finally, some volatile or highly dispersed drugs can be taken by **inhalation** (breathed in through the nose or mouth) while non-polar compounds are often administered **transdermally** (applied to the skin in the form of patches, ointments, or therapeutic baths).

Therapeutic effects of pharmaceutical drugs depend on the chemical structure and the route of administration of the drug. Pharmaceutical drugs can affect the physiological state (including metabolism, consciousness, activity level, and coordination) of the body, alter mood and emotions, or change the perception of sensory information. Certain drugs may have little or no effect on the patient but instead target specific pathogenic organisms within the patient's body, or perform purely diagnostic functions (for example, biologically inert barium sulfate used for X-ray examination of the gastrointestinal tract).

In some cases, the desired therapeutic effect can be achieved by assisting the body in its natural healing process. This may be done through counselling or administering a biologically inert substance known as a **placebo**. Although the exact mechanisms of such apparently successful treatments are not fully understood, there is strong experimental evidence that the body can sometimes be deceived into healing itself without receiving any help in the form of medical drugs.

The placebo effect and clinical trials

The therapeutic action of placebo, known as the **placebo effect**, must be taken into account during clinical trials of pharmaceutical drugs. In a typical experiment, laboratory animals or human volunteers are separated into two groups of equal size, one of which receives the drug while the other is given a placebo. To reduce the possibility of conscious

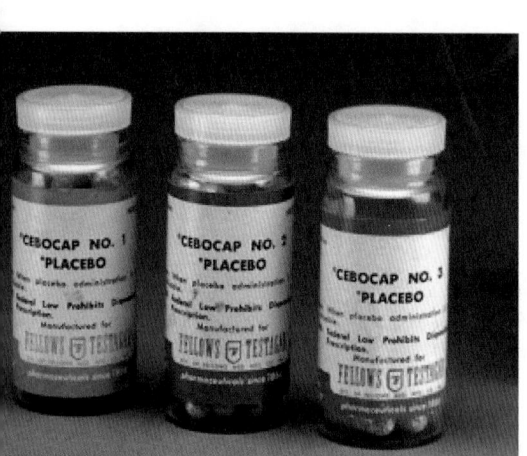

▲ Figure 1 Placebos are produced for clinical use in a range of different shapes and colours

or subconscious bias in the interpretation of the experimental results, neither the researchers directly observing the patients nor the patients themselves know who is given the real drug and who receives placebo, so this type of experiment is known as a **double-blind test**. At the end of the trial the therapeutic effects in the two groups are compared, and any difference in results is attributed to the pharmacological action of the drug.

Side effects

Pharmaceutical drugs interfere with biological processes so no drug is completely safe or free from non-beneficial effects on the human body, known as **side effects**. For example, aspirin (sub-topic D.2) increases the risk of gastrointestinal bleeding while opiates (sub-topic D.3) are addictive and often become substances of abuse. Any drug can become a poison if taken in excess. Overdoses of paracetamol, the most common analgesic in the world, often cause kidney, liver, and brain damage, which in severe cases can be fatal. At the same time, insufficient doses or irregular use of antibiotics can lead to antibiotic resistance (sub-topic D.2), so every pharmaceutical drug must be administered with caution and only in the recommended amounts.

Effectiveness and safety

The effectiveness and safety of a pharmaceutical drug can be expressed using its **therapeutic index** (*TI*), which is determined as the ratio between the therapeutic dose and the toxic (or lethal) dose of the drug. The **effective dose** (ED_{50}) is usually defined as the minimum dose of the drug that produces the desired therapeutic effect in 50% of laboratory animals or human patients. Similarly, the **lethal dose** (LD_{50}) of the drug is the dose that causes death in 50% of laboratory animals. The LD_{50} value for humans is not determined for obvious ethical reasons; instead, the **toxic dose** (TD_{50}) of the drug is measured as the dose that causes toxicity (an unacceptable adverse effect) in 50% of patients. Therefore, the therapeutic index of a drug can be defined as follows:

$$TI \text{ (in animals)} = \frac{LD_{50}}{ED_{50}} \qquad TI \text{ (in humans)} = \frac{TD_{50}}{ED_{50}}$$

The greater the therapeutic index, the "safer" the drug. For example, an overdose of a drug with $TI = 100$ occurs when the patient takes 100 times more drug than prescribed while a drug with $TI = 5$ becomes dangerous when the recommended dose is exceeded only five times. Pharmaceutical drugs available over the counter usually have high *TI* values, which reduces the risk of overdose in patients who take these drugs without obtaining medical advice. At the same time, certain drugs with therapeutic indices as low as 2 can still be used safely if administered by qualified medical personnel.

Therapeutic window and bioavailability

Another important characteristic of a drug, the **therapeutic window**, is the range of doses where the drug provides the desired therapeutic effect without causing unacceptable adverse effects in most patients

Risks and benefits

Comparing the risks versus the benefits of pharmaceutical drugs is the central problem in medicinal chemistry. Before a drug is made commercially available it must go through a variety of tests that determine its efficiency, stability, side effects, and the potential for abuse. Many other factors, such as the environmental impact of the drug's synthesis, administration, and disposal, must be also considered. After the tests are complete, drugs are classified into several categories which determine the form and extent of their release to the market. However, local regulations vary greatly, so the same drug may be available over the counter in some countries but require a prescription or even be completely banned in other parts of the world. Such differences restrict international trade and raise many ethical questions, such as the balance between the freedom of individuals and the right of public bodies to protect the health of their citizens.

(figure 2). In contrast to the therapeutic index, the term "therapeutic window" is not strictly defined and serves only as a general indication of the recommended drug dosages. Typically, the therapeutic window "opens" below the ED_{50} (where some patients can still be provided with minimal beneficial effect) and "closes" below the TD_{50} (where only a small percentage of patients might experience significant adverse effects). Similar to drugs with low *TI*, drugs with narrow therapeutic windows must be administered with great care and often require constant monitoring of their actual levels in the patient's body.

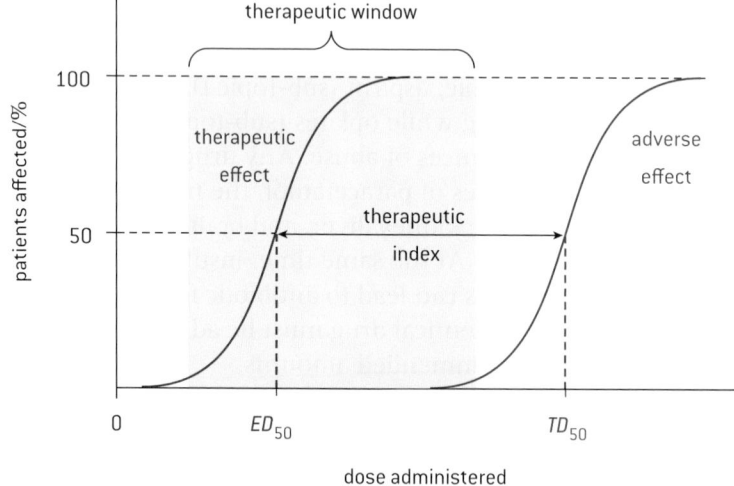

▲ Figure 2 Therapeutic index and therapeutic window

The effective and toxic doses of the drug depend on its route of administration. In order to reach the target organ or part of the body, most drugs have to pass into the bloodstream, which may be problematic if a drug has limited solubility in water or has a slow absorption rate from the gastrointestinal tract when it is administered orally. The fraction of the administered dose that is absorbed into the bloodstream is known as the **drug bioavailability**. By definition, when a drug is injected intravenously its bioavailability is 100%. Other routes of drug administration generally decrease its bioavailability (figure 3) due to incomplete absorption, decomposition, and many other factors including physiological differences in individual patients.

The bioavailability of pharmaceutical drugs depends on their solubility, polarity, and the presence of certain functional groups. Polar molecules containing hydroxyl, carboxyl, and amino groups are usually soluble in water and are therefore quickly absorbed from the gastrointestinal tract into the bloodstream. However, such molecules cannot easily pass through hydrophobic cell membranes, which in many cases reduces their biological activity. The effects of specific functional groups on the bioavailability and activity of pharmaceutical drugs will be discussed in sub-topics D.2 and D.3.

Tolerance and addiction

Regular administration of certain drugs may reduce the body's response to specific medications or classes of pharmaceutical drugs due to accelerated drug metabolism or changes in cellular functions. This phenomenon,

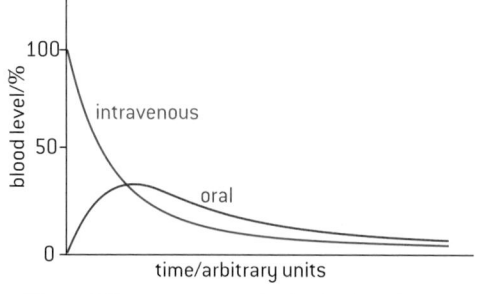

▲ Figure 3 Drug concentration in the bloodstream as a function of time

known as **drug tolerance**, is typical for opiates (sub-topic D.3) and other narcotic drugs, where drug users need progressively higher doses of the drug to obtain the desired therapeutic effect. Increased doses lead to more pronounced side effects, which may eventually become unacceptable and "close" the therapeutic window for some patients.

Another adverse effect of certain pharmaceutical drugs, known as **drug addiction**, is the compulsive desire of the user to take the drug regardless of the health problems it might cause. Addiction may be purely psychological but it often involves some degree of physiological dependence that leads to withdrawal symptoms when the drug use is reduced or interrupted. Drug addiction becomes particularly dangerous when combined with drug tolerance, which is the case for opiates and many illegal drugs. Patients addicted to such drugs require higher and higher doses, which soon exceed the toxic level and can lead to irreversible physiological changes or death.

Together with other adverse effects, the risks of drug tolerance and addiction must be taken into account when the drug becomes commercially available or is prescribed to a patient. In many cases, addictive properties of drugs outweigh their medical benefits and prevent their release to the market. However, even the most addictive drugs are sometimes used as painkillers in life-threatening situations or for patients with incurable diseases, where the high risk of side effects is less important than the therapeutic result.

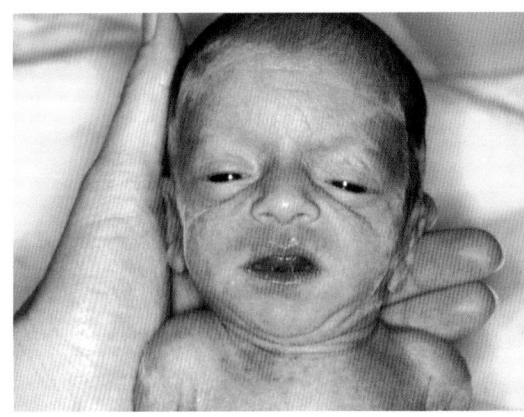

▲ Figure 4 A premature baby born to a cocaine addict is suffering from withdrawal symptoms

Drug action and development of new drugs

At the molecular level, pharmaceutical drugs interact with the binding sites of **enzymes** or cellular **receptors**, which are proteins composed of 2-amino acids (sub-topic B.2). In binding to enzymes most drugs act as **inhibitors**, reducing the activity of enzymes via competitive or non-competitive mechanisms (sub-topic B.7). If a drug binds to a cellular receptor, the cell responds to this chemical message by altering its state or allowing specific molecules to pass through the cell membrane.

The type and efficiency of drug–receptor interactions depend on the chemical structures of the drug and the binding site. Ideally, the functional groups of the drug and receptor should be complementary to one another and have correct orientations that allow them to form dipole-dipole interactions, hydrogen bonds or ionic bonds (sub-topics 4.1 and 4.4). Alkyl chains and phenyl groups of the drug molecule can also interact with non-polar groups of the receptor via London forces (sub-topic 4.4). Drug–receptor interactions can involve any types of chemical bonds, some of which are shown in figure 5.

Although the structures of real drugs and their target receptors or enzymes do not match exactly, efficient binding can be achieved by slight conformational changes of both the binding site and the drug molecule (as in the "induced fit" theory for enzymes (see sub-topic B.7). At the same time, the nature and strength of binding can be affected by chemical modification of certain functional groups of the drug. The analysis of pharmacological activity in a series of similar compounds provides some information on the structure of the binding site.

Ethanol and nicotine are common substances of abuse that have many side effects, including toxicity and high addiction potential. Each of these substances causes more deaths around the globe than all illegal drugs combined. Nevertheless, alcoholic beverages and tobacco are available in most countries to any adult over a certain age. This fact raises many questions about the roles of traditional culture and scientific evidence in drug legislation.

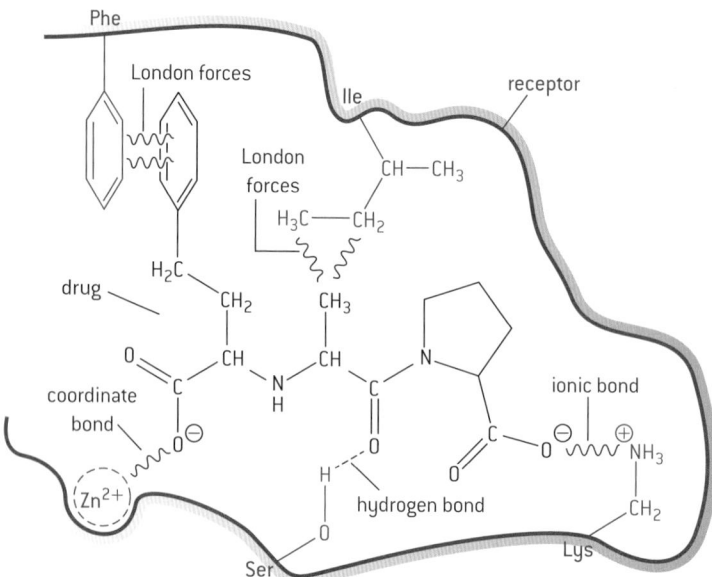

▲ Figure 5 Drug—receptor interactions

In turn, this information can be used for further modification of the drug and optimization of its activity.

The development of new pharmaceutical drugs is a long and complex process that often involves fundamental research and requires close collaboration of specialists from various disciplines. In addition pharmaceutical drugs have to satisfy many practical, legal, and ethical requirements, which must be considered at every stage of the drug development process.

The first step of a drug development is the identification of a **lead compound** that shows any kind of promising activity towards a specific biological target. The lead compound, also known as a **new chemical entity** (**NCE**), can be isolated from natural products with known therapeutic effects or synthesized in the laboratory and screened against cell cultures, bacteria, or animals. This approach, known as **drug discovery**, is a slow, expensive, and inefficient process, which often fails to identify the lead compound with satisfactory pharmacological activity.

An alternative approach, **drug design**, relies on knowledge about drug–receptor interactions. If the chemical composition and three-dimensional structure of a particular biological target are known, a small molecule with a complementary structure can be designed using computer modelling techniques. The designed molecule is then synthesized and tested on a cell culture or isolated enzyme in order to determine its pharmacological activity. Any differences between actual and predicted activities can be used to refine the computer model, which eventually allows identification of the lead compound and, at the same time, better understanding of the drug–receptor interactions.

Once the lead compound has been identified, a series of similar compounds is synthesized, characterized, and subjected to **preclinical trials**. Each compound is rated according to its activity, toxicity, chemical stability, solubility in water and lipids, preparation cost, and many other properties that might be desirable for a potential pharmaceutical drug. In addition, the best candidates must have minimal activity towards unrelated biological targets, which can be responsible for side effects. Finally, the potential drug must be accessible (able to be synthesized with high yield or easily isolated from a natural source) and have minimal environmental impact (sub-topic D.6).

If all the above tests are successful, information about the new drug is submitted to regulatory authorities and, with their approval, the drug is tested on humans in a series of **clinical trials** (table 1). Most clinical tests involve double-blind experiments in which the patients are randomly given the drug or placebo. Any clinical trials can be carried out only with the full and informed consent of all participating patients or their legal representatives.

Phase	Subjects	Test results
I	small number of healthy volunteers	toxicity and safety dosage (TD_{50}), side effects
II	small number of patients	effectiveness and effective dosage (ED_{50}), safety and side effects
III	large number of patients	comparison with other available drugs, drug compatibility, further data on effectiveness, safety and side effects

▲ Table 1 Clinical (human) trials

If the drug successfully passes all clinical trials, it is approved by regulatory authorities for marketing and general use. However, the study of effectiveness and safety of the drug continues during the whole period of its commercial use, which is known as **post-clinical studies** or **phase IV trials**. Post-clinical studies are particularly important for determining the long-term effects and chronic toxicity of the drug, including its carcinogenic properties and the effects on the immune system, fertility, and reproductive functions.

Many potential drugs fail to pass clinical trials due to their toxicity, low efficiency, or unacceptable risk-to-benefit ratios. It is estimated that only one in 10 000 compounds synthesized by pharmaceutical companies is approved for medical use. In addition, some drugs are removed from the market during post-clinical trials, usually because of newly discovered side effects or the development of more efficient alternatives.

Questions

1 a) Explain the meanings of the terms *lethal dose* (LD_{50}), *toxic dose* (TD_{50}), and *effective dose* (ED_{50}).

   b) Explain how the above doses can be determined in animal and human studies.

2 Medicines and drugs alter the physiological state of the body including consciousness and coordination.

   a) State one other effect of medicines and drugs on the body. [1]

   b) Explain the meaning of the following terms: (i) therapeutic window; (ii) tolerance. [2]

   IB, May 2009

3 Describe how computers can be used to predict how changes to the structure of a drug might affect its activity. [2]

   IB, May 2012

4 Drugs can be prescribed for treating various diseases and assisting in healing the human body. However, any drug presents potential risks. The properties of three drugs are summarized in table 2.

Drug	Physiological effect	Side-effects	Therapeutic window
A	high	severe	medium
B	moderate	moderate	narrow
C	low	minimal	wide

▲ Table 2

Suggest which drug (**A**, **B**, or **C**) could be:

   a) considered safe enough to be taken by patients without supervision [1]

   b) administered only by qualified staff [1]

   c) used only in a medical emergency. [1]

   IB, May 2010

5 Drugs are most commonly taken orally.

   a) State one advantage and one disadvantage of this. [2]

   b) List three methods, other than orally, that can be used for the administration of a drug. [2]

   IB, May 2012

6 a) The effectiveness of a drug depends on the method of administration. One method of injecting drugs into the body results in the drug having a very rapid effect. State the method and explain its rapid action. [2]

   b) List the **two** other methods which can be used to inject drugs into the body. [1]

   c) Identify the method of administration used to treat respiratory diseases such as asthma. [1]

   IB, November 2009

7 Medicines and drugs are natural or synthetic substances used for their effects on the body.

   a) List two general effects of medicines and drugs on the functioning of the body.

   b) Explain the meaning of the term *side effect*.

   c) Describe the *placebo effect* and state its importance in drug development.

8 Creating a new pharmaceutical product is a long and complex process. Outline the main stages of this process in the correct order.

9 Describe briefly how pharmaceutical drugs can interact with receptors and enzymes.

10 The same drug can be identified by different names. Discuss whether the names of drugs are only labels, or whether they can influence our knowledge and perception.

11 ⊘ All drugs carry risks and benefits, which can be assessed differently by public bodies and individuals. Discuss the right of the government to protect the health of society and the right of individuals to make their choices about the use and abuse of drugs.

D.2 Aspirin and penicillin

Understandings

Aspirin

→ Mild analgesics function by intercepting the pain stimulus at the source, often by interfering with the production of substances that cause pain, swelling, or fever.

→ Aspirin is prepared from salicylic acid.

→ Aspirin can be used as an anticoagulant, in prevention of the recurrence of heart attacks and strokes, and as a prophylactic.

Penicillin

→ Penicillins are antibiotics produced by fungi.

→ A beta-lactam ring is a part of the core structure of penicillins.

→ Some antibiotics work by preventing cross-linking of the bacterial cell walls.

→ Modifying the side-chain results in penicillins that are more resistant to the penicillinase enzyme.

Applications and skills

Aspirin

→ Description of the use of salicylic acid and its derivatives as mild analgesics.

→ Explanation of the synthesis of aspirin from salicylic acid, including yield, purity by recrystallization, and characterization using IR and melting point.

→ Discussion of the synergistic effects of aspirin with alcohol.

→ Discussion of how aspirin can be chemically modified into a salt to increase its aqueous solubility and how this facilitates its bioavailability.

Penicillin

→ Discussion of the effects of chemically modifying the side-chain of penicillins.

→ Discussion of the importance of patient compliance and the effects of the over-prescription of penicillin.

→ Explanation of the importance of the beta-lactam ring on the action of penicillin.

Nature of science

→ Serendipity and scientific discovery – the discovery of penicillin by Sir Alexander Fleming.

→ Making observations and replication of data – many drugs need to be identified, isolated, and modified from natural sources. For example, salicylic acid from bark of willow tree for relief of pain and fever.

Natural products in medicine

Natural products have been used in traditional medicine for thousands of years. Even today about a quarter of all pharmaceutical drugs are derived from plants, animal tissues, and minerals. However, natural medicines have many disadvantages, including low efficiency, variable composition, instability, and numerous side effects caused by the presence of many bioactive substances in the same material. Therefore scientists and medical practitioners work to isolate, identify, and modify the chemical substances responsible for the therapeutic properties of natural products.

Aspirin

One of the first active ingredients, salicylic (2-hydroxybenzoic) acid, was isolated from the bark of willow tree in the first half of the nineteenth century and used as a pharmaceutical drug for pain and fever relief. However, pure salicylic acid caused severe digestive problems such as stomach irritation, bleeding, and diarrhoea. These side effects could be significantly reduced by the use of chemically modified salicylic acid, known as **acetylsalicylic acid** or **aspirin**:

salicylic acid (2-hydroxybenzoic acid)   ethanoic anhydride   aspirin (acetylsalicylic acid)   ethanoic acid

An alternative synthetic route to aspirin involves ethanoyl chloride and a base catalyst:

The first reaction can be used in a school laboratory for the preparation of aspirin. In a typical experiment, salicylic acid is mixed with excess ethanoic anhydride and several drops of catalyst (concentrated phosphoric acid). The mixture is heated for a short time, then diluted with water, and allowed to cool down slowly, producing crystals of aspirin. The obtained product is usually impure, so it needs to be recrystallized from hot ethanol. The identity of the product can be confirmed by IR spectroscopy (sub-topics 11.3 and 21.1) (figure 1) and by determining its melting point (sub-topic 10.2).

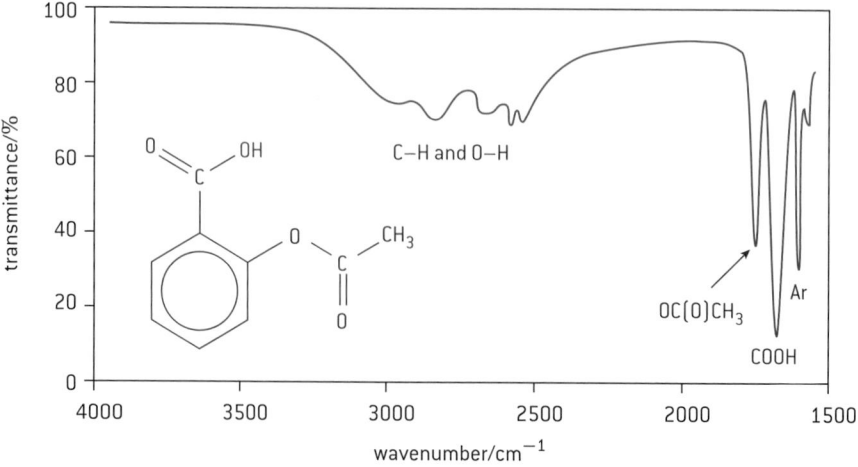

▲ Figure 1 Part of the IR spectrum of aspirin. Ar = aromatic ring

Worked example

Two students prepared samples of aspirin using the reaction conditions shown in table 1.

Sample	Masses of reactants and products / g			Melting point of product / °C	Product isolation
	Salicylic acid	Ethanoic anhydride	Aspirin		
1	2.57	2.85	2.11	134–135	Filtering, recrystallizing from ethanol, and drying for 24 hours
2	2.06	4.49	3.42	124–126	Filtering, washing with water, and drying for 10 minutes

▲ Table 1 Reaction conditions for the synthesis of two samples of aspirin

a) Calculate the amounts, in mol, of reactants used by both students and deduce the limiting reactant in each case.

b) Calculate the theoretical yields, in g, of aspirin in both cases.

c) Calculate the percentage yield of aspirin obtained by each student.

d) The melting point of pure aspirin is 136 °C. Deduce, referring to percentage yields and melting points, which sample of aspirin is likely to be more pure.

Solution

a) The molecular masses of salicylic acid ($C_7H_6O_3$) and ethanoic anhydride ($C_4H_6O_3$) are 138.13 and 102.10 g mol^{-1}, respectively. The first student used $2.57/138.13 \approx 0.0186$ mol of salicylic acid and $2.85/102.10 \approx 0.0279$ mol of ethanoic anhydride, so salicylic acid was the limiting reactant. Similarly, the second student used $2.06/138.13 \approx 0.0149$ mol of salicylic acid and $4.49/102.10 \approx 0.0440$ mol of ethanoic anhydride, so salicylic acid was again the limiting reactant.

b) The molecular mass of aspirin ($C_9H_8O_4$) is 180.17 g mol^{-1}. The theoretical yield depends on the amount of the limiting reactant, so the theoretical yield of aspirin was 0.0186 mol × 180.17 g mol^{-1} ≈ 3.35 g in the first case and 0.0149 mol × 180.17 g mol^{-1} ≈ 2.68 g in the second case.

c) The percentage yield of aspirin obtained by the first student was (2.11/3.35) × 100% ≈ 63.0%. In the second case, the percentage yield appears to be greater than 100%: (3.42/2.68) × 100% ≈ 128%.

d) The percentage yield above 100% indicates that sample 2 contains some impurities. In addition, the melting point of sample 2 (124–126 °C) is much lower than that of pure aspirin (136 °C), which also suggests the presence of impurities, probably water, ethanoic acid, and unreacted salicylic acid. In contrast, the percentage yield of sample 1 of aspirin is below 100% (probably because some of the product was lost during the recrystallization step), and its melting point (134–135 °C) is very close to the expected value (136 °C). Therefore, sample 1 of aspirin is likely to be more pure than sample 2.

The effects of aspirin

Aspirin and salicylic acid belong to the class of **mild analgesics**, also known as **non-narcotic analgesics** and **non-steroidal anti-inflammatory drugs** (**NSAIDs**). In addition to pain-relieving (analgesic) and fever-reducing (antipyretic) properties, these drugs can also reduce inflammation caused by irritation, infection, or physical damage to cell tissues. In contrast to strong analgesics (sub-topic D.3), mild analgesics affect the nervous system by intercepting the pain stimulus at the source. In particular, aspirin irreversibly binds to the enzyme cyclooxygenase and suppresses the production of prostaglandins, which are responsible for fever, swelling, and the transmission of pain impulses from the site of injury to the brain.

Two other mild analgesics, paracetamol (acetaminophen) and ibuprofen (figure 2), are commonly used for relieving pain and fever. In many countries these drugs are preferred to aspirin because they have less pronounced side effects, particularly in young children. However, paracetamol has a relatively narrow therapeutic window (sub-topic D.1) and in high doses can cause permanent damage to the brain, liver, and kidneys.

Similar to aspirin, ibuprofen increases the risk of stomach bleeding when taken with alcohol.

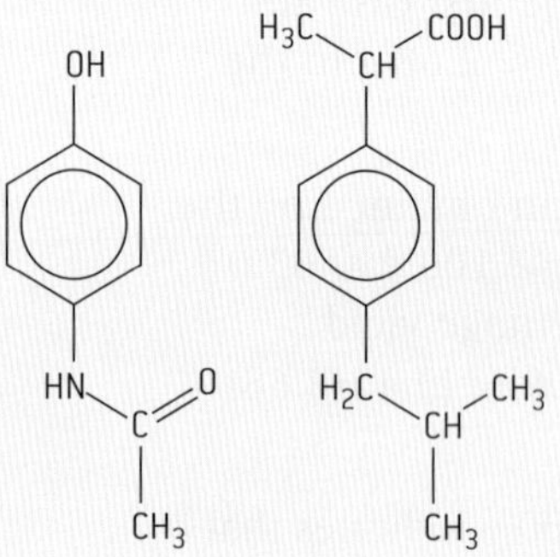

▲ Figure 2 The structures of paracetamol (left) and ibuprofen (right)

TOK

Although Fleming's discovery of penicillin is often described as serendipitous, the significance of his observations would have been missed by non-experts or less inquisitive scientists. In fact, the ability of mould to inhibit the growth of bacteria had been observed and reported in the early 1900s, with a conclusion that "the only thing to do now is to throw the culture away". This is a good example of the importance of a flexible and prepared mind in understanding the significance of observations.

Prostaglandins are also involved in the production of thromboxanes, which stimulate the aggregation of platelets (thrombocytes) and blood clotting. By inhibiting cyclooxygenase, aspirin prevents the formation of thromboxanes and acts as an anticoagulant, reducing the risk of strokes and heart attacks. At the same time, the anticlotting action of aspirin can lead to excessive bleeding and ulceration of the stomach. The risk of stomach bleeding significantly increases when aspirin is taken together with alcohol (ethanol) or other anticoagulants. This synergistic side effect is an example of a **drug interaction**, which must be taken into account when several drugs are prescribed to the same patient. Other side effects of aspirin include allergies, acidosis (decreased pH of the blood caused by salicylic acid), and Reye's syndrome in young children (potentially fatal liver and brain damage).

Soluble aspirin

Because aspirin is almost insoluble in water, its bioavailability (sub-topic D.1) is limited. The solubility and bioavailability of pharmaceutical drugs can be increased by converting them into ionic salts. In the case of aspirin, the carboxyl group can be neutralized with sodium hydroxide, producing the water-soluble sodium salt of acetylsalicylic acid (known as "soluble aspirin"):

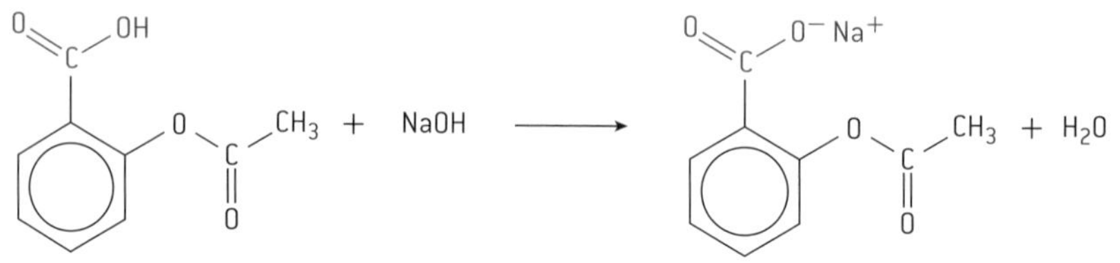

In aqueous solution the sodium salt of acetylsalicylic acid dissociates completely into sodium cations and acetylsalicylate anions, which form multiple ion–dipole interactions and hydrogen bonds with water. However, the sodium salt is immediately converted back into aspirin by the reaction with hydrochloric acid in the stomach, so the bioavailability of soluble aspirin is only slightly higher than that of plain aspirin.

Many drugs contain amino groups, which can also be converted into more soluble ionic salts by reactions with acids. For example, the common antidepressant fluoxetine is almost insoluble in water while its salt fluoxetine hydrochloride (Prozac™) is water soluble and can be administered orally.

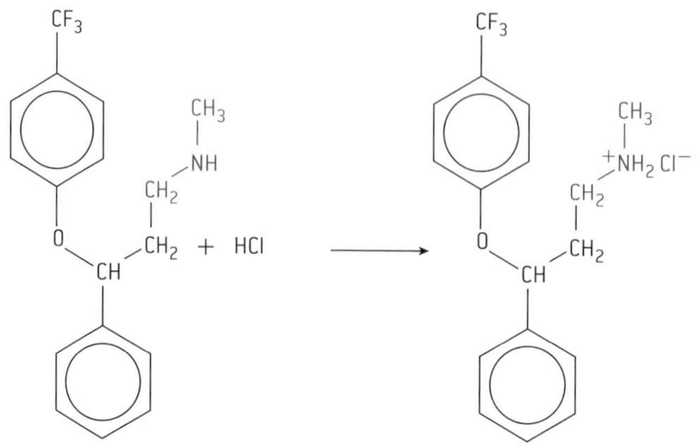

fluoxetine   fluoxetine hydrochloride

Penicillin

In 1928 the Scottish bacteriologist Alexander Fleming noticed that a Petri dish with a bacterial culture had been mistakenly left open. The dish became contaminated with a blue-green mould that inhibited the growth of bacteria. Fleming concluded that the mould produced a substance that was toxic to the bacteria and prevented them from developing normally. He grew a culture of the mould, determined its type (*Penicillium*), and named the unknown antibacterial substance "**penicillin**". Although Fleming published his observations, he could not isolate a pure sample of penicillin and did not pursue his discovery any further.

The development of penicillin into a drug

In 1938 Howard Florey and Ernest Chain read Fleming's reports on penicillin and decided to continue his research. Very soon they managed to concentrate penicillin and show that it was harmless to mice and effective *in vivo* against certain infectious diseases. In 1941 they used penicillin on their first patient who was suffering from a fatal blood infection. Within a day of treatment the patient started recovering, but later relapsed and died because the researchers ran out of penicillin. Nevertheless, the initial improvement in the patient's condition was dramatic, so Florey and Chain continued their studies. In 1943 Andrew Moyer and Margaret Rousseau developed a technology for the large-scale production of penicillin by growing *Penicillium* mould in large tanks filled with corn steep liquor. Since that time penicillin has become the most widely used antibiotic, and has saved more lives across the globe than any other pharmaceutical drug.

The term "penicillins" is now used as a collective name for a group of structurally similar natural and synthetic substances (figure 3). The chemical structure of the first penicillin, known as **benzylpenicillin** or **penicillin G**, was determined by Dorothy Hodgkin in 1945. The prefix "benzyl" refers to the **side-chain** (**R**) of benzylpenicillin, which is $-CH_2-C_6H_5$ in this particular compound but varies in other penicillins. For example, the side-chain in ampicillin contains an additional amino group [$R = -CH(NH_2)-C_6H_5$].

The mechanism of action of penicillin

A distinctive structural feature of penicillins, the four-membered **beta-lactam ring**, is responsible for the antibacterial properties of these drugs. The bond angles (sub-topic 4.3) of the carbon and nitrogen atoms in this ring are approximately 90° (instead of 109° and 120° for sp^3- and sp^2-hybridized atoms, respectively; see sub-topic 14.2 for more details). Such bond angles create significant ring strain and make the amide group in the beta-lactam ring very reactive. Once in bacteria the beta-lactam ring opens and irreversibly binds to the enzyme **transpeptidase**, which is responsible for cross-linking of bacterial cell walls. This weakens the cell walls in multiplying bacteria and makes them more permeable to water. The osmotic pressure causes water to enter the bacteria until they burst open and die. Human and other animal cells do not have cell walls and therefore are not affected by penicillin.

The discovery of penicillin has dramatically reduced the occurrence and severity of bacterial infections caused by surgical procedures and common diseases. In the 1950s and 1960s, when benzylpenicillin became

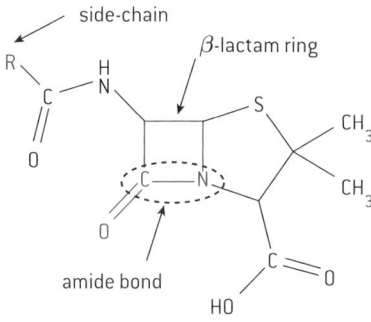

▲ Figure 3 The general structure of penicillins

Antibiotic resistance

Penicillin resistance is caused not only by the over-prescription of penicillin but also by the failure of some patients to complete their course of antibacterial treatment. Many patients stop taking medications soon after the symptoms of the disease disappear, which allows some of the most resistant bacteria to survive, multiply, and pass their resistance to the next generations. Another factor contributing to penicillin resistance is the use of antibiotics in agriculture, where penicillins are commonly given to healthy animals to prevent infectious diseases (sub-topic D.6). These antibiotics are eventually consumed by humans in the meat and dairy products, accelerating the development of resistant bacteria.

readily available around the world, it was routinely prescribed for treating minor illnesses or even as a **prophylactic** medicine. As a result, certain bacteria mutated and developed varying degrees of **antibiotic resistance** due to increased production of the enzyme **penicillinase**. This enzyme was able to deactivate benzylpenicillin and prevent it from binding to transpeptidase. Over time, bacteria with high levels of penicillinase became the dominant species and therefore greatly reduced the effectiveness of benzylpenicillin against many common diseases.

To overcome this bacterial resistance, new penicillins with modified side-chains were developed. Initially these penicillins could not be deactivated by penicillinase and were effective against a wider range of bacterial infections. In addition, some modified penicillins were stable in the acidic environment of the stomach and thus could be administered orally. However, new strands of constantly mutating bacteria became resistant to most penicillins (figure 4). Therefore scientists had to create new classes of antibacterial drugs which in turn triggered the development of **multidrug resistance** (**MDR**) in bacteria. The treatment of infectious diseases caused by MDR bacteria requires the use of a "cocktail" of different antibiotics and strict patient compliance to medical procedures. The problem of multidrug resistance is one of the major challenges of the twenty-first century and can be resolved only by the collective efforts of the international scientific community.

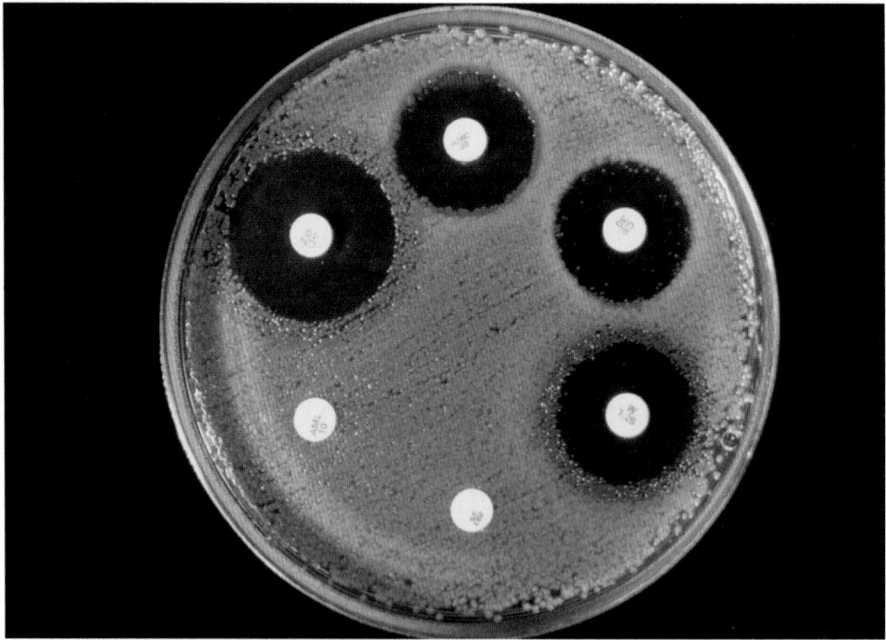

▲ Figure 4 A Petri dish with a bacterial culture (grey) and six different antibiotics (white pellets). Four antibiotics inhibit the bacterial growth (dark circles around the pellets). The remaining two pellets are surrounded by bacteria that are resistant to these drugs

Questions

1 a) Aspirin is thought to interfere with the production of prostaglandins. Explain how this produces an analgesic effect. [1]

b) State one important use for aspirin other than the relief of pain and fever. [1]

IB, May 2010

2 Acetylsalicylic acid (aspirin) can be synthesized from salicylic (2-hydroxybenzoic) acid.

a) Deduce the equation of the reaction of salicylic acid with ethanoic anhydride.

b) State the type of this reaction.

c) "Extra strength" aspirin tablets contain 500 mg of acetylsalicylic acid. Calculate the mass of salicylic acid needed to produce a pack of 10 "extra strength" aspirin tablets if the reaction yield is 60%.

3 Two examples of mild analgesics are aspirin and paracetamol (acetaminophen). Paracetamol is often used as an alternative to aspirin. State one advantage and one disadvantage of the use of paracetamol. [2]

IB, November 2010

4 Physiological effects of drugs can be significantly reduced, enhanced, or altered by other drugs or foods. The problem of drug interactions is particularly important for patients who consume excessive amounts of ethanol. State one possible adverse effect of consuming ethanol together with aspirin. [1]

IB, November 2012

5 Drugs such as fluoxetine and aspirin can be converted into salts.

a) Identify the functional group present in each of fluoxetine and aspirin which allows them to be converted into a salt. Suggest a reagent required for each conversion. [2]

b) Explain the advantage of converting drugs such as fluoxetine and aspirin into salts. [2]

IB, May 2011

6 The discovery of penicillin by Alexander Fleming in 1928 is often given as an example of serendipity in science.

a) Describe the chance event that led to Alexander Fleming's discovery of penicillin. [1]

b) Outline the work of Florey and Chain in developing penicillin. [3]

c) Describe what happens to bacteria when they come into contact with penicillin. [2]

d) The structure of a particular type of penicillin called dicloxacillin is shown in figure 5. State the name of the functional group in dicloxacillin, circled below. [1]

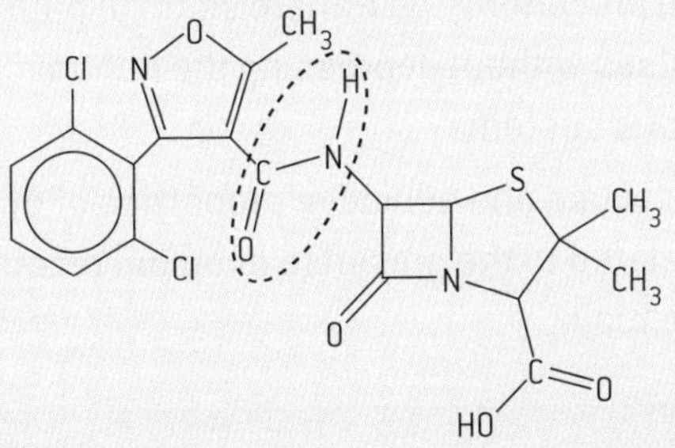

▲ Figure 5

e) Identify the β-lactam ring by drawing a circle around it and explain why the β-lactam ring is so important in the mechanism of the action of penicillin. [1]

f) Comment on the fact that many bacteria are now resistant to penicillins. [2]

IB, May 2012

7 The efficiency of certain drugs is strongly dependent on the frequency and regularity of their administration. Explain the importance of patient compliance when the patient is treated with antibacterials.

8 "In the field of observation, chance favours the prepared mind." – *Louis Pasteur*. Using the discovery of penicillin as an example, discuss the influence of an open-minded attitude on our perceptions.

D.3 Opiates

Understandings

→ The ability of a drug to cross the blood–brain barrier depends on its chemical structure and solubility in water and lipids.

→ Opiates are natural narcotic analgesics that are derived from the opium poppy.

→ Morphine and codeine are used as strong analgesics. Strong analgesics work by temporarily binding to receptor sites in the brain, preventing the transmission of pain impulses without depressing the central nervous system.

→ Medical use and addictive properties of opiates are related to the presence of opioid receptors in the brain.

Applications and skills

→ Explanation of the synthesis of codeine and diamorphine from morphine.

→ Description and explanation of the use of strong analgesics.

→ Comparison of the structures of morphine, codeine, and diamorphine (heroin).

→ Discussion of the advantages and disadvantages of using morphine and its derivatives as strong analgesics.

→ Discussion of side effects and addiction to opiate compounds.

→ Explanation of the increased potency of diamorphine compared to morphine based on their chemical structure and solubility.

Nature of science

→ Data and its subsequent relationships – opium and its many derivatives have been used as a painkiller in a variety of forms for thousands of years. One of these derivatives is diamorphine.

Opium and opiates

Opium and its derivatives have been used as painkillers for thousands of years. The primary bioactive ingredient of opium, morphine (figure 1), is a natural analgesic that belongs to the group of **alkaloids** – naturally occurring chemical compounds containing basic nitrogen atoms. Although morphine can be synthesized in the laboratory it is usually extracted from the opium poppy, which is a common plant around the world (figure 2).

▲ Figure 2 The opium poppy (*Papaver somniferum*) exuding opium sap from shallow cuts in the fresh seed pod

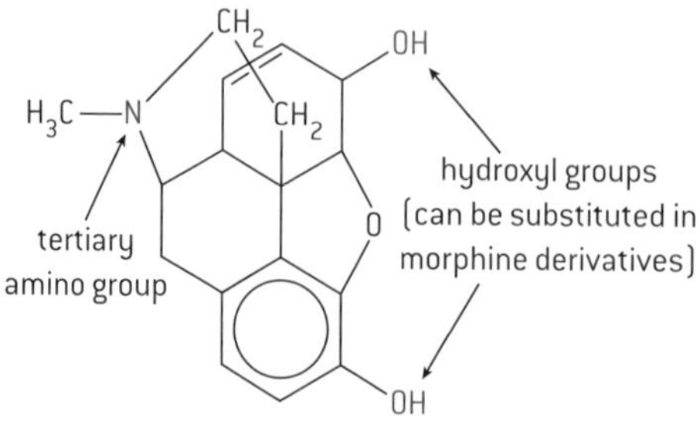

▲ Figure 1 The chemical structure of morphine

Morphine and its derivatives (**opiates**) are **strong analgesics**, which are used to relieve severe pain caused by injury, surgical procedures, heart attack, or chronic diseases such as cancer. In contrast to mild analgesics (sub-topic D.2), strong analgesics block the transmission of pain impulses by temporarily binding to **opioid receptors** (topic D.1) in the brain. Although strong analgesics act as depressants of the central nervous system (CNS), they do not significantly affect perception, attention, or coordination when taken in low to moderate doses. However, high doses of opiates affect all functions of the CNS and can lead to drowsiness, confusion, and potentially fatal asphyxia caused by respiratory depression.

Opiates are also known as **narcotic analgesics** because of their specific effects on the human body. In addition to their painkilling properties, large doses of opiates cause a strong feeling of euphoria, provide relief from all forms of distress, and stimulate sociability. As a result morphine and other opiates have a very high potential for misuse, which often leads to drug addiction. Non-medical use of opiates quickly leads to psychological dependence and tolerance (sub-topic D.1), forcing the user to take constantly increasing doses of the drug to achieve the desired effect. This affects the metabolic processes in the body and leads to physiological dependence, further increasing the required dose of the drug and the risk of adverse effects. Therefore the production and use of opiates in most countries is strictly regulated by the law and limited to the most severe cases of pain and suffering.

Crossing the blood–brain barrier

The physiological activity of opiates strongly depends on their ability to cross the so-called **blood-brain barrier**: a series of lipophilic cell membranes (sub-topic B.3) that coat the blood vessels in the brain and prevent polar molecules from entering the CNS. The presence of one amino and two hydroxyl groups (sub-topic 10.2) in the morphine molecule makes it sufficiently polar to be soluble in water but at the same time reduces its solubility in lipids and therefore limits its ability to reach the opioid receptors in the brain.

The polarity of morphine can be reduced by chemical modification of one or both hydroxyl groups in its molecule. In **codeine**, the phenolic –OH group is replaced with the less polar ether group, $-OCH_3$:

Codeine readily crosses the blood–brain barrier but does not bind to the opioid receptor because of the steric effect of the ester group. However,

Side effects and withdrawal symptoms

Short-term adverse effects of opiates include decreased breathing and heart rates, nausea and vomiting (in first-time users); high doses can lead to coma and death. Common long-term effects include constipation, loss of sex drive, disrupted menstrual cycle, and poor appetite. Illegal drug users face an increased risk of AIDS, hepatitis, and other diseases transmitted through shared needles, as well as acute poisoning caused by contaminants in street drugs. In addition the high cost of opiates causes many social problems such as theft and prostitution.

Drug addiction is a serious health condition that usually requires long-term medical and psychological treatment. When the drug intake is stopped or significantly reduced, most drug addicts experience **withdrawal symptoms**. In the case of opiates, withdrawal symptoms include perspiration, diarrhoea, cramps, and acute feelings of distress. Without medical treatment these effects can last from several days to a few weeks or even months. Certain medical drugs such as methadone can be used to alleviate withdrawal symptoms. These drugs are structurally similar to morphine and bind to opioid receptors in the brain without producing the euphoria craved by addicts.

codeine is slowly metabolized into morphine, which is ultimately responsible for its pharmaceutical properties. As a result, codeine is approximately 10 times less potent an analgesic than morphine. Its low activity, wide therapeutic window (sub-topic D.1), and limited potential for abuse makes codeine the most widely used opiate in the world. In some countries, codeine is available over the counter as a component of cough syrups or in combination with mild analgesics (sub-topic D.2).

The development of synthetic opiates

Systematic observations of opium users allowed scientists to establish certain patterns in the physiological and psychological effects of this drug on the human body. These data stimulated the study of opiates and eventually led to the isolation of morphine from the opium poppy. Further studies of morphine allowed its structure and reactivity to be established, producing a broad range of opiates with greater potency or specific types of pharmaceutical activity. In turn, clinical studies of various opiates led to better understanding of the basic functions of the CNS and the development of new generations of pharmaceutical drugs.

Study tip

The structures of morphine, diamorphine (heroin), and codeine are given in the *Data booklet*, which will be available during the examination.

Diamorphine

In another derivative of morphine, **diamorphine**, both hydroxyl groups are substituted with ester groups which greatly reduces the polarity of the molecule. Diamorphine can be prepared from morphine in the same way as aspirin is prepared from salicylic acid and ethanoic anhydride (sub-topic D.2):

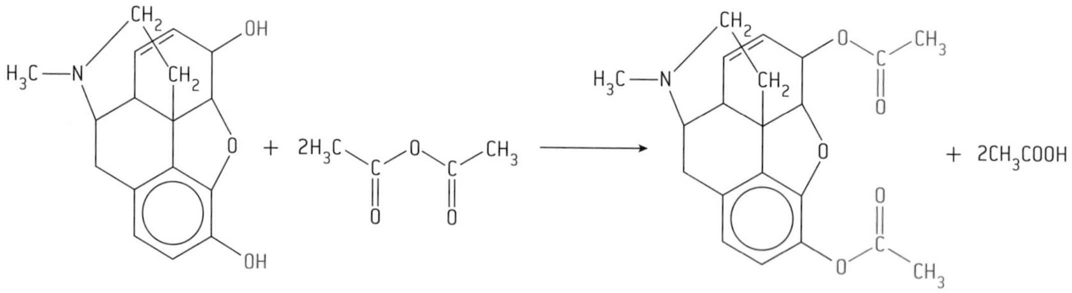

Similar to codeine, diamorphine is soluble in lipids and can easily cross the blood–brain barrier. In the brain diamorphine is quickly metabolized into morphine, which binds to the opioid receptor. This mechanism of action makes diamorphine about five times more potent an analgesic than morphine. At the same time diamorphine has more severe side effects, including tolerance, addiction, and CNS depression. Under the street name **"heroin"** diamorphine is one of the most dangerous substances of abuse; it is responsible for nearly 50% of all drug-related deaths around the globe. In most countries the use of diamorphine is either banned or restricted to terminally ill patients with certain forms of cancer or CNS disorders.

Cultural views on drugs

Morphine, heroin, and many other substances of abuse are illegally produced in a small number of countries and then distributed globally by criminal organizations. According to the UN World Drug Report over 80% of illicit opiates are produced in a single country, Afghanistan, with less than 2% of these drugs consumed locally and the remaining 98% exported to Europe, Asia, Africa, and North America. This situation reflects differences in cultural and economic viewpoints on the production and sale of non-medical drugs around the world. The problem of drug abuse can be resolved only by recognizing and addressing these differences, primarily through education, economic development, and international cooperation.

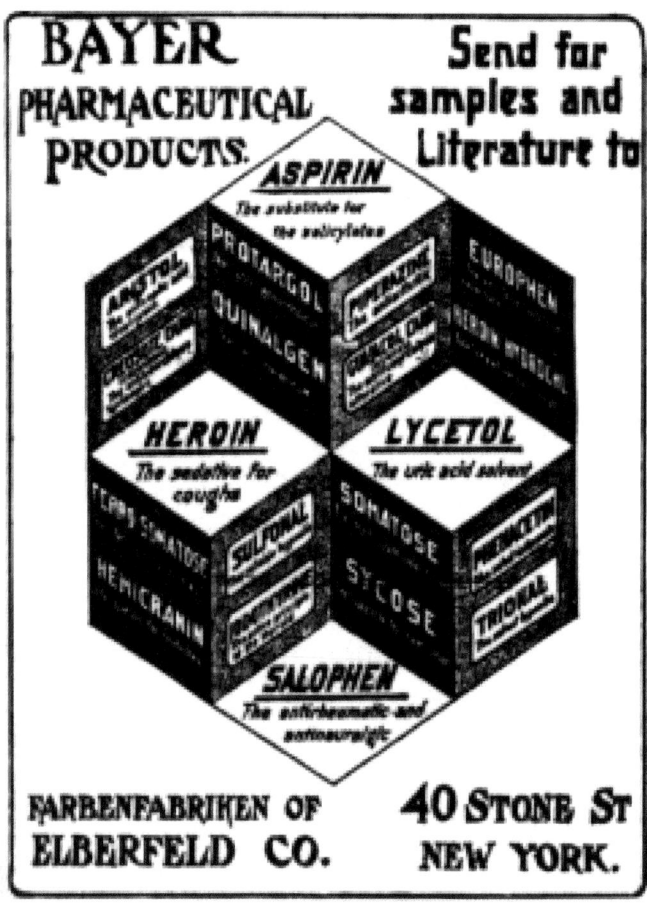

▲ Figure 3 From 1898 to 1910, diamorphine was available over the counter in many countries under the trademark name Heroin

Questions

1 Examples of strong analgesics are morphine, codeine, and diamorphine (heroin). Their structures are shown in the *Data booklet*.

   a) Identify two functional groups present in all three of these analgesics. [2]

   b) Identify one functional group present in morphine, but not in diamorphine. [1]

   c) State the name of the type of chemical reaction which is used to convert morphine into diamorphine. [1]

   IB, November 2010

2 Mild analgesics such as aspirin, and strong analgesics such as opiates, differ not only in their potency but also in the ways they act on the central nervous system.

   a) Describe how mild and strong analgesics provide pain relief. [2]

   b) Discuss two advantages and two disadvantages of using morphine and other opiates for pain relief. [4]

   c) Explain why heroin is a more potent drug than morphine. [2]

   IB, May 2010

3 Aspirin, morphine, and diamorphine (heroin) are painkillers. Their structures are given in the *Data booklet*.

   a) Other than the phenyl group, state the name of one other functional group that is common to both aspirin and diamorphine. [1]

   b) Suggest a reagent that could be used to convert morphine into diamorphine and state the name of the type of reaction taking place. [2]

   IB, May 2010

4 Diamorphine (heroin) is often administered as an ionic salt, diamorphine hydrochloride.

   a) State the name of the functional group in diamorphine that can be protonated by strong acids.

   b) Deduce the equation for the reaction of diamorphine with hydrogen chloride.

   c) Suggest how the bioavailability of diamorphine will be affected by its conversion into an ionic salt.

5 Morphine, diamorphine, and codeine are strong analgesics. Their solubility in water and lipids depends on the nature of the functional groups present in their molecules.

   a) Suggest which of these three drugs will be most soluble in water.

   b) Explain, with reference to intermolecular interactions, how the drug named in (a) will interact with water in solutions.

   c) Suggest which of the three drugs will be most soluble in lipids.

6 Methadone is an analgesic that is commonly used in the treatment of opioid dependence. The structure of methadone is given in figure 4.

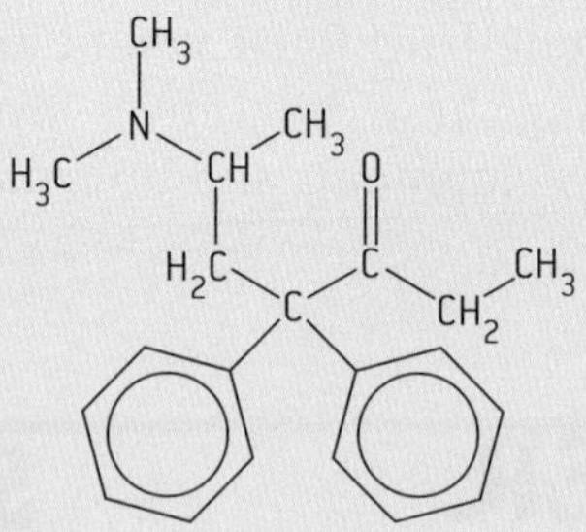

▲ Figure 4 Methadone

   a) State the names of two different functional groups in the molecule of methadone.

   b) Identify, by marking it with an asterisk (*) on a copy of figure 4, the chiral carbon atom in methadone.

   c) Deduce the equation for the reaction of methadone with hydrogen chloride.

   d) Suggest which drug (methadone or methadone hydrochloride) will be more soluble in water, and which one will have higher bioavailability.

   e) Methadone binds to the opioid receptor in the same way as morphine but does not produce the euphoric effect of opiates. Deduce whether methadone is a strong analgesic or a mild analgesic.

   f) Suggest, by comparing the structures of methadone and morphine, which functional groups in their molecules are likely to be involved in binding to the opioid receptor.

7 ⊘ Views on the problem of illegal drug production and trafficking are very different across the globe. Discuss whether it is ever appropriate for one ethnic group or nation to impose change on another.

D.4 pH regulation of the stomach

Understandings

→ Non-specific reactions, such as the use of antacids, are those that work to reduce excess stomach acid.

→ Active metabolites are the active forms of a drug after it has been processed by the body.

Applications and skills

→ Explanation of how excess acidity in the stomach can be reduced by the use of different bases.

→ Construction and balancing of equations for neutralization reactions and the stoichiometric application of these equations.

→ Solving buffer problems using the Henderson–Hasselbalch equation.

→ Explanation of how compounds such as ranitidine (Zantac) can be used to inhibit stomach acid production.

→ Explanation of how compounds like omeprazole (Prilosec) and esomeprazole (Nexium) can be used to suppress acid secretion in the stomach.

Nature of science

→ Collecting data through sampling and trialling – one of the symptoms of dyspepsia is the overproduction of stomach acid. Medical treatment of this condition often includes the prescription of antacids to instantly neutralize the acid, or H2-receptor antagonists or proton pump inhibitors which prevent the production of stomach acid.

Stomach acid

The process of digestion involves a series of catabolic reactions (sub-topic B.1) that transform food nutrients into small molecules. Many of these reactions take place in the stomach, where the food is mixed with a digestive fluid. This fluid, also known as **gastric juice**, is composed of water, salts (mostly KCl and NaCl), hydrochloric acid (HCl), and enzymes (pepsins), which are secreted by the cells in the stomach lining. These enzymes are primarily responsible for the breakdown of proteins into peptides and individual amino acids (sub-topic B.2). Other cells produce hydrogencarbonate ions (HCO_3^-) and gastric mucus to buffer the acid (sub-topic 18.3) and prevent the gastric juice from digesting the stomach tissues.

The concentration of hydrochloric acid in the stomach varies from approximately 0.003 to 0.1 mol dm^{-3} (0.01–0.4%), which corresponds to a pH range of 1.0 to 2.5 (sub-topic 8.3). Although the acid itself does not break down food molecules, it denatures proteins and provides an optimum pH (sub-topic B.7) for pepsins and other enzymes in the gastric juice. In addition, hydrochloric acid acts as a disinfectant, killing nearly all harmful microorganisms that are ingested with the food.

Worked example

Hypochlorhydria is a health condition caused by insufficient production of gastric acid. A 20.0 cm³ sample of gastric juice with a density of 1.03 g cm⁻³ was taken from a patient suffering from hypochlorhydria and titrated with a 0.0215 mol dm⁻³ solution of sodium hydroxide to pH = 7.0. The volume of the titrant used was 1.47 cm³. Calculate:

a) the molar concentration of hydrogen chloride in the sample

b) the pH of the sample, to two significant figures

c) the mass percentage of hydrogen chloride in the sample.

Solution

a) The amount of NaOH is 0.00147 dm³ × 0.0215 mol dm⁻³ ≈ 3.16 × 10⁻⁵ mol. Since the neutralization of HCl requires an equal amount of NaOH, the amount of HCl in the original sample was the same, 3.16 × 10⁻⁵ mol. Therefore, the molar concentration of HCl in

the sample was 3.16×10^{-5} mol / 0.0200 dm³ = 1.58×10^{-3} mol dm⁻³.

b) Hydrogen chloride is a strong acid and dissociates completely in aqueous solutions:

$$HCl(aq) \rightarrow H^+(aq) + Cl^-(aq)$$

Therefore:

$$pH = -\log [H^+(aq)] = -\log c(HCl)$$
$$= -\log (1.58 \times 10^{-3}) = 2.8$$

This value is higher than the typical pH range of gastric juice (1.0–2.5), which confirms the case of hypochlorhydria.

c) The molar mass of hydrogen chloride is $35.45 + 1.01 = 36.46$ g mol⁻¹, so the mass of hydrogen chloride in the original sample was 36.46 g mol⁻¹ × 3.16×10^{-5} mol = 1.15×10^{-3} g.

The mass of gastric juice sample 20.0 cm³ × 1.03 g cm⁻³ = 20.6 g. Therefore, the mass percentage of HCl in the sample was (1.15×10^{-3} g/20.6 g) × 100% ≈ 5.58×10^{-3}%.

Antacids

Excessive production of hydrochloric acid in the stomach is commonly associated with indigestion (also known as dyspepsia), gastritis, and peptic ulcer disease. It is often accompanied by abdominal pain, heartburn, bloating, nausea, and other unpleasant feelings, which can be alleviated by neutralizing excess acid or reducing its secretion. Certain pharmaceutical drugs known as **antacids** can quickly increase the pH of gastric juice by reacting with hydrochloric acid. Common antacids are hydroxides, carbonates, and hydrogencarbonates of calcium, magnesium, aluminium, and sodium, which act as weak Brønsted–Lowry bases (sub-topics 8.1 and 8.2), for example:

$$Al(OH)_3(s) + 3HCl(aq) \rightarrow AlCl_3(aq) + 3H_2O(l)$$

$$CaCO_3(s) + 2HCl(aq) \rightarrow CaCl_2(aq) + CO_2(g) + H_2O(l)$$

$$NaHCO_3(s) + HCl(aq) \rightarrow NaCl(aq) + CO_2(g) + H_2O(l)$$

The ionic equations for the above processes clearly show that antacids reduce the concentration of H⁺(aq) ions and therefore increase the pH of gastric juice:

$$Al(OH)_3(s) + 3H^+(aq) \rightarrow Al^{3+}(aq) + 3H_2O(l)$$

$$CaCO_3(s) + 2H^+(aq) \rightarrow Ca^{2+}(aq) + CO_2(g) + H_2O(l)$$

$$NaHCO_3(s) + H^+(aq) \rightarrow Na^+(aq) + CO_2(g) + H_2O(l)$$

The discovery of gastric acid

The presence of acid in the gastric juice was first described in 1838 by surgeon William Beaumont, who was observing a patient with a gastric fistula (an unhealed hole in the stomach) left by a gunshot. By taking samples of gastric juice and using them to "digest" food in glass containers, Beaumont discovered that digestion was a chemical rather than mechanical process.

Further experiments revealed the negative effects of excess stomach acid, which led to the development of antacids. Finally, the study of digestion at the cellular level led to the creation of new pharmaceutical drugs such as ranitidine and omeprazole (see below), which regulate the acidity of the stomach by suppressing the secretion of hydrochloric acid.

Worked example

An antacid tablet contains 350 mg of magnesium hydroxide and 650 mg of sodium hydrogencarbonate.

a) State the equations for the reactions of these antacids with hydrochloric acid.

b) Deduce which of the two antacids can neutralize the greater amount of the stomach acid.

Solution

a) Magnesium hydroxide:

molecular equation:
$$Mg(OH)_2(s) + 2HCl(aq) \rightarrow MgCl_2(aq) + 2H_2O(l)$$

ionic equation:
$$Mg(OH)_2(s) + 2H^+(aq) \rightarrow Mg^{2+}(aq) + 2H_2O(l)$$

The equations for sodium hydrogencarbonate are given in the text.

b) The amounts of $Mg(OH)_2$ and $NaHCO_3$ in the tablet are $0.35\ g / 58.32\ g\ mol^{-1} \approx 0.0060$ mol and $0.65\ g / 84.01\ g\ mol^{-1} \approx 0.0077$ mol, respectively. One mole of $Mg(OH)_2$ reacts with two moles of HCl, so 0.0060 mol of $Mg(OH)_2$ can neutralize $0.0060 \times 2 = 0.012$ mol of HCl. One mole of $NaHCO_3$ reacts with one mole of HCl, so 0.0077 mol of $NaHCO_3$ can neutralize 0.0077 mol of HCl. Therefore, 350 mg of $Mg(OH)_2$ can neutralize more stomach acid than 650 mg of $NaHCO_3$.

As with any pharmaceutical drugs, antacids may have various side effects (sub-topic D.1) and must be taken with care. For example, aluminium hydroxide reduces the concentration of phosphates in the body fluids (due to the precipitation of aluminium phosphate) while carbonates and hydrogencarbonates produce carbon dioxide, which causes bloating and belching. In addition, excessive intake of calcium, magnesium, and sodium ions affects the electrolyte balance in the body and can lead to various conditions, ranging from diarrhoea and constipation to kidney stones and heart failure.

Antacids are often combined with anti-foaming agents and alginates. Anti-foaming agents such as organosilicon polymers (dimethicone) relieve bloating by allowing the bubbles of carbon dioxide to coalesce and leave the body via belching and flatulence. Alginates produce a protective layer that floats on the stomach contents and prevents heartburn, which is caused by gastric juice rising up the esophagus.

Regulation of acid secretion

The acidity of gastric juice can be controlled at the cellular level by targeting the biochemical mechanisms of acid production. The secretion of acid in the stomach is triggered by histamine (a derivative of amino

Study tip

The structures of ranitidine (Zantac), omeprazole (Prilosec), and esomeprazole (Nexium) are given in the *Data booklet*, which will be available during the examination.

acid histidine) that binds to **H2-histamine receptors** in the cells of the gastric lining. Certain pharmaceutical drugs such as ranitidine (Zantac) block H2-histamine receptors and reduce the secretion of stomach acid. Ranitidine and other H2-histamine receptor inhibitors provide short-term relief from the symptoms of indigestion and usually require frequent administration (two to four times a day).

Another group of pharmaceutical drugs including omeprazole (Prilosec) and esomeprazole (Nexium) reduce the production of stomach acid by inhibiting a specific enzyme, known as the **gastric proton pump**, which is directly responsible for secreting $H^+(aq)$ ions into the gastric juice. In contrast to ranitidine, the action of proton pump inhibitors reduces the secretion of stomach acid for prolonged periods (up to three days).

Indigestion

Dyspepsia or indigestion is a common problem that affects up to 40% of the global population. However, the occurrence and symptoms of indigestion differ around the world. Culture, diet, lifestyle, and genetics are among the main factors affecting the pH of the stomach and therefore the risk of indigestion and other gastric disorders. In many cases, indigestion is related to excessive consumption of alcohol and fizzy drinks, smoking, stress, spicy or heavy food, and irregular eating patterns. Positive changes in lifestyle and dietary habits often relieve the symptoms of indigestion and reduce the need for medical treatment.

Omeprazole and esomeprazole

Omeprazole and esomeprazole have the same molecular formula ($C_{17}H_{19}N_3O_3S$) and differ only in their stereoisomeric structure (sub-topic 20.3). Due to the presence of three different substituents and a lone pair at the sulfur atom, these compounds are chiral and can exist as two enantiomers (figure 1). Omeprazole is a racemic mixture of both enantiomers while esomeprazole is a single enantiomer.

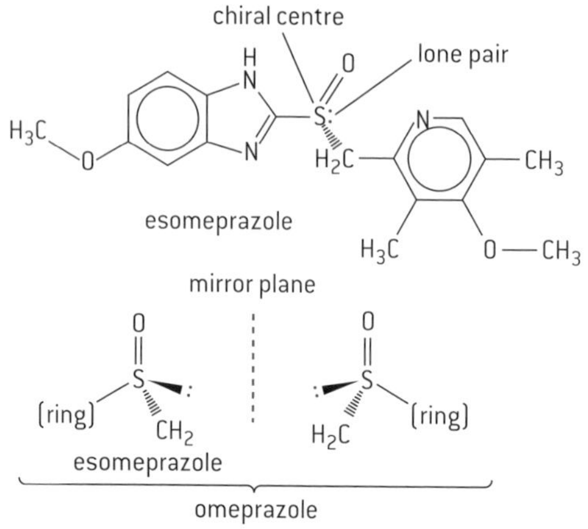

▲ Figure 1 The structures of esomeprazole (top) and chiral centres in omeprazole (bottom)

In contrast to many other drugs, both enantiomers of omeprazole show very similar pharmacological activity (sub-topic D.7). In their original form they are inactive and do not interact with the gastric proton pump directly. Due to their low polarity, omeprazole and esomeprazole readily cross cell membranes (sub-topic D.1) and enter the intracellular compartments containing hydrochloric acid. In this acidic environment both enantiomers undergo a series of chemical transformations and produce the same **active metabolites**, which bind to the proton pump enzymes and inhibit the secretion of stomach acid. This mechanism of action increases the efficiency of both drugs and allows a reduced frequency of administration.

Acid–base buffers

In contrast to gastric juice, where the concentration of acid varies by a factor of 100, the pH of other biological fluids remains relatively constant. This is achieved by the action of **acid–base buffers** (topic 18.3), which can neutralize small amounts of strong acids and bases without significantly changing their pH. Each acid–base buffer system contains two molecular or ionic species which differ by a single proton (H^+). Such species are known as **conjugate acid–base pairs**, where the more protonated species is the **conjugate acid** and the less protonated species is the **conjugate base** (sub-topic 8.1). For example, an acetate buffer consists of ethanoic (acetic) acid, CH_3COOH and ethanoate (acetate) anions, CH_3COO^-. The CH_3COOH molecule contains one more proton than the CH_3COO^- anion, so ethanoic acid is the conjugate acid while ethanoate anion is the conjugate base.

In buffer solutions both the conjugate acid and the conjugate base are weak and exist in equilibrium, for example:

$$CH_3COOH(aq) \rightleftharpoons CH_3COO^-(aq) + H^+(aq)$$
conjugate acid   conjugate base

The acid–base equilibrium is characterized by the **dissociation constant** (K_a) of the conjugate acid or, more commonly, its negative logarithm (pK_a, see sub-topic 18.2 and table 1 below):

$$K_a = \frac{[\text{conjugate base}][H^+]}{[\text{conjugate acid}]} \qquad pK_a = -\log K_a \qquad pK_a = -\log \frac{[\text{conjugate base}][H^+]}{[\text{conjugate acid}]}$$

Since $pH = -\log [H^+]$, the pK_a expression can be transformed into the **Henderson–Hasselbalch equation**:

$$pH = pK_a + \log \frac{[\text{conjugate base}]}{[\text{conjugate acid}]}$$

The Henderson–Hasselbalch equation allows us to calculate the pH of a buffer solution with known acid–base composition, or the concentrations of the conjugate acid and base in a solution with known pH. For example, if $pH = pK_a$, $\log ([\text{conjugate base}]/[\text{conjugate acid}]) = 0$ and therefore [conjugate base] = [conjugate acid]. According to table 1, an acetate buffer solution prepared from equal amounts of ethanoic acid and sodium ethanoate will have a pH of 4.76.

Buffer	Conjugate acid	Conjugate base	pK_a
acetate (ethanoate)	CH_3COOH	CH_3COO^-	4.76
ammonia	NH_4^+	NH_3	9.25
hydrogencarbonate (bicarbonate)	H_2CO_3 or $CO_2 \cdot H_2O$	HCO_3^-	6.36
carbonate	HCO_3^-	CO_3^{2-}	10.3
dihydrogen phosphate	H_3PO_4	$H_2PO_4^-$	2.12
hydrogen phosphate	$H_2PO_4^-$	HPO_4^{2-}	7.20
phosphate	HPO_4^{2-}	PO_4^{3-}	12.3

▲ Table 1. Common acid–base buffers

Worked example

An ammonia buffer is commonly used in biochemical experiments when high pH is required.

a) Calculate the pH of an aqueous solution that contains 0.040 mol dm⁻³ ammonium chloride and 0.16 mol dm⁻³ ammonia.

b) State the equations that show the buffer action of the solution in (a) when a small amount of hydrochloric acid is added and when a small amount of sodium hydroxide is added.

Solution

a) Ammonium chloride is an ionic salt (sub-topic 4.1) that dissociates completely in aqueous solutions:

$$NH_4Cl(aq) \rightarrow NH_4^+(aq) + Cl^-(aq)$$

Therefore, the concentration of $NH_4^+(aq)$ (the conjugate acid) will be the same as the concentration of ammonium chloride (0.040 mol dm⁻³). According to table 1, $pK_a(NH_4^+) = 9.25$, so:

$$pH = 9.25 + \log(0.16/0.040)$$
$$\approx 9.25 + 0.60 = 9.85$$

b) The conjugate base of the buffer system, NH_3, will neutralize the strong acid, HCl. This reaction can be represented by molecular and ionic equations:

$$NH_3(aq) + HCl(aq) \rightarrow NH_4Cl(aq)$$
$$NH_3(aq) + H^+(aq) \rightarrow NH_4^+(aq)$$

Similarly, the conjugate acid of the buffer system will neutralize the strong base:

$$NH_4Cl(aq) + NaOH(aq) \rightarrow NH_3(aq) + NaCl(aq) + H_2O(l)$$
$$NH_4^+(aq) + OH^-(aq) \rightarrow NH_3(aq) + H_2O(l)$$

Study tip

The same ionic or molecular species *in a particular acid–base buffer* cannot neutralize *both* the strong acid *and* the strong base. If you attempt to use the same species (such as hydrogencarbonate ion) in both neutralization reactions, in one case you will produce a species that cannot exist in this particular buffer solution and will immediately react with another component of the buffer system to give the original ion or molecule. Therefore, before writing any equations you should identify the conjugate acid–base pair and make sure that only these two species are used as reactants or formed as products in each neutralization reaction.

Hydrogencarbonate and carbonate buffers

The primary acid–base buffer system in the human body consists of carbon dioxide and hydrogencarbonate ions. Carbon dioxide is soluble in water and forms unstable carbonic acid, H_2CO_3, which is usually represented as $CO_2 \cdot H_2O$. The equilibrium between carbon dioxide and hydrogencarbonate ions is characterized by the first dissociation constant of carbonic acid:

$$CO_2 \cdot H_2O \rightleftharpoons HCO_3^-(aq) + H^+(aq) \qquad pK_{a1} = 6.36$$

conjugate acid   conjugate base

At high pH a hydrogencarbonate ion can lose the second proton and produce a carbonate buffer. The equilibrium between carbonate and hydrogencarbonate ions is characterized by the second dissociation constant of carbonic acid:

$$HCO_3^-(aq) \rightleftharpoons CO_3^{2-}(aq) + H^+(aq) \qquad pK_{a2} = 10.3$$

conjugate acid   conjugate base

Therefore, depending on the solution pH, hydrogencarbonate ions can form two different buffer systems and play the role of either the conjugate acid (at low pH) or the conjugate base (at high pH). This situation is similar to that of amino acid buffers (sub-topic B.7).

Worked example

A hydrogencarbonate buffer was prepared by slow addition of 20.0 cm³ of 0.100 mol dm⁻³ hydrochloric acid to 80.0 cm³ of a 0.200 mol dm⁻³ solution of sodium hydrogencarbonate.

a) Calculate the pH of this buffer solution. Assume that the densities of all solutions are 1.00 kg dm⁻³ and all carbon dioxide stays in the solution.

b) Calculate the pH change after the addition of 0.0200 g of solid sodium hydroxide to this buffer solution. Assume that the addition of NaOH does not affect the volume of the solution.

Solution

a) The initial amounts of HCl and $NaHCO_3$ are 0.0200 dm³ × 0.100 mol dm⁻³ = 0.00200 mol and 0.0800 dm³ × 0.200 mol dm⁻³ = 0.0160 mol, respectively. Hydrochloric acid reacts with sodium hydrogencarbonate to produce unstable carbonic acid, $CO_2 \cdot H_2O$:

$$NaHCO_3(aq) + HCl(aq) \rightarrow CO_2 \cdot H_2O + NaCl(aq)$$

or, in ionic form,

$$HCO_3^-(aq) + H^+(aq) \rightarrow CO_2 \cdot H_2O$$

initial amount:	0.0160	0.00200	—
amount change:	−0.00200	−0.00200	+0.00200
final amount:	0.0140	—	0.00200

Since the volume of the final solution is 0.0200 + 0.0800 = 0.100 dm³, the concentrations of $CO_2 \cdot H_2O$ (conjugate acid) and HCO_3^- (conjugate base) in the buffer solution will be 0.0140/0.100 = 0.140 mol dm⁻³

and 0.00200/0.100 = 0.0200 mol dm⁻³, respectively. The equilibrium between $CO_2 \cdot H_2O$ and HCO_3^- is characterized by $pK_{a1} = 6.36$ (table 1). Using the Henderson–Hasselbalch equation, pH = 6.36 + log (0.140/0.0200) ≈ 6.36 + 0.85 = 7.21.

b) The amount of NaOH is 0.0200 g/40.00 g mol⁻¹ = 0.000500 mol. Since NaOH is a strong base it will dissociate completely to produce 0.000 50 mol of hydroxide ions, which will be neutralized by the conjugate acid of the buffer solution, $CO_2 \cdot H_2O$:

$$CO_2 \cdot H_2O + OH^-(aq) \rightarrow HCO_3^-(aq)$$

initial amount:	0.00200	0.00050	0.0140
amount change:	−0.00050	−0.00050	+0.00050
final amount:	0.00150	—	0.0145

The concentrations of the $CO_2 \cdot H_2O$ and HCO_3^- in the final solution will be 0.00150/0.100 = 0.0150 mol dm⁻³ and 0.0145/0.100 = 0.145 mol dm⁻³, respectively. Therefore, the pH of the final solution will be 6.36 + log (0.145/0.0150) ≈ 6.36 + 0.99 = 7.35, and ΔpH = 7.35 − 7.21 = 0.14.

As you can see, the addition of a strong base to a buffer solution caused a very small change in pH. If the same amount of NaOH (0.00050 mol) were added to 100 cm³ of pure water, the pH change would be much greater, approximately 4.7 units (you can calculate it using the formulae from sub-topic 8.3).

Buffer pH range

The ability of acid–base buffers to resist pH changes is limited and depends on the concentrations and ratios of the conjugate acid and base in the solution. At pH = pK_a, an acid–base buffer reaches its maximum efficiency and can neutralize the greatest amounts of strong acids or bases before any significant pH change occurs. According to the Henderson–Hasselbalch equation, the ratio between the components of a conjugate acid–base pair increases or decreases 10 times when the pH of the solution changes by one unit. Therefore an acid–base buffer can be used from pH = pK_a − 1 to pH = pK_a + 1. For example, a hydrogencarbonate buffer with pK_a = 6.36 (table 1) works efficiently between pH = 5.36 and pH = 7.36. Outside this range the concentration of one of the buffer components becomes too low and the buffer loses its ability to maintain a constant pH of the solution.

Questions

1   Hydrochloric acid is primarily responsible for the acidity of gastric juice. Calculate the concentration, in mol dm^{-3}, and mass percentage of hydrochloric acid in the sample of gastric juice with pH 1.5 and density 1.03 kg dm^{-3}.

2   A well-known brand of antacids contains 0.160 g of aluminium hydroxide and 0.105 g of magnesium carbonate in each tablet.

   a)   State the separate equations for the reactions of aluminium hydroxide and magnesium carbonate with hydrochloric acid.   [2]

   b)   Determine which of the two components of the tablet will neutralize the most acid.   [2]

   c)   The tablets also contain alginic acid and sodium hydrogencarbonate. The function of the sodium hydrogencarbonate is to react with the alginic acid to form sodium alginate. State the function of the sodium alginate produced.   [1]

   IB, May 2012

3   A suspension of magnesium hydroxide in water, known as "milk of magnesia", is a common antacid. A 2.00 cm^3 sample of the suspension has a density of 1.15 kg dm^{-3} and can neutralize 15.8 cm^3 of 0.400 mol dm^{-3} hydrochloric acid. Calculate the mass percentage of magnesium hydroxide in the suspension.

4   Two substances commonly used in antacid tablets are magnesium hydroxide and aluminium hydroxide.

   a)   Suggest why compounds such as sodium hydroxide or potassium hydroxide cannot be used as antacids.   [1]

   b)   Explain why alginates and dimethicone are often included in antacid tablets.   [2]

   IB, May 2011

5   The acidity of gastric juice can be temporarily reduced by antacids or controlled at the cellular level by certain drugs, such as ranitidine (an H2-receptor antagonist), omeprazole, and esomeprazole (proton pump inhibitors). Each of these methods has benefits and disadvantages. Discuss how we choose between different approaches that can be utilized to solve the same problem.

6   An acetate buffer was prepared from 500 cm^3 of 0.100 mol dm^{-3} ethanoic acid (pK$_a$ = 4.76) and 16.4 g of solid sodium acetate.

   a)   Assuming that the addition of sodium acetate does not affect the solution volume, calculate the pH of this buffer solution.

   b)   State the pH range in which acetate buffers can be used.

   c)   Deduce molecular and ionic equations that show the buffer action of this solution when a small amount of hydrochloric acid is added and when a small amount of sodium hydroxide is added.

7   Phosphoric acid (pK$_{a1}$ = 2.12, pK$_{a2}$ = 7.20, pK$_{a3}$ = 12.3) and its anions can produce several acid–base buffer systems that exist at different pH.

   a)   Identify the conjugate acid and conjugate base in the buffer solution with pH = 6.8 prepared from phosphoric acid and sodium hydroxide.

   b)   Calculate the mole ratio of the conjugate acid and conjugate base in this solution.

   c)   Deduce molecular and ionic equations that show the buffer action of this solution.

   d)   Suggest how the ratio from (b) will change when the buffer solution is diluted with an equal volume of water.

8   An ammonia buffer with pH = 8.8 was prepared by dissolving solid ammonium chloride in 0.100 dm^3 of a 0.200 mol dm^{-3} solution of ammonia. The pK$_a$ for ammonium ion is 9.25. Calculate the mass of solid ammonium chloride that was used to prepare this buffer solution. Assume that the solution volume did not change when ammonium chloride was added.

9   Calculate the volumes, in cm^3, of 0.100 mol dm^{-3} solutions of sodium carbonate and sodium hydrogencarbonate that need to be mixed together to prepare 300 cm^3 of a buffer solution with pH 10.0. The pK$_{a2}$ for carbonic acid is 10.3. Assume that the volume of the final solution is equal to the sum of volumes of initial solutions.

10   The buffer solution from question 9 was mixed with 50.0 cm^3 of 10.0 mmol dm^{-3} hydrochloric acid. Calculate the pH of the final solution. Assume that the volume of the final solution is equal to the sum of volumes of the initial solutions.

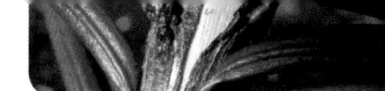

D.5 Antiviral medications

Understandings

→ Viruses lack a cell structure and so are more difficult to target with drugs than bacteria.

→ Antiviral drugs may work by altering the cell's genetic material so that the virus cannot use it to multiply. Alternatively, they may prevent the viruses from multiplying by blocking enzyme activity within the host cell.

 ## Applications and skills

→ Explanation of the different ways in which antiviral medications work.

→ Description of how viruses differ from bacteria.

→ Explanation of how oseltamivir (Tamiflu) and zanamivir (Relenza) work as preventative agents against flu viruses.

→ Comparison of the structures of oseltamivir and zanamivir.

→ Discussion of the difficulties associated with solving the AIDS problem.

Nature of science

→ Scientific collaboration – recent research in the scientific community has improved our understanding of how viruses invade our systems.

Viruses

The discovery of penicillin (sub-topic D.2) and other antibiotics has dramatically improved the chances of success in the treatment of bacterial infections. However, antibiotics are completely ineffective against **viruses**, which differ from bacteria in many ways. While bacteria are living cells that can feed, excrete, grow, and multiply, viruses lack cellular structure and do not have their own metabolism. Therefore viruses are not considered to be life forms but rather very complex chemical compounds, which can be synthesized in the laboratory and isolated in crystalline form (figure 1). The sizes of individual viruses are intermediate between those of bacteria and large biomolecules (figure 2).

Most viruses are nucleoproteins containing a nucleic acid (RNA or DNA) surrounded by a protein coat. This coat, known as a **capsid**, consists of multiple protein units (**capsomeres**) arranged in helical or polyhedral structures (figure 3).

Although viruses can exist outside living organisms, they cannot perform any biological functions on their own. Viruses use the machinery and metabolism of host cells for creating multiple copies of themselves. In order to do this the capsid proteins of the virus bind to receptors on the host cell surface (sub-topic D.1) and then either cross the cell membrane or inject their genome (RNA or DNA) into the cell. The virus genome is interpreted by the cell as a set of instructions for synthesizing proteins

▲ Figure 1 A crystal of satellite tobacco mosaic virus grown on the Mir space station in 1998

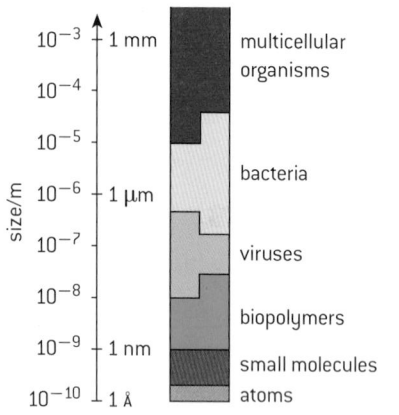

▲ Figure 2 Relative sizes of life forms, viruses, and biopolymers

and nucleic acids, which self-assemble into new copies of the virus. Finally the replicated viruses are released from the host cell, usually by lysis (breaking of the cell membrane) that destroys the cell.

The lack of cellular structure and metabolism makes viruses very difficult to target with pharmaceutical drugs. Most viral diseases have no cure and can be treated only symptomatically (by reducing pain, fever, and the probability of secondary infections). For many years the best defence against specific types of virus has been immunization, which in some cases was particularly successful. For example, smallpox (figure 4), a viral disease responsible for nearly 500 million deaths in the twentieth century, was eradicated in 1979 after several decades of worldwide vaccination. The occurrences of other viral diseases such as measles and polio have been significantly reduced by the vaccination programmes coordinated by the World Health Organization (WHO).

▲ Figure 3 A computer model of Pariacoto virus. The protein capsid is cut in half to show the virus RNA

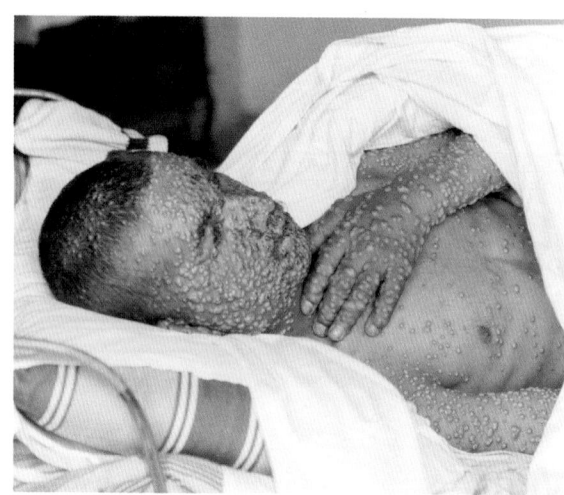

▲ Figure 4 A patient infected with smallpox

Antiviral drugs

In recent years several kinds of antiviral medication have been developed. Similar to antibiotics, antiviral drugs target specific types or classes of viruses. Since viruses are not alive, they cannot be "killed" by drugs; instead antivirals interfere with different stages of the virus replication cycle, including:

- attachment of the virus to a host cell

- uncoating of the virus and injection of viral RNA or DNA into the cell

- biosynthesis of viral components by the cell machinery

- release of viruses from the cell.

During the first stage antivirals can bind to the cell receptors or capsid proteins, preventing the attachment of the virus to the cell. The development of such drugs is a slow and expensive process, which so far has not led to any commercial products.

In the second stage, antivirals can inhibit the uncoating of the virus and the injection of its genetic material into the cell. This strategy was utilized in **amantadine** and **rimantadine** (figure 5), drugs designed for treating influenza and the common cold. However, nearly all

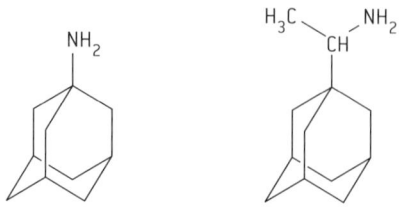

▲ Figure 5 The structures of amantadine (left) and rimantadine (right)

viral strains have now developed resistance (sub-topic D.2) to both amantadine and rimantadine, which greatly decreased the efficiency of these drugs.

The third stage, the biosynthesis of viral components by the host cell, is targeted by antivirals that mimic the structures of nucleotides (sub-topic B.8). These drugs include **acyclovir** and **zidovudine** (figure 6), which are effective against herpes and human immunodeficiency virus (HIV) (see below). In the host cell acyclovir and zidovudine undergo phosphorylation and produce non-standard nucleotides, which are mistakenly incorporated into RNA and DNA sequences. The enzymes produced from these altered nucleic acids are inactive and cannot be used for replicating viral components.

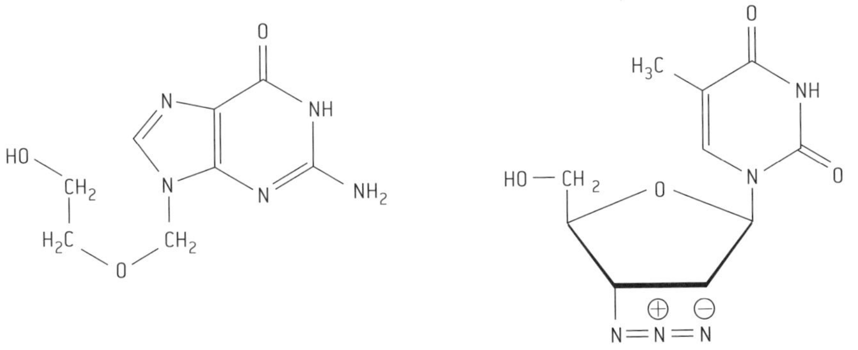

▲ Figure 6 The structures of acyclovir (left) and zidovudine (right)

The final stage of the virus replication cycle can also be targeted by antivirals. Two such drugs, **oseltamivir** (Tamiflu) and **zanamivir** (Relenza), prevent the release of virus copies from the cell by inhibiting certain viral enzymes called neuraminidases. These enzymes trigger the process of budding, which allows viruses to bulge through the outer membrane of the host cell. The inhibition of neuraminidases keeps viruses trapped within the cell and slows their spread around the body.

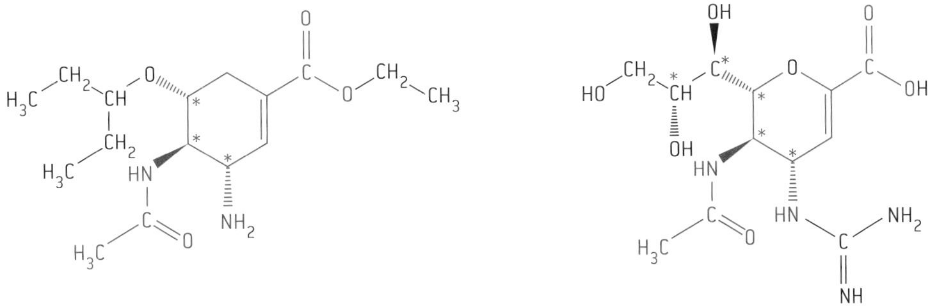

▲ Figure 7 The structures of oseltamivir (left) and zanamivir (right). The chiral carbon atoms are marked with asterisks; common structural features are shown in red

Both oseltamivir and zanamivir target the same enzymes and their structures have many similarities (figure 7). Both molecules contain a six-membered ring with three chiral carbon atoms (marked with asterisks in figure 7). However, the side-chains in oseltamivir and zanamivir contain different functional groups, which affect the pharmacological properties of these drugs. In particular, the presence of an ester group makes oseltamivir inactive in its original form. In the body the ester group is hydrolysed into a carboxyl group, producing an active metabolite (sub-topic D.4) with enhanced antiviral activity. The zanamivir molecule already has a carboxyl group so it is active in its original form.

Oseltamivir and zanamivir are used in many countries for the treatment and prevention of influenza. Both drugs show varying degrees of efficiency against all strains of influenza viruses, including potentially fatal H1N1 (swine flu) and H5N1 (bird flu). Over the years some viral strains have developed significant resistance to oseltamivir while cases of zanamivir resistance are still very rare.

The significance of antiviral drugs

The emergence of antivirals over recent decades is the result of scientific collaboration and exchange of information on a global scale. The availability of protein, DNA, and RNA sequences, crystal structures of biomolecules, and extensive medical data via public databases has greatly expanded our knowledge of the interactions between viruses and host organisms on the molecular level. Better understanding of the structure and functions of viruses leads to the development of new drugs that target viral infections at all stages of the virus replication cycle. The progress in antiviral therapy has already changed the way of treatment of many viral infections and will probably have the same effect on modern medicine as the discovery of antibiotics in the twentieth century.

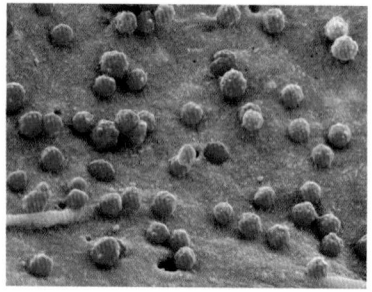

▲ Figure 8 A scanning electron microphotograph of HIV particles (red) budding from an infected lymphocyte (brown)

HIV and AIDS

Despite progress in antiviral therapy many viruses use various methods to evade the action of medicinal drugs and the immune response of the host organism. One such virus, the **human immunodeficiency virus (HIV)**, is responsible for **acquired immunodeficiency syndrome (AIDS)**, which is characterized by progressive failure of the immune system and the development of life-threatening opportunistic infections and cancers. Due to its fast replication cycle and high mutation rate, HIV can produce up to 10^{10} new copies per day and is often present in several modifications within the same organism. In addition, HIV infects the very cells (certain types of lymphocytes or white blood cells) that are responsible for fighting viral and bacterial infections (figure 8). Finally, HIV is able to incorporate itself into the host DNA, where it can remain dormant for many years. Such behaviour makes HIV extremely difficult to eradicate and to prevent from multiplying and infecting other cells.

HIV belongs to the class of **retroviruses**, which use **reverse transcriptase enzymes** (sub-topic B.8) to produce DNA strands from their RNA genomes. This process is the reverse of normal transcription, where RNA copies are produced from DNA templates using transcriptase enzymes. Since reverse transcriptase is used only by retroviruses, its inhibition does not affect normal cells but significantly reduces the ability of viruses to multiply. Certain antiviral drugs such as zidovudine (see above) use this technique to combat AIDS and prevent HIV transmission (for example, from mother to child during birth). However, zidovudine cannot eliminate HIV completely, allowing the virus to become resistant to this drug over time. Therefore zidovudine is often used in combination with other reverse transcriptase inhibitors, which slows down the development of resistance and increases the overall efficiency of HIV/AIDS therapy.

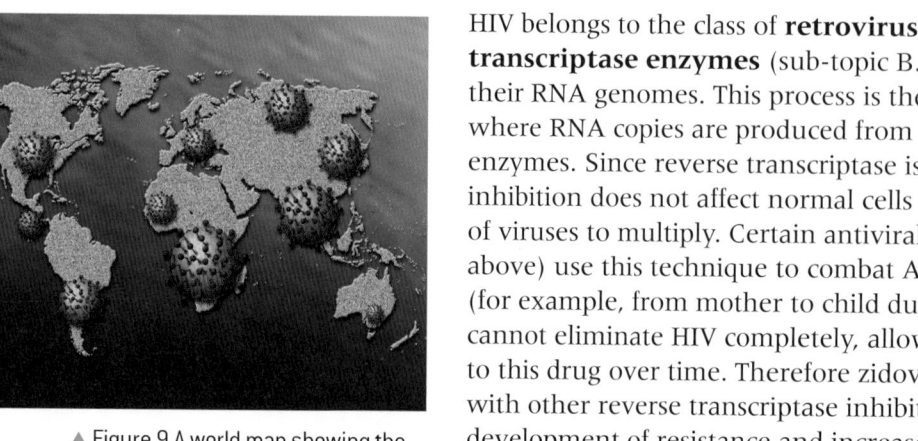

▲ Figure 9 A world map showing the proportional distribution of HIV/AIDS

The control and treatment of HIV/AIDS is further complicated by a lack of health care, poor education, and sociocultural issues. In many countries the cost of anti-retroviral treatment exceeds the average income of patients while governments provide little or no financial support to people with HIV/AIDS. A significant proportion of HIV-positive people are unaware of their infection and therefore do not seek medical help and continue spreading the disease. The most efficient protective measure against HIV, the use of condoms, is rejected in certain societies due to economic or religious reasons. At the same time, illegal drug use, prostitution, and casual sexual contacts also increase the risk of HIV and AIDS. Finally, HIV/AIDS patients are often stigmatized and suffer various forms of discrimination, ranging from avoidance to physical violence. All these factors contribute to the global pandemic of HIV/AIDS, which now affects over 35 million people worldwide (figure 9).

Since its discovery in the early 1980s HIV has killed 30 million people around the world. About two-thirds of all HIV cases and AIDS-related deaths have occurred in Sub-Saharan Africa, where 5% of the population is now HIV positive. As a result, the life expectancy in that region has fallen sharply (figure 10), which has had a significant social and economic impact on many African countries. South Africa is the worst hit country, with over 10% of the population HIV positive and 1.2 million "AIDS orphans", who generally depend on the state for care and financial support. Recently the situation has been slowly improving, mostly due to internationally supported programmes in healthcare and education. However, much more needs to be done before the HIV/AIDS pandemic can be reversed.

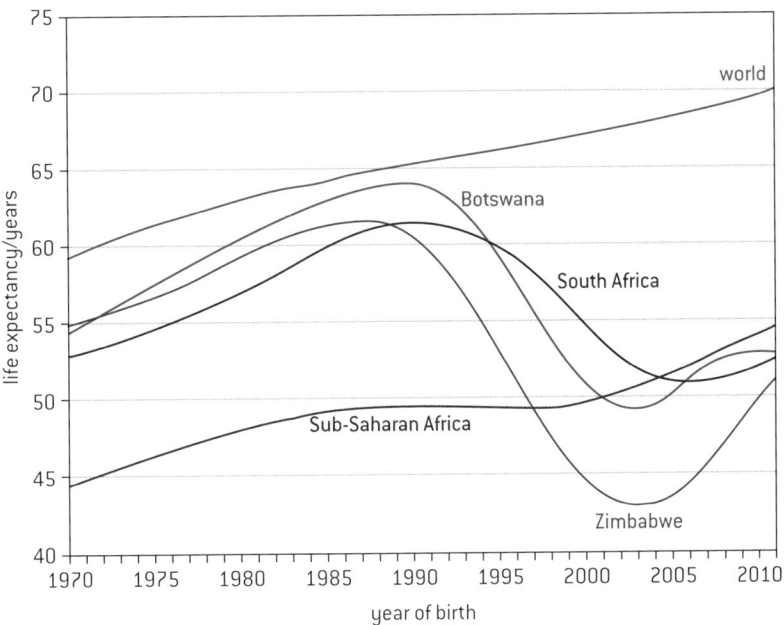

▲ Figure 10 Life expectancy at birth for some sub-Saharan countries. The sharp fall in the 1990s was primarily due to the HIV/AIDS pandemic. Data from http://data.worldbank.org/indicator/SP.DYN.LE00.IN

Questions

1 a) State two differences in structure between viruses and bacteria. [2]

b) Describe two ways in which antiviral drugs work. [2]

c) Discuss two difficulties associated with the development of drugs for the effective treatment of AIDS. [2]

IB, May 2011

2 The structures of two antiviral drugs, amantadine and rimantadine, are given in the text.

a) Deduce the molecular formula of amantadine.

b) Deduce the number of primary, secondary, tertiary, and quaternary carbon atoms in the molecule of rimantadine.

c) State whether the amino groups in amantadine and rimantadine are primary, secondary, or tertiary.

d) Indicate with asterisks (*) the chiral centres in amantadine and rimantadine (if any).

e) Explain why viral infections are so difficult to treat.

3 An antiviral drug, acyclovir, can alleviate some symptoms of the common cold. The structure of acyclovir is given in the text.

a) Draw the structure of acyclovir and identify the amido group by drawing a circle around it.

b) Explain why acyclovir is more soluble in dilute acids than in water.

c) Many drugs including acyclovir can be administered orally. However, some other drugs must be injected directly into the bloodstream. Suggest two reasons why certain drugs cannot be taken orally.

4 Oseltamivir (Tamiflu®) and zanamivir (Relenza®) are antiviral drugs. Their structures are given in the text.

a) State the names of two functional groups that are present in oseltamivir but not in zanamivir.

b) State the names of two functional groups that are present in zanamivir but not in oseltamivir.

c) State the names of two functional groups that are present in both drugs.

d) Predict and explain which of the two drugs is likely to be more soluble in water.

e) In the human body oseltamivir undergoes hydrolysis, producing ethanol and an active metabolite. (i) State the meaning of the term "active metabolite". (ii) Draw the structural formula of the active metabolite of oseltamivir.

5 Acquired immunodeficiency syndrome (AIDS) is a disease caused by human immunodeficiency virus (HIV). Zidovudine is an antiretroviral drug used in the treatment of AIDS.

a) The structure of zidovudine is given in the text. State the number of chiral carbon atoms in a molecule of zidovudine.

b) State the meaning of the term "retrovirus".

c) Outline how zidovudine slows down the replication of HIV.

d) Zidovudine is often used in combination with other antiviral drugs. This approach is similar to the treatment of tuberculosis, where a "cocktail" of antibacterials is used. State the reason why more than one drug is needed in both cases.

e) Discuss the social and economic impacts of the HIV/AIDS pandemic.

6 AIDS (acquired immune deficiency syndrome) has resulted in millions of deaths worldwide since it was first recorded in 1981. The control and treatment of HIV is made worse by the high price of anti-retroviral agents and sociocultural issues. Discuss one sociocultural difficulty facing society today associated with solving this global problem. [3]

IB, November 2010

7 The 1918–1919 pandemic of influenza killed more people in just one year than HIV/AIDS in 25 years. Discuss whether this fact can justify the claim that influenza viruses are more dangerous to the global population than is HIV.

D.6 Environmental impact of some medications

Understandings

→ High-level waste (HLW) is waste that gives off large amounts of ionizing radiation for a long time.

→ Low-level waste (LLW) is waste that gives off small amounts of ionizing radiation for a short time.

→ Antibiotic resistance occurs when microorganisms become resistant to antibacterials.

 ## Applications and skills

→ Description of the environmental impact of medical nuclear waste disposal.

→ Discussion of environmental issues related to left-over solvents.

→ Explanation of the dangers of antibiotic waste from improper drug disposal and animal waste, and the development of antibiotic resistance.

→ Discussion of the basics of green chemistry (sustainable chemistry) processes.

→ Explanation of how green chemistry was used to develop the precursor for Tamiflu (oseltamivir).

 ## Nature of science

→ Ethical implications and risks and problems – the scientific community must consider both the side effects of medications on the patient and the side effects of the development, production, and use of medications on the environment (i.e. disposal of nuclear waste, solvents, and antibiotic waste).

Medical waste and the environment

For many years the environmental impact of medical waste has been largely ignored as scientists concentrated on well known contaminants generated by the agricultural and industrial sectors (sub-topic B.6). **Pharmacologically active compounds** (**PACs**) used in medicine and biochemical studies have not been treated as potentially toxic and have been routinely released to the environment. However, prolonged exposure to PACs causes significant changes in the metabolism and behaviour of various organisms. In particular, uncontrolled release of antibiotics to the environment leads to the development of resistant bacteria (sub-topic D.2) while other drugs can act as endocrine disruptors, increasing the risk of cancer and reproductive disorders in humans and other animals.

Another type of environmental pollutant is radioactive materials used in medical treatment and diagnostics (sub-topic D.8). Although the activity of these materials is usually very low, they are often disposed of as common waste and add to radiation levels in local ecosystems. Certain radioisotopes can undergo bioaccumulation and biomagnification, increasing the risk of radiation exposure for predators at the tops of food chains.

Environmental xenobiotics

Environmental xenobiotics are artificial bioactive compounds that are found as pollutants in the natural environment. Along with industrial products, environmental xenobiotics include various PACs such as antibiotics, analgesics, cytostatics (chemotherapy drugs), disinfectants, steroids, and hormones. Most PACs easily pass through waste-water treatment plants which are not designed to manage this type of pollutant. In 2012 over a million tonnes of PACs were released to the environment worldwide.

751

The production, storage, and distribution of pharmaceutical drugs also contribute to environmental pollution through the release of greenhouse gases (sub-topic C.5), ozone-depleting substances (sub-topic 14.1), and toxic materials including left-over solvents and biologically active by-products of organic synthesis. These negative effects can be greatly reduced by the introduction of sustainable industrial processes or **green chemistry**, which will be discussed later in this sub-topic.

Antibiotic resistance

The widespread use of penicillin and other antibiotics in the second half of the twentieth century led to the development of **antibiotic resistance** (sub-topic D.2) in many strains of harmful bacteria. As a result the efficiency of traditional antibiotics against common diseases has significantly decreased, so scientists need to create new drugs in order to combat bacterial infections. However, it becomes progressively more difficult as bacteria constantly evolve and become resistant to increasing numbers of antibiotics (figure 1).

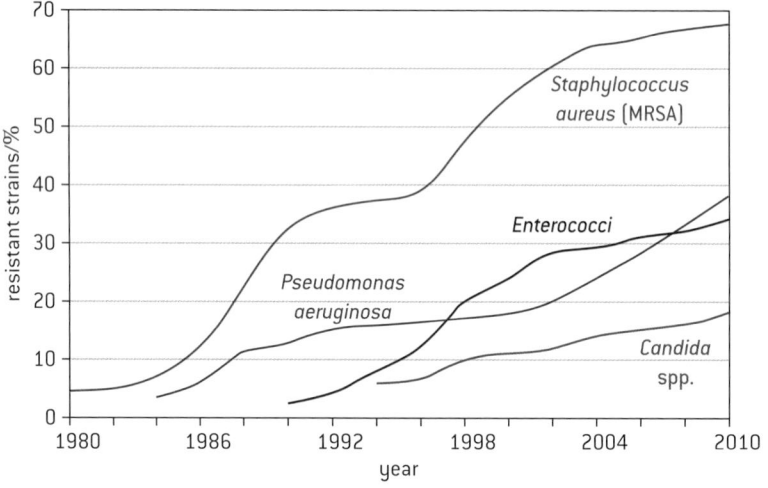

▲ Figure 1 Antibiotic-resistant strains of common bacteria

Antibiotic resistance in bacteria is caused by several factors, including the over-prescription of antibacterials, non-compliance of patients in finishing a course of treatment, the use of antibacterials in agriculture, and the release of antibacterial waste by hospitals and the pharmaceutical industry. In all cases, exposure to low levels of antibiotics allows some bacteria to survive and mutate, eventually developing the ability to tolerate higher and higher concentrations of the drug. Such bacteria pass their resistance to new generations, gradually replacing non-resistant strains. This process can take place both in individual patients and in the environment. In the latter case, exposure to antibacterials increases the antibiotic resistance of the whole bacterial population.

Over the past two decades the use of antibiotics in agriculture has nearly doubled and now contributes to 50–60% of global consumption. Most of these drugs are given to healthy animals to prevent infectious diseases and promote livestock growth. Although this practice allows increased output and reduced prices in agricultural production, it is also the primary source of antibiotic waste in the environment. Since antibiotics are never completely metabolized in animal organisms, a significant percentage of each drug is excreted in unchanged form and released into the ground water or absorbed by other organisms. Some of these antibiotics are eventually consumed by humans with meat, dairy products, and water, further accelerating the development of resistant bacteria.

Restrictions on the use of antibiotics

Since the late 1990s the use of antibiotics as growth promoters in agriculture has been banned in the European Union and some other countries. However, these measures had no immediate effect on bacterial resistance in humans while the rates of death and disease in animals increased significantly. Apparently, several decades of excessive antibiotic intake have weakened the immune systems of animals and made them more susceptible to infections. There is strong evidence that similar changes have taken place in the human population, so the problem of antibiotic resistance has much broader implications than was initially thought.

It is now obvious that antibiotic therapy should be restricted to the most severe cases of bacterial infections while non-medical use of antibacterial drugs should be banned completely. At the same time, the amount of antibiotic waste from hospitals and the pharmaceutical industry must be reduced to a minimum and thoroughly processed before being released into the environment. In addition, new antibacterial drugs must be produced and used under strict control to prevent the development of antibiotic resistance. To be effective, these measures need to be taken by all countries and coordinated at the international level.

Nuclear waste

Many medical procedures involve the use of **radionuclides** – unstable isotopes of certain elements that undergo spontaneous radioactive decay (sub-topic D.8). Some of these isotopes are administered to patients as water-soluble salts or radiopharmaceutical drugs (sub-topic D.8) while other radionuclides are used in medical equipment as sources of ionizing radiation. During medical procedures radionuclides and ionizing radiation come into contact with various materials that also become radioactive. These materials, together with left-over radionuclides, produce **nuclear waste**, which must be disposed of in accordance with specific procedures.

Most radionuclides used in hospitals and medical research centres have very low activity and short half-life times (sub-topic D.8). The waste containing such radionuclides is known as **low-level waste (LLW)** and typically consists of contaminated syringes, tools, swabs, paper, and protective clothing. Such waste has limited environmental impact and is usually suitable for shallow land burial or incineration. Some types of LLW, such as concentrated solutions of radionuclides, must be stored for several days or weeks in shielded containers until most of the radioactive isotopes have decayed and the radiation level has dropped below a safe limit.

Medical equipment for radiotherapy may contain large quantities of radioactive isotopes such as Co-60 and Cs-137. These radionuclides remain active for many years and produce very high levels of ionizing

In some cases resistant bacteria can be passed directly from domestic animals to humans, causing serious diseases. A recent study showed that 75–80% of strains of *Salmonella* bacteria found in chicken and turkey were resistant to at least one antibiotic while nearly 50% were resistant to three or more drugs. Certain types of *Salmonella* bacteria cause typhoid fever, which is responsible for 200 000 deaths in developing countries each year. Therefore this finding is particularly worrying because an outbreak of multidrug-resistant typhoid fever can be very difficult to treat.

Goiânia accident

In 1987 a Cs-137 radiation source was stolen from an abandoned hospital site in Goiânia (Brazil) and disassembled at a local scrapyard. As a result four people died of radiation sickness, including a six-year-old girl who was fascinated by the deep-blue glow of the source and applied some of the radioactive material to her body. Another 249 people received varying doses of radiation and needed medical treatment. Several houses had to be demolished and topsoil removed from contaminated areas. According to the *International Atomic Energy Agency (IAEA)*, it was one of the world's worst radiological incidents to date.

radiation. Although Co-60 and Cs-137 are classified as LLW, they cannot be released to the environment and are usually recycled or stored in underground repositories (figure 2).

High-level waste (**HLW**) is produced in nuclear reactors and contains a mixture of nuclear fission products (sub-topic C.3) with unused nuclear fuel. Many radionuclides in HLW have very long half-lives (from several decades to billions of years) while other isotopes are short lived but highly active. Due to ongoing nuclear reactions, concentrated HLW releases heat and must be constantly cooled with water for up to several years. When the radioactivity level decreases, HLW can be reprocessed and partly recycled. The remaining waste is fused with glass ("**vitrified**") or immobilized in certain minerals ("**Synroc**" or "synthetic rock" technology), producing water-resistant and chemically stable solid materials. These materials are encased in steel cylinders, covered with concrete, and buried deep underground in geologically stable locations.

The treatment, transportation, and disposal of nuclear waste present serious risks due to possible release of radionuclides to the environment. In high doses ionizing radiation is harmful to all living organisms, causing extensive cellular and genetic damage. Low doses of radiation increase the number of mutations and the probability of developing certain diseases such as cancer, birth defects, and reproductive disorders. In addition, ionizing radiation weakens the immune system by triggering apoptosis (programmed cell death) in lymphocytes and rapidly dividing bone marrow cells. As a result, organisms exposed to radiation are more likely to contract infectious diseases and develop complications.

The effects of ionizing radiation and other environmental pollutants can be cumulative. For example, radioactive materials discarded together with antibiotic waste can increase the mutation rate in bacteria and accelerate the development of drug-resistant strains. A personal injury caused by contaminated hypodermic needles or broken glass can introduce such bacteria directly into the bloodstream and lead to a serious disease. Therefore each kind of medical waste must be disposed of separately and always treated as a potential environmental hazard.

Waste products from the pharmaceutical industry

Many pharmaceutical drugs are produced on an industrial scale using a wide range of technological processes. Most of these processes involve the use of toxic chemicals that have to be recycled or disposed of after the synthesis is complete. **Organic solvents** used in the pharmaceutical industry constitute a significant proportion of chemical waste. Most solvents are toxic to living organisms, primarily affecting nervous and respiratory systems, certain internal organs (liver and kidneys), and the reproductive organs. Some solvents such as benzene (C_6H_6) and chloroform ($CHCl_3$) increase the risk of cancer in humans and other animals. In addition, many solvents are highly flammable while their vapours contribute to the greenhouse effect (sub-topic C.5).

Chlorinated solvents such as carbon tetrachloride (tetrachloromethane, CCl_4), chloroform ($CHCl_3$), dichloromethane (CH_2Cl_2), trichloroethene ($Cl_2C=CHCl$), and tetrachloroethene ($Cl_2C=CCl_2$) present specific

▲ Figure 2 An underground storage area for low-level nuclear waste (Fontenay-aux-Roses, France)

environmental hazards. Due to low bond enthalpies (sub-topic 5.3) of the C–Cl bonds, these compounds act as ozone-depleting agents (sub-topic 14.1) and contribute to the formation of "photochemical smog" in large industrial cities. Some chlorinated solvents have limited biodegradability (sub-topic B.6) and may accumulate in the groundwater, causing long-term damage to local ecosystems.

The disposal of chlorinated solvents is an expensive and complex process. Chlorine-containing compounds cannot be incinerated together with common organic waste because their incomplete combustion could produce highly toxic phosgene ($COCl_2$) and dioxins. To minimize the formation of such by-products, chlorinated solvents must be oxidized separately at very high temperatures or recycled by distillation.

▲ Figure 3 Chlorinated and non-chlorinated chemical waste must be kept separately for correct disposal or recycling

Supercritical fluids

For every substance there is a certain combination of temperature and pressure (the "**critical point**") where all differences between gaseous and liquid phases disappears. Above that point the substance behaves as a **supercritical fluid**, which can pass through porous solids like a gas and dissolve other substances like a liquid. **Supercritical carbon dioxide** is an excellent solvent that is increasingly used in the pharmaceutical industry for extraction, recrystallization and purification of various compounds. In contrast to common organic solvents it is non-toxic, non-flammable, and can easily be removed from the solution by reducing the pressure. In food processing supercritical carbon dioxide is used for making decaffeinated coffee and tea. The extracted caffeine is used as a component of pharmaceutical drugs and soft drinks. An anticancer drug Taxol (sub-topic D.7) is also extracted from plant material using supercritical carbon dioxide.

Another supercritical fluid, water, is used as a solvent for the oxidation of hazardous materials such as polychlorinated biphenyls (PCBs) and certain types of LLW. These materials cannot be destroyed by incineration because they release toxic combustion products. In supercritical water saturated with oxygen these products are oxidized and hydrolysed into hydrochloric acid, carbon dioxide, and inorganic compounds that can easily be separated and recycled. Similar to carbon dioxide, supercritical water is an excellent solvent but can exist only at very high pressures and temperatures.

Green chemistry

The efficiency of a synthetic procedure in traditional chemistry is measured in terms of the product yield and the cost of raw materials. In contrast, the primary goal of **green chemistry** is to reduce the environmental impact of technological processes by minimizing the use and generation of hazardous chemicals. Common practices of green chemistry include aqueous or solvent-free reactions, renewable starting materials, mild reaction conditions, regio- and stereoselective catalysis (sub-topic 20.1), and the utilization of any by-products formed during the synthesis.

Atom economy

One of the key concepts of green chemistry, **atom economy**, expresses the efficiency of a synthetic procedure as the ratio between the molecular mass of the isolated target product and the combined molecular masses of all starting materials, catalysts, and solvents used in the reaction. The problems involving atom economy are discussed in sub-topic B.6.

Origins of "green chemistry"

The term "green chemistry" was coined in 1991 by Paul Anastas and John Warner, who formulated 12 principles that explain their approach to chemical technology. These principles emphasize the benefits of non-hazardous chemicals and solvents, efficient use of energy and reactants, reduction of waste ("the best form of waste disposal is not to create it in the first place"), choice of renewable materials, and prevention of accidents. The philosophy of green chemistry has been adopted by many companies and eventually passed into national and international laws, encouraging the development of environmentally friendly technologies.

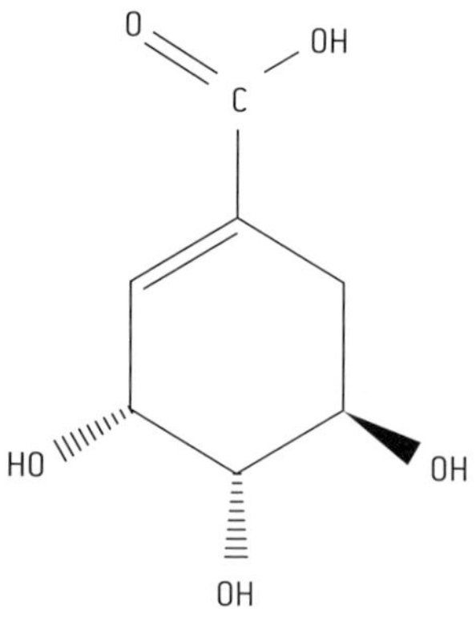

▲ Figure 4 Shikimic acid

Another important field of green chemistry is the use of biotechnologies and bioengineering in organic synthesis. Enzyme-catalysed biochemical reactions are highly selective, efficient, and proceed in aqueous solution under mild conditions. Similar to penicillin (sub-topic D.2), many pharmaceutical drugs or synthetic intermediates can be produced from renewable materials by genetically modified organisms. One such intermediate, **shikimic acid** (figure 4), is a precursor to the antiviral drug oseltamivir, which is also known under the trade name Tamiflu (sub-topic D.5).

For many years shikimic acid was extracted from Chinese star anise in a ten-stage process that took a year to complete. In 2005 an outbreak of "bird flu" (sub-topic D.5) increased the demand for oseltamivir and led to a worldwide shortage of this drug due to a limited supply of star anise. Modern biosynthetic technologies allow shikimic acid to be produced on an industrial scale by genetically modified *E. coli* bacteria, which effectively prevents any shortages of oseltamivir in the future.

The industrial use of natural products leads to various ecological and social issues such as the extinction of plant species (sub-topic D.7) and rising food prices. At the same time some non-hazardous substances branded as "green" or "environmentally friendly" still require toxic chemicals or large amounts of energy for their production. Therefore the criteria used in assessing the "greenness" of a substance or technological process must include all direct and indirect environmental implications, which remains one of the most controversial problems in green chemistry.

> Standards and practices in the pharmaceutical industry vary greatly around the world. Increasing adoption of green technological processes in developed countries has significantly reduced the emissions of many hazardous chemicals such as chlorinated solvents and greenhouse gases. Although green technologies often involve expensive equipment and recycling facilities, they reduce the costs of environmental remediation, waste management, and energy consumption, making green chemistry a commercially attractive and sustainable alternative to traditional organic synthesis.

Questions

1 a) State one difference between viruses and bacteria. [1]

   b) Discuss three human activities that have increased the resistance to penicillin in bacteria populations. [3]

   IB, November 2010

2 ⊘ In the case of antibacterial treatment, the short-term benefits to the patient must be weighed against the long-term individual and environmental risks. Discuss how we balance ethical concerns that appear to be at odds with one another when trying to formulate a solution to the problem.

3 High-level and low-level wastes are two types of radioactive waste. Compare the half-lives and the methods of disposal of these two types of waste.[3]

   IB, November 2009

4 a) State the characteristics and sources of low-level nuclear waste. [2]

   b) The disposal of nuclear waste in the sea is now banned in many countries. Discuss **one** method of storing high-level nuclear waste and two problems associated with it. [3]

   IB, May 2010

5 Disposal of radioactive waste is a major ecological concern.

   a) State one source of low-level radioactive waste and one source of high-level radioactive waste. [2]

   b) Consider the following types of radioactive waste (table 1).

Type	Waste	Isotopes	Half-life	Emissions
A	syringes and other disposable materials used in radiotherapy	^{90}Y	64 hours	$\beta-$
B	diluted aqueous solution of cobalt-60 complexes	^{60}Co	5.3 years	$\beta-, \gamma$
C	partially processed solid materials from a nuclear reactor	U, Pu, Am and other actinides	10^3-10^9 years	α, γ

▲ Table 1

Identify which method can be used for the disposal of radioactive wastes **A**, **B**, and **C**:

   (i) vitrification followed by long-term underground storage [1]

   (ii) storage in a non-shielded container for two months followed by disposal as normal (non-radioactive) waste [1]

   (iii) ion-exchange and adsorption on iron(II) hydroxide, storage in a shielded container for 50 years, then mixing with concrete and shallow land burial. [1]

   IB, May 2010

6 Caffeine is a mild stimulant that can be extracted from plant material such as coffee beans or tea leaves. State three advantages and one disadvantage of using supercritical carbon dioxide instead of traditional organic solvents for caffeine extraction.

7 Many technological processes of green chemistry involve the use of supercritical carbon dioxide as solvent, hydrogen peroxide as oxidant, and molecular hydrogen as reducing agent. Explain how these compounds reduce the environmental impact of the chemical industry.

8 Shikimic acid is used as an intermediate in the synthesis of the antiviral drug oseltamivir (Tamiflu). The structure of shikimic acid is given in figure 5.

   a) Identify two different named functional groups in the molecule of shikimic acid.

   b) Deduce the number of stereoisomers of shikimic acid (assume no *E/Z* isomerism in this compound).

   c) Shikimic acid can be extracted from plant material or produced by genetically modified bacteria. Discuss the impact of these two methods on the environment.

9 ⊘ Pharmaceutical companies use different approaches to spending funds on research projects. Discuss how the philosophy of green chemistry has affected the ethics of drug development and production.

D.7 Taxol – a chiral auxiliary case study (AHL)

Understandings

→ Taxol is a drug that is commonly used to treat several different forms of cancer.

→ Taxol naturally occurs in yew trees but is now commonly synthetically produced.

→ A chiral auxiliary is an optically active substance that is temporarily incorporated into an organic synthesis so that it can be carried out asymmetrically with the selective formation of a single enantiomer.

Applications and skills

→ Explanation of how Taxol (paclitaxel) is obtained and used as a chemotherapeutic agent.

→ Description of the use of chiral auxiliaries to form the desired enantiomer.

→ Explanation of the use of a polarimeter to identify enantiomers.

Nature of science

→ Advances in technology – many of these natural substances can now be produced in laboratories in high enough quantities to satisfy the demand.

→ Risks and problems – the demand for certain drugs has exceeded the supply of natural substances needed to synthesize these drugs.

▲ Figure 1 Pacific yew tree (*Taxus brevifolia*), the source of Taxol

The discovery of paclitaxel

The discovery and development of the anticancer drug **paclitaxel** (**Taxol®**) illustrates the challenges faced by researchers when an unknown substance with useful pharmaceutical activity needs to be isolated from natural sources. At the same time it clearly shows the importance of collaboration between scientists from different disciplines and the environmental implications of drug production on an industrial scale.

In 1960 the American National Cancer Institute (NCI) initiated an antitumour screening programme that involved the analysis of 650 samples of plant material. Among those samples were the stem and bark of the Pacific yew tree, *Taxus brevifolia* (figure 1). In 1964 samples of Pacific yew were studied by a team of scientists led by Monroe Wall. Approximately 12 kg of air-dried stem and bark were extracted with ethanol and the solution was concentrated and partitioned between water and chloroform. The organic layer yielded 146 g of semi-solid material that showed good activity against a certain type of cancer, Walker-256 solid tumour.

The obtained material was fractionated using multi-step partitioning between various solvents (figure 2). The activity of each fraction was determined as the degree of tumour inhibition in laboratory animals. The degree of inhibition was recorded as a *T/C* value:

$$T/C = \frac{\text{mean tumour mass of treated animals}}{\text{mean tumour mass of control animals}} \times 100\%$$

After each step all active fractions were combined and the process was repeated using different solvents and extraction conditions. In total several hundred fractions were analysed, which took two and a half years to complete.

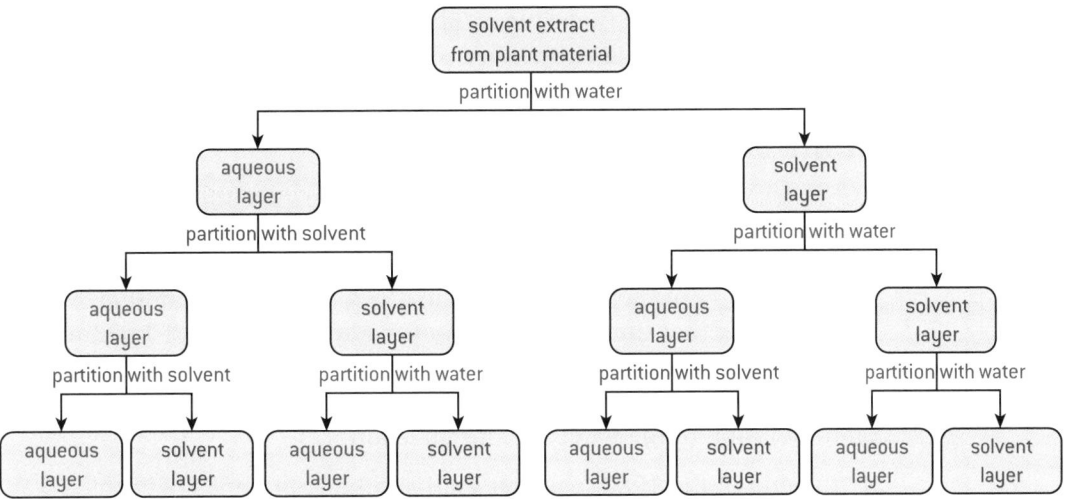

▲ Figure 2 Multi-step liquid–liquid extraction. Automatic extractors can process and analyse hundreds of fractions, discarding empty extracts and combining similar fractions for further separation

Each extraction step produced material with progressively higher anticancer activity (table 1). The final extraction afforded 0.5 g of pure Taxol with an overall yield of only 0.004%. Four years later, in 1971, the structure of Taxol (figure 3) was determined by Mansukh Wani using a combination of chemical degradation and X-ray crystallography.

Extraction series	Mass of active extract / g	T/C / %	Dose / mg kg^{-1}	Relative anticancer activity
1	146	31	100	0.04
2	41	30	45	0.09
3	14	30	23	0.17
4	2.4	16	15	0.50
5	0.5*	24	5.0	1.00

▲ Table 1 Extraction of Taxol from Pacific yew

*Pure Taxol

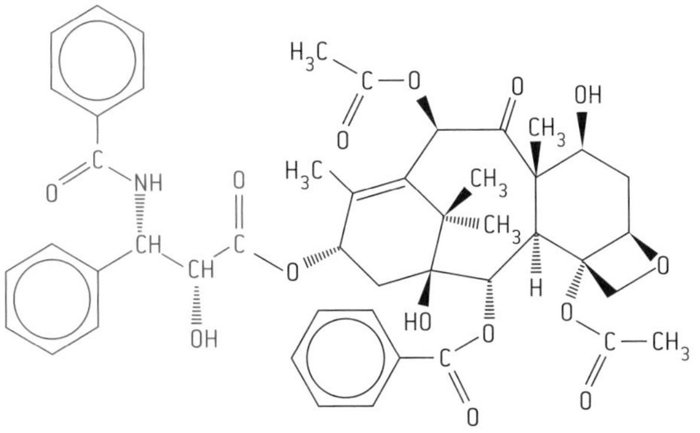

▲ Figure 3 The structure of the anticancer drug paclitaxel (Taxol). The side-chain (red) can be synthesized using chiral auxiliaries (see page 761)

Study tip

The structure of Taxol is given in the *Data booklet*, which will be available during the examination.

Further development of the drug was hindered by the high cost of extraction, low yield of final product, and limited supply of Pacific yew bark, the only known natural source of Taxol. In addition, Taxol was found to be almost insoluble in water and therefore unsuitable for intravenous administration. Finally, the presence of 11 chiral carbon centres in the molecule of Taxol made the synthesis of this drug extremely difficult and expensive.

Semi-synthetic production

In 1979 it was discovered that Taxol destroyed cancerous cells in a unique way, by binding to certain proteins (tubulins) and interfering with the process of cell division. This discovery allowed clinical trials (sub-topic D.1) of the drug to begin in 1983, which took another six years. During that time the problem of the low solubility of Taxol was also resolved. For intravenous administration a mixture of the drug with chemically modified castor oil and ethanol was diluted with normal saline solution immediately before injection.

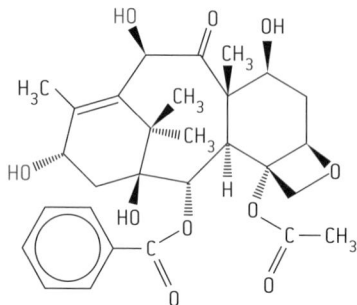

▲ Figure 4 The structure of 10-deacetylbaccatin, a precursor of Taxol. The synthesis of Taxol from 10-deacetylbaccatin requires chemical modification of two hydroxyl groups (red)

By the end of the 1980s the first semi-synthetic methods of Taxol production were developed. A precursor of Taxol, 10-deacetylbaccatin (figure 4), was isolated from the leaves of European yew (*Taxus baccata*) with a yield of 0.2%, which was 50 times higher than the yield of Taxol (0.004%). The molecule of 10-deacetylbaccatin can be converted into Taxol in several synthetic steps, which involve condensation reactions and the use of organometallic reagents.

🧬 Environmental considerations of Taxol production

The environmental impact of drug research and development is one of the major problems faced by the pharmaceutical industry. Although anticancer drugs save lives, the isolation of active ingredients from natural sources put certain species at risk of extinction. To produce 1 g of Taxol using traditional technologies, three 100-year old Pacific yew trees had to be destroyed, which was completely unacceptable from the ecological perspective. Therefore the extraction of Taxol was replaced by its semi-synthetic production, where the natural precursor (10-deacetylbaccatin) was obtained from the leaves of European yew. In contrast to slow-growing and rare Pacific yew, European yew is a common plant that can easily be cultivated. The leaves harvested from the tree are quickly regenerated, providing sufficient supply of 10-deacetylbaccatin to meet the increasing demand for anticancer drugs. Recent studies suggest that Taxol precursors can also be synthesized by plant cell cultures or by genetically engineered organisms such as *E. coli* and yeast.

Clinical use

Between 1992 and 1995, after three decades of research and development, Taxol was finally approved for clinical use in the USA, Europe, and other countries. In 1994 the total synthesis of Taxol was performed by two groups of scientists led by Robert Holton and Kyriacos Nicolaou. However this synthetic drug was too expensive, so nearly all Taxol in the world is produced by semi-synthetic methods from 10-deacetylbaccatin and other natural precursors. Small amounts of Taxol are still isolated from Pacific yew using advanced techniques such as extraction with supercritical carbon dioxide (sub-topic D.6).

The availability of 10-deacetylbaccatin and advances in chemical technology satisfied the global demand for Taxol and created new anticancer drugs with a wide range of activity. One such drug, docetaxel, is known under the trade name Taxotere (figure 5). Docetaxel is slightly more active than Taxol and more soluble in water, which makes it more suitable for intravenous administration. It also remains in the cancer cells for a longer time than Taxol, reducing the effective dose and leading to fewer side effects. However, the cost of anticancer therapy with docetaxel and Taxol remains high ($4000–6000 per course), which limits the availability of these drugs in many developing countries.

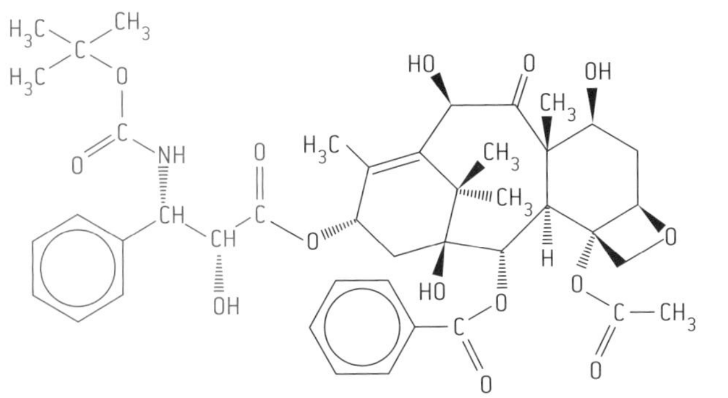

▲ Figure 5 The structure of docetaxel (Taxotere). The side-chain (red) is synthesized using chiral auxiliaries (see below) and combined with 10-deacetylbaccatin (figure 4)

Chiral auxiliaries

To produce Taxol or docetaxel from their precursors, the side-chains of these drugs need to be synthesized in the laboratory. Because these chains contain two chiral carbon centres their synthesis from non-chiral starting materials is problematic because it would lead to a mixture of several stereoisomers (sub-topic 20.3). Therefore both side-chains are synthesized using **chiral auxiliaries** – readily available chiral reagents that can be temporarily introduced to the starting material and easily removed when the synthesis is complete. This process involves three steps:

$$ S \xrightarrow{+A^*} S\!-\!A^* \xrightarrow{reagent} P^*\!-\!A^* \xrightarrow{-A^*} P^* $$

| substrate (non-chiral) | intermediate 1 (single enantiomer) | intermediate 2 (single diastereomer) | product (single enantiomer) |

In the first step the auxiliary A* is combined with a non-chiral substrate S, producing a chiral intermediate S–A*. When another chiral centre in the substrate is created, its configuration is affected by the configuration of the existing chiral centre in the auxiliary. As a result the second step of the reaction usually produces only one of the two possible diastereomers P*–A*. In the last step the auxiliary A* is removed, producing the desired enantiomer P*. To some extent this scheme is similar to biochemical reactions (sub-topic B.7) in which enzymes temporarily bind to substrates and play the roles of biological chiral auxiliaries.

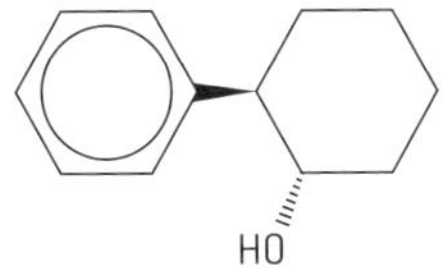

The chiral auxiliary used in the synthesis of Taxol and docetaxel is *trans*-2-phenylcyclohexanol (figure 6). It is a large molecule with two chiral centres, which strongly favour the formation of specific diastereomers in the subsequent steps of the synthesis. At the end of the synthesis the

▲ Figure 6 The structure of the chiral auxiliary *trans*-2-phenylcyclohexanol

761

The Thalidomide disaster

From 1957 to 1962, a new sedative drug was aggressively marketed worldwide under the trade names Thalidomide and Contergan. In many countries it was available without prescription and routinely taken by pregnant women to relieve the symptoms of morning sickness. Despite numerous reports of adverse side effects, sales of Thalidomide kept increasing until 1961, when it was proven to be teratogenic (causing malformations in embryos). By that time over 10 000 children with missing or deformed limbs had been born in 46 countries. Most of those children, known as "thalidomide babies", died within a few months after birth while others remained disabled for the rest of their lives.

The molecule of thalidomide contains a chiral carbon atom and can exist as two enantiomers. Initially, it was thought that only one enantiomer was teratogenic while the other enantiomer provided the desired sedative effect. However, later studies have shown that both enantiomers can interconvert in the human body and therefore are equally dangerous to unborn children.

Surprisingly, thalidomide returned to the market soon after its ban in 1962. However, this drug is now used under strict control and prescribed to patients with certain forms of cancer, leprosy, and AIDS complications (sub-topic D.5). Once again, the story of thalidomide demonstrates the risks associated with drug development and the importance of rigorous testing of any substance intended for medical use.

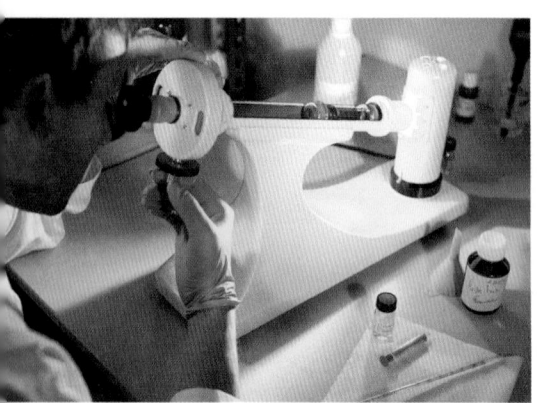

▲ Figure 7 A researcher using a polarimeter to test the purity of pharmaceutical products

auxiliary is removed and recycled, reducing the cost and environmental impact of the drug's production.

Although the use of chiral auxiliaries allows specific stereoisomers to be synthesized, small quantities of the other isomers always form along with the target product. Since the configuration of the chiral centre or centres in the auxiliary is fixed, all unwanted isomers will be diastereomers of the target product and therefore will have different physico-chemical properties (such as solubility, melting point, etc.). Unwanted diastereomers can be removed from the mixture by crystallization, extraction, or chromatography (sub-topic B.2). However, no separation is perfect, so the purity of the final product must be confirmed by laboratory tests.

The identity and purity of chiral compounds can be determined using a polarimeter (figure 7). This instrument measures the angle of rotation of plane-polarized light caused by optically active molecules. The angle depends on the nature and concentration of chiral compounds in the studied solution. Under identical conditions, two enantiomers of the same compound will rotate plane-polarized light by the same angle but in opposite directions (topic 20.3). Each optically active isomer has a unique rotation angle. Therefore, a pure isomer of an unknown compound can be identified by its rotation angle. At the same time, any change in the rotation angle of a known compound will indicate the presence of some impurities. For example, a racemic mixture of two enantiomers (50% purity with respect to each isomer) will be **optically inactive** (will have a rotation angle of 0°). Other proportions of enantiomers in the mixture will produce rotation angles from $+A°$ to $-A°$, where $+A$ and $-A$ are the rotation angles of pure enantiomers.

Optical isomers of pharmaceutical drugs can have very different physiological activities. In some drugs, one isomer may be responsible for the therapeutic effect while other isomers may be less active, inactive, or even harmful to the patient. However, clinical studies of all possible isomers can be very expensive, take a long time, and unnecessarily put patients at risk. Therefore nearly all new drugs contain only a single isomer of the active compound while the levels of other stereoisomers are rigorously controlled and kept as low as possible.

Questions

1 Paclitaxel (Taxol) is an anticancer drug that can be extracted from the bark of Pacific yew tree (*Taxus brevifolia*) or produced semi-synthetically using extracts from the leaves of European yew tree (*Taxus baccata*).

   a) State what is meant by the term "semi-synthetic".

   b) Discuss the advantages and disadvantages of extraction and semi-synthetic production of Taxol.

   c) Since 1994 the total synthesis of Taxol has been reported by several research groups in different countries. Suggest why total synthesis is not used for producing Taxol on an industrial scale.

2 Chirality plays an important role in the action of drugs.

   a) Using an asterisk (*), identify the chiral carbon atom in a copy of the structure of thalidomide (figure 8). [1]

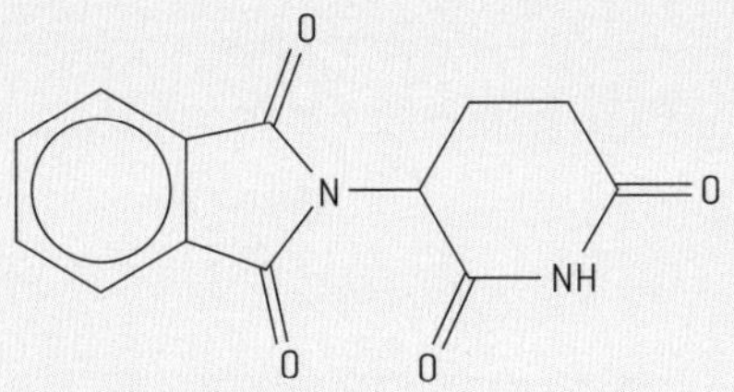

▲ Figure 8

   b) Describe the composition of a racemic mixture. [1]

   c) Discuss the importance of chirality in drug action. [2]

IB, November 2011

3 Paroxetine, whose structure is shown in figure 9, is a drug prescribed to people suffering from mental depression.

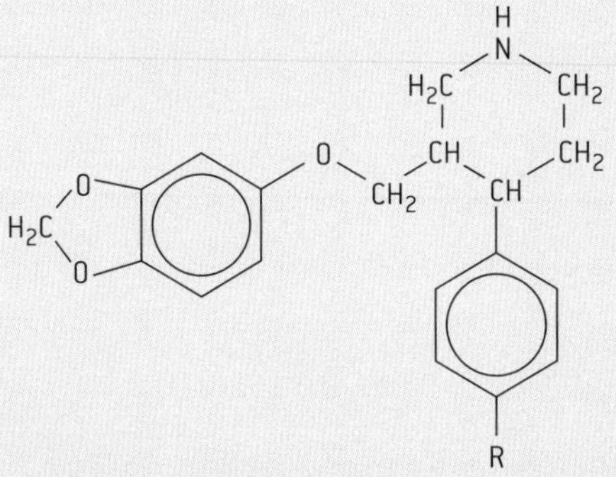

▲ Figure 9

   a) Identify the two chiral carbon atoms in a copy of figure 9 with an asterisk (*). [2]

   b) Describe the use of chiral auxiliaries to synthesize the desired enantiomer of a drug. [2]

IB, May 2010

4 Taxotere (docetaxel) is an anticancer drug that can be synthesized using chiral auxiliaries. A fragment of its structure is shown in figure 10.

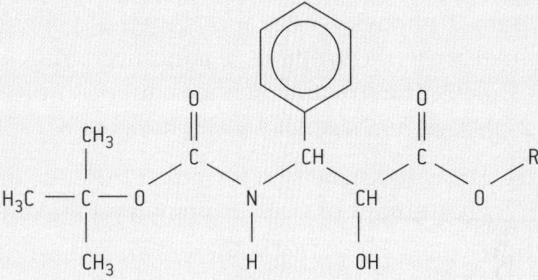

▲ Figure 10

   a) On a copy of figure 10, identify with asterisks (*) two chiral centres in this structural fragment.

   b) Deduce the number of possible stereoisomers of this structural fragment.

   c) Suggest how the presence of unwanted stereoisomers in a drug might affect its pharmacological activity.

5 Baccatin III is the name of a biologically active compound that can be isolated from the Pacific yew tree, *Taxus brevifolia*. Together with 10-deacetylbaccatin, it is a precursor of the anticancer drug Taxol. Baccatin III can be converted into 10-deacetylbaccatin by the following reaction:

   a) State the type of reaction shown above.

   b) State the names of the two circled functional groups.

   c) Suggest why baccatin III cannot be synthesized with a reasonable yield by the reaction of 10-deacetylbaccatin with ethanoic acid.

   d) Deduce the number of chiral carbon centres in the molecule of baccatin III.

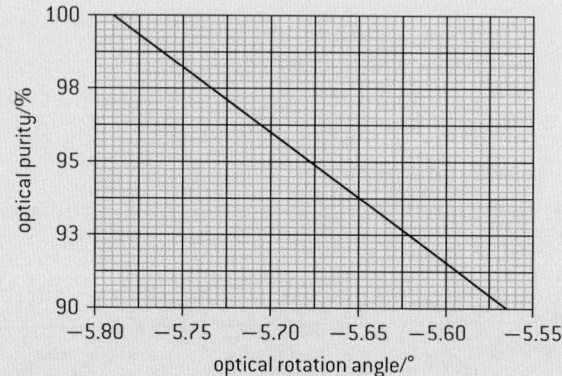

6 Trans-2-phenylcyclohexanol is used as a chiral auxiliary in the synthesis of anticancer drugs such as Taxol. The structure of one enantiomer of *trans-2-phenylcyclohexanol* is given in figure 6.

   a) Draw the structural formula of the second enantiomer of *trans-2-phenylcyclohexanol*.

   b) Explain how a polarimeter can be used to identify enantiomers.

   c) A solution of *trans-2-phenylcyclohexanol* was analysed by polarimetry. At a certain concentration the rotation angle of the solution was −5.73°. Using the calibration curve in figure 11, determine the optical purity of the sample.

▲ Figure 11

D.8 Nuclear medicine (AHL)

Understandings

→ Alpha, beta, gamma, proton, neutron, and positron emissions are used for medical treatment.

→ Magnetic resonance imaging (MRI) is an application of NMR technology.

→ Radiotherapy can be internal and/or external.

→ Targeted Alpha Therapy (TAT) and Boron Neutron Capture Therapy (BNCT) are two methods which are used in cancer treatment.

Nature of science

→ Risks and benefits – it is important to try and balance the risk of exposure to radiation with the benefit of the technique being considered.

Applications and skills

→ Discussion of common side effects of radiotherapy.

→ Explanation of why technetium-99m is the most common radioisotope used in nuclear medicine based on its half-life, emission type, and chemistry.

→ Explanation of why lutetium-177 and yttrium-90 are common isotopes used for radiotherapy based on the type of radiation emitted.

→ Balancing nuclear equations involving alpha and beta particles.

→ Calculating the percentage and amount of radioactive material decayed and remaining after a certain period of time using the nuclear half-life equation.

→ Explanation of TAT and how it might be used to treat diseases that have spread throughout the body.

Radionuclides in nuclear medicine

Nuclear medicine uses radioactive materials in the diagnosis and treatment of diseases. These materials contain **radionuclides** – unstable isotopes of certain elements that undergo spontaneous radioactive decay and emit ionizing radiation. In some cases radionuclides are administered to patients in the form of water-soluble salts or complexes (sub-topic 13.2) that are distributed around the body by the blood. This method is commonly used in diagnostics, where nuclear emissions from the body are detected by radiation sensors and processed by a computer to produce two- or three-dimensional images of internal organs (figure 1).

Unstable isotopes can be combined with biologically active compounds, producing **radiopharmaceuticals** – drugs that deliver radionuclides to specific tissues or cellular receptors. In **brachytherapy**, also known as **internal radiotherapy**, radiation sources are inserted into the patient's body in the form of metal wires or pellets that deliver radiation directly to the site of the disease. More powerful sources of ionizing radiation such as particle accelerators or large quantities of radioisotopes are used in **external radiotherapy**, in which cancerous cells are destroyed by precisely directed beams of gamma rays, protons, electrons, or neutrons (sub-topic 2.1).

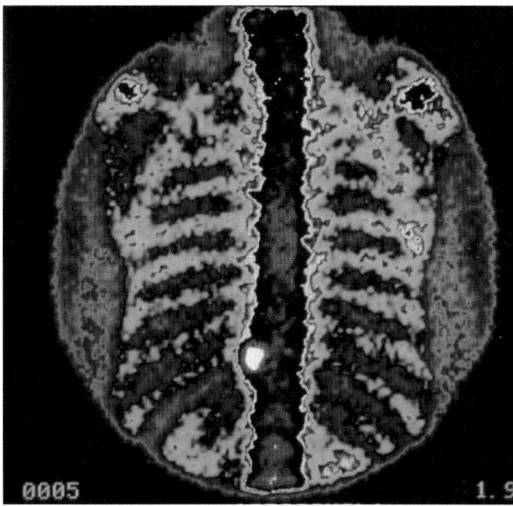
▲ Figure 1 Bone scintigram (gamma-ray photograph) of spine cancer. The tumour appears as a "hot spot" (white area near the bottom of the image)

Ionizing radiation

Ionizing radiation is dangerous to living organisms as it can damage cells, cause mutations, and increase the probability of developing cancer. However, cancerous cells are more sensitive to nuclear emissions so a carefully selected dose of radiation can destroy these cells without causing unacceptable damage to healthy tissues. Over time normal cells will regenerate while the development of the cancer will be slowed down or reversed. Still, radiotherapy is often traumatic to patients and produce severe side effects so is used only in life-threatening situations, where the benefits of the treatment outweigh the risks of radiation exposure.

Radiotherapy

The primary use of radiotherapy is the treatment of cancer. Along with other physiological effects, ionizing radiation induces errors in DNA sequences (sub-topic B.8), which can be passed to other cells through division. Rapidly dividing cancer cells are particularly sensitive to genetic damage because they accumulate DNA errors and this eventually limits their ability to grow and multiply. In addition, a reduced ability of cancer cells to repair their genetic material makes them more likely than normal cells to die from radiation exposure. However, normal dividing cells are also sensitive to induced DNA errors. Hair loss is a common side effect of radiotherapy, caused by damage to hair follicles which contain one of the fastest-growing cells in the human body. In contrast to chemotherapy (sub-topic D.7), the hair loss caused by ionizing radiation is often irreversible.

Other side effects of radiotherapy include skin and nail damage, nausea, fatigue, and sterility. Most of these effects are also caused by DNA errors in dividing cells (such as epidermal cells in the skin or germ cells in the reproductive organs), although some may be a result of psychological stress. A long-term risk of radiotherapy is the development of secondary cancers, which may occur several years or decades after the treatment.

Types of radiation

Radionuclides used in medicine produce various types of ionizing radiation. The three most common types of radiation (alpha particles, beta particles, and gamma rays) were discovered at the end of the nineteenth century and named after the first letters of the Greek alphabet. **Alpha particles** (α or $^4_2$He) are nuclei of helium-4 containing two protons and two neutrons (sub-topic 2.1); **beta particles** (β^- or e^-) are high-energy electrons emitted from atomic nuclei; and gamma rays (γ) are photons with very short wavelengths (sub-topic 2.2). Later it was found that radionuclides can emit other subatomic particles including **protons** (p), **neutrons** (n), and **positrons** (positively charged electrons, β^+ or e^+). The properties and sources of various kinds of nuclear emission are summarized in table 1.

Common name	Particle	Symbol	Charge, e*	Mass, u**	Common sources
alpha particle	helium-4 nucleus	α, $^4_2$He	+2	4.0	^{212}Pb, ^{225}Ac
beta particle	electron	β^-, e^-	−1	5.5×10^{-4}	^{90}Y, ^{131}I, ^{177}Lu, ^{192}Ir
positron emission	positron	β^+, e^+	+1	5.5×10^{-4}	^{11}C, ^{13}N, ^{15}O, ^{18}F
proton beam	proton	p, $^1_1$p, $^1_1$H	+1	1.0	particle accelerators
neutron beam	neutron	n, $^1_0$n	0	1.0	bombardment of ^{9}Be with protons or alpha particles
gamma ray	photon	γ	0	0	^{60}Co, ^{99m}Tc, ^{131}I, ^{137}Cs
X-ray***	photon	—	0	0	X-ray tubes

▲ Table 1 Types and sources of ionizing radiation used in medicine

* 1 e ≈ 1.6×10^{-19} C; ** 1 u ≈ 1.7×10^{-27} kg; *** not emitted by radionuclides

Ionizing radiation is produced by nuclear reactions or by the spontaneous decay of unstable isotopes, which can be represented by **nuclear equations**. In nuclear equations radioactive emissions are identified by their common symbols (table 1) while atomic nuclei are shown using the symbol for the chemical element with two additional numbers ($_Z^A X$, sub-topic 2.1). The **mass number** A shows the total number of protons and neutrons in the nucleus while the **atomic number** Z, also known as the **nuclear charge**, shows the number of protons in the nucleus. For example, a nucleus of carbon-11 containing 6 protons and 5 neutrons is written as $_6^{11} C$. An alpha particle containing 2 protons and 2 neutrons is a nucleus of helium-4, so it can be represented as either α or $_2^4 He$. Similarly, a proton is a nucleus of hydrogen-1 so can be written as p or $_1^1 H$. The mass numbers and/or charges of nuclear emissions can be also shown with symbols (for example, $_2^4 \alpha$ or $_1^1 p$).

The simplest kind of nuclear transformation, **radioactive decay**, is similar to decomposition reactions in chemistry, where a single species (radioactive nucleus) produces two or more other species (nuclei or elementary particles). For example, a nucleus of the radioactive isotope lead-212 ($_{82}^{212} Pb$) emits a beta particle, β^- and produces a nucleus of bismuth-212 ($_{83}^{212} Bi$):

$$_{82}^{212} Pb \rightarrow {}_{83}^{212} Bi + \beta^-$$

In the nucleus of lead-212, one neutron decays into a proton and an electron. The extra proton remains in the nucleus and increases the atomic number by one unit (from 82 to 83), so lead-212 (the **parent nucleus**) becomes bismuth-212 (the **daughter nucleus**). The electron is expelled from the nucleus as a beta particle while the mass number (212) of the nucleus does not change.

 ## Worked example

The nucleus of bismuth-212 produced in the above reaction is radioactive and emits either an alpha or a beta particle. The daughter nuclei in both cases undergo further decays and produce the same stable isotope, lead-208. Deduce the nuclear equations for the radioactive decay of bismuth-212 and its daughter nuclei.

Solution

In alpha decay the parent nucleus emits an alpha particle, $_2^4 He$, which contains 2 protons and 2 neutrons. The loss of 2 protons reduces the atomic number of $_{83}^{212} Bi$ by 2 units ($83 - 2 = 81$), so bismuth, $_{83} Bi$ will become thallium, $_{81} Tl$. At the same time the mass number of the parent nucleus will decrease by 4 units, from 212 to 208. Therefore the alpha decay of bismuth-212 will produce thallium-208:

$$_{83}^{212} Bi \rightarrow {}_{81}^{208} Tl + {}_2^4 He$$

Beta decay increases the atomic number of the parent nucleus by one unit so bismuth, $_{83} Bi$ will become polonium, $_{84} Po$. The mass number does not change, so polonium-212 will be produced:

$$_{83}^{212} Bi \rightarrow {}_{84}^{212} Po + \beta^-$$

We know that both $_{81}^{208} Tl$ and $_{84}^{212} Po$ produce $_{82}^{208} Pb$, so we can deduce their decay types by comparing the mass numbers and charges of parent and daughter nuclei. The mass numbers of thallium-208 and lead-208 are the same while their atomic numbers differ by one unit, which indicates a beta decay:

$$_{81}^{208} Tl \rightarrow {}_{82}^{208} Pb + \beta^-$$

Similarly, the mass numbers of $_{84}^{212} Po$ and $_{82}^{208} Pb$ differ by 4 units while their atomic numbers differ by 2 units, so polonium-212 undergoes an alpha decay:

$$_{84}^{212} Po \rightarrow {}_{82}^{208} Pb + {}_2^4 He$$

The **decay chain** (sequence of radioactive transformations) of lead-212 can be represented by a single scheme (figure 2).

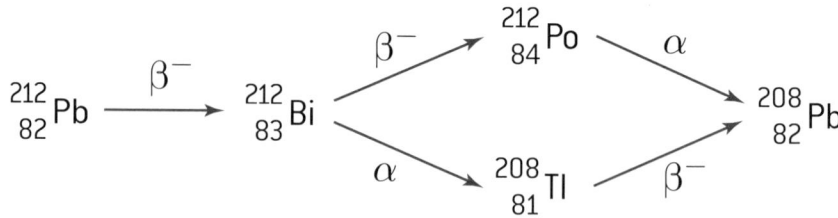

▲ Figure 2 The decay chain of lead-212

Techniques in nuclear medicine

In the human body alpha particles cause more damage to cellular tissues than any other form of radiation. However, these particles have very low penetrating power and are completely absorbed within a short range (0.05–0.1 mm) of their emission. This property is used in **targeted alpha therapy** (**TAT**) for treating leukaemia and other dispersed cancers. Controlled amounts of alpha emitters such as lead-212 (figure 2) or actinium-225 can be delivered by a carrier drug or protein directly to the targeted cancer cells, which will be selectively destroyed by radiation without significant damage to surrounding tissues. At the same time the collisions of alpha and beta particles with atomic nuclei produce secondary gamma radiation, which can be detected and used for mapping the distribution of cancer cells in the body.

Pure beta emitters such as yttrium-90 and lutetium-177 are also used in radiotherapy. These nuclides decay in one step and produce stable isotopes of zirconium and hafnium, respectively:

$$^{90}_{39}\text{Y} \rightarrow \,^{90}_{40}\text{Zr} + \beta^-$$

$$^{177}_{71}\text{Lu} \rightarrow \,^{177}_{72}\text{Hf} + \beta^-$$

Yttrium-90 is a common radiation source for cancer brachytherapy and palliative treatment of arthritis. Lutetium-177 produces low-energy beta particles with reduced tissue penetration, which is very useful in the targeted therapy of small tumours. In addition, lutetium-177 emits just enough gamma rays for visualizing tumours and monitoring the progress of their treatment.

Many kinds of ionizing radiation are produced not by the radioactive decay of individual nuclei but by **nuclear reactions**, where a target nucleus is bombarded with elementary particles or other nuclei. For example, neutrons can be generated by collisions of protons or alpha particles with beryllium-9:

$$^{9}_{4}\text{Be} + \,^{1}_{1}\text{p} \rightarrow \,^{9}_{5}\text{B} + \,^{1}_{0}\text{n}$$

$$^{9}_{4}\text{Be} + \,^{4}_{2}\text{He} \rightarrow \,^{12}_{6}\text{C} + \,^{1}_{0}\text{n}$$

High-intensity neutron beams are used in **boron neutron capture therapy** (**BNCT**), which utilizes the ability of boron-10 to absorb neutrons. After capturing a neutron the nucleus of boron-10 transforms into boron-11, which immediately undergoes alpha decay:

$$^{10}_{5}\text{B} + \,^{1}_{0}\text{n} \rightarrow [^{11}_{5}\text{B}] \rightarrow \,^{7}_{3}\text{Li} + \,^{4}_{2}\text{He}$$

The availability of nuclear medicine

The use of nuclear technology in medicine varies greatly from country to country. The main problem is the high cost of radiotherapeutic equipment, which in certain cases can exceed $100 million per unit. Sources of ionizing radiation are also expensive and require qualified staff for handling and maintenance. Another problem is the limited life span of many radionuclides, some of which can be stored for only a few days, while others must be produced in nuclear reactors or particle accelerators immediately before administration to patients. All these factors, together with cultural traditions and beliefs, significantly reduce the availability of radiodiagnostics and radiotherapy in many parts of the world.

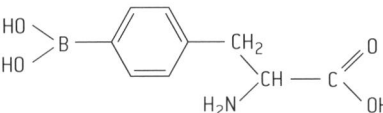

Both lithium-7 ions and alpha particles cause extensive cellular damage in a very limited range, 0.005–0.01 mm, which is approximately the size of a single cell. Therefore tumours can be destroyed by BNCT if they accumulate sufficient boron-10. This isotope can be administered to the patient by intravenous injection of certain organoboron compounds such as boronophenylalanine (BPA, figure 3).

BPA is structurally similar to amino acids used in protein synthesis so it is accumulated in all growing tissues including tumours. Certain types of cancer cell absorb BPA at levels sufficient for BNCT treatment. This kind of radiotherapy is still under development, with clinical trials taking place in many countries around the world.

Proton beam therapy (**PBT**) is another experimental technique of nuclear medicine. The protons are produced by a particle accelerator and released towards the tumour target. In contrast to other types of ionizing radiation, the absorption of protons by cellular tissues reaches a maximum within a narrow range, deep inside the patient's body (figure 4). This phenomenon, known as the **Bragg's peak effect** (figure 4), allows the proton beam to be focused on the tumour with minimal radiation damage to healthy tissues.

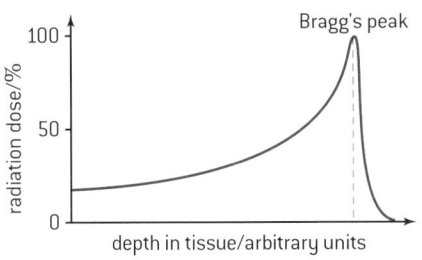

▲ Figure 3 The structure of boronophenylalanine, used to deliver boron-10 to cancer cells in the body

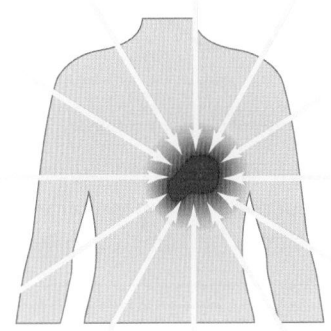

▲ Figure 4 Absorption of protons by cellular tissues

Gamma radiation

Many radionuclides used in medicine emit gamma radiation – high-energy photons that easily penetrate the human body and damage cellular tissues along their path. A series of low-intensity gamma rays can be used to deliver the maximum radiation dose to cancer cells (figure 5). These rays are focused on the tumour and destroy the cells within a small area while other parts of the body are exposed to relatively low levels of gamma radiation. Alternatively, a single gamma ray can be fired at the tumour many times from different angles, producing the same therapeutic effect.

An array of gamma emitters known as the **gamma knife** (figure 6) is a common tool for treating brain tumours. A typical gamma knife consists of 200 cobalt-60 sources mounted on a heavily shielded helmet. Each source emits a narrow ray of gamma radiation, which can be focused on a specific area of the brain. All the rays penetrate the skull and converge on the tumour, producing a very high local effect but sparing normal brain cells from extensive damage. Gamma knife treatment has very few side effects and can be used for almost any kind of brain tumour.

Radiodiagnostics

An important area of nuclear medicine is radiodiagnostics in which ionizing radiation is used to visualize internal organs, tumours, or physiological processes within the body. X-ray imaging, once the most common method of radiodiagnostics, has now been largely replaced with advanced techniques which allow the creation of three-dimensional images and animations of body parts, blood circulation or CNS activity. In **computed tomography** (**CT**), cross-sections of biological objects are generated by a computer from multiple two-dimensional X-ray scans taken at various angles. The source of X-rays, the cathode tube, does not contain radioactive materials and therefore can be switched on and off at any time.

▲ Figure 5 Multi-beam radiotherapy. Gamma rays (yellow) intersect at the target area (pink) and deliver most damage to the tumour (red)

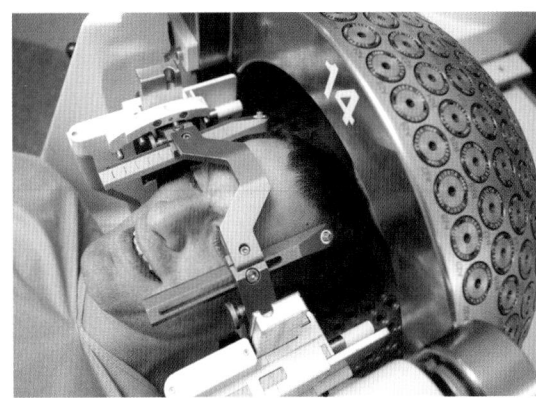

▲ Figure 6 Treatment of a brain tumour with a gamma knife

Another imaging technique detects the emissions of radionuclides inside the patient's body. These radionuclides, also known as **radiotracers,** are administered to the patient shortly before the scan and either absorbed in the blood or concentrated in certain organs or tumours (figure 1). For example, iodine-131 accumulates in the thyroid gland, producing sharp images of this organ even at extremely low doses. Higher doses of iodine-131 are used in radiotherapy for treating thyroid hyperfunction or malformations.

Positron emission tomography

Physiological processes in the body can be examined by **positron emission tomography** (PET). Many positron emitters are isotopes of macrobioelements (see table 1 above) so they can be chemically incorporated into any biologically active molecule. The most common substance used in PET is 2-fluoro-2-deoxyglucose (FDG) containing a radiotracer, fluorine-18 (figure 7).

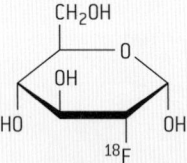

▲ Figure 7 FDG with a fluorine-18 radiotracer

When FDG is injected into the circulation it is distributed around the body in the same way as normal glucose. Positrons (β^+) emitted by fluorine-18 collide with electrons (e^-) and annihilate, producing pairs of high-energy photons (gamma rays) moving in opposite directions:

$$^{18}_{9}F \rightarrow\ ^{18}_{8}O + \beta^+$$

$$\beta^+ + e^- \rightarrow 2\gamma$$

These pairs of photons can be detected by a gamma camera and processed by a computer in the same way as X-rays are processed in CT scanning, producing a three-dimensional image of the body. The intensity of the detected radiation is proportional to the concentration of FDG, which in turn depends on the metabolic activity of cellular tissues. Any unusual variation in such activity may indicate a pathological process such as cancer, brain disease, or developing heart problems. Modern instruments can perform PET and CT scans simultaneously, greatly increasing the efficiency of both techniques.

Technetium-99m

Over 80% of diagnostic procedures in modern nuclear medicine rely on a single radionuclide, **technetium-99m** ($^{99m}_{43}Tc$). The letter "m" means that the nucleus of technetium-99m is metastable and can exist only for a short period of time. Similar to exited electrons in atoms and molecules (sub-topic 2.2), metastable nuclei eventually return to a lower-energy state by emitting electromagnetic radiation:

$$^{99m}_{43}Tc \rightarrow\ ^{99}_{43}Tc + \gamma$$

The photons produced by technetium-99m have approximately the same wavelength as X-rays, so they can be detected using traditional X-ray equipment. At the same time, the energy of these photons is relatively low which reduces the radiation dose received by the patient and medical personnel. Finally, technetium has several stable oxidation states (+3, +4, +7) and readily forms complexes with various ligands, which can be administered by injection and delivered to specific organs or tissues.

One of the major problems of nuclear medicine is the very nature of radionuclides, many of which decay quickly and therefore can be used only within a short time period. Kinetically, radioactive decay is a first order process (sub-topic 16.1) so the activity of a radionuclide decreases exponentially with time (figure 8). The time required for half of the initial amount of radionuclide to decay is known as its **half-life period** or simply **half-life** $(t_{1/2})$.

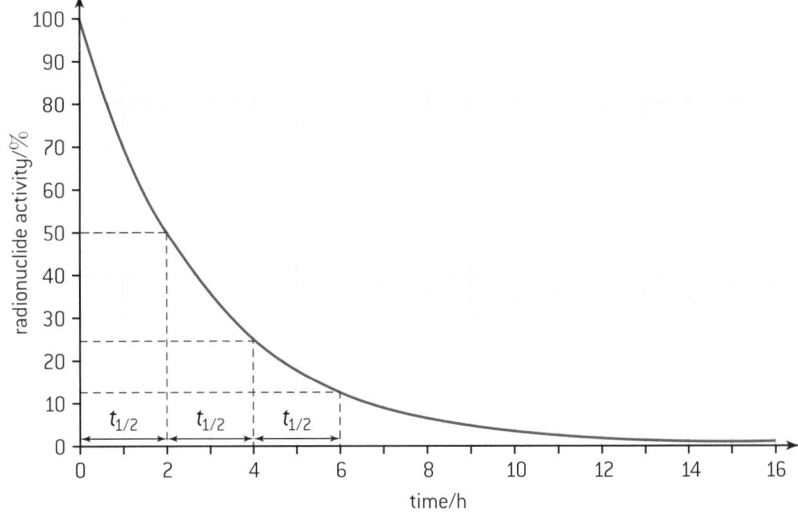

▲ Figure 8 Radioactive decay of a nuclide with $t_{1/2} = 2$ h. After each half-life period, the activity of the nuclide has decreased to half the previous level

Each radionuclide has a specific half-life which can vary from nanoseconds to billions of years (table 2). Half-life is inversely proportional to the nuclide activity, so more active radionuclides decay faster and have shorter half-lives than less active but longer-lived isotopes.

Technetium-99m has a half-life of 6.0 hours, which makes it ideal for medical imaging. A very small amount of this nuclide (typically 10^{-14} to 10^{-13} mol) administered to a patient in a single injection produces enough gamma radiation for most diagnostic procedures. After the gamma scan is complete nearly all the injected radionuclide decays within 2 days, minimizing the patient's exposure to radiation. At the same time the half-life of technetium-99m is long enough to prepare various complexes of this radionuclide with biologically active ligands.

Nuclide	Half-life	Decay type	Medical applications
^{18}F	110 min	β^+	positron emission tomography (PET)
^{60}Co	5.3 years	β^-, γ	external radiotherapy including "gamma knife"; sterilization of medical instruments
^{90}Y	64 h	β^-	cancer brachytherapy; palliative treatment of arthritis
^{99m}Tc	6.0 h	γ	imaging of tumours, internal organs, bone, muscle, brain, and biological fluids
^{131}I	8.0 days	β^-, γ	internal radiotherapy of thyroid hyperfunction and cancer; imaging of the thyroid and internal organs
^{137}Cs	30 years	β^-, γ	external radiotherapy
^{177}Lu	6.6 days	β^-, γ	targeted therapy and imaging of small tumours
^{192}Ir	74 days	β^-, γ	cancer brachytherapy
^{212}Pb	10.6 h	α, β^-	targeted alpha therapy (TAT) of cancer
^{225}Ac	10 days	α	targeted alpha therapy (TAT) of cancer

▲ Table 2 Half-lives of common radionuclides used in medicine

Worked example

Unused injection solutions and other materials containing technetium-99m ($t_{1/2} = 6.0$ h) are classified as low-level nuclear waste (sub-topic D.6), which must be stored in shielded containers for several days before disposal. Calculate the percentage of the initial amount of technetium-99m left in the container after 3 days of storage.

Solution

After each half-life period the amount of technetium-99m will have decreased by a half, so after 6 hours, 50% of the isotope will remain. After another 6 hours (total 12 hours), the remaining percentage will be 25% of the original, and so on. This process will continue as shown in table 3.

Therefore after 3 days (72 hours) only 0.02% of the initial amount of technetium-99m will remain in the container.

The same result could be obtained by another method. Since the amount of a radionuclide decreases to half the current level after each half-life period, after n half-life periods this amount will halve n times. So in 72 h (after $\frac{72}{6.0} = 12$ half-life periods), the amount of technetium-99m will fall to $\left(\frac{1}{2}\right)^{12} = \frac{1}{4096} \approx 0.0002$ (0.02%) of the initial value.

Time/h	0	6	12	18	24	30	36	42	48	54	60	66	72
Number of $t_{1/2}$	0	1	2	3	4	5	6	7	8	9	10	11	12
Nuclide left/%	100	50	25	12.5	6.25	3.13	1.56	0.78	0.39	0.20	0.10	0.05	0.02

▲ Table 3

Decay constant

Along with the half-life, the activity of a radionuclide can be characterized by its decay constant (λ), which is related to the half-life as follows:

$$\lambda = \frac{\ln 2}{t_{1/2}} \approx \frac{0.693}{t_{1/2}}$$

If the initial quantity (N_0) of the radionuclide is known, the remaining quantity (N) of this nuclide after any given period of time (t) can be found:

$$N = N_0 e^{-\lambda t}$$

It is also possible to find the time required for a certain fraction of the radionuclide to decay:

$$t = \frac{\ln \frac{N_0}{N}}{\lambda}$$

These calculations are particularly important when a short-lived radionuclide is administered to a patient. The activity of such a nuclide can change significantly during the medical procedure, which must be taken into account when interpreting the diagnostics results or determining the dose and duration of the treatment.

Worked example

In a typical PET examination, a dose of FDG containing radioactive fluorine-18 ($t_{1/2} = 110$ min) is administered to a patient 1 hour before the scan, which takes 40 minutes to complete. Calculate the number of fluorine-18 atoms that will decay inside the patient's body during the scan if the amount of ^{18}F in the injected FDG was 1.5×10^{-8} mol.

Solution

Substituting in the formula for λ above, $\lambda \approx \frac{0.693}{110}$ ≈ 0.0063 min^{-1}, so after 1 hour (60 min) the amount of ^{18}F will be:

$$1.5 \times 10^{-8} \times e^{-0.0063 \times 60} \approx 1.0 \times 10^{-8} \text{ mol}$$

Since each mole contains $N_A \approx 6.0 \times 10^{23}$ atoms, the number of ^{18}F atoms in the body before the scan will be $1.0 \times 10^{-8} \times 6.0 \times 10^{23}$ $= 6.0 \times 10^{15}$.

If the scan takes 40 minutes the number of ^{18}F atoms will decrease further to $6.0 \times 10^{15} \times e^{-0.0063 \times 40}$ $\approx 4.7 \times 10^{15}$. Therefore, $6.0 \times 10^{15} - 4.7 \times 10^{15} =$ 1.3×10^{15} atoms of fluorine-18 will decay inside the patient's body during the scan.

Magnetic resonance imaging

Magnetic resonance imaging (**MRI**) is a medical application of nuclear magnetic resonance (NMR, sub-topics 11.3 and 21.1). Modern MRI scanners use superconductive magnets (sub-topic A.8) to create powerful magnetic fields (up to 100 000 times stronger than the magnetic field of the Earth). The instrument also produces electromagnetic radiation of low frequency and long wavelength (radio waves). When a patient is placed inside the magnet the protons (^{1}H) in the body constantly change their states, absorbing and emitting radio waves of certain frequency. These radio waves are detected by the scanner and processed on a computer. By focusing the scanner on different parts of the body, two- or three-dimensional images of internal organs or body parts can be created.

MRI produces more detailed images of the human body than CT or PET scanning techniques. The protons in water, lipids, carbohydrates, and proteins have different chemical environments, which can be easily distinguished by ^{1}H NMR chemical shifts (sub-topic 11.3). Because the concentrations of these compounds in various tissues are different, MRI provides highly detailed images of the brain, heart, muscles, and body fluids. The technique does not use ionizing radiation so can be used repeatedly without increasing the risk of cancer to the patient. The only drawbacks of MRI are the high cost of the equipment and the interaction of magnetic fields with metal body implants such as prosthetics and heart pacemakers.

Multinuclear MRI

As well as proton NMR, modern MRI instruments can detect other nuclei including carbon-13, sodium-23, and phosphorus-31. Multinuclear MRI studies are particularly useful for the imaging of organs that have insufficient contrast in ^{1}H NMR. For example, images of lungs can be obtained by ^{3}He or ^{129}Xe NMR, where a noble gas (helium or xenon, respectively) is inhaled by the patient during the MRI scan. Another nucleus, naturally occurring ^{31}P, can provide important information on the structure of bone tissues and brain functions.

Questions

1 Define the terms "nuclear medicine", "radionuclide", "half-life", "radiopharmaceutical", "brachytherapy", and "external radiotherapy".

2 Radionuclides produce ionizing radiation such as alpha and beta particles, positrons, and gamma rays.

   a) Explain how ionizing radiation can be used in medical diagnostics and the treatment of diseases.

   b) Discuss common side effects of radiotherapy.

3 In theory, it would take an infinite time for all the unstable nuclei in a sample of a radionuclide to decay. However, the activity of radionuclides decreases sharply within 5–10 periods of their half-lives. Calculate the percentage of a radionuclide that will remain after: (a) 5 half-life periods (b) 10 half-life periods.

4 The activity of a radionuclide has been measured every 6 hours and recorded in table 4.

Time/h	0	6	12	18	24	30	36	42	48
Activity /%	100	78.3	61.3	48.0	37.6	29.4	23.0	18.0	14.1

▲ Table 4

   a) Draw a graph of activity versus time on graph paper.
   b) Determine the half-life period of the radionuclide from the plot.
   c) Calculate the half-life period of the same radionuclide using the data from the table and the formulae given in the text.

5 Actinium-225 ($t_{1/2} = 10$ days) is an alpha emitter used in targeted alpha therapy (TAT).

   a) Deduce the nuclear equation for the decay of actinium-225.
   b) Explain how TAT can be used for treating cancers that have spread around the body.
   c) Suggest why alpha particles are particularly effective in cancer treatment.

6 Beta emitters such as yttrium-90 and lutetium-177 are commonly used in nuclear medicine.

   a) Explain why these radionuclides are administered directly to the patient's body rather than used for external radiotherapy.

   b) Calculate how much of a 7.0 mg sample of lutetium-177 ($t_{1/2} = 6.6$ days) would remain after 30 days.

7 Boron neutron capture therapy (BNCT) and proton beam therapy (PBT) are advanced nuclear medicine techniques.

   a) Explain how BNCT can be used to target cancer cells.
   b) Explain why PBT is more effective in treating cancers than traditional methods of external radiotherapy.

8 Nitrogen-13 ($t_{1/2} = 10$ min) is a radioactive tracer used in positron emission tomography (PET).

   a) Deduce the nuclear equation for the decay of nitrogen-13.
   b) To deliver nitrogen-13 to a specific organ the tracer must be chemically incorporated into a biologically active compound. The synthesis of a particular compound with a ^{13}N tracer takes 40 min, followed by 5 min for the preparation of the injection solution. Calculate the percentage of ^{13}N that will decay before the compound can be administered to a patient.
   c) Other than the cost of radionuclides and equipment, suggest one factor that limits the availability of PET in remote medical centres.

9 The radionuclide cobalt-60 ($t_{1/2} = 5.3$ years) is used in external radiotherapy. It emits a beta particle and a gamma ray, producing a stable isotope of another element.

   a) Deduce the nuclear equation for the decay of cobalt-60.
   b) Calculate how many times the activity of a ^{60}Co source will decrease in 10 years.
   c) Decommissioned ^{60}Co sources must be stored in protected areas until most of the radionuclide has decayed into non-radioactive materials. Calculate the time needed for the decay of 99.99% of cobalt-60.

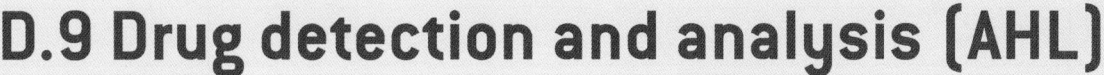

D.9 Drug detection and analysis (AHL)

Understandings

→ Organic structures can be analysed and identified through the use of infrared spectroscopy, mass spectroscopy, and proton NMR.

→ The presence of alcohol in a sample of breath can be detected through the use of either a redox reaction or a fuel cell type of breathalyzer.

 ## Nature of science

→ Advances in instrumentation – modern analytical techniques (IR, MS, and NMR) have assisted in drug detection, isolation, and purification.

 ## Applications and skills

→ Interpretation of a variety of analytical spectra to determine an organic structure including infrared spectroscopy, mass spectroscopy, and proton NMR.

→ Description of the process of extraction and purification of an organic product. Consider the use of fractional distillation, Raoult's law, the properties on which extractions are based, and explaining the relationship between organic structure and solubility.

→ Description of the process of steroid detection in sport utilizing chromatography and mass spectroscopy.

→ Explaining how alcohol can be detected with the use of a breathalyzer.

Analytical techniques

A variety of analytical techniques is used for the detection and analysis of pharmaceutical drugs. Some of these techniques, including chromatography, electrophoresis (sub-topics B.2 and B.8), nuclear magnetic resonance (NMR) and infrared (IR) spectroscopy (sub-topics 11.3 and 21.1), mass spectrometry (MS), and X-ray crystallography (sub-topic 21.1) have been discussed earlier. Analysed drugs or other compounds often need to be isolated and purified by crystallization, distillation, or extraction (sub-topics 10.2 and 21.1). In this sub-topic we shall discuss how spectroscopic data can be related to the molecular structure of a drug and how a target compound can be separated from a mixture with other substances.

Spectroscopic identification of drugs

Many pharmaceutical drugs are relatively simple organic molecules containing various functional groups (topic 10). The presence or absence of these groups in pharmaceutical products can be determined by IR, NMR, and mass spectroscopy. For example, all the functional groups in the molecule of aspirin (sub-topic D.2) have characteristic absorptions in the IR spectrum (figure 1 in sub-topic D.2). Additional information can be obtained from the ^{1}H NMR spectrum of aspirin, where the protons in different chemical environments produce signals with specific chemical shifts and splitting patterns (figure 1 and table 1 on the next page).

 ## Advances in analytical techniques

Recent advances in instrumentation have dramatically improved the sensitivity and accuracy of drug analysis in medical studies, forensic science, and the pharmaceutical industry. Modern analytical techniques can detect trace amounts of illegal substances in the human body, distinguish between stereoisomers of biologically active compounds, or confirm the identity and purity of pharmaceutical products. These technological changes improve the quality of our lives and protect society from the consequences of substance abuse. At the same time, an increasing number of people are now legally required to provide samples of their blood or urine for routine drug tests, which limits their personal freedom and affects the ethical choices of individuals.

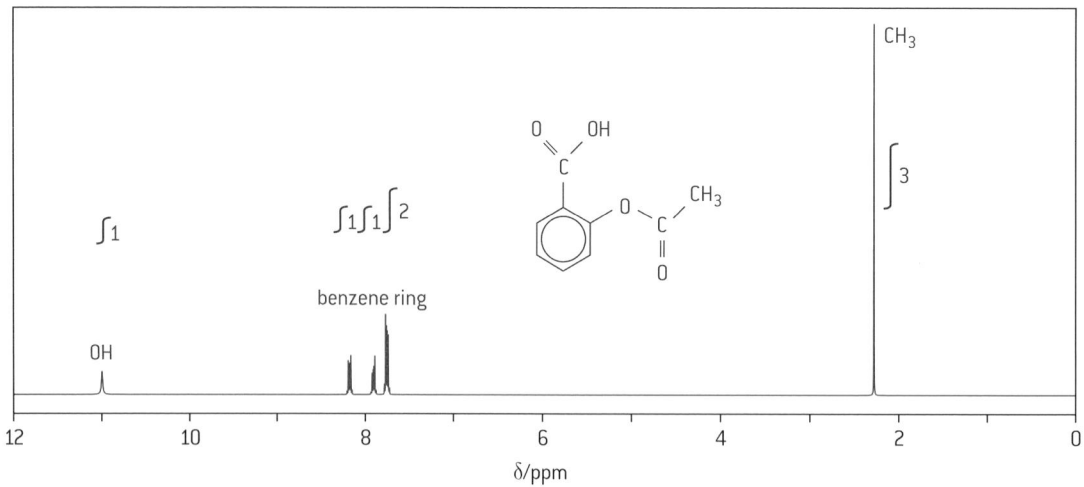

▲ Figure 1 ¹H NMR spectrum of aspirin

Chemical environment	Chemical shift / ppm	Number of protons (integration)	Number of adjacent protons	Splitting pattern
CH₃	2.3	3	0	none (singlet)
C₆H₄ (benzene ring)	7.7, 7.9, and 8.2	4 (2 + 1 + 1)	—	multiplets*
OH	11.0	1	0	none (singlet)

▲ Table 1 Chemical shifts and splitting patterns of protons in the molecule of aspirin

* The splitting pattern of protons in the benzene ring will not be assessed

In addition, the structure of aspirin can be confirmed by its mass spectrum (figure 2). Certain structural fragments such as CH_3^+ ($m/z = 15$) and CH_3CO^+ ($m/z = 43$) produce stable cations that can be directly observed in the mass spectrum. A cation with $m/z = 163$ is formed by the loss of a hydroxyl radical (HO•, $M_r = 17$) from the molecular ion $M^{•+}$ ($m/z = 180$). Other species ($m/z = 92$, 120, and 138) are produced by further fragmentation and rearrangements of these cations.

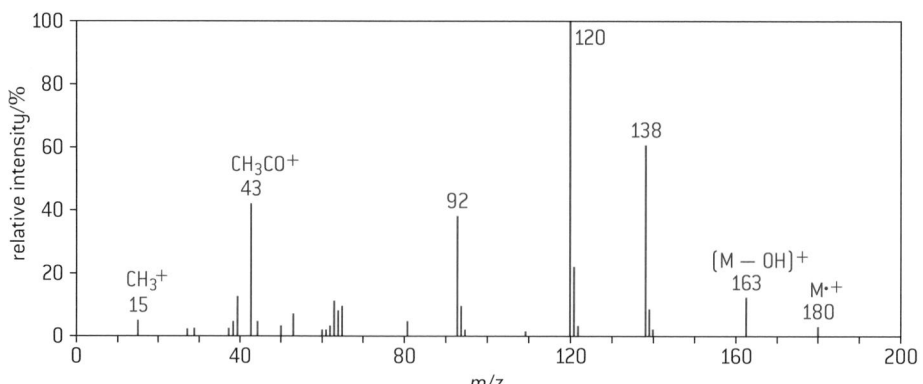

▲ Figure 2 Mass spectrum of aspirin

Identifying unknown compounds

The most common task for a pharmaceutical chemist is the identification of a drug or other organic molecule from various analytical data. If some information about the drug (molecular mass, elemental composition, retention factor (R_f) in a chromatogram) is already known, the molecule can be identified by comparison with a library of known compounds. Otherwise, its molecular mass can be determined from its mass spectrum (assuming that the peak with the greatest m/z value belongs to the molecular ion). The functional groups in a molecule can be identified by IR and ^{1}H NMR spectroscopy and then matched to the MS fragmentation pattern to confirm the identity of the compound.

 Worked example

Methamphetamine (N-methyl-1-phenylpropan-2-amine), colloquially known as "meth", is a stimulant drug and a common substance of abuse. Depending on the manufacturing method it can contain various impurities, including ephedrine, methcathinone, and N-benzylpropan-2-amine (figure 3).

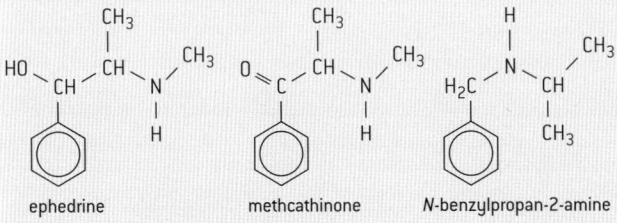

▲ Figure 3 Impurities commonly found in methamphetamine

One of these impurities has been isolated from a sample of illicit methamphetamine and analysed by ^{1}H NMR, IR, and MS.

a) Deduce the number of chemical environments of protons in the side-chains of ephedrine, methcathinone, and N-benzylpropan-2-amine (ignore the protons of the benzene ring).

b) The ^{1}H NMR spectrum of the impurity is given in figure 4. Identify the splitting patterns of signals in this spectrum.

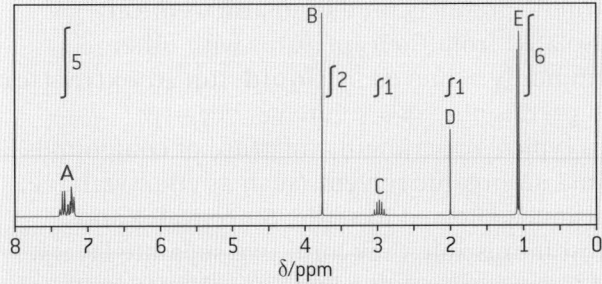

▲ Figure 4 ^{1}H NMR spectrum of the impurity

c) Identify the impurity using its ^{1}H NMR spectrum (figure 4) and IR spectrum (figure 5).

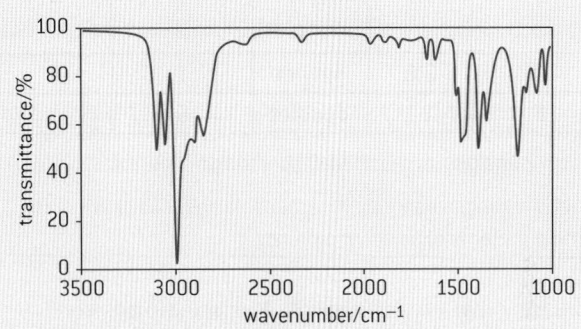

▲ Figure 5 IR spectrum of the impurity

d) The mass spectrum of the same impurity is given in figure 6. Identify the cationic species responsible for all labelled peaks in this mass spectrum.

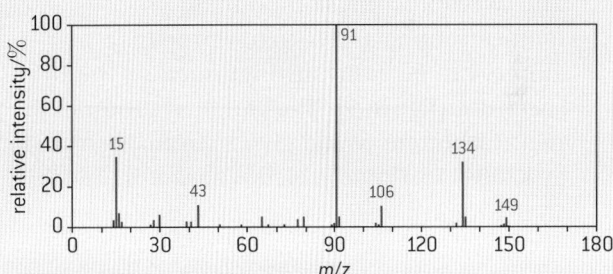

▲ Figure 6 Mass spectrum of the impurity

Solution

a) The protons in the side-chain of ephedrine have six different chemical environments (one OH, one NH, two different CH, and two different CH_3 groups). The side-chain in methcathinone has four different chemical environments (one NH, one CH, and two different CH_3 groups). The side-chain in N-benzylpropan-2-amine also has four different chemical environments (one NH, one CH, one CH_2, and two identical CH_3 groups).

b) In addition to the splitting patterns, the integrations and numbers of adjacent protons are shown in table 2.

Chemical environment	Chemical shift / ppm	Splitting pattern	Number of protons (integration)	Number of adjacent protons*
A	7.2–7.4	multiplet	5	—
B	3.8	singlet	2	0
C	2.9	septet (multiplet)	1	6
D	2.0	singlet	1	0
E	1.1	doublet	6	1

▲ Table 2 Analysis of the ^{1}H NMR spectrum in figure 4

* Due to hydrogen bonding, NH groups do not usually affect the splitting patterns of adjacent protons.

c) Chemical environment **A** corresponds to the protons of the phenyl group (see *Data booklet*), so the protons of the side-chain have four different chemical environments (signals **B–E**). Therefore this spectrum cannot belong to ephedrine, which has a side-chain with six different chemical environments.

The two remaining compounds, methcathinone and *N*-benzylpropan-2-amine, can be easily distinguished by the IR spectrum (figure 5). The carbonyl group in methcathinone would give a strong absorption at 1700–1750 cm^{-1} which is absent in figure 5, so the impurity is *N*-benzylpropan-2-amine.

The same conclusion could be reached by analysing the integrations and splitting patterns in the ^{1}H NMR spectrum. The protons in the side-chain of methcathinone would give the integration ratio of 1 : 1 : 3 : 3. However, in figure 4 the integration ratio is 1 : 1 : 2 : 6, which corresponds to *N*-benzylpropan-2-amine. Similarly, the septet (a multiplet with seven components) at 2.9 ppm could only be produced by the CH proton of an isopropyl group, $-CH(CH_3)_2$, which is absent in methcathinone but present in *N*-benzylpropan-2-amine.

d) Typical fragmentations of the molecule of *N*-benzylpropan-2-amine are shown in figure 7.

▲ Figure 7 MS fragmentations of *N*-benzylpropan-2-amine

Therefore the first five labelled *m/z* peaks in figure 6 belong to cations CH_3^+ (15), $CH(CH_3)_2^+$ (43), $C_7H_7^+$ (91), $C_6H_5CH_2NH^+$ or $(M - C_3H_7)^+$ (106), and $C_6H_5CH_2NHCHCH_3^+$ or $(M - CH_3)^+$ (134). The last peak (*m/z* = 149) belongs to the molecular ion, $M^{•+}$, which is a radical cation.

▲ Figure 8 Partition of a yellow dye between an organic solvent (top) and water (bottom). The dye can be isolated by collecting the organic layer and evaporating the solvent

Extraction and purification of organic products

Many natural and synthetic products used in pharmaceutical chemistry have to be isolated from their mixtures with other compounds. This is commonly achieved by **liquid–liquid extraction**, a process that involves partitioning of a solute between two immiscible liquids. In a typical experiment a mixture of compounds is shaken with water and an organic solvent (such as ethoxyethane) and the resulting emulsion is allowed to settle. Since water and ethoxyethane are almost immiscible they form two separate layers. Polar compounds tend to be more soluble in polar solvents (such as water) and therefore stay in the aqueous layer while non-polar substances dissolve in the organic layer. Each layer can be run into a different beaker using a separation funnel (figure 8). The organic solvent and water can be evaporated from the separated layers, leaving the components of the original mixture.

For complex mixtures the separation process can be repeated many times using the same or different solvents. In the case of the anticancer drug Taxol (sub-topic D.7), the isolation of the target compound required several hundred extractions and took over two years to complete.

The partition of a solute between two immiscible liquids can be described as a heterogeneous equilibrium (sub-topics 7.1 and 17.1) between different states of the same compound. For example, when molecular iodine, I_2 is partitioned between water (designated as "aq") and an organic solvent ("org"), the following equilibrium takes place:

$$I_2(aq) \rightleftharpoons I_2(org)$$

The constant of this equilibrium is known as the **partition coefficient**, P_c:

$$P_c = \frac{[I_2(org)]}{[I_2(aq)]}$$

Similar to K_c (sub-topic 7.1), the partition coefficient depends on the nature of the participating species and the temperature of the mixture. At 25 °C the partition coefficient of iodine in ethoxyethane/water is 760, which is typical for non-polar molecules. In contrast, polar compounds are more soluble in polar solvents, so their partition coefficients in ethoxyethane/water are usually less than 1.

 Worked example

Extraction is commonly used in drug analysis. In one experiment a steroidal hormone X was extracted from 0.10 dm³ of urine using 5.0 cm³ of hexane. The hormone concentration in hexane was found to be 120 nmol dm⁻³. Calculate the hormone concentrations, in nmol dm⁻³, in the urine sample before and after the extraction if $P_c(X)$ in hexane/water is 250.

Solution

$P_c(X) = \frac{[X(org)]}{[X(aq)]}$. After the extraction

$[X(aq)] = \frac{[X(org)]}{P_c(X)} = \frac{120}{250} = 0.48$ nmol dm⁻³

The amounts of X(org) and X(aq) are 0.48 × 0.10 = 0.048 nmol and 120 × 0.0050 = 0.60 nmol, respectively. Before the extraction all the hormone (0.048 + 0.60 ≈ 0.65 nmol) was dissolved in the urine, so its initial concentration was $\frac{0.65}{0.10}$ = 6.5 nmol dm⁻³.

This example shows the importance of extraction techniques in medicine. A relatively simple experiment allowed the extraction of $\frac{0.060}{0.065}$ × 100% ≈ 92% of the hormone and its concentration in the solution to be increased $\frac{120}{6.5}$ ≈ 18 times, enhancing the sensitivity of further laboratory analyses.

Fractional distillation

The pharmacological properties of a drug depend largely on its polarity. Polar (**hydrophilic**) molecules tend to stay in the blood plasma while non-polar (**lipophilic**) drugs accumulate in lipid tissues. In medicine the polarity of a drug is often represented by the logarithm of its partition coefficient (log P) between octan-1-ol and water. For example, the log P values for morphine and diamorphine are 0.9 and 1.58, respectively, which explains the greater ability of diamorphine to cross the blood brain barrier (sub-topic D.3) and produce a stronger analgesic effect.

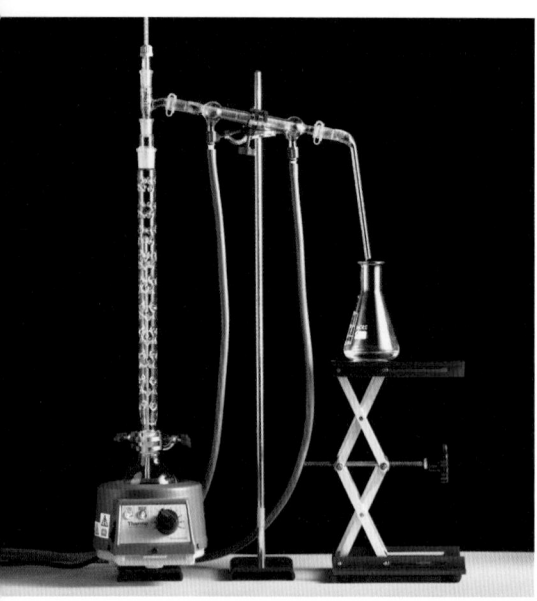

▲ Figure 9 Fractional distillation

Fractional distillation is another common method of isolation and purification of organic compounds (sub-topics 10.2 and 21.1). According to **Raoult's law**, the vapour pressure of a volatile substance A is proportional to the mole fraction of A in the mixture:

$$p(A) = p^*(A) \cdot x(A)$$

where

- $p(A)$ is the vapour pressure of A over the mixture (also known as the **partial pressure**) at a given temperature,

- $p^*(A)$ is the vapour pressure over a pure sample of A at the same temperature,

- $x(A)$ is the mole fraction of A, which is the ratio of the amount of A to the sum of the amounts of all components in the mixture.

In a boiling mixture of several substances, the more volatile compounds will have higher vapour pressures and evaporate faster than other components of the mixture. If a sufficiently long distillation column (figure 9) is used, vapours of different components will partly condense and evaporate again at different heights. Each cycle of condensation and evaporation will enrich the mixture with more volatile components, increasing their mole fractions and therefore partial pressures. As a result, the vapours of more volatile components will move up the column while less volatile substances will stay as liquids and fall back into the flask. Eventually the most volatile compound will reach the top of the column, pass through the water-cooled condenser, and flow into the receiver flask, producing the first fraction of the distillate. Other components of the mixture will form subsequent fractions, which can be collected in different flasks. If the separation is incomplete, each fraction can be distilled again until individual compounds are obtained.

In the pharmaceutical industry fractional distillation is often used as a continuous process, with the mixture constantly being added to the distillation apparatus while different fractions are collected at various column heights. Industrial distillation columns can be over 100 m high and produce several cubic metres of distillate every hour.

Drug detection in sports and forensic studies

The misuse of performance-enhancing substances in sports is a serious international problem. The most common type of these substances, anabolic steroids, accelerate the synthesis of proteins and cellular growth, especially in the muscle and bone tissues. Anabolic steroids are banned by most sports organizations including the International Olympic Committee. Athletes are regularly required to provide urine and blood samples for laboratory analyses in which steroids and their metabolites can be detected by a combination of gas chromatography (GC) or high performance liquid chromatography (HPLC) (sub-topic B.2) with mass spectrometry (MS) (sub-topics 11.3 and 21.1).

Anabolic steroids are predominantly non-polar compounds, so they can be extracted from biological materials with organic solvents and concentrated for further studies. Each steroid produces a characteristic mass spectrum (figure 10) which can be compared with a library of known compounds. Modern GC/MS and HPLC/MS instruments can detect anabolic steroids and their metabolites at concentrations as low as $1 \ ng \ cm^{-3}$ ($3 \times 10^{-9} \ mol \ dm^{-3}$), giving positive results for many weeks or even months after the use of these drugs has been discontinued.

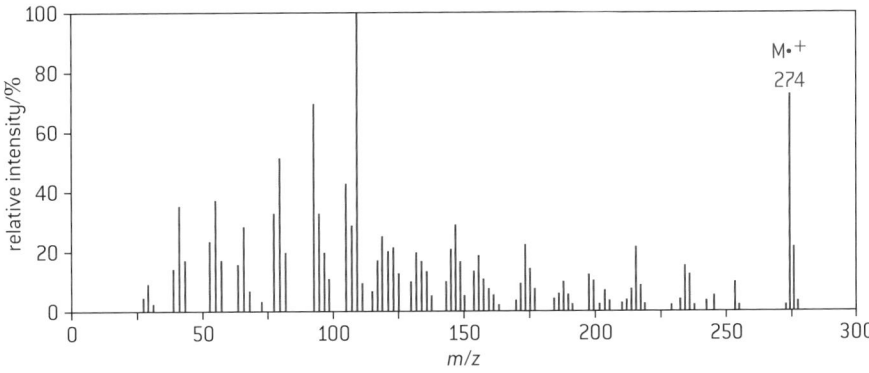

▲ Figure 10 Mass spectrum of the anabolic steroid nandrolone ($M_r = 274$)

Alcohol (ethanol) is the most common substance of abuse in the world. Excessive consumption of alcohol impairs judgement, concentration, and motor skills, often causing road accidents and violent behaviour. In many countries there is a legal limit for the **blood alcohol concentration** (BAC) that must not be exceeded by drivers or people operating heavy machinery. A motorist suspected of being under the influence of alcohol may be stopped by the police and asked to take an alcohol test on a portable device known as a **breathalyzer**. Instead of measuring BAC directly the breathalyzer determines the concentration of alcohol in the breath, which is roughly proportional to the BAC.

The simplest breathalyzer consists of a glass tube filled with acidified crystals of potassium dichromate(VI). When an intoxicated person blows into the tube the orange crystals turn green, as dichromate(VI) ions are reduced by ethanol in the breath to chromium(III) ions:

$$Cr_2O_7^{2-}(s) + 14H^+(aq) + 6e^- \rightarrow 2Cr^{3+}(aq) + 7H_2O(l)$$
orange   green

Depending on the reaction conditions, ethanol in a breathalyzer is oxidized to ethanoic acid or ethanal, for example:

$$C_2H_5OH(g) + H_2O(l) \rightarrow CH_3COOH(aq) + 4H^+(aq) + 4e^-$$

Another type of breathalyzer uses a fuel cell (sub-topic C.6) in which ethanol is oxidized by atmospheric oxygen on the surface of platinum electrodes. When a suspect exhales air into the fuel cell, ethanol in the breath is oxidized at the anode (the same reaction as above) while oxygen is reduced at the cathode:

$$O_2(g) + 4H^+(aq) + 4e^- \rightarrow 2H_2O(l)$$

The electric current produced by the fuel cell is proportional to the concentration of ethanol in the breath, which can be related to the BAC.

Portable breathalyzers are relatively simple instruments, so the results of roadside alcohol tests are not very reliable and cannot be used in court. An accurate measurement of the alcohol concentration in the breath or blood can be performed in a laboratory using IR spectroscopy,

GC, or HPLC. An IR spectrometer detects the presence of alcohol in the breath by the absorption of infrared light at certain wavelengths, which is caused by the C−H and C−O bonds in ethanol. A beam of IR radiation alternately passes through two identical chambers, one of which contains a breath sample while another is filled with atmospheric air. The difference in absorption between the sample and reference chambers can be converted into the concentration of ethanol in the breath using the Beer–Lambert law (sub-topic B.7).

GC and HPLC techniques are used for direct measurement of the BAC. When a blood sample containing alcohol is injected into a GC instrument, ethanol evaporates and passes into a column containing a non-volatile liquid (the stationary phase) and a carrier gas (the mobile phase). As the ethanol travels along the column it constantly evaporates and condenses, producing a narrow band of vapour and liquid. When this band leaves the column it passes through a detector that converts the absorption of IR or UV radiation by ethanol into electric current. Most instruments can also produce a chromatogram, in which the analysed compounds appear as peaks of different sizes (figure 11). The presence of ethanol in the blood can be confirmed by its **retention time** (the time between the injection and detection). The amount of ethanol is proportional to the area under the peak, which can be converted to BAC using a calibration curve.

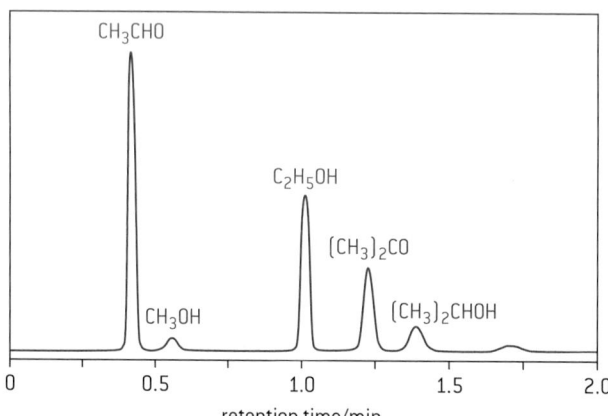

▲ Figure 11 A typical gas chromatogram used in BAC analysis. Ethanol, C_2H_5OH and its primary metabolite ethanal, CH_3CHO are shown in red

An HPLC instrument works in a similar way to GC except that the blood sample is not evaporated but mixed with a liquid mobile phase and injected into a column containing a solid or liquid stationary phase. The components of the blood are partitioned between the stationary and mobile phases and move through the column at different speeds according to their polarities and affinities to each phase. Similar to GC, the presence and concentration of ethanol in the blood sample are determined by its retention time and the area under the peak on the chromatogram.

Questions

1 The ^{1}H NMR spectrum of an intermediate compound formed during the synthesis of the painkiller ibuprofen is shown in figure 12. The peaks labelled **A** to **G** are not fully expanded to show the splitting but the integration trace for each peak is included.

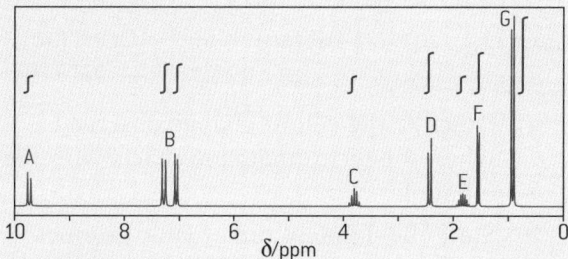

▲ Figure 12

The peak labelled **A** is a doublet. The two peaks labelled **B** centred at 7.1 ppm are due to the four hydrogen atoms on the benzene ring. The expansions to show the splitting for the other five peaks are shown in figure 13.

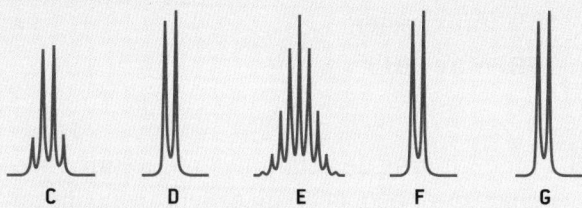

▲ Figure 13

The structure of the intermediate compound is given in figure 14, with seven hydrogen atoms labelled.

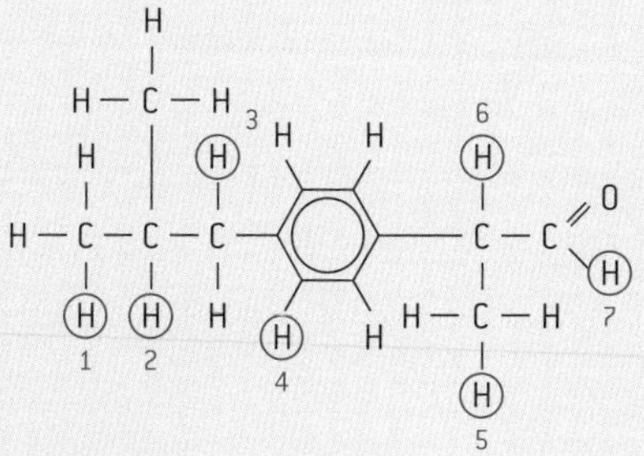

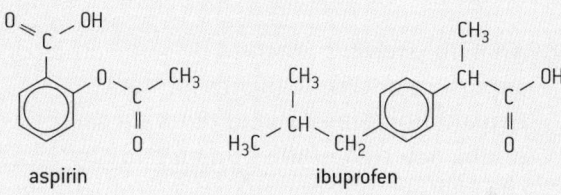

▲ Figure 14

Deduce which labelled hydrogen atoms are responsible (wholly or in part) for each of the peaks and complete a copy of table 3. [6]

Peak	A	B	C	D	E	F	G
Hydrogen atom responsible		4					

▲ Table 3

IB, May 2013

2 Aspirin and ibuprofen are painkillers. The structures of aspirin and ibuprofen are shown in figure 15:

aspirin   ibuprofen

▲ Figure 15

a) State the number of peaks in the ^{1}H NMR spectrum of aspirin (ignore the peaks due to the hydrogen atoms on the benzene ring and the reference sample). [1]

b) Describe the splitting pattern for each of the peaks given in (a). [1]

c) State how the infrared spectra of aspirin and ibuprofen will differ in the region 1700–1750 cm^{-1}. [2]

IB, May 2013

3 Pharmacological properties of drugs depend on their polarities. The partition coefficient of a certain drug between cellular tissues and blood plasma is 125. Calculate the concentration, in µmol dm^{-3}, this drug in tissues if its concentration in the blood plasma is maintained at 0.60 µmol dm^{-3} by continuous injection.

4 Extraction is an important technique in medicinal chemistry.

a) Outline how a mixture of two organic compounds with different polarities can be separated by extraction.

b) The partition of a pharmaceutical drug (X) between water and an organic solvent can be represented by the following equation: $X(aq) \rightleftharpoons X(org)$. Deduce the equation for the partition coefficient of X.

c) An aqueous solution with $c(X) = 0.46$ mol dm^{-3} was extracted with an equal volume of octan-1-ol. After the extraction, the concentration of X in the aqueous phase decreased to 0.012 mol dm^{-3}. Calculate the concentration of X in the organic phase and the log P value for this drug.

5 Anabolic steroids are used by some athletes as performance-enhancing substances. Explain how steroids and other illegal drugs can be detected in the human body by chromatography and mass spectrometry.

6 Ethanol is sufficiently volatile to pass into the lungs from the bloodstream. The roadside breathalyzer uses potassium dichromate(VI), which reacts with ethanol present in the breath.

a) Deduce the oxidation and reduction half-equations that occur in the breathalyzer.

b) State and explain, in terms of electron transfer and oxidation number change, whether chromium in potassium dichromate(VI) is oxidized or reduced.

c) Explain how the concentration of ethanol can be determined by the use of a fuel cell and IR spectroscopy.

7 The presence of ethanol in the breath can be detected by blowing into a "bag" through a tube with acidified potassium dichromate(VI). The half-equation for the dichromate reaction is:

$$Cr_2O_7^{2-}(aq) + 14H^+(aq) + 6e^- \rightarrow 2Cr^{3+}(aq) + 7H_2O(l)$$

a) Describe the colour change observed when the dichromate ion reacts with the ethanol. [1]

b) State the name of the organic product formed during the reaction. [1]

c) In order to quantify exactly how much ethanol is present in the blood, a person may be required to give a blood sample or may be asked to blow into an intoximeter. Explain the chemistry behind the techniques for determining the ethanol content in a blood sample and by using an intoximeter. [4]

IB, May 2013

8 Modern drug detection techniques increase the chances of people being caught using illegal substances. Discuss how changes in technology influence our ethical choices.

INTERNAL ASSESSMENT

Introduction

In this chapter you will discover the important role of experimental work in chemistry. It guides you through the expectations and requirements of an independent investigation called the internal assessment (IA).

Advice on the internal assessment

Understandings

→ theory and experiment

→ internal assessment requirements

→ internal assessment guidance

→ internal assessment criteria

Nature of science

Empirical evidence is a key to objectivity in science. Evidence is obtained by observation, and the details of observation are embedded in experimental work. Theory and experiment are two sides of the same coin of scientific knowledge.

Applications and skills

→ appreciation of the interrelationship of theory and experiment

→ ability to plan your internal assessment

→ understanding of teacher guidance

→ appreciation of the formal requirements of an internal assessment

→ critical awareness of academic honesty

Theory and experiment

The sciences use a wide variety of methodologies and there is no single agreed scientific method. However, all sciences are based on evidence obtained by experiment. Evidence is used to develop theories, which then form laws. Theories and laws are used to make predictions that can be tested in experiments. Science moves in a cycle that moves between theory and experiment. Observations inform theory. Observations help us determine a theory, but a theory equally can re-focus our observations. Experimentation allows us to have confidence that a theory is not merely pure speculation.

Consider a famous analogy used by Albert Einstein and Leopold Infeld of a man trying to understand the mechanism of a pocket watch. The following quote illustrates that our scientific knowledge can be tested against reality. It shows that we can confirm or deny a theory by experiment, but we can never know reality itself. There is a continual dance between theory and experiment.

> "Physical concepts are free creations of the human mind, and are not, however it may seem, uniquely determined by the external world. In our endeavor to understand reality we are somewhat like a man trying to understand the mechanism of a closed watch. He sees the face and the moving hands, even hears it ticking, but he has no way of opening the case; if he is ingenious he may form some picture of the mechanism which could be responsible for all the things he observes, but he may never be quite sure his picture is the only one which could explain his observations. He will never be able to compare his picture with the real mechanism and he cannot even imagine that possibility of the meaning of such a comparison. But he certainly believes that, as his knowledge increases, his picture of reality will become simpler and simpler and will explain a wider and wider range of his sensuous impressions. He may also believe in the existence of the ideal limit of knowledge and that it is approached by the human mind. He may call this ideal limit the objective truth."

> —*Albert Einstein and Leopold Infeld, "The Evolution of Physics."*

The internal assessment requirements

Experimental work is not only an essential part of the dynamic of scientific knowledge, it also plays a key role in the teaching and learning of chemistry. Experimental work should be an integral and regular part of your chemistry lessons consisting of demonstrations, hands-on group work, and individual investigations. It may also include computer simulations, molecular modelling, and online database resources. It is only natural then that time should be allocated to you in order to formulate, design, and implement your own experimental project. You will produce a single investigation that is called an **internal assessment**. Your teacher will assess your report using IB criteria, and the IB will externally moderate your teacher's assessment.

Your investigation will involve:

- selecting an appropriate topic
- researching the scientific content of your topic
- defining a workable research question
- adapting or designing a methodology
- obtaining, processing, and analysing data
- identifying errors, uncertainties, and limits of data
- writing a scientific report 6–12 A4 pages long
- receiving continued guidance from your teacher.

Planning and guidance

After the idea of an internal assessment investigation is introduced, you will have an opportunity to discuss your investigation topic with your teacher. Through dialogue with your teacher you can select an appropriate topic, define an appropriate research question, and begin conducting research into what is already known about your topic. You will not be penalized for seeking advice.

It is your **teacher's responsibility** to provide you with a clear description of the IA guidelines.

Your teacher will:

- provide you with continued guidance at all stages of your work

- help you focus on a topic, discuss your chosen research question with you, and assist in your selection of an appropriate methodology

- provide guidance as you work and read a draft of your report, making general suggestions for improvements or completeness.

It is not the role of the teacher to edit your report nor give you a tentative grade or achievement level for your project until it is finally completed. Once your report is completed and formally submitted you are not allowed to make any changes.

As the student, it is **your responsibility** to appreciate the meaning of academic honesty, especially authenticity and the respect of intellectual property. You are also responsible for initiating your research question, seeking assistance when in doubt, and demonstrating independence of thought and initiative in the design and implementation of your investigation. You are also responsible for meeting the deadlines set by your teacher.

The internal assessment report

There is no prescribed format for your investigative report. However, the IA criteria encourage a logical and justified approach, one that demonstrates personal involvement and exhibits sound scientific work.

The style and form of your report for the IA investigation should model a scientific journal article. You should be familiar with a number of chemistry journal articles. For example, journals and magazines like *Education in Chemistry, Chemistry International, The Australian Journal of Education in Chemistry,* and *The Journal of Chemical Education* often have articles that are appropriate for high school work. Moreover, many of these articles can provide good ideas for an investigation.

There is no prescribed narrative mode, and your teacher will direct you to the style that they wish you to use. However, because a report describes what you have carried out in your investigation, it is appropriate to write in the past tense. Descriptions are always clearer to understand if you avoid the use of pronouns (usually 'it') and refer specifically to the relevant noun ('the beaker', 'the voltmeter', 'a pipette', etc.).

Academic honesty

The IB learner profile (see page iv) describes the IB student as ideally possessing many qualities, including that of being "principled". This means that you act with integrity and honesty, with a strong sense of fairness and justice, and that you take responsibility for your actions and their consequences. The IA is your responsibility, and it is your work. Plagiarism and copying others' work is not permissible. You must clearly distinguish between your own words and thoughts and those of others by the use of quotation marks (or another method like indentation) followed by an appropriate citation that denotes an entry in the bibliography. In fact, your IA report is strengthened when you demonstrate that you have the skills to research relevant information and incorporate these references into your report, ensuring its academic integrity.

Although the IB does not prescribe referencing style or in-text citation, certain styles may prove most commonly used; you are free to choose a style that is appropriate. It is expected that the minimum information included is: name of author, date of publication, title of source, and page numbers as applicable.

Types of investigations

After you have covered a number of topics and performed a number of hands-on experiments in class, you will be required to research, design, perform, and write up your own investigation. The IA accounts for 20% of your final grade and requires you to spend 10 hours performing laboratory work, during which time you will be engaged in constant dialogue with your teacher. The time required for you to write your report cannot be included in the 10 hours and this should be compiled outside of the classroom period.

The variety and range of possible investigations is large, you could choose from:

- **Traditional hands-on experimental work.** You may want to estimate the level of organic pollution in water by measuring the biological oxygen demand (BOD), determine the percentage of iron in a medication such as iron supplements, or synthesize a drug such as aspirin and characterise it using a variety of analytical techniques.

- **Database investigations.** You may obtain data from scientific websites and process and analyse the information for your investigation. Perhaps find a correlation between the frequency of cancer cases in your community and the level of harmful substances in the atmosphere released by local industries, or you might be interested in structural systematics, an emerging area of chemistry that looks at structure-property relationships in chemical compounds.

- **Spreadsheet.** You can make use of a spreadsheet with data from any type of investigation. You can process the data, graph the results or design a simple model to compare theoretical values with your experimental values.

- **Simulations.** It may not be feasible to perform some investigations in the classroom, but you may be able to utilise a computer simulation. The data from a simulation may be processed and presented in such a way to reveal some novel aspect of the scientific work.

A combination of these alternatives is possible. The subject matter of your investigation is a personal decision and may be situated inside or outside the boundaries of the IB chemistry syllabus. The depth of understanding should be, however, commensurate with the course you are taking. Your knowledge of IB Chemistry (either SL or HL) will enable you to achieve the maximum mark when your report is assessed.

The assessment criteria

Your IA consists of a single investigation with a report 6–12 pages long. The report should have an academic and scholarly presentation, and demonstrate scientific rigor commensurate with the course. There is the expectation of personal involvement and a sound understanding of chemistry. You must clearly identify the current scientific understanding of your chosen topic, thereby establishing a point of departure for your scientific inquiry.

There are five assessment criteria, ranging in weight from 8–25% of the total possible marks. Each criterion reflects a different aspect of your investigation and are applied equally to SL and HL students.

Criterion	Marks	Weight
Personal engagement	0–2	8%
Exploration	0–6	25%
Analysis	0–6	25%
Evaluation	0–6	25%
Communication	0–4	17%
Total	**0–24**	**100%**

PERSONAL ENGAGEMENT. *This criterion assesses the extent to which you engage with the investigation and make it your own. Personal engagement may be recognized in different attributes and skills. These include thinking independently and/or creatively, addressing personal interests, and presenting scientific ideas in your own way.*

For maximum marks under the personal engagement criterion, you must provide clear evidence that you have contributed significant thinking, initiative, or insight to your investigation: that you take the responsibility for ownership of your investigation. Your research question could be based upon something covered in class or an extension of your own interest.

For example you may be a keen athlete and your teacher may have demonstrated various analytical techniques for the detection of prohibited substances in sports as specified by the World Anti-Doping Agency (WADA). You might be very interested in the role of the analytical chemist in drug testing in sports and decide to design and perform an investigation based on performance-enhancing drugs. Personal significance, interest, and curiosity are expressed here.

You can demonstrate personal engagement through personal input and initiative in the design, implementation, or presentation of the investigation. Perhaps you designed a novel method for the synthesis of a particular drug, resulting in a greater yield of product or devised an improved method for the analysis of data. You are not to simply perform a recipe-like experiment.

The key here is to be involved in your investigation, to contribute something that makes it your own.

EXPLORATION. *This criterion assesses the extent to which you establish the scientific context for your work, state a clear and focused research question, and use concepts and techniques appropriate to the course you are studying. Where appropriate, this criterion also assesses awareness of safety, environmental, and ethical considerations.*

For maximum marks under the exploration criterion, your topic must be appropriately identified and a relevant and fully focused research question developed. Background information about your investigation must be relevant, and the methodology appropriate to enable your research question to be addressed. Moreover, for maximum marks, your research must identify significant factors that may influence the relevance, reliability, and sufficiency of your data. Finally, your work must be safe and it must demonstrate a full awareness of relevant environmental and ethical issues. Safety plays a fundamental role in any wet laboratory based experimental work and the environmental aspects associated with the field of **Green Chemistry** continues to be a growth area in science.

The key here is your ability to select, develop, and apply appropriate methodology and produce a solid, scientific piece of work.

ANALYSIS. *This criterion assesses the extent to which your report provides evidence that you have selected, processed, analysed, and interpreted the data in ways that are relevant to the research question and can support a conclusion.*

For maximum marks under the analysis criterion, your investigation must include sufficient raw data to support a detailed and valid conclusion to your research question. Your processing of the data must be carried out with sufficient accuracy. Experimental uncertainties need to be identified and the propagation of these random errors will enable you to demonstrate their impact on the final result. For maximum marks, you must correctly interpret your data, so that completely valid and detailed conclusions to the proposed research question can be deduced.

EVALUATION. *This criterion assesses the extent to which your report provides evidence of evaluation of the investigation and results with regard to the research question and the wider world.*

For maximum marks under the evaluation criterion, you must describe a detailed and justified conclusion that is entirely relevant to the research question, and fully supported by your analysis of the data presented. You should make a comparison to the accepted scientific context if relevant. The strengths and weaknesses of your investigation, such as the limitations of data and sources of uncertainty, must be discussed and you will need to provide evidence of a clear understanding of the scientific methodology involved in establishing your conclusion. You should discuss realistic and relevant improvements and propose possible extensions to your investigation.

The focus of evaluation is to incorporate the methodology used and set the results within a a wider scientific context while making reference to your initial research question.

COMMUNICATION. *This criterion assesses whether the investigation is presented and reported in a way that supports effective communication of the investigation's focus, process, and outcomes.*

For maximum marks under the communication criterion, your report must be clear and well structured, focus on the necessary information, and the process and outcomes must be presented in a logical and coherent manner. Your text must be relevant and avoid wandering off onto tangential issues. Your use of specific chemistry terminology and conventions must be appropriate and correct. Graphs, tables, and images must all be well presented.

The IA represents a unique opportunity for you to take ownership of your chemistry learning by investigating something that matters to you. It is an opportunity for you to work independently and to follow your own scientific instincts. You should be prepared to research your topic independently and approach your teacher full of ideas and suggestions.

INDEX

Page numbers in *italics* refer to question sections.

benzylpenicillin 729–30
  beta-lactam ring 729
  development of penicillin into a drug 729
  mechanism of action of penicillin 729–30
  penicillin G 729
  side-chain (R) 729
penicillinase 729
pentoses 581
peptides 548
  C-terminal 555
  dipeptides 555
  example of peptide 555
  how many peptides can we make? 555
  hydrolysis of peptides 556
  N-terminal 555
  naming peptides 555
  peptide bonds 554
  peptide linkage 554
  peptides in the human body 556
  properties of peptides 556
percentage yield 20, 23
perfumes 455
periodic law 68–9, 69
periodic table 67, 68–9, 76
  actinoids 72–3, 302
  d-block elements 72–3, 303
  electron configurations and the periodic table 73–4
  f-block elements 72–3, 302, 303
  lanthanoids 72–3, 302
  main-group elements 72–3
  metals, non-metals and metalloids 71–2
  p-block elements 72–3
  period number 67, 71
  periodic table today 70–1
  periods 71
  s-block elements 72–3
  transition elements 72, 302–3
periodic trends 75, *91–2*
  periodic trends in atomic radius 78–9
  periodic trends in electron affinity 82
  periodic trends in electronegativity 84
  periodic trends in ionic radius 79–80
  periodic trends in ionization energy 80–2, 296–9
  periodic trends in metallic and non-metallic character 75, 85–7
periodicity 67, 68
perms 559
pH curves 404–8
pH scale 191, 197, 198
  calculating pH 198–9
  ionization of water 199
  pH and acid–base titrations 200
  pH changes 75
pharmaceutical drugs 717, 718, *724*
  administration 718
  clinical trials 718–19, 720, 723
  drug action and development of new drugs 721–3
  drug design 722
  effective dose 717, 719
  effectiveness and safety 719–20
  lead compounds 722
  lethal dose 717, 719
  new chemical entities (NCEs) 722
  phase IV trials 723
  placebo effect 718–19
  post-clinical studies 723
  preclinical trials 722–3
  receptors and inhibitors 717, 721
  risks and benefits 719
  side effects 719
  social implications of the pharmaceutical industry 752
  therapeutic effects 718
  therapeutic index 717, 719
  therapeutic window and bioavailability 717, 719–20
  tolerance and addiction 717, 720–1
  toxic dose 717, 719
  waste products 754–5, 746

pharmaceutically active compounds (PACs) 600, 751
pharmacodynamics 456
pharmacokinetics 456
phenalymine 447
phenol–methanal plastics 530
philosophy 307
phlogiston 2, 192
phonons 519
phospholipids 565, 573–5
  amphiphilic 574
photochromic lenses 212
photons 52, 53, 292
photosynthesis 540, 543, 544–5, 658, *678*
  harnessing solar energy by chlorophyll 675–7
  photosynthesis, respiration, and the atmosphere 544–5
photosystems 635
photovoltaic cells 710, 712–13
pi bonds 113, 334–6
picometre (pm) 41
placebos 718
  double-blind tests 719
  placebo effect and clinical trials 718–19
Planck's equation 279–80
plasma 480–2
  Inductively Coupled Plasma (ICP) 480–2
plasticizers 494
  plasticizers and chlorine-free plastics 511
  PVC and the use of plasticizers 496–7
plastics 509, *515*
  biodegradable plastics 597, 602
  effect of plastic waste and POPs on wildlife 510
  phenol–methanal plastics 530
  plastics and polymers 602
  recycling of plastics 511–13
  resin identification code (RIC) 512–13
  sorting plastics 514
  see also polymers
polarizability 125
Poliakoff, Martyn 76
polychloroethene *see* PVC
polyethene, 494, 496
polymers 248, 472, 494, *500*
  addition polymerization 254, 529
  atom economy of polymerization reactions 499, 529
  condensation polymerization 528–9
  cross-linking polymer chains 531
  elastomers 496
  high density and low density polyethene 495
  identifying monomers 498–9
  ion implantation 531
  isotactic, atactic, and syndiotactic addition polymers 497–8
  modifying polymers 531–2
  plastics and polymers 602
  polymerization of alkenes 254–5
  polymers in society 496
  polystyrene 497
  PVC and the use of plasticizers 496–7
  thermoplastics and thermosets 495–6
  vulcanization 531
  see also plastics
polypeptides 548
polysaccharides 586–7
polystyrene 497
polyurethanes 530–1
polyvinyl chloride *see* PVC
POPs (persistent organic pollutants) 510, 598
  dioxins and PCBs 510–11
porphyrins 631–2
  porphin 631
positron emission tomography (PET) 45, 770
postulates 24, 38
potassium permanganate 308–9
potential energy profile 167, 171
powers 15
precision 263–4, 267
predictions 201
pressure 20, 25

effect of pressure on reactions in the gas phase 187
  partial pressure 780
primary compounds 245–6
primary solutions 33
primary standards 16
principal quantum number 57
probability density 56
products 1
proligands 312
propagation 251

prophylactics 730
prosthetic groups 559, 613
proteins 213, 547, 548, *563–4*, 606, *617–18*

  acid–base of proteins 560
  acid–base properties of 2-amino acids 606–8
  amino acids and peptides 548–9
  amino acids as zwitterions 550
  biuret test 560, 616
  central role of proteins in biochemistry 547–8
  denatured proteins 559
  enzymes 560–3
  fibrous proteins 558–9
  gel electrophoresis 551, 610
  globular proteins 558
  isoelectric focusing 560, 610
  native states/structures 559
  predicting secondary structures 557
  primary structure 556
  prosthetic groups 559, 613
  protein assay 614–17
  protein deficiency 561
  protein sequencing 556
  protein subunits 559
  proteins and heredity 547
  proteins as biological buffers 610–11
  quaternary structure 559
  scleroproteins 559
  secondary structure 556–7
  tertiary structure 558–9
proteomics 556, 560
proton beam therapy (PBT) 769
proton nuclear magnetic resonance spectroscopy (1H NMR) 261, 283–5, 462
  chemical shift 283
  high resolution 1H NMR spectroscopy 462–4
  integration trace 285
  magnetic resonance imaging (MRI) 285
  tetramethylsilane 465–9
protons 42, 766
Proust, Joseph 38
Proust's law of constant composition 6
PTFE (polytetrafluoroethene) 244
puckering 453
pure sciences 472
pure substances 5
purines 620, 621
PVC 496–7, 509
pyrimidines 620

qualitative analysis 18, 262
quantitative analysis 18, 33, 262
quantitative measurements 1
  mole 13–14
  si units, 3, 12–13
quantization 52–4
  quantization and atomic structure 55–8
quantum mechanics 187
  quantum mechanical model of the atom 56–8
quantum numbers 58
  azimuthal quantum number 58
  magnetic quantum number 58
  principal quantum number 58
  spin magnetic quantum number 58
quartz 120

racemic mixtures 456